Stem Cell Biology and Regenerative Medicine

Series Editor
Kursad Turksen, Ph.D.
kturksen@ohri.ca

For other titles published in this series, go to
http://www.springer.com/series/7896

Vinagolu K. Rajasekhar, M.Sc., M.Phil., Ph.D.
Mohan C. Vemuri, Ph.D.
Editors

Regulatory Networks in Stem Cells

Humana Press

Editors
Vinagolu K. Rajasekhar, MSc, MPhil, PhD
Memorial Sloan-Kettering
Cancer Center, New York
NY, USA
vinagolr@mskcc.org

Mohan C. Vemuri, PhD
Invitrogen Corporation
Frederick, MD, USA
mohan.vemuri@invitrogen.com

ISBN: 978-1-4939-5696-8 ISBN: 978-1-60327-227-8 (eBook)
DOI 10.1007/978-1-60327-227-8

Cover illustration: Neurons derived from hESC were labeled with post mitotic neuronal marker of HuC/D(Green) and of neuronal filament of DCX (Red). Courtesy Soojung Shin

Printed on acid-free paper

springer.com

Dedication

I (V.K. Rajasekhar) sincerely dedicate this book to:

My beloved family: my wife, Birgit Baur; my children, Julia and Jessica; my parents, Suseela and Krishnamachari; my sister, Late Bhanu; my brothers, Veerabrahmam, Suresh, and Dasaradharam; my immediate family members, Erika and Hugo Baur, Late Veerabrahmam uncle, Hari Gopal uncle, Raghunath uncle, Jaya Aunty, and Venkatachari uncle.
My great friends: James and Loretta Laplander, Julie Cerrato, Christine Geiger, Bela Shah, Guyu Ho, Ravindranathreddy, Zaki Qureshi, Pinakin Patel, S.V. Prasad, S. Gopi, A. Jayakumar, Toshiro Shigaki, Martin Begemann, and Sandra Cohen.
My invaluable teachers and advisors: Drs. Sudhir Sopory, Hans Mohr, Michael Mulligan, Howard I. Scher, Lorenz Studer, Mark Ptashne, and Nahum Sonenberg.

Mohan C. Vemuri dedicates this book to his parents Vemuri Lakshmanachari and Vemuri Nagarathnamma.

Preface

Stem cells appear to be fundamental cellular units associated with the origin of multicellular organisms and have evolved to function in safeguarding the cellular homeostasis in organ tissues. The characteristics of stem cells that distinguish them from other cells have been the fascinating subjects of stem cell research. The important properties of stem cells, such as *maintenance of quiescence, self-renewal capacity, and differentiation potential,* have propelled this exciting field and presently form a common theme of research in developmental biology and medicine. The derivation of pluripotent embryonic stem cells, the prospective identification of multipotent adult stem cells, and, more recently, the induced pluripotent stem cells (popularly called iPS) are important milestones in the arena of stem cell biology. Complex networks of transcription factors, different signaling molecules, and the interaction of genetic and epigenetic events constantly modulate stem cell behavior to evoke programming and reprogramming processes in normal tissue homeostasis during development. In any given cellular scenario, the regulatory networks can pose considerable complexity and yet exert an orderly control of stem cell differentiation during normal development. An aberration in these finely tuned processes during development usually results in a spectrum of diseases such as cancers and neurological disorders. This underscores the imminent need for a more complete understanding of molecular mechanisms underlying the regulatory circuitries required for stem cell maintenance.

Over the past 3–5 years, a diverse group of bench and physician scientists have prospectively enhanced our knowledge of stem cell biology. These studies are unveiling many unrecognized or previously unknown fundamentals of developmental biology. Furthermore, they are also opening up new horizons in clinical medicine. Some of the basic questions that are being currently pursued include: H*ow are the cellular context-dependent and cell type-specific gene expression patterns controlled by the cross-talk between intracellular signaling pathways and extracellular signals? What are the genetic and epigenetic controls associated with stem cells, and how do they operate during their developmental transitions and in the maintenance of the resulting phenotype? What are the functional roles of the recently emerging new layers of control in gene expression at the post-transcriptional/translational/post-translational levels in stem cells? Most importantly, can stem cells be engineered with fidelity for a particular application, and would the application have to be patient specific?* Attempts to meaningfully address many of these questions face challenges from biological complexities such as the potential of multiple stem cell types in a given cellular context. Understanding these stem-cell-specific regulatory networks is therefore the key to successfully harvesting the fruits of stem cell research.

Regulatory Networks in Stem Cells is an initial attempt to decipher the key factors involved in stem cell pluripotency, maintenance, and directed differentiation toward specific cell lineages and stem cell types. The presentation of the contents is such that upper-grade undergraduates, graduate students, postgraduates, and basic research as well as clinical research scientists are provided with accessible information about recent advances in the stem cell field. This book also covers the necessary basic concepts of developmental biology in order to facilitate the reader's understanding and also to anticipate potential applications in this fast-growing field.

We and the authors have worked together to ensure that the chapters address current topics for advanced, inquisitive researchers as well. For the convenience of readers, chapters are grouped into the following sections: (A) Molecular Regulation in Stem Cells, (B) Regulation by Stem Cell Niches, (C) Epigenetic Mechanisms in Stem Cells, (D) Signaling and Regulation in Select Stem Cell Types, and (E) Disease Paradigms and Stem Cell Therapeutics. Stem cell regulatory networks are only just beginning to emerge in the field that is regularly inundated with a myriad of novel developments.

We hope that this edition enables readers to gain crucial insights into the field of stem cell research and provides a framework for planning stem cell research with the goal of improving human health.

V.K. Rajasekhar and Mohan C. Vemuri

Contents

Part I Molecular Regulation in Stem Cells

The Molecular Basis of Embryonic Stem Cell Self-Renewal 3
Stephen Dalton

Asymmetric Behavior in Stem Cells 13
Bridget M. Deasy

Determinants of Pluripotency in Mouse and Human Embryonic Stem Cells 27
Leon M. Ptaszek and Chad A. Cowan

Maintenance of Embryonic Stem Cell Pluripotency by Nanog-Mediated Dedifferentiation of Committed Mesoderm Progenitors 37
Atsushi Suzuki, Ángel Raya, Yasuhiko Kawakami, Masanobu Morita, Takaaki Matsui, Kinichi Nakashima, Fred H. Gage, Concepción Rodríguez-Esteban and Juan Carlos Izpisúa Belmonte

Human Embryonic Stem Cells and Germ Cell Development 55
Nina J. Kossack, Joerg Gromoll, and Renee A. Reijo Pera

Genomic Stability in Stem Cells 67
Irene Riz and Robert G. Hawley

Genetic Manipulation of Human Embryonic Stem Cells 75
Dimitris G. Placantonakis, Mark J. Tomishima, Fabien G. Lafaille, and Lorenz Studer

Transcriptional Networks Regulating Embryonic Stem Cell Fate Decisions 87
Emily Walker and William L. Stanford

Use of Zebrafish to Dissect Gene Programs Regulating Hematopoietic Stem Cells 101
Colleen E. Albacker and Leonard I. Zon

HOXB4 in Hematopoietic Stem Cell Regulation 111
Mohan C. Vemuri

Telomere and Telomerase for the Regulation of Stem Cells 123
Eiso Hiyama and Keiko Hiyama

The Role of Mitochondria in Stem Cell Biology 135
Claudia Nesti, Livia Pasquali, Michelangelo Mancuso, and Gabriele Siciliano

Part II Regulation by Stem Cell Niches

Stem Cells and Stem Cell Niches in Tissue Homeostasis: Lessons from the Expanding Stem Cell Populations of *Drosophila* 147
Yukiko M. Yamashita

Extrinsic and Intrinsic Control of Germline Stem Cell Regulation in the *Drosophila* Ovary 155
Nian Zhang and Ting Xie

The Niche Regulation of Hematopoietic Stem Cells 165
Hiroko Iwasaki and Toshio Suda

Environmental Signals Regulating Mesenchymal Progenitor Cell Growth and Differentiation 175
Meirav Pevsner-Fischer and Dov Zipori

Microenvironmental Regulation of Adult Mesenchymal Stem Cells 185
Thomas P. Lozito, Catherine M. Kolf and Rocky S. Tuan

Stem Cells, Hypoxia and Hypoxia-Inducible Factors 211
Suzanne M. Watt, Grigorios Tsaknakis, Sinead P. Forde and Lee Carpenter

Part III Epigenetic Mechanisms in Stem Cells

Stem Cell Epigenetics 235
Joyce E. Ohm and Stephen B. Baylin

Epigenetic Signature of Embryonal Stem Cells: A DNA Methylation Perspective ... 247
Monther Abu-Remaileh and Yehudit Bergman

Epigenetic Basis for Differentiation Plasticity in Stem Cells 257
Philippe Collas, Sanna Timoskainen and Agate Noer

Role of DNA Methylation and Epigenetics in Stem Cells 269
Bhaskar Thyagarajan and Mahendra Rao

DNA Methylation and the Epigenetic Program in Stem Cells 277
Laurie Jackson-Grusby

Polycomb Group Protein Homeostasis in Stem Cell Identity – A Hypothetical Appraisal 285
Vinagolu K. Rajasekhar

Part IV Signaling and Regulation in Select Stem Cell Types

Signaling Pathways in Embryonic Stem Cells 293
D. Reynolds, Ludovic Vallier, Zhenzhi Chng and Roger Pedersen

Regulation of Stem Cell Systems by PI3K/Akt Signaling 309
Tohru Kimura and Toru Nakano

Endothelial Ontogeny During Embryogenesis: Role of Cytokine Signaling Pathways 319
Daylon James, Marco Seandel and Shahin Rafii

Signaling Networks in Mesenchymal Stem Cells 329
Vivek M. Tanavde, Lailing Liew, Jiahao Lim and Felicia Ng

Single-Cell Approaches to Dissect Cellular Signaling Networks 337
Weijia Wang and Julie Audet

Hematopoietic Stem Cells 347
Malcolm A.S. Moore

Renal Stem Cells and Kidney Regeneration 379
Takashi Yokoo, Akira Fukui, Kei Matsumoto and Tetsuya Kawamura

The Endometrium: A Novel Source of Adult Stem/Progenitor Cells 391
Caroline E. Gargett and Kjiana E. Schwab

Epithelial Stem Cells and the Development of the Thymus, Parathyroid, and Skin 405
Chew-Li Soh, Joanna M.C. Lim, Richard L. Boyd and Ann P. Chidgey

Hepatic Stem Cells and Liver Development 439
Nalu Navarro-Alvarez, Alejandro Soto-Gutierrez and Naoya Kobayashi

Part V Disease Paradigms and Stem Cell Therapeutics

The Idea and Evidence for the Tumor Stemness Switch 473
Bikul Das, Rika Tsuchida, Sylvain Baruchel, David Malkin and Herman Yeger

The Role of the Tumor Suppressor Fhit in Cancer-Initiating Cells 489
Hideshi Ishii

History of Cancer Stem Cells 495
Stewart Sell

Immune Responses to Stem Cells and Cancer Stem Cells 505
Xiao-Feng Yang and Hong Wang

Leukemic Stem Cells: New Therapeutic Targets? 519
Dominique Bonnet

Solid Tumor Stem Cells – Implications for Cancer Therapy 527
Tobias Schatton, Natasha Y. Frank and Markus H. Frank

Therapeutic Approaches to Target Cancer Stem Cells 545
Lisa R. Rogers and Maxs Wicha

Preclinical Evidence for Cellular Therapy as a Treatment for Neurological Disease 561
Matthew T. Harting, Charles S. Cox and Stephen G. Hall

Improving Memory with Stem Cell Transplantation 575
Mathew Blurton-Jones, Tritia R. Yamasaki and Frank M. LaFerla

Index 585

Contributors

Monther Abu-Remaileh The Hubert Humphrey Center for Experimental Medicine and Cancer Research, Hebrew University Medical School, Ein Kerem Jerusalem, 91120 Israel, Email: monther@geneinfo.net

Colleen E. Albacker Division of Hematology/Oncology, Harvard Medical School and Howard Hughes Medical Institute, Children's Hospital Boston, One Blackfan Circle, Boston, MA 02115, USA, Email: zon@enders.tch.harvard.edu Tel: 617-919-2069,
Fax: 617-730-0222

Julie Audet Institute of Biomaterials and Biomedical Engineering, Terrence Donnelly Centre for Cellular and Biomolecular Research, University of Toronto, 164 College St., Room 407, Toronto, ON, Canada, M5S 3G9, Email: julie.audet@utoronto.ca Tel: 416-946-0209,
Fax: 416-978-4317

Sylvain Baruchel, MD Division of Hematology & Oncology, The Hospital for Sick Children, Faculty of Medicine, Institute of Medical Science, University of Toronto, Toronto, ON, Canada M5S 3G9

Stephen B. Baylin Cancer Biology Division, The Sidney Kimmel Comprehensive Cancer Center, The Johns Hopkins University Medical Institutions, Baltimore, MD 21231, USA Email: sbaylin@jhmi.edu Tel: 410-955-8506, Fax: 410-614-9884,

Juan Carlos Izpisúa Belmonte Gene Expression Laboratory, Salk Institute for Biological Studies, 10010 N. Torrey Pines Rd., La Jolla, CA 92037, USA

Yehudit Bergman The Hubert Humphrey Center for Experimental Medicine and Cancer Research Hebrew University Medical School Ein Kerem Jerusalem, 91120 Israel, Email: yehudit.bergman@huji.ac.il

Mathew Blurton-Jones, PhD Department of Neurobiology and Behavior and Institute for Brain Aging and Dementia, University of California, Irvine. Irvine, CA 92697-4545, USA

Dominique Bonnet Cancer Research UK, London Research Institute, Haematopoietic Stem Cell Laboratory, Cancer Research UK, London Research Institute, 44 Lincoln's Inn Fields, London, WC2A 3PX, UK, Email: dominique.bonnet@cancer.org.uk

Richard L. Boyd Monash Immunology and Stem Cell Laboratories (MISCL), Level 3, Building 75, Monash University, Wellington Road, Clayton, Victoria 3800, Australia
Tel: +613 9905 0628; Fax: +613 9905 0680

Lee Carpenter Stem Cell Research Laboratory, Nuffield Department of Clinical Laboratory Sciences, University of Oxford, Oxford, UK

Ann P. Chidgey Monash Immunology and Stem Cell Laboratories (MISCL), Level 3, Building 75, Monash University, Wellington Road, Clayton, Victoria 3800, Australia,
Email: n.chidgey@med.monash.edu.au Tel: +613 9905 0628; Fax: +613 9905 0680

Zhenzhi Chng Department of Surgery and Cambridge Institute for Medical Research (CIMR), Addenbrookes Hospital, University of Cambridge, Hills Road, Cambridge, CB2 0XY, UK

Philippe Collas Institute of Basic Medical Sciences, Department of Biochemistry, Faculty of Medicine, University of Oslo, Norway, Email: philippe.collas@medisin.uio.no
Tel: 47-22851066, Fax 47-22851058

Chad A. Cowan Harvard Stem Cell Institute, 42 Church Street, Cambridge, MA 02138, USA; Stowers Medical Institute, Cardiovascular Research Center and Center for Regenerative Medicine, Massachusetts General Hospital, 185, Cambridge Street CPZN 4265-A, Boston, MA 02114 Tel: 617.643.3569, Fax: 617.643.0674

Charles S. Cox, Jr MD University of Texas Medical School at Houston, Department of Pediatric Surgery; 6431 Fannin St. MSB 5.228, Houston, TX 77030, USA,
Email: charles.s.cox@uth.tmc.edu

Stephen Dalton Department of Biochemistry and Molecular Biology, Paul D. Coverdell Center for Biomedical and Health Sciences, 500 DW Brooks Drive, Athens, GA 30602, USA,
Email: sdalton@uga.edu Tel: 706-583-0480

Bikul Das M.B.B.S., PhD Stem Cell and Developmental Biology program, Research Institute, and Division of Hematology & Oncology, The Hospital for Sick Children, Toronto; Faculty of Medicine, Institute of Medical Science, University of Toronto, Toronto, ON, Canada,
Email: bikul.das@utoronto.ca Tel: 416-813-5937/5977,

Bridget M Deasy Departments of Orthopedic Surgery and Bioengineering, University of Pittsburgh, Stem Cell Research Center, McGowan Institute for Regenerative Medicine, University of Pittsburgh Medical Center, Pittsburgh, PA 15219, USA,
Email: deasybm@upmc.edu Tel: 412-692-3223

Sinead P. Forde Stem Cell Research Laboratory, Nuffield Department of Clinical Laboratory Sciences, University of Oxford, Oxford, UK, Stem Cells and Immunotherapies, NHS Blood and Transplant, Oxford, UK.

Markus H. Frank, MD Transplantation Research Center, Children's Hospital Boston & Brigham and Women's Hospital, Harvard Medical School, 300 Longwood Avenue, Boston, MA 02115, USA, Email: mfrank@rics.bwh.harvard.edu

Natasha Y. Frank, MD Department of Medicine, VA Boston Healthcare System, 1400 VFW Parkway, West Roxbury, MA 02132, USA

Akira Fukui MD Project Team for Kidney Regeneration, Institute of DNA Medicine, Division of Nephrology and Hypertension, Department of Internal Medicine, The Jikei University School of Medicine, 3-25-8, Nishi-Shimbashi, Minato-ku, Tokyo, 105-8461 Japan

Fred H. Gage Laboratory of Genetics, Salk Institute for Biological Studies, 10010 N. Torrey Pines Rd., La Jolla, CA 92037, USA

Caroline E. Gargett, PhD Centre for Women's Health Research, Monash Institute of Medical Research and Monash University, Department of Obstetrics and Gynaecology, Monash Medical Centre, 246 Clayton Road, Clayton, Victoria, 3168, Australia, Email: caroline.gargett@med.monash.edu.au Tel: +61 3 9594 5392, Fax: +61 3 9594 6389,

Joerg Gromoll Institute of Reproductive Medicine, Westphalian Wilhelms University, Muenster, Germany

Stephen G. Hall, PhD AlphaGenix, Inc, Department of Regenerative Medicine, 2329 North Career Avenue, Sioux Falls, SD 57107, USA, Email: hallsg@alphagenix.com Tel: 605-274-2268

Matthew T. Harting MD, University of Texas Medical School at Houston, Department of Pediatric Surgery; 6431 Fannin St. MSB 5.228, Houston, TX 77030, USA, Email: matthew.t.harting@uth.tmc.edu

Robert G. Hawley Department of Anatomy and Regenerative Biology, The George Washington University, School of Medicine and Health Sciences, 2300 I Street NW, Washington, DC 20037, USA Tel: 202-994-2763/3511, Fax: 202-994-8886

Eiso Hiyama Natural Science Center for Basic Research and Development, Hiroshima University, 1-2-3, Kasumi, Minami-ku, Hiroshima, 734-8551, Japan, Email: eiso@hiroshima-u.ac.jp Tel: 81-82-257-5951, Fax: 81-82-257-5416

Keiko Hiyama Research Institute for Radiation Biology and Medicine, Hiroshima University, 1-2-3, Kasumi, Minami-ku, Hiroshima, 734-8551, Japan, Email: khiyama@hiroshima-u.ac.jp Tel: 81-82-257-5841, Fax: 81-82-256-7105

Hideshi Ishii Center for Molecular Medicine, Jichi Medical University, Tochigi 329-0498 Japan. Email: hishii@ms.jichi.ac.jp

Hiroko Iwasaki Department of Cell Differentiation, Keio University School of Medicine, 35 Shinanomachi, Shinjuku-ku, Tokyo 160-8582 Japan, Email: di065022@sc.itc.keio.ac.jp Tel & Fax: +81-3-5363-3475

L. Jackson-Grusby Children's Hospital Boston, Harvard Stem Cell Institute, Harvard Medical School, Boston, MA 02115, USA, Email: grusby@childrens.harvard.edu/laurie.jackson

Daylon James, PhD Howard Hughes Medical Institute, Ansary Center for Stem Cell Therapeutics, Department of Genetic Medicine, Division of Hematology-Oncology, Weill Cornell Medical College, 1300 York Avenue, Room A-863, New York, NY 10021, USA

Yasuhiko Kawakami Gene Expression Laboratory, Salk Institute for Biological Studies, 10010 N. Torrey Pines Rd., La Jolla, CA 92037, USA

Tetsuya Kawamura, MD, PhD Project Team for Kidney Regeneration, Institute of DNA Medicine, Division of Nephrology and Hypertension, Department of Internal Medicine, The Jikei University School of Medicine, 3-25-8, Nishi-Shimbashi, Minato-ku, Tokyo, 105-8461 Japan

Tohru Kimura Department of Pathology, Graduate School of Medicine, Osaka University, 2-2 Yamada-oka, Suita, Osaka, Japan 565-0871, Email: tkimura@patho.med.osaka-u.ac.jp Tel: +81-6-6879-3722, Fax: +81-6-6879-3729

Naoya Kobayashi Department of Surgery, Okayama University Graduate School of Medicine and Dentistry, 2-5-1 Shikata-cho, Okayama 700-8558, Japan,
Email: immortal@md.okayama-u.ac.jp Tel & Fax: 81-86-235-7485

Catherine M. Kolf Cartilage Biology and Orthopaedics Branch, National Institute of Arthritis, and Musculoskeletal & Skin Diseases, National Institutes of Health Building 50, Room 1523, 50 South Drive, MSC 8022, Bethesda, MD 20892-8022, USA Tel: 301-451-6854. Fax: 301-435-8018

Nina J. Kossack Center for Human Embryo and Embryonic Stem Cell Research and Education; Institute for Stem Cell Biology and Regenerative Medicine, Stanford University, Palo Alto, CA 94304-5542, USA

Fabien G. Lafaille Program in Developmental Biology, Memorial Sloan-Kettering Cancer Center, New York, NY

Frank M. LaFerla Department of Neurobiology and Behavior and Institute for Brain Aging and Dementia; University of California, Irvine, Irvine, CA 92697-4545, USA,
Email: laferla@uci.edu

Lailing Liew Bsc Biomolecular Function Discovery Group, Bioinformatics Institute, Singapore

Jiahao Lim Bsc Genome & Gene Expression Data Analysis Group, Bioinformatics Institute, Singapore

Joanna M.C. Lim Monash Immunology and Stem Cell Laboratories (MISCL), Level 3, Building 75, Monash University, Wellington Road, Clayton, Victoria 3800, Australia
Tel: +613 9905 0628, Fax: +613 9905 0680

Thomas P. Lozito Cartilage Biology and Orthopaedics Branch, National Institute of Arthritis, and Musculoskeletal & Skin Disease,s National Institutes of Health Building 50, Room 1523, 50 South Drive, MSC 8022, Bethesda, MD 20892-8022, USA Tel: 301-451-6854,
Fax: 301-435-8017

David Malkin MD Divisions of Hematology/Oncology, The Hospital for Sick Children, Toronto, Canada; Medical Biophysics, University of Toronto, Toronto Canada

Michelangelo Mancuso Department of Neurosciences, Section of Neurology, University of Pisa, Via Roma 67, 56126 Pisa, Italy

Takaaki Matsui Gene Expression Laboratory, Salk Institute for Biological Studies, 10010 N. Torrey Pines Rd., La Jolla, CA 92037, USA; Gene Regulation Research, Graduate School of Biological Sciences, Nara Institute of Science and Technology, 8916-5 Takayama, Ikoma 630-0101, Japan

Kei Matsumoto MD Project Team for Kidney Regeneration, Institute of DNA Medicine, Division of Nephrology and Hypertension, Department of Internal Medicine, The Jikei University School of Medicine, 3-25-8, Nishi-Shimbashi, Minato-ku, Tokyo, 105-8461 Japan

Malcolm A.S. Moore Enid A. Haupt Professor of Cell Biology, Cell Biology Program, Memorial Sloan Kettering Cancer Center, New York, NY 10065, USA,
Email: m-moore@ski.mskcc.org Tel: 212-639-7090, Fax: 212-717-3618

Masanobu Morita Gene Expression Laboratory, Salk Institute for Biological Studies, 10010 N. Torrey Pines Rd., La Jolla, CA 92037

Toru Nakano Department of Pathology, Graduate School of Medicine, Graduate School of Frontier Biosciences, Osaka University, 2-2 Yamada-oka, Suita, Osaka, Japan 565-0871
Tel: +81-6-6879, Fax: +81-6-6879-3729

Kinichi Nakashima Laboratory of Genetics, Salk Institute for Biological Studies, 10010 N. Torrey Pines Rd., La Jolla, CA 92037; Laboratory of Molecular Neuroscience, Graduate School of Biological Sciences, Nara Institute of Science and Technology, 8916-5 Takayama, Ikoma 630-0101, Japan

Nalu Navarro-Alvarez Department of Surgery, Okayama University Graduate School of Medicine and Dentistry, 2-5-1 Shikata-cho, Okayama 700-8558, Japan

Claudia Nesti Center for the Clinical Use of Stem Cells, University of Pisa, Via Roma 67, 56126, Pisa, Italy, E-mail: cla_nesti@yahoo.it Tel: 0039-050-993191, Fax: 0039-050-992748

Felicia Ng, MSc Genome & Gene Expression Data Analysis Group, Bioinformatics Institute, Singapore

Agate Noer, Institute of Basic Medical Sciences, Department of Biochemistry, Faculty of Medicine, University of Oslo, PO Box 1112, Blindern, 0317 Oslo, Norway

Joyce E. Ohm Cancer Biology Division, The Sidney Kimmel Comprehensive Cancer Center, The Johns Hopkins University Medical Institutions, Baltimore, MD 21231

Livia Pasquali Department of Neurosciences, Section of Neurology, University of Pisa, Via Roma 67, 56126, Pisa, Italy

Roger Pedersen Department of Surgery and Cambridge Institute for Medical Research (CIMR), Addenbrookes Hospital, University of Cambridge, Hills Road, Cambridge, CB2 1DQ, United Kingdom

Meirav Pevsner-Fischer Department of Molecular Cell Biology, Weizmann Institute of Science, Rehovot, 76100, Israel, Email: meirav.pevsner@weizmann.ac.il
Tel: 972-8-9343550, Fax: 972-8-9344125

Dimitris G. Placantonakis Department of Neurosurgery, Memorial Sloan-Kettering Cancer Center, New York, NY; Department of Neurological Surgery, Weill Cornell Medical College, NewYork – Presbyterian Hospital, New York, NY

Leon M. Ptaszek Harvard Stem Cell Institute, 42 Church Street,Cambridge, MA 02138, USA; Stowers Medical Institute, Cardiovascular Research Center and Center for Regenerative Medicine, Massachusetts General, Hospital, 185 Cambridge Street, CPZN 4265-A, Boston, MA 02114, USA, Email: lptaszek@partners.org, Tel: 617-643-3569; Fax: 617-643-0674

Shahin Rafii, MD Howard Hughes Medical Institute, Ansary Center for Stem Cell Therapeutics, Department of Genetic Medicine, Division of Hematology-Oncology, Weill Cornell Medical College, 1300 York Avenue, Room A-863, New York, NY 10023, USA, Email: srafii@med.cornell.edu.

Vinagolu K. Rajasekhar Program in Developmental Biology, Memorial Sloan-Kettering Cancer Center, New York, NY 10021, USA, Email: vinagolr@mskcc.org

Mahendra Rao Invitrogen Corporation, 5781 Van Allen Way, Carlsbad, CA 92008, USA Email: mahendra.rao@invitrogen.com

Ángel Raya Gene Expression Laboratory, Salk Institute for Biological Studies, 10010 N. Torrey Pines Rd., La Jolla, CA 92037, USA; Center of Regenerative Medicine in Barcelona, Dr. Aiguader 88, 08029 Barcelona, Spain; ICREA, 5CIBER-BBN, Gene Regulation Research, Graduate School of Biological Sciences, Nara Institute of Science and Technology, 8916-5 Takayama, Ikoma 630-0101, Japan

Renee A. Reijo Pera Center for Human Embryo and Embry**onic** Stem Cell Research and Education; Institute for Stem Cell Biology and Regenerative Medicine, Stanford University, Palo Alto, CA 94304-5542, USA, Email: reneer@stanford.edu Tel: 650-725-3803,
Fax: 650-736-2961

Daniel Reynolds University of Cambridge Department of Surgery, Laboratory for Regenerative Medicine, West Forvie Building, Forvie Site, Robinson Way, Cambridge, CB2 0SZ, Email: d.reynolds.00@cantab.net

Irene Riz Department of Anatomy and Regenerative Biology, The George Washington University, School of Medicine and Health Sciences, 2300 I Street NW, Washington, DC 20037, USA. Tel: 202-994-2763/3511, Fax: 202-994-8885

Concepción Rodríguez-Esteban Gene Expression Laboratory, Salk Institute for Biological Studies, 10010 N. Torrey Pines Rd., La Jolla, CA 92037, USA

Lisa R. Rogers DO, Department of Neurology, University of Michigan Medical Center, 1500 E. Medical Center Drive, 1920F Taubman Center, Ann Arbor, MI 48109-53161, USA
Email: lisaroge@umich.edu Tel: (734) 936-7910, Fax: (734) 936-8763

Tobias Schatton, PharmD Transplantation Research Center, Children's Hospital Boston & Brigham and Women's Hospital, Harvard Medical School, 300 Longwood Avenue, Boston, MA 02115, USA

Kjiana E. Schwab B. Biomed. Sci. (Hons), Centre for Women's Health Research, Monash Institute of Medical Research and Monash University, Department of Obstetrics and Gynaecology, Monash Medical Centre, 246 Clayton Road, Clayton, Victoria, 3168, Australia

Marco Seandel, PhD Department of Medicine/Medical Oncology, Box 8, Memorial Sloan Kettering Cancer Center, 1275 York Avenue, New York, NY 10065, USA,
Email: seandelm@mskcc.org

Stewart Sell, MD Wadsworth Center, Ordway Research Institute and University at Albany; Empire State Plaza; Room C-551; Albany, NY 12201, USA,
Email: ssell@wadsworth.org

Gabriele Siciliano Department of Neurosciences, Section of Neurology, University of Pisa, Via Roma 67, 56126, Pisa, Italy

Chew-Li Soh Monash Immunology and Stem Cell Laboratories (MISCL), Level 3, Building 75, Monash University, Wellington Road, Clayton, Victoria 3800, Australia Tel: +613 9905 0628, Fax: +613 9905 0680

Alejandro Soto-Gutierrez Department of Surgery, Okayama University Graduate School of Medicine and Dentistry, 2-5-1 Shikata-cho, Okayama 700-8558, Japan

William L. Stanford Institute of Biomaterials and Biomedical Engineering, Institute of Medical Science, Department of Chemical Engineering and Applied Chemistry, University of Toronto, Toronto, Ontario, Canada; 164 College Street. Toronto, ON, Canada, M5S 3G9, Email: william.stanford@utoronto.ca Tel: 416-946-8379, Fax: 416-978-4317

Lorenz Studer Department of Neurosurgery and Program in Developmental Biology, Memorial Sloan-Kettering Cancer Center, New York, NY, Cancer Center, New York, NY

Toshio Suda Department of Cell Differentiation, Keio University School of Medicine, 35 Shinanomachi, Shinjuku-ku, Tokyo 160-8582 Japan, Email: sudato@sc.itc.keio.ac.jp
Tel & Fax: +81-3-5363-3475

Atsushi Suzuki Gene Expression Laboratory, Salk Institute for Biological Studies, 10010 N. Torrey Pines Rd., La Jolla, CA 92037, USA; Research Unit for Organ Regeneration, Center for Developmental Biology, 2-2-3 Minatojima-minamimachi, Chuo-ku, Kobe, Hyogo 650-0047, Japan

Vivek M. Tanavde, PhD Genome & Gene Expression Data Analysis Group, Bioinformatics Institute, Singapore; 30 Biopolis St., Matrix #07-01, A* STAR Singapore 138671,
Email: vivek@bii.a-star.edu.sg Tel: (65)64788383 Fax: (65)64789047

Bhaskar Thyagarajan Invitrogen Corporation, 5781 Van Allen Way, Carlsbad, CA 92008, USA, Email: bhaskar.thyagarajan@invitrogen

Sanna Timoskainen Institute of Basic Medical Sciences, Department of Biochemistry, Faculty of Medicine, University of Oslo, PO Box 1112 Blindern, 0317 Oslo, Norway

Mark J. Tomishima SKI Stem Cell Research Facility, Program in Developmental Biology, Memorial Sloan-Kettering Cancer Center, New York, NY

Grigorios Tsaknakis Stem Cell Research Laboratory, Nuffield Department of Clinical Laboratory Sciences, University of Oxford, Oxford, UK; Stem Cells and Immunotherapies, NHS Blood and Transplant, Oxford, UK

Rika Tsuchida, MD, PhD Divisions of Hematology/Oncology, The Hospital for Sick Children, Toronto, Canada

Rocky S. Tuan Cartilage Biology and Orthopedics Branch National Institute of Arthritis, and Musculoskeletal & Skin Diseases, National Institutes of Health, Building 50, Room 1523, 50 South Drive, MSC 8022, Bethesda, MD 20892-8022, USA, Email: Tuanr@mail.nih.gov
Tel: 301-451-6854; Fax: 301-435-8017

Ludovic Vallier Department of Surgery and Cambridge Institute for Medical Research (CIMR), Addenbrookes Hospital, University of Cambridge, Hills Road, Cambridge, CB2 0XY, UK

Mohan C. Vemuri Stem Cells and Regenerative Medicine, Invitrogen Corporation, 7335 Executive Way, Frederick, MD 21704, USA, Email: mohan.vemuri@invitrogen.com
Tel: 716-774-6908, Fax: 716-774-6996

Emily Walker Institute of Biomaterials and Biomedical Engineering; University of Toronto, Toronto, Ontario, Canada; 164 College Street. Toronto, ON, Canada, M5S 3G9
Tel: 416-946-8379, Fax: 416-978-4317

Hong Wang Department of Pharmacology, Temple University School of Medicine, 3420 North Broad Street, MRB 325, Philadelphia, PA 19140, USA

WeiJia Wang Institute of Biomaterials and Biomedical Engineering, Terrence Donnelly Centre for Cellular and Biomolecular Research, University of Toronto, 164 College St., Room 407, Toronto, ON, Canada, M5S 3G10

Suzanne M. Watt Stem Cell Research Laboratory, Nuffield Department of Clinical Laboratory Sciences, University of Oxford, Oxford, UK; Stem Cells and Immunotherapies, NHS Blood and Transplant, Oxford, UK

Max Wicha, MD Distinguished Professor of Oncology Director, University of Michigan Comprehensive Cancer Center, 1500 E. Medical Center Drive, Ann Arbor, MI 48109, USA

Ting Xie Stowers Institute for Medical Research, 1000 East 50th Street, Kansas City, MO 64111, USA

Tritia R. Yamasaki Department of Neurobiology and Behavior and Institute for Brain Aging and Dementia; University of California, Irvine. Irvine, CA 92697-4545, USA

Yukiko M. Yamashita Center for Stem Cell Biology, Life Sciences Institute, and Department of Cell and Developmental Biology, University of Michigan, Ann Arbor, 210 Washtenaw Avenue Rm 5403, Ann Arbor, MI 48104, USA, Email: yukikomy@umich.edu Tel 734-615-8508, Fax 734-615-5520

Xiao-Feng Yang, MD, PhD Department of Pharmacology, Temple University School of Medicine, 3420 North Broad Street, MRB 325, Philadelphia, PA 19140, USA, Email: xfyang@temple.edu

Herman Yeger, PhD Stem Cell and Developmental Biology Program, Research Institute, and Department of Pediatric Laboratory Medicine and Pathobiology, Hospital for Sick Children. Faculty of Medicine, Department of Laboratory Medicine and Pathobiology, University of Toronto, Toronto, ON MSG 1X8, Canada, Email: herman.yeger@sickkids.ca

Takashi Yokoo, MD, PhD Project Team for Kidney Regeneration, Institute of DNA Medicine, Division of Nephrology and Hypertension, Department of Internal Medicine, The Jikei University School of Medicine, 3-25-8, Nishi-Shimbashi, Minato-ku, Tokyo, 105-8461 Japan, Email: tyokoo@jikei.ac.jp Tel: +813-3433-1111, Fax: +813-3433-4297

Nian Zhang Stowers Institute for Medical Research, 1000 East 50th Street, Kansas City, MO 64110, USA

Dov Zipori, PhD Department of Molecular Cell Biology, Weizmann Institute of Science, Rehovot, 76100, Israel, Email: dov.zipori@weizmann.ac.il Tel: 972-8-9342484, Fax: 972-8-9344125

Leonard I. Zon, Division of Hematology/Oncology, Harvard Medical School and Howard Hughes Medical Institute, Children's Hospital Boston, One Blackfan Circle, Boston, MA 02115, USA, Email: zon@enders.tch.harvard.edu Tel: 617-919-2069, Fax: 617-730-0223

Part I
Molecular Regulation in Stem Cells

The Molecular Basis of Embryonic Stem Cell Self-Renewal

Stephen Dalton

Abstract Peri-implantation stage embryos are a source of pluripotent cells that can be cultured indefinitely in vitro as a stable, self-renewing population. By definition, these cell populations have the capacity to differentiate into all cell types of the adult. Consequently, these cells are of special interest to developmental biologists and have significant potential in the area of cell replacement therapy. In this chapter we discuss the hallmarks of pluripotent cells derived from murine and human embryos and compare signaling pathways and transcription factor networks required for the self-renewing, pluripotent state. Maintenance of pluripotent cells derived from murine and human embryos requires different culture conditions for their in vitro maintenance, indicative of distinct differences at the molecular level and developmental nonequivalence. This chapter will evaluate the literature in terms of what is critical, from a signal transduction perspective, for maintenance of pluripotency and will highlight common themes that exist between embryonically derived stem cell populations. Recent findings describing epiblast stem cells (EpiScs), a self-renewing pluripotent cell type derived from post-implantation stage embryos, will be discussed with respect to embryonic stem cells and primitive ectoderm. EpiScs seem to be more closely related to hESCs than mESCs, posing some interesting questions as to the developmental equivalence of hESCs and mESCs. Finally, the new revolution of reprogramming from a differentiated state to an induced pluripotent stem (iPS) cell state will be discussed in relation to what we know about self-renewal regulatory networks and how this technology promises to revolutionize stem cell–based regenerative therapy.

Keywords Embryonic stem cell · Pluripotency · Self-renewal

S. Dalton (✉)
Department of Biochemistry and Molecular Biology,
Paul D. Coverdell Center for Biomedical and Health Sciences,
500 DW Brooks Drive, Athens, GA 30606, USA
e-mail: sdalton@uga.edu

1 Pluripotent Cells from Peri-implantation Stage Mammalian Embryos

Pre-implantation and early post-implantation stage embryos contain a population of pluripotent cells that have the capacity to differentiate into the three embryonic germ layers and, thus, contribute to all of the adult cell lineages. Pluripotency refers to the developmental potency of a cell, however, although pluripotent cells exist throughout pre-gastrula development, they are continually changing as part of a developmental continuum. Cells of the inner cell mass (ICM) from which murine ESCs are derived represent one particular stage of pluripotent cell development. While pluripotent cells from various stages of pregastrulation development have definable developmental potency, they are not necessarily equivalent and can not be compared directly. Primitive ectoderm, for example, is an Oct4+ pluripotent population that is distinctly different from its predecessors from the ICM. In short, pluripotent cells exist in a variety of different states during early mammalian development [1]. This is true for all mammalian embryos even though the configuration of the early embryo varies considerable in different species [2]. The significance of this will become apparent throughout this chapter, when different in vitro pluripotent populations are discussed, including human embryonic stem cells (hESCs), murine embryonic stem cells (mESCs), epiblast-like stem cells (EpiSCs), early primitive ectoderm-like (EPL) cells, and induced-pluripotent stem (IPS) cells.

1.1 Developmental Origins of Murine Pluripotent Cells

The mammalian embryo is comprised of three predominant cell populations during the early stage of development. First, the trophectoderm (TE), which gives rise to extraembryonic tissues, such as the placenta. These extraembryonic tissues are critical for supporting embryonic development

V.K. Rajasekhar, M.C. Vemuri (eds.), *Regulatory Networks in Stem Cells*, Stem Cell Biology and Regenerative Medicine,
DOI 10.1007/978-1-60327-227-8_1, © Humana Press, a part of Springer Science+Business Media, LLC 2009

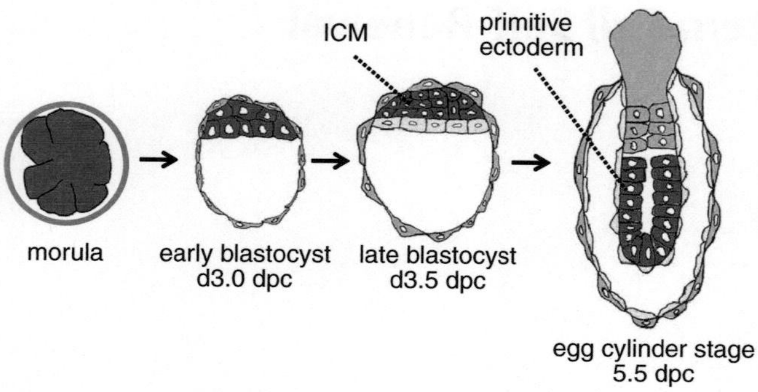

Fig. 1 The peri-implantation mouse embryo illustrating stages of pluripotent cell development including the inner cell mass (ICM) from blastocyst stage embryos and primitive ectoderm from post-implantation stage (egg cylinder) embryos

by facilitating exchange of nutrients and oxygen with the mother [3, 4]. The second, is a population of approximately twenty cells known as the inner cell mass (ICM). Cells of the ICM are pluripotent and amplify rapidly during the epiblast stage of development. Subsequently, they differentiate into the three embryonic germ layers (ectoderm, mesoderm, and definitive endoderm) and, therefore, are the founders of all adult tissues. The ICM is surrounded by a third cell type known as primitive endoderm (PrEn). Formation of PrEn around the ICM denotes the epiblast stage of development, which extends until gastrulation [4]. As the epiblast develops past the ICM stage to the egg cylinder stage a cavity forms in the core of the ball of the pluripotent cells, resulting in the formation of a single layer of pseudostratified epithelia, known as primitive ectoderm (PrEct) (Fig. 1). This cavity represents a classic example of how cavities are sculpted within tissues through a developmentally regulated apoptotic mechanism [5]. Primitive ectoderm is pluripotent but represents a different stage of the pluripotent continuum that exists during pregastrulation development. As pluripotent cells commit toward one of the three germ layers, they lose pluripotency and, just prior to or coinciding with this, the germ cells are segregated away into the allantois, where they will preserve the germline by migrating to the genital ridge. How a subpopulation of pluripotent cells is selected to be segregated away into the germ cell compartment is an area of intense, ongoing investigation.

1.2 Embryonic Stem Cells from the Mouse

In 1981 two papers published independently by Gail Martin and Sir Martin Evans described the first isolations of mESCs from blastocyst stage embryos [6, 7]. mESCs exhibit several remarkable features in culture. Under the appropriate conditions, they can be cultured over extended periods of time as a genetically stable, self-renewing population where at every cell division both mother and daughter cells retain stem cell identity following a symmetric cell division [8]. This immortalized phenotype allows the stem cell state to be maintained over extended periods of time. Upon differentiation, this feature is lost and progeny succumb to cellular aging mechanisms (Hayflick limit), as has been well documented for all other nontransformed primary cells. This self-renewing phenomenon seems to be developmentally regulated, but it is not clear whether it is inherently tied into the pluripotent state (this will be addressed in more detail later). A second feature of more developmental relevance is that, during extended culture, mESCs retain their pluripotency and can differentiate into the same range of cell types as those formed in the embryo from the ICM (Fig. 2). For the purposes of this discussion, we describe pluripotency as being the ability to generate all adult cell types and totipotency as the ability to form all adult, germline, and extra-embryonic tissues. The latter definition is usually reserved for fertilized eggs since mESCs cannot contribute to TE or PrEn when injected into blastocysts [9] but do differentiate into other lineages [10, 11]. A stringent test of the developmental potential for mESCs is their ability to contribute to the germline and all tissues of an adult animal following injection into recipient blastocysts. This technology enables genetically engineered mESCs to be introduced into the blastocyst for subsequent germline transmission and production

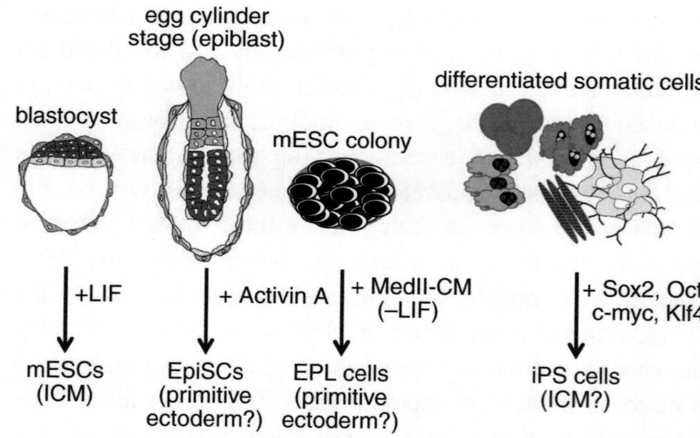

Fig. 2 Different in vitro populations of pluripotent cells and their developmental origins. mESCs are derived directly from the ICM of blastocyst stage embryos and can be maintained on mouse embryo fibroblast feeder layers or under feeder-free conditions in the presence of LIF. EpiSCs are isolated from post-implantation mouse embryos and require factors such as Fgf2 and Activin A to be maintained in a self-renewing, pluripotent state. EPL cells are isolated directly from mESCs in the absence of LIF, but in the presence of conditioned media from HepG2 cells (MedII-CM), and represent an in vitro population of primitive ectoderm. The relationship between EpiSCs and EPL cells has not been thoroughly investigated. iPS cells are generated from differentiated somatic cells by retroviral-mediated expression of Sox2, Oct4, c-myc, and Klf4. Since these cells can contribute to the germline they would appear to more closely resemble cells of the ICM than primitive ectoderm

of genetically modified animals. Sir Martin Evans, Oliver Smithies, and Mario Capecchi shared the Nobel Prize for Medicine in 2007 for their contributions in this area.

A common alternate assay for ESC potency, particularly where embryo transfer is not practical, is to inject ESCs into immunocompromised mice where they form mixed cell tumors known as teratomas. The ability of injected ESCs to generate a tumor comprising mesoderm, ectoderm, and endoderm lineages is indicative of their multipotency. As ESCs differentiate, however, they lose their tumorigenic potential. The basis behind this is not clearly understood but it is noteworthy that some tumor cells also express markers previously thought to be specifically associated with the pluripotent state. It should be pointed out that pluripotent cells of embryonic origin express high levels of potentially oncogenic factors such as c-myc [12] and lack many tumor suppressors normally involved in restraining cell growth [13]. It is ironic that we begin life as a ball of latent tumor cells that lose tumorigenic potential as a function of development.

While functional analysis of ESC developmental potential is the gold standard for ESC analysis, molecular markers are often used as readouts for the stem cell state because of practical issues. Many of these markers are transcription factors expressed in the ICM and by ESCs that have functional roles in their development/maintenance. The best-characterized examples include the POU domain transcription factor Oct3/4 [14], the homeodomain transcription factor, Nanog [15, 16], and the HMG protein Sox2 [17]. Transcription factor networks involved in ESC pluripotency have been reviewed elsewhere [18, 19]. Other signatures include high TERT expression and presentation of characteristic cell surface antigens such as the glycomarker SSEA1 [20], the tetraspanin CD9 [21], and the carbohydrate epitope N-acetylgalactosamine [22].

Most of the transcription factor regulatory networks responsible for maintenance of pluripotency appear to have been conserved between human and mouse ESCs and will be considered in greater detail later in this chapter. Striking differences do emerge, however, when cell surface markers of ESCs are compared. hESCs do not exhibit high SSEA1 reactivity but instead are identified based on elevated SSEA3,4 and TRA-1-60, TRA-1-81 antigens [23–25]. In contrast to mouse ESCs, N-acetylgalactosamine epitopes recognized by the lectin DBA are not a feature of hESCs (SD, unpublished observation). The central question again arises: are human and mouse ESCs representative of different stages of development, or are mouse and human pluripotent cells different in many respects? While human and murine ESCs pack tightly together and individual cells exhibit a high nuclear to cytoplasmic volume ratio, gross morphological differences exist between the structures of ESC colonies. For example, mESCs grow as three-dimensional, dome-shaped colonies whereas hESCs grow in colonies as thin layers, often monolayers. Self-renewal signals required for hESCs and mESCs maintenance are quite different and will be considered later in this review.

1.3 Primitive Ectoderm and Epiblast Stem Cells

Isolation of pluripotent cells from mammalian embryos has clearly focused on the pre-implantation (ICM) stage of development where clear success has been obtained, resulting in the successful isolation of ESC-like cells from several species. Successful isolation of self-renewing populations from post-implantation stages was not successful until recently, when two groups reported the isolation of stem cells from the murine late-stage epiblast [26, 27]. Establishment of epiblast stem cells (EpiSCs) does not require the same cocktail of media components as for mESCs. Although EpiSCs can be maintained on MEFs (as can mESCs and hESCs), a major difference between mESCs and EpiSCs is that under feeder-free conditions, mESCs have a requirement for interleukin (IL)6 family member cytokines such as leukemia inhibitory factor (LIF) to maintain self-renewal and pluripotency [28]. This is not the case for EpiScs, where LIF is not required. Moreover, the colony morphology of EpiSCs is more reminiscent of hESCs than mESCs since they grow as flat, epithelial colonies. In common with hESCs, EpiSCs have a requirement for Activin/Nodal signaling to promote self-renewal and they respond to BMP4 by differentiating into trophectoderm [26, 27]. Although EpiSCs can differentiate into the three embryonic germ layers, their ability to contribute to the germline has not been tested and so the differentiation capacity of these cells has yet to be fully characterized. These are very exciting findings and indicate that the ICM is not the only embryonic stage from which self-renewing stem cells can be isolated. It should be pointed out that successful attempts have been made to isolate mESCs and hESCs from different stages of preimplantation stages of development [29, 30] but not from post-implantation stages. The parallels between EpiSCs and hESC did not escape the authors' attention [26, 27] and the possibility that they represent a developmentally equivalent cell types was raised.

Another interesting pluripotent ectoderm-like population has also been described, but this time they were created directly from mESCs using HepG2 conditioned media [31]. This cell type, known as early primitive ectoderm-like (EPL) cells, expresses markers characteristic of primitive ectoderm from post-implantation stage embryos as denoted by the expression of markers such as Fgf5 but not ICM-specific markers such as Rex1 and Gbx2 [31]. Remarkably, EPL cells can be reverted back to an ESC state by replacement of conditioned media with LIF. These studies illustrate the concept that multiple pluripotent cell types exist in early embryonic development, but more surprisingly, indicate that there is developmental plasticity in that these closely related cell types can interconvert. The relationship between EpiSCs and EPL cells has not yet been evaluated and it will be interesting to establish whether they represent equivalent or alternate stages of post-implantation development.

2 Murine ESC Self-Renewal Signaling Pathways

As mentioned already, the signaling requirements for maintenance of human and murine ESCs differ considerably. Why should this be? We will discuss the literature and try to reconcile these differences by considering known signaling pathways implicated in ESC self-renewal. The best-characterized effector of mESC self-renewal is leukemia inhibitory factor (LIF). LIF is a member of the IL6 family of cytokines that plays a key role in maintaining mESC self-renewal and functions by engaging the LIF/gp130 heterodimeric receptor, thereby recruiting and activating STAT3, a transcription factor that translocates to the nucleus and regulates genes required for "stemness" [32–34]. While LIF can activate JAK-STAT3 and Ras-MAPK pathways in mESCs, studies in mice indicate that genetic inactivation of LIF signaling has no major effect on development [35]. This may be due to compensation by other IL6 family members, such as CNTF, which can also signal through LIF/gp130 receptors [36]. LIF/STAT3 does seem to be important for blastocyst maintenance during delayed implantation [25], although this does not appear to be relevant to human development [36].

Several efforts have been made to understand the mechanism of STAT3-dependent self-renewal in mESCs [32–34]. One of the most promising targets identified is the proto-oncogene c-myc, a helix-loop helix transcription factor that is a direct transcriptional target of STAT3 [12]. Following LIF withdrawal, c-myc transcript levels decrease due to inactivation of STAT3. Maintenance of myc levels using inducible transgenes can maintain self-renewal in the absence of LIF indicating that myc is a major target of the LIF-STAT3 self-renewal pathway in mESCs [12]. A second pathway that controls myc levels involves the serine/threonine protein kinase, glycogen synthase kinase 3 beta (GSK3β). When LIF signaling ceases, GSK3β is rapidly activated and phosphorylates c-myc on threonine 58 (T58), triggering its ubiquitination and proteosome dependent degradation. How GSK3β activity is suppressed in mESCs is unclear but is likely to involve PI3K activity either directly or indirectly as a consequence of LIF signaling. Another intriguing connection between GSK3β and self-renewal was made when the efficiency of mESC derivation was shown to be markedly enhanced in the presence of BIO, a chemical inhibitor of GSK3β [37]. Hence, low GSK3β activity could be an absolute requirement for pluripotency and mESC self-renewal.

A second pathway implicated in mESC self-renewal involves BMP signaling. Although BMP is generally not added as a recombinant factor, as in the case of LIF, BMP in fetal calf serum appears to have a pro-maintenance effect at least under some culture conditions [38, 39]. Under these conditions, BMP acts by promoting Id gene expression, which serves to block neural differentiation. The report by Ying and colleagues [38] was the first to seriously raise the issue that self-renewal must be a coordinated series of events that involves maintenance of the pluripotent state and the blockade of differentiation pathways. In the case of BMP signaling, ectoderm specification is inhibited. By this model, other factors would work in collaboration with BMP to restrict differentiation pathways for mesoderm and endoderm. Qi et al. [39] have alternative explanations for how BMP impacts on mESC self-renewal. In their experiments they show that BMP blocks differentiation by suppressing p38 MAP kinase. Since different laboratories use different culture conditions, including fetal calf serum, which is a huge variable, it seems likely that BMP contributes to suppression of differentiation by context dependent mechanisms. The main outcome, however, is to suppress pro-differentiation signaling pathways.

Phosphatidylinositol 3 kinase (PI3K) is involved in many aspects of cell behavior such as proliferation, apoptosis, and differentiation [40]. A major effector of PI3K signaling is protein kinase B (PKB)/AKT1. There is a large body of evidence demonstrating that PI3K signaling is crucial for mESC self-renewal [41, 42]. Inhibition of PI3K signaling by small molecule inhibitors such as LY294002 promotes differentiation even in the presence of LIF [43]. As mESCs differentiate AKT activity declines, consistent with PI3K signaling being important for self-renewal. Sustained AKT activity, achieved by ectopic expression of a constitutively active mutant, significantly delays differentiation of murine and monkey ESCs [44]. Although PI3K/AKT seems to be crucial for mESC self-renewal, factors promoting their activity have not been clearly defined. Candidates include serum components such as IGF or even LIF, a known activator of PI3K signaling through LIF-gp130 receptor complexes. Mechanistically, PI3K/AKT may function by suppressing GSK3β, a known antagonist of pro-self-renewal regulators such as c-myc [12, 45, 46].

Increased ERK activity is thought to be correlated with early differentiation of mESCs following LIF withdrawal, and suppression of its activity by addition of PD98059 reduces the level of LIF required to maintain mESC self-renewal [47]. LIF itself promotes ERK activity but this is balanced by self-renewal signals generated at the LIF-gp130 receptor. Recent work indicates that Fgf4 secreted by mESCs primes cells for differentiation by acting through ERK and that suppression of this signal compromises differentiation [48].

In summary, LIF-STAT3 is critical for mESCs self-renewal. In conjunction with additional signals in serum, self-renewal is promoted. Besides LIF, PI3K/AKT appears to be most critical and may be activated as part of the LIF signaling pathway or from other factors in media (insulin, IGF, for example). The absence of defined media formulations has compounded the definition of self-renewing signaling pathways in mESCs. Part of the problem relates back to the different culture conditions used by laboratories in the field and the nemesis of many tissue culture systems, variability in batches of fetal calf serum.

3 hESC Self-Renewal Signaling Pathways

It did not take long for the field to realize that culture conditions required for mESC self-renewal are quite different from that required for hESC maintenance. Although human and murine ESCs can be maintained on MEF feeder layers in fetal calf serum (which has to be carefully batch tested), differences clearly emerge under feeder-free conditions. LIF is clearly not required for hESC self-renewal [36] but several other factors have been identified instead, such as Fgf2, Activin A, and activators of PI3K signaling such as IGF/insulin.

From the onset of discussions relating to hESC culture, it should be clearly stated that first generation feeder-free conditions utilized MEF-CM, a complex mixture of secreted factors and fetal calf serum/synthetic serum replacement formulations. The complexity of MEF-CM again raises question in terms of defining the critical factors. For example, Xu et al. [49] identified BMP activity in serum replacement media associated with serum albumin. Clearly, there is a great need to progress toward defined media that can sustain hESC self-renewal independently of MEFs, serum, and other undefined media components. From the early experiments using MEF-CM it was clear that supplementation with Fgf2 had profound effects on hESC stability [50, 51]. Since then Fgf2 (basic Fgf; bFGF) has been consistently used in both MEF-CM–based and defined media formulations for hESCs. Fgf2 may promote hESC self-renewal in two ways. First, by directly activating signaling pathways required for self-renewal, perhaps through transcriptional networks (to be discussed below). Second, it could work indirectly by stimulating autocrine effects. Since Fgf2 is added at the time of MEF media conditioning, it may serve to promote secretion of factors from MEFs [52]. Since Fgf2 is a key component of defined media, where MEFs and MEF-CM are absent, the first possibility certainly seems likely.

Members of the TGFβ family such as Activin A also seem to play a role in maintaining hESC self-renewal, perhaps in collaboration with Fgf2 [53, 54]. This may involve a mechanism where Activin A signals directly through Smads to promote transcription of genes encoding transcription factors required for self-renewal such as Nanog and Oct3/4 [55]. In contrast, Activin A does not appear to be involved in mESC self-renewal [56] and would probably interfere with BMP-dependent self-renewal pathways in this system because of reciprocal antagonism [57]. In contrast to the situation in mESCs, BMP promotes differentiation into hESCs [58] and antagonism of BMP signaling by GDF3 can promote hESC self-renewal [59].

Evidence is emerging that PI3K signaling is crucial for hESCs self-renewal. We previously showed that for specification signals such as Activin A to promote hESCs differentiation, PI3K signaling must first be inactivated [60]. Although the mechanism for this has not been resolved, it appears that PI3K antagonizes signaling pathways required for cell fate commitment as well as by promoting self-renewal regulatory circuits. In the case of Activin A-dependent definitive endoderm specification, PI3K can be suppressed by chemical inhibitors such as LY 29402 or by removing/reducing FCS or serum supplements [60, 61]. Removal of insulin/IGF type molecules seems to be important for reducing PI3K signaling in this context. Addition of PI3K agonists such as IGF and insulin to defined media formulations would therefore play two roles: i) promoting self-renewal by suppressing differentiation and ii) promoting cell survival. Another anticipated outcome of PI3K signaling would be to suppress GSK3 activity. This is consistent with reports from Sato et al. [46], who showed that suppression of GSK3 is central to hESCs self-renewal in short-term assays. In contrast to what has been described for mESCs, ERK activity is inhibited during hESCs differentiation when cells are cultured in MEF-CM [62]. This raises questions about the generality of ERK in self-renewal/differentiation and it is unclear if these differences represent differences between species, culture conditions, or whether this can be attributed to hESCs and mESCs representing different phases of development.

Now that defined media formulations are being widely used to propagate hESCs, the key growth factors and signaling pathways are now being revealed. Activin A, Fgf2, and insulin/IGF seem to be the consensus players revealing key roles for Smad, PI3K signaling, and possibly ERK signaling in hESC self-renewal. This clearly portrays a different picture from that which has emerged from studies in mESCs [63, 64].

4 Transcriptional Networks that Control Pluripotency

Several transcription factors have been identified that, in an embryonic context, are important for development of pluripotent cells in the murine epiblast. These include Sox2,

UTF1, FoxD3, Oct4, and Nanog [65]. Oct4 is a POU domain transcription factor that is restricted to pluripotent cells including the germline. Although constitutive expression of Oct4 cannot prevent differentiation, its overexpression promotes differentiation in a manner similar to that seen following LIF withdrawal [66]. In contrast, reducing the level of Oct4 expression promotes formation of trophectoderm from murine and human ESCs [66, 67], which, together with the previous observation, indicates that Oct4 levels are critical in determining cell fate. More recently, the homeodomain transcription factor Nanog, which is also expressed in pluripotent cells of the embryonic epiblast, has been shown to perform a key role in ES cell identity and cell lineage specification [15, 16]. This comes from evidence showing that overexpression of Nanog in ES cells can support self-renewal in the absence of LIF signaling and by genetic approaches showing that inactivation of Nanog triggers differentiation into parietal/visceral endoderm [16]. This observation is recapitulated in the embryo where genetic ablation of Nanog causes a failure in the specification of pluripotent cells from the embryonic epiblast, leading to a visceral/parietal endoderm fate [15]. Since Nanog expression is independent of Oct4 status and because Nanog fails to block differentiation following loss of Oct4, it would appear that Oct4 and Nanog work in concert to promote self-renewal and pluripotency [65]. A major deficiency in our knowledge is that we have no understanding of how these transcription factors connect with the self-renewal signaling pathways. Global chromatin immunoprecipitation assays (ChIP on chip) indicates additional intimate relationships between these factors since they seem to co-occupy many pluripotent and differentiation specific genes. One model to emerge from this type of study predicts that Oct4, Sox2, and Nanog not only promote transcription associated with the pluripotent state but also suppress the activity of genes associated with differentiation. A second series of studies with a biochemical emphasis has shown that Oct4, Nanog, and Sox2 can associate together in complexes. This is consistent with observations that these factors co-occupy target promoters. The power of mass spectrometry is now beginning to dissect the types of functional multiprotein complexes found in ESCs and has identified additional factors, such as Zfp281, not previously implicated in pluripotent cell biology [68].

Myc transcription factors have also emerged as being critical for ESC self-renewal [12]. The Myc family was originally identified through their homology to viral transforming genes and later, through their role in a wide variety of cancers [69]. The best characterized members of the Myc family, c-, L-, and N-myc have roles in a variety of cellular functions including cell proliferation, cell transformation, growth, differentiation, and apoptosis [70]. In mESCs, sustained c-myc activity can relieve the requirement for leukemia inhibitory factor and the myc gene itself is a direct target for STAT3. A requirement for sustained myc activity is the suppression of GSK3β activity. Activation of GSK3β seems to be associated with myc degradation and commitment to differentiation. This is consistent with other reports showing that inhibitors of GSK3β can promote self-renewal of ESCs [46].

5 Induced Pluripotent Stem (iPS) Cell

Takahashi et al. [71] recently showed that co-expression of four transcription factors by retroviral transduction could dedifferentiate MEFs back to an ESC-like state [71]. The factors required were Sox2, Oct3/4, the Kruppel-like factor 4 (Klf4), and c-myc. Nanog was not required for dedifferentiation but its expression was reestablished by the four named factors. These observations have been repeated by several laboratories in human cells [72, 73] to a level of robustness that makes this approach an attractive alternative to the use of hESCs for research and therapeutic applications.

How could these factors participate in reestablishment of an ESC-like state? c-myc, for example, is known to be a global gene regulator and plays an active role in gene activation and repression [74]. In particular, it is known to promote global acetylation of chromatin [75]. Establishing an open chromatin state by enhanced histone acetylation could then facilitate complexes of Oct3/4, Sox2, and Klf4 to bind target genes required for establishment and maintenance of pluripotency. Similar experiments could not reproduce this effect in human cells however. In summary, genetic and biochemical analysis has established a role for Oct3/4, Nanog, Sox2, c-myc, and Klf4 in ESC cell identity. However, how the transcription factors cooperate on reprogramming event needs to be addressed.

6 Comments and Perspectives

Understanding mechanisms that promote self-renewal of pluripotent cells is crucial if we are to harness their full potential in a therapeutic context. Throughout this chapter, the identity and characteristics of different pluripotent cell populations has been discussed. Although they each represent useful models for embryonic development, it is unclear which will eventually have most utility for cell therapeutic applications. Understanding differences between these pluripotent populations at the molecular level will be critical in establishing how to control their behavior in vitro and to apply them for practical purposes.

For over a decade, investigators have been committed to solving the question of how ESCs retain their self-renewing capacity, but several major questions still need to be addressed. It is still not understood how cell-signaling pathways involved in this process integrate with the core self-renewal machinery. The solution to this issue has been confounded by variations in culture conditions and the use of nondefined media formulations. How does LIF signaling impact on Oct4 expression in mESCs and how does Fgf2 impact on Nanog in hESCs? In the mouse, LIF-STAT3 signaling targets c-myc, but beyond this there are few validated targets of this pathway that could give clues as to how self-renewal is maintained. In human ESCs, Activin A seems to regulate Nanog transcription but a full picture of self-renewal at the molecular level has yet to be obtained. Hints as to how the transcription network in ESCs functions on a global scale, however, are now beginning to emerge. Genome-wide studies point toward interactions between Sox2, Nanog, and Oct3/4 on target promoters, involving context dependent activation or repression. The full details of this still need to be rigorously evaluated and much of the global analysis needs to be functionally validated. This and the area of epigenetic regulation in ESCs are fertile ground that needs to be explored further.

Perhaps the major factor hampering the field relates back to variability in culture conditions. The complexity of signaling components reflects the true nature of self-renewal in culture – a "balancing act" where different signaling pathways need to be juggled. There appear to be several ways of balancing the self-renewal equation but, ultimately, culture conditions must suppress differentiation and promote continued cell division. In the case of hESCs, many of these problems will be circumvented by the use of defined media where purified, recombinant growth factors are being used.

One of the most provocative observations made over the last year is that hESCs may reflect a later stage of embryonic development than the ICM. Similarities between EpiSCs and hESCs confirm suspicions that many in the field have held for a long time – that while pluripotent, hESCs are not the developmental equivalent of mESCs. This may explain many of the differences between mouse and human ESCs that inexplicably have accumulated in the literature over the last 5 years. Perhaps this will lead to new efforts directed toward derivation of new hESC cells lines representing alternate stages of pre-implantation development.

Finally, the ability of several laboratories to generate iPS cells flags a potential turning point for the field where an alternative source of human pluripotent cells is available for research and therapeutic applications. Although it is early days, all of the signs indicate that what was once though to be impossible is now a reality. What then the future of hESCs as a therapeutic platform? At this stage they still represent the best option for development of cell therapeutics.

Acknowledgments We thank members of the Dalton laboratory for useful discussions. This work was supported by grants to SD from NIH-NICHD, the Georgia Cancer Coalition, and the Georgia Research Alliance.

References

1. Hogan B, Beddington R, Constantini F, et al. Manipulating the mouse embryo. 2nd ed. Cold Spring Harbor, NY: Cold Spring Harbor Laboratory Press; 1994.
2. Stern CD. Gastrulation: from cells to embryo. Cold Spring Harbor, NY: Cold Spring Harbor Laboratory Press; 2004.
3. Rossant J. Stem cells and lineage development in the mammalian blastocyst. Reprod Fertil Dev. 2007;19:111–8.
4. Nagy A, Gertsenstein M, Vintersten K, et al. Manipulating the mouse embryo. 3rd ed. Cold Spring Harbor Laboratory Press; 2003. pp. 131–40.
5. Coucavanis E, Martin GR. Signals for death and survival: a two-step mechanism for cavitation in the vertebrate embryo. Cell. 1995;83:279–87.
6. Martin GR. Isolation of a pluripotent cell line from early mouse embryos cultured in medium conditioned by teratocarcinoma stem cells. Proc Natl Acad Sci U S A. 1981;78:7634–8.
7. Evans MJ, Kaufman MH. Establishment in culture of pluripotential cells from mouse embryos. Nature. 1981;292:154–6.
8. Zwaka TP, Thomson JA. Differentiation of human embryonic stem cells occurs through symmetric cell division. Stem Cells. 2005;23:146–9.
9. Beddington RS, Robertson EJ. An assessment of the developmental potential of embryonic stem cells in the midgestation mouse embryo. Development. 1989;105:733–7.
10. Shimosato D, Shiki M, Niwa H. Extra-embryonic endoderm cells derived from ES cells induced by GATA factors acquire the character of XEN cells. BMC Dev Biol. 2007;7:80.
11. Fujikura J, Yamato E, Yonemura S, et al. Differentiation of embryonic stem cells is induced by GATA factors. Genes Dev. 2002;16:784–9.
12. Cartwright P, McLean C, Sheppard A, et al. LIF/STAT3 controls ES cell self-renewal and pluripotency by a Myc-dependent mechanism. Development. 2005;132:885–96.
13. White J, Dalton S. Cell cycle control of embryonic stem cells. Stem Cell Rev. 2005;1:131–8
14. Nichols J, Zevnik B, Anastassiadis K, et al. Formation of pluripotent stem cells in the mammalian embryo depends on the POU transcription factor Oct4. Cell. 1998;95:379–91.
15. Mitsui K, Tokuzawa Y, Itoh H, et al. The homeoprotein Nanog is required for maintenance of pluripotency in mouse epiblast and ES cells. Cell. 2003;113:631–42.
16. Chambers I, Colby D, Robertson M, et al. Functional expression cloning of Nanog, a pluripotency sustaining factor in embryonic stem cells. Cell. 2003;113:643–55.
17. Avilion AA, Nicolis SK, Pevny LH, et al. Multipotent cell lineages in early mouse development depend on SOX2 function. Genes Dev. 2003;17:126–40.
18. Niwa H. How is pluripotency determined and maintained? Development 2007;134:635–46.
19. Boiani M, Scholer HR. Regulatory networks in embryo-derived pluripotent stem cells. Nat Rev Mol Cell Biol. 2005;6:872–84.
20. Solter D, Knowles BB. Monoclonal antibody defining a stage-specific mouse embryonic antigen (SSEA-1). Proc Natl Acad Sci U S A. 1978;75:5565–9.
21. Oka M, Tagoku K, Russell TL, et al. CD9 is associated with leukemia inhibitory factor-mediated maintenance of embryonic stem cells. Mol Biol Cell. 2002;13:1274–81.
22. Nash R, Neves L, Faast R, et al. The lectin Dolichos biflorus agglutinin recognizes glycan epitopes on the surface of murine embryonic stem cells: a new tool for characterizing pluripotent cells and early differentiation. Stem Cells. 2007;25:974–82.

23. Brimble SN, Sherrer ES, Uhl EW, et al. The cell surface glycosphingolipids SSEA-3 and SSEA-4 are not essential for human ESC pluripotency. Stem Cells. 2007;25:54–62.
24. Thomson JA, Itskovitz-Eldor J, Shapiro SS, et al. Embryonic stem cell lines derived from human blastocysts. Science. 1998;282:1145–7.
25. Reubinoff BE, Pera MF, Fong CY, et al. Embryonic stem cell lines from human blastocysts: somatic differentiation in vitro. Nat Biotechnol. 2000;18:399–404.
26. Brons IG, Smithers LE, Trotter MW, et al. Derivation of pluripotent epiblast stem cells from mammalian embryos. Nature. 2007;448:191–5.
27. Tesar PJ, Chenoweth JG, Brook FA, et al. New cell lines from mouse epiblast share defining features with human embryonic stem cells. Nature. 2007;448:196–9.
28. Smith AG, Heath JK, Donaldson DD, et al. Inhibition of pluripotential embryonic stem cell differentiation by purified polypeptides. Nature. 1988;336:688–90.
29. Wakayama S, Hikichi T, Suetsugu R, et al. Efficient establishment of mouse embryonic stem cell lines from single blastomeres and polar bodies. Stem Cells. 2007;25:986–93.
30. Klimanskaya I, Chung Y, Becker S, et al. Derivation of human embryonic stem cells from single blastomeres. Nat Protoc. 2007;2:1963–72.
31. Stead E, White J, Faast R, et al. Pluripotent cell division cycles are driven by ectopic Cdk2, cyclin A/E and E2F activities. Oncogene. 2002;21:8320–33.
32. Matsuda T, Nakamura T, Nakao K, et al. STAT3 activation is sufficient to maintain an undifferentiated state of mouse embryonic stem cells. EMBO J. 1999;18:4261–9.
33. Niwa H, Burdon T, Chambers I, et al. Self-renewal of pluripotent embryonic stem cells is mediated via activation of STAT3. Genes Dev. 1998;12:2048–60.
34. Raz R, Lee CK, Cannizzaro LA, et al. Essential role of STAT3 for embryonic stem cell pluripotency. Proc Natl Acad Sci U S A. 1999;96:2846–51.
35. Nichols J, Chambers I, Taga T, et al. Physiological rationale for responsiveness of mouse embryonic stem cells to gp130 cytokines. Development. 2001;128:2333–9.
36. Humphrey RK, Beattie GM, Lopez AD, et al. Maintenance of pluripotency in human embryonic stem cells is STAT3 independent. Stem Cells. 2004;22:522–30.
37. Umehara H, Kimura T, Ohtsuka S, et al. Efficient derivation of embryonic stem cells by inhibition of glycogen synthase kinase-3. Stem Cells. 2007;25:2705–11.
38. Ying QL, Nichols J, Chambers I, et al. BMP induction of Id proteins suppresses differentiation and sustains embryonic stem cell self-renewal in collaboration with STAT3. Cell. 2003;115: 281–92.
39. Qi X, Li TG, Hao J, et al. BMP4 supports self-renewal of embryonic stem cells by inhibiting mitogen-activated protein kinase pathways. Proc Natl Acad Sci U S A. 2004;101:6027–32.
40. Takahashi K, Murakami M, Yamanaka S. Role of the phosphoinositide 3-kinase pathway in mouse embryonic stem (ES) cells. Biochem Soc Trans. 2005;33:1522–5.
41. Storm MP, Bone HK, Beck CG, et al. Regulation of Nanog expression by phosphoinositide 3-kinase-dependent signaling in murine embryonic stem cells. J Biol Chem. 2007;282:6265–73.
42. Paling NR, Wheadon H, Bone HK, et al. Regulation of embryonic stem cell self-renewal by phosphoinositide 3-kinase-dependent signaling. J Biol Chem. 2004;279:48063–70.
43. Welham MJ, Storm MP, Kingham E, et al. Phosphoinositide 3-kinases and regulation of embryonic stem cell fate. Biochem Soc Trans. 2007;35:225–8.
44. Watanabe S, Umehara H, Murayama K, et al. Activation of Akt signaling is sufficient to maintain pluripotency in mouse and primate embryonic stem cells. Oncogene. 2006;25:2697–707.
45. Cross DA, Alessi DR, Cohen P, et al. Inhibition of glycogen synthase kinase-3 by insulin mediated by protein kinase B. Nature. 1995;378:785–9.
46. Sato N, Meijer L, Skaltsounis L, et al. Maintenance of pluripotency in human and mouse embryonic stem cells through activation of Wnt signaling by a pharmacological GSK-3-specific inhibitor. Nat Med. 2004;10:55–63.
47. Burdon T, Stracey C, Chambers I, et al. Suppression of SHP-2 and ERK signalling promotes self-renewal of mouse embryonic stem cells. Dev Biol. 1999;210:30–43.
48. Kunath T, Saba-El-Leil MK, Almousailleakh M, et al. FGF stimulation of the Erk1/2 signalling cascade triggers transition of pluripotent embryonic stem cells from self-renewal to lineage commitment. Development. 2007;134: 2895–902
49. Xu RH, Peck RM, Li DS, et al. Basic FGF and suppression of BMP signaling sustain undifferentiated proliferation of human ES cells. Nat Methods. 2005;2:185–90.
50. Amit M, Carpenter MK, Inokuma MS, et al. Clonally derived human embryonic stem cell lines maintain pluripotency and proliferative potential for prolonged periods of culture. Dev Biol. 2000;227:271–8.
51. Levenstein ME, Ludwig TE, Xu RH, et al. Basic fibroblast growth factor support of human embryonic stem cell self-renewal. Stem Cells. 2006;24:568–74.
52. Greber B, Lehrach H, Adjaye J. Fibroblast growth factor 2 modulates transforming growth factor beta signaling in mouse embryonic fibroblasts and human ESCs (hESCs) to support hESC self-renewal. Stem Cells. 2007;25:455–64.
53. Wang G, Zhang H, Zhao Y, et al. Noggin and bFGF cooperate to maintain the pluripotency of human embryonic stem cells in the absence of feeder layers. Biochem Biophys Res Commun. 2005;330:934–42.
54. Vallier L, Alexander M, Pedersen RA. Activin/Nodal and FGF pathways cooperate to maintain pluripotency of human embryonic stem cells. J Cell Sci. 2005;118:4495–509.
55. Babaie Y, Herwig R, Greber B, et al. Analysis of Oct4-dependent transcriptional networks regulating self-renewal and pluripotency in human embryonic stem cells. Stem Cells. 2007;25: 500–10.
56. Ogawa K, Saito A, Matsui H, et al. Activin-Nodal signaling is involved in propagation of mouse embryonic stem cells. J Cell Sci. 2007;120:55–65.
57. Herpin A, Cunningham C. Cross-talk between the bone morphogenetic protein pathway and other major signaling pathways results in tightly regulated cell-specific outcomes. FEBS J. 2007;274:2977–85.
58. Xu RH, Chen X, Li DS, et al. BMP4 initiates human embryonic stem cell differentiation to trophoblast. Nat Biotechnol. 2002;20:1261–64.
59. Levine AJ, Brivanlou AH. GDF3, a BMP inhibitor, regulates cell fate in stem cells and early embryos. Development. 2006;133: 209–16
60. McLean AB, D'Amour KA, Jones KL, et al. Activin A efficiently specifies definitive endoderm from human embryonic stem cells only when phosphatidylinositol 3-kinase signaling is suppressed. Stem Cells. 2007;25:29–38.
61. D'Amour KA, Agulnick AD, et al. Efficient differentiation of human embryonic stem cells to definitive endoderm. Nat Biotechnol. 2005;23:1534–41.
62. Li J, Wang G, Wang C, et al. MEK/ERK signaling contributes to the maintenance of human embryonic stem cell self-renewal. Differentiation. 2007;75:299–307.
63. Ludwig TE, Levenstein ME, Jones JM, et al. Derivation of human embryonic stem cells in defined conditions. Nat Biotechnol. 2006;24:185–7.

64. Wang L, Schulz TC, Sherrer EC, et al. Self-renewal of human embryonic stem cells requires insulin-like growth factor-1 receptor and ERBB2 receptor signaling. Blood. 2007;110:4111–9.
65. Calaveri F, Scholer HR. Nanog: a new recruit to the embryonic stem cell orchestra. Cell. 2003;113:551–2.
66. Niwa H, Miyazaki J, Smith, AG. Quantitative expression of Oct-3/4 defines differentiation, dedifferentiation or self-renewal of ES cells. Nat Genet. 2000;24:372–6.
67. Matin MM, Walsh JR, Gokhale PJ, et al. Specific knockdown of Oct4 and beta2-microglobulin expression by RNA interference in human embryonic stem cells and embryonic carcinoma cells. Stem Cells. 2004;22:659–68.
68. Wang J, Rao S, Chu J, et al. A protein interaction network for pluripotency of embryonic stem cells. Nature. 2006;444:364–8.
69. Lutz W, Leon J, Eilers M. Contributions of Myc to tumorigenesis. Biochim Biophys Acta. 2002;1602:61–71.
70. Ryan KM, Birnie GD. Myc oncogenes: the enigmatic family. Biochem J. 1996;314:713–21.
71. Takahashi K, Yamanaka S. Induction of pluripotent stem cells from mouse embryonic and adult fibroblast cultures by defined factors. Cell. 2006;126:663–76.
72. Yu J, Vodyanik MA, Smuga-Otto K, et al. Induced pluripotent stem cell lines derived from human somatic cells. Science. 2007;318:1917–20.
73. Meissner A, Wernig M, Jaenisch R. Direct reprogramming of genetically unmodified fibroblasts into pluripotent stem cells. Nat Biotechnol. 2007;25:1177–81.
74. Fernandez PC, Frank SR, Wang L, et al. Genomic targets of the human c-Myc protein. Genes Dev. 2003;17:1115–29.
75. Knoepfler PS, Zhang XY, Cheng PF, et al. Myc influences global chromatin structure. EMBO J. 2006;25:2723–34.

Asymmetric Behavior in Stem Cells

Bridget M. Deasy

Abstract Asymmetry in the stem cell niche refers to the notion that daughter cells are different from each other. There is significant evidence that many stem cell divisions result in one daughter cell that is similar to the parent cell and, hence, necessarily allows for self-renewal of the stem cell phenotype, whereas the other daughter cell is a differentiated or committed cell type. In this chapter we will discuss the role of asymmetry in stem cell divisions and the evidence that supports different asymmetric scenarios in different model systems. We first present the early asymmetric divisions that have been described in first divisions of the zygote and in gametogenesis. Next, we will discuss evidence of asymmetry in postnatal stem cells. Here we will describe two systems in particular – the hematopoietic system and muscle stem cells. Lastly, we will present a theory of the immortal strand hypothesis in which the role of DNA strand segregation is discussed as it relates to asymmetry in cell divisions and the protection of the self-renewing stem cell.

Keywords Polarized · Polarity · Niche · Microenvironment · Immortal strand · Cancer · Cell expansion · Cell therapy · Lineage · Division history

1 Stem Cell Asymmetry

Asymmetry in stem cell behavior refers to the notion that daughter cells are different from each other (Fig. 1A). It has been shown that some stem cell divisions result in one daughter cell that is similar to the parent cell and, hence, necessarily allows for self-renewal of the stem cell phenotype, whereas the other daughter cell is a differentiated or committed cell type. Asymmetry in the phenotype of the daughter cells can occur in theory from two different mechanisms. First, there may be directed or random events occurring within the cytoplasm that result in asymmetric partitioning of cytoplasmic contents and, hence, distinct daughter cell phenotypes (Fig. 1B). Alternatively, the event of cell division may yield daughter cells that are equivalent at birth, and the cells then respond to extrinsic cues that prompt one cell to differentiate while the other does not. This scenario implies that the equivalent daughter cells are positioned in the microenvironment such that they receive different cues and, hence, the result is two different phenotypes (Fig. 1C).

Studies of asymmetry in stem cell biology must specifically define the aspect in which the resulting daughter cells differ. For example, to one investigator, asymmetric division may refer to a behavioral parameter such as cell division activity – one daughter cell is actively dividing and one daughter cell is nondividing (quiescent, terminally differentiated, or senescent). To another investigator, asymmetry may mean that one daughter cell maintains its location, while one daughter cell is physically moved to a new position. Asymmetry may also mean that one daughter cell expresses a specific transcription factor, and one daughter cell does not express that transcription factor. Therefore, the point in time at which an investigator can identify differences in the daughter cells, and hence recognize asymmetry, will vary with the parameter that is being investigated.

Here, we will discuss the role of asymmetry in stem cell divisions and the evidence that supports each of these scenarios in different model systems. We first present the early asymmetric divisions that have been described in first divisions of the zygote and in gametogenesis. Next, we will discuss evidence of asymmetry in postnatal stem cells. Here we will describe two systems in particular – the hematopoietic system and muscle stem cells. Lastly, we will present a theory of the immortal strand hypothesis in which the role of DNA strand segregation is discussed as it relates to asymmetry in cell divisions and the protection of the self-renewing stem cell.

B.M. Deasy (✉)
Departments of Orthopaedic Surgery and Bioengineering, University of Pittsburgh, Stem Cell Research Center, Children's Hospital of Pittsburgh of UPMC, McGowan Institute of Regenerative Medicine, University of Pittsburgh Medical Center; 5113 Rangos Research Center, 3705 Fifth Avenue, Pittsburgh, PA 15213
e-mail: deasybm@upmc.edu

V.K. Rajasekhar, M.C. Vemuri (eds.), *Regulatory Networks in Stem Cells*, Stem Cell Biology and Regenerative Medicine,
DOI 10.1007/978-1-60327-227-8_2, © Humana Press, a part of Springer Science+Business Media, LLC 2009

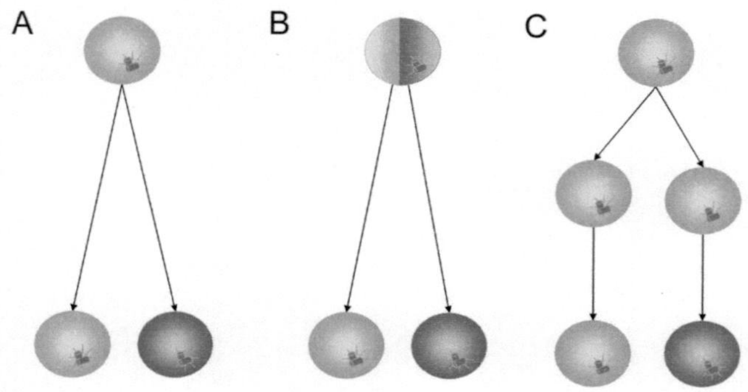

Fig. 1 Asymmetry in cell division gives rises to daughter cells with unique properties. (**A**) Asymmetry. General concept of asymmetric division with unique daughter cells. Observations of this general asymmetric scenario are often made without an understanding of underlying mechanism. Intrinsic or extrinsic factors play a role in whether the daughter cells are different due to internal cues or environmental cues. Asymmetry may occur in theory by two mechanisms. (**B**) Asymmetric Division. The parent cell may have cytoplasmic asymmetry, or the process of division that involves the centrosomes and mitotic spindle alignment may result in unequal partitioning of molecular determinants. The result is two unique daughter cells. (**C**) Asymmetric Fate. Asymmetry may arise from a parent cell that gives rise to two equal daughter cells at the time of division, but the daughter cells respond differentially to the microenvironment and adopt different phenotypes or undergo different developmental programs. The result is two unique daughter cells. The ability to distinguish which of these two patterns may be occurring in a given system depends on the spatial and temporal resolution of the experimental analysis, and the asymmetric parameter of interest. In addition, we show in this chapter, that some stem cell niches involve both mechanisms to maintain the stem cell phenotype and permit cell differentiation

2 Asymmetry in Embryonic and Germ Cells

2.1 *Zygote First Division*

Clearly, the most potent of stem cells is the zygote, having totipotent capability to give rise to all cell types of the organism and support development of extraembryonic tissues (e.g., the placenta). The notion that the first division of the zygotic cell establishes two cells with unique fates appears contradictory to the established finding that, in many systems, all cells of the 4-, 8-, or 16-cell stage have potential to give rise to all cell types [1–3]. Totipotency in the early cell stages was first shown by Hans Spemann in the newt salamander, and later, others showed that loss of totipotency in mammals spanned a range from the 2-cell stage up to nuclei totipotency of sheep embryos at the 64-cell stage [2]. Blastomeres that are separated at the two-cell stage show equal potential to become viable organisms, yet, more recent findings also support the occurrence of asymmetry in the first two blastomeres, or daughter cells, of the zygote.

Asymmetry has been examined comprehensively in embryonic development of the nematode *Caenorhabditis elegans*, and in the insect model *Drosophila melanogaster.* Both undergo an extensive number of asymmetric divisions during development. In particular, all 959 cells of the worm have been traced (from 671 divisions) though the work of J. Sulston and colleagues [4, 5]. In particular, the first division of the *C. elegans* zygote involves *par* (partitioning) genes that lead to two cells of different developmental pathways; ne cell develops to the ectodermal lineage and one cell to the endodermal and mesodermal lineages. The par genes are highly conserved. In many species, the membrane around the sperm entry position (SEP) is marked by a fertilization cone that consists of cytoplasmic elements including par proteins [6].

Studies with mammalian embryos, predominantly mouse embryos, show that the asymmetry of the first zygote division also may be established by environmental cues [7, 8]. First, the primary cleavage of the zygote that results in two cells with bilateral symmetry appears to be oriented with respect to the sperm entry position [9, 10]. It has also been demonstrated that the cleavage axis for the first division can be predicted with high probability by the SEP markers [9]. Further, the daughter cell that receives the SEP marker also has a tendency to divide before its sister cell. In *C. elegans*, the par proteins will accumulate near the SEP. Tracking the lineage of the cell membrane also showed that this earlier dividing, SEP-inheriting cell contributes preferentially to the embryonic part of the mouse blastocyst (Fig. 2). Another environmental cue that appears to play a role in polarity of the blastocyst is the polar body of the second meiotic division [11]. The final step in gametogenesis (also discussed below) yields a smaller haploid cell, or polar body, associated with the larger zygote. Gardner et al. [11] have shown that the location of the polar body has a tendency to be aligned with the boundary between the embryonic and extraembryonic regions. This axis relates to the animal-vegetal pole – the axis of bilateral symmetry is normally aligned with the animal-vegetal axis of the zygote and the embryonic-extraembryonic axis is orthogonal to it. Lineage analysis again shows that the cells that are adjacent to the polar body give rise to cells of the animal pole [12]. In sum, the SEP and polar body location appear to predict the first cleavage plane and these environmental cues may be the earliest signals that direct lineage fate of the blastocyst in the mouse blastocyst [7, 9] (Fig. 2).

2.2 *Gametogenesis*

Another clear pattern for asymmetric divisions is demonstrated in germ cell differentiation. Primordial germ cells (PGCs) are the embryonic precursors to the gametes. Primordial germ cells (PGCs) are the embryonic precursors to the gametes (also see Chapter 5). The point in embryogenesis at which germ lineage determination is made differs among species. In insects, nematodes, and some amphibians, for

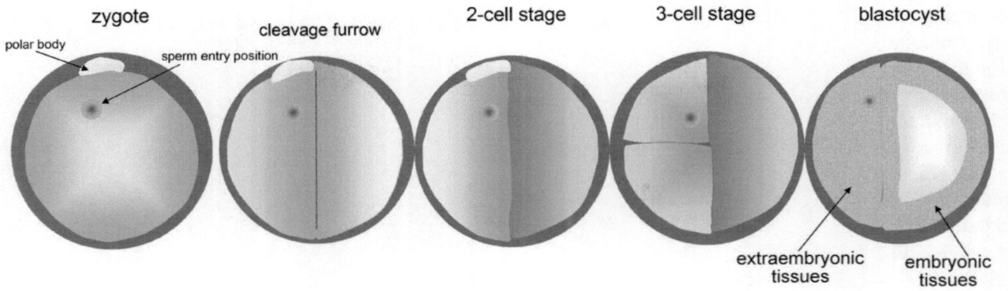

Fig. 2 Asymmetry in first cell cleavage or division of the mouse embryo. The sperm entry position (SEP) and the polar body appear to have a role in directing polarization of the early division of the zygote. In many species, the SEP is associated with positioning of the developing embryo. Later, the cell lineage shows partitioning between the embryonic and embryonic tissues. This figure is adapted from *Zernicka-Goetz, M., Development, 2002, 129:815* [7]. One of the challenges to understanding asymmetry at this early time point involves reconciling asymmetry with the demonstrated equal developmental plasticity of the early blastomeres

example, specific maternal cytoplasm of the zygote, called germ plasm, is responsible for signaling germline differentiation [13, 14]. Germ plasm is comprised of RNA, protein, and polar granules, or electron-dense structures that are associated with mitochondria [15]. The first division of the nematode *Ascaris* zygote, for example, results in two cells with different developmental potential; with one cell being committed entirely to somatic cells while the other may give rise to both germ cells and somatic cells [16, 17] (Fig. 3). In the *Drosophila* model, a number of maternal genes in the oocyte play a role in specifying germ cell fate. Here the nuclei destined to become germ cells are located at one pole of the developing syncytium. Nuclei associated with this cytoplasm are the first to form a unique cell membrane or cellularize to form cells of a distinct cell fate [18].

In mammals and other amphibians, germ cell differentiation appears to be signaled much later through cell-cell interactions of gastrulation [19]. In studies of the developing mouse embryo, it appears that cell interactions associated with gastrulation induce germ cell specification [20] and that the process is mediated by secreted factors of the bone morphogenetic protein (BMP) family [21]. Alkaline phosphatase expression has been classically used to identify PGCs. Cell divisions that give rise to cells of specific fates have not been identified here. Rather, cells appear to adopt distinct fates based on positional information. Whatever the mechanism of specification, once primordial germs cells are specified, the process of differentiation to mature gametes again involves asymmetry.

The well-described differentiation of the *Drosophila* male primordial germ cells to sperm provides a clear example of stem cell asymmetry and the role of the stem cell niche. Like other species, the *Drosophila* testes contain compartments of cells at the various stages of spermatogenesis. A cluster of post-mitotic somatic cells, termed hub cells, resides at the apical tip of the fly testis and this hub is surrounded by the germline stem cells [22] (Fig. 4A). Upon cell division, the male germ cell gives rise to 1 cell which will remain adjacent to the hub, and retain the stem cell phenotype, and 1 cell which is physically displaced from the hub and is no longer in direct physical contact with the hub. The displaced cell, termed the gonialblast, gives rise to transiently amplifying cells and spermatogonia. The apical hub cells express the ligand Unpaired (Upd), which activates the Janus kinase-signal transducer and activator of transcription (JAK-STAT) pathway in adjacent germ cells [23, 24]. This pathway is required for self-renewal of the germ cells [24, 25]. Further, this local acting ligand appears to have limited diffusion [26] and may therefore act on the adjacent germ cells to signal self-renewal, while cells further from the hub initiate differentiation [27]. In addition to the hub cells, signals from the cyst progenitor cells also regulate germ cell differentiation; the epidermal growth factor receptor pathway acting within the cyst cell plays a role in inducing differentiation and regulating amplification in the germ cells [28, 29].

It has been proposed that the male germ cell niche of *Drosophila* also requires the function of adherens junctions and specific orientation of mitotic spindles to ensure that one daughter cell self-renews and remains within the niche and the other daughter cell is displaced [27, 30]. Yamashita et al. [30] showed that dividing germline stem cells use mechanisms involving centrosome activity and a cortically localized protein to orient the mitotic spindles perpendicular to the hub cells of the niche. The high concentration of the E-cadherin homolog (Shg) at the interface of hub cells and germ cells, and the architectural association of the adheren with intracellular APC of germ cells, would facilitate an asymmetric division in which one daughter cell remains in the niche and self-renews and the other is displaced and initiates differentiation [30].

An asymmetric pattern also is observed in many species of female germ cell development. The differentiation of female primordial germ cells to oocytes involves two steps of asymmetric meiosis (Fig. 4B). The daughter cells differ from

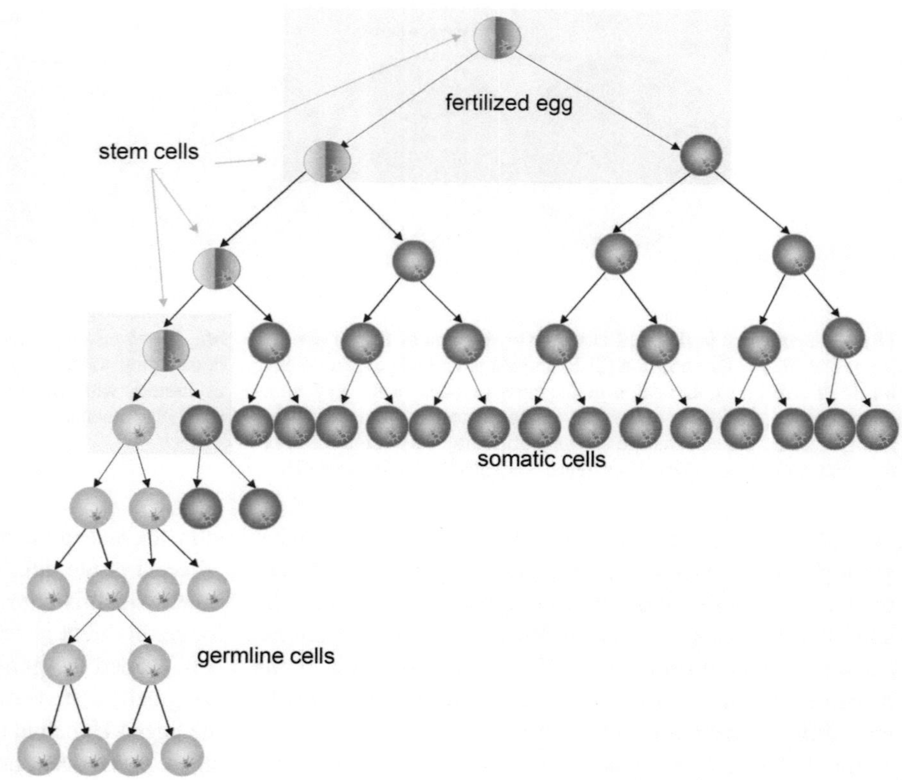

Fig. 3 Asymmetry in cell cleavage of *Ascaris*. A unique mode of asymmetric lineage development is observed in the invertebrate *Ascaris* nematode. Asymmetry results from portions of the genome being lost in some daughter cells. The somatic cells have reduced chromatin content, while cells of the germ line retain a full chromosome complement. Adapted from *Ham, Mechanisms of Development, 1980, Mosby Publishers, St. Louis, MO* [16]

each other mainly in cytoplasmic volume. In the process, the germ cell gives rise to the diploid oogonium, which may undergo symmetric divisions to give rise to more oogonia or may mature to an oocyte. In the first asymmetric division of meiosis, the primary oocyte gives rise to a secondary oocyte and a polar body. Both daughters receive a second complement of chromosomes; however, one of the daughters, termed a polar body, randomly receives a much smaller portion of the cytoplasm. In the second meiotic event, the secondary oocyte gives rise to the haploid ootid, which will mature to the oovum, and another smaller polar body.

Further asymmetry is observed in the epigenetic characteristics of parental genomes of the fertilized egg. Imprinting during gametogenesis gives rise to differential developmental roles for the maternal and paternal genomes in embryonic and extraembryonic tissues [31, 32]. In mammals and a number of other species, the higher degree of methylation of the maternal DNA and histones, as compared to the paternal DNA and histones methylation, is responsible for epigenetic asymmetry. A rapid loss of methylation occurs in the hours following zygote formation and some regions are resistant to demethylation [33–35]. The mechanism responsible for the methylation differences is not clearly understood. It may be that the high level of methylation at the maternal zygote protects against the demethylase activity of the zygote. Or the differential may be due to increased targeting of the paternal genome by the demethylases. The results of Nakamora et al. [35] suggest that the maternal factor called PGC7/Stella protects the maternal genome from demethylation after it localizes to the nucleus, where it maintains the methylation of several imprinted genes. Additional epigenetic asymmetry is observed during development. The DNA and histone methylation and polycomb gene silencing are asymmetric in the embryonic (deriving from the inner cell mass) versus extraembryonic tissues (mainly deriving from the trophoectoderm) [36, 37]. X chromosome inactivation is random generally in the embryonic and somatic tissue but imprinted in the extraembryonic placental and umbilical tissues.

In relation to the current interest in stem cells for therapeutics, the potential of ESCs and PGCs has been examined. The potential of ESCs has been widely discussed. In vitro and in vivo studies have shown that PGCs may give rise to pluripotent stem cells that are capable of giving rise to cells of multiple lineages. However, transplantation of PGCs to the mouse blastocyst showed that the cells did not contribute to either germ cells or somatic cells [38]. Regulatory molecular mechanisms that control development of the mammalian EC and germline cells are the focus of ongoing studies. This will contribute to both the potential use of these cells and an understanding of the role of division patterns in basic biology of early ECs and PGCs.

2.3 Neurogenesis

Neurogenesis during embryonic development has been well characterized using the *Drosophila* model system. Key in

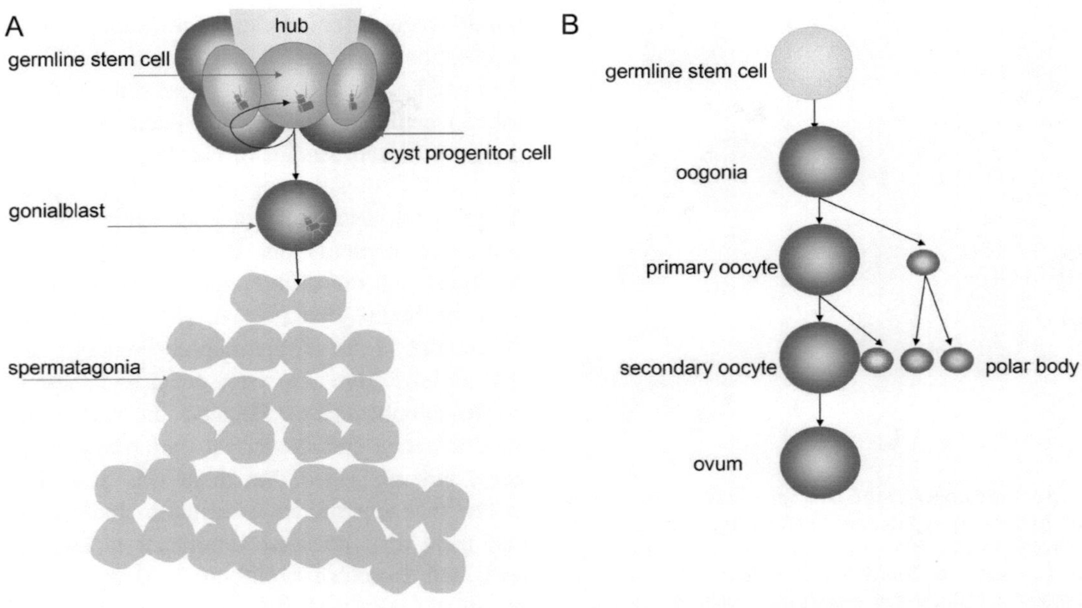

Fig. 4 Gametogenesis. (**A**) A cluster of post-mitotic somatic cells resides at the apical tip of the *Drosophila* testis and this hub is surrounded by the germline stem cells. The male germ cell gives rise to one cell that will remain adjacent to the hub, and retain the stem cell phenotype, and one cell, the gonialblast, which is physically displaced from the hub and is no longer in direct physical contact with the hub. The gonialblast gives rise to transiently amplifying cells, through four divisions, and spermatogonia. The hub cells express the ligand Unpaired, which activates the JAK-STAT pathway in adjacent germ cells, and is required for self-renewal the germ cells. Signals from the cyst progenitor cells also regulate germ cell differentiation. Finally, dividing germline stem cells use mechanisms involving the centrosomes and a cortically localized protein to orient the mitotic spindles perpendicular to the hub and facilitate an asymmetric division in which one daughter cell remains in the niche and self-renews and the other is displaced and initiates differentiation. (**B**) The coordinated process of oogenesis in several mammalian species creates one ovum and three smaller polar bodies. The daughter cells differ from each other mainly in cytoplasmic volume. The primordial germ cell gives rise to the diploid oogonium, the oogonia, and then the oocyte. In the first asymmetric division of meiosis, the primary oocyte gives rise to a secondary oocyte and a polar body. While both daughters receive a second complement of chromosomes, one of the daughters, termed a polar body, will randomly receive a much smaller portion of the cytoplasm. The second meiotic event, whose timing varies among species, will give rise to the mature oovum, and another smaller polar body. In humans, the second meiotic division occurs after fertilization

this process is Numb, a membrane-bound intracellular protein that directs fate specification of neuron and sheath cells (cells that form a sheath around the dendrite of the neuron) and other cells associated with the external sensory organ. Asymmetry related to Numb and neurogenesis has been extensively described elsewhere, for *Drosophila* and mammals [39–43], and will only be highlighted here.

The lineage of the sensory organ precursor (SOP) cell eventually gives rise to five cells of the *Drosophila* external sensory organ. Rhyu et al. [44] first showed that Numb protein segregates asymmetrically and this event is required for the fate specification of the daughter cells. Spindle orientation also plays a role in the asymmetric divisions as it orients the plane of cell cleavage. The crescent-shaped surface localization of Numb on the cell correlates with the mitotic spindle arrangement. The first SOP division, occurring along the anterior-posterior axis, results in pIIa and pIIb; pIIb subsequently divides along the apical-basal axis to give rise to a glial cell and pIIIb, which again divides apical-basal to yield a sheath cell and a neuron. All divisions in the lineage are asymmetric (Fig. 5).

A number of other cytoplasmic factors interact with or inhibit Numb and affect cell fate specification [45–47]. Notable among these factors is Notch. Notch signaling was shown to inhibit neuronal differentiation in *Drosophila* and other species [48]. Numb is an inhibitor of Notch signaling [49]; it prevents nuclear translocation of Notch and antagonizes its activity. Morrison et al showed that a transient activation of Notch was sufficient to cause an irreversible loss of neurogenic differentiation potential; accelerated glial differentiation was also observed following Notch activation [50]. In sum, interactions between Numb, Notch [and Numblike (d-Numb homolog)] play important roles in controlling asymmetry and directing neuronal cell fate specification.

3 Asymmetry in Postnatal Cells

As the germ layers are formed from the ESCs, and cell determination and organogenesis evolve during embryonic development, it is believed that asymmetry may generate stem cells that maintain the stem cell pool specific for

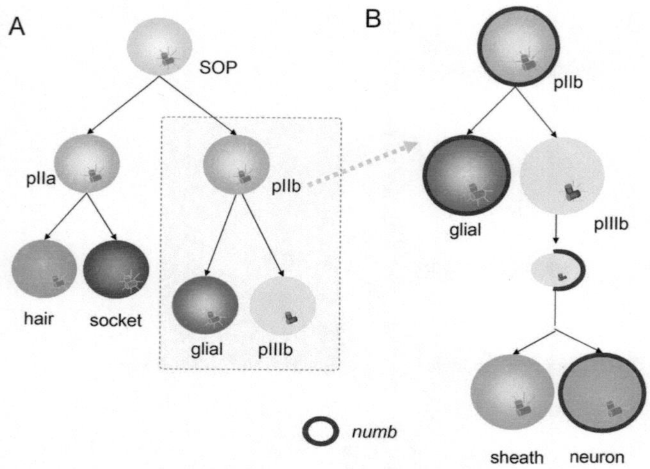

Fig. 5 Asymmetry in neurogenesis of *Drosophila* external sensory organ. The SOP (sensory organ progenitor) cell lineage gives rise to five cells of the external sensory organ – a hair cell, a socket cell, a glial cell, a sheath cell and a neuron. (**A**) The SOP cell divides to give rise to pIIa and pIIb. The progenitor pIIa yields the hair and socket cells, while the progenitor pIIb yields a glial cell and a pIIIb cell. The progenitor pIIIb undergoes an additional division to yield the sheath cell and neuron. (**B**) Numb expression regulates cell fate determination. Other important factors (not shown here) include the niche, the mitotic alignment, and the presence or absence of notch, inscuteable, and delta. All divisions are asymmetric

different organs and tissues, and other cells that initiate the process of differentiation through transiently amplifying stages and become the progenitors of somatic cells. Postnatal or adult stem cells are resident tissue-specific stem cells that are responsible for tissue homeostasis and tissue repair.

3.1 Hematopoietic Stem Cells

Stem cells of the blood tissue have been the model system for studying adult-derived stem cells. Hematopoietic stem cells (HSCs) give rise to all blood cell types, which fall into two general categories: myeloid lineages – monocytes, macrophages, neutrophils, basophils, eosinophils, erythrocytes, megakaryocytes, platelets, dendritic cells; and lymphoid lineages – T-cells, B-cells, NK-cells, and dendritic cells. In addition, postnatal HSCs from cord blood, peripheral blood, and bone marrow have been used successfully in therapy to treat blood disorders and some types of cancer. (See Chapters 9, 10, 15 and 30 for additional discussions of HSCs.)

Early clonal assays revealed that single HSCs and progenitors are capable of giving rise to colonies of mixed progenies that include, for example, macrophages eosinophils, neutrophils, basophils, erythrocytes, and megakaryocytes [50–52]. These studies provided evidence for a single cell origin with multilineage potential, and also showed that a structurally intact, or physical, microenvironment was not necessary for multilineage differentiation. The findings opened the door to questions regarding the cellular mechanisms that lead to the mixed colonies. In subsequent studies, the question was asked whether asymmetry may occur in the originating cell division. Indeed, studies of paired daughter cells that result from a HSC division suggested that there is asymmetry in daughter cell developmental potential [53–56]. Using single-cell micromanipulation, daughter cells were physically separated and the differentiation fates of the cell progeny were examined. Suda et al. [54] found that there were differences in the differentiation directions of the progeny that derived from the sister cells, and these were termed nonhomologous pairs. Because the cells were in similar environmental conditions, yet they produced different progeny, this suggested that there was a stochastic element in cell fate determination. In these studies, some daughter cell pairs also revealed significant differences in colony size, and therefore proliferation rates [53, 54, 56]. The results of Leary et al. [55, 56] supported the findings as they reported that sister multipotent progenitors had differences in colony-forming potential. Asymmetric cell phenotypes, here determined by asymmetry in differentiation fate, were observed in up to 17% of the paired progenitor cells of human umbilical cord blood [57]; the fate did not appear to be affected by cytokines, which again supports the idea of a stochastic component. Other results performed on clones and subclones, rather than sister cells, provided further support for the notion of asymmetry in that they demonstrate heterogeneity and intrinsic control in cell fate [58].

Studies using defined phenotypes have also demonstrated asymmetry in HSCs' fate. HSCs isolated on the basis of CD34 expression, a surface glycoprotein that functions in hematopoiesis and hematopoietic cell adhesion [59, 60], were examined by time-lapsed microscopy and it was observed that HSC divisions resulted in some daughter cells remaining as quiescent cells while other daughter cells underwent extensive proliferation [61]. Other groups subsequently showed that the CD34 cells that gave rise to myeloid-lymphoid initiating cells had slower division times and were associated with asymmetry more so than CD34 cells that gave rise to colony-forming units [62]. They also showed that contact with supporting cells, which may mimic the microenvironment, caused an increase in daughter cell asymmetry [63]. In these studies, approximately 30% of the defined CD34 cells gave rise to daughter cells with mixed proliferation rates [61, 62]. Geibel et al. [64] also showed that both primitive and more committed cells gave rise to differentially specified daughters. However, it was within the primitive compartment that the majority of cells appeared to diverge asymmetrically, while the majority of divisions of the committed cells led to symmetric cell expansions.

It has been shown clearly that stem and progenitors of the hematopoietic compartment can give rise to daughter cells whose progeny have different cell fates. However, it has not been shown that there is asymmetry in intracellular determinants at the time of HSC cell division. There is still a focus to identify segregation of molecular determinants within the cell such that unique phenotypes between the daughter cells is apparent at the time of cell division. A recent study indeed suggests that an interaction between CD34 cells affects the cleavage plane of cell division, and may subsequently result in unequal distribution of Notch-1 to the daughter cells [65]. Certainly, although one type of asymmetry has been shown in an in vitro culture setting, it is not known whether HSCs divide symmetrically or asymmetrically in vivo.

3.2 Skeletal Muscle Stem Cells

The skeletal muscle cell compartment of adult tissues, like the blood cells, includes a variety of cell types. The stem cell that is described classically in skeletal muscle is the satellite cell, which fuses to form the mature multinucleated muscle fiber. Satellite cells, which appear to be committed precursor cells, were first described based on their location and morphology [66]. Satellite cells surround the mature functional cell of skeletal muscle; the specific niche for satellite cells is in between the sarcolemma and the basal lamina of the muscle fiber. In adult muscle, satellite cells remain quiescent until external stimuli trigger re-entry into the cell cycle. Their progeny, myoblasts, fuse to form new multinucleated myofibers [67–71]. Cell surface markers associated with the in situ satellite stem cell phenotype, either in the quiescent or activated state, include M-cadherin, c-met, CD34, Pax7, and CD56 [70, 72–77]. These cells have been described as having multilineage differentiation potential [78] and have been examined as candidates in cell therapy for muscle repair [79–83].

More recently, a number of other stem-cell-like populations have been identified from the adult skeletal muscle tissue. These phenotypes include side population or SP cells [81, 84–89], mesoangioblasts [90–92], pericytes [93–95], and endothelium-related cells such as AC133 cells [96, 97], preplate muscle-derived cells [98–100], and myo-endothelial cells [101]. The developmental origins and relationships among these cells are still being investigated (for review, see [102]). However it is generally believed that the satellite cell is downstream of the other cell types, which often do not express the Pax7 transcription factor that appears to induce satellite cell specification [76]. The mix of cell types present in an adult muscle biopsy have led to similar questions regarding the role of asymmetry in muscle cell population heterogeneity and hierarchy.

The general notion of muscle stem cell self-renewal implies that the cell division results in one daughter cell that maintains the stem cell phenotype and one daughter cell that is committed to the myogenic lineage. While it has not been demonstrated conclusively that asymmetry of this sort occurs with adult muscle stem cells, there is growing evidence that supports this idea. Olguin and Olwin [103] examined clonal cultures, initiated with 500 cells, and found heterogeneity within individual clones – both differentiated progeny and cells that regained quiescent phenotype markers (Pax-7+ and MyoD$^-$ or myogenin$^-$). Clonal and subclonal cultures of muscle-derived stem cells have also demonstrated mixed phenotypes in terms of both marker expression (CD34, Sca-1, and myogenic markers) [99, 104] and proliferative behavior [105]. Although it was not reported that these clonal cultures were explicitly initiated with single cells [99, 104, 105], a separate study observed proliferative heterogeneity in a single cell colony, from a population of muscle stem cells, which was tracked using time-lapsed imaging [106]. Zammit et al. [107] also proposed a model for asymmetry in cell fates of daughters of satellite cells based on their observations of cell clusters on muscle fibers. Cell clusters were heterogeneous in the expression of Pax7 and MyoD [107]; these results were also supported by findings of Pax7+/MyoD– cells in chicken muscle cell cultures, initiated with 10 cells/plate, which showed both Pax7+/MyoD– and Pax7+/MyoD+ progeny in the cultures [108]. Actively dividing satellite cells (BrdU or PCNA+) also showed asymmetric cellular localization of Numb [109], an inhibitor of Notch signaling, and a determinant of asymmetry in *Drosophila* neurogenesis [44]. Numb was asymmetrically localized to one pole of the cell, and Numb+ and Numb– progenitors showed different patterns of expression of myogenic genes. Adjacent cells that may represent daughter cells showed Numb+/Pax3– and Numb–/Pax3+ cell pairs, Numb+/Myf5+ and Numb–/Myf5– pairs, and Numb+/desmin+ and Numb–/desmin– pairs [109].

In vivo experiments for muscle regeneration have shown that there may be only a subset of satellite cells or myoblasts that contribute to new myofibers; these studies support the notion of self-renewal [110–113] A specific subpopulation of slowly dividing cells (refractory to [^{3}H]thymidine uptake in culture) appeared to survive intramuscular transplantation and proliferate well in vivo [110]. In studies in which intact myofibers, with associated satellite cells, were transplanted to muscles of immunocompromised, dystrophic (*mdx*-nude) mice, investigators observed the generation of large numbers of donor-derived functional Pax7 satellite cells, which supports the concept of stem cell self-renewal in satellite cells [112]. These reports, however, were not designed to identify asymmetric events that yielded these potentially self-renewing subpopulations.

Kuang et al. [114] performed in situ examination of adjacent satellite cells that appear as the daughter cells of a satellite cell division. The orientation of the mitotic spindle within the stem cell niche appears to influence divisional symmetry. Asymmetric division may occur when the mitotic spindle is oriented perpendicular to the fiber axis and cytokinesis gives rise to 1 Myf5− self-renewing cell that remains in contact with the basal lamina and 1 Myf5+ committed cell that is adjacent to the plasma membrane but does not contact the basal lamina. They propose that symmetric divisions may occur parallel to the axis of the myofiber, and give rise to either two self-renewing cells or two committed myogenic cells – both daughter cells contact the basal lamina and the plasma membrane [114, 115] (Fig. 5). The proposed asymmetry will be strengthened by additional studies that include temporal analysis to determine whether the cells are different at the time of cell division or if the cells adopt these different fates.

Broad heterogeneity has been described for satellite cells and other muscle stem cell populations [112, 116–121]. As illustrated for other examples of asymmetry and shown in Fig. 6, clonally derived mixed populations could arise from unequal partitioning of cytoplasmic components or the cells may stochastically adopt unique cell fates. While some data suggest that population heterogeneity derives from asymmetric divisions, additional clonal studies utilizing single cells will strengthen the understanding of this stem cell activity in adult muscle stem cells.

4 Immortal Strand Hypothesis

As stem cells are responsible for the long-term health and maintenance of tissue throughout the adult life of the organism, it is necessary for these cells to have a mechanism to resist the accumulation of replication errors that would occur during normal tissue repair. A proposed mechanism by which cells protect themselves from DNA damage could also give rise to asymmetric divisions in stem cell self-renewal and differentiation. In 1975, John Cairns hypothesized that stem cell division may involve segregation of new and old DNA strands [122] (Fig. 7). If most spontaneous mutations arise during DNA replication, and since DNA is replicated semi-conservatively, Carins hypothesized that the strand that acquires the mutation would be the stand passed to the progeny of stem cells. The nonmutated stand, in theory, could be to retained by the self-renewed stem cell, that is, the immortal daughter cell. In this way, an immortal strand would be maintained through successive divisions while the mutation(s) would accumulate in the mortal daughter that would become the differentiated tissue cells or senescence in time. Further, this analysis can be extended to show how heterogeneity could result in the expanding population. Distinct phenotypes can be categorized based on the DNA template and

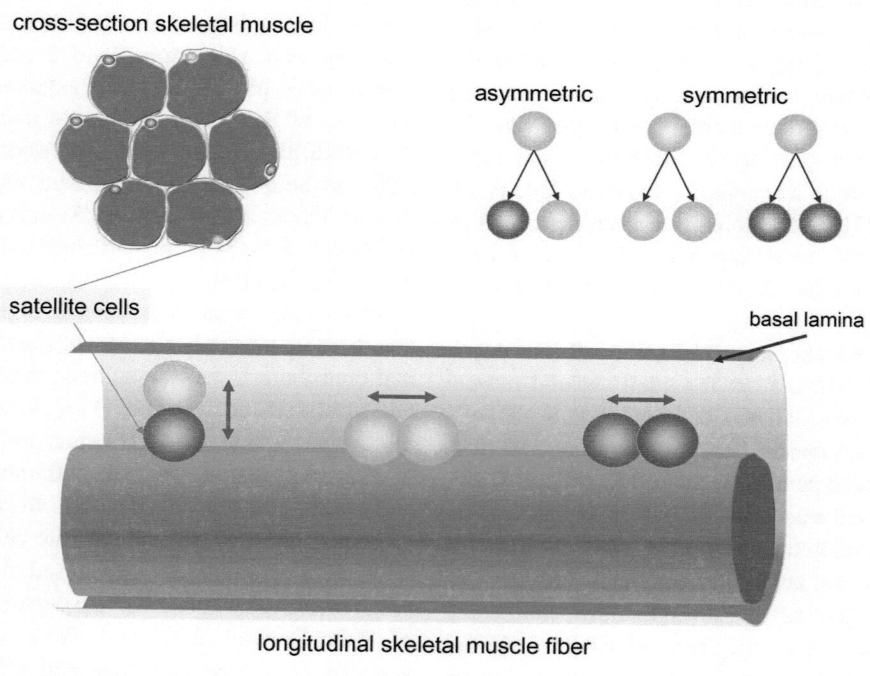

Fig. 6 Asymmetric division of skeletal muscle satellite cells. The orientation of the mitotic spindle within the stem cell niche may influence divisional (a)symmetry. Satellite cells reside in between the basal lamina and the plasma membrane of the muscle fiber. Asymmetric division may occur when the mitotic spindle is oriented perpendicular to the fiber axis and cytokinesis gives rise to one self-renewing cell (Pax7+/Myf5−) that remains in contact with the basal lamina and one committed cell (Pax7+/Myf5+) that is adjacent to the plasma membrane but does not contact the basal lamina. Symmetric divisions may occur parallel to the axis of the myofiber, and appear to give rise to either two self-renewing cells or two committed myogenic cells – both daughter cells contact the basal lamina and the plasma membrane. Image based on *Cossu and Tajbakhsh Cell, 2007, 129:859* [115] and the work of *Kuang S et al, Cell, 2007,129: 999* [114]

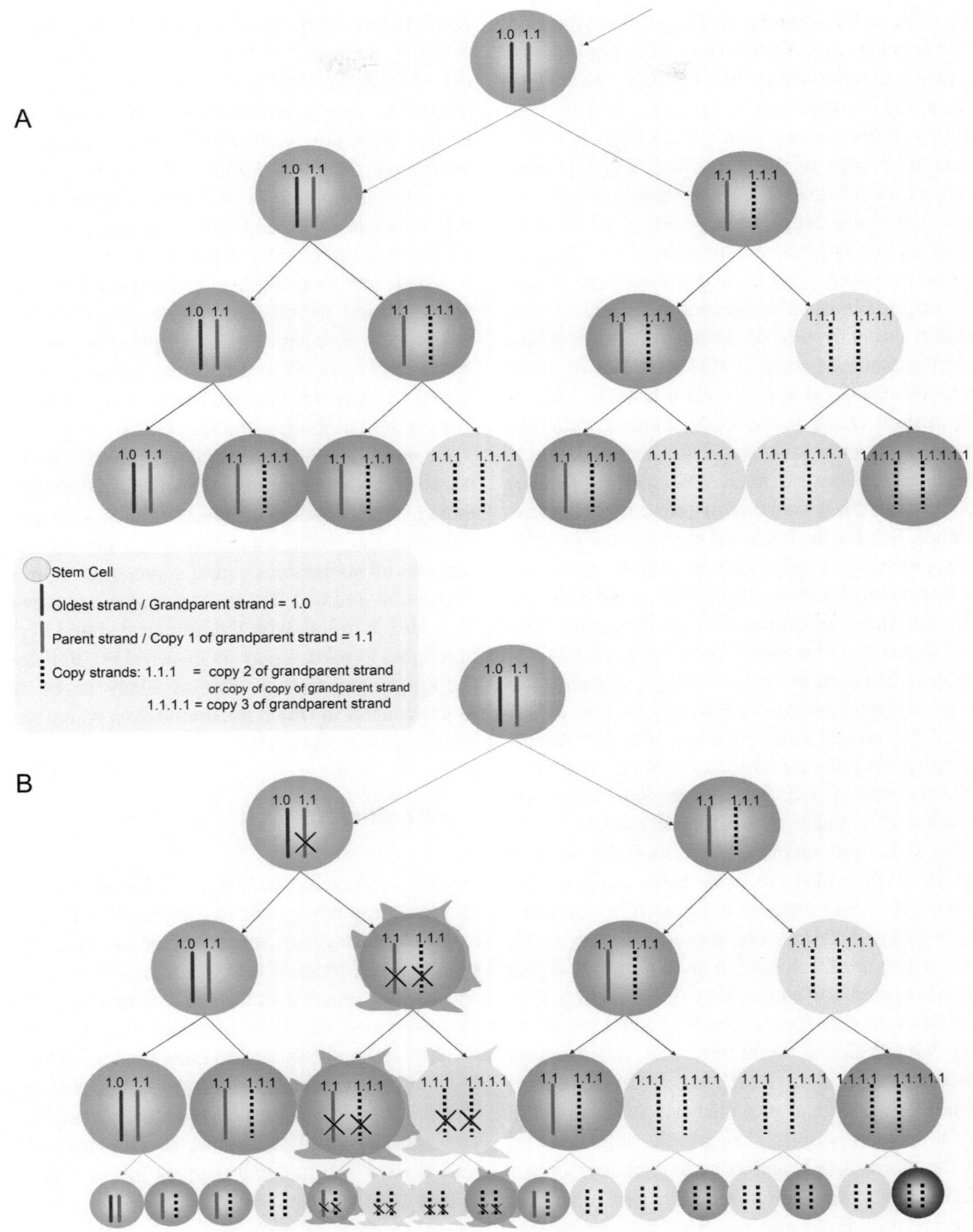

Fig. 7 Immortal strand hypothesis. (**A**) Asymmetry in strand segregation would allow for stem cell self-renewal. The (*blue*) stem cell would retain the oldest DNA strands. This cartoon shows segregation of one chromosome: the oldest/grandparent strand is blue and designated 1.0, and the parent strand (a copy of the grandparent strand) is red and designated 1.1. All other copies are *dashed lines* and designated copy numbers are 1.1.1, 1.1.1.1, 1.1.1.1.1 etc. If nonrandom strand segregation occurs among all chromatids in the cell, the result is asymmetric divisions and self-renewal of the stem cell. We extend Cairns analysis to show here how heterogeneity would result in the expanding population. If distinct phenotypes occur based on the DNA strand copy numbers, then these phenotypes can be categorized based on the template and copy number. For example, after the three divisions shown in the lineage tree above, there would be one stem cell (1.0 and 1.1 strands, p0 phenotype), three pI cells would have (1.1/1.1.1 stands), three pII cells (1.1.1/1.1.1.1), and one pIII cell (1.1.1.1/1.1.1.1.1). (**B**) Further, as Cairns hypothesis showed, a mutation, X, which is heterozygous in the chromosome strands, would segregate and stem cells would be protected against duplication errors. Nonrandom segregation would be required to maintain the immortal strand. This figure also shows how heterogeneity could occur in cancer cells that may develop from the mutation. Some of the cells that derive from the original mutation would also have higher strand-copy numbers that increase the probability of errors in the DNA code (e.g., the green cells with X)

copy number (Fig. 6A). For example, after the three divisions shown in the lineage tree in Fig. 6A, there would be one stem cell that contains a grandparent strand (1.0) and a parent (1.1) strand, there would be three cells that have a parent strand (1.1) and a copy of the parent strand (1.1.1), there would be three cells that have copy of the parent strand (1.1.1) and a copy of a copy of the parent (1.1.1.1) and there would be one cell that has a copy of a copy of the parent (1.1.1.1) and a copy of a copy of a copy of the parent strand (1.1.1.1.1). As the lineage tree or colony grows, the phenotypic difference between the categories would become more distinct.

The immortal strand hypothesis assumes that there is minimal sister chromatid exchange. If this were not the case, the stem cell would not be able to retain a strain that did not have replication errors in the code. Further, cells that preserve immortal strands could avoid the accumulation of errors if they inhibit pathways for DNA repair [123]. Such pathways could potentially cause error-prone resynthesis of damaged strands. Finally, the immortal strands would need to be marked in some way in order for nonrandom segregation to occur. Cairns recognized that the centromeres need to be able to distinguish the sister chromatids and the centromeres would need to behave in a co-coordinated fashion [122].

Some evidence has been presented to support this hypothesis, although several questions regarding its plausibility remain [124, 125]. Several years prior to Carins hypothesis, nonrandom segregation of sister chromatids was reported for mouse embryonic cells [126]. The investigators examined the incorporation of a pulse of tritiated thymidine in the grand-daughter cells and quantitatively observed unequal label distribution. There also are some early studies using lower organisms that may support the possibility of DNA strand co-segregation [124]. More recently, Potten et al. [127] examined the mouse epithelial in the crypts of the small intestinal mucosa for nonrandom strand segregation. The template DNA strands (or regenerating cells) were first pulse-labeled with tritiated thymidine, and subsequently received a bromodeoxyuridine (BrdU) pulse. Co-expression of these two DNA markers provided evidence that the cells were actively dividing. Long-term retention of the tritiated thymidine label, and concomitant loss of the BrdU label, illustrated that an immortal (label-retaining) strand was actively dividing and the label was segregated nonrandomly. A similar study of mouse neural stem cells used a bromodeoxyuridine label alone to show the long-term retention of (BrdU) in the actively dividing cells [128].

Two studies in skeletal muscle also support the idea of asymmetry in DNA strand segregation. Conboy et al. [129] also showed that mouse myogenic progenitors behaved similarly. In the in vivo studies, cells were first labeled with 5-chloro-2-deoxyuridine (CldU) and then a short pulse of 5-iodo-2-deoxyuridine (IdU). Asymmetric inheritance of CldU was evident by all of the detected label being identified in only one daughter. Lastly, Shinin observed selective template-DNA strand segregation during satellite cell mitosis in vivo, and in culture; this provides strong indication that genomic DNA strands are nonequivalent [130]. Interestingly, this study also showed that Numb, previously described for its role in asymmetry, undergoes selective partitioning to one daughter cell. They also found that template DNA and Numb co-segregated in long-term label-retaining cells that express Pax7 [130].

There are some reports that appear to counter the immortal strand hypothesis. Studies that used mouse HSCs showed that co-labeling of BrdU (pulse 1) and halogenated 2-deoxyuridines (CldU or IdU, pulse 2) indicated that all HSCs segregate their chromosomes randomly; both in vivo and in vitro results supported this idea [131].

Overall, there is increasing support for the immortal strand hypothesis; there is also further development of the theory of the function of nonrandom strand segregation. For example, the silent sister hypothesis distinguishes that the purpose of nonrandom strand segregation is to direct gene expression and cell fate in stem and progenitor cells [125]. This idea is in line with the immortal strand hypothesis and highlights important players involved in cell determination – the epigenetic factors. This is likely to be the exciting future context in which the nonrandom strand segregation is investigated in stem cells.

5 Conclusions

Stem cells function to balance self-renewal with differentiation during embryonic development, and in adult tissue, to maintain tissue homeostasis. One mechanism to maintain stem cell self-renewal is asymmetric cell division, in which one cell self-renews while one cell initiates differentiation. The stem cell niche, or microenvironment, provides both biochemical and biophysical components for these regulated stem cell activities. An increased understanding of the extrinsic cues and intrinsic cues, including nonrandom strand segregation, will allow for the development of methods to control stem cell fate and perhaps increase the use of stem cells in cell therapeutics.

References

1. Spemann H. Embryonic development and induction. New Haven, London: Yale University Press; H. Milford, Oxford University Press; 1938.
2. Cibelli JB. Principles of cloning. Amsterdam; Boston: Academic Press; 2002.
3. Sell S. Stem cells handbook. Totowa, NJ: Humana Press; 2004.
4. Sulston JE, Schierenberg E, White JG, Thomson JN. The embryonic cell lineage of the nematode Caenorhabditis elegans. Dev Biol. 1983;100(1):64–119.

5. Schnabel R, Hutter H, Moerman D, Schnabel H. Assessing normal embryogenesis in Caenorhabditis elegans using a 4D microscope: variability of development and regional specification. Dev Biol. 1997;184(2):234–65.
6. Betschinger J, Knoblich JA. Dare to be different: asymmetric cell division in Drosophila, *C. elegans* and vertebrates. Curr Biol. 2004;14(16):R674–85.
7. Zernicka-Goetz M. Patterning of the embryo: the first spatial decisions in the life of a mouse. Development. 2002;129(4):815–29.
8. Beddington RS, Robertson EJ. Axis development and early asymmetry in mammals. Cell. 1999;96(2):195–209.
9. Piotrowska K, Zernicka-Goetz M. Role for sperm in spatial patterning of the early mouse embryo. Nature. 2001;409(6819): 517–21.
10. Plusa B, Piotrowska K, Zernicka-Goetz M. Sperm entry position provides a surface marker for the first cleavage plane of the mouse zygote. Genesis. 2002;32(3):193–8.
11. Gardner RL. The early blastocyst is bilaterally symmetrical and its axis of symmetry is aligned with the animal-vegetal axis of the zygote in the mouse. Development. 1997;124(2):289–301.
12. Ciemerych MA, Mesnard D, Zernicka-Goetz M. Animal and vegetal poles of the mouse egg predict the polarity of the embryonic axis, yet are nonessential for development. Development. 2000;127(16):3467–74.
13. Saffman EE, Lasko P. Germline development in vertebrates and invertebrates. Cell Mol Life Sci. 1999;55(8–9):1141–63.
14. Raz E. Primordial germ-cell development: the zebrafish perspective. Nat Rev Genet. 2003;4(9):690–700.
15. Mahowald AP. Polar granules of Drosophila. 3. The continuity of polar granules during the life cycle of Drosophila. J Exp Zool. 1971;176(3):329–43.
16. Ham RG, Veomett MJ. Mechanisms of development. St. Louis: Mosby; 1980.
17. Wilson EB. The cell in development and heredity. 3rd ed. New York: Macmillan; 1925.
18. Underwood EM, Caulton JH, Allis CD, Mahowald AP. Developmental fate of pole cells in Drosophila melanogaster. Dev Biol. 1980;77(2):303–14.
19. Wylie C. Germ cells. Cell. 1999;96(2):165–74.
20. Tam PP, Zhou SX. The allocation of epiblast cells to ectodermal and germ-line lineages is influenced by the position of the cells in the gastrulating mouse embryo. Dev Biol. 1996;178(1):124–32.
21. Lawson KA, Dunn NR, Roelen BA, et al. Bmp4 is required for the generation of primordial germ cells in the mouse embryo. Genes Dev. 1999;13(4):424–36.
22. Hardy RW, Tokuyasu KT, Lindsley DL, Garavito M. The germinal proliferation center in the testis of Drosophila melanogaster. J Ultrastruct Res. 1979;69(2):180–90.
23. Kiger AA, Jones DL, Schulz C, Rogers MB, Fuller MT. Stem cell self-renewal specified by JAK-STAT activation in response to a support cell cue. Science. 2001;294(5551):2542–5.
24. Tulina N, Matunis E. Control of stem cell self-renewal in Drosophila spermatogenesis by JAK-STAT signaling. Science. 2001;294(5551):2546–9.
25. Kiger AA, White-Cooper H, Fuller MT. Somatic support cells restrict germline stem cell self-renewal and promote differentiation. Nature. 2000;407(6805):750–4.
26. Harrison DA, McCoon PE, Binari R, Gilman M, Perrimon N. Drosophila unpaired encodes a secreted protein that activates the JAK signaling pathway. Genes Dev. 1998;12(20): 3252–63.
27. Yamashita YM, Fuller MT, Jones DL. Signaling in stem cell niches: lessons from the Drosophila germline. J Cell Sci. 2005;118(Pt 4):665–72.
28. Tran J, Brenner TJ, DiNardo S. Somatic control over the germline stem cell lineage during Drosophila spermatogenesis. Nature. 2000;407(6805):754–7.
29. Schulz C, Wood CG, Jones DL, Tazuke SI, Fuller MT. Signaling from germ cells mediated by the rhomboid homolog stet organizes encapsulation by somatic support cells. Development. 2002;129(19):4523–34.
30. Yamashita YM, Jones DL, Fuller MT. Orientation of asymmetric stem cell division by the APC tumor suppressor and centrosome. Science. 2003;301(5639):1547–50.
31. Barton SC, Surani MA, Norris ML. Role of paternal and maternal genomes in mouse development. Nature. 1984;311(5984):374–6.
32. Reik W, Santos F, Mitsuya K, Morgan H, Dean W. Epigenetic asymmetry in the mammalian zygote and early embryo: relationship to lineage commitment? Philos Trans R Soc Lond B Biol Sci. 2003;358(1436):1403–9; discussion 9.
33. Lane N, Dean W, Erhardt S, et al. Resistance of IAPs to methylation reprogramming may provide a mechanism for epigenetic inheritance in the mouse. Genesis 2003;35(2):88–93.
34. Haaf T. Methylation dynamics in the early mammalian embryo: implications of genome reprogramming defects for development. Curr Top Microbiol Immunol. 2006;310:13–22.
35. Nakamura T, Arai Y, Umehara H, et al. PGC7/Stella protects against DNA demethylation in early embryogenesis. Nat Cell Biol. 2007;9(1):64–71.
36. Chapman V, Forrester L, Sanford J, Hastie N, Rossant J. Cell lineage-specific undermethylation of mouse repetitive DNA. Nature. 1984;307(5948):284–6.
37. Kalantry S, Mills KC, Yee D, Otte AP, Panning B, Magnuson T. The Polycomb group protein Eed protects the inactive X-chromosome from differentiation-induced reactivation. Nat Cell Biol. 2006;8(2):195–202.
38. Donovan PJ. The germ cell-the mother of all stem cells. Int J Dev Biol. 1998;42(7):1043–50.
39. Zhong W. Diversifying neural cells through order of birth and asymmetry of division. Neuron. 2003;37(1):11–4.
40. Roegiers F, Younger-Shepherd S, Jan LY, Jan YN. Two types of asymmetric divisions in the Drosophila sensory organ precursor cell lineage. Nat Cell Biol. 2001;3(1):58–67.
41. Wang H, Chia W. Drosophila neural progenitor polarity and asymmetric division. Biol Cell. 2005;97(1):63–74.
42. Jan YN, Jan LY. Asymmetric cell division in the Drosophila nervous system. Nat Rev Neurosci. 2001;2(11):772–9.
43. Zhong W, Feder JN, Jiang MM, Jan LY, Jan YN. Asymmetric localization of a mammalian numb homolog during mouse cortical neurogenesis. Neuron. 1996;17(1):43–53.
44. Rhyu MS, Jan LY, Jan YN. Asymmetric distribution of numb protein during division of the sensory organ precursor cell confers distinct fates to daughter cells. Cell. 1994;76(3):477–91.
45. Orgogozo V, Schweisguth F, Bellaiche Y. Lineage, cell polarity and inscuteable function in the peripheral nervous system of the Drosophila embryo. Development. 2001;128(5):631–43.
46. Berdnik D, Torok T, Gonzalez-Gaitan M, Knoblich JA. The endocytic protein alpha-Adaptin is required for numb-mediated asymmetric cell division in Drosophila. Dev Cell. 2002;3(2):221–31.
47. Kraut R, Chia W, Jan LY, Jan YN, Knoblich JA. Role of inscuteable in orienting asymmetric cell divisions in Drosophila. Nature. 1996;383(6595):50–5.
48. Artavanis-Tsakonas S, Rand MD, Lake RJ. Notch signaling: cell fate control and signal integration in development. Science. 1999;284(5415):770–6.
49. Zhong W, Jiang MM, Weinmaster G, Jan LY, Jan YN. Differential expression of mammalian Numb, Numblike and Notch1 suggests distinct roles during mouse cortical neurogenesis. Development. 1997;124(10):1887–97.
50. Johnson GR, Metcalf D. Pure and mixed erythroid colony formation in vitro stimulated by spleen conditioned medium with no detectable erythropoietin. Proc Natl Acad Sci U S A. 1977;74(9):3879–82.

51. Fauser AA, Messner HA. Granuloerythropoietic colonies in human bone marrow, peripheral blood, and cord blood. Blood. 1978;52(6):1243–8.
52. Suda T, Suda J, Ogawa M. Single-cell origin of mouse hemopoietic colonies expressing multiple lineages in variable combinations. Proc Natl Acad Sci U S A. 1983;80(21):6689–93.
53. Suda J, Suda T, Ogawa M. Analysis of differentiation of mouse hemopoietic stem cells in culture by sequential replating of paired progenitors. Blood. 1984;64(2):393–9.
54. Suda T, Suda J, Ogawa M. Disparate differentiation in mouse hemopoietic colonies derived from paired progenitors. Proc Natl Acad Sci U S A. 1984;81(8):2520–4.
55. Leary AG, Ogawa M, Strauss LC, Civin CI. Single cell origin of multilineage colonies in culture. Evidence that differentiation of multipotent progenitors and restriction of proliferative potential of monopotent progenitors are stochastic processes. J Clin Invest. 1984;74(6):2193–7.
56. Leary AG, Strauss LC, Civin CI, Ogawa M. Disparate differentiation in hemopoietic colonies derived from human paired progenitors. Blood. 1985;66(2):327–32.
57. Mayani H, Dragowska W, Lansdorp PM. Lineage commitment in human hemopoiesis involves asymmetric cell division of multipotent progenitors and does not appear to be influenced by cytokines. J Cell Physiol. 1993;157(3):579–86.
58. Brummendorf TH, Dragowska W, Zijlmans J, Thornbury G, Lansdorp PM. Asymmetric cell divisions sustain long-term hematopoiesis from single-sorted human fetal liver cells. J Exp Med. 1998;188(6):1117–24.
59. Healy L, May G, Gale K, Grosveld F, Greaves M, Enver T. The stem cell antigen CD34 functions as a regulator of hemopoietic cell adhesion. Proc Natl Acad Sci U S A. 1995;92(26):12240–4.
60. Hu MC, Chien SL. The cytoplasmic domain of stem cell antigen CD34 is essential for cytoadhesion signaling but not sufficient for proliferation signaling. Blood. 1998;91(4):1152–62.
61. Huang S, Law P, Francis K, Palsson BO, Ho AD. Symmetry of initial cell divisions among primitive hematopoietic progenitors is independent of ontogenic age and regulatory molecules. Blood. 1999;94(8):2595–604.
62. Punzel M, Zhang T, Liu D, Eckstein V, Ho AD. Functional analysis of initial cell divisions defines the subsequent fate of individual human CD34(+)CD38(–) cells. Exp Hematol. 2002;30(5):464–72.
63. Punzel M, Liu D, Zhang T, Eckstein V, Miesala K, Ho AD. The symmetry of initial divisions of human hematopoietic progenitors is altered only by the cellular microenvironment. Exp Hematol. 2003;31(4):339–47.
64. Giebel B, Zhang T, Beckmann J, et al. Primitive human hematopoietic cells give rise to differentially specified daughter cells upon their initial cell division. Blood. 2006;107(5):2146–52.
65. Bullock TE, Wen B, Marley SB, Gordon MY. Potential of CD34 in the regulation of symmetrical and asymmetrical divisions by hematopoietic progenitor cells. Stem Cells. 2007;25(4):844–51.
66. Mauro A. Satellite cells of skeletal muscle fibers. J Biochem Biophys Cytol. 1961;9:493–8.
67. Leblond CP. Classification of cell populations on the basis of their proliferative behavior. Natl Cancer Inst Monogr. 1964;14:119–50.
68. Cossu G, Zani B, Coletta M, Bouche M, Pacifici M, Molinaro M. In vitro differentiation of satellite cells isolated from normal and dystrophic mammalian muscles. A comparison with embryonic myogenic cells. Cell Differ. 1980;9(6):357–68.
69. Bischoff R. The satellite cell and muscle regeneration. In: Engel AG, Franzini-Armstrong C, eds. Myology: basic and clinical. 2nd ed. New York: McGraw-Hill; 1994. pp. 97–118.
70. Cornelison DD, Wold BJ. Single-cell analysis of regulatory gene expression in quiescent and activated mouse skeletal muscle satellite cells. Dev Biol. 1997;191(2):270–83.
71. Yablonka-Reuveni Z, Rivera AJ. Temporal expression of regulatory and structural muscle proteins during myogenesis of satellite cells on isolated adult rat fibers. Dev Biol. 1994;164(2): 588–603.
72. Beauchamp JR, Heslop L, Yu DS, et al. Expression of CD34 and Myf5 defines the majority of quiescent adult skeletal muscle satellite cells. J Cell Biol. 2000;151(6):1221–34.
73. Yoshida N, Yoshida S, Koishi K, Masuda K, Nabeshima Y. Cell heterogeneity upon myogenic differentiation: down-regulation of MyoD and Myf-5 generates "reserve cells". J Cell Sci. 1998;111(Pt 6):769–79.
74. Miller JB, Schaefer L, Dominov JA. Seeking muscle stem cells. Curr Top Dev Biol. 1999;43:191–219.
75. Seale P, Rudnicki MA. A new look at the origin, function, and "stem-cell" status of muscle satellite cells. Dev Biol. 2000;218(2):115–24.
76. Seale P, Sabourin LA, Girgis-Gabardo A, Mansouri A, Gruss P, Rudnicki MA. Pax7 is required for the specification of myogenic satellite cells. Cell. 2000;102(6):777–86.
77. Covault J, Sanes JR. Distribution of N-CAM in synaptic and extrasynaptic portions of developing and adult skeletal muscle. J Cell Biol. 1986;102(3):716–30.
78. Asakura A, Komaki M, Rudnicki M. Muscle satellite cells are multipotential stem cells that exhibit myogenic, osteogenic, and adipogenic differentiation. Differentiation. 2001;68(4–5):245–53.
79. Partridge TA, Morgan JE, Coulton GR, Hoffman EP, Kunkel LM. Conversion of mdx myofibres from dystrophin-negative to -positive by injection of normal myoblasts. Nature. 1989;337 (6203):176–9.
80. Huard J, Acsadi G, Jani A, Massie B, Karpati G. Gene transfer into skeletal muscles by isogenic myoblasts. Hum Gene Ther. 1994;5(8):949–58.
81. Gussoni E, Soneoka Y, Strickland CD, et al. Dystrophin expression in the mdx mouse restored by stem cell transplantation. Nature. 1999;401(6751):390–4.
82. Kinoshita I, Vilquin JT, Guerette B, Asselin I, Roy R, Tremblay JP. Very efficient myoblast allotransplantation in mice under FK506 immunosuppression. Muscle Nerve. 1994;17(12):1407–15.
83. Vilquin JT, Wagner E, Kinoshita I, Roy R, Tremblay JP. Successful histocompatible myoblast transplantation in dystrophin-deficient mdx mouse despite the production of antibodies against dystrophin. J Cell Biol. 1995;131 (4):975–88.
84. McKinney-Freeman SL, Jackson KA, Camargo FD, Ferrari G, Mavilio F, Goodell MA. Muscle-derived hematopoietic stem cells are hematopoietic in origin. Proc Natl Acad Sci U S A. 2002;99(3):1341–6.
85. Asakura A, Seale P, Girgis-Gabardo A, Rudnicki MA. Myogenic specification of side population cells in skeletal muscle. J Cell Biol. 2002;159(1):123–34.
86. Montanaro F, Liadaki K, Schienda J, Flint A, Gussoni E, Kunkel LM. Demystifying SP cell purification: viability, yield, and phenotype are defined by isolation parameters. Exp Cell Res. 2004;298(1):144–54.
87. Jackson KA, Mi T, Goodell MA. Hematopoietic potential of stem cells isolated from murine skeletal muscle. Proc Natl Acad Sci U S A. 1999;96(25):14482–6.
88. Liadaki K, Kho AT, Sanoudou D, et al. Side population cells isolated from different tissues share transcriptome signatures and express tissue-specific markers. Exp Cell Res. 2005;303(2):360–74.
89. Schienda J, Engleka KA, Jun S, et al. Somitic origin of limb muscle satellite and side population cells. Proc Natl Acad Sci U S A. 2006;103(4):945–50.

90. Galvez BG, Sampaolesi M, Brunelli S, et al. Complete repair of dystrophic skeletal muscle by mesoangioblasts with enhanced migration ability. J Cell Biol. 2006;174(2):231–43.
91. Minasi MG, Riminucci M, De Angelis L, et al. The meso-angioblast: a multipotent, self-renewing cell that originates from the dorsal aorta and differentiates into most mesodermal tissues. Development. 2002;129(11):2773–83.
92. Sampaolesi M, Torrente Y, Innocenzi A, et al. Cell therapy of alpha-sarcoglycan null dystrophic mice through intra-arterial delivery of mesoangioblasts. Science. 2003;301(5632): 487–92.
93. Dellavalle A, Sampaolesi M, Tonlorenzi R, et al. Pericytes of human skeletal muscle are myogenic precursors distinct from satellite cells. Nat Cell Biol. 2007;9(3):255–67.
94. Andreeva ER, Pugach IM, Gordon D, Orekhov AN. Continuous subendothelial network formed by pericyte-like cells in human vascular bed. Tissue Cell. 1998;30(1):127–35.
95. Di Rocco G, Iachininoto MG, Tritarelli A, et al. Myogenic potential of adipose-tissue-derived cells. J Cell Sci. 2006;119(Pt 14):2945–52.
96. Gavina M, Belicchi M, Rossi B, et al. VCAM-1 expression on dystrophic muscle vessels has a critical role in the recruitment of human blood-derived CD133+ stem cells after intra-arterial transplantation. Blood. 2006;108(8):2857–66.
97. Torrente Y, Belicchi M, Sampaolesi M, et al. Human circulating AC133(+) stem cells restore dystrophin expression and ameliorate function in dystrophic skeletal muscle. J Clin Invest. 2004;114(2):182–95.
98. Sarig R, Baruchi Z, Fuchs O, Nudel U, Yaffe D. Regeneration and transdifferentiation potential of muscle-derived stem cells propagated as myospheres. Stem Cells. 2006;24(7):1769–78.
99. Lee JY, Qu-Petersen Z, Cao B, et al. Clonal isolation of muscle-derived cells capable of enhancing muscle regeneration and bone healing. J Cell Biol. 2000;150(5):1085–100.
100. Winitsky SO, Gopal TV, Hassanzadeh S, et al. Adult murine skeletal muscle contains cells that can differentiate into beating cardiomyocytes in vitro. PLoS Biol. 2005;3(4):e87.
101. Zheng B, Cao B, Crisan M, et al. Prospective identification of myogenic endothelial cells in human skeletal muscle. Nat Biotechnol. 2007;25(9):1025–34.
102. Peault B, Rudnicki M, Torrente Y, et al. Stem and progenitor cells in skeletal muscle development, maintenance, and therapy. Mol Ther. 2007;15(5):867–77.
103. Olguin HC, Olwin BB. Pax-7 up-regulation inhibits myogenesis and cell cycle progression in satellite cells: a potential mechanism for self-renewal. Dev Biol. 2004;275(2):375–88.
104. Qu-Petersen Z, Deasy B, Jankowski R, et al. Identification of a novel population of muscle stem cells in mice: potential for muscle regeneration. J Cell Biol. 2002;157(5):851–64.
105. Deasy BM, Qu-Peterson Z, Greenberger JS, Huard J. Mechanisms of muscle stem cell expansion with cytokines. Stem Cells. 2002;20(1):50–60.
106. Deasy BM, Jankowski RJ, Payne TR, et al. Modeling stem cell population growth: incorporating terms for proliferative heterogeneity. Stem Cells. 2003;21(5):536–45.
107. Zammit PS, Golding JP, Nagata Y, Hudon V, Partridge TA, Beauchamp JR. Muscle satellite cells adopt divergent fates: a mechanism for self-renewal? J Cell Biol. 2004;166(3):347–57.
108. Halevy O, Piestun Y, Allouh MZ, et al. Pattern of Pax7 expression during myogenesis in the posthatch chicken establishes a model for satellite cell differentiation and renewal. Dev Dyn. 2004;231(3):489–502.
109. Conboy IM, Rando TA. The regulation of Notch signaling controls satellite cell activation and cell fate determination in postnatal myogenesis. Dev Cell. 2002;3(3):397–409.
110. Beauchamp JR, Morgan JE, Pagel CN, Partridge TA. Dynamics of myoblast transplantation reveal a discrete minority of precursors with stem cell-like properties as the myogenic source. J Cell Biol. 1999;144(6):1113–22.
111. Baroffio A, Hamann M, Bernheim L, Bochaton-Piallat ML, Gabbiani G, Bader CR. Identification of self-renewing myoblasts in the progeny of single human muscle satellite cells. Differentiation. 1996;60(1):47–57.
112. Collins CA, Olsen I, Zammit PS, et al. Stem cell function, self-renewal, and behavioral heterogeneity of cells from the adult muscle satellite cell niche. Cell. 2005;122(2):289–301.
113. Collins CA. Satellite cell self-renewal. Curr Opin Pharmacol. 2006;6(3):301–6.
114. Kuang S, Kuroda K, Le Grand F, Rudnicki MA. Asymmetric self-renewal and commitment of satellite stem cells in muscle. Cell. 2007;129(5):999–1010.
115. Cossu G, Tajbakhsh S. Oriented cell divisions and muscle satellite cell heterogeneity. Cell. 2007;129(5):859–61.
116. Mitchell PO, Mills T, O'Connor RS, Graubert T, Dzierzak E, Pavlath GK. Sca-1 negatively regulates proliferation and differentiation of muscle cells. Dev Biol. 2005;283(1): 240–52.
117. Molnar G, Ho ML, Schroedl NA. Evidence for multiple satellite cell populations and a non-myogenic cell type that is regulated differently in regenerating and growing skeletal muscle. Tissue Cell. 1996;28(5):547–56.
118. Schultz E. Satellite cell proliferative compartments in growing skeletal muscles. Dev Biol. 1996;175(1):84–94.
119. Deasy BM, Li Y, Huard J. Tissue engineering with muscle-derived stem cells. Curr Opin Biotechnol. 2004;15(5):419–23.
120. Zammit P, Beauchamp J. The skeletal muscle satellite cell: stem cell or son of stem cell? Differentiation. 2001;68(4–5): 193–204.
121. Wagers AJ, Conboy IM. Cellular and molecular signatures of muscle regeneration: current concepts and controversies in adult myogenesis. Cell. 2005;122(5):659–67.
122. Cairns J. Mutation selection and the natural history of cancer. Nature. 1975;255(5505):197–200.
123. Cairns J. Somatic stem cells and the kinetics of mutagenesis and carcinogenesis. Proc Natl Acad Sci U S A. 2002;99(16): 10567–70.
124. Rando TA. The immortal strand hypothesis: segregation and reconstruction. Cell. 2007;129(7):1239–43.
125. Lansdorp PM. Immortal strands? Give me a break. Cell. 2007;129(7):1244–7.
126. Lark KG, Consigli RA, Minocha HC. Segregation of sister chromatids in mammalian cells. Science. 1966;154(753): 1202–5.
127. Potten CS, Owen G, Booth D. Intestinal stem cells protect their genome by selective segregation of template DNA strands. J Cell Sci. 2002;115(Pt 11):2381–8.
128. Karpowicz P, Morshead C, Kam A, et al. Support for the immortal strand hypothesis: neural stem cells partition DNA asymmetrically in vitro. J Cell Biol. 2005;170(5): 721–32.
129. Conboy MJ, Karasov AO, Rando TA. High incidence of non-random template strand segregation and asymmetric fate determination in dividing stem cells and their progeny. PLoS Biol. 2007;5(5):e102.
130. Shinin V, Gayraud-Morel B, Gomes D, Tajbakhsh S. Asymmetric division and cosegregation of template DNA strands in adult muscle satellite cells. Nat Cell Biol. 2006;8(7):677–87.
131. Kiel MJ, He S, Ashkenazi R, et al. Haematopoietic stem cells do not asymmetrically segregate chromosomes or retain BrdU. Nature. 2007;449(7159):238–42.

Determinants of Pluripotency in Mouse and Human Embryonic Stem Cells

Leon M. Ptaszek and Chad A. Cowan

Abstract Embryonic stem cells, derived from the inner cell mass of blastocyst stage embryos prior to implantation, remain pluripotent and self-renewing due to both their inherent properties and the culture conditions in which they are propagated. Recent study of the genetic and epigenetic mechanisms that underlie pluripotency in embryonic stem cells has revealed that mouse and human embryonic stem cells have a number of key features in common; however, our knowledge of this area is incomplete. Detailed analyses of mouse and human embryonic stem cells have revealed a number of differences whose significance is not yet understood. An improved knowledge of the molecular underpinnings of embryonic stem cell properties will be required if these cells are to be utilized as part of cell-based therapies. This chapter offers a review of the current understanding of the molecular mechanisms of pluripotency in human and mouse embryonic stem cells. We also describe insights produced by the use of alternate strategies for production of pluripotent cells, such as somatic cell nuclear transfer and direct reprogramming of terminally differentiated somatic cells.

Keywords Embryonic stem cell · Pluripotency · Epigenetics · Reprogramming · Development

1 Introduction

The mammalian zygote is totipotent, as it gives rise to both embryonic and extraembryonic tissues. As cell division proceeds during early preimplantation embryonic development (PED), the resultant daughter cells progressively lose developmental potential. By the time the embryo reaches the blastocyst stage, the multicellular structures that lead to embryonic and extraembryonic lineages are present. Cells in the external envelope, or trophectoderm, of the blastocyst give rise to extraembryonic tissues such as the amniotic sac and the placenta (Fig. 1). Cells in the inner cell mass (ICM) of the blastocyst embryo give rise to all embryonic structures, and are defined as being pluripotent. The traditional definition of pluripotency imputes ability of a cell to give rise to all three germ layers present in the developed organism (endoderm, mesoderm, and ectoderm), as well as germ cells. After implantation of the blastocyst stage embryo, cells from the ICM amplify and give rise to the epiblast, which in turn gives rise to the three distinct germ layers. As cells commit to germ layer lineages, they lose both pluripotency and capacity for self-renewal. By the time gastrulation occurs, true pluripotent cells are no longer thought to be present in the embryo, with the exception of spermatogonial stem cells.

Mouse pluripotent cells were first successfully cultivated in vitro by two independent groups in 1981. Both groups took cells from the ICM of pre-implantation, blastocyst stage embryos [1, 2]. These ICM cells were then propagated in culture, using different methods: Evans and Kaufman utilized feeder layers of mouse embryonic fibroblasts, whereas Martin utilized medium conditioned with teratocarcinoma stem cells. The resultant pluripotent cells, which can be cultured indefinitely without diminution of their pluripotency or ability to self-renew, are termed embryonic stem (ES) cells. Over time, feeder layer culture became the generally accepted technique for mouse ES cell propagation.

Derivation of human ES cells from blastocyst embryos (using mouse fibroblast feeder layer culture) was first published in 1998 by Thomson and co-workers [3]. Since these initial derivations, both mouse and human ES cells have been successfully propagated in culture by numerous independent research groups. Mouse ES cells have also been derived from the epiblast (primitive endoderm) of peri-implantation blastocysts [4]. Multiple ES lines have been established for both mouse and human. Our group has reported 17 distinct human ES lines that were derived and maintained using a single protocol [5]. Recently, several groups have attempted to put

L.M. Ptaszek (✉)
Harvard Stem Cell Institute, 42 Church Street, Cambridge, MA 02138; Stowers Medical Institute, Cardiovascular Research Center and Center for Regenerative Medicine, Massachusetts General Hospital, 185 Cambridge Street CPZN 4265-A, Boston, MA 02114,
e-mail: lptaszek@partners.org

V.K. Rajasekhar, M.C. Vemuri (eds.), *Regulatory Networks in Stem Cells*, Stem Cell Biology and Regenerative Medicine,
DOI 10.1007/978-1-60327-227-8_3, © Humana Press, a part of Springer Science+Business Media, LLC 2009

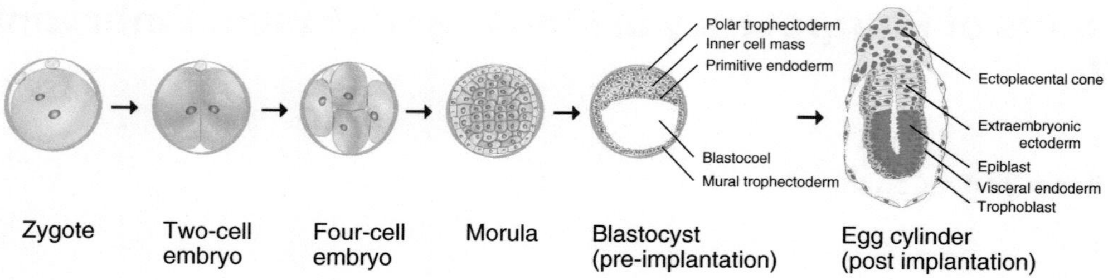

Fig. 1 Summary of embryonic development. The zygote is formed upon fertilization. Cellular activities at this stage of development are directed by maternally deposited transcripts. Paternal transcripts start to contribute at the time of zygote genome activation (ZGA). In mouse, this occurs at the two-cell stage, and in human, this occurs at the four-cell stage. Transcription of genes associated with pluripotency starts at the time of ZGA [95]. Symmetric cleavage divisions continue until the eight-cell stage. After the eight-cell stage, not all cells have an equal proportion of cell surface facing the outside of the embryo and some degree of cell polarity is observed. It is thought that cell fate decisions occur at this time, with external cells being more likely to contribute to the trophectoderm [96]. At this point, the embryo undergoes compaction and eventually forms the morula. Soon after morula formation, a cavity forms in the embryo. This cavity, termed the blastocoel, separates the inner cell mass (ICM) from the mural trophectoderm. Cells of the ICM are pluripotent and are generally the source for mammalian ES cells. The blastocyst then hatches from the zona pellucida and implants into the uterine wall. After implantation, cells from the ICM form the primitive epiblast in the egg cylinder, a structure from which stem cells can be derived [17, 18]. This structure subsequently differentiates into the three germ layers

forward a standardized set of ES derivation conditions [5–8]. In addition, the International Stem Cell Initiative recently published a thorough verification and comparison of a group of 49 human ES lines from 17 laboratories worldwide [9]. Although these lines were propagated using different protocols, all appeared pluripotent and exhibited unlimited capacity for self-renewal in culture. Pluripotency of many of these lines was verified using the "gold standard" assay for determining the pluripotency of human ES cells: assessment of the ability of the ES cell to form tissues from all three germ layers. In general, this is determined by the ability of ES cells injected into immunodeficient mice to form teratomas [10].

2 Phenotypic Differences Between Mouse and Human ES Cells

Analysis of protein expression patterns in ES cells reveals some variation between mouse and human: previously described molecular markers of pluripotency were present in all cell lines studied, but significant differences in expression of several lineage markers were observed. Both mouse and human ES cells express high levels of alkaline phosphatase and telomerase, as well as the cell surface marker CD133 and members of the SSEA receptor family (SSEA1 in mouse, SSEA3/4 in human). Tra1-60/81 receptors, on the other hand, are only expressed on human ES cells. A comprehensive list of markers used to identify ES cells can be found on the NIH stem cell resource web site (http://stemcells.nih.gov/info/scireport/appendixe.asp#eii) as noted previously [11].

Proposed explanations for phenotype differences between mouse and human ES cells include interspecies variability and differences in techniques of pluripotent cell procurement from embryos and propagation of ES cells in culture. Since ES cells are the artificial products of in vitro culture systems, it is not possible to attribute observed differences solely to interspecies variability. Cells from the ICM, when cultured in vitro, are highly responsive to changes in media conditions. It is therefore possible that some of the observed differences in cellular properties are due to culture artifacts rather than true interspecies differences. Culture conditions required to maintain mouse and human ES cells are notably different. Perhaps the most salient difference is the dependence of mouse ES cell growth and pluripotency on the presence of myeloid leukemia inhibitory factor, or LIF [12]. It is noteworthy that LIF, a member of the interleukin (IL)-6 family, supports mouse ES cell growth only in the context of serum, suggesting that other signaling molecules are also necessary.

It may also be the case that mouse and human ES cells are different because they are derived from cells with subtle differences in pluripotency. This is a possible explanation for the differences between human and mouse ES cells with respect to their ability to differentiate into extraembryonic tissues. While human ES cells can be directed to differentiate into trophoblastic cells, the same is not generally true for mouse ES cells [13]. It should be noted that an exception exists: mouse ICM cells and mouse ES cells, under specific culture conditions, have been shown to give rise to trophectodermal tissues [14, 15]. Only mouse ES cells have been shown to contribute to extraembryonic mesoderm [16].

The possibility that currently available mouse and human stem cells have been derived from cells with slightly different developmental potential is supported by recent reports of successful creation of stem cells from post-implantation embryos [17, 18]. In these studies, epiblast cells from mouse and rat embryos were used to create stem cells. These epiblast stem cells (EpiSC) are pluripotent, but likely represent a transition between pluripotent status and differentiation into specific tissue lineages. Perhaps the most striking property of the mouse and rat EpiSCs is the absence of LIF requirement for propagation in culture. In fact, mouse EpiSCs could not be derived in the presence of LIF or BMP4 signal [18]. Instead, much like human ES cells, mouse EpiSCs can be maintained in chemically defined medium in which Activin/Nodal signaling is activated. The absence of a LIF requirement suggests that the overall phenotype of mouse/rat EpiSCs is similar to human ES cells, at least in some respects: this hypothesis is supported by the results of transcriptional profiling and whole genome ChIP-on-chip promoter arrays [18]. Similar comparisons of EpiSCs to epiblast cells, ICM cells, and mouse ES cells reveals greater similarity between EpiSCs and epiblast cells than between EpiSCs and ICM and ES cells [17]. Although these data are intriguing, the relative contributions of interspecies variability and culture artifact remain incompletely defined.

Due to these uncertainties, much energy has recently been placed into efforts to improve our knowledge of the basic properties of ES cells. One point of focus in recent research is further refinement of the definition of the molecular phenotype of a stem cell. We review the current understanding of the genetic and epigenetic features that are unique to stem cells.

3 Genetic Determinants of Pluripotency in Embryonic Stem Cells

The transcriptional networks responsible for maintenance of pluripotency appear to be highly conserved between human and mouse. These similarities are not surprising, as pluripotency appears to be a fundamental and conserved biological property. The expression of these common factors appears to reflect a specific transcriptional state associated with pluripotency. Despite these similarities, stimulation of these common factors via cell surface receptors appears to be quite different between mouse and human [19]. As noted above, our incomplete understanding of the properties of ES cells limits our ability to ascribe this variability solely to interspecies differences. We start our discussion here with a review of the distinct sets of cell surface inputs associated with pluripotency/self-renewal in mouse and human ES cells. This is followed by a discussion of the transcription factors known to be expressed in both mouse and human ES cells.

3.1 *Signaling in Mouse ES Cells*

As stated above, the generally accepted technique for mouse ES cell culture involves the use of mouse embryonic fibroblast (MEF) feeder culture with fetal calf serum–containing medium. All published protocols for mouse ES propagation in serum-based conditions indicate that the presence of additional LIF is required [8]. The presence of LIF activates several distinct signaling pathways.

LIF activation of STAT3 has been shown to be critical for mouse ES cell self-renewal [20–22]. LIF engages the LIF/gp130 heterodimeric receptor on the ES cell surface (Fig. 2). Thus bound, this receptor activates signaling through the JAK-STAT pathway through the recruitment of STAT3: binding-activated STAT3 translocates from the perimembrane cytoplasm to the nucleus, leading to activation of multiple effectors of the JAK-STAT pathway, including the transcription factor c-myc [23]. Activation of c-myc has been linked to the increased transcription of many genes. C-myc transcription appears to also be increased by signaling through the PI3K pathway [24, 25]. LIF signaling also maintains the deactivation of glycogen synthase kinase 3β (GSK3β), apparently through Akt [26]. In fact, Akt activation is sufficient to maintain for mouse and primate ES cell self-renewal in the absence of LIF [26]. In its activated state, GSK3β phosphorylates c-myc, thus targeting it for ubiquitination and consequent degradation [27]. Direct inhibition of GSK3β by application of BIO can improve the efficiency of mouse ES culture, but cannot replace LIF altogether [28]. LIF has been shown to activate the Ras-MAPK pathway, although the downstream effectors of this pathway relevant to pluripotency/self-renewal are not well described [21–23]. Removal of LIF from mouse ES cell culture leads to cellular differentiation, likely through the effects of Fgf4 activation of ERK. Interestingly, LIF is capable of increasing ERK activity, but it appears that direct activation of ERK by LIF is counterbalanced by the other pro-pluripotency effects of LIF [29, 30]. The role of LIF in native mouse development is less clear. LIF may act to maintain the blastocyst in the event that implantation is delayed, but the absence of LIF does not affect development under normal circumstances [31]. In addition, LIF has no clear role in human development: the discrepancy between the prominence of LIF in vitro versus in vivo may be due to the presence of other redundant factors in vivo [32].

Proteins in the WNT family have established roles in determining differentiation and organogenesis at several phases of development. In mouse, several Wnt proteins, including Wnt5a and Wnt11, are expressed late in preimplantation development and at the time of uterine implantation [33]. Activation of the Frizzled receptor by extracellular Wnt protein leads to inhibition of GSK3β. The specific role of Wnt in determining pluripotency is not yet fully understood, although there is some evidence that signaling through the Wnt pathway is involved in the maintenance of ES cell pluripotency. Direct inhibition of GSK3β by use of the BIO molecule can increase the amount of pluripotency-associated factors

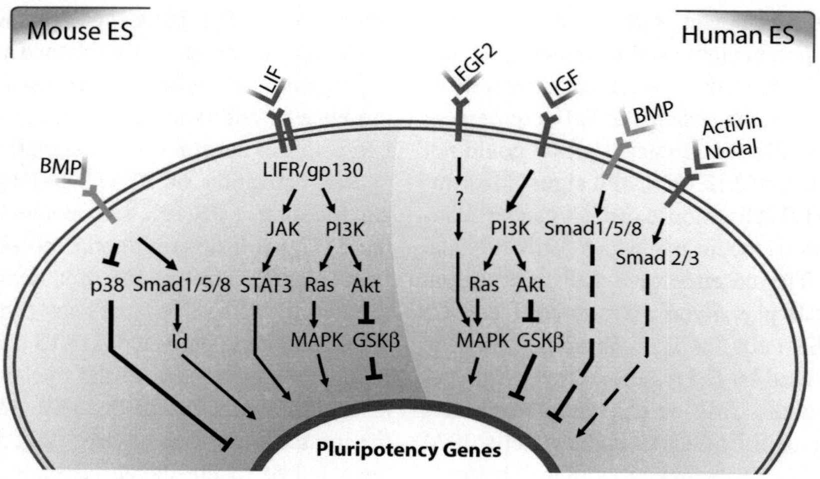

Fig. 2 Signaling pathways in mouse and human ES cells. Unlike human ES cells, mouse ES cells in culture require the presence of LIF to retain pluripotency. LIF appears to maintain pluripotency in mouse ES cells by stimulating signaling through the JAK-STAT and the PI3K pathways. It has been noted that PI3K activates both Ras and Akt. Ras serves to stimulate signaling through MAPK, which subsequently upregulates transcription of pluripotency-related factors through poorly characterized intermediates. Akt activation through PI3K leads to inhibition of GSK3β. This inhibition releases the inhibition of pluripotency-related transcription factors, including c-myc [20–27]. In human ES cells, as in mouse ES cells, PI3K and Activin/Nodal-related signaling plays critical roles in the maintenance of the ES cell state [39, 61]. In human, Activin/Nodal signaling is critical for the maintenance of the pluripotent state. In contrast, Activin/Nodal signaling has been implicated in maintenance of self-renewal, but not pluripotency, in mouse ES cells [36–38]. Several other notable differences exist between mouse and human ES cells. For example, BMP signaling appears to play distinct roles in mouse and human ES cells. In mouse ES cells, BMP has been noted to be required for maintenance of pluripotency. BMP receptor signaling leads to the inhibition of p38 MAPK signaling, which releases the transcription of pluripotency-related genes from inhibition. BMP signaling also increases Id-related signaling through the Smad pathway, which upregulates the transcription of STAT3 and other transcription-related genes [35]. In human ES cells, BMP signaling appears to drive differentiation into trophectoderm.

expressed in cultured mouse ES cells, although Wnt proteins themselves have not yet been directly implicated [28]. Recent screening studies have implicated Wnt5a and Wnt6 as factors that inhibit differentiation of mouse ES cells; however, it has been found that the Wnt-related effects on pluripotency (such as upregulation of STAT3) can be recapitulated by stimulation of downstream factors in the Wnt pathway, such as β-catenin [34].

Members of the Transforming Growth Factor β (TGFβ) family, including the Bone Morphogenic Protein (BMP) family and Activin/Nodal, are also critical to self-renewal in mouse ES cells. The primary action of BMPs in mouse ES cells appears to be the inhibition of differentiation [35, 36]. BMPs have been shown to act through several pathways. Inhibition of p38 MAPK appears to maintain pluripotency by blocking the pro-differentiation effects of p38 MAPK [29], but the specific transcription factors involved have not been well defined. BMPs also induce Id genes via the Smad 1/5/8 pathway [35], leading to maintenance of the pluripotent state. Interactions between BMP-mediated signaling intermediates and Nanog have also been implicated in the maintenance of pluripotency through inhibition of differentiation signals [37]. Ample amounts of BMPs are present in fetal calf serum, so addition of supplemental BMP (extracted or recombinant) is not necessary for propagation of mouse ES cells using fetal calf serum-containing media; however, addition of BMP to some conditions may improve the efficiency of pluripotency maintenance in ES culture [35].

Signaling through the Transforming Growth Factor β (TGFβ) family has also been shown to promote self-renewal of mouse ES cells, without affecting pluripotency. Inhibition of signaling through Activin/Nodal (by increase in Smad7 level or use of chemical inhibitor SB-431542) decreases mouse ES cell division in culture, likely through reduction of Smad2/3 signaling [38]. This decrease in cell division is not associated with a change in ES cell pluripotency, as measured by the expression of molecular markers.

3.2 Signaling in Human ES Cells

Most investigators use MEF feeder layer culture with fetal calf serum–containing media for growth of human ES cells. Detailed analysis of the factors present in serum has revealed a number of factors necessary for pluripotency/self-renewal of human ES cells. Although signaling through the core pathways involving PI3K, Activin/Nodal, and BMPs are present in human and mouse, the upstream effectors and effects of signaling frequently differ [39, 40].

As stated above, cultured human ES cells do not require LIF. Rather, human ES cells appear to require FGF2: a recombinant form of FGF2 does not need to be added to serum-containing medium, but removal of this factor from serum-free media decreases the stability of human ES cells [39, 41]. Blocking the FGF pathway in human ES cells can lead to differentiation [42]. Ding and co-workers [43, 44] have reported several chemical compounds that can replace serum, at least for finite periods of time, presumably through replacement of FGF2 activity and other activities. Even in standard MEF/serum culture conditions, the true effects of FGF2 are somewhat difficult to pinpoint, and FGF2 has been shown to stimulate release of TGFβ from MEFs [45]. Another noteworthy difference between human and mouse is signaling through BMPs. In cultured mouse ES cells, presence of BMP signaling is thought to contribute to the maintenance of pluripotency. In contrast, enhanced BMP-mediated signaling in human ES cells has been shown to drive differentiation toward trophectoderm, and removal of BMP-mediated signaling can restore the pluripotent phenotype [46, 47].

Up-regulation of ERK signaling may have opposing effects in mouse and human ES cells. In mouse ES cells, ERK signaling may promote differentiation. In experiments performed using human ES cells cultured in serum-free conditions, ERK signaling acts to preserve the pluripotent phenotype [48]. As these effects have been observed using different culture conditions, it is not certain that this represents a true difference between species: it remains a formal possibility that this difference is at least partially the result of a culture artifact.

There are several notable similarities between self-renewal/pluripotency signals in mouse and human ES cells (Fig. 2). For example, signaling through PI3K is necessary in both species. Important agonists for PI3K signaling in cultured human ES cells include insulin/insulin-like growth factor (IGF) [49]. Human ES cells express insulin-like growth factor receptor type 1 (IGFR1), which appears to be responsible for mediating insulin/IGF signaling. Blocking of IGFR1 leads to decreased pluripotency of human ES cells [42]. Decreased IGFR1 signaling leads to a decrease in PI3K-mediated signaling. This leads to suppression of Akt, leading to a presumed decrease in suppression of GSK3β activity. Downstream effectors in this pathway have not been well-defined. As in mouse ES cells, an increase in GSK3β activity is associated with increase in ES cell differentiation [50, 51].

Members of the TGFβ family are critical to maintenance of pluripotency in human ES cells, as is the case in mouse ES cells. In particular, Activin/Nodal has been shown to increase the transcription of pluripotency-associated genes, such as Oct4 and Nanog [38]. It has been shown that the critical products produced by MEF feeder layers include Activin A and FGF2 [52]. The effects of Activin A are thought to be mediated through Smad2/3, and may also be up-regulated by FGF2 [38, 53, 54]. In fact, addition of FGF2 to ES cell culture medium allows for the propagation of pluripotent ES cells in the absence of a MEF feeder layer [53, 55]. Stimulation of signaling through Activin A is also thought to be associated with an increase in expression in the FGF, Wnt, and Hedgehog pathways. Activin A signal is also associated with suppression of the differentiating effects of the BMP signal [52]. Activin A/Nodal signaling is also implicated in differentiation, under specific circumstances. For example, Activin A/Nodal is necessary for differentiation of ES cells into endoderm, but only when PI3K and IGFR-mediated signaling is repressed [50].

4 Control of the Cell Cycle in Pluripotent Cells

The cell cycle in mouse and human pre-gastrulation embryos is significantly faster than the cell cycle post-gastrulation in which three distinct germ layers are present. For mouse, the molecular underpinnings of this shift are fairly well understood, as reviewed previously [19]. The difference in total cell cycle time in mouse is associated with a shift in the relative proportion of time spent in S phase: the rapid cell cycle of pluripotent cells is thought to be largely the result of a very short G1 phase and an extended S phase. With differentiation, the length of time spent in G1 increases, leading to a lengthening of the cell cycle and decreased proportion of time spent in S phase. This shift, in mouse, has been attributed to a change in differential Cdk activity. In mouse ES cells, only Cdk1/cyclin B is under cell cycle control. With cellular differentiation, Cdk2/cyclin E comes under cell cycle control as a result of changes in expression of several Cdk inhibitors, such as $p16^{INK4a}$, $p21^{cip1}$, and $p27^{Kip1}$ [56, 57]. While this process is not as well understood in human ES cells, it is thought that a similar series of processes occur, as the shift in cell cycle timing is very similar in mouse and human. None of these cell cycle changes has yet been linked directly to the pluripotency-related cell signaling events discussed here.

5 Pluripotency-Related Transcription Factors in Mouse and Human ES Cells

Extensive study of both mouse and human embryos during PED revealed a core group of transcription factors necessary for the maintenance of pluripotency in both species: Oct3/4, Sox2, and Nanog [58–62]. These factors appear to work in concert to control expression of the complement of genes that ultimately determines the phenotype of the pluripotent cell [59]. As yet, the full complement of the affected genes has not been defined, but much has been learned about the interactions between these transcription factors [11, 19, 63]. In the following paragraphs, we review the current knowledge of the function of Oct3/4, Sox2, Nanog, Klf4, and c-myc in the context of pluripotent cells.

Oct3/4, also referred to as Pou5f1, is a member of the Pou family of transcription factors expressed by pluripotent cells in both mouse and human embryos during PED [64]. Oct3/4 has also been shown to be an important regulator of pluripotency. Oct3/4-deficient mouse embryos do not successfully complete preimplantation development: these embryos develop to a quasi-blastocyst stage, but no true ICM forms [65]. Artificial suppression of Oct3/4 leads to differentiation of pluripotent cells into trophectoderm. Interestingly, artificial overexpression of Oct3/4 also leads to differentiation, in this case to primitive endoderm [66]. There are many identified downstream effectors of Oct3/4. One noteworthy example is Cdx2, which promotes trophectoderm formation: Cdx2 is down-regulated by Oct3/4, leading to maintenance of pluripotency. Suppression of Oct3/4 leads to elevation in Cdx2 levels [54].

Oct3/4 activity is thought to be modulated by an extensive group of factors. For example, Oct3/4 overexpression is not the only known means of driving pluripotent cells toward the primitive endoderm phenotype (this effect is not dependent on Oct3/4 DNA binding activity). The same effect can be achieved through overexpression of Stat3, suggesting the presence of a Stat3-activated cofactor that modulates Oct3/4 levels. This cofactor could be inhibited by overexpression of Oct3/4, leading to stimulation of Stat3 activity [67]. Oct3/4 has been shown to work in concert with other transcriptional factors in the maintenance of pluripotency, notably Sox2 and Nanog [59].

Sox2 is a member of the HMG family of DNA binding proteins that binds with Oct1 and Oct3/4. The Sox2-Oct3/4 complex has been shown to bind to enhancer elements of the Fgf4 gene [68]. In the mouse embryo, Sox2 is expressed in early embryonic development, with decreasing levels noted with tissue differentiation. Sox2 appears to be necessary for formation of primitive and extra-embryonic ectoderm: $Sox2^{-/-}$ embryos are defective in primitive ectoderm [69]. Together, Oct3/4 and Sox2 regulate a number of important, pluripotency-related factors, such as the transcription factor Zinc Finger Protein 206 (ZFP 206) in mouse, or ZNF 206 in human [70]. Sox2 is also important for the regulation of Oct3/4 expression (Fig. 3) but not absolutely required for the maintenance of mouse ES cell pluripotency: ES cells from Sox2 null mice can be rescued by overexpression of Oct3/4 [71].

Nanog is a homeodomain-containing transcription factor (Nk2 class) that contributes to the maintenance of pluripotency in specific cell populations in the mammalian embryo [72]. As with Oct3/4 and Sox2, Nanog mRNA is present

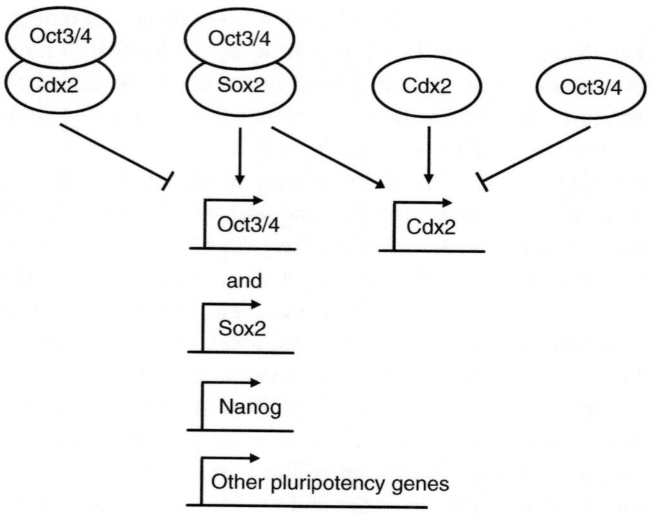

Fig. 3 Regulation of Oct3/4 expression. Oct3/4 acts as a transcription factor that promotes the transcription of a number of genes, including itself and other transcription factors, notably Sox2, and Nanog. Oct3/4 also upregulates the transcription of a number of other genes associated with pluripotency. The means by which the effects of Oct3/4 are suppressed illustrates the mechanisms by which the forces promoting pluripotency and differentiation are kept in balance. Oct3/4 works in concert with Sox2 to promote the transcription of a number of genes, including Cdx2 (promotes trophectoderm formation). The Cdx2 protein works in concert with Oct3/4 to down-regulate Oct3/4 transcription, thus modulating the effects of Oct3/4. This effect is reciprocal, as the Oct3/4 protein down-regulates Cdx2 transcription. Comparable regulatory mechanisms for Nanog have been proposed, but none have yet been experimentally verified [54, 64, 65]

in early embryonic development. Unlike Oct3/4 and Sox2, Nanog expression appears to persist after tissue differentiation [73]. Nanog can block the differentiation of mouse ES cells into primitive endoderm when LIF and BMP are removed (in serum-free conditions), but Nanog is not necessary for the maintenance of pluripotent cells in the developing embryo [74]. In fact, ES cells can propagate in the absence of Nanog. It has been suggested that Nanog can help maintain the pluripotent state by preventing a shift toward transcriptional programs that favor differentiation [75]. Although Nanog is dispensable for the maintenance of somatic pluripotency, it is required for formation of germ cells.

Several groups have recently discovered that simultaneous lentiviral overexpression of a small group of transcription factors (Oct3/4, Sox2, Klf4, and c-myc) is sufficient to reprogram terminally differentiated, adult skin fibroblasts from mouse and human into inducible pluripotent stem (iPS) cells whose phenotype cannot be distinguished from that of embryonic stem cells [76–80]. Expression of Nanog, while not necessary for somatic cell reprogramming, is induced in iPS cells produced by expression of Oct3/4, Sox2, Klf4, and c-myc. This striking result suggests that "direct reprogramming" with these transcription factors produces a highly conserved transcriptional program that maintains pluripotency in mammalian embryos. The full complement of genes that carry out the reprogramming of the somatic cell nucleus is not yet known in detail.

As several of these "reprogramming" factors have been associated with changes in histone modifications, it may be that a change in chromatin structure is central to the reprogramming reaction. For example, c-myc produces an increase in histone acetylation [81], but there is no clear evidence linking the abovementioned factors with discrete changes in chromatin structure and epigenetic state of pluripotent cells. Given the high likelihood that chromatin structure plays a key role in maintenance of pluripotency, further investigation into the epigenetics of pluripotent cells is warranted.

6 Epigenetic Determinants of Pluripotency in Embryonic Stem Cells

It has been noted that the promoter/enhancer regions for the Oct3/4, Sox2, and Nanog genes contain binding sites for all three proteins: therefore, it is possible that these factors govern pluripotency through a self-regulating network [63, 82]. Even if this is the case, the transcriptional effects of these pluripotency-related factors on other genes are undoubtedly influenced by epigenetic factors. For example, consider the recently reported iPS cells. The phenotype of the pluripotent cells produced by this technique, as assessed through teratoma formation assays and measurement of molecular markers, is not distinguishable from the ES cell phenotype. In addition, no differences between mouse iPS and ES cells were noted in global transcriptional profiling and histone modification assays. It has been suggested that pluripotent cells have a stereotypic epigenetic state. The specific epigenetic factors that contribute to pluripotency in embryonic development and embryonic stem cells are not known in detail; however, pluripotent cells (iPS and ES cells) do bear the epigenetic hallmarks of transcriptionally active chromatin.

For the purposes of this discussion, we consider epigenetics to describe the regulation of genome function through chemical modifications of DNA and histone proteins [83]. The epigenetic phenomena most relevant to a discussion of pluripotency include DNA CpG methylation and chemical modifications of histones. Cytosine methylation of DNA is associated with varied effects, including: tissue differentiation and X chromosome inactivation. Similarly, histone modification can lead to a wide array of effects.

ES cells are essentially a culture artifact: therefore, there is no exact epigenetically analogous state during PED. Even so, it stands to reason that the broader forces guiding pluripotency apply in both ES cells and embryonic cells. Specifically, as ES cells are generally procured from the ICM or the epiblast, the epigenetic mechanisms that are operational at the time of implantation are likely the most relevant to cultured ES cells.

Therefore, we begin our discussion with epigenetics of PED. The early phases of PED are associated with several, dynamic epigenetic transitions, as reviewed recently by Surani and co-workers [84]. Immediately after fertilization, the paternal genome is in a condensed, heavily methylated state. Before the paternal and maternal genomes join to form the zygotic genome, the paternal genome undergoes several modifications. Protamine-histone exchange occurs in the male pronucleus and the paternal genome is then actively demethylated. The maternal genome is demethylated passively over multiple cell divisions [85]. The epigenetic state of the totipotent zygote and early blastomeres is "permissive" in that many genes are available to key transcription factors such as Oct3/4 and Cdx2 [15, 86]. These transcription factors modulate formation of the inner cell mass and other structures present in the blastocyst.

The epigenetic state of the blastocyst is quite different from that of the zygote. In the blastocyst, the embryonic genome undergoes a rapid increase in methylation: a differential pattern is observed between the inner cell mass and the trophectoderm, reflecting the difference in pluripotency of the cells in each structure. Differential patterns in histone modifications are also observed during PED. In female embryos, the paternal X chromosome is preferentially silenced in extraembryonic tissues during PED. In the context of mouse SCNT, the paternal X chromosome is preferentially inactivated in extraembryonic tissues as well [87]. This epigenetic "memory" ends with the formation of the ICM at the blastocyst stage: after this point, either X chromosome is equally likely to be silenced in cells that will form embryonic tissues. Therefore, the oocyte is capable of DNA demethylation, but it cannot modify the histone modification involved in X inactivation (trimethylation of Lys 27 of histone 3, or H3K27me3, produced by the PcG protein Ezh2). The blastocyst stage embryo is not limited in this manner [83, 88, 89].

In ES cells, gene silencing is governed by two predominant histone methylation events: H3K27me3, usually associated with gene repression, and H3K4me3, usually associated with active chromatin [83]. These types of methylation events can be catalyzed by several factors, including: Ezh2, Eset, and members of the NuRD complex such as MBD [90, 91]. It is interesting to note that the documented effects of these factors on development occur primarily during the peri-implantation period, the point of development whose epigenetic state is likely to be the most relevant to cultured ES cell epigenetics [84, 92]. The significance of the simultaneous presence of "opposing" histone methylation events in ES cells is not completely understood. Pluripotent cells produced through SCNT exhibit dramatic changes in histone methylation in the somatic cell genome. Genes that are ordinarily repressed are associated with enrichment of H3K27me3 events, and genes that are ordinarily expressed are associated with enrichment of H3K4me3 events [93].

Epigenetic modifications appear to be critically important in iPS cells. Indirect evidence comes from the kinetics of the reprogramming of mouse somatic cells. Initial reports of mouse iPS cells included those cells that were selected early. These cells, while they are pluripotent and express the molecular markers generally associated with pluripotency, did not appear identical to mouse ES cells. Comparison of global transcriptional profiles revealed small, but significant, differences [76]. Later reports included mouse iPS cells that were propagated after a longer period of selection. Comparison of these "later" iPS cells to ES cells did not reveal any discernible differences with respect to pluripotency, expression of molecular markers, transcription profiling, or histone modification [80]. The difference between early versus late selection of iPS cells may be attributed to an incomplete "reset" of the epigenetic state of the terminally differentiated somatic cell utilized for iPS. It therefore appears that epigenetic modifications brought about by transcriptional events must play an important role in resolving the pluripotent cell state.

So what role is played the genetic factors in the maintenance of the presumably unique epigenetic state of ES cells? Only one direct link has been suggested in the literature, as pointed out by Niwa [82]. The transcription of the histone demethylases Jumanji family (Jmjd1a and Jmjd2c, which can demethylate H3K9 events) is under the control of Oct3/4. While these factors may represent a link between the genetic and epigenetic regulators of pluripotency, the activity of Jmjd1a and Jmjd2c in ES cells has not been established [94].

7 Conclusions

Much has been learned recently about the genetic and epigenetic contributors to pluripotency in ES cells. Of note, we are now capable of using terminally differentiated somatic cells to create pluripotent cells that appear very similar to ES cells, albeit with low efficiency. Subtle differences in the properties of different populations of pluripotent cells, such as ICM cells, iPS cells, and ES cells produced from the ICM or SCNT have yielded some insights into the mechanism responsible for pluripotency and self-renewal. Detailed analysis of the components of the cell culture systems in which ES cells are propagated continues to improve our knowledge of the signaling factors required for ES cell pluripotency. While our understanding of specific aspects of ES cell biology, such as the role of key transcription factors such as Oct3/4 and Sox2, is much improved, we do not yet have a grasp of how genetic and epigenetic factors collaborate to produce a pluripotent cell. Comparison of various pluripotent cells using several experimental methods, such as microRNA profiling, global transcriptional profiling, and ChIP, holds promise for the discovery of links between the genetic and epigenetic determinants of pluripotency.

Acknowledgments This work was supported by a grant from the Stowers Medical Institute (C.A.C.). The authors thank K. Brennand for assistance with Figure 1 and L. Fenno for assistance with Figure 2. The authors would like to emphasize that every attempt was made to make this review as comprehensive as possible. Given space constraints, some omissions were necessary.

References

1. Evans M, Kaufman M. Establishment in culture of pluripotential cells from mouse embryos. Nature. 1981;292:154–6.
2. Martin G. Isolation of a pluripotent cell line from early mouse embryos cultured in medium conditioned by teratocarcinoma stem cells. PNAS. 1981;78:7634–8.
3. Thomson J, Itskovitz-Eldor J, Shapiro S, et al. Embryonic stem cell lines derived from human blastocysts. Science. 1998;282:1145–7.
4. Rossant J. Stem cells from the Mammalian blastocyst. Stem Cells. 2001;19:477–82.
5. Cowan C, Klimanskaya I, McMahon J, et al. Derivation of embryonic stem-cell lines from human blastocysts. NEJM. 2004;350:1353–6.
6. Sullivan S, Egli D, Akutsu H, Melton D, Eggan K, Cowan C. Derivation of human embryonic stem cell lines. In: Sullivan S, Cowan C, Eggan K, editors. Human embryonic stem cells: the practical handbook. Chichester: Wiley and Sons; 2007. pp. 35–51.
7. Akutsu H, Cowan C, Melton D. Human embryonic stem cells. Method Enzymol. 2006;418:78–92.
8. Bryja V, Bonilla S, Arenas E. Derivation of mouse embryonic stem cells. Nat Protoc. 2006;1:2082–7.
9. Initiative ISC, Adewumi O, Aflatoonian B, et al. Characterization of human embryonic stem cell lines by the International Stem Cell Initiative. Nat Biotechnol. 2007;25:803–16.
10. Blum B, Benvenisty N. Clonal analysis of human embryonic stem cell differentiation into teratomas. Stem Cells. 2007;25:1924–30.
11. Boiani M, Scholer H. Regulatory networks in embryo-derived pluripotent stem cells. Nat Rev Mol Cell Biol. 2005;6:872–84.
12. Williams R, Hilton D, Pease S, et al. Myeloid leukaemia inhibitory factor maintains the developmental potential of embryonic stem cells. Nature. 1988;336:684–7.
13. Soares M, Wolfe M. Human embryonic stem cells assemble and fulfill their developmental destiny. Endocrinology. 2004;145:1514–6.
14. Pierce G, Arechaga J, Muro C, Wells R. Differentiation of ICM cells into trophectoderm. Am J Pathol. 1988;132:356–64.
15. Niwa H, Toyoka Y, Shimosato D, et al. Interaction between Oct3/4 and Cdx2 determines trophectoderm differentiation. Cell. 2005;123:917–29.
16. Beddington R, Robertson E. An assessment of the developmental potential of embryonic stem cells in the midgestation mouse embryo. Development. 1989;105:733–7.
17. Brons I, Smithers L, Trotter M, et al. Derivation of pluripotent epiblast stem cells from mammalian embryos. Nature. 2007;448:191–5.

18. Tesar P, Chenoweth J, Brook F, et al. New cell lines from mouse epiblast share defining features with human embryonic stem cells. Nature. 2007;448:196–9.
19. Ohtsuka S, Dalton S. Molecular and biologic properties of pluripotent embryonic stem cells. Gene Therapy. 2008;15:74–81.
20. Matsuda T, Nakamura T, Nakao K, et al. STAT3 activation is sufficient to maintain an undifferentiated state of mouse embryonic stem cells. EMBO J. 1999;18:4261–9.
21. Niwa H, Burdon T, Chambers I, Smith A. Self-renewal of pluripotent embryonic stem cells is mediated via activation of STAT3. Genes Dev. 1998;12:2048–60.
22. Raz R, Lee C, Cannizzaro L, d'Eustachio P, Levy D. Essential role of STAT3 for embryonic stem cell pluripotency. PNAS. 1999;96:2846–51.
23. Cartwright P, McLean C, Sheppard A, Rivett D, Jones K, Dalton S. LIF/STAT3 controls ES cell self-renewal and pluripotency by a Myc-dependent mechanism. Development. 2005;132:885–96.
24. Storm M, Bone H, Beck C, et al. Regulation of Nanog expression by phosphoinositide 3-kinase-dependent signaling in murine embryonic stem cells. J Biol Chem. 2007;282:6265–73.
25. Paling N, Wheadon H, Bone H, Welham M. Regulation of embryonic stem cell self-renewal by phosphoinositide 3-kinase-dependent signaling. J Biol Chem. 2004;279:48063–70.
26. Watanabe S, Umehara H, Murayama K, Okabe M, Kimura T, Nakano T. Activation of Akt signaling is sufficient to maintain pluripotency in mouse and primate embryonic stem cells. Oncogene. 2006;25:2697–707.
27. Umehara H, Kimura T, Ohtsuka S, et al. Efficient derivation of embryonic stem cells by inhibition of glycogen synthase kinase-3. Stem Cells. 2007;25:2705–11.
28. Sato N, Meijer L, Skaltsounis L, Greengard P, Brivanlou A. Maintenance of pluripotency in human and mouse embryonic stem cells through activation of Wnt signaling by a pharmacological GSK-3-specific inhibitor. Nat Med. 2004;10:55–63.
29. Kunath T, Saba-El-Leil M, Almousailleakh M, Wray J, Meloche S, Smith A. FGF stimulation of the Erk1/2 signalling cascade triggers transition of pluripotent embryonic stem cells from self-renewal to lineage commitment. Development. 2007;134:2895–902.
30. Burdon T, Stracey C, Chambers I, Nichols J, Smith A. Suppression of SHP-2 and ERK signalling promotes self-renewal of mouse embryonic stem cells. Dev Biol. 1999;210:30–43.
31. Nichols J, Chambers I, Taga T, Smith A. Physiological rationale for responsiveness of mouse embryonic stem cells to gp130 cytokines. Development. 2001;128:2333–9.
32. Humphrey R, Beattie G, Lopez A, et al. Maintenance of pluripotency in human embryonic stem cells is STAT3 independent. Stem Cells. 2004;22:522–30.
33. Mohamed O, Dufort D, Clarke H. Expression and estradiol regulation of Wnt genes in the mouse blastocyst identify a candidate pathway for embryo-maternal signaling at implantation. Biol Reprod. 2004;71:417–24.
34. Hao J, Li T, Qi X, Zhao D, Zhao G. WNT/beta-catenin pathway up-regulates Stat3 and converges on LIF to prevent differentiation of mouse embryonic stem cells. Dev Biol. 2006;290:81–91.
35. Ying Q, Nichols J, Chambers I, Smith A. BMP induction of Id proteins suppresses differentiation and sustains embryonic stem cell self-renewal in collaboration with STAT3. Cell. 2003;31:281–92.
36. Qi X, Li T, Hao J, et al. BMP4 supports self-renewal of embryonic stem cells by inhibiting mitogen-activated protein kinase pathways. PNAS. 2004;101:6027–32.
37. Suzuki A, Raya A, Kawakami Y, et al. Nanog binds to Smad1 and blocks bone morphogenetic protein-induced differentiation of embryonic stem cells. PNAS. 2006;103:10294–9.
38. Ogawa K, Saito A, Matsui H, et al. Activin-Nodal signaling is involved in propagation of mouse embryonic stem cells. J Cell Sci. 2007;120:55–65.
39. Ludwig T, Levenstein M, Jones J, et al. Derivation of human embryonic stem cells in defined conditions. Nat Biotechnol. 2006;24:185–7.
40. Wang L, Schulz T, Sherrer E, et al. Self-renewal of human embryonic stem cells requires insulin-like growth factor-1 receptor and ERBB2 receptor signaling. Blood. 2007;110:4111–9.
41. Levenstein M, Ludwig T, Xu R, Llanas R, VanDenHeuvel-Kramer K, Manning D, Thomson J. Basic fibroblast growth factor support of human embryonic stem cell self-renewal. Stem Cells. 2006;24:568–74.
42. Bendall S, Stewart M, Menendez P, et al. IGF and FGF cooperatively establish the regulatory stem cell niche of pluripotent human cells in vitro. Nature. 2007;448:1015–21.
43. Chen S, Do J, Zhang Q, et al. Self-renewal of embryonic stem cells by a small molecule. PNAS. 2006;103:17266–71
44. Yao S, Chen S, Clark J, et al. Long-term self-renewal and directed differentiation of human embryonic stem cells in chemically defined conditions. PNAS. 2006;103:6907–12.
45. Greber B, Lehrach H, Adjaye J. Fibroblast growth factor 2 modulates transforming growth factor beta signaling in mouse embryonic fibroblasts and human ESCs (hESCs) to support hESC self-renewal. Stem Cells. 2007;25:455–64.
46. Xu R, Chen X, Li D, et al. BMP4 initiates human embryonic stem cell differentiation to trophoblast. Nat Biotechnol. 2002;20:1261–4.
47. Levine A, Brivanlou A. GDF3, a BMP inhibitor, regulates cell fate in stem cells and early embryos. Development. 2006;133:209–16.
48. Feng G. Shp2-mediated molecular signaling in control of embryonic stem cell self-renewal and differentiation. Cell Res. 2007;17:37–41.
49. Wang L, Schulz T, Sherrer E, et al. Self-renewal of human embryonic stem cells requires insulin-like growth factor-1 receptor and ERBB2 receptor signaling. Blood. 2007;110:4111–9.
50. McLean A, D'Amour K, Jones K, et al. Activin a efficiently specifies definitive endoderm from human embryonic stem cells only when phosphatidylinositol 3-kinase signaling is suppressed. Stem Cells. 2007;25:29–38.
51. D'Amour K, Agulnick A, Eliazer S, Kelly O, Kroon E, Baetge E. Efficient differentiation of human embryonic stem cells to definitive endoderm. Nat Biotechnol. 2005;23:1534–41.
52. Xiao L, Yuan X, Sharkis S. Activin A maintains self-renewal and regulates fibroblast growth factor, Wnt, and bone morphogenic protein pathways in human embryonic stem cells. Stem Cells. 2006;24:1476–86.
53. Vallier L, Alexander M, Pedersen R. Activin/Nodal and FGF pathways cooperate to maintain pluripotency of human embryonic stem cells. J Cell Sci. 2005;118:4495–509.
54. Babaie Y, Herwig R, Greber B, et al. Analysis of Oct4-dependent transcriptional networks regulating self-renewal and pluripotency in human embryonic stem cells. Stem Cells. 2007;25:500–10.
55. Beattie G, Lopez A, Bucay N, et al. Activin A maintains pluripotency of human embryonic stem cells in the absence of feeder layers. Stem Cells. 2005;23:489–95.
56. White J, Dalton S. Cell cycle control of embryonic stem cells. Stem Cell Rev. 2005;1:131–8.
57. White J, Stead E, Faast R, Conn S, Cartwright P, Dalton S. Developmental activation of the Rb-E2F pathway and establishment of cell cycle-regulated cyclin-dependent kinase activity during embryonic stem cell differentiation. Mol Biol Cell. 2005;16:2018–27.
58. Hyslop L, Stojkovic M, Armstrong L, et al. Downregulation of NANOG induces differentiation of human embryonic stem cells to extraembryonic lineages. Stem Cells. 2005;23:1035–43.
59. Boyer L, Lee T, Cole M, et al. Core transcriptional regulatory circuitry in human embryonic stem cells. Cell. 2005;122:947–56.

60. Loh Y, Wu Q, Chew J, et al. The Oct4 and Nanog transcription network regulates pluripotency in mouse embryonic stem cells. Nat Genet. 2006;38:431–40.
61. Wang J, Rao S, Chu J, et al. A protein interaction network for pluripotency of embryonic stem cells. Nature. 2006;444: 364–8.
62. Ivanova N, Dobrin R, Lu R, et al. Dissecting self-renewal in stem cells with RNA interference. Nature. 2006;442:533–8.
63. Niwa H. How is pluripotency determined and maintained? Development. 2007;134:635–46.
64. Rosner M, Vigano M, Ozato K, et al. A POU-domain transcription factor in early stem cells and germ cells of the mammalian embryo. Nature. 1990;345:686–92.
65. Nichols J, Zevnik B, Anastassiadis K, et al. Formation of pluripotent stem cells in the mammalian embryo depends on the POU transcription factor Oct4. Cell. 1998;95:379–91.
66. Niwa H, Miyazaki J, Smith A. Quantitative expression of Oct-3/4 defines differentiation, dedifferentiation or self-renewal of ES cells. Nat Genet. 2000;24:372–6.
67. Niwa H. Molecular mechanism to maintain stem cell renewal of ES cells. Cell Struct Funct. 2001;26:137–48.
68. Yuan H, Corbi N, Basilico C, Dailey L. Developmental-specific activity of the FGF-4 enhancer requires the synergistic action of Sox2 and Oct-3. Genes Dev. 1995;9:2635–45.
69. Avilion A, Nicolis S, Pevny L, Perez L, Vivian N, Lovell-Badge R. Multipotent cell lineages in early mouse development depend on SOX2 function. Genes Dev. 2003;17:126–40.
70. Wang Z, Kueh J, Teh C, et al. Zfp206 is a transcription factor that controls pluripotency of embryonic stem cells. Stem Cells. 2007;25:2173–82.
71. Masui S, Nakatake Y, Toyooka Y, et al. Pluripotency governed by Sox2 via regulation of Oct3/4 expression in mouse embryonic stem cells. Nat Cell Biol. 2007;9:625–35.
72. Mitsui K, Tokuzawa Y, Itoh H, et al. The homeoprotein Nanog is required for maintenance of pluripotency in mouse epiblast and ES cells. Cell. 2003;113:631–42.
73. Hart A, Hartley L, Ibrahim M, Robb L. Identification, cloning and expression analysis of the pluripotency promoting Nanog genes in mouse and human. Dev Dyn. 2004;230:187–98.
74. Chambers I, Colby D, Robertson M, et al. Functional expression cloning of Nanog, a pluripotency sustaining factor in embryonic stem cells. Cell. 2003;113:643–55.
75. Chambers I, Silva J, Colby D, et al. Nanog safeguards pluripotency and mediates germline development. Nature. 2007;450:1230–4.
76. Maherali N, Sridharan R, Xie W, et al. Directly reprogrammed fibroblasts show global epigenetic remodeling and widespread tissue contribution. Cell Stem Cell. 2007;1:55–70.
77. Takahashi K, Yamanaka S. Induction of pluripotent stem cells from mouse embryonic and adult fibroblast cultures by defined factors. Cell. 2006;126:663–76.
78. Takahashi K, Tanabe K, Ohnuki M, et al. Induction of pluripotent stem cells from adult human fibroblasts by defined factors. Cell. 2007;131:861–72.
79. Okita K, Ichisaka T, Yamanaka S. Generation of germline-competent induced pluripotent stem cells. Nature. 2007;448:313–7.
80. Wernig M, Meissner A, Foreman R, et al. In vitro reprogramming of fibroblasts into a pluripotent ES-cell-like state. Nature. 2007;448:318–24.
81. Fernandez P, Frank S, Wang L, et al. Genomic targets of the human c-Myc protein. Genes Dev. 2003;17:1115–29.
82. Niwa H. Open conformation chromatin and pluripotency. Genes Dev. 2007;21:2671–6.
83. Bernstein B, Meissner A, Lander E. The mammalian epigenome. Cell. 2007;128:669–81.
84. Surani M, Hayashi K, Hajkova P. Genetic and epigenetic regulators of pluripotency. Cell. 2007;128:747–62
85. Arney K, Erhardt S, Drewell R, Surani M. Epigenetic reprogramming of the genome–from the germ line to the embryo and back again. Int J Dev Biol. 2001;45:533–40.
86. Zernicka-Goetz M. Cleavage pattern and emerging asymmetry of the mouse embryo. Nat Rev Mol Cell Biol. 2005;6:919–28.
87. Eggan K, Akutsu H, Hochedlinger K, Rideout W, Yanagimachi R, Jaenisch R. X-Chromosome inactivation in cloned mouse embryos. Science. 2000;290:1578–81.
88. Lepikhov K, Walter J. Differential dynamics of histone H3 methylation at positions K4 and K9 in the mouse zygote. BMC Dev Biol. 2004;4:12.
89. Santos F, Zakhartchenko V, Stojkovic M, et al. Epigenetic marking correlates with developmental potential in cloned bovine preimplantation embryos. Curr Biol. 2003;13:1116–21.
90. Kaji K, Caballero I, MacLeod R, Nichols J, Wilson V, Hendrich B. The NuRD component Mbd3 is required for pluripotency of embryonic stem cells. Nat Cell Biol. 2006;8:285–92.
91. Kaji K, Nichols J, Hendrich B. Mbd3, a component of the NuRD co-repressor complex, is required for development of pluripotent cells. Development. 2007;134:1123–32.
92. Erhardt S, Su I, Schneider R, et al. Consequences of the depletion of zygotic and embryonic enhancer of zeste 2 during preimplantation mouse development. Development. 2003;130: 4235–48.
93. Kimura H, Tada M, Nakatsuji N, Tada T. Histone code modifications on pluripotential nuclei of reprogrammed somatic cells. Mol Cell Biol. 2004;24:5710–20.
94. Klose R, Kallin E, Zhang Y. JmjC-domain-containing proteins and histone demethylation. Nat Rev Genet. 2006;7:715–27.
95. Schier A. The maternal-zygotic transition: death and birth of RNAs. Science. 2007;316:406–7.
96. Zernicka-Goetz M. The first cell-fate decisions in the mouse embryo: destiny is a matter of both chance and choice. Curr Opin Genet Dev. 2006;16:406–12.

Maintenance of Embryonic Stem Cell Pluripotency by Nanog-Mediated Dedifferentiation of Committed Mesoderm Progenitors

Atsushi Suzuki, Ángel Raya, Yasuhiko Kawakami, Masanobu Morita, Takaaki Matsui, Kinichi Nakashima, Fred H. Gage, Concepción Rodríguez-Esteban and Juan Carlos Izpisúa Belmonte

Abstract Embryonic stem (ES) cells can be propagated indefinitely in culture while retaining the ability to differentiate into any cell type in the organism. The molecular and cellular mechanisms underlying ES cell pluripotency are, however, poorly understood. Here, we characterize a population of early mesoderm-committed (EM) progenitors that is generated from mouse ES cells by bone morphogenetic protein (BMP) stimulation. We further show that EM progenitors are actively dedifferentiated to ES cells by the action of Nanog, which, in turn, is directly up-regulated in EM progenitors by the combined action of leukemia inhibitory factor (LIF) and the early mesoderm transcription factor T/Brachyury. Finally, we demonstrate that this negative feedback mechanism contributes to the maintenance of ES cell pluripotency. These findings uncover specific roles of LIF, Nanog, and BMP in the self-renewal of ES cells and provide novel insights into the cellular bases of ES cell pluripotency.

Keywords Pluripotency · T (Brachyury) · Self-renewal · Mesoderm differentiation · Leukemia inhibitory factor

1 Introduction

Mouse embryonic stem (ES) cells are permanent cell lines derived from pre-implantation embryos [1, 2] that display the peculiarities of combining unlimited self-renewal and pluripotency abilities while retaining a normal karyotype. In practical terms, these peculiarities mean that mouse ES cells can be maintained in culture for indefinite periods of time while conserving their ability to differentiate into any cell type if the appropriate context is provided, either in vivo or in vitro [3].

Strict culture conditions must be followed in order to maintain the self-renewal of pluripotent mouse ES cells. Two extrinsic culture requirements, a feeder layer of fibroblasts and the addition of fetal bovine serum, have been identified to be necessary to sustain proliferation of undifferentiated mouse ES cells and their activities pinpointed to specific molecules [4]. Thus, self-renewal of mouse ES cells can be sustained in feeder-free conditions by supplementing the culture media with the cytokine leukemia inhibitory factor (LIF) [5, 6]. In the absence of LIF, ES cell colonies flatten and form epithelial-like sheets [5, 6]. More recently, the self-renewal-promoting activity of animal serum has been identified to be mediated by ligands of specific families of the transforming growth factor-β (TGFβ)superfamily, including the bone morphogenetic protein (BMP) family members BMP2 and BMP4, and the growth and differentiation factor (GDF) family member GDF6 [7]. In the absence of BMP/GDF signals, LIF is not sufficient to prevent the neural differentiation of ES cells, whereas the absence of both BMP/GDF and LIF stimulation results in a flattened cell phenotype similar to that of LIF withdrawal [7].

The intracellular signaling cascades initiated by both LIF and BMP/GDF that sustain self-renewal of mouse ES cells have been worked out in a significant degree of detail [4]. In summary, binding of LIF to its cognate LIF receptor results in the recruitment of gp130 and the formation of a ternary complex that catalyzes the tyrosine phosphorylation, dimerization, and nuclear translocation of the downstream signal transducer STAT3. BMP/GDF, in turn, promotes ES cell self-renewal by inducing the expression of members of the Inhibitor of differentiation (Id) family of negative transcriptional modulators, most likely mediated by activation of the TGFβ downstream signal transducer Smad1 [7].

In addition to extrinsic requirements, the pluripotency of mouse ES cells has been shown to depend on intrinsic determinants, such as the expression of the POU transcription factor Oct4 [8] and the divergent homeodomain-containing

J.C.I. Belmonte (✉)
Gene Expression Laboratory, Salk Institute for Biological Studies, 10010 N. Torrey Pines Rd., La Jolla, CA 92037; Center of Regenerative Medicine in Barcelona, Dr. Aiguader 88, 08029 Barcelona, Spain
e-mail: belmonte@salk.edu

V.K. Rajasekhar, M.C. Vemuri (eds.), *Regulatory Networks in Stem Cells*, Stem Cell Biology and Regenerative Medicine, DOI 10.1007/978-1-60327-227-8_4,

factor Nanog [9, 10]. Both factors are absolutely required for ES cells to maintain their pluripotent identity. Thus, the lack [8] or down-regulation [11] of *Oct4* expression induces trophoectoderm differentiation, whereas ES cells lacking *Nanog* function differentiate to endoderm lineages [9]. The relationships between extrinsic and intrinsic determinants of ES cell identity are only recently beginning to be understood. The maintenance of pluripotent ES cell self-renewal by Oct4 requires functional LIF/STAT3 and BMP/GDF/Id signaling cascades [7, 11], but the function of LIF/STAT3 does not appear to be the maintenance of *Oct4* expression [11]. Overexpression of *Nanog*, in turn, circumvents the necessity of either LIF or BMP/GDF stimulation [7, 10], although synergism between Nanog function and LIF/STAT3 signaling has been noted [10].

The manipulation of mouse ES cells is currently a standard tool in many laboratories, inasmuch as it allows the generation of mice carrying targeted gene mutations for the direct analysis of a gene's function. Moreover, the ability of mouse ES cells to give rise in vitro to virtually any cell type of the organism has been exploited to gain insights into the molecular and cellular mechanisms of cell differentiation to specific lineages. However, most of the recent interest in ES cell research results from the successful derivation of human pluripotent cell lines [12, 13], which created new prospects for future cell replacement therapies. Many basic questions about the biology of these promising cells need be answered if their potential is to be realized. Human ES cells share with their mouse homonyms the peculiarities of self-renewal and pluripotency. The molecular mechanisms by which self-renewal and pluripotency are maintained in human and mouse ES cells, however, appear to differ [14]. Moreover, the cellular bases of pluripotency of either mouse or human ES cells are largely unknown. How intrinsic and extrinsic determinants of ES cell identity crosstalk to maintain cell pluripotency, whether the symmetric self-renewal of pluripotent ES cells depends on a truly symmetrical cell division, on a particular resistance of ES cells to undergo cell differentiation – yet retaining the ability to do so, or on the reversal of early steps of cell differentiation, and what kinds of mechanisms operate in ES cells so they maintain pluripotency are still open questions.

Here we show that mouse ES cells cultured on feeders in the presence of LIF and serum contain committed mesoderm progenitors, the number of which is dependent on the amount of LIF in the culture medium. By clonal analyses, we show that, in the presence of LIF, the commitment of these cells to mesoderm fate can be reverted, so that they give rise to fully pluripotent ES cells. We further demonstrate that the process of dedifferentiation of committed mesoderm progenitors is important to maintain the pluripotency of mouse ES cells over long-term cultures, and that this process is regulated by *Nanog*. Specifically, we show that *Nanog* expression is up-regulated in mesoderm progenitor cells by the combinatorial action of STAT3 and the mesoderm-specific transcription factor T/Brachyury. Finally, we provide evidence from gain- and loss-of-function experiments demonstrating that Nanog prevents the progression of BMP-induced mesoderm differentiation of ES cells by directly binding to Smad1 and interfering with the recruitment of co-activators, thus blocking the transcriptional activation of downstream targets, including that of *T/Brachyury*.

2 Results

To characterize the early steps of mouse ES cell differentiation toward mesoderm lineages, we generated transgenic ES cell lines expressing enhanced green fluorescent protein (eGFP) under the regulatory sequences of *T/Brachyury* (*T*), which encodes one of the earliest markers of mesoderm differentiation [15, 16]. The expression of eGFP in nine independent *T-eGFP* ES cell lines faithfully recapitulated that of endogenous *T*, as assayed by the presence of *T* transcripts in *T-eGFP*-positive [T(+)] cells sorted from embryoid bodies differentiated in vitro, and by their absence in *T-eGFP*-negative [T(–)] cells (Fig. 1A). Interestingly, colonies of undifferentiated *T-eGFP* ES cells grown under standard culture conditions (on a fibroblast feeder layer in culture medium containing serum and 1000 μ/mL of LIF) contained T(+) cells. These cells were found in small numbers (1–3 cells per colony) in colonies of otherwise undifferentiated morphology (6.7%, n = 120, Fig. 1B), and no colonies formed exclusively by T(+) were ever detected under these

Fig. 1 (continued) 400 μ/mL of LIF. Bar shows mean ± SD (n = 4). (**E**, **F**, **G**) Colony formation from single T(–) or T(+) cells isolated from *T-eGFP* ES cells obtained at P20 with 1000 μ/mL of LIF (**E**), at P20 with 400 u/ml of LIF (**F**), or at P4 without LIF (**G**). Scale bar = 10 μm. *Graphs* show the percentage of colonies containing only T(+) or T(–) cells, or both types of cells (mosaic colony) formed from purified T(–) or T(+) cells. *Bar* shows mean ± SD (n = 3; 40 colonies were examined in each dish). (**H, I**) ES cells cultured with 400 μ/mL of LIF expressed the mesoderm marker *T*, but not ectoderm or endoderm markers. RT-PCR analyses of ectoderm (*Sox1*), mesoderm (*T*), and endoderm (*HNF4α*) marker genes in *T-eGFP* ES cells obtained at P20 (1000 and 400 μ/mL of LIF) or P4 (without LIF) (**H**). Western blotting analysis of T in *T-eGFP* ES cells obtained at P20 (1000 and 400 μ/mL of LIF) or P4 (without LIF) (**I**). (**J**) Western blotting analysis of tyrosine-phosphorylated STAT3 (STAT3-P) in *T-eGFP* ES cells obtained at P20 (1000 and 400 μ/mL of LIF) or P4 (without LIF). Total STAT3 levels are shown as a control. Panels A-B and H-J are reproduced with permission of Nature Publishing Group [48]

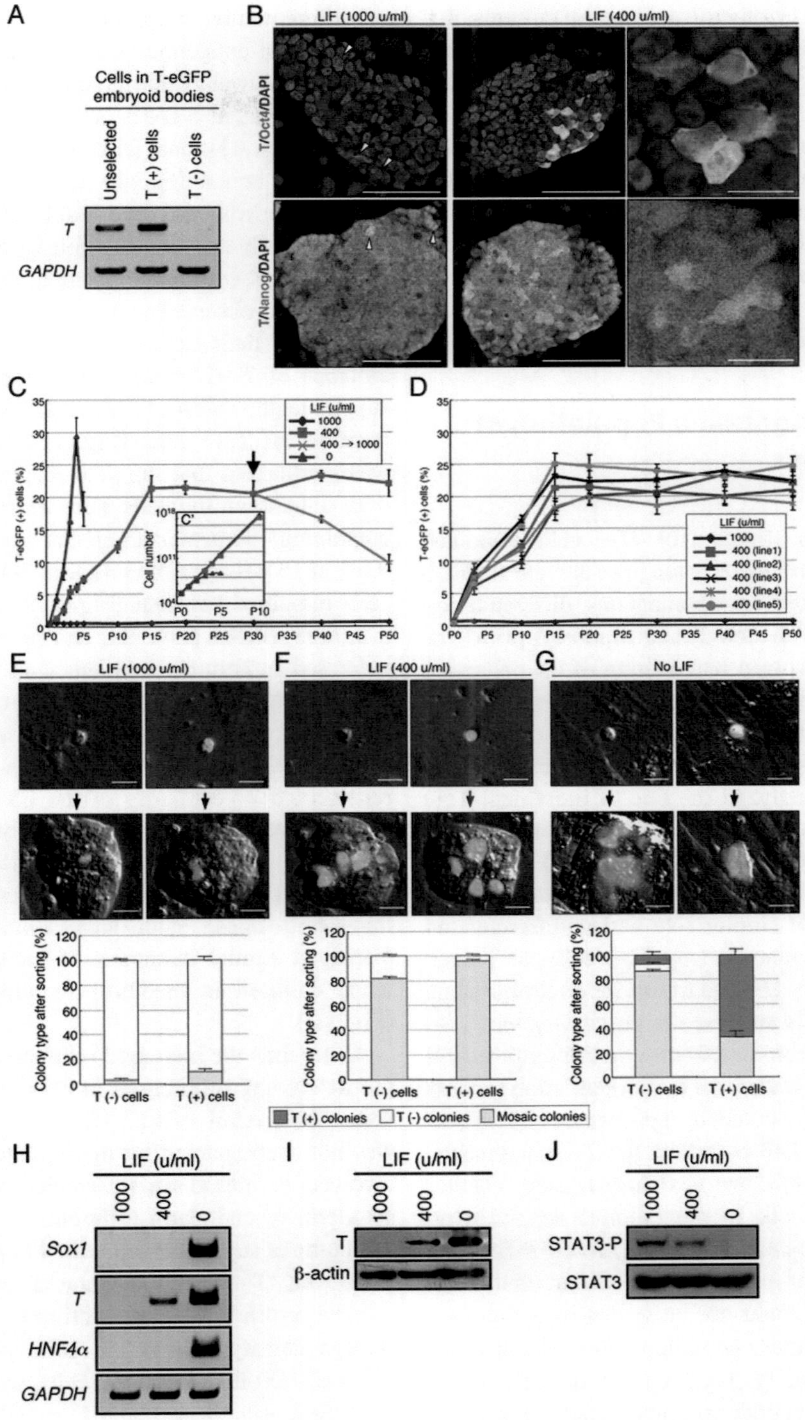

Fig. 1 LIF regulates the percentage of *T*-expressing cells in mouse ES cell cultures. (**A**) RT-PCR analysis of *T* expression in *T-eGFP* embryoid bodies (unselected cells), and in T(+) or T(–) cells isolated from *T-eGFP* embryoid bodies. *T-eGFP* expression recapitulated endogenous *T* expression. (**B**) Fluorescent images of *T* (eGFP)/Oct4 and *T* (DsRed2)/*Nanog* (eGFP) expression in mouse ES cell colonies at passage (P) 25 in culture with 1000 or 400 μ/mL of LIF. DAPI staining identifies individual cells in each field. T(+) cells were observed not only in colonies formed in culture with 400 μ/mL of LIF (*right* and *center* panels), but also in colonies formed in culture with 1000 μ/mL of LIF (*left* panels, *arrowheads*). Scale bar = 50 μm (*left* and *center* panels) or 10 μm (*right* panels). (**C**) Flow-cytometric analysis of T(+) cells in *T-eGFP* ES cell cultures with various LIF concentrations (1000, 400, and 0 μ/mL). A portion of ES cells cultured with 400 μ/mL of LIF was retreated with 1000 μ/mL of LIF from P30 (shown by *arrow*). Bar shows mean ± SD (n = 4). (**C'**) Growth of *T-eGFP* ES cells expressed by the number of cells in culture with 1000, 400, or 0 μ/mL of LIF (n = 4). (**D**) Flow-cytometric analysis of T(+) cells in five independent ES cell lines carrying the *T-eGFP* reporter, cultured with 1000 or 400 μ/mL of LIF. Cell line 1, 2, 3, 4, and 5 corresponds to J1, SAT1, SAT2, SAT6, and SAT11, respectively. Five independent ES cell lines exhibited similar dynamics of T(+) cell-accumulation in culture with

culture conditions. Flow cytometric analyses of cultures of *T-eGFP* ES cells revealed a 0.59 ± 0.05% of T(+) cells (n = 4), a fraction that remained virtually invariable after more than 50 passages (Fig. 1C). These findings are consistent with the widely known (though frequently overlooked) fact that mouse ES cells, even when maintained under optimal culture conditions, exhibit some degree of spontaneous differentiation. Indeed, this very fact has been recognized as a hallmark of a "good" ES cell line [3]. Our observation is also consistent with the identification of *T* as a transcript selectively enriched in undifferentiated mouse ES cells [9, 17].

3 Mouse ES Cells Contain a Population of *T*-Expressing Cells

We reasoned that, if the fraction of T(+) cells remained constant at 0.5% after continuous passaging, the rate at which T(+) cells were produced by spontaneous differentiation must be in equilibrium with their disappearance. A priori, the most likely scenarios involved lengthening of the cell cycle and eventual mitotic arrest of T(+) cells, selective cell death, the progression of differentiation with loss of *T* expression, or a combination of these mechanisms. To directly address these possibilities, we analyzed the fate of individual T(+) cells isolated by fluorescence activated cell sorting (FACS) and plated at clonal density. Surprisingly, T(+) cells gave rise to colonies that contained only a few or no T(+) cells, and which were of similar size and undifferentiated characteristics as those generated by T(–) cells, as judged by their morphology (Fig. 1E) and transcription profile (data not shown). These results suggest the possibility that T(+) cells are not selectively eliminated from undifferentiated ES cell cultures; rather, they undergo a dedifferentiation process that gives rise to undifferentiated T(–) progeny.

Since the addition of LIF is critical for the maintenance of undifferentiated ES cells, we next investigated whether the percentage of T(+) cells in our cultures depended on the amount of LIF in the culture medium. *T-eGFP* ES cells cultured in the absence of exogenous LIF differentiated extensively over time and could not be grown over five passages (Fig. 1C'). Under these conditions, the percentage of T(+) cells rose exponentially (Fig. 1C) and markers of ectoderm, mesoderm, and endoderm lineages were expressed (Fig. 1H), indicating that the majority of ES cells underwent spontaneous differentiation. T(+) cells sorted from cultures maintained for three to four passages in the absence of LIF displayed a dramatically reduced ability to generate colonies containing T(–) cells, when compared with cells grown in medium containing 1000 μ/mL of LIF (Fig. 1G). In contrast, reduction of exogenous LIF to 400 μ/mL resulted in colonies of normal ES cell morphology (Fig. 1B) that could be maintained in culture for over 50 passages with no signs of differentiation or crisis. This finding is consistent with the range of exogenous LIF concentrations reported to sustain self-renewal of pluripotent mouse ES cells [6, 18], with the activation of STAT3 phosphorylation under these conditions (Fig. 1J), and with the fact that ES cells cultured in medium containing 400 μ/mL of LIF for extended periods of time, when injected into blastocysts, result in degrees of chimerism and contribution to the germline not different from ES cells maintained in 1000 μ/mL of LIF (our unpublished observations).

Despite their apparently undifferentiated morphology, colonies of *T-eGFP* ES cells adapted to grow in medium supplemented with 400 μ/mL of LIF contained large numbers of T(+) cells (Fig. 1B). The adaptation process was gradual, so that the percentage of T(+) cells increased over time when the cells were switched to culture medium containing 400 μ/mL of LIF and reached a plateau of 21.2± 2.0% at 15 passages (n = 4), after which the fraction of T(+) cells remained constant (Fig. 1C). Also, the expression of *T* at both mRNA (Fig. 1H) and protein (Fig. 1I) levels was increased in cultures supplemented with 400 μ/mL of LIF when compared with *T-eGFP* ES cells grown in medium containing 1000 μ/mL. To rule out the possibility that the increased number of T(+) cells under these conditions represented a peculiarity of the ES cell line used in these experiments (J1), we established four independent ES cell lines from a different genetic background (C57BL/6 × 129/TerSv) that were used to generate additional *T-eGFP* transgenic lines. With small variations, all four lines displayed similar dynamics of accumulation of T(+) cells when cultured in medium containing 400 μ/mL of LIF (Fig. 1D).

Other than the increased numbers of T(+) cells, we could not detect any differences in cultures maintained with 400 versus 1000 μ/mL of LIF. Thus, the increase in *T* expression was not accompanied by up-regulation of other transcripts involved in mesoderm differentiation, and no markers of ectoderm or endoderm differentiation were detected in ES cell cultures supplemented with 400 μ/mL of LIF (Fig. 1H). Moreover, T(+) cells generated under these conditions showed proliferation rates similar to T(–) cells when plated at high density (0.2×10^6 cells yielded $4.18 \pm 0.27 \times 10^6$ and $4.00 \pm 0.31 \times 10^6$ cells, respectively, after 7 days in culture, n = 3; see also Fig. 1C'), displayed an ability to give rise to T(–) cells comparable to that of T(–) cells when plated at clonal density (Fig. 1F), co-expressed the pluripotency-associated markers *Oct4*, *Nanog*, and *Rex1* (Figs. 1B, 3A), and stained positive for alkaline phosphatase (not shown). Importantly, the number of T(+) cells in cultures supplemented with 400 μ/mL of LIF declined progressively after the concentration of the cytokine was increased to 1000 μ/mL (Fig. 1C). Taken together, our results indicate that, in cultures of mouse ES cells, a population of T(+) cells exists whose size is controlled by the amount of LIF present in the culture medium. Importantly, lowering the concentration of LIF to 400 u/ml did not appear to be detrimental for the self-renewal of mouse ES cells, even though T(+) cells formed up to 20% of the cells in these culture conditions. For these reasons, and since the size of the T(+) fraction was more amenable to analysis, we continued our studies in culture medium containing 400 μ/mL of LIF.

4 Reversal of Mesoderm Commitment of ES Cells by LIF

To characterize the identity of T(+) cells, we first analyzed their ability to generate differentiated progeny. For this purpose, we performed in vitro differentiation assays of bulk *T-eGFP* ES cells as well as of sorted populations of T(+) and T(–) cells. In these assays, unsorted ES cells and T(–) cells behaved similarly, giving rise to differentiated cells that expressed markers of ectoderm, mesoderm, and endoderm fates (Fig. 2A). In contrast, embryoid bodies formed from T(+) cells differentiated exclusively into cells expressing markers of mesoderm lineages (Fig. 2A), indicating that the T(+) cells present in our cultures of *T-eGFP* ES cells were lineage-committed. Thus, we termed this population of T(+) cells "early mesoderm-committed" (EM) progenitors.

In our experiments of colony formation from sorted cells plated at clonal density (Fig. 1E, 1F), EM progenitors gave rise to large numbers of T(–) cells. In light of the mesoderm commitment of T(+) cells, two possible scenarios could account for these findings: T(–) cells generated in these conditions could represent (i) a further step of mesoderm differentiation of EM progenitors, in which *T* was no longer expressed [16]; or (ii) undifferentiated ES cells dedifferentiated from EM progenitors. To investigate these possibilities, we analyzed the differentiation potential of T(–) cells generated from EM progenitors (Fig. 2B). In vitro differentiation assays revealed that T(–) cells derived from EM progenitor cells were able to give rise to cells of ectoderm, mesoderm, and endoderm lineages (Fig. 2C). Moreover, T(–) cells readily generated beating cardiomyocytes that stained positive for myosin, albumin-positive cells after prolonged periods under differentiation-promoting conditions, and Tuj1-positive cells after treatment with retinoic acid (Fig. 2D). We also analyzed the ability of T(–) cells to colonize embryo lineages in vivo after their introduction into mouse blastocysts. To trace the progeny of the injected cells, a constitutively expressed LacZ reporter was introduced into *T-eGFP* ES cells. We recovered 11 embryos from 2 independent injections of T(+)-derived T(–) cells into mouse blastocysts, of which 5 were overtly chimeric. In these embryos LacZ-positive cells contributed to organs of ectoderm, mesoderm, and endoderm lineages (Fig. 2F–2K). As a control for our in vitro and in vivo differentiation assays of T(–) cells, we analyzed T(+) cells generated from EM progenitors. In all cases, T(+) cells behaved the same as T(+) cells isolated from bulk cultures of *T-eGFP* ES cells, displaying in vitro differentiation potential restricted to mesoderm lineages (Fig. 2C) and failing to contribute to embryogenesis in vivo (we could not detect any signs of chimerism in eight embryos recovered; Fig. 2E). Our results show that T(+) cells represent a population of EM progenitors that, in the presence of LIF, are able to recover the abilities of self-renewal, pluripotency, and chimera contribution characteristics of mouse ES cells.

5 Nanog Regulates the Dedifferentiation of EM Progenitors

In our characterization of the transcriptional profile of EM progenitors, we did not detect changes in the expression of the pluripotency-associated markers *Oct4* and *Rex1* or in the levels of *Gbx2*, *Fgf5*, or *Lif* expression (see Discussion below). However, we detected a clear up-regulation in the expression of the pluripotency-associated marker *Nanog*, when compared to that of ES T(–) cells (Fig. 3A). Indeed, aside from the expression of *T* itself, these two cell populations only appeared to differ in the level of *Nanog* expression, raising the possibility that Nanog function is mechanistically linked to the transition of ES cells to EM progenitors, or vice versa. We investigated this possibility by directly manipulating the level of *Nanog* expression in *T-eGFP* ES cells and analyzing the consequences of such manipulations in the size and characteristics of the T(+) and T(–) cell populations. The introduction of a *Nanog* expression transgene driven by a strong constitutive promoter resulted in sustained overexpression of *Nanog* transcripts in *T-eGFP* ES cells (Fig. 3B). In these conditions, the expression of *T* was down-regulated (Fig. 3B), and the transition of EM progenitors to ES cells was facilitated, as evaluated by the ~3-fold reduction of T(+) cells (Fig. 3F) and the ~17-fold increase in the percentage of colonies composed exclusively by T(–) cells generated by EM progenitors overexpressing *Nanog* (Fig. 3H, 3J), when compared to mock-transfected *T-eGFP* ES cells (Fig. 3E, 3H, 3I).

For the converse experiment, down-regulation of Nanog function, we assayed the efficiency of short-hairpin RNAs (shRNAs) to induce partial *Nanog* silencing, since complete loss of Nanog function is incompatible with the pluripotent phenotype of ES cells [9]. Introduction of *Nanog-shRNA* into *T-eGFP* ES cells resulted in a marked decrease in the levels of Nanog, as evaluated by RT-PCR (Fig. 3D) and immunoblotting with specific antibodies (Fig. 3C). In these

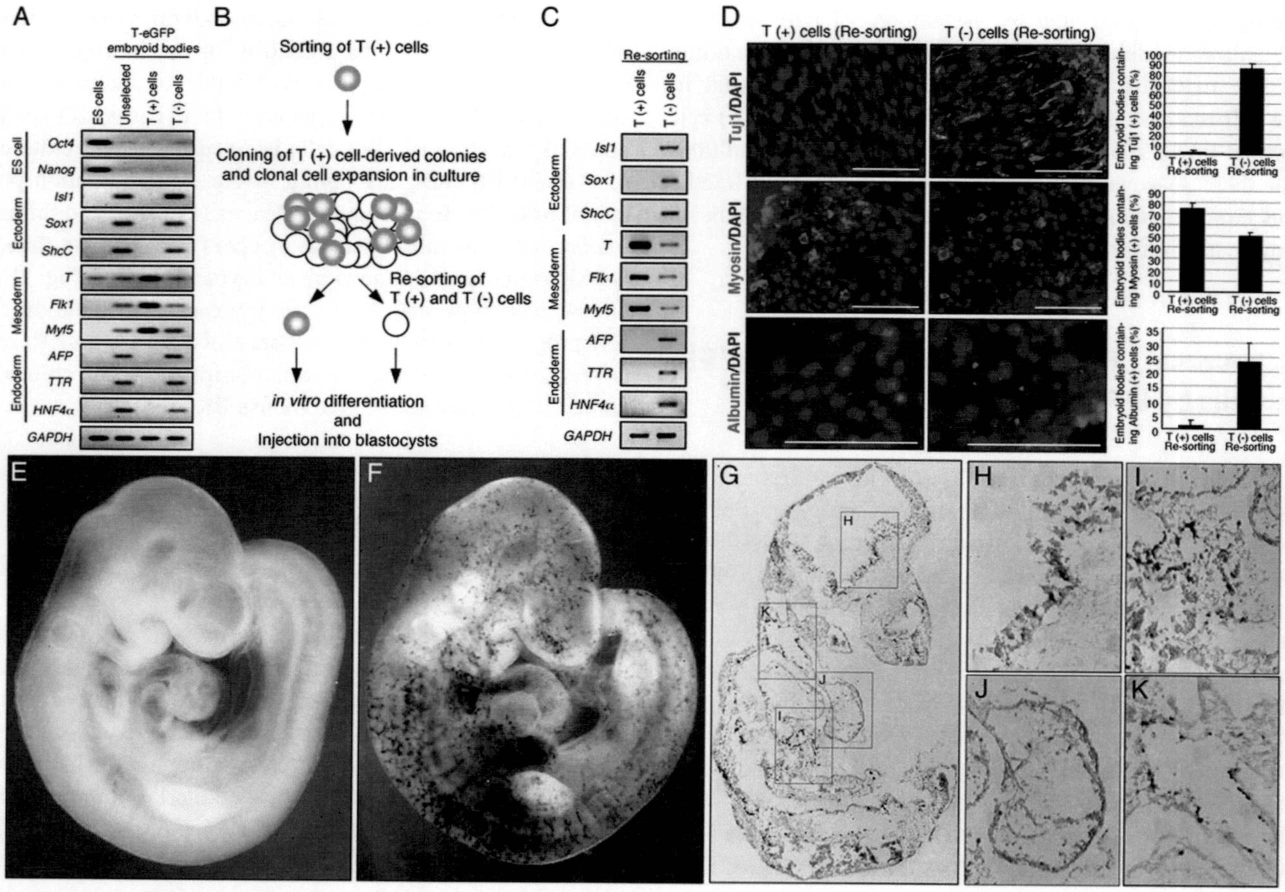

Fig. 2 Dedifferentiation of lineage-committed mesoderm progenitors into pluripotent stem cells. (**A**) RT-PCR analysis of ES cell, ectoderm, mesoderm, and endoderm marker gene expression in *T-eGFP* ES cells obtained at P20 (400 μ/mL of LIF), and in embryoid bodies derived from unselected *T-eGFP* ES cells, and from purified T(+) or T(–) cells. (**B**) Schematic representation of the experimental procedure for characterizing T(+) and T(–) cells derived from a single T(+) cell (EM progenitor cell). A single T(+) cell formed a mosaic colony in clonal density cultures with 400 μ/mL of LIF. Each mosaic colony was independently picked up and expanded. Then, T(+) and T(–) cells originally derived from a single T(+) cell were re-sorted by FACS to analyze their differentiation potential in vitro (**C, D**), and in vivo by injecting into blastocysts (**E, F, G, H, I, J, K**). (**C**) RT-PCR analysis for ectoderm, mesoderm, and endoderm marker gene expression in embryoid bodies derived from re-sorted T(+) or T(–) cells. (**D**) Immunocytochemical analysis of Tuj1 (neuronal lineage), myosin (muscle lineage), and albumin (hepatic lineage) production in embryoid bodies derived from re-sorted T(+) or T(–) cells. Scale bar = 100 μm. *Graphs* show the percentage of embryoid bodies containing antigen-positive cells. *Bar* shows mean ± SD (n = 3; 30 embryoid bodies were examined in each dish). (**E, F**) Chimeric analysis demonstrating contribution of re-sorted T(+) (**E**) or T(–) (**F**) cells in E10 mouse embryos. β-galactosidase activity was used to visualize the contribution of re-sorted T(+) or T(–) cells in chimeric embryos. Images are *right* lateral view with the anterior to the *top*. (**G, H, I, J, K**) Histological images of the embryos shown in panel (**F**). A sagittal section counter stained with eosin (**G**) and close ups of different areas (**H, I, J, K**): (**H**) neuroepithelium; (**I**) gut tube and hepatic primordia within septum transversum; (**J**) myocardial wall; (**K**) artery. Panels A-D are reproduced with permission of Nature Publishing Group [48]

conditions, cell colonies of undifferentiated ES cell morphology formed, and no signs of endoderm differentiation were apparent (not shown). However, the number of T(+) cells generated from EM progenitors expressing *Nanog-shRNA* more than doubled that of mock-transfected *T-eGFP* ES cells maintained under similar culture conditions (Fig. 3G). Also, EM progenitors expressing *Nanog-shRNA* generated ~20% of colonies composed exclusively of T(+) cells when plated at clonal density (Fig. 3H, 3K), indicating that the transition of EM progenitors to ES cells was impaired upon down-regulation of Nanog function. Together with the results of our gain-of-function experiments, these findings demonstrate that Nanog controls the dedifferentiation of EM progenitors to ES cells.

6 Positive Regulation of *Nanog* Expression by T and LIF/STAT3

We next investigated the mechanism by which *Nanog* expression is up-regulated in EM progenitors. Based on our observations that *Nanog* expression is found to be increased

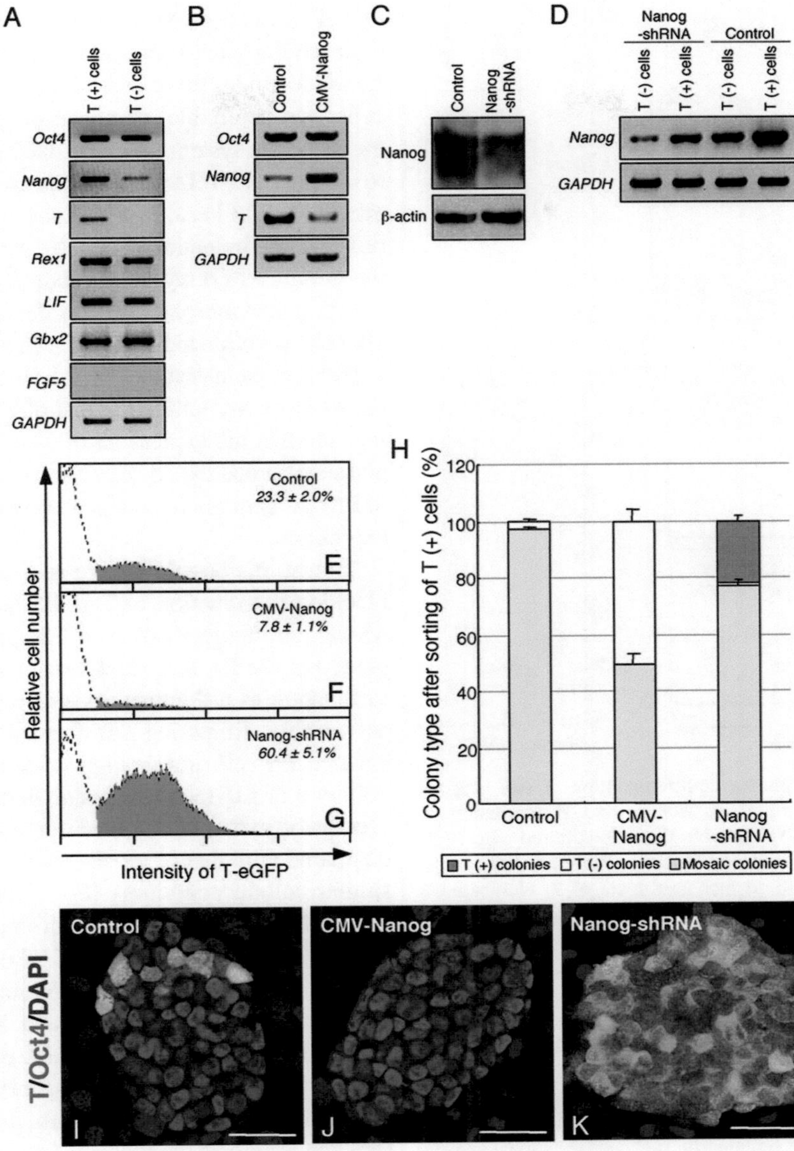

Fig. 3 *Nanog*is sufficient and necessary for dedifferentiation of EM progenitors into pluripotent stem cells. (A) RT-PCR analysis of gene expression in T(+) or T(–) cells isolated from *T-eGFP* ES cells. T(+) cells exhibited a higher level of *Nanog* expression. (**B**) RT-PCR analysis of *Oct4*, *Nanog*, and *T* expression in *T-eGFP* ES cells carrying control or CMV-*Nanog* constructs. Overexpression of *Nanog* is associated with down-regulation of *T* expression. (**C**) Western blotting analysis of Nanog in *T-eGFP* ES cells carrying control or *Nanog-shRNA* constructs. *Nanog-shRNA* efficiently down-regulates the level of Nanog protein in ES cells. (**D**) RT-PCR analysis of *Nanog* expression in T(+) or T(–) cells isolated from *T-eGFP* ES cells carrying control or *Nanog-shRNA* constructs. (**E–G**) Flow-cytometric analysis of T(+) cells produced from purified T(+) cells carrying control (**E**), CMV-*Nanog* (**F**), or *Nanog-shRNA* (**G**) constructs. *T-eGFP* ES cells were transfected with each construct, then T(+) cells were isolated, selected with puromycin, and cultured for 7 days with 400 μ/mL of LIF. Percentages of T(+) cells are shown in each panel (n = 4, mean ± SD). (**H**) Percentage of types of colonies in cultures of (**E–G**) before FACS analysis. *Bar* shows mean ± SD (n = 4; 40 colonies were examined in each dish). (**I–K**) Fluorescent images of *T* (eGFP)/Oct4 expression in colonies formed in cultures of experiments shown in panels (**E–G**). DAPI staining identified individual cells in each field. Scale bar = 20 μm. Panels A-D and I-K are reproduced with permission of Nature Publishing Group [48]

in cell populations with high levels of *T* expression (Fig. 3A), we first asked whether T could induce *Nanog* expression. For this purpose, we attempted to overexpress *T* in ES cells by transfecting a full-length cDNA driven by a constitutive promoter. Under these conditions, ES cells underwent massive differentiation and could not be propagated, even in culture medium supplemented with 1000 μ/mL of LIF (not shown). As an alternative approach, we used a doxycycline-inducible conditional expression system. ES cells expressing the tetracycline-inducible transcriptional activator (rtTA)

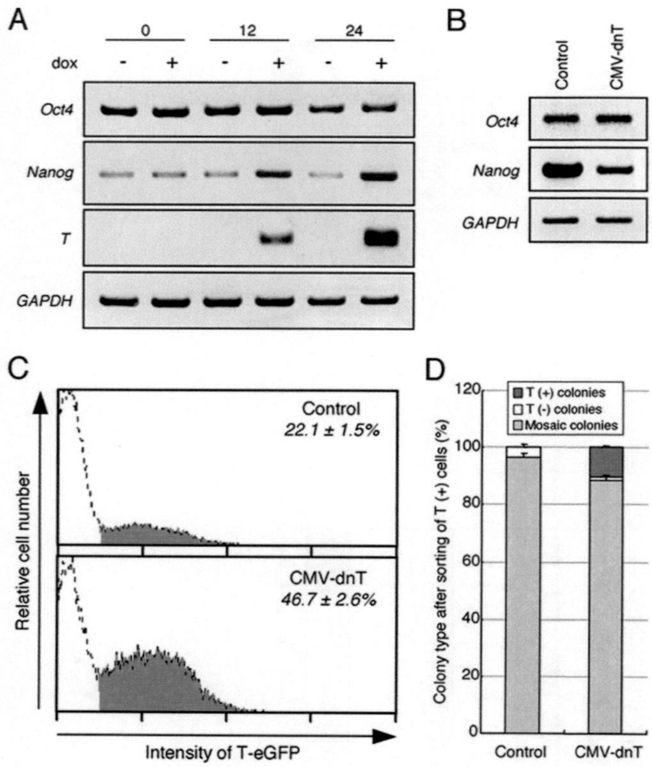

Fig. 4 T controls Nanog-dependent dedifferentiation of EM progenitors. (**A**) RT-PCR analysis of *Oct4*, *Nanog*, and *T* expression in doxicycline- (dox) dependent *T*-inducible ES cells cultured with 1000 μ/mL of LIF. *T* expression was induced by dox, and *Nanog* up-regulation was associated with an increased level of *T*. Numbers on the *top* indicate induction hours with dox. (**B**) RT-PCR analysis of *Oct4* and *Nanog* expression in *T-eGFP* ES cells carrying control or CMV-*dnT* constructs. dnT down-regulated *Nanog* expression without affecting *Oct4* expression in ES cells cultured with 400 μ/mL of LIF. (**C, D**) Blockage of T function impaired transition of T(+) cells to T(–) cells. Flow-cytometric analysis of T(+) cells produced from purified T(+) cells in the presence or absence of down-regulation of T by CMV-*dnT* (**C**). *T-eGFP* ES cells were transfected with each construct, then T(+) cells were isolated, selected with puromycin, and cultured for 7 days with 400 μ/mL of LIF. Percentages of T(+) cells are shown in each panel (n = 4, mean ± SD). Percentage of types of colonies in cultures of (**C**) before FACS analysis (**D**). *Bar* shows mean ± SD (n = 4; 40 colonies were examined in each dish). Panels A-B are reproduced with permission of Nature Publishing Group [48]

[19] were transfected with a construct containing the full-length *T* cDNA driven by a tetracycline-responsive promoter (T-rtTA ES cells). In the absence of doxycycline, T-rtTA ES cells cultured in medium supplemented with 1000 μ/mL of LIF formed colonies of undifferentiated ES cells, with a morphology indistinguishable from parental untransfected ES cells. Under these conditions,*T* expression in T-rtTA ES cells was negligible, demonstrating the tight regulation of the inducible system (Fig. 4A). Upon addition of doxycycline to cultures of T-rtTA ES cells, *T* expression was induced progressively and became detectable by RT-PCR after 12 h of doxycycline induction (Fig. 4A). T-rtTA ES cells cultured in the presence of doxycycline started to acquire the characteristic flattened morphology of differentiated colonies 48–72 h after doxycycline induction (not shown). Importantly, *Nanog* expression increased in T-rtTA ES cells after 12 h of doxycycline induction, paralleling that of *T* (Fig. 4A). These results indicate that the expression of *Nanog* can be regulated by T and suggest that the up-regulation of *Nanog* expression found in EM progenitors may depend on their increased levels of *T* expression.

To test this possibility, we analyzed the consequences of blocking the function of T in ES cells in the expression of *Nanog* and the generation of EM progenitors. For this purpose, we used a truncated version of T previously shown to function as a dominant-negative (dnT) [20]. *T-eGFP* ES cells stably expressing *dnT* formed colonies with undifferentiated ES cell morphology in culture medium containing 400 u/ml of LIF (see Discussion below). The level of *Nanog* expression in these cells, however, was down-regulated when compared with that of mock-transfected *T-eGFP* cells under similar culture conditions (Fig. 4B). Moreover, the blockade of T function resulted in an impaired transition of EM progenitors to ES cells (Fig. 4C, 4D), consistent with the reduced levels of *Nanog* expression found in these conditions. Thus, our results from gain- and loss-of-function experiments identify a negative feedback mechanism by which increased *T* expression in EM progenitors up-regulates the expression of *Nanog*, which, in turn, down-regulates *T* expression and promotes the regeneration of an ES cell phenotype.

To gain insights into the regulation of *Nanog* expression by T, we analyzed the mouse *Nanog* gene in the search of regulatory sequences. At 4.91 kb upstream of the translation start site of *Nanog*, we identified a 20-bp sequence forming an imperfect palindrome that shared homology with the proposed binding site for T [21] (Fig. 5A). We tested the ability of T to bind to oligonucleotides representing this sequence, but not to mutated versions thereof, by performing

Fig. 5 (continued) maintained with 400 μ/mL of LIF (**E**). Both the T- and STAT3-binding sites were required for activation of the *Nanog* EM enhancer activity in ES cells cultured with 400 μ/mL LIF (**F**). WT: –5203 to –4192 bp, MUT/T, MUT/S and MUT/TS indicate mutation in T-, STAT3- or both T- and STAT3-binding sites, respectively, in the *Nanog* EM enhancer. Bars show mean ± SD (n = 4). (**G**) T and STAT3 physically interact inside cells. T and STAT3 were co-immunoprecipitated when STAT3 was activated by LIF. (**H**) Fluorescent images of *T*-eGFP and *Nanog* EM enhancer-DsRed2 expression in ES cell colonies formed in culture with 400 μ/mL of LIF. Co-expression of eGFP with DsRed2 in wild-type, but not when T- and/or STAT3-binding sites are mutated, indicates that the activity of the *Nanog* EM enhancer in EM progenitor cells is regulated by T and STAT3. Modified figure reproduced with permission from ref. [49]

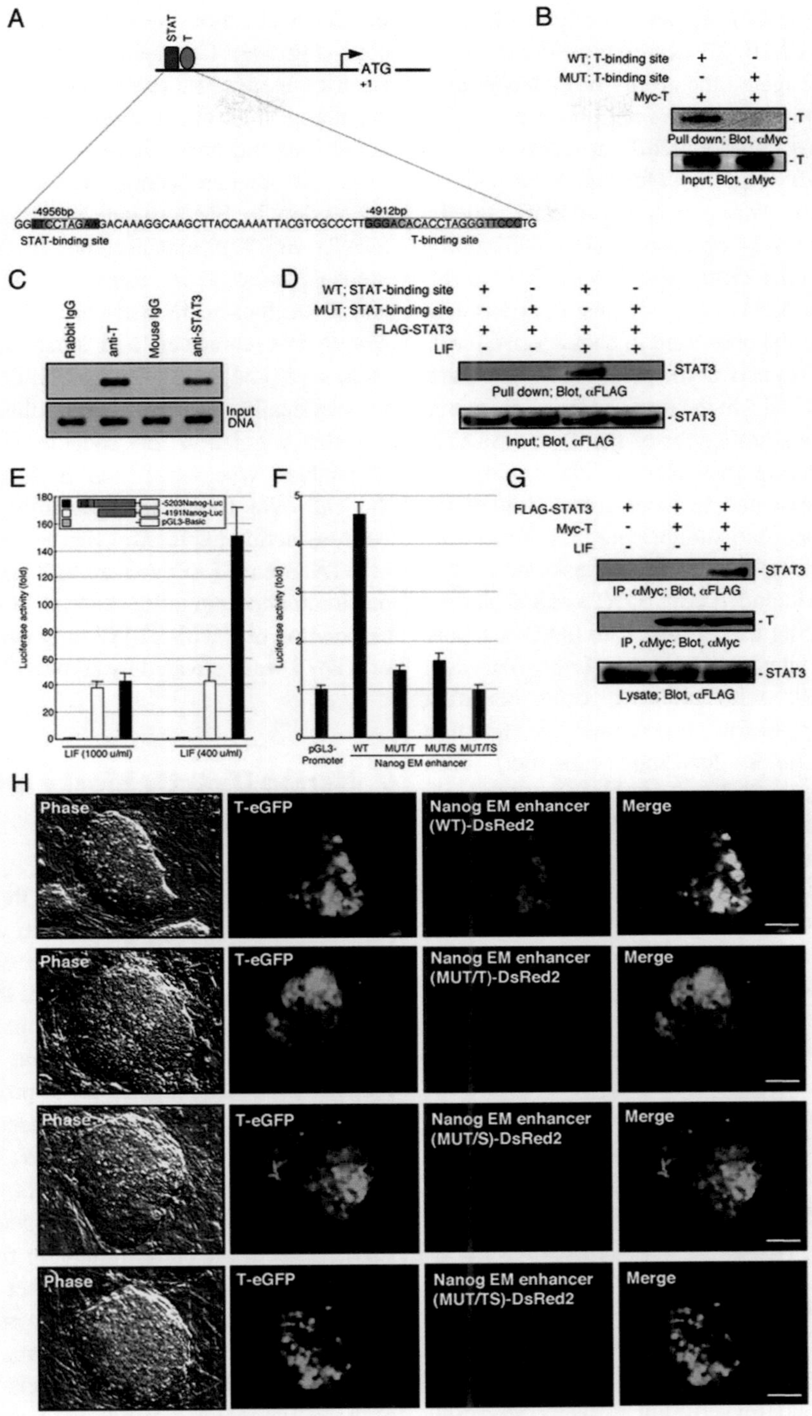

Fig. 5 Binding of STAT3 and T on *Nanog*EM enhancer is required for up-regulation of *Nanog*expression in EM progenitors. (**A**) Schematic representation of the 5′-upstream regulatory region of the mouse *Nanog* gene. Putative STAT- and T-binding sites are indicated. (**B**) T bound to the putative T-binding site in the *Nanog* regulatory region, as shown by pull-down assays. Wildtype (WT) and mutated (MUT) versions of double-strand oligonucleotides representing the putative T-binding site were used as probes. Input lysates were also blotted with anti-Myc antibody. (**C**) ChIP assay for the putative T- and STAT-binding sites in the *Nanog* regulatory region showed specific binding of T and STAT3 to the regulatory region. The *lower panel* is a PCR-amplification of input DNA prior to immunoprecipitation. (**D**) LIF-dependent binding of STAT3 to the putative STAT-binding site in the *Nanog* regulatory region. Wild-type and mutated versions of double-strand oligonucleotides for the putative STAT-binding site were used as probes. Input lysates were also blotted with anti-FLAG antibody. (**E, F**) Analysis of transcriptional activities of the *Nanog* regulatory region by luciferase reporter assay in mouse ES cells (1000 or 400 μ/mL of LIF). Both –5203Nanog-Luc and –4191Nanog-Luc showed a similar activation with 1000 μ/mL of LIF, whereas –5203Nanog-Luc activity was further increased in cultures

in vitro pull-down assays of biotin-labeled oligonucleotides incubated with lysates of NIH3T3 expressing Myc-tagged T (Fig. 5B). We also investigated the ability of endogenous T to bind the region of interest in the *Nanog* promoter in vivo by chromatin immunoprecipitation (ChIP) assays of ES cells with a T-specific antibody (Fig. 5C). Our search for putative regulatory elements in the *Nanog* promoter also identified a predicted STAT-binding site 44 bp upstream of the T-binding site (Fig. 5A). We tested the ability of STAT3 to bind to this site in vitro (Fig. 5D) and in vivo (Fig. 5C) using experimental approaches similar to the ones used to characterize the T-binding site. These results uncover the presence of functional binding sites for T and STAT3 in the mouse *Nanog* promoter.

We next analyzed the significance of the T- and STAT3-binding sites in the *Nanog* promoter for the biology of EM progenitors. We generated two constructs driving the expression of luciferase, one comprising 5.2 kb of the *Nanog* genomic sequence upstream of the translation start (–5203Nanog-Luc, which included both STAT3 and T-binding sites), and the other lacking the 5′-most 1 kb (and thus, both STAT3 and T-binding sites, –4191Nanog-Luc). Transient transfection of ES cells with either reporter construct resulted in a similar ~40-fold transcriptional induction (compared to a promoterless luciferase construct) when ES cells were cultured in medium containing 1000 μ/mL of LIF (Fig. 5E), a condition in which EM progenitors are generated at very low frequency (Fig. 1C). These results indicate that the regulatory elements responsible for the constitutive expression of *Nanog* in ES cells are located in the first 4.2 kb of the mouse *Nanog* gene upstream of the translation start. Importantly, the transcriptional activity of the –5203Nanog-Luc was increased by ~4-fold with respect to that of –4191Nanog-Luc in ES cells adapted to grow in medium supplemented with 400 μ/mL of LIF (Fig. 5E), in which the EM progenitor population represents ~20% of the culture (Fig. 1C). These findings suggest that the enhancer element responsible for *Nanog* up-regulation in EM progenitors (*Nanog* EM enhancer) is located between –5203 and –4192 bp upstream of the translation start of the mouse *Nanog* gene, a region containing the functional STAT3- and T-binding sites.

We then generated a luciferase reporter construct driven by the *Nanog* EM enhancer and a minimal promoter. This enhancer element increased transcription levels by ~4.5-fold when transiently transfected into ES cells cultured with 400 μ/mL of LIF (Fig. 5F). Moreover, the activity of the *Nanog* EM enhancer was lost when either or both the STAT3- and the T-binding sites were mutated (Fig. 5F). To visualize the activity of the *Nanog* EM enhancer in specific cells, we used it to drive the expression of a red fluorescent protein (DsRed2) reporter in ES cells. *T-eGFP* ES cells stably expressing this reporter showed activity of the *Nanog* EM enhancer only in EM progenitors, as evaluated by the co-localization of eGFP and DsRed2 signals in these cells (Fig. 5H). Consistent with the results of the luciferase reporter assays, mutation of either or both STAT3- and T-binding sites in the *Nanog* EM enhancer abrogated the activity of this reporter in EM progenitors (Fig. 5H). Our results so far demonstrate that the up-regulation of *Nanog* expression in EM progenitors depends on the binding of STAT3 and T to specific sites in the EM enhancer in the mouse *Nanog* gene. Since the binding sites for STAT3 and T are located in close proximity to one another in the *Nanog* EM enhancer, and since both T-box transcription factors [22–24] and STAT3 [25, 26] have been described to physically interact with other transcription factors for the regulation of specific promoters, we decided to analyze whether T and STAT3 could interact inside the cell. We tested this possibility in NIH3T3 cells by co-transfecting expression vectors encoding tagged versions of STAT3 and T (FLAG-STAT3 and Myc-T) and carrying out immunoprecipitation assays. Interestingly, we found an association of T with STAT3 only when nuclear translocation of STAT3 was activated by stimulation with LIF (Fig. 5G).

7 Nanog Directly Blocks Mesoderm Induction by BMPs

Our findings so far demonstrate that *Nanog* expression is up-regulated in EM progenitors in the presence of LIF by the combined action of activated STAT3 and T, and that increased Nanog function promotes the transition of EM progenitors to ES cells. We reasoned that a likely mechanism of Nanog action in this process could be preventing the generation or blocking the effects of pro-differentiation factors. BMPs are potent inducers of mesoderm differentiation in the context of embryo development [27, 28], as well as in mouse ES cells [29–31]. EM progenitors, in turn, are generated rapidly during the first ~15 passages after reducing the LIF supplement in the culture medium from 1000 to 400 μ/mL (Fig. 1C). Thus, we tested whether the generation of EM progenitors in these conditions was modified by increasing or decreasing BMP signaling in cultures of ES cells. After three passages in medium containing 400 μ/mL of LIF, the size of the EM progenitor population reached ~6% in cultures of *T-eGFP* ES cells (Fig. 6A, see also Fig. 1C). This percentage almost doubled when cells were incubated in the presence of recombinant BMP2, BMP4, or BMP7 and was reduced by half upon incubation with noggin (Fig. 6A), a secreted factor that blocks BMP signaling [32, 33]. We then tested whether BMP signaling was also regulating the maintenance of EM progenitors. When pure populations of EM progenitors were plated in culture medium containing 400 μ/mL of LIF, ~75% of the resulting cells underwent a transition

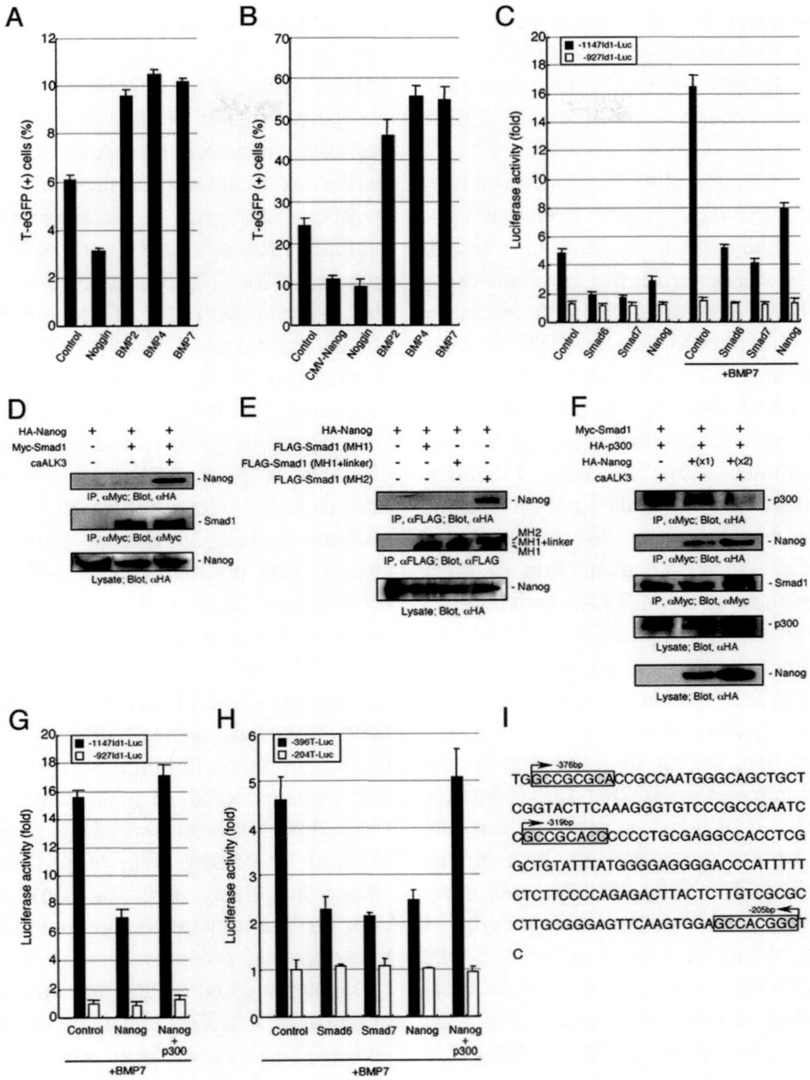

Fig. 6 Nanog interferes with BMP signaling at the level of transcription activity of Smad. (**A**) Flow-cytometric analysis of T(+) cells in *T-eGFP* ES cells cultured for three passages with 400 μ/mL of LIF under conditions of inhibition (noggin) or activation (BMP2, BMP4, and BMP7) of BMP signaling. *Bar* shows mean ± SD (n = 4). (**B**) Flow-cytometric analysis of T(+) cells produced from purified T(+) cells cultured with 400 μ/mL of LIF under conditions of inhibition or activation of BMP signaling. Inhibition of endogenous BMP signaling by noggin decreased the percentage of T(+) cells at a similar level of Nanog overexpression, whereas BMP activation increased the percentage of T(+) cells. *Bar* shows mean ± SD (n = 4). (**C**) A reporter construct of –1147Id1-Luc containing the Smad-binding sites, but not –927Id1-Luc, was activated in a BMP-dependent manner in ES cells cultured with 400 μ/mL of LIF. Nanog and inhibitory Smads (Smad6 and Smad7) down-regulated –1147Id1-Luc activity in a similar manner. Bars show mean ± SD (n = 4). (**D–F**) Co-immunoprecipitation assays of the physical interaction between Nanog and BMP-responsive Smad1. Nanog interacted with activated Smad1 (**D**). Nanog interacted with the MH2 domain of Smad1 (**E**). Nanog interfered with the interaction between activated Smad1 and p300 by competitively binding to Smad1 in a dose-dependent manner (**F**). The relative p300:Nanog ratio was 1:1 or 1:2. (**G**) Overexpression of p300 rescued the down-regulation of –1147Id1-Luc activity induced by *Nanog*. Bars show mean ± SD (n = 4). (**H**) The –396T-Luc reporter construct, but not –204T-Luc, was activated in a BMP-dependent manner in ES cells cultured with 400 u/ml of LIF. Nanog and inhibitory Smads (Smad6 and Smad7) down-regulated –396T-Luc activity in a similar manner. The down-regulation of –396T-Luc activity induced by Nanog was rescued by overexpression of p300. Bars show mean ± SD (n = 4). (**I**) Sequence of the 5'-upstream regulatory region of the mouse *T* gene. Three putative BMP-responsive Smad-binding sites are indicated with boxes. Please, add the following text: Panels A-C and G-I are reproduced with permission of The National Academy of Sciences of the United States of America [49]

to ES cells, whereas the remaining ~25% maintained EM progenitor identity (Fig. 6B, see also Figs. 3E, 3I, 4C). When the cultures were supplemented with BMPs, the maintenance of EM progenitors increased by ~2-fold, whereas it was decreased by half upon incubation with noggin (Fig. 6B). Interestingly, overexpression of *Nanog* in EM progenitors resulted in a decrease in their maintenance similar to that induced by noggin (Fig. 6B). These results indicate that the generation and maintenance of EM progenitors depends, at least in part, on the differentiation-promoting activity of BMPs and suggest that Nanog's ability to reduce the numbers of EM progenitors may depend on the blockade of BMP signaling.

Signaling by BMPs is intracellularly transduced by receptor-regulated Smads (Smad1, 5, and 8) and the co-mediator Smad4 and is antagonized by inhibitory Smads (Smad6 and 7) [34]. To characterize the mechanism by which Nanog blocks BMP signaling, we first analyzed the effects of *Nanog* overexpression in the BMP-induced transcriptional activation of *Id1*. *Id1* is a well-characterized transcriptional target of BMP signaling [35], for which the Smad-binding elements have been mapped to a specific region in the *Id1* promoter [36]. We used luciferase reporter constructs containing (–1147Id1-Luc) or lacking (–927Id1-Luc) the Smad-binding sites [36] and analyzed their activity in ES cells. Transient transfection of these reporters in ES cells resulted in a ~4.5-fold activation of the –1147Id1-Luc reporter when compared to –927Id1-Luc (Fig. 6C), indicating the existence of a significant level of endogenous BMP signaling associated with our culture conditions (see Discussion below). Addition of BMP to the culture medium resulted in a strong up-regulation of the –1147Id1-Luc reporter compared to –927Id1-Luc (Fig. 6C). That the activation of the –1147Id1-Luc reporter was due to BMP signaling was further confirmed by the fact that co-transfection of ES cells with cDNAs encoding inhibitory Smads drastically reduced the transcriptional activity of the reporter induced by endogenous or exogenous BMPs (Fig. 6C). Interestingly, *Nanog* overexpression in ES cells closely mimicked the effect of inhibitory Smads (Fig. 6C), suggesting that Nanog may block BMP signaling by interfering with the formation of activated Smad complexes.

Inhibitory Smads negatively regulate BMP signaling by binding to activated receptor-regulated Smads, hence limiting their availability to form transcriptionally active complexes with Smad4 and/or other nuclear cofactors [34]. To address whether Nanog blocked BMP signaling by a similar mechanism, we first analyzed its ability to interact with the receptor-regulated Smad1 inside the cell. Co-immunoprecipitation assays in NIH3T3 cells revealed that Nanong was indeed able to bind Smad1 only when the latter was activated by co-transfection of a constitutively active ALK3 (caALK3, Fig. 6D). Next, we mapped the interaction domain of Smad1 with Nanog. The different Smads contain two conserved domains, the N-terminal Mad homology (MH) 1 and the C-terminal MH2 domain, separated by a poorly conserved linker. The interaction of receptor-regulated Smads with Smad4 and other transcription factors and cofactors, as well as with inhibitory Smads, occurs through the MH2 domain [34]. In cells co-transfected with Nanog and expression constructs encoding the individual MH1, MH1+linker, or MH2 domains of Smad1, interaction with Nanog was found exclusively with the MH2 domain (Fig. 6E). These results are consistent with a negative role of Nanog on BMP signaling by interfering with the interaction of receptor-activated Smads with Smad4 and/or additional nuclear factors.

The paralogous transcriptional coactivators CREB-binding protein (CBP) and p300 are nuclear cofactors important for TGFβ signaling, including that of BMPs, that interact with the MH2 domain of receptor-regulated Smads and Smad4 [37–39]. To gain further insights into the mechanism of Nanog-mediated down-regulation of BMP signaling, we tested whether Nanog interfered with the recruitment of p300 to the complexes of activated Smads. For this purpose, Myc-tagged Smad1, HA-tagged p300, and caALK3 were expressed in NIH3T3 cells with or without HA-tagged Nanog. Immunoprecipitations of cell lysates were performed with anti-Myc antibodies followed by Western blotting utilizing anti-HA antibodies. In the absence of Nanog, Smad1 efficiently co-immunoprecipitated p300 (Fig. 6F). In the presence of co-expressed Nanog, the amount of p300 bound to Smad1 decreased in a Nanog dose-dependent manner (Fig. 6F). The functional significance of these findings was further verified by the fact that overexpression of p300 completely rescued the down-regulation in the transcriptional activity of the *Id* promoter induced by Nanog (Fig. 6G). These results indicate that Nanog negatively regulates BMP signaling by interfering with the recruitment of the co-activator p300 to the Smad transcriptional complex.

Finally, the finding that the expression of the *Xbra*, the homologue of *T* in *Xenopus*, is regulated by TGFβ signals [40] prompted us to investigate whether *T* could be a transcriptional target of BMP signaling in ES cells, and, if so, whether Nanog could directly block the induction of *T* by BMPs. In a preliminary analysis, we identified a BMP-responsive element in the ~1.2-kb region upstream of the translation initiation site of the mouse *T* promoter (data not shown). We then generated a series of luciferase reporter constructs covering this region. We transfected these constructs into ES cells cultured in medium containing 400 μ/mL of LIF and supplemented with BMP7, and further mapped the BMP-responsive element to a region located between –396 and –204 bp of the mouse *T* gene (Fig. 6H). Under these conditions, the activity of the –396T-Luc reporter was ~5-fold that of –204T-Luc and decreased by half upon co-expression of inhibitory Smads or Nanog (Fig. 6H). Interestingly, the down-regulation of –396T-Luc activity induced by Nanog could be completely rescued by co-expression of p300 (Fig. 6H). The analysis of this region in the mouse *T* promoter detected three motifs with homology to the reported consensus of BMP-responsive Smad-binding sites [41]. These results indicate that *T* is a direct transcriptional target of BMP signaling, and that Nanog down-regulates *T* expression by inhibiting BMP signaling at the level of the formation of active Smads/p300 complexes.

8 Discussion

Mouse ES cells, consistent with their developmental origin in the embryo epiblast, have the ability to give rise to derivatives of all three primary germ layers. However, unlike cells in the epiblast, in which pluripotency is very transient, mouse ES cells can be maintained in culture indefinitely in a pluripotent state. The mechanism(s) whereby the adaptation to culture conditions releases epiblast cells from the loss of pluripotency remain an outstanding question in the biology of ES cells.

9 The Transient EM Progenitor Population

In this study, we identify a population of EM progenitors normally present in cultures of mouse ES cells. The commitment of EM progenitors to mesoderm fates is evident upon LIF withdrawal, which results in differentiation restricted to mesoderm lineages and by their failure to contribute to embryogenesis in vivo (Fig. 2A, 2C, 2E). In the presence of LIF, however, EM progenitors are phenotypically indistinguishable from ES cells, as both populations co-exist in colonies of undifferentiated morphology and both maintain pluripotency over extended periods of time in culture (Fig. 1C). Moreover, EM progenitors maintain the expression of pluripotency-associated markers such as *Oct4*, *Nanog*, and *Rex1* and have high levels of alkaline phosphatase activity (Figs. 1B, 3A, and data not shown). Indeed, in the presence of LIF, EM progenitors and ES cells interchange their identities at a rate that depends, precisely, on the amount of LIF.

In this sense, the regeneration of a pluripotent ES cell phenotype from EM progenitors is reminiscent of the reversion to ES cells of early primitive ectoderm-like (EPL) cells [42]. EPL cells are generated in vitro by culturing ES cells in medium conditioned by HepG2 cells with or without LIF [42]. Similar to the population of EM progenitors characterized in this study, EPL cells express markers of pluripotency at levels comparable to ES cells, do not contribute to embryonic lineages upon injection into mouse blastocysts, can be reverted to an ES cell phenotype in the presence of LIF upon withdrawal of HepG2-conditioned medium [42], and differentiate in vitro preferentially (though not exclusively) into mesoderm-derived lineages [43]. EPL cells can be differentiated from ES cells based on their characteristic expression profile, which includes high levels of *Fgf5* expression and down-regulation of *Rex1* and *Gbx2* expression [42]. In contrast, EM progenitors display expression levels of *Rex1* and *Gbx2* comparable to ES cells and do not express *Fgf5* (Fig. 3A). Thus, despite some similarities, EM progenitor cells are distinct from EPL cells and are more closely related to pluripotent ES cells, at least based on their respective transcriptional profiles. In addition, two important characteristics of the EM progenitor population make its analysis especially relevant for our understanding of ES cell pluripotency: (i) EM progenitors are generated from ES cells under standard culture conditions, not by addition of ill-characterized conditioned media, and (ii) in the same culture conditions, EM progenitors undergo a dedifferentiation process that gives rise to pluripotent ES cells.

10 Differentiation-Promoting Activity of BMPs

The results from our analyses indicate that the generation of EM progenitors from ES cells depends on the direct mesoderm-inducing ability of BMP stimulation (Fig. 6). This finding is consistent with the reported roles of BMP signaling during embryo development [27, 28], and with previous studies of ES cell differentiation in vitro [29–31]. However, the mesoderm-differentiating activity of BMPs seems to be at odds with their role in maintaining the self-renewal of pluripotent ES cells [7]. Indeed, BMP signaling appears to have contrasting effects in the maintenance of ES cell pluripotency. On the one hand, BMPs are necessary to prevent ES cell differentiation toward neural fates [7, 30, 44]. On the other hand, signaling by BMPs results in loss of ES cell pluripotency by promoting their differentiation toward non-neural fates such as mesoderm-derived lineages [29, 31] (and this study). These opposing effects of BMPs can be partially explained by differences in the experimental conditions used in those studies. Thus, in the absence of LIF, low concentrations of BMPs ($\sim$0.25–10 ng/mL) promote mesoderm differentiation [29–31] at the expense of neural fates [30]. In the presence of LIF, however, similar low concentrations of BMPs prevent neural differentiation of ES cells [7, 44] and maintain their pluripotency with no signs of mesoderm differentiation [7]. Consistent with this notion, we did not detect increased generation of EM progenitors with BMP concentrations below 100 ng/mL in the presence of LIF (data not shown). Thus, LIF appears to render ES cells refractory to the mesoderm-inducing activity of BMPs. Our studies demonstrate that this resistance is, at least in part, dependent on the negative feedback mechanism mediated by Nanog.

11 A Negative Feedback that Blocks Mesoderm Differentiation

While investigating the role of LIF in the maintenance of mouse ES cell pluripotency, Rathjen and colleagues [45] found that the expression of *Lif* itself is up-regulated in the

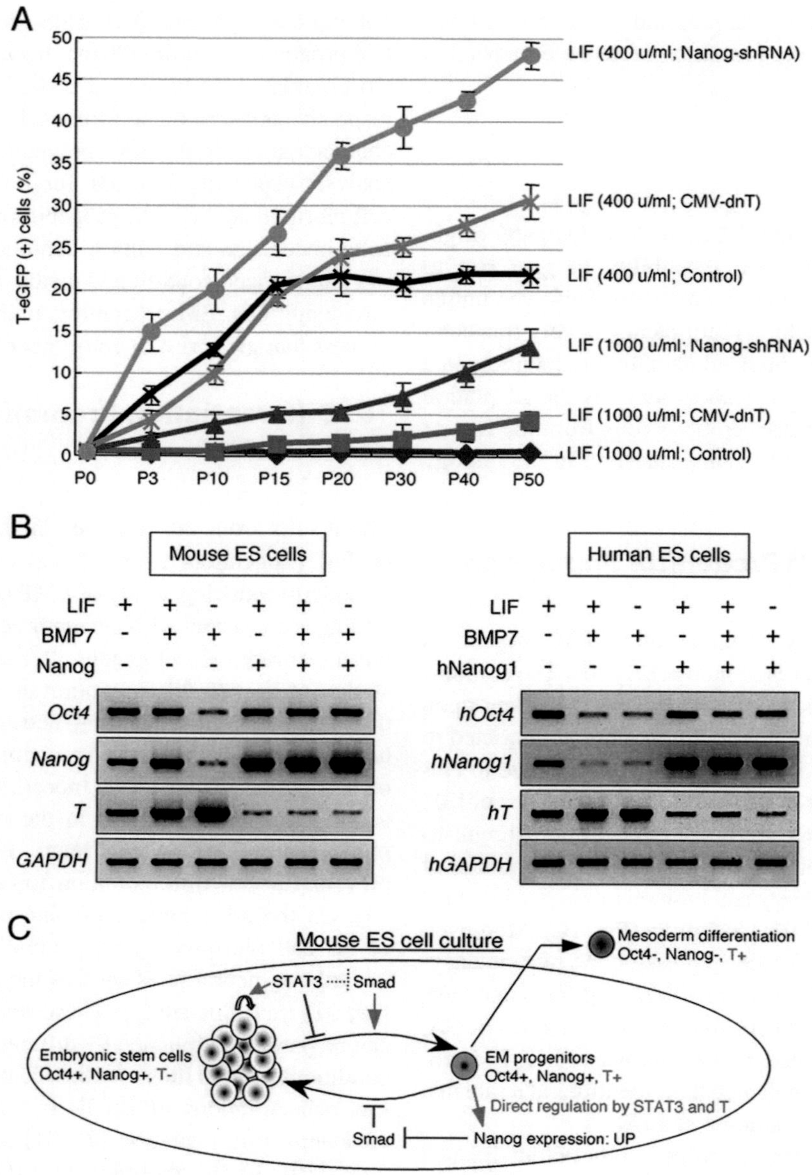

Fig. 7 Nanog-mediated dedifferentiation of EM progenitors is required for maintaining mouse ES cells. (**A**) Flow-cytometric analysis of T(+) cells in *T-eGFP* ES cells (1000 or 400 u/ml of LIF), after down-regulation of T or Nanog activities with *dnT* or *Nanog-shRNA*. (**B**) Comparison of expression profiles of mouse and human ES cells. Mouse and human ES cells were transfected with mouse *Nanog* and *hNanog1*, respectively, or a control vector, then treated with LIF and/or BMP7. In human ES cells, *hNanog1* was down-regulated in response to BMP7 stimulation, whereas *T* expression increased. In contrast, in mouse ES cells, these genes were up-regulated by LIF and BMP7. (**C**) Schematic representation of the mechanism for the maintenance of pluripotent mouse ES cells. EM progenitors are generated in ES cell cultures depending on the balance between the level of LIF/STAT3 and BMP/Smad activation. *Nanog* expression increases in EM progenitors by the combinatorial action of T and activated STAT3. As a result, Nanog inhibits the differentiation signal of BMP by interfering with the formation of activated Smad/p300 complexes, and promotes the dedifferentiation of EM progenitors into pluripotent ES cells. See main text for details

early phases of ES cell differentiation. This mechanism provides a negative feedback that may limit the progression of ES cell differentiation and contribute to the self-renewal of pluripotent ES cells [45]. The transition of EM progenitors to ES cells does not appear to depend on such a mechanism, since *Lif* expression is not noticeably up-regulated in EM progenitors (Fig. 3A). In contrast, EM progenitors do up-regulate the expression of *Nanog* (Fig. 3A). Our results also show that, in the presence of LIF, *Nanog* over-expression is sufficient to accelerate the transition of EM progenitors to ES cells (Fig. 3F, 3H, 3J). More importantly, down-regulation of Nanog function results in impaired dedifferentiation of EM progenitors to ES cells (Fig. 3G, 3H, and 3K). Thus, we identify Nanog as a critical component of a negative feedback mechanism that blocks the progression of ES cell differentiation toward mesoderm fates (Fig. 7C). In this mechanism, mesoderm differentiation of ES cells is initiated by BMP signaling. Possible sources of BMP activity in our culture conditions include fetal calf serum [29], fibroblast feeder layer, and/or ES cells themselves [7]. Consistent with this, we detect a significant activation of the –1147Id1-Luc reporter even in the absence of exogenous BMP supplements (Fig. 6C). ES cells that initiate mesoderm differentiation express the early mesoderm marker *T*. In the presence of LIF, activated STAT3 cooperates with T to directly up-regulate the expression of *Nanog* (Fig. 5), which, in turn, provides a negative feedback that down-regulates *T* expression and eventually leads to the regeneration of ES cells from EM progenitors. In the absence of LIF/STAT3 signaling, T is not sufficient to up-regulate the expression of *Nanog*, and mesoderm differentiation proceeds. The relevance of this negative feedback mechanism for maintaining the pluripotency of ES cells is evident in long-term cultures. Thus, the size of the EM population cannot be maintained in ES cells in which either Nanog or T function is experimentally down-regulated (Fig. 7A). In such conditions, the cultures progressively accumulate EM progenitors and eventually lose pluripotency (not shown).

The existence of such a negative feedback mechanism in mouse ES cells contributes to explaining previous observations on the function of Nanog. For instance, the fact that overexpression of *Nanog* bypasses the need for LIF/STAT3 signaling [10] and BMP stimulation [7] to maintain self-renewal of pluripotent ES cells is easily understood in light of this mechanism. Since the final outcome of BMP activity is the up- regulation of *Nanog* expression, which is mediated by LIF/STAT3 signaling, the experimental up-regulation of Nanog would obviate the need for both BMP and LIF. It is clear, however, that the functions of LIF and Nanog in the maintenance of ES cell pluripotency are not restricted to participating in the negative feedback mechanism characterized in this study. Thus, the complete lack of Nanog function promotes differentiation of ES cells to endoderm lineages [9], indicating the existence of additional roles of Nanog other than that of preventing mesoderm differentiation. Indeed, the up-regulation of *Nanog* expression by T and STAT3 only takes place in EM progenitors, whereas the constitutive expression of *Nanog* in ES cells is regulated by more proximal regions of the *Nanog* promoter (Fig. 5E).

The requirement of LIF/STAT3 signaling for the maintenance of mouse ES cells has been related to the ability of pre-implantation mouse embryos to arrest development when implantation is prevented (a phenomenon known as dispause) [46]. In keeping with this idea, it appears reasonable that cells in the inner cell mass of mouse blastocysts evolve specific mechanisms to prevent unwanted cell differentiation during dispause. The mechanism described in this study could very well serve this purpose. In contrast, ES cells derived from human embryos, in which dispause does not occur, do not depend on LIF/STAT3 signaling to maintain pluripotency [47]. Interestingly, the Nanog-mediated negative feedback mechanism characterized in this study does not appear to be operative in human ES cells. First, the overall conservation of the mouse *Nanog* EM-enhancer in the human *Nanog* gene is very poor, and no T-binding site is present (not shown). Second, unlike mouse ES cells, human ES cells do not up-regulate *Nanog* expression in response to LIF and BMP stimulation, even though *T* expression is induced under these conditions (Fig. 7B). The absence of a functional negative feedback mechanism mediated by T, LIF/STAT3, and Nanog in human ES cells provides additional mechanistic insights into the reasons why LIF is dispensable for the self-renewal of human ES cells.

Taken together, our results uncover a mechanism underlying mouse ES cell pluripotency, by which committed mesoderm progenitors undergo an active process of dedifferentiation mediated by the combined action of the extrinsic cytokine LIF and the intrinsic pluripotency factor Nanog. These findings contribute to unravel the complex network of molecular interactions required to maintain the self-renewal of ES cells and shed light on the cellular bases of ES cell pluripotency. Furthermore, the possibility of reverting the differentiation status of committed cells offers new ways to approach the generation of pluripotent cells for future therapeutic interventions of regenerative medicine.

Acknowledgments We thank Robert Benezra, Senyon Choe, Neil G. Copeland, Richard Eckner, Douglas Melton, Kohei Miyazono, Gustavo Tiscornia, and Shinya Yamanaka for sharing reagents, Dirk Buscher, Chris Kintner, Isao Oishi, Junichiro Sonoda, and Ayumu Tashiro for helpful suggestions, Harley Pineda, Timothy Chapman, and Henry Juguilon for excellent technical assistance, and May-Fun Schwarz for help in the preparation of this manuscript. AS was partially supported by JSPS Research Fellowships for Young Scientists, Japan; AS, TM, and KN are partially supported by JSPS Postdoctoral Fellowships for Research Abroad, Japan; AR and CRE are partially supported by postdoctoral fellowships from Fundación Inbiomed, Spain. The authors are indebted to the Salk Institute administration for the establishment of a non-NIH core in the Stem Cell Research Center through support of institutional funds, the Lookout Fund, and the G. Harold and Leila Y. Mathers Charitable Foundation. Additional funding for mouse ES cell work in JCIB's laboratory was from the G. Harold and Leila Y. Mathers Charitable Foundation and the NIH.

References

1. Martin GR. Isolation of a pluripotent cell line from early mouse embryos cultured in medium conditioned by teratocarcinoma stem cells. Proc Natl Acad Sci U S A. 1981;78:7634–8.
2. Evans MJ, Kaufman MH. Establishment in culture of pluripotential cells from mouse embryos. Nature. 1981;292:154–6.

3. Smith AG. Embryo-derived stem cells: of mice and men. Annu Rev Cell Dev Biol. 2001;17:435–62.
4. Chambers I, Smith A. Self-renewal of teratocarcinoma and embryonic stem cells. Oncogene. 2004;23:7150–60.
5. Smith AG, Heath JK, Donaldson DD, et al. Inhibition of pluripotential embryonic stem cell differentiation by purified polypeptides. Nature. 1988;336:688–90.
6. Williams RL, Hilton DJ, Pease S, et al. Myeloid leukaemia inhibitory factor maintains the developmental potential of embryonic stem cells. Nature. 1988;336:684–7.
7. Ying QL, Nichols J, Chambers I, et al. BMP induction of Id proteins suppresses differentiation and sustains embryonic stem cell self-renewal in collaboration with STAT3. Cell. 2003;115: 281–92.
8. Nichols J, Zevnik B, Anastassiadis K, et al. Formation of pluripotent stem cells in the mammalian embryo depends on the POU transcription factor Oct4. Cell. 1998;95:379–91.
9. Mitsui K, Tokuzawa Y, Itoh H, et al. The homeoprotein Nanog is required for maintenance of pluripotency in mouse epiblast and ES cells. Cell. 2003;113:631–42.
10. Chambers I, Colby D, Robertson M, et al. Functional expression cloning of Nanog, a pluripotency sustaining factor in embryonic stem cells. Cell. 2003;113:643–55.
11. Niwa H, Miyazaki J, Smith AG. Quantitative expression of Oct-3/4 defines differentiation, dedifferentiation or self-renewal of ES cells. Nat Genet. 2000;24:372–6.
12. Shamblott MJ, Axelman J, Wang S, et al. Derivation of pluripotent stem cells from cultured human primordial germ cells. Proc Natl Acad Sci U S A. 1998;95:13726–31.
13. Thomson JA, Itskovitz-Eldor J, Shapiro SS, et al. Embryonic stem cell lines derived from human blastocysts. Science. 1998;282:1145–7.
14. Rao M. Conserved and divergent paths that regulate self-renewal in mouse and human embryonic stem cells. Dev Biol. 2004;275: 269–86.
15. Herrmann BG, Labeit S, Poustka A, et al. Cloning of the T gene required in mesoderm formation in the mouse. Nature. 1990;343:617–22.
16. Wilkinson DG, Bhatt S, Herrmann BG. Expression pattern of the mouse T gene and its role in mesoderm formation. Nature. 1990;343:657–9.
17. Ramalho-Santos M, Yoon S, Matsuzaki Y, et al. "Stemness": transcriptional profiling of embryonic and adult stem cells. Science. 2002;298:597–600.
18. Raz R, Lee CK, Cannizzaro LA, et al. Essential role of STAT3 for embryonic stem cell pluripotency. Proc Natl Acad Sci U S A. 1999;96:2846–51.
19. Gossen M, Freundlieb S, Bender G, et al. Transcriptional activation by tetracyclines in mammalian cells. Science. 1995;268: 1766–9.
20. Hoffmann A, Czichos S, Kaps C, et al. The T-box transcription factor Brachyury mediates cartilage development in mesenchymal stem cell line C3H10T1/2. J Cell Sci. 2002;115:769–81.
21. Kispert A, Herrmann BG. The Brachyury gene encodes a novel DNA binding protein. EMBO J. 1993;12:3211–20.
22. Hiroi Y, Kudoh S, Monzen K, et al. Tbx5 associates with Nkx2-5 and synergistically promotes cardiomyocyte differentiation. Nat Genet. 2001;28:276–80.
23. Stennard FA, Costa MW, Elliott DA, et al. Cardiac T-box factor Tbx20 directly interacts with Nkx2-5, GATA4, and GATA5 in regulation of gene expression in the developing heart. Dev Biol. 2003;262:206–24.
24. Garg V, Kathiriya IS, Barnes R, et al. GATA4 mutations cause human congenital heart defects and reveal an interaction with TBX5. Nature. 2003;424:443–7.
25. Zhu M, John S, Berg M, et al. Functional association of Nmi with Stat5 and Stat1 in IL-2- and IFNgamma-mediated signaling. Cell. 1999;96:121–30.
26. Collum RG, Brutsaert S, Lee G, et al. A Stat3-interacting protein (StIP1) regulates cytokine signal transduction. Proc Natl Acad Sci U S A. 2000;97:10120–5.
27. Dale L, Howes G, Price BM, et al. Bone morphogenetic protein 4: a ventralizing factor in early Xenopus development. Development. 1992;115:573–85.
28. Jones CM, Lyons KM, Lapan PM, et al. DVR-4 (bone morphogenetic protein-4) as a posterior-ventralizing factor in Xenopus mesoderm induction. Development. 1992;115:639–47.
29. Johansson BM, Wiles MV. Evidence for involvement of activin A and bone morphogenetic protein 4 in mammalian mesoderm and hematopoietic development. Mol Cell Biol. 1995;15: 141–51.
30. Finley MF, Devata S, Huettner JE. BMP-4 inhibits neural differentiation of murine embryonic stem cells. J Neurobiol. 1999;40: 271–87.
31. Czyz J, Wobus A. Embryonic stem cell differentiation: the role of extracellular factors. Differentiation. 2001;68:167–74.
32. Holley SA, Neul JL, Attisano L, et al. The Xenopus dorsalizing factor noggin ventralizes Drosophila embryos by preventing DPP from activating its receptor. Cell. 1996;86:607–17.
33. Zimmerman LB, De Jesus-Escobar JM, Harland RM. The Spemann organizer signal noggin binds and inactivates bone morphogenetic protein 4. Cell. 1996;86:599–606.
34. Shi Y, Massague J. Mechanisms of TGF-beta signaling from cell membrane to the nucleus. Cell. 2003;113:685–700.
35. Hollnagel A, Oehlmann V, Heymer J, et al. Id genes are direct targets of bone morphogenetic protein induction in embryonic stem cells. J Biol Chem. 1999;274:19838–45.
36. Nakashima K, Takizawa T, Ochiai W, et al. BMP2-mediated alteration in the developmental pathway of fetal mouse brain cells from neurogenesis to astrocytogenesis. Proc Natl Acad Sci U S A. 2001;98:5868–73.
37. Feng XH, Zhang Y, Wu RY, et al. The tumor suppressor Smad4/DPC4 and transcriptional adaptor CBP/p300 are coactivators for smad3 in TGF-beta-induced transcriptional activation. Genes Dev. 1998;12:2153–63.
38. Janknecht R, Wells NJ, Hunter T. TGF-beta-stimulated cooperation of smad proteins with the coactivators CBP/p300. Genes Dev. 1998;12:2114–9.
39. Pearson KL, Hunter T, Janknecht R. Activation of Smad1-mediated transcription by p300/CBP. Biochim Biophys Acta. 1999;1489:354–64.
40. Latinkic BV, Umbhauer M, Neal KA, et al. The Xenopus Brachyury promoter is activated by FGF and low concentrations of activin and suppressed by high concentrations of activin and by paired-type homeodomain proteins. Genes Dev. 1997;11: 3265–76.
41. Kim J, Johnson K, Chen HJ, et al. Drosophila Mad binds to DNA and directly mediates activation of vestigial by Decapentaplegic. Nature. 1997;388:304–8.
42. Rathjen J, Lake JA, Bettess MD, et al. Formation of a primitive ectoderm like cell population, EPL cells, from ES cells in response to biologically derived factors. J Cell Sci. 1999;112 (Pt 5): 601–12.
43. Lake J, Rathjen J, Remiszewski J, et al. Reversible programming of pluripotent cell differentiation. J Cell Sci. 2000;113(Pt 3): 555–66.
44. Tropepe V, Hitoshi S, Sirard C, et al. Direct neural fate specification from embryonic stem cells: a primitive mammalian neural stem cell stage acquired through a default mechanism. Neuron. 2001;30: 65–78.

45. Rathjen PD, Nichols J, Toth S, et al. Developmentally programmed induction of differentiation inhibiting activity and the control of stem cell populations. Genes Dev. 1990;4:2308–18.
46. Nichols J, Chambers I, Taga T, et al. Physiological rationale for responsiveness of mouse embryonic stem cells to gp130 cytokines. Development. 2001;128:2333–9.
47. Daheron L, Opitz SL, Zaehres H, et al. LIF/STAT3 signaling fails to maintain self-renewal of human embryonic stem cells. Stem Cells. 2004;22:770–8.
48. Suzuki A, Raya A, Kawakami Y, et al. Maintenance of embryonic stem cell pluripotency by Nanog-mediated reversal of mesoderm specification. Nat Clin Pract Cardiovasc Med. 2006;3 Suppl 1:S114–22.
49. Suzuki A, Raya A, Kawakami Y, et al. Nanog binds to Smad1 and blocks bone morphogenetic protein-induced differentiation of embryonic stem cells. Proc Natl Acad Sci U S A. 2006;103: 10294–9.

Human Embryonic Stem Cells and Germ Cell Development

Nina J. Kossack, Joerg Gromoll, and Renee A. Reijo Pera

Abstract Embryonic stem cells (ESCs) are derived from the inner cell mass (ICM) of blastocysts and are characterized by the ability to differentiate into the three primary germ layers. Evidence shows, however, that the cells of the ICM and derived ESCs are not identical. Expression of early germ cell–specific markers in undifferentiated ESCs and the ability of ESCs to differentiate into functional germ cells in vitro suggest that early germ cells and ESCs may be closely related cell types. Proteins such as Dazl, Pumilio, and Nanos are essential for specification, maintenance, and maturation of the germ cell population and are conserved from invertebrates to vertebrates. Homologs of these RNA-binding proteins have recently been identified in human germ cells as well as in human ESCs, suggesting a role in differentiation of ESCs towards the germ cell lineage. This review summarizes properties of ESCs and germ cells and highlights the importance of protein complex formation in differentiation of ESCs towards the germ cell lineage.

Keywords Human embryonic stem cells · Germ cell specification · Germ plasm · Primordial germ cells · Germ cells · Protein/protein interactions · RNA/protein complexes · DAZL · NANOS · PUMILIO

1 Embryonic Stem Cells

The derivation of human embryonic stem cells (hESCs) was first described by Thomson et al. [1]. As described, for the derivation of hESC lines, embryos are cultured to the blastocyst stage in vitro. Subsequently, the inner cell mass (ICM) of the blastocyst is removed and plated on mitotically inactivated mouse embryonic fibroblast cells (MEFs). The resulting pluripotent hESCs are characterized by their ability to proliferate indefinitely and by their potential to differentiate into derivatives of all three embryonic germ layers. Other characteristics of hESCs include a high ratio of nucleus to cytoplasm, increased telomerase activity, and the ability to form teratomas following the injection into severe combined immunodeficient (SCID) mice [1].

2 Germ Cells

One of the first events during mammalian development is the formation or specification of the germ cell lineage. There are two apparently divergent developmental programs of germ cell specification. In nonmammalian model organisms such as *Drosophila melanogaster* (*D. melanogaster*), *Caenorhabditis elegans* (*C. elegans*), and *Xenopus laevis* (*X. laevis*), the germ cells are formed from maternally synthesized germ plasm [2, 3]. This germ plasm is characterized by the presence of ribosome-rich structures, specific RNAs, and RNA-binding proteins. Those cells that receive a portion of the germ plasm during the progressing cell divisions differentiate to become germ cells, while those cells lacking germ plasm develop into somatic cells [2, 3].

In contrast to germ plasm-dependent germ cell specification, germ cell formation in mammalian species is initiated by cell-cell induction. Human germ cell specification takes place after the implantation of the embryo and has not therefore been studied in depth, unlike mouse germ cell specification and development. In the mouse embryo germ cell specification in the proximal epiblast has been shown to be induced by bone morphogenetic proteins (BMPs) 4 and 7, which are secreted from the extraembryonic ectoderm. Therefore, it is the position of the respective cells that determines their fate as germ cells rather than their origin. Mouse germ cells can first be identified at 7.2 days post coitum (dpc) as an extraembryonic cell cluster that is localized at the base of the allantois. These so-called primordial germ cells

R.A. Reijo Pera (✉)
Center for Human Embryonic Stem Cell Research and Education, Institute for Stem Cell Biology & Regenerative Medicine, Department of Obstetrics and Gynecology, Stanford University School of Medicine, Palo Alto, CA 94304-5542
e-mail: reneer@stanford.edu

V.K. Rajasekhar, M.C. Vemuri (eds.), *Regulatory Networks in Stem Cells*, Stem Cell Biology and Regenerative Medicine,
DOI 10.1007/978-1-60327-227-8_5,

(PGCs) subsequently migrate through the hindgut to the dorsal body wall and finally into the genital ridges where they become nonmigratory gonocytes [4].

After migration, the gonocytes in the female enter meiosis, thereby becoming primary oocytes and follicles. The gonocytes in the male, on the other hand, arrest in the G0/G1 stage of mitosis as prospermatogonia and keep their proliferative potential. These prospermatogonia migrate to the basement membrane of the seminiferous tubules after birth and differentiate into spermatogonial stem cells. The spermatogonial stem cells form the basis for spermatogenesis throughout male adult life [5].

Embryonic germ cells (EGCs) are the in vitro manifestation of pluripotent migratory germ cells and it has been noted that mouse EGCs are highly similar to mouse ESCs [6].

Although EGCs and ESCs share important characteristics, one hallmark of migrating germ cells is that the somatic status of imprinted genes is gradually erased [7]. The methylation pattern of imprinted genes will therefore be different depending on the time point of EGC isolation.

Independent of the mode of germ cell specification, the expression of certain germ cell specific genes is highly conserved from invertebrates to vertebrates. The RNA binding proteins Pumilio and Nanos, for instance, are important for the germ cell survival and migration in *Drosophila* as well as in the mouse [8].

3 In Vitro Differentiation of Mouse ESCs into Germ Cells

Mouse ESCs can differentiate into germ cells in vitro as demonstrated by several reports. Hubner et al. [9] demonstrated the ability of mouse ESCs to produce oocytes. To visualize and enrich for the developing germ cells, a germ cell-specific *Oct-4*-promoter-driven green fluorescent protein (GFP) reporter gene was used. Under differentiating culture conditions, the formation of oocytes, which were able to develop into blastocyst-like structures, was observed. More recently, Lacham-Kaplan et al. [10] have demonstrated the positive effect of conditioned media collected from testicular cell cultures on the differentiation of ESCs into oocytes. However, it has not yet been determined whether these in vitro generated oocytes can be fertilized and give rise to progeny.

Additional studies have demonstrated that mouse ESCs can differentiate into PGCs that can differentiate further to form haploid male germ cells. As in the study described above, an *Oct-4*-promoter-driven GFP reporter gene was used to identify PGCs. Fluorescence-activated cell sorting (FACS) analysis was applied to isolate haploid cells, which were then used to fertilize mature oocytes via intracytoplasmatic sperm injection (ICSI). Approximately 20% of these fertilized oocytes progressed to the blastocyst stage. Even though the formation of haploid male germ cells was inefficient, this study demonstrated the ability of mouse ESCs to form male haploid germ cells with the characteristics and the potential of round spermatids [11]. Toyooka et al. [12] likewise succeeded in differentiating mouse ESCs into male germ cells using a knock-in of the *GFP/LacZ* gene into the germ cell-specific *Vasa* gene. Vasa-positive cells were enriched and transplanted into mouse testis, resulting in the production of sperm [12]. In a more recent study, it has been demonstrated that male gametes which have been derived from mouse ESCs can be used to successfully generate offspring, thereby providing proof of functional gametogenesis in vitro [13].

4 In Vitro Differentiation of hESCS into Germ Cells

Human ESCs have also been investigated for their potential to differentiate into germ cells in vitro [14]. Surprisingly, however, it was noted that undifferentiated hESCs express early germ cell markers such as DAZL, STELLAR, NANOS1, and PUM2. Closer analysis using immunohistochemical methods showed that the DAZL and STELLAR proteins are present in most of the cells of undifferentiated ESC colonies. The expression of later germ cell markers such as VASA, BOULE, and SYCP3, on the other hand, was not detected. Upon in vitro differentiation of hESCs into embryoid bodies (EBs), the expression of the gonocyte marker VASA, the meiotic marker SYCP1, and even the postmeiotic markers GDF9 and TEKT1 was induced [14].

The expression of several germ cell-specific genes in the ICM was investigated to determine whether DAZL expression was an indicator for the spontaneous differentiation of the hESCs towards the germ cell lineage [14]. Analysis of the gene expression in ICM cells revealed the expression of *STELLAR* and *NANOS1* at high levels. However, the expression of *DAZL* could not be detected, indicating that the removal of the ICM and the subsequent culture of hESCs may lead to the spontaneous differentiation towards the germ cell lineage. This data underlines the differences between the ICM and hESCs and indicates that hESCs might be more closely related to epiblast cells or EGCs [14].

Another study has investigated the effect of BMPs on the differentiation of hESCs into germ cells in vitro [15] since it has been shown that BMPs are essential for the specification and maintenance of germ cells in mice [16]. To study the influence of BMPs on the differentiation of hESCs towards the germ cell lineage the expression of two germ cell-specific markers, *VASA* and *SYCP3*, was measured. The results demonstrated that the expression of *VASA* in particular was increased by the addition of BMP4 to the differentiation media. While BMP7 and BMP8b alone do not have an effect on the expression of *VASA*, the combined use of BMP4, BMP7, and BMP8b leads to a distinct increase of *VASA* expression. One explanation for this observation might be that BMP4 is essential for the initial induction of PGC formation while BMP7 and BMP8b might play a role in the self renewal or the proliferation of PGCs. However, it is also possible that the presence of BMP4 induces the expression of receptors that are necessary for the action of BMP7 and BMP8b. Although the role of BMPs in the formation of human germ cells has yet to be determined, the study showed that the number of hESCs differentiating towards the germ cell lineage in vitro was increased by the addition of BMPs [15].

In vitro differentiation of human ESCs to VASA-expressing germ cells has been successfully shown, although the efficient completion of human meiosis in vitro has not been established [14].

5 ESC-Like Properties of Germ Cells

A recent study has shown that mouse spermatogonial stem cells can be isolated selecting for the spermatogonial-specific marker Stra8 [17]. It was also demonstrated that the isolated spermatogonial stem cells can acquire embryonic stem cell properties under certain culture conditions. The pluripotency of these cells was determined by their ability to spontaneously differentiate into derivatives of the three primary germ layers and their ability to form teratomas after injection into SCID/beige mice. In addition, the developmental potential was investigated by injecting the spermatogonial stem cells into 3.5-day-old blastocysts that were transferred into the uterus of pseudopregnant mice and resulted in the birth of several chimeric animals [17]. Pluripotency therefore seems to be a shared characteristic of hESCs, the epiblast, PGCs [1] and, under certain conditions, of spermatogonial stem cells.

6 Genes that Play a Role in Germ Cell Development Are Highly Conserved

As noted above, there are two distinct developmental programs of germ cell specification. Nonetheless, despite these different programs, several genes that are important for the establishment and the maintenance of the germ cell population are conserved from invertebrates to vertebrates.

7 The DAZ Gene Family

The *DAZ* gene family has three members, *BOULE*, *DAZ* (*Deleted in Azoospermia*), and *DAZL* (*DAZ-Like*), which are characterized by three features: First, their expression is germ cell-specific [18]. Second, all three proteins contain a highly conserved RNP-type RNA-recognition motif (RRM) [19]. Third, each member of the *DAZ* gene family contains at least one DAZ repeat coding for 24 amino acid residues. Although the function of these DAZ repeats has not been clarified yet, it is assumed that they are involved in protein-protein interactions [19–22]. While *BOULE* is the ancestral gene of the *DAZ* family and is conserved from flies to humans, *DAZL* orthologues are only found in vertebrates. It is thought that the autosomal *DAZL* gene arose via duplication of the *Boule* gene before the divergence of vertebrates and invertebrates [18]. The Y chromosomal *DAZ* genes arose from an ancestral *DAZL* gene through transposition and repeated amplification. In contrast to *BOULE* and *DAZL*, the *DAZ* genes can only be found on the Y chromosome of Old World monkeys and great apes [21–23]. An overview of the evolutionary development of the *DAZ* gene family is shown (Fig. 1).

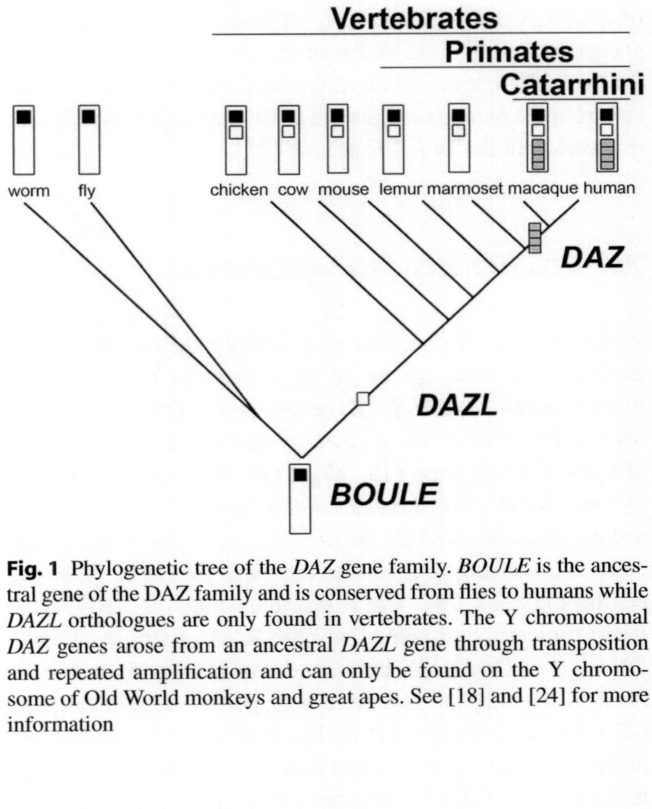

Fig. 1 Phylogenetic tree of the *DAZ* gene family. *BOULE* is the ancestral gene of the DAZ family and is conserved from flies to humans while *DAZL* orthologues are only found in vertebrates. The Y chromosomal *DAZ* genes arose from an ancestral *DAZL* gene through transposition and repeated amplification and can only be found on the Y chromosome of Old World monkeys and great apes. See [18] and [24] for more information

7.1 BOULE

The human BOULE protein was originally identified using a DAZ/DAZL construct as bait for a yeast two-hybrid screen [18]. Sequence analysis showed that its DNA sequence is highly conserved from flies to humans. No mutations within the coding region of the human *BOULE* gene have been described, indicating a strong functional constraint and suggesting that variations within the *BOULE* gene might interfere with human reproduction. The *BOULE* gene is localized on chromosome 2 at position 2q33 and consists of 11 exons [18]. Apart from that, the *BOULE* gene contains one DAZ repeat and encodes a protein containing a RNA-binding domain with specific ribonucleoprotein-1 (RNP-1) and RNP-2 motifs.

Unlike DAZ, BOULE is only expressed in the testis and cannot be found in the early embryo or in primordial germ cells. Within the testis BOULE protein can first be detected in germ cells of the first meiotic division and then reaches its highest expression level in pachytene spermatocytes [18, 24, 25]. Expression levels decrease from the diplotene stage and are not detectable in later spermatogenic stages such as round and elongated spermatids [18]. It has been shown that BOULE regulates the translation of the *CDC25A* phosphatase mRNA. This CDC25A phosphatase then activates the M phase-promoting factor, which is essential for the G2/M transition of meiosis. While the onset of CDC25A phosphatase protein expression is identical to that of BOULE, CDC25A is also expressed at later stages of spermatogenesis. The assumption that BOULE is essential for the meiotic transition is underlined by the fact that a large subgroup of patients diagnosed with meiotic arrest lacks the expression of the BOULE protein [25].

7.2 DAZ (Deleted in Azoospermia)

A screen for Y-chromosomal abnormalities that cause azoospermia (no sperm in the ejaculate) led to the identification of the *DAZ* genes [19]. The *DAZ* genes encode for RNA-binding proteins and are deleted in about 10% of infertile men [19, 26]. Four *DAZ* genes, arranged in two clusters are located on the human Y chromosome and are characterized by the presence of 7–24 DAZ repeats, each of them coding for 24 amino acids rich in N, Y, and Q residues [19, 27, 28]. The N-terminus contains a RNP-type RRM typical for proteins that bind RNA or single-stranded DNA [19]. This RNA binding domain suggests a role of DAZ in the translational control of germ cell specific gene [25]. Because of the sequence heterogeneity of the members of the *DAZ* family, it is likely that the proteins interact with different sets of RNA substrates and proteins [19, 29].

DAZ protein can be detected in male germ cells prenatally in the gonocytes of the fetus as well as in adult spermatogonia and the spermatocytes. DAZ expression in postmeiotic germ cells is very low [30]. While DAZ can be detected in the nucleus and the cytoplasm in gonocytes, the localization is mostly nuclear in spermatogonial stem cells and is finally restricted to the cytoplasm in later stages of spermatogenesis. This expression pattern indicates that the *DAZ* genes are important for the early stages of germ cell development [30]. This assumption is underlined by studies showing that men who have deletions in the DAZ gene cluster have a disrupted spermatogenesis [31]. It has been shown that deletions of the *DAZ* genes cause oligozoospermia (<20 million sperm/ml of ejaculate) and azoospermia (no sperm in the ejaculate) [19, 32] demonstrating that DAZ is important, but not essential for the completion of spermatogenesis [18].

7.3 DAZL (Deleted in Azoospermia-Like)

DAZL is an autosomal homolog of the *DAZ* genes and has been implicated in human male and female fertility. The human *DAZL* gene is located on chromosome 3 and consists of 11 exons [21, 26, 33, 34]. In the embryo, DAZL proteins are expressed in the nucleus and in the cytoplasm of primordial germ cells. Immunohistochemical analysis has shown that the intracellular location of the DAZL protein in the adult changes during the process of germ cell maturation. While DAZL is expressed in the nucleus and the cytoplasm of spermatogonia it can only be detected in the cytoplasm of meiotic spermatocytes and round and elongating spermatids. This expression pattern suggests that the ancestral *DAZL* gene has evolved a function in late germ cell development [30]. In contrast to the Y-chromosomal *DAZ* genes, the expression of DAZL can also be detected in the cytoplasm of oocytes [18, 30]. It has also been shown that the DAZL protein is not only expressed in male and female germ cells, but can also be detected in undifferentiated hESCs [14].

Many single nucleotide polymorphisms (SNPs) and mutations have been described for the *DAZL* gene. Interestingly those polymorphisms do not affect male and female reproduction in the same way. On the contrary, several SNPs have been described that are associated with a decreased age at menopause in women but with increased sperm counts in men [31, 35].

Expression patterns and the results from null mutants suggest that DAZ and DAZL play an important role in the development of PGCs and in germ cell differentiation and maturation, whereas the role of BOULE appears to be restricted to meiosis [36]. The ability to genetically dissect the role of these genes in human germ cell development may be realized by hESC differentiation to the germ cell lineage in vitro.

7.4 DZIP (DAZ-Interacting Protein)

The DZIP protein was identified as a protein that interacts with DAZ and DAZL [37]. Closer analysis revealed that the *DZIP* gene is located on chromosome 13q31 and encodes a protein with a C2H2 zinc-finger domain. This binding motif has been demonstrated to mediate interactions with

RNA, DNA or protein domains. Further analysis has shown that *DZIP* encodes three different isoforms: *DZIPb* (*DZIP brain*), *DZIPt1* (*DZIP testis1*), and *DZIPt2* (*DZIP testis2*). The transcript of the splice variant *DZIPt1* is characterized by the inclusion of two alternative exons. These exons encode for 46 additional amino acids that are located on the N-terminus of the protein and their inclusion into the transcript leads to the truncation of the protein at the carboxyterminal end. The *DZIPt2* transcript, on the other hand, contains only one alternative exon, which codes for 19 additional amino acids that are located downstream of the C2H2 zinc-finger domain. Although the function of DZIP remains unclear, the presence of two testis specific splice variants suggests that they may be involved in the regulation of different stages of germ cell maturation [37].

The expression of *DZIP* was detected in fetal testis, fetal ovary, and in brain tissue. Interestingly, *DZIP* is also expressed in hESCs and, in contrast to *DAZL*, the expression levels remain constant during differentiation. In the adult, *DZIP* mRNAs can be detected in the testis and at a lower expression level in skeletal muscle and in the ovary. The localization of the DZIP protein is very similar to that of DAZL. DZIP protein can be detected in the nucleus and the cytoplasm of PGCs, whereas in the adult testis, the DZIP protein is localized mainly in the nucleus but can also be detected in the cytoplasm of spermatogonia. While DZIP has not been observed in primary spermatocytes, it is present primarily in the nuclei of secondary spermatocytes. The expression pattern within germ cells is therefore similar to that of DAZL [37].

7.5 NANOG

The *NANOG* gene was named after the mythological Celtic land of the ever young, 'Tir nan Og', underlining its function in ESC self-renewal [38]. It is localized on chromosome 12 and consists of 4 exons [39]. The NANOG protein contains a highly conserved DNA-binding homeodomain, suggesting its function as a transcriptional regulator [38]. Two transactivator modules have been identified at the C-terminus of the NANOG protein, which is therefore considered the functionally dominant domain of the protein [40]. The expression of *NANOG* has been shown to be regulated by the binding of OCT-4 and SOX2 to the *NANOG* promoter region and it has been suggested that other pluripotency factors are also able to regulate the expression of NANOG [38, 41]. Thus, together with other proteins, NANOG functions as a transcriptional regulator to maintain pluripotency [40].

During early mouse embryonic development, *Nanog* expression can be detected through the blastocyst stage where it is then down-regulated prior to implantation. In addition to that, *Nanog* is expressed in PGCs, undifferentiated ESCs, EGCs, and embryonic carcinoma (EC) cells [42]. Human NANOG protein can be detected in the fetal and in the adult ovary and in the testis, albeit at low levels. NANOG is also highly expressed in hESC lines and its expression has been shown to decrease upon EB differentiation [39].

7.6 NOS1 (NANOS1) NOS2 (NANOS2), and NOS3 (NANOS3)

The *Nanos* gene family has been shown to play an important role in germ cell formation, maintenance, and maturation. The human *NOS1* gene is localized on chromosome 10 and consists of only one exon and a long 3′ untranslated region (UTR), typical of developmental genes [8]. *NOS1* encodes a protein with a C-terminal zinc finger RNA-binding domain indicating a function as a translational regulator [8]. This RNA-binding domain is highly conserved between species, showing a 62% amino acid identity with *Drosophila* Nanos, for instance [43].

The NOS1 protein has been shown to be expressed during male germ cell development. It can be detected in spermatogonia, spermatocytes, and in post-meiotic male germ cells. Within these cells, NOS1 is not equally distributed but is present in a perinuclear subregion of the cytoplasm. The mRNA expression of *NOS1* in fetal ovaries indicates that the protein may be present in premeiotic female germ cells, however, neither the mRNA nor the protein can be detected in mature postmeiotic oocytes. The expression pattern of *NOS1* in the adult testis suggests a role in the differentiation of the germ cells. *NOS1* expression can also be detected in mouse and human ESCs [8].

Nos2 and *Nos3* have been studied in model organisms and were shown to be essential for germ cell development. While the disruption of *Nos2* leads to impaired spermatogenesis in male mice, it does not influence oogenesis in female mice. Nos3, on the other hand, seems to be essential for earlier stages of germ cell development, thereby influencing male and female gametogenesis. It has been shown that the PGC population in mice with a disrupted *Nos3* gene was not maintained leading to the migration of a limited number of germ cells that finally enter the genital ridges [44]. The importance of NANOS2 and NANOS3 for human germ cell development remains to be elucidated.

7.7 OCT-4 (Octamer-Binding Transcription Factor 4)

OCT-4 is a member of a family of transcription factors and is characterized by the presence of a highly conserved POU-DNA binding domain [45–50]. This DNA binding domain specifically binds to the sequence 5′-ATGCAAAT-3′ [51]. The OCT-4 protein is expressed in pluripotent cells like in hESCs, the cells of the ICM, and the epiblast indicating its function in early mammalian development [52] and in maintaining cells in an undifferentiated state [53]. OCT-4 expression has been detected in PGCs [49, 52, 54] and it has been shown that Oct-4 is essential for the viability of the mammalian germline [55].

The *OCT-4* gene is located on chromosome 6 and consists of five exons. Two splice variants of the human OCT-4 protein have been described, which result from the alternative usage of exon 1a and exon 1b, respectively. In contrast to OCT-4A, the OCT-4B protein does not contain the sequence that is necessary for transactivation and is therefore unable to stimulate transcription from OCT-4-dependent promoters. Furthermore, OCT-4B is mostly localized in the cytoplasm and is not able to maintain stem cell self-renewal [56].

7.8 PUM1 (Pumilio 1) and PUM2 (Pumilio 2)

The *PUM1* gene is located on chromosome 1 and consists of 22 exons while the *PUM2* gene is located on chromosome 2 and consists of 20 exons. Apart from the presence of the first two additional exons in the *PUM1* gene, *PUM1* and *PUM2* are characterized by highly conserved exon sizes and exon/intron boundaries indicating that they evolved through the duplication of a common ancestral gene [57]. PUM2 protein contains 8 C-terminal RNA-binding PUF (Pumilio/FBF) repeats consisting of 36 amino acids [57, 58]. PUM1 and PUM2 are RNA-binding proteins that specifically bind to sequences in the 3′-UTR of their target mRNAs [59] leading to the inhibition of translation and destabilization of the mRNA by de-adenylation [60–62]. Via this mechanism, the PUMILIO proteins act as translational repressors during development and differentiation [57].

PUM2 mRNA is highly expressed in ESCs, in the fetal and adult ovary and fetal and adult testis, whereas *PUM1* is expressed ubiquitously [59]. The expression of PUM2 protein can be detected in primordial germ cells of the fetal testis and the fetal ovary. During adult male germ cell development, PUM2 protein is present in the cytoplasm and the nucleus of spermatogonia and also in the cytoplasm of early and late spermatocytes. In the female, PUM2 protein can be detected in the cytoplasm of oocytes [58].

Like the *BOULE* gene, *PUM2* is conserved from flies and worms to mice and humans [18, 58]. The *Drosophila* RNA-binding domain is 80% identical to the human PUM2 RNA-binding domain, indicating strong functional constraints [63].

Studies on model organisms suggest that the Pumilio proteins not only function as translational regulators of embryogenesis but that they are also involved in the regulation of germ cell development and differentiation. It is also hypothesized that Pumilio plays a role in the self-renewal of mammalian stem cells [64, 65].

7.9 STELLAR

STELLAR was identified in a study aimed at finding genes with a similar expression pattern as the pluripotency marker *OCT-4,* which might therefore be involved in the regulation of pluripotency as well [39]. In the adult, *OCT-4* is highly expressed in human germ cell tumors. Since several studies have shown that structural chromosomal changes within chromosome 12p are involved in the formation of these germ cell tumors, genes located on chromosome 12p were characterized in more detail. One of those genes is *STELLAR,* which consists of four exons and codes for a protein with 159 amino acids and a mass of 17.9 kD. STELLAR is a nuclear protein that does not contain any functional motifs. Besides *STELLAR*, *NANOG,* and *GDF3* were identified as genes that also localize to human chromosome 12p. *STELLAR*, *NANOG,* and*GDF3* are expressed at very low levels in the fetal ovary, the adult ovary and the germ cells of the adult testis. In addition to that, *STELLAR*, *NANOG,* and *GDF3* are also expressed in hESCs and their expression levels decrease upon differentiation of ESCs [39].

8 Formation of Protein Complexes in Germ Cells Is Conserved from Invertebrates to Vertebrates

Proteins like Vasa, Dazl, Pumilio, and Nanos all localize to the germ plasm in lower organisms. Two common features of these proteins are that they contain RNA-binding domains and that they function in a ribonucleoprotein complex in the germ plasm of lower organisms. By regulating the translation of specific mRNAs, these proteins play an important role in the development and the maintenance of early germ cells [43, 66–71].

In recent years, homologs of Vasa, Dazl, Pumilio, and Nanos have been identified in human germ cells as well as in hESCs [8, 30, 58, 72, 73], leading to the assumption that these proteins might function in a ribonucleoprotein complex in humans as well. In fact, the formation of those protein complexes has been reported in recent studies.

9 Interaction of DAZ and PUM2

It has been suggested that the RNA-binding DAZ proteins can interact with other proteins and thereby regulate the translation of distinct mRNAs. Coimmunoprecipitation experiments have shown that DAZ and PUM2 are able to form a stable complex. Further analysis has demonstrated that the RNA-binding region of PUM2, which contains eight PUF repeats [74, 75] and the linker region of DAZ, located between the RNA-binding DAZ repeat domains are required for the interaction of these two proteins [58]. The co-localization of DAZ and PUM2 in the germ cell lineage underlines the possible importance of this complex.

In addition, it was shown that DAZL and PUM2 are able to form a stable complex, which can bind to the *Drosophila* Nanos responsive element (NRE) RNA and thereby regulates its transcription. It has therefore been suggested that PUM2 interacts with the DAZ/DAZL proteins and functions as a translational regulator in germ cells and ESCs [58] indicating its importance for the maintenance of cells that give rise to mature germ cells.

10 Binding Specificity of PUM2 and DAZL

Multiple studies have shown that the DAZ/DAZL proteins can form a stable complex with the PUM2 protein. It has also been demonstrated that these protein complexes can bind to the same mRNA [37, 58]. Human PUM2 has been shown to bind to the *Drosophila* NRE sequence. Several nucleotides of this sequence were mutagenized to identify the minimal sequence 'GNNNNNNNNNNNUGUA' required for PUM2 binding. The fact that DAZL and DAZ are not able to bind to this sequence demonstrates the specificity of this RNA-protein interaction. The search for sequences that contain the minimal binding sequence lead to the identification of *SDAD1*, which is expressed in ESCs, fetal and adult tissues [76]. The function of the SDAD1 protein has been closely investigated in several model organisms. In *Saccharomyces cerevisiae*, SDAD1 plays a role in the regulation of the cell cycle, whereas it is essential for the maintenance of the cells in the central nervous system in the zebrafish [77–79]. Closer analysis of the PUM2 and DAZL binding sites within the 3′UTR revealed that DAZL can bind to a region close to the stop codon and to another fragment that also binds PUM2. The DAZL consensus sequence has been determined to consist of 10 essential nucleotides and to be rich in uridines [76].

PUM2 specifically binds to the 3′UTR region within the *SDAD1* gene, and the RNA sequence UNUUANNUGUA within the *SDAD1* mRNA has been shown to be essential for the binding of PUM2. This sequence was therefore termed PUM2 binding element I (hPBE1). The PUM2 binding site is characterized by an UGUA motif that is also present in the *Drosophila* NRE, the FBF binding elements of *C. elegans,* and the murine PUM2 binding element [80]. An additional binding sequence (UAUANNUAGU) was identified in the *SDAD1* gene which was named PUM2 binding element 2 (hPBE2) [76].

11 Interaction of NOS1 and PUM2

It has also been demonstrated that human PUM2 and NOS1 can form a stable complex [8]. For the interaction of these two proteins the RNA-binding domain plus 155 amino acids upstream of PUM2, and the zinc finger domain of NOS1 are required. The interaction of these two proteins seems to be independent of the presence of the target RNAs to which the complex binds. In addition, experiments designed to identify proteins that interact with PUM2 showed that PUM2 can form homodimers. To homodimerize, the presence of the N-terminal amino acids 261–550 is required. This N-terminal interaction still allows the binding of NOS1 proteins and suggests the presence of a NOS-1-PUM2/PUM2-NOS-1 mRNA complex, which might be involved in early germ cell development and the differentiation of germ cells in the adult [8]. This assumption is supported by the fact that PUM2 [58] and NOS1 protein are both expressed in a perinuclear subregion of the cytoplasm of meiotic and pre-meiotic germ cells [8].

12 Interaction of BOULE and PUM2

Using the yeast two hybrid assay, it was shown that human BOULE can homodimerize and interact with PUM2. BOULE has been demonstrated to be evolutionarily conserved from flies to humans and therefore studies have been performed investigating the ability of human and *Drosophila* proteins to interact. *Drosophila* Boule is able to homodizerize and to interact with the human BOULE protein, but it cannot interact with the human PUM2 protein. In contrast, *Drosophila* Nanos is not able to interact with the human BOULE protein or human PUM2, but is able to form homodimers [59].

Further analysis of the BOULE regions required for homodimerization has shown that the region between residues 161 and 185 is essential for the interaction of two BOULE proteins. This region corresponds to the location of the single DAZ repeat within the BOULE protein, which is also essential for the interaction between BOULE and DAZL proteins. In contrast, most of the BOULE protein needs to be present for a successful interaction between BOULE and PUM2 [59].

PUF repeats 2, 4, and 8 within the PUM2 protein are essential for the interaction of PUM2 and DAZL proteins. The interaction of BOULE and PUM2, on the other hand, does not require the presence of these PUF repeats suggesting a different interacting mechanism [59].

The Nanos response element (NRE) has been shown to be the *Drosophila* Pumilio RNA target sequence. It has also been demonstrated that PUM2, the PUM2/DAZL complex, and the PUM2/BOULE complex can interact with the NRE. Further analysis of the PUM2 complexes indicated that the formation of specific PUM2 complexes depends on the target RNA [59].

The ability of PUM2 to interact with different proteins suggests that those complexes may play different roles during germ cell development. This assumption is supported by the fact that DAZL and BOULE are expressed at different stages during spermatogenesis. In addition, it has been shown that DAZL and BOULE do not interact with the same regions of the PUM2 protein and that the PUM2/DAZL complex binds to different RNA targets than the PUM2/BOULE complex.

These findings indicate that the function of human PUM2 depends on the interacting protein. It is possible that the binding of PUM2 to a certain target RNA requires a specific PUF domain. However, it is also likely that the interaction of PUM2 with another protein induces conformational changes within the protein complex and thereby influences the specificity for certain target RNAs. It might also be that the interaction of PUM2 with a specific target RNA influences the formation of a specific complex with another protein [59].

13 Interaction of DAZ and DZIP

DAZ protein has also been shown to interact with the DZIPt1 or DZIPt2 proteins. Multiple regions of the DAZ protein seem to be important for the formation of a complex with DZIPt1, a region close to the RNA-binding domain (amino acids 90–123), the linker region (124–173), and the DAZ repeat. However, deletion analysis revealed that the region between amino acids 90 and 123 of the DAZ protein is sufficient for the interaction of DAZ with DZIPt1. The interaction of DAZ and DZIPt1 therefore requires the same region that is necessary for the formation of DAZ homo- and heterodimers. The DZIPt1 protein, on the other hand, only interacts via the C2H2 zinc-finger region with the DAZ protein.

The interaction of DAZ and DAZL with several different proteins suggests that certain protein complexes are able to bind and to regulate the expression of different target RNAs [37].

14 Interaction of PUM2 and DAZL in ESCs and Germ Cells

Fluorescence resonance energy transfer (FRET) was used to investigate the interaction of PUM2 and DAZL in ESCs, germ cells, and EGCs [81]. The results show that DAZL protein is located in the nuclei and in the cytoplasm of mouse and human ESCs. In contrast, PUM2 was only detected in the cytoplasm of these cells. Neither homo- nor heterodimerization of DAZL and PUM2 was observed and, even in the presence of DAZL, the PUM2 proteins were only detected in the cytoplasm of the ESCs. Heterodimerization and the movement of PUM2 into the nucleus of EGCs was observed, although homodimerization of DAZL and PUM2 was not detected. In mouse premeiotic germ cells, DAZL and PUM2 proteins are able to form homo- and heterodimers that can be found in the cytoplasm and in the nucleus [81]. These results strongly indicate that the presence of PUM2 in the nucleus depends on the heterodimerization with the DAZL protein. Finally, the results for somatic cells showed neither homo- nor heterodimerization of DAZL and PUM2 nor the redistribution of PUM2 to the nucleus of the cells. This data indicates the presence of an RNA-binding protein complex that plays a role in the post-transcriptional regulation of gene expression in germ cells. These results also show the distinct differences between germ cells and ESCs concerning the homo- and heterodimerization of DAZL and PUM2 [81].

It has been demonstrated that protein complexes from homologs of germ plasm components are formed in mammalian germ cells [81]. It has also been shown that ESCs express early germ cell markers and that in vitro differentiation of ESCs leads to the expression of late germ cell markers [14]. Based on this data, a model of germ cell formation from ESCs has been suggested. ESCs express proteins that are essential for the formation or maintenance of germ cells, then extracellular signals result in the formation of protein complexes committing the ESCs to the germ cell lineage [81]. An outline of this model is shown in Fig. 2.

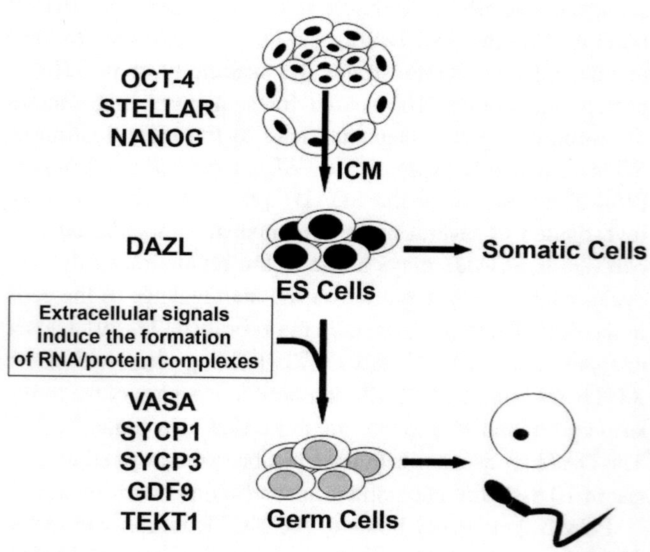

Fig. 2 Model of germ cell specification from ESCs based on the previous publications [37], [58], [14], and [81]. The cells of the inner cell mass express early germ cell markers such as STELLAR and NANOG in addition to the pluripotency marker OCT-4. The expression of DAZL has not been demonstrated. In contrast to that, hESCs do express DAZL and upon differentiation towards the germ cell lineage the expression of later germ cell markers such as VASA, SYCP1, SYCP3, GDF9, and TEKT1 is induced. This figure *underlines* the ability of hESCs to differentiate into somatic cells as well as germ cells and *highlights* the importance of RNA/protein complexes for the formation of germ cells

15 Similarities and Differences Between ESCs and Germ Cells

EGCs are the in vitro manifestation of pluripotent migratory germ cells and it has been noted that mouse EGCs are highly similar to mouse ESCs [6]. Even though EGCs and ESCs share important characteristics, one hallmark of migrating germ cells is that the somatic status of imprinted genes is gradually erased [7]. The methylation pattern of imprinted genes will therefore be different depending on the time point of EGC isolation. In contrast, mouse ESCs have been shown to have a somatic imprinting pattern [11]. If ESCs were derived from germ cells, then based on the methylation pattern, the ESCs must have been derived from germ cells prior to E9.5 [82].

Differentiation experiments from ESCs towards the germ cell lineage have been impeded by the similarity of these cell types concerning their morphology, mRNA, and protein expression prior to the erasure and sex-specific establishment of imprinting. Even genes like *DAZL, PUM2,* and *DZIP* that had previously been thought to be germ cell-specific can be detected in hESCs [14, 37, 58, 83]. The expression of later germ cell markers such as *VASA*, *BOULE,* and *SYCP3*, on the other hand, were not shown. Upon in vitro differentiation of hESCs into EBs, the expression of the gonocyte marker *VASA*, the meiotic marker *SYCP1,* and even of the postmeiotic markers *GDF9* and *TEKT1* were induced [14].

To determine whether the DAZL expression was an indicator for the spontaneous differentiation of hESCs towards the germ cell lineage, DAZL expression in the ICM was investigated. However, the expression of *DAZL* could not be detected indicating that the removal of the ICM and the subsequent culture of ESCs may lead to the spontaneous differentiation towards the germ cell lineage. This data underlines the differences between the ICM and hESCs and indicates that hESCs might be more closely related to epiblast cells or EGCs.

The expression of germ cell specific genes in ESCs supports the hypothesis that early germ cells are the closest in vivo equivalent to ESCs [82]. This idea, however, raises the question as to why the differentiation of ESCs to mature germ cells in vitro is so highly inefficient. Apart from that, PGCs do not share the ESC property of contributing to chimeras after the injection into blastocysts, thus demonstrating that ESCs and PGCs are highly similar in terms of marker expression but they are not functionally identical [82].

The derivation of hESCs has opened the door to extensive studies of human developmental genetics. For example, future research may focus on the comparison of the transcriptomes from the ICM, ESCs, and early germ cells, thereby enhancing the understanding of the identity of hESCs and germ cells, their common characteristics and differences. This research will help to clarify whether ESCs represent an in vivo cell type or whether their expression pattern is the result of ESC culture conditions [82]. Other studies may focus on the genetics of infertility. Although 10–15% of those couples wishing to have children are infertile, the reasons for infertility are still poorly understood. *DAZ* deletions have been described in men with impaired spermatogenesis and rare variants in the *DAZL* gene have been described in women with ovarian failure. Nonetheless, the biological consequences of these variations in human germ cell development remain to be elucidated.

References

1. Thomson J, Itskovitz-Eldor J, Shapiro S, et al. Embryonic stem cell lines derived from human blastocysts. Science. 1998;282: 1145–7.
2. Wylie C. Germ cells. Curr Opin Genet Dev. 2000;10:410–13.
3. Zhou Y, King M. Sending RNAs into the future: RNA localization and germ cell fate. IUBMB Life. 2004;56:19–27.
4. McLaren A. Primordial germ cells in the mouse. Dev Biol. 2003;262:1–15.
5. Brinster RL. Germline stem cell transplantation and transgenesis. Science. 2002;296:2174–6.
6. Donovan PJ, de Miguel MP. Turning germ cells into stem cells. Curr Opin Genet Dev. 2003;13:463–71.
7. Yamazaki Y, Mann MRW, Lee SS, et al. Reprogramming of primordial germ cells begins before migration into the genital ridge, making these cells inadequate donors for reproductive cloning. Proc Natl Acad Sci U S A. 2003;100:12207–12.
8. Jaruzelska J, Kotecki M, Kusz K, Spik A, Firpo M, Reijo Pera RA. Conservation of a Pumilio-Nanos complex from *Drosophila* germ plasm to human germ cells. Dev Genes Evol. 2003;213:120–6.
9. Hubner K, Fuhrmann G, Christenson L, et al. Derivation of oocytes from mouse embryonic stem cells. Science. 2003;300:1251–6.
10. Lacham-Kaplan O, Chy H, Trounson A. Testicular cell conditioned medium supports differentiation of embryonic stem cells into ovarian structures containing oocytes. Stem Cells. 2006;24:266–73.
11. Geijsen N, Horoschak M, Kim K, Gribnau J, Eggan K, Daley GQ. Derivation of embryonic germ cells and male gametes from embryonic stem cells. Nature. 2004;427:148–54.
12. Toyooka Y, Tsunekawa N, Akasu R, Noce T. Embryonic stem cells can form germ cells in vitro. Proc Natl Acad Sci U S A. 2003;100:11457–62.
13. Nayernia K, Nolte J, Michelmann H, et al. In vitro-differentiated embryonic stem cell give rise to male gametes that can generate offspring mice. Dev Cell. 2006;11:125–32.
14. Clark AT, Bodnar MS, Fox MS, et al. Spontaneous differentiation of germ cells from human embryonic stem cells in vitro. Hum Mol Genet. 2004;13:727–39.
15. Kee K, Gonsalves J, Clark A, Reijo Pera RA. Bone morphogenetic proteins induce germ cell differentiation from human embryonic stem cells. Stem Cells Dev. 2006;15:831–7.
16. Fujiwara T, Dunn NR, Hogan BL. Bone morphogenetic protein 4 in the extraembryonic mesoderm is required for allantois development and the localization and survival of primordial germ cells in the mouse. Proc Natl Acad Sci U S A. 2001;98:13739–44.
17. Guan K, Nayernia K, Maier L, et al. Pluripotency of spermatogonial stem cells from adult mouse testis. Nature. 2006;440: 1199–203.
18. Xu EY, Moore FL, Reijo Pera RA. A gene family required for human germ cell development evolved from an ancient meiotic gene conserved in all metazoans. Proc Natl Acad Sci U S A. 2001;98:7414–9.

19. Reijo RA, Lee TY, Salo P, et al. Diverse spermatogenic defects in humans caused by Y chromosome deletions encompassing a novel RNA-binding protein gene. Nat Genet. 1995;10:383–93.
20. Tsui S, Dai T, Roettger S, Schempp W, Salido EC, Yen PH. Identification of two novel proteins that interact with germ-cell specific RNA-binding proteins DAZ and DAZL1. Genomics. 2000;65:266–73.
21. Saxena R, Brown LG, Hawkins T, et al. The*DAZ* gene cluster on the human Y chromosome arose from an autosomal gene that was transposed, repeatedly amplified and pruned. Nat Genet. 1996;14:292–9.
22. Gromoll J, Weinbauer GF, Skaletsky H, et al. The Old World monkey *DAZ* (*Deleted in AZoospermia*) gene yields insights into the evolution of the *DAZ* gene cluster on the human Y chromosome. Hum Mol Genet. 1999;8:2017–24.
23. Shan Z, Hirschmann P, Seebacher T, et al. A *SPGY* copy homologous to the mouse gene *Dazla* and the *Drosophila* gene *boule* is autosomal and expressed only in the human male gonad. Hum Mol Genet. 1996;5:2005–11.
24. Tung J, Luetjens C, Wistuba J, Xu E, Reijo Pera RA, Gromoll J. Evolutionary comparison of the reproductive genes, *DAZL* and *BOULE*, in primates with and without *DAZ*. Dev Genes Evol. 2006 216:158–68.
25. Luetjens CM, Xu EY, Reijo Pera RA, Kamischke A, Nieschlag E, Gromoll J. Association of meiotic arrest with lack of BOULE protein expression in infertile men. J Clin Endocrinol Metab. 2004;89:1926–33.
26. Yen PH, Chai NN, Salido EC. The human autosomal gene *DAZLA*: testis specificity and a candidate for male infertility. Hum Mol Genet. 1996;5:2013–7.
27. Yen PH, Chai NN, Salido EC. The human *DAZ* genes, a putative male infertility factor on the Y chromosome, are highly polymorphic in the DAZ repeat regions. Mamm Genome. 1997;8:756–9.
28. Saxena R, Vries JWA, Repping S, et al. Four *DAZ* genes in two clusters found in the *AZFc* region of the human Y chromosome. Genomics. 2000;67:256–67.
29. Ruggiu M, Speed R, Taggart M, et al. The mouse *Dazla* gene encodes a cytoplasmic protein essential for gametogenesis. Nature. 1997;389:73–7.
30. Reijo RA, Dorfman DM, Slee R, et al. DAZ family proteins exist throughout male germ cell development and transit from nucleus to cytoplasm at meiosis in humans and mice. Biol Reprod. 2000;63:1490–6.
31. Tung J, Rosen M, Nelson L, et al. Novel missense mutations of the *Deleted-in-AZoospermia-Like* (*DAZL*) gene in infertile women and men. Reprod Biol Endocrinol. 2006;4:40.
32. Reijo RA, Alagappan RK, Patrizio P, Page DC. Severe oligospermia resulting from deletions of the *Azoospermia Factor* gene on the Y chromosome. Lancet. 1996;347:1290–3.
33. Cooke HJ, Lee M, Kerr S, Ruggiu M. A murine homologue of the human DAZ gene is autosomal and expressed only in male and female gonads. Hum Mol Genet. 1996;5:513–6.
34. Reijo RA, Seligman J, Dinulos MB, et al. Mouse autosomal homolog of *DAZ*, a candidate male sterility gene in humans, is expressed in male germ cells before and after puberty. Genomics. 1996;35:346–52.
35. Tung J, Rosen M, Nelson L, et al. Variants in *Deleted in AZoospermia-Like* (*DAZL*) are correlated with reproductive parameters in men and women. Hum Genet. 2006;118:730–40.
36. Yen PH. Putative biological functions of the *DAZ* family. Int J Androl. 2004;27:125–9.
37. Moore FL, Jaruzelska J, Dorfmann D, Reijo Pera RA. Identification of a novel gene, *DZIP* (*DAZ Interacting Protein*), that encodes a protein that interacts with DAZ and DAZL (Deleted in AZoospermia and DAZ-Like) and is expressed in embryonic stem cells and germ cells. Genomics. 2004;83:834–43.
38. Chambers I, Colby D, Robertson M, et al. Functional expression cloning of Nanog, a pluripotency sustaining factor in embryonic stem cells. Cell. 2003;113:643–55.
39. Clark AT, Rodriguez R, Bodnar M, et al. Human *STELLAR*, *NANOG*, and *GDF3* genes are expressed in pluripotent cells and map to chromosome 12p13, a hot-spot for teratocarcinoma. Stem Cells. 2004;22:169–79.
40. Pan G, Thomson JA. Nanog and transcriptional networks in embryonic stem cell pluripotency. Cell Res. 2007;17:42–9.
41. Kuroda T, Tada M, Kubota H, et al. Octamer and Sox elements are required for transcriptional *cis* regulation of *Nanog* gene expression. Mol Cell Biol. 2005;25:2475–85.
42. Chambers I, Smith A. Self-renewal of teratocarcinoma and embryonic stem cells. Oncogene. 2004;23:7150–60.
43. Wang C, Lehmann R. Nanos is the localized posterior determinant in *Drosophila*. Cell. 1991;68:1177.
44. Tsuda M, Sasaoka Y, Kiso M, et al. Conserved role of nanos proteins in germ cell development. Science. 2003;301:1239–41.
45. Hansis C, Grifo JA, Krey LC. Oct-4 expression in inner cell mass and trophectoderm of human blastocysts. Mol Hum Reprod. 2000;6:999–1004.
46. Scholer H, Dressler G, Balling R, Rohdewohld H, Gruss P. Oct-4: a germ line specific transcription factor mapping to the mouse t-complex. EMBO J. 1990;9:2185–95.
47. Burdon T, Smith A, Savatier P. Signalling, cell cycle and pluripotency in embryonic stem cells. Trends Cell Biol. 2002;12:432–8.
48. Rosner MH, Vigano MA, Ozato K, et al. A POU-domain transcription factor in early stem cells and germ cells of the mammalian embryo. Nature. 1990;345:686–92.
49. Goto T, Adjaye J, Rodeck CH, Monk M. Identification of genes expressed in human primordial germ cells at the time of entry of the female germ line into meiosis. Mol Hum Reprod. 1999;5:851–60.
50. Okamoto K, Okazawa H, Okuda A, Sakai M, Muramatsu M, Hamada H. A novel octamer binding transcription factor is differentially expressed in mouse embryonic cells. Cell. 1990;60:461–72.
51. Scholer HR. Octamania: the POU factors in murine development. Trends Genet. 1991;7:323–9.
52. Pesce M, Wang X, Wolgemuth DJ, Scholer H. Differential expression of the Oct-4 transcription factor during mouse germ cell differentiation. Mech Dev. 1998;71:89–98.
53. Niwa H, Miyazaki J, Smith AG. Quantitative expression of *Oct-3/4* defines differentiation, dedifferentiation or self-renewal of ES cells. Nat Genet. 2000;24:372–6.
54. Adjaye J, Bolton V, Monk M. Developmental expression of specific genes detected in high-quality cDNA libraries from single human preimplantation embryos. Gene. 1999;237:373–83.
55. Kehler J, Tolkunova E, Koschorz B, et al. Oct4 is required for primordial germ cell survival. EMBO Rep. 2004;5:1078–83.
56. Lee J, Kim HK, Rho JY, Han YM, Kim J. The human OCT-4 isoforms differ in their ability to confer self-renewal. J Biol Chem. 2006;281:33554–65.
57. Spassov DS, Jurecic R. Mouse *Pum1* and *Pum2* genes, members of the Pumilio family of RNA-binding proteins, show differential expression in fetal and adult hematopoietic stem cells and progenitors. Blood Cells Mol Dis. 2003;30:55–69.
58. Moore FL, Jaruzelska J, Fox MS, et al. Human Pumilio-2 is expressed in embryonic stem cells and germ cells and interacts with DAZ (Deleted in AZoospermia) and DAZ-Like proteins. Proc Natl Acad Sci U S A. 2003;100:538–43.
59. Urano J, Fox M, Reijo Pera RA. Interaction of the conserved meiotic regulators, BOULE (BOL) and PUMILIO-2 (PUM2). Mol Reprod Dev. 2005;71:290–8.
60. Murata Y, Wharton RP. Binding of Pumilio to maternal *hunchback* mRNA is required for posterior patterning in *Drosophila* embryos. Cell. 1995;80:747–56.

61. Sonoda J, Wharton RP. Recruitment of Nanos to hunchback mRNA by Pumilio. Genes Dev. 1999;13:2704–12.
62. Chagnovich D, Lehmann R. Poly(A)-independent regulation of maternal *hunchback* translation in the *Drosophila* embryo. Proc Natl Acad Sci U S A. 2001;98:11359–64.
63. Wharton RP, Sonoda J, Lee T, Patterson M, Murata Y. The Pumilio RNA-binding domain is also a translational regulator. Mol Cell. 1998;1:863–72.
64. Crittenden SL, Bernstein D, Bachorik J, et al. FBF and control of germline stem cells in *Caenorhabditis elegans*. Nature. 2002;417:660–3.
65. Wickens M, Bernstein DS, Kimble J, Parker R. A PUF family portrait: 3'UTR regulation as a way of life. Trends Genet. 2002;18:150–7.
66. Lehmann R, Nusslen-Volhard C. Involvement of the *Pumilio* gene in the transport of an abdominal signal in the *Drosophila* embryo. Nature. 1987;329:167–70.
67. Mosquera L, Forristall C, Zhou Y, King ML. A mRNA localized to the vegetal cortex of *Xenopus* oocytes encodes a protein with a nanos-like zinc finger domain. Development. 1993;117: 377–86.
68. Houston DW, Zhang J, Maines JZ, Wasserman SA, King ML. A *Xenopus DAZ-like* gene encodes an RNA component of germ plasm and is a functional homologue of *Drosophila boule*. Development. 1998;125:171–80.
69. Maegawa S, Yasuda K, Inoue K. Maternal mRNA localization of zebrafish *DAZ-like* gene. Mech Dev. 1999;81:223–6.
70. Subramaniam K, Seydoux G. *nos-1* and *nos-2,* two genes related to *Drosophila nanos,* regulate primordial germ cell development and survival in *Caenorhabditis elegans*. Development. 1999;126: 4861–71.
71. Houston DW, King ML. A critical role for *Xdazl,* a germ plasm-localized RNA, in the differentiation of primordial germ cells in *Xenopus.* Development. 2000;127:447–56.
72. Dorfman DM, Genest DR, Pera RAR. Human *DAZL1* encodes a candidate fertility factor in women that localizes to the prenatial and postnatal germ cells. Hum Reprod. 1999;14:2531–6.
73. Tanaka SS, Toyooka Y, Akasu R, et al. The mouse homolog of *Drosophila Vasa* is required for the development of male germ cells. Genes Dev. 2000;14:841–53.
74. Wang X, Zamore PD, Hall TMT. Crystal structure of a Pumilio homology domain. Mol Cell. 2001;7:855–65.
75. Edwards TA, Pyle SE, Wharton RP, Aggarwal AK. Structure of Pumilio reveals similarity between RNA and peptide binding motifs. Cell. 2001;105:281–9.
76. Fox M, Urano J, Reijo Pera RA. Identification and characterization of RNA sequences to which human PUMILIO-2 (PUM2) and deleted in Azoospermia-Like (DAZL) bind. Genomics. 2005;85:92–105.
77. Zimmerman ZA, Kellog DR. The Sda1 protein is required for passage through start. Mol Biol Cell. 2001;12:201–19.
78. Buscemi G, Saracino F, Masnada D, Carbone ML. The *Saccharomyces cerevisiae SDA1* gene is required for actin cytoskeleton organization and cell cycle progression. J Cell Sci. 2000;113: 1199–211.
79. Golling G, Amsterdam A, Sun Z, et al. Insertional mutagenesis in zebrafish rapidly identifies genes essential for early vertebrate development. Nat Genet. 2002;31:135–40.
80. White E, Moore-Jarrett T, Ruley H. Pum2, a novel murine puf protein, and its consensus RNA-binding site. RNA. 2001;7:1855–66.
81. Fox MS, Clark AT, El Majdoubi M, et al. Intermolecular interactions of homologs of germ plasm components in mammalian germ cells. Dev Biol. 2007;301:417–31.
82. Zwaka T, Thomson J. A germ cell origin of embryonic stem cells? Development. 2005;132:227–33.
83. Abeyta M, Clark AT, Rodriguez R, Bodnar M, Reijo Pera RA, Firpo M. Unique gene expression signatures of independently-derived human embryonic stem cell lines. Hum Mol Genet. 2004;13:601–8.

Genomic Stability in Stem Cells

Irene Riz and Robert G. Hawley

Abstract Accumulated data suggest that the unique function of stem cells to self-renew is under the strong control of a sensor system detecting potential threats to genomic integrity. Reactive oxygen species, the most significant mutagens in stem cells, when elevated, activate the protective mechanisms blocking self-renewal and at the same time serve as a signal stimulating stem cell differentiation. Based on studies performed primarily on hematopoietic stem cells and embryonic stem cells, we outline the signaling networks connecting the ATM, MAPK14 (p38) and FRAP1 (mTOR) protein kinases, the p53 tumor suppressor, the PTEN phosphatase, and the TEL, NFκB, FoxO, and HIF transcription factor families as a potential stress-controlled differentiation mechanism regulating stem cell fate decisions. An intriguing observation to come from these studies is that stem cell self-renewal capacity appears to be pharmacologically preserved by anti-oxidative agents.

Keywords Stem cells · Genomic stability · Oxidative stress · Self-renewal · p53 tumor suppressor · PTEN phosphatase · ATM protein kinase

1 Introduction

The main function of stem cells is to store the genetic information that makes it possible to correctly replenish cellular components of the organism throughout its life span and in case of germ cells allows creation of a new organism. Since the number of divisions of a eukaryotic cell is limited and associated with increased mutations, it is imperative for the existence of the multicellular organism that specialized cells store the genetic information. When the genetic information is needed, there must be a relatively rare event of asymmetrical division where one daughter cell is destined to keep the information and the other is directed to expand and to replenish those constantly exhausting cellular components of the body.

Adult stem cells divide infrequently and primarily reside in the G_0 phase of the cell cycle [1–3], helping to minimize cell division-associated mutations [4]. In contrast, embryonic stem (ES) cells continuously divide in culture. Despite this dramatic difference, the mutation frequency in ES cells is still about 100 times lower in comparison to their more differentiated descendants [5, 6], suggesting that stem cells have specialized mechanisms to preserve the genome. For example, asymmetrical self-renewal has been proposed to preserve older DNA strands within certain stem cells (the "immortal strand hypothesis") preventing replication-induced errors [7–12]. The major source of DNA damage in stem cells besides that associated with cell proliferation is endogenously generated genotoxic agents produced as signaling molecules or metabolic byproducts, among which reactive oxygen species are the most common threat. In comparison to their differentiated descendants, stem cells express increased amounts of specific proteins such as anti-oxidative or anti-alkylating enzymes that neutralize genotoxic agents or members of the ATP-binding cassette family of transporters that could efflux them out of the cell [13–18]. In addition, stem cells are less exposed to environmental stress due to the protective effect of their specialized microenvironments or niches [19, 20]. To prevent the accumulation of mutations, stem cells have evolved efficient mechanisms to repair or eliminate those cells with damaged DNA [6, 21, 22]. However, even with superior protective mechanisms, accumulation of DNA damage upon acute exposure to genotoxic agents or during aging of stem cells does occur [23–25].

In the current chapter, we focus on the molecular network regulating stem cell fate in response to oxidative stress. The data we discuss suggest that the stress response may reversibly block self-renewal potential in order to protect the genetic integrity of the stem cells and that stress sensors may

R.G. Hawley (✉)
Department of Anatomy and Regenerative Biology, The George Washington University, 2300 I Street NW, Washington, DC 20037
e-mail: rghawley@gwu.edu

V.K. Rajasekhar, M.C. Vemuri (eds.), *Regulatory Networks in Stem Cells*, Stem Cell Biology and Regenerative Medicine,
DOI 10.1007/978-1-60327-227-8_6, © Humana Press, a part of Springer Science+Business Media, LLC 2009

redirect stem cell fate outcomes from self-renewal to differentiation, senescence or apoptosis depending on the levels of stress and associated DNA damage.

2 Reactive Oxygen Species Are a Major Threat to Stem Cell DNA

Stem cells are mainly protected from environmental mutagens such as UV light and ionizing radiation by their localization within the niche. Major sources of DNA damage in stem cells are endogenous factors associated with cell division and metabolism. Some of the most significant endogenous mutagens are reactive oxygen species (ROS) including superoxide anions, hydrogen peroxide, organic peroxides, and hydroxyl radicals. ROS are continuously generated by normal metabolic processes such as oxidative phosphorylation and nucleotide catabolism or by specific plasma membrane oxidases in response to growth factors or cytokines. If not inactivated in a timely manner by cellular anti-oxidative systems, ROS may cause DNA damage. Based on mathematical modeling of oxygen distributions in the bone marrow, Chow and colleagues [26, 27] suggested that hematopoietic stem cells (HSCs) are located within the most anoxic region to prevent oxygen radicals from damaging these important cells. The hypothesis was experimentally corroborated for murine HSCs by Parmar and collaborators [28]. The authors measured the levels of intracellular uptake of Hoechst 33342 intravenously injected into animals as an indicator of oxygen perfusion. Hematopoietic precursor cell subsets showed a steep gradient in dye distribution according to their bone marrow localization. As predicted, the cells from the least dye accessible regions of bone marrow showed the highest long-term HSC repopulating potential and exhibited the highest staining for the reductive 2-nitroimidazole compound pimonidazole, a chemical marker for hypoxia.

The energy metabolism of HSCs is based mostly on glycolysis and very low levels of oxidative phosphorylation. Interestingly, half of the oxygen consumption in human $CD34^+$ HSCs is extra-mitochondrial and attributed to the constitutively active plasma membrane NADPH oxidase (NOX) which produces superoxide anions [29]. This finding underscores the significance of ROS in stem cell biology. In a subsequent report, the Capitanio group showed that human HSCs express the NOX1, NOX2, and NOX4 catalytic isoforms along with the complete set of their regulatory subunits and the NOX2s spicing variant [30]. Moreover, they suggested that NOX1 and NOX2 can be placed downstream and upstream of ROS signaling with the constitutively active NOX4 functioning as an oxygen sensor, signaling to the cell when a change in the environmental oxygen concentration is experienced, integrating redox signaling to determine HSC fate.

In other work, murine ES cells were shown to be highly proficient in antioxidant defense, withstanding supra-physiological levels of oxygen, an ability that decreases during the early steps of differentiation into embryoid bodies [14]. Accordingly, intracellular ROS levels were shown to increase after differentiation and a search for gene expression changes associated with ES cell differentiation revealed decreased mRNA levels of ROS inactivating enzymes [14]. An increase in ROS levels was detected in various models of ES cell differentiation (e.g., human ES cells [31], cardiovascular differentiation of mouse ES cells [32]), and associated with differentiation of mesenchymal and neural stem cells [33, 34]. A quantitative proteomic screen comparing murine long term versus short term repopulating HSCs (lineage negative Sca^+Kit^+ cells versus more differentiated lineage negative Sca^+Kit^- cells) revealed hypoxia-related changes in proteins controlling metabolism and oxidative protection [17]. Thus, differentiation of both embryonic and adult stem cells showed decreased expression of genes regulating synthesis of natural antioxidants and those providing protection against oxidative damage: superoxide dismutases (which convert reactive oxygen species to H_2O_2), glutathione transferases, and peroxiredoxins (which inactivate H_2O_2). Therefore, a high activity of antioxidant enzymes appears to be an important feature of self-renewing stem cells.

The significance of down-regulation of antioxidative activity associated with differentiation can be better understood when considered in the context of a signaling function for ROS (for reviews, see refs. [35, 36]). Cells with elevated antioxidants may be refractory to signals from many growth factors and cytokines. For example, IL-3 and SCF survival/proliferation signals are mediated in part by ROS [37]. However, by facilitating the signaling function of ROS, stem cells could be subjected to a higher risk of DNA damage, suggesting the existence of regulatory mechanisms sensing ROS and controlling stem cell quiescence and self-renewal. It is reasonable to propose that the increased levels of ROS, regardless of whether they are induced by the signals promoting differentiation or occur as a result of environmental fluctuation, might inhibit self-renewal. As part of the mechanism protecting stem cells from the genotoxic effects of ROS, cells that fail to inactivate ROS in a timely manner might be destined to exit stem cell compartments. Depending on the levels of ROS and associated DNA damage, stem cell fate outcomes may vary – with the options being differentiation, senescence, or apoptosis. Since the hypothesis implies the existence of a common regulatory mechanism controlling ROS defense and self-renewal ability of stem cells, we next discuss the major known regulators of HSC fate and their proposed role in ROS sensing or inactivation.

3 Molecular Sensors of ROS and Their Influence on Stem Cell Fate Decisions

In an elegant study performed on murine HSCs, Ito and colleagues [38] suggested that self-renewal capacity of HSCs depends on ATM-mediated inhibition of oxidative stress. The authors found that murine $Atm^{-/-}$ HSCs lack long-term repopulating capacity. ATM, ataxia telangiectasia mutated, is a member of the PI3 kinase-like protein kinase family. Since ATM is a known key regulator of oxidative and DNA damage-induced stress [39], the investigators compared the levels of ROS in the lineage negative $Sca^{+}Kit^{+}$ populations in mutant versus wild-type animals and found the ROS levels to be significantly elevated in ATM-deficient cells. Treatment with a permeant anti-oxidative agent thiol *N*-acetyl-l-cysteine (NAC) restored the self-renewal potential of this population as measured in vivo by a hematopoietic repopulating assay. In a follow-up study, the authors demonstrated that elevated levels of ROS activate HSC-specific phosphorylation of the p38 MAP kinase accompanied by a defect in maintenance of quiescence [40]. Similar to ROS scavengers, treatment with the p38 inhibitor or inactivation of MAP3K5, an upstream activator of p38 MAP kinase, restored self-renewal of $Atm^{-/-}$ HSCs. Although it is not clear whether the treatment reduced ROS levels or acted on the downstream signaling pathways, the work emphasized the role of p38 MAP kinase as a regulator of stem cell oxidative stress and differentiation.

What are the downstream targets of p38 MAP kinase in stem cells? Some interesting examples are the TEL/Etv6, p53, NFκB, and hypoxia-inducible factor (HIF) signaling pathways [41–43]. From a stem cell biology perspective, one of the most fascinating direct targets of the p38 MAP kinase is the transcription factor TEL/Etv6, since the repressor function of TEL is necessary for HSC self-renewal [44]. p38 MAP kinase binds, phosphorylates, and inhibits the repressor function of TEL [45]. Inactivation of p38 MAP kinase was also associated with expansion of repopulation competent HSCs in Lnk-defficient mice [46]. In order to gain a better understanding of the stress/differentiation control in stem cells, it is important to learn more about p38 MAP kinase targets.

The tumor suppressor p53 safeguards the genome by preventing proliferation and survival of cells with damaged DNA. Apropos, ATM kinase directly activates p53 and, via Mdm2 and Chk2, stabilizes its protein levels [47–51]. Recently, p53 – in the absence of a stress response – was shown to transcriptionally activate genes encoding anti-oxidative proteins neutralizing intracellular ROS [52, 53]. Downregulation of p53 caused an increase of intracellular ROS levels and resulted in excessive oxidation of DNA, increased mutation rates, and karyotypic instability. NAC treatment prevented the effect. It is possible that in stem cells p53 plays a similar role. An interesting and intriguing study performed by the Sherley group showed that restoration of wild type p53 function in immortalized embryonic fibroblasts or adult mammary epithelium led to activation of asymmetrical cell divisions with preservation of the conserved DNA strands – modeling self-renewal divisions of stem cells [54, 55]. The effect was inhibited by IMP dehydrogenase, a key enzyme in the de novo pathway of purine biosynthesis, or its product xanthosine. Since nucleotide metabolism – and oxidation of xanthosine specifically – may elevate intracellular ROS levels, it would be informative to compare the levels of ROS in these cells; the prediction is that active IMP dehydrogenase or an excess of xanthosine will cause elevation of ROS that in turn will override the self-renewal function of p53. In light of the studies discussed above, it is important to directly ask whether p53 controls self-renewal and DNA strand conservation in stem cells similar to that in model cell lines. Can the tumor suppressor function of p53 be separated from self renewal function [56, 57]? In this regard, do p53 mutants exhaust their stem cell pools in a manner comparable to p21 mutant mice in response to stress [58]?

Several years ago, Donehower and colleagues generated a mutant form of p53 consisting of exons 7–11 encoding an N-terminally truncated protein missing both transcriptional activation domains and part of the DNA-binding domain. The resulting 24 kDa mutant protein was not functional as a transactivator as measured by a p21 promoter-driven reporter gene assay. Mice carrying one copy of this truncated protein ($p53^{+/m}$) developed a premature aging phenotype that included decreased proliferative and homing capacity of HSCs, reduced longevity, organ mass and cellularity, and heightened tumor resistance compared with their wild-type counterparts [59]. The authors interpreted the $p53^{+/m}$ phenotype as "apparently hyperactive" and suggested that premature exhaustion of the stem cell pool in $p53^{+/m}$ mice may be caused by the lack of efficient self-renewal. Transgenic mice expressing increased levels of a naturally occurring short 44 kDa isoform of p53 missing the first transactivation domain encoded by the first four exons also showed signs of premature aging and a lower incidence of tumors [60]. Although not characterized at the HSC level, the observed premature infertility associated with "catastrophic loss" of sperm-producing cells suggested failure in germ cell self-renewal. Both endogenous and mutant N-termini truncated isoforms require a wild type p53 allele for their effects. Interestingly, $p53^{+/-}$ mice do not show any of the signs of premature aging or HSC exhaustion, suggesting that interaction of wild-type and truncated forms of p53 may specifically affect the self-renewal function of p53.

The balance between short and long isoforms of p53 was suggested to control the function of the lipid phosphatase PTEN, with the p44 form being inhibitory [60]. PTEN is critical for self-renewal of HSCs [61]. Mice conditionally deficient for PTEN showed rapid depletion of long-term repopulation-competent HSCs. Importantly, PTEN belongs to the family of cysteine-based phosphatases that contain nucleophilic catalytic cysteine residues that are highly susceptible to oxidation by ROS, a property that permits redox regulation of enzyme activity [62, 63]. One might speculate that elevated levels of ROS indicating potential threat to genomic integrity of stem cells would inhibit PTEN, in turn preventing self-renewal and accumulation of mutated stem cells.

The PI3K/Akt axis is the major downstream signaling pathway from PTEN. PI3K/Akt is a known integrator of survival and oxidative stress signaling [64]. PTEN inhibition leads to stabilization of the active phoshorylated form of Akt. Since the mammalian target of rapamycin (mTOR) is one of the Akt targets, Morrison and colleagues [61] evaluated whether HSC self-renewal in PTEN-deficient mice can be restored by inhibition of mTOR function. Indeed, the effect was reversed by rapamycin treatment. Similar to ATM, mTOR belongs to the PI3 kinase-like protein kinase family, mediating cellular responses to stresses such as DNA damage and nutrient deprivation [65]. In addition to the PTEN/PI3kinase/Akt pathway, mTOR is targeted by other factors, among them the inhibitory action of hypoxia is noteworthy since self-renewing HSCs exist under hypoxic conditions [28, 66].

Akt also controls the stability of the FoxO family of transcription factors [67]. This family of transcription factors was shown to induce target genes inactivating ROS while concurrently supporting quiescence and survival of murine HSCs. The phenotypes of PTEN and FoxO deficiencies are nearly identical. Triple deficiency for FoxO1/3/4 blocked long-term repopulating activity of the lineage negative $Sca^{+}Kit^{+}$ population and was associated with increased levels of ROS in this hematopoietic compartment. Again, treatment with NAC restored HSC self-renewal potential.

In other studies, using an intracellular ROS indicator 2′-7′-dichlorofluorescene diacetate (DCF-DA), Jang and Sharkis [68] found that ROS levels are remarkably heterogeneous within the lineage negative $Sca^{+}Kit^{+}$population of mouse HSCs. $ROS^{(high)}$ cells had significantly lower self-renewal capacity than $ROS^{(low)}$ cells, and the difference could be completely abolished by ROS scavengers treatment. The authors suggested that an increase of intracellular ROS associated with transition of the cells from the osteoblastic niche to the more oxygenated vascular niche of bone marrow and decrease of self-renewal capacity may be one of the early differentiation events of HSCs. As predicted from previous studies of ATM and PTEN [40, 61], treatment with p38 or mTOR inhibitors restored the self-renewal capacity of the $ROS^{(high)}$ population similar to that observed following NAC treatment. Congruent with data describing an antioxidant function of p53 and work suggesting a role for p53 in stem cell self-renewal [52–54], the $ROS^{(low)}$ population exhibited higher levels of p53 mRNA. An important conclusion from this study is that a reversible shift from self-renewal to differentiation divisions may be controlled by sensors of intracellular ROS. If the ROS levels saturate the sensors, dividing HSC are only allowed to exit stem cell pool. This mechanism could be part of a normal developmental process associated with transition of HSCs to a more highly oxygenated niche, as was suggested by Jang and Sharkis. Alternatively, or in addition, it may reflect a stress response signaling network that safeguards the genome of the stem cell pool, since increased ROS may cause DNA damage. When no DNA damage is detected and ROS levels return to normal, the self-renewal divisions could be activated again. Thus, to begin to understand the signaling network that executes oxygen-sensitive ROS-mediated control over stem cell fate decisions, the following interactive components should be considered: ATM-p38 MAP kinase interaction with potential downstream targets TEL, NFκB, and HIF; and ATM-p53/p44 and PTEN-Akt interaction with downstream targets mTOR and FoxO.

4 Lessons from Differentiated Cells

Next we will discuss several other well known ubiquitous redox sensitive regulators whose function, in our opinion, is needed to be further characterized in order to better understand the oxidative stress control of stem cell self-renewal capacity. One such regulator is APE/Ref-1 [69]. It is a ubiquitously expressed protein with a dual function: one as an apurinic/apyrimidinic endonuclease involved in DNA repair and another as a redox regulator of transcription. The redox function of APE/Ref-1 was documented to control the hematopoietic differentiation potential of ES cells [70]. Among the downstream targets of APE/Ref-1 are stress-inducible transcription factors such as AP1, ATF, CREB, p53, NFκB, and HIF-1α In this context, NFκB is a well-recognized stress-inducible redox regulated transcription factor, which has been documented to support the survival and development of HSC and progenitor cells. For example, NFκB was shown to be present in a latent form in human CD34 positive hematopoietic precursors and induced by stress [71]. Survival and repopulation potential of HSCs is supported by NFκB targets such as MCL1 and BCL2 [72–74]. In addition, NFκB exhibits a strong antioxidant activity mediated by its downstream target genes such as Mn-superoxide dismutase (MnSOD), glutathione peroxidase, and TRX [75, 76]. Nakata et al. [75] showed that ROS scavenger treatment is sufficient to prevent the apoptotic effect of a dominant negative mutant of NFκB and suggested that activation of antioxidant proteins makes a more important contribution to NFκB-mediated HSC survival mechanisms than BCL2 family members. Thus, the accumulated data suggest that NFκB may play an important role in protecting HSCs from ROS fluctuations. Importantly, NFκB is subject to complex regulation by ROS: negatively via oxidation of cysteine residues in IκB kinase which prevents NFκB activation [77] or positively via ROS-induced phosphorylation [78]. This raises the possibility that the functional outcome of ROS-NFκB interaction in stem cells may depend on the levels of ROS: if the levels are too high and associated with signs of oxidative damage, NFκB might become inactivated and the cells destined to undergo apoptosis. How might this balance be regulated in stem cells?

Hypoxia-inducible factor-1 (HIF-1) is another ROS-sensitive transcription factor [79]. HIF-1 is a heterodimeric transcription factor composed of HIF-1α, a labile subunit that becomes stabilized under low oxygen conditions, and the arylhydrocarbon receptor nuclear translocation, ARNT, a constitutively expressed subunit. HIF-1 activates a broad spectrum of genes adapting cells to low oxygen conditions. Its transcriptional targets encode more than 70 different proteins such as enzymes of glycolysis as well as regulators of proliferation and survival. Recently, HIF-1α was shown to induce transcription of FoxO3a [80] as well as hTERT [81, 82]. Regulation of HIF-1 is cell dependent and stress specific, and it is well appreciated that HIF-1α is targeted for degradation by high ROS in normoxic conditions [79]. Under hypoxic conditions within the stem cell niche, ROS may activate HIF-1 by analogy to the processes that take place in differentiated cells during hypoxia: potential regulators being NOX1, mTOR, p38 kinase, Ref-1/thyredoxin, and NFκB [41, 83–86].

Under hypoxic conditions, HIF-1α activates transcription of Notch target genes by stabilizing activated Notch and forming a HIF-1α/NotchIC heterodimer [87]. Notch is a strong inducer of self-renewal pathways in neural and hematopoietic stem cells [88, 89]. Is HIF-1/Notch interaction involved in the control of self renewal? Further studies directly assessing activation levels of HIF-1 in stem cells promise to be enlightening. An increase in HIF-1α mRNA levels was documented during differentiation of ES cells [14]. One might suggest that activation of HIF-1 in stem cells may be a general event associated with differentiation onset: while still under hypoxic conditions, stem cells would exit quiescence, increase their energy metabolism by activation of glycolysis in preparation for differentiation. Recent data suggest that the relationship between HIF-1 and p53 is complex [79]. It would be interesting to follow the interaction of these redox stress-sensitive components in stem cells and explore the functional implications as regards p53-mediated quiescence or self-renewal of stem cells.

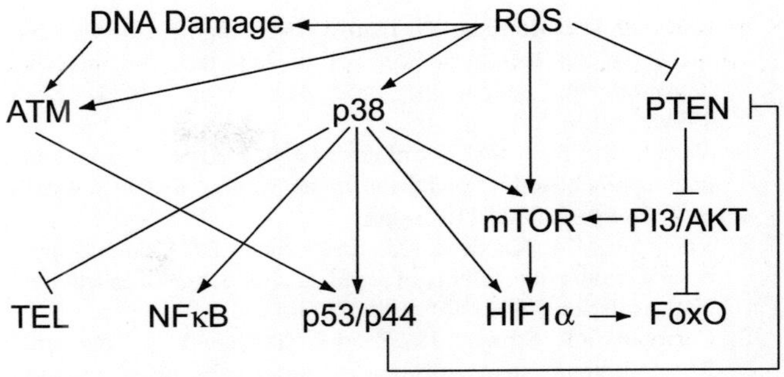

Fig. 1 Signaling components controlling stem cell self-renewal capacity in response to oxidative stress. Stem cell fate outcomes shift from self-renewal to differentiation, senescence or apoptosis depending on the levels of ROS and DNA damage. See text for details

5 Conclusions

Based on studies performed mainly on HSCs and extrapolating data obtained in other cell types, we outline several signaling pathways connecting the cellular regulators discussed above (Fig. 1). In this scheme, to safeguard the genome, stem cells have ROS sensors and express highly active anti-oxidative enzymes. Mechanisms controlling stem cell fate adjust to ROS fluctuations via several key redox sensitive components: ATM, p53, NFκB, and FoxO directly contribute to antioxidant defense while regulating survival and self-renewal of HSCs. PTEN, NFκB, HIF, and p53 are examples of regulators directly sensing ROS levels and controlling self-renewal and survival of stem cells. Thus, charged with the unique tasks to preserve the genome throughout the life span of the organism and to replenish damaged tissues and organs, stem cells integrate the stress and differentiation signals at the level of ROS. They self-renew when ROS levels are low and no danger to the genome is detected. They may undergo differentiation divisions when ROS levels are higher to produce differentiated precursors with much less stringent requirements for the integrity of the genome. If ROS levels are elevated long enough, in case of a deficiency in anti-oxidant defense mechanisms, the stem cell pool might be completely depleted. Finally, stem cells might die by apoptosis if ROS levels are too high such that they become DNA damaging and overwhelm the cells intrinsic DNA repair capacity.

Acknowledgments This work was supported in part by National Institutes of Health grants R01HL65519, R01HL66305, and R24RR16209, and by the King Fahd Endowment Fund and an Elaine H. Snyder Cancer Research Award (The George Washington University Medical Center).

References

1. Bradford GB, Williams B, Rossi R, Bertoncello I. Quiescence, cycling, and turnover in the primitive hematopoietic stem cell compartment. Exp Hematol. 1997;25:445–53.

2. Cheshier SH, Morrison SJ, Liao X, Weissman IL. In vivo proliferation and cell cycle kinetics of long-term self-renewing hematopoietic stem cells. Proc Natl Acad Sci U S A. 1999;96:3120–5.
3. Vickers M, Brown GC, Cologne JB, Kyoizumi S. Modelling haemopoietic stem cell division by analysis of mutant red cells. Br J Haematol. 2000;110:54–62.
4. Vilenchik MM, Knudson AG. Endogenous DNA double-strand breaks: production, fidelity of repair, and induction of cancer. Proc Natl Acad Sci U S A. 2003;100:12871–6.
5. Cervantes RB, Stringer JR, Shao C, Tischfield JA, Stambrook PJ. Embryonic stem cells and somatic cells differ in mutation frequency and type. Proc Natl Acad Sci U S A. 2002;99: 3586–90.
6. Hong Y, Cervantes RB, Tichy E, Tischfield JA, Stambrook PJ. Protecting genomic integrity in somatic cells and embryonic stem cells. Mutat Res. 2007;614:48–55.
7. Cairns J. Cancer and the immortal strand hypothesis. Genetics. 2006;174:1069–72.
8. Potten CS, Owen G, Booth D. Intestinal stem cells protect their genome by selective segregation of template DNA strands. J Cell Sci. 2002;115:2381–8.
9. Karpowicz P, Morshead C, Kam A, et al. Support for the immortal strand hypothesis: neural stem cells partition DNA asymmetrically in vitro. J Cell Biol. 2005;170:721–32.
10. Shinin V, Gayraud-Morel B, Gomes D, Tajbakhsh S. Asymmetric division and cosegregation of template DNA strands in adult muscle satellite cells. Nat Cell Biol. 2006;8:677–87.
11. Conboy MJ, Karasov AO, Rando TA. High incidence of non-random template strand segregation and asymmetric fate determination in dividing stem cells and their progeny. PLoS Biol. 2007;5:e102.
12. Kiel MJ, He S, Ashkenazi R, et al. Haematopoietic stem cells do not asymmetrically segregate chromosomes or retain BrdU. Nature. 2007;449:238–42.
13. Bunting KD, Hawley RG. Integrative molecular and developmental biology of adult stem cells. Biol Cell. 2003;95:563–78.
14. Saretzki G, Armstrong L, Leake A, Lako M, von ZT. Stress defense in murine embryonic stem cells is superior to that of various differentiated murine cells. Stem Cells. 2004;22:962–71.
15. Eaker SS, Hawley TS, Ramezani A, Hawley RG. Detection and enrichment of hematopoietic stem cells by side population phenotype. Methods Mol Biol. 2004;263:161–80.
16. Hawley RG, Ramezani A, Hawley TS. Hematopoietic stem cells. Methods Enzymol. 2006;419:149–79.
17. Unwin RD, Smith DL, Blinco D, et al. Quantitative proteomics reveals posttranslational control as a regulatory factor in primary hematopoietic stem cells. Blood. 2006;107:4687–94.
18. Saretzki G, Walter T, Atkinson S, et al. Downregulation of multiple stress defense mechanisms during differentiation of human embryonic stem cells. Stem Cells. 2008;26:455–64.
19. Suda T, Arai F, Hirao A. Hematopoietic stem cells and their niche. Trends Immunol. 2005;26:426–33.
20. Zhang W, Remenyik E, Zelterman D, Brash DE, Wikonkal NM. Escaping the stem cell compartment: sustained UVB exposure allows p53-mutant keratinocytes to colonize adjacent epidermal proliferating units without incurring additional mutations. Proc Natl Acad Sci U S A. 2001;98:13948–53.
21. Khanna KK, Jackson SP. DNA double-strand breaks: signaling, repair and the cancer connection. Nat Genet. 2001;27:247–54.
22. Aladjem MI, Spike BT, Rodewald LW, et al. ES cells do not activate p53-dependent stress responses and undergo p53-independent apoptosis in response to DNA damage. Curr Biol. 1998;8:145–55.
23. Donahue SL, Lin Q, Cao S, Ruley HE. Carcinogens induce genome-wide loss of heterozygosity in normal stem cells without persistent chromosomal instability. Proc Natl Acad Sci U S A. 2006;103:11642–6.
24. Nijnik A, Woodbine L, Marchetti C, et al. DNA repair is limiting for haematopoietic stem cells during ageing. Nature. 2007;447:686–90.
25. Rossi DJ, Bryder D, Seita J, Nussenzweig A, Hoeijmakers J, Weissman IL. Deficiencies in DNA damage repair limit the function of haematopoietic stem cells with age. Nature. 2007;447:725–9.
26. Chow DC, Wenning LA, Miller WM, Papoutsakis ET. Modeling pO(2) distributions in the bone marrow hematopoietic compartment. I. Krogh's model. Biophys J. 2001;81:675–84.
27. Chow DC, Wenning LA, Miller WM, Papoutsakis ET. Modeling pO(2) distributions in the bone marrow hematopoietic compartment. II. Modified Kroghian models. Biophys J. 2001;81: 685–96.
28. Parmar K, Mauch P, Vergilio JA, Sackstein R, Down JD. Distribution of hematopoietic stem cells in the bone marrow according to regional hypoxia. Proc Natl Acad Sci U S A. 2007;104: 5431–6.
29. Piccoli C, Ria R, Scrima R, et al. Characterization of mitochondrial and extra-mitochondrial oxygen consuming reactions in human hematopoietic stem cells. Novel evidence of the occurrence of NAD(P)H oxidase activity. J Biol Chem. 2005;280: 26467–76.
30. Piccoli C, D'Aprile A, Ripoli M, et al. Bone-marrow derived hematopoietic stem/progenitor cells express multiple isoforms of NADPH oxidase and produce constitutively reactive oxygen species. Biochem Biophys Res Commun. 2007;353:965–72.
31. Cho YM, Kwon S, Pak YK, et al. Dynamic changes in mitochondrial biogenesis and antioxidant enzymes during the spontaneous differentiation of human embryonic stem cells. Biochem Biophys Res Commun. 2006;348:1472–8.
32. Schmelter M, Ateghang B, Helmig S, Wartenberg M, Sauer H. Embryonic stem cells utilize reactive oxygen species as transducers of mechanical strain-induced cardiovascular differentiation. FASEB J. 2006;20:1182–4.
33. Madhavan L, Ourednik V, Ourednik J. Increased "vigilance" of antioxidant mechanisms in neural stem cells potentiates their capability to resist oxidative stress. Stem Cells. 2006;24:2110–9.
34. Wang N, Xie K, Huo S, Zhao J, Zhang S, Miao J. Suppressing phosphatidylcholine-specific phospholipase C and elevating ROS level, NADPH oxidase activity and Rb level induced neuronal differentiation in mesenchymal stem cells. J Cell Biochem. 2007;100:1548–57.
35. Finkel T. Redox-dependent signal transduction. FEBS Lett. 2000;476:52–4.
36. Poulsen HE, Jensen BR, Weimann A, Jensen SA, Sorensen M, Loft S. Antioxidants, DNA damage and gene expression. Free Radic Res. 2000;33 Suppl:S33–9.
37. Sattler M, Winkler T, Verma S, et al. Hematopoietic growth factors signal through the formation of reactive oxygen species. Blood. 1999;93:2928–35.
38. Ito K, Hirao A, Arai F, et al. Regulation of oxidative stress by ATM is required for self-renewal of haematopoietic stem cells. Nature. 2004;431:997–1002.
39. Barzilai A, Rotman G, Shiloh Y. ATM deficiency and oxidative stress: a new dimension of defective response to DNA damage. DNA Repair (Amst). 2002;1:3–25.
40. Ito K, Hirao A, Arai F, et al. Reactive oxygen species act through p38 MAPK to limit the lifespan of hematopoietic stem cells. Nat Med. 2006;12:446–51.
41. Emerling BM, Platanias LC, Black E, Nebreda AR, Davis RJ, Chandel NS. Mitochondrial reactive oxygen species activation of p38 mitogen-activated protein kinase is required for hypoxia signaling. Mol Cell Biol. 2005;25:4853–62.

42. Sanchez-Prieto R, Rojas JM, Taya Y, Gutkind JS. A role for the p38 mitogen-acitvated protein kinase pathway in the transcriptional activation of p53 on genotoxic stress by chemotherapeutic agents. Cancer Res. 2000;60:2464–72.
43. Ryan S, McNicholas WT, Taylor CT. A critical role for p38 map kinase in NF-kappaB signaling during intermittent hypoxia/reoxygenation. Biochem Biophys Res Commun. 2007;355:728–33.
44. Hock H, Meade E, Medeiros S, et al. Tel/Etv6 is an essential and selective regulator of adult hematopoietic stem cell survival. Genes Dev. 2004;18:2336–41.
45. Arai H, Maki K, Waga K, et al. Functional regulation of TEL by p38-induced phosphorylation. Biochem Biophys Res Commun. 2002;299:116–25.
46. Seita J, Ema H, Ooehara J, et al. Lnk negatively regulates self-renewal of hematopoietic stem cells by modifying thrombopoietin-mediated signal transduction. Proc Natl Acad Sci U S A. 2007;104:2349–54.
47. Bae IH, Kang SW, Yoon SH, Um HD. Cellular components involved in the cell death induced by cisplatin in the absence of p53 activation. Oncol Rep. 2006;15:1175–80.
48. Caspari T. How to activate p53. Curr Biol. 2000;10:R315–7.
49. Chen C, Shimizu S, Tsujimoto Y, Motoyama N. Chk2 regulates transcription-independent p53-mediated apoptosis in response to DNA damage. Biochem Biophys Res Commun. 2005;333:427–31.
50. Khanna KK, Keating KE, Kozlov S, et al. ATM associates with and phosphorylates p53: mapping the region of interaction. Nat Genet. 1998;20:398–400.
51. Khosravi R, Maya R, Gottlieb T, Oren M, Shiloh Y, Shkedy D. Rapid ATM-dependent phosphorylation of MDM2 precedes p53 accumulation in response to DNA damage. Proc Natl Acad Sci U S A. 1999;96:14973–7.
52. Budanov AV, Sablina AA, Feinstein E, Koonin EV, Chumakov PM. Regeneration of peroxiredoxins by p53-regulated sestrins, homologs of bacterial AhpD. Science. 2004;304:596–600.
53. Sablina AA, Budanov AV, Ilyinskaya GV, Agapova LS, Kravchenko JE, Chumakov PM. The antioxidant function of the p53 tumor suppressor. Nat Med. 2005;11:1306–13.
54. Merok JR, Lansita JA, Tunstead JR, Sherley JL. Cosegregation of chromosomes containing immortal DNA strands in cells that cycle with asymmetric stem cell kinetics. Cancer Res. 2002;62:6791–5.
55. Rambhatla L, Ram-Mohan S, Cheng JJ, Sherley JL. Immortal DNA strand cosegregation requires p53/IMPDH-dependent asymmetric self-renewal associated with adult stem cells. Cancer Res. 2005;65:3155–61.
56. Meletis K, Wirta V, Hede SM, Nister M, Lundeberg J, Frisen J. p53 suppresses the self-renewal of adult neural stem cells. Development. 2006;133:363–9.
57. Dumble M, Moore L, Chambers SM, et al. The impact of altered p53 dosage on hematopoietic stem cell dynamics during aging. Blood. 2007;109:1736–42.
58. Cheng T, Rodrigues N, Shen H, et al. Hematopoietic stem cell quiescence maintained by p21cip1/waf1. Science. 2000;287:1804–8.
59. Tyner SD, Venkatachalam S, Choi J, et al. p53 mutant mice that display early ageing-associated phenotypes. Nature. 2002;415:45–53.
60. Maier B, Gluba W, Bernier B, et al. Modulation of mammalian life span by the short isoform of p53. Genes Dev. 2004;18:306–19.
61. Yilmaz OH, Valdez R, Theisen BK, et al. Pten dependence distinguishes haematopoietic stem cells from leukaemia-initiating cells. Nature. 2006;441:475–82.
62. Cho SH, Lee CH, Ahn Y, et al. Redox regulation of PTEN and protein tyrosine phosphatases in H(2)O(2) mediated cell signaling. FEBS Lett. 2004;560:7–13.
63. Salmeen A, Barford D. Functions and mechanisms of redox regulation of cysteine-based phosphatases. Antioxid Redox Signal. 2005;7:560–77.
64. Barthel A, Klotz LO. Phosphoinositide 3-kinase signaling in the cellular response to oxidative stress. Biol Chem. 2005;386:207–16.
65. Abraham RT, Gibbons JJ. The mammalian target of rapamycin signaling pathway: twists and turns in the road to cancer therapy. Clin Cancer Res. 2007;13:3109–14.
66. Brugarolas J, Lei K, Hurley RL, et al. Regulation of mTOR function in response to hypoxia by REDD1 and the TSC1/TSC2 tumor suppressor complex. Genes Dev. 2004;18:2893–904.
67. Tothova Z, Kollipara R, Huntly BJ, et al. FoxOs are critical mediators of hematopoietic stem cell resistance to physiologic oxidative stress. Cell. 2007;128:325–39.
68. Jang YY, Sharkis SJ. A low level of reactive oxygen species selects for primitive hematopoietic stem cells that may reside in the low-oxygenic niche. Blood. 2007;110:3056–63
69. Fritz G, Grosch S, Tomicic M, Kaina B. APE/Ref-1 and the mammalian response to genotoxic stress. Toxicology. 2003;193:67–78.
70. Zou GM, Luo MH, Reed A, Kelley MR, Yoder MC. Ape1 regulates hematopoietic differentiation of embryonic stem cells through its redox functional domain. Blood. 2007;109:1917–22.
71. Pyatt DW, Stillman WS, Yang Y, Gross S, Zheng JH, Irons RD. An essential role for NF-kappaB in human CD34(+) bone marrow cell survival. Blood. 1999;93:3302–8.
72. Opferman JT, Iwasaki H, Ong CC, et al. Obligate role of anti-apoptotic MCL-1 in the survival of hematopoietic stem cells. Science. 2005;307:1101–4.
73. Domen J, Weissman IL. Hematopoietic stem cells need two signals to prevent apoptosis; BCL-2 can provide one of these, Kitl/c-Kit signaling the other. J Exp Med. 2000;192:1707–18.
74. Domen J, Cheshier SH, Weissman IL. The role of apoptosis in the regulation of hematopoietic stem cells: overexpression of Bcl-2 increases both their number and repopulation potential. J Exp Med. 2000;191:253–64.
75. Nakata S, Matsumura I, Tanaka H, et al. NF-kappaB family proteins participate in multiple steps of hematopoiesis through elimination of reactive oxygen species. J Biol Chem. 2004;279:55578–86.
76. Tanaka H, Matsumura I, Ezoe S, et al. E2F1 and c-Myc potentiate apoptosis through inhibition of NF-kappaB activity that facilitates MnSOD-mediated ROS elimination. Mol Cell. 2002;9:1017–29.
77. Korn SH, Wouters EF, Vos N, Janssen-Heininger YM. Cytokine-induced activation of nuclear factor-kappa B is inhibited by hydrogen peroxide through oxidative inactivation of IkappaB kinase. J Biol Chem. 2001;276:35693–700.
78. Kamata H, Manabe T, Oka S, Kamata K, Hirata H. Hydrogen peroxide activates IkappaB kinases through phosphorylation of serine residues in the activation loops. FEBS Lett. 2002;519:231–7.
79. Bardos JI, Ashcroft M. Negative and positive regulation of HIF-1: a complex network. Biochim Biophys Acta. 2005;1755:107–20.
80. Bakker WJ, Harris IS, Mak TW. FOXO3a is activated in response to hypoxic stress and inhibits HIF1-induced apoptosis via regulation of CITED2. Mol Cell. 2007;28:941–53.
81. Nishi H, Nakada T, Kyo S, Inoue M, Shay JW, Isaka K. Hypoxia-inducible factor 1 mediates upregulation of telomerase (hTERT). Mol Cell Biol. 2004;24:6076–83.
82. Yatabe N, Kyo S, Maida Y, et al. HIF-1-mediated activation of telomerase in cervical cancer cells. Oncogene. 2004;23:3708–15.
83. Bonello S, Zahringer C, BelAiba RS, et al. Reactive oxygen species activate the HIF-1alpha promoter via a functional NFkappaB site. Arterioscler Thromb Vasc Biol. 2007;27:755–61.

84. Hudson CC, Liu M, Chiang GG, et al. Regulation of hypoxia-inducible factor 1alpha expression and function by the mammalian target of rapamycin. Mol Cell Biol. 2002;22:7004–14.
85. Wartenberg M, Gronczynska S, Bekhite MM, et al. Regulation of the multidrug resistance transporter P-glycoprotein in multicellular prostate tumor spheroids by hyperthermia and reactive oxygen species. Int J Cancer. 2005;113:229–40.
86. Goyal P, Weissmann N, Grimminger F, et al. Upregulation of NAD(P)H oxidase 1 in hypoxia activates hypoxia-inducible factor 1 via increase in reactive oxygen species. Free Radic Biol Med. 2004;36:1279–88.
87. Gustafsson MV, Zheng X, Pereira T, et al. Hypoxia requires notch signaling to maintain the undifferentiated cell state. Dev Cell. 2005;9:617–28.
88. Suzuki T, Chiba S. Notch signaling in hematopoietic stem cells. Int J Hematol. 2005;82:285–94.
89. Shi Y, Sun G, Zhao C, Stewart R. Neural stem cell self-renewal. Crit Rev Oncol Hematol. 2008;65:43–53.

Genetic Manipulation of Human Embryonic Stem Cells

Dimitris G. Placantonakis, Mark J. Tomishima, Fabien G. Lafaille, and Lorenz Studer

Abstract The ability to genetically manipulate stem cells is central in our effort to harness their potential. Various genetic approaches are now being implemented in human embryonic stem cells with the goal of understanding basic regulatory mechanisms, as well as modifying them toward potential therapeutic applications. This chapter will review genetic strategies available for the modification of human embryonic stem cells.

Keywords Transgenesis · Human · Embryonic · Stem cell · Genetic

1 Introduction

Human embryonic stem cells (hESCs) have received enormous attention in recent years by scientists and clinicians alike due to their theoretically unlimited potential to provide all cell types in the body. Human ESCs, which are derived from the inner cell mass of the blastocyst [1], possess two critical properties that uniquely define them. First, they have the ability to self-renew while maintaining their undifferentiated state. Second, they are pluripotent in that they can differentiate into cell types of all three germ layers (ectoderm, mesoderm, and endoderm), under appropriate conditions. Over the past decade, reliable techniques have been developed for culturing hESCs in vitro, as well as for directing their differentiation in vitro by mimicking specific developmental pathways. The explosive rate of development of new technologies involving hESCs has been unparalleled and has revolutionized the fields of stem cell biology, human developmental biology, and regenerative medicine.

The ability to genetically manipulate hESCs is central in our efforts to harness their enormous potential. Fortunately, such manipulations are now becoming feasible with hESCs thanks to prior expertise with murine ESCs. This chapter will discuss recent advances in the genetic manipulation of hESCs using conventional transfection techniques and small transgenic constructs, viral transduction, homologous recombination, and protein transduction. In addition, we will discuss a novel technology that we recently developed in our laboratory involving the transgenesis of hESCs with bacterial artificial chromosomes (BACs). We will also describe two relatively novel approaches: episomal transgenesis and targeted transgenesis. Finally, along with the description of different modes of transgenesis we will revisit the long-standing issues of silencing and insertional mutagenesis associated with them.

2 Nonviral Transfection of Small DNA Constructs

Perhaps the simplest approach to genetic modification of hESCs is their nonviral transfection with small DNA constructs, such as plasmids [2, 3]. A variety of transfection strategies have been successfully used in hESCs, including electroporation, lipofection, and nucleofection, the latter being a modified type of electroporation that increases the efficiency of nuclear transport of the transfected nucleic acid.

An important issue surrounding the transfection of hESCs is their poor clonogenic ability when they are dissociated to single cells, as opposed to their murine counterparts. For reasons that are not clear, hESCs grow in clumps or colonies and undergo massive apoptosis as single cells, with approximately only 1% of the cells surviving and proliferating. The poor clonogenicity of hESCs represents a major obstacle to optimizing the efficiency of their genetic manipulation, since conceivably single cells are more amenable to transfection. Recently, however, a report by Watanabe et al.

L. Studer (✉)
Department of Neurosurgery and Program in Developmental Biology, Memorial Sloan-Kettering Cancer Center, 1275 York Avenue, Box 256, New York, NY 10021
e-mail: studerl@mskcc.org

V.K. Rajasekhar, M.C. Vemuri (eds.), *Regulatory Networks in Stem Cells*, Stem Cell Biology and Regenerative Medicine,
DOI 10.1007/978-1-60327-227-8_7, © Humana Press, a part of Springer Science+Business Media, LLC 2009

[4] showed that the clonogenic potential of single hESCs can be dramatically improved when the Rho-associated kinase (ROCK) is pharmacologically inhibited. This manipulation may prove to be beneficial in enhancing transfection efficiencies.

An additional issue surrounding the transfection of hESCs involves the environment in which they are cultured. Human ESCs have traditionally required the presence of FGF2 in the media, as well as a feeder substrate, in order to proliferate while maintaining their undifferentiated state [1, 5]. However, the feeders, commonly mitotically inactivated mouse embryonic fibroblasts, compete with hESCs for the uptake of transfected nucleic acids, thus reducing the overall transfection efficiency. The use of feeder-free culture systems based on Matrigel (BD Biosciences) substrate and feeder-conditioned media has facilitated the efforts to optimize transfection rates, without compromising the undifferentiated identity of the cells [6–8].

Transfected DNA is only transiently expressed in the vast majority of the transfected cells, and stable integration into the host genome is a rare event. Thus, when stable transgenesis is the goal, selection is required to ensure the integration of the transfected DNA into the genome. Transgene integration entails replication and proper segregation after cell division, while episomal DNA is lost through degradation and cell division.

2.1 Electroporation

Electroporation is the application of a brief electric field across cells, which is thought to cause transient pores in the cell membrane and concurrent transfer of charged molecules, such as nucleic acids, into the cytosol. Electroporation is well tolerated by mouse ESCs and, thus, has been widely used for their transfection with different types of DNA [9], including BACs [10, 11]. In contrast, electroporation has received mixed reviews in the case of hESCs. A study by Eiges et al. [12] compared electroporation to various lipofection reagents and found it to be of marginal efficacy when reporter plasmids were used. This finding was corroborated by the experiments performed by Liew et al. [13]. However, Siemen et al. [14] electroporated dissociated hESCs and reported a transient transfection rate of 40%, again using a small reporter construct expressing GFP under the control of the CMV promoter. Electroporation was successfully used by Nolden et al. [15] to stably transfect hESCs with a Cre recombinase reporter construct. Finally, as mentioned below, electroporation was used by Zwaka and Thomson [16] to introduce linearized targeting vectors into hESCs for homologous recombination. Our experience with electroporation has been disappointing, for reasons that are not clear.

2.2 Lipofection

Lipofection is a transfection technique based on the coating of DNA with either charged or neutral lipophilic polymers. The polymer shell binds to and fuses with the cell membrane, thus releasing the nucleic acid content into the cytosol. The first demonstration that lipofection is an effective transfection modality in hESCs was by Eiges et al. [12]. In that study, ExGen 500 (Fermentas), a cationic reagent that coats DNA and binds cell surface proteoglycans, was found to be the most efficacious transfection agent when compared to the lipid-based polymers FuGENE (Roche Molecular Biochemicals) and LipofectAMINE (Invitrogen), as well as conventional electroporation. However, the transient transfection efficiency with ExGen 500 was only approximately 10%. Using the same reagent, the group developed hESC lines stably transfected with a GFP reporter driven by a 700 bp fragment of the mouse *Rex1* promoter, a transcription factor expressed in undifferentiated hESCs.

A study by Liew et al. [13] corroborated the above findings. They also found that ExGen 500 produced superior transfection rates compared to the cationic lipids Superfect (QIAGEN) and Lipofectin (Invitrogen). However, Vallier et al. [17] found LipofectAMINE to be 4 times more efficient than ExGen 500 at producing stably transfected hESC lines. In another comparison of electroporation to lipofection, using either FuGENE or LipofectAMINE, showed the former to be more efficient by approximately 40% [14]. Singh Roy et al. [18] used FuGENE-mediated lipofection of hESCs to transiently express GFP in hESC-derived motor neurons using a motor neuron-specific promoter.

Lipofection has been extensively used for the stable transfection of hESCs with plasmid constructs, using a variety of reagents [4, 17, 19, 20]. Gerrard et al. [19] and Thyagarajan et al. [20], in particular, utilized the technique to generate stably transfected hESC lines that expressed GFP under the control of a human *Oct4* promoter, thus selectively labeling only undifferentiated ES cells. In conclusion, lipofection represents an easy and convenient approach to the genetic modification of hESCs, especially given its relatively low toxicity.

2.3 Nucleofection

Nucleofection represents a novel transfection strategy that provides a variety of electroporation and buffer conditions that suit many different cell types and promote nuclear DNA transport, rather than simple cytosolic delivery. The first demonstration of nucleofection in hESCs was performed by Siemen et al. [14], who succeeded in obtaining both robust transient transfection rates (70%) with excellent viability, as well as stably transfected hESC lines using reporter

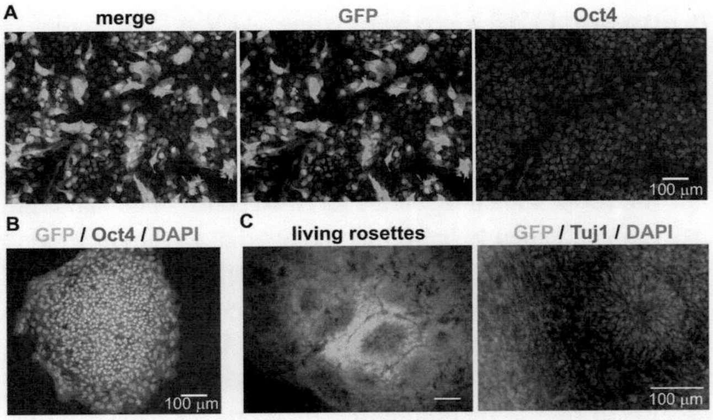

Fig. 1 Transient and stable nucleofection of hESCs with the *EF1α::GFP* reporter plasmid. (**A**) Immunofluorescent micrographs of hESCs transiently transfected with the reporter plasmid. The cells were fixed 2 d after the nucleofection and immunostained for GFP (*green*) and Oct4 (*red*). The transient nucleofection efficiency with such constructs is approximately 75% in our experiments. (**B**) This photomicrograph shows a fixed hESC colony stably nucleofected with the *EF1α::GFP* reporter plasmid. Note the colocalization of the GFP and Oct4 immunoreactivities in the undifferentiated hESCs. (**C**) These photomicrographs depict living neural precursors (rosettes) derived from the hESC line shown in (**B**) (*left*), and immunofluorescent staining of such precursors for both GFP (*green*) and neuronal β-tubulin (Tuj1; *red*). Note that GFP expression persists in the differentiated state (*see* also Color Insert)

plasmids. In that study, nucleofection was found to be superior to both electroporation and lipofection in transient transfection comparisons.

We have reproduced the findings by Siemen et al. [14] using reporter plasmids expressing GFP and the H9 (WA-09) hESC line. In our hands, nucleofection results in a 75% transient transfection rate, and provides stable GFP labeling of hESCs using the *EF1α* promoter after selection (Fig. 1A, 1B). Such lines continue to express GFP over several passages (over 40) and after differentiation to neuronal lineages without obvious attenuation of GFP expression, suggesting that there is no transcriptional silencing (Fig. 1C).

More recently, we adapted the nucleofection technology to BAC transgenesis of mouse and human ESCs. We generated mouse ESC lines that stably express GFP engineered into a BAC under the control of the *Oct4* promoter (Fig. 2). Oct4 is a master regulator of pluripotency and, as such, drives GFP expression in ESCs but not their differentiated progeny. In conjunction, we adapted the nucleofection technology toward the stable transfection of hESCs with BACs, a recent breakthrough that will be discussed below. Importantly, with regard to the BAC transgenesis of hESCs, nucleofection succeeded where electroporation and lipofection had failed. This suggests that nucleofection may be a powerful addition to our arsenal of bioengineering tools toward the genetic modification of hESCs. Moreover, its efficacy is paralleled by its relative lack of toxicity, as opposed to the harsher conditions associated with electroporation.

3 Viral Transduction

One of the most robust techniques for transgenesis of hESCs is viral transduction [3]. A variety of viral vectors have been utilized for the expression of transgenes in hESCs. The utility of modified viruses as gene delivery vehicles requires the following critical properties:

1. Viral vectors need to be replication-incompetent.
2. Their virion has to provide space for the insertion of transgenes.
3. They must be able to infect the target cell.
4. Usually it is preferred that they produce long-term, rather than transient, transduction of the target cells, although in certain situations that may not be the case.

In this chapter, we will focus on retroviruses, lentiviruses, adenoviruses, adeno-associated viruses (AAV), and baculoviruses. All of these viral vectors have been successfully tested with hESCs, however, they differ with regard to the above properties.

3.1 Retroviruses

Retroviruses are single-stranded RNA viruses with a 10 kb genome that contains at least three genes: *gag*, which encodes core proteins that help assemble viral particles; *pol*, which encodes reverse transcriptase; and *env*, which encodes the viral envelope protein. Engineered retroviruses used for gene transfer are replication-incompetent vectors. Upon infection of dividing target cells, the viral RNA is reverse

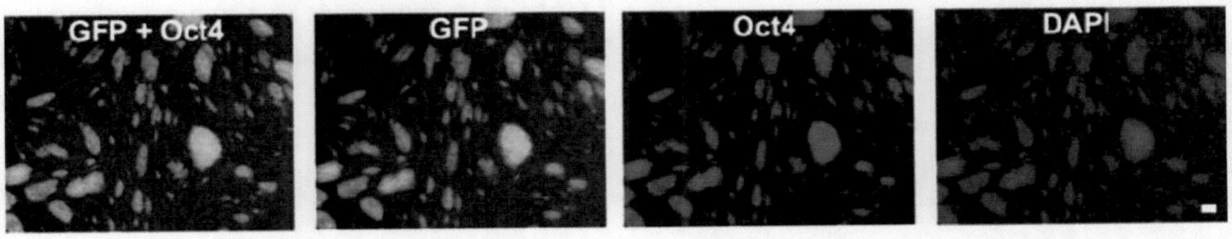

Fig. 2 Mouse ESCs stably nucleofected with the *mOct4::GFP* BAC. The photomicrographs show immunofluorescent staining for GFP (*green*) and Oct4 (*red*). Nuclei are counterstained with DAPI. Scale bar, 100 μm (*see* also Color Insert)

transcribed into DNA, which is then stably integrated into the genome. Gammaretroviruses, such as the oncoretroviruses used as gene transfer vehicles, can only infect dividing cells because the nuclear envelope dissolution during mitosis is required for the viral genome to enter the nucleus. A subclass of retroviruses, the lentiviruses, have evolved molecular machinery that allows the nuclear import of the viral genome without mitosis and can therefore infect postmitotic cells as well. The retroviral DNA has a predilection for transcriptionally active genomic regions as targets for integration [21, 22], likely because transcription occurs in euchromatic areas where unraveled genomic DNA is more conducive to integration. In particular, murine leukemia virus (MLV), a widely used retroviral vector, was found to preferentially integrate around the start of transcriptional units, especially within promoter regions [21].

The fact that retroviral vectors permanently modify the host cell genome raises the question of insertional mutagenesis. Indeed, one potential pitfall linked to retroviruses is that their genomic insertion can either inactivate tumor suppressor genes or activate proto-oncogenes, thus promoting cellular transformation and tumorigenesis. This phenomenon occurred recently in a retroviral gene therapy trial [23, 24].

The experience with retroviruses in the context of hESCs is limited, likely due to the more recent development of lentiviral vectors, as well as the limitations of retroviral silencing. In one study, a vesicular stomatitis virus G protein (VSV-G) – pseudotyped retrovirus expressing GFP under the control of the murine stem cell virus (MSCV) long terminal repeat (LTR), an endogenous retroviral promoter that has been engineered to be more resistant to silencing compared to wild-type LTRs, successfully transduced a high percentage of hESCs, without cytopathic effects [25].

One type of retrovirus successfully adapted to hESCs is the foamy virus. These viruses are replication-incompetent and not pathogenic in humans. They also have the advantage of transferring approximately 9 kb of foreign DNA, similar to lentiviruses. Moreover, foamy viruses are less likely to integrate near transcriptional start sites compared to other retroviruses, hence lowering the probability of insertional mutagenesis [26]. Gharwan et al. [27] showed that foamy viruses can efficiently transduce hESCs. When the cells were infected with a virus carrying the reporter GFP driven by the phosphoglycerate kinase (*PGK*) promoter, approximately 30% of the cells were transduced.

Our laboratory has recently adapted another retroviral gene transfer system to hESCs, the RCAS/TVA system previously used to model tumor formation in murine cells [28]. The premise of this system lies in the use of subgroup A avian leukosis virus (ALV-A), which normally infects only avian cells, since they express the cognate avian-specific cell surface receptor TVA. Dividing mammalian cells become susceptible to infection by ALV-A when they are engineered to express TVA. Moreover, while ALV-A can replicate in avian cells (hence RCAS, which stands for replication competent ALV-A), it becomes replication-incompetent in mammalian cells. The viral genome can be engineered to accommodate small transgenes, which are usually placed under the control of the endogenous LTR. Recently, this paradigm has been extended to lentiviral vectors that are engineered to utilize the envelope protein of ALV-A and, therefore, require the presence of TVA to infect the target cell [29].

In order to implement the RCAS/TVA system in hESCs, we generated hESC lines stably expressing the TVA receptor driven by the ubiquitous *EF1α* promoter. However, we found no GFP-positive cells after infection of undifferentiated *EF1α::TVA* hESCs with a GFP-expressing ALV-A vector. This suggests either that hESCs are not permissive to infection by ALV-A, or that transgene expression driven by the LTR is rapidly and efficiently suppressed. However, neural precursors derived from *EF1α::TVA* hESCs are permissive to infection and transgene expression. We are currently using the system to develop brain tumor models derived from hESCs, taking advantage of cell type-specific TVA expression to achieve targeted viral transduction with oncogenes.

3.2 Lentiviruses

Among retroviruses, lentiviruses have the unique ability to infect non-dividing cells due to sophisticated molecular machinery that mediates nuclear transport of the viral genome [30–32]. Lentiviral vectors can accommodate transgenes with sizes up to 9 kb. Like other retroviruses, reverse-transcribed lentiviral genomes integrate into the host cell's genome, thus supporting long-term transgene expression. HIV-derived lentiviral vectors have undergone several levels of engineering to ensure their safety. In particular, they are replication-incompetent, lack virulence factors encoded by accessory genes that have been implicated in HIV pathogenesis, are packaged with a conditional packaging system that relies on viral nucleic acid and protein assembly in *trans*, and are self-inactivating (SIN) [33–36]. The self-inactivation feature refers to a deletion in the U3 region of the 3′ endogenous LTR causing loss of promoter function in the 5′ LTR following reverse transcription of the viral genome. Thus, such vectors rely on exogenous promoters to drive expression of the transgene. The efficiency of transduction is increased by the presence of the central polypurine tract (cPPT) [32], which facilitates transport of the viral genome into the host cell's nucleus. Transgene expression is enhanced in the presence of the woodchuck hepatitis virus post-transcriptional regulatory element (WPRE) placed 3′ to the transgene [37]. Additional exogenous elements that may be used to enhance transgene expression include the scaffold attachment region (SAR) of the human interferon-β gene, which promotes euchromatic organization of the genome around the integration site and decreases transgene methylation [38], as well as the chicken β-globin chromatin insulator, which prevents nucleoprotein condensation [39].

Lentiviruses, like gammaretroviruses, allow for a fair amount of flexibility regarding the choice of the envelope protein. Pseudotyping the vector with VSV-G confers upon it extremely broad tropism, since it is presumed that VSV-G directly interacts with the lipid bilayer to mediate endocytosis, envelope fusion to endosomal membranes, and release of the viral contents devoid of envelope into the cytosol [40, 41]. Recent efforts by Jang et al. [42] led to the development of lentiviral vectors pseudotyped with either the gibbon ape leukemia virus or the RD114 feline virus envelope proteins that specifically transduce hESCs, but not the mouse fibroblast feeder cells.

Lentiviral vectors sustain stable transgene expression in hESCs [43]. There exist several examples of highly efficient transduction of hESCs with lentiviruses expressing reporters such as GFP [42, 44–49], with sustained transgene expression over several passages, as well as after differentiation [3]. Overall, lentiviral vectors are the most widely used viral gene delivery vehicles in the context of hESCs.

Transgene silencing is a concern for both lentiviral and gammaretroviral vectors. Gammaretroviruses undergo transgene silencing and variegation in mouse and human ESCs [22, 50], as well as in human hematopoietic stem cells [51, 52]. While early reports claimed that lentiviruses were silenced less than gammaretroviruses [53], more recently it has become recognized that lentiviruses are also subject to epigenetic modifications and positional effects. Such silencing events are expected to be more pronounced when host cells are transduced with a single copy of the lentiviral genome, since multiple integrations confer statistically an improved chance of avoiding transcriptional suppression. The genomic integration patterns differ between lentiviruses and gammaretroviruses [21]. While the former have a predilection for euchromatic genomic regions [54] and, in particular, transcriptionally active open reading frames, the latter are more likely to integrate around transcriptional start sites and promoters. Such differences may account for different degrees of epigenetic modifications and potentially silencing. Moreover, modifications in the 5′ LTR, such as the SIN feature of lentiviral vectors, may also play a crucial role in modulating the suppression of transgene expression.

Silencing, however, depends not only on the gene delivery vehicle, but also on the promoter used for transgene expression. Xia et al. [49] recently showed that lentiviral transgene silencing in hESCs is highly promoter dependent. Expression of a GFP transgene under the *PGK* promoter yielded a 38.1% expression maintenance rate against 18.6% under the *EF1α*, 3.3% under the CAG (cytomegalovirus immediate-early enhancer/chicken β-actin hybrid) and only 1.1% under the cytomegalovirus promoters. These data obtained with viral vectors are concordant with studies on promoter-specific silencing using plasmids and lipofection reagents [13], and suggest that viral promoters may be more prone to silencing than eukaryotic or hybrid promoters. An alternative explanation is that silencing is dependent on transcriptional activity, such that overactive promoters, as occurs with cytomegalovirus promoter constructs, are silenced. Interestingly, transcriptional suppression appears to be more powerful in hESCs, as opposed to their murine counterparts [49].

The mechanisms underlying the transcriptional silencing are not well understood. Early experiments showed that retroviral transcription in infected mouse embryonal carcinoma cells is effectively suppressed via mechanisms that depend on a *cis* viral DNA element termed the repressor binding site (RBS) at the 5′ end of the genome, as well as via DNA methylation [55, 56]. A recent study by Wolf and Goff [57] identified the transcriptional silencer TRIM28 as a major component of the complex that binds RBS and brings about silencing via recruitment of a number of histone modification enzymes. Understanding the molecular processes that underlie transcriptional silencing is essential to our improving the approaches to the genetic manipulation of stem cells.

3.3 Adeno-Associated Viruses

Adeno-associated viruses (AAV) are DNA viruses that are replication-incompetent, unless a helper virus, such as adenovirus or herpes simplex virus [58], coinfects the host cell. As such, they lend themselves as gene transfer vectors, although they lack the capacity for large transgene insertion. AAV has been shown to integrate into transcriptionally active regions of the host genome [59]. In human hosts, the AAV genome usually integrates at a specific site on chromosome 19q [60]. However, modified AAV vectors tend to remain episomal. When tested with hESCs, AAV serotype 2 was the only AAV subtype that succeeded at transducing hESCs, albeit at very low efficiencies [61].

3.4 Adenoviruses

Adenoviruses (Ad) are DNA viruses whose genome remains episomal in the host cell. Adenoviral vectors are replication-incompetent and can accommodate up to 8 kb of foreign transgene DNA. However, because the adenoviral genome is episomal and does not integrate into the host genome, Ad can be used in circumstances when transient transgene expression is needed. In the context of hESCs, Smith-Arica et al. [61] tested Ad type 5 expressing the β-galactosidase reporter and found that Ad5 infected a small fraction of hESCs, resulting in transgene expression.

3.5 Baculoviruses

Baculoviral vectors have traditionally been used in the context of large-scale protein expression, because they support high levels of gene expression without significant cytopathic effects. Moreover, they can accommodate large transgenic inserts, up to 30 kb, and are replication-incompetent in mammalian cells. However, baculovirus-mediated transgenesis is episomal and, therefore, transient. A recent study by Zeng et al. [62] showed that hESCs can be transiently transduced with baculoviral vectors expressing the GFP reporter at efficiencies up to 80%. Moreover, Zeng et al. [62] engineered a hybrid baculovirus-AAV vector that was able to stably integrate into the AAV cognate integration site on chromosome 19q. Such viral strategies are promising, since baculovirus allows for insertion of large transgenes.

4 Homologous Recombination

The use of homologous recombination in mouse ES cells [9, 63] revolutionized the field of mouse genetics by allowing the dissection of genetic pathways using either knock-out or knock-in approaches. Over the past few years, homologous recombination has also been achieved in hESCs, although such experiments are hampered by the generally inferior transfection efficiencies of hESCs and their intrinsically poor clonogenic ability as single cells, as discussed above. Zwaka and Thomson [16] were the first group to succeed in targeting specific genes in hESCs by homologous recombination, by using electroporation of cell clumps and subsequent plating at high density. The efficiency of obtaining correct targeting approximated 0.5×10^{-6}. Using the above protocol, they were able to target the *HPRT1* and *Oct4* loci independently, thus demonstrating the reproducibility of the technique.

A recent demonstration of homologous recombination involved the targeting of the human *ROSA26* locus in hESCs [64], using a modification of the protocol developed by Zwaka and Thomson [16], in which single dissociated hESCs, rather than clumps, were electroporated with the targeting vector. The discovery of the human *ROSA26* locus is a significant contribution to our arsenal of genetic tools within the hESC field. The *ROSA26* gene is ubiquitously expressed and does not undergo silencing, thereby representing an ideal site for directed integration of transgenes.

The availability of homologous recombination in hESCs allows a great deal of flexibility in designing basic genetic experiments. Moreover, homologous recombination can serve as a powerful approach in correcting gene defects in a therapeutic context. A recent publication offers an example of such an application. Hanna et al. [65] used an established genetic protocol for reprogramming mouse somatic cells harboring the human sickle cell hemoglobin mutation into induced pluripotent embryonic stem-like cells (iPS cells) [66, 67]. Homologous recombination was then used to correct the gene defect, and the genetically corrected iPS cells were differentiated into hematopoietic progenitors that were transplanted into syngeneic hosts, thus rescuing the disease phenotype. Such strategies may soon be feasible in human iPS cells [68, 69] derived from patients with specific gene defects.

5 Protein Transduction

Over the past decade, protein transduction has been used as a strategy to introduce peptides and nucleic acids into target cells [70, 71]. The premise of protein transduction lies in the use of certain peptides bearing protein transduction domains (PTDs) that interact with the cell membrane and are subsequently internalized, a phenomenon first observed with the HIV Tat protein by Frankel and Pabo [72], as well as Green and Loewenstein [73]. Other native proteins known to contain PTDs include several homeodomain proteins, such as Antennapedia [74, 75], HoxA-5 [76], Engrailed [77, 78], and the maize homeoprotein Knotted-1 [78]; and the herpes simplex virus protein VP22 [79]. The Tat protein's PTD, in particular, a stretch of 11 basic amino acids, has been widely used as a vehicle for protein transduction in the form of fusions with proteins of interest. It is thought that the positive charges on the Tat PTD electrostatically interact with the negatively charged moieties on the cell membrane, thus facilitating endocytosis via lipid raft-dependent macropinocytosis [80]. The utility of Tat as a mediator of protein transduction and potentially protein therapy was demonstrated in vivo by Schwarze et al. [81], who injected mice intraperitoneally with a fusion of Tat to the β-galactosidase reporter and found robust transduction throughout the body, including the brain.

The mechanism underlying protein transduction suggests that the approach is endowed with wide target cell tropism. Indeed, protein transduction has been successfully tested in hESCs. Kwon et al. [82] used the Tat PTD as a vehicle to transduce hESCs with Pdx1, a transcription factor related to pancreatic development. Interestingly, Antennapedia was not an adequate carrier in the context of hESCs, a discrepancy that may be explained by different internalization mechanisms between Tat and Antennapedia. More recently, Nolden et al. [15] used the Tat PTD as vehicle to transduce hESCs with Cre recombinase, which catalyzed the *loxP*-dependent mutagenesis of a Cre reporter construct stably transfected into the cells. These encouraging first steps demonstrate that the protein transduction strategy is not only applicable to hESCs, but that it may facilitate their genetic manipulation as well.

6 BAC Transgenics

BACs are large genomic fragments (150–250 kb) from any organism of interest contained in a low-copy plasmid. The plasmid backbone allows maintenance of the genomic DNA in bacteria, making it easy to prepare large amounts of the DNA. In addition, BACs are less prone to rearrangements than yeast artificial chromosomes or cosmids [83]. These advantages led to the use of BAC libraries as the foundation for most genome sequencing projects, and hundreds of libraries are now available for many different species.

Functional BAC gene expression was first used to clone the circadian Clock gene through in vivo complementation [84]. Since then, the experience with BACs as gene expression and transgenesis tools has demonstrated several advantages over conventional small plasmids. First, it has become clear that BAC promoter-driven transgenes usually provide better expression fidelity compared to conventional plasmids [85–89]. One explanation for the increased specificity is that the large size of BACs insulates the transgene from the effects of the surrounding chromatin. Another possibility is that the additional promoter and/or enhancer elements present in the BAC help maintain expression fidelity.

Second, in the past molecular cloning techniques limited conventional transgenic construct size to less than 20 kb. Due to the large size of mammalian genes and their promoters, promoter analysis was necessary to identify smaller, more manageable sequences necessary for correct transgene expression. After discovery of a suitable sequence, the identified pieces of the promoter were assembled together with the reporter's open reading frame before use. BAC transgenics obviate the need for promoter mapping, since BACs spanning the entire gene and large regions of flanking sequence are available for most genes of interest (from sources such as BACPAC Resource Center [http://bacpac.chori.org/] at Children's Hospital Oakland Research Institute).

One of the caveats of BACs as genetic tools, however, is their large size, which renders direct BAC modification in vitro using restriction enzymes and ligases impractical. However, recent developments in BAC bioengineering allow rapid modification with base pair precision using homologous recombination in bacteria. Two main techniques are used: (1) *allelic exchange* and (2) *recombineering*. Allelic exchange relies on bacterial recA for homologous recombination. In this method, a shuttle vector is constructed using conventional cloning. The shuttle vector contains the in vitro engineered allele flanked by two regions of homology. This vector is introduced into bacteria containing the BAC of interest where it undergoes a single homologous crossover event creating a cointegrate. After cointegration, the direct repeats in the BAC allow for a resolution to occur that might contain the allele of interest, depending on the recombination that took place [88, 90]. Recombineering methods exploit the conditional expression of phage-mediated recombination systems that permit homologous recombination between as little as 30 bp of homology [91, 92]. In practical terms, recombineering allows the use of a PCR product containing an allele of interest flanked by regions of homology built directly into the PCR primers as a targeting construct. The hybrid PCR product can be directly inserted into a BAC through two crossover events. Either method can modify BACs in a variety of ways, including subtle single base pair modifications.

A direct product of the exciting advances in BAC technology has been the construction of the GENSAT library (www.gensat.org). GENSAT (Gene Expression Nervous System Atlas) is an NIH-funded project aimed at producing a library of GFP- and Cre-expressing BAC transgenic mice. To date, over 2000 BACs have been engineered and constructed, and multiple transgenic mouse lines from each BAC have been created and quality controlled to visualize cells (and their projections) expressing a gene of interest (personal communication, Shiaoching Gong). Ultimately, the project aims to have around 5000 CNS-expressed genes [93]. Reagents created through this project are readily available to the research community: BACs are distributed through ATCC (www.atcc.org), while the mouse lines are available from the MMRRC (www.mmrrc.org).

GENSAT was originally intended for mouse development in vivo, but we demonstrated that this BAC library can be used in mouse ESCs as well [11]. BAC transgenic mice are made by injecting BAC DNA into a fertilized egg (pronuclear injection). Pronuclear injection is efficient enough to produce 10–40% transgenic pups without drug selection [94]. Unlike pronuclear injection, ESC transgenesis requires drug selection because very few cells integrate the BAC DNA into the genome after electroporation. Therefore, we modified the GENSAT BACs by inserting a selection cassette into a *loxP* site in the BAC backbone, using a technique termed BAC retrofitting [95]. This permits the selection of rare ESCs that have stably integrated BAC DNA. Depending on the BAC in question, between 5 and 40% of the drug-resistant ESC colonies expressed GFP after differentiation into the appropriate cell type. The clones that express GFP expressed it correctly, at least within the limits of our assays [11]. In our laboratories, we have only observed two possible clones out

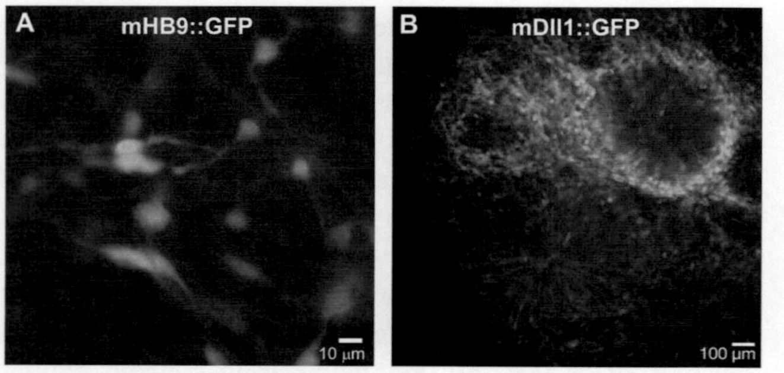

Fig. 3 Stable BAC transgenesis of hESCs via nucleofection. (**A**) Living GFP-positive motor neurons derived from hESCs stably transfected with the *mHb9::GFP* BAC. (**B**) Living rosettes derived from hESCs transfected with the *mDll1::GFP* BAC (*see* also Color Insert)

of many hundreds that obviously expressed GFP incorrectly (unpublished observations).

In addition to mouse ESCs, our groups have recently found that GENSAT BACs function correctly in hESCs (Being reviewed for publication – Fig. 3). We created hESC lines stably transfected with the retrofitted *mHb9::GFP* and *mDll1::GFP* BACs, thereby labeling motor neurons and neuroblasts/neurons, respectively, with GFP. The ability to rapidly create BAC-transgenic hESCs should help transplantation biology, drug and gene screening, and allow directed microarray interrogation that focuses on specific cells in a complex milieu.

7 Episomal Maintenance

Genome integration endows a transgene with two essential properties: the transgene is replicated with the genome, and it is segregated during cell division. As a result, the transgene is stably inherited without the need for selection, assuming that the transgene does not slow cell division or affect cell viability. While these traits are desirable, there are potential costs associated with integration. Illegitimate integration can cause inappropriate activation or inactivation of endogenous genes. Such insertional mutagenesis was vividly underscored in a gene therapy trial that used retroviral vectors to treat SCID children [23]. In this case, a subset of patients developed leukemia, likely due to retrovirus-mediated transcriptional activation of a proto-oncogene in lymphoid precursors that acts to block their differentiation and promote their overgrowth [24]. In addition to changing gene expression, transgene integration has frequently been observed near major chromosomal changes [96]. The cause and effect of this observation is not well understood. One possibility is that transgenes are frequently inserted at a site of damage, such as a double-stranded DNA break that subsequently causes a translocation. Once selected, the transgenic population now contains not only the transgene, but also the chromosomal damage that initially allowed transgene insertion. Finally, the host chromatin can affect how the transgene is expressed, changing the expression strength and fidelity. Transgene integration and subsequent clonal selection, therefore, can have drastic consequences to gene expression within the host cell and vice versa.

Episomal maintenance can avoid these problems by maintaining the transgene outside of the genome. To allow for episomal maintenance, two functions have to be supplied by the transgene: an origin of replication to allow replication of the transgene, and segregation functions so that both cells receive a copy of the transgene after cell division. This latter function can be dispensed with if the copy number of the transgene is high enough to guarantee that each daughter cell receives the transgene. One method of keeping DNA episomal has been co-opted from Epstein-Barr Virus (EBV), a large DNA virus that evolved mechanisms to maintain the viral genome in dividing cells. One drawback of such systems is the need to express a viral protein, called EBNA-1, in order to carry out these functions. Mouse ESCs with EBV-based expression systems contain between 10 and 30 episomal copies per cell [97]. Such "superfection" systems have been extensively used in maintaining small, conventional transgenes in mouse ESCs, typically for functional screening [97–108]. Initial work in hESCs demonstrated that cells expressing EBNA-1 were easier to transfect, and that the hESCs expressed GFP from the episome for short periods of time. However, after one week, there was severe silencing in one report [109]. A second report claimed sustained episomal GFP expression as long as selection was maintained [110]. These investigators used a different promoter in the episome, and they expressed a bicistronic reporter containing GFP-IRES-puromycin. Presumably, the reporter selection was required to overcome the silencing observed in the initial paper, but further work is needed to clarify the discrepancies noted between these two reports in hESC-based episomal systems. BAC-based episomal systems have been described in many mammalian cell culture systems, but not yet in ESCs [111, 112]. It is possible that BAC-based episomes would escape the silencing observed with small episomal constructs.

Newer artificial chromosome engineering technologies do not require the expression of viral proteins [113]. Instead, this technique aims to exploit cellular strategies for artificial chromosome segregation. This is accomplished largely by incorporating centromeric sequences to assure artificial chromosome segregation. Origins of replication are not needed for episomes containing large genomic sequences since they occur naturally approximately every few kb [114]. So far, no reports have demonstrated the utility of this approach in ESCs.

8 Targeted Transgenesis

As mentioned above, transgenesis is associated with undesired effects both on the host cells' physiology brought about by insertional mutagenesis, as well as on transgene expression, due to transcriptional silencing. Both of these caveats could potentially be addressed if transgenes were to integrate into specific regions of the genome, where the insertion would not disrupt the endogenous gene regulation and, likewise, the surrounding genome would be conducive to transgene expression. Targeted transgenesis does occur, of course, in the case of knock-in constructs where homologous recombination directs transgenes to a targeted locus. However, such strategies are not always possible and often interfere with endogenous gene function. Below we will describe two novel and promising approaches that may allow nonrandom, targeted transgenesis of hESCs.

The first strategy makes use of the phage phiC31 integrase to target transgenes bearing an *attB* motif into pseudo-*attP* sites of the hESC genome [20]. PhiC31 integrase, unlike Cre and Flp, can mediate recombination between non-identical but homologous DNA sequences, such as the *attB* and pseudo-*attP* sequences. The human genome contains multiple pseudo-*attP* sites, that are often located within introns, a safety feature that reduces the possibility of alterations in coding sequences. Using *attB*-flanked transgenes and exogenous transient expression of phiC31 integrase, Thyagarajan et al. [20] were able to obtain stably transfected hESC lines expressing GFP under the control of either the human *Oct4* or the *EF1α* promoter. An additional benefit of this approach is that phiC31 integrase may also increase the rate of stable integration when compared to conventional strategies where integration is random.

The second approach involves the use of zinc-finger nucleases (ZFNs) to specifically engineer targeted sequences of the hESC genome [115]. More specifically, the C_2H_2 class of zinc-finger binding domains can be engineered to recognize a particular target sequence. Fusion of such a zinc-finger protein to the *FokI* endonuclease can then produce a ZFN that will cleave a cognate DNA sequence. Due to FokI enzymatic requirements, two such ZFN molecules are required to recognize the two complementary DNA stands, so as to produce a double-stranded DNA break. The most faithful repair modality of such a break is termed homology-directed repair, which utilizes homologous recombination from an identical template to restore structure. Lombardo et al. [115] were able to use this technology to insert transgenes into the *CCR5* gene, a nonessential gene in hESCs, after transducing the cells with integrase-defective lentiviral vectors expressing the appropriate ZFNs. Such lentiviral vectors cannot integrate into the genome and, thus, remain temporarily episomal, eventually disappearing from the host cells after a few cell divisions. This is an exciting new technology, since it can allow the targeted integration of transgenes into specific regions of the genome, where the risk of insertional mutagenesis is minimized by the dispensability of the targeted gene.

9 Conclusion

We are undoubtedly experiencing the coming of a new era in biology and medicine, where stem cells, and in particular hESCs, will play a major role in our understanding of human development and disease, as well as in our efforts to treat disease. Fortunately, the explosive rate of growth of the hESC field is paralleled by equally impressive advancements in genetic engineering. We now have in our possession a powerful arsenal of tools that will allow elaborate genetic manipulation of hESCs, an essential element in the effort to harness their potential.

We believe that the development of BAC-mediated tissue-specific transgenesis in hESCs will allow us to pursue new avenues in the biology of hESCs and their derivatives. By being able to genetically label specific hESC-derived populations with high fidelity and, thus, identify them from other cell types, we can begin to ask specific questions about the genetic and developmental pathways that govern their biology; their transcriptome and how it relates to the transcriptomes of other cell types; their complex interactions with neighboring cells and how such interplay leads to the formation of larger scale cellular networks and systems; and, finally, the changes associated with disease processes and potentially their therapeutic reversal in drug screens. From the technical point of a view, we now feel confident that we have available to us full array of genetic manipulations that are as applicable to hESCs as they have been to their mouse counterparts. Conceivably in the near future, we will be able to routinely use targeted transgenesis and homologous recombination in hESCs to perform sophisticated genetic experiments, including knock-ins and conditional knock-outs, knock-downs, and transgenics.

References

1. Thomson JA, Itskovitz-Eldor J, Shapiro SS, Waknitz MA, Swiergiel JJ, Marshall VS, Jones JM. Embryonic stem cell lines derived from human blastocysts. Science. 1998;282:1145–7.
2. Yates F, Daley GQ. Progress and prospects: gene transfer into embryonic stem cells. Gene Ther. 2006;13:1431–9.
3. Strulovici Y, Leopold PL, O'Connor TP, Pergolizzi RG, Crystal RG. Human embryonic stem cells and gene therapy. Mol Ther. 2007;15:850–66.
4. Watanabe K, Ueno M, Kamiya D, Nishiyama A, Matsumura M, Wataya T, et al.. A ROCK inhibitor permits survival of dissociated human embryonic stem cells. Nat Biotechnol. 2007;25:681–6.
5. Amit M, Carpenter MK, Inokuma MS, Chiu CP, Harris CP, Waknitz MA, et al. Clonally derived human embryonic stem

cell lines maintain pluripotency and proliferative potential for prolonged periods of culture. Dev Biol. 2000;227:271–8.
6. Xu C, Inokuma MS, Denham J, Golds K, Kundu P, Gold JD, et al. Feeder-free growth of undifferentiated human embryonic stem cells. Nat Biotechnol. 2001;19:971–4.
7. Xu C, Rosler E, Jiang J, Lebkowski JS, Gold JD, O'Sullivan C, et al. Basic fibroblast growth factor supports undifferentiated human embryonic stem cell growth without conditioned medium. Stem Cells. 2005;23:315–23.
8. Levenstein ME, Ludwig TE, Xu RH, Llanas RA, Van DenHeuvel-Kramer K, Manning D, et al. Basic fibroblast growth factor support of human embryonic stem cell self-renewal. Stem Cells. 2006;24:568–74.
9. Thomas KR, Capecchi MR. Site-directed mutagenesis by gene targeting in mouse embryo-derived stem cells. Cell. 1987;51: 503–12.
10. Xian HQ, Werth K, Gottlieb DI. Promoter analysis in ES cell-derived neural cells. Biochem Biophys Res Commun. 2005;327:155–62.
11. Tomishima MJ, Hadjantonakis AK, Gong S, Studer L. Production of green fluorescent protein transgenic embryonic stem cells using the GENSAT bacterial artificial chromosome library. Stem Cells. 2007;25:39–45.
12. Eiges R, Schuldiner M, Drukker M, Yanuka O, Itskovitz-Eldor J, Benvenisty N. Establishment of human embryonic stem cell-transfected clones carrying a marker for undifferentiated cells. Curr Biol. 2001;11:514–8.
13. Liew CG, Draper JS, Walsh J, Moore H, Andrews PW. Transient and stable transgene expression in human embryonic stem cells. Stem Cells. 2007;25:1521–8.
14. Siemen H, Nix M, Endl E, Koch P, Itskovitz-Eldor J, Brüstle O. Nucleofection of human embryonic stem cells. Stem Cells Dev. 2005;14:378–83.
15. Nolden L, Edenhofer F, Haupt S, Koch P, Wunderlich FT, Siemen H, et al. Site-specific recombination in human embryonic stem cells induced by cell-permeant Cre recombinase. Nat Methods. 2006;3:461–7.
16. Zwaka TP, Thomson JA. Homologous recombination in human embryonic stem cells. Nat Biotechnol. 2003;21:319–21.
17. Vallier L, Rugg-Gunn PJ, Bouhon IA, Andersson FK, Sadler AJ, Pedersen RA. Enhancing and diminishing gene function in human embryonic stem cells. Stem Cells. 2004;22:2–11.
18. Singh Roy N, Nakano T, Xuing L, Kang J, Nedergaard M, Goldman SA. Enhancer-specified GFP-based FACS purification of human spinal motor neurons from embryonic stem cells. Exp Neurol. 2005;196:224–34.
19. Gerrard L, Zhao D, Clark AJ, Cui W. Stably transfected human embryonic stem cell clones express OCT4-specific green fluorescent protein and maintain self-renewal and pluripotency. Stem Cells. 2005;23:124–33.
20. Thyagarajan B, Liu Y, Shin S, Lakshmipathy U, Scheyhing K, Xue H, et al. Creation of engineered human embryonic stem cell lines using phiC31 integrase. Stem Cells. 2008;26:119–26.
21. Wu X, Li Y, Crise B, Burgess SM. Transcription start regions in the human genome are favored targets for MLV integration. Science. 2003;300:1749–51.
22. Ellis J, Yao S. Retrovirus silencing and vector design: relevance to normal and cancer stem cells? Curr Gene Ther. 2005;5: 367–73.
23. Hacein-Bey-Abina S, Von Kalle C, Schmidt M, McCormack MP, Wulffraat N, Leboulch P, et al. LMO2-associated clonal T cell proliferation in two patients after gene therapy for SCID-X1. Science. 2003;302:415–9.
24. Pike-Overzet K, de Ridder D, Weerkamp F, Baert MR, Verstegen MM, Brugman MH, et al. Ectopic retroviral expression of LMO2, but not IL2Rgamma, blocks human T-cell development from CD34+ cells: implications for leukemogenesis in gene therapy. Leukemia. 2007;21:754–63.
25. Lebkowski JS, Gold J, Xu C, Funk W, Chiu CP, Carpenter MK. Human embryonic stem cells: culture, differentiation, and genetic modification for regenerative medicine applications. Cancer J. 2001;7 Suppl 2:S83–93.
26. Trobridge GD, Miller DG, Jacobs MA, Allen JM, Kiem HP, Kaul R, et al. Foamy virus vector integration sites in normal human cells. Proc Natl Acad Sci U S A. 2006;103:1498–503.
27. Gharwan H, Hirata RK, Wang P, Richard RE, Wang L, Olson E, et al. Transduction of human embryonic stem cells by foamy virus vectors. Mol Ther. 2007;15:1827–33.
28. Holland EC, Varmus HE. Basic fibroblast growth factor induces cell migration and proliferation after glia-specific gene transfer in mice. Proc Natl Acad Sci U S A. 1998;95:1218–23.
29. Lewis BC, Chinnasamy N, Morgan RA, Varmus HE. Development of an avian leukosis-sarcoma virus subgroup A pseudotyped lentiviral vector. J Virol. 2001;75:339–44.
30. Naldini L, Blömer U, Gallay P, Ory D, Mulligan R, Gage FH, et al. In vivo gene delivery and stable transduction of nondividing cells by a lentiviral vector. Science. 1996;272:263–7.
31. Blömer U, Naldini L, Kafri T, Trono D, Verma IM, Gage FH. Highly efficient and sustained gene transfer in adult neurons with a lentivirus vector. J Virol. 1997;71:6641–9.
32. Zennou V, Petit C, Guetard D, Nerhbass U, Montagnier L, Charneau P. HIV-1 genome nuclear import is mediated by a central DNA flap. Cell. 2000;101:173–85.
33. Zufferey R, Nagy D, Mandel RJ, Naldini L, Trono D. Multiply attenuated lentiviral vector achieves efficient gene delivery in vivo. Nat Biotechnol. 1997;15:871–5.
34. Dull T, Zufferey R, Kelly M, Mandel RJ, Nguyen M, Trono D, et al. A third-generation lentivirus vector with a conditional packaging system. J Virol. 1998;72:8463–71.
35. Mochizuki H, Schwartz JP, Tanaka K, Brady RO, Reiser J. High-titer human immunodeficiency virus type 1-based vector systems for gene delivery into nondividing cells. J Virol. 1988;72: 8873–83.
36. Zufferey R, Dull T, Mandel RJ, Bukovsky A, Quiroz D, Naldini L, et al. Self-inactivating lentivirus vector for safe and efficient in vivo gene delivery. J Virol. 1998;72:9873–80.
37. Zufferey R, Donello JE, Trono D, Hope TJ. Woodchuck hepatitis virus posttranscriptional regulatory element enhances expression of transgenes delivered by retroviral vectors. J Virol. 1999;73:2886–92.
38. Dang Q, Auten J, Plavec I. Human beta interferon scaffold attachment region inhibits de novo methylation and confers long-term, copy number-dependent expression to a retroviral vector. J Virol. 2001;74:2671–8.
39. Emery DW, Yannaki E, Tubb J, Stamatoyannopoulos G. A chromatin insulator protects retrovirus vectors from chromosomal position effects. Proc Natl Acad Sci U S A. 2000;97:9150–5.
40. Fredericksen BL, Whitt MA. Vesicular stomatitis virus glycoprotein mutations that affect membrane fusion activity and abolish virus infectivity. J Virol. 1995;69:1435–43.
41. Aiken C. Pseudotyping human immunodeficiency virus type 1 (HIV-1) by the glycoprotein of vesicular stomatitis virus targets HIV-1 entry to an endocytic pathway and suppresses both the requirement for Nef and the sensitivity to cyclosporin A. J Virol. 1997;71:5871–7.
42. Jang JE, Shaw K, Yu XJ, Petersen D, Pepper K, Lutzko C, et al. Specific and stable gene transfer to human embryonic stem cells using pseudotyped lentiviral vectors. Stem Cells Dev. 2006;15:109–17.
43. Gropp M, Reubinoff B. Lentiviral vector-mediated gene delivery into human embryonic stem cells. Methods Enzymol. 2006;420:64–81.

44. Gropp M, Itsykson P, Singer O, Ben-Hur T, Reinhartz E, Galun E, et al. Stable genetic modification of human embryonic stem cells by lentiviral vectors. Mol Ther. 2003;7:281–7.
45. Ma Y, Ramezani A, Lewis R, Hawley RG, Thomson JA. High-level sustained transgene expression in human embryonic stem cells using lentiviral vectors. Stem Cells. 2003;21:111–7.
46. Xiong C, Tang DQ, Xie CQ, Zhang L, Xu KF, Thompson WE, et al. Genetic engineering of human embryonic stem cells with lentiviral vectors. Stem Cells Dev. 2005;14:367–77.
47. Xue T, Cho HC, Akar FG, Tsang SY, Jones SP, Marbán E, et al. Functional integration of electrically active cardiac derivatives from genetically engineered human embryonic stem cells with quiescent recipient ventricular cardiomyocytes: insights into the development of cell-based pacemakers. Circulation. 2005;111:11–20.
48. James D, Noggle SA, Swigut T, Brivanlou AH. Contribution of human embryonic stem cells to mouse blastocysts. Dev Biol. 2006;295:90–102.
49. Xia X, Zhang Y, Zieth CR, Zhang SC. Transgenes delivered by lentiviral vector are suppressed in human embryonic stem cells in a promoter-dependent manner. Stem Cells Dev. 2007;16: 167–76.
50. Ellis J. Silencing and variegation of gammaretrovirus and lentivirus vectors. Hum Gene Ther. 2005;16:1241–6.
51. Guenechea G, Gan OI, Dorrell C, Dick JE. Distinct classes of human stem cells that differ in proliferative and self-renewal potential. Nat Immunol. 2001;2:75–82.
52. Pannell D, Ellis J. Silencing of gene expression: implications for design of retrovirus vectors. Rev Med Virol. 2001;11:205–17.
53. Pfeifer A, Ikawa M, Dayn Y, Verma IM. Transgenesis by lentiviral vectors: lack of gene silencing in mammalian embryonic stem cells and preimplantation embryos. Proc Natl Acad Sci U S A. 2002;99:2140–5.
54. Schröder AR, Shinn P, Chen H, Berry C, Ecker JR, Bushman F. HIV-1 integration in the human genome favors active genes and local hotspots. Cell. 2002;110:521–9.
55. Niwa O, Yokota Y, Ishida H, Sugahara T. Independent mechanisms involved in suppression of the Moloney leukemia virus genome during differentiation of murine teratocarcinoma cells. Cell. 1983;32:1105–13.
56. Loh TP, Sievert LL, Scott RW. Evidence for a stem cell-specific repressor of Moloney murine leukemia virus expression in embryonal carcinoma cells. Mol Cell Biol. 1990;10:4045–57.
57. Wolf D, Goff SP. TRIM28 mediates primer binding site-targeted silencing of murine leukemia virus in embryonic cells. Cell. 2007;131:46–57.
58. Buller RML, Janik JE, Sebring ED, Rose JA. Herpes simplex virus types 1 and 2 completely help adenovirus-associated virus replication. J Virol. 1981;40:241–247.
59. Nakai H, Montini E, Fuess S, Storm TA, Grompe M, Kay MA. AAV serotype 2 vectors preferentially integrate into active genes in mice. Nat Genet. 2003;34:297–302.
60. Kotin RM, Linden RM, Berns KI. Characterization of a preferred site on human chromosome 19q for integration of adeno-associated virus DNA by non-homologous recombination. EMBO J. 1992;11:5071–8.
61. Smith-Arica JR, Thomson AJ, Ansell R, Chiorini J, Davidson B, McWhir J. Infection efficiency of human and mouse embryonic stem cells using adenoviral and adeno-associated viral vectors. Cloning Stem Cells. 2003;5:51–62.
62. Zeng J, Du J, Zhao Y, Palanisamy N, Wang S. Baculoviral vector-mediated transient and stable transgene expression in human embryonic stem cells. Stem Cells. 2007;25:1055–61.
63. Doetschman T, Maeda N, Smithies O. Targeted mutation of the Hprt gene in mouse embryonic stem cells. Proc Natl Acad Sci U S A. 1988;85:8583–7.
64. Irion S, Luche H, Gadue P, Fehling HJ, Kennedy M, Keller G. Identification and targeting of the ROSA26 locus in human embryonic stem cells. Nat Biotechnol. 2007;25:1477–82.
65. Hanna J, Wernig M, Markoulaki S, Sun CW, Meissner A, Cassady JP, et al. Treatment of sickle cell anemia mouse model with iPS cells generated from autologous skin. Science. 2007;318:1920–3.
66. Takahashi K, Yamanaka S. Induction of pluripotent stem cells from mouse embryonic and adult fibroblast cultures by defined factors. Cell. 2006;126:663–76.
67. Wernig M, Meissner A, Foreman R, Brambrink T, Ku M, Hochedlinger K, et al. In vitro reprogramming of fibroblasts into a pluripotent ES-cell-like state. Nature. 2007;448:318–24.
68. Takahashi K, Tanabe K, Ohnuki M, Narita M, Ichisaka T, Tomoda K, et al. Induction of pluripotent stem cells from adult human fibroblasts by defined factors. Cell. 2007;131:861–72.
69. Yu J, Vodyanik MA, Smuga-Otto K, Antosiewicz-Bourget J, Frane JL, Tian S, et al. Induced pluripotent stem cell lines derived from human somatic cells. Science. 2007;318:1917–20.
70. Wadia JS, Dowdy SF. Protein transduction technology. Curr Opin Biotechnol. 2002;13:52–6.
71. Chauhan A, Tikoo A, Kapur AK, Singh M. The taming of the cell penetrating domain of the HIV Tat: myths and realities. J Control Release. 2007;117:148–62.
72. Frankel AD, Pabo CO. Cellular uptake of the tat protein from human immunodeficiency virus. Cell. 1988;23:1189–93.
73. Green M, Loewenstein PM. Autonomous functional domains of chemically synthesized human immunodeficiency virus tat trans-activator protein. Cell. 1988;55:1179–88.
74. Perez F, Joliot A, Bloch-Gallego E, Zahraoui A, Triller A, Prochiantz A. Antennapedia homeobox as a signal for the cellular internalization and nuclear addressing of a small exogenous peptide. J Cell Sci. 1992;102:717–22.
75. Bolton SJ, Jones DN, Darker JG, Eggleston DS, Hunter AJ, Walsh FS. Cellular uptake and spread of the cell-permeable peptide penetratin in adult rat brain. Eur J Neurosci. 2000;12:2847–55.
76. Chatelin L, Volovitch M, Joliot AH, Perez F, Prochiantz A. Transcription factor hoxa-5 is taken up by cells in culture and conveyed to their nuclei. Mech Dev. 1996;55:111–7.
77. Joliot A, Maizel A, Rosenberg D, Trembleau A, Dupas S, Volovitch M, et al. Identification of a signal sequence necessary for the unconventional secretion of Engrailed homeoprotein. Curr Biol. 1998;8:856–63.
78. Balayssac S, Burlina F, Convert O, Bolbach G, Chassaing G, Lequin O. Comparison of penetratin and other homeodomain-derived cell-penetrating peptides: interaction in a membrane-mimicking environment and cellular uptake efficiency. Biochemistry. 2006;45:1408–20.
79. Elliott G, O'Hare P. Intercellular trafficking and protein delivery by a herpesvirus structural protein. Cell. 1997;88:223–33.
80. Wadia JS, Stan RV, Dowdy SF. Transducible TAT-HA fusogenic peptide enhances escape of TAT-fusion proteins after lipid raft macropinocytosis. Nat Med. 2004;10:310–5.
81. Schwarze SR, Ho A, Vocero-Akbani A, Dowdy SF. In vivo protein transduction: delivery of a biologically active protein into the mouse. Science. 1999;285:1569–72.
82. Kwon YD, Oh SK, Kim HS, Ku SY, Kim SH, Choi YM, et al. Cellular manipulation of human embryonic stem cells by TAT-PDX1 protein transduction. Mol Ther. 2005;12:28–32.
83. Kelley JM, Field CE, Craven MB, Bocskai D, Kim UJ, Rounsley SD, et al. High throughput direct end sequencing of BAC clones. Nucleic Acids Res. 1999;27:1539–46.
84. Antoch MP, Song EJ, Chang AM, Vitaterna MH, Zhao Y, Wilsbacher LD, et al. Functional identification of the mouse circadian clock gene by transgenic BAC rescue. Cell. 1997;89:655–67.
85. Jessen JR, Meng A, McFarlane RJ, Paw BH, Zon LI, Smith GR, et al. Modification of bacterial artificial chromosomes through

chi-stimulated homologous recombination and its application in zebrafish transgenesis. Proc Natl Acad Sci U S A. 1998;95: 5121–6.
86. Lee EC, Yu D, Martinez de Velasco J, Tessarollo L, Swing DA, Court DL, et al. A highly efficient Escherichia coli-based chromosome engineering system adapted for recombinogenic targeting and subcloning of BAC DNA. Genomics. 2001;73: 56–65.
87. Reizis B, Leder P. The upstream enhancer is necessary and sufficient for the expression of the pre-T cell receptor alpha gene in immature T lymphocytes. J Exp Med. 2001;194:979–90.
88. Yang XW, Model P, Heintz N. Homologous recombination based modification in Escherichia coli and germline transmission in transgenic mice of a bacterial artificial chromosome. Nat Biotechnol. 1997;15:859–65.
89. Zhang XM, Ng AH, Tanner JA, Wu WT, Copeland NG, Jenkins NA, et al. Highly restricted expression of Cre recombinase in cerebellar Purkinje cells. Genesis. 2004;40:45–51.
90. Gong S, Yang XW, Li C, Heintz N. Highly efficient modification of bacterial artificial chromosomes (BACs) using novel shuttle vectors containing the R6Kgamma origin of replication. Genome Res. 2002;12:1992–8.
91. Copeland NG, Jenkins NA, Court DL. Recombineering: a powerful new tool for mouse functional genomics. Nat Rev Genet. 2001;2:769–79.
92. Court DL, Sawitzke JA, Thomason LC. Genetic engineering using homologous recombination. Annu Rev Genet. 2002;36:361–88.
93. Heintz N. Gene expression nervous system atlas (GENSAT). Nat Neurosci. 2004;7:483.
94. Palmiter RD, Brinster RL. Germ-line transformation of mice. Annu Rev Genet. 1986;20:465–99.
95. Wang Z, Engler P, Longacre A, Storb U. An efficient method for high-fidelity BAC/PAC retrofitting with a selectable marker for mammalian cell transfection. Genome Res. 2001;11:137–42.
96. Wurtele H, Little KC, Chartrand P. Illegitimate DNA integration in mammalian cells. Gene Ther. 2003;10:1791–9.
97. Gassmann M, Donoho G, Berg P. Maintenance of an extrachromosomal plasmid vector in mouse embryonic stem cells. Proc Natl Acad Sci U S A. 1995;92:1292–6.
98. Abe M, Sato Y. Puromycin insensitive leucyl-specific aminopeptidase (PILSAP) is required for the development of vascular as well as hematopoietic system in embryoid bodies. Genes Cells. 2006;11:719–29.
99. Adachi K, Soeta-Saneyoshi C, Sagara H, Iwakura Y. Crucial role of Bysl in mammalian preimplantation development as an integral factor for 40S ribosome biogenesis. Mol Cell Biol. 2007;27: 2202–14.
100. Fujikura J, Yamato E, Yonemura S, Hosoda K, Masui S, Nakao K, et al. Differentiation of embryonic stem cells is induced by GATA factors. Genes Dev. 2002;16:784–9.
101. Jackson M, Baird JW, Cambray N, Ansell JD, Forrester LM, Graham GJ. Cloning and characterization of Ehox, a novel homeobox gene essential for embryonic stem cell differentiation. J Biol Chem. 2002;277:38683–92.
102. Nakagawa T, Abe M, Yamazaki T, Miyashita H, Niwa H, Kokubun S, et al. HEX acts as a negative regulator of angiogenesis by modulating the expression of angiogenesis-related gene in endothelial cells in vitro. Arterioscler Thromb Vasc Biol. 2003;23:231–7.
103. Niwa H, Burdon T, Chambers I, Smith A. Self-renewal of pluripotent embryonic stem cells is mediated via activation of STAT3. Genes Dev. 1998;12:2048–60.
104. Niwa H, Masui S, Chambers I, Smith AG, Miyazaki J. Phenotypic complementation establishes requirements for specific POU domain and generic transactivation function of Oct-3/4 in embryonic stem cells. Mol Cell Biol. 2002;22:1526–36.
105. Pritsker M, Ford NR, Jenq HT, Lemischka IR. Genomewide gain-of-function genetic screen identifies functionally active genes in mouse embryonic stem cells. Proc Natl Acad Sci U S A. 2006;103:6946–51.
106. Shimoda M, Kanai-Azuma M, Hara K, Miyazaki S, Kanai Y, Monden M, et al. Sox17 plays a substantial role in late-stage differentiation of the extraembryonic endoderm in vitro. J Cell Sci. 2007;120:3859–69.
107. Suzuki H, Watabe T, Kato M, Miyazawa K, Miyazono K. Roles of vascular endothelial growth factor receptor 3 signaling in differentiation of mouse embryonic stem cell-derived vascular progenitor cells into endothelial cells. Blood. 2005;105:2372–9.
108. Tokuzawa Y, Kaiho E, Maruyama M, Takahashi K, Mitsui K, Maeda M, et al. Fbx15 is a novel target of Oct3/4 but is dispensable for embryonic stem cell self-renewal and mouse development. Mol Cell Biol. 2003;23:2699–708.
109. Kameda T, Smuga-Otto K, Thomson JA. A severe de novo methylation of episomal vectors by human ES cells. Biochem Biophys Res Commun. 2006;349:1269–77.
110. Ren C, Zhao M, Yang X, Li D, Jiang X, Wang L, et al. Establishment and applications of epstein-barr virus-based episomal vectors in human embryonic stem cells. Stem Cells. 2006;24:1338–47.
111. Magin-Lachmann C, Kotzamanis G, D'Aiuto L, Wagner E, Huxley C. Retrofitting BACs with G418 resistance, luciferase, and oriP and EBNA-1 – new vectors for in vitro and in vivo delivery. BMC Biotechnol. 2003;3:2–13.
112. Wade-Martins R, Frampton J, James MR. Long-term stability of large insert genomic DNA episomal shuttle vectors in human cells. Nucleic Acids Res. 1999;27:1674–82.
113. Monaco ZL, Moralli D. Progress in artificial chromosome technology. Biochem Soc Trans. 2006;34:324–7.
114. Gilbert DM. Making sense of eukaryotic DNA replication origins. Science. 2001;294:96–100.
115. Lombardo A, Genovese P, Beausejour CM, Colleoni S, Lee YL, Kim KA, et al. Gene editing in human stem cells using zinc finger nucleases and integrase-defective lentiviral vector delivery. Nat Biotechnol. 2007;25:1298–306.

Transcriptional Networks Regulating Embryonic Stem Cell Fate Decisions

Emily Walker and William L. Stanford

Abstract A comprehensive understanding of the transcriptional regulation of embryonic stem cell (ESC) fate decisions will provide the key to their successful manipulation for therapeutic purposes as well as provide insight into the process of early embryogenesis. Traditional molecular and genetic approaches have been successful in identifying several essential regulators of pluripotency, notably *Oct4*, *Nanog,* and *Sox2*. However, these approaches will not be sufficient to understand the global regulatory control of transcriptional networks. Genome-wide work in model organisms such as *Escherichia coli*, yeast, and sea urchin reveal that transcriptional networks can be broken down into a small set of evolutionarily conserved network motifs, each with its own biological function. Initial genome-wide studies in ESCs reveal the presence of these same network motifs, providing mechanistic explanations of cell fate decisions. Thus, as is being performed in lower organisms, the drafting of a comprehensive transcriptional network controlling ESC fate will require systematic characterisation of the functional targets of each ESC-expressed transcription factor.

Keywords Embryonic stem cells (ESCs) · Transcription factor · Network motif · Transcriptional network · Microarray · Promoter occupancy

1 Introduction

Embryonic stem cells (ESCs) are unspecialized cells that have the ability to self-renew, producing daughter cells with equivalent developmental potential, or to differentiate into more specialized cells. They are derived from the inner cell mass of the pre-implantation embryo and are pluripotent, as they are able to differentiate in vivo into all cell types of the adult organism, but not into extraembryonic tissue.

Exogenous control of the pluripotent state can be achieved by a limited number of factors. When grown in fetal bovine serum (FBS)-containing medium and in the presence of murine embryonic fibroblast feeder cells [1, 2] or the cytokine leukemia inhibitory factor (LIF) [3–6], mouse ESCs remain undifferentiated. LIF activates gp130 signalling through binding to LIFRβ. LIFRβ dimerizes with gp130 and transduces the signal through the JAK-STAT pathway. While STAT3 plays an important role in self-renewal of ESCs, *Stat3*$^{-/-}$ embryos can undergo gastrulation, suggesting the existence of a STAT3-independent pathway for ESC self-renewal [7]. Another factor, BMP4, provided by the serum, functions in the presence of LIF to maintain pluripotency by inducing phosphorylation and nuclear localization of Smad1, followed by upregulation of Id proteins that block neural differentiation [8]. Despite similarities to mouse ESCs, human ESCs do not require LIF to maintain their undifferentiated state [9] and BMP4 initiates differentiation towards the trophoblast lineage [10]. Instead it has been shown that activin/nodal [11], basic FGF [12], and IGF [13] signalling are required to maintain human ESC pluripotency. Recent reports have shown that pluripotent mouse cell lines can also be derived from the post-implantation epliblast [14, 15]. These cells, termed epiblast stem cells (EpiSCs), can be maintained in activin and FGF, have a similar morphology, gene expression, and epigenetic profile to human ESCs. This suggests that the disparity between signalling responsiveness may not be due to species differences but rather to the developmental origin (early or late epiblast) of the cells.

Despite differences in extrinsic factors and possibly the developmental origins of mouse and human ESCs, self-renewal of both human and mouse ESCs is controlled by the key transcription factors *Oct4* [16], *Nanog* [17, 18], and *Sox2* [19]. These factors form a core transcriptional network module.

Because ESCs can be expanded indefinitely in their pluripotent state and are also able to differentiate into cell

W.L. Stanford (✉)
Institute of Biomaterials and Biomedical Engineering; Institute of Medical Science; Department of Chemical Engineering and Applied Chemistry, University of Toronto, 164 College Street, Toronto, ON, Canada, M5S 3G9
e-mail: william.stanford@utoronto.ca

V.K. Rajasekhar, M.C. Vemuri (eds.), *Regulatory Networks in Stem Cells*, Stem Cell Biology and Regenerative Medicine, DOI 10.1007/978-1-60327-227-8_8, © Humana Press, a part of Springer Science+Business Media, LLC 2009

types of all three germ layers in vitro, they are an ideal system for transcriptional network analysis. Of additional value for analysis, they can be genetically manipulated to stably or inducibly express shRNAs, transgenes, or reporter systems. In general, it is important to understand transcriptional regulation in ESCs because they represent both a tractable model of early embryogenesis and somatic stem cells and a potential source of cells for therapeutic applications via tissue engineering and regenerative medicine [20].

In this chapter, we will first define a transcriptional network, its constituent motifs and their predicted biological functions. We will discuss the current understanding of individual transcriptional regulators in ESCs and the initial genome-wide studies in ESCs. We will describe an integrative approach that could extend our understanding of the complex transcriptional cascades responsible for defining ESC fate. Finally, we will describe the model of ESC maintenance that has emerged from these works.

2 Transcriptional Networks

Cells respond to external stimuli and carry out cellular processes by regulating gene expression. *Transcription factors* are proteins that bind to the promoter region of their target gene through interaction with DNA binding domains and function to regulate expression by either increasing (activation) or decreasing (repression) the rate of gene transcription. More than 2000 human genes are classified as transcription factors, mainly through characterization of common structural elements known to be involved in DNA binding [21].

Transcription factors regulate complex processes through *transcriptional regulatory networks*. It has been shown in many organisms, including the bacteria *E. coli* [22], the budding yeast *Saccharomyces cerevisiae* [23], the plant *Arabidopsis* [24], mouse hematopoietic stem cells [25], and human pancreatic cells [26], that these networks are made up of *network motifs*. Network motifs are the simplest recurring patterns of interactions between transcription factors and their gene targets. They have been more rigorously defined as patterns of interconnections that recur in many different parts of a network at a frequency much higher than those found in randomized networks [22]. The recurrence of the same patterns amongst widely varying organisms suggests that these motifs comprise the basic building blocks of any transcriptional regulatory network. It is proposed that each network motif confers a specific regulatory capacity to the network and it is hoped that the dynamics of an entire transcriptional network can be understood, at least in part, through an understanding of the dynamics of each component network motif. Although not discussed here, many other proteins play important roles in transcriptional regulation, such as coactivators or corepressors or through the epigenetic effects of chromatin remodelling proteins, histone acetylases, histone deacetylases and histone and DNA methyltransferases.

2.1 Transcriptional Motifs and Their Predicted Biological Functions

2.1.1 Negative Autoregulation and Negative Feedback

Autoregulation occurs when a transcription factor binds to the promoter of its own gene [27, 28]. It has been reported that in *E. coli*, a large percentage of genes are autoregulated (52–74%) [22, 29], whereas in *S. cerevisiae*, about 10% of genes are autoregulated [28]. There are two instances of autoregulation, positive and negative, each with a unique function.

Negative autoregulation occurs when a transcription factor represses the expression of its own gene [27]. The goal of this motif is to rapidly produce a steady-state protein concentration. The three advantages of the negative autoregulatory motif over an unregulated system (Fig. 1A) are increased stability of gene expression levels [30], reduced noise [31], and reduced response time [32]. The response time or rise-time is defined as the delay from the initiation of transcript production until half maximal product concentration is reached [32]. To understand how this motif confers stability, consider the example shown in Fig. 1B. Transcription factor “B” is initially produced until it reaches a repression threshold. It is then able to repress transcription of its own gene, “gene b,” and the production of transcription factor “B” is then reduced. Once it is reduced below the repression threshold, transcription of “gene b” resumes. This cycle continues, fixing the concentration of “B” into a steady-state level, hovering around its repression threshold. In comparison, an unregulated system, in which a transcription factor activates the expression of a gene other than its own (Fig. 1A), must rely on alternative methods of controlling concentration of the transcription factor, such as promoter binding affinities or transcription factor degradation. Such a system is slow to achieve steady-state and thus results in large cell-cell variability. An unregulated motif can only reach a steady-state equal to the negative autoregulatory motif when the unregulated gene is controlled by a weak promoter and the autoregulated gene is controlled by a strong promoter. The use of the strong promoter is advantageous because it dramatically increases the rise-time of the system. It has been experimentally demonstrated in *E. coli* that the rise-time of an unregulated system is approximately one cell-cycle compared to the negative autoregulatory system with a rise-time of 1/5th a cell cycle [32]. Additional factors influencing the response of these motifs include protein degradation

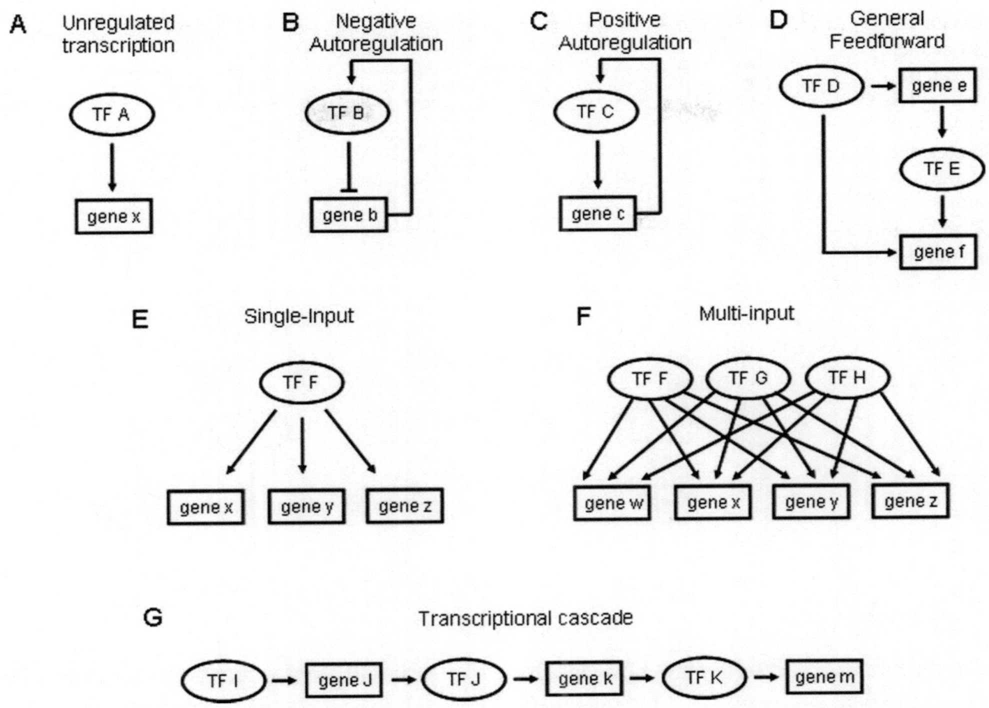

Fig. 1 Six basic network motifs. For all figures, an *oval* represents a protein product and in all cases shown it is a transcription factor (TF). A *rectangle* represents the promoter of the gene. An *arrowhead* represents an activating interaction and a perpendicular line represents a repressive interaction. (**A**) In unregulated transcription, transcription factor (TF) A binds to "gene x," causing activation or repression. (**B**) In negative autoregulation, "TF B" represses the expression of its own gene, "gene b." (**C**) In positive autoregulation, "TF C" activates the expression of its own gene, "gene c." (**D**) In the general feedforward motif, "TF D" regulates the expression of "gene e" and both "TF D" and "TF E" regulate the expression of "gene f." (**E**) In the single-input motif, "TF F" regulates a group of target genes. (**F**) In the multi-input motif, a set of transcription factors (TF F, G, and H) all control a set of target genes (w, x, y, z). (**G**) In a transcriptional cascade, "TF I" regulates "gene j." "TF J" regulates "gene k." 'TF K' regulates "gene m" and so on

rates, mRNA production and degradation rates, delays in the formation of protein from mRNA, strength of promoter activation and/or repression and cooperativity with other regulators. Mathematical models to account for these factors have been developed and outside of extreme inputs, correlate well with the theory and experimental evidence. Details of these models can be found in Becskei and Serrano [30] and Rosenfeld et al. [32].

Negative feedback is a variation of negative autoregulation and occurs when the gene target of a transcription factor represses the activating transcription factor. It has the same effects on response-time, noise reduction, and gene expression stability. In addition, there is evidence to suggest that it significantly increases the production of the downstream gene target [31]. A mathematical simulation of a negative feedback motif can be found in Dublanche et al. [31].

2.1.2 Positive Autoregulation and Positive Feedback

Positive autoregulation occurs when a transcription factor increases the expression of its own gene. The effects of this motif are to slow response time [33] and to decrease the stability of gene expression levels [34, 35]. To understand how this motif slows response times, consider the example in Fig. 1C. Transcription factor "C" is initially produced, but while levels of "C" remain low, transcription rates are also low. Once "C" reaches an activation threshold, it is able to increase the rate of transcription of its own gene, "gene c." This motif results in an increase in cell-cell variability when in some cells, the activation threshold of "gene c" has not been reached and the concentration of "C" remains low, while in other cells, the activation threshold is surpassed and the concentration of "C" becomes dramatically higher [34]. Once the concentration of "C" becomes high, it tends to stay high, even for a time after the loss of the activating signal. It has been suggested that this mechanism confers "memory" to the cell since the response delay can last longer than one cell cycle [33, 35]. In some systems, positive autoregulation can result in a bimodal distribution in which there are essentially two populations of cells – those in which "C" is off (low expression) and those is which "C" is on (activated expression) [34].

Positive feedback is a variation of positive autoregulation and occurs when the gene target of a transcription factor activates the activating transcription factor. It has similar effects

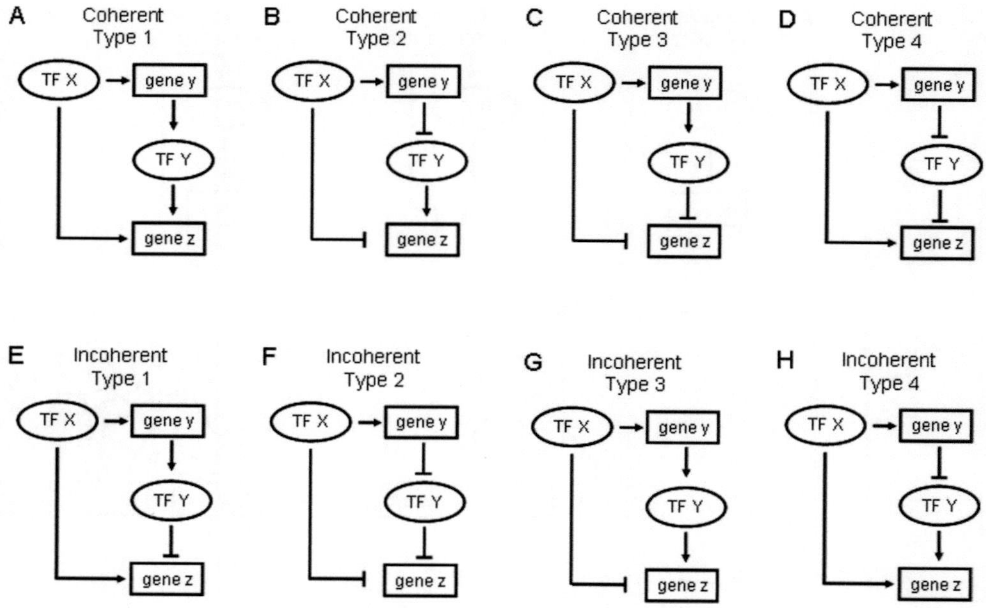

Fig. 2 Feed-forward motifs. Since each regulatory element in a feed-forward motif can be either an activator or a repressor, there are eight possible motif structures. (**A–D**) In coherent feed-forward motifs, the sign of the direct path from transcription "TF X" to "gene z" is the same as the overall sign of the indirect path through "TF Y." (**E–H**) In incoherent feed-forward motifs, the two paths have opposite signs

on response-time, gene expression stability and bimodal distribution [33]. A mathematical simulation of a positive feedback motif can be found in Maeda et al. [33].

2.1.3 Feed-Forward

The feed-forward motif consists of three elements, two transcription factors and one effector gene. As depicted in Fig. 1D, the first transcription factor ("D") regulates the expression of the second ("E") and both "D" and "E" regulate the expression of the effector gene "gene f." Since each regulatory element can be acting as either an activator or a repressor, there are eight possible feed-forward motif structures (Fig. 2) [27]. As such, this is diverse class of motifs and has been shown to have multiple functions [36]. It has been broken down into two subclasses, coherent feed-forward motifs (Fig. 2A–2D) in which the sign of the direct path from transcription factor "X" to "gene z" is the same as the overall sign of the indirect path through transcription factor "Y," and incoherent feed-forward motifs (Fig. 2E–2H) in which the two paths have opposite signs [27].

It has been proposed that coherent feed-forward loops can provide a delayed response to the appearance/disappearance of an input [27, 37, 38]. The biological function of these motifs is to act as a filter that can protect the target gene from fluctuations in input. Consider the motif in Fig. 2A in which transcription factor "X" initially induces the expression of transcription factor "Y." If expression of "gene z" requires both "X" and "Y," there will be a delay in "gene z" expression until "Y" accumulates to its activation threshold. Thus, its expression is initially resistant to fluctuations in either "X" or "Y," as it will require constant activation of both for a defined period before it becomes activated. However, once the signal causing activation of either "X" or "Y" is removed, expression of "gene z" will be rapidly deactivated, as it requires both inputs. This same motif can also work in the opposite way. Consider that only one of either "X" or "Y" is required for activation of "gene z." In this case, upon expression of "X," "gene z" will be quickly expressed. At the same time, "X" will cause production of "Y" which will eventually also enhance "gene z." When stimulation by either "X" or "Y" is removed, there will be a delay in the deactivation of "gene z" because if "X" is shut down, the presence of "Y" will continue until it is degraded and even if "Y" is shut down, "X" can still be active. Thus, shutting off this motif requires absence of both inputs for a defined length of time but will otherwise be able to filter out brief fluctuations in "X" and "Y" and maintain stable levels of "gene z."

In contrast, the incoherent feed-forward motif has been shown to accelerate the response time of the target [39]. Consider the motif in Fig. 2E in which transcription factor "X" initially induces the expression of "gene z." At the same time, transcription factor "X" induces the expression of transcription factor "Y." Once "Y" reaches the repressive threshold of "gene z," it will begin to repress "gene z." In the case that "Y" does not entirely repress its target, "gene z" will become expressed at a steady-state level, balanced by activation through "X" and repression through "Y." Mathematical and experimental models demonstrate that this will occur approximately three times more rapidly than in an unregulated system [39]. Note that the function of this motif is similar to the autoregulatory motif except that the effects are felt upon the "gene z" target that is not necessarily a transcription factor itself.

Mathematical models of various feed-forward motifs can be found in Mangan et al. [37], Wall et al. [36], and Mangan et al. [39].

2.1.4 Single-Input Motif

A single-input motif occurs when a transcription factor regulates a group of target genes. These target genes are not regulated by any other transcription factor. The predicted function of this motif is to allow coordinated expression of a group of genes with a common function [27]. A more refined function of this motif is to create a temporal cascade of target gene expression, defined by the activation threshold for each target gene. Considering the motif in Fig. 1E, as the concentration of "F" rises over time, it will cross the activation threshold of each target gene – x, y, and z, turning on its targets in a defined order. Alternatively, if "F" is a repressor, as "F" rises over time, the repressive threshold of each target will be crossed, turning off its targets in a defined order. This system is important to prevent protein production before it is necessary and also to coordinate developmental processes, for example controlling the activation of the endomesodermal zygotic control apparatus in sea urchin [40].

2.1.5 Multi-Input Motif

A multi-input motif, shown in Fig. 1F, occurs when a set of transcription factors all control a set of target genes. Generally, the targets will all co-ordinate an essential cellular function [27]. It has been proposed that this motif allows for the regulation of a set of genes under different biological conditions, depending on which transcription factors are present [28].

2.1.6 Transcriptional Cascades

A transcriptional cascade, such as shown in Fig. 1G, occurs when one transcription factor binds to the promoter of a second transcription factor, which in turn binds a third transcription factor and so on. This cascade passes a signal on a slow timescale, perhaps one cell generation at each cascade step making it an appropriate system for executing a developmental program. Developmental programs often use repressor cascades which can be more robust to noise in protein-production rates than those of activator cascades. For example, in sea urchin development, a cascade of transcription factors is activated. Each is stabilized by association with an autoregulatory or feedback loop and then proceeds to activate the next gene in the cascade. Eventually, through the stabilization acquired at each stage, the initial signal is no longer required to propagate the cascade [40].

3 Core Transcriptional Regulators in ES Cells

Considerable work has been performed to identify and characterize transcription factors crucial for ESC fate decisions. Because most of this work has been on targeted, single-gene studies, it has not been possible to recognize network motifs in the context of all possible network connections, as they have been defined in lower organisms. However, each observed transcriptional interaction can be placed into one of these predefined motifs. Assuming that these motifs are conserved across species and are the building blocks of all transcriptional networks, it is possible to infer the biological function of these interactions in maintaining the identity of the ESC. Studies on a genome-wide scale are required to confirm that the above described motifs are in fact conserved in ESC transcriptional networks. Genome-wide studies supporting this will be discussed in the next section, but first we will describe the detailed understanding of key transcriptional interactions maintaining ESC pluripotency.

3.1 Oct4 and Sox2

Oct4 (*Pou5f1*) is a POU-domain transcription factor expressed exclusively in totipotent and pluripotent ES cells and germ cells [41]. It has a highly conserved role in maintaining pluripotent cell populations [42, 43] and its expression level dictates ESC fate [16]. If *Oct4* expression remains within ±50% of normal diploid expression, the ESC will maintain the undifferentiated phenotype. If *Oct4* expression is increased above 150%, ESCs will differentiate into primitive endoderm and mesoderm and finally, if *Oct4* expression is reduced below 50%, cells will dedifferentiate into the trophectoderm lineage. *Oct4*-null embryos die at the blastocyst stage because although they form the trophoblast lineage, they cannot form the inner cell mass [42].

OCT4 is known to bind an octomeric sequence, ATGCAAAT, and associates with HMG-containing transcription factor, SOX2, which binds to a neighboring sox element [41]. It is proposed that the OCT4-SOX2 complex is necessary to co-operatively activate target genes which contain a sox-oct regulatory element [44, 45]. Genes with regulatory regions containing sox-oct elements include *Fgf4*, *Utf1*, *Lefty1*, *Fbxo15*, and *Nanog*. In addition, OCT4 regulates the transcription of *Sox2* and the OCT4-SOX2 complex activates both *Oct4* and *Sox2* expression [46–48]. This autoregulatory interaction is shown in Fig. 3A.

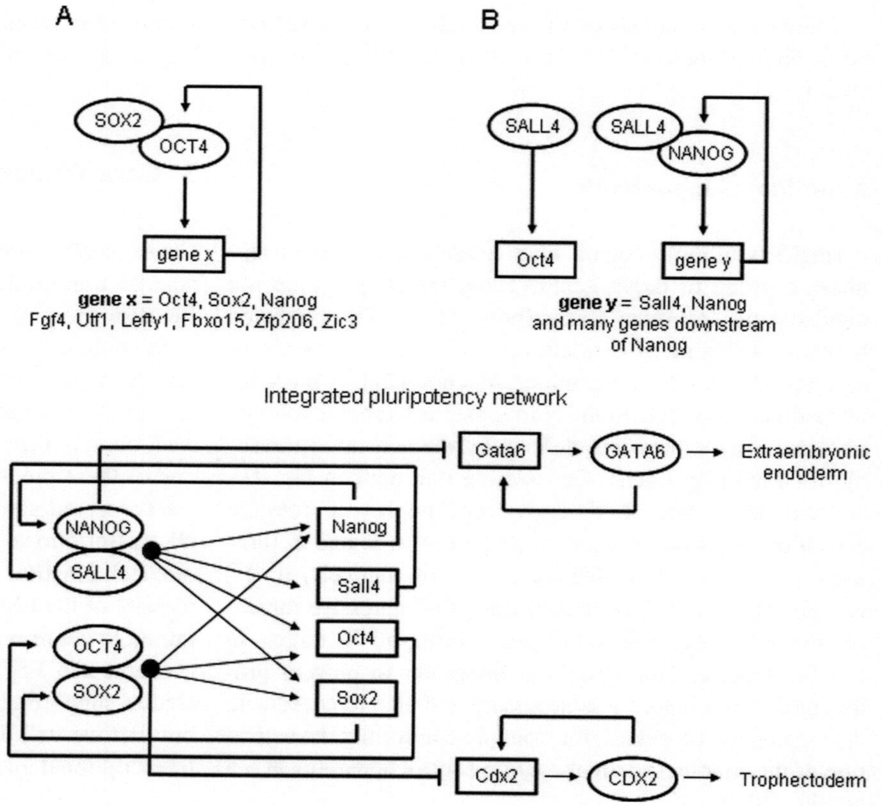

Fig. 3 ESC-specific transcriptional networks. (**A**) OCT4 and SOX2 function in a complex to activate many targets genes. Activation of *Oct4* and *Sox2* themselves constitute one important ESC autoregulatory motif. (**B**) SALL4 and NANOG function in a complex to activate many targets genes while SALL4 alone activates *Oct4*. Activation of *Nanog* and *Sall4* themselves constitutes another important ESC autoregulatory motif. (**C**) An integrated pluripotency network showing the intricate assembly of autoregulatory motifs controlling *Oct4*, *Nanog*, *Sox2*, and *Sall4* expression. *Black dots* represent binding of a complex (either the NANOG/SALL4 or OCT4/SOX2 complex) to the indicated promoter. *Gata6* and *Cdx2* are repressed by NANOG and OCT4/SOX2, respectively, while also participating in autoregulatory loops to enhance their own expression once activated

In more recent work, *Sox2*-null ESCs were used to show that although *Sox2* is essential for the maintenance of pluripotency, it was not essential for the activation of the sox-oct elements in several predicted targets, as its presence could be compensated for by other Sox factors: SOX4, SOX11, or SOX15. Microarray analysis following induction of *Sox2* loss showed up-regulation of negative regulators of *Oct4* (*Nr2f2*) and differentiation inducers (*Eomes* and *Esx1*), and down-regulation of positive regulators (*Nr5a2*) suggesting that the essential function of *Sox2* may be to regulate the expression of *Oct4* through regulation of these other transcription factors [49].

3.2 Nanog

Nanog is a homeobox-containing transcription factor expressed prior to blastocyst formation in the inner cells of the morula and then restricted to the inner cell mass in the blastocyst. *Nanog* expression is not detectable at implantation but reappears in the proximal epiblast at E6 and then remains restricted to the epiblast [50]. Forced overexpression of *Nanog* maintains pluripotency and OCT4 levels in ESCs, even in the absence of LIF [17, 18]. *Nanog* expression is regulated through a sox-oct element in its proximal promoter [50, 51]. Conflicting results report that this element is either occupied by OCT4 and SOX2 [50] or OCT4 and some other sox-element binding transcription factor [51]. The *Nanog* promoter also contains a NANOG-bound region, but it has not been shown that this association is functional [52].

3.3 Sall4

Sall4 is a spalt-like zinc finger transcription factor identified through mass spectrometry analysis as interacting with NANOG [53]. SALL4 functionally binds to an upstream enhancer of *Nanog*. NANOG and SALL4 bind to an intronic enhancer of *Sall4* [52]. In addition, SALL4 binds to many genomic sites that are also known to be bound by NANOG [53]. This suggests that there exists a NANOG-SALL4 complex that functions to activate transcription, similar to the OCT4-SOX2 complex.

Sall4 is essential for ESC self-renewal as well as being involved in many developmental processes. In human, SALL4 mutations have been linked to Okihiro syndrome, characterized by limb deformities and eye movement abnormalities and *Sall4*-null mice are embryonic lethal following a failure to properly form the epiblast [54]. Outgrowths of *Sall4*-deficient embryos show that *Sall4* is essential for the expansion of the inner cell mass, but is dispensable for trophectoderm development. Despite attempts, it has not been possible to generate *Sall4*-null ESCs [54] but in one study, *Sall4* shRNA knockdown experiments in ESCs resulted in reduced proliferation due to decreased in S-phase and increased G-1 phase, but no altered colony morphology, *Oct4* expression or up-regulation of differentiation markers.

In another study [55], *Sall4* was targeted because of the similarity between its expression and *Oct4* expression during early embryonic development [56]. Contrary to the previous study, shRNA knockdown of*Sall4* led to reductions in the expression of *Oct4*, *Nanog*, *Sox2*, increased colony differentiation and up-regulation of trophectodermal markers. Overexpression of *Sall4* led to the specific up-regulation of *Oct4*, *Nanog,* and *Sox2*. It is important to note that the second shRNA study was performed in feeder-free conditions, while the first study was performed in the presence of feeders, suggesting that factors produced by the feeder population were compensating for the loss of *Sall4*, masking the effect on differentiation.

Finally, there is a direct interaction between SALL4 and the *Oct4* promoter that is responsible for the activation of *Oct4* in ESCs [55]. These data suggest that *Sall4* functions to both regulate the expression of *Oct4*, preventing the aberrant development of trophectoderm and to regulate the expression of *Nanog*, while also forming a complex with NANOG to activate additional downstream targets. These interactions are summarized in Fig. 3B.

3.4 Zfp206

Zfp206 is a SCAN-zinc finger transcription factor originally identified because of its highly specific expression in undifferentiated ESCs [57]. It is rapidly down-regulated upon treatment with differentiation agent retinoic acid (RA) and overexpression and shRNA knockdown studies show that it has a positive influence on the ability of the ESC to remain undifferentiated [57], even following treatment with RA [58]. In in vitro experiments, ZFP206 is both able to enhance its own transcription and also enhance the transcription of *Oct4* and *Nanog* [58], although it has not been established that ZFP206 acts directly on the promoters of either *Oct4* or *Nanog*. It has also been shown that the first intron of *Zfp206* is directly bound by both OCT4 and SOX2 through a sox-oct element that confers transcriptional activation of *Zfp206* [52, 59, 60].

3.5 Zic3

Zic3 is a Gli-family zinc finger transcription factor first studied in ESCs because its promoter region was found to be occupied by OCT4, NANOG, and SOX2 in both mouse and human ESCs [52, 60]. It is highly expressed in undifferentiated ESCs and its expression is down-regulated following induction of retinoic acid (RA) differentiation [61]. *Zic3* is dynamically expressed during the embryonic development of mouse [62], *Xenopus* [63], chick [64], and zebrafish [65]. In mouse it is primarily involved in the development of ectoderm and mesoderm during gastrulation [66], during neurulation and is exclusively expressed in the cerebellum of the adult [62]. In *Zic3* mutant mice, left-right patterning is disrupted [67], 50% of null animals die in utero and 30% die perinatally due to congenital heart defects, pulmonary isomerism, and CNS defects [68]. These data suggest that *Zic3* is an important mediator of mesoderm and ectoderm development.

In ESCs, shRNA knockdown of *Oct4* or *Sox2* reduced *Zic3* transcripts to 25% of control and knockdown of *Nanog* reduced *Zic3* to 70% of the control. Likewise, overexpression of *Nanog* sustained *Zic3* expression during RA differentiation, together suggesting that observed OCT4, SOX2, and NANOG promoter occupancy may result in functional regulation [61]. Colony-forming assays with *Zic3* shRNA and overexpression cells show that it has a positive influence on pluripotency, however, *Oct4*, *Sox2,* and *Nanog* show very little change in expression in the *Zic3* knockdowns, suggesting that the effect of *Zic3* is downstream of these key pluripotency factors [61]. Finally, *Zic3* knockdown, both transient and stable, induced expression of a panel of endodermal markers (*Sox17*, *PDGFRA*, *Gata6*, *Gata4*, *Foxa2*) while having no effect on markers for mesoderm, ectoderm, or trophectoderm [61]. This suggests that the specific role of *Zic3* in undifferentiated ESCs is to repress expression of endodermal genes, thus positively influencing the development of mesoderm and ectoderm in the developing embryo.

3.6 Induced Pluripotent Stem (iPS) Cells

Three groups reported that it is possible to induce mouse fibroblasts, either embryonic or adult tail-tip, to acquire pluripotent characteristics including chimera formation and, in one case, germline transmission through the overexpression of only four transcription factors: *Oct4*, *Sox2*, *Klf4,* and *c-Myc* [69–71]. Given the importance of *Oct4* and *Sox2* in maintaining pluripotency, their involvement is not surprising, but *Klf4* and *c-Myc* must also play crucial roles in either setting the stage for pluripotency or for maintaining it. Interestingly, ectopic expression of *Nanog* was not required as *Nanog* expression was induced by reprogramming.

c-Myc is a basic helix-loop-helix transcription factor that is a well-known accelerator of cell cycle [72]. It is a direct target of STAT3 and overexpression can maintain self-renewal of ESCs in the absence of LIF [73]. Its role in reprogramming may be due to its association with histone acetyltransferase (HAT) complexes [74]. By inducing global histone acetylation, the chromatin structure may be opened so that *Oct4* and *Sox2* can access necessary targets.

Klf4 is a zinc-finger containing transcription factor that has been characterized as an oncogene [75] and has many cell cycle targets [76]. *Klf4* has been shown to repress *p53* [77], which may be beneficial for reprogramming considering evidence that *p53* suppresses *Nanog* [78]. It has also been shown that KLF4 cooperates with OCT4 and SOX2 to activate at least one target gene, *Lefty1*, and may similarly activate other *Oct4*-target genes [79]. Mice derived from iPS cells developed tumors with age, likely due to the overexpression of the oncogenes *c-Myc* and *Klf4* [80].

3.7 Transcription Factors Regulating Commitment to Specific Lineages

Since down-regulation of *Oct4* leads to trophectoderm specification and down-regulation of *Nanog* leads to extra-embryonic or primitive endoderm specification, it has been proposed that one mechanism by which ESCs prevent commitment is through the repression of transcription factors that drive lineage specific differentiation. Thus far, very few direct transcriptional interactions have been identified.

Cdx2 is a caudal-type homeodomain transcription factor that is specifically expressed in the trophectoderm at the blastocyst stage [81]. *Cdx2* homozygous mutant mice die around the time of implantation due to a failure to maintain the blastocoel [82, 83]. While it is possible to generate ESCs from $Cdx2^{-/-}$ embryos [84], indicating that the ICM is not affected, it is not possible to form trophectoderm stem (TS) cells [85]. Activation of *Cdx2* results in differentiation towards the trophectoderm lineage [86] and loss of CDX2 leads to aberrant expression of *Oct4* and *Nanog* outside the cells of the ICM. These data indicate that CDX2 is required for the formation of the trophoblast lineage and is also required for the restriction of Oct4 and Nanog to the ICM. It has also been reported that *Cdx2* is autoregulated and that OCT4 and CDX2 interact with one another to form a repressive complex that acts to shut down *Cdx2* expression is ESCs and *Oct4* expression in trophectoderm [86].

Gata6 is a zinc-finger transcription factor expressed in extraembryonic and primitive endoderm lineages [87]. *Gata6* null homozygous mice die between embryonic day 5.5 and 7.5 due to defects in visceral endoderm formation [88]. The activation of *Gata6* is sufficient for the induction of ESCs to the extraembryonic endoderm lineage [89]. This is similar to the phenotype observed following down-regulation of *Nanog*, leading to the suggestion that ESCs prevent endodermal specification through *Nanog*-mediated repression of *Gata6* [90]. There is currently no direct evidence to support this suggestion.

4 Genome-Wide Analyses to Extend Transcriptional Networks

In general, traditional molecular, genetic and developmental biology approaches are indispensable for confirming, defining and fully characterizing specific interactions but will not be sufficient to reveal the global architecture of the pluripotency network. One problem so far is that all the associated pluripotency transcription factors have been identified through their interaction with one of either *Oct4* or *Nanog*. Remaining questions that must be addressed on a genome-wide scale include: (1) What is the nature of the targets of OCT4, NANOG and SOX2 in ESCs? (2) Are there be additional regulators of pluripotency working outside the *Oct4*, *Nanog*, *Sox2* network? (3) What are the overall mechanisms of pluripotency in ESCs? Genome-wide strategies currently being employed in ESC research include gene expression studies, with both microarray and high-throughput sequencing, protein-protein interaction studies such as mass spectrometry, and promoter occupancy studies using chromatin immunoprecipitation (ChIP) combined with either promoter array analysis (ChIP-chip) or high-throughput sequencing (ChIP-PET). Each of these techniques can yield valuable information about the overall structure of ESC transcriptional networks.

4.1 Microarray Expression Studies

Expression analysis can be studied either along the time course of an important biological event, such as ESC commitment to differentiation [91, 92], or following perturbation of a specific transcription factor of interest by shRNA knockdown or overexpression [92, 93]. The idea is to identify genes that are up or down-regulated in response to the changing conditions. Time course studies are more likely to identify new transcriptional targets involved in the process under study, while genome-wide expression analysis following genetic perturbation of known transcription factors will identify downstream targets of that transcription factor.

4.1.1 Common Stem Cell Genes

Initial microarray studies tested the hypothesis that genes commonly expressed between different stem cell populations (ESCs, hematopoietic stem cells, and neural stem cells) would constitute a set of genes that defined stem cell identity [94, 95]. Despite testing the same three types of stem cells and using controls to eliminate the identification of housekeeping genes, there was an almost nonexistent correlation between the lists of genes generated by these two groups, demonstrating that it is not possible to characterize all stem cells with a common list of genes.

4.1.2 Undifferentiated Vs. Differentiated

Instead, it may be more useful to identify the differences in expression between undifferentiated and differentiated populations of ESCs. This has been studied in mouse and human ESCs using large-scale EST sequencing [96, 97]. In this strategy, ESTs over-represented or under-represented in undifferentiated ESCs compared to differentiated populations were identified. The differentiated populations in one study were: (1) day 8 embryoid bodies (EBs), (2) day 5 under DMSO differentiation (known to differentiate towards hepatocytes), and (3) 7 days under RA differentiation (known to differentiate towards neuroectoderm). The second study compared undifferentiated hESCs and mESCs to (1) day 4 mouse EBs, and (2) day 12 human EBs. The major contribution of these studies was to identify differences between signalling pathways essential to mESC and hESC growth.

In terms of using this data to extend transcriptional networks, one study, for example, successfully identified, *Oct4*, *Sox2*, *Znf206* (*Zfp206* in mouse), and *Zic3* but not other important pluripotency factors, such as *Nanog* [96]. This suggests that the cell populations under study were still too divergent to be able to identify the critical factors influencing pluripotency. Forming traditional EBs introduces large gene expression changes due to cell-cell interactions and microenvironmental stimuli within the EB. In addition, the time points studied were too far along their respective differentiation pathways. Given the large number of genes that would have shown differential expression between these populations, it was impossible to identify the subtle changes occurring during the initial commitment decision. Construction of network connections requires observation of step-by-step changes in gene expression that can be directly associated with the change in the regulating transcription factor.

4.1.3 Time Courses of Differentiation

Thus, a third strategy is to identify the genes that are dynamically regulated during differentiation of ESCs by performing microarrays at incremental time points throughout the early stages of differentiation. Each time point represents a snapshot of gene expression that can be used to construct the temporal profile of transcriptional repression or activation of each transcription factor in the system. This strategy has been attempted in mouse ESCs [91, 92].

One study employed RA induced differentiation of mouse ESCs in monolayer culture for six days with a microarray performed at one-day intervals [92]. Genes whose expression changed by at least 2-fold were hypothesized to be regulators of ESC fate. This study revealed three genes, *Tbx3*, *Esrrb*, and *Tcl1*, which were shown to have previously unknown effects on ESC pluripotency. This group also performed microarrays on shRNA knockdowns and overexpressors for *Oct4*, *Nanog*, *Sox2* as well as *Tbx3*, *Esrrb*, and *Tcl1*. Analysis of genes regulated following these manipulations reveal that OCT4 represses trophectodermal genes, NANOG represses genes of endoderm, trophectoderm and epliblast origin, SOX2 represses trophectoderm and epiblast genes, ESRRB and TBX3 repress mesoderm, endoderm, and neural crest genes, and TCL1 represses a subset of neural crest genes.

A second study employed two modes of differentiation, one following the simple withdrawal of LIF and the other following the withdrawal of LIF supplemented by RA. Mouse ESCs were cultured in monolayer for five days with microarrays performed at two-day intervals [91]. Genes whose expression changed gradually and consistently under both differentiation scenarios were hypothesized to be the regulators of ESC fate. Importantly for the construction of transcriptional networks, a step-wise reduction in levels of *Oct4*, *Nanog*, and *Sox2* expression in the LIF time course was observed, although in the RA time course these genes were already dramatically depleted by the second day of culture. These step-wise changes facilitated the construction of a temporal cascade of transcription factors, outlining the order in which these factors are activated or repressed. In general, down-regulated genes were enriched for transcription factors and chromatin remodeling proteins while the up-regulated genes were enriched for transcription factors responsible for differentiation to all germ layers.

4.1.4 shRNA Knockdown and Overexpression

A strategy to more directly assess targets of key transcription factors is to perform expression microarray studies on shRNA and overexpression constructs of those key transcription factors. In general, if a gene is up-regulated in the knockdown, it is predicted that the transcription factor was responsible for its repression in the undifferentiated state. Conversely, if a gene is down-regulated, it is predicted that the transcription factor was responsible for its activation. Likewise, if a gene is up-regulated or down-regulated in the overexpressor, it will be predicted that the transcription factor was responsible for its activation or repression, respectively. Knockdown experiments have been done for *Oct4*, *Nanog*, and *Sox2* as well as other predicted pluripotency genes [91,

92, 98–102]. Knockdown of these key genes results in expression changes in over 1000 genes. In general, it has been confirmed that *Oct4* knockdown results in up-regulation of a trophectodermal markers while *Nanog* knockdown results in up-regulation of endodermal markers [98, 99]. In addition, many regulated genes are involved in epigenetics and chromatin remodelling, further suggesting this as important mediator of stem cell pluripotency [100]. This gives important clues as to the nature of the downstream targets of the pluripotency genes. An improvement on this technique is to combine expression analysis of knockdowns and overexpressors to determine genes that are oppositely regulated, as was done to predict the targets of OCT4 [93] and ZFP206 [57]. The limitation of this approach is that microarray expression analysis cannot distinguish between direct and indirect regulation.

4.2 Promoter Occupancy Studies

Another strategy to identify new regulators of ESC fate decisions is to study the promoter occupancy of known regulators of pluripotency. Two major studies have taken advantage of the established importance of the transcription factors OCT4, NANOG, and SOX2. The hypothesis is that genes bound by any of these three factors, and especially in combination, would be the additional key regulators in ESCs [52, 60].

In human ESCs, chromatin immunoprecipitation experiments for OCT4, NANOG and SOX2 were combined with promoter array analysis [60]. The arrays used for this study contained oligonucleotide probes ranging from 8 kb upstream to 2 kb downstream of the start site of 17,917 annotated human genes, at a resolution of 60 bases [60]. OCT4, NANOG, and SOX2 co-occupied many target genes. This may be expected considering the evidence that OCT4 and SOX2 co-operate to activate target genes, but interestingly, NANOG was also present on the promoters of over 90% of target genes co-occupied by OCT4 and SOX2. In addition, these three factors were often found in close proximity to one another, suggesting that they may function together. OCT4, NANOG, and SOX2 occupied the promoters of both transcriptionally active and inactive genes, suggesting that they can act as activators or repressors. Active genes occupied by all three included several known pluripotency regulators, including *Oct4*, *Nanog*, *Sox2*, *Stat3,* and *Zic3*. Gene ontology (GO) analysis revealed that many of the inactive genes were transcription factors involved in development of all three germ layers.

In mouse ESCs, chromatin immunoprecipitation for OCT4 and NANOG was combined with high throughput sequencing of DNA fragments, thus achieving an unbiased, genome-wide survey of OCT4 and NANOG binding [52]. In general, binding sites frequently occurred much further from the start site of genes than was studied in the human experiments, suggesting that the human dataset only constitutes a portion of the total number of binding sites. There were a large number of genes co-occupied by OCT4 and NANOG, including *Oct4*, *Sox2,* and *Nanog*. Additionally, the predominant transcription factor binding motif occupied by OCT4 included the sox element (approximately 70%), supporting evidence that OCT4 targets genomic sequences in co-operation with SOX2, through the sox-oct element. *Rif1* and *Essrb* were identified as two downstream targets of OCT4 and NANOG that were reduced following shRNA knockdown of *Oct4* and *Nanog* and also displayed altered colony phenotypes when knocked down in ESCs.

4.3 Future Directions: Systematic Integration of Expression and Promoter Occupancy Data

Expression analysis and promoter occupancy studies are the two prominent genome-wide methods for determining transcriptional interactions. Despite the enormous amount of data generated from the studies discussed above, there has been little progress towards constructing a comprehensive transcriptional network controlling ESC fate decisions. Part of the difficulty lies in the limitations of using either of these experimental techniques in isolation.

The limitation of expression analysis of either a time course or following perturbation of a specific gene is that it is impossible to determine which genes are direct targets and which are downstream or indirect targets. In addition, there will likely be a large number of false positive predictions of genes that are changing due to secondary effects such as culture conditions.

Promoter occupancy studies do demonstrate a direct biochemical interaction between a transcription factor and the promoters of all genes available on the microarray chip. One major complication is that even if a gene can be bound by a transcription factor, it may not necessarily be functionally regulated. In addition, it may not be functional under the studied conditions. This could be due, for example, to low affinity of the binding site, absence of critical co-factors or unfavorable spatial orientation of the chromatin. Another issue is that the regulatory regions controlling genes in the mammalian system are very large and can be 5′, 3′, or even intergenic [52]. Thus, to capture the complete profile of binding sites for a given transcription factor, a truly genome-wide experimental technique must be employed. For some transcription factors, for which the transcription factor binding site is known, it is possible to use a purely bioinformatic approach to determine predicted binding sites by searching the genome for these conserved binding site motifs, but this approach will unavoidably identify many more sites than are regulated, or even accessible.

The next logical step in extending ESC transcriptional networks is to combine expression data following perturbations in a system of interest, with genome-wide promoter occupancy studies, as has been done in yeast [101, 102]. The strictest criteria, designed to eliminate false positive predictions, would be to only accept interactions that included both binding data and expression correlation. Ensuring both direct biochemical interaction and regulation of expression provides more confidence in the identification of a functional transcriptional interaction. To achieve complete characterization, the functional targets of each expressed transcription factor must be defined. In this way, the basic transcriptional control of each gene can be appreciated.

Two studies have employed this strategy to narrow down the list of OCT4, NANOG, and SOX2 targets. The first used expression analysis of *Oct4* and *Nanog* shRNA knockdown cell lines to determine which targets found by ChIP-PET were regulated [52]. A total of 776 genes were found to be bound by OCT4, and 4711 genes were regulated followed by *Oct4* knockdown but only 394 were both bound and regulated. In the same study, 1802 genes were bound by NANOG, 2264 were regulated following knockdown of *Nanog* but only 475 were both bound and regulated.

A second study combined a time course of ESC differentiation with ChIP-chip and ChIP-PET data for OCT4, NANOG, and SOX2 from published sources to reveal the temporal architecture of gene regulation during early commitment [91]. Many of the regulated genes were directly bound by OCT4, NANOG, and SOX2, but many more were not, revealing that they were either independently regulated or indirectly regulated by OCT4, NANOG, or SOX2. Further, microarray analysis of *Oct4* and *Sox2* knockdowns was used to confirm that those genes predicted to be both bound and regulated during commitment were regulated following enforced down-regulation of *Oct4* or *Sox2*. Finally, additional ChIP-chip data for two genes that were both bound by OCT4 and NANOG and down-regulated in the time course was incorporated to build two new nodes in the network. Both of these genes were polycomb group (PcG) proteins, known to repress developmental targets through trimethylation of lysine 27 on histone 3. These genes had a set of targets that were commonly bound by OCT4 and NANOG and also up-regulated during differentiation. Together, these connections resulted in the prediction of 43 targets of coherent type 3 feed-forward motifs (Fig. 2C), many of which were transcription factors controlling development of specific germ layers. One of these targets was *Gata6*, whose importance in defining endoderm has been described above. Thus, this integrative approach predicted many new targets that could play similar roles to *Cdx2* and *Gata6*.

A third study combined microarray analysis of under and over-expression of *Oct4* to determine a list of OCT4 targets [93]. Their data was then combined with published ChIP-chip and ChIP-PET studies of OCT4 to distinguish primary targets from secondary or tertiary targets.

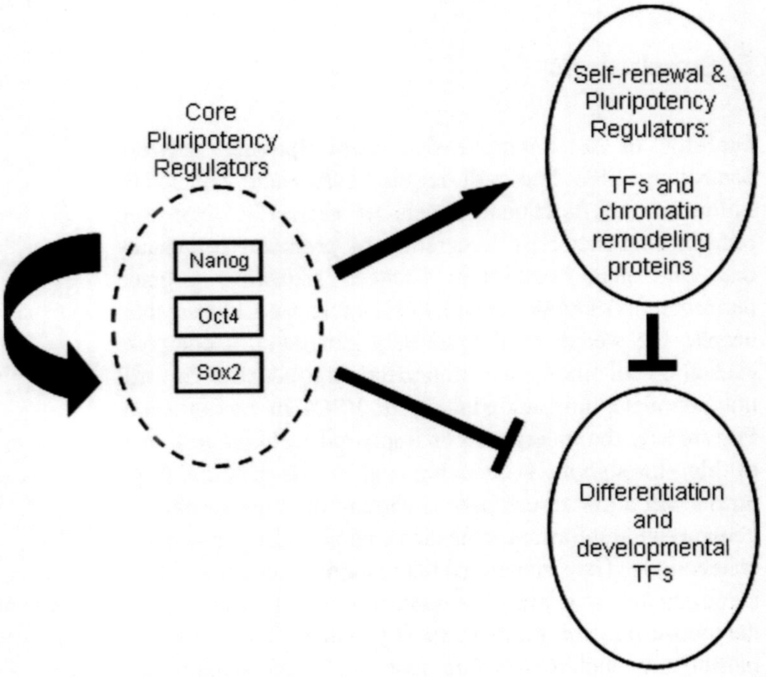

Fig. 4 A general model of ESC maintenance. OCT4, NANOG, and SOX2 function to both activate and repress a large set of target genes, including themselves. Activated genes are transcription factors and chromatin remodeling genes that co-operate with pluripotency genes to maintain repression of transcription factors controlling development of tissues derived from all three germ layers

In these studies, the intersection of these two experiments represents a higher confidence dataset of true OCT4 and NANOG targets to construct the first level of a transcriptional network controlling ESCs. The second study reveals how performing analysis on new nodes will expand the transcriptional network and also provide new insights into the overall mechanism of ESC maintenance. These data reveal that the action of OCT4 and NANOG is widespread and acts to both activate pluripotency genes and repress developmental genes.

4.4 A General Model of ESC Maintenance

Together, the genome-wide studies surveying both the regulation of genes during commitment and the influences of *Oct4*, *Nanog,* and *Sox2* have suggested a general model of stem cell maintenance in which OCT4, NANOG, and SOX2 function to both activate and repress a large set of target genes. Activated genes are transcription factors and chromatin remodeling genes while repressed genes are transcription factors controlling development of tissues derived from all three germ layers. Thus, a stem cell remains undifferentiated while expression of *Oct4*, *Nanog*, and *Sox2* as well as many other transcription factors and chromatin remodeling proteins remain high. The elevated expression of these genes results in the strict repression of developmental regulators. Once perturbed by the signal to differentiate, these factors are down-regulated, gradually releasing their repression and allowing the expression of key developmental regulators (Fig. 4).

5 Conclusions

Drafting of a comprehensive transcriptional network controlling ESC fate will require systematic characterisation of the functional targets of each ESC-expressed transcription factor by integration of promoter occupancy data and gene expression changes following targeted perturbations of these transcription factors. Considering that, despite the wealth of data already generated, a complete evaluation of the yeast transcriptional network is still not complete, this undertaking in ESCs in monumental. Fortunately, the emergence of improved technologies such as high-throughput sequencing and whole-genome tiling arrays will allow researchers to capture the high-resolution, genome-wide information needed to build accurate network connections. The structure of the emerging network and the identification of relevant network motifs will help identify the transcriptional mechanisms by which ESCs maintain pluripotency and control fate decisions. Understanding the transcriptional control of pluripotency will enable deliberate and intelligent manipulation of ESCs for therapeutic uses.

References

1. Evans MJ, Kaufman MH. Establishment in culture of pluripotential cells from mouse embryos. Nature. 1981;292(5819):154–6.
2. Martin GR. Isolation of a pluripotent cell line from early mouse embryos cultured in medium conditioned by teratocarcinoma stem cells. Proc Natl Acad Sci U S A. 1981;78(12):7634–8.
3. Smith AG, Heath JK, Donaldson DD, et al. Inhibition of pluripotential embryonic stem cell differentiation by purified polypeptides. Nature. 1988;336(6200):688–90.
4. Smith AG, Hooper ML. Buffalo rat liver cells produce a diffusible activity which inhibits the differentiation of murine embryonal carcinoma and embryonic stem cells. Dev Biol. 1987;121(1):1–9.
5. Williams RL, Hilton DJ, Pease S, et al. Myeloid leukaemia inhibitory factor maintains the developmental potential of embryonic stem cells. Nature. 1988;336(6200):684–7.
6. Nichols J, Evans EP, Smith AG. Establishment of germ-line-competent embryonic stem (ES) cells using differentiation inhibiting activity. Development. 1990;110(4):1341–8.
7. Takeda K, Noguchi K, Shi W, et al. Targeted disruption of the mouse Stat3 gene leads to early embryonic lethality. Proc Natl Acad Sci U S A. 1997;94(8):3801–4.
8. Ying QL, Nichols J, Chambers I, Smith A. BMP induction of Id proteins suppresses differentiation and sustains embryonic stem cell self-renewal in collaboration with STAT3. Cell. 2003;115(3):281–92.
9. Daheron L, Opitz SL, Zaehres H, et al. LIF/STAT3 signaling fails to maintain self-renewal of human embryonic stem cells. Stem Cells. 2004;22(5):770–8.
10. Xu RH, Chen X, Li DS, et al. BMP4 initiates human embryonic stem cell differentiation to trophoblast. Nat Biotechnol. 2002;20(12):1261–4.
11. Vallier L, Alexander M, Pedersen RA. Activin/Nodal and FGF pathways cooperate to maintain pluripotency of human embryonic stem cells. J Cell Sci. 2005;118(Pt 19):4495–509.
12. Xu RH, Peck RM, Li DS, Feng X, Ludwig T, Thomson JA. Basic FGF and suppression of BMP signaling sustain undifferentiated proliferation of human ES cells. Nat Methods. 2005;2(3):185–90.
13. Bendall SC, Stewart MH, Menendez P, et al. IGF and FGF cooperatively establish the regulatory stem cell niche of pluripotent human cells in vitro. Nature. 2007;448(7157):1015–21.
14. Tesar PJ, Chenoweth JG, Brook FA, et al. New cell lines from mouse epiblast share defining features with human embryonic stem cells. Nature. 2007;448(7150):196–9.
15. Brons IG, Smithers LE, Trotter MW, et al. Derivation of pluripotent epiblast stem cells from mammalian embryos. Nature. 2007;448(7150):191–5.
16. Niwa H, Miyazaki J, Smith AG. Quantitative expression of Oct-3/4 defines differentiation, dedifferentiation or self-renewal of ES cells. Nat Genet. 2000;24(4):372–6.
17. Mitsui K, Tokuzawa Y, Itoh H, et al. The homeoprotein Nanog is required for maintenance of pluripotency in mouse epiblast and ES cells. Cell. 2003;113(5):631–42.
18. Chambers I, Colby D, Robertson M, et al. Functional expression cloning of Nanog, a pluripotency sustaining factor in embryonic stem cells. Cell. 2003;113(5):643–55.
19. Avilion AA, Nicolis SK, Pevny LH, Perez L, Vivian N, Lovell-Badge R. Multipotent cell lineages in early mouse development depend on SOX2 function. Genes Dev. 2003;17(1):126–40.

20. Beddington RS, Robertson EJ. An assessment of the developmental potential of embryonic stem cells in the midgestation mouse embryo. Development. 1989;105(4):733–7.
21. Brivanlou AH, Darnell JE, Jr. Signal transduction and the control of gene expression. Science. 2002;295(5556):813–8.
22. Shen-Orr SS, Milo R, Mangan S, Alon U. Network motifs in the transcriptional regulation network of Escherichia coli. Nat Genet. 2002;31(1):64–8.
23. Guelzim N, Bottani S, Bourgine P, Kepes F. Topological and causal structure of the yeast transcriptional regulatory network. Nat Genet. 2002;31(1):60–3.
24. Saddic LA, Huvermann B, Bezhani S, et al. The LEAFY target LMI1 is a meristem identity regulator and acts together with LEAFY to regulate expression of CAULIFLOWER. Development. 2006;133(9):1673–82.
25. Swiers G, Patient R, Loose M. Genetic regulatory networks programming hematopoietic stem cells and erythroid lineage specification. Dev Biol. 2006;294(2):525–40.
26. Odom DT, Zizlsperger N, Gordon DB, et al. Control of pancreas and liver gene expression by HNF transcription factors. Science. 2004;303(5662):1378–81.
27. Alon U. Network motifs: theory and experimental approaches. Nat Rev. 2007;8(6):450–61.
28. Lee TI, Rinaldi NJ, Robert F, et al. Transcriptional regulatory networks in Saccharomyces cerevisiae. Science. 2002;298 (5594):799–804.
29. Thieffry D, Huerta AM, Perez-Rueda E, Collado-Vides J. From specific gene regulation to genomic networks: a global analysis of transcriptional regulation in Escherichia coli. Bioessays. 1998;20(5):433–40.
30. Becskei A, Serrano L. Engineering stability in gene networks by autoregulation. Nature. 2000;405(6786):590–3.
31. Dublanche Y, Michalodimitrakis K, Kummerer N, Foglierini M, Serrano L. Noise in transcription negative feedback loops: simulation and experimental analysis. Mol Syst Biol. 2006;2:41.
32. Rosenfeld N, Elowitz MB, Alon U. Negative autoregulation speeds the response times of transcription networks. J Mol Biol. 2002;323(5):785–93.
33. Maeda YT, Sano M. Regulatory dynamics of synthetic gene networks with positive feedback. J Mol Biol. 2006;359(4): 1107–24.
34. Becskei A, Seraphin B, Serrano L. Positive feedback in eukaryotic gene networks: cell differentiation by graded to binary response conversion. EMBO J. 2001;20(10):2528–35.
35. Xiong W, Ferrell JE, Jr. A positive-feedback-based bistable 'memory module' that governs a cell fate decision. Nature. 2003;426(6965):460–5.
36. Wall ME, Dunlop MJ, Hlavacek WS. Multiple functions of a feed-forward-loop gene circuit. J Mol Biol. 2005;349(3):501–14.
37. Mangan S, Zaslaver A, Alon U. The coherent feedforward loop serves as a sign-sensitive delay element in transcription networks. J Mol Biol. 2003;334(2):197–204.
38. Hayot F, Jayaprakash C. A feedforward loop motif in transcriptional regulation: induction and repression. J Theor Biol. 2005;234(1):133–43.
39. Mangan S, Itzkovitz S, Zaslaver A, Alon U. The incoherent feed-forward loop accelerates the response-time of the gal system of Escherichia coli. J Mol Biol. 2006;356(5):1073–81.
40. Davidson EH, Rast JP, Oliveri P, et al. A genomic regulatory network for development. Science. 2002;295(5560):1669–78.
41. Pesce M, Scholer HR. Oct-4: gatekeeper in the beginnings of mammalian development. Stem Cells. 2001;19(4):271–8.
42. Nichols J, Zevnik B, Anastassiadis K, et al. Formation of pluripotent stem cells in the mammalian embryo depends on the POU transcription factor Oct4. Cell. 1998;95(3):379–91.
43. Morrison GM, Brickman JM. Conserved roles for Oct4 homologues in maintaining multipotency during early vertebrate development. Development. 2006;133(10):2011–22.
44. Yuan H, Corbi N, Basilico C, Dailey L. Developmental-specific activity of the FGF-4 enhancer requires the synergistic action of Sox2 and Oct-3. Gene Dev. 1995;9(21):2635–45.
45. Ambrosetti DC, Basilico C, Dailey L. Synergistic activation of the fibroblast growth factor 4 enhancer by Sox2 and Oct-3 depends on protein-protein interactions facilitated by a specific spatial arrangement of factor binding sites. Mol Cell Biol. 1997;17(11): 6321–9.
46. Okumura-Nakanishi S, Saito M, Niwa H, Ishikawa F. Oct-3/4 and Sox2 regulate Oct-3/4 gene in embryonic stem cells. J Biol Chem. 2005;280(7):5307–17.
47. Tomioka M, Nishimoto M, Miyagi S, et al. Identification of Sox-2 regulatory region which is under the control of Oct-3/4-Sox-2 complex. Nucleic Acids Res. 2002;30(14):3202–13.
48. Chew JL, Loh YH, Zhang W, et al. Reciprocal transcriptional regulation of Pou5f1 and Sox2 via the Oct4/Sox2 complex in embryonic stem cells. Mol Cell Biol. 2005;25(14):6031–46.
49. Masui S, Nakatake Y, Toyooka Y, et al. Pluripotency governed by Sox2 via regulation of Oct3/4 expression in mouse embryonic stem cells. Nature Cell Biol. 2007;9(6):625–35.
50. Rodda DJ, Chew JL, Lim LH, et al. Transcriptional regulation of nanog by OCT4 and SOX2. J Biol Chem. 2005;280(26): 24731–7.
51. Kuroda T, Tada M, Kubota H, et al. Octamer and Sox elements are required for transcriptional cis regulation of Nanog gene expression. Molecular Cell Biol. 2005;25(6):2475–85.
52. Loh YH, Wu Q, Chew JL, et al. The Oct4 and Nanog transcription network regulates pluripotency in mouse embryonic stem cells. Nat Genet. 2006;38(4):431–40.
53. Wu Q, Chen X, Zhang J, et al. Sall4 interacts with Nanog and co-occupies Nanog genomic sites in embryonic stem cells. J Biol Chem. 2006;281(34):24090–4.
54. Sakaki-Yumoto M, Kobayashi C, Sato A, et al. The murine homolog of SALL4, a causative gene in Okihiro syndrome, is essential for embryonic stem cell proliferation, and cooperates with Sall1 in anorectal, heart, brain and kidney development. Development. 2006;133(15):3005–13.
55. Zhang J, Tam WL, Tong GQ, et al. Sall4 modulates embryonic stem cell pluripotency and early embryonic development by the transcriptional regulation of Pou5f1. Nat Cell Biol. 2006;8(10): 1114–23.
56. Hamatani T, Carter MG, Sharov AA, Ko MS. Dynamics of global gene expression changes during mouse preimplantation development. Dev Cell. 2004;6(1):117–31.
57. Zhang W, Walker E, Tamplin OJ, Rossant J, Stanford WL, Hughes TR. Zfp206 regulates ES cell gene expression and differentiation. Nucleic Acids Res. 2006;34(17):4780–90.
58. Wang ZX, Kueh JL, Teh CH, et al. Zfp206 is a transcription factor that controls pluripotency of embryonic stem cells. Stem Cells. 2007;25(9):2173–82.
59. Wang ZX, Teh CH, Kueh JL, Lufkin T, Robson P, Stanton LW. Oct4 and Sox2 directly regulate expression of another pluripotency transcription factor, Zfp206, in embryonic stem cells. J Biol Chem. 2007;282(17):12822–30.
60. Boyer LA, Lee TI, Cole MF, et al. Core transcriptional regulatory circuitry in human embryonic stem cells. Cell. 2005;122(6): 947–56.
61. Lim LS, Loh YH, Zhang W, et al. Zic3 is required for maintenance of pluripotency in embryonic stem cells. Mol Biol Cell. 2007;18(4):1348–58.
62. Herman GE, El-Hodiri HM. The role of ZIC3 in vertebrate development. Cytogenet Genome Res. 2002;99(1–4):229–35.

63. Nakata K, Nagai T, Aruga J, Mikoshiba K. Xenopus Zic family and its role in neural and neural crest development. Mech Dev. 1998;75(1–2):43–51.
64. Warner SJ, Hutson MR, Oh SH, Gerlach-Bank LM, Lomax MI, Barald KF. Expression of ZIC genes in the development of the chick inner ear and nervous system. Dev Dyn. 2003;226(4): 702–12.
65. Grinblat Y, Sive H. zic Gene expression marks anteroposterior pattern in the presumptive neurectoderm of the zebrafish gastrula. Dev Dyn. 2001;222(4):688–93.
66. Kitaguchi T, Mizugishi K, Hatayama M, Aruga J, Mikoshiba K. Xenopus Brachyury regulates mesodermal expression of Zic3, a gene controlling left-right asymmetry. Dev. Growth Differ. 2002;44(1):55–61.
67. Ware SM, Peng J, Zhu L, et al. Identification and functional analysis of ZIC3 mutations in heterotaxy and related congenital heart defects. Am J Hum Genet. 2004;74(1):93–105.
68. Purandare SM, Ware SM, Kwan KM, et al. A complex syndrome of left-right axis, central nervous system and axial skeleton defects in Zic3 mutant mice. Development. 2002;129(9):2293–302.
69. Takahashi K, Yamanaka S. Induction of pluripotent stem cells from mouse embryonic and adult fibroblast cultures by defined factors. Cell. 2006;126(4):663–76.
70. Meissner A, Wernig M, Jaenisch R. Direct reprogramming of genetically unmodified fibroblasts into pluripotent stem cells. Nat Biotechnol. 2007;25(10):1177–81.
71. Wernig M, Meissner A, Foreman R, et al. In vitro reprogramming of fibroblasts into a pluripotent ES-cell-like state. Nature. 2007;448(7151):318–24.
72. Hooker CW, Hurlin PJ. Of Myc and Mnt. J Cell Sci. 2006;119(Pt 2):208–16.
73. Cartwright P, McLean C, Sheppard A, Rivett D, Jones K, Dalton S. LIF/STAT3 controls ES cell self-renewal and pluripotency by a Myc-dependent mechanism. Development. 2005;132(5):885–96.
74. McMahon SB, Van Buskirk HA, Dugan KA, Copeland TD, Cole MD. The novel ATM-related protein TRRAP is an essential cofactor for the c-Myc and E2F oncoproteins. Cell. 1998;94(3):363–74.
75. Rowland BD, Peeper DS. KLF4, p21 and context-dependent opposing forces in cancer. Nat Rev Cancer. 2006;6(1):11–23.
76. Chen X, Whitney EM, Gao SY, Yang VW. Transcriptional profiling of Kruppel-like factor 4 reveals a function in cell cycle regulation and epithelial differentiation. J Mol Biol. 2003;326(3): 665–77.
77. Rowland BD, Bernards R, Peeper DS. The KLF4 tumour suppressor is a transcriptional repressor of p53 that acts as a context-dependent oncogene. Nat Cell Biol. 2005;7(11):1074–82.
78. Lin T, Chao C, Saito S, et al. p53 induces differentiation of mouse embryonic stem cells by suppressing Nanog expression. Nat Cell Biol. 2005;7(2):165–71.
79. Nakatake Y, Fukui N, Iwamatsu Y, et al. Klf4 cooperates with Oct3/4 and Sox2 to activate the Lefty1 core promoter in embryonic stem cells. Mol Cell Biol. 2006;26(20):7772–82.
80. Okita K, Ichisaka T, Yamanaka S. Generation of germline-competent induced pluripotent stem cells. Nature. 2007;448(7151): 313–7.
81. Beck F, Erler T, Russell A, James R. Expression of Cdx-2 in the mouse embryo and placenta: possible role in patterning of the extra-embryonic membranes. Dev Dyn. 1995;204(3):219–27.
82. Chawengsaksophak K, James R, Hammond VE, Kontgen F, Beck F. Homeosis and intestinal tumours in Cdx2 mutant mice. Nature. 1997;386(6620):84–7.
83. Strumpf D, Mao CA, Yamanaka Y, et al. Cdx2 is required for correct cell fate specification and differentiation of trophectoderm in the mouse blastocyst. Development. 2005;132(9):2093–102.
84. Chawengsaksophak K, de Graaff W, Rossant J, Deschamps J, Beck F. Cdx2 is essential for axial elongation in mouse development. Proc Natl Acad Sci U S A. 2004;101(20):7641–5.
85. Tanaka S, Kunath T, Hadjantonakis AK, Nagy A, Rossant J. Promotion of trophoblast stem cell proliferation by FGF4. Science. 1998;282(5396):2072–5.
86. Niwa H, Toyooka Y, Shimosato D, et al. Interaction between Oct3/4 and Cdx2 determines trophectoderm differentiation. Cell. 2005;123(5):917–29.
87. Morrisey EE, Ip HS, Lu MM, Parmacek MS. GATA-6: a zinc finger transcription factor that is expressed in multiple cell lineages derived from lateral mesoderm. Dev Biol. 1996;177(1):309–22.
88. Morrisey EE, Tang Z, Sigrist K, et al. GATA6 regulates HNF4 and is required for differentiation of visceral endoderm in the mouse embryo. Genes Dev. 1998;12(22):3579–90.
89. Fujikura J, Yamato E, Yonemura S, et al. Differentiation of embryonic stem cells is induced by GATA factors. Genes Dev. 2002;16(7):784–9.
90. Niwa H. How is pluripotency determined and maintained? Development. 2007;134(4):635–46.
91. Walker E, Ohishi M, Davey RE, et al. Prediction and testing of novel transcriptional networks regulating embryonic stem cell self-renewal and commitment. Cell Stem Cell. 2007;1(1):16.
92. Ivanova N, Dobrin R, Lu R, et al. Dissecting self-renewal in stem cells with RNA interference. Nature. 2006;442(7102):533–8.
93. Matoba R, Niwa H, Masui S, et al. Dissecting Oct3/4-regulated gene networks in embryonic stem cells by expression profiling. PLoS ONE. 2006;1:e26.
94. Ramalho-Santos M, Yoon S, Matsuzaki Y, Mulligan RC, Melton DA. 'Stemness': transcriptional profiling of embryonic and adult stem cells. Science. 2002;298(5593):597–600.
95. Ivanova NB, Dimos JT, Schaniel C, Hackney JA, Moore KA, Lemischka IR. A stem cell molecular signature. Science. 2002;298(5593):601–4.
96. Brandenberger R, Wei H, Zhang S, et al. Transcriptome characterization elucidates signaling networks that control human ES cell growth and differentiation. Nat Biotechnol. 2004;22(6): 707–16.
97. Wei CL, Miura T, Robson P, et al. Transcriptome profiling of human and murine ESCs identifies divergent paths required to maintain the stem cell state. Stem Cells. 2005;23(2):166–85.
98. Zaehres H, Lensch MW, Daheron L, Stewart SA, Itskovitz-Eldor J, Daley GQ. High-efficiency RNA interference in human embryonic stem cells. Stem Cells. 2005;23(3):299–305.
99. Hough SR, Clements I, Welch PJ, Wiederholt KA. Differentiation of mouse embryonic stem cells after RNA interference-mediated silencing of OCT4 and Nanog. Stem Cells. 2006;24(6):1467–75.
100. Babaie Y, Herwig R, Greber B, et al. Analysis of Oct4-dependent transcriptional networks regulating self-renewal and pluripotency in human embryonic stem cells. Stem Cells. 2007;25(2): 500–10.
101. Bar-Joseph Z, Gerber GK, Lee TI, et al. Computational discovery of gene modules and regulatory networks. Nat Biotechnol. 2003;21(11):1337–42.
102. Gao F, Foat BC, Bussemaker HJ. Defining transcriptional networks through integrative modeling of mRNA expression and transcription factor binding data. BMC Bioinformatics. 2004;5:31.

Use of Zebrafish to Dissect Gene Programs Regulating Hematopoietic Stem Cells

Colleen E. Albacker and Leonard I. Zon

Abstract Hematopoietic stem cells (HSCs) are responsible for creating each cellular component of the vertebrate blood system. However, HSCs must also self-renew to maintain their own population so that the blood system always has the capacity to be reconstituted. This balance of HSCs producing more HSCs as well as differentiated blood cells is regulated by several extrinsic pathways. The *Cdx/Hox* pathway has been shown to have a role in regulating embryonic stem cells, with the zebrafish *caudal* genes *cdx4* and *cdx1a* acting upstream of the *Hox* clusters to aid in the formation of blood. The *Notch* pathway has been shown in mice and zebrafish to be a positive regulator of HSC self-renewal; experiments in fish revealed that this pathway signals through the HSC transcription factor *runx1* to do so. Lastly, the action of prostaglandin E2 was found to positively regulate HSCs, and treatment with this compound leads to increased recovery kinetics of the hematopoietic system upon acute injury or transplantation. This knowledge is currently being applied in the clinic as a means to increase the success rates of umbilical cord blood transplantations in adult patients.

Keywords Zebrafish · Hematopoietic stem cells · Caudal · Hox · Notch · Prostaglandin

1 Introduction to Hematopoiesis and Zebrafish

All vertebrates require blood flowing through their circulatory systems that carries oxygen and other nutrients to cells of all tissues of the body and that provides defense against foreign pathogens. Hematopoiesis, or the differentiation of the mature blood lineages from immature progenitors, occurs throughout the lifetime of an organism to regularly replenish the different cellular components of the blood. It is ultimately supported by the hematopoietic stem cell (HSC). The HSC sits atop the entire hematopoietic hierarchy as it has the ability to differentiate into each member of every blood lineage; it also has the capacity to self-renew, maintaining a population of undifferentiated, immature HSCs.

Vertebrate hematopoiesis occurs in the following two successive waves: primitive, to support the developing embryo, and definitive, to provide the organism with longterm HSCs to last its entire lifetime. The primitive and definitive waves produce different cell types from different locations, with the primitive typically limited to nucleated erythrocytes containing embryonic globins and macrophages for nonspecific defense. The definitive wave produces long term HSCs (LT-HSCs), which give rise to a multipotent progenitor; this produces a common lymphoid progenitor, which will become T lymphocytes, B lymphocytes, dendritic cells, and natural killer cells, and a common myeloid progenitor, which will become monocytes, neutrophils, eosinophils, basophils, thrombocytes, and erythrocytes [1, 2].

Danio rerio, commonly known as the zebrafish, is a member of the teleost family, bony fish with rayed fins. Technically easy to work with, the zebrafish is an excellent vertebrate model organism for studying hematopoiesis. Zebrafish eggs are fertilized externally, development occurs entirely ex vivo, and embryos are optically clear. Additionally, zebrafish perform well as a genetic model organism; they are competent for large-scale genetic screens since they require less space than mammalian systems, females can produce 100–200 eggs every week, and their generation time is short. The zebrafish genome is fully sequenced, and a series of large-scale mutagenesis screens have been completed, leading to the production of numerous genetic mutants that are available for study [3, 4].

Zebrafish development occurs rapidly over the course of a few days. Immediately after fertilization, the zygote remains in the one cell stage for 45 minutes, facilitating genetic manipulations of the embryo as nucleic acid injections can be performed at this stage. Gastrulation commences

L.I. Zon (✉)
Division of Hematology/Oncology, Harvard Medical School and Howard Hughes Medical Institute, Children's Hospital Boston, One Blackfan Circle, Boston, MA 02115
e-mail: zon@enders.tch.harvard.edu

V.K. Rajasekhar, M.C. Vemuri (eds.), *Regulatory Networks in Stem Cells*, Stem Cell Biology and Regenerative Medicine,
DOI 10.1007/978-1-60327-227-8_9, © Humana Press, a part of Springer Science+Business Media, LLC 2009

around 5 hours post fertilization (hpf), as morphogenetic movements form the three primary germ layers. Somites and organs begin to form at 10 hpf. Shortly after 24 hpf, circulation and pigmentation commence; by 48 hpf the zebrafish begin hatching out of their clear chorions and by 72 hpf are swimming freely [5].

2 Stages of Hematopoiesis

2.1 Primitive Hematopoiesis

In vertebrates, the first wave of hematopoiesis, the primitive wave, produces erythrocytes and myeloid cells for the developing embryo. In the zebrafish, the primitive erythropoiesis wave occurs within the intermediate cell mass (ICM); this hematopoietic site is functionally equivalent to the extraembryonic yolk sac blood islands in mammals and birds. The ICM is located in the trunk ventral to the notochord. Following gastrulation, the developing primitive HSCs are marked by expression of the genes *stem cell leukemia (scl)* and *LIM domain only 2 (lmo2)*, first seen by the 2 somite stage (10 hpf). They exist in two bilateral stripes in the embryo until merging into the ICM at the 18 somite stage (18 hpf). Around the 6 somite stage (12 hpf), expression of *gata1*, an erythrocyte marker, commences in a subset of *scl*-positive cells in the ICM as the first primitive red blood cells are produced (Fig. 1A). About 300 primitive erythrocytes will be produced by this hematopoietic wave. These cells are first seen in circulation by about 24 hpf and last until about 4 days post fertilization (dpf). The rest of the *scl*-positive cells, those that do not express *gata1*, will differentiate into vasculature, marked by expression of the *vascular endothelial growth factor receptor 2* (*vegfr2/flk1*) [1, 6, 7].

Primitive myelopoiesis occurs concomitantly with primitive erythropoiesis. However, whereas primitive erythropoiesis takes place in the ICM, primitive myelopoiesis occurs instead in the rostral blood islands (RBI) in the anterior mesoderm. The RBI are also marked by expression of *scl* and *lmo2*. Similar to how a proportion of cells in the ICM will express *gata1,* a subset of these cells in the RBI will additionally express *pu.1*, a myeloid-specific transcription factor. These cells will develop into primitive macrophages and granulocytes, which are functional and circulating in the bloodstream by 26 hpf [1].

The appropriate proportion of erythroid to myeloid cells is maintained during the primitive wave by an equilibrium between expression of *gata1* and *pu.1* in the ICM. ICM cells have a bipotentiality, being able to form either cell lineage, so this regulation plays an important role in the embryo. Loss of *gata1* results in an expansion of myeloid precursors at the expense of erythroid cells, with *pu.1* expression persisting longer than usual; conversely, loss of *pu.1* leads to ectopic *gata1* expression and an expansion of erythroid cells [8, 9]. This reciprocal regulation suggests that each marker promotes development of one lineage while actively suppressing the other.

2.2 Definitive Hematopoiesis

The definitive wave of hematopoiesis provides an animal with LT-HSCs, which, due to their ability to undergo self-renewal, will support the generation of all blood lineages for the life of the organism. These differentiated lineages include not only erythroid and myeloid cells like the primitive wave, but also lymphocytes, thrombocytes, and a larger variety of myeloid cells. In mammals, the first anatomic site

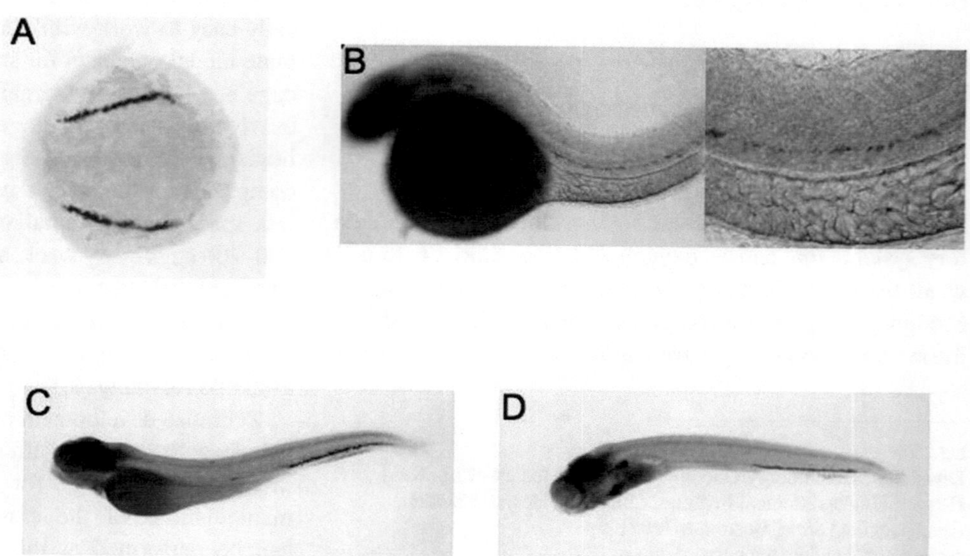

Fig. 1 In situ hybridizations of blood markers during the various stages of zebrafish hematopoiesis. (**A**) *Gata1*, an erythrocyte marker, at the 5 somite stage (12 hpf). The two bilateral stripes form the ICM and will merge together by 18 somites (18 hpf) (X. Bai). (**B**) *C-myb* and *runx1*, markers of definitive stem cells, at 36 hpf in the ventral wall of the dorsal arota (T. North). (**C**) *Scl*, an HSC and progenitor marker, is seen expressed in the CHT at 4 dpf (T. Bowman). (**D**) *Scl* at 6.5 dpf; additional expression near the head is the kidney marrow (T. Bowman)

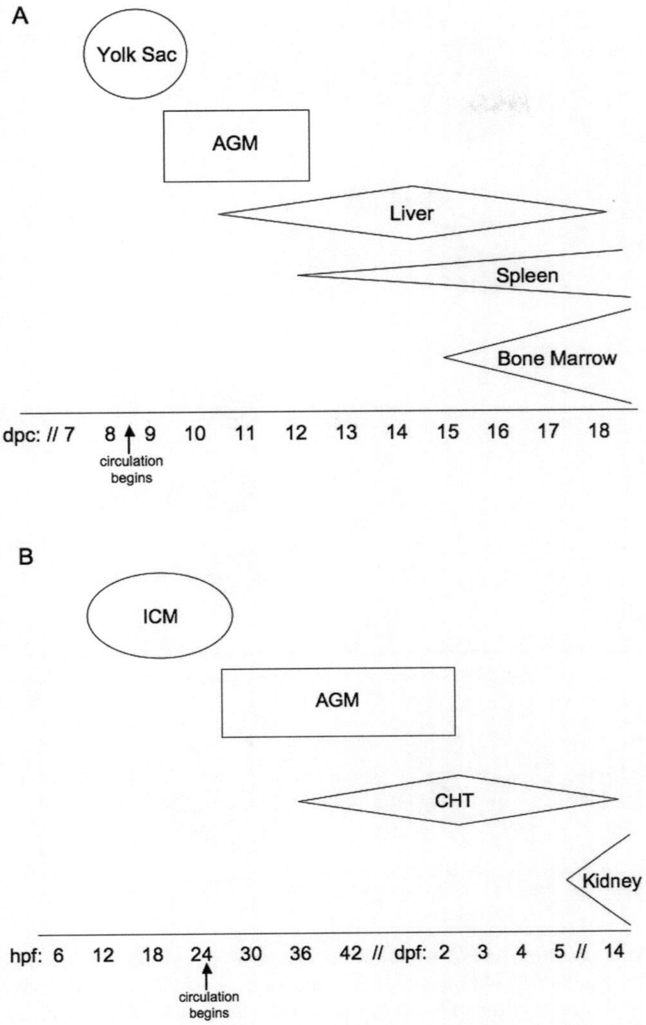

Fig. 2 Approximate temporal and anatomical locations of hematopoietic activity, based on current knowledge (adapted from [7]). (**A**) Mouse. (**B**) Zebrafish

of definitive hematopoiesis is the aorta-gonad-mesonephros region (AGM; Fig. 2). The AGM equivalent in zebrafish is the ventral wall of the dorsal aorta. It is here at 31 hpf that expression of the first definitive HSC markers, the transcription factors *c-myb* and *runx1*, arises. In situ hybridizations at 36 hpf for these two genes reveal clusters of HSCs on the floor of the dorsal aorta (Fig. 1B). They then likely bud off from the epithelium to enter circulation and migrate to their next location [1, 6, 7]. Additionally, in mammals, a parallel population of definitive HSCs can be found in the placenta [10].

In mammalian systems, HSCs are next found in the fetal liver; the equivalent intermediate larval site of hematopoietic expansion in the zebrafish is the caudal hematopoietic tissue (CHT; Figs. 1C, 2). Lineage tracing of HSCs forming in the vascular plexus below the floor of the dorsal aorta shows that the cells formed by 2 dpf will seed the CHT, later becoming myeloid cells and lymphoid precursors. The CHT develops in association with the caudal vein, consisting of endothelial stroma that harbors the hematopoietic cells [11].

Finally, in mammals, HSCs seed the bone marrow, their permanent location. In zebrafish, the permanent niche for HSCs is the kidney marrow, in the reticular stromal cells of the kidney between the renal tubules and blood vessels (Figs. 1D, 2). While most blood cell progenitors are also found in the kidney marrow, T lymphocyte progenitors, after being born in the kidney, can additionally be found in the thymus. HSCs formed at 2 dpf go through the CHT on their way to the kidney marrow and thymus. Those HSCs formed at 3 dpf will seed the kidney with fewer cells heading for the CHT and thymus; these are presumed to be the definitive HSCs [11, 12]. By 5 dpf, nearly all HSCs can be found in the kidney marrow and will inhabit there for the remainder of the zebrafish's life. All mature blood cell lineages originate from the kidney, with only T lymphocytes progenitors then migrating to the bilateral thymi to be educated. Kidney marrow cells can be analyzed by flow cytometry; separation by forward scatter and side scatter, which are measures of cell size and granularity, respectively, reveals four distinct populations that correspond to lymphoid, myeloid, erythroid, and precursor cells [13; plots on Fig. 3].

The definitive wave of erythropoiesis to replace the primitive red blood cells commences around 5 dpf. Erythrocyte differentiation is highly conserved between zebrafish and mammals, with the exception that mature erythrocytes in zebrafish retain their nuclei and have an elliptical shape, as opposed to mammalian erythroid cells that have an enucleated discoid-shape. Myelopoiesis results in two granulocyte lineages, one resembling mammalian neutrophils and the other resembling a combination of mammalian eosinophils and basophils. They are found in the kidney by 7 dpf. The zebrafish immune system is supported by B cells that develop from the kidney by 19 dpf and by T cells that originate in the kidney and mature in the bilateral thymi beginning around 3–4 dpf. Thrombopoiesis in zebrafish yields small round nucleated thrombocytes that are equivalent to mammalian platelets, which appear beginning at 36 hpf [1].

3 Hematopoietic Assays

Hematopoiesis can be studied in vitro through a series of colony-forming unit-culture (CFU-C) assays. Hematopoietic progenitors are grown in culture, consisting of a mixture of media, growth factors, and methylcellulose. When the progenitors proliferate and differentiate, a colony forms, which can be quantified and characterized [14, 15]. Variations in the media conditions promote the growth of specific kinds of colonies with varying potentials.

In the murine system, the classic in vivo assay to determine the presence of HSCs is the colony-forming unit-spleen

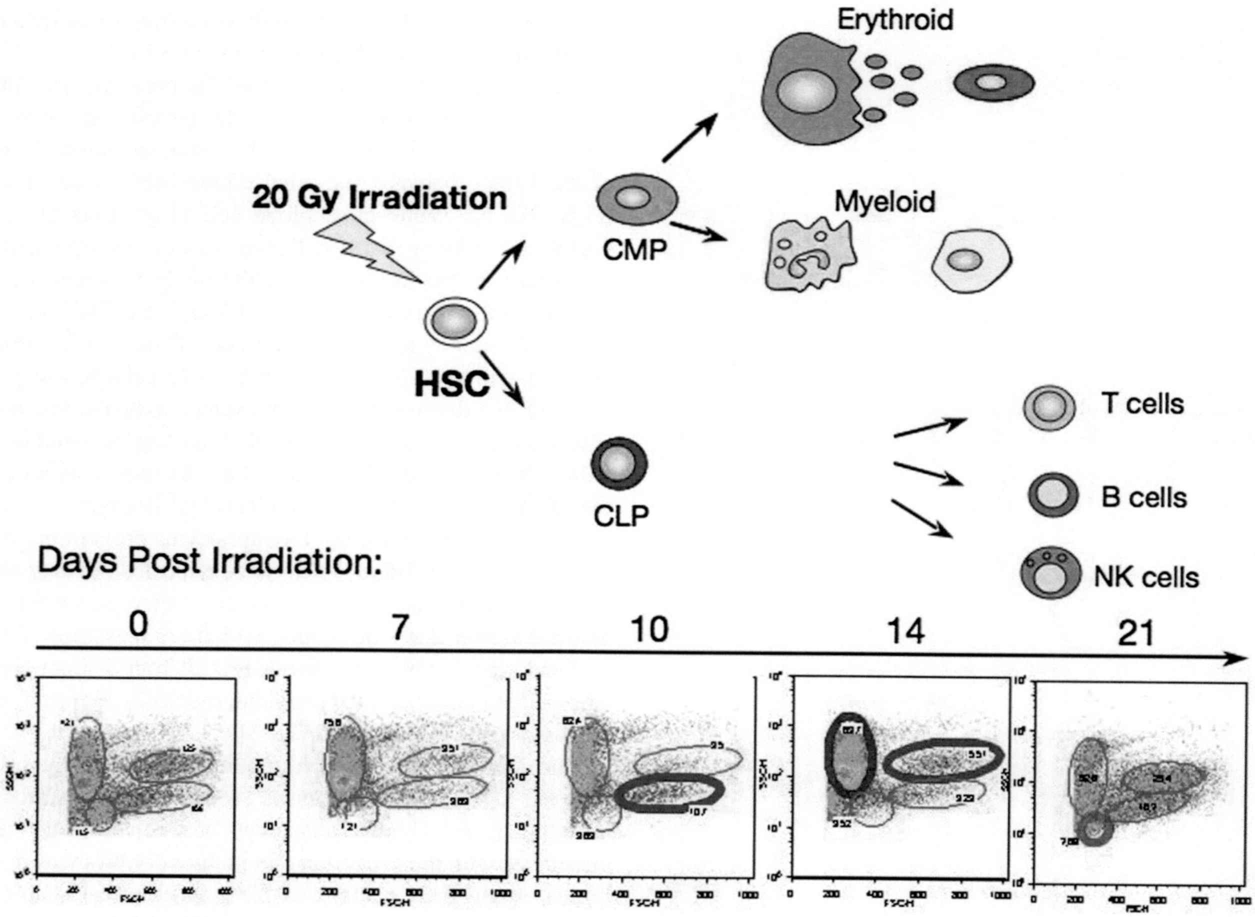

Fig. 3 Schematic of the acute injury recovery assay. By treating zebrafish with a sublethal dose of irradiation, the blood system undergoes injury and recovers over time. By day 7 post-irradiation, the HSCs will begin to produce cells that will differentiate into the blood system. Common lymphoid and myeloid progenitors recover by day 10 (lower left population on FACS plot). More mature erythroid and myeloid cells recover by day 14 (upper left and upper right populations, respectively) with mature lymphocytes returning by day 21 (lower left population). Within a few weeks, levels of all populations are back to and can even surpass pre-irradiation levels (T. Bowman) (*see* also Color Insert)

(CFU-S) assay. After lethal irradiation, progenitor cells from the bone marrow will form nodules of hematopoietic cells on the spleen. Those that appear relatively quickly, by 8 days post-irradiation, likely arise from short-term HSCs or precursor cells; nodules forming later, by 12 or 14 days, are from the long-term HSCs [16].

HSCs can be sorted out in mammalian systems based on the antigens on their cell surface; recognition of these antigens by monoclonal antibodies facilitates the sorting process. In the human, the HSCs are marked by the presence of CD34 and the absence of Lineage markers. In the mouse, the long-term HSCs are enriched in the Lineage-negative, Sca-positive, c-kit-positive population [17].

HSC presence can also be noted by performing an acute injury-recovery experiment, where the hematopoietic system is injured and assessed for recovery. In mice, the injury is traditionally 5-fluorouracil (5-FU) injection; then the recovery kinetics of the whole bone marrow are monitored. A similar assay has been developed for the zebrafish. The fish are sublethally irradiated, then analyzed over a time course by forward scatter-side scatter FACS analysis. In this way, the recovery of the different blood populations, erythroid, myeloid, lymphoid, and precursor, can be monitored over time [13, 18; Fig. 3].

Transplantation is another classic assay to definitively determine the presence of HSCs in a cell population. This can be executed in both mouse and zebrafish [17, 13]. Transplanting donor marrow cells into an irradiated recipient allows for the assessment of long-term donor engraftment, which can only be fulfilled by long-term HSCs. If 5% of the recipient's blood is found to originate from the donor, either by cell surface antigens in the mouse or fluorescent markers in the zebrafish, the marrow cells are considered to have engrafted. The actual percentage of donor cells found in the recipient's blood is known as the percent chimerism. Competitive transplants are performed to test the desired cell population against a secondary population, serving as the challenger. The two cell populations are mixed, and co-transplanted into the same animal; if one engrafts better than the other, this will be noted in the recipient. Serial transplants are performed to assess self-renewal. A population containing HSCs is transplanted into a recipient; at a later time point, marrow cells from this first recipient are then transplanted into another animal, a secondary recipient. A population containing a greater number of HSCs or HSCs enhanced for self-renewal will sustain more rounds of serial transplants; those with fewer HSCs or HSCs impaired in self-renewal will undergo hematopoietic exhaustion, essentially running out of HSCs to repopulate the recipient's blood.

Alongside these traditional assays, recent work has combined pathway analysis, mutants, and transgenic animals to study hematopoietic stem cells. Understanding of the extrinsic pathways regulating HSCs has benefited greatly from this work. In the paragraphs that follow, three of these pathways will be discussed: Cdx/Hox, Notch, and Prostaglandins.

4 Regulatory Pathways for HSC Development

4.1 Cdx/Hox Pathway

Many aspects of embryonic patterning and development, including hematopoiesis, are determined by the homeobox transcription factors, which are encoded by the *Hox* genes. In vertebrates, there exist thirty-eight *Hox* genes clustered into four areas of the genome; these clusters are each oriented in the same transcriptional direction and are known as *Hoxa, Hoxb, Hoxc,* and *Hoxd* [19]. Within these clusters, there is a direct correlation between the order of genes along the chromosome and the gene's function according to the anterior-posterior (AP) axis of the organism. This colinearity refers both to the location a gene is expressed along the AP axis as well as the time it is expressed in development. For example, the genes positioned at the 3′ end of a *Hox* cluster tend to be expressed first and the most anterior; the more 5′ genes in the cluster are expressed at later time points and spatially more posterior in the embryo. Any gene mutations or other perturbations in the *Hox* clusters typically lead to changes in cell fate and embryonic patterning. Loss of function in the most 3′ region of a cluster will cause the embryo to become posteriorized, losing some anterior structures; conversely, gain of function in this genomic region would result in anteriorization of the embryo and loss of certain posterior structures [19].

Vertebrates also have other homeobox genes that are not members of the main four *Hox* clusters. Among these genes include the *caudal* paralogues, *Cdx1, Cdx2,* and *Cdx4*. A proposed role for such genes, being that they are in a different genomic location from the *Hox* clusters, is that they affect AP patterning by directly regulating the *Hox* genes. Promoters of many of the clustered *Hox* genes contain binding sites with consensus sequences for *Cdx1, Cdx2,* and *Cdx4* [20]. Several studies of murine mutants for *Cdx1* and *Cdx2* have shown that these mutations led to shifts in the expression patterns and boundaries of several *Hox* genes, resulting in aberrant vertebral column and tail development [21, 22].

Being that the Cdx/Hox genes play key roles in development, it was not surprising that some of these genes would participate in hematopoietic development. A zebrafish mutant referred to as *kugelig* has a mutation in the *cdx4* gene. It has abnormal AP patterning, and several genes of the *Hoxa* and *Hoxb* clusters display altered expression patterns. The *cdx4–/–* mutant presents with severe anemia by 1 dpf that partially recovers by 5 dpf; pleiotropic defects lead to the mutant's early death by 7–10 dpf. *Kugelig* also has reduced expression of hemoglobin, *scl*, and *gata1* with a complete lack of *runx1*. Overexpressing *hoxb7a* or *hoxa9a* in the *cdx4–/–* mutant rescues levels of *gata1* expression. However, overexpressing *scl* in the *cdx4–/–* mutant was unable to rescue these erythroid precursors, suggesting that *scl*'s role in determining hematopoietic cell fate is *cdx4*-dependent and further downstream. Overexpression analysis of *cdx4* was also performed, which showed a posteriorization of the zebrafish embryo and ectopic expression of *scl* and *gata1* [23]. These observations showed that *cdx4* is upstream of *Hox* gene expression.

A follow-up study was performed involving one of the other *caudal* genes, *cdx1a*, since it was noted that the *cdx4–/–* mutant was not completely bloodless, likely due to redundancy. Knocking down *cdx1a* by injection of an antisense morpholino oligonucleotide led to an increase in expression of the gene, implying that it normally negatively regulates itself. Injection of this same *cdx1a* morpholino into the *cdx4–/–* mutant zebrafish embryos yielded loss of posterior structures and development of fewer somites; it appears that the two genes normally act simultaneously to form the posterior tissues of the animal. *Cdx1a*-morpholino; *cdx4+/–* embryos had fewer *scl*-positive cells in the ICM as well as a reduction in *gata1*-positive cells; these phenotypes were more severe in *cdx1a*-morpholino; *cdx4–/–* embryos, suggesting a dose-dependent effect of *cdx4*. The *cdx1a*-morpholino; *cdx4–/–* embryos also suffered from a complete failure to specify the ICM blood precursors; these also lacked in forming definitive HSCs, as noted by a reduction in *runx1a* expression. However, these effects could be rescued by co-injection of *hoxa9a* mRNA [24]. The *cdx* pathway plays a major role in primitive hematopoiesis in the ICM, as this region was particularly sensitive to the gene dosage of *cdx4*; the *cdx-hox* pathway is also required to form definitive HSCs in the AGM. These studies confirmed the idea that the *caudal* genes act upstream of and regulate the *Hox* clusters, playing an important role in blood formation.

Other studies of *Cdx4* have confirmed the pathway's role among HSCs. In the mouse, *Cdx4* is expressed in adult stem and progenitor population yet is significantly reduced in differentiated blood cells [25]. Induction of *Cdx4* in embryoid bodies promoted formation of CD41-positive and c-kit-positive presumptive HSCs in a colony-forming assay. Additionally, an increase in blood markers was observed upon *Cdx4* induction, including *lmo2*, *scl*, *gata1*, embryonic and adult globins, *c-myb*, and *runx1* expression. Similar to *cdx* regulation in zebrafish, *Hox* genes were found to be the downstream mediators of *cdx4* in vitro [26, 27]. These observations show that the *Cdx* genes play a role in regulating blood stem and progenitor stems through the *Hox* genes.

4.2 Notch Signaling

The Notch pathway is an evolutionarily conserved signaling pathway that is known for playing important roles in development, and it has been shown to indeed participate in the formation of HSCs. The Notch proteins themselves are type-1, single-pass transmembrane glycoprotein receptors that activate downstream transcriptional targets upon binding to their transmembrane ligands, Jagged and Delta. Receptor-ligand interactions occur when a cell expressing the ligand is juxtaposed to a cell expressing the Notch receptor. This interaction results in an external cleavage of Notch, caused by an ADAM metalloprotease, which splits the receptor into the Notch extracellular domain and an activated membrane-bound form of Notch. A second cleavage then occurs within the activated transmembrane domain by presenilin-dependent gamma-secretase; this releases the Notch intracellular domain (NICD), which translocates to the nucleus and participates as a member of various transcriptional activation complexes [28].

Examination of live Notch expression in HSCs was facilitated by the creation of a transgenic Notch reporter mouse. The transgene contained an upstream Notch response element as well as four binding sites for CBF-1, a Notch target gene, driving enhanced green fluorescent protein (GFP). In the transgenic mice, HSCs, designated by the marker c-kit, were also GFP-positive, showing in vivo that Notch signaling is active in HSCs. Analysis of more mature blood cells revealed significantly less GFP expression, suggesting that Notch activation decreases with differentiation. This was confirmed in vitro by comparing the amount of CFU-S12 colonies formed by HSC-enriched GFP-positive and GFP-negative cells; those expressing GFP, and thus having active Notch signaling, tended to form more CFU-S12 colonies and thus had more HSC function [29].

Results of overexpression studies of Notch are consistent with these conclusions. Mice containing an expanded population of osteoblasts in the murine bone marrow, which serve as a stem cell niche, possessed increased numbers of Notch-expressing HSCs. The osteoblast niche presents the appropriate Notch ligands to HSCs, leading to an increase in NICD within HSCs and a concomitant increase in HSC number, presumably reflecting an increase in HSC self-renewal [30]. If primary murine bone marrow cells enriched for stem cells are retrovirally transduced with the NICD, differentiation is inhibited, the cells appear primitive, and more colonies form in the CFU-C in vitro assay, implying there are more HSCs present than in controls. Upon competitive transplantation of the NICD-transduced bone marrow cells into irradiated mice, there are more primitive cells and fewer mature cells present, as the Notch gain of function appears to impede exit from the undifferentiated state of the HSCs. Secondary recipients in serial transplants do not experience the hematopoietic exhaustion that control cells undergo, possibly due to enhanced self-renewal creating a larger HSC pool [31]. Forced expression of NICD in irradiated zebrafish adults enhanced recovery of multi-lineage precursor cells, and thus all the blood lineages; HSCs were likely increased as well, as shown by increased *runx1* and *lmo2* expression [32]. Similarly, retroviral transduction of Hairy-Enhancer of Split (HES-1), a Notch target gene encoding a basic helix-loop-helix transcription factor, cDNA into fetal liver and bone marrow HSCs maintained the cells as immature, undergoing self-renewal. Upon transplantation, the HES-1 transduced HSCs also yielded a higher proportion of immature cells in the recipient mouse as well as a higher donor chimerism [33].

Loss of function studies have also helped elucidate the role of Notch in definitive HSC induction. A zebrafish known as *mindbomb*, harboring a mutation in an E3 ligase required for Delta ligand trafficking, lacks expression of *runx1* and *c-myb* in the AGM; this suggests that Notch signaling is required for the definitive wave of hematopoiesis [32]. To assess the effect of Notch1 on AGM HSC formation, AGM cells from day E9.5 embryonic Notch1–/– mice were removed and cultured in vitro, then assayed for functional implications. The cells were found to have impaired ability to produce functional HSCs in a colony-forming assay or to reconstitute an irradiated mouse. These effects could not be rescued by the addition of signaling factors belonging to other pathways in the environment, suggesting that HSCs have a cell autonomous requirement for Notch [34].

The zebrafish has proved particularly useful in the elucidation of a genetic pathway involved in definitive HSC formation. Crossing two fish, one carrying the yeast transactivator Gal4 driven by the zebrafish heat-shock promoter (*hsp70:gal4*) and the other with the Gal4-responsive *upstream activating sequence* (*uas*) driving the zebrafish NICD (*uas:NICD*), led to a fish harboring heat shock inducible NICD. A pulse of heat shock at 14 hpf led to NICD induction, which caused ectopic *runx1* and *c-myb* expression at 36 hpf. This included an expansion of these definitive HSC markers into the aortic roof and vein. Similar results were found by overexpressing *runx1* by injection of *runx1* mRNA into single cell stage embryos. Conversely, knocking down *runx1* by morpholino injection led to a lack of definitive HSC formation; both this and the *mindbomb* mutant could not be rescued by NICD induction by heat shock or overexpression of *runx1* mRNA. This implied that *runx1* and Notch acted together in a pathway to induce HSC formation. This pathway was teased out by knocking down either *mindbomb* or *runx1* in the heat shock inducible NICD fish. Forced NICD expression showed Notch gain of function effects when *mindbomb* was knocked down, thus showing that Notch is downstream of *mindbomb*; however, with *runx1* knocked down and NICD induced, the fish still had a lack of HSCs, thus displaying that *runx1* is downstream of Notch. These zebrafish experiments enabled the finding that Notch is necessary to establish HSC fate, sufficient to expand HSCs in vivo, and signals through the transcriptional factor *runx1* to perform these effects [32].

4.3 Prostaglandin Signaling

Prostaglandins (PG) are a type of eicosanoid that have a variety of physiological effects at low concentrations. They are derived from 20-carbon unsaturated fatty acids, the most common being arachidonic acid. Arachidonic acid is stored in the cell membrane and is turned into PGH_2 by cyclooxygenases (COX). COX1 is expressed in most tissues constitutively whereas COX2 expression is regulated within specific tissues; aspirin inhibits both COX enzymes. The PGH_2 intermediate then gives rise to a variety of PGs through PG synthases; the most common PG is PGE_2. The PGs bind to G-protein coupled receptors to mediate their actions [35].

A few studies have shown the role of PGE_2 in promoting blood cell proliferation. Adding PGE_2 to a murine bone marrow suspension stimulated cell proliferation at concentrations comparable to those found in human serum. A colony-forming assay was performed and revealed a higher proportion of colonies were in S phase than the controls. These results implied that PGE_2 prompts quiescent HSCs to move into the cell cycle [36]. A similar effect was shown in the zebrafish thymi. Embryos treated with PGE_2 contained increased expression of the lymphocyte-specific *recombination activation gene* (*rag1*) in the thymi whereas those treated with COX inhibitors or antagonist of the PGE_2 receptor lacked *rag1* expression. Additionally, PGE_2 treatment could rescue the COX2 inhibitor phenotype. These effects of PGE_2 were shown to be due to increased cell proliferation as shown by the presence of the proliferating cell nuclear marker (PCNA) that was missing in controls [37].

Recently, a major study on PGE_2 and HSC development was completed in the zebrafish. It was shown that treatment of zebrafish embryos with PGE_2 increased AGM expression of *runx1* and *c-myb*, while exposure to COX1 and COX2 inhibitors gave a decrease in HSC number. Knockdown of COX1/2 by morpholino injection similarly decreased *runx1* and *c-myb* expression; this effect was rescued by addition of dimethylPGE_2 (dmPGE_2), a long-acting form of PGE_2, showing the loss of AGM HSCs was a result of the loss of PG signaling. COX1 is expressed in the endothelial cells of the AGM region, yet both COX1 and COX2 are expressed in CD41-positive cells, likely the HSCs themselves. This implies that PGE_2 could be regulating both the HSC and the niche. In adult zebrafish, exposure to dmPGE_2 enhanced the recovery kinetics of the kidney marrow following irradiation, leading to faster repopulation of the precursor cells [38].

Follow-up studies in mammalian systems confirmed that the regulation of HSCs by PGE_2 was evolutionarily conserved. Addition of dmPGE_2 to murine embryonic stem cells increased the amount of hematopoietic colonies formed in a dose-dependent manner. Similar to zebrafish AGM analysis, treatment with an inhibitor of COX1/2 prevented colony growth, and this was rescued by dmPGE_2 treatment. Murine bone marrow was exposed *ex vivo* to dmPGE_2 and then transplanted into an irradiated recipient. This led to an increase in the number of CFU-S8 and CFU-S12 colonies formed, indicating an increase in HSC and precursor populations, respectively; complementary results were obtained upon treatment with COX inhibitors. Similar results were obtained when these experiments were performed on isolated HSCs. dmPGE_2-treated bone marrow showed an increase in the frequency of repopulating cells in a limiting dilution competitive transplant assay. Since no difference in homing to the bone marrow was observed, these results indicate an increase in HSC number [38]. COX2–/– mice have normal hematopoiesis and HSC formation due to maternal and sibling contribution of the enzyme. However, upon 5-FU injury treatment, these mice recover slowly, have significantly lower numberse of cells in all blood lineages by day 12 post-injury and fail to repopulate the bone marrow with hematopoietic cells [39]. These findings conclude that PGE_2, produced by COX2 in particular, plays a key role in the creation of HSCs.

5 Clinical Applications

Umbilical cord blood transplantation is a common therapy used when an immunological match donor cannot be found for a patient. However, cord blood contains few HSCs and typically takes longer to engraft than bone marrow transplantations. It has been successful for the pediatric population, but expansion of HSCs is necessary to be more useful for adult patients [40]. Understanding the extrinsic pathways that impact HSC development can lead to the development of an ex vivo treatments that increase HSC frequency in cord blood samples in the clinic.

Researchers have made several attempts to take knowledge of the above pathways to expand HSC populations in cord blood. Similar to overexpression of the *Hox* genes in zebrafish and mouse studies, by treating cord blood *ex vivo* with a peptide containing *Hox* sequences, the amount of HSCs and multipotent progenitors doubled; additionally, the cord blood reconstituted the blood system when transplanted into irradiated mice more efficiently and with faster kinetics [41]. Since Notch signaling seems to be required for HSC formation, adding Notch ligands at appropriate concentrations to cord blood has been shown to increase the number of immature precursor cells [42]. Whereas lower concentrations of the ligands result in better repopulation of irradiated mice in comparison to no ex vivo treatment, higher concentrations appear to cause too much apoptosis and hence a decreased popultion [43]. Similar trials are underway for prostaglandin E2, since overexpression has also been shown to increase HSC numbers and functionality in zebrafish and mice [North TE, Zon LI. Personal communication]. This is just one example of how work in model organisms, especially the zebrafish, can be translated into therapies for human disease.

Acknowledgments We thank Teresa Bowman, Narie Storer, and Lee Albacker for helpful advice and critical reading of the manuscript and Teresa Bowman, Xiaoying Bai, and Trista North for their contributions to the figures.

References

1. deJong JLO, Zon LI. Use of the zebrafish system to study primitive and definitive hematopoiesis. Annu Rev Genet. 2005;39:481–501.
2. Zon LI. Developmental biology of hematopoiesis. Blood. 1995;86:2876–91.
3. Driever W, Fishman MC. The zebrafish: heritable disorders in transparent embryos. J Clin Invest. 1996;97:1788–94.
4. Haffter P, Granato M, Brand M, Mullins MC, Hammerschmidt M, Kane DA, Odenthal J, van Eeden FJ, Jiang YJ, Heisenberg CP, Kelsh RN, Furutani-Seiki M, Vogelsang E, Beuchle D, Schach U, Fabian C, Nusslein-Volhard C. The identification of genes with unique and essential functions in the development of the zebrafish, Danio rerio. Development. 1996;123:1–36.
5. Kimmel CB, Ballard WW, Kimmel SR, Ullmann B, Schilling TF. Stages of embryonic development of the zebrafish. Dev Dyn. 1995;203:253–310.
6. Davidson AJ, Zon LI. The definitive (and primitive) guide to zebrafish hematopoiesis. Oncogene. 2004;23:7233–46.
7. Galloway JL, Zon LI. Ontogeny of hematopoiesis: examining the emergence of hematopoietic cells in the vertebrate embryo. Curr Top Dev Biol. 2003;53:139–58.
8. Galloway JL, Wingert RA, Thisse C, Thisse B, Zon LI. Loss of gata1 but not gata2 converts erythropoiesis to myelopoiesis in zebrafish embryos. Dev Cell. 2005;8:109–16.
9. Rhodes J, Hagen A, Hsu K, Deng M, Liu TX, Look AT, Kanki JP. Interplay of pu.1 and gata1 determines myelo-erythroid progenitor cell fate in zebrafish. Dev Cell. 2005;8:97–108.
10. Mikkola HK, Gekas C, Orkin SH, Dieterlen-Lievre F. Placenta as a site for hematopoietic stem cell development. Exp Hematol. 2005;33:1048–54.
11. Murayama E, Kissa K, Zapata A, Mordelet E, Briolat V, Lin HF, Handin RI, Herbomel P. Tracing hematopoietic precursor migration to successive hematopoietic organs during zebrafish development. Immunity. 2006;25:963–75.
12. Burns CE, Zon LI. Homing sweet homing: odyssey of hematopoietic stem cells. Immunity. 2006;25:859–62.
13. Traver D, Paw BH, Poss KD, Penberthy WT, Lin S, Zon LI. Transplantation and in vivo imaging of multilineage engraftment in zebrafish bloodless mutants. Nat Immunol. 2003;4:1238–46.
14. Bradley TR, Metcalf D. The growth of mouse bone marrow cells in vitro. Aust J Exp Biol Med Sci. 1966;44:287–99.
15. Ichikawa Y, Pluznik DH, Sachs L. In vitro control of the development of macrophage and granulocyte colonies. Proc Natl Acad Sci U S A. 1966;56:488–95.
16. Magli MC, Iscove NN, Odartchenko N. Transient nature of early hematopoietic spleen colonies. Nature. 1982;295:527–9.
17. Shizuru JA, Negrin RS, Weissman IL. Hematopoietic stem and progenitor cells: clinical and preclinical regeneration of the hepatolymphoid system. Annu Rev Med. 2005;56:509–38.
18. Traver D, Winzeler A, Stern HM, Mayhall EA, Langenau DM, Kutok JL, Look AT, Zon LI. Effects of lethal irradiation in zebrafish and rescue by hematopoietic cell transplantation. Blood. 2004;104:1298–1305.
19. Krumlauf R. Hox genes in vertebrate development. Cell 1994;78:191–201.
20. Charite J, de Graaff W, Consten D, Reijnen MJ, Korving J, Deschamps J. Transducing positional information to the Hox genes: critical interaction of cdx gene products with position-sensitive regulatory elements. Development. 1998;125:4349–58.
21. Subramanian V, Meyer BI, Gruss P. Disruption of the murine homeobox gene Cdx1 affects axial skeletal identities by altering the mesodermal expression domains of Hox genes. Cell. 1995;83:641–53.
22. van den Akker E, Forlani S, Chawengsaksophak K, de Graaff W, Beck F, Meyer BI, Deschamps J. Cdx1 and Cdx2 have overlapping functions in anteroposterior patterning and posterior axis elongation. Development. 2002;129:2181–93.
23. Davidson AJ, Ernst P, Wang Y, Dekens MP, Kingsley PD, Palis J, Korsmeyer SJ, Daley GQ, Zon LI. cdx4 mutants fail to specify blood progenitors and can be rescued by multiple hox genes. Nature. 2003;425:300–6.
24. Davidson AJ, Zon LI. The caudal-related homeobox genes cdx1a and cdx4 act redundantly to regulate hox gene expression and the formation of putative hematopoietic stem cells during zebrafish embryogenesis. Dev Biol. 2006;292:506–18.
25. Bansal D, Scholl C, Frohling S, McDowell E, Lee BH, Dohner K, Ernst P, Davidson AJ, Daley GQ, Zon LI, Gilliland DG, Huntly BJ. Cdx4 dysregulates Hox gene expression and generates acute myeloid leukemia alone and in cooperation with Meis1a in a murine model. Proc Natl Acad Sci U S A. 2006;103:16924–9.
26. Wang Y, Yates F, Naveiras O, Ernst P, Daley GQ. Embryonic stem cell-derived hematopoietic stem cells. Proc Natl Acad Sci U S A. 2005;102:19081–6.
27. Lengerke C, McKinney-Freeman S, Naveiras O, Yates F, Wang Y, Bansal D, Daley GQ. The cdx-hox pathway in hematopoietic

stem cell formation from embryonic stem cells. Ann N Y Acad Sci. 2007;1106:197–208.
28. Maillard I, Adler SH, Pear WS. Notch and the immune system. Immunity. 2003;19:781–91.
29. Duncan AW, Rattis FM, DiMascio LN, Congdon KL, Pazianos G, Zhao C, Yoon K, Cook JM, Willert K, Gaiano N, Reya T. Integration of Notch and Wnt signaling in hematopoietic stem cell maintenance. Nat Immunol. 2005;6:314–22.
30. Calvi LM, Adams GB, Weibrecht KW, Weber JM, Olson DP, Knight MC, Martin RP, Schipani E, Divieti P, Bringhurst FR, Milner LA, Kronenberg HM, Scadden DT. Osteoblastic cells regulate the haematopoietic stem cell niche. Nature. 2003;425: 841–6.
31. Stier S, Cheng T, Dombkowski D, Carlesso N, Scadden DT. Notch1 activation increases hematopoietic stem cell self-renewal in vivo and favors lymphoid over myeloid lineage outcome. Blood. 2002;99:2369–78.
32. Burns CE, Traver D, Mayhall E, Shepard JL, Zon LI. Hematopoietic stem cell fate is established by the Notch-Runx pathway. Genes Dev. 2005;19:2331–42.
33. Kunisato A, Chiba S, Nakagami-Yamaguchi E, Kumano K, Saito T, Masuda S, Yamaguchi T, Osawa M, Kageyama R, Nakauchi H, Nishikawa M, Hirai H. HES-1 preserves purified hematopoietic stem cells ex vivo and accumulates side population cells in vivo. Blood. 2003;101:1777–83.
34. Kumano K, Chiba S, Kunisato A, Sata M, Saito T, Nakagami-Yamaguchi E, Yamaguchi T, Masuda S, Shimizu K, Takahashi T, Ogawa S, Hamada Y, Hirai H. Notch1 but not Notch2 is essential for generating hematopoietic stem cells from endothelial cells. Immunity. 2003;18:699–711.
35. Voet D, Voet JG. Lipid metabolism: eicosanoid metabolism. In: Biochemistry: Biomolecules, mechanisms of enzyme action, and metabolism, 3rd ed., vol. 1. Hoboken, NJ: John Wiley and Sons, Inc; 2004. pp. 959–63.
36. Feher I, Gidali J. Prostaglandin E2 as stimulator of haemopoietic stem cell proliferation. Nature. 1974;247:550–1.
37. Villablanca EJ, Pistocchi A, Court FA, Cotelli F, Bordignon C, Allende ML, Traversari C, Russo V. Abrogation of prostaglandin E2/EP4 signaling impairs the development of rag1+ lymphoid precursors in the thymus of zebrafish embryos. J Immunol. 2007;179:357–64.
38. North TE, Goessling W, Walkley CR, Lengerke C, Kopani KR, Lord AM, Weber GJ, Bowman TV, Jang IH, Grosser T, Fitzgerald GA, Daley GQ, Orkin SH, Zon LI. Prostaglandin E2 regulates vertebrate haematopoietic stem cell homeostasis. Nature. 2007;447:1007–11.
39. Lorenz M, Slaughter HS, Wescott DM, Carter SI, Schnyder B, Dinchuk JE, Car BD. Cyclooxygenase-2 is essential for normal recovery from 5-fluorouracil-induced myelotoxicity in mice. Exp Hematol. 1999;27:1494–502.
40. Hofmeister CC, Zhang J, Knight KL, Le P, Stiff PJ. Ex vivo expansion of umbilical cord blood stem cells for transplantation: growing knowledge from the hematopoietic niche. Bone Marrow Transplant. 2007;39:11–23.
41. Tanaka H, Matsumura I, Itoh K, Hatsuyama A, Shikamura M, Satoh Y, Heike T, Nakahata T, Kanakura Y. HOX decoy peptide enhances the ex vivo expansion of human umbilical cord blood CD34+ hematopoietic stem cells/hematopoietic progenitor cells. Stem Cells. 2006;24:2592–2602.
42. Varnum-Finney B, Brashem-Stein C, Bernstein ID. Combined effects of Notch signaling and cytokines induce a multiple log increase in precursors with lymphoid and myeloid reconstituting ability. Blood. 2003;101:1784–89.
43. Delaney C, Varnum-Finney B, Aoyama K, Brashem-Stein C, Bernstein ID. Dose-dependent effects of the Notch ligand Delta1 on ex vivo differentiation and in vivo marrow repopulating ability of cord blood cells. Blood. 2005;106:2693–9.

HOXB4 in Hematopoietic Stem Cell Regulation

Mohan C. Vemuri

Abstract HOXB4 overexpression has been previously shown to increase hematopoietic stem cell renewal and enhance murine HSC competitive repopulation without compromising differentiation or homeostatic regulation of HSC pool size. However, recent studies on human HSC have indicated that forced expression of HOXB4 may dysregulate lymphoid and myeloid differentiation. In this study the effects of HOXB4 forced expression were examined on the distribution and lineage specification of HOXB4 expressing cells in a congenic competitive repopulation assay. Retroviral HOXB4 overexpression in hematopoietic cells (HSCs) resulted in increased frequency of long-term culture initiating cells (LTC-IC) in vitro, indicative of selective growth advantage but with no significant difference in mean CFC/LTC-IC ratio, an indicator of balanced differentiation. Competitive in vivo repopulation assays in congenic transplants with different combinations of HOXB4- and GFP-transduced bone marrow cells showed long-term engraftment and a selective growth advantage of HOXB4 transduced cells in bone marrow, thymus, blood, and spleen. Engraftment levels were disproportionately high in bone marrow and thymus relative to spleen and peripheral blood, while multilineage analysis of bone marrow showed a predominance of myeloid lineages. Thymic analysis revealed a HOXB4-dependent increase in thymic size due to increased numbers of mature CD4 and CD8 T cells consistent with an abnormality of release of lymphocytes from the thymus, and the abnormalities persisted in secondary transplants. These results support the potential of HOXB4 forced expression to alter the balance and distribution of hematopoiesis and suggest fundamental alterations induced by HOXB4 in thymic T-cell trafficking. These results emphasize the need for a careful evaluation of clinical gene therapy applications with HOXB4.

Keywords HoxB4 · HSC · Hematopoietic stem cells · Thymic regulation · Transduction · Differentiation · Gene therapy

M.C. Vemuri (✉)
Stem Cells and Regenerative Medicine, Invitrogen Corporation, 7335 Executive Way, Frederick, MD 21704, USA
e-mail: mohan.vemuri@invitrogen.com

1 Introduction

Hematopoietic stem cells (HSCs) are attractive targets for gene therapy because of their unique ability to reconstitute long-term hematopoiesis after transplantation. However, HSC directed gene therapy has been hampered by inefficient gene transfer and limited expansion capacity of genetically modified HSCs [1–3]. Thus, an important achievement for facilitation of HSC directed gene therapy would be the ability to manipulate HSCs to enhance in vitro expansion and improve their in vivo competitive capacity. Although the genetic and molecular mechanisms that regulate HSC self-renewal and differentiation are not completely understood, there are a number of promising genetic approaches to expand HSCs while maintaining balanced differentiation and self-renewal [4–6]. Perhaps the most promising candidate gene to achieve regulated expansion of HSCs in vivo and in vitro is the homeobox transcription factor HOXB4. Retroviral overexpression of HOXB4 enhances HSC self-renewal and induces a 1000-fold in vivo expansion [4, 7, 8]. Forced expression of HOXB4 also induces rapid ex vivo expansion of transduced HSCs [9] and, in more recent studies, ex vivo expansion is directly achieved using soluble HOXB4 recombinant protein [10, 11]. In vivo studies of retroviral mediated HOXB4 gene transfer into HSCs demonstrated expansion of HSCs with normal lineage development and without causing hematological abnormalities or induction of leukemia [4, 8]. Remarkably, homeostatic regulation of HSC pool size is maintained, suggesting that HSC that overexpress HOXB4 remain responsive to homeostatic regulatory signals governing HSC proliferation [4, 8]. However, there are questions that need to be addressed prior to clinical application. In contrast to the studies cited above, there is evidence of impaired lympho-myeloid differentiation when HOXB4 is overexpressed in human $CD34^+$cells, although the cells retained their in vivo competitive advantage [12, 13]. HOXB4-mediated proliferation is dose-dependent and high levels of ectopic HOXB4 expression results in substantial skewing of myelo/erythroid differentiation [14, 15]. These studies indicate a risk of inadvertent effects of HOXB4, an important concern since vector-mediated expression of other

V.K. Rajasekhar, M.C. Vemuri (eds.), *Regulatory Networks in Stem Cells*, Stem Cell Biology and Regenerative Medicine,
DOI 10.1007/978-1-60327-227-8_10,

HOX family members such as *Hoxb8* or *Hoxa10* have been shown to cause leukemia in mouse models [16–18].

In the course of these studies intended to generate highly competitive HSC, some hematopoietic abnormalities transplant in were noted recipient mice. The purpose of the present study was to systematically investigate the regulatory role of HOXB4 gene in long-term effects of retroviral overexpression of HOXB4 in vivo in the murine model. HOXB4-transduced cells were transplanted into lethally irradiated congenic recipients, and the effects of HOXB4 expression on competitive engraftment, long-term hematopoiesis in primary and secondary recipients, and on compartmental distribution and phenotype of transduced cells were assessed. These results confirm the competitive repopulating advantage conferred by HOXB4, but document aberrations in myelo- and lymphopoiesis resulting in a skewed distribution of HOXB4 expressing cells in hematopoietic compartments, most notably the thymus.

2 Results

2.1 Retroviral Transduction Efficiency with GFP and HOXB4-GFP Vectors

In independent experiments, bone marrow cells (BMCs) were prestimulated and transduced with GFP- or HOXB4-GFP retroviral vector supernatants. Gene transfer rates were determined by FACS analysis of GFP fluorescence following 36 h of exposure to the vectors and 48 h post-transduction expansion (Fig. 1). A mean transduction efficiency of 49.8 ± 7.1% and 47.9 ± 3.7% was obtained with the GFP (n = 6) and HOXB4-GFP (n = 6) viral vectors, respectively.

2.2 Immunophenotypic Profile of Transduced Cells

The reactivity of GFP and HOXB4-GFP retroviral transduced cells with the primary antibody panel is shown in Fig. 2A. CD45R/B220 and CD11b were the most predominantly expressed antigens in both GFP- and HOXB4-GFP–transduced cells (Fig. 2A). Transduced cells were negative for CD25, CD34, and NK 1.1 antigens. The cells also did not express T-cell-associated antigens CD3, CD4, and CD8 (< than 1%) indicating that the culture conditions were least favorable for T-cell survival and growth.

2.3 HOXB4 Overexpression Enhances Progenitor Cell-Derived Colony Formation

To determine how the forced HOXB4 expression influences the formation of myeloid and erythroid progenitor cells, standard colony-forming cell culture assays were performed in methylcellulose media with GFP-, HOXB4-GFP–transduced cells, and mock-infected BMCs and colonies were scored after 10 days (Fig. 2B). In cultures with HOXB4-transduced cells, the total number of colony-forming cells was always high relative to either mock-infected cells (147%, $p < 0.001$) or GFP-transduced cells (148%, $p < 0.001$). A cross-sectional analysis of different types of colonies

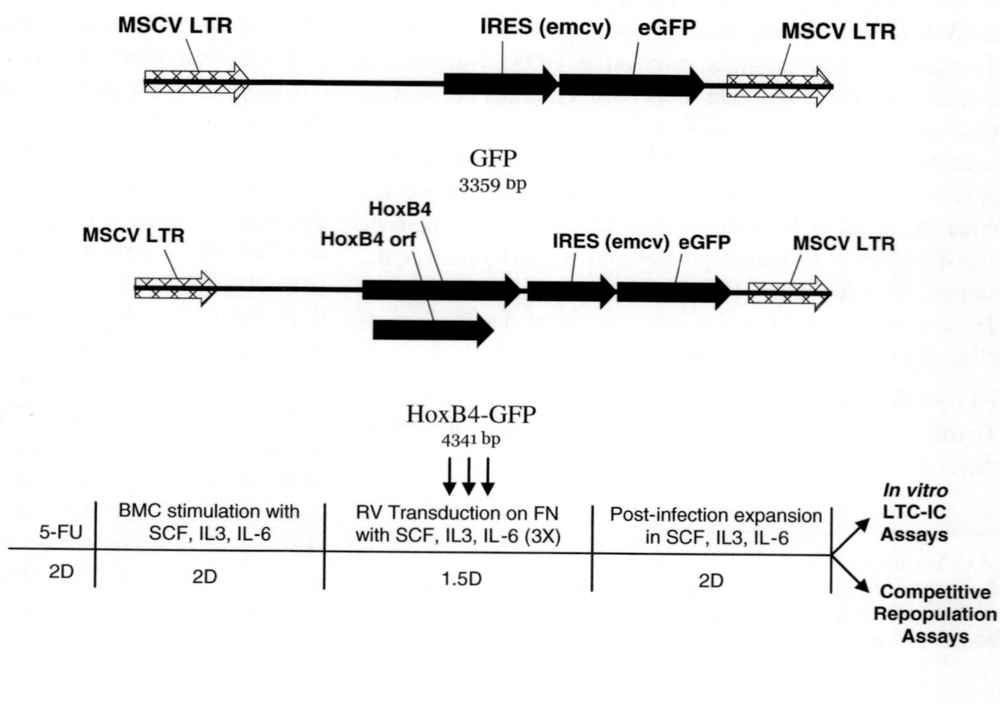

Fig. 1 Retrovirus-mediated overexpression of HOXB4. Structure of integrated MSCV-GFP (*top*) MSCV-HOXB4-GFP vector (*middle*) and schematic time frame of experiments (*bottom*). Plasmid, pM-HOXB4-IG, was constructed by inserting the hu HOXB4 cDNA molecule, a kind gift from Guy Sauvageau[4] upstream of the internal ribosomal entry site (IRES) of MiGR1, a kind gift from WS. Pear[28]

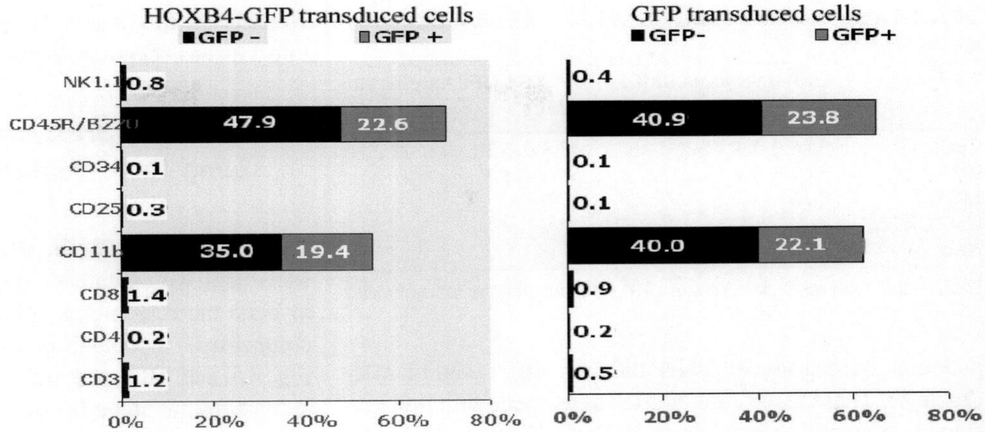

Fig. 2A Immunophenotypic analysis of BMC cultures transduced with HOXB4-GFP or GFP-vector

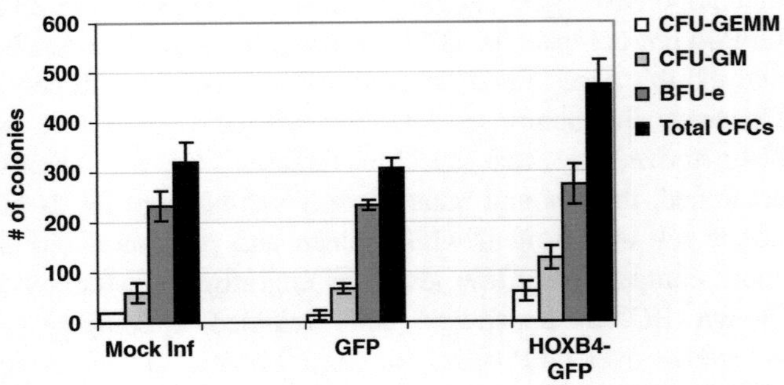

Fig. 2B Frequency and distribution of colony forming units (CFU) derived from total BMCs transduced with HOXB4-GFP or GFP control vector or mock infected cells

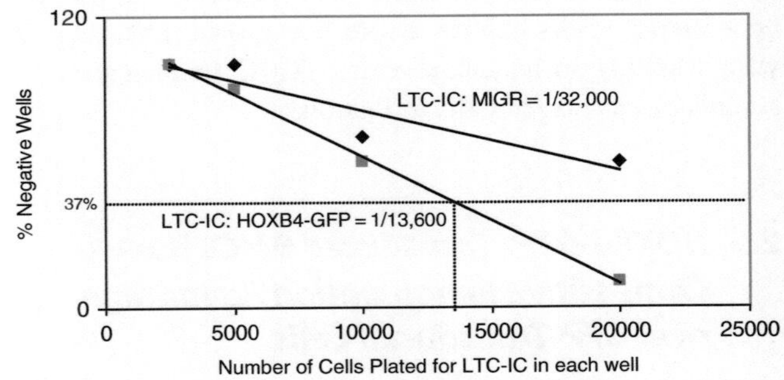

Fig. 2C Limiting dilution analysis of LTC-IC frequency in GFP and HOXB4-GFP transduced bone marrow cells from CD 45.2 donor mice. Transduced BMCs were grown on irradiated feeder stromal layers for 5 weeks and subjected to CFC cultures for 2 weeks in methylcellulose media supplemented with mSCF, IL-3, IL-6 and erythropoietin. The LTC-IC frequency corresponding to the cell dose at which 37% of the cultures/wells yield a negative response was derived, using Poisson statistics and linear regression polynomial curve fitting. GFP transduced BMC showed an LTC-IC frequency of 1 in 32,000 cells (at 95% confidence interval: 1 in 42,733 cells, $p < 0.26$) while HOXB4-GFP transduced cells showed an LTC-IC frequency of 1 in 13,600 (at 95% confidence interval: 1 in 16,266, $p < 0.01$). The frequency levels were calculated using L-calc software from Stem Cell Technologies

also revealed enhanced numbers of granulocyte macrophage erythrocyte mixed CFUs (CFU-GEMM, 200%, $p < 0.001$), granulocyte-macrophage CFU (CFU-GM, 100%, $p < 0.001$) and erythroid burst-forming units (BFU-e, 17%, $p < 0.01$) when compared with the colony forming abilities of GFP vector-transduced cells.

To analyze the specific effect of HOXB4 overexpression on progenitor derived CFUs, fluorescence positive and negative colonies were scored from colony-forming methylcellulose cell culture assays. HOXB4-GFP-vector–transduced cells always produced a significantly increased number of GFP+ colonies (61%, $p < 0.04$), in contrast to the colonies from control GFP vector–transduced cells (44%) (Fig. 2C). The number of GFP+ colonies derived from HOXB4 vector-transduced cells was significantly higher in all the progenitor colony types: CFU-GEMM (64% versus 25%, $p < 0.08$), CFU-GM (58% versus 30%, $p < 0.005$) and BFU-e (59% versus 43%, $p < 0.06$).

Together, the in vitro results suggest that forced HOXB4 expression can increase the proliferation and differentiation of all progenitor cell types, although we observed a significant enhancement in the formation of the most primitive and early progenitor colonies (CFU-GEMM) relative to mature and differentiated colonies (BFU-e).

2.4 HOXB4 Overexpression in HSCs Results in Increased Frequency of Long-Term Culture Initiating Cells (LTC-IC) In Vitro

Analysis of LTC-IC frequency in transduced bone marrow cell populations corresponding to the cell dose at which 37% of cultures yield a negative response showed an LTC-IC of 1 per 13,600 ($p < 0.01$, 95% confidence interval) with HOXB4-GFP–transduced cells compared to 1 per 32,000 cells ($p < 0.26$, 95% confidence interval) with GFP–transduced cells, a 2.5-fold increase in LTC-IC frequency due to HOXB4 overexpression. However, the mean CFC/LTC-IC ratio is nearly the same in cells transduced either with GFP (8.16) or HOXB4 (8.5), indicating that the clonogenic output and differentiation

Table 1 Frequencies of CFCs and LTC-IC from GFP and HOXB4-GFP transduced BMCs

Analysis	GFP-transduced BMCs	HOXB4-GFP transduced BMCs
CFCs	30.6 ± 2.1	47.3 ± 5.3
LTC-ICs	3.67 ± 1.2	5.56 ± 1.3
Mean CFC/LTC-IC ratio	8.16	8.50

Each of the analyses is derived from 10^4 transduced bone marrow cells.

potential of transduced cells did not differ significantly (Table 1). These results are in good agreement with previous studies of forced expression of HOXB4 in murine HSC [4, 8] and demonstrate that HOXB4 overexpression in primary bone marrow cells confers a selective growth advantage in vitro resulting in an increase in LTC-IC frequency and a secondary increase in CFC formation.

2.5 HOXB4-GFP–Transduced BMCs Have a Competitive Repopulation Advantage over GFP-Transduced Cells

To assess the in vivo hematopoietic reconstitution potential of HOXB4, HOXB4 and GFP control vector transduced cells were evaluated for their ability to engraft and repopulate the marrow of mice in competitive repopulation assays. Lethally irradiated recipient mice were transplanted with various combinations of GFP-transduced (CD45.1) and HOXB4-GFP–transduced (CD45.2) cells and the relative contribution to reconstituted hematopoiesis by the two cell populations was monitored over time (Fig. 3). Keeping the total transplant dose constant, competition groups were set up consisting of either equal numbers or a 9:1 ratio of GFP- and HOXB4-GFP–transduced cells respectively. A third group received HOXB4 transduced cells alone without competition from GFP CD45.1 cells. FACS profiles of recipient peripheral blood chimerism as early as 6 weeks after transplantation showed the presence of both donor cell types (Fig. 3). As we did not sort the GFP+ cells prior to transplantation the HOXB4-expressing cells were also competing against the nontransduced CD45.2 cells. The results demonstrate a competitive advantage for HOXB4 expressing cells over nonexpressing GFP cells but not over nontransduced CD45.2 cells. In addition, the nontransduced CD45.2 cells were more competitive than the nontransduced CD45.1 cells. This suggests that there may be a collateral enhancement of the competitive capacity of nontransduced cells by transduced cells, perhaps related to HOXB4 secretion during the expansion phase of our protocol. Overall, these results support a competitive advantage rendered by HOXB4 in this study, although perhaps not as robust an effect as has been reported in previous studies [8, 20].

2.6 HOXB4 Expression Leads to Defects in Long-Term Hematopoiesis

To determine whether the HOXB4-induced repopulation was sustained, we examined long-term donor cell engraftment in bone marrow, blood, spleen, and thymus from the three competitive repopulation groups. The data presented in Fig. 4A and 4B represent a summary of the hematopoietic reconstitution potential of CD45.2+ HOXB4-GFP+ cells (Fig. 4A) versus CD45.1+ GFP+ cells (Fig. 4B) in different hematopoietic organs. HOXB4-transduced cells (CD45.2+, GFP+) out-compete the GFP control vector-transduced cells and exhibit a sustained long-term donor cell engraftment. However, the nature of HOXB4-induced engraftment is characterized by two specific features. Of the tissues examined, thymus and bone marrow exhibit high levels of donor cell engraftment while spleen and peripheral blood show comparatively low levels of engraftment. Moreover, though HOXB4-transduced cells retained a competitive advantage over GFP cells, the engraftment is qualitatively and quantitatively different from the control group of lethally irradiated recipient mice (CD45.1) transplanted with nontransduced donor cells (CD45.2). Several conclusions may be drawn from this data. First, HOXB4 expression results in a compartmental imbalance in hematopoiesis, with high percentage representation maintained in the thymus and bone marrow but minimal representation in the peripheral blood and spleen. Second, HOXB4 expression did not confer a sustained competitive advantage over nonexpressing cells, as one would expect progressive conversion to complete engraftment if that were the case.

2.7 HOXB4 Overexpression Disturbs Lymphoid-Myeloid Differentiation In Vivo

In view of the observed compartmental imbalance in hematopoietic reconstitution, whether overexpression of HOXB4 would disturb the generation of specific hematopoietic lineages was examined. FACS analysis of bone marrow cells was performed to assess donor lineage distribution and engraftment in long-term reconstituted primary mice. Analysis for the lymphoid (CD19), myeloid (CD11b), and T-cell (CD3) lineages showed that HOXB4 overexpression can cause a disturbance in lymphoid-myeloid lineage generation (Fig. 5). In particular, HOXB4 expression promotes myeloid differentiation in vivo an observation consistent with previous observations with human CD34$^+$ CB cells [15].

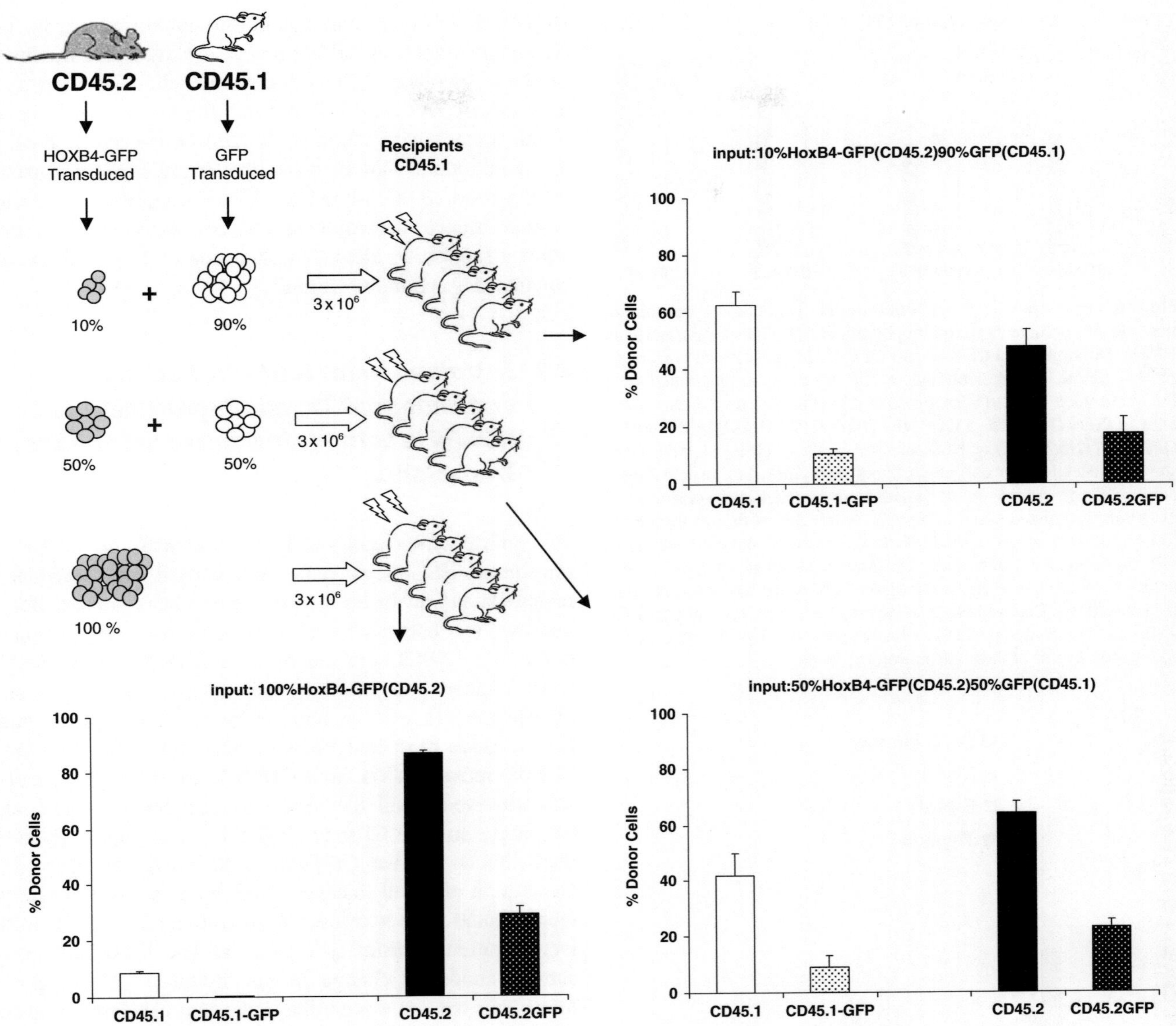

Fig. 3 Competitive repopulation assay with HOXB4-GFP and GFP-transduced BMCs in lethally irradiated recipients. Six weeks after transplantation, GFP+ donor derived CD 45.2 or CD 45.1 cells are identified in peripheral blood. Increasing number of HOXB4-GFP transduced test cells from CD 45.2 mice are mixed with proportionately decreasing number of GFP transduced BMCs from CD 45.1 mice that differ from a congenic marker, to derive 10%HOXB4-GFP+90%GFP or 50%HOXB4-GFP+50%GFP or 100% HOXB4-GFP transduced cell combinations. Fixed number of these combined cell populations, in this case, 3×10^6 per mouse are injected to groups of lethally irradiated (500cGy × 2) recipient mice (CD 45.1) and analyzed for HOXB4-GFP conferred competitive advantage and long term donor chimerism. An additional group of lethally irradiated CD 45.1 mice reconstituted with untransduced CD 45.2 donor BMCs served as control in repopulation assay

2.8 Forced HOXB4 Expression Enables High Levels of Thymic Engraftment In Vivo with Associated Thymic Enlargement

A detailed examination of the bone marrow, peripheral blood, and spleen of reconstituted mice did not reveal any abnormalities. However, mice that were reconstituted with a complete transplant dose of HOXB4-transduced cells displayed enlarged thymuses (3 out of 4 mice) and showed high cell numbers ($15.13 \pm 2.04 \times 10^6$) compared to the control thymuses ($1.4 \pm 0.02 \times 10^6$) from age-matched lethally irradiated CD45.1 mice reconstituted with non-transduced BMCs from CD45.2 donor mice. In the groups of mice that received equal numbers of HOXB4- and GFP-transduced cells or in the group injected with 10-fold greater number of GFP cells, the thymuses were normal. In

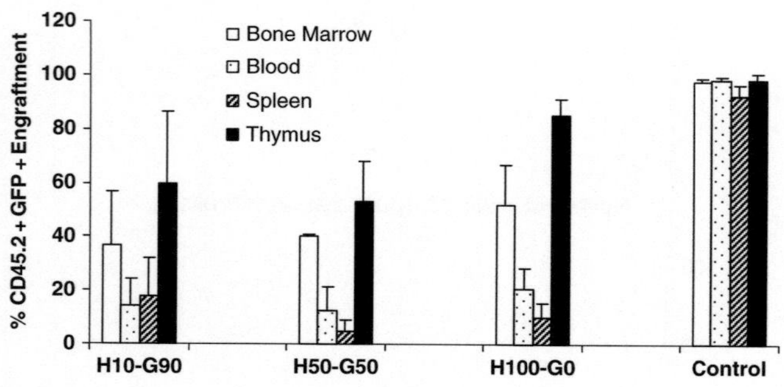

Fig. 4A Hematopoietic reconstitution after 5 months of transplantation and competitive engraftment potential of HOXB4-GFP transduced BMCs. Donor derived CD45.2 and GFP double positive cells from peripheral blood, bone marrow, spleen and thymus of reconstituted mice. Recipient mice injected with mixed combinations of transduced BMCs with HOXB4-GFP 10% + GFP 90% (*H10–G90*), HOXB4-GFP 50% + GFP 50% (*H50–G50*) or HOXB4-GFP 100% + GFP 0% (*H100–G0*) show enhanced donor cell engraftment in thymus, but an imbalanced donor cell engraftment in all tissues compared with the control group of lethally irradiated CD 45.1 recipients injected with non-transduced CD 45.2 BMCs which showed balanced chimerism and complete donor cell engraftment in all tissues. The final read-out in control group is only CD 45.2 positive and do not have GFP, since the donor cells are not transduced. Data represent the average ± s.e.m. percentage of CD 45.2 and GFP double positive cells in experimental groups or only CD 45.2 positive cells in the control group (N = 4)

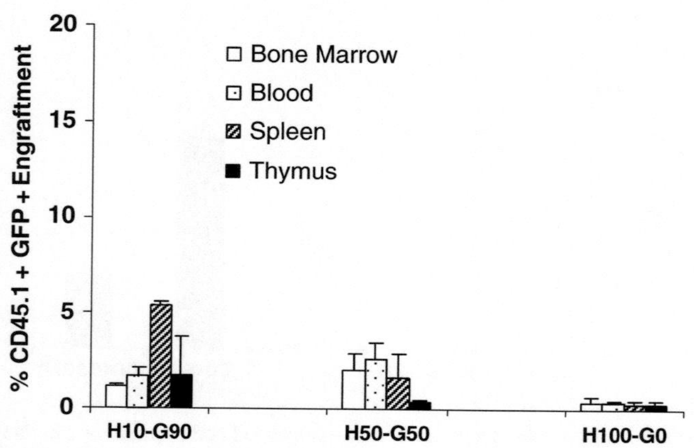

Fig. 4B Engraftment potential of donor derived CD 45.1 and GFP double positive cells from peripheral blood, bone marrow, spleen and thymus of reconstituted mice after 5 months of transplantation in competitive repopulation assays. Details of assays are same as in the legend of figure 4A, except that the analysis is with congenic marker CD 45.1 in this figure. The scale on Y-axis is intentionally expanded to show up the difference and distinct failure of GFP transduced CD 45.1 donor cells in competitive assays. Data represent the average ± s.e.m. percentage of CD 45.1 and GFP double positive cells ($N = 4$)

view of the thymic morphological anomalies in reconstituted mice, T-cell populations were further analyzed. FACS analysis of T-cell lineages with $CD4^+$ and $CD8^+$ markers revealed dramatically increased numbers of $CD4^+$ and $CD8^+$ cells derived from $CD45.2^+$ HOXB4-GFP^+ donor cells (Fig. 6). Interestingly, though the CD4+ and CD8+ cell numbers increased significantly, the cells were maintained at 1:1 ratio in every animal similar to control mice (Fig. 7). Histological analysis did not reveal any overt abnormalities in these thymuses. Thus, the combination of increased numbers of HOXB4-GFP–derived single positive T cells in the thymus with minimal HOXB4-GFP–derived T cells in the peripheral blood suggest that lymphocyte precursors arising from HOXB4-transduced HSC populations undergo normal thymic development but are sequestered in the thymus supporting altered mechanisms of T-cell release or attenuated T-cell mobilization.

2.9 *Sustained Hematopoietic Recovery and Enhanced Thymic Engraftment by HOXB4 in Lethally Irradiated Secondary Transplants*

To confirm that engraftment was secondary to HOXB4-transduced HSC transplants were carried into secondary recipients and analyzed for long-term hematopoiesis. Bone marrow cells harvested from a 5-month-old primary transplant with HOXB4-transduced cells (CD45.2 → CD45.1) were injected into lethally irradiated secondary congenic mice (CD45.1) and the bone marrow, blood, spleen, and the thymuses were analyzed 4 months after transplantation. Thymic reconstitution from $CD45.2^+$ HOXB4-GFP^+ cells was observed in all the secondary recipients. In general, the single positive CD4 or CD8 cells were slightly higher than double positive CD4+8+ cells in thymus (Fig. 8). Generation of T-cell lineages in the thymuses of secondary reconstituted mice was less robust compared to the thymus reconstitution in primary recipients. The CD4/CD8 ratio showed moderate change in the thymuses of secondary transplants but that was similar to the control transplant mice which received nontransduced CD45.2 donor cells (Fig. 9). Assessment of the distribution of HOXB4-GFP–derived hematopoiesis in secondary recipients demonstrated an identical pattern to that in primary recipients with increased numbers of HOXB4-GFP cells in the BM and thymus and relatively low levels of PB and splenic chimerism (Fig. 10). These results of secondary transplants confirmed that chimerism from HOXB4-GFP–transduced cells was due to engraftment of HSC and that though less pronounced in secondary recipients, the abnormalities of distribution of hematopoiesis persist.

3 Discussion

HOXB4 is one of the important molecules implicated in regulating hematopoietic stem cell renewal, but the mechanism remains unclear [18]. Retroviral overexpression of HOXB4

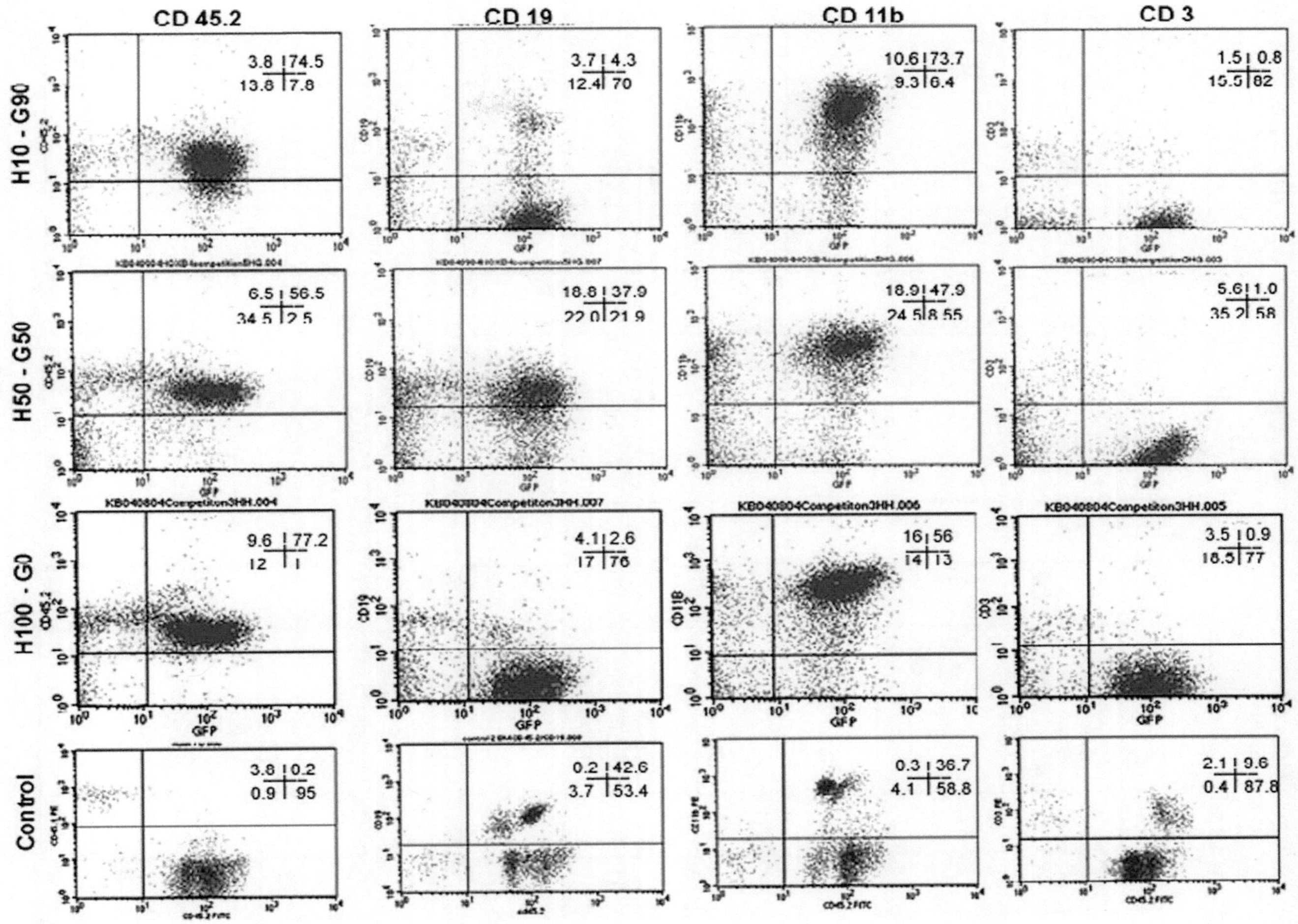

Fig. 5 Multilineage contribution by HOXB4-GFP transduced cells was analyzed in recipient BMCs by flow cytometry on gated cell populations of donor type congenic marker CD 45.2 and GFP using myeloid (CD11b), lymphoid (CD19) and T cells (CD3) lineage specific markers. Percentages of cells are indicated in quadrants. Each panel is a representative of data from at least 4 different recipients. Panel *H10–G90* represents recipient mice injected with mixed combinations of HOXB4-GFP 10% + GFP 90% transduced BMCs, panel *H50–G50* represents data from mice that received HOXB4-GFP 50% + GFP 50% and panel *H100–G0* represents data from mice that received HOXB4-GFP 100% + GFP 0%. Panel *Control* represents data from age-matched lethally irradiated CD 45.1 mice that received nontransduced BMCs from CD 45.2 mice

in HSCs mediates a profound expansion of gene-modified hematopoietic stem and precursor populations in vitro and in vivo, a property that could be valuable for gene therapy applications. Though several studies confirmed the unique role and utility of HOXB4 in hematopoietic stem cell expansion [4, 8, 11, 21, 22], a few recent studies point out skewed differentiation patterns [12–15] that underscore the need for further assessment of HOXB4 modified HSCs. Results presented in this study show for the first time that hematopoietic cells overexpressed with retroviral HOXB4 can significantly increase thymic cellularity with pronounced thymic enlargement and variable levels of engraftment in different hematopoietic organs of long-term engrafted congenic transplanted mice. This observation was made, despite features of our transduced cells that were consistent with previous reports [4, 8], that is, HOXB4-induced expansion in culture, superior proliferative self-renewing capacity, and an in vivo competitive advantage for repopulation of lethally irradiated recipients.

The ability of HOXB4 to confer growth advantage to HSCs and modulate stem cell responses in a coordinated manner of self-renewal and differentiation has important implications in gene therapy and transplantation biology. LTC-ICs reflect a close approximation to transplantable stem cells, since they are distinguished by their ability to generate multilineage colonies over extended periods of culture [23]. Analysis of HOXB4-transduced HSC populations showed 2.5-fold greater LTC-IC frequency. Additionally, the colony-forming ability of clonogenic progenitors was not significantly altered as indicated by mean CFC/LTC-IC ratio. These in vitro results are in agreement with previous reports and support the in vitro capacity of HOXB4-transduced cells to

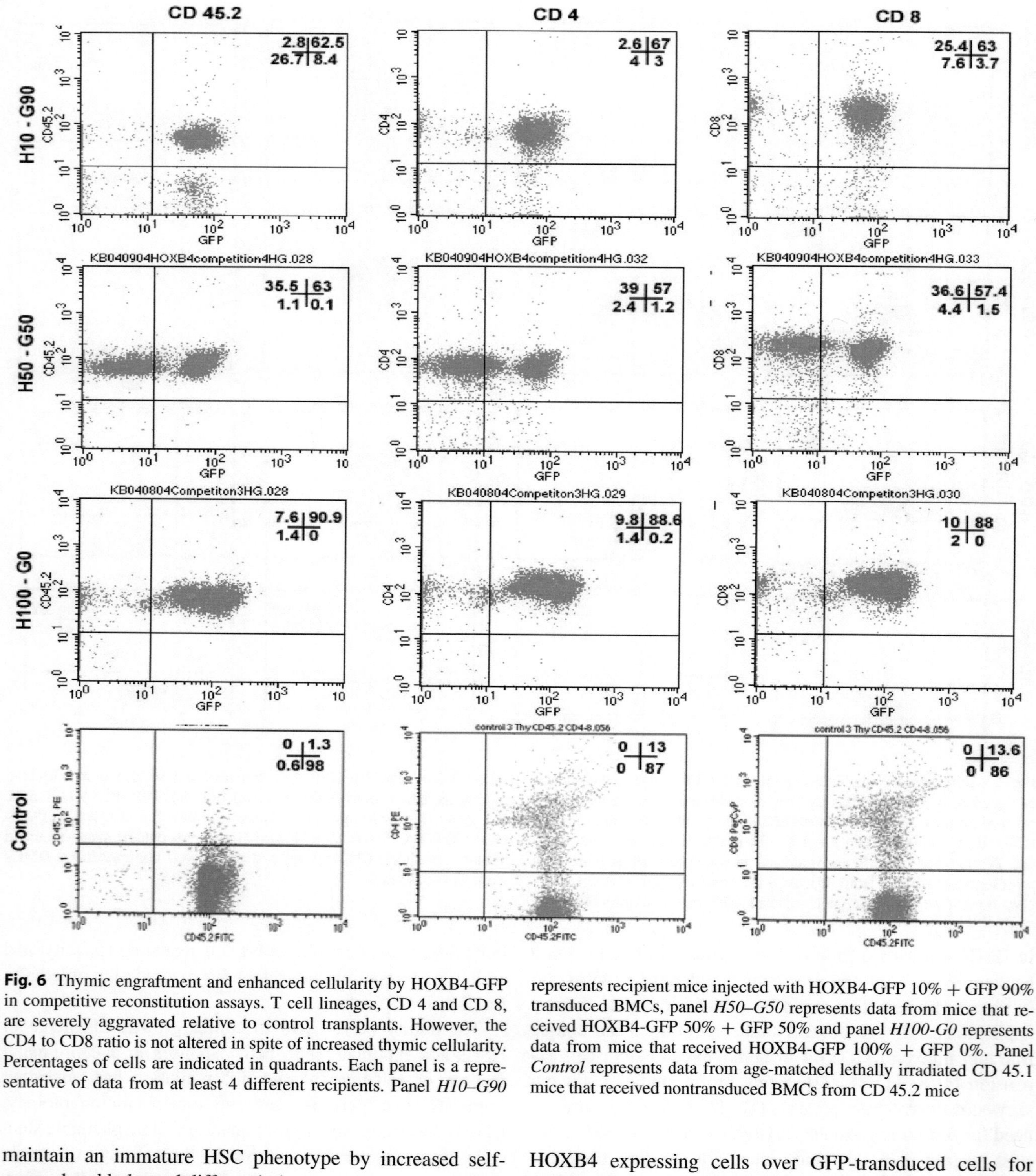

Fig. 6 Thymic engraftment and enhanced cellularity by HOXB4-GFP in competitive reconstitution assays. T cell lineages, CD 4 and CD 8, are severely aggravated relative to control transplants. However, the CD4 to CD8 ratio is not altered in spite of increased thymic cellularity. Percentages of cells are indicated in quadrants. Each panel is a representative of data from at least 4 different recipients. Panel *H10–G90* represents recipient mice injected with HOXB4-GFP 10% + GFP 90% transduced BMCs, panel *H50–G50* represents data from mice that received HOXB4-GFP 50% + GFP 50% and panel *H100-G0* represents data from mice that received HOXB4-GFP 100% + GFP 0%. Panel *Control* represents data from age-matched lethally irradiated CD 45.1 mice that received nontransduced BMCs from CD 45.2 mice

maintain an immature HSC phenotype by increased self-renewal and balanced differentiation.

However, in vivo data differs from previous murine studies. Skewed distribution of engraftment and a striking abnormality in thymopoiesis related to the HOXB4-transduced cells relative to control transplanted mice were noticed in this study. The competitive repopulation assays, in all tissues analyzed, demonstrate a marked competitive advantage of HOXB4 expressing cells over GFP-transduced cells for engraftment. Interestingly, the improved competitive capacity was also apparent in the nontransduced cells that were co-cultured with HOXB4-expressing cells. These cells were able to compete equally with HOXB4-expressing cells and could out compete nontransduced cells in the GFP-transduced population. These results are in agreement with reports of improved competitive capacity of cells

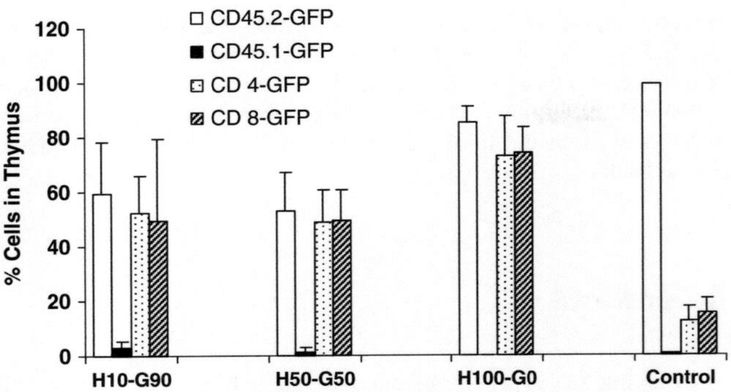

Fig. 7 Summary of and T-cell lineages from donor derived GFP positive cells in the thymus in different groups of mice from competitive repopulation assays

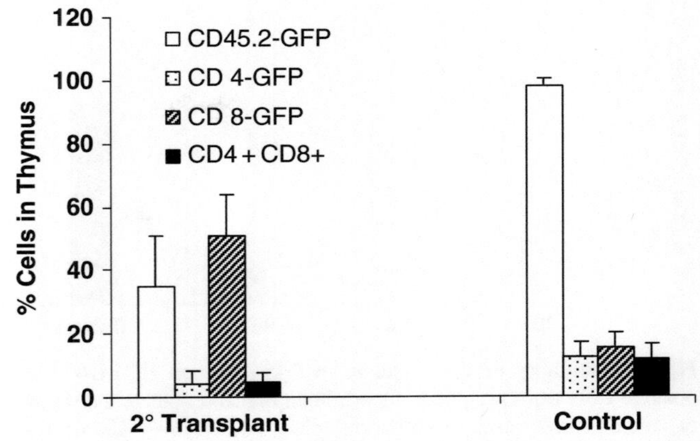

Fig. 9 Summary of and T-cell lineages from donor derived GFP positive cells in the thymus of secondary transplant mice

exposed to HOXB4 protein prior to transplantation. The aberrations in engraftment distribution and thymopoiesis, however, appear to be due solely to HOXB4-expressing cells suggesting that the ongoing intracellular production of HOXB4 is related to the deregulation. Multilineage analysis of BM demonstrated skewing toward myelopoiesis, a finding that differs from some previous studies that showed balanced differentiation [8] but adds to the evidence that at least in some circumstances, forced HOXB4 expression results in alterations in the balance of lineage specification [12, 15]. The mechanisms of HOXB4-induced variations in lymphomyeloid differentiation are poorly understood but appear to be dose-dependent. Following a two-fold increase in HOXB4 mRNA, growth advantage effects of HOXB4 were achieved with balanced differentiation of blood cell lineages [8], while high-level ectopic HOXB4 expression resulted in impaired lymphomyeloid differentiation [12]. In contrast, enforced adenoviral HOXB4 expression in CD34+ cells resulted in myeloid differentiation [15]. This supports the likely need for regulation of HOXB4 expression in clinical strategies utilizing this approach.

An additional finding in this study was the abnormal distribution of HOXB4 expressing cells. Specifically, the relatively high levels in the BM and thymus compared to PB and Spleen. As PB engraftment normally reflects the level of BM engraftment, this suggests abnormal sequestration of presumably myeloid cells in the BM. More impressive were the abnormalities in the thymus in mice receiving higher doses of HOXB4-transduced cells. The presence of thymic enlargement due to increased numbers of HOXB4

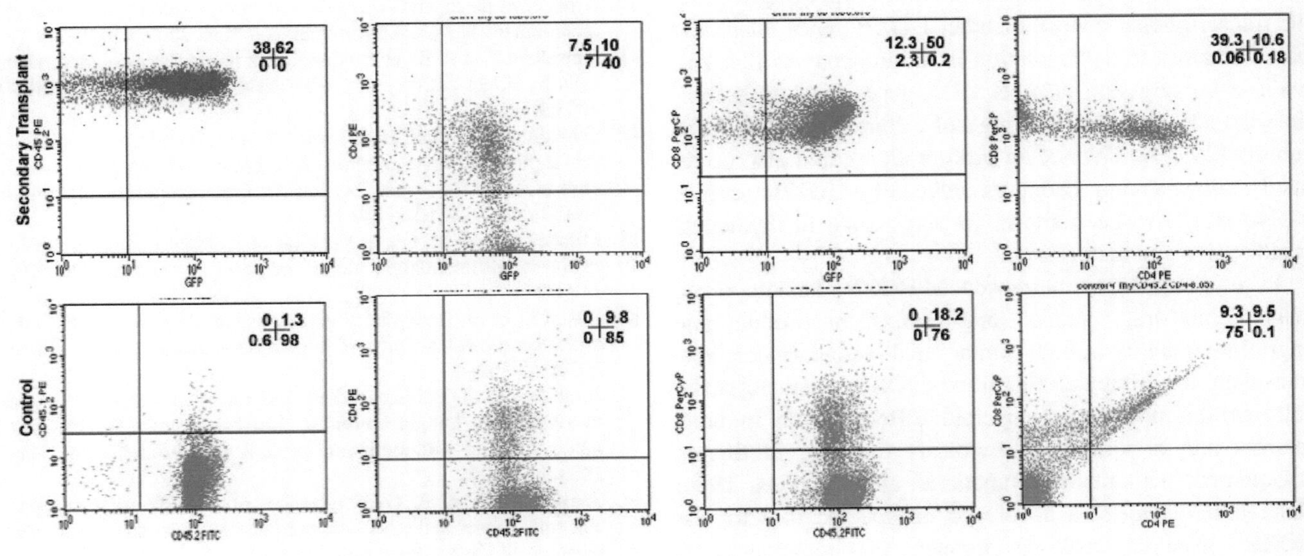

Fig. 8 Thymic engraftment and enhanced cellularity by HOXB4 in Secondary Transplant mice. T cell lineages, CD 4 and CD 8, continue to be severely aggravated similar to primary transplants. CD8 cellularity relative to CD4 is increased causing an imbalanced thymic cellularity. Percentages of cells are indicated in quadrants

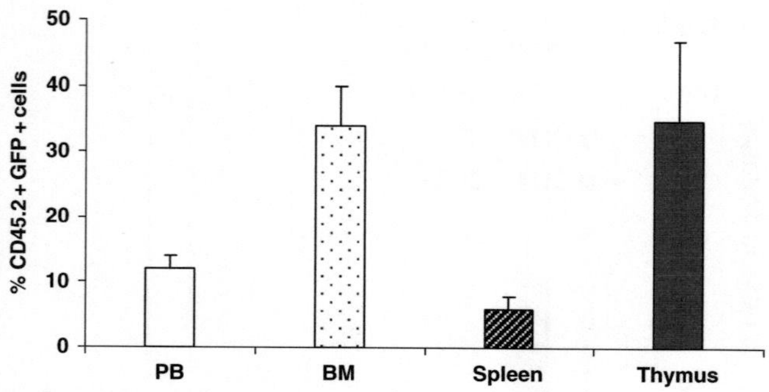

Fig. 10 Summary of donor derived CD45.2 + and HOXB4-GFP+ positive cells from peripheral blood, bone marrow, spleen and thymus of congenic secondary transplant recipient (CD45.2 → CD 45.1 → CD45.1), 4 months after of transplantation

expressing lymphocytes with minimal levels of HOXB4-transduced lymphocytes in the PB also suggests failure of release of transduced cells from the thymus. This is further supported by the phenotypic analysis of these cells, which were predominantly single positive (SP) CD4 or CD8 lymphocytes. These cells normally transit through the medullary zone of the thymus, after positive and negative selection that occurs predominantly in the double positive (DP) phase of thymocyte development. The accumulation of SP cells, in greater numbers than DP cells is distinctly abnormal and suggests that HOXB4 expression somehow interferes with one or more of the many steps required for lymphocytes to leave the thymus. T-cell defects continue to persist even in secondary transplants that are long term engrafted suggesting that the genetically manipulated HOXB4 hematopoietic cells retain the dysregulated thymic repopulation property. Previous studies suggest that enforced HOXB4 might affect the transcriptional control of other HOXB genes that could possibly result in dysregulation of T lymphocytes [24, 25], while overexpression of other HOX genes resulted in multiple defects including blockage of differentiation of T-cell subsets [26, 27]. Thymocyte proliferation could also be related to downstream mediators induced by HOXB4 such as c-Myc that have been shown to play a role in thymocyte proliferation and maturation.

In summary, retroviral transduced HOXB4 hematopoietic cells demonstrate stable long-term reconstitution and engraftment ability in both primary and secondary congenic transplant experiments. Engrafted HOXB4 hematopoietic cells exhibit preferential myeloid differentiation in bone marrow and abnormal thymopoiesis resulting in thymic enlargement without overt functional abnormalities. These observations suggest a novel and unexpected role for the HOXB4 gene in T-cell development that remains to be defined. These results reinforce the need for regulated HOXB4 gene expression to avoid potential complications in gene therapy and transplantation applications.

Acknowledgments Mohan C. Vemuri thanks Dr. Alan W. Flake for the facilities to carry out the work. The author would also like to acknowledge the support from Kirsten J. Beggs, Stephanie C. Filice, William H. Peranteau, and Phillip W. Zoltick at the Children's Institute of Surgical Science, Children's Hospital of Philadelphia, Philadelphia, Pennsylvania.

References

1. Bunting KD, et al. Transduction of murine bone marrow cells with an MDR1 vector enables ex vivo stem cell expansion, but these expanded grafts cause a myeloproliferative syndrome in transplanted mice. Blood. 1998;92:2269–79.
2. Varnum-Finney B, et al. Pluripotent, cytokine-dependent, hematopoietic stem cells are immortalized by constitutive Notch1 signaling. Nat Med. 2000;6:1278–81.
3. Reya T, et al. A role for Wnt signalling in self-renewal of haematopoietic stem cells. Nature. 2003;423:409–14.
4. Sauvageau G, et al. Overexpression of HOXB4 in hematopoietic cells causes the selective expansion of more primitive populations in vitro and in vivo. Genes Dev. 1995;9:1753–65.
5. Stier S, et al. Notch1 activation increases hematopoietic stem cell self-renewal in vivo and favors lymphoid over myeloid lineage outcome. Blood. 2002;99:2369–78.
6. Reya T. Regulation of hematopoietic stem cell self-renewal. Recent Prog Horm Res. 2003;58:283–95.
7. Thorsteinsdottir U, Sauvageau G, Humphries RK. Enhanced in vivo regenerative potential of HOXB4-transduced hematopoietic stem cells with regulation of their pool size. Blood. 1999;94: 2605–12.
8. Antonchuk J, Sauvageau G, Humphries RK. HOXB4 overexpression mediates very rapid stem cell regeneration and competitive hematopoietic repopulation. Exp Hematol. 2001;29: 1125–34.
9. Antonchuk J, Sauvageau G, Humphries RK. HOXB4-induced expansion of adult hematopoietic stem cells ex vivo. Cell. 2002;109:39–45.
10. Krosl J, et al. In vitro expansion of hematopoietic stem cells by recombinant TAT-HOXB4 protein. Nat Med. 2003;9:1428–32.
11. Amsellem S, et al. Ex vivo expansion of human hematopoietic stem cells by direct delivery of the HOXB4 homeoprotein. Nat Med. 2003;9:1423–7.
12. Schiedlmeier B, et al. High-level ectopic HOXB4 expression confers a profound in vivo competitive growth advantage on human cord blood CD34+ cells, but impairs lymphomyeloid differentiation. Blood. 2003;101:1759–68.
13. Milsom MD, et al. Overexpression of HOXB4 confers a myelo-erythroid differentiation delay in vitro. Leukemia. 2005;19: 148–53.
14. Buske C, et al. Deregulated expression of HOXB4 enhances the primitive growth activity of human hematopoietic cells. Blood. 2002;100:862–8.
15. Brun AC, et al. Enforced adenoviral vector-mediated expression of HOXB4 in human umbilical cord blood cd34(+) cells promotes myeloid differentiation but not proliferation. Mol Ther. 2003;8:618–28.
16. Sauvageau G, et al. Overexpression of HOXB3 in hematopoietic cells causes defective lymphoid development and progressive myeloproliferation. Immunity. 1997;6:13–22.
17. Thorsteinsdottir U, et al. Overexpression of HOXA10 in murine hematopoietic cells perturbs both myeloid and lymphoid differentiation and leads to acute myeloid leukemia. Mol Cell Biol. 1997;17:495–505.

18. Thorsteinsdottir U, Sauvageau G, Humphries RK. Hox homeobox genes as regulators of normal and leukemic hematopoiesis. Hematol Oncol Clin North Am. 1997;11:1221–37.
19. Vemuri MC, Campagnoli C, Beggs KJ, Ashizuka S, Zoltick PW, Flake AW. HOXB4 confers enhanced stem cell regeneration and allogenic hematopoietic reconstitution. Exp Hematol. 2005: [abstract of conference proceedings].
20. Schmittwolf C, et al. HOXB4 confers a constant rate of in vitro proliferation to transduced bone marrow cells. Oncogene. 2005;24:561–72.
21. Krosl J, et al. The competitive nature of HOXB4-transduced HSC is limited by PBX1: the generation of ultra-competitive stem cells retaining full differentiation potential. Immunity. 2003;18: 561–71.
22. Thorsteinsdottir U, et al. The oncoprotein E2A-Pbx1a collaborates with Hoxa9 to acutely transform primary bone marrow cells. Mol Cell Biol. 1999;19:6355–66.
23. Coulombel L. Identification of hematopoietic stem/progenitor cells: strength and drawbacks of functional assays. Oncogene. 2004;23:7210–22.
24. Almeida AR, Borghans JA, Freitas AA. T cell homeostasis: thymus regeneration and peripheral T cell restoration in mice with a reduced fraction of competent precursors. J Exp Med. 2001;194: 591–9.
25. Care A, et al. Coordinate expression and proliferative role of HOXB genes in activated adult T lymphocytes. Mol Cell Biol. 1994;14:4872–7.
26. Izon DJ, et al. Loss of function of the homeobox gene Hoxa-9 perturbs early T-cell development and induces apoptosis in primitive thymocytes. Blood. 1998;92:383–93.
27. Taghon T, et al. Homeobox gene expression profile in human hematopoietic multipotent stem cells and T-cell progenitors: implications for human T-cell development. Leukemia. 2003;17: 1157–63.

Telomere and Telomerase for the Regulation of Stem Cells

Eiso Hiyama and Keiko Hiyama

Abstract Telomeres, guanine-rich tandem DNA repeats of chromosomal ends, provide chromosomal stability, and cellular replication causes their loss. In somatic cells, the activity of telomerase, a reverse transcriptase that can elongate telomeric repeats, is usually diminished after birth so that the telomere length is gradually shortened with cell divisions and triggers cellular senescence. In embryonic stem cells, telomerase is activated and maintains telomere length and cellular immortality. On the other hand, in adult stem cells, the level of telomerase activity is low and insufficient to maintain telomere length. Thus, even in stem cells, except for embryonic stem cells and cancer stem cells, telomere shortening occurs during replicative aging, possibly at a slower rate than that in normal somatic cells. In the past few years, the importance of telomere maintenance in human stem cells has been highlighted by the studies on dyskeratosis congenita, aplastic anemia, and idiopathic pulmonary fibrosis, a part of which are genetic disorders in the human telomerase component and are characterized by premature loss of tissue regeneration with stem cell dysfunction. The regulation of telomere length and telomerase activity is a complex and dynamic process that is tightly linked to cell cycle regulation in human stem cells. Here we review the role of telomeres and telomerase in human stem cells.

Keywords Telomere · Telomerase · Stem cell · Cancer stem cell · Dyskeratosis congenita · Aplastic anemia · Idiopathic pulmonary fibrosis

Stem cells are characterized by the capacity to self-renew and/or differentiate and this capacity appears to be finite except for embryonic stem cells. Through symmetric cell division, stem cells repopulate and expand their numbers. Through asymmetric division arising from unequal partition of cellular components, a stem cell can produce a daughter stem cell and a progenitor cell, which ultimately gives rise to a differentiated cell. Since stem cell compartments undergo low but constant rates of cell turnover during their lifetime, mechanisms for maintaining their lifespan are necessary.

The ends of chromosomes, known as telomeres, are formed by a special chromatin structure, which is essential to protect chromosome ends from chromosomal fusion, recombination, and terminal DNA degradation [1]. During DNA replication in cell division, telomeres gradually shorten as a result of the incomplete replication of linear chromosomes, the so-called "end-replication problem." In certain cell types the telomere loss is counteracted by telomere elongation, catalyzed by the enzyme telomerase [2, 3]. This enzyme is inactivated in most somatic cells and, however, only expressed in a few cell types in humans [4–6]. The role of telomerase in maintaining stem cell self-renewal is one key distinguishing feature between stem cells and non-stem cells.

In the past few years, the specific role of telomerase in different kinds of adult stem cells has been elucidated, especially in well-characterized stem cells such as hematopoietic stem cells (HSC) [5, 7–12], epidermal stem cells (ESC) [13, 14], and neural stem cells (NSC) [15]. HSCs derived from humans and mice lose telomeric DNA with age despite the presence of detectable telomerase activity [9–12]. By the recent advances to elucidate the dyskeratosis congenita (DKC) or related diseases, telomerase activity and/or maintenance of telomere length are necessary to maintain the function of some kinds of tissue stem cells [16]. This chapter will review recent data on the importance of telomere biology for regulation of stem cells.

1 Telomere and Telomerase

1.1 Telomere

The ends of chromosomes, known as telomeres, are formed by a special chromatin structure, which is essential to

E. Hiyama (✉)
Natural Science Center for Basic Research and Development, Hiroshima University, 1-2-3, Kasumi, Minami-ku, Hiroshima, 734-8551, Japan
e-mail: eiso@hiroshima-u.ac.jp

V.K. Rajasekhar, M.C. Vemuri (eds.), *Regulatory Networks in Stem Cells*, Stem Cell Biology and Regenerative Medicine, DOI 10.1007/978-1-60327-227-8_11,

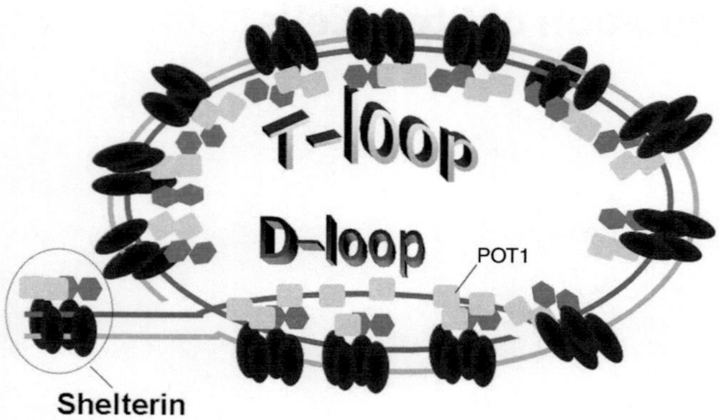

Fig. 1 Telomere structure. Mammalian telomeres consist of the TTAGGG repeats that are bound to the shelterin complex [20]. The T-loop structure is formed by invasion of the 3′ single-strand overhang into an adjacent duplex telomeric repeat array, forming a D-loop. The size of the loop is variable and the length of the telomere is likely to be regulated by shelterin complex

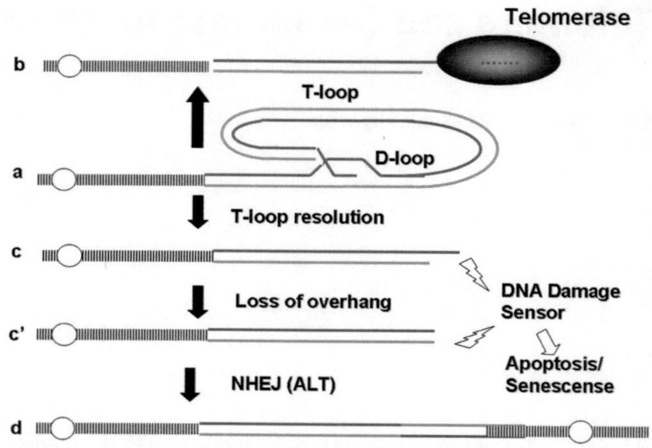

Fig. 2 Telomere elongation mechanism. When the T-loop structure is formed by invasion of the 3′ single-strand overhang into an adjacent duplex, telomerase is unable to bind the telomeric ends (**a**). Once telomere is opened, telomerase is able to elongate the telomere (**b**). Telomerase is the main mechanism for telomere elongation in human stem cells. But, the telomeric structure regulated by shelterin complex or other related substances is important to the length of the telomere. In the replication process of stem cells as well as normal somatic cells, telomere lengths gradually shorten and then T-loop resolution or loss of 3′ overhang strand activates the DNA damage sensors, including p53, leading to apoptosis or senescence (**c, c'**). The alternative lengthening of telomeres (ALT) mechanism instead of telomerase is shown to involve homologous recombination events between telomeric sequences (**d**) [28–30]. ALT might occur in some human normal stem cells

protect chromosome ends from degradation and DNA repair activities (Fig. 1) [1]. Telomere repeats consist of short guanine-rich repeat sequences in all eukaryotes with linear chromosomes, and range from 10 to 15 Kb in human somatic cells and 25 to 40 Kb in mice [17, 18]. During DNA replication in cell division, telomeres gradually shorten as a result of the incomplete replication of linear chromosomes, the so-called "end-replication problem." This progressive telomere shortening is one of the molecular mechanisms underlying aging, because critically short telomeres trigger chromosome senescence and apoptosis (Fig. 2) [18, 19]. The end of the telomere forms a 3′ overhang of the G-rich strand, which is generated by the postreplicative processing of the C-rich strand and folds back into the D-loop of the duplex telomeric DNA to form a protective "T-loop" to prevent degradation by exonucleases or processing as damaged DNA [20]. This loop hides the 3′ end from the telomerase and degradation activities as a primitive mechanism for telomere protection. The telomeric repeats are bound to a large multiprotein complex recently named "shelterin" [20]. This complex is proposed to regulate both telomere length and telomere protection [20]. In humans as well as higher animals, there is, an inherent problem with chromosome capping structure, namely that the telomere repeats become shorter and shorter with each cell cycle [21–23]. This is due to the fact that the replication machinery cannot replicate the most distal parts of the chromosomal DNA. A consequence of this is that 100–150 base pairs of DNA are lost from the telomere repeat block at each replication [21]. In addition to shelterin, telomeres are bound by nucleosome arrays, which show histone modifications characteristic of constitutive heterochromatin domains [18, 24]. Constitutive heterochromatin is generally found at transcriptionally inactive ("silenced") genomic regions of repetitive DNA, such as pericentric satellite repeats. Similar to pericentric chromatin, telomeres are bound by the heterochromatin protein 1 (HP1) and high levels of trimethylated H3-K9 and H4-K20, two histone modifications carried out by the histone methyltransferases (HMTases), suppressor of variegation 3–9 homolog (Suv39h) and suppressor of variegation 4-20 homolog (Suv4-20h), respectively [25–27]. The retinoblastoma proteins are also required for efficient H4-K20 trimethylation at both telomeres and centromeres through direct interaction with the Suv4-20h HMTases [25, 27]. The heterochromatic nature of telomeres, therefore, suggests that chromosome ends are in a compacted and "silenced" chromatin conformation, which has to be finely regulated in order to properly control telomere length [25].

1.2 Telomerase and Other Mechanisms for Telomere Elongation

Telomerase is the main mechanism for telomere elongation (Fig. 2). Telomerase is a complex of a reverse transcriptase protein encoded by the *TERT* (telomerase reverse transcriptase) gene and a template RNA TERC (telomerase RNA component). Telomerase can add telomeric repeats onto the chromosome ends, and prevents the replication-dependent loss of telomere and cellular senescence in highly proliferative cells of the germline and in majority of cancers [18]. Thus, telomerase activity and telomere maintenance are associated with immortality of cancer cells, germ-line cells, and embryonic stem cells.

Some immortal cell lines and tumors are able to maintain or elongate their telomeres without telomerase activity. This mechanism, called alternative lengthening of telomeres (ALT) [28, 29], has been shown to involve homologous recombination events between telomeric sequences (Fig. 2) [28–30]. ALT-positive cells are characterized by heterogeneous telomere lengths, with both very short and very long telomeres being present at the same time, and by the colocalization of telomeres with a specific type of promyelocytic leukemia (PML) body—so-called ALT-associated PML bodies (APBs) [28–30]. ALT mechanisms are also activated in *Terc*-deficient mice in cultured mouse embryonic fibroblasts and embryonic stem cells [31–33], and in vivo during germinal center formation [34], which indicates that ALT mechanisms can also be selected in nontumoral settings. ALT is mostly restricted to *Terc*-deficient mice, as well as immortal cell lines and tumors, indicating the existence of mechanisms that actively repress ALT in normal cells. Recent data indicate that components of shelterin complex such as Pot1 and TRF2, or TRF2-interacting proteins such as WRN, can influence telomere recombination [35–37] and are potential regulators of ALT. Similarly, subtelomeric DNA methylation [38, 39] and histone methylation at telomeres [39, 40] are potent repressors of telomere recombination and ALT activation. ALT can rescue the viability of telomerase-deficient yeast strains [30], but ALT mechanisms cannot rescue the viability of *Terc*-deficient mice, which suggests that ALT mechanisms do not operate to rescue survival of most higher organisms.

The third mechanism of telomere regulation was reported that telomeric repeats are transcribed by DNA-dependent RNA polymerase II, which, in turn, interacts with the TRF1 shelterin protein. Telomeric RNAs (TelRNAs) contain UUAGGG repeats, are polyadenylated and are transcribed from the telomeric C-rich strand. Transcription of mammalian telomeres is regulated by several mechanisms, including developmental status, telomere length, cellular stress, and tumor stage and chromatin structure. Using RNA-fluorescent in situ hybridization (FISH), TelRNAs are novel structural components of telomeric chromatin. Importantly, the evidence that TelRNAs block the activity of telomerase in vitro suggests that TelRNAs may regulate telomerase activity at chromosome ends [41]. TelRNAs may be a novel component of mammalian telomeres, which are anticipated to be fundamental for understanding telomere biology in stem cells and telomere-related diseases.

2 Telomere and Telomerase in Stem Cells

2.1 *Telomere and Telomerase in Germ Cells and Embryonic Stem Cells*

Telomerase activity is high in male germline and embryonic stem cells, but is low or absent in mature oocytes and cleavage stage embryos, and then high again in blastocysts [42]. The mechanism to reset telomere length in embryo remains poorly understood. Oocytes actually have shorter telomeres than somatic cells, but after fertilization their telomeres lengthen remarkably during early cleavage development [43]. Thus the capacity to elongate telomeres must reside within oocytes themselves. Moreover, telomeres are also elongated in the early cleavage embryos of telomerase-null mice, demonstrating that telomerase is unlikely to be responsible for the abrupt lengthening of telomeres in these cells (Fig. 3).

Coincident with telomere lengthening, extensive telomere sister-chromatid exchange (T-SCE) and co-localization of the DNA recombination proteins Rad50 and Trf1 were observed in early cleavage embryos [43]. Recently, telomeric repeats have been shown to be transcribed by DNA-dependent RNA polymerase II, which, in turn, interacts with the TRF1 shelterin protein in developmental stage [41]. Moreover, TelRNAs block the activity of telomerase in vitro, suggesting that TelRNAs may regulate telomerase activity at chromosome ends.

Embryonic stem (ES) cells are cultured stem cells derived from blastcyst and likely immortal and pluripotent having indefinite self-renewal capacity together with the ability to differentiate into all three germ layers. ES cells display high levels of telomerase activity and hTERT expression, both of which are rapidly down-regulated during differentiation [44] and much lower or absent in somatic cells including stem cells in self-renewal tissues (Fig. 3). The down-regulation of telomerase activity in differentiating EC cells was reported to be tightly correlated with histone deacetylation and DNA methylation of the *TERT* gene [45]. Moreover, increased telomerase activity enhanced the self-renewal ability, proliferation, and differentiation efficiency in the Tert-overexpressing ES cells [44]. High telomerase activity or expression of TERT can therefore be regarded as a marker of undifferentiated ES cells.

In cloned animals originated from adult nuclei with shortened telomeres, telomere length in somatic cells has been found to be comparable with that in age-matched normal animals originated from a fertilized egg with long telomeres, except for Dolly [46]. This finding indicates that the enucleated oocyte has an ability to reset the telomere length of the nucleus derived from a donor adult somatic cell by the elongation of telomeres [47]. Elucidation of this "reset"

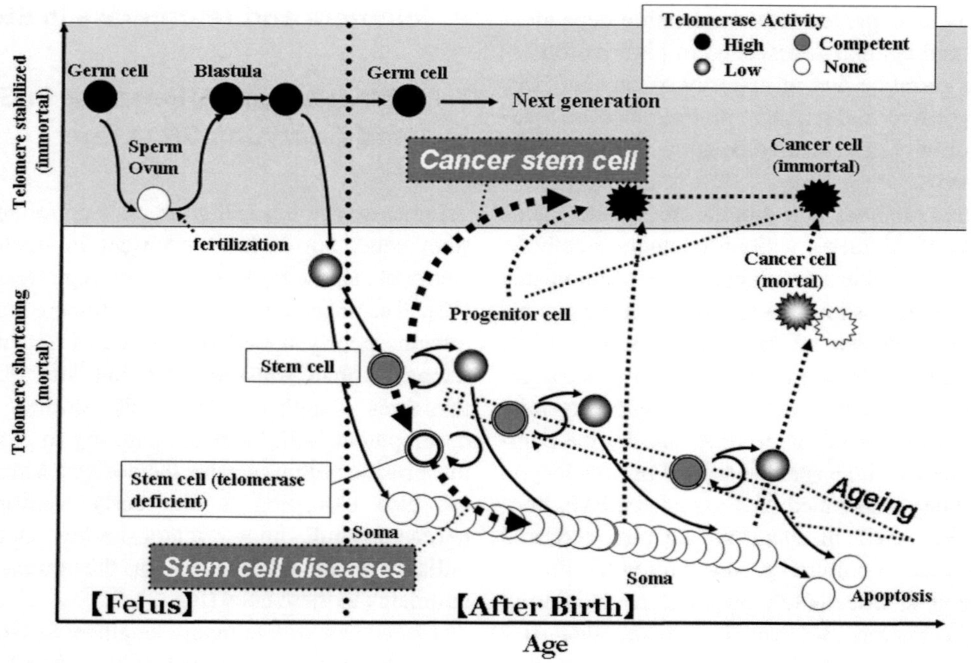

Fig. 3 Telomere and telomerase dynamics in human stem cells. Germ cells and embryonic stem cells have high levels of telomerase activity during rapid proliferation. While the telomerase activity is diminished in nonproliferating sperm and ovum, it is highly activated after fertilization and maintained in blastula and germ cells of the next generation. In the developmental stage, telomerase activity gradually decreases and diminishes in most somatic cells after birth. In adult stem cells, the level of telomerase activity is low or undetectable and up-regulated in committed progenitor cells that have high reproducible activity in each tissue, but it is insufficient to stably maintain their telomere length. Thus, normal stem cells are considered to be mortal and finally senesce by telomere shortening, indicating aging process. In telomerase-deficient adult stem cells, progressive telomere shortening leads to premature senescence and decreased regeneration of the stem cells, which consequently results in stem cell diseases. Cancer stem cells can be derived from normal stem cells, progenitor cells, or possibly somatic cells and might be immortal, having a capacity for indefinite self-renewal and proliferation

mechanism in oocyte will make a technical breakthrough in developing cloned animals. This reset may be the same mechanism in the elongation of telomere of germ cells. Both T-SCE and DNA recombination proteins decrease in blastocyst stage embryos, whereas telomerase activity increases and telomeres elongate only slowly. Telomeres may be elongated during the early cleavage cycles following fertilization through a recombination-based mechanism and, from the blastocyst stage onwards, telomerase only maintains the telomere length established by this alternative mechanism.

2.2 Telomere Length and Stem Cell Behaviors

Since adult stem cells usually express very low levels of telomerase, the telomeres of these cells slowly shorten. In agreement with this, it has been recently shown that human adult stem cells, including hematopoietic stem cells (hHSCs) and mesenchymal stem cells (hMSCs), diminish the mean telomere length during the long-term culturing period (Fig. 4) [48]. In addition, the telomere distribution on individual chromosomes in hMSCs is similar to the profile found in lymphocytes using a PNA–FISH method [49]. A similar comparison between telomerase-immortalized hMSC and lymphocytes revealed that this pattern is similar at both early and late phases of cellular expansion [48, 49]. This long conservation of a profile of telomere distribution, which is in accordance with the results from lymphocytes, clearly suggests that in both lymphocytes and in hMSCs there is a low degree of random fluctuation in the telomere dynamics, or that the telomerase activity – in addition to elongating the telomeres – has the effect of conserving the profile.

In the hematopoietic system, the telomerase-expressing cells are stem cells with self-renewal potential and their early descendants [5, 50, 51]. Age-dependent loss of telomeric DNA has been observed in both lymphocytes and neutrophils [52] as well as in other tissues [13, 17]. The length of telomere in adult blood leukocytes was reported to be shorter than that in germline cells from the same donor [53]. The subset of stem cells isolated from adult bone marrow showed shorter telomeres than the similar subset isolated from fetal liver or cord blood, suggesting that a progressive decline in telomere length with age occurs in HSCs [52]. It is possible that transplanted HSCs derived from aged donors may reach their proliferative limit during the lifetime of the recipient. In

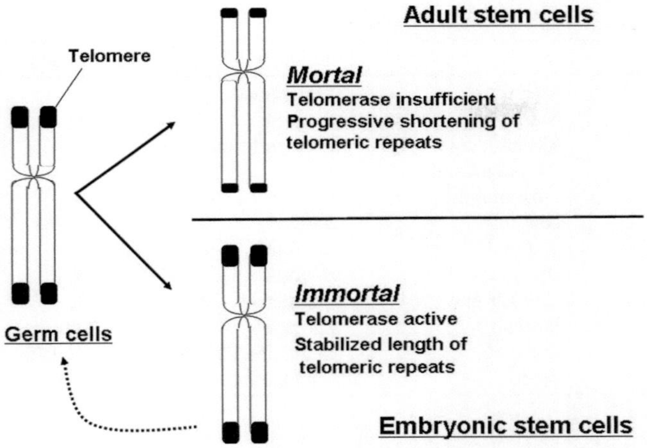

Fig. 4 Telomere and telomerase dynamics in human stem cells Since adult stem cells usually express very low levels of telomerase, the telomeres of these cells slowly shorten. Human adult stem cells, including hematopoietic stem cells (hHSCs) and mesenchymal stem cells (hMSCs), diminish the mean telomere length during the long-term culturing period, consequently going into senescence or apoptosis. However, the low level of telomerase is very important to maintain adult stem cell function, especially to regenerate tissue and organs. On the other hand, telomerase activity is high in male germline and embryonic stem cells in order to maintain the telomere length

fact, several cases of graft failure after HSC transplant have been reported to have unusual telomere shortening [54].

Low levels of telomerase activity are readily detectable in adult stem cells, including HSCs and in some of their differentiated progeny [5, 51]. Nevertheless, the age-related loss of telomeric DNA in these cells suggests that the telomerase activity in stem cells is insufficient to completely prevent telomere loss (Fig. 4). Telomere length alterations have also been noted in myelodysplastic syndromes (MDS) and chronic myeloid leukemia, where telomere loss in HSCs may be associated with cellular senescence and ineffective hematopoiesis and also contribute to genomic instability and subsequent leukemic progression [55]. These data support the idea that the replicative potential of adult stem cells could be limited by progressive telomere shortening and that telomere length can be used as an indirect indicator of stem cell proliferative history and potential (Fig. 3).

In human mesenchymal stem cells (hMSCs), one of the human nonhematopoietic stem cells, the telomere length gradually shorten at rate similar to non-stem cells (30–120 bp/PD), and telomerase activity is usally undetectable [56]. Several growth factors enhanced the mitotic potential of hMSCs, which maintained long telomeres without up-regulation of telomerase activity for more than 100 population doublings under culture with basic FGF [57]. Telomerase activity is lower or undetectable in hMSCs, and there may be a mechanism of telomere maintenance other than telomerase, such as ALT (alternative lengthening of telomeres), in hMSCs.

2.3 Telomerase for Regulation of Stem Cell Function

In mammals, stem cells of the germline are the only cells that maintain the link through animal generations. Since the phenomenon of self-renewal is unique to stem cells, allowing for the generation of at least one daughter cell with equivalent developmental potential, this ensures an unlimited supply of stem cells, irrespective of the demands put on the system by animal aging or stress caused by injury or disease. Unlike germ cells, adult stem cells such as those of the hematopoietic system are incapable of sustaining an indefinite line of generation either in vitro using cell culture or in vivo through serial transplantation (Fig. 4). The limited lifespan of HSCs, as manifested during serial transplantation, suggest that stem cells age. This correlates with indicators of cellular aging such as telomere shortening and a reduction in telomerase activity. This contrasts with embryonic stem cells and germ cells in which the telomere lengths are maintained, regardless of serial passage in vitro and animal age. The temporary activation or low expression of telomerase in adult stem cells may partially account for the resistance to telomere erosion of stem cells in general but the telomere lengths of these cells were gradually eroded, consequently yielding the aging of stem cells. Progenitor cells and adult stem cells possess some levels of telomerase, but these are insufficient to prevent telomere shortening associated with continuous tissue renewal with increasing age (Table 1).

Telomerase activity appears to be up-regulated in response to cytokine-induced proliferation and cell cycle activation in primitive HSCs and progenitor cells, and progressively down-regulated in more mature subsets [58], except for mitogenically activated lymphocytes (Fig. 3) [5]. Stem cells are capable of generating a very large number of committed progenitors and their descendants during a small number of self-renewal divisions. As such, individual stem cell turnover at any given point would be minimal and up-regulation of telomerase activity in stem cells could be minimal, preventing the generation of overly long telomeres [19]. Transient activation of telomerase activity induced during periods of rapid cellular proliferation could reduce telomere loss in lineage-committed progenitor cells. In more mature cells, telomerase activity is repressed regardless of their proliferation. These findings suggest that one important function of telomerase in HSCs is to reduce the rate of telomere loss during the period of rapid cell divisions, preventing premature critical shortening of telomeres and loss of telomere function. Such a role of telomerase would be critically important in case of increased proliferative demand, such as infections or blood loss [11].

Telomerase is up-regulated in cells that undergo rapid expansion in highly proliferative tissues, such as lymphocytes or keratinocytes, and in various stem cell

Table 1 Telomere and telomerase in human stem cells

Cell type	Lineage	Growth	Telomerase	Telomere	Reference
Embryonic	All tissue	Unlimited	High	Maintained	
Hematopoietic	Multipotent	Limited	Absent/low	Shorten	[95]
Mesenchymal	Multipotent/Skeletal tissue	Limited/Unlimited	Absent/low Upregulated by stimuli	Shorten/Maintained	[57, 96, 97]
Skin	Epidermis	Limited	Low	Shorten	[69, 98]
Hair follicle	Hair	Limited	Low	Shorten	
Intestinal crypt	Intestinal mucosa	Limited	Low	Shorten	[13]
Neuronal*	Several neuron	Limited	Low/Absent	Shorten	[99]
Pancreatic*	Ductal epithelial	Limited	Absent	Shorten	[100]
Liver epithelial	Bi-potential	Limited (durable)	Low	Maintained/Shorten	[101, 102]
Cancer stem cell	Multipotent	Unlimited	High	Maintained	[103]

*These were hMSCs derived *in vitro* cultured cells.

compartments [5, 7, 59]. In normal somatic cells, telomerase activity is mainly restricted to stem cells, suggesting that telomerase levels in these cells may be a determinant for maintenance of organism. Indeed, mutations in the telomerase core components, TERT and TERC, are present in patients suffering from several diseases such as aplastic anemia and dyskeratosis congenita (DKC) [60, 61]. During the past few years the specific role of telomerase in different stem cell compartments has been elucidated, mostly in well-characterized stem cell subtypes such as HSCs, epidermal stem cells (ESCs), and neural stem cells (NSCs). In particular, HSCs derived from human and mice lose telomeric DNA with age despite the presence of detectable telomerase activity [9, 10]. This progressive telomere shortening is proposed to act as a developmental barrier for HSCs, which may limit hematopoietic regeneration (Fig. 3). In support of this notion, HSCs from telomerase-deficient mice with short telomeres show a reduced ability to repopulate irradiated mice [12, 62]. Interestingly, stabilization of telomere length in these cells by Tert overexpression throughout the hematopoietic system is not sufficient to extend their transplantation capacity, suggesting that additional telomere-independent barriers limit HSC regeneration capacity [12].

In hMSCs, forced telomerase expression led to an extended life span and enhanced differentiation potential [63]. Thus, most supposedly, telomerase is required for both cell replication and differentiation. This implies that hMSCs may express at least a minimum level of telomerase activity to bring about regenerative capacity and differentiation potential.

Finally, it is important to note that the effects of telomere length and telomerase activity on different stem cell compartments (ESC, HSC, and adult NSC) are cell-autonomous, as demonstrated using in vitro clonogenicity assays [14, 15, 62]. This fact is relevant for designing potential therapeutic strategies based on telomerase reactivation, since it indicates that the effects of telomerase and telomere length on stem cell behavior are intrinsic to the stem cells and do not depend on physiological niche micro-environments. However, differentiation of stem cells occurs concomitantly with a diminishment of telomerase activity, indicating that stem cells are characterized by positive telomerase activity.

2.4 Animal Models for Telomere and Telomerase in Stem Cells

The use of loss-of-function and gain-of-function mouse models for telomerase has also served to establish the role of telomere length and telomerase activity in stem cell behavior. In successive generations of mice lacking telomerase RNA (*Terc*), late-generation animals manifested defective spermatogenesis due to increased apoptosis and reduced proliferation in the testis, as well as reduced proliferative capacity of hematopoietic cells in the bone marrow and spleen [8]. In primary cultures of mouse mesenchymal stem cells obtained from telomerase knockout mice, these cells lose multipotency and the capacity to differentiate into adipocytes and chondrocytes [8, 64]. Telomere shortening in the $Terc^{-/-}$ mice has been shown to result in decreased functionality of their skin stem cell compartment [14]. Since mice have much longer telomeres that cannot be seriously shortened within their short lifetime, this situation contrasts markedly with *Terc* knockout mice, where completely absent telomerase activity is tolerated without abnormality in the first few generations. In fact, proliferation and migration of epidermal stem cells out of the hair follicle niche upon mitogen-induced proliferation are partially inhibited in G1 $Terc^{-/-}$ mice with a slight reduction in telomere length and strongly inhibited in G3 $Terc^{-/-}$ mice with critically short telomeres [14]. After the fourth generation, *Terc* knockout mice begin to exhibit abnormalities similar to the phenotype of human DKC [62]. The immediate consequences of such mobilization defects are lower rates of proliferation in the hair follicle and in the adjacent transient-amplifying compartments, resulting in defective hair growth and a stunted hyperplastic response [14]. Besides the skin, other tissues with a high cell turnover, such as bone marrow, intestine, and testis, show atrophies in telomerase-deficient mice with critically short telomeres [65, 66], supporting the notion that telomere length is a determinant for tissue maintenance in the wide context of the organism. Wild-type mice derived from late generations of *Terc* +/– mice with short telomeres and positive telomerase activity displayed a similar phenotype of DKC, indicating that short telomeres themselves, rather than lack of telomerase activity, cause stem cell failure [67]. Recently, *Cdkn1a* deletion improved stem cell function and lifespan of mice with telomere dysfunction, indicating that up-regulation of p21 (encoded by *Cdkn1a*) in response to shortened telomeres impairs repopulation capacity of stem cells in age-related diseases and senescence [68].

On the other hand, the reverse phenomenon was observed in transgenic mice with constitutive Tert overexpression in the epidermis, including the stem cell compartment (K5-*mTert* mice) that presents increased epidermal stem cell mobilization upon proliferation stimuli. This increased mobilization in epidermal stem cells is concomitant with increased keratinocyte proliferation, enhanced hair growth, and augmented skin hyperplasia [14]. Similar activation in epidermal stem cells and hair growth has been reported by a different group using a transgenic mouse in which Tert is overexpressed in a conditional manner [69]. Interestingly, in the later report, the hair-growth-promoting effects of Tert were found to be independent of the telomerase RNA component as well as telomerase activity, suggesting an unclassical role of Tert in addition to its known role in telomere synthesis. However, the potential involvement of Tert independent of Terc in other in vivo proliferative responses is still unclear, since it has been recently shown that the absence of Terc abrogates the enhanced skin tumorigenesis and wound healing responses shown by transgenic mice that constitutively overexpress Tert in the skin [70]. These different requirements for Terc in epidermal growth versus hair growth may be explained by the existence of distinct cell populations involved in these processes. Indeed, recent data indicate the existence of distinct stem cell populations within the epidermis, which are separately involved in regenerating either the hair follicles or the epidermis [71–73].

Another important implication from these experiments was that overexpression of the telomerase catalytic subunit *hTERT* is obviously required but not sufficient to transform normal cells into malignant cells. Hahn et al. [74] reported that the ectopic expression of *hTERT*, in combination with the simian virus 40 large-T antigen and oncogenic H-ras, led to the creation of tumorigenic cells from normal human fibroblasts. Because telomerase expression is found in 80–90% of human cancers [75–77], it has become an attractive target for the treatment of neoplastic tumors. Indeed, recent observations have demonstrated that the expression of a mutant catalytic subunit of human telomerase resulted in complete inhibition of telomerase activity, reduction of telomere length, and death of tumor cells [78].

These insights into the central role of telomeres and telomerase in cellular senescence and immortality came from studies on telomerase-deficient mice and hTERT transgenic mice and have contributes to elucidate the stem cell regulation by telomere and telomerase.

3 Stem Cell Diseases

3.1 *Aging and Cancer as Stem Cell Diseases*

The diseases due to deficiency in telomerase activity and short telomeres are characterized by skin abnormalities and bone marrow failure, the latter resulting from defects in maintaining the hematopoietic stem cell compartment [60, 79, 80]. In the adult, stem cells are essential for homeostasis, growth, and repair. In humans, some parts of diseases such as aplastic anemia and DKC, which are characterized by skin abnormalities and bone marrow failure, are caused by mutations in telomerase components TERT and TERC [60, 80]. The skin and bone marrow are well-characterized major stem cell compartments [81]. Mutations in these genes have also been described in familial idiopathic pulmonary fibrosis. The loss of telomerase activity is postulated to be the cause of premature loss of tissue renewal and premature death. Therefore, there is a presumptive connection between the role of telomerase in mediating stem cell cycling for the purpose of tissue renewal and its loss, which leads to organ malfunction. Telomerase can be regarded as a case of antagonistic pleiotropy, and it facilitates chromosomal duplication and cellular division through the maintenance of telomere length and preventing premature senescence. On the other hand, prolonged replications and longevity would subject cells to the accumulation of mutations leading to transformation of stem cells to cancerous cells. It has been proposed that cellular senescence in aging acts primarily as an anticancer mechanism, where the number of cell divisions allowed would represent the balance between the advantage gained from cancer prevention and the disadvantages that arise due to limiting cellular turnover required for homeostasis [82]. Moreover, cancer and aging, two biological processes in which telomerase activity has been implicated, are increasingly seen as stem cell diseases [18, 83]. In particular, cancer may often originate from the transformation of normal stem cells, while aging has been associated with a progressive decline in the number and/or functionality of certain stem cells [18, 84]. Thus, aging and cancer can be classified as stem cell diseases.

3.2 Diseases Due to Lack of Telomerase in Stem Cells: Dyskeratosis Congenita (DKC) and Aplastic Anemia

The critical importance of telomerase activity in the human stem cells has been highlighted as the etiology of dyskeratosis congenita (DKC). In one type of this disease, a defect of the telomerase RNA template gene results in absence of telomerase activity and premature telomere shortening, resulting in development of bone marrow failure, intestinal disorders, or malignancy, typically in patients less than 50 years old. In other types, positional cloning identified a gene affected in many patients with X-linked DKC as the human homolog of yeast CBF5, termed *DKC1*, encoding a protein dyskerin. The association of the telomerase complex with dyskerin suggested that DKC phenotype may be the result of altered telomerase activity [19]. Then, the discovery of a 3′ deletion in the gene encoding TERC in a single large family with autosomal dominant DKC confirmed the importance of telomerase deficiency as an etiology for the DKC and highlighted the importance of telomerase in maintaining cellular lifespan and replicative potential in human organs. After the fourth generation, the *Terc* knockout mice begin to exhibit abnormalities similar to the phenotype of human DKC [62]. Several types of DKC have been summarized [85].

Aplastic anemia results from the failure of bone marrow to produce sufficient quantities of all hematopoietic lineages. The etiology of this disease is unclear in most cases, but is generally thought to be the result of HSC damage or loss. Telomere length in peripheral blood granulocytes and monocytes in patients with aplastic anemia or related disorders was significantly shorter than that in age-matched controls and correlated with disease duration [86]. An inverse correlation between age-adjusted telomere length and peripheral blood counts was also observed in aplastic anemia [87]. As in DKC-associated aplastic anemia, there is a report that found mutations in *TERC* gene in patients with aplastic anemia [88].

3.3 Cancer Stem Cells

Currently, there is evidence that cancer originates from the transformation of normal stem cells, while aging has been associated with progressive decline in number and/or functionality of certain stem cells. According to the cancer stem cell hypothesis, cancer arises from the transformation of rare adult tissue stem cells into cancer stem cells [89]. The growth and expansion of a tumor is supported by a small pool of such cells capable of self-renewal, and asymmetric division producing a daughter cancer stem cell and another progeny capable of differentiation. The transformation of the adult stem cell is attributed to the acquisition of mutations due to the exceptional longevity of stem cells. Hematopoietic malignancies provide some of the most convincing evidence as the diseases have been shown to originate from and be sustained by cancer stem cells [90]. These malignancies can be divided into premalignant, chronic, and acute stages, the last being the most advanced and malignant. It is known that the premalignant stage, such as myelodysplasia (MDS), and the chronic stage, such as chronic myeloid or lymphoid leukemia (CML or CLL), can progress into more malignant and acute stage of the disease, for example, acute lymphoid or myeloid leukemia (AML or ALL) [91]. In this progression, telomere shortening in HSCs is considered to be a risk factor that contributes to the development of chromosomal instability and malignant transformation. In agreement with this hypothesis, telomere shortening in patients with DKC correlates with an increased risk of hematological neoplasia [60]. Similar to hematopoietic malignancies, several kinds of solid cancers, especially breast and brain tumors [92, 93], are considered to be generated and/or propagated by cancer stem cells.

To maintain their characteristic replicative ability, cancer stem cells (CSCs) should have telomere-lengthening mechanisms. However, since identification and purification of CSCs are difficult in most solid tumors, the telomere and telomerase dynamics in CSCs remain unclear. No matter from what the CSCs derived – normal stem cells, committed progenitor cells, or differentiated cells [94] – CSCs might have acquired immortality by mutational events in telomere-lengthening mechanisms, typically activation of telomerase [55], considering the fact that telomere shortening and cellular senescence are inevitable in all of these original cells (Figs. 3, 4). Theoretically, inhibition of telomerase in CSCs to limit proliferation capacity and induce apoptosis, maintenance of telomere length in premalignant lesions to prevent activation of telomerase and transformation to cancer stem cells, and identification of specific marker proteins in cancer stem cells in each organ are likely to optimize screening and therapeutic interventions as novel anticancer strategies.

3.4 Future Prospects

Recent research has been focused on the telomere and its role in regulation of stem cells. However, much remains to be learned, and what is clear is the complexity of telomere and telomerase interactions in cell cycle progression and cell signaling in both normal somatic cells and stem cells.

Both elements have a key role in determining the proliferative capacity in all human stem cells. Telomere shortening occurs in most human somatic cells and triggers DNA damage responses that mediate cell cycle arrest or apoptosis, while human stem cells can escape this trigger by employing a telomerase-dependent telomere lengthening mechanism in replication. This process in normal adult stem cells, however, appears to have only limited potential in extending the replicative capacity, because the telomere shortening occurs slowly with aging. Recently, decrease of telomerase activity has been found in the stem cell diseases such as dyskeratosis, aplastic anemia, and idiopathic pulmonary fibrosis. On the contrary, cancer stem cells have complete telomere-lengthening mechanisms that provide indefinite proliferation capacity. Research on the telomere-telomerase dynamics in stem cells could lead to the development of novel etiology-based radical therapies in stem cell diseases and cancers.

References

1. Blackburn EH. Switching and signaling at the telomere. Cell. 2001;106:661–73.
2. Blackburn EH, Greider CW, Henderson E, Lee MS, Shampay J, Shippen-Lentz D. Recognition and elongation of telomeres by telomerase. Genome. 1989;31:553–60.
3. Greider CW, Blackburn EH. The telomere terminal transferase of Tetrahymena is a ribonucleoprotein enzyme with two kinds of primer specificity. Cell. 1987;51:887–98.
4. Dhaene K, Van Marck E, Parwaresch R. Telomeres, telomerase and cancer: an up-date. Virchows Arch. 2000;437:1–16.
5. Hiyama K, Hirai Y, Kyoizumi S, et al. Activation of telomerase in human lymphocytes and hematopoietic progenitor cells. J Immunol. 1995;155:3711–5.
6. Smogorzewska A, de Lange T. Regulation of telomerase by telomeric proteins. Annu Rev Biochem. 2004;73:177–208.
7. Counter CM, Gupta J, Harley CB, Leber B, Bacchetti S. Telomerase activity in normal leukocytes and in hematologic malignancies. Blood. 1995;85:2315–20.
8. Lee HW, Blasco MA, Gottlieb GJ, Horner JW, 2nd, Greider CW, DePinho RA. Essential role of mouse telomerase in highly proliferative organs. Nature. 1998;392:569–74.
9. Vaziri H, Dragowska W, Allsopp RC, Thomas TE, Harley CB, Lansdorp PM. Evidence for a mitotic clock in human hematopoietic stem cells: loss of telomeric DNA with age. Proc Natl Acad Sci U S A. 1994;91:9857–60.
10. Allsopp RC, Cheshier S, Weissman IL. Telomere shortening accompanies increased cell cycle activity during serial transplantation of hematopoietic stem cells. J Exp Med. 2001;193:917–24.
11. Allsopp RC, Morin GB, DePinho R, Harley CB, Weissman IL. Telomerase is required to slow telomere shortening and extend replicative lifespan of HSCs during serial transplantation. Blood. 2003;102:517–20.
12. Allsopp RC, Morin GB, Horner JW, DePinho R, Harley CB, Weissman IL. Effect of TERT over-expression on the long-term transplantation capacity of hematopoietic stem cells. Nat Med. 2003;9:369–71.
13. Hiyama E, Hiyama K, Tatsumoto N, Kodama T, Shay JW, Yokoyama T. Telomerase activity in human intestine. Int J Oncol. 1996;9:453–8.
14. Flores I, Cayuela ML, Blasco MA. Effects of telomerase and telomere length on epidermal stem cell behavior. Science. 2005;309:1253–6.
15. Ferron S, Mira H, Franco S, et al. Telomere shortening and chromosomal instability abrogates proliferation of adult but not embryonic neural stem cells. Development. 2004;131:4059–70.
16. Vulliamy T, Marrone A, Goldman F, et al. The RNA component of telomerase is mutated in autosomal dominant dyskeratosis congenita. Nature. 2001;413:432–5.
17. Wright WE, Shay JW. Telomere biology in aging and cancer. J Am Geriatr Soc. 2005;53:S292–4.
18. Blasco MA. Telomeres and human disease: ageing, cancer and beyond. Nat Rev Genet. 2005;6:611–22.
19. Collins K, Mitchell JR. Telomerase in the human organism. Oncogene. 2002;21:564–79.
20. de Lange T. Shelterin: the protein complex that shapes and safeguards human telomeres. Genes Dev. 2005;19:2100–10.
21. Harley CB, Futcher AB, Greider CW. Telomeres shorten during ageing of human fibroblasts. Nature. 1990;345:458–60.
22. Harley CB, Vaziri H, Counter CM, Allsopp RC. The telomere hypothesis of cellular aging. Exp Gerontol. 1992;27:375–82.
23. Hastie ND, Dempster M, Dunlop MG, Thompson AM, Green DK, Allshire RC. Telomere reduction in human colorectal carcinoma and with ageing. Nature. 1990;346:866–8.
24. Blasco MA. Mice with bad ends: mouse models for the study of telomeres and telomerase in cancer and aging. EMBO J. 2005;24:1095–103.
25. Flores I, Benetti R, Blasco MA. Telomerase regulation and stem cell behaviour. Cur Op Cell Biol. 2006;18:254–60.
26. Garcia-Cao M, O'Sullivan R, Peters AH, Jenuwein T, Blasco MA. Epigenetic regulation of telomere length in mammalian cells by the Suv39h1 and Suv39h2 histone methyltransferases. Nat Genet. 2004;36:94–9.
27. Gonzalo S, Garcia-Cao M, Fraga MF, et al. Role of the RB1 family in stabilizing histone methylation at constitutive heterochromatin. Nat Cell Biol. 2005;7:420–8.
28. Muntoni A, Reddel RR. The first molecular details of ALT in human tumor cells. Hum Mol Genet. 2005;14 Spec No. 2: R191–6.
29. Dunham MA, Neumann AA, Fasching CL, Reddel RR. Telomere maintenance by recombination in human cells. Nat Genet. 2000;26:447–50.
30. Lundblad V. Telomere maintenance without telomerase. Oncogene. 2002;21:522–31.
31. .Hande MP, Samper E, Lansdorp P, Blasco MA. Telomere length dynamics and chromosomal instability in cells derived from telomerase null mice. J Cell Biol. 1999;144:589–601.
32. Chang S, Khoo CM, Naylor ML, Maser RS, DePinho RA. Telomere-based crisis: functional differences between telomerase activation and ALT in tumor progression. Genes Dev. 2003;17:88–100.
33. Niida H, Shinkai Y, Hande MP, et al. Telomere maintenance in telomerase-deficient mouse embryonic stem cells: characterization of an amplified telomeric DNA. Mol Cell Biol. 2000;20:4115–27.
34. Herrera E, Martinez AC, Blasco MA. Impaired germinal center reaction in mice with short telomeres. EMBO J. 2000;19:472–81.
35. Blanco R, Munoz P, Flores JM, Klatt P, Blasco MA. Telomerase abrogation dramatically accelerates TRF2-induced epithelial carcinogenesis. Genes Dev. 2007;21:206–20.
36. Wu L, Multani AS, He H, et al. Pot1 deficiency initiates DNA damage checkpoint activation and aberrant homologous recombination at telomeres. Cell. 2006;126:49–62.

37. Laud PR, Multani AS, Bailey SM, et al. Elevated telomere-telomere recombination in WRN-deficient, telomere dysfunctional cells promotes escape from senescence and engagement of the ALT pathway. Genes Dev. 2005;19:2560–70.
38. Gonzalo S, Jaco I, Fraga MF, et al. DNA methyltransferases control telomere length and telomere recombination in mammalian cells. Nat Cell Biol. 2006;8:416–24.
39. Benetti R, Garcia-Cao M, Blasco MA. Telomere length regulates the epigenetic status of mammalian telomeres and subtelomeres. Nat Genet. 2007;39:243–50.
40. Benetti R, Gonzalo S, Jaco I, et al. Suv4-20h deficiency results in telomere elongation and derepression of telomere recombination. J Cell Biol. 2007;178:925–36.
41. Schoeftner S, Blasco MA. Developmentally regulated transcription of mammalian telomeres by DNA-dependent RNA polymerase II. Nat Cell Biol. 2008;10:228–36.
42. Wright WE, Piatyszek MA, Rainey WE, Byrd W, Shay JW. Telomerase activity in human germline and embryonic tissues and cells. Dev Genet. 1996;18:173–9.
43. Liu L, Bailey SM, Okuka M, et al. Telomere lengthening early in development. Nat Cell Biol. 2007;9:1436–41.
44. Armstrong L, Saretzki G, Peters H, et al. Overexpression of telomerase confers growth advantage, stress resistance, and enhanced differentiation of ESCs toward the hematopoietic lineage. Stem Cells. 2005;23:516–29.
45. Lopatina NG, Poole JC, Saldanha SN, et al. Control mechanisms in the regulation of telomerase reverse transcriptase expression in differentiating human teratocarcinoma cells. Biochem Biophys Res Commun. 2003;306:650–9.
46. Morton O, Williams N. First Dolly, now headless tadpoles. Science. 1997;278:798.
47. Meerdo LN, Reed WA, White KL. Telomere-to-centromere ratio of bovine clones, embryos, gametes, fetal cells, and adult cells. Cloning Stem Cells. 2005;7:62–73.
48. Serakinci N, Christensen R, Graakjaer J, et al. Ectopically hTERT expressing adult human mesenchymal stem cells are less radiosensitive than their telomerase negative counterpart. Exp Cell Res. 2007;313:1056–67.
49. Graakjaer J, Christensen R, Kolvraa S, Serakinci N. Mesenchymal stem cells with high telomerase expression do not actively restore their chromosome arm specific telomere length pattern after exposure to ionizing radiation. BMC Mol Biol. 2007;8:49.
50. Lansdorp PM. Lessons from mice without telomerase. J Cell Biol. 1997;139:309–12.
51. Morrison SJ, Prowse KR, Ho P, Weissman IL. Telomerase activity in hematopoietic cells is associated with self-renewal potential. Immunity. 1996;5:207–16.
52. Vaziri H, Schachter F, Uchida I, et al. Loss of Telomeric DNA during aging of normal and trisomy 21 human lymphocytes. Am J Hum Genet. 1993;52:661–7.
53. Cooke HJ, Smith BA. Variability at the telomeres of the human X/Y pseudoautosomal region. Cold Spring Harb Symp Quant Biol. 1986;51 Pt 1:213–9.
54. Wynn RF, Cross MA, Hatton C, et al. Accelerated telomere shortening in young recipients of allogeneic bone-marrow transplants. Lancet. 1998;351:178–81.
55. Ohyashiki JH, Sashida G, Tauchi T, Ohyashiki K. Telomeres and telomerase in hematologic neoplasia. Oncogene. 2002;21:680–7.
56. Fehrer C, Lepperdinger G. Mesenchymal stem cell aging. Exp Gerontol. 2005;40:926–30.
57. Yanada S, Ochi M, Kojima K, Sharman P, Yasunaga Y, Hiyama E. Possibility of selection of chondrogenic progenitor cells by telomere length in FGF-2-expanded mesenchymal stromal cells. Cell Prolif. 2006;39:575–84.
58. Chiu CP, Dragowska W, Kim NW, et al. Differential expression of telomerase activity in hematopoietic progenitors from adult human bone marrow. Stem Cells. 1996;14:239–48.
59. Harrington L. Does the reservoir for self-renewal stem from the ends? Oncogene. 2004;23:7283–9.
60. Mason PJ, Wilson DB, Bessler M. Dyskeratosis congenita – a disease of dysfunctional telomere maintenance. Curr Mol Med. 2005;5:159–70.
61. Ball SE, Gibson FM, Rizzo S, Tooze JA, Marsh JC, Gordon-Smith EC. Progressive telomere shortening in aplastic anemia. Blood. 1998;91:3582–92.
62. Samper E, Fernandez P, Eguia R, et al. Long-term repopulating ability of telomerase-deficient murine hematopoietic stem cells. Blood. 2002;99:2767–75.
63. Simonsen JL, Rosada C, Serakinci N, et al. Telomerase expression extends the proliferative life-span and maintains the osteogenic potential of human bone marrow stromal cells. Nat Biotechnol. 2002;20:592–6.
64. Liu L, Saldanha SN, Pate MS, Andrews LG, Tollefsbol TO. Epigenetic regulation of human telomerase reverse transcriptase promoter activity during cellular differentiation. Genes Chromosomes Cancer. 2004;41:26–37.
65. Lee DH, Yang SC, Hong SJ, Chung BH, Kim IY. Telomerase: a potential marker of bladder transitional cell carcinoma in bladder washes. Clin Cancer Res. 1998;4:535–8.
66. Rudolph KL, Chang S, Lee HW, et al. Longevity, stress response, and cancer in aging telomerase-deficient mice. Cell. 1999;96:701–12.
67. Hao LY, Armanios M, Strong MA, et al. Short telomeres, even in the presence of telomerase, limit tissue renewal capacity. Cell. 2005;123:1121–31.
68. Choudhury AR, Ju Z, Djojosubroto MW, et al. Cdkn1a deletion improves stem cell function and lifespan of mice with dysfunctional telomeres without accelerating cancer formation. Nat Genet. 2007;39:99–105.
69. Sarin KY, Cheung P, Gilison D, et al. Conditional telomerase induction causes proliferation of hair follicle stem cells. Nature. 2005;436:1048–52.
70. Cayuela ML, Flores JM, Blasco MA. The telomerase RNA component Terc is required for the tumour-promoting effects of Tert overexpression. EMBO Rep. 2005;6:268–74.
71. Ghazizadeh S, Taichman LB. Multiple classes of stem cells in cutaneous epithelium: a lineage analysis of adult mouse skin. EMBO J. 2001;20:1215–22.
72. Levy V, Lindon C, Harfe BD, Morgan BA. Distinct stem cell populations regenerate the follicle and interfollicular epidermis. Dev Cell. 2005;9:855–61.
73. Ito M, Liu Y, Yang Z, et al. Stem cells in the hair follicle bulge contribute to wound repair but not to homeostasis of the epidermis. Nat Med. 2005;11:1351–4.
74. Hahn WC, Counter CM, Lundberg AS, Beijersbergen RL, Brooks MW, Weinberg RA. Creation of human tumor cells with defined genetic elements. Nature. 1999;400:464–8.
75. Hiyama E, Hiyama K. Clinical utility of telomerase in cancer. Oncogene. 2002;21:643–9.
76. Hiyama E, Yokoyama T, Hiyama K, et al. Alteration of telomeric repeat length in adult and childhood solid neoplasias. Int J Oncol. 1995;6:13–6.
77. Kim NW, Piatyszek MA, Prowse KR, et al. Specific association of human telomerase activity with immortal cells and cancer. Science. 1994;266:2011–5.
78. Hahn WC, Stewart SA, Brooks MW, et al. Inhibition of telomerase limits the growth of human cancer cells. Nat Med. 1999;5:1164–70.
79. Greenwood MJ, Lansdorp PM. Telomeres, telomerase, and hematopoietic stem cell biology. Arch Med Res. 2003;34:489–95.

80. Maciejewski JP, Risitano A. Hematopoietic stem cells in aplastic anemia. Arch Med Res. 2003;34:520–7.
81. Armanios MY, Chen JJ, Cogan JD, et al. Telomerase mutations in families with idiopathic pulmonary fibrosis. N Engl J Med. 2007;356:1317–26.
82. Wright WE, Shay JW. Time, telomeres and tumours: is cellular senescence more than an anticancer mechanism? Trends Cell Biol. 1995;5:293–7.
83. Bell DR, Van Zant G. Stem cells, aging, and cancer: inevitabilities and outcomes. Oncogene. 2004;23:7290–6.
84. McEachern MJ, Krauskopf A, Blackburn EH. Telomeres and their control. Annu Rev Genet. 2000;34:331–58.
85. de la Fuente J, Dokal I. Dyskeratosis congenita: advances in the understanding of the telomerase defect and the role of stem cell transplantation. Pediatr Transplant. 2007;11: 584–94.
86. Lee JJ, Kook H, Chung IJ, et al. Telomere length changes in patients with aplastic anaemia. Br J Haematol. 2001;112: 1025–30.
87. Brummendorf TH, Maciejewski JP, Mak J, Young NS, Lansdorp PM. Telomere length in leukocyte subpopulations of patients with aplastic anemia. Blood. 2001;97:895–900.
88. Vulliamy T, Marrone A, Dokal I, Mason PJ. Association between aplastic anaemia and mutations in telomerase RNA. Lancet. 2002;359:2168–70.
89. McCulloch EA. Stem cells in normal and leukemic hemopoiesis (Henry Stratton Lecture, 1982). Blood. 1983;62:1–13.
90. Lapidot T, Sirard C, Vormoor J, et al. A cell initiating human acute myeloid leukaemia after transplantation into SCID mice. Nature. 1994;367:645–8.
91. Greenberg P, Cox C, LeBeau MM, et al. International scoring system for evaluating prognosis in myelodysplastic syndromes. Blood. 1997;89:2079–88.
92. Al-Hajj M, Clarke MF. Self-renewal and solid tumor stem cells. Oncogene. 2004;23:7274–82.
93. Singh SK, Clarke ID, Terasaki M, et al. Identification of a cancer stem cell in human brain tumors. Cancer Res. 2003;63:5821–8.
94. Clarke MF, Fuller M. Stem cells and cancer: two faces of eve. Cell. 2006;124:1111–5.
95. Brummendorf TH, Balabanov S. Telomere length dynamics in normal hematopoiesis and in disease states characterized by increased stem cell turnover. Leukemia. 2006;29:273–5.
96. Zimmermann S, Voss M, Kaiser S, Kapp U, Waller CF, Martens UM. Lack of telomerase activity in human mesenchymal stem cells. Leukemia. 2003;17:1146–9.
97. Izadpanah R, Trygg C, Patel B, et al. Biologic properties of mesenchymal stem cells derived from bone marrow and adipose tissue. J Cell Biochem. 2006;99:1285–97.
98. Sugihara M, Ohshima K, Nakamura H, et al. Decreased expression of telomerase-associated RNAs in the proliferation of stem cells in comparison with continuous expression in malignant tumors. Int J Oncol. 1999;15:1075–80.
99. Wright LS, Prowse KR, Wallace K, Linskens MH, Svendsen CN. Human progenitor cells isolated from the developing cortex undergo decreased neurogenesis and eventual senescence following expansion in vitro. Exp Cell Res. 2006;312:2107–20.
100. Moriscot C, de Fraipont F, Richard MJ, et al. Human bone marrow mesenchymal stem cells can express insulin and key transcription factors of the endocrine pancreas developmental pathway upon genetic and/or microenvironmental manipulation in vitro. Stem Cells. 2005;23:594–603.
101. Dan YY, Riehle KJ, Lazaro C, et al. Isolation of multipotent progenitor cells from human fetal liver capable of differentiating into liver and mesenchymal lineages. Proc Natl Acad Sci U S A. 2006;103:9912–7.
102. Oh BK, Lee CH, Park C, Park YN. Telomerase regulation and progressive telomere shortening of rat hepatic stem-like epithelial cells during in vitro aging. Exp Cell Res. 2004;298:445–54.
103. Hertzog RG. Ancestral telomere shortening: a countdown that will increase mean life span? Med Hypotheses. 2006;67:157–60.

The Role of Mitochondria in Stem Cell Biology

Claudia Nesti, Livia Pasquali, Michelangelo Mancuso, and Gabriele Siciliano

Abstract This chapter summarizes the current data on the role of mitochondrial DNA mutations and mitochondrial metabolism in stem cell biology. Topics include oxidative stress in normal and in cancer stem cells, subcellular mitochondrial localization, and oxygen consumption related to the maintenance of stem cells properties and differentiation capacity. The possible uses of stem cells as a therapeutic tool in mitochondrial disorders are also reported.

Keywords Mitochondria · Adult stem cells · Embryonic stem cells · Oxidative stress

1 Introduction

Mitochondria are cytoplasmic organelles found in all eukaryotic cells, responsible for converting nutrients into the energy-yielding molecule adenosine triphosphate (ATP) to fuel the activities of the cells. This function, known as aerobic respiration, is the reason why mitochondria are frequently referred to as the powerhouse of the cell. Furthermore, they play essential roles in processes such as steroid metabolism, calcium homeostasis, apoptosis, and cellular proliferation.

Stem cells can be defined as cells having high proliferative potential with the ability to self-renew. They are undifferentiated, but can generate daughters of more than one distinct phenotype. Stem cells have potential for numerous biomedical applications, including therapeutic cell replacement to repair damaged body organs, as tools for studying genetic defects and testing drugs, and as models for studying cell differentiation and early development. Considering the potential uses of stem cells in regenerative medicine, until now the large majority of studies on both adult and embryonic stem cells have been focused on the assessment of their genetic stability in culture rather than considering functional cellular markers such as mitochondrial properties.

Due to their critical functions in all cell types, the role of mitochondria must also be crucial for stem cells. However, despite the enormous production of studies on stem cell viability, proliferation, and differentiation capacity, their functional activity, in particular their energy status, has been virtually neglected thus far.

2 Mitochondrial Localization in Stem Cells

It has recently been hypothesized that stem cell competence (i.e., stem cell stability and pluripotency) can be verified using functional mitochondrial characteristics [1, 2]. Lonergan and colleagues [1] found that several mitochondrial properties, such as subcellular localization and metabolic activity, change with increasing passage number in culture and therefore can be considered markers of "stemness." In fact, it is known that in the oocytes of different species (rodent and primate) mitochondria are randomly distributed, while after fertilization their localization becomes peripronuclear [3]. Considering that embryonic stem cells (ESCs) are derived from the blastomeres, it has been shown that a perinuclear arrangement of mitochondria is maintained in stem cells, and that a more scattered cytoplasmic distribution is acquired with differentiation and senescence (culture passages). Mitochondrial metabolic activity has been correlated to the degree of cell differentiation: ATS cells (an adult primate stromal cell line derived from adipose tissue) showed a lower ATP/cell content and a high rate of oxygen consumption during the first culture passages, compared to cells cultured for longer periods [1].

Furthermore, cultures with a higher percentage of stem cells, as judged by the predominantly perinuclear arrangement of mitochondria, had a 10-fold higher rate of oxygen consumption compared to the more differentiated cells. Collectively, these data indicate that the perinuclear arrangement of mitochondria, accompanied by a high metabolic rate but a low ATP/cell content, may be valid indicators of stem cell

C. Nesti (✉)
CUCCS, Center for the Clinical Use of Stem Cells, University of Pisa, Via Roma 67, 56126 – Pisa, Italy
e-mail: cla_nesti@yahoo.it

V.K. Rajasekhar, M.C. Vemuri (eds.), *Regulatory Networks in Stem Cells*, Stem Cell Biology and Regenerative Medicine,
DOI 10.1007/978-1-60327-227-8_12,

differentiation competence, whereas divergence from this profile indicates that cells are differentiating or perhaps becoming senescent. Therefore, it can be assumed that periodic measurement of the percentage of cells with a perinuclear mitochondrial arrangement might serve as a method to monitor the maintenance of "stemness" in stem cell populations, at least in adult stem cell lines.

However, these observations were only partially confirmed by Piccoli and co-workers [4], since they found a predominantly perinuclear mitochondrial arrangement accompanied by a low mitochondrial oxygen consumption rate and a low amount of mitochondrial respiratory chain complexes in human CD34+ hematopoietic stem cells. The fact that these cells seem to be a poor oxidative phosphorylating cell type may be explained by their normally quiescent state in the stromal niche, where oxygen tension is supposed to be minimal.

Human ESCs derived from pre-implantation blastocysts also grow well in hypoxic conditions. In this environment, ESCs are resistant to spontaneous differentiation and stably remain pluripotent [5]; they present only few immature mitochondria, but with the differentiation, the number of mitochondria increases and their morphology becomes that of mature cells, with a higher number of cristae and dense matrix [6]. In fact, fully differentiated individual cell types can be characterized by the number of mitochondria they possess and the number of copies of mitochondrial genome per organelle [7].

Mitochondrial clustering is frequently correlated with function [8], and alterations of this pattern have been associated with changes in cell activities: treatment of cultured embryonic hippocampal neurons with ethidium bromide (EtBr, an agent that depletes mitochondrial DNA), prior to the establishment of cell polarity, prevents axon formation by altering mitochondrial ultrastructure and subcellular localization [9].

From a methodological point of view, the functional activity of mitochondria in cultured neural precursor cells can serve as a criterion of optimum conditions for culturing and isolation of a population of neural stem cells (NSCs) with high proliferative activity and differentiation potential. Since high energy potential of some cells in neurospheres has been associated with the expression of nestin, an early marker of stem cells, a relationship between mitochondrial function and the status of NSCs [10] has been suggested.

3 Mitochondrial DNA Alterations in Stem Cells

Mitochondrial DNA (mtDNA) is a small (16.6-kb) self-replicating molecule that encodes 13 essential proteins of the mitochondrial oxidative phosphorylation complexes, 2 rRNA and 22 tRNA genes. Within cells, there are thousands of mitochondria, each containing multiple copies of mtDNA. MtDNA mutations are random and are thought to occur because of the presence of free-radical generating enzymes, potentiated by poor DNA repair mechanisms and the lack of protective histones [11, 12]. These mutations can affect all copies of the mitochondrial genome (homoplasmy) or a portion of them (heteroplasmy). For a mutated mitochondrial genotype to result in a mutated cellular phenotype, homoplasmy or high levels of heteroplasmy must be present.

Although heteroplasmy is usually observed in post-mitotic tissues characterized by high energy demand, such as brain and skeletal muscle, mtDNA mutations also accumulate in dividing cells, including adult stem cells such as epithelial cells of the mouth and of the colonic crypts [13, 14]. Considering that one colonic crypt stem cell with a mtDNA mutation has the potential to expand and replace the entire colonic crypt stem cell population, the mutation load in these cells can be high. It has been suggested that a somatic mutation accumulates in the cell through "clonal expansion," that is, accumulation of the progeny of the single initially mutated DNA molecule. However, this mechanism may play a minor role in other stem cell types that are normally quiescent and are activated only in response to injury or damage, making difficult the accumulation of mutations to levels sufficient to result in a biochemical defect. Quiescence is critical for protecting the stem cell compartment: analysis of bone marrow side-population (SP) cells (G_0 phase) under 5-fluorouracil (5-FU)-induced myelosuppressive conditions showed that these cells were resistant to bone marrow suppression induced by 5-FU [15]. Since 5-FU induces apoptosis of actively cycling cells, this result indicates that SP cells are quiescent and resistant to apoptosis.

Several studies looked for changes in telomerase activity and in the expression of cell surface markers and transcription factors in stem cells. The expression of antigens and transcription factors (such as OCT-4, hTERT, SOX2, and Cripto), telomerase activity, telomere length, and karyotype appear stable for the ES cell lines analyzed after continuous culture for over 1 year [16, 17]. By contrast, in adult mesenchymal stem cells (MSCs) tested for the expression of telomerase activity, human telomerase reverse transcriptase (hTERT) transcripts, and alternative lengthening of telomere (ALT) mechanism at different passages, a progressive decrease in proliferative capacity until reaching senescence was observed [18].

The incidence of aneuploidy and chromosomal alterations has also been evaluated in stem cell cultures. Draper and collaborators [19] reported karyotypic changes involving the gain of chromosome 17q in three independent human ESC lines on five independent occasions, meanwhile the gain of chromosome 12 was seen occasionally: it may be speculated that increased dosage of chromosome 17q and 12 gene(s) provides a selective advantage for the propagation of undifferentiated human ESCs, at least in in vitro condition. Other studies confirmed the tendency of ESCs to acquire karyotypic abnormalities [20], which are accompanied by a proliferative advantage and a noticeable shortening in the population doubling time [21]. Therefore, it can be stated that evaluating the genetic assessment of stem cells is a particularly important issue when working with embryonic stem cells, which are highly heterogeneous and likely to present abnormalities having derived from the inner cell mass of stored embryos. Besides, ESCs derived from in vitro–produced embryos are prone to carry mtDNA mutations because of the gonadotropin stimulation [22]; this proves that mitochondrial replication must be occurring during oocyte maturation, contrary to current beliefs. Accumulation of mitochondrial mutations with multiple culture passages may be a problem for cell lines derived from embryonic as well as adult tissues. However a recent report analyzing rhesus macaque adult bone marrow stromal stem cells (BMSC) showed high levels of the mtDNA common deletion, regardless of passage [23]. Therefore, existing stem cell lines, whether generated from in vitro– or in vivo–derived embryos, are not suitable for therapeutic strategies and perhaps not for drug testing as well. It is imperative that the cell lines generated for these purposes be attentively characterized. On the other hand, adult stem cells seem to be more genetically stable: karyotypic analysis of several rat adult pluripotent stem cell lines was normal [24], human MSCs expanded in vitro did not show chromosomal abnormalities [18], adult NSCs propagated in vitro for over a year maintain a stable profile with regard to self-renewal, differentiation, growth factor dependence, karyotype, and molecular profiling [25]. This evidence, together with the absence of spontaneous differentiation, suggest that stem cells derived from adults may have a higher degree of regulatory control than embryonic stem cells, rendering them suitable for cell therapy approaches.

It is known that murine oocytes possess 10^5 mtDNA copies immediately prior to fertilization and there is no increase in mtDNA copy number until post-implantation; therefore, each newly divided blastomere within the pre-implantation embryo will possess fewer copies of the genome at each stage of post-fertilization cell division [26]. Consequently, undifferentiated human ESCs, derived from the inner cell mass of the blastocyst, have very low numbers of mitochondria, but their differentiation should result in the expansion of mitochondrial number [6], which is likely to be matched by increased mtDNA copy number. MtDNA replication is regulated by the nuclear-encoded mitochondrial transcription factor A (TFAM) and the mitochondrial-specific DNA polymerase gamma, the latter consisting of a catalytic (POLG) and an accessory (POLG2) subunit. Recently, Facucho-Oliveira and colleagues [27] observed the up-regulation of *Polg*, *Polg2*, and *Tfam* in spontaneously differentiating murine ESCs, but this event was not accompanied by a significant increase in mtDNA replication. The authors speculate that the increase in expression could be related to repair of the replicated mtDNA, because POLG has been shown to replace single nucleotide gaps generated by the apurinic/apyrimidinic endonucleases [28]. This event taking place in the early days (days 1–5) of fetal development could be crucial to ensure that the cells – future precursors of all other lineages – do not harbor large number of rearrangements possibly leading to the onset of severe mtDNA-type diseases. This would also be crucial for the continued viability of ESCs, especially because these are prone to rearrangement as passage number increases [29]. After ESCs commit to a specific lineage, mtDNA is extensively replicated, leading to the enrichment of the mitochondrial content, which provides the higher levels of ATP required for further differentiation.

4 Stem Cells, Oxygen, and Oxidative Stress

Oxygen is an important signal in all major aspects of stem cell biology including proliferation and tumorigenesis, cell death and differentiation, self-renewal and migration. Oxygen can also exert different effects on different cell types (Table 1). For the cultivation of stem cells, more oxygen is not necessarily better: virtually all stem cells cultivated in lower oxygen conditions proliferate more than in traditional 20% O_2 environments. This may reproduce the physiological in vivo conditions: hematopoietic stem cell populations normally reside in a particularly low oxygen niche of bone marrow, and oxygen has been suggested to be a primary determinant of stem cell mobilization from marrow in disease processes [30]. Low oxygen culture conditions have an anti-apoptotic effect on different types of stem cells [31, 32], probably because fewer reactive oxygen species are generated in culture in low vs. high oxygen conditions. Mesoderm-derived stem cells show remarkable differentiation changes in response to lower oxygen in culture. Both skeletal muscle stem cells and mesenchymal stem cell lines are more likely to undergo myogenesis rather than adipogenesis in lower oxygen culture than in 20% O_2 [33]. Also, central and peripheral nervous system stem cells in culture show remarkable changes in differentiation patterns when exposed to low oxygen in culture: central nervous system (CNS) stem cells in low oxygen differentiated toward a dopaminergic neuron phenotype, and the serotonergic neuron differentiation is enhanced as well [31]. Stem cells are characterized by the capacity to migrate to the sites of injury or inflammation; among the growth factors known to enhance mobilization

Table 1 Tabular summary of the effects of oxygen and reactive oxygen species (ROS) on different stem cell types

Cell type	Effects of the variation of O_2 and ROS levels on stem cell biology	Refs.
Hematopoietic stem cells	↑ O_2: mobilization from bone marrow	[30]
CNS precursors and stem cells	↓ O_2: anti-apoptotic effect, preferential dopaminergic differentiation	[31]
Megakaryocytes	↓ O_2: anti-apoptotic effect	[32]
Mesoderm-derived stem cells	↓ O_2: preferential myogenic differentiation	[33]
Differentiating ESCs	↑ ROS: ↑ antioxidant enzymes expression	[37]
Mouse embryonic, neural, hematopoietic stem cells	↑ ROS: up-regulation of detoxifier system genes	[39, 40]
Grafted neural stem cells	↑ ROS: protection of host neurons from oxidative stress	[41]
ESCs overexpressing hTERT	↑ resistance to oxidative stress	[43]
ESCs	Chronic ↑ ROS: inhibition of cardiomyogenesis and vasculogenesis	[46]
ESCs	Low level of ROS: enhanced cardiomyogenesis and vasculogenesis	[46]
Rodent NSCs	↑ ROS: mitochondria-mediated apoptosis	[50]
ESCs	Chronic ↑ ROS: activation of NF-κB pathway and possible transformation to tumor stem cells	[62]

of some marrow stem cells into the circulation are vascular endothelial growth factor (VEGF) [34] and erythropoietin, both classically oxygen-regulated [35]. Since injuries that induce marrow mobilization often involve hypoxia or oxidative stress, oxygen is likely fundamental to this aspect of stem cell biology as well.

Oxidative stress, defined as a disturbance in the prooxidant–antioxidant balance (Fig. 1), has been implicated in a variety of physiological and pathological processes, including DNA damage, proliferation, cell adhesion, and cell survival. The generation of reactive oxygen species (ROS) can arise from toxic insults or normal metabolic processes as the mitochondrial electron transport chain. Under normal physiological conditions, enzymes like superoxide dismutases, SOD1 and SOD2, glutathione (GSH) peroxidase, and catalase prevent formation/accumulation of oxygen metabolites. However, overproduction of oxidants and/or dysfunction of endogenous antioxidant defences may occur and result in oxidative stress-induced injury with damage to all the major classes of biological macromolecules, such as nucleic acids, proteins, lipids and carbohydrates.

It has been shown that in differentiating cells there is an increase in mitochondrial mass, accompanied by increased ATP production and, consequently, by higher production of ROS. Therefore, differentiating cells should activate effective antioxidant systems, such as catalase, GSH peroxidase, peroxiredoxins (Prx), and others [36]. Cho and colleagues [37] found that the antioxidant enzymes expression showed a dramatic change during differentiation, with significant increase in the expression of Cu/Zn-superoxide dismutase (Cu/Zn-SOD), Prx1, Prx2, and GSH peroxidase. Nonetheless, intracellular ROS amount was found to be increased

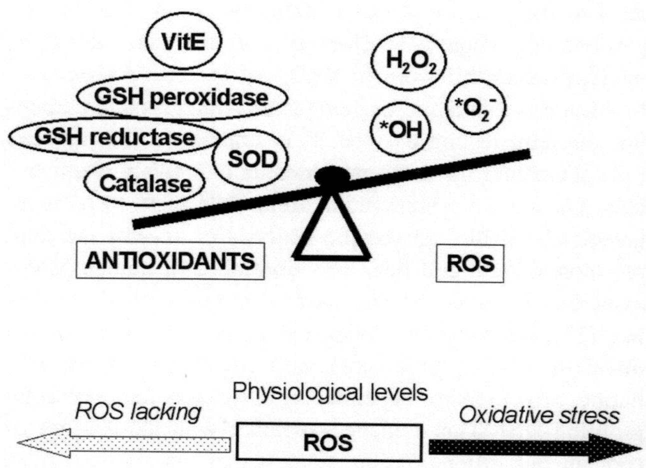

Fig. 1 Reactive oxygen species (ROS) are formed during a variety of biochemical reactions and cellular functions, such as mitochondrial metabolism. Under normal conditions, the formation of pro-oxidants (free radicals) is balanced by their consumption by antioxidants, both occurring at similar rate. Antioxidants are molecules or compounds that act as free radical scavengers, forming innocuous end-products. Antioxidants include intracellular enzymes such as superoxide dismutase (SOD), glutathione (GSH) peroxidase, catalase, as well as endogenous molecules (such as GSH and coenzyme Q10), essential nutrients (vitamin C, vitamin E, selenium, etc.) and dietary compounds (such as bioflavonoids). Oxidative stress results from an imbalance between formation and neutralization of pro-oxidants

H_2O_2: hydrogen peroxide; $^*O_2^-$: superoxide anion; *OH: hydroxyl radical.

with differentiation, suggesting a possible role of these unstable molecules in cell signalling, proliferation, and differentiation [38].

Transcriptional profiling of mouse embryonic, adult neural, and hematopoietic stem cells revealed a characteristic pattern of expression of "stemness" genes that distinguishes stem cells from mature cells: in particular, stem cells showed a higher resistance to stress, with up-regulation of DNA repair, protein folding, ubiquitin system, and detoxifier system genes [39]. Redox state is itself modulated by cell-extrinsic signaling molecules that alter the balance between self-renewal and differentiation: growth factors that promote self-renewal cause progenitor cells to become more reduced, while exposure to signaling molecules that promote differentiation causes progenitors to become more oxidized. Thus, ROS may be involved in the balance between self-renewal and differentiation [40]. At the same time, the redox balance changes as cells mature, suggesting that antioxidant enzymes could play a major role in the preservation of "stemness," and that at least some of the up-regulated genes provide resistance against environmental stresses. Interestingly, grafted NSCs have also been proved to metabolically protect compromised host neurons from oxidative-mediated damage. Madhavan and co-workers [41] showed that mice protected by intrastriatal grafting of NSCs prior to 3-nitropropionic acid (3-NP, an oxidative stress-causing agent) treatment expressed none or only mild movement symptoms, and that the grafted mice recovered substantially faster than the control animals. Moreover, the majority of preserved neurons in these brains remained of host origin while donor NSCs remained mostly undifferentiated, suggesting that the integrity of the host cells had been preserved from 3-NP intoxication.

Oxidative stress shortens the life span of stem and progenitor cells, which can then no longer repair tissue and organ damage, leading to an accumulation of damage and a shortened life span. There is usually a good correlation between the metabolic rate of an animal and the level of ROS generated by its mitochondria. ROS accelerate aging through random and sequential damage to cell components. One of the mechanisms by which the life span is shortened is impairment of telomerase, which repairs DNA damage due to oxidation. Murasawa and collaborators [42] reported that overexpression of hTERT by adenovirus-mediated gene delivery could delay senescence and recover/enhance the regenerative properties of endothelial progenitor cells. Furthermore, in embryonic stem cells overexpressing hTERT, an enhanced self-renewal ability, improved resistance to apoptosis, and increased proliferation have been shown [43]. In addition, these cells accumulate lower concentrations of peroxides than the wild-type cells, suggesting greater resistance to oxidative stress [43]. These results are conflicting with other data: it has been shown that hTERT has a functional N-terminal mitochondrial targeting sequence, and that ectopic hTERT expression in human fibroblasts correlates with increase in mtDNA damage after hydrogen peroxide (H_2O_2) treatment [44]. However, cells carrying mutated N-terminal mitochondrial leader sequence have significantly reduced levels of mtDNA damage following H_2O_2 treatment, without showing any loss of viability or cell growth. Thus, localization of hTERT in the mitochondria renders cells more susceptible to oxidative stress-induced mtDNA damage and subsequent cell death, whereas nuclear-targeted hTERT, in the absence of mitochondrial localization, is associated with diminished mtDNA damage, increased cell survival, and protection against cellular senescence [45]. These findings suggest that nuclear- and mitochondrial-localized hTERT may be differently regulated in stem cells and in adult somatic cells, explaining the different results found in the literature.

The role of ROS in cardiovascular differentiation of embryonic stem cells appears to be antagonistic. While continuous exposure to ROS results in inhibition of cardiomyogenesis and vasculogenesis, the pulse chase exposure to low-level ROS enhances differentiation toward the cardiomyogenic as well as vascular cell lineage [46]. Elevated intracellular ROS generation may provide the differentiation stimulus that directs transplanted stem cells into the cardiovascular lineage. Therefore, it is not surprising that mitochondrial oxidative metabolism is a critical regulator of the cardiac differentiation of stem cells: any disruption of the respiratory chain functions prevents mitochondrial organization and compromises the energetic infrastructure, causing deficient sarcomerogenesis and contractile malfunction [47].

Mitochondria also play a central role as regulators of apoptosis: studies on NSCs from the subventricular zone of adult rat brain have shown that these cells undergo mitochondria-mediated cell death in response to apoptotic stimuli, while the Fas pathway may function as a mediator of growth rather than death in NSCs [48, 49]. Rodent NSCs exposed to neurotoxic compounds (methyl-mercury and 2,3-dimethoxy-1,4-naphthoquinone) known to induce oxidative stress exhibited typical apoptotic morphology with nuclear condensation, and release of cytochrome *c* from the mitochondria with subsequent caspase activation [50].

There is increasing evidence that prenatal events may predispose to diseases later in life [51]. In support of this hypothesis, several studies have pointed out that prenatal exposure to high levels of ROS reduced the antioxidant activity of catalase [52], increasing the susceptibility of the whole nervous system and of NSCs to oxidative stress later in life.

5 Cancer Stem Cells and Oxidative Stress

Currently, there is increasing evidence that mtDNA content directly affects the metabolic state of cells [53, 54]. Moreover, a possible role of mtDNA mutations in tumor progression has been suggested by the finding that the trans-mitochondrial hybrids (cybrid) cell line carrying mutations in ATPase6 gene, encoded by mtDNA, exhibited increased tumorigenicity and reduced apoptosis [55, 56]. An increasing number of studies show that proliferating tumors in both human and animal models contain mutated or deleted mtDNA and/or dysfunctional mitochondria [56, 57].

Stem cells and cancer cells share several traits, including unlimited self-renewal capabilities and the ability to generate a diverse range of other cell types. However, while proliferation of stem cells is highly concerted and controlled, in cancer, proliferation is uncontrolled and detrimental. The presence of stem cell populations has now been identified in a number of malignancies of the blood, breast, colon, pancreas, and brain [58, 59]. This has many implications for the diagnosis and treatment of cancers, since the most effective way to eliminate the disease would be to target cancer stem cells for the complete eradication of the tumor.

Several signaling pathways have been implicated in both cancer and stem cells (recently reviewed by Dreesen and Brivanlou [60]). In particular, the nuclear factor-kappa-B (NF-κB) pathway regulates genes involved in key cellular processes such as proliferation, stress response, innate immunity, and inflammation [61]. NF-κB signaling is activated by a variety of extracellular factors including oxidative stress. Such stimuli allow NF-κB dimers to translocate to the nucleus and activate transcription of target genes. It has been found that in about 50% of Hodgkin's lymphoma, one of the NF-κB transcription factors is amplified, and that NF-κB signaling is necessary to maintain pluripotency in human ESCs. These findings might support the hypothesis that stem cells might undergo transformation into cancer stem cells under prolonged oxidative stress, probably due to molecular modifications such as hyperoxia-induced NF-κB [62]. However, other recent findings indicate that normal and malignant cells of various origins differ in their sensitivity to oxidative stress: the growth of malignant mesenchymal stem cells treated with aldehyde 4-hydroxynonenal (HNE), known as a second messenger of free radicals and a signaling molecule, results more affected by the compound than the normal cells [63]. Therefore, it remains to be fully elucidated how the ROS status relates to cancer stem cells and their ability to self-renew and escape death signals.

6 Neural Stem Cells, Irradiation, and Oxidative Stress

Radiation therapy is an important adjuvant treatment of various primary or metastatic tumors that occur in the central nervous system (CNS). Several reports, however, suggest that irradiation of the brain can cause dysfunctions of the normal CNS. It has been suggested that the cellular events leading to radiation injury can be explained in terms of a loss in the reproductive capacity of cells regarded as targets within the CNS. Since the discovery of regions of active neurogenesis in mammal adult brain [64], that is, the subventricular zone (SVZ) of the lateral ventricles and the dentate subgranular zone (SGZ) of the hippocampus, the proliferating NSCs could be considered those targets. Irradiation also appears to be surprisingly selective toward certain types of proliferating cells: neuronal precursors have proved to be more sensitive to radiation than glial precursors [65, 66]. The effects on neurogenesis observed thus far seem irreversible, and no direct effects of radiation on the structure or function of the mature neurons have been noted [67].

Radiation has been found to cause a delayed and/or persistent induction of ROS and apoptosis (induced by the release of cytochrome *c* from the mitochondria) in various cell lines, including neural precursors cells [68]. It has been hypothesized that oxidative stress plays a major role in affecting neurogenesis and subsequent cognitive decline after radiation-induced cell injury [69]. In vitro and in vivo studies regarding hippocampal precursor survival and differentiation after irradiation showed that local microenvironmental factors, such as oxidative stress, may play inhibitory roles in post-irradiation precursor cell function and recovery [70, 71]. This evidence suggests that antioxidant treatment strategies focused on reducing radiation-induced ROS could conceivably be used to restore neurogenesis and perhaps to ameliorate specific adverse effects of irradiation. In support of this hypothesis it has been recently reported that edaravone (the free-radical scavenger 3-methyl-1-phenyl-2-pyrazolin-5-one) protects human NSCs in vitro from cell death and restores their differentiation ability after irradiation; however, the protective effect of edaravone is not observed in human glioma cell lines [72].

7 Mitochondrial Diseases and Stem Cells

The idea that almost every tissue contains stem-like progenitor cells that repair tissues after damage is currently accepted [73]; local repair can be enhanced by stem-like progenitor cells from the bone marrow. Stem/progenitor cells repair tissues by differentiating to replace lost cells, providing cytokines and growth factors, and by cell fusion with endogenous cells.

A variety of heritable and acquired diseases is linked to mitochondrial dysfunction. In a recent report the possibility that stem cells may repair cells with non-functional mitochondria has been investigated. Spees and colleagues [74] cocultured ρ^0 cells (cells depleted of mtDNA and therefore incapable of aerobic respiration) with either adult nonhematopoietic stem/progenitor cells from human bone marrow (MSCs) or skin fibroblasts. In both cases cocultures produced clones of rescued ρ^0 cells with functional mitochondria, proving that mitochondria or mtDNA can move between cells. The possibility that the cells underwent fusion followed by selective loss of the donor cell nuclei is unlikely because transfer of genetic markers was never detected. The transfer of mitochondria appeared to involve an active cellular process rather than passive uptake of cellular fragments or organelles, since coculture with platelets or isolated mitochondria did not rescue the aerobic respiration. The existence of mitochondrial transfer in other cell types has been very recently confirmed by the work of Plotnikov and co-workers [75], which revealed mitochondrial transfer from MSCs to cardiomyocytes in a coculture system. Transport in the opposite direction was never observed, probably because of the type of cross-talking cells. This mechanism may be important to restore the energetic status in conditions in which mitochondria are damaged or dysfunctional. Despite these intriguing results, the frequency of such transfers in vivo is difficult to estimate, even if it may partially explain the beneficial effects registered after either in vivo mobilization of MSCs or administration of large numbers of human MSCs in animal models of spinal cord injury, stroke, and heart disease [76–78].

To our knowledge, the first mitochondrial disease in which stem cell treatment has been conducted so far is mitochondrial neurogastrointestinal encephalomyopathy (MNGIE). This multisystemic autosomal recessive disorder is due to mutations in the *ECGF1* gene encoding thymidine phosphorylase (TP). Patients with MNGIE present severe reduction of TP activity, resulting in systemic accumulation of its substrates, thymidine (dThd) and deoxyuridine (dUrd) [79, 80]. Excess of these nucleosides leads to deoxynucleotide pool imbalances, which cause mtDNA instability [79]. To reduce circulating nucleosides, allogenic stem cell transplantations (alloSCTs) have been performed in two patients [81]. In the first case, in which the source of stem cells was umbilical cord blood from an unrelated donor, alloSCT failed to engraft, but the second patient (donor cells were from the HLA-identical patient's brother) achieved mixed donor chimerism, which partially restored TP activity and lowered plasma nucleosides. Thus, alloSCT can correct biochemical abnormalities in the blood of patients with MNGIE.

8 Conclusions

In summary, the use of mitochondria has proved to be an important tool in investigating stem cell properties. In particular, the significance of mitochondrial intracellular arrangement as an indicator of "stemness" and the role of oxidative stress in stem cells and in cancer stem cells have been discussed. Regarding DNA alterations, we think that it is of fundamental importance to check for the presence of mtDNA mutations, as well as for nuclear DNA integrity, in embryonic and adult stem cell lines before they can be used for either experimental or therapeutic purposes.

Moreover, even if the long-term effects and the clinical efficacy of the studies conducted so far should be verified, the potential applications of stem cells in the treatment of mitochondrial disorders may open new strategies for the cure of these almost untreatable diseases.

References

1. Lonergan T, Brenner CA, Bavister BD. Differentiation-related changes in mitochondrial properties as indicators of stem cell competence. J Cell Physiol. 2006;208:149–53.
2. Bavister BD. The mitochondrial contribution to stem cell biology. Reprod Fertil Dev. 2006;18:829–38.
3. Barnett DK, Kimura J, Bavister BD. Translocation of active mitochondria during hamster preimplantation embryo development studied by confocal laser scanning microscopy. Dev Dyn. 1996;205:64–72.
4. Piccoli C, Ria R, Scrima R, et al. Characterization of mitochondrial and extra-mitochondrial oxygen consuming reactions in human hematopoietic stem cells. J Biol Chem. 2005;280:26467–76.
5. Ezashi T, Das P, Roberts RM. Low O2 tensions and the prevention of differentiation of hES cells. Proc Natl Acad Sci U S A. 2005;102:4783–8.
6. St John JC, Ramalho-Santos J, Gray HL, et al. The expression of mitochondrial DNA transcription factors during early cardiomyocyte in vitro differentiation from human embryonic stem cells. Cloning Stem Cells. 2005;7:141–53.
7. St John JC, Amaral A, Bowles E, et al. The analysis of mitochondria and mitochondrial DNA in human embryonic stem cells. Methods Mol Biol. 2006;331:347–74.
8. Yaffe M. The machinery of mitochondrial inheritance and behaviour. Science. 1999;283:1493–7.
9. Mattson MP, Partin J. Evidence for mitochondrial control of neuronal polarity. J Neurosci Res. 1999;56:8–20.
10. Plotnikov EY, Marei MV, Podgornyi OV, Aleksandrova MA, Zorov DB, Sukhikh GT. Functional activity of mitochondria in cultured neural precursor cells. Bull Exp Biol Med. 2006;141:142–6.
11. Carew JS, Huang P. Mitochondrial defects in cancer. Mol Cancer. 2002;1:9.
12. Taylor RW, Turnbull DM. Mitochondrial DNA mutations in human disease. Nat Rev Genet. 2005;6:389–402.
13. Nekhaeva E, Bodyak ND, Kraytsberg Y, et al. Clonally expanded mtDNA point mutations are abundant in individual cells of human tissues. Proc Natl Acad Sci U S A. 2002;99:5521–6.
14. Taylor RW, Barron MJ, Borthwick GM, et al. Mitochondrial DNA mutations in human colonic crypt stem cells. J Clin Invest. 2003;112:1351–60.
15. Arai F, Hirao A, Ohmura M, et al. Tie2/angiopoietin-1 signaling regulates hematopoietic stem cell quiescence in the bone marrow niche. Cell. 2004;118:149–61.
16. Rosler ES, Fisk GJ, Ares X, et al. Long-term culture of human embryonic stem cells in feeder-free conditions. Dev Dyn. 2004;229:259–74.
17. Oh SK, Kim HS, Ahn HJ, et al. Derivation and characterization of new human embryonic stem cell lines: SNUhES1, SNUhES2, and SNUhES3. Stem Cells. 2005;23:211–9.
18. Bernardo ME, Zaffaroni N, Novara F, et al. Human bone marrow derived mesenchymal stem cells do not undergo transformation after long-term in vitro culture and do not exhibit telomere maintenance mechanisms. Cancer Res. 2007;67:9142–9.

19. Draper JS, Smith K, Gokhale P, et al. Recurrent gain of chromosomes 17q and 12 in cultured human embryonic stem cells. Nat Biotechnol. 2004;22:53–4.
20. Eggan K, Rode A, Jentsch I, et al. Male and female mice derived from the same embryonic stem cell clone by tetraploid embryo complementation. Nat Biotechnol. 2002;20:455–9.
21. Cowan CA, Klimanskaya I, McMahon J, et al. Derivation of embryonic stem-cell lines from human blastocysts. N Engl J Med. 2004;350:1353–6.
22. Gibson TC, Kubisch HM, Brenner CA. Mitochondrial DNA deletions in rhesus macaque oocytes and embryos. Mol Hum Reprod. 2005;11:785–9.
23. Gibson TC, Pei Y, Quebedeaux TM, Brenner CA. Mitochondrial DNA deletions in primate embryonic and adult stem cells. Reprod Biomed Online. 2006;12:101–6.
24. Henson NL, Heaton ML, Holland BH, et al. Karyotypic analysis of adult pluripotent stem cells. Histol Histopathol. 2005;20: 769–84.
25. Foroni C, Galli R, Cipelletti B, et al. Resilience to transformation and inherent genetic and functional stability of adult neural stem cells ex vivo. Cancer Res. 2007;67:3725–33.
26. Shoubridge EA, Wai T. Mitochondrial DNA and the mammalian oocyte. Curr Top Dev Biol. 2007;77:87–111.
27. Facucho-Oliveira JM, Alderson J, Spikings EC, Egginton S, St John JC. Mitochondrial DNA replication during differentiation of murine embryonic stem cells. J Cell Sci. 2007;120:4025–34.
28. Graziewicz MA, Longley MJ, Copeland WC. DNA polymerase gamma in mitochondrial DNA replication and repair. Chem Rev. 2006;106:383–405.
29. Maitra A, Arking DE, Shivapurkar N, et al. Genomic alterations in cultured human embryonic stem cells. Nat Genet. 2005;37: 1099–1103.
30. Antoniou ES, Sund S, Homsi EN, Challenger LF, Rameshwar P. A theoretical simulation of hematopoietic stem cells during oxygen fluctuations: prediction of bone marrow responses during hemorrhagic shock. Shock. 2004;22:415–22.
31. Studer L, Csete M, Lee SH, et al. Enhanced proliferation, survival, and dopaminergic differentiation of CNS precursors in lowered oxygen. J Neurosci. 2000;20:7377–83.
32. Mostafa SS, Miller WM, Papoutsakis ET. Oxygen tension influences the differentiation, maturation and apoptosis of human megakaryocytes. Br J Haematol. 2002;111:879–89.
33. Csete M, Walikonis J, Slawny N, et al. Oxygen-mediated regulation of skeletal muscle satellite cell proliferation and adipogenesis in culture. J Cell Physiol. 2001;189:189–96.
34. Rabbany SY, Heissig B, Hattori K, Rafii S. Molecular pathways regulating mobilization of marrow-derived stem cells for tissue revascularization. Trends Mol Med. 2003;9:109–17.
35. Kessinger A, Sharp JG. The whys and hows of hematopoietic progenitor and stem cell mobilization. Bone Marrow Transplant. 2003;31:319–29.
36. Rhee SG, Kang SW, Chang TS, Jeong W, Kim K. Peroxiredoxin, a novel family of peroxidases. IUBMB Life. 2001;52:35–41.
37. Cho YM, Kwon S, Pak YK, et al. Dynamic changes in mitochondrial biogenesis and antioxidant enzymes during the spontaneous differentiation of human embryonic stem cells. Biochem Biophys Res Commun. 2006;348:1472–8.
38. Sauer H, Wartenberg M, Hescheler J. Reactive oxygen species as intracellular messengers during cell growth and differentiation. Cell Physiol Biochem. 2001;11:173–86.
39. Ramalho-Santos M, Yoon S, Matsuzaki Y, Mulligan RC, Melton DA. "Stemness": transcriptional profiling of embryonic and adult stem cells. Science. 2002;298:597–600.
40. Noble M, Smith J, Power J, Mayer-Pröschel M. Redox state as a central modulator of precursor cell function. Ann N Y Acad Sci. 2003;991:251–71.
41. Madhavan L, Ourednik V, Ourednik J. Grafted neural stem cells shield the host environment from oxidative stress. Ann NY Acad Sci. 2005;1049:185–8.
42. Murasawa S, Llevadot J, Silver M, Isner JM, Losordo DW, Asahara T. Constitutive human telomerase reverse transcriptase expression enhances regenerative properties of endothelial progenitor cells. Circulation. 2002;106:1133–9.
43. Armstrong L, Saretzki G, Peters H, et al. Overexpression of telomerase confers growth advantage, stress resistance, and enhanced differentiation of ESCs toward the hematopoietic lineage. Stem Cells. 2005;23:516–29.
44. Santos JH, Meyer JN, Skorvaga M, Annab LA, Van Houten B. Mitochondrial hTERT exacerbates free-radical-mediated mtDNA damage. Aging Cell. 2004;3:399–411.
45. Santos JH, Meyer JN, Van Houten B. Mitochondrial localization of telomerase as a determinant for hydrogen peroxide-induced mitochondrial DNA damage and apoptosis. Hum Mol Genet. 2006;15:1757–68.
46. Sauer H, Wartenberg M. Reactive oxygen species as signaling molecules in cardiovascular differentiation of embryonic stem cells and tumor-induced angiogenesis. Antioxid Redox Signal. 2005;7:1423–34.
47. Chung S, Dzeja PP, Faustino RS, Perez-Terzic C, Behfar A, Terzic A. Mitochondrial oxidative metabolism is required for the cardiac differentiation of stem cells. Nat Clin Pract Cardiovasc Med. 2007;4(S1):S60–7.
48. Ceccatelli S, Tamm C, Sleeper E, Orrenius S. Neural stem cells and cell death. Toxicol Lett. 2004;149:59–66.
49. Tamm C, Robertson JD, Sleeper E, et al. Differential regulation of the mitochondrial and death receptor pathways in neural stem cells. Eur J Neurosci. 2004;19:2613–21.
50. Tamm C, Duckworth J, Hermanson O, Ceccatelli S. High susceptibility of neural stem cells to methylmercury toxicity: effects on cell survival and neuronal differentiation. J Neurochem. 2006;97: 69–78.
51. Barker DJ. In utero programming of chronic disease. Clin Sci. 1998;95:115–28.
52. Ahlbom E, Gogvadze V, Chen M, Celsi G, Ceccatelli S. Prenatal exposure to high levels of glucocorticoids increases the susceptibility of cerebellar granule cells to oxidative stress-induced cell death. Proc Natl Acad Sci U S A. 2000;97:14726–30.
53. Meierhofer D, Mayr JA, Foetschl U, et al. Decrease of mitochondrial DNA content and energy metabolism in renal cell carcinoma. Carcinogenesis. 2004;25:1005–10.
54. Singh KK, Kulawiec M, Still I, Desouki MM, Geradts J, Matsui S. Inter-genomic cross talk between mitochondria and the nucleus plays an important role in tumorigenesis. Gene. 2005;354:140–6.
55. Petros JA, Baumann AK, Ruiz-Pesini E, et al. mtDNA mutations increase tumorigenicity in prostate cancer. Proc Natl Acad Sci U S A. 2005;102:719–24.
56. Shidara Y, Yamagata K, Kanamori T, et al. Positive contribution of pathogenic mutations in the mitochondrial genome to the promotion of cancer by prevention from apoptosis. Cancer Res. 2005;65:1655–63.
57. Parrella P, Xiao Y, Fliss M, et al. Detection of mitochondrial DNA mutations in primary breast cancer and fine-needle aspirates. Cancer Res. 2001;61:7623–6.
58. Huntly BJ, Gilliland DG. Cancer biology: summing up cancer stem cells. Nature. 2005;435:1169–70.
59. Clarke MF, Fuller M. Stem cells and cancer: two faces of eve. Cell. 2006;124:1111–5.
60. Dreesen O, Brivanlou AH. Signaling pathways in cancer and embryonic stem cells. Stem Cell Rev. 2007;3:7–17.
61. Courtois G, Gilmore TD. Mutations in the NfkappaB signaling pathway: implications for human disease. Oncogene. 2006;25:6831–43.

62. Orzechowski A. Possible implications of redox-sensitive tumour cell transformation; lessons from cell culture studies. Pol J Vet Sci. 2007;10:123–6.
63. Borovic S, Cipak A, Meinitzer A, et al. Differential sensitivity to 4-hydroxynonenal for normal and malignant mesenchymal cells. Redox Rep. 2007;12:50–4.
64. Reynolds BA, Weiss S. Generation of neurons and astrocytes from isolated cells of the adult mammalian central nervous system. Science. 1992;255:1707–10.
65. Mizumatsu S, Monje ML, Morhardt DR, Rola R, Palmer TD, Fike JR. Extreme sensitivity of adult neurogenesis to low doses of X-irradiation. Cancer Res. 2003;63:4021–7.
66. Snyder JS, Hong N, McDonald RJ, Wojtowicz JM. A role for adult hippocampal neurogenesis in spatial long-term memory. Neuroscience. 2005;130:843–52.
67. Wojtowicz JM. Irradiation as an experimental tool in studies of adult neurogenesis. Hippocampus. 2006;16:261–6.
68. Limoli CL, Giedzinski E, Rola R, Otsuka S, Palmer TD, Fike JR. Radiation response of neural precursor cells: linking cellular sensitivity to cell cycle checkpoints, apoptosis and oxidative stress. Radiat Res. 2004;161:17–27.
69. Limoli CL, Rola R, Giedzinski E, Mantha S, Huang TT, Fike JR. Cell-density-dependent regulation of neural precursor cell function. Proc Natl Acad Sci U S A. 2004;101:16052–7.
70. Monje ML, Mizumatsu S, Fike JR, Palmer TD. Irradiation induces neural precursor-cell dysfunction. Nat Med. 2002;8:955–62.
71. Monje ML, Toda H, Palmer TD. Inflammatory blockade restores adult hippocampal neurogenesis. Science. 2003;302:1760–1765.
72. Ishii J, Natsume A, Wakabayashi T, et al. The free-radical scavenger edaravone restores the differentiation of human neural precursor cells after radiation-induced oxidative stress. Neurosci Lett. 2007;423:225–30.
73. Prockop DJ, Gregory CA, Spees JL. One strategy for cell and gene therapy: harnessing the power of adult stem cells to repair tissues. Proc Natl Acad Sci U S A. 2003;100 Suppl 1:11917–23.
74. Spees JL, Olson SD, Whitney MJ, Prockop DJ. Mitochondrial transfer between cells can rescue aerobic respiration. Proc Natl Acad Sci U S A. 2006;103:1283–8.
75. Plotnikov EY, Khryapenkova TG, Vasileva AK, et al. Cell-to-cell cross-talk between mesenchymal stem cells and cardiomyocytes in coculture. J Cell Mol Med. 2007 Dec 14; [Epub ahead of print].
76. Hofstetter CP, Schwarz EJ, Hess D, et al. Marrow stromal cells form guiding strands in the injured spinal cord and promote recovery. Proc Natl Acad Sci U S A. 2002;99:2199–204.
77. Chen J, Li Y, Katakowski M, et al. Intravenous bone marrow stromal cell therapy reduces apoptosis and promotes endogenous cell proliferation after stroke in female rat. J Neurosci Res. 2003;73:778–86.
78. Gnecchi M, He H, Liang OD, et al. Paracrine action accounts for marked protection of ischemic heart by Akt-modified mesenchymal stem cells. Nat Med. 2005;11:367–8.
79. Spinazzola A, Marti R, Nishino I, et al. Altered thymidine metabolism due to defects of thymidine phosphorylase. J Biol Chem. 2002;277:4128–33.
80. Marti R, Nishigaki Y, Hirano M. Elevated plasma deoxyuridine in patients with thymidine phosphorylase deficiency. Biochem Biophys Res Commun. 2003;303:14–8.
81. Hirano M, Marti R, Casali C, et al. Allogeneic stem cell transplantation corrects biochemical derangements in MNGIE. Neurology. 2006;67:1458–60.

Part II
Regulation by Stem Cell Niches

Stem Cells and Stem Cell Niches in Tissue Homeostasis: Lessons from the Expanding Stem Cell Populations of *Drosophila*

Yukiko M. Yamashita

Abstract New stem cell populations and their niches have been discovered continuously in *Drosophila*, changing the view that the adult *Drosophila* body consists primarily of post-mitotic organs and is not suited for the study of dynamic tissue homeostasis. With a century of genetics history as well as many sophisticated tools for the genetic analysis of cellular and developmental biology, *Drosophila* has now emerged as an impressive model system in which to study stem cell biology in great detail.

Keywords Niche · Centrosome · Asymmetric division · Cell polarity · Spindle orientation

1 Stem Cells and Tissue Homeostasis

Adult stem cells continuously supply highly differentiated but short-lived cells, such as those found in blood, skin, intestinal epithelium, and testis, throughout life. In the absence of adult stem cells, it is estimated that we would lose our complete gut in two days, all of our skin in three weeks, and our blood cells in four months. Therefore, adult stem cells are critically important for tissue homeostasis. However, because of their importance, stem cells can be the source of many problems that organisms encounter during their lives.

1.1 Stem Cells and Tumorigenesis

Features of stem cell identity, such as proliferative capacity and maintenance in a relatively undifferentiated state, can be easily modified or exploited to lead to tumorigenesis. Indeed, it is now suggested that many tumors arise from stem cell populations, and that tumors maintain "seeding cells," or cancer stem cells (which share many characteristics with normal tissue stem cells), that can initiate tumorigenesis [1]. Their low abundance, resistance to therapeutic treatments, and ability to generate a whole new tumor hamper the efficient treatment of cancer. Many treatments that significantly reduce tumor size often fail to cure patients due to a failure in targeting such cancer stem cells. While essential, stem cells can be very dangerous to the maintenance of our bodies. Therefore, multiple mechanisms have evolved to limit stem cell expansion while ensuring the maintenance of the stem cell population.

One great evolutionary process is the prevention of uncontrolled proliferation of stem cells by the stem cell niche. The stem cell niche is a microenvironment that specifies stem cell identity. It provides essential signals to its resident stem cells for the maintenance of stem cell identity. In this manner, stem cells rely on external sources for proliferation [2, 3]. The separation of potentially dangerous, undifferentiated stem cells from mitogens and other proliferation factors from tissue ensures that stem cells do not overproliferate automatically. Interestingly, many niche-composing cells are known to be post-mitotic, thereby ensuring a limited availability of essential signals for stem cell proliferation.

1.2 Stem Cells and Tissue Aging

Another type of biological problem caused by stem cell dysfunction/malfunction is the opposite of cancer. That is, while cancer can be initiated by the abnormal overproliferation of stem cells, a failure in stem cell maintenance/proliferation can lead to tissue degeneration. Also, because of their fundamental involvement in tissue homeostasis, a decline in stem cell number/function has been proposed to contribute to age-related changes to tissues, such as decreased regenerative capacity and tissue hypoplasia [4, 5]. It has been proposed that tissue aging is a consequence of tumor suppression mechanism(s) that increase the survival of organisms in an

Y.M. Yamashita (✉)
Center for Stem Cell Biology, Life Sciences Institute and Department of Cell and Developmental Biology, University of Michigan,
Ann Arbor, 210 Washtenaw Avenue, Rm 5403, Ann Arbor, MI 48104
e-mail: yukikomy@umich.edu

V.K. Rajasekhar, M.C. Vemuri (eds.), *Regulatory Networks in Stem Cells*, Stem Cell Biology and Regenerative Medicine,
DOI 10.1007/978-1-60327-227-8_13, © Humana Press, a part of Springer Science+Business Media, LLC 2009

earlier stage of life [4]. While limiting stem cell proliferation can help organisms survive at an early age (by preventing tumorigenesis), such a mechanism can result in the inability of stem cells to proliferate, thus depleting stem cells and/or their progenies (differentiated cells) when aging tissues/organs need additional differentiated cells. However, the underlying mechanisms that connect stem cell function to tissue aging are poorly understood. For example, it has been shown that the cell cycle inhibitor Ink4A accumulates in stem cells with age, contributing to a decline in stem cell function [6–8]. However, how the expression of this cell cycle inhibitor is controlled in stem cells is not understood. Furthermore, it has yet to be elucidated whether Ink4A signaling is the strategy for stem cells to prevent tumorigenesis or whether it is the consequence of cell cycle arrest that is caused by other mechanisms.

2 Asymmetric vs. Symmetric Stem Cell Division

The balance between stem cell self-renewal and differentiation is of critical importance since an imbalance in this process may lead to either tumorigenesis (caused by overproliferation of stem cells) or tissue degeneration (caused by depletion of stem cells). One way to maintain this critical balance is by asymmetric stem cell divisions that give rise to one stem cell and one differentiating cell, thereby maintaining both populations. Alternatively, total stem cell number may be monitored at either the tissue or organism level and each stem cell division may be symmetric or asymmetric [9]. During tissue homeostasis, maintaining the ratio of stem cell self-renewal and differentiation would be favored and, thus, asymmetric stem cell division might be an ideal way to do so. In contrast, when stem cell populations require expansion, such as during embryonic development where organ size is expanding as well as during recovery after injury, stem cell expansion would be favored. While the former example allows a stricter control over stem cell number, advantageous in tumor suppression, the latter example has more flexibility in stem cell number control, advantageous in adapting to the ever-changing requirements for tissue homeostasis. Therefore, the choice of cell fates, as well as the choice of the mode of stem cell division (symmetric vs. asymmetric), must be appropriately controlled to respond to the demand of tissue development, homeostasis, and regeneration.

3 Signaling in the Stem Cell Niche

Although conceptual components of stem cells and their niches are conserved in many stem cell systems, signaling pathways involved in each stem cell/niche compartment appears to vary to a large extent [10]. Here I would like to focus on several examples from *Drosophila* where stem cells and their niches are well documented, illustrating the diversity of stem cell-niche signaling.

3.1 *Drosophila* Female Germline Stem Cells (GSCs)

In adult *Drosophila* females, each ovary consists of approximately 15 ovarioles. At the very anterior tip of each ovariole, somatic cells (terminal filament and cap cells) comprise the stem cell niche for the female GSCs, in which 2–3 GSCs reside. Female GSCs directly attach to cap cells via adherens junctions [11], which ensure the transduction of the dpp (TGFβ) ligand from cap cells to GSCs in which TGFβ signaling maintains stem cell identity (Fig. 1A). In response to dpp signaling from the terminal filament and cap cells, female GSCs repress the expression of *bam*, which is a master regulator of the differentiation program [12, 13]. Once a daughter of a GSC leaves the stem cell niche, *bam* expression is induced, initiating the differentiation program. Such control of stem cell identity and initiation of differentiation appears to solely depend on signals from the niche, since germ cells that committed to the differentiation program can revert (or dedifferentiate) to regain stem cell identity once they are placed back into the niche [14]. A similar phenomenon is observed in male GSCs, which suggests a general dominance of niche signaling [15].

In addition to GSCs, another stem cell population called escort stem cells (ESCs) was reported to be located close to the terminal filament/cap cell niche. ESCs are somatic stem cells that provide a support for GSCs and their progenies [16]. In contrast to male GSCs (see below), female GSCs do not depend on the JAK-STAT signaling pathway for the maintenance of stem cell identity. Instead, ESCs are maintained by the JAK-STAT pathway (Fig. 1A). The source of the ligand involved in the JAK-STAT pathway to maintain ESCs has not yet been identified.

Drosophila ovaries maintain yet another population of stem cells; follicle stem cells (FSCs) are located in the germarium of each ovariole [17]. Progenies of FSCs proliferate and encase developing germline cells (nurse cells and oocytes), replacing escort cells. Proliferation of FSCs depends on Hedgehog (Hh) [18] and the maintenance of FSCs requires adherens junctions [19]. Recently, Nystrul and Spradling [20] have shown that this FSC population resides in an acellular niche (i.e., no cellular components that constitute a stem cell niche exist). Exactly two FSCs exist in each ovariole and, interestingly, daughters of these two FSCs can migrate across the germarium and replace

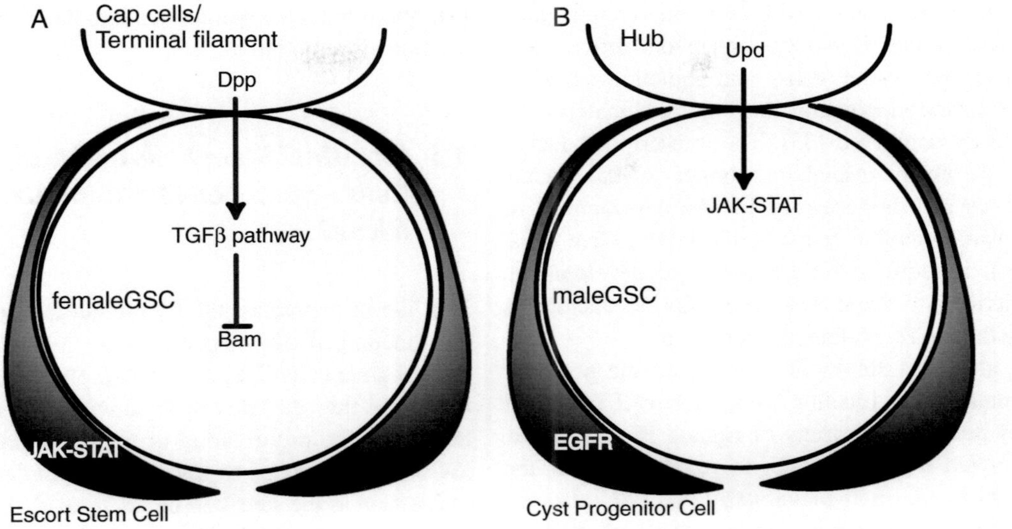

Fig. 1 Signaling in *Drosophila* germline stem cells and their niche. (**A**) Female germline stem cells are maintained by the TGFβ signaling pathway. Niche supporting cells, cap cells and terminal filament, secrete signaling ligand, Dpp, to activate the TGFβ pathway within the germ line stem cells. The TGFβ pathway in turn represses expression of Bam gene, which induces differentiation of germ cells. Another stem cell population, somatic escort stem cell, is maintained by the JAK-STAT pathway. (**B**) Male germline stem cells are maintained by the JAK-STAT pathway, as well as cyst progenitor cells, somatic stem cells. The hub cells secrete signaling ligand, Upd, to activate the JAK-STAT pathway in the male germline stem cells. EGFR signaling between germline stem cells and cyst progenitor cells (and/or between differentiating germ cells and cyst cells, progenies of germline stem cells and cyst progenitor cells, respectively) is required for the encapsulation of germ cells by somatic cells. Such germ cell-somatic interaction is essential for proper germ cell development

FSCs in the opposite niche. The biological significance of such FSC behavior, such as whether this migration plays a role in the maintenance of the stem cell population, has yet to be determined.

3.2 Male Germline Stem Cells

The adult testis has a coiled-tubular structure that is filled with cells from all stages of spermatogenesis [21]. The apical end of the testis is a blind end, where stem cells reside in the stem cell niche. The male GSC niche is composed of the hub, which is composed of the tightly clustered post-mitotic cells (hub cells), and possibly cyst progenitor cells, a pair of which encapsulates each GSC. Hub cells secrete the signaling ligand, Unpaired (Upd), which activates the JAK-STAT pathway within GSCs and thereby maintains stem cell identity [22, 23] (Fig. 1B). The ectopic expression of Upd in early germ cells releases GSCs from dependence on the niche, leading to a tumor-like phenotype with testes filled with early germ cells (stem cells and transit-amplifying cells [spermatogonia]). Similar to female GSCs, male GSCs are attached to hub cells via adherens junctions [24]. Since the range of Upd signaling appears to be small (probably due to its glycosylation-mediated attachment to the hub cell surface [25]), physical attachment to hub cells appears to be essential for the maintenance of GSC identity.

Once stem cell daughters leave the niche, they first become gonialblasts and then spermatogonia, undergoing transit-amplifying divisions. In *Drosophila*, the transit-amplifying divisions occur four times, generating 16 spermatogonia, which are interconnected to one another by incomplete cytokinesis. BMP (bone morphogenetic protein) signaling has been suggested to control GSC self-renewal and/or the number of transit-amplifying divisions [26–28].

In addition to the above signaling process, cyst progenitor cells and cyst cells (which encapsulate GSCs and gonialblasts/spermatogonia, respectively) regulate stem cell identity/differentiation [29] (Fig. 1B). A mutation in the Egf (epithelial growth factor) receptor leads to overproliferation of stem cells and spermatogonia. This Egf signaling pathway between germ cells and somatic support cells is also essential for proper encapsulation of germ cells by somatic cells [30]. The compromised function in the Egf signaling pathway results in GSCs that are not encapsulated by somatic cells as well as a failure in stem cell differentiation into gonialblast/spermatogonia [30, 31].

3.3 Blood Stem Cells

Drosophila have only a few blood cell types, which function similarly to vertebrate myeloid lineages. *Drosophila* blood cells (hemocytes) emerge from the head mesoderm during the embryonic stage, then from the lymph gland during the larval stage. Hemocyte precursors differentiate into plasmatocytes, crystal cells, and lamellocytes [32]. Interestingly, transcription factors that control hematopoiesis are evolutionarily conserved [33]. The transcription factor, Lozenge (Lz), which resembles human AML1 (acute myeloid leukemia-1), is necessary for the development of crystal cells, while another transcription factor, Gcm (glial cells missing), is required for plasmatocyte development. The transcription of these two transcription factors is controlled by the GATA protein, Serpent (Srp).

In lymph glands, a subset of cells called the posterior signaling center (PSC) function as a stem cell niche for hemocytes by providing a variety of signals. First, PSC expresses the Notch ligand, Serrate, to regulate Lz for the specification of crystal cell precursors [34]. PSC also expresses Hedgehog (Hh) protein, the loss of which leads to the premature differentiation of hemocyte precursors [35]. In addition, PSC appears to maintain hemocyte precursors via the JAK-STAT pathway [36]. Further studies will be required to address how these multiple signaling pathways are integrated into stem cell (hemocyte precursors) maintenance and lineage specification.

3.4 Neuroblasts

Drosophila neuroblasts divide asymmetrically in a stem cell-like manner, giving rise to another neuroblast and a ganglion mother cell that undergoes differentiation. *Drosophila* neuroblasts do not rely on the stem cell niche for their asymmetric division. Instead, they segregate intrinsic fate determinants asymmetrically, thereby controlling the differential fates of daughter cells. Key determinants that regulate differentiation are localized to the basal cortex of the neuroblast prior to cell division and, due to an apicobasally oriented and asymmetrically positioned mitotic spindle, are sequestered into the nascent GMC [37]. Numb functions as a primary fate determinant that directs cells toward differentiation, and Numb loss of function leads to the overproliferation of neuroblasts [38]. Numb's function as a fate determinant is exerted at least partly by its ability to modulate Notch signaling [39].

Asymmetric localization of fate determinants to the basal cortex depends on an evolutionarily conserved mechanism based on the Par6, Baz (Par 3), aPKC cell polarity complex, which localizes to the apical side of the neuroblast cortex prior to division. The apical Par complex also serves as a binding platform for proteins that control orientation and positioning of the mitotic spindle. The coordination of the position of cell fate determinants with spindle orientation (which depends on Pins and G protein signaling) means that determinants localized to the basal cortex of the dividing neuroblast are reliably inherited into the GMC, whereby they promote differentiation.

3.5 Intestinal Stem Cells and Malpighian Stem Cells – Stem Cell Maintenance Without the Niche?

Just like in mammals, the fly midgut is also maintained by the function of tissue-specific stem cells (multi-potent, intestinal stem cells [ISCs]) [40–42]. Notch signaling appears to control the cell fates of ISCs with high Notch signaling leading to the proliferation of undifferentiated progenitors. ISCs are not attached to any cellular components that appear to function as the stem cell niche. Instead, differential Notch signaling within the ISCs controls daughter cell fates, suggesting that the ISC itself functions as a sort of "niche" that signals to its daughters to dictate their fates [42].

More recently, it has been found that fly Malpighian tubules (fly counterparts of kidneys) are also maintained by the function of resident stem cells [43]. This stem cell population, RNSCs (renal and nephric stem cells), is multipotent, giving rise to multiple cell types within the Malpighian tubules. It was also shown that the autocrine JAK-STAT signaling pathway is responsible for the stem cell identity of RNSCs; all components of the JAK-STAT pathway, including the ligand (Upd) and the receptor (Dome), are expressed in the RNSCs, suggesting that stem cells do not require extrinsic signals for proliferation. It would be interesting to address how stem cell number is controlled in normal Malpighian tubules (since all signaling components are present within one cell), why they do not overproliferate, and whether any other extrinsic signal contributes to the regulation of stem cell number.

In summary, *Drosophila* have multiple tissue-specific stem cell populations, far more than once was believed. Some of these populations are maintained by signals from the cellular niche, others are maintained by autocrine signaling, and still others are maintained in an acellular niche. It has yet to be determined whether stem cell populations that do not reside in the niche are really independent of any environmental influence.

4 Stem Cell Niche and Asymmetric Stem Cell Division – Translation of Extracellular Polarity into Intracellular Polarity

For stem cells that reside in the stem cell niche, asymmetric division essentially depends on the asymmetric placement of daughter cells inside and outside of the influence of the niche signaling pathway. In many cases, such asymmetric placement of stem cell daughters is accomplished by an oriented mitotic spindle.

4.1 Drosophila Male GSCs

Drosophila male GSCs always divide asymmetrically by orienting the mitotic spindle perpendicularly toward the niche (hub cells). The orientation of the mitotic spindle occurs during interphase by the positioning of centrosomes, which become spindle poles in mitosis. One centrosome within a GSC always locates very close to the hub. After centrosome duplication, one centrosome stays close to the hub, while the other moves toward the opposite side of the GSC, thereby setting up the orientation of the mitotic spindle [24]. We have recently shown that difference(s) between mother and daughter centrosomes underlie the stereotyped positioning of the centrosomes; the mother centrosome that contains the oldest centriole remains anchored to the hub, while the daughter centrosome that contains younger centrioles moves away from the hub (Fig. 2A). As a result, the mother centrosome is inherited by the stem cell, and the daughter by the gonialblast, which is destined to differentiate [44].

Although it is tempting to speculate that mother and/or daughter centrosomes also harbor certain fate determinants that direct stem cell identity and/or differentiation, no such factors have been identified so far. It has been reported that in mollusk early embryos, fate-determining RNAs associate with one centrosome and are inherited by only one daughter cell [45]. However, it has not been shown whether the difference(s) between mother and daughter centrosomes are utilized in such asymmetric cell divisions.

The anchoring of the mother centrosome to the hub-GSC interface depends on microtubules and their linkages to the cell cortex between the hub and GSCs (Fig. 2A). First, the mother centrosome harbors many astral microtubules throughout the cell cycle [44]. Mutations in cnn (centrosomin, a component of pericentriolar material that is responsible for astral microtubule anchoring to centrosomes) lead to a failure in typical centrosome orientation and randomization of mother-daughter centrosome positioning. Then, such microtubules emanating from the mother centrosome appear to be "captured" by the adherens junctions formed between the hub and GSC. This microtubule capture is mediated by a linker protein, apc2 (adenomatous polyposis coli homolog 2). Apc2 co-localizes to adherens junctions from the GSC side (unpublished data) and the loss of apc2 function leads to a failure in centrosome positioning, similar to that of cnn mutants. Interestingly, such centrosome-cortex interactions, as a mechanism of spindle orientation, might be evolutionarily conserved since budding yeast orient their mitotic spindles in a very similar manner [46].

4.2 Drosophila Neuroblasts

A second example of asymmetric behavior of centrosomes during asymmetric stem cell division was recently demonstrated in *Drosophila* larval neuroblasts [47, 48]. Although *Drosophila* neuroblasts do not rely on the stem cell niche, asymmetric distribution of fate determinants into daughter cells plays a critical role in stem cell self-renewal and asymmetric stem cell division as described above (Section 3.4) [37]. Therefore, the control of the oriented spindle in the

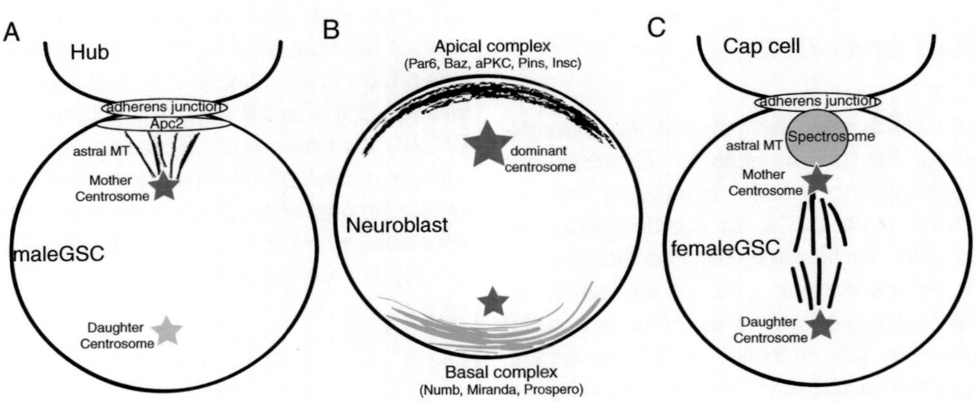

Fig. 2 Mechanisms of spindle orientation. (**A**) Male GSC: Mother centrosome is captured to the adherens-junction/apc2 cortical complex between the hub and GSC. (**B**) Larval neuroblast: Dominant centrosome consistently locates to the same side as the apical protein complex containing Par6, Baz (Par3), aPKC, Pins, Insc, while the dormant centrosome moves away to the opposite side, where the basal complex, containing Numb, Miranda, Prospero, locates. (**C**) Female GSC: female GSC only orients toward the cap cell in mitosis, by anchoring one centrosome to the spectrosome next to the cap cell-GSC junction. During interphase, centrosomes are not consistently oriented toward the cap cell

context of the cell axis is crucially important in neuroblasts as in *Drosophila* male GSCs.

In larval neuroblasts, one centrosome stays close to the apical side of the cell and maintains robust microtubule arrays throughout the cell cycle (Fig. 2B). On the other hand, the other centrosome is inactivated immediately after centrosome duplication, loses activity to anchor astral microtubules, and remains dormant until it is reactivated prior to the onset of mitosis. When this dormant centrosome becomes activated, it is already located on the basal side of the neuroblast and therefore spindle orientation is already established. It has yet to be determined whether differences between mother and daughter centrosomes underlie centrosome behavior in neuroblasts.

Interestingly, centrosomes in embryonic neuroblasts behave quite differently. After duplication, two centrosomes stay close to the apical surface as in larval neuroblasts. However, both centrosomes leave this position when they begin to separate from each other, establishing a mitotic spindle that is parallel to the apical surface [49, 50]. After entering mitosis, the spindle undergoes a programmed rotation, acquiring a desired orientation that is perpendicular to the apical surface (namely, parallel to apical-basal axis). This is an elegant example that stem cell populations of the same lineage behave differently during development, and such control might be responsible for the regulated behavior of stem cells depending on the demand of the tissue (flexibility to divide asymmetrically/symmetrically vs. tighter control to maintain asymmetric stem cell division). Indeed, it has been reported that the spindle orientation of mouse skin stem cells shift from parallel to perpendicular (to skin surface) during development, causing a shift from symmetric, expanding stem cell division to asymmetric stem cell division [51].

4.3 *Drosophila* Female GSCs

Drosophila female GSCs also normally divide asymmetrically by orienting the mitotic spindle. However, the underlying mechanism for spindle orientation appears to be different from male GSCs. In female GSCs, the oriented mitotic spindle is achieved by the association of one spindle pole with the spectrosome, which is a membranous subcellular organelle, located close to the niche (cap cells) [52]. Upon progression into mitosis, one of the spindle poles is captured by the spectrosome, thereby anchoring the mitotic spindle in an oriented manner (Fig. 2C). However, in female GSCs, interphase centrosomes are not oriented with respect to the cap cells and asymmetric stem cell divisions are unaffected by the absence of centrosomes [53], suggesting that orientation of the mitotic spindle and regulation of asymmetric stem cell division are carried out by distinct mechanisms in male and female GSCs. Although male GSCs also contain spectrosomes, they are not oriented with respect to the hub cells throughout the cell cycle, while centrosomes are consistently oriented toward the hub [24].

Interestingly, female GSCs can undergo symmetric stem cell divisions to replace the empty niche after stem cell loss [54]. In contrast, male GSCs rarely undergo symmetric stem cell division (unpublished results), implying that rigid stem cell orientation throughout the cell cycle may lead to the inability of symmetric stem cell division.

In other stem cell populations found in *Drosophila* as described in Section 3, the detailed examination of stem cell division is lacking. A comprehensive description of anatomy/cellular architecture of stem cell residence and signaling pathways in future studies will be required to fully understand how asymmetric/symmetric stem cell divisions are controlled and what determines the nature of the cellular machinery that is responsible for such stem cell divisions.

5 Stem Cells and Tissue Aging in Fly GSCs

Drosophila is an ideal model system to study tissue aging because of its short life span (a mean life span of $\sim$40 days [55]). Recently, a decline in GSC numbers in aged flies was reported [56, 57]. Interestingly, in both male and female GSCs, a decline in cell adhesion between the niche and GSCs appears to be, at least in part, responsible for the loss of stem cells with age. Furthermore, it has been suggested that signaling from the niche decreases with age in the testis, possibly accounting for stem cell decline with age [56].

However, the actual rate of the decline in stem cell number is significantly lower than the rate calculated from the half-life of each stem cell [58], suggesting that the tissue possesses a mechanism to compensate for stem cell loss during aging. Indeed, a decline in stem cell number is only approximately 25% after 50 days of age both in male and female GSCs when tissue aging is already dramatic, implying that additional mechanism(s) must exist that cause overall tissue aging. Our unpublished data suggest that intrinsic changes in stem cells, which lead to cell cycle arrest of stem cells, start as early as 5 days of age.

6 Summary

Stem cells are the source of highly differentiated but short-lived cells throughout life, and, therefore, the dysfunction/malfunction of stem cells has a tremendous impact on organisms, including tumorigenesis and tissue aging. To be able to treat any disease or symptom related to stem cell function, the list of which is growing fast, it is essential to understand the behavior of stem cells in great detail. For example, it has been imagined that understanding the mechanism of tissue aging may lead to a treatment that extends life span for a prolonged time period. However, to be able to approach such a goal, we may need to understand the mechanisms that are responsible for tumor suppression and tissue aging. More practically, understanding the mechanism of asymmetric stem cell division may lead to manipulation of stem cells to adopt symmetric division, which may allow us to expand tissue specific stem cells in culture. If realized, we may not need to look for bone marrow donors for transplantation.

As discussed here, *Drosophila* contains a wide variety of stem cell populations, with different types of signaling pathways, niches, and asymmetric/symmetric cell divisions. Given the powerful genetics, as well as the cell and developmental biology, of *Drosophila*, this tiny organism will continue to contribute to the understanding of stem cell behavior during both tissue homeostasis and aging.

References

1. Clarke MF, Fuller M. Stem cells and cancer: two faces of eve. Cell. 2006;124(6):1111–5.
2. Spradling A, Drummond-Barbosa D, Kai T. Stem cells find their niche. Nature. 2001;414(6859):98–104.
3. Fuchs E, Tumbar T, Guasch G. Socializing with the neighbors: stem cells and their niche. Cell. 2004;116(6):769–78.
4. Kirkwood TB. Understanding the odd science of aging. Cell. 2005;120(4):437–47.
5. Campisi J, d'Adda di Fagagna F. Cellular senescence: when bad things happen to good cells. Nat Rev Mol Cell Biol. 2007;8(9):729–40.
6. Molofsky AV, Slutsky SG, Joseph NM, et al. Increasing p16INK4a expression decreases forebrain progenitors and neurogenesis during ageing. Nature. 2006;443(7110):448–52.
7. Janzen V, Forkert R, Fleming HE, et al. Stem-cell ageing modified by the cyclin-dependent kinase inhibitor p16INK4a. Nature. 2006;443(7110):421–6.
8. Krishnamurthy J, Ramsey MR, Ligon KL, et al. p16INK4a induces an age-dependent decline in islet regenerative potential. Nature. 2006;443(7110):453–7.
9. Morrison SJ, Kimble J. Asymmetric and symmetric stem-cell divisions in development and cancer. Nature. 2006;441(7097):1068–74.
10. Moore KA, Lemischka IR. Stem cells and their niches. Science. 2006;311(5769):1880–5.
11. Song X, Zhu CH, Doan C, Xie T. Germline stem cells anchored by adherens junctions in the Drosophila ovary niches. Science. 2002;296(5574):1855–7.
12. Chen D, McKearin D. Dpp signaling silences bam transcription directly to establish asymmetric divisions of germline stem cells. Curr Biol. 2003;13(20):1786–91.
13. Chen D, McKearin DM. A discrete transcriptional silencer in the bam gene determines asymmetric division of the Drosophila germline stem cell. Development. 2003;130(6):1159–70.
14. Kai T, Spradling A. Differentiating germ cells can revert into functional stem cells in Drosophila melanogaster ovaries. Nature. 2004;428(6982):564–9.
15. Brawley C, Matunis E. Regeneration of male germline stem cells by spermatogonial dedifferentiation in vivo. Science. 2004;304(5675):1331–4.
16. Decotto E, Spradling AC. The Drosophila ovarian and testis stem cell niches: similar somatic stem cells and signals. Dev Cell. 2005;9(4):501–10.
17. Margolis J, Spradling A. Identification and behavior of epithelial stem cells in the Drosophila ovary. Development. 1995;121(11):3797–807.
18. Forbes AJ, Lin H, Ingham PW, Spradling AC. Hedgehog is required for the proliferation and specification of ovarian somatic cells prior to egg chamber formation in Drosophila. Development. 1996;122(4):1125–35.
19. Song X, Xie T. DE-cadherin-mediated cell adhesion is essential for maintaining somatic stem cells in the Drosophila ovary. Proc Natl Acad Sci U S A. 2002; 99(23):14813–8.
20. Nystul T, Spradling A. An Epithelial Nihce in the Drosophila ovary undergoes long-range stem cell replacement. Cell Stem Cell. 2007;1(3):277–85.
21. Fuller MT. Spermatogenesis. In: Bate M, Martinez-Arias, A. editors. The development of drosophila melanogaster. New York: Cold Spring Harbor Laboratory Press; 1993. pp. 71–147.
22. Tulina N, Matunis E. Control of stem cell self-renewal in Drosophila spermatogenesis by JAK-STAT signaling. Science. 2001;294(5551):2546–9.
23. Kiger AA, Jones DL, Schulz C, Rogers MB, Fuller MT. Stem cell self-renewal specified by JAK-STAT activation in response to a support cell cue. Science. 2001;294(5551):2542–5.
24. Yamashita YM, Jones DL, Fuller MT. Orientation of asymmetric stem cell division by the APC tumor suppressor and centrosome. Science. 2003;301(5639):1547–50.
25. Harrison DA, McCoon PE, Binari R, Gilman M, Perrimon N. Drosophila unpaired encodes a secreted protein that activates the JAK signaling pathway. Genes Dev. 1998;12(20):3252–63.
26. Schulz C, Kiger AA, Tazuke SI, et al. A misexpression screen reveals effects of bag-of-marbles and TGFbeta class signaling on the Drosophila male germ-line stem cell lineage. Genetics. 2004;167(2):707–23.
27. Kawase E, Wong MD, Ding BC, Xie T. Gbb/Bmp signaling is essential for maintaining germline stem cells and for repressing bam transcription in the Drosophila testis. Development. 2004;131(6):1365–75.
28. Shivdasani AA, Ingham PW. Regulation of stem cell maintenance and transit amplifying cell proliferation by tgf-Beta signaling in Drosophila spermatogenesis. Curr Biol. 2003;13(23):2065–72.
29. Kiger AA, White-Cooper H, Fuller MT. Somatic support cells restrict germline stem cell self-renewal and promote differentiation. Nature. 2000;407(6805):750–4.
30. Schulz C, Wood CG, Jones DL, Tazuke SI, Fuller MT. Signaling from germ cells mediated by the rhomboid homolog stet organizes encapsulation by somatic support cells. Development. 2002;129(19):4523–34.
31. Sarkar A, Parikh N, Hearn SA, Fuller MT, Tazuke SI, Schulz C. Antagonistic roles of Rac and Rho in organizing the germ cell microenvironment. Curr Biol. 2007;17(14):1253–8.
32. Evans CJ, Hartenstein V, Banerjee U. Thicker than blood: conserved mechanisms in Drosophila and vertebrate hematopoiesis. Dev Cell. 2003;5(5):673–90.
33. Lebestky T, Chang T, Hartenstein V, Banerjee U. Specification of Drosophila hematopoietic lineage by conserved transcription factors. Science. 2000;288(5463):146–9.

34. Lebestky T, Jung SH, Banerjee U. A Serrate-expressing signaling center controls Drosophila hematopoiesis. Genes Dev. 2003;17(3):348–53.
35. Mandal L, Martinez-Agosto JA, Evans CJ, Hartenstein V, Banerjee U. A Hedgehog- and Antennapedia-dependent niche maintains Drosophila haematopoietic precursors. Nature. 2007;446(7133):320–4.
36. Krzemien J, Dubois L, Makki R, Meister M, Vincent A, Crozatier M. Control of blood cell homeostasis in Drosophila larvae by the posterior signalling centre. Nature. 2007;446(7133):325–8.
37. Yu F, Kuo CT, Jan YN. Drosophila neuroblast asymmetric cell division: recent advances and implications for stem cell biology. Neuron. 2006;51(1):13–20.
38. Lee CY, Andersen RO, Cabernard C, et al. Drosophila Aurora-A kinase inhibits neuroblast self-renewal by regulating aPKC/Numb cortical polarity and spindle orientation. Genes Dev. 2006;20(24):3464–74.
39. Berdnik D, Torok T, Gonzalez-Gaitan M, Knoblich JA. The endocytic protein alpha-Adaptin is required for numb-mediated asymmetric cell division in Drosophila. Dev Cell. 2002;3(2):221–31.
40. Micchelli CA, Perrimon N. Evidence that stem cells reside in the adult Drosophila midgut epithelium. Nature. 2006;439(7075):475–9.
41. Ohlstein B, Spradling A. The adult Drosophila posterior midgut is maintained by pluripotent stem cells. Nature. 2006;439(7075):470–4.
42. Ohlstein B, Spradling A. Multipotent Drosophila intestinal stem cells specify daughter cell fates by differential notch signaling. Science. 2007;315(5814):988–92.
43. Singh SR, Liu W, Hou SX. The adult Drosophila malpighian tubules are maintained by multipotent stem cells. Cell Stem Cell. 2007;1(2):191–203.
44. Yamashita YM, Mahowald AP, Perlin JR, Fuller MT. Asymmetric inheritance of mother versus daughter centrosome in stem cell division. Science. 2007;315(5811):518–21.
45. Lambert JD, Nagy LM. Asymmetric inheritance of centrosomally localized mRNAs during embryonic cleavages. Nature. 2002;420(6916):682–6.
46. Bienz M. Spindles cotton on to junctions, APC and EB1. Nat Cell Biol. 2001;3(3):E67–8.
47. Rusan NM, Peifer M. A role for a novel centrosome cycle in asymmetric cell division. J Cell Biol. 2007;177(1):13–20.
48. Rebollo E, Sampaio P, Januschke J, Llamazares S, Varmark H, Gonzalez C. Functionally unequal centrosomes drive spindle orientation in asymmetrically dividing Drosophila neural stem cells. Dev Cell. 2007;12(3):467–74.
49. Kaltschmidt JA, Davidson CM, Brown NH, Brand AH. Rotation and asymmetry of the mitotic spindle direct asymmetric cell division in the developing central nervous system. Nat Cell Biol. 2000;2(1):7–12.
50. Siegrist SE, Doe CQ. Extrinsic cues orient the cell division axis in Drosophila embryonic neuroblasts. Development. 2006;133(3):529–36.
51. Lechler T, Fuchs E. Asymmetric cell divisions promote stratification and differentiation of mammalian skin. Nature. 2005;437(7056):275–80.
52. Deng W, Lin H. Spectrosomes and fusomes anchor mitotic spindles during asymmetric germ cell divisions and facilitate the formation of a polarized microtubule array for oocyte specification in Drosophila. Dev Biol. 1997;189(1):79–94.
53. Stevens NR, Raposo AA, Basto R, St Johnston D, Raff JW. From stem cell to embryo without centrioles. Curr Biol. 2007;17(17):1498–503.
54. Xie T, Spradling AC. A niche maintaining germ line stem cells in the Drosophila ovary. Science. 2000;290(5490):328–30.
55. Helfand SL, Rogina B. Molecular genetics of aging in the fly: is this the end of the beginning? Bioessays. 2003;25(2):134–41.
56. Boyle M, Wong C, Rocha M, Jones DL. Decline in self-renewal factors contributes to aging of the stem cell niche in the Drosophila testis. Cell Stem Cell. 2007;1(4):470–8.
57. Pan L, Chen S, Weng C, et al. Stem cell aging is controlled both intrinsically and extrinsically in the Drosophila ovary. Cell Stem Cell. 2007;1(4):458–69.
58. Wallenfang MR, Nayak R, DiNardo S. Dynamics of the male germline stem cell population during aging of Drosophila melanogaster. Aging Cell. 2006;5(4):297–304.

Extrinsic and Intrinsic Control of Germline Stem Cell Regulation in the *Drosophila* Ovary

Nian Zhang and Ting Xie

Abstract Germline stem cells (GSCs) in the *Drosophila* ovary represent one of the best studied adult stem cell types, while their regulatory microenvironment or niche is also one of the best defined ones. Due to the availability of powerful genetic tools and a large number of mutants, much progress has been made in understanding the molecular mechanisms underlying intrinsic and extrinsic controls of GSC regulation within the past decade. Since several excellent reviews have been written on these stem cells, this review primarily discusses progress within the last couple of years and important questions remaining to be answered.

Keywords *Drosophila* ovary · Germline stem cells · Self-renewal · Differentiation · Niche · Asymmetric cell division

Germline stem cells are truly immortal because they transmit their genetic information from generation to generation and their progeny ultimately differentiate into gametes. In human males, genetic mutations and chemotherapy for treating childhood cancer can cause depletion of GSCs, leading to infertility. The infertility caused by chemotherapy can be restored by transplantation of the GSCs that are preserved before chemotherapy. The molecular mechanisms for controlling GSC regulation are critical for understanding how certain genetic mutations cause GSC loss and for expanding GSCs in vitro for treating the infertility caused by chemotherapy. Since GSCs in different organisms share many similarities and different stem cell types also share many regulatory mechanisms and strategies [1, 2], the knowledge gained from the studies on GSCs in the *Drosophila* ovary should be useful in understanding their counterparts as well as other adult stem cell types in mammals. The structural simplicity of the *Drosophila* ovary and the wealth of genetic materials and tools have made the *Drosophila* ovary an ideal system to study stem cell biology in vivo at the molecular and cellular level [3]. Since GSCs and their niche in the *Drosophila* ovary are well defined [4–6], great strides have been made in understanding the cellular and molecular mechanisms regulating niche functions and GSC behaviors. The studies on the *Drosophila* ovarian stem cells have provided mechanistic insights into, and conceptual framework for, stem cell/niche interaction and stem cell function and behaviors [2].

1 Easily Identifiable GSCs and Their Well-Defined Physical Surroundings Make the *Drosophila* Ovary an Attractive System for Studying Stem Cell Biology

The *Drosophila* ovary consists of 12–16 parallel ovarioles, which represent individual egg production units [7]. The simple tubular structure at the tip of the ovariole is known as the germarium, where three types of stem cells reside. At the most distal end of the germarium, there is a stack of disc-like terminal filament (TF) cells, the most posterior one of which directly contacts a group of five to seven tightly packed cap cells [7]. Two or three large and round GSCs lie posteriorly and are anchored to cap cells [8, 9], while four to six thin escort stem cells (ESCs) contact cap cells anteriorly and also wrap around GSCs laterally [10]. Cap cells and ESCs are arranged in a cap-like structure that encases GSCs, providing the physical support and signals for their development and functioning as the niche [11]. When a GSC divides to generate two daughter cells, the daughter remaining within the niche self-renews as a stem cell, while the one moving away from the niche differentiates into a cystoblast. Similarly, an ESC divides to generate a self-renewing stem cell that remains in contact with cap cells and a differentiated escort cell. Then, the differentiated escort cells wrap around the differentiated cystoblasts and support their further proliferation

T. Xie (✉)
Stowers Institute for Medical Research, 1000 East 50th Street, Kansas City, MO 64110
e-mail: tgx@stowers-institute.org

V.K. Rajasekhar, M.C. Vemuri (eds.), *Regulatory Networks in Stem Cells*, Stem Cell Biology and Regenerative Medicine,
DOI 10.1007/978-1-60327-227-8_14,

and differentiation. The cystoblasts undergo four rounds of synchronous divisions with incomplete cytokinesis to produce 16-cell cysts that are still encased by escort cells. After the cysts move to the middle of the germarium, they change their round shape to a lens-like shape, and at the same time the lens-shaped cysts shed the surrounded escort cells and become covered by a layer of epithelial follicle cells. The follicle cells are produced by two follicular stem cells that are located in the middle of the germarium [12, 13]. Finally, after a cyst is completely covered by a follicular epithelium, it then forms an egg chamber that is separated from the germarium by a somatic stalk. The linear arrangement of GSCs and their differentiated progeny in a simple tubular structure greatly facilitates analysis of gene functions in stem cell development.

In addition, different cells types in the germarium can be reliably identified by their specific molecular markers, distinct morphologies, and anatomical locations. TF cells and cap cells can be identified by high expression of Lamin-C on their nuclear membranes and a *hedgehog* (*hh*)*-lacZ* line [11, 14], while they can also be distinguished from each other by their location (TF: the most anterior somatic cells; cap cells: between TF and GSCs) and morphology (TF: disc-like; cap cells: small and round) [11]. Escort stem cells and their progeny (differentiated escort cells) can be identified by their molecular markers such as c587-drived UAS-GFP expression and a protein trap line fax-GFP [10, 15]. Two or three GSCs stand out in the germarium based on a combination of their location (surrounded by cap cells and ESCs and the most anterior germ cells), molecular markers (germline markers such as Vasa), size (the largest cell in the germarium), and anteriorly localized spectrosomes. Spectrosomes in GSCs and cystoblasts and branched fusomes in differentiated germline cysts are the same germ cell-specific organelles, which are rich in the membrane skeletal protein spectrin and Hu-li tai-shao (Hts) [16, 17]. Therefore, GSCs and their surrounding cells can be studied at the single cell level.

2 The GSC Niche is Composed of Adjacent Somatic Cells

As indicated earlier, GSCs are in direct contact with cap cells and ESCs. A series of genetic and cell biological studies have shown that cap cells and ESCs form the niche for GSCs (Fig. 1). First, TF and cap cells express Yb and Piwi, which are essential for controlling GSC maintenance [18–20].

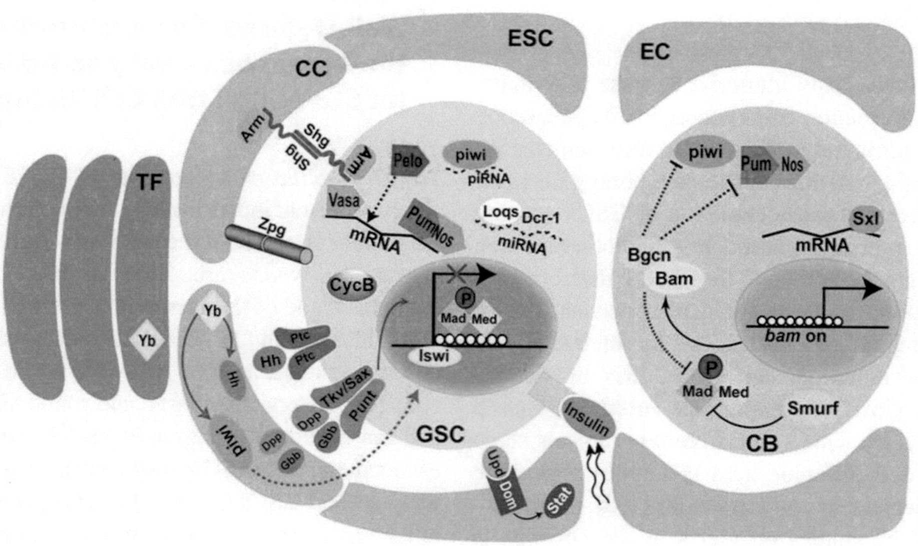

Fig. 1 Niche structure and major signaling pathways involved in GSC self-renewal and differentiation in the *Drosophila* ovary. GSCs are anchored to their niche by DE-cadherin (Shg)–mediated cell adhesions. The niche cells (TF/cap cell) express BMPs (Dpp and Gbb), Yb, Piwi, and Hh, which are required for GSC self-renewal and proliferation. In the GSC, BMP receptors (Tkv/Sax and Punt) transmit BMP signal to downstream effectors Mad and Med, which repress *bam*. ISWI also repress *bam* expression. Bam is required for cystoblast differentiation but is repressed in GSCs by Mad/Med complexes. However, this repression of Bam is relieved due to distance to the cap cell and through degradation of pMAD by Smurf. The GSC intrinsic factors Pum/Nos and Vasa are required for their self-renewal, while Piwi regulates GSC division. JAK-STAT signaling in ESCs regulates of GSCs. Gap junctions formed by Zpg are important for GSC maintenance and CB differentiation. The niche-independent insulin signal regulates GSC proliferation
TF: terminal filament cell; CC: cap cell; ESC: escort stem cell; EC: escort cell; GSC: germline stem cell; CB: cystoblast.

Disruption of their functions in somatic cells can lead to rapid GSC loss. Second, ESCs and cap cells express BMPs, Dpp, and Gbb, which are essential for controlling GSC self-renewal and proliferation [15, 21]. Removal of Gbb and Dpp function from the niche or blockage of BMP signaling inside GSCs causes them to lose their self-renewing ability. Third, when a normally differentiating GSC daughter is placed in direct contact with cap cells and ESCs, it can develop into a functional GSC. This provides direct evidence that cap cells and ESCs provide a regulatory microenvironment or niche sufficient for maintaining GSC self-renewal [11]. Fourth, the anchorage of GSCs to cap cells through E-cadherin–mediated cell adhesion is absolutely essential for long-term GSC self-renewal, indicating that proximity to cap cells is critical for GSC self-renewal [9]. Fifth, a signal controlled by JAK-STAT signaling in ESCs is also essential for controlling GSC self-renewal [10]. The observation that the absence of active JAK-STAT signaling in ESCs leads to rapid GSC loss underscores the importance of ESCs in the regulation of GSCs. Finally, an increase in cap cell number can result in niche size expansion and thus an increase in GSC number, while a decrease in cap cell number results in reduced GSC number [11, 22]. Together, these pieces of experimental evidence demonstrate that cap cells and ESCs form a functional niche for GSCs.

The niche exhibits an obvious structural asymmetry [11, 23]. This asymmetry allows the newly generated stem cell daughters to stay in different physical environments, permitting them to receive different signals or different levels of the same signals to adopt different cell fates. When a GSC divides to generate two daughters, the anterior daughter remains in direct contact with cap cells and ESCs (the niche) to continue self-renewal, while the other daughter directly interacts with differentiated escort cells, but not cap cells and ESCs, and undergoes differentiation. Neuroblasts in the *Drosophila* embryo undergo asymmetric division to generate two daughters that are intrinsically distinct [24]. Unlike neuroblasts, immediately following GSC division, the cell fates of the two daughters are not intrinsically determined since, if they both remain in the niche, they develop into GSCs [11]. The niche can reprogram already differentiated germline cysts back into the GSC state in both *Drosophila* ovary and testis when they are put in a functional niche [25, 26]. Together, cap cells and ESCs function as a niche, which has a unique ability to sufficiently maintain stem cell self-renewal and reprogram differentiated cells back into stem cells. Further revelation of the signals originated from the niche should help understand how the niche controls GSC functions and expand stem cells in vitro for future cell therapy.

Although much has been learned about how the niche controls GSC self-renewal and proliferation, relatively little is known about how stem cell niche formation and maintenance are controlled. Ectopically activated Notch signaling can sufficiently induce formation of extra cap cells, which then support more GSCs [22, 27]. Reduction of Notch signaling decreases the number of cap cells and thus the number of GSCs [22]. During the late third-instar larval stage, newly formed terminal filament cells express the Notch ligand Delta and activate Notch signaling in the adjacent somatic cells, which develop into cap cells [22]. Interestingly, adult GSCs require Delta and Serrate to keep GSCs in the niche [27], while Notch signaling is required for the survival of niche cells [22]. Therefore, Notch signaling is required for controlling niche formation and maintenance.

3 The BMP Signal and the Piwi-Mediated Signal from the Niche Control GSC Self-Renewal by Repressing Differentiation

The best-studied niche signal is the BMP signal, which is composed of two BMP-like ligands, Dpp (BMP2/4) and Gbb(BMP5/8) [15, 21]. Both *dpp* and *gbb* mRNAs can be detected in the somatic cells surrounding GSCs but not in GSCs themselves [15]. Loss-of-function mutations on either gene leads to slow proliferation and precocious differentiation leading to complete GSC loss. Conversely, when overexpressed, *dpp*, but not *gbb*, leads to the complete suppression of cystoblast differentiation. Although the cause for the difference between *dpp* and *gbb* overexpression remains unknown at the present time, it has become clear that both proteins act directly on the receptors on the surface of GSCs to activate its BMP signaling and control GSC self-renewal and proliferation. Removal of BMP receptors (TKV, Sax, and Punt) and downstream transducers (Mad and Medea) specifically from GSCs can lead to their premature differentiation and loss. In *Drosophila* imaginal discs, Dpp functions as a long-range morphogen to pattern cells in the distance of many cell diameters [28]. Interestingly, BMP signaling activity, which is detected by expression of pMad and Dad-lacZ, is primarily restricted to GSCs, indicating that BMP is a short-range niche signal. This finding can explain how Dpp and Gbb can maintain GSCs and at the same time allow cystoblasts lying next to GSCs to differentiate.

In addition to the BMP pathway, functions of *fs(1)Yb* (*Yb*) and *piwi* in niche cells are also indispensable for GSC self-renewal since their loss-of-function mutations cause rapid GSC loss and their overexpression increases GSC-like or cystoblast-like germ cells [18–20, 29, 30]. *Yb* encodes a novel protein, while *piwi* encodes a highly conserved protein involved in regulation of piRNA production and function. *Yb* is required for the expression of *piwi* and *hedgehog* (*hh*) in the TF and cap cell; the latter only plays a redundant role in niche function. Piwi has been recently shown to control production of a class of small RNAs known as piRNAs; disruption of a gene cluster encoding piRNAs can suppress *piwi* mutant GSC loss phenotype, indicating that piRNAs are indeed responsible for mediating the function of Piwi in controlling GSC self-renewal [31]. Likely, piRNAs in niche cells epigenetically or translationally control the production of a signal important for GSC self-renewal [31, 32]. However, the key niche signal regulated by *piwi* remains to be identified. Recent studies have given some clues about the Piwi-regulated niche signal. The GSC loss phenotype caused by *piwi* mutation can be rescued by a loss-of-function mutation of *smurf*, which encodes a E3 ubiquitin ligase known to target pMad for degradation [33], while overexpression of *dpp* neither produces similar effects as does *piwi* nor rescues the *piwi* phenotype [20]. These findings suggest that the Piwi-regulated niche signal is integrated with the BMP signal in GSCs by regulating pMad levels.

Interestingly, the functions of the Yb/Piwi-mediated signal and BMPs are to prevent GSCs from expressing differentiation-promoting genes, such as *bam* [15, 33,–35]. *bam* is necessary and sufficient for cystoblast differentiation since mutations in *bam* completely prevent cystoblast differentiation and forced *bam* expression in GSCs can sufficiently cause them to differentiate [36, 37]. In the absence of either BMP signaling or Piwi function, *bam* is up-regulated in GSCs [15, 33–35]. In addition, mutations in *bam* can also block GSC loss caused by defective *piwi* and BMP signaling. At least for BMP signaling, downstream transcription factors, Mad and Medea, can bind to a silencer element of *bam* in vitro, while the silencer element is necessary and sufficient for its repression in GSCs [15, 34, 38]. Finally, in order for cystoblast differentiation to proceed normally, the residue BMP signaling activity in newly formed cystoblasts is effectively removed by a *bam*-mediated negative feedback loop mechanism [39]. Therefore, BMP and Yb/Piwi-mediated signaling pathways are coordinated in GSCs to maintain an undifferentiated state and thus control their continuous self-renewal by preventing expression of differentiation genes.

4 Global Signals and Nutritional Status Also Regulate GSC Proliferation

Stem cell proliferation must be tightly controlled by systemic signals and nutrient status of an organism in order to better serve a role in maintaining tissue homeostasis. By controlling the amount of protein-rich nutrients in the diet, the rate of egg production can vary 60-fold due to reduced proliferation rates of GSCs and follicular stem cells, as well as their progeny [40]. However, the number of active GSCs does not appear to change in response to different nutritional conditions. α-Endosulfine, which is a small protein that regulates insulin secretion, is expressed in the ovary, the brain, and certain regions of the intestine, and is also required for proliferative response of GSCs to nutritional changes [41]. This requirement is non-cell autonomous, which is consistent with a role of α-endosulfine in secretion of *Drosophila* insulin-like peptides (DILPs). Furthermore, neural-derived DILPs directly regulate GSC division rate through activation of insulin pathways in stem cells [42]. This finding demonstrates that signals mediating the ovarian response to nutritional input can modify stem cell activity in a niche-independent manner. Therefore, systemic signals mediated by nutritional status and physiological conditions can modulate stem cell proliferation but not their self-renewal ability.

5 Interactions Between GSCs and Their Niche Are Important for Stem Cell Self-Renewal

In addition to secreted growth factors, direct cell-cell interactions also play important roles in controlling GSC self-renewal and proliferation. Indeed, homophilic adhesion protein E-cadherin and its intraceullar partner Armadillo (Arm) are expressed in the stem cell-niche interface to form adherens junctions [9]. This anchorage is essential for GSC self-renewal since removal of E-cadherin and Arm from GSCs leads to their rapid loss. In addition, E-cadherin–mediated cell adhesion is required for recruiting GSC precursors to their niche to become GSCs. Since E-cadherin–mediated cell adhesion is not directly required for GSC self-renewal, its function is mainly to ensure GSCs in close contact with the niche cells to receive maximal niche signals for their continuous self-renewal. Therefore, E-cadherin-mediated cell adhesion keeps GSCs in the niche for their long-term self-renewal.

Gap junctions allow contacting cells to transport small molecules, such as ions and small peptides, which can then regulate cellular functions. In addition to adherens junctions, gap junctions are also found between GSCs and cap cells, and between the differentiated cystoblast and IGS cells. Interestingly, when *zero population growth* (*zpg*), encoding a gap junction protein in *Drosophila*, is mutated, GSCs are gradually lost, and mutant germ cells do not differentiate properly [43, 44]. Thus, gap junctions are important for the maintenance of GSCs and differentiation of their progeny. However, it remains unknown which small molecules passing through gap junctions are responsible for GSC maintenance and differentiation.

6 Escort Cells May Form a Differentiation-Promoting Niche for the GSC Lineage

As cystoblasts moving away from the niche continue to divide and differentiate, they are encased by cellular processes of escort cells during the differentiation process until further surrounded by follicle cells. In the *Drosophila* testis, the counterparts of the escort cells, known as somatic cyst cells, are critical for differentiation of GSC daughters [45, 46]. Similarly, *stet*, encoding a Rhomboid-like membrane protease that cleaves and releases an EGF-like ligand, is expressed in escort cells and its mutations disrupt the differentiation of cystoblast differentiation causing the accumulation of cystoblast-like single germ cells [47]. *stet* mutations reduce the cellular processes of the escort cells, and the amount of active MAP kinase is reduced in *stet* mutant escort cells, indicating that there is a signal loop between early differentiated germ cells and escort cells, which is mediated by EGF signaling. These findings support the theory that escort cells form a differentiation niche for the GSC lineage. Future elucidation of the differentiation signals from escort cells will confirm the theory.

7 Different Classes of Intrinsic Factors Are Also Critical for Controlling GSC Self-Renewal

Within GSCs, different classes of intrinsic factors operate in concert in response to the extrinsic environmental signals to dictate whether to undergo self-renewal or differentiation. As expected, BMP downstream components, such as the BMP receptor and the Smads complex, are required in GSCs themselves to control their response to the BMP niche signal and maintain self-renewal [21]. Since transcriptional regulation often requires coordinated actions of chromatin remodeling factors and transcriptional factors, two chromatin remodeling factors, imitation switch (ISWI) and Stonewall (Stwl), have recently been identified for their essential roles in repressing GSC differentiation [48, 49]. Interestingly, ISWI, a SNF2-like ATP-dependent chromatin remodeling factor that regulates nucleosome sliding along DNA, is expressed in GSCs to repress *bam* expression but is not required for normal BMP signaling since pMad expression is normal but bam expression is upregulated in *iswi* mutant GSCs [48]. This finding suggests that it may help Mad-Medea transcriptional complexes bind to the *bam* silencer to repress its expression. By contrast, the requirement of Stwl for repressing GSC differentiation is not through suppressing *bam* expression since mutant *stwl* GSCs are lost prematurely but do not upregulate *bam* expression [49]. The results from gene profiling show that *stwl* is required for controlling expression of many genes in GSCs, including self-renewing factor *nanos* (*nos*). These findings also indicate that multiple differentiation pathways need to be simultaneously suppressed in order for GSCs to self-renew.

Four translational regulators, namely *pumilio* (*pum*), *nanos* (*nos*), *pelota* (*pelo*), and *vasa* (*vas*), have been identified so far for their essential roles in maintaining GSCs. In the germline, *pum* and *nos* are required in early GSC precursors and adult GSCs to prevent precocious differentiation, and their mutations cause premature differentiation and thus rapid GSC loss [30, 50–52]. Since Pum and Nos proteins are shown to form complexes in *Drosophila* embryos to repress RNA translation [53, 54], it is likely that they are also involved in repressing expression of differentiation-promoting genes in GSCs. Vas, a germline-specific translation initiation factor eIF4A, functions in GSCs to control their survival since *vas* mutant germaria contain only degenerated or growth-arrested germ cells [55]. Pelo has been recently shown to be a potential RNase by structural analysis [56]. In the *pelo* mutant germaria, the GSCs divide at a retarded rate and are quickly lost due to differentiation, indicating that Pelo is also involved in degrading mRNAs encoding differentiation-promoting proteins [57]. Genetic and cell biological studies indicate that Pum/Nos and Pelo are not involved in regulating *bam* transcription, translation, or protein function [33, 35, 57]. Therefore, these studies indicate that translational regulation plays a critical role in GSC self-renewal by repressing differentiation pathways independent of Bam function. However, target mRNAs regulated by these translation regulators remain to be identified. The identification of these target mRNAs will surely provide significant insight into the molecular mechanisms underlying GSC regulation.

Another class of molecules essential for GSC renewal and division are the proteins involved in small RNA production. In addition to its niche function in controlling GSC renewal, Piwi is also required intrinsically to stimulate GSC proliferation since its mutant GSCs divide much slower than wild-types [18]. Since it has been recently shown that Piwi interacts with heterochromatin protein 1 (HP1) to epigenetically control gene expression and is also involved in the production of piRNAs, the Piwi function inside GSCs could be mediated through one or both of the mechanisms [31, 32]. Two double-strand RNA binding proteins, *Dicer-1* (*Dcr-1*) and *Loquacious* (*Loqs*, also known as R3D1), have been recently implicated in controlling GSC division and self-renewal [58–60]. These two proteins form a complex to process and generate 19–21 nucleotide long single-stranded RNAs for repressing translation and/or degrading target mRNAs [61, 62]. *dcr-1* mutant GSCs divide significantly slower than wild-types, possibly due to up-regulating a p21-like

CDK inhibitor encoded by *decapo* and thus retarding the G1-to-S transition of the cell cycle [58]. *dcr-1* and *loqs* mutant GSCs are lost much faster than wild-types due to differentiation but not increased apoptosis, indicating that Dcr-1 and Loqs control GSC self-renewal [59, 60]. *bam* transcription and protein levels are not up-regulated in these mutant GSCs, whereas a mutation in *bam* cannot suppress the GSC loss phenotype of *dcr-1* or *loqs* mutants, indicating that Dcr-1 controls GSC self-renewal by repressing differentiation pathways independent of Bam function. Small RNA pathways are important for controlling GSC self-renewal and proliferation. It will be essential to identify responsible small RNAs and their target mRNAs in GSCs for a better understanding of GSC self-renewal mechanisms.

Interestingly, the cycle regulator Cyclin B (CycB) is also required for GSC division and self-renewal [51]. In *cycB* mutant female gonads, GSC precursors, also known as PGCs, differentiate prematurely, while in the mutant adult ovaries, GSCs are lost precociously and divide slower than wild-types. Since CycB is a key cell cycle regulator, it raises an interesting possibility that cell cycle progression is tightly linked to the self-renewal process. However, it remains unclear how CycB controls GSC self-renewal through regulating cell cycle progression or a cell cycle-dependent function. The answer to this question will surely help us understand functional relationships between proliferation and self-renewal of GSCs.

8 Multiple Differentiation Pathways Control GSC Differentiation

Many differentiation factors that promote GSC differentiation have also been identified, including *bam* (*bag of marbles*), *bgcn* (*benign gonial cell neoplasm*), *otu* (*ovarian tumor*), *Orb,* and *Sxl* (*Sex lethal*) [37, 63–67]. Although all these mutant germaria accumulate GSC-like or cystoblast-like single germ cells, there are significant differences. Mutations in *bam* or *bgcn* can completely block GSC differentiation, resulting in the accumulation of single germ cells only, indicating that they are critical for GSC differentiation [37]. Forced *bam* expression in GSCs can drive them to differentiate because *bgcn* is normally expressed in GSCs and early differentiated germ cells [36]. In addition, *bam* cannot drive GSCs to differentiation without functional *bgcn*, and they genetically enhance each other's mutant phenotypes [66]. These findings indicate that *bam* and *bgcn* function in the same genetic pathway. By contrast, mutations in *otu*, *orb,* and *Sxl* can only slow down, but not completely block, GSC differentiation, causing the accumulation of mixed single germ cells and germ cell cysts, indicating that these genes play regulatory roles in the control of GSC differentiation [63, 64, 67–69]. Interestingly, *bam* encodes a novel protein, while *bgcn*, *Sxl*, *otu,* and *orb* encode RNA-binding proteins, suggesting that these differentiation factors function in differentiated germ cells to control mRNA splicing, stability, and/or translation regulation of differentiation-promoting genes [63, 64, 67–69]. In addition, *bam* mutant GSCs can still differentiate in *pum* and *pelo* mutant backgrounds, demonstrating the existence of differentiation pathways independent of Bam function [33, 35]. Thus, multiple differentiation pathways must exist to execute differentiation of GSC daughters.

9 Stem Cell Aging Is Controlled Extrinsically and Intrinsically

Functional decline of aged stem cells has been widely postulated to contribute to organismal aging, but it remains unknown how stem cell aging itself is controlled. Recently, it has been shown that the numbers of GSCs and niche cells decline with age and GSC proliferation also decreases with age [70]. By manipulating the dosage of BMP signaling and E-cadherin that are known to be important for GSC function, age-dependent decline in BMP signaling and E-cadherin contributes to stem cell aging. Strengthening BMP signaling and E-cadherin-mediated cell adhesion from the niche or wthin GSCs can prolong GSC lifespan. One recent study in the *Drosophila* testis has also shown that declined niche signaling and adhesion between stem cells and their niche also contribute to stem cell aging [71]. In addition, overexpression of SOD, an enzyme that helps remove reactive oxygen species (ROS), in either the niche or GSCs can significantly prolong GSC lifespan, indicating that ROS-induced damage contributes to aging of GSCs and their niche [70]. Therefore, stem cell aging is controlled extrinsically and intrinsically. In the future, it will be important to determine what triggers decline in niche signaling and adhesion and the increase in ROS-induced cellular damage in stem cells and the niche.

10 Conclusions and Future Directions

The hallmark of stem cells is their ability to self-renew, while the central question of stem cell biology is how self-renewal is controlled at the cellular and molecular level. The studies on *Drosophila* ovarian GSCs have provided the conceptual framework that guides stem cell research in other systems, notably the niche theory, and also have provided a thorough understanding of the cellular and molecular mechanisms underlying stem cell regulation. The findings that have been made in *Drosophila* ovarian GSCs have important implications for stem cell biology in general: (1) the niche structure exhibits an obvious asymmetry in such a way that following stem cell division the daughter remaining in the niche renews as a stem cell, while the other positioning away from the niche differentiates; (2) the signals from the niche only function in a distance of one cell diameter so that only the daughter in the niche receives enough signaling to repress differentiation and remain undifferentiated, while the other daughter just outside the niche fails to do so; (3) stem cells are anchored to the niche in order to maintain long-term self-renewal; (4) different classes of intrinsic self-renewing factors are required to control self-renew by repressing different branches of differentiation pathways; (5) extrinsic niche signaling and intrinsic factors work coordinately to repress differentiation and thereby maintain self-renewal; and (6) stem cell aging is controlled extrinsically and intrinsically. These general principles are also shared by stem cells in other systems, including the *Drosophila* testis.

Although rapid progress has been made in understanding stem cell regulation in the *Drosophila* ovary, there are many important questions remaining to be answered. These questions include how stem cell niche formation and maintenance are controlled, whether and how stem cells in the same niche or tissue interact with one another, how stem cell aging is regulated, how multiple signals from the niche are integrated in GSCs to control their self-renewal, how niche signaling and intrinsic self-renewing factors work together to maintain GSCs, and how different classes of intrinsic factors work together to repress different differentiation pathways to control self-renewal. The combination of sophisticated genetics, elegant cell biology, and powerful genomics will continue to make *Drosophila* ovarian stem cells a productive system to answer important questions and reveal novel insights into stem cell biology.

References

1. Wong MD, Jin Z, Xie T. Molecular mechanisms of germline stem cell regulation. Annu Rev Genet. 2005;39:173–95.
2. Li L, Xie T. Stem cell niche: structure and function. Annu Rev Cell Dev Biol. 2005;21:605–31.
3. Xie T, Spradling A. The Drosophila ovary: an n vivo stem cell system. In: Marshak DR, Gardner RL, Gottlieb D, editors. Stem cell biology. Cold Spring Harbor, NY: Cold Spring Harbor Laboratory Press; 2001. pp.129–48.
4. Kirilly D, Xie T. The Drosophila ovary: an active stem cell community. Cell Res. 2007;17(1):15–25.
5. Ohlstein B, Kai T, Decotto E, Spradling A. The stem cell niche: theme and variations. Curr Opin Cell Biol. 2004;16(6):693–9.
6. Lin H. The stem-cell niche theory: lessons from flies. Nat Rev Genet. 2002;3(12):931–40.
7. Spradling AC. Developmental genetics of oogenesis. In: Bate M, Martinez Arias A, editors. The development of *Drosophila melanogaster*. 1st ed. Cold Spring Harbor, NY: Cold Spring Harbor Laboratory Press; 1993. pp. 1–71.
8. Lin H, Spradling AC. Germline stem cell division and egg chamber development in transplanted Drosophila germaria. Dev Biol. 1993;159(1):140–52.
9. Song X, Zhu CH, Doan C, Xie T. Germline stem cells anchored by adherens junctions in the Drosophila ovary niches. Science. 2002;296(5574):1855–7.
10. Decotto E, Spradling AC. The Drosophila ovarian and testis stem cell niches: similar somatic stem cells and signals. Dev Cell. 2005;9:501–10.
11. Xie T, Spradling AC. A niche maintaining germ line stem cells in the Drosophila ovary. Science. 2000;290(5490):328–30.
12. Nystul T, Spradling A. An epithelial niche in the Drosophila ovary undergoes long-range stem cell replacement. Cell Stem Cell. 2007;1(3):277–85
13. Margolis J, Spradling A. Identification and behavior of epithelial stem cells in the Drosophila ovary. Development. 1995;121(11):3797–807.
14. Forbes AJ, Lin H, Ingham PW, Spradling AC. Hedgehog is required for the proliferation and specification of ovarian somatic cells prior to egg chamber formation in Drosophila. Development. 1996;122(4):1125–35.
15. Song X, Wong MD, Kawase E, et al. Bmp signals from niche cells directly repress transcription of a differentiation-promoting gene, bag of marbles, in germline stem cells in the Drosophila ovary. Development. 2004;131(6):1353–64.
16. Lin H, Yue L, Spradling AC. The Drosophila fusome, a germline-specific organelle, contains membrane skeletal proteins and functions in cyst formation. Development. 1994;120(4):947–56.
17. de Cuevas M, Lee JK, Spradling AC. Alpha-spectrin is required for germline cell division and differentiation in the Drosophila ovary. Development. 1996;122(12):3959–68.
18. Cox DN, Chao A, Lin H. Piwi encodes a nucleoplasmic factor whose activity modulates the number and division rate of germline stem cells. Development. 2000;127(3):503–14.
19. Cox DN, Chao A, Baker J, Chang L, Qiao D, Lin H. A novel class of evolutionarily conserved genes defined by piwi are essential for stem cell self-renewal. Genes Dev. 1998;12(23):3715–27.
20. King FJ, Szakmary A, Cox DN, Lin H. Yb modulates the divisions of both germline and somatic stem cells through piwi- and hh-mediated mechanisms in the Drosophila ovary. Mol Cell. 2001;7(3):497–508.
21. Xie T, Spradling AC. Decapentaplegic is essential for the maintenance and division of germline stem cells in the Drosophila ovary. Cell. 1998;94(2):251–60.
22. Song X, Call GB, Kirilly D, Xie T. Notch signaling controls germline stem cell niche formation in the Drosophila ovary. Development. 2007;134(6):1071–80.
23. Song X, Xie T. DE-cadherin-mediated cell adhesion is essential for maintaining somatic stem cells in the Drosophila ovary. Proc Natl Acad Sci U S A. 2002;99(23):14813–8.
24. Jan YN, Jan LY. Asymmetric cell division in the Drosophila nervous system. Nat Rev Neurosci. 2001;2(11):772–9.
25. Kai T, Spradling A. Differentiating germ cells can revert into functional stem cells in Drosophila melanogaster ovaries. Nature. 2004;428(6982):564–9.

26. Brawley C, Matunis E. Regeneration of male germline stem cells by spermatogonial dedifferentiation in vivo. Science. 2004;304(5675):1331–4.
27. Ward EJ, Shcherbata HR, Reynolds SH, Fischer KA, Hatfield SD, Ruohola-Baker H. Stem cells signal to the niche through the Notch pathway in the Drosophila ovary. Curr Biol. 2006;16(23):2352–8.
28. Nellen D, Burke R, Struhl G, Basler K. Direct and long-range action of a DPP morphogen gradient. Cell. 1996;85(3):357–68.
29. King FJ, Lin H. Somatic signaling mediated by fs(1)Yb is essential for germline stem cell maintenance during Drosophila oogenesis. Development. 1999;126(9):1833–44.
30. Lin H, Spradling AC. A novel group of pumilio mutations affects the asymmetric division of germline stem cells in the Drosophila ovary. Development. 1997;124(12):2463–76.
31. Yin H, Lin H. An epigenetic activation role of Piwi and a Piwi-associated piRNA in Drosophila melanogaster. Nature. 2007; 450(7167):304–8.
32. Brower-Toland B, Findley SD, Jiang L, et al. Drosophila PIWI associates with chromatin and interacts directly with HP1a. Genes Dev. 2007;21(18):2300–11.
33. Chen D, McKearin D. Gene circuitry controlling a stem cell niche. Curr Biol. 2005;15:179–84.
34. Chen D, McKearin D. Dpp signaling silences bam transcription directly to establish asymmetric divisions of germline stem cells. Curr Biol. 2003;13(20):1786–91.
35. Szakmary A, Cox DN, Wang Z, Lin H. Regulatory relationship among piwi, pumilio, and bag-of-marbles in Drosophila germline stem cell self-renewal and differentiation. Curr Biol. 2005;15: 171–8.
36. Ohlstein B, McKearin D. Ectopic expression of the Drosophila Bam protein eliminates oogenic germline stem cells. Development. 1997;124(18):3651–62.
37. McKearin D, Ohlstein B. A role for the Drosophila bag-of-marbles protein in the differentiation of cystoblasts from germline stem cells. Development. 1995;121(9):2937–47.
38. Chen D, McKearin DM. A discrete transcriptional silencer in the bam gene determines asymmetric division of the Drosophila germline stem cell. Development. 2003;130(6):1159–70.
39. Casanueva MO, Ferguson EL. Germline stem cell number in the Drosophila ovary is regulated by redundant mechanisms that control Dpp signaling. Development. 2004;131(9):1881–90.
40. Drummond-Barbosa D, Spradling AC. Stem cells and their progeny respond to nutritional changes during Drosophila oogenesis. Dev Biol. 2001;231(1):265–78.
41. Drummond-Barbosa D, Spradling AC. Alpha-endosulfine, a potential regulator of insulin secretion, is required for adult tissue growth control in Drosophila. Dev Biol. 2004;266:310–21.
42. LaFever L, Drummond-Barbosa D. Direct control of germline stem cell division and cyst growth by neural insulin in Drosophila. Science. 2005;309(5737):1071–3.
43. Tazuke SI, Schulz C, Gilboa L, et al. A germline-specific gap junction protein required for survival of differentiating early germ cells. Development. 2002;129(10):2529–39.
44. Gilboa L, Forbes A, Tazuke SI, Fuller MT, Lehmann R. Germ line stem cell differentiation in Drosophila requires gap junctions and proceeds via an intermediate state. Development. 2003;130(26):6625–34.
45. Kiger AA, White-Cooper H, Fuller MT. Somatic support cells restrict germline stem cell self-renewal and promote differentiation. Nature. 2000;407(6805):750–4.
46. Tran J, Brenner TJ, DiNardo S. Somatic control over the germline stem cell lineage during Drosophila spermatogenesis. Nature. 2000;407(6805):754–7.
47. Schultz C, Wood CG, Jones DL, Tazuke SI, Fuller MT. Signaling from germ cells mediated by the rhomboid homolog stet organizes encapsulation by somatic support cells. Development. 2002;129:4523–34.
48. Xi R, Xie T. Stem cell self-renewal controlled by chromatin remodeling factors. Science. 2005;310:1487–9.
49. Maines JZ, Park JK, Williams M, McKearin DM. Stonewalling Drosophila stem cell differentiation by epigenetic controls. Development. 2007;134(8):1471–9.
50. Forbes A, Lehmann R. Nanos and Pumilio have critical roles in the development and function of Drosophila germline stem cells. Development. 1998;125(4):679–90.
51. Wang Z, Lin H. The division of Drosophila germline stem cells and their precursors requires a specific cyclin. Curr Biol. 2005;15: 328–33.
52. Gilboa L, Lehmann R. Repression of primordial germ cell differentiation parallels germ line stem cell maintenance. Curr Biol. 2004;14(11):981–6.
53. Murata Y, Wharton RP. Binding of pumilio to maternal hunchback mRNA is required for posterior patterning in Drosophila embryos. Cell. 1995;80(5):747–56.
54. Barker DD, Wang C, Moore J, Dickinson LK, Lehmann R. Pumilio is essential for function but not for distribution of the Drosophila abdominal determinant Nanos. Genes Dev. 1992;6(12A):2312–26.
55. Styhler S, Nakamura A, Swan A, Suter B, Lasko P. Vasa is required for GURKEN accumulation in the oocyte, and is involved in oocyte differentiation and germline cyst development. Development. 1998;125(9):1569–78.
56. Lee HH, Kim YS, Kim KH, et al. Structural and functional insights into Dom34, a key component of no-go mRNA decay. Mol Cell. 2007;27(6):938–50.
57. Xi R, Doan C, Liu D, Xie T. Pelota controls self-renewal of germline stem cells by repressing a Bam-independent differentiation pathway. Development. 2005;132:5365–74.
58. Hatfield SD, Shcherbata HR, Fischer KA, Nakahara K, Carthew RW, Ruohola-Baker H. Stem cell division is regulated by the microRNA pathway. Nature. 2005;435:974–8.
59. Jin Z, Xie T. Dcr-1 maintains Drosophila ovarian stem cells. Curr Biol. 2007;17(6):539–44.
60. Park JK, Liu X, Strauss TJ, McKearin DM, Liu Q. The miRNA pathway intrinsically controls self-renewal of drosophila germline stem cells. Curr Biol. 2007;17(6):533–8.
61. Forstemann K, Tomari Y, Du T, et al. Normal microRNA maturation and germ-line stem cell maintenance requires Loquacious, a double-stranded RNA-binding domain protein. PLoS Biol. 2005;3:e236.
62. Jiang F, Ye X, Liu X, Fincher L, McKearin D, Liu Q. Dicer-1 and R3D1-L catalyze microRNA maturation in Drosophila. Genes Dev. 2005;19:1674–9.
63. Bopp D, Horabin JI, Lersch RA, Cline TW, Schedl P. Expression of the Sex-lethal gene is controlled at multiple levels during Drosophila oogenesis. Development 1993;118(3): 797–812.
64. Christerson LB, McKearin DM. Orb is required for anteroposterior and dorsoventral patterning during Drosophila oogenesis. Genes Dev. 1994;8(5):614–28.
65. Lantz V, Chang JS, Horabin JI, Bopp D, Schedl P. The Drosophila orb RNA-binding protein is required for the formation of the egg chamber and establishment of polarity. Genes Dev. 1994;8(5):598–613.
66. Lavoie CA, Ohlstein B, McKearin DM. Localization and function of Bam protein require the benign gonial cell neoplasm gene product. Dev Biol. 1999;212(2):405–13.
67. Ohlstein B, Lavoie CA, Vef O, Gateff E, McKearin DM. The Drosophila cystoblast differentiation factor, benign gonial cell neoplasm, is related to DExH-box proteins and interacts genetically with bag-of-marbles. Genetics. 2000;155: 1809–19.

68. McKearin DM, Spradling AC. Bag-of-marbles: a Drosophila gene required to initiate both male and female gametogenesis. Genes Dev. 1990;4(12B):2242–51.
69. Mulligan PK, Mohler JD, Kalfayan LJ. Molecular localization and developmental expression of the otu locus of Drosophila melanogaster. Mol Cell Biol. 1988;8(4):1481–8.
70. Pan L, Chen S, Weng C, et al. Stem cell aging is controlled both intrinsically and extrinsically in the *Drosophila* ovary. Cell Stem Cell. 2007;1:458–69.
71. Boyle M, Wong C, Rocha M, Jones DL. Decline in self-renewal factors contributes to aging of the stem cell niche in the Drosophila testis. Cell Stem Cell. 2007;1:470–8.

The Niche Regulation of Hematopoietic Stem Cells

Hiroko Iwasaki and Toshio Suda

Abstract Stem cells' major capabilities, that is, pluripotency and self-renewal, are key to sustaining the lifelong functionality of organs. Stem cells reside in the special microenvironment called the niche. The niche and stem cells adhere to each other via adhesion molecules and exchange the molecular signals that maintain stem cell features. It has been suggested that tumor tissue also contains such type of cells.

In this chapter, the hematopoietic system is used as an example to show the interaction between the stem cell and its niche. Different kinds of niches and regulatory mechanisms are explained. Furthermore, the niche's involvement in cancer stem regulation, tumor invasion, and metastasis, and novel therapeutic approaches used in association with the cancer stem cell niche are also described.

Keywords Niche · Centrosome · Asymmetric division · Cell polarity · Spindle orientation

1 Introduction

Virtually every organ has special cells called stem cells. These stem cells possess two major roles – pluripotency, which gives rise to the mature cells that compose the specific organ or tissue, and self-renewal capability, which supplies an adequate number of cells to maintain the organ's function. These stem cells tend to be found in specific areas in the organ, where a special microenvironment, called the *niche*, maintains the stem cell functions. The stem cell and the niche cells adhere to each other via adhesion molecules. They exchange molecular signals that maintain the stem cell features.

The concept of the stem cell niche was proposed by Schofield [1] in 1978 for the hematopoietic stem cell (HSC) in bone marrow. Schofield hypothesized that "the cellular environment which retains the stem cell I shall call a stem cell 'niche'. As long as the stem cell remains fixed its further maturation is prevented and it continues indefinitely to replicate as a stem cell, i.e., it exhibits immortality[1]." Since this initial proposal, not only the HSCs but various other kinds of stem cells have also been identified, and their niches and the molecular mechanisms for stemness maintenance have been revealed.

Stem cells are not only for healthy normal organs or tissues. Recent findings suggest that there exists a group of cells in tumor that behave in a similar way to regular stem cells. Such cells are called cancer stem cells and are considered to be deeply involved in tumor proliferation, invasion, and metastasis. Their niche plays a regulatory role for cancer stem cell maintenance.

To date, the most advanced studies in the stem cell niche field can be found in the hematopoietic system. In this chapter, the interaction between the HSC and its niche will be discussed. Furthermore, the niche involvement in cancer stem regulation, tumor invasion, and metastasis, and novel therapeutic approaches used in association with the cancer stem cell niche will be described.

2 Stem Cell and Niche

Stem cell research in terms of the maintenance of pluripotency and self-renewal capacity made a great advancement through an experiment using *Drosophila* and *Caenorhabditis elegans*. In 1997, the group of Deng and Lin [2] demonstrated that the female *Drosophila* has cells named "cap

H. Iwasaki (✉)
Department of Cell Differentiation, The Sakaguchi Laboratory of Developmental Biology, Keio University School of Medicine; 35 Shinanomachi, Shinjuku, Tokyo 160-8582 Japan
e-mail: di065022@sc.itc.keio.ac.jp

[1] This article was published in Blood Cells, 4, Schofield R, The relationship between the spleen colony-forming cell and the haematopoietic stem cell, 7–25, Copyright (1978).

V.K. Rajasekhar, M.C. Vemuri (eds.), *Regulatory Networks in Stem Cells*, Stem Cell Biology and Regenerative Medicine, DOI 10.1007/978-1-60327-227-8_15, © Humana Press, a part of Springer Science+Business Media, LLC 2009

cells" at the tip of the ovary, to which the germline stem cells adhere. In mitosis, daughter stem cells that divided in such a way that they no longer adhered to the cap cells started to differentiate, growing into cystoblasts In contrast, those that kept the adhesion to the cap cells after the cell division remained as the germline stem cells. These results suggested that cap cells provide the special environment for the germline stem cell and play a role as the stem cell niche in the *Drosophila* ovary.

Xie and Spradling [3] demonstrated in 2000 that the cells newly introduced to the *Drosophila* ovary perform as the germline stem cell as long as they adhere to the cap cells, thus proving that the cap cells function as the true stem cell niche. They showed that the germline stem cells require a signal mediated by dpp, a member of the TGF-β super family, in order to maintain adhesion to the cap cells (i.e., in order to maintain the stem cell) and control the frequency of the cell division. Additional signals, such as hedgehog, wingless, and armadillo, were also suggested to be involved in the regulation of the germline stem cells.

3 Interaction Between Hematopoietic Stem Cell and Niche

3.1 Osteoblastic Niche

In human hematopoiesis in bone marrow, the osteoblasts offer the microenvironment as the niche, that is, the osteoblastic niche, for the HSCs. The HSC and the osteoblast bind each other via adhesion molecules, such as N-cadherin (Fig. 1). Two independent research groups, led by Li and Scadden, respectively, revealed this new function of the osteoblast as the HSC niche in 2003. Both groups approached the issue in the investigation of the regulatory system for the number of the HSCs. Li's group reported that the bone morphogenetic protein (BMP) signaling pathway, through BMP receptor type IA (BMPRIA) expressed in osteoblasts, controls the number of HSCs by regulating the niche size [4]. Scadden and his colleagues showed that the osteoblastic cells, stimulated by activated PTH (parathyroid hormone)/PTHrP (parathyroid hormone-related peptide) receptors (PPRs), increased in number, produced high levels of the Notch ligand jagged 1, and supported the increase in the number of HSCs [5].

It is widely known that the stem cell in general is in a quiescent state (G0 phase in the cell cycle) and that this quiescence prevents the stem cells from entering into the cell cycle and differentiation. Arai et al. [6] demonstrated that angiopoietin-1 (Ang1) expressed in the osteoblast interacts with Tie-2, a type of receptor tyrosine kinase, expressed in HSCs in bone marrow, and that this enhanced adhesion between the niche cell and the stem cell contributes to maintenance of the quiescence of the stem cell.

Osteopontin (Opn) expressed by osteoblasts is also reported as a key component in the HSC niche [7]. Nilsson and colleagues demonstrated that the LSK (lineage$^-$ Sca-1$^+$c-Kit$^+$) cells isolated from bone marrow of Opn knockout mice after a continuous 4-week BrdU treatment were 100% BrdU positive, meaning that the HSCs divide at least once in 4 weeks in bone marrow of the Opn knockout mice. On the other hand, approximately 60% of the collected LSK cells were BrdU$^+$ in the case of wild-type mice. They also showed

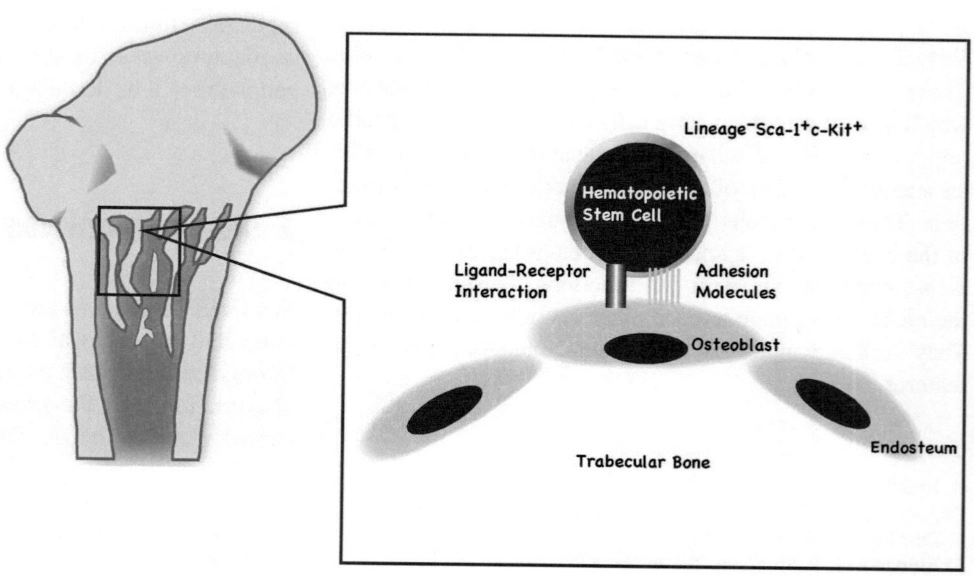

Fig. 1 Schematic model for the hematopoietic stem cell (HSC) and the osteoblastic niche. The osteoblastic niche lodges HSC via adhesion molecules, such as N-cadherin. Osteoblasts and HSCs exchange signals in order to maintain stem cell function

that CD34$^+$ human hematopoietic progenitors were inhibited to bind to Opn when treated with the specific β_1 integrin-blocking antibody prior to the cell culture. Also, LSK cells were observed to be rather randomly distributed within bone marrow after transplantation into Opn knockout mice, while they were likely to be located in the endosteal region, where osteoblasts reside, in the case of wild-type mice. These data suggest that Opn expressed by osteoblasts contributes to the adhesion between the HSC and the osteoblastic niche and negatively regulates HSC proliferation, contributing to the maintenance of the stem cell quiescence.

c-Myc is another important regulator of the HSC fate. Trumpp and colleagues demonstrated, using conditional c-Myc–deficient mice, that the number of HSCs in bone marrow was significantly higher, and that HSC differentiation was prevented due to the up-regulated adhesion molecules between HSCs and the niche, such as N-cadherin and various integrins, resulting in the severe cytopenia. Conversely, overexpression of c-Myc led to loss of self-renewal activity and augmentation of differentiation of HSCs. The endogenous c-Myc mRNA level appeared to be consistent with these results, that is, a lower expression level in self-renewing long-term HSCs and a higher level in differentiating short-term HSCs. c-Myc controls the balance between HSC self-renewal and differentiation by adjusting the adhesion between HSCs and the niche [8].

3.2 ECM Contribution to Osteoblastic Niche

Aside from these factors expressed in osteoblasts, various extracellular matrix (ECM) proteins in bone marrow are also confirmed to be involved in regulation of the osteoblastic niche environment. Glycosaminoglycan hyaluronic acid (HA), for example, is one of the key components of the niche regulators. HA is the major component of the ECM and is found ubiquitously, including in bone marrow. Its receptor, CD44, is a multifunctional transmembrane protein expressed by a wide variety of cells, including the HSCs and progenitors. Matrosova et al. [9] reported that mice treated with 5-fluorouracil (5-FU) for bone marrow suppression exhibited quicker recovery in terms of the white blood cell and platelet counts when administered with HA during the recovery period, compared to phosphate-buffered saline as control. In order to examine the location of injected HA in vivo, FITC-labeled HA was traced and found concentrated in bone marrow, binding to CD44. This suggests that HA plays a role as a regulator of supportive function of the niche in bone marrow for hematopoietic stem cells.

Another report by Lapidot and colleagues [10] demonstrated that the xenografting of human CD34$^+$ hematopoietic stem/progenitor cells into the bone marrow of nonobese diabetic/severe combined immunodeficient (NOD/SCID) mice was significantly impaired by masking the HSCs' cell surface with anti-CD44 antibody or HA, indicating that CD44-HA interaction is an important factor for homing and engraftment of the HSC to the niche.

Because osteoblasts are located at the endosteal area, where higher calcium concentration is expected due to its active bone modeling/remodeling, Scadden and co-workers [11] hypothesized that the sensor for calcium must be expressed on HSCs and contribute to their homing and lodgment to bone marrow. They revealed that calcium-sensing receptor, CaR, was expressed on normal HSCs and that CaR-deficient mice had a normal number of primitive hematopoietic cells in the circulation and spleen, but very few of them were found in bone marrow. HSCs isolated from fetal liver of CaR-deficient mice could not localize to the endosteal niche in wild-type recipient mice because of the defective adhesion to the collagen I, which is one of the major components of the ECM and is secreted by osteoblasts. Thus, it was suggested that the HSCs preferentially localize the calcium-rich endosteum and lodge to the niche via adhesion to the ECM surrounding osteoblast.

3.3 Vascular Niche

As described above, adhesion molecules and their signaling play a fundamental role in HSC maintenance of the osteoblastic niche. There has recently been found another kind of niche, existing along the endothelial cells of the sinusoidal vessels in bone marrow or spleen. Such a microenvironment is called a vascular niche. This new type of niche was identified through a recent study to distinguish the stem and progenitor cells in bone marrow using the SLAM family receptors. The most purified HSCs were fractioned as CD150$^+$CD244$^-$CD48$^-$ cells, and the majority of these cells were found to be associated with sinusoidal endothelium [12].

Sugiyama et al. [13] showed that HSCs were found specifically adjacent to cells expressing a high level of CXCL12 (also called stromal cell-derived factor-1, or SDF-1) surrounding the sinusoidal endothelial cells. They named these cells CXCL12-abundant reticular (CAR) cells. They demonstrated that depletion of CXCR4 led to reduction of the HSC pool and poor survival rate after 5-FU-induced bone marrow suppression, thus suggesting that CXCL12-CXCR4 chemokine signaling plays an essential role in maintaining the quiescent HSC pool. The connection between CAR cells and HSCs is also observed in the endosteum. It appears to be the universal component of the hematopoietic stem cell niche.

The integrin very late antigen 4 (VLA-4) and its ligand vascular cell adhesion molecule 1 (VCAM-1) were proposed to be the regulator for the homing and lodgment mechanism to the vascular niche [14]. Pretreatment of donor cells with either anti- VLA-4 or VCAM-1 antibody prior to transplantation resulted in impaired lodgment of HSCs to bone marrow and an increased number of the hematopoietic progenitors in circulating blood and spleen.

The functional difference between these osteoblastic and vascular niches is yet to be elucidated, however, it may be reasonable to approach this issue from the physiological aspect. One of the major differences between these two microenvironments is oxygen level. In the vascular niche, a higher oxygen level is expected than in the osteoblastic niche. In such a microenvironment, the cell cycle of the stem cell would resume [15]. Thus, a model of the dynamics of the hematopoiesis can be hypothesized: HSCs in the G0 state located at the osteoblastic niche under regional hypoxia would move to the vascular niche at a certain time and undergo differentiation and supply the required pool of the mature cells into the peripheral blood stream [16]. Once the necessary supply of mature cells is fulfilled, the HSCs at the vascular niche would move back to the osteoblastic niche, where they are maintained in the G0 state again. The logistics of shuttling the HSCs between these two types of niches may be the key to well-balanced hematopoiesis (Fig. 2).

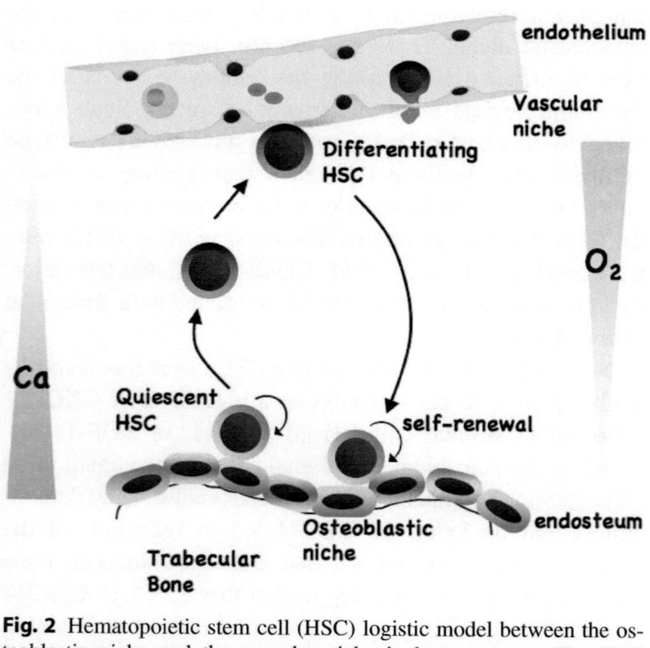

Fig. 2 Hematopoietic stem cell (HSC) logistic model between the osteoblastic niche and the vascular niche in bone marrow. The HSC stays in quiescence in the osteoblastic niche and activates the cell cycle in leaving for the vascular niche, where differentiation is initiated. HSCs are guided by multiple factors, such as local calcium or oxygen concentration gradient between these two different kinds of niches

4 Seed or Soil? Niche Disruption and Disease

If a stem cell comes out of the niche, its differentiation program resumes and mature cells are supplied to the organ. If a stem cell niche becomes corrupted for some reason, more stem cells will be forced to leave. This may lead to the ideal situation for the organ, at least temporarily, given plenty of healthy mature cells to sustain the function. However, what if a majority of the stem cell niche is permanently destroyed? If a disease is developed, is it because the niche is bad, or because the stem cell itself is no longer healthy due to the niche malfunction?

Walkley et al. [17] delivered the answer to this question by introducing retinoic acid receptor γ (RARγ)–deficient mice. These mice exhibited the myeloproliferative syndromes (MPS), with significantly increased granulocyte/macrophage progenitors and granulocytes in bone marrow, peripheral blood, and spleen. The phenotypes became more severe with age. A significant reduction of trabecular bone was observed, resulting in depletion of the osteoblasts, thus destroying the niche. In order to determine whether this disease is caused by a bad seed (the HSCs) or the soil (the niche), the transplantation of wild-type bone marrow (BM) cells into lethally irradiated RARγ overexpressed and deficient mice was performed, revealing the result that the onset of MPS occurred exclusively in the RARγ-deficient mice. These data clearly proved that the MPS was not due to the altered HSC but the malfunctioning niche.

Walkley et al. also investigated the effect of the *Rb* gene, a major cell cycle regulator, using the interferon-inducible *Mx*-Cre transgene and $pRb^{fl/fl}$ mice [18]. When *Rb* was widely inactivated in the transgenic mice, including the BM and HSCs, the mice exhibited profound myeloproliferation, loss of HSC from BM, differentiation, and extramedullary hematopoiesis. Substantial loss of trabecular bone was also observed after the *Rb* inactivation. However, when either the transplanted HSC or recipient is wild type, no such phenotype was observed.

The effect of *Rb* deletion was also examined using Lysozyme-M-Cre mice, by which the gene deletion specific to the myeloid lineage, that is, granulocytes, macrophages, and osteoclasts, is made possible. Myeloproliferation occurred when the *Rb*-inactivated hematopoietic cells, including the unaffected HSCs, harvested from the Lysozyme-M-Cre mice were transplanted into the widely *Rb*-inactivated mice, but not when transplanted into the wild-type mice. Collectively, this *Rb* deletion model concluded that the myeloproliferative disease is observed only when *Rb*-inactivated myeloid-derived cells were interacted with *Rb*-inactivated microenvironment, regardless of the condition of the HSCs.

Both of these studies exhibited loss of the osteoblastic niche and myeloproliferation, however, the causes were not identical. The status of the residual microenvironment, the recovery system for the lost niche, or the relevance of the gene to hematopoiesis and microenvironment may explain the differences in the observations between these two cases. The bottom line is that disruption of the niche or microenvironment can trigger the system disorder, even if the stem cells are healthy. Understanding the pathological involvement of the niche can bring about novel therapeutic strategies with enhanced effectiveness, and further studies in this area are awaited.

5 Is There a Niche for Cancer Stem Cells?

It has been reported lately by various studies that there is a functional microenvironment to support cancer stem cells as well. This should also be considered a niche, and is thus called the cancer stem cell niche. A representative example is AML (acute myeloid leukemia) and its niche in bone marrow. Dick and co-workers [19] showed that anti-CD44 antibody-treated NOD/SCID mice transplanted with AML cells exhibited a significantly lower rate of disease onset. Also, Van Etten and colleagues [20] showed impaired induction of CML (chronic myeloid leukemia)-like myeloproliferative disease among the recipient mice when transplanted with *BCR-ABL1*–transduced CML progenitors from CD44-mutant donors. These results indicate that, in both AML and CML cases, CD44 is essential for the homing and engraftment of cancer stem cells to the niche. In another words, CD44-expressing leukemic stem cells adhere to the niche, binding to its ligand hyaluronic acid expressed by cells on the surface of the sinusoidal endothelium or endosteum in bone marrow; this binding is crucial for the niche's maintenance of the stem cells. Interestingly, this molecular mechanism resembles that of healthy HSC and the vascular niche described earlier. Cancer and healthy stem cells have much in common in the maintenance system in the niche.

Interestingly, cancer stem cells can evolve according to the microenvironment and initiate different types of cancer. Dick [21] demonstrated that human hematopoietic cells infected with a retrovirus encoding mixed-lineage leukemia (MLL)-eleven-nineteen leukemia (ENL) fusion gene cause acute lymphoid leukemia (ALL) when transplanted into sublethally irradiated immunodeficient mice, but acute myeloid leukemia (AML) when cultured in a myeloid-promoting culture condition prior to transplantation for a prolonged period. This indicates that a cancer stem cell with the *MLL* fusion gene can initiate either ALL or AML or both, depending on the microenvironment.

In contrast to the hematopoietic system, stem cell research in the solid cancer field is relatively new. A study of stems cells in brain tumors has made a significant progress. Gilbertson and colleagues [22] showed that the brain tumor cells co-expressing Nestin and CD133, the fraction believed to contain the cancer stem cell, are located close to the capillaries in the brain tumor. When these cells were co-cultured, the cancer stem cells selectively adhered to the endothelial cells. This suggested that the endothelial cells secreted factors necessary to maintain the cancer stem cells. Furthermore, the CD133-positive cells derived from human medulloblastoma developed into brain tumor only when xenografted to the brain of a recipient nude mouse with endothelial cells. These data collectively suggest that the cancer stem cells of the brain tumor rely on the endothelial cells, which form a vascular niche, for the maintenance of their self-renewal, differentiational, and proliferative capacity.

6 Roles of the Niche Against Development, Maintenance, and Proliferation of Cancer

As described above, the niche has the ability to control the pool, function, and even the fate of stem cells. It is indeed an important aspect of the niche to maintain the stem cells in a quiescent state. The niche also has to drive an adequate number of stem cells into the proliferation and differentiation paths at the same time, however, as far as maintenance of the organ is concerned. This two-way signal of the niche against stem cells must be carefully regulated (Fig. 3). For example, in the osteoblastic niche for HSCs in bone marrow, Tie2/Ang1 adhesion or BMP (bone morphogenetic protein), a member of the TGF-β (transforming growth factor beta) super family, are involved in maintaining quiescence and proliferation suppression, while Wnt or Notch signaling promote self-renewal, proliferation, and differentiation (Fig. 4) [23].

It is obvious that cancer is a highly proliferative disease. Does the stem cell regulatory system in the niche for cancer then lean more toward the proliferative side than that for regular stem cells? Li and co-workers [24] approached this problem via the dependence of stem cells on the niche. They hypothesized that the behavior of cancer and regular stem cells is regulated by the niche to different degrees. The cancer stem cell is derived through intrinsic mutation that leads to its highly proliferative activity. This highly proliferative state itself alters the signaling balance between niche and stem cell. Namely, the niche function of quiescence maintenance

Fig. 3 Dynamics of two-way interaction between the stem cell and the niche: a comparison between normal and cancer stem cells. In the normal stem cells and the niche, signaling towards stem cell quiescence and self-renewal is dominant, while proliferative signaling is more dominant between the cancer stem cell and its niche

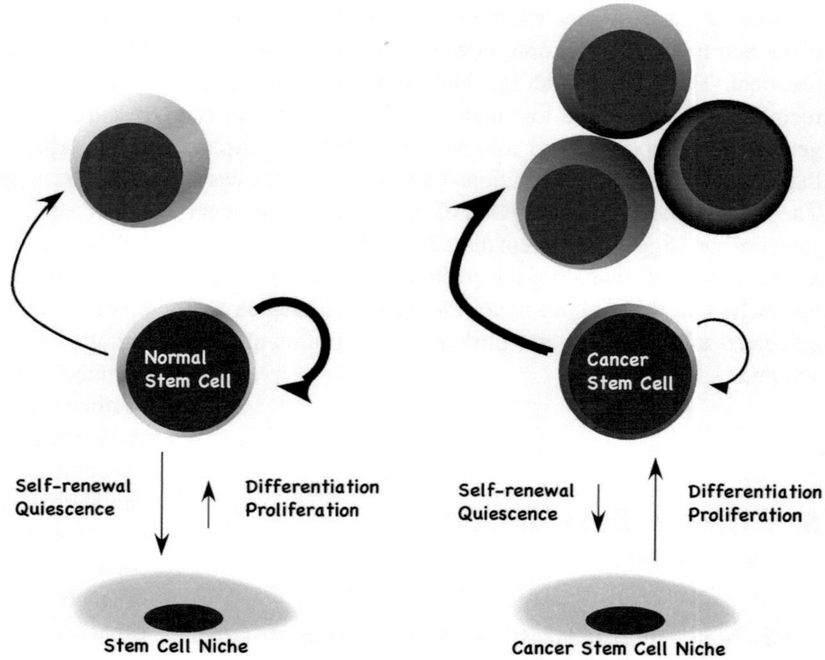

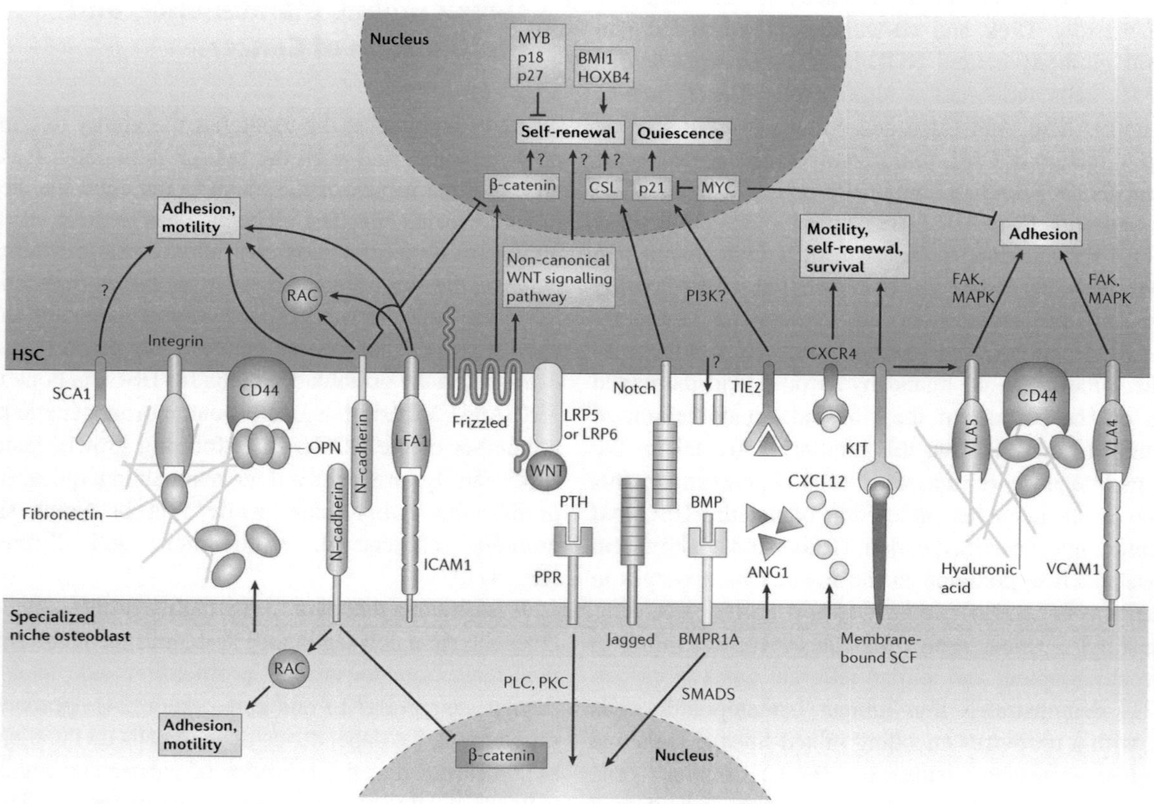

Fig. 4 A number of signaling pathways between HSCs and the niche are involved in stem cell regulation. Components in extracellular matrix are also important factors in niche formation. Normal and cancer stem cells share a majority of this regulatory signaling but show different kinds of phenotypes

becomes relatively ineffective, thus the functions to support proliferation and differentiation become more dominant (Fig. 3). This model is supported by some clinical symptoms, one of which is the blast crisis of CML. The Wnt/β-catenin signal, a causative pathway for self-renewal, differentiation, and proliferation, is enhanced in the mutated nucleus of granulocyte-macrophage progenitors in CML patients due to the higher concentration of β-catenin and irregularly activated TCF/LEF (T-cell factor/lymphocyte-enhancement factor) transcription activity [25]. The blast crisis is thought to be triggered by this irregular TCF/LEF activity, leading to excess transcription of Wnt target gene products. Similarly, the relevance of irregular Wnt/β-catenin activation has been reported in colon cancer or melanoma, in which cases the β-catenin itself has the mutation. It is important to note that many signaling pathways involved in the regular stem cell and niche interaction are also found between the cancer stem cell and its niche, playing a role as the promoter of tumorigenesis and cancer proliferation. Identical sets of proteins under slightly different conditions can deliver totally different results.

7 Mechanism of Cancer Metastasis Regulated by the Niche

The niche is not only for cradling the existing cancer stem cell but also for the incoming cancer stem cell in the future. It is passively sending an inviting signal to remote cancer stem cells.

MMPs (matrix metalloproteinases) are well-known factors, not only for their contribution to the repair of inflammation or wounds but also for their involvement in cancer invasion and metastasis. A model of the molecular mechanism for remote metastasis and invasion in association with MMPs has been proposed for the lung [26]. In this model, VEGF (vascular endothelial growth factor) secreted by primary cancer cells induces the MMP9 expression, specifically in lung endothelial cells and macrophages, via VEGFR (vascular endothelial growth factor receptor) tyrosine kinase, resulting in the formation of the cancer stem cell niche. This means that cancer cells can produce their own favorable microenvironment, the future cancer stem cell niche, from a distance by secreting factors to influence the protein composition at that site.

This niche creation mechanism by cancer cells themselves has been verified for various other kinds of metastasis as well: bone metastasis of prostate cancer has been shown to be supported by uPA (urokinase-type plasminogen activator) or PSA (prostate-specific antigen) secreted by the prostate cancer cells through alteration of the growth factors in the bone microenvironment, thus enhancing the proliferation of the osteoblasts that serve as the cancer stem cell niche [27]. Lung metastasis of breast cancer via SPARC (secreted protein acidic and rich in cysteine, osteonectin) or MMP2 has also been found to be based on this mechanism [28].

The factors existing in the ECM (extracellular matrix) can also be a part of the cancer mediator. In cancer stem cells in hypoxia, the oncogene *MET* is up-regulated. MET binds to its ligand HGF (hepatocyte growth factor) existing in the ECM and enhances the transcription of PAI-1 (plasminogen activator inhibitor type 1) and COX2 (cyclooxygenase 2). This leads to enhanced blood coagulation and fibrin deposition. The fibrin deposition (fibrin nest) induces vasculogenesis in the surrounding area, supporting the homing of cancer stem cells and their proliferation, thus serving as the cancer stem cell niche. The migration of the vascular vessel triggered by the fibrin nest also facilitates metastasis to other site [29].

Lyden and colleagues [30] recently reported a new type of cancer stem cell niche created by a third party. They demonstrated this mechanism using lethally irradiated mice, transplanted with BMDCs (bone marrow-derived cell), followed by lung cancer or melanoma cells. The transplanted BMDCs expressing VEGFR1 (vascular endothelial growth factor receptor-1) gather at a site, such as lung, creating a group of cells. This microenvironment is highly receptive for cancer metastasis, and they call this microenvironment the "premetastatic niche." The formation of the premetastatic niche has been found for breast, lung, and gastrointestinal cancer in humans. BMDCs creating the premetastatic niche are a type of hematopoietic progenitor expressing VLA-4 (integrin $\alpha_4\beta_1$) and/or Id3 (inhibitor of DNA binding 3). Primary cancer cells secrete factors (such as cytokines) that positively regulate the expression of fibronectin (a ligand of VLA-4) of fibroblast located at the site of the future premetastatic niche, inducing the homing of VEGFR1-expressing BMDCs. The mice transplanted with VEGFR1-depleted BMDCs or treated with anti-VEGFR1 antibody did not show the formation of either the premetastatic niche or metastasis, proving this new metastatic mechanism induced by cancer cells involving the third party.

8 Novel Cancer Therapy Targeting the Cancer Stem Cell and Its Niche

Most current anticancer drugs act as an inhibitor of DNA synthesis or cell division of the cancer cell, consequently suppressing the growth of tumor. Therefore, the cancer cells must be in the cell cycle, not in quiescence, in order to see the positive effect of the chemotherapy. Cancer stem cells as well as regular stem cells are supposed to be in the quiescent state. Even if the tumor size is decreased after the chemotherapy, it is simply because the regular cell-cycling cancer cells are killed. One must expect the quiescent, tumorigenic cancer stem cells to survive chemotherapy.

A report by Graham et al. [31] proved the insensitivity of cancer stem cells by demonstrating that most of the *BCR-ABL* expressing blood cells of patients with chronic myeloid leukemia arrested after cell culture with addition of STI571 (Gleevec®); however, those in the quiescent state, namely the leukemic stem cells, survived. This suggests that degenerating the quiescent cancer stem cell is one of the most promising therapeutic methods in the next generation of cancer treatment aiming for complete cure. Various attempts based on this concept have been investigated so far, including alternation of the cancer stem cell characteristics, manipulation of the niche environment to induce a resumption of the cell cycle, and differentiation of cancer stem cells. From the viewpoint of cancer stem cell control, complete inhibition of cell division of cancer stem cells, even if it happens minimally, may also have potential.

Parthenolide (PTL) is a sesquiterpene lactone and a major component in the herbal medicine feverfew. Recently, PTL has been found to have a unique characteristic against tumors, including inhibition of DNA synthesis, cancer cell proliferation, and NF-κB activation, and to increase intracellular reactive oxygen species (ROS). Jordan and co-workers [32] successfully demonstrated that PTL induces stem cell–specific apoptosis among leukemic cells in primary acute myeloid leukemia (AML) and blast crisis chronic myeloid leukemia (bcCML) without affecting the normal hematopoietic cells. This action was identified as a result of multiple effects of PTL, such as inhibition of up-regulated NF-κB expression in tumor cells, proapoptotic activation of p53, and increased intracellular ROS.

Glioblastoma is considered one of the most malignant cancers and is extremely proliferative and resistant to not only chemotherapy but also radiotherapy. The prognosis is very poor, with an average survival duration of less than a year. Vescovi et al. [33] showed a successful result for a novel therapy for glioblastoma based on the above-mentioned concept. They discovered that BMPs, especially BMP4, can induce the differentiation of CD133-positive glioblastoma cells, which are considered to be the cancer stem cells, and ultimately decrease the cancer stem cell pool. Moreover, xenograft of human glioblastoma cells to mice showed a significant suppression of tumor growth and an improvement in survival rate, which would have been 100% lethal otherwise, when combined with BMP4. In this case, BMP4 or other BMPs did not kill the glioblastoma cells but led to their extinction by inducing temporary differentiation and proliferation through alteration of stem cell characteristics. This positive experimental result should stimulate exploration of this novel therapeutic method targeting the cancer stem cell or its niche.

For niche-targeting therapy, Gilbertson's group suggested that alternation of the vascular niche is effective in brain tumor treatment [22]. Medulloblastoma cells express a high level of ERBB2 (erythroblastic leukemia viral oncogenic homolog 2) and also, consequently, VEGF. ERBB2-overexpressing mice bearing brain tumor were treated with either the ERBB2 inhibitor erlotinib (Tarceva®) or anti-VEGF antibody bevacizumab (Avastin®). The treatment significantly inhibited tumor vasculature and suppressed tumor growth. Cancer stem cells expressing Nestin and CD133, the self-renewing cancer cells, were not obtained from the mice after either treatment. Similar treatment was applied to mice bearing glioma, and the result also showed significant suppression of vasculogenesis and tumor proliferation, as well as a decrease of the Nestin^{+}CD133^{+} cancer stem cell pool. The small number of surviving Nestin^{+}CD133^{+} cancer stem cells were all found in association with the tumor vessels. These data collectively suggested that suppression of the vascular niche by inhibiting vasculogenesis has potential as a novel treatment of medulloblastoma or glioma.

9 Summary

The stem cell and its niche influence each other to maintain their capabilities and control the fate of, and ultimately sustain, healthy organs and tissue. Besides the interactions between stem cell and niche, some intrinsic factors of stem cells have been also revealed as regulators of the stem cell capacity. A good example may be Bmi-1, which is a member of the polycomb group of genes and is involved in maintenance of transcriptional repression of target genes. HSCs derived from Bmi-1–deficient mice exhibited defective self-renewal phenotype while maintaining the multilineage differentiation potential. The microenvironment in Bmi-1–deficient mice was found to be intact. The forced expression of Bmi-1 led to enhanced repopulation capacity in vivo and augmented expansion of progenitors ex vivo. These data collectively suggest that the self-renewal capacity of HSCs is controlled by the intrinsic expression level of Bmi-1 [34].

Stem cell regulation and maintenance, whether intrinsic or extrinsic, is not a simple single-factor phenomenon, and a large part of this complicated molecular mechanism is yet to be elucidated. Understanding the regulatory system could lead to novel therapeutic methods that directly target stem cells (normal or cancer) or niches. As described in this chapter, the latest drugs targeting the cancer stem cell or its microenvironment employ this state-of-the-art strategy and are expected to minimize complications and improve patient quality of life at the same time. From both scientific and clinical viewpoints, the biology of the stem cell and its niche is the one of the most promising research fields in the next few decades.

References

1. Schofield R. The relationship between the spleen colony-forming cell and the haematopoietic stem cell. Blood Cells. 1978;4:7–25.
2. Deng W, Lin H. Spectrosomes and fusomes anchor mitotic spindles during asymmetric germ cell divisions and facilitate the formation of a polarized microtubule array for oocyte specification in *Drosophila*. Dev Biol. 1997;189:79–94.
3. Xie T, Spradling AC. A niche maintaining germ line stem cells in the *Drosophila* ovary. Science. 2000;290:328–330.
4. Zhang J, Niu C, Ye L, et al. Identification of the haematopoietic stem cell niche and control of the niche size. Nature. 2003;425:836–841.
5. Calvi LM, Adams GB, Weibrecht KW, et al. Osteoblastic cells regulate the haematopoietic stem cell niche. Nature. 2003;425: 841–846.
6. Arai F, Hirao A, Ohmura M, et al. Tie2/angiopoietin-1 signaling regulates hematopoietic stem cell quiescence in the bone marrow niche. Cell. 2004;118:149–161.
7. Nilsson SK, Johnston HM, Whitty GA, et al. Osteopontin, a key component of the hematopoietic stem cell niche and regulator of primitive hematopoietic progenitor cells. Blood. 2005;15: 1232–1239.
8. Wilson A, Murphy MJ, Oskarsson T, et al. c-Myc controls the balance between hematopoietic stem cell self-renewal and differentiation. Genes Dev. 2004;18:2747–2763.
9. Matrosova VY, Orlovskaya IA, Serobyan N, Khaldoyanidi SK. Hyaluronic acid facilitates the recovery of hematopoiesis following 5-Fluorouracil administration. Stem Cells. 2004;22:544–555.
10. Avigdor A, Goichberg P, Shivtiel S, et al. CD44 and hyaluronic acid cooperate with SDF-1 in the trafficking of human $CD34^+$ stem/progenitor cells to bone marrow. Blood. 2004;15:2981–2989.
11. Adams GB, Chabner KT, Alley IR, et al. Stem cell engraftment at the endosteal niche is specified by the calcium-sensing receptor. Nature. 2006;439:599–603.
12. Kiel MJ, Yilmaz OH, Iwashita T, Yilmaz OH, Terhorst C, Morrison SJ. SLAM family receptors distinguish hematopoietic stem and progenitor cells and reveal endothelial niches for stem cells. Cell. 2005;121:1109–1121.
13. Sugiyama T, Kohara H, Noda M, Nagasawa T. Maintenance of the hematopoietic stem cell pool by CXCL12-CXCR4 chemokine signaling in bone marrow stromal cell niches. Immunity. 2006;25:977–988.
14. Papayannopoulou T, Craddock C, Nakamoto B, Priestley GV, Wolf NS. The VLA_4/VCAM-1 adhesion pathway defines contrasting mechanism of lodgement of transplanted murine hemopoietic progenitors between bone marrow and spleen. Proc Natl Acad Sci U S A. 2007;92:9647–9651.
15. Parmar K, Mauch P, Vergilio JA, Sackstein R, Down JD. Distribution of hematopoietic stem cells in the bone marrow according to regional hypoxia. Proc Natl Acad Sci U S A. 2007;104: 5431–5436.
16. Heissig B, Hattori K, Dias S, et al. Recruitment of stem and progenitor cells from the bone marrow niche requires MMP-9 mediated release of kit-ligand. Cell. 2002;109:625–637.
17. Walkley CR, Olsen GH, Dworkin S, et al. A microenvironment-induced myeloproliferative syndrome caused by retinoic acid receptor γ deficiency. Cell. 2007;129:1097–1110.
18. Walkley CR, Shea JM, Sims NA, Purton LE, Orkin SH. Rb regulates interactions between hematopoietic stem cells and their bone marrow microenvironment. Cell. 2007;129:1081–1095.
19. Jin L, Hope KJ, Zhai Q, Smadja-Joffe F, Dick JE. Targeting of CD44 eradicates human acute myeloid leukemic stem cells. Nat Med. 2006;12:1167–1174.
20. Krause DS, Lazarides K, von Andrian UH, Van Etten RA. Requirement for CD44 in homing and engraftment of BCR-ABL-expressing leukemic stem cells. Nat Med. 2006;12:1175–1180.
21. Barabe F, Kennedy JA, Hope KJ, Dick JE. Modeling the initiation and progression of human acute leukemia in mice. Science. 2007;316:600–604.
22. Calabrese C, Poppleton H, Kocak M, et al. A perivascular niche for brain tumor stem cells. Cancer Cell. 2007;11:69–82.
23. Wilson A, Trumpp A. Bone-marrow haematopoietic-stem-cell niches. Nat Rev Immunol. 2006;6:93–106.
24. Li L, Neaves WB. Normal stem cells and cancer stem cells: The Niche matters. Cancer Res. 2006;66:4553–4557.
25. Jamieson CH, Ailles LE, Dylla SJ, et al. Granulocyte-macrophage progenitors as candidate leukemic stem cells in blast-crisis CML. N Engl J Med. 2004;351:657–667.
26. Hiratsuka S, Nakamura K, Iwai S, et al. MMP9 induction by vascular endothelial growth factor receptor-1 is involved in lung-specific metastasis. Cancer Cell. 2002;2:289–300.
27. Logothetis CJ, Lin SH. Osteoblasts in prostate cancer metastasis to bone. Nat Rev Cancer. 2005;5:21–28.
28. Minn AJ, Gupta GP, Siegel PM, et al. Genes that mediate breast cancer metastasis to lung. Nature. 2005;426:518–524.
29. Boccaccio C, Sabatino G, Medico E, et al. The *MET* oncogene drives a genetic programme linking cancer to haemostasis. Nature. 2005;434:396–400.
30. Kaplan RN, Riba RD, Zacharoulis S, et al. VEGFR1-positive haematopoietic bone marrow progenitors initiate the pre-metastatic niche. Nature. 2005;438:820–827.
31. Graham SM, Jørgensen HG, Allan E, et al. Primitive, quiescent, Philadelphia-positive stem cells from patients with chronic myeloid leukemia are insensitive to STI571 in vitro. Blood. 2002;99: 319–325.
32. Guzman ML, Rossi RM, Karnischky L, et al. The sesquiterpene lactone parthenolide induces apoptosis of human acute myelogenous leukemia stem and progenitor cells. Blood. 2005;105: 4163–4169.
33. Piccirillo SG, Reynolds BA, Zanetti N, et al. Bone morphogenetic proteins inhibit the tumorigenic potential of human brain tumour-initiating cells. Nature. 2006;444:761–765.
34. Iwama A, Oguro H, Negishi M, Kato Y, Nakauchia H. Epigenetic regulation of hematopoietic stem cell self-renewal by polycomb group genes. Int J Hematol. 2005;81:294–300.

Environmental Signals Regulating Mesenchymal Progenitor Cell Growth and Differentiation

Meirav Pevsner-Fischer and Dov Zipori

Abstract Mesenchymal progenitor cells are widespread in the organism and are implicated in a variety of physiological and pathological processes. As such, these cells should be able to respond to microenvironmental signals. Here we review some of the conditions that modulate the biological functions of mesenchymal progenitors, particularly during inflammation and stress.

Keywords Multipotent stromal cells · Hemopoietic stem cells · Microenvironments · Toll-like receptors · Inflammation · Stress

1 Introduction

Multipotent stromal cells (MSC) are progenitors that give rise to multiple mesodermal derivatives such as bone, muscle, and fat. They are often referred to as mesenchymal stem cells. We prefer, however, the former designation since the in vivo ability of MSC to self-renew, divide asymmetrically, and reside within specific niches at a single cell level is not well established. MSC are assumed to perform in vivo many cellular functions such as participation in the formation of the hematopoietic stem cell (HSC) niches, regulation of immune responses, and maintenance and repair of damaged or worn out tissues [1–13]. A vast body of in vitro evidence supports the idea that MSC possess the above functions. One major downside of current MSC research is the lack of specific markers shared by cultured MSC and their in vivo counterparts. On one hand, this may be only a technical issue and such markers may eventually be found. On the other hand, it is equally possible that MSC, like other progenitors that lack lineage markers, are in a state in which the genome is widely and nonspecifically expressed. This may be a part of the mechanism that permits multipotent progenitors to rapidly up-regulate specific gene sets upon encountering the appropriate environmental cues [14, 15]. The lack of specific MSC markers limits the ability to follow endogenous MSC, both in animal models and in human tissues. This is the major reason that most factors, discussed below, that effect MSC functions, were studied ex vivo. One major challenge for future research would therefore be the identification of MSC or their precursors in situ.

The localization of MSC in vivo relates to the question of the nature of the MSC niche. In the bone marrow, hematopoietic stem cells (HSC) reside, at least partially, adjacent to the endosteum. This niche has been suggested to include osteoblasts [16], which are one possible differentiated product of MSC. It is not known whether MSC are a part of this niche. If they are, MSC may share their niche with HSC. In the absence of precise knowledge on the nature of the MSC niche in vivo, one way to learn about the possible structure of this niche would be to study the survival and growth requirements of MSC in vitro and the molecules that regulate their functions. This approach may implicate specific domains within tissues and organs that eventually could be directly tested as candidate niches for MSC in vivo. The inability to detect and follow endogenous MSC in vivo has lead to the use of models in which cultured MSC are transplanted into inflicted animals. It was demonstrated that the timing of MSC administration relative to the tissue damage infliction is crucial; in nondamaged tissues, the turnover of cells is slow. Therefore, MSC engraftment is less extensive than engraftment into damaged sites at the time of acute inflammation or cell destruction induced chemically, mechanically, or following radio-damage; MSC have been shown to engraft into injured neural tissue, myocardium, hepatic, and muscle sites as well as into sites of graft rejection and heart infraction [17–27].

For the finesse regulation of these functions, MSC must sense multiple aspects of their surrounding microenvironment. MSC have been shown to express chemokine receptors, growth factors receptors, cytokine receptors, adhesion proteins, and immune system related proteins.

D. Zipori (✉)
Department of Molecular Cell Biology, Weizmann Institute of Science, Rehovot 76100, Israel
e-mail: dov.zipori@weizmann.ac.il

V.K. Rajasekhar, M.C. Vemuri (eds.), *Regulatory Networks in Stem Cells*, Stem Cell Biology and Regenerative Medicine,
DOI 10.1007/978-1-60327-227-8_16, © Humana Press, a part of Springer Science+Business Media, LLC 2009

MSC respond to microenvironmental signals including external factors such as mechanical forces and pathogen molecules. These cells also respond to endogenous physiological cues such as growth factors, cytokines, and chemokines. The growing receptor repertoire expressed by MSC may implicate the factors that affect MSC functions. Indeed, following their discovery by Friedenstein [1], MSC functions, primarily their differentiation, was shown to be regulated by cytokines and growth factors that induce the activation of lineage specific genes [28, 29].

2 Effects of Inflammation Related-Cytokines, Chemokines, and Growth Factors on Migration and Mobilization of MSC

Inflammation following pathogen infection or tissue damage recruit MSC into the inflicted sites where MSC possibly participate in tissue regeneration and modulation of immune responses [30–32]. This section reviews the conditions and factors that may be involved in the attraction of MSC into injured tissues (Fig. 1).

Cytokines are small, secreted proteins that mediate and regulate hematopoiesis, immunity, and inflammation. Cytokines have been proposed to regulate MSC migration into damaged and inflamed tissues. Transforming growth factor (TGF)-β, interleukin-1 $(IL\text{-}1)_{\beta}$, tumor necrosis factor (TNF)-α, and the chemokine stromal derived factor (SDF)-1 induce human MSC chemotactic migration, as was demonstrated in an in vitro transwell chamber system coated with human ECM proteins. Such migration was dependent (with the exception of SDF-1 that will be discussed below) on Matrix metalloproteinase-2 (MMP-2), MMP-9, and membrane-type-1 matrix metalloproteinase (MT1-MMP) [33]. Other inflammatory cytokines and chemokines, such as MPC-1, MIP-1, and IL-8, were shown to enhance the migration of hMSC, in a microchemotaxis chamber system [24]. Contradicting data is available regarding MSC mobilization following granulocyte-colony stimulating factor (G-CSF) administration. While some investigators detected small numbers of MSC in peripheral blood, others did not (reviewed in [34]). The effects of cytokines on MSC induced migration, detailed above, were assayed predominantly using in vitro migration chamber systems. Therefore, direct evidence of migratory MSC response toward cytokines in vivo is still missing.

Chemokines are secreted low-molecular-weight proteins that are defined by their ability to induce directed migration. MSC express many chemokine receptors (reviewed in [35]). The most studied are CXCR4 and CX3CR1, the receptors for SDF-1 and fractalkine, respectively. These chemokines stimulate MSC migration in vitro and in vivo [36–42]. SDF-1 and fractalkine are both inflammatory chemokines that recruit leukocytes to damaged sites in response to physiological stresses, such as tissue damage, inflammation, hypoxia or total body irradiation (reviewed in [43, 44]). While some studies show that all of the cells within the MSC population express CXCR4 [40, 45], others show that only a small subpopulation of human MSC (hMSC) express extracellular CXCR4 and CX3CR1 [38, 39]. This later population was proposed to be a distinct rapidly self-renewing MSC (RS-MSC), that have a greater potential for multi-lineage differentiation and show better in vivo engraftment compared to the slowly renewing MSC (SR-MSC) [37, 46]. A combination of cytokines was proposed as a mechanism for up-regulation of CXCR4 expression on 83–98% of MSC [36]. In studies demonstrating MSC responsiveness to SDF-1, these cells were infused either to injured [40] or irradiated [36] mice causing SDF-1 and fractalkine expression and subsequent MSC recruitment.

Hepatocyte growth/scatter factor (HGF/SF) acts as a potent mitogen, motogen, and morphogen on different cell types [47, 48]. HGF levels are increased in damaged and inflamed tissues [49, 50]. There are some indications that HGF acts as a mobilizing agent to stimulate MSC migration and recruitment from tissues of residence to the inflamed sites;

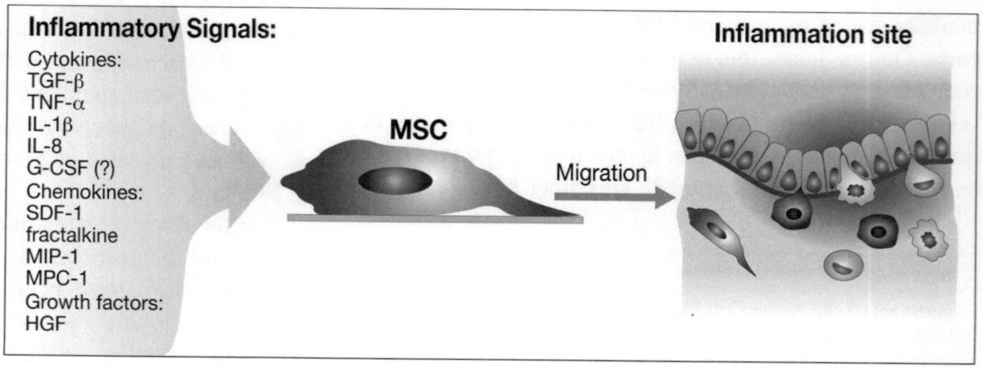

Fig. 1 Inflammatory signals directing MSC to inflicted sites

Mouse and human MSC were shown to express both HGF and its receptor c-Met. HGF induced human bone marrow (BM)- or cord blood (CB)-derived MSC [45] chemotaxis, while inhibiting proliferation. HGF also induced the expression of myocyte-specific genes in mouse MSC [45, 51, 52].

3 Regulation of MSC Functions by Toll-Like Receptor (TLR) Activation

An additional way by which MSC can detect infection or danger signals in their surrounding microenvironment is by sensing pathogen molecules or stress-released intracellular constituents through TLR. These receptors play a role in *Drosophila* development [53] and immune responses [54–58]. In mammalians, the TLR family is of importance in the innate immune system for the recognition of pathogen-associated molecular patterns (PAMP). TLR activation initiates primary responses towards invading pathogens and the subsequent recruitment of the adaptive immune response [54, 59–62]. TLR can be activated by pathogen components as well as by mammalian endogenous molecules such as heat-shock proteins and extracellular matrix (ECM) breakdown products such as HSP60 and fibronectin, respectively [63–65]. Cultured mouse MSC express TLR molecules 1–8, but not TLR-9. TLR expression was also demonstrated in human adipose tissue and bone marrow–derived MSC [66]. In the latter study, TLR expression was altered by hypoxia.

Activation of MSC by TLR ligands induced interleukin-6 (IL-6) secretion and nuclear factor (NF)-κB translocation from the cytoplasm to the nucleus. Pam3Cys, a prototypic ligand for TLR-2, induced proliferation of MSC and inhibited their differentiation (Fig. 2). These findings suggest that TLR ligands may control the balance between MSC proliferation versus differentiation. In addition to this implied role, TLR may also be involved in the generation of MSC during development; MyD88, an adaptor protein that signals downstream from most TLR molecules [67], was required for the ability of MSC to differentiate into osteocytes or chondrocytes under the conditions that were permissive for differentiation of their normal counterparts. TLR signaling may therefore be required for acquisition of MSC multipotency [68].

4 Regulators of the MSC Immune-Suppressive Activity

In view of the above, it is likely that MSC are mobilized and subsequently localize to inflamed and damaged tissues by multiple signals originating from the injured sites. When arriving at the inflamed/damaged sites, MSC may perform multiple functions related to tissue repair by differentiating and replacing damaged cells, secreting cytokines and survival factors, and, as discussed below, by regulating the immune response.

The mechanisms by which MSC exert immune regulation have not been completely elaborated. Both contact-dependent and -independent mechanisms have been suggested. These involve soluble factors secreted by MSC, such as interleukin-10 (IL-10), TGF-β, prostaglandin E2 (PGE2), nitric oxide (NO), HGF, and indoleamine 2,3-dioxygenase (IDO) (reviewed in [12]). MSC were also reported to induce a regulatory phenotype in T-cells [69, 70]. The pro-inflammatory cytokines IFN-γ and TNF-α can be found at site of ongoing inflammatory response. MSC exposure to these inflammatory cytokines was shown to either promote or revert MSC immunosuppressive activity. In vitro administration of IFN-γ and TNF-α to MSC cultures increased expression levels of PGE-2, IDO, HGF, and TGF-β [71–73]. However, TNF-α could revert MSC inhibition of T-cell proliferation [74]. IFN-γ was suggested to be involved in MSC pro- versus anti-inflammatory behavior depending

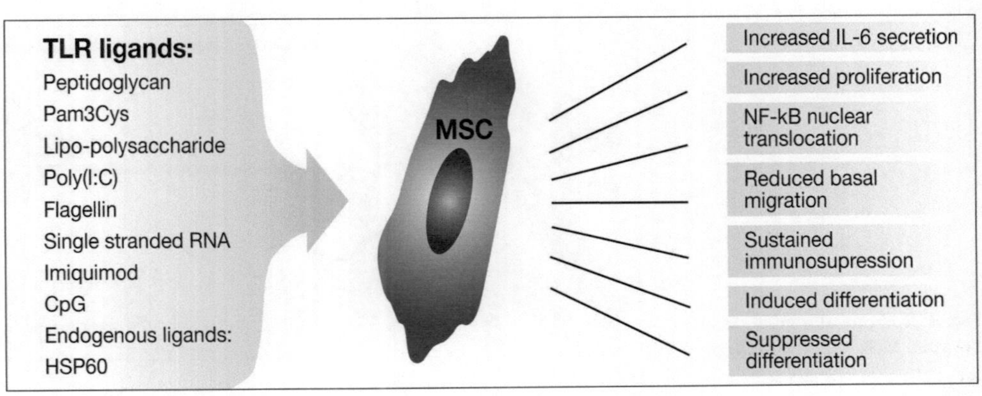

Fig. 2 TLR ligands regulate MSC functions

on its concentration; while high IFN-γ concentrations promoted MSC immunosuppressive activity [75], low IFN-γ doses stimulated MSC to act as T-cell activators: it promoted both mouse and human MSC MHC-II–mediated APC function in vivo and in vitro [76] and increased MSC MHC-II expression and antigen processing. Additional in vivo data showed that ovalbumin-pulsed IFN-γ-treated MSC were fully able to generate antigen-specific CD8pos cytotoxic T cells and to protect mice transplanted with ovalbumin expressing tumors [76].

There is still some controversy regarding MSC being immune privileged and eliciting suppressive activity. Some studies show that allogeneic and syngeneic MSC can engraft and suppress immune responses in vivo. Others show that MSC were rejected following allo-transplantation and did not prevent tissue rejection such as graft versus host disease (GvHD) in a mouse model of allogeneic BM transplantation [77–81]. It seems that MSC immune functions can be modulated depending on microenvironmental conditions, such as cytokine levels.

5 Interactions of MSC with the Extracellular Matrix

Integrins are a major family of receptors that facilitate the interactions between cells and the surrounding ECM. These receptors activate intercellular signaling pathways and control gene expression leading to many cellular processes, including actin cytoskeleton re-arrangement, proliferation, differentiation, survival, motility, and apoptosis. During damage infliction and inflammation, various stimuli alter the composition of ECM from a latent to an activated form. New combinations of ECM moieties are formed, and mediators such as inflammatory cytokines, chemokines, and growth factors bind the ECM [82, 83]. ECM remodeling by MSC can result either from synthesis and deposition of newly formed components, or due to degradation by proteases such as MMP. MSC express MMP and their inhibitors TIMP. MMP and TIMP were shown to be involved in MSC differentiation (reviewed in [84]), migration [45], angiogenesis [85], and invasive behavior [86]. Thus, MSC may remodel ECM that, in turn, affects MSC functions (Fig. 3). Interactions between cells and the ECM are important for wound repair and tissue regeneration, processes in which MSC apparently participate. Therefore, it is not surprising that MSC have been shown to express integrin receptors, which when activated, can control differentiation, proliferation, survival, adhesion, and migration. Human bone marrow MSC express many integrin molecules including receptors for fibronectin, collagen type I, laminin, and vitronectin (reviewed in [87]). ECM affects MSC osteogenic differentiation; rat MSC cultured with type I collagen showed high ALP activity, mineralization, collagen synthesis, and up-regulation of osteoblastic gene expression. Osteoblastic differentiation was dependent on α2β1 receptor, which is the functional receptor for type I collagen in osteoblasts [88, 89]. Seeding human MSC on purified vitronectin and collagen type I, was sufficient to induce osteogenic differentiation either with or without the addition of induction medium. Moreover, blocking the β_1 integrin subunit diminished osteogenic differentiation of MSC [90, 91]. ECM deposited on titanium scaffolds induced osteoblastic differentiation of MSC in the absence of osteogenic supplements [92]. Other ECM components that have been shown to promote MSC osteogenic differentiation include laminin-5 [93] and phosphophorin [94]. ECM components can also regulate chondrogenic differentiation of MSC. Collagen type I and II were suggested to promote chondrogenic differentiation of human, rabbit, and bovine MSC [95, 96]. In addition to regulation of MSC differentiation, ECM components including fibronectin, vitronectin, and collagen type I were shown to induce chemotaxis and haptotaxis of human and rabbit MSC [97]. This might indicate that ECM-integrin interaction is one of the mechanisms involved in the recruitment of MSC into tissue repair sites.

The ECM serves as a specialized reservoir of factors that promote cell proliferation, differentiation, activation,

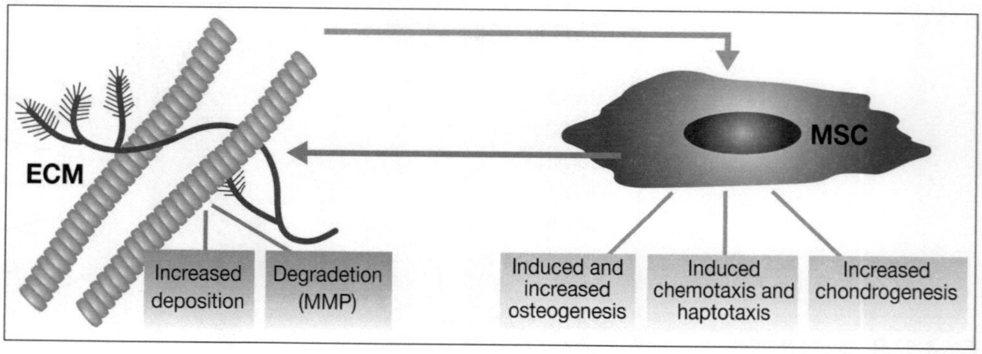

Fig. 3 Reciprocal regulation between MSC and the ECM

and migration. ECM moieties such as heparan sulfate, fibronectin, and laminin among others were shown to bind cytokines such as bFGF, TGF-β, IFN-γ, and TNF-α and growth factors such as IGF [98]. These factors can be either released enzymatically by MMP secreted by cells including MSC, or may act while bound to ECM. The ECM-bound growth factors and cytokines might target immune cells and cells such as epithelial cells, hematopoietic stem cells, keratinocytes, and possibly MSC (reviewed in [99]). Cytokines and growth factors in association with ECM create high local concentration, thereby increasing cellular responses and restricting the response to a limited site. In addition, combination of cytokines, growth factors, and ECM signaling might induce combinatorial signals, altering MSC responses.

6 The Role of Oxygen Pressure in the Regulation of MSC Functions

Oxygen pressure affects the function of various stem cell types such as neural, embryonic, and muscle stem cells [100–102]. Under hypoxic conditions, a key transcription factor called hypoxia-inducible factor (HIF) is expressed by the cells. Under normoxia, the α subunit of the heterodimeric transcription factors HIF-1, -2, and -3 is unstable, but upon exposure to hypoxic conditions, it is rapidly stabilized. Following heterodimerization with the constitutively expressed β subunit, HIF activates the transcription of a large number of genes involved in maintaining oxygen homeostasis, as well as other genes involved in cellular functions, such as cell-migration, proliferation, cytokine secretion, and differentiation [103]. HSC and the supporting cells of the stem cell niche are predominantly located at the lowest end of an oxygen gradient in the bone marrow [104], suggesting that oxygen pressure regulates differentiation versus self-renewal of HSC. MSC were proposed to reside within the HSC niches and therefore might be affected as well by oxygen pressure. A large body of in vitro evidence discussed below supports these assumptions.

Hypoxia develops following severe bone or cartilage injury as a result of disrupted vascular supply, forming hypoxic gradient in the wound. Hypoxia is therefore an important signal by which MSC can "sense" their microenvironment. Since MSC have been suggested to contribute to wound healing, a hypoxic condition might contribute to MSC mobilization, migration, and differentiation [100, 101, 103, 104]. Hypoxia augmented cytokine and growth factor secretion by human MSC [105–107]. In vitro systems were used to examine the effect of hypoxia on MSC: low oxygen (5%) concentration, increased rat MSC proliferation and osteogenic differentiation, as well as the ability to form colonies [108]. Increased proliferation in response to low oxygen was reported in additional studies; hypoxic conditions increased proliferation of mouse MSC that exhibited changes in cell morphology, increased adipogenic differentiation, and diminished Oct4 expression [109]. Hypoxia also increased human MSC proliferation, ECM formation and organization, HIF-2α and Oct4 transcription, and connexin-43 expression [110].

In contrast to the above consensus relating to the effects of hypoxia on MSC proliferation, conflicting data on MSC differentiation has been reported; increased [111–114] versus decreased [115–117] differentiation of mouse and human MSC into adipocytes, osteoblasts, and chondrocytes were demonstrated. These contradictions might be explained by differences in the species used or the tissue of origin of MSC, as well as by differences in the experimental conditions (degree of hypoxia and the use of primary cells versus cell lines) used in these studies. Regulation of MSC by hypoxia was further demonstrated in animal models where rat MSC grown in low O_2 concentrations were more likely to differentiate into osteoblasts and chondrocytes when transplanted in vivo than MSC grown in 20% O_2 [108]. Hypoxia was proposed to affect MSC migration and mobilization. It induced expression of the chemokine receptors CXCR4 and CX3CR1. The migration of MSC towards SDF-1 or fractalkine in vitro increased after hypoxic conditioning and resulted in increased engraftment into early chick embryos. In a rat model of chronic hypoxia MSC were mobilized into the peripheral blood while maintaining differentiation potential [118]. Whether in these experiments the BM was the source of the mobilized MSC, is unclear. A number of mechanisms might be responsible for hypoxia induced MSC mobilization, such as an increase of chemokines that target MSC, changes in the MSC niche due to hypoxia or increased vascular permeability.

In conditions such as inflammation and tissue ischemia, release of growth factors such as SDF-1 and VEGF promote angiogenesis [119]. SDF-1 expression is under the control of HIF-1 in endothelial cells [120] and hypoxia-induced angiogenesis is mediated by increasing VEGF production 121]. Transplanted mouse MSC were shown to induce angiogenesis via VEGF and to differentiate in vivo into CD31 positive and VEGF producing cells that contribute to blood vessel formation. This process was suggested to be mediated by hypoxia, since VEGF is hypoxia responsive [122]. Indeed, other studies indicated that low oxygen pressure induced MSC migration and capillary-like tube formation. MSC improved myocardium function after ischemic injury through VEGF production [123, 124]. MSC up-regulated VEGF receptor 1 expression after exposure to hypoxia through HIF-1 up-regulation [125]. These cells were further found to secrete a variety of angiogenic factors such as VEGF, TGF-β1, and bFGF following hypoxia and therefore may contribute to vascularization leading to protection of ischemic tissues. This process was suggested to be mediated by hypoxia [126–128].

7 The Effect of Mechanical Force on MSC Functions

Mechanical loading plays an important role in bone remodeling. Lack of mechanical load leads to decreased bone matrix protein production, mineral content, and bone formation through increased bone resorption. Indeed, lack of exercise and gravitational forces leads to atrophied bone [129, 130]. MSC have been recently shown to respond to mechanical forces. Systems applying mechanical load and shear flow were used to simulate physiological conditions of locomotion or stresses such as injuries. Different mechanical stresses that effect MSC differentiation are discussed in this section (Fig. 4).

Mechanical strain simulating the mechanical forces existing in the vascular environment was shown to promote MSC differentiation into smooth muscle cells (SMC) and to have a potential role in directing mesenchymal cells for repair of damaged vascular walls. MSC were subjected, either to equiaxial or uniaxial stretch, simulating the mechanical forces existing in the vascular environment. The stretch drove MSC differentiation toward SMC. Equiaxial stretch reduced expression of SMC related genes smooth muscle (SM) α-actin and SM 22α. Uniaxial stretch transiently increased the expression of these genes, as well as the expression of collagen type I, thereby demonstrating differential responses of MSC to different mechanical strains [131]. Additional studies showed that mechanical forces promote MSC differentiation toward SMC, thus contributing to vascular remodeling [132–136]. Mechanical strain mimicking the forces to which a ligament is exposed during physiological function was applied on MSC loaded on collagen gels. These MSC up-regulated expression of collagen type I and III and tenascin-C and, in addition, the cells formed oriented collagen fibers characteristic of ligament cells [137]. Cyclic mechanical strain was shown to inhibit MSC proliferation but to increase osteogenic differentiation through ERK1/2 activation [138]. An additional study [139] showed that combined mechanical strain and attachment to collagen type I increases human MSC osteogenic differentiation while reducing gene expression of chondrogenic, adipogenic, and neurogenic lineages. Another form of mechanical stress that cells may encounter is shear flow. The later was shown to increase osteogenic differentiation of rat, human, and goat MSC [140–144]. In the C3H10T1/2 cell-line, shear flow induced endothelial differentiation, as detected by expression of the endothelial markers CD31, vWF, and vascular endothelial–cadherin. More importantly, endothelial functions of the shear flow–treated cells, such as formation of capillary-like structures on Matrigel, were enhanced [145]. Integrin receptors are capable of transducing mechanical strain into changes in gene expression by different cell types. This mechanotransduction can be initiated through signaling pathways induced by adhesion to ECM [146, 147]. As mentioned previously, ECM components can regulate MSC differentiation, proliferation, morphology, and migration. ECM activation of integrins is one of the suggested mechanisms by which mechanic stress activates cellular responses. Other mechanisms by which mechanical stimulation might be translated into cellular responses are mobilization of second messengers such as calcium and nitric oxide and activation of kinase cascades including the mitogen-activated protein (MAP) kinase and protein kinase C (PKC) pathways.

8 Conclusions

MSC have been proposed to be involved in a plethora of biological processes, from the regulation of HSC self-renewal and bone formation to wound healing and tumor development, to mention just a few examples. The study of MSC responses to microenvironmental signals in

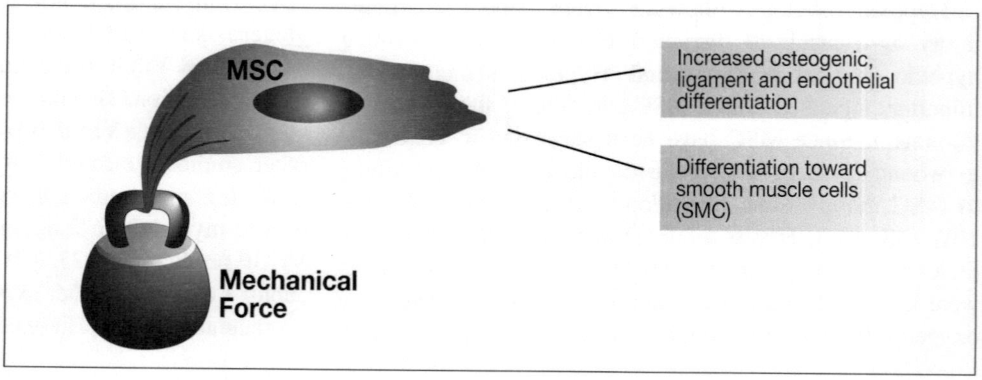

Fig. 4 Mechanical force drives MSC differentiation

vitro and also in vivo strongly supports the role of these cells in repair mechanisms. The knowledge gained by the analysis of isolated molecules, or otherwise, environmental conditions, already provides powerful tools by which MSC can be manipulated to perform better. This may open novel ways to the translation of MSC methodologies into therapeutic modalities.

References

1. Friedenstein AJ, Chailakhyan RK, Latsinik NV, Panasyuk AF, Keiliss-Borok IV. Stromal cells responsible for transferring the microenvironment of the hemopoietic tissues. Cloning in vitro and retransplantation in vivo. Transplantation. 1974;17:331–40.
2. Dexter TM, Allen TD, Lajtha LG. Factors controlling the proliferation of haemopoietic stem cells in vitro. Symp Soc Dev Biol. 1978:149–60.
3. Zipori D, Bol S. The role of fibroblastoid cells and macrophages from mouse bone marrow in the in vitro growth promotion of haemopoietic tumour cells. Exp Hematol. 1979;7:206–18.
4. Zipori D, Sasson T. Adherent cells from mouse bone marrow inhibit the formation of colony stimulating factor (CSF) induced myeloid colonies. Exp Hematol. 1980;8:816–7.
5. Zipori D, Sasson T. Myelopoiesis in the presence of stromal cells from mouse bone marrow: II. Mechanism of glucose dependent colony formation. Exp Hematol. 1981;9:663–74.
6. Zipori D, Sasson T, Frenkel A. Myelopoiesis in the presence of stromal cells from mouse bone marrow: I. Monosaccharides regulate colony formation. Exp Hematol. 1981;9:656–63.
7. Zipori D, Duksin D, Tamir M, Argaman A, Toledo J, Malik Z. Cultured mouse marrow stromal cell lines. II. Distinct subtypes differing in morphology, collagen types, myelopoietic factors, and leukemic cell growth modulating activities. J Cell Physiol. 1985;122:81–90.
8. Zipori D, Toledo J, von der Mark K. Phenotypic heterogeneity among stromal cell lines from mouse bone marrow disclosed in their extracellular matrix composition and interactions with normal and leukemic cells. Blood. 1985;66:447–55.
9. Zipori D, Tamir M, Toledo J, Oren T. Differentiation stage and lineage-specific inhibitor from the stroma of mouse bone marrow that restricts lymphoma cell growth. Proc Natl Acad Sci U S A. 1986;83:4547–51.
10. Zipori D, Lee F. Introduction of interleukin-3 gene into stromal cells from the bone marrow alters hemopoietic differentiation but does not modify stem cell renewal. Blood. 1988;71:586–96.
11. Barda-Saad M, Rozenszajn LA, Globerson A, Zhang AS, Zipori D. Selective adhesion of immature thymocytes to bone marrow stromal cells: relevance to T cell lymphopoiesis. Exp Hematol. 1996;24:386–91.
12. Uccelli A, Moretta L, Pistoia V. Immunoregulatory function of mesenchymal stem cells. Eur J Immunol. 2006;36:2566–73.
13. Prockop DJ, Gregory CA, Spees JL. One strategy for cell and gene therapy: harnessing the power of adult stem cells to repair tissues. Proc Natl Acad Sci U S A. 2003;100 Suppl 1:11917–23.
14. Zipori D. The nature of stem cells: state rather than entity. Nat Rev Genet. 2004;5:873–8.
15. Zipori D. The stem state: plasticity is essential, whereas self-renewal and hierarchy are optional. Stem Cells. 2005;23:719–26.
16. Calvi LM, Adams GB, Weibrecht KW, et al. Osteoblastic cells regulate the haematopoietic stem cell niche. Nature. 2003;425:841–6.
17. Azizi SA, Stokes D, Augelli BJ, DiGirolamo C, Prockop DJ. Engraftment and migration of human bone marrow stromal cells implanted in the brains of albino rats–similarities to astrocyte grafts. Proc Natl Acad Sci U S A. 1998;95:3908–13.
18. Francois S, Bensidhoum M, Mouiseddine M, et al. Local irradiation not only induces homing of human mesenchymal stem cells at exposed sites but promotes their widespread engraftment to multiple organs: a study of their quantitative distribution after irradiation damage. Stem Cells. 2006;24:1020–9.
19. Lu D, Li Y, Wang L, Chen J, Mahmood A, Chopp M. Intraarterial administration of marrow stromal cells in a rat model of traumatic brain injury. J Neurotrauma. 2001;18:813–9.
20. Schwarz EJ, Alexander GM, Prockop DJ, Azizi SA. Multipotential marrow stromal cells transduced to produce L-DOPA: engraftment in a rat model of Parkinson disease. Hum Gene Ther. 1999;10:2539–49.
21. Sato Y, Araki H, Kato J, et al. Human mesenchymal stem cells xenografted directly to rat liver are differentiated into human hepatocytes without fusion. Blood. 2005;106:756–63.
22. Chopp M, Zhang XH, Li Y, et al. Spinal cord injury in rat: treatment with bone marrow stromal cell transplantation. Neuroreport. 2000;11:3001–5.
23. Shake JG, Gruber PJ, Baumgartner WA, et al. Mesenchymal stem cell implantation in a swine myocardial infarct model: engraftment and functional effects. Ann Thorac Surg. 2002;73:1919–25; discussion 1926.
24. Wang L, Li Y, Chen X, et al. MCP-1, MIP-1, IL-8 and ischemic cerebral tissue enhance human bone marrow stromal cell migration in interface culture. Hematology. 2002;7:113–7.
25. Kawada H, Fujita J, Kinjo K, et al. Nonhematopoietic mesenchymal stem cells can be mobilized and differentiate into cardiomyocytes after myocardial infarction. Blood. 2004;104: 3581–7.
26. Mackenzie TC, Flake AW. Human mesenchymal stem cells persist, demonstrate site-specific multipotential differentiation, and are present in sites of wound healing and tissue regeneration after transplantation into fetal sheep. Blood Cells Mol Dis. 2001;27:601–4.
27. Barbash IM, Chouraqui P, Baron J, et al. Systemic delivery of bone marrow-derived mesenchymal stem cells to the infarcted myocardium: feasibility, cell migration, and body distribution. Circulation. 2003;108:863–8.
28. Minguell JJ, Erices A, Conget P. Mesenchymal stem cells. Exp Biol Med (Maywood). 2001;226:507–20.
29. Gregory CA, Prockop DJ, Spees JL. Non-hematopoietic bone marrow stem cells: molecular control of expansion and differentiation. Exp Cell Res. 2005;306:330–5.
30. Le Blanc K. Mesenchymal stromal cells: tissue repair and immune modulation. Cytotherapy. 2006;8:559–61.
31. Caplan AI, Dennis JE. Mesenchymal stem cells as trophic mediators. J Cell Biochem. 2006;98:1076–84.
32. Stagg J, Galipeau J. Immune plasticity of bone marrow-derived mesenchymal stromal cells. Handb Exp Pharmacol. 2007;180: 45–66.
33. Ries C, Egea V, Karow M, Kolb H, Jochum M, Neth P. MMP-2, MT1-MMP, and TIMP-2 are essential for the invasive capacity of human mesenchymal stem cells: differential regulation by inflammatory cytokines. Blood. 2007;109:4055–63.
34. Levesque JP, Winkler IG, Larsen SR, Rasko JE. Mobilization of bone marrow-derived progenitors. Handb Exp Pharmacol. 2007;180:3–36.
35. Fox JM, Chamberlain G, Ashton BA, Middleton J. Recent advances into the understanding of mesenchymal stem cell trafficking. Br J Haematol. 2007;137:491–502.
36. Shi M, Li J, Liao L, et al. Regulation of CXCR4 expression in human mesenchymal stem cells by cytokine treatment:

role in homing efficiency in NOD/SCID mice. Haematologica. 2007;92:897–904.
37. Lee RH, Hsu SC, Munoz J, et al. A subset of human rapidly self-renewing marrow stromal cells preferentially engraft in mice. Blood. 2006;107:2153–61.
38. Wynn RF, Hart CA, Corradi-Perini C, et al. A small proportion of mesenchymal stem cells strongly expresses functionally active CXCR4 receptor capable of promoting migration to bone marrow. Blood. 2004;104:2643–5.
39. Sordi V, Malosio ML, Marchesi F, et al. Bone marrow mesenchymal stem cells express a restricted set of functionally active chemokine receptors capable of promoting migration to pancreatic islets. Blood. 2005;106:419–27.
40. Ji JF, He BP, Dheen ST, Tay SS. Interactions of chemokines and chemokine receptors mediate the migration of mesenchymal stem cells to the impaired site in the brain after hypoglossal nerve injury. Stem Cells. 2004;22:415–27.
41. Zhu W, Boachie-Adjei O, Rawlins BA, et al. A novel regulatory role for stromal-derived factor-1 signaling in bone morphogenic protein-2 osteogenic differentiation of mesenchymal C2C12 cells. J Biol Chem. 2007;282:18676–85.
42. Parameswaran R MV, Larrone A, Rousso-Noori L, Shani N, Naffar-Abu-Amara S, Zipori D. Targeting the bone marrow with Activin A overexpressing embryonic multipotent stromal cells specifically modified B lymphopoiesis. Stem Cells Dev. 2008;17:93–106.
43. Kucia M, Ratajczak J, Reca R, Janowska-Wieczorek A, Ratajczak MZ. Tissue-specific muscle, neural and liver stem/progenitor cells reside in the bone marrow, respond to an SDF-1 gradient and are mobilized into peripheral blood during stress and tissue injury. Blood Cells Mol Dis. 2004;32:52–7.
44. Umehara H, Bloom ET, Okazaki T, Nagano Y, Yoshie O, Imai T. Fractalkine in vascular biology: from basic research to clinical disease. Arterioscler Thromb Vasc Biol. 2004;24:34–40.
45. Son BR, Marquez-Curtis LA, Kucia M, et al. Migration of bone marrow and cord blood mesenchymal stem cells in vitro is regulated by stromal-derived factor-1-CXCR4 and hepatocyte growth factor-c-met axes and involves matrix metalloproteinases. Stem Cells. 2006;24:1254–64.
46. Colter DC, Sekiya I, Prockop DJ. Identification of a subpopulation of rapidly self-renewing and multipotential adult stem cells in colonies of human marrow stromal cells. Proc Natl Acad Sci U S A. 2001;98:7841–5.
47. Matsumoto K, Nakamura T. Emerging multipotent aspects of hepatocyte growth factor. J Biochem (Tokyo). 1996;119:591–600.
48. Zarnegar R, Michalopoulos GK. The many faces of hepatocyte growth factor: from hepatopoiesis to hematopoiesis. J Cell Biol. 1995;129:1177–80.
49. Yoshida S, Yamaguchi Y, Itami S, et al. Neutralization of hepatocyte growth factor leads to retarded cutaneous wound healing associated with decreased neovascularization and granulation tissue formation. J Invest Dermatol. 2003;120:335–43.
50. Cowin AJ, Kallincos N, Hatzirodos N, et al. Hepatocyte growth factor and macrophage-stimulating protein are upregulated during excisional wound repair in rats. Cell Tissue Res. 2001;306: 239–50.
51. Neuss S, Becher E, Woltje M, Tietze L, Jahnen-Dechent W. Functional expression of HGF and HGF receptor/c-met in adult human mesenchymal stem cells suggests a role in cell mobilization, tissue repair, and wound healing. Stem Cells. 2004;22:405–14.
52. Forte G, Minieri M, Cossa P, et al. Hepatocyte growth factor effects on mesenchymal stem cells: proliferation, migration, and differentiation. Stem Cells. 2006;24:23–33.
53. Kambris Z, Hoffmann JA, Imler JL, Capovilla M. Tissue and stage-specific expression of the Tolls in Drosophila embryos. Gene Expr Patterns. 2002;2:311–7.
54. Hoebe K, Janssen E, Beutler B. The interface between innate and adaptive immunity. Nat Immunol. 2004;5:971–4.
55. Beutler B, Hoebe K, Du X, Ulevitch RJ. How we detect microbes and respond to them: the Toll-like receptors and their transducers. J Leukoc Biol. 2003;74:479–85.
56. Beutler B, Jiang Z, Georgel P, et al. Genetic analysis of host resistance: Toll-like receptor signaling and immunity at large. Annu Rev Immunol. 2006;24:353–89.
57. Gangloff M, Weber AN, Gibbard RJ, Gay NJ. Evolutionary relationships, but functional differences, between the Drosophila and human Toll-like receptor families. Biochem Soc Trans. 2003;31:659–63.
58. Ferrandon D, Imler JL, Hoffmann JA. Sensing infection in Drosophila: Toll and beyond. Semin Immunol. 2004;16:43–53.
59. Krieg AM. Antitumor applications of stimulating toll-like receptor 9 with CpG oligodeoxynucleotides. Curr Oncol Rep. 2004;6:88–95.
60. Lund JM, Alexopoulou L, Sato A, et al. Recognition of single-stranded RNA viruses by Toll-like receptor 7. Proc Natl Acad Sci U S A. 2004;101:5598–603.
61. Meier A, Kirschning CJ, Nikolaus T, Wagner H, Heesemann J, Ebel F. Toll-like receptor (TLR) 2 and TLR4 are essential for Aspergillus-induced activation of murine macrophages. Cell Microbiol. 2003;5:561–70.
62. O'Neill LA. Therapeutic targeting of Toll-like receptors for inflammatory and infectious diseases. Curr Opin Pharmacol. 2003;3:396–403.
63. Cohen-Sfady M, Nussbaum G, Pevsner-Fischer M, et al. Heat shock protein 60 activates B cells via the TLR4-MyD88 pathway. J Immunol. 2005;175:3594–602.
64. Ohashi K, Burkart V, Flohe S, Kolb H. Cutting edge: heat shock protein 60 is a putative endogenous ligand of the toll-like receptor-4 complex. J Immunol. 2000;164:558–61.
65. Okamura Y, Watari M, Jerud ES, et al. The extra domain A of fibronectin activates Toll-like receptor 4. J Biol Chem. 2001;276:10229–33.
66. Hwa Cho H, Bae YC, Jung JS. Role of toll-like receptors on human adipose-derived stromal cells. Stem Cells. 2006;24:2744–52.
67. Kaisho T, Akira S. Dendritic-cell function in Toll-like receptor- and MyD88-knockout mice. Trends Immunol. 2001;22:78–83.
68. Pevsner-Fischer M, Morad V, Cohen-Sfady M, et al. Toll-like receptors and their ligands control mesenchymal stem cell functions. Blood. 2007;109:1422–32.
69. Maccario R, Podesta M, Moretta A, et al. Interaction of human mesenchymal stem cells with cells involved in alloantigen-specific immune response favors the differentiation of CD4+ T-cell subsets expressing a regulatory/suppressive phenotype. Haematologica. 2005;90:516–25.
70. Aggarwal S, Pittenger MF. Human mesenchymal stem cells modulate allogeneic immune cell responses. Blood. 2005;105: 1815–22.
71. Krampera M, Cosmi L, Angeli R, et al. Role for interferon-gamma in the immunomodulatory activity of human bone marrow mesenchymal stem cells. Stem Cells. 2006;24:386–98.
72. Ryan JM, Barry F, Murphy JM, Mahon BP. Interferon-gamma does not break, but promotes the immunosuppressive capacity of adult human mesenchymal stem cells. Clin Exp Immunol. 2007;149:353–63.
73. English K, Barry FP, Field-Corbett CP, Mahon BP. IFN-gamma and TNF-alpha differentially regulate immunomodulation by murine mesenchymal stem cells. Immunol Lett. 2007;110: 91–100.
74. Djouad F, Fritz V, Apparailly F, et al. Reversal of the immunosuppressive properties of mesenchymal stem cells by tumor necrosis factor alpha in collagen-induced arthritis. Arthritis Rheum. 2005;52:1595–603.

75. Chan JL, Tang KC, Patel AP, et al. Antigen-presenting property of mesenchymal stem cells occurs during a narrow window at low levels of interferon-gamma. Blood. 2006;107:4817–24.
76. Stagg J, Pommey S, Eliopoulos N, Galipeau J. Interferon-gamma-stimulated marrow stromal cells: a new type of nonhematopoietic antigen-presenting cell. Blood. 2006;107:2570–77.
77. Nauta AJ, Westerhuis G, Kruisselbrink AB, Lurvink EG, Willemze R, Fibbe WE. Donor-derived mesenchymal stem cells are immunogenic in an allogeneic host and stimulate donor graft rejection in a nonmyeloablative setting. Blood. 2006;108: 2114–20.
78. Zappia E, Casazza S, Pedemonte E, et al. Mesenchymal stem cells ameliorate experimental autoimmune encephalomyelitis inducing T-cell anergy. Blood. 2005;106:1755–61.
79. Le Blanc K, Rasmusson I, Sundberg B, et al. Treatment of severe acute graft-versus-host disease with third party haploidentical mesenchymal stem cells. Lancet. 2004;363:1439–41.
80. Eliopoulos N, Stagg J, Lejeune L, Pommey S, Galipeau J. Allogeneic marrow stromal cells are immune rejected by MHC class I- and class II-mismatched recipient mice. Blood. 2005;106: 4057–65.
81. Sudres M, Norol F, Trenado A, et al. Bone marrow mesenchymal stem cells suppress lymphocyte proliferation in vitro but fail to prevent graft-versus-host disease in mice. J Immunol. 2006;176:7761–7.
82. Vaday GG, Franitza S, Schor H, et al. Combinatorial signals by inflammatory cytokines and chemokines mediate leukocyte interactions with extracellular matrix. J Leukoc Biol. 2001;69: 885–92.
83. Raghow R. The role of extracellular matrix in postinflammatory wound healing and fibrosis. FASEB J. 1994;8:823–31.
84. Mannello F, Tonti GA, Bagnara GP, Papa S. Role and function of matrix metalloproteinases in the differentiation and biological characterization of mesenchymal stem cells. Stem Cells. 2006;24:475–81.
85. Ghajar CM, Blevins KS, Hughes CC, George SC, Putnam AJ. Mesenchymal stem cells enhance angiogenesis in mechanically viable prevascularized tissues via early matrix metalloproteinase upregulation. Tissue Eng. 2006;12:2875–88.
86. Alexander CM, Werb Z. Targeted disruption of the tissue inhibitor of metalloproteinases gene increases the invasive behavior of primitive mesenchymal cells derived from embryonic stem cells in vitro. J Cell Biol. 1992;118:727–39.
87. Docheva D, Popov C, Mutschler W, Schieker M. Human mesenchymal stem cells in contact with their environment: surface characteristics and the integrin system. J Cell Mol Med. 2007;11:21–38.
88. Mizuno M, Fujisawa R, Kuboki Y. Type I collagen-induced osteoblastic differentiation of bone-marrow cells mediated by collagen-alpha2beta1 integrin interaction. J Cell Physiol. 2000;184:207–13.
89. Mizuno M, Kuboki Y. Osteoblast-related gene expression of bone marrow cells during the osteoblastic differentiation induced by type I collagen. J Biochem (Tokyo). 2001;129:133–8.
90. Salasznyk RM, Williams WA, Boskey A, Batorsky A, Plopper GE. Adhesion to vitronectin and collagen I promotes osteogenic differentiation of human mesenchymal stem cells. J Biomed Biotechnol. 2004;2004:24–34.
91. Gronthos S, Simmons PJ, Graves SE, Robey PG. Integrin-mediated interactions between human bone marrow stromal precursor cells and the extracellular matrix. Bone. 2001;28: 174–81.
92. Datta N, Holtorf HL, Sikavitsas VI, Jansen JA, Mikos AG. Effect of bone extracellular matrix synthesized in vitro on the osteoblastic differentiation of marrow stromal cells. Biomaterials. 2005;26:971–7.
93. Klees RF, Salasznyk RM, Kingsley K, Williams WA, Boskey A, Plopper GE. Laminin-5 induces osteogenic gene expression in human mesenchymal stem cells through an ERK-dependent pathway. Mol Biol Cell. 2005;16:881–90.
94. Jadlowiec J, Koch H, Zhang X, Campbell PG, Seyedain M, Sfeir C. Phosphophoryn regulates the gene expression and differentiation of NIH3T3, MC3T3-E1, and human mesenchymal stem cells via the integrin/MAPK signaling pathway. J Biol Chem. 2004;279:53323–30.
95. Bosnakovski D, Mizuno M, Kim G, Takagi S, Okumura M, Fujinaga T. Chondrogenic differentiation of bovine bone marrow mesenchymal stem cells (MSCs) in different hydrogels: influence of collagen type II extracellular matrix on MSC chondrogenesis. Biotechnol Bioeng. 2006;93:1152–63.
96. Chen CW, Tsai YH, Deng WP, et al. Type I and II collagen regulation of chondrogenic differentiation by mesenchymal progenitor cells. J Orthop Res. 2005;23:446–53.
97. Thibault MM, Hoemann CD, Buschmann MD. Fibronectin, vitronectin, and collagen I induce chemotaxis and haptotaxis of human and rabbit mesenchymal stem cells in a standardized transmembrane assay. Stem Cells Dev. 2007;16:489–502.
98. Jones JI, Gockerman A, Busby WH, Jr., Camacho-Hubner C, Clemmons DR. Extracellular matrix contains insulin-like growth factor binding protein-5: potentiation of the effects of IGF-I. J Cell Biol. 1993;121:679–87.
99. Vaday GG, Lider O. Extracellular matrix moieties, cytokines, and enzymes: dynamic effects on immune cell behavior and inflammation. J Leukoc Biol. 2000;67:149–59.
100. Csete M. Oxygen in the cultivation of stem cells. Ann N Y Acad Sci. 2005;1049:1–8.
101. Zhu LL, Wu LY, Yew DT, Fan M. Effects of hypoxia on the proliferation and differentiation of NSCs. Mol Neurobiol. 2005;31:231–42.
102. Ezashi T, Das P, Roberts RM. Low O2 tensions and the prevention of differentiation of hES cells. Proc Natl Acad Sci U S A. 2005;102:4783–8.
103. Wenger RH. Mammalian oxygen sensing, signalling and gene regulation. J Exp Biol. 2000;203:1253–63.
104. Parmar K, Mauch P, Vergilio JA, Sackstein R, Down JD. Distribution of hematopoietic stem cells in the bone marrow according to regional hypoxia. Proc Natl Acad Sci U S A. 2007;104:5431–6.
105. Kinnaird T, Stabile E, Burnett MS, et al. Marrow-derived stromal cells express genes encoding a broad spectrum of arteriogenic cytokines and promote in vitro and in vivo arteriogenesis through paracrine mechanisms. Circ Res. 2004;94:678–85.
106. Wang M, Zhang W, Crisostomo P, et al. STAT3 mediates bone marrow mesenchymal stem cell VEGF production. J Mol Cell Cardiol. 2007;42:1009–15.
107. Becker MI, De Ioannes AE, Leon C, Ebensperger LA. Females of the communally breeding rodent, Octodon degus, transfer antibodies to their offspring during pregnancy and lactation. J Reprod Immunol. 2007;74:68–77.
108. Lennon DP, Edmison JM, Caplan AI. Cultivation of rat marrow-derived mesenchymal stem cells in reduced oxygen tension: effects on in vitro and in vivo osteochondrogenesis. J Cell Physiol. 2001;187:345–55.
109. Ren H, Cao Y, Zhao Q, et al. Proliferation and differentiation of bone marrow stromal cells under hypoxic conditions. Biochem Biophys Res Commun. 2006;347:12–21.
110. Grayson WL, Zhao F, Bunnell B, Ma T. Hypoxia enhances proliferation and tissue formation of human mesenchymal stem cells. Biochem Biophys Res Commun. 2007;358:948–53.
111. Khan WS, Adesida AB, Hardingham TE. Hypoxic conditions increase hypoxia-inducible transcription factor 2alpha and enhance chondrogenesis in stem cells from the infrapatellar fat pad of osteoarthritis patients. Arthritis Res Ther. 2007;9:R55.

112. Grayson WL, Zhao F, Izadpanah R, Bunnell B, Ma T. Effects of hypoxia on human mesenchymal stem cell expansion and plasticity in 3D constructs. J Cell Physiol. 2006;207:331–9.
113. Wang DW, Fermor B, Gimble JM, Awad HA, Guilak F. Influence of oxygen on the proliferation and metabolism of adipose derived adult stem cells. J Cell Physiol. 2005;204:184–91.
114. Robins JC, Akeno N, Mukherjee A, et al. Hypoxia induces chondrocyte-specific gene expression in mesenchymal cells in association with transcriptional activation of Sox9. Bone. 2005;37:313–22.
115. Malladi P, Xu Y, Chiou M, Giaccia AJ, Longaker MT. Effect of reduced oxygen tension on chondrogenesis and osteogenesis in adipose-derived mesenchymal cells. Am J Physiol Cell Physiol. 2006;290:C1139–46.
116. Salim A, Nacamuli RP, Morgan EF, Giaccia AJ, Longaker MT. Transient changes in oxygen tension inhibit osteogenic differentiation and Runx2 expression in osteoblasts. J Biol Chem. 2004;279:40007–16.
117. Hung SC, Pochampally RR, Hsu SC, et al. Short-term exposure of multipotent stromal cells to low oxygen increases their expression of CX3CR1 and CXCR4 and their engraftment in vivo. PLoS ONE. 2007;2:e416.
118. Rochefort GY, Delorme B, Lopez A, et al. Multipotential mesenchymal stem cells are mobilized into peripheral blood by hypoxia. Stem Cells. 2006;24:2202–8.
119. Ferrara N, Kerbel RS. Angiogenesis as a therapeutic target. Nature. 2005;438:967–74.
120. Ceradini DJ, Kulkarni AR, Callaghan MJ, et al. Progenitor cell trafficking is regulated by hypoxic gradients through HIF-1 induction of SDF-1. Nat Med. 2004;10:858–64.
121. Yancopoulos GD, Davis S, Gale NW, Rudge JS, Wiegand SJ, Holash J. Vascular-specific growth factors and blood vessel formation. Nature. 2000;407:242–8.
122. Al-Khaldi A, Eliopoulos N, Martineau D, Lejeune L, Lachapelle K, Galipeau J. Postnatal bone marrow stromal cells elicit a potent VEGF-dependent neoangiogenic response in vivo. Gene Ther. 2003;10:621–9.
123. Tang YL, Zhao Q, Zhang YC, et al. Autologous mesenchymal stem cell transplantation induce VEGF and neovascularization in ischemic myocardium. Regul Pept. 2004;117:3–10.
124. Uemura R, Xu M, Ahmad N, Ashraf M. Bone marrow stem cells prevent left ventricular remodeling of ischemic heart through paracrine signaling. Circ Res. 2006;98:1414–21.
125. Okuyama H, Krishnamachary B, Zhou YF, Nagasawa H, Bosch-Marce M, Semenza GL. Expression of vascular endothelial growth factor receptor 1 in bone marrow-derived mesenchymal cells is dependent on hypoxia-inducible factor 1. J Biol Chem. 2006;281:15554–63.
126. Heppenstall RB, Grislis G, Hunt TK. Tissue gas tensions and oxygen consumption in healing bone defects. Clin Orthop Relat Res. 1975:357–65.
127. Kinnaird T, Stabile E, Burnett MS, et al. Local delivery of marrow-derived stromal cells augments collateral perfusion through paracrine mechanisms. Circulation. 2004;109:1543–49.
128. Annabi B, Lee YT, Turcotte S, et al. Hypoxia promotes murine bone-marrow-derived stromal cell migration and tube formation. Stem Cells. 2003;21:337–47.
129. Meyers VE, Zayzafoon M, Gonda SR, Gathings WE, McDonald JM. Modeled microgravity disrupts collagen I/integrin signaling during osteoblastic differentiation of human mesenchymal stem cells. J Cell Biochem. 2004;93:697–707.
130. Zayzafoon M, Gathings WE, McDonald JM. Modeled microgravity inhibits osteogenic differentiation of human mesenchymal stem cells and increases adipogenesis. Endocrinology. 2004;145:2421–32.
131. Park JS, Chu JS, Cheng C, Chen F, Chen D, Li S. Differential effects of equiaxial and uniaxial strain on mesenchymal stem cells. Biotechnol Bioeng. 2004;88:359–68.
132. Riha GM, Wang X, Wang H, et al. Cyclic strain induces vascular smooth muscle cell differentiation from murine embryonic mesenchymal progenitor cells. Surgery. 2007;141:394–402.
133. Hamilton DW, Maul TM, Vorp DA. Characterization of the response of bone marrow-derived progenitor cells to cyclic strain: implications for vascular tissue-engineering applications. Tissue Eng. 2004;10:361–9.
134. Gojo S, Gojo N, Takeda Y, et al. In vivo cardiovasculogenesis by direct injection of isolated adult mesenchymal stem cells. Exp Cell Res. 2003;288:51–9.
135. Kurpinski K, Chu J, Hashi C, Li S. Anisotropic mechanosensing by mesenchymal stem cells. Proc Natl Acad Sci U S A. 2006;103:16095–100.
136. Mangi AA, Noiseux N, Kong D, et al. Mesenchymal stem cells modified with Akt prevent remodeling and restore performance of infarcted hearts. Nat Med. 2003;9:1195–201.
137. Altman GH, Horan RL, Martin I, et al. Cell differentiation by mechanical stress. FASEB J. 2002;16:270–2.
138. Simmons CA, Matlis S, Thornton AJ, Chen S, Wang CY, Mooney DJ. Cyclic strain enhances matrix mineralization by adult human mesenchymal stem cells via the extracellular signal-regulated kinase (ERK1/2) signaling pathway. J Biomech. 2003;36: 1087–96.
139. Ward DF, Jr., Salasznyk RM, Klees RF, et al. Mechanical strain enhances extracellular matrix-induced gene focusing and promotes osteogenic differentiation of human mesenchymal stem cells through an extracellular-related kinase-dependent pathway. Stem Cells Dev. 2007;16:467–80.
140. Datta N, Pham QP, Sharma U, Sikavitsas VI, Jansen JA, Mikos AG. In vitro generated extracellular matrix and fluid shear stress synergistically enhance 3D osteoblastic differentiation. Proc Natl Acad Sci U S A. 2006;103:2488–93.
141. Sikavitsas VI, Bancroft GN, Holtorf HL, Jansen JA, Mikos AG. Mineralized matrix deposition by marrow stromal osteoblasts in 3D perfusion culture increases with increasing fluid shear forces. Proc Natl Acad Sci U S A. 2003;100:14683–8.
142. Kim SH, Choi YR, Park MS, et al. ERK 1/2 activation in enhanced osteogenesis of human mesenchymal stem cells in poly(lactic-glycolic acid) by cyclic hydrostatic pressure. J Biomed Mater Res A. 2007;80:826–36.
143. Knippenberg M, Helder MN, Doulabi BZ, Semeins CM, Wuisman PI, Klein-Nulend J. Adipose tissue-derived mesenchymal stem cells acquire bone cell-like responsiveness to fluid shear stress on osteogenic stimulation. Tissue Eng. 2005;11: 1780–8.
144. Li YJ, Batra NN, You L, et al. Oscillatory fluid flow affects human marrow stromal cell proliferation and differentiation. J Orthop Res. 2004;22:1283–9.
145. Wang H, Riha GM, Yan S, et al. Shear stress induces endothelial differentiation from a murine embryonic mesenchymal progenitor cell line. Arterioscler Thromb Vasc Biol. 2005;25:1817–23.
146. Saha S, Ji L, de Pablo JJ, Palecek SP. Inhibition of human embryonic stem cell differentiation by mechanical strain. J Cell Physiol. 2006;206:126–37.
147. Matsuda N, Yokoyama K, Takeshita S, Watanabe M. Role of epidermal growth factor and its receptor in mechanical stress-induced differentiation of human periodontal ligament cells in vitro. Arch Oral Biol. 1998;43:987–97.

Microenvironmental Regulation of Adult Mesenchymal Stem Cells

Thomas P. Lozito, Catherine M. Kolf and Rocky S. Tuan

Abstract This chapter aims to present the importance of the cell-extrinsic factors that influence the identity and activities of adult mesenchymal stem cells (MSCs). The focus is on microenvironmental cues that represent a concert of signals coming from the niche in which the MSC finds itself. These signals are generated by contact with surrounding cells, by soluble factors that can be autocrine, paracrine, or endocrine in nature, and by the extracellular matrix surrounding the cell. Even fragments of matrix molecules and small molecules like oxygen can influence the behavior of an MSC. In vivo, these signals work together in intricate harmony to regulate the regenerative abilities of MSCs. By learning more about the mechanisms of action, specificities and interactions of each member of this environmental orchestra, it will be possible to manipulate MSCs for therapeutic applications to diseases and injuries.

Keywords Niche · Microenvironment · Mesenchymal stem cells · Extracellular matrix · Cell-cell interactions · Signaling molecules

1 Introduction

The activities of an adult stem cell are intimately dependent on and regulated by its microenvironment. The purpose of this chapter is to present the properties and interactive nature of the environmental influences that regulate adult stem cells, specifically mesenchymal stem cells (MSCs). We have focused on cell-cell interactions, soluble signaling factors, and the extracellular matrix (ECM), and the interactions among these key components (Fig. 1). In addition, examples of developmental morphogenesis (Fig. 2) involving environmental regulation of cell differentiation are presented to illustrate the exquisite and intricate nature of these networks in controlling cellular activities.

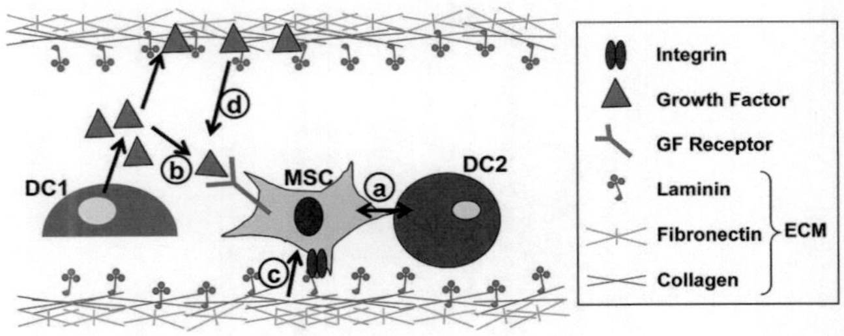

Fig. 1 Extracellular signals acting on mesenchymal stem cells. (**a**) Cell-cell interaction mediated by molecules such as cadherins and connexins regulate the differentiation of MSCs. (**b**) Secreted growth factors (VEGF, FGF, IGF, Wnts, TGF-β, etc.) produced by MSCs and/or other cells bind specific receptors on MSCs and regulate differentiation. (**c**) Molecules of the extracellular matrix (ECM), such as collagens and laminins, also provide differentiation signals by interacting with cell surface receptors, e.g., integrins, on MSCs. (**d**) Furthermore, some ECM molecules can bind growth factors, modulating their activities and regulating how they interact with MSCs.
MSC, mesenchymal stem cell; DC1, differentiated cell type 1; DC2, differentiated cell type 2; GF, growth factor; ECM, extracellular matrix

2 Mesenchymal Stem Cells

MSCs are adult stem cells that can be isolated from many adult tissues of mesodermal origin on the basis of plastic-adherence and clonogenicity, for example, the formation of colony-forming units (CFUs) [1]. They have been shown to be multipotent, becoming mesodermally derived bone, cartilage, and adipose cells [2]. They are less frequently differentiated into other mesodermal tissues like muscle, tendon, and stromal cells that support hematopoiesis [3]. Recent studies have also suggested that certain subpopulations of MSCs might have the ability to become cells of nonmesodermal origin (e.g., nerve, liver), implicating that MSCs could some

The authors declare that they have no competing interests.

R.S. Tuan (✉)
Cartilage Biology and Orthopaedics Branch, National Institute of Arthritis, and Musculoskeletal and Skin Diseases, National Institutes of Health, Department of Health and Human Services, Bethesda, MD, USA; Building 50, Room 1523, MSC 8022, Bethesda, MD, 20892-8022, USA
e-mail: tuanr@mail.nih.gov

V.K. Rajasekhar, M.C. Vemuri (eds.), *Regulatory Networks in Stem Cells*, Stem Cell Biology and Regenerative Medicine,
DOI 10.1007/978-1-60327-227-8_17, © Humana Press, a part of Springer Science+Business Media, LLC 2009

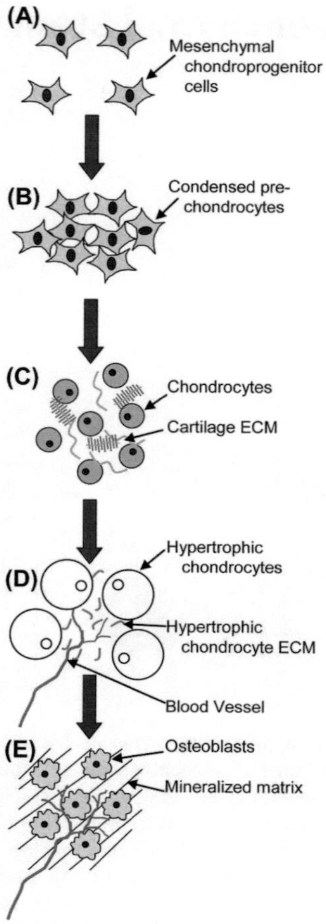

Fig. 2 Endochondral ossification exemplifies each aspect of the environmental regulation of mesenchymal stem cells. (**A**) At the start of endochondral ossification, chondroprogenitor cells produce matrix with collagen type IIA (Col IIA), which binds and inactivates TGF-β, and B+A+ fibronectin (FN), which promotes chondrocyte condensation. (**B**) As prechondrocytes proliferate, their cell-cell interactions, mediated by proteins such as N-cadherin, increase and the cells condense. They also begin secreting matrix metalloproteases (MMPs) that degrade Col IIA, causing the release of TGF-β. The clearance of Col IIA coincides with a switch in expression to Col IIB, which does not bind TGF-β. (**C**) Free TGF-β and Col IIB continue to drive the deposition of cartilage extracellular matrix (ECM) and to complete chondrocyte maturation. TGF-β also promotes a switch to expression of B+A- FN, which promotes hypertrophy of the cartilage cells. (**D**) Hypertrophic chondrocytes revert to secreting Col IIA, which binds BMP. They also secrete Col X as a part of a temporary matrix that facilitates the matrix turnover that occurs between cartilage and bone. Blood vessels, made up of endothelial cells (ECs), begin to invade, perhaps as a result of VEGF secretion in response to hypoxic conditions. (**E**) Osteoblasts are carried in by blood vessels. ECs secrete BMP-2, which helps to stimulate osteogenesis. Osteoblasts also degrade Col IIA, freeing more BMP. Osteoblasts begin to secrete matrix, typified by Col I, which gradually gives rise to the mineralized bone matrix

day be defined as pluripotent stem cells [4]. These studies also point to the fact that current isolation techniques are not capable of achieving a homogenous cell population but, rather, produce a heterogeneous mix of progenitor cells, which have varying tendencies to mature along each differentiation pathway.

MSCs are like most adult stem cells in that they can proliferate extensively, but not indefinitely, in culture [5]. One of the most exciting features of these cells, one that might result in many successful therapies, is their immunomodulatory activity (reviewed in [6]). Their apparent ability to regulate and evade the immune system (e.g., in [7]) will greatly enhance our ability to transplant them – and their progeny – into patients, without requiring life-long immunosuppressive drug regimens. In addition to their ability to regulate their environment, MSCs are also under intimate control by their environment, the subject of this chapter.

The normal microenvironment of a stem cell has been termed its "niche" [8]. This word, taken from ecology, describes the natural habitat of the stem cell. It is that specialized location in the body, made of mature cells, extracellular matrix, and soluble factors, where a given type of stem cell normally resides. There it is able to divide asymmetrically, both maintaining its own population and creating daughter/progenitor cells capable of proliferating and differentiating into the type(s) of cells the body needs at the moment. A salient characteristic of the niche is its ability to maintain the "stemness" of the resident stem cells. That is, the niche helps to maintain the naïveté and plasticity of the stem cell and prevent its differentiation. Understanding the mechanisms behind this "niche effect" will provide the tools needed to manipulate adult stem cells, including extended expansion and targeted differentiation into therapeutically relevant cell types.

3 Regulation via Cell-Cell Interactions

Regardless of their location, MSCs are in the proximity of and in direct contact with cells of other types, particularly non-stem cells. Many studies have characterized the interactions between MSCs and these cells.

3.1 Cardiomyocytes

Cardiomyocytes are often studied in connection with MSCs because exogenous MSCs have been successfully used in therapy for myocardiac infarcts [9]. A persistent question for researchers is how the MSCs provide the observed benefits. Some argue that recipient cardiomyocytes fuse with MSCs, gaining renewed healing potential. Others believe that MSCs differentiate into cardiomyocytes due to the environmental cues they are receiving. Another current hypothesis is that MSCs do not differentiate, but instead secrete cytokines that help rejuvenate the injured cardiac tissue. Rangappa et al. [10] found that direct contact of MSCs with cardiomyocytes induced their differentiation into cardiomyocytes while exposure to cardiomyocyte-conditioned medium (CM) did not. This work was confirmed by Yoon et al. [11], but with a caveat. These investigators found that the developmental status of the cardiomyocytes was an important factor in determining the behavior of the MSCs. Specifically, MSCs were able to differentiate into cardiomyocytes only when in direct contact with neonatal cardiomyocytes, not with adult cardiomyocytes. This effect is most likely mediated by a cell surface component because the cardiomyocytes could be fixed before coculture and still produced the same effect.

3.2 Skeletal Muscle

While skeletal muscle has its own reservoir of stem cells for repair (known as satellite cells), MSCs also can differentiate into skeletal muscle in vitro [12]. Under conditions of injury, injected MSCs can even be recruited to muscle to become satellite cells and then multinucleated myofibers [13]. The mechanism for this is not yet known but very likely depends on some level of direct contact with muscle cells. An in vitro study reports that MSC-myoblast contact induces myotube formation and dystrophin expression [14]. However, it is possible that the MSCs did not actually differentiate into muscle cells before fusing with the myoblasts to form myotubes. Two other studies, though, convincingly show that fusion with myoblasts is not necessary for MSCs to differentiate into skeletal muscle in vitro [15] and in vivo [16]. Therapeutically, the exact mechanism may not be relevant, since when MSCs were injected into mice suffering from muscle degeneration, both in vivo studies reported improvements [14, 16].

The De Bari in vivo study [16] pointed to a further issue in the regulation of MSCs by their environment. When they injured one tibialis anterior muscle but not the other, MSCs injected into the bloodstream were more likely to be found in the injured muscle. This indicates that either the injured muscle was secreting a homing molecule or it was expressing a cell-binding protein that allowed for better engraftment of the MSCs.

3.3 Epithelial Cells

MSCs also respond to injured epithelial cells. When cocultured with heat-shocked small airway epithelial cells, about 4% of GFP-labeled MSCs assume the morphology and genetic markers of the epithelial cells [17]. The caveat is that, like the muscle study above, some of this "differentiation" was due to fusion since about one-fourth of the GFP-positive epithelial cells were multinucleated. Whether the fusion mechanism is of significance in the therapeutic application of MSCs for tissue injury remains to be studied.

3.4 Neural Cells

Neural cells represent another differentiated cell type that can affect the ability of MSCs to differentiate into similar cell types. After being induced to express nestin, the marker for neuronal precursors, MSCs from rat bone marrow can then be cocultured on fixed mouse cerebellar granule neurons in CM from those same cells. Under these conditions, they can differentiate into neuron-like cells that are either GFAP (astroglial) or NeuN (neuronal) positive [18]. What is most impressive about these neuron-like cells is that they are electrically excitable and react to neurotransmitters, although no spontaneous synaptic events were observed. It would be interesting to see what effect coculture with live granule neurons would have on MSCs.

3.5 Osteoblasts

Osteoblasts have been shown to tightly bind hematopoietic stem cells (HSCs) and inhibit their differentiation [19]. It is therefore plausible that they maintain a similar relationship with MSCs; however, our recent studies (Kolf CM, Song L, and Tuan RS, unpublished data) suggest that osteoblasts do not inhibit, but instead enhance MSC differentiation, at least along the osteogenic lineage.

3.6 Endothelial Cells

Many studies have strongly shown a perivascular niche for MSCs [20], suggesting that perhaps MSCs are not normally in contact with the cell types discussed above but are most often interacting with endothelial cells (ECs) encapsulating blood vessels. (Note: Interestingly, endothelial niches have also been proposed for HSCs [21], suggesting the possibility that HSCs and MSCs share the same niche and interact with one another.)

Several studies have focused on the effect of ECs on the osteogenic differentiation of MSCs, since the development of blood vessels and bone is tightly connected in vivo. Using human umbilical cord vascular endothelial cells (HUVEC) and bone marrow MSCs, Villars et al. [22] found that direct coculture, and only direct coculture, caused an increase in alkaline phosphatase (ALP) activity in the MSCs (an indication of osteogenic differentiation). This study was further confirmed using human dermal microvascular ECs in vitro and in vivo [23]. The effect was found to be somewhat dependent on the presence and function of gap junctions between the two cell types [22], which explains why non-contact HUVEC/MSC transwell cultures and HUVEC matrix alone did not produce the same effect as direct coculture [24].

Other coculture studies of MSCs and ECs have focused on MSC smooth muscle differentiation. Bovine aortic ECs have been shown to induce migration of mouse embryonic C3H10T1/2 cells, an MSC-like cell line [25]. When contact between the cells occurred, the MSCs changed their morphology and gene expression to mimic that of smooth muscle cells. Since a similar result was obtained by simply adding TGF-β1 to the MSCs, it is possible that direct contact between the two cell types is not necessary to induce smooth muscle cell differentiation. Alternatively, contact may be necessary to initiate secretion of TGF-β1 by ECs. In either case, the MSC-derived smooth muscle cells were capable of incorporating into newly formed blood vessels in mice, an important verification of the functionality of these cells.

That coculture of MSCs with ECs could give rise to either a smooth muscle or an osteoblastic phenotype also illustrates the variability associated with differences in tissue sources, developmental stages, and media supplements as studied in different laboratories.

3.7 MSC-MSC Interactions

Homotypic interactions among MSCs are also critical in regulating MSC biology as evidenced by the dependence of various MSC differentiation regimens on cell plating density. Chondrogenesis of MSCs displays the most dramatic requirement for high-density contact. Robust chondrogenesis requires the MSCs to be seeded in high-density micromass or pellet; only minimal differentiation is seen in monolayer, suggesting that MSCs need contact with each other to become chondrocytes [26]. The high-density mediated cell-cell interactions likely mimic conditions during embryonic development in which homotypic N-cadherin interactions are needed for chondrogenesis (Fig. 2) [27]. This is further supported by data showing that the chondro-inductive growth factor, TGF-β1, induces chondrogenesis through the regulation of N-cadherin by Wnt7a and MAP kinases [28]. How do cell adhesion molecules, such as cadherins, mediate these critical cell-cell interactions in MSC differentiation?

3.7.1 Adherens Junctions/Cadherins

Cadherins are transmembrane, Ca^{2+}-dependent proteins that bind to other cadherins on neighboring cells, mostly in a homotypic manner. In addition to mediating cell-cell adhesion, cadherins also play other functional roles. Specifically, through their cytoplasmic domains, cadherins bind β-catenin, a critical component in the canonical Wnt signaling pathway. The interactions of cadherins, β-catenin, and Wnts are summarized below (also see review [29]).

β-Catenin is a key regulator of many transcription factors involved in differentiation. Canonical Wnt signaling, via binding of Frizzled receptors, blocks phosphorylation of β-catenin by GSK-3β, reducing subsequent ubiquitination and proteosomal degradation of β-catenin and increasing its cytoplasmic levels. Build-up of cytoplasmic β-catenin also occurs upon activation of receptor tyrosine kinases, which results in a loss of cadherin-mediated cell-cell adhesion and an increase in cell migration. Nuclear translocation of β-catenin from the cytoplasm allows it to interact with the transcription factor LEF/TCF (lymphoid enhancer factor/T-cell factor), which sets into motion many changes in gene expression. In the absence of Wnt signaling, this nuclear translocation is prevented since β-catenin is either bound to cadherins at the cell membrane, or is degraded following phosphorylation by the GSK/Axin/APC complex.

Several cadherins have been specifically implicated in niche studies. N-Cadherin is functional in the interaction between HSCs and osteoblasts, which are thought to be an integral part of the HSC niche [19]. When MSCs interact with epithelial cells, the presence of E-cadherin may be important for their ability to differentiate [17]. In *Drosophila*, too, the loss of Wnt or β-catenin/cadherin complexes results in the loss of stem cells in the ovary [30]. It is also possible that cadherin complexes help to orient division asymmetrically, giving rise to a stem cell and a progenitor that will go on to differentiate [31]. These are just a few of the many stem cell processes that are influenced by cadherins.

3.7.2 Gap Junctions/Connexins

The other major class of proteins responsible for cell-cell interactions is the connexins, which form the hexameric units of gap junctions that act as channels to allow molecules of specific sizes to flow from cell to cell, the basis of cell-cell communication and juxtacrine signaling. MSCs express connexins 32, 40, 43, and 45 [10]. Among the many cell types that could interact with them through these connexins are ECs and cardiomyocytes. In fact, the osteo-inductive effect of HUVECs on MSCs depends in part on gap junctions formed by connexin 43 [22]. Connexins 40 and 43 are similarly responsible for the ability of MSCs to couple with one another and to form channels with canine ventricular cardiomyocytes [32].

4 Regulation Via Soluble Factors

In addition to direct contact with neighboring cells, stem cells are also affected by the soluble factors in their microenvironment. These factors are most often proteins that originate from the stem cells themselves, from nearby cells, from the endocrine system of the organism, or from degradation of nearby extracellular matrix. In addition, the effects of non-proteinaceous factors, including oxygen (and ions, such as Ca^{2+}), must also be taken into consideration.

4.1 Endothelial Cells

While direct contact with HUVECs increases osteogenesis in MSCs as assayed by ALP activity, conditioned medium (CM) from HUVECs increases MSC proliferation and decreases their osteogenic differentiation [24]. This same result was obtained by another group, who also monitored matrix mineralization and several other markers of MSC osteogenesis [33]. They also found that HUVECs had a more potent inhibitory effect on MSCs after having been stimulated with vascular endothelial growth factor (VEGF). Since VEGF itself was not found in the CM nor was it effective when added exogenously to MSCs, it is thought that it stimulates HUVEC secretion of other molecules that exert the inhibitory effect seen.

ECs are known to secrete many growth factors that affect bone biology [34]. Among those with well-studied functions is bone morphogenetic protein-2 (BMP-2) (Fig. 2). For example, the increase in ALP seen in direct cocultures with ECs can be partly abrogated when BMP-2 production in ECs is down-regulated via siRNA [23]. Since there was no increase in ALP seen in MSCs cultured with CM from ECs, direct contact with MSCs might be necessary for ECs to begin producing BMP-2. This contact-initiated alteration of cellular secretion profiles is important to keep in mind when designing similar studies.

4.2 Skeletal Muscle

As seen in Section 3.2, tissue injury is a key variable in considering MSC activity. It seems that injured cells secrete different signals than healthy ones. In an interesting study from Santa Maria et al. [35], CM from undamaged muscle had no effect on MSCs, while that from damaged muscle induced their proliferation in a dose-dependent manner. It was also able to induce MSC myogenesis as supported by the expression of late gene markers of myogenesis, myotube formation and spontaneous twitching of myofibers.

4.3 Platelets

Platelets are one of the earliest cell types recruited to an injury site, and it is reasonable to speculate that they may influence the activity of tissue progenitor cells during the repair process. The supernatants of leukocyte-depleted thrombin-activated platelets have been shown to increase proliferation and migration of MSCs, but to decrease their osteogenesis [36]. Once the MSCs have migrated to the site of injury, it is possible that other environmental cues would override these inhibitory effects.

4.4 Chondrocytes

Another physiological state to consider is skeletal development. During endochondral ossification (Fig. 2), chondrogenesis precedes osteogenesis, and it is reasonable to expect an active relationship between chondrocytes and the osteoprecursor cells, MSCs. Indeed, Transwell culture of C3H10T1/2 cells with chick chondrocytes induced in the former proliferation, ALP activity, and the expression of Cbfa1 and osteocalcin mRNAs, indicating an enhancement of MSC osteogenesis [37]. The identity of the active factor(s) is not known, although BMP-7 has been ruled out since its addition to the MSCs did not mimic the transculture results.

4.5 Osteoblasts

When the same C3H10T1/2 cells are cultured in Transwells with embryonic chick osteoblasts, their proliferation is also greatly increased while their ALP activity is totally inhibited [37]. These results suggest that an osteoblast-secreted factor is active in maintaining MSC stemness. (Note: Although indirect coculture with osteoblasts also caused a slight up-regulation of the osteogenic genes Cbfa-1 and osteocalcin in the MSCs, this was only analyzed at the RNA level and is therefore less compelling than the aforementioned evidence.) Identification of the secreted factor(s) could greatly enhance our ability to control MSCs.

4.6 MSCs (Autocrine Regulation)

MSCs themselves secrete a large number of regulatory proteins. Many of these are known to affect other cells (as reviewed in [6]) but they may also contribute to MSC self-regulation. For example, MSCs concomitantly express several Wnts and their Frizzled receptors [38, 39] and MSC osteogenesis is clearly influenced by several members of the Wnt pathway [39]. In addition, Wnt signaling via canonical and noncanonical pathways appears to act in a cross-talking manner to provide both autocrine and paracrine regulation of MSC activities [40, 41].

When freshly isolated, MSCs also express high levels of stromal-derived factor-1 (SDF-1) together with its receptor CXCR4 [42]. SDF-1 can enhance the ability of platelet-derived growth factor-BB (PDGF-BB) to increase colony-forming activity in MSCs. SDF-1 expression is normally down-regulated upon osteo-induction; however, its overexpression in MSCs enhances the formation of new bone when they are injected into NOD/SCID mice [42]. Since the same is not seen in vitro, SDF-1 must be acting indirectly, possibly through increasing MSC proliferation and resistance against apoptosis. Alternatively, SDF-1 could promote the migration of the transplanted MSCs to the correct microenvironment [43]. Interestingly, SDF-1 is also expressed by endothelial, perivascular, myelosupportive, and B-lymphosupportive cells within the bone marrow and at the endosteal surface [42], presenting itself as another candidate protein involved in modulating cell-cell interactions.

Another protein that MSCs secrete and to which they also respond is VEGF. VEGF is best known for inducing angiogenesis. Indeed, an increase in the proliferation of endothelial and smooth muscle cells in vitro can be partly ascribed to its presence in MSC-CM [44]. VEGF secretion is also increased during MSC osteogenesis [24, 45, 46], thus accounting for the infiltration of blood vessels during bone development [47]. While VEGF has been reported to decrease the synthesis of osteocalcin (a marker of osteoblast maturation) in osteogenic MSCs when it is added to HUVEC-CM [24], it has also been observed to increase mineralization [45] and to be regulated by proteins involved in osteoblast differentiation [46], suggesting that it usually acts as an enhancer of both angiogenesis and osteogenesis.

4.7 Specific Signaling Molecules

While CM and Transwell cultures, like those reported above, are helpful for discerning the effects of a given cell type on MSCs, they do not identify the effective molecules. Often, educated guesses are made and a single protein is added to the MSCs and their differentiation is assayed. A large number of studies have been done using this approach, and have been reviewed in the literature. A summary is provided below.

4.7.1 TGF-βs

The TGF-β superfamily has long been implicated in development and MSC differentiation (reviewed in [48]). TGF-βs themselves are added to several differentiation cocktails and affect the differentiation of MSCs into chondrocytes, osteoblasts, adipocytes, myoblasts, and smooth muscle cells. Proteomic profiling has been used successfully to identify changes in MSC protein expression due to TGF-β stimulation [49]. This method would be very useful if applied to MSCs stimulated with other proteins, such as BMPs, the other major group of proteins in the TGF-β superfamily. BMPs were first discovered for their role in bone development but are now known to control a host of other differentiation pathways [50]. What still needs to be determined is whether or not they have roles in MSC biology that are similar to the roles of their homologues in *Drosophila*, where they have been implicated in several niche functions, such as stem cell maintenance [51, 52] and asymmetric division [53]. Elucidation of these mechanisms in human stem cells is essential to our understanding of the regulation of their identity.

4.7.2 Wnts

Another family of proteins critical in MSC biology is the Wnt family of signaling molecules. Wnts are highly conserved proteins that are essential to limb development and musculoskeletal morphogenesis in vertebrates [54]. At least Wnts 2, 4, 5a, 11, and 16 are expressed by MSCs along with several Wnt receptors (Frizzleds 2-6) [38]. Wnts are very active in determining MSC patterns of proliferation and differentiation (e.g., [55] and [39]) and they also play a role in cell adhesion and migration through their indirect interactions with the cadherin pathway (see Section 3.7.1 above and [29]).

4.7.3 Other Signaling Factors

Another major group of proteins important for MSC functions are growth factors that signal via the receptor tyrosine kinases. These include many major growth factor families, including epithelial growth factor (EGF), fibroblast growth factors (FGFs), platelet-derived growth factor (PDGF), VEGF, and others. FGFs have received recent attention in the stem cell field because of many observations suggesting that they can increase their proliferation and plasticity. FGF-2, for example, increases colony size and proliferation rate while also enhancing the ability of MSCs to form bone once FGF is removed [56]. It has also been shown to increase telomere length, total population doublings, and chondrogenic potential [57].

PDGF also affects the proliferation, differentiation, and migration of MSCs. Bovine aortic ECs secrete a factor(s) in vitro that causes C3H10T1/2 cells to migrate towards them. Anti-PDGF-BB antibodies completely inhibit this effect [25]. PDGF-BB can also increase CFU-F numbers, especially when combined with SDF-1 [58]. It is also active in the supernatants of leukocyte-depleted thrombin-activated platelets. When exposed to these supernatants, MSCs exhibit increased mitogenic and migratory activity and decreased osteogenesis [36].

Finally, a recent report [59] suggests that insulin-like growth factor (IGF) may play a critical role in stem cell maintenance within the niche. Human embryonic stem cells (ESCs) cannot usually survive without a feeder layer of fibroblasts; in this study, these fibroblasts were given bFGF, which induced them to secrete TGF-β and IGF-II. These two growth factors, in turn, helped to keep the ESCs in their stem state. Interestingly, ESCs themselves also appear to secrete IGF-II in an autocrine manner. During osteogenesis, rat osteoblasts secrete increasing amounts of IGF-I; IGF-II, however, has not been detected [60]. Assuming IGF-II has a niche effect on MSCs similar to its effect on ESCs, expression of low levels of IGF-II during MSC differentiation would be consistent with the ESC data. Additional work is needed to decipher the interactive roles among these different signaling molecules.

4.8 Hypoxia

While it may seem counterintuitive, low levels of oxygen actually enhance MSC function. Hypoxia not only increases the number and proliferation of CFU-Fs but may also increase their plasticity since the ESC stemness genes *oct4* and *rex1* are up-regulated in MSCs under hypoxic conditions [61]. These changes in gene regulation are potentially caused by activation of hypoxia-induced factor-1α (HIF-1α). Hypoxia also affects the expression of other proteins in MSCs and MSC neighbors. VEGF-A and its receptor VEGFR-1, for example, were up-regulated in MSCs under hypoxic (2%) conditions in undifferentiated trabecular bone cells [45]. Hypoxia has also been reported to increase TGF-β secretion from MSCs [44, 62] and increase BMP-2 production in endothelial cells [23]. The exact role of oxygen tension in MSC biology awaits further investigations.

5 Extracellular Matrix Regulation of MSCs

In addition to direct cell-cell interactions and those mediated by soluble factors, a major extracellular regulatory network, which is complex and tightly controlled, is the extracellular matrix (ECM). The ECM contains many molecules, and increasing numbers are being identified. Many of these molecules are multifunctional, with multiple domains, each with their own "activity," ranging from structural to signaling (Tables 1 and 2). This leads to a seemingly endless web of interactions involving ECM molecules; they all bind each other, not to mention the growth factors and other cell-surface receptors. The end result is a complex (often confusing) and interesting network with multiple outputs on cell behavior. Out of this chaos, however, arise some clear examples of both positive and negative regulatory mechanisms that act to guide MSC differentiation. The goal of this section is to highlight the best-studied examples of these regulatory networks.

5.1 Tissue Specificity of Matrix Regulation or "When in Rome..."

The first two examples of ECM regulation of MSC differentiation will focus, quite appropriately, on collagen types I and II (Col I and Col II), the most abundant collagen types. Studies involving these molecules provide evidence of the following trend – the prevailing ECM molecule of a particular tissue induces differentiation of MSCs into the cell type of that tissue. The tissue-specific cell lineage often goes on to secrete more of the tissue-specific ECM molecule, thereby effectively driving and maintaining MSC differentiation. This positive feedback may act to ensure terminal differentiation and tissue maturation once it has begun.

5.1.1 Collagen Type I

As the most abundant organic component of bone (90%), Col I is of particular importance when it comes to analyzing extracellular signals seen by MSCs [63, 64]. Col I has been shown to promote adhesion and maintenance of mature osteoblasts and osteocytes [63]. As one of the ~26 known collagen isoforms, Col I contains trimeric triple-helical domains [64, 65]. These domains are characterized by a unique amino acid motif (Gly-X-Y), where the X- and Y-positions are usually hydroxylysine or hydroxyproline [65]. This motif aids in the formation of the characteristic triple helix; three domains of three single collagen chains come together to form a triple helix. The small side chain of glycine packs in the center.

Col I belongs to the "fibrillar" subclass of collagen due to is capacity to form fibers [65]. The Col I triple helix is a heterotrimer consisting of two identical α1(I) chains and one α2(I) chain [64]. Like many other ECM molecules, it contains an RGD domain(s). The RGD sequence serves as a cell attachment site as it is recognized by nearly half of the over 20 known integrin matrix receptors on the cell surface [66] (Table 1). This includes integrins α1β1 and αvβ3, both of which are found on MSCs [66, 67]. Col I interaction with these integrins has been shown to enhance osteogenesis in MSCs [68]. In fact, Col I interaction with these integrins

Table 1 Integrin-ECM ligand interactions*

Subunits		Ligands	Domains/sequences	Effects	Refs.
β1	α1	Collagens			[139]
		Col I	RGD	Enhances osteogenesis	[68]
		Col I	GFPGER	Enhances osteogenesis?	[70]
		Col II	ND	Enhances chondrogenesis	[73]
		Col IV	RGD of Arrestin	ND	[65, 122, 123]
		Laminins		ND	[139]
	α2	Collagens			[139]
		Col I	DGEA	Enhances osteogenesis	[69]
		Col I	GFPGER	Enhances osteogenesis?	[70]
		Col II	ND	Enhances	[73]
		Laminins		Chondrogenesis	[139]
		Laminin γ2	Short arm	ND	[81]
	α3	Collagens			[139]
		Col IV	Canstatin	ND	[65, 124]
		Laminins		ND	[139]
		Laminin α5	Domain VI	Suppresses chondrogenesis	[77]
		Laminin α3	LG3		[80]
		Fibronectin			[139]
		Entactin			[139]
	α4	Fibronectin	LDV of IIICS-1	ND	[113, 139]
		Fibronectin	REDV of IIICS-5	ND	[113, 139]
		VCAM-1			[139]
	α5	Fibronectin	RGD	Inhibits adipogenesis	[111, 139]
		Laminin α5	RGD of Domain IVa	ND	[77]
	α6	Laminins			[139]
		Col IV	Tumstatin	ND	[65, 126, 127]
	αv	Fibronectin			[139]
		Vitronectin	RGD	Enhances osteogenesis	[139]
β2	αL	ICAM-1, -2, -3			[139]
β3	αv	Fibrinogen			[139]
		Fibronectin			[139]
		vWF			[139]
		Vitronectin	RGD	Enhances osteogenesis	[68, 139]
		Thrombospondin			[139]
		Osteopontin			[139]
		Collagens			[139]
		Col I	RGD	Enhances osteogenesis	[68]
		Col IV	HUIV26 epitope	ND	[128]
		Col IV	Canstatin	ND	[65, 124]
		Col IV	Tumstatin	ND	[65, 126, 127]
		Laminin α5	RGD of Domain IVa	ND	[77]
β4	α6	Laminin α3	LG3	ND	[80, 139]
β5	αv	Vitronectin			[139]
		Fibronectin			[139]
		Col IV	Canstatin	ND	[65, 124]
		Laminin α5	RGD of Domain	ND	[77]

*Only integrins reported to be expressed by MSCs are included. When possible, the ligand domains involved and the effects of binding on MSC differentiation are given.
ND, not determined.

Table 2 Domains of ECM molecules with known effects on cell behavior*

Molecule	Domain/ sequences	Receptor	Effect	Refs.
Collagen type I	ND	ND	Osteoblast/osteocytes adhesion and maintenance	[63]
Collagen type I	RGD	Integrins $\alpha1\beta1$, $\alpha v\beta3$	Promotes MSC osteogenesis	[68]
Collagen type I	DGEA	Integrin $\alpha2\beta1$	Promotes MSC osteogenesis	[69]
Collagen type I	GFPGER	Integrins $\alpha1\beta1$, $\alpha2\beta1$ (A-domains)	Promotes MSC osteogenesis?	[70]
Collagen type I	P-15 (FN-binding domain)	ND	Osteoblast attachment/activity, BMP secretion	[71]
Collagens	Helical domain	DDR1, DDR2	MMP-1 expression	[72]
Collagen type I	ND	ND	Downregulates TGF-β1 secretion	[73]
Vitronectin	RGD	Integrin $\alpha v\beta3$	Promotes MSC osteogenesis, Col 1 expression	[68]
Collagen type II	ND	Integrins $\alpha1\beta1$, $\alpha2\beta1$, $\alpha10\beta1$	Promotes MSC chondrogenesis, enhance TGF-β1 signaling	[73]
Collagen type II	ND	ND	Downregulates TGF-β1 secretion	[73]
ND, EC matrix	ND	ND	Enhances MSC growth, proliferative life-span, and differentiation potential	[63]
ND, Matrigel	ND	ND	Enhances MSC expansion efficiency	[76]
Laminin α chain	LG domains	ND, cell-surface HSPGs	Cell adhesion	[77]
Laminin $\alpha5$	RGD (Domain IVa)	Integrins $\alpha5\beta1$,$\alpha v\beta3$, $\alpha v\beta5$	Cell Adhesion	[77]
Laminin $\alpha5$	Domain VI	Integrin $\alpha3\beta1$	Cell adhesion	[77]
Laminin $\alpha3$	LG3	Integrin $\alpha3\beta1$	Cell adhesion	[80]
Laminin $\alpha3$	LG3	Integrin $\alpha6\beta4$	Forms hemidesmosomes	[80]
Laminin $\gamma2$	Short arm	Integrin $\alpha2\beta1$	Keratinocyte migration	[81]
Laminin $\alpha3$	LG4 A3G75aR	Syndecan-2,-4	Cell adhesion	[82]
Laminin $\beta1$	YIGSR	Laminin receptor	Mediates EC binding	[140, 141]
Laminin $\beta1$	PDSGR	ND	Mediates EC binding	[140,141]
Fibronectin	GRGDS	Integrin $\alpha5\beta1$	Inhibits adipogenesis in ST-13 preadipocytes	[111]
Fibronectin	ND	ND	Increases expression of FN-degrading enzymes	[114]
Collagen type IV	Helical domain	$\beta1$ integrins	Cell adhesion	[122]
Collagen type IV $\alpha2$, $\alpha3$, $\alpha6$ chains	NC1 domains	αv (including $\alpha v\beta3$) and $\beta1$ integrins	Enhances EC binding, angiogenesis	[122]
Collagen type IV $\alpha3$ chain	RGD of NC1 domain	Integrin $\alpha1\beta1$	Cell adhesion	[122]

*Where possible, the cell receptors mediating the interaction are given.
ND, not determined.

was reported to drive osteogenesis alone (without the use of osteogenic medium). Enhancement of osteogenesis may be due to activation of ERK 1/2 in response to integrin-binding to Col I. These pro-osteogenic properties were enhanced by vitronectin, another RGD-containing matrix molecule, which also interacts with integrin $\alpha v\beta3$ [68]. It appears that Col I stimulates MSC expression of osteogenic genes directly, while contact with vitronectin may stimulate expression of Col I. This fits nicely with a positive feedback loop concerning osteogenesis and bone-ECM molecules. However, other RGD-containing, integrin-binding ECM molecules did not enhance osteogenesis in MSCs. Thus, other signals must be involved, most likely involving other domains on Col I.

Indeed, Col I contains other non-RGD domains that interact with cell-surface integrins. For example, the peptide domain DGEA interacts with integrin α2β1, and this interaction has also been shown to mediate osteogenesis in MSCs [69]. Furthermore, the GFPGER sequence at residues 502–508 bind the von Willebrand Factor A-like domains (A-domains) of integrins α1β1 and α2β1 [70]. In fact, GER and GER-like motifs are present in all of the integrin-binding domains of Col I.

Col I contains yet another cell-binding domain that has been shown to induce osteogenesis in MSCs, known as the cell-binding domain P-15 [63]. Located on the α1 (I) chains, the P-15 sequence (766 GTPGPQGIAGQRDVV 780) exhibits a conformation characterized by two stretched β-strands flanking a β-bend [71]. While the specific cell-surface receptor that binds P-15 has yet to be identified, the P-15 sequence falls within a known fibronectin-binding domain; thus, it is possible that MSCs indirectly interact with the P-15 sequence with the help of other ECM molecules [70]. Interestingly, synthetic P-15 has been shown to enhance cell attachment and osteoblastic activity in osteoblasts [71]. P-15 has been shown to increase the gene expression of BMP-2 and BMP-7 – osteogenic factors that have been reported to stimulate osteogenic activity. In turn, these growth factors stimulate the synthesis of P-15-containing Col I. The P-15 domain may thus act in the positive feedback loop that cycles between collagen stimulation of osteogenesis and collagen synthesis in MSCs.

Surprisingly, triple-helical collagens, including Col I, have been shown to bind and activate both discoidin domain receptor tyrosine kinases (DDR1 and DDR2) [72]. Evidence suggests that DDR binding to collagen occurs in the GXY repeat region. Triple helical conformation is required for receptor activation, and the glycosylation state of collagen may also regulate binding/activation. Collagen activation of DDR1 induces phosphorylation of a docking site for the Shc phosphotyrosine binding domain. Shc is also recruited to integrin α1β1 complexes; thus, stimulated DDR1 and activated α1β1 may converge on the same pathways. However, DDR1-mediated signaling in response to binding collagens has been shown to be independent from β1 integrins. Interestingly, one of the results of this signaling is increased expression of MMP-1, which degrades Col I. It is not known whether MSCs express the DDRs, but this could add an element of negative feedback to collagen-MSC signaling. Also of note, collagen-mediated stimulation of DDR follows a much delayed time course relative to conventional growth factor receptors. Thus, DDR receptors may work as monitors of the ECM environment rather than as mediators of acute signaling responses. In this way, only prolonged Col I signaling through DDR would result in MMP-1 expression. This acts as a balance to the positive feedback described between Col I and MSC osteogenesis.

5.1.2 Collagen Type II

Col II, another fibrillar collagen, has also been shown to affect MSC differentiation. Col II is a major ECM component of hyaline cartilage and is capable of maintaining the chondrocytic phenotype [73]. While Col I has also been reported to induce primarily osteogenesis, Col II biases toward chondrogenesis [69, 73, 74]. In fact, Col II hydrogels promoted chondrogenesis in MSCs under serum-free conditions without growth factor stimulation [73]. Integrins α1β1, α2 β1, and α10 β1 are reported to bind Col II. These are all found on chondrocytes, while undifferentiated MSCs express all except α10 [67, 73], raising the interesting possibility that perhaps integrin α10 is involved in the maintenance of chondrocytic phenotype.

As discussed above, TGF-β1 is a well-documented potent chondrogenic factor. TGF-β1 interacts with a heterotrimeric receptor complex (type I and II) and leads to activation of the Sox9 transcription mechanisms through the Smad pathway [73]. Col II, which induces chondrogenesis by itself, was reported to greatly enhance TGF-β effects [73]. Interestingly, it was found that cells treated with TGF-β1 produce endogenous TGF-β1, but Col II (and I) is reported to downregulate production of endogenous TGF- β1 [73]. This is a clear example of negative feedback regulation of cell production of TGF-β1, which is highly regulated in other ways, as illustrated below.

5.2 Role of ECM in Niche and Stemness

The findings on these first two examples of Col I and Col II begin to suggest the trend that the prevailing ECM molecule of a particular tissue may induce differentiation of MSCs into the cell type of that tissue (i.e., Col I, the main component of bone matrix, promotes osteogenesis, while Col II, a major cartilage ECM molecule, induces chondrogenesis). The corollary question can thus be posed: "Do ECM molecules also act to prevent differentiation, thus maintaining the stemness of MSCs in their undifferentiated state?" In this context, the issue thus addresses the stem cell niche, the collection of microenvironmental cues that maintain the stemness state of the MSCs. Specifically, does the notion of "tissue specificity" of the ECM also hold true in terms of ECM molecules of the niche acting to promote stemness? The answer to this question is – "possibly".

5.2.1 Endothelial Cell Matrix/Matrigel

It has been reported that MSCs occupy a perivascular niche, thereby interacting with a perivascular matrix. This matrix, also known as basement membrane, is produced, in part, by ECs, the principal blood vessel cell type. Matrix produced by ECs (EC matrix) was found to enhance the growth and proliferative life-span (period before the cells reach growth arrest) of MSCs [63]. Furthermore, MSCs expanded on EC matrix retained tri-lineage potential throughout many mitotic divisions [75]. The EC matrix is composed of laminin, heparan sulfate, entactin, and collagen type IV (Col IV) [75]. Some studies involving a similar vascular matrix, Matrigel, have yielded similar results. Matrigel is a high-salt urea extract of Englebreth-Holm-Swarm (EHS) mouse carcinoma consisting primarily of laminin (60%) and Col IV, with lesser amounts of other matrix molecules, including fibronectin, entactin, and the heparan sulfate proteoglycan (HSPG), perlecan [74]. High coating densities of Matrigel (50 $\mu g/cm^2$) were found to enhance MSC expansion efficiency [76].

5.2.2 Laminin

Because both EC matrix and Matrigel are a mixture of ECM molecules, efforts to pinpoint the specific matrix factor responsible for promoting stemness in MSCs has proved difficult. However, as the principal component of both EC matrix and Matrigel, laminin is a good candidate. Laminin refers to a group of cross-shaped heterotrimeric glycoproteins, each composed of an α, β, and γ chain [77]. Five distinct α chains, four β chains, and three γ chains have been identified that can combine to form 15 different isoforms (laminin 1-15). Laminins α, β, and γ chains share homologous structures that are divided into short and long arms. The N-terminal short arms include globular domains (domains IV and VI) and rod-like domains containing EGF-repeats (domains III and V). Additional rod-like and globular domains (IIIa and IVa) are found in the short arm regions of the α chains [142]. The C-terminal long arms contain domains that form the α-helical coiled-coil of the molecule (domains I and II) [77]. In addition, the long arms of all laminin α chains contain a large COOH-terminal globular (G) domain with five internal repeat motifs (LG1-5). The current view is that domains IV and VI are essential to self-assembly and, therefore, incorporation into the basement membrane. Laminin has the ability to self-associate, forming large aggregated structures. This activity is mediated by the globular regions at the tips of the short arms (domain VI), and is Ca^{2+}-dependent [142]. These globular domains have also been shown to cross-link the laminin network to the Col IV network [77]. There are also exceptions: laminin forms a particularly stable complex with nidogen/entactin, which can be extracted from mouse EHS tumors as a 1:1 complex [142]. Nidogen binds to the cysteine-rich, EGF-like repeats (domain III) of both β and γ chains. Also, a nidogen/entactin-domain has been identified on the γ1 chain [78]. Also, the G domain has been shown to bind Col IV. Interestingly, nidogen has also been demonstrated to bind Col IV, thereby acting as a bridge between laminin and Col IV.

The major cell binding domains are located in the α chain G domains [77]. There are also exceptions: laminin α5 (found in laminins 10, 11), which has an exceptionally long N-terminal short arm due to the presence of a large IVb domain and additional EGF repeats in domains IIIa, IIIb, and V, is the only laminin α chain that carries exposed RGD-cell binding sites in the short arm (two RGD binding sites occur in domain IVa and adjacent EGF-like repeats and support integrin α5β1, αvβ3, and αvβ5-mediated binding). Although such RGD sequences occur in other laminins, they are normally exposed only after proteolytic cleavage and do not represent major cell binding sites. Another exception is that domain VI of laminin α5 binds integrin α3β1. Laminins also bind heparin found in both cell-surface and matrix HSPGs. A major binding site for heparan sulfate/heparin is the α chain long arm G domain.

Laminin 5

While most laminins are found in basement membranes, not all are localized to the vascular matrix and MSC niche. One laminin isoform that has been linked to the MSC bone marrow microenvironment is laminin 5 [79]. Like all laminins, it exists as a heterotrimer of an α, β, and γ chain [77]. Along with laminins 6, 7, and 13, laminin 5 contains the α3 chain. However, it is the only source of the β3 and γ2 chains. The LG3 module of its α chain G domain contains a binding site for integrin α3β1 [80]. It also binds the hemidesmosomal integrin α6β4. There is also evidence to suggest that integrin α2β1 binds to the short arm of the γ2 [81]. Interestingly, the LG3 and LG4 domains of the α3 chain bind heparin [82]. The LG4 in particular contains an A3G75aR sequence that is a binding site for heparin and HSPGs, such as syndecan-2 and -4 on the surface of cells.

The activities of laminin 5 are heavily regulated by proteolytic cleavage of its chains. For example, the α3 chain is cleaved in the LG4 domain by BMP-1 [78]. The loss of at least part of LG4 and all of LG5 restricts the ability of laminin 5 to bind heparin and HSPGs. The cleaved fragment has not yet been identified, but it may retain bioactivity if it still binds syndecan. The processed α3 chain sill possesses the LG3 domain, and retains cell-binding capacity. Interestingly, the short arm of α3 is also processed by BMP-1 [78]. This cleavage produces a new N-terminal sequence within EGF-like repeat 2 of domain IIIa, near the domain II–IIIa border, thus eliminating the bulk of the short arm. The effects of this processing event on the biological activity of laminin 5, and those of the resultant fragment, have yet to be identified.

The short arm of the $\gamma 2$ chain is also processed. It contains cleavage sites for MMP-2, MT1-MMP, as well as BMP-1 [78, 83, 84]. One site occurs between amino acid residues 413–414, and is acted on by BMP-1 and MT1-MMP. The other occurs between residues 587–586 and is acted on by MMP-2 and MT1-MMP. The BMP-1/MT1-MMP 413–414 cleavage site is located within the second EGF repeat of domain III. The sequence at this site matches the consensus sequence for the cleavage of type I, II, and III procollagens by BMP-1, also known as type I procollagen C-proteinase. [78]. Cleavage at this site results in a fragment containing domains IV and V [83]. Interestingly, a fragment comprising domains IV and V has been found to mediate integration of laminin 5 into the ECM and to promote cell adhesion. The MMP-2/MT1-MMP 587–586 cleavage site immediately precedes two closely spaced cysteines in domain III, which are thought to be involved in joining the $\alpha 3$, $\beta 3$, $\gamma 2$ chains at the center of the "cross" architecture of laminin [83, 84]. Cleavage at this site results in liberation of the entire $\gamma 2$ short arm (domains III, IV, and V) [83]. Processing at this site in particular has been shown to alter the properties of the molecule dramatically. The loss of most of the short arm means processed laminin no longer binds $\alpha 2\beta 1$ [81]. Processed laminin still binds integrins $\alpha 3\ \beta 1$ and $\alpha 6\ \beta 4$, however. This has particular ramifications in the cell attachment and motility properties of laminin 5. Interestingly, processing at both sites, either by MT1-MMP alone or a combination of both MMP-2 and BMP-1, produces fragments of domain III separate from the domain IV/V fragment.

The fragmented domain III domains are of particular interest because of the EGF-like repeats [83]. Liberated domain III (but neither intact laminin 5 nor $\gamma 2$ domain III–V) binds EGFR and stimulates downstream signaling (MAP kinases), MMP-2 gene expression, and cell migration. In the case of enhanced MMP-2 expression, this is a good example of positive feedback. Cleavage by MMP-2 at the 587–586 site on the $\gamma 2$ also reveals a cryptic cell motility-stimulating site on the $\alpha 3$ chain (maybe the M1G1 epitope of the $\alpha 3$ chain) [84]. Thus, $\gamma 2$ cleavage at this site is closely linked to cell migration; both the liberated fragment and the molecule left behind have been implicated as migration factors.

The modulation of laminin-5 domains through cleavage controls its role in ECM assembly. Laminin 5 does not self-assemble/interact with ECM molecules to the degree of other laminins. The $\alpha 3$ and $\gamma 3$ chains of laminin 5 lack the laminin-binding domain IV and the $\gamma 3$ chain lacks the nidogen/entactin-binding site found on $\gamma 1$ [78]. Plus, its early-processed $\alpha 3$ chain loses the ability to bind HSPGs. Further processing, however, works to enhance incorporation of laminin 5 into the ECM. For example, processing by BMP-1 of the short arms of the $\gamma 2$ and $\alpha 3$ chains opens access of unpaired cysteine residues within the fourth EGF-like repeat of $\alpha 3$ domain IIIa and domain VI of $\beta 3$. This allows covalent association by disulfide bonds with laminins 6 and 7. The laminin 5–6/7 complex binds nidogen due to the $\gamma 1$ chain, thereby incorporating the complex into the ECM. Interestingly, only monomeric laminin 5 links hemidesmosomes (through integrin $\alpha 6\beta 5$ with fibril protein type VII collagen). This further suggests that BMP-1 activity facilitates membrane assembly, but not hemidesmosome assembly.

These characteristics of laminin 5 suggest a possible involvement in their regulation of MSC differentiation. Activation of EGFR has been reported to induce cell expansion and reversibly prevents multilineage differentiation of MSCs [85]. Since laminin 5 produces a fragment that activates EGFR, does this imply that laminin 5 may prevent multilineage differentiation of MSCs? One recent study [86] reports that MSCs incubated with laminin 5 proliferated more but retained differentiation potential. Laminin-5 potently suppressed chondrogenic differentiation but had no effect on osteogenic differentiation; however, these authors link this effect to integrin $\alpha 3\beta 1$, but not domain III-EGFR. These findings support the hypothesis that laminin 5 promotes naivety in MSCS. In addition, laminin $\gamma 2$ chain is a marker for the bone marrow microenvironment [79], and may in fact reflect some aspects of the niche of the bone marrow stromal cells, which contain MSCs.

However, other studies have reported that laminin 5 enhances osteogenic differentiation [68, 87, 88]; it was not clear whether the laminin used was in its unprocessed or processed form.

5.3 Extracellular Matrix Regulation of Growth Factors

The studies described above suggest that laminin, or some other component(s) of EC matrix/basement membrane/Matrigel, promotes stemness of MSCs. However, in considering this postulate, the fact that the ECM avidly binds growth factors must also be taken into consideration. For example, while some studies indicate that EC matrix/Matrigel promotes stemness, other studies indicate that Matrigel stimulates neuronal differentiation and chondrogenesis [74, 76]. These differentiation properties could not be attributed only to laminin or Col IV, the main structural proteins found in Matrigel. Furthermore, the stemness maintenance properties of EC matrix were probably not solely supported by laminin; laminin exhibited the same maintenance abilities as Col IV, and neither of these molecules separately gave the same maintenance abilities as the EC matrix mixture [75]. These facts strongly indicate the involvement of other matrix-associated factors.

Matrigel, like most ECMs, binds and sequesters growth factors. For example, soluble factors including TGF-β, FGF, and IGF and PDGF have been identified in Matrigel [74]. On the other hand, it should be noted that the chondrogenic properties of Matrigel could not be attributed solely to TGF-β or FGF, Matrigel isolated from EHS tumors grown in young mice worked better than that from old mice, but no differences were found in the relative levels of the major matrix molecules or TGF-β and FGF.

5.3.1 Transforming Growth Factor

Binding of growth factors by the ECM often regulates growth factor activity (Table 3). For example, binding of GFs (here, the term GF is used loosely to describe growth, differentiation, survival factors, and cytokines) can serve to stabilize or increase local concentration of the GFs, modulate the interaction of GFs with their receptors, alter the rates of GF diffusion through the matrix, or to protect the GF from proteolytic degradation [89, 90]. These characteristics are exemplified by TGF-β. TGF-β is one of the most potent signaling molecules for MSCs, and is thus also the most tightly controlled. TGF-β is secreted in an inactive form as part of a complex with its propeptide, also known as latency-associated protein (LAP) [91]. LAP noncovalently binds TGF-β with high affinity and prevents TGF-β from interacting with its receptors. In addition, before secretion, disulfide linkages are formed between cysteine residues of LAP and specific cysteine residues in the latent-TGF-β-binding protein (LTBP). LTBP is part of the LTBP/fibrillin protein family. Interestingly, these proteins also contain multiple EGF-like repeats. This complex comprised of TGF-β, LAP, and LTBP is referred to as the large latent complex (LLC). The N-terminal region of LTBP-1 is covalently cross-linked to as yet unidentified ECM proteins by transglutaminase. In this way, inactive TGF-β is sequestered in the ECM. LAP-bound, inactive TGF-β can be released from the matrix through protease-sensitive regions on LTBP. However, TGF-β must be freed from LAP to become activated. A diverse group of activators has been identified, including proteases (plasmin, MMP-2, MMP-9), TSP-1, the integrin αvβ6, reactive oxygen species, and low pH.

5.3.2 Small Leucine-Rich Proteoglycans

Other ECM molecules have also been shown to bind TGF-β. One such group of molecules is the small leucine-rich proteoglycans (SLRPs) [92] These small chondroitin sulfate-dermatan sulfate proteoglycans are abundant in bone and cartilage matrix and include biglycan, decorin, and fibromodulin. In addition to TGF-β, SLRPs also bind other matrix molecules, including collagen, fibronectin (FN), and thrombospondin, further stabilizing the complex. In this way, SLRPS allow for the association of TGF-β with a wide range of matrices and matrix molecules. The core protein of both decorin and biglycan have been shown to bind and inhibit TGF-β [93]. The binding site might be the leucine-rich repeat. Biglycan is found pericellularly, while decorin is located further away, indicating an additional spatial factor in the regulated distribution of growth factors. Interestingly, TGF-β induces the synthesis of SLRPs and other ECM chondroitin/dermatan sulfate proteoglycans, suggesting a negative feedback system for regulating TGF-β and its cellular effects [93, 94]. In this way, availability of active TGF-β is tightly regulated by the ECM.

Similarly, biglycan can also bind BMP-4, another member of the TGF-β superfamily [95]. BMPs have been shown to be very potent regulators of MSC differentiation, and their activity and availability are also regulated by extracellular factors. The principal BMP antagonists are made up of two families: the cysteine knot family (Noggin and Dan family proteins) and the cysteine-rich repeat proteins exemplified by chordin [96]. Noggin binds with various affinity BMP-2, -4, -5, -6, and -7 and growth and differentiation factor (GDF)-5, -6 and Vg1, but not other members of the TGF-β family [97]. Noggin inhibits BMP signaling by binding with its cysteine-knot domain [96, 98]. This interaction blocks Type I and Type II receptor binding

Table 3 Effects (activation or inhibition) of binding of Growth Factors by ECM molecules

Molecule	Domain*	GF	Activate/inhibit	Refs.
LTBP-1	CR Domain	TGF-β/LAP	Inhibit	[91]
Biglycan	Leucine-rich repeat?	TGF- β, BMP-4	Inhibit	[93, 95]
Decorin	Leucine-rich repeat?	TGF- β	Inhibit	[93]
Fibromodulin	Leucine-rich repeat?	TGF- β	Inhibit	[93]
Procollagen type IIA	CR domain, N-terminal propeptide	TGF-β, BMP-2	Inhibit	[103, 104]
IGFBP-5	C- and N-terminal domains	IGF-I	Activate	[90, 106]
Perlecan (HSPGs)	HS chain, domain I	FGF, Wnt	Activate	[108, 109]
Collagen type IV	ND	TGF- β1, BMP-1, BMP-4	ND	[129, 130, 131]

*When possible, the exact ECM molecule domain that mediates the binding is given.
ND, not determined.

sites, thereby preventing BMP from binding to its receptors. In mice, Noggin is expressed in condensing cartilage and immature chondrocytes [99]. Noggin expression in osteoblasts is limited, but it is induced following exposure to BMP-2, -4, or -6, suggesting that this may be another negative feedback system/protective mechanism to prevent excessive exposure of skeletal cells to BMPs [97]. Similar protective mechanisms may exist in chondrocytes in which the expression of Noggin and chordin is up-regulated by Indian hedgehog (Ihh), an inducer of endochondral ossification [97].

Chordin contains four cysteine-rich (CR) domains of about 70 amino acids each [99]. These domains, particularly CR1 and CR3, determine chordin's ability to bind BMPs [97]. Chordin binds BMPs specifically, preventing their receptor signaling, and does not bind to other members of the TGF-β family of peptides. Chordin may be cleaved by Tolloid/Xolloid metalloprotease, releasing bound BMP [99]. Xolloid cleaves chordin at two conserved aspartic acids just after CR1 and CR3 [100]. This may be modulated by *twisted gastrulation* (tsg), which binds and stabilizes the BMP/chordin complex and which makes chordin susceptible to degradation by Tolloid [99]. Tsg also binds BMP with a CR domain [97]. Tsg is expressed by osteoblasts, but it is not regulated by BMPs, and it is not known whether it is also expressed by chondrocytes. Chordin is more predominantly expressed by chondrocytes, and Tolloid is expressed by osteoblasts and chondrocytes. Thus, the molecules that act to release TGF-β are also highly regulated.

Being soluble, Noggin and chordin are also highly regulated in terms of their localization, consistent with the necessity of tight control of the highly active BMPs. Both Noggin and chordin bind cell-surface HSPGs (chordin, at least, has been shown to not bind matrix HSPGs [101]). The localization of these molecules thus focuses the regulation of growth factor activities to the crucial pericellular space [101, 102].

While Noggin and chordin are technically not ECM molecules, chordin-like CR domains are found in other matrix molecules that interact with MSCs, including most forms of fibrillar procollagens (types I, II, III, and V) [100].

5.3.3 Collagen Type II

Col II is another interesting GF-regulating matrix molecule. As stated above, Col II is the most abundant protein in hyaline cartilage ECM [64]. Like other fibrillar collagens, it is initially synthesized as a procollagen that includes both an NH_2- and COOH-terminal propeptide [103]. Procollagen type II exhibits some intrinsic qualities that make it particularly interesting. First, unlike other collagens, procollagen type II transcripts exist in two alternatively spliced forms, type IIA and type IIB [103]. The translated proteins from these alternatively spliced transcripts differ in several important ways. The NH_2-terminal propeptide of type IIA procollagen contains an additional 69-amino acid CR domain, not found in type IIB procollagen. Thus, procollagen type IIA has been shown to bind TGF-β1 and BMP-2, while procollagen type IIB does not. This binding inactivates members of the TGF-β superfamily [104].

Procollagen type II exhibits other intrinsic qualities that make it a particularly efficient, compared to other collagens TGF-β/BMP binder. First, it is a heterotrimer of three identical α chains, each containing the CR domain [104]. Type I procollagen, on the other hand, only contains a single CR domain on its α1(I)chain. This feature may affect the structure and the TGF-β/BMP binding capabilities of the trimeric NH_2-propeptide. Next, type II procollagen contains a triple-helical domain (minor helix) in its NH_2-propeptide that is almost twice as long as that in procollagen type I. This may provide additional extension of the CR domain from the procollagen type II fibril, allowing more effective binding of TGF-β/BMP. Finally, the type II propeptide contains proteinase cleavage sites not found in type I procollagen. This last point becomes interesting when one considers how bound, inactive TGF-β/BMP is released.

All procollagens are processed by proteinases that cleave off the NH_2- and COOH-terminal propeptides, resulting in mature collagen molecules [104]. In the case of the NH_2-propeptide of procollagen type II, both type IIA and type IIB have been found to be susceptible to cleavage by N-proteinase, and MMPs 3, 7, 8, 9, 12, and 14. Processing by these enzymes occur at four cleavage sites (A–D). Cleavage at sites C and D do not disturb the minor helix and result in a freed triple-helical trimeric propeptide that retains the ability to bind and inactivate BMPs. These sites are found in both the IIA and IIB propeptides and are susceptible to MMP-3, 9, and 14 and N-proteinase cleavage. (Note: Cleavage at site D interferes with the cross-linking of the major helical domain of the rest of the collagen type II fibril. Thus, cleavage at this site makes the collagen type II susceptible to structural disruption, thereby playing a role in matrix turnover and compositional change. This feature becomes interesting and important when one considers the sequential replacement of IIA with IIB during cartilage development.) However, cleavage at sites A and B (B is found in both the IIA and IIB propeptides, while A is only in the IIA propeptide) release monomeric CR domains with reduced growth factor binding efficiency. Processing at these sites, which are cleaved by MMP-13 and 7, would result in freed, active BMP.

5.3.4 Insulin-Like Growth Factor

Members of the TGF-β family are not the only growth factors to be bound by matrix molecules. Another highly regulated growth factor is IGF. Many of the principles of regulation are similar to those involving TGF-β but, in this scheme, binding of the GF by the ECM activates signaling. IGF is a major anabolic growth factor in articular cartilage [105]. It is bound and modulated by IGFBPs [90]. IGFBP-5 binds Col III and IV, laminin, and fibronectin, and there is evidence to suggest that this is through an ionic interaction. A more recent report indicates that IGFBP-5 has carboxyl-terminal basic motifs incorporating heparin-binding and additional basic residues that interact with various glycosaminoglycans, osteopontin, thrombospondin, and hydroxyapatite [106]. ECM-bound IGFBP-5 strengthened IGF signaling and was protected from degradation [90]. Conversely, soluble IGFBP-5 was degraded to a 22 kD fragment and had no effect on IGF-signaling. Interestingly, ECM-bound IGFBP-5 has less affinity for IGF than soluble IGFBP-5 [90]. The general idea is that IGFBP-5 acts to localize IGF to the receptors, where the GF is free to interact with the receptor due to the lower affinity matrix binding. As previously mentioned, IGF is a very potent stimulator of MSC differentiation, particularly chondrogenesis [105]. Thus, ECMs that bind IGF, such as cartilage and bone matrices, may support enhanced MSC differentiation through novel, IGF-depended mechanisms. Interestingly, it would seem that IGFBP-5 also has osteogenic IGF-independent properties, a fact made more interesting when one considers the fact that it binds hydroxyapatite and would be found in bone matrix [90].

5.3.5 Fibroblast Growth Factor

In addition to IGF and members of the TGF-β family, many other growth factors are bound by ECM molecules. It is noteworthy that most of these growth factors bind not the protein portions of the matrix molecules, but to heparan sulfate-containing glycosaminoglycans, and that these interactions generally work to strengthen the growth factor signal. Thus, an interesting pattern arises; binding of growth factors to the core protein of matrix molecules – such as TGF-β/BMP binding to the CR domains of chordin, biglycan, and decorin – usually causes inactivation of the growth factor, while binding to heparan sulfate glycosaminoglycan chains usually activates growth factors. A prime example of the latter case involves the growth factors Wnt and fibroblast growth factor (FGF), specifically basic FGF (bFGF, FGF-2). bFGF is present in ECM and associates with HSPGs both in the matrix and on the cell surface [107]. In fact, binding to free heparin/heparan sulfate or HSPGs is actually required for bFGF to be able to interact with the FGF receptor. This may be due to a conformational change in FGF induced by binding these glycosaminoglycans that allows FGF to bind its receptor. In addition, certain HSPGs serve as low affinity/high abundance reservoir of bFGF and work to concentrate the GF in the vicinity of receptors. As in the case of IGF, it is important that FGF is not bound to these HSPGs so tightly so that it is unable to interact with its receptors. For example, heparan sulfate chains attached to the N-terminal domain I of perlecan, a large basal lamina HSPG, function as low affinity, accessory receptors for bFGF, capable of facilitating potent FGF signaling [108, 109]. Other HSPGs bind bFGF with higher affinity and protect the growth factors from proteolytic cleavage [107]. In this way, HSPG binding creates a matrix-bound reservoir of FGFs from which active FGF-HSPG complexes can be liberated by degradation of either the HSPG core protein the heparan sulfate components. Finally, FGF receptors also require HSPGs to function; interaction between heparin/heparin sulfate and the extracellular domain of the FGF tyrosine kinase receptor have been shown to be required for FGF binding [108]. Thus, the regulation of FGF signaling is highly controlled, making use of HSPGs with specific binding strengths to regulate the activation of FGF and its receptors. In terms of regulation of MSC differentiation, FGF, as previously stated, promotes maintenance in an undifferentiated state. Thus, it becomes interesting that matrices with HSPGs that have been shown to sequester FGF – such as EC matrix/Matrigel, which contains perlecan – are also the matrices that have been implicated in promoting stemness in MSCs.

In concluding our discussion on the interplay between matrix and growth factors, we would like to highlight once more the many layers of negative feedback that work to prevent uncontrolled growth factor signaling. Cases in point are the many examples of matrix molecules binding growth factors that have already been discussed. Recent studies have also suggested that growth factors can actually affect the levels of these matrix molecules available to bind them. For example, TGF-β regulates the transcription of a wide spectrum of ECM proteins, increasing their production while decreasing their proteolysis [81]. As these ECM molecules bind and inactivate TGF-β, negative feedback systems arise. Various examples of this have already been discussed, including TGF-β/BMP stimulation of Col II and SLRPs. Furthermore, several studies also indicate the reverse situation, that matrix molecules regulate production of growth factors. For example, the expression of endogenous TGF-β1 is down-regulated in a dose-dependent manner by the presence of Col I and Col II in the ECM [73]. Given that TGF-β1 promotes the secretion of cartilage matrix molecules, including Col II, and that Col II itself enhances chondrogenesis, Col II down-regulation of TGF-β1 secretion adds yet another element of negative feedback to the regulatory network. Finally, matrix molecules have been found to modulate cellular responses to growth factors. For example, FN augments chondrogenesis-inducing effects of IGF in prechondrocytes in alginate culture, but inhibits them for cells in monolayer [105]. Combinations of these regulatory mechanisms result in very tightly controlled grow factor activities, and, thus, very tightly controlled differentiation in stem cells like MSCs.

5.4 Cryptic Domains and Bioactive, ECM-Molecule Derived Fragments

The final regulatory network linking ECM to MSCs to be discussed here involves cryptic domains and bioactive fragments. A growing number of studies have revealed the susceptibility of matrix molecules to processing that results in exposure of hidden domains and/or release of fragments with biological activities (Tables 4 and 5). This processing often includes proteolytic degradation, but can also involve other, non-enzymatic means. And when MSCs themselves are the source of this processing, the results are interesting feedback networks in which stem cells alter the very microenvironmental signals acting upon them.

Examples of this kind of network that have already been discussed include laminin 5 and Col II, where processing of matrix proteins by MMPs secreted by MSCs results in protein fragments with very different biological activities. In our next example, FN, this scheme is taken another step further with the inclusion of cryptic domains.

5.4.1 Fibronectin

FN is a dimeric glycoprotein found in plasma as well as most ECMs [105]. Each monomer consists of multiple copies of three homologous units, the type I, II, and III repeats [110]. FN is a multifunctional molecule that exhibits multiple domains that interact with a diverse array of other molecules, including collagen, fibrin, heparin, integrins, as well as other FN molecules [110]. Many of these domains are cryptic, only becoming exposed and/or active following some processing of the FN molecule. Actually, most of the molecule can be considered cryptic; studies of the secondary and tertiary structures of FN suggest that the molecule folds back on itself under physiological conditions [111]. This results in the "hiding" of several domains that affect how FN associates with other FN molecules. Thus, changes to FN expose these domains, thereby regulating the formation and properties of FN matrix. Almost all of these cryptic self-association sites are localized in FN type III (FN III) repeating modules, each consisting of seven β strands arranged in two β sheets [112]. FN III modules contain no disulfide bonds, allowing reversible unfolding and full extension in response to cell-generated tension, which in turn exposes cryptic self-association sites that can lead to FN polymerization. Thus, FN is like a spring that can be "stretched" by interacting with cells such as MSCs. This stretching allows FN to bind other FN molecules, in turn causing them to "stretch" and resulting in the propagation of FN matrix assembly. Conformational changes in FN can also expose other cryptic domains, including several found in the type I modules (FN I) of FN. For example, partially cryptic, intrinsic disulfide-isomerase activity has been assigned to the Cys-X-X-Cys motif of the C-terminal twelfth type I module (FN I12) [112]. It allows cross-linking of disulfide bonds, resulting in the rearrangement of FN molecules into stable, rigid matrices during their incorporation into the ECM.

FN cryptic domains can also be exposed through proteolytic processing. For example, the first type III module (FN III1) contains a cryptic heparin-binding domain [112]. Cleavage at Ile 597 in this domain increases affinity for heparin at physiological ionic strength. Furthermore, another cryptic

Table 4 Cryptic domains found on ECM molecules and their cellular effects

Cryptic site	Molecule	Mechanism of uncovering*	Receptor	Effects	Refs.
RGD	Laminins	Proteolysis	RGD-binding integrins	Cell adhesion	[77]
M1G1 epitope	Laminin α3 chain	MMP-2 cleavage at 587–586 site on the laminin γ2	ND	Directly stimulates cell motility	[84]
Heparin-binding domain	Fibronectin III1	Cleavage at Ile 597	Cell-surface HSPGs?	Cell adhesion?	[112]
FN III1 cryptic site	Fibronectin III1	ND	ND	Regulates cytoskeletal organization and cell growth	[112]
REDV	Fibronectin IIICS-5	ND	α4β1, α4β7	EC attachment	[113]
LDV	Fibronectin IIICS-1	ND	α4β1, α4β7	EC attachment	[113]
HUIV26 epitope	Collagen type IV, helical domain	Proteolysis	αvβ3	Replacement of integrin α1β1 binding with αvβ3	[128]

*When possible, the mechanism that leads to uncovering of the domain, as well as the cell-surface receptor the domain binds, is given.
ND, not determined.

Table 5 Bioactive fragments produced from ECM molecules and their cellular effects*

Fragment	Derived from	Mechanism of fragmentation	Receptor	Effects	Refs.
Laminin α3 short arm	Laminin α3	BMP-1 processing	ND	ND	[78]
Domains IV/V	Laminin γ2 short arm	BMP-1/MT1-MMP processing	ND	Laminin 5 matrix integration, cell adhesion	[83]
Domain III	Laminin γ2 short arm	BMP-1 + MMP-2 and/or MT1-MMP processing processing	EGFR	MAPK signaling, MMP-2 expression, cell migration	[83]
				Promotes MSC ‘stemness”	[85, 86]
				Enhances MSC osteogenesis	[87, 88]
FN fragment	Fibronectin	ND	ND	Inhibit growth of ECs	[111]
29, 50 kDa	Fibronectin, N-terminal gelatin binding domain	ND	ND	Chondrolysis	[114]
24/28/29 kDa Fib I	Fibronectin, N-terminal Fib 1 domain of Fibronectin	Thermolysin degradation	Cation-dependent receptor?	Stimulate adipogenesis	[111, 115]
Anastellin	Fibronectin FN III1	Proteolysis	ND	Anti-angiogenesis, interfered ERK signaling, inhibited proliferation, G1 arrest in ECs	[116]
Endostatin	Col XVIII α1 NC1 domain	Cathespins, elastase, MMP processing	HSPGs	Inhibition of FGF-induced angiogenesis, proliferation, migration in ECs	[117, 119]
Restin	Col XV α1 NC1 domain	Proteolysis?	ND	Anti-angiogenic	[117, 118]
Endorepellin	Perlecan LG3 of Domain V	MMP processing	ND	Anti-angiogenic, inhibition of migration, tube formation, and adhesion in ECs	[109, 121]
Arrestin	Col IV α1 NC1 domain	Cathespins, elastase, MMP processing	αvβ3	Anti-angiogenic, inhibition of VEGF- and FGF-induced proliferation and migration in ECs,	[65, 123]
Canstatin	Col IV α2 NC1 domain	Cathespins, elastase, MMP processing	αvβ3, αvβ5, α3β1	Anti-angiogenic, inhibition proliferation and migration and induction of apoptosis in ECs	[65, 124]
Tumstatin	Col IV α3 NC1 domain	Cathespins, elastase, MMP processing	αvβ3, α6β1,	Anti-angiogenic, inhibition of proliferation protein synthesis, and tube formation and induction of apoptosis and G1 arrest in ECs	[65, 124]

*When possible, the mechanisms that lead to release of the fragment, as well as the cell-surface receptor the fragment binds, are given. ND, not determined.

site in FN III1 may be involved in regulating cytoskeleton organization and cell growth.

The exposure of cryptic domains and their effects on MSCs become particularly interesting when they involve the cell-binding domains of FN. For example, the CS-1 and CS-5 regions of the FN variable spliced type III connecting segment (IIICS) domain contains the integrin $\alpha4$ binding sites LDV and REDV, [113]. Peptides with these sequences bind integrins $\alpha4\beta1$ and $\alpha4\beta7$ on ECs, and perhaps on MSCS, whereas full-length FN does not. This suggests these sequences are cryptic in intact FN.

FN is also susceptible to proteolytic processing that results in the release of bioactive fragments [111]. Many of these fragments have very different bioactivities from intact FN [111]. This makes sense given the structure of FN and the fact that most of its domains are hidden. Proteolytic cleavage, therefore, may be an efficient trigger to liberate cryptic activity. For example, FN fragments are reported to inhibit growth of bovine aortic ECs, while intact FN does not. Also, amino terminal 29- and 50-kD gelatin-binding FN fragments cause chondrolysis of bovine articular cartilage slices in culture, whereas intact FN does not [114]. Interestingly, these findings suggest possibly important roles for FN fragments in the pathogenesis of osteoarthritis and other degenerative joint diseases.

Another example of full length FN and FN fragments differing drastically in bioactivity involves the effect of FN on adipocyte differentiation. This would make sense because, as adipogenesis progresses, differentiating adipocytes reduce their interactions with the ECM [111]. Thus, studies suggest that molecules that enhance cell adhesion interfere with adipogenesis. For example, intact rat plasma FN inhibits ST-13 preadipocyte differentiation [111]. This inhibition involves RGDS-integrin $\alpha5\beta1$ interaction. Thermolysin degradation results in 24- and 28-kD fragments relating to the Fib1 domain that is a potent stimulator of adipogenesis. These stimulatory effects are independent of the RGDS-integrin $\alpha5\beta1$ interaction. A putative receptor is divalent cation-dependent [115]. That these adipo-stimulating interactions are not facilitated by intact FN may be attributed to the fact that the Fib 1 domain is hidden by a fold in intact FN [111]. The amino-terminal region of FN also plays a role in FN matrix assembly. (Indeed, it is bound by the cryptic FN III1 domain.) It is possible, therefore, that the N-terminal fragment also functions through interfering with assembly of FN assemblies. Interestingly, FN also causes increased expression of FN-cleaving enzymes, providing another possible feedback system [114].

5.4.2 Anti-angiogenic Fragments

We will conclude our discussion of cryptic domains with several interesting examples of fragments produced from matrix molecules found in the basement membranes of blood vessels. The fragments have proved to be powerful inhibitors of angiogenesis and have created a great deal of interest as anti-tumor therapies. Thus, while most of the studies involving these fragments have focused on their effects on ECs and cancer cells, they provide valuable paradigms for the eventual investigation into interactions between anti-angiogenic fragments and MSCs. As previously stated, MSCs have been found in the perivascular niche. Thus, it is highly likely that MSCs are in constant contact with these fragments, either as targets of their signals or, equally interesting, as regulators of their production.

Anastellin

As previously mentioned, FN is susceptible to proteolytic processing that results in the release of fragments that can differ substantially in bioactivities from those of intact FN. One of these fragments, known as anastellin, is derived from the first type III repeat (FN III1) [116]. Anastellin has been shown to inhibit angiogenesis through interfering with ERK signaling and expression of cell cycle regulatory proteins in ECs. This results in inhibited proliferation and eventual G_1 arrest in exposed ECs.

Endostatin and Restin

Perhaps the most widely studied of these anti-angiogenic fragments are endostatin and restin derived from collagen type XVIII (Col XVIII) and collagen type XV (Col XV), respectively. Col XVIII and Col XV belong to a sub-family of nonfibrillar collagens known as multiplexins [117]. Structurally, multiplexins are characterized by multiple interruptions in the central collagenous triple-helical domain and the presence of a noncollagenous domain (NC1) at the C-terminus. Both are highly glycosylated; Col XVIII is an HSPG, while Col XV is a chondroitin sulfate proteoglycan (CSPG). Both are found in the vascular basement membranes, as well as other locations. The noncollagenous domain of Col XVIII ((XVIII)NC1) consists of three segments: an N-terminal association region that aids in the self-assembly of Col XVIII homotrimers, a central protease sensitive hinge region, and a 20 kD C-terminal domain, termed endostatin. The noncollagenous domain of Col XV ((XV)NC1) also contains these regions, and its C-terminal domain, termed restin, shows ~60% sequence homology to endostatin [117, 118].

(XVIII)NC1 fragments containing endostatin are released following processing of Col XVIII by cathepsins, elastase, and MMPs [119]. As previously mentioned, processing occurs in the hinge region of (XVIII)NC1. Different proteases cleave at different regions, resulting in fragments of varying sizes. Interestingly, (XVIII)NC1 fragments that include portions of the association region are able to form trimers while those that lack this region remain as monomers [117]. Furthermore, several of these proteases, such as the cathepsins, also degrade the endostatin region [119]. Only monomeric fragments containing intact endostatin have proved to have anti-angiogenic properties. For example, mouse endostatin has been shown to inhibit the proliferation and migration of ECs in response to bFGF [120]. Structurally, endostatin exhibits a compact globular fold with an exposed face rich in Arg residues. This localization of positive charge allows endostatin to bind HSPGs. As previously mentioned, FGF also binds HSPGs, and the interaction is required for FGF to bind to its receptor. Thus, endostatin may compete with FGF for HSPG binding sites, thereby inhibiting FGF-induced angiogenesis. Interestingly, endostatin may not affect VEGF-induced angiogenesis, because the clustering of VEGF receptors 1 and 2 that precedes signaling does not require HSPG. Interestingly, several studies indicate that endostatin also acts as a protease inhibitor, interfering with the activation of proMMP-2 and reducing the catalytic activity of mature MMP-2 [117]. Given that MMP-2 is responsible for the degradation of the perivascular ECM, including the processing of Col XVIII that results in the release of endostatin, this is another example of an autoregulator feedback loop.

Like endostatin, the (XV)NC1 fragment known as restin is also anti-angiogenic [117, 118]. However, despite its high homology to endostatin, it differs in several important aspects that perhaps contribute to its comparatively reduced potency. First, the hinge region of Col XV NC1 is less sensitive to proteolysis than that of Col XVIII. This may result in less efficient release of restin than endostatin. Furthermore, restin lacks a heparin-binding domain. Since such a domain has been attributed to the ability of endostatin to inhibit FGF-induced angiogenesis by competing with FGF for HSPG binding sites, this may result in weaker bioactivity for restin. Indeed, in contrast to endostatin, restin lacks anti-migratory and anti-tube formation activity.

Endorepellin

Another example of a vascular matrix molecule that gives rise to anti-angiogenic fragments following processing is perlecan [109]. As previously mentioned, perlecan is a large basement membrane HSPG capable of binding growth factors such as Wnt and FGF with its amino-terminal domain I. Studies have also identified a fragment derived from the C-terminal domain V that is released following processing of perlecan by MMPs [121]. This fragment, known as endorepellin, inhibits EC migration, tube formation, and adhesion [109]. Most of the anti-angiogenic activity of endorepellin has been attributed to its C-terminal laminin-like globular (LG3) domain [121]. Interestingly, endorepellin can be further processed by members of the BMP-1/Tolloid family of metalloproteases, resulting in the generation of LG3 fragments. Cleaving of the LG3 has been shown to modulate endorepellin activity, and may represent a further level of regulation of anti-angiogenic signaling. Furthermore, endorepellin has been shown to bind endostatin and interfere with its anti-angiogenic activities [121]. This binding has been attributed to the LG1 and LG2 domains of endorepellin, but not the LG3. Thus, BMP-1/Tolloid-processed endorepellin may retain its endostatin-regulating abilities despite losing its angiostatic activity due to the cleaving of LG3. Clearly the regulatory networks involving these anti-angiogenic fragments may be even more complex and tightly controlled than initially realized.

Collagen Type IV

The final matrix molecule to be used as an example of cryptic domains and bioactive fragments is Col IV. As previously mentioned, Col IV, together with laminin, are the main ECM components of the basal lamina of blood vessels [122]. Like other collagens, Col IV exists as a trimer of three α chains. In the case of Col IV, there are six homologous forms, α1(IV)–α6(IV). Each chain has a 7S domain at the amino terminus, a long (approximately 1400 residues) triple helical collagenous domain, and a noncollagenous domain ((IV)NC1) of approximately 230 residues at the carboxyl terminus. Three α(IV) chains assemble into a triple helical protomer, such as $(\alpha 1)_2\ \alpha 2$ Col IV. Found in most basal lamina, including those of blood vessels, this is the most widely distributed protomer. The distribution of the other combinations of α(IV) chains exists in a tissue-specific manner. Col IV is unique among collagens in its intermolecular arrangements, and is also known as the network-forming collagen [65]. The NC1 domain of one Col IV protomer can bind to a specific region within the cartilaginous domain of another, giving rise to lateral associations among molecules [122]. In addition, dimers form when two protomers interact through their NC1 domains, while tetramers can from when four protomers interact via their 7S domains. These groups of protomers can then come together to form supramolecular networks as protomers continue to interact via their 7S and NC1 domains. Interestingly, varying the specific chains that make up the protomers among the six α chains means that each network has a distinct composition.

Like other collagens, Col IV binds cells by interacting with integrins. The central helical collagenous domain promotes cellular interactions through β1 integrins [122]. Studies have shown that (IV)NC1 also interacts with integrins, although the specific integrins and the exact sequences they interact with are still being investigated

[122]. For example, the NC1 domains of α2, α3, and α6(IV) chains interact with ECs by binding αV and β1 integrins. Surprisingly, integrin αVβ3, traditionally thought of as a vitronectin/RGD receptor, has been shown to interact with α2, α3, and α6(IV)NC1 domains. While α3(IV)NC1 contains an RGD sequence, both α2 and α6(IV)NC1 do not, suggesting that these α(IV)NC1 domains contain novel, non-RGD-dependent αVβ3 binding sites. Interestingly, EC interactions with αvβ3 integrin-binding sites on NC1 domains of intact Col IV have been shown to enhance angiogenesis.

Col IV α chains are susceptible to processing by MMPs, elastase, and cathepsins, leading to the release of (IV)NC1 fragments [65]. Three of these fragments were found to have anti-angiogenic and anti-tumor properties and were named arrestin (α1(IV)NC1), canstatin (α2(IV)NC1), and tumstatin (α2(IV)NC1). The α6(IV)NC1 fragment has also been shown to inhibit EC proliferation, but has yet to be named. However, while each appears to inhibit angiogenesis, there is evidence to suggest that they function through different mechanisms. For example, arrestin has been shown to inhibit EC proliferation, migration, tube formation, and neovascularization by interacting with integrin α1β1 [123]. Studies suggest that binding of arrestin to α1β1 down-regulates VEGF- and FGF-induced proliferation and migration of ECs. Canstatin, on the other hand, inhibits EC proliferation and migration and induces apoptosis through interactions with integrins αvβ3, αvβ5, and α3β1 [65, 124]. Studies suggest that binding of canstatin to these integrins does not affect signaling events activated by VEGF and FGF [125]. Apoptosis induced by canstatin appears to be caused by a down-regulation of the anti-apoptotic protein FLIP [125]. Tumstatin inhibits EC proliferation and tube formation and induces apoptosis and G_1 cell cycle arrest through interactions with integrins αvβ3 and α6β1 [65, 126]. Tumstatin binding to integrin αvβ3 was also found to inhibit cap-dependent protein synthesis in ECs [127].

These soluble (IV)NC1 fragments may also function through interference with Col IV network assembly; soluble NC1 has been shown to bind to the central collagenous domain of Col IV, thereby disrupting the lateral associations that normally occur between the central domain and the NC1 domains of other intact collagen IV protomers [117]. Such a breakdown in the Col IV network may act to reverse the pro-angiogenic effect observed with intact Col IV. Another possibility is that the anti-angiogenic effects of soluble NC1 are due to competitive disruption of cell binding to intact Col IV, as ECs normally bind NC1 domain via integrin αvβ3. Soluble NC1 may disrupt this interaction by competing for binding of integrin αvβ3 with the NC1 domains of intact Col IV. Furthermore, a β1 integrin binding site is located approximately 100 nm from the amino terminus near the region in collagen IV where soluble NC1 domain has been suggested to bind. Interestingly, soluble NC1 domains have also been shown to inhibit the activities of proteases, including MMP-2 and MMP-3 [117]. Once again, this would present a means of regulating the release of these fragments via negative feedback. Also of interest, Col IV has also been found to contain a cryptic site that, when exposed, potentiates angiogenesis [128]. This site, known as the HUIV26 epitope, is located within the helical domain of Col IV and is exposed following proteolysis. Exposure of HUIV26 was associated with a replacement of integrin α1β1 binding with αvβ3 integrin binding. How these activities affect the integrin-binding sites associated with the (IV)NC1 remains unclear, however.

Finally, Col IV has been reported to bind TGF-β1, BMP-3 and BMP-4 [129, 130, 131]. The specific binding domain(s) on Col IV is not known, and it is unclear whether these binding events inhibit or activate the growth factors involved, but it appears that other matrix molecules, such as laminin and FN, do not interfere with the binding. For Col IV, the combination of these binding sites and the cryptic domains suggests the complexity of the regulatory networks involving this molecule.

5.5 Role of Regulatory Networks Involving Matrix Molecules in Development

In concluding this discussion of the role of ECM in MSC differentiation, we refer to endochondral ossification as a particularly elegant example of how matrix regulation of cell behavior can give rise to the tightly controlled development of a tissue (Fig. 2). In endochondral ossification, the components of the axial skeleton are formed first as cartilage then later change into bone. It begins as mesenchymal prechondrocytes proliferate and condense before differentiating into mature chondrocytes. Thus, chondrogenesis begins when dividing prechondrocytes express ECM molecules that cause them to condense. Afterwards, the cells become chondrocytes and begin secreting chondrocyte-specific matrix. Once this cartilaginous "model" is formed, the chondrocytes begin to undergo hypertrophy. The matrix becomes susceptible to invasion by blood vessels. As the cartilage matrix is further degraded, hypertrophic chondrocytes die, and osteoblasts carried in by the blood vessels begin to secrete bone matrix [47].

Each stage of endochondral ossification is typified by a particular matrix that has a profound impact on the environmental signals reaching the cells at that time. As previously mentioned, Col II is a major component of cartilage matrix, and there are two alternatively spliced forms, type IIA and IIB, that differ in their procollagen forms and by the fact that only type IIA is able to bind/inactivate TGF-β and BMP.

These molecules come into play during endochondral ossification as Col IIA is expressed by chondroprogenitor cells and hypertrophic chondrocytes, while Col IIB is synthesized by mature chondrocytes [103]. Thus, before chondrogenesis, prechondrocytes produce IIA, which inactivates TGF-β. Then, during these early stages of chondrogenesis, MMPs 3, 9, and 14 are synthesized, which cleave Col IIA (sites C and D), thereby causing the disruption and facilitating the removal of Col IIA. Synthesis then switches over to Col IIB, replacing IIA in the matrix. MMP 13 is also secreted at this time, and the IIA propeptide is cleaved at sites A and B, thus causing the release of TGF-β, which further drives chondrogenesis.

The role of Col II continues to come into play as skeletal development proceeds to endochondral ossification. Col IIA has been localized to the hypertrophic region of the growth plate [103]. Also, during hypertrophy, chondrocytes again secrete MMPs 13 and 14. This results in both the removal of Col II and the release of bound BMP. Removal of Col II facilitates matrix turnover from cartilage to bone matrix and vascularization, and the release of active BMP drives osteogenesis. Endochondral skeletal development, therefore, illustrates how a regulatory network involving ECM and bound growth factors can coordinate development and tissue morphogenesis.

In addition to Col II, another alternatively spliced matrix molecule, FN, is also involved. Like Col II, FN exists as different splice variants derived from a common transcript. In the case of FN, alternative splicing of three regions (exons IIIB, IIIA, and V) results in multiple splice forms [110]. Again similar to Col II, these splice variants differ in their temporal and spatial localizations, as well as in their cellular effects, particularly as they relate to chondrogenesis and chondrocytes [132]. For example, FN mRNA from mesenchymal pro-chondrocytes contains both exons IIIB and IIIA, resulting in the FN isoform that includes extra domain (ED) B and EDA. This isoform, known as B^+A^+ FN, is expressed through condensation, at which point splicing out of exon IIIA begins, resulting in the isoform lacking EDA [110]. This isoform, known as B^+A^- FN, is the splice form secreted by mature chondrocytes. Conversely, B^-A^+ FN is the isoform expressed by most cells, while B^-A^- FN is the splice form found in plasma. In addition to different localizations, these splice forms also exhibit varying cell-adhesion properties. As previously mentioned, FN contains domains that interact with integrins. Indeed, FN contains multiple integrin-binging sequences, including an RGD-containing loop, and most integrins bind FN, including α5β1 (*the* FN integrin) [105]. However, while cartilage B^+A^- FN, B^-A^+ FN, and plasma B^-A^- FN induce cell spreading in a variety of cell types, mesenchymal B^+A^+ FN does not support spreading in prechondrocytes [110]. Furthermore, B^+A^+ FN promotes condensation and enhances chondrogenesis in prechondrocytes, while the other isoforms inhibit condensation and chondrogenesis. These opposing effects on differentiation may be due to differences in cell-binding strengths between B^+A^+FN and the other isoforms; cell adhesion to the B^+A^-, B^-A^+, and B^-A^- isoforms may be too powerful to allow for the migration required to form cellular condensations. B^+A^+ FN, on the other hand, may be of the appropriate adhesive strength to support the type of cellular migration required for condensation and chondrogenesis.

The switch in expression from B^+A^+ to B^+A^- FN in prechondrocytes has been linked to TGF-β exposure and to the splicing factor SRp40 [133]. Treatment with TGF-β1 causes a shift in the form of SRp40 expressed by prechondrocytes, from a short form to a long form. Where the short form of SRp40 stimulates the retention of exon IIIA, the long form of SRp40 (SRp40LF) inhibits this inclusion. Thus, TGF-β1 exposure shifts expression of SRp40 toward SRp40LF, which causes splicing of exon IIIA and results in expression of B^+A^- FN instead of B^+A^+.

Combined with what we know about Col II, these results present an elegant example of how interplay between matrix molecules whose biological properties are affected by alternative splicing can modulate the development of a tissue. As previously mentioned, the maturing cartilage matrix is characterized by a switch in Col II profile from the IIA to the IIB isoform, and by a switch in FN profile from the B^+A^+ to the B^+A^- splice form. Col IIA binds and inactivates TGF-β, while Col IIB does not. TGF-β induces splicing of FN exon IIIA, causing expression to shift to B^+A^- FN. Thus, prechondrocytes begin by expressing Col IIA and FN B^+A^+. The low cell adhesion of B^+A^+ FN allows the prechondrocytes to migrate and condense, while Col IIA binds and inactivates TGF-β, preventing FN splicing to shift to the B^+A^- isoform. After condensation, however, prechondrocytes continue to mature. Expression of Col II shifts to IIB while IIA is cleared by proteolytic degradation. TGF-β is no longer bound and is free to induce the shift to B^+A^- FN. This FN isoform supports increased cell attachment and spreading, and may support the next stage of endochondral ossification, the invasion of blood vessels and osteoblasts. B^+A^- FN also binds decorin and biglycan, which in turn bind TGF-β [105]. This may work to further halt chondrogenesis, allowing chondrocytes to hypertrophy and osteogenesis to begin.

The final example of the regulatory role of matrix in endochondral ossification involves collagen type X (Col X). Col X secretion is restricted to hypertrophic chondrocytes in regions undergoing endochondral ossification, such as the epiphyseal growth plate [134, 135]. In fact, it is the major protein secreted by hypertrophic chondrocytes, accounting for 80–90% of newly synthesized protein. Col X acts to physically stabilize the hypertrophic zone, which is otherwise experiencing a reduction in ECM. It is a member of the "short chain" collagen family, a distinct ECM protein family which also includes Col VIII. Structurally, Col X exists as trimers with a single short triple-helical domain flanked at both ends by non-triple-helical regions. These trimers form hexagonal lattice supramolecular assemblies distinct from other collagenous proteins. Col X is very susceptible to collagenases, which break down its unique lattice assembly. Interestingly, hypertrophic chondrocytes actively secrete a number of collagenases. This is important because production of Col X by hypertrophic chondrocytes precedes both vascular invasion and mineralization of the matrix. Thus, this collagenase susceptibility of Col X has been associated with degradation of the cartilage matrix, preparation of the microenvironment for vascular invasion, and mineralization. Col X may be considered to act like a temporary fix to fill the "void" left from the high matrix turnover that occurs during cartilage hypertrophy. Since it is meant to be temporary, only to provide structural support until it is replaced by bone matrix, Col X comes with a "built-in removal option"; it is easily cleared by collagenases, secreted by the hypertrophic chondrocytes.

6 Conclusion

As discussed, there are many variables that play a role in microenvironmental regulation of MSCs. The most salient feature of this interaction is its dynamism. Throughout development and the mature life of an organism, the MSC microenvironment is continually changing in a tightly coordinated way. The ability of the MSC to respond to changes in its microenvironment is absolutely critical for the MSC to fulfill its role in the body. Failure of these regulatory networks can have severe and deleterious consequences for an organism. For example, there is increasing evidence suggesting that cancers may be formed by stem cells that are no longer responsive to their microenvironments [136].

The complexity of the environmental regulation of MSCs begins with the diversity of the types of cells with which an MSC comes in contact. As an organism develops, the cells surrounding an MSC are continuously differentiating and migrating, resulting in a highly dynamic milieu that is difficult to characterize [137]. Even after the major developmental processes are complete, MSCs are found in a large number of tissue types throughout the body. One possible scenario is that a similar microenvironment (e.g., a perivascular one) is present in each of these tissues. Alternatively, MSCs may in fact be at home in several different microenvironments, or are at least capable of maintaining their identity while being exposed to different microenvironments.

To understand the mechanisms responsible for microenvironmental regulation of MSCs, cues from cells, from ECM molecules and their fragments, from secreted and small molecules, including oxygen and ions, must be taken into consideration. Each of these signaling mechanisms is tightly regulated while also being inseparably intertwined. Elucidating this tapestry of environmental regulatory networks depends critically on deciphering the individual and interactive action(s) of each of these components. A striking example of successful manipulation of cell-cell and cell-matrix interactions was just reported. Beginning with a decellularized rat heart, researchers were able to bioengineer a new, functional heart merely by seeding endothelial and cardiac cells on this natural scaffold [138]. Understanding the regulatory networks at work here, and in other systems, will provide the tools needed to manipulate MSCs through their microenvironment for the development of therapeutic approaches to diseases and tissue injuries.

Acknowledgments Supported by the Intramural Research Program of NIAMS, NIH (Z01 AR41131).

References

1. Friedenstein AJ, Chailakhjan RK, Lalykina KS. The development of fibroblast colonies in monolayer cultures of guinea-pig bone marrow and spleen cells. Cell Tissue Kinet. 1970;3:393–403.
2. Pittenger MF, Mackay AM, Beck SC, et al. Multilineage potential of adult human mesenchymal stem cells. Science. 1999;284: 143–7.
3. Tuan RS, Boland G, Tuli R. Adult mesenchymal stem cells and cell-based tissue engineering. Arthritis Res Ther. 2003;5:32–45.
4. Herzog EL, Chai L, Krause DS. Plasticity of marrow-derived stem cells. Blood. 2003;102:3483–93.
5. Bruder SP, Jaiswal N, Haynesworth SE. Growth kinetics, self-renewal, and the osteogenic potential of purified human mesenchymal stem cells during extensive subcultivation and following cryopreservation. J Cell Biochem. 1997;64:278–94.
6. Caplan AI, Dennis JE. Mesenchymal stem cells as trophic mediators. J Cell Biochem. 2006;98:1076–84.
7. Spaggiari GM, Capobianco A, Becchetti S, Mingari MC, Moretta L. Mesenchymal stem cell-natural killer cell interactions: evidence that activated NK cells are capable of killing MSCs, whereas MSCs can inhibit IL-2-induced NK-cell proliferation. Blood. 2006;107:1484–90.
8. Schofield R. The relationship between the spleen colony-forming cell and the haemopoietic stem cell. Blood Cells. 1978;4:7–25.
9. Pittenger MF, Martin BJ. Mesenchymal stem cells and their potential as cardiac therapeutics. Circ Res. 2004;95:9–20.
10. Rangappa S, Entwistle JW, Wechsler AS, Kresh JY. Cardiomyocyte-mediated contact programs human mesenchymal stem cells to express cardiogenic phenotype. J Thorac Cardiovasc Surg. 2003;126:124–32.
11. Yoon J, Shim WJ, Ro YM, Lim DS. Transdifferentiation of mesenchymal stem cells into cardiomyocytes by direct cell-to-cell contact with neonatal cardiomyocyte but not adult cardiomyocytes. Ann Hematol. 2005;84:715–21.
12. Wakitani S, Saito T, Caplan AI. Myogenic cells derived from rat bone marrow mesenchymal stem cells exposed to 5-azacytidine. Muscle Nerve. 1995;18:1417–26.
13. LaBarge MA, Blau HM. Biological progression from adult bone marrow to mononucleate muscle stem cell to multinucleate muscle fiber in response to injury. Cell. 2002;111:589–601.

14. Saito T, Dennis JE, Lennon DP, Young RG, Caplan AI. Myogenic expression of mesenchymal stem cells within myotubes of mdx mice in vitro and in vivo. Tissue Eng. 1995;1:327–43.
15. De Bari C, Dell'Accio F, Tylzanowski P, Luyten FP. Multipotent mesenchymal stem cells from adult human synovial membrane. Arthritis Rheum. 2001;44:1928–42.
16. De Bari C, Dell'Accio F, Vandenabeele F, Vermeesch JR, Raymackers JM, Luyten FP. Skeletal muscle repair by adult human mesenchymal stem cells from synovial membrane. J Cell Biol. 2003;160:909–18.
17. Spees JL, Olson SD, Ylostalo J, et al. Differentiation, cell fusion, and nuclear fusion during ex vivo repair of epithelium by human adult stem cells from bone marrow stroma. Proc Natl Acad Sci U S A. 2003;100:2397–402.
18. Wislet-Gendebien S, Hans G, Leprince P, Rigo JM, Moonen G, Rogister B. Plasticity of cultured mesenchymal stem cells: switch from nestin-positive to excitable neuron-like phenotype. Stem Cells. 2005;23:392–402.
19. Zhang J, Niu C, Ye L, et al. Identification of the haematopoietic stem cell niche and control of the niche size. Nature. 2003;425:836–41.
20. Shi S, Gronthos S. Perivascular niche of postnatal mesenchymal stem cells in human bone marrow and dental pulp. J Bone Miner Res. 2003;18:696–704.
21. Kiel MJ, Yilmaz OH, Iwashita T, Yilmaz OH, Terhorst C, Morrison SJ. SLAM family receptors distinguish hematopoietic stem and progenitor cells and reveal endothelial niches for stem cells. Cell. 2005;121:1109–21.
22. Villars F, Guillotin B, Amedee T, et al. Effect of HUVEC on human osteoprogenitor cell differentiation needs heterotypic gap junction communication. Am J Physiol Cell Physiol. 2002;282:C775–85.
23. Kaigler D, Krebsbach PH, West ER, Horger K, Huang YC, Mooney DJ. Endothelial cell modulation of bone marrow stromal cell osteogenic potential. FASEB J. 2005;19:665–7.
24. Villars F, Bordenave L, Bareille R, Amedee J. Effect of human endothelial cells on human bone marrow stromal cell phenotype: role of VEGF? J Cell Biochem. 2000;79:672–85.
25. Hirschi KK, Rohovsky SA, D'Amore PA. PDGF, TGF-beta, and heterotypic cell-cell interactions mediate endothelial cell-induced recruitment of 10T1/2 cells and their differentiation to a smooth muscle fate. J Cell Biol. 1998;141:805–14.
26. Mackay AM, Beck SC, Murphy JM, Barry FP, Chichester CO, Pittenger MF. Chondrogenic differentiation of cultured human mesenchymal stem cells from marrow. Tissue Eng. 1998;4:415–28.
27. Oberlender SA, Tuan RS. Expression and functional involvement of N-cadherin in embryonic limb chondrogenesis. Development. 1994;120:177–87.
28. Tuli R, Tuli S, Nandi S, et al. Characterization of multipotential mesenchymal progenitor cells derived from human trabecular bone. Stem Cells. 2003;21:681–93.
29. Nelson WJ, Nusse R. Convergence of Wnt, beta-catenin, and cadherin pathways. Science. 2004;303:1483–7.
30. Song X, Zhu CH, Doan C, Xie T. Germline stem cells anchored by adherens junctions in the Drosophila ovary niches. Science. 2002;296:1855–7.
31. Yamashita YM, Jones DL, Fuller MT. Orientation of asymmetric stem cell division by the APC tumor suppressor and centrosome. Science. 2003;301:1547–50.
32. Valiunas V, Doronin S, Valiuniene L, et al. Human mesenchymal stem cells make cardiac connexins and form functional gap junctions. J Physiol. 2004;555:617–26.
33. Meury T, Verrier S, Alini M. Human endothelial cells inhibit BMSC differentiation into mature osteoblasts in vitro by interfering with osterix expression. J Cell Biochem. 2006;98:992–1006.
34. Collin-Osdoby P. Role of vascular endothelial cells in bone biology. J Cell Biochem. 1994;55:304–9.
35. Santa Maria L, Rojas CV, Minguell JJ. Signals from damaged but not undamaged skeletal muscle induce myogenic differentiation of rat bone-marrow-derived mesenchymal stem cells. Exp Cell Res. 2004;300:418–26.
36. Gruber R, Karreth F, Kandler B, et al. Platelet-released supernatants increase migration and proliferation, and decrease osteogenic differentiation of bone marrow-derived mesenchymal progenitor cells under in vitro conditions. Platelets. 2004;15:29–35.
37. Gerstenfeld LC, Cruceta J, Shea CM, Sampath K, Barnes GL, Einhorn TA. Chondrocytes provide morphogenic signals that selectively induce osteogenic differentiation of mesenchymal stem cells. J Bone Miner Res. 2002;17:221–30.
38. Etheridge SL, Spencer GJ, Heath DJ, Genever PG. Expression profiling and functional analysis of wnt signaling mechanisms in mesenchymal stem cells. Stem Cells. 2004;22:849–60.
39. Boland GM, Perkins G, Hall DJ, Tuan RS. Wnt 3a promotes proliferation and suppresses osteogenic differentiation of adult human mesenchymal stem cells. J Cell Biochem. 2004;93:1210–30.
40. Baksh D, Boland GM, Tuan RS. Cross-talk between Wnt signaling pathways in human mesenchymal stem cells leads to functional antagonism during osteogenic differentiation. J Cell Biochem. 2007;101:1109–24.
41. Baksh D, Tuan RS. Canonical and non-canonical Wnts differentially affect the development potential of primary isolate of human bone marrow mesenchymal stem cells. J Cell Physiol. 2007;212:817–26.
42. Kortesidis A, Zannettino A, Isenmann S, Shi S, Lapidot T, Gronthos S. Stromal-derived factor-1 promotes the growth, survival, and development of human bone marrow stromal stem cells. Blood. 2005;105:3793–801.
43. Song L, Young NJ, Webb NE, Tuan RS. Origin and characterization of multipotential mesenchymal stem cells derived from adult human trabecular bone. Stem Cells Dev. 2005;14:712–21.
44. Kinnaird T, Stabile E, Burnett MS, et al. Local delivery of marrow-derived stromal cells augments collateral perfusion through paracrine mechanisms. Circulation. 2004;109:1543–9.
45. Mayer H, Bertram H, Lindenmaier W, Korff T, Weber H, Weich H. Vascular endothelial growth factor (VEGF-A) expression in human mesenchymal stem cells: autocrine and paracrine role on osteoblastic and endothelial differentiation. J Cell Biochem. 2005;95:827–39.
46. Deckers MM, Karperien M, van der Bent C, Yamashita T, Papapoulos SE, Lowik CW. Expression of vascular endothelial growth factors and their receptors during osteoblast differentiation. Endocrinology. 2000;141:1667–74.
47. Shum L, Coleman CM, Hatakeyama Y, Tuan RS. Morphogenesis and dysmorphogenesis of the appendicular skeleton. Birth Defects Res C Embryo Today. 2003;69:102–22.
48. Roelen BA, Dijke P. Controlling mesenchymal stem cell differentiation by TGFBeta family members. J Orthop Sci. 2003;8:740–8.
49. Wang D, Park JS, Chu JS, et al. Proteomic profiling of bone marrow mesenchymal stem cells upon transforming growth factor beta1 stimulation. J Biol Chem. 2004;279:43725–34.
50. Granjeiro JM, Oliveira RC, Bustos-Valenzuela JC, Sogayar MC, Taga R. Bone morphogenetic proteins: from structure to clinical use. Braz J Med Biol Res. 2005;38:1463–73.
51. Xie T, Spradling AC. decapentaplegic is essential for the maintenance and division of germline stem cells in the Drosophila ovary. Cell. 1998;94:251–60.

52. Song L, Tuan RS. Transdifferentiation potential of human mesenchymal stem cells derived from bone marrow. FASEB J. 2004;18:980–2.
53. Gronthos S, Chen S, Wang CY, Robey PG, Shi S. Telomerase accelerates osteogenesis of bone marrow stromal stem cells by upregulation of CBFA1, osterix, and osteocalcin. J Bone Miner Res. 2003;18:716–22.
54. Yang Y. Wnts and wing: Wnt signaling in vertebrate limb development and musculoskeletal morphogenesis. Birth Defects Res Part C Embryo Today. 2003;69:305–17.
55. Gregory CA, Singh H, Perry AS, Prockop DJ. The Wnt signaling inhibitor dickkopf-1 is required for reentry into the cell cycle of human adult stem cells from bone marrow. J Biol Chem. 2003;278:28067–78.
56. Martin I, Muraglia A, Campanile G, Cancedda R, Quarto R. Fibroblast growth factor-2 supports ex vivo expansion and maintenance of osteogenic precursors from human bone marrow. Endocrinology. 1997;138:4456–62.
57. Bianchi G, Banfi A, Mastrogiacomo M, et al. Ex vivo enrichment of mesenchymal cell progenitors by fibroblast growth factor 2. Exp Cell Res. 2003;287:98–105.
58. Kortesidis A, Zannettino A, Isenmann S, Shi S, Lapidot T, Gronthos S. Stromal derived factor-1 promotes the growth, survival and development of human bone marrow stromal stem cells. Blood. 2005;105:3793–801.
59. Bendall SC, Stewart MH, Menendez P, et al. IGF and FGF cooperatively establish the regulatory stem cell niche of pluripotent human cells in vitro. Nature. 2007;448:1015–21.
60. Birnbaum RS, Bowsher RR, Wiren KM. Changes in IGF-I and -II expression and secretion during the proliferation and differentiation of normal rat osteoblasts. J Endocrinol. 1995;144:251–9.
61. Grayson WL, Zhao F, Izadpanah R, Bunnell B, Ma T. Effects of hypoxia on human mesenchymal stem cell expansion and plasticity in 3D constructs. J Cell Physiol. 2006;207:331–9.
62. Kinnaird T, Stabile E, Burnett MS, et al. Marrow-derived stromal cells express genes encoding a broad spectrum of arteriogenic cytokines and promote in vitro and in vivo arteriogenesis through paracrine mechanisms. Circ Res. 2004;94:678–85.
63. Heng BC, Cao T, Stanton LW, Robson P, Olsen B. Strategies for directing the differentiation of stem cells into the osteogenic lineage in vitro. J Bone Miner Res. 2004;19:1379–94.
64. Gelse K, Poschl E, Aigner T. Collagens-structure, function, and biosynthesis. Adv Drug Deliv Rev. 2003;55:1531–46.
65. Kalluri R. Basement membranes: structure, assembly and role in tumour angiogenesis. Nat Rev Cancer. 2003;3:422–33.
66. Ruoslahti E. RGD and other recognition sequences for integrins. Annu Rev Cell Dev Biol. 1996;12:697–715.
67. Majumdar MK, Keane-Moore M, Buyaner D, et al. Characterization and functionality of cell surface molecules on human mesenchymal stem cells. J Biomed Sci. 2003;10:228–41.
68. Salasznyk RM, Williams WA, Boskey A, Batorsky A, Plopper GE. Adhesion to vitronectin and collagen I promotes osteogenic differentiation of human mesenchymal stem cells. J Biomed Biotechnol. 2004;2004:24–34.
69. Mizuno M, Fujisawa R, Kuboki Y. Type I collagen-induced osteoblastic differentiation of bone-marrow cells mediated by collagen-alpha2betal integrin interaction. J Cell Physiol. 2000; 184:207–13.
70. Di Lullo GA, Sweeney SM, Korkko J, Ala-Kokko L, San Antonio JD. Mapping the ligand-binding sites and disease-associated mutations on the most abundant protein in the human, type I collagen. J Biol Chem. 2002;277:4223–31.
71. Nguyen H, Qian JJ, Bhatnagar RS, Li S. Enhanced cell attachment and osteoblastic activity by P-15 peptide-coated matrix in hydrogels. Biochem Biophys Res Commun. 2003;311:179–86.
72. Vogel W, Gish GD, Alves F, Pawson T. The discoidin domain receptor tyrosine kinases are activated by collagen. Mol Cell. 1997;1:13–23.
73. Bosnakovski D, Mizuno M, Kim G, Takagi S, Okumura M, Fujinaga T. Chondrogenic differentiation of bovine bone marrow mesenchymal stem cells (MSCs) in different hydrogels: influence of collagen type II extracellular matrix on MSC chondrogenesis. Biotechnol Bioeng. 2006;93:1152–63.
74. Bradham DM, Passaniti A, Horton WE, Jr. Mesenchymal cell chondrogenesis is stimulated by basement membrane matrix and inhibited by age-associated factors. Matrix Biol. 1995;14: 561–71.
75. Matsubara T, Tsutsumi S, Pan H, et al. A new technique to expand human mesenchymal stem cells using basement membrane extracellular matrix. Biochem Biophys Res Commun. 2004;313: 503–8.
76. Qian L, Saltzman WM. Improving the expansion and neuronal differentiation of mesenchymal stem cells through culture surface modification. Biomaterials. 2004;25:1331–7.
77. Hallmann R, Horn N, Selg M, Wendler O, Pausch F, Sorokin LM. Expression and function of laminins in the embryonic and mature vasculature. Physiol Rev. 2005;85:979–1000.
78. Amano S, Scott IC, Takahara K, et al. Bone morphogenetic protein 1 is an extracellular processing enzyme of the laminin 5 gamma 2 chain. J Biol Chem. 2000;275:22728–35.
79. Siler U, Rousselle P, Muller CA, Klein G. Laminin gamma2 chain as a stromal cell marker of the human bone marrow microenvironment. Br J Haematol. 2002;119:212–20.
80. Shang M, Koshikawa N, Schenk S, Quaranta V. The LG3 module of laminin-5 harbors a binding site for integrin alpha3beta1 that promotes cell adhesion, spreading, and migration. J Biol Chem. 2001;276:33045–53.
81. Decline F, Rousselle P. Keratinocyte migration requires alpha2beta1 integrin-mediated interaction with the laminin 5 gamma2 chain. J Cell Sci. 2001;114:811–23.
82. Utani A, Nomizu M, Matsuura H, et al. A unique sequence of the laminin alpha 3 G domain binds to heparin and promotes cell adhesion through syndecan-2 and -4. J Biol Chem. 2001;276: 28779–88.
83. Schenk S, Hintermann E, Bilban M, et al. Binding to EGF receptor of a laminin-5 EGF-like fragment liberated during MMP-dependent mammary gland involution. J Cell Biol. 2003;161: 197–209.
84. Giannelli G, Falk-Marzillier J, Schiraldi O, Stetler-Stevenson WG, Quaranta V. Induction of cell migration by matrix metalloprotease-2 cleavage of laminin-5. Science. 1997;277: 225–8.
85. Krampera M, Pasini A, Rigo A, et al. HB-EGF/HER-1 signaling in bone marrow mesenchymal stem cells: inducing cell expansion and reversibly preventing multilineage differentiation. Blood. 2005;106:59–66.
86. Hashimoto J, Kariya Y, Miyazaki K. Regulation of proliferation and chondrogenic differentiation of human mesenchymal stem cells by laminin-5 (laminin-332). Stem Cells. 2006;24: 2346–54.
87. Klees RF, Salasznyk RM, Vandenberg S, Bennett K, Plopper GE. Laminin-5 activates extracellular matrix production and osteogenic gene focusing in human mesenchymal stem cells. Matrix Biol. 2007;26:106–14.
88. Klees RF, Salasznyk RM, Kingsley K, Williams WA, Boskey A, Plopper GE. Laminin-5 induces osteogenic gene expression in human mesenchymal stem cells through an ERK-dependent pathway. Mol Biol Cell. 2005;16:881–90.
89. Streuli C. Extracellular matrix remodelling and cellular differentiation. Curr Opin Cell Biol. 1999;11:634–40.

90. Jones JI, Gockerman A, Busby WH, Jr., Camacho-Hubner C, Clemmons DR. Extracellular matrix contains insulin-like growth factor binding protein-5: potentiation of the effects of IGF-I. J Cell Biol. 1993;121:679–87.
91. Annes JP, Munger JS, Rifkin DB. Making sense of latent TGFbeta activation. J Cell Sci. 2003;116:217–24.
92. Young MF. Bone matrix proteins: more than markers. Calcif Tissue Int. 2003;72:2–4.
93. Yamaguchi Y, Mann DM, Ruoslahti E. Negative regulation of transforming growth factor-beta by the proteoglycan decorin. Nature. 1990;346:281–4.
94. Bassols A, Massague J. Transforming growth factor beta regulates the expression and structure of extracellular matrix chondroitin/dermatan sulfate proteoglycans. J Biol Chem. 1988;263:3039–45.
95. Chen XD, Fisher LW, Robey PG, et al. The small leucine-rich proteoglycan biglycan modulates BMP-4-induced osteoblast differentiation. Faseb J. 2004;18:948–58.
96. Groppe J, Greenwald J, Wiater E, et al. Structural basis of BMP signaling inhibition by Noggin, a novel twelve-membered cystine knot protein. J Bone Joint Surg Am. 2003;85-A Suppl 3:52–8.
97. Canalis E, Economides AN, Gazzerro E. Bone morphogenetic proteins, their antagonists, and the skeleton. Endocr Rev. 2003;24:218–35.
98. Groppe J, Greenwald J, Wiater E, et al. Structural basis of BMP signalling inhibition by the cystine knot protein Noggin. Nature. 2002;420:636–42.
99. Yanagita M. BMP antagonists: their roles in development and involvement in pathophysiology. Cytokine Growth Factor Rev. 2005;16:309–17.
100. Garcia Abreu J, Coffinier C, Larrain J, Oelgeschlager M, De Robertis EM. Chordin-like CR domains and the regulation of evolutionarily conserved extracellular signaling systems. Gene. 2002;287:39–47.
101. Jasuja R, Allen BL, Pappano WN, Rapraeger AC, Greenspan DS. Cell-surface heparan sulfate proteoglycans potentiate chordin antagonism of bone morphogenetic protein signaling and are necessary for cellular uptake of chordin. J Biol Chem. 2004;279: 51289–97.
102. Paine-Saunders S, Viviano BL, Economides AN, Saunders S. Heparan sulfate proteoglycans retain Noggin at the cell surface: a potential mechanism for shaping bone morphogenetic protein gradients. J Biol Chem. 2002;277:2089–96.
103. Zhu Y, Oganesian A, Keene DR, Sandell LJ. Type IIA procollagen containing the cysteine-rich amino propeptide is deposited in the extracellular matrix of prechondrogenic tissue and binds to TGF-beta1 and BMP-2. J Cell Biol. 1999;144:1069–80.
104. Fukui N, McAlinden A, Zhu Y, et al. Processing of type II procollagen amino propeptide by matrix metalloproteinases. J Biol Chem. 2002;277:2193–201.
105. Martin JA, Buckwalter JA. Effects of fibronectin on articular cartilage chondrocyte proteoglycan synthesis and response to insulin-like growth factor-I. J Orthop Res. 1998;16:752–7.
106. Firth SM, Baxter RC. Cellular actions of the insulin-like growth factor binding proteins. Endocr Rev. 2002;23:824–54.
107. Ruoslahti E, Yamaguchi Y. Proteoglycans as modulators of growth factor activities. Cell. 1991;64:867–9.
108. Aviezer D, Hecht D, Safran M, Eisinger M, David G, Yayon A. Perlecan, basal lamina proteoglycan, promotes basic fibroblast growth factor-receptor binding, mitogenesis, and angiogenesis. Cell. 1994;79:1005–13.
109. Mongiat M, Sweeney SM, San Antonio JD, Fu J, Iozzo RV. Endorepellin, a novel inhibitor of angiogenesis derived from the C terminus of perlecan. J Biol Chem. 2003;278:4238–49.
110. White DG, Hershey HP, Moss JJ, Daniels H, Tuan RS, Bennett VD. Functional analysis of fibronectin isoforms in chondrogenesis: full-length recombinant mesenchymal fibronectin reduces spreading and promotes condensation and chondrogenesis of limb mesenchymal cells. Differentiation. 2003;71: 251–61.
111. Fukai F, Iso T, Sekiguchi K, Miyatake N, Tsugita A, Katayama T. An amino-terminal fibronectin fragment stimulates the differentiation of ST-13 preadipocytes. Biochemistry. 1993;32: 5746–51.
112. Schenk S, Quaranta V. Tales from the crypt[ic] sites of the extracellular matrix. Trends Cell Biol. 2003;13:366–75.
113. Rodenberg EJ, Pavalko FM. Peptides derived from fibronectin type III connecting segments promote endothelial cell adhesion but not platelet adhesion: implications in tissue-engineered vascular grafts. Tissue Eng. 2007;13:2653–66.
114. Homandberg GA, Meyers R, Xie DL. Fibronectin fragments cause chondrolysis of bovine articular cartilage slices in culture. J Biol Chem. 1992;267:3597–604.
115. Limper AH, Quade BJ, LaChance RM, Birkenmeier TM, Rangwala TS, McDonald JA. Cell surface molecules that bind fibronectin's matrix assembly domain. J Biol Chem. 1991;266:9697–702.
116. Ambesi A, Klein RM, Pumiglia KM, McKeown-Longo PJ. Anastellin, a fragment of the first type III repeat of fibronectin, inhibits extracellular signal-regulated kinase and causes G(1) arrest in human microvessel endothelial cells. Cancer Res. 2005;65:148–56.
117. Marneros AG, Olsen BR. The role of collagen-derived proteolytic fragments in angiogenesis. Matrix Biol. 2001;20:337–45.
118. Ramchandran R, Dhanabal M, Volk R, et al. Antiangiogenic activity of restin, NC10 domain of human collagen XV: comparison to endostatin. Biochem Biophys Res Commun. 1999;255:735–9.
119. Ferreras M, Felbor U, Lenhard T, Olsen BR, Delaisse J. Generation and degradation of human endostatin proteins by various proteinases. FEBS Lett. 2000;486:247–51.
120. Sasaki T, Larsson H, Kreuger J, et al. Structural basis and potential role of heparin/heparan sulfate binding to the angiogenesis inhibitor endostatin. EMBO J. 1999;18:6240–8.
121. Gonzalez EM, Reed CC, Bix G, et al. BMP-1/Tolloid-like metalloproteases process endorepellin, the angiostatic C-terminal fragment of perlecan. J Biol Chem. 2005;280:7080–7.
122. Petitclerc E, Boutaud A, Prestayko A, et al. New functions for non-collagenous domains of human collagen type IV. Novel integrin ligands inhibiting angiogenesis and tumor growth in vivo. J Biol Chem. 2000;275:8051–61.
123. Colorado PC, Torre A, Kamphaus G, et al. Anti-angiogenic cues from vascular basement membrane collagen. Cancer Res. 2000;60:2520–6.
124. Magnon C, Galaup A, Mullan B, et al. Canstatin acts on endothelial and tumor cells via mitochondrial damage initiated through interaction with alphavbeta3 and alphavbeta5 integrins. Cancer Res. 2005;65:4353–61.
125. Panka DJ, Mier JW. Canstatin inhibits Akt activation and induces Fas-dependent apoptosis in endothelial cells. J Biol Chem. 2003;278:37632–6.
126. Maeshima Y, Manfredi M, Reimer C, et al. Identification of the anti-angiogenic site within vascular basement membrane-derived tumstatin. J Biol Chem. 2001;276:15240–8.
127. Maeshima Y, Sudhakar A, Lively JC, et al. Tumstatin, an endothelial cell-specific inhibitor of protein synthesis. Science. 2002;295:140–3.
128. Xu J, Rodriguez D, Petitclerc E, et al. Proteolytic exposure of a cryptic site within collagen type IV is required for angiogenesis and tumor growth in vivo. J Cell Biol. 2001;154:1069–79.

129. Paralkar VM, Vukicevic S, Reddi AH. Transforming growth factor beta type 1 binds to collagen IV of basement membrane matrix: implications for development. Dev Biol. 1991;143: 303–8.
130. Paralkar VM, Nandedkar AK, Pointer RH, Kleinman HK, Reddi AH. Interaction of osteogenin, a heparin binding bone morphogenetic protein, with type IV collagen. J Biol Chem. 1990;265:17281–4.
131. Paralkar VM, Weeks BS, Yu YM, Kleinman HK, Reddi AH. Recombinant human bone morphogenetic protein 2B stimulates PC12 cell differentiation: potentiation and binding to type IV collagen. J Cell Biol. 1992;119:1721–8.
132. Bennett VD, Pallante KM, Adams SL. The splicing pattern of fibronectin mRNA changes during chondrogenesis resulting in an unusual form of the mRNA in cartilage. J Biol Chem. 1991;266:5918–24.
133. Han F, Gilbert JR, Harrison G, et al. Transforming growth factor-beta1 regulates fibronectin isoform expression and splicing factor SRp40 expression during ATDC5 chondrogenic maturation. Exp Cell Res. 2007;313:1518–32.
134. Warman ML, Abbott M, Apte SS, et al. A type X collagen mutation causes Schmid metaphyseal chondrodysplasia. Nat Genet. 1993;5:79–82.
135. Iyama K, Ninomiya Y, Olsen BR, Linsenmayer TF, Trelstad RL, Hayashi M. Spatiotemporal pattern of type X collagen gene expression and collagen deposition in embryonic chick vertebrae undergoing endochondral ossification. Anat Rec. 1991;229: 462–72.
136. Neth P, Ries C, Karow M, Egea V, Ilmer M, Jochum M. The Wnt signal transduction pathway in stem cells and cancer cells: influence on cellular invasion. Stem Cell Rev. 2007;3:18–29.
137. Gilbert SF, Tuan RS. New vistas for developmental biology. J Biosci. 2001;26:293–8.
138. Ott HC, Matthiesen TS, Goh S, et al. Perfusion-decellularized matrix: using nature's platform to engineer a bioartificial heart. Nat Med. 2008;14:213–21.
139. Docheva D, Popov C, Mutschler W, Schieker M. Human mesenchymal stem cells in contact with their environment: surface characteristics and the integrin system. J Cell Mol Med. 2007;11:21–38.
140. Hubbell JA, Massia SP, Desai NP, Drumheller PD. Endothelial cell-selective materials for tissue engineering in the vascular graft via a new receptor. Biotechnology (NY). 1991;9: 568–72.
141. Nomizu M, Kim WH, Yamamura K, et al. Identification of cell binding sites in the laminin alpha 1 chain carboxyl-terminal globular domain by systematic screening of synthetic peptides. J Biol Chem. 1995;270:20583–90.
142. Beck K, Hunter I, Engel J. Structure and function of laminin: anatomy of a multidomain glycoprotein. Faseb J. 1990;4:148–60.

Stem Cells, Hypoxia and Hypoxia-Inducible Factors

Suzanne M. Watt, Grigorios Tsaknakis, Sinead P. Forde and Lee Carpenter

Abstract Oxygen is a critical environmental factor that regulates the fate of stem cells. In this review, our aims are twofold: (i) to consider the contribution of oxygen tension to the environmental niches in which stem cells and their progeny find themselves and which have a role in determining their fate, and (ii) to define the regulatory networks that control the response of stem/progenitor cells to hypoxia, particularly those that affect early embryonic stem cell specification and hematopoietic stem cell development and function.

Keywords Hypoxia · Hypoxia-inducible factors · Stem cells · Regulatory networks · Fate determination

1 Introduction

Oxygen homeostasis is critical to all organisms that rely on aerobic respiration, yet there is often a lack of clarity in the published literature regarding the partial pressures of oxygen to which cells and tissues are exposed. This is in part due to the difficulty in accurately measuring or defining normal oxygen tensions and oxygen consumption of individual cells and tissues. These vary with cell type, location in the body or tissue, proliferative state, distance from the blood supply, oxygen diffusion rates through cells and matrices, relative proportions of mitochondria within a cell, exertion rate, the tissue or organ architecture, and the stage of development or differentiation. It is therefore evident that cells will have oxygen-sensing mechanisms to detect changes in oxygen tensions. One mechanism that has been identified is the oxygen-dependent hydroxylation of amino acids on specific α subunits of the hypoxia-inducible transcription factors (HIFs). This post-translational modification controls the stability and activity of these HIF-α protein subunits, eventually allowing their targeting for proteosomal degradation. In hypoxic conditions as oxygen levels drop below ~8–10% and this hydroxylation is inhibited, the HIF-α subunits are stabilized, bind to hypoxia response elements (HRE) in target genes, and, in association with transcriptional co-activators, turn these genes on. Genes activated by HIFs include those encoding proteins involved in cell proliferation, self-renewal, survival, energy metabolism, motility, and invasion. In the following sections, we will focus principally on organogenesis and the development of hematopoiesis, examining the molecular mechanisms that regulate the stem cell response to hypoxia in specific microenvironmental niches and under such stress conditions as transplantation. The majority of studies described relate to human and murine systems.

2 Defining Normoxia and Hypoxia

Oxygen is the terminal electron acceptor for the electron transport chain in mitochondria during aerobic respiration. It is essential for the conversion of ADP to ATP during this process and hence for cell metabolism, which is inhibited at ~3 mmHg (~0.5% O_2) oxygen partial pressure (reviewed in [1] and references therein). Hypoxia is strictly defined as a reduction in normal oxygen partial pressures for the body as a whole or for specific tissues or cells. This may occur under certain physiological and pathological conditions or stresses, which include exposure to low oxygen environments due to high altitudes or obstructed airways (hypoxic hypoxia), anemia, ischemia, infection, tumor growth, histoxic hypoxia (e.g., cyanide poisoning), and tissue injury or remodeling. Oxygen tensions in the atmosphere are ~159 mmHg (~21%) at sea level, while at 10,000 ft. above sea level these decrease to ~60 mmHg (~8%). Very few tissues or cells in mammals are exposed to atmospheric oxygen levels, exceptions being the outer layers of the skin and of parts of the gastrointestinal and respiratory

S.M. Watt (✉)
Stem Cells and Immunotherapies, NHS Blood and Transplant, John Radcliffe Hospital, Headington, Oxford, OX3 9DU, UK
e-mail: suzanne.watt@nbs.nhs.uk

V.K. Rajasekhar, M.C. Vemuri (eds.), *Regulatory Networks in Stem Cells*, Stem Cell Biology and Regenerative Medicine,
DOI 10.1007/978-1-60327-227-8_18,

tracts, and the cornea of the eye. In alveolar air, oxygen tensions are reported to be on average ~100 mmHg (~13.2%) and these drop to ~95 mmHg (~12.6%) in arterial blood and to ~40 mmHg (~5.3%) in venous blood (reviewed in [1]). This will vary at rest, with exertion, and with a number of other factors including the partial pressure of atmospheric oxygen. The natural or physiological oxygenated environment for most mammalian tissues and cells not exposed to atmospheric oxygen is thought to range from ~1 to ~13%, levels that may be considered hypoxic in comparison with atmospheric oxygen levels at sea level (reviewed in [1, 2]). Throughout this review, we will follow convention and refer to these tissue oxygen tensions as "physiological hypoxia" and normal atmospheric oxygen partial pressures at sea level as "normoxic." Hypoxia can develop locally when oxygen demand exceeds supply in pathological conditions such as cancers, ischemia, infections, and inflammation. The outcome can be the stimulation of neovascularization, increased neutrophil and macrophage survival, and increased neutrophil diapedesis to sites of inflammation. This "pathological hypoxia" has been well described recently by others and will not be described in this review (reviewed in [3–7]).

3 Physiological Hypoxia In Vivo in the Early Embryo Prior to Implantation

Embryogenesis and organogenesis in mammals take place in environments that are physiologically hypoxic. Such an environment prevails in the female reproductive organs and tract, where oocytes and the developing embryo may be exposed to oxygen levels ranging from ~1 to 5.5% [1, 8–18]. Prior to implantation and in the hypoxic environment of the female reproductive tract, the zygote undergoes cleavage. By the four-cell stage, at least in mice, epigenetic differences between blastomeres occur and this affects their fate. Blastomeres with a high content of arginine methylation on histone H3 contribute to polar trophectoderm and the inner cell mass, and those with low levels of arginine methylation of histone H3 contribute to mural trophectoderm [19], although this is also dependent on the order of cleavage divisions and the orientation of the cleaved cells in relation to the animal and vegetal axes of the egg [20, 21] Overt differentiation of the embryo occurs with compaction of the morula and the development of the blastocyst. The resulting segregation of cells into the inner cell mass (which gives rise to the embryo proper) and the trophectoderm (which surrounds the inner cell mass and fluid filled blastocoel and gives rise to the placental tissues) unfolds again under physiological hypoxia.

4 The Effects of Hypoxia on Embryonic Stem Cells Derived In Vitro from the Preimplantation Blastocyst

Embryonic stem cells (ES) can be derived from the inner cell mass prior to blastocyst implantation. Since the blastocyst develops in a physiological hypoxic environment, it is not surprising that hypoxic environments have been shown to be crucial for optimal early embryonic development in vitro in a variety of species. In fact, several studies have used oxygen tensions of ~5% to demonstrate these effects [22–30]. Michalska et al. [31] have more recently reported that ~1–4% oxygen levels contribute to the maintenance of human embryonic stem (hES) cells in an undifferentiated and pluripotent state in vitro, while normoxic conditions (21% O_2) promote their differentiation. Furthermore, low oxygen concentrations (5% O_2) have also been reported to increase hES cell cloning efficiency.

Hypoxia may also contribute to the prevention of in vitro chromosomal instability observed in culture adapted hES cells over time by reducing microenvironmental stress [32].

5 Hypoxia, Implantation and Placentation

Placentation also takes place in a physiologically hypoxic environment. It is initiated after implantation of the hatched blastocyst (at ~5–6 days postfertilization in the human) into the uterine endometrium, the surface of which is reported to have an oxygen partial pressure of ~17.9 mmHg (~2.5% O_2) (reviewed in [1]; [11, 18, 33, 34]). The placental villi that subsequently form contain a mononuclear cytotrophoblast stem cell layer surrounding a mesenchymal connective tissue core overlaid with the syncytiotrophoblast layer. The villous cytotrophoblast stem cells can either give rise to extravillous trophoblasts, which can differentiate into invasive trophoblasts, or fuse to form the syncytiotrophoblast again under physiological hypoxic conditions. The invasive extravillous trophoblasts invade and modify the maternal spiral arteries, allowing them to respond to the need for an increased blood supply in the second and third trimesters of pregnancy. In early pregnancy, it is generally accepted that the invading extravillous trophoblasts form plugs that block the maternal spiral arteries and prevent the flow of maternal blood through the intervillous space, hence maintaining the hypoxic environment. By about week 10 of gestation in the human (reviewed in [18] and references therein), maternal blood flow to the fetus begins as the extravillous trophoblast plugs loosen. Before 11 weeks of gestation, the oxygen partial pressure of the developing placenta is ~17.9 mmHg (~2.5% O_2), approximately doubling to ~39.6

mmHg by ~12–13 weeks of gestation, the level found in the decidua ([11, 33]; reviewed in [18]). The oxygen partial pressures found in the fetal arterial and venous blood once the placental blood system becomes functional are reported to rarely exceed 30–40 mmHg (see review in [1]). In vitro studies on the effects of hypoxia in enhancing trophoblast proliferation versus invasion are contradictory, with the experimental evidence being recently well reviewed [1, 18].

6 Hematopoietic Development During Early Embryonic Development in Hypoxic Environments

Although recently challenged [35], it has generally been accepted over the past decade that hematopoiesis first develops extra-embryonically in the yolk sac blood islands and then, independently, in the embryo proper from splanchnopleural mesoderm (reviewed in [36 , 37]).

In the human, hematopoiesis in the blood islands of the yolk sac generates mainly nucleated erythroid and to a lesser extent other myeloid cells (macrophages and primitive megakaryocytes). It begins at approximately day 16 of gestation and completely disappears by day 60 (reviewed in [36–40]).

Experimental evidence initially demonstrated that hematopoiesis originating in the yolk sac provided precursors for fetal and adult hematopoiesis [41]. This was supplanted by the view that (i) hematopoiesis develops independently within the embryo from the splanchnopleural mesoderm, which becomes hemogenic by day 19 of gestation, 3 days before the onset of the circulation, and (ii) that the potential of the human embryo proper to produce blood cells capable of generating both erythromyeloid and lymphoid lineages is not realized until approximately ~day 27 of gestation. At this time, HSC/HPC are found associated with the ventral floor of the dorsal aorta and the vitelline artery (termed hematopoietic intra-aortic clusters; HIACs) and in subaortic patches (SAPs) and may derive from $BB9^{+}$ hemogenic endothelium [42]. This activity then ceases by ~day 40 of gestation (reviewed in [36, 37, 43]). From these sites, HSC/HPC or their progeny migrate to and seed other organs, such as the fetal liver, thymus, spleen and eventually the bone marrow (reviewed in [36, 37, 42]). Recent studies in mice have also suggested that (i) hematopoietic cells colonising other organs in the embryo may be in part derived from the yolk sac [35] and that (ii) the allantois/chorion/placenta may act as an additional and perhaps independent extra-embryonic niche of HSC/HPC generation or accumulation during these early stages of development (reviewed in [37, 44, 45]).

Until this is resolved, current indications are that a first wave of hepatic hematopoiesis, which is characterized by the presence of erythromyeloid cells within developing sinusoids, is observed in the human fetal liver at around 23 day of gestation, following their putative migration from the yolk sac (reviewed in [36 , 37]). This precedes a second wave of hepatic colonization by HSC/HPC at about day 30–32. Here, hematopoiesis predominates until about month 7 of gestation, although it can persist in the liver until shortly after birth (reviewed in [36, 40]). Splenic hematopoiesis occurs mainly between the 3rd and 6th months of gestational age. Lymphopoiesis, which finds its full expression in such specific lymphoid organs, is reviewed elsewhere [37, 46] and will not be discussed further in this review. The first bone marrow to form in the human is in the clavicles, which are the first bones to ossify (reviewed in [40]). Bone marrow hematopoiesis begins by ~week 10.5 of intrauterine development following migration of cells from the fetal liver. It is preceded, at least in developing long bones, by the appearance of $CD68^{+}$ cells at the chondrolysis stage during formation/remodeling of the bone rudiment, and then by the formation of a vascular bed. The latter is associated with a period of granulopoiesis [47], implicating vascular niches in this process. In the young adult human, the bone marrow is largely confined to the skull, vertebrae, ribs, clavicles, sternum, pelvis, and the proximal ends of the humeri and femora (reviewed in [40]).

As well as hematopoietic precursors, mesenchymal stem cells (MSC) have been found to circulate in the blood during embryonic development, having their highest concentration and potentiality within the first trimester in human fetal blood and gradually declining in the circulation with age [48 , 49]. Interestingly, they can be sourced from human amniotic fluid taken at the time of amniocentesis (17–22 weeks of gestational age) [50 , 51]. Several studies (reviewed in [52]) suggest that MSCs originate from a subset of perivascular cells or pericytes. In the postnatal bone marrow, these cells develop into osteoblasts/stromal reticular cells, and along with osteoclasts and vascular endothelial cells, are organized into stem cell niches, which regulate HSC/HPC positioning and fate (reviewed in [53]).

Thus, it is apparent that in vivo (i) hematopoietic development occurs at a variety of transient sites during ontogeny, (ii) HSC/HPC or their progeny circulate and migrate in a temporal manner during development to these sites before finally colonizing the developing bone marrow, (iii) multiple sites of origin for definitive hematopoiesis remain a matter for debate, and (iv) the colonizing cells in the fetus, placenta, and amniotic fluid are likely to be exposed to oxygen partial pressures that do not exceed ~30–40 mmHg (4–5%), with each of these hematopoietic organs or regions providing specific inductive microenvironments for the production of blood cells or maintenance of HSC/HPC [54].

7 Hypoxic Regions Within the Postnatal Bone Marrow Hematopoietic Stem Cell Niche

The bone marrow was first postulated as the major site of hematopoiesis in the adult human by Neumann and Bizzozero more than a century ago (reviewed in [40]; [55]). Since then, the bone marrow has been demonstrated experimentally to be the main postnatal residence of the human HSC, with limited numbers of these cells circulating in the peripheral blood and exiting and entering the bone marrow in adult life. A switch in the cell cycle status of hematopoietic stem/progenitor cells (HSC/HPC) has been proposed to occur in the bone marrow within the first few weeks to years post natally, depending on the species studied. In mice, HSCs are actively cycling until ~3 weeks after birth, when they abruptly switch, within a further week, to become a predominantly quiescent population within the bone marrow [56 , 57].

This highlights the importance of both intrinsic stem cell programming and the microenvironmental influences of the stem cell niche, as well as the changes that can occur in its regulation of hematopoietic stem cell function during ontogeny.

8 The Postnatal Bone Marrow Hematopoietic Stem Cell Niches

It is generally accepted that, within the post-natal bone marrow compartment, HSC/HPC reside in microenvironmental niches either in close proximity to the endosteal bone surface in the "osteoblastic niche" or near the bone marrow sinusoids in the "vascular niche" (reviewed in [53, 58]). Indeed, it has been suggested that these niches restrict or control hematopoietic stem cell numbers, thereby protecting the organism from the excessive production of these cells that occur in malignancies (reviewed in [59]). Despite this, the cellular composition, anatomical location and importance of these niches are not completely resolved and remain a matter of some debate.

The endosteal niche as a site for primitive hematopoiesis was first proposed by Maloney and Patt [60] in 1969. These studies were developed further by others, who proposed that HSC/HPC resided close to bony trabeculae in association with osteoblasts and osteoclasts (reviewed in [53]). Although not fully supported by osteoblast ablation studies, it has been hypothesized that HSC/HPC residing in the "osteoblastic niches" are maintained in an immature quiescent state, while the cells in the vascular niche lack the factors sustaining quiescence and instead proliferate and differentiate. The quiescence-inducing factors are not entirely understood, although the bone morphogenetic proteins (BMPs), Notch ligands, Ang-1/Tie-2, parathyroid hormone-related protein (PTHrP), collagen I, fibronectin, and/or high Ca^{++} levels for example have been suggested to play a role (reviewed in [59]). Other recent studies in mice have placed many of the phenotypically identifiable murine $CD150^{+}CD48^{+}$ HSC subset in close association with the sinusoidal vasculature [61]. Live cell imaging of murine calvaria in vivo, gene marking, and the in vivo transplantation studies suggest that different HSC subsets may either lodge within vascular niches or engraft closer to the bone, with both being associated with osteoblast/stromal reticular cells [59, 62]. Whether the bone marrow hematopoietic stem cell niche consists principally of osteoblasts/stromal reticular cells associated with vasculature or includes osteoblasts and osteoclasts associated with bone remains to be resolved, but there is experimental evidence in the mouse to demonstrate that, in both niches, HSC are found closely associated with stromal reticular (CAR) cells that abundantly express the hypoxia-regulated gene, CXCL12, and that this cell type plays a central role in HSC localization and retention in the bone marrow (reviewed in [53]; [63]).

9 Oxygen Tensions in HSC/HPC Bone Marrow Niches

Low oxygen partial pressures are a physiologically important component of the microenvironmental niches occupied by stem cells and their progeny (reviewed in [6, 64, 65] and references therein). It is believed that low oxygen concentrations in specific niches protect HSC and other stem cells from the potential damaging effects of oxygen free radicals [66]. Cells must either be able to resist these physiological hypoxic environments or, when necessary, respond rapidly to sudden changes in oxygen levels in order to survive and maintain energy-dependent activities. Therefore, depending on their type and stage of differentiation, cells have adaptive mechanisms for responding to hypoxia and reoxygenation. These may be global responses in the body's ability to enhance oxygenation of tissues, such as by vasodilation, enhanced blood vessel formation, and enhanced erythropoiesis, or cell type–specific responses to local hypoxic stimuli, that is, increased glycolysis brought about by the action of HIF-α and discussed below (reviewed in [5, 67]).

Although the bone marrow compartment has been identified as one of the few naturally hypoxic tissues in the healthy adult mammal, measuring the oxygen levels of the bone marrow in vivo has proven difficult (reviewed in [6]). Cells consume oxygen at different rates and may respond differently to changes in oxygen tensions (reviewed in [2, 66, 68–70]). The Kroghian model, which predicts oxygen distribution in solid tissues (see [66]) has been used to estimate the average steady state oxygen consumption of normal murine bone marrow to be of the order of 1.64 μmole per femur per hour [66, 71]. However, the bone marrow contains multiple stem, precursor and maturing cell types, which, like HSC, are spatially and dynamically distributed within specific niches [66, 72]. It has been predicted that oxygen can diffuse 100–150 μm from the blood supply at least into cancerous tissues [73] and that oxygen tensions in tissues reduce in proportion to the distance from the blood supply and the composition both in terms of cell type and numbers and extracellular matrix components of the intervascular space (see discussions in [66, 68] and references therein). It thus follows that oxygen consumption will depend on the stem/progenitor cell type, its associated microenvironmental niche, and whether such cells are quiescent, proliferating or at a particular stage of differentiation. Indeed, it has been proposed that HSCs are maintained in an undifferentiated state in the more hypoxic niches in the bone marrow and that the expansion and differentiation of the HSC/HPC occurs along an oxygen gradient towards the sinusoids [66]. The distribution of the vasculature and the anatomical location of HSC niches in the bone marrow compartment are therefore important considerations (reviewed in [53]).

Mathematical modeling of pO_2 distributions in bone marrow [66] and gas analyses of bone marrow aspirates taken from healthy human volunteers [74] and patients with chronic pulmonary disease [75] have suggested bone marrow oxygen partial pressures are equivalent to between ~1–6%. More recently, experiments using chemical markers for hypoxia have allowed the in vivo identification of these areas within murine bone marrow. When administered in vivo, pimonidazole, a 2-nitroimidazole compound, forms stable adducts in hypoxic regions ($\leq$10 mmHg; $\leq$1.3% O_2). Using this compound, Parmar et al. [76] found that murine bone marrow cells in the high Hoechst 33,342 dye efflux side population that were enriched for HSC could bind pimonidazole at much higher levels than other bone marrow cells. HSC from the side population were also selectively depleted following in vivo administration of tirapazamine, a hypoxia-associated cytotoxic compound that once reduced in hypoxic conditions, leads to double-stranded breaks in DNA. Jang and Sharkis [77] have additionally enriched more primitive and quiescent murine HSCs resident in low oxygen tension bone marrow niches and expressing TERT, Bcrp, p53, p21, N-cadherin, CaR, and Notch-1 based on differential reactive oxygen species (ROS) activity. Interestingly, raised ROS levels due to oxidative stress that follows deficiencies in the Atm cell cycle regulator, or in FoxO1, FoxO2, and FoxO3 forkhead transcription factors have been shown to result in defective murine hematopoiesis [78–80].

In addition, a direct analysis of hypoxic regions in the bone marrow of mice has demonstrated high levels of pimonidazole staining along the endosteum which decreased rapidly within the first 50 μm of the bone marrow stroma towards the centre of the bone marrow compartment [71] suggesting that certain regions of the bone marrow are very hypoxic. The release of HSC/HPC into the circulation and homing back to the bone marrow is recapitulated and enhanced under conditions of hematological stress and transplantation, both of which can generate n hypoxic bone marrow environment. In response to stress or administration of the mobilizing agents granulocyte-colony stimulating factor (G-CSF), cyclophosphamide, or the CXCR4 antagonist AMD3100, HSC/HPC egress from the bone marrow into the peripheral circulation (reviewed in [53, 81]). Following G-CSF administration into mice, larger areas of hypoxia were shown to be visible throughout the bone marrow as a result of massive cell proliferation [71].

The result was an increase in the synthesis of VEGF-A throughout the bone marrow, but particularly at the bone marrow endothelial sinuses, with a concomitant increase in vascular permeability and cell egress into the peripheral blood. Interestingly, exposure to hypoxic conditions also allows mesenchymal stem cells and endothelial progenitors to mobilize into the peripheral blood [82–85].

These data support the view that (i) at least murine HSC/HPC with the characteristics described above are organized along an oxygen gradient within distinct bone marrow microenvironmental niches, (ii) those HSC/HPC associated with the endosteal niche are likely to experience lower oxygen tensions than those associated with vascular niches, (iii) the oxygen gradient ranges from the equivalent of ~1 to $\geq$6% oxygen, (iv) the bone marrow microenvironment becomes significantly more hypoxic under such stress conditions as those experienced with both low atmospheric oxygen tensions and following administration of mobilizing agents, and (v) the hypoxic environment may maintain HSCs in a quiescent state providing protection from exposure to oxygen free radicals.

10 The Response of HSC/HPC to Hypoxia In Vitro

Oxygen tensions affect HSC/HPC function in vitro, particularly in terms of cell survival and proliferation or "self-renewal" as opposed to differentiation. Hematopoietic precursors subjected to hypoxic conditions in vitro have been analyzed using specific clonogenic/proliferation (CFU, pre-CFU) assays or transplanted into mice in order to examine their repopulating potential. The original observations revealed that the production of myeloid progenitors in mice was enhanced in 5% O_2 (38 mmHg pO_2) and that more immature hematopoietic progenitors were more resistant to chronic exposure to low oxygen tensions (equivalent to 1% O_2 for 5 days) (see [66, 86] and references therein). More recently, both acute (1.5% O_2 for 24 h) and chronic (3% O_2 for 7 days) hypoxia have been found to enhance CFU or pre-CFU growth of human umbilical cord blood HPC, while limited expansion of human bone marrow CFU-G and SCID repopulating cells (4–6 fold SRC) has been observed in a continuous 1.5% oxygen environment for 4 days [3, 87–89].

In this latter study, the G_2/M fraction of cycling cells was reduced in hypoxia, while a much higher fraction of the G_0/G_1 peak was actively cycling. Hermitte et al. [90] subsequently demonstrated that very low oxygen levels (0.1%) favor the return of dividing human umbilical cord blood $CD34^+$ cells to G_0. More recent studies by Dao et al. [91] illustrate that the differential effects of low O_2 tensions in vitro may also depend in part on specific factors within culture media and hence on other microenvironmental influences.

Taken together, these studies suggest that the response of HSC/HPC to low oxygen tensions in vitro is variable and depends on the oxygen tension, acute or chronic exposure to low oxygen tensions, and the composition of the culture media. It would appear that overall, the more immature hematopoietic progenitors are more resistant to low oxygen tensions and that, if using strictly defined conditions, culturing HSC/HPC in hypoxia (<6% O_2) in vitro can maintain or enhance their clonogenic and reconstituting potential. The effects of chronic and acute hypoxic exposure of other stem/progenitor cells from hematopoietic tissues (e.g., MSCs, endothelial progenitors) in vitro in terms of cell survival, proliferation, differentiation, migration, and engraftment have also been assessed to varying degrees and are described elsewhere [81, 89, 92–94].

11 Molecular Networks that Regulate the Response to Hypoxia: The Hypoxia-Inducible Transcription Factors (HIFs)

There are multiple pathways that allow cells to adapt or respond to hypoxia. This review will concentrate exclusively on those involving the hypoxia inducible factors (HIFs), a family of transcription factors that act as master regulators for oxygen homeostasis within the cell (reviewed in [4, 5, 68, 70, 95]).

These HIFs are members of the basic helix-loop-helix (bHLH)-Per Arnt Sim (PAS; periodic circadian protein-aryl hydrocarbon receptor-aryl hydrocarbon receptor nuclear translocator-single minded protein) family of transcription factors. Three HIF-α subunits have been identified, HIF-1α, HIF-2α, and HIF-3α (reviewed in [4, 70, 96, 97]). The specific roles of each of these molecules are not fully understood and this remains a major research challenge in this field. HIFs can exist as heterodimers comprising the HIF-α subunit and a constitutively expressed nuclear HIF-β (ARNT) subunit. HIF-1α, HIF-2α, and certain HIF-3α isoforms can dimerize with HIF-1α (ARNT), with a much more limited ability to bind to ARNT2 (reviewed in [96, 98] and references therein). A third structurally similar ARNT molecule, ARNT3, has also been described (reviewed in [96]). Compared with the other HIF-α paralogs, less is known about HIF-3α, although it exists as multiple isoforms (HIF-3α1 to 3α6) and at least one of its isoforms functions as an inhibitor of transcription, the so-called HIF-3α2 or inhibitory PAS (IPAS) domain protein (reviewed in [96]; [99]).

12 HIF-α Subunit Structure-Function Relationships

Comparative structures of the HIF-1α HIF-2α, and the longest HIF-3α isoform, HIF-3α1, are illustrated in Fig. 1. All three HIF-α subunits shown contain a bHLH domain, the function of which is to mediate DNA binding and HIF-β dimerization [100], a PAS domain that acts as a second dimerization interface, a von Hippel-Lindau VHL-targeted ODD (oxygen dependent degradation) domain that is essential for proteolytic regulation of HIF-α, and an N-terminal transactivation domain (N-TAD). HIF-1α and HIF-2α, but none of the HIF-3α splice variants, contain a C-terminal transactivation domain (C-TAD), which is responsible for transcriptional co-activator binding [98, 101, 102]. Notably, HIF-1α, HIF-2α, and all HIF-3α splice variants (with the exception of HIF-3α1) lack the LZIP domain, while the HIF-3α3, 5, and 6 splice variants lack the bHLH domain and HIF-3α4, 5, and 6 lack the ODD/N-TAD domains (reviewed in [96]). In comparison, HIF-1β contains both the bHLH and PAS domains, but not the ODD/N-TAD/C-TAD. The HIF-1α and HIF-2α subunits share 48% amino acid sequence identity with each other, and substantially less with the HIF-3α1 variant.

13 The Tissue and Cell Distribution of HIF-α Subunits. Are HIF-1α and HIF-2α Expressed in Stem Cells?

It is self-evident that the activity of HIF-α subunits will depend not only on their cell and tissue distribution, but also on the distribution and activity of the HIF interacting partners.

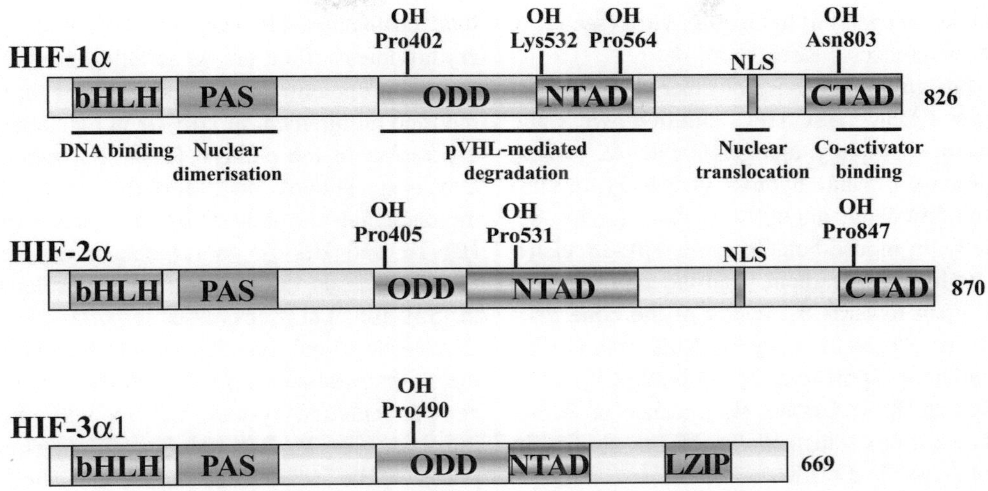

Fig. 1 Schematic representation of structural domains of HIF-1α, HIF-2α and HIF-3α1 subunits. Functional domains are illustrated as boxes of various shades of grey bHLH, basic helix-loop-helix; PAS, Per/Arnt/Sim; N-TAD, N-terminal transactivation domain; C-TAD, C-terminal transactivation domain; ODD, oxygen-dependent degradation domain; NLS, nuclear localization signal; LZIP, leucine zipper. The total number of amino acids of each subunit is marked at the end of the domain structure

Despite this, the detailed distribution of these molecules has not been completely determined. Original tissue screening studies demonstrated that HIF-1α mRNA was expressed ubiquitously throughout development, with the exception of blood leukocytes. In the mouse, HIF-1α mRNA was found to be expressed highly in early heart, brain spinal cord, trigeminal ganglion, branchial arches, hepatic primordia, and primitive gut of the E9.5 embryo [103].

At E13.5, HIF-1α mRNA expression was further reported in the primitive pituitary, neural and olfactory epithelium, developing thymus, and ventricular wall of the heart and kidneys. In later stages of development, expression remained ubiquitous, but at lower levels than observed earlier.

HIF-2α mRNA was found to be highly expressed in vascular endothelial cells in the mesenchyme of the lung, neural crest derivatives during embryonic development, bone marrow macrophages, kidney interstitial cells, cardiac myocytes, and liver parenchymal cells (reviewed in [95]; [103–107]). Compared with the other HIF-α homologues, less is known about HIF-3α mRNA distributions (reviewed in [4, 97]), although, probing with exons 3–6, which are present in all HIF-3α1-6 isoforms, has revealed strong mRNA expression pattern in human heart, placenta and skeletal muscle tissues and weaker mRNA expression in lung, liver, and kidney [99].

HIF-3α2 or IPAS mRNA has also been observed to be highly expressed in murine corneal epithelium and to a lesser extent the Purkinje cells of the cerebellum [108 , 109].

Since HIF-1α and HIF-2α are both stabilized by hypoxia at the post-translational protein level, the tissue and cell distributions of these proteins, rather than their mRNAs, are likely to provide more important information. In 2000, Talks et al. [110] analyzed selected adult human tissues and cancers for HIF-1α and HIF-2α protein expression and demonstrated that neither protein was detectable in the normal human tissues screened with the exception of bone marrow macrophages, which were HIF-2α positive. This is in contrast to studies on murine HIF-1α, where Stroka et al. [111] demonstrated HIF-1α protein expression in certain tissues (brain, kidney, liver, and heart) and its weak induction upon exposure of animals to atmospheric hypoxia. Subsequently, Wiesener et al. [107] examined HIF-2α protein expression in perfused tissues of rats before and after exposure to low atmospheric oxygen tensions (6 h with 8% O_2) and showed that HIF-2α was not expressed under normoxic conditions but was widely upregulated in all normal tissues investigated. The level of up-regulation was found to vary among tissues and with the time and degree of hypoxic exposure. In some tissues such as the liver and kidney, HIF-2α expression was clearly associated with tissue oxygen gradients.

Comparing these in vivo rodent studies [107, 111] suggests similarities and differences in the regulation of these two proteins, at least in those tissues examined, in response to atmospheric hypoxia, viz. (i) HIF-2α protein persisted longer than HIF-1α, (ii) HIF-2α was induced at higher oxygen tensions than HIF-1α, (iii) HIF-2α, but not HIF-1α, was induced in the lung (in 6% O_2), (iv) HIF-1α and HIF-2α were both induced in myocardial endothelial cells, cardiomyocytes and hepatocytes, and (v) while HIF-1α was induced in neuronal cells and renal tubular epithelial, HIF-2α was found in renal glomerular and peritubular endothelial cells and fibroblasts and in brain endothelial and glial cells.

As described earlier, embryonic and placental derived stem/progenitor cells as well as stem/progenitor cells from hematopoietic tissues respond to hypoxia in vitro (see [1, 6, 18, 112–116] and references therein).

Studies from our group and those of others indicate that human CD133$^+$ or CD34$^+$ HSC/HPC, cultured MSCs and endothelial progenitors rapidly upregulate HIF-1α protein expression in response to acute hypoxia (1.5% O_2) in vitro (G Tsaknakis and SM Watt, unpublished data; [3, 88, 89, 113]). Additionally, in mouse bone marrow, HIF-1α distribution follows the hypoxic gradient from the endosteum to approximately 50 μm towards the center of the bone marrow in mice and this expression is upregulated with G-CSF administration and hematopoietic cell proliferation [71].

Further studies on the functional significance of HIF-α subunits have come from transgenic/knock-out studies in mice and these are described below.

14 HIF Deficiency In Vivo and In Vitro

The importance of HIFs and the functional effects of the HIF subunits have been studied by examining the effects of HIF deficiency. In the mouse, disruption of the HIF-1α and HIF-2α loci results in severe embryonic malformations, with a range of defects suggesting functionally distinct roles while retaining some degree of overlap, and most often lethality at around E9.5-E11. HIF-1α null murine embryos show morphological defects by E8.0 with a lack of cephalic vascularization, neural fold defects, and reduced number of somites [114].

These defects progress to lethality by E10.5 with death ensuing from cardiac and vascular malformation, failure of neural tube closure, and cell death in the cephalic mesenchyme [115]. Knock-out studies for HIF-2α indicate a role in placental morphogenesis and trophoblast lineage determination [116], reduced hematopoiesis [117 , 118], and other less severe phenotypes (depending on mouse background), such as impaired lung development in perinatal mice [106]. Multiorgan failure and mitochondrial dysfunction can also result in embryonic and postnatal lethality in HIF-2α–deficient mice [117].

The few viable adults generated from HIF-2α–deficient mice were pancytopenic and had hypocellular bone marrows, with quantitatively less multilineage hematopoietic maturation [117 , 118].

The defect appeared to be (i) within the microenvironmental niche since reciprocal transplantation of HIF-2α–/– bone marrow into wild-type irradiated recipients or of wild-type bone marrow into HIF-2α–/– recipients revealed normal hematopoiesis in the former but not the latter, (ii) VEGF and VEGFR independent, and (iii) dependent on adhesion molecules such as fibronectin and VCAM-1. Further conditional knock-out studies [119] have revealed that, postnatally, HIF-2α, but not HIF-1α, is required for erythropoiesis (but not generally for myelopoiesis), and for erythropoietin production. This is in contrast to the reported requirement of HIF-1α in regulating erythropoietin expression in the embryo [120]. Interestingly, conditional knock-out studies have revealed that HIF-1α in limb mesenchyme is required for early skeletogenesis and that HIF-1α and HIF-2α in osteoblasts couples angiogenesis to osteogenesis during long bone development (reviewed in [52]; [65, 121]), processes important for the subsequent development and functioning of the HSC/HPC niche in these long bones. Using a RAG-2–deficient blastocyst complementation system, HIF-1α was also shown to be important for B, but not T, cell development in mice [122], while conditional knock-out studies of HIF-1α, VHL, and VEGF in mice have revealed that HIF-1α is required for the energy metabolism in myeloid cell-mediated inflammation in a VEGF–independent but VHL- and β2 integrin–dependent fashion, as well as for myeloid cell survival ([123 , 124]; reviewed in [7, 125]). To address the level of redundancy between HIF-1α and HIF-2α and whether HIF-2α could rescue a HIF-1α null mutation, the HIF-2α gene has been knocked-into the HIF-1α locus. Interestingly, the "knock-in" embryos were neither rescued nor displayed a HIF-1α null phenotype, but instead showed resorption commencing at day E3.5 with those surviving embryos showing severe defects and lethality by day E7.5. Recovered embryos showed aberrant patterning of embryonic and extra-embryonic tissues, and elevated yet heterogeneous expression patterns for Oct-4 [126 , 127].

Additionally, the HIF-2α gene knockin (into the HIF-1α locus) day 9 embryoid body in vitro showed severely perturbed hematopoietic differentiation with reduced CFU-GEMM, -GM, -G, -M and -E formation. This activity was restored with Oct-4 shRNA, suggesting a reduced ability for commitment to the hematopoietic lineage [127].

HIF-1β is a common subunit for several transcription factors including HIF-1α, HIF-2α, the Ah receptor (AHR) [128], and singleminded-2 [129] and some of the defects observed with the disruption of HIF-1β can recapitulate phenotypes observed for both HIF-1α and HIF-2α deficiency. Indeed, HIF-1β (ARNT) deficient murine embryos have been reported to display many aspects of both HIF-1α and HIF-2α mutations (i.e., delayed development, cardiovascular defects, reduced yolk sac vascularization and VEGF expression, failure of neural tube closure and lethality from placental defects) at E10.5 [130–132].

In addition, the number of hematopoietic progenitors in the yolk sac was decreased [132 , 133].

In terms of hematopoietic development, hypoxia itself has been used as a method to drive hematopoietic cell fate and increase hematopoietic cell numbers from mES cells by promoting early expression of the mesoderm markers

Brachyury, Flk-1, and Bmp-4, and enhancing hemangioblast specification during early development [112, 132 , 133].

In HIF-1β (ARNT) –/– mES cells, fewer flk-1^{+} and BL-CFCs were found to be produced, although the efficiency of mesoderm formation was similar to their wild-type counterparts. This defect was rescued when ARNT–/– mES cells were co-cultured with wild-type mES cells. This suggests extrinsic factors secreted by cells within the embryoid body feedback to promote hemangioblast formation, with this feedback loop being promoted by hypoxia. VEGF+/– and –/– mES cells produce normal flk-1^{+} cells but fewer BL-CFCs. Additional factors regulated by HIF (other than VEGF) therefore appear to be involved in the early events of hemangioblast commitment.

Thus, HIFs play a role in both the pluripotent stem cell and in hematopoietic cell production and end cell function, but the HIF-α subunits involved appear to have differing roles during embryonic and postnatal development.

15 Mechanisms of HIF Regulation

HIF regulation occurs at three levels: (i) the regulation of HIF stability, (ii) the regulation of HIF functional activity, and (iii) the regulation of the HIF regulators.

The regulation of HIF stability and activity is known to occur principally through its α subunits, which are translated into proteins and then degraded rapidly under oxygen tensions equivalent to >8–10%. The HIF-α subunits, however, accumulate as the level of hypoxia increases. Although the oxygen sensing mechanisms in hypoxia have not been fully determined and remain controversial, it is generally agreed that specific propyl or asparaginyl hydroxylases, which hydroxylate the HIF-α subunits, are key hypoxic/metabolic sensors (reviewed in [95]). Under normoxic conditions, oxygen-dependent propyl hydroxylation of HIF-α subunits leads to their ubiquitination via the von Hippel-Lindau (pVHL) protein-elongin C-E3 ubiquitin ligase complex and their subsequent 26S proteosomal degradation (reviewed in [95]). This rate limiting oxygen-dependent HIF-α hydroxylation reaction has been shown to be linear over a range of physiologically relevant oxygen tensions (reviewed in [68, 95]).

To describe these processes in more detail, the prolyl hydroxylase domain-containing enzymes 1–3 (PHDs), which are dioxygenases and are oxygen, iron, 2-oxoglutarate (2-OG), and ascorbate dependent, hydroxylate specific proline residues within the ODD domains of the HIF-α subunits in normoxic conditions. As shown in Fig. 2, key proline residues in the oxygen-dependent degradation (ODD) domains are hydroxylated on Pro402 and Pro564 for HIF-1α [134], Pro405 and Pro531 for HIF-2α [135 , 136], and Pro490 for HIF-3α1-3 [99]. It is of interest that PHD1-3 hydroxylate Pro564 in HIF-1α, while PHD1 and PHD2 hydroxylate Pro402 (reviewed in [4]). The hydroxylated proline residues interact with the von Hippel-Lindau tumor suppressor protein (pVHL), which is the recognition component of a multiprotein complex comprising elongins B/C, Cul-2, and Rbx1 and which in turn functions as an E3 ubiquitin ligase complex tagging appropriate HIF-α subunits for 26S proteosomal degradation [137]. All three PHD isoforms show distinct patterns of tissue expression and subcellular localization (reviewed in [95]).

Both PHD2 and PHD3, but not PHD1, gene transcription is induced under decreased O_2 tensions by HIF itself, thus constituting a potential negative feedback loop which controls HIF-α degradation upon tissue reoxygenation and allows individual cells to adapt to oxygen thresholds [138–140].

It is PHD2 that is principally responsible for maintaining low HIF-1α levels in those oxygenated cells tested [138], although it has been shown that PHD3 can act as a negative regulator of HIF-1α induction under moderate hypoxic conditions [141].

Appelhoff et al. [142] have reported that PHD3 may have relatively more control over HIF-2α than HIF-1α, since inhibition of PHD2 expression by RNAi had a smaller effect on the induction of HIF-2α levels than on HIF-1α levels. It is of interest that PHD2 knock-out is embryonic lethal in mice, while PHD1 and 3 deficiency is not [143].

Several other factors modulate HIF-α subunit degradation either positively or negatively (reviewed in [2, 4, 70, 96]). Some examples are listed below. The OS-9 protein interacts with both HIF-1α and PHD2, and positively regulates the PHD2 hydroxylation of HIF-1α [144].

Acetylation of the Lys532 in the HIF-1α subunit by the murine acetyltransferase ARD1 was reported to promote its interaction with pVHL and subsequent degradation [145], but this has not been confirmed for human ARD1 [146], perhaps indicating differences between species in HIF-α subunit regulation (reviewed in [4]). SSAT2 acetyl transferase stabilizes the interaction of VHL and Elongin C with HIF-1α and promotes the ubiquitination of the latter [147].

In addition, the Siah1a and 2 E3 ubiquitin ligases negatively regulate the hydroxylation process by targeting PHD1 and 3 for proteosomal degradation in hypoxia (reviewed in [141]), while VHL deubiquitinating enzyme 2 stabilizes HIF-1α by deubiquitinating it [148].

As oxygen levels decrease below 8–10% and become limiting, the activity of these propyl hydroxylases is inhibited and the relevant HIF-α subunits are increasingly stabilized, thereby avoiding proteosomal degradation. Here, the actual mechanisms are a matter of debate. Some researchers (for details see [5, 149] and references therein) suggest that anoxic conditions (0% O_2) inhibit PHD activity,

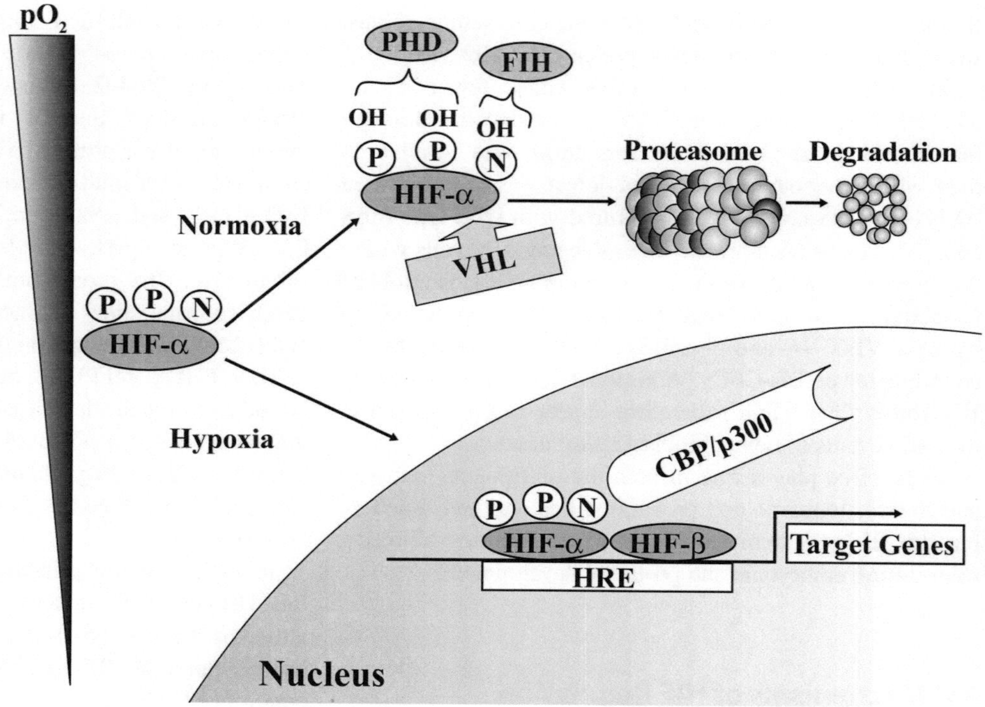

Fig. 2 The mechanism of HIF-1α regulation under normoxia and hypoxia. In normoxia, HIF-1α is rapidly degraded by the hydroxylation of proline residues by proline hydroxylases. This hydroxylated HIF-1α is recognized by pVHL, which tags it for proteasomal degradation. However, in hypoxia, HIF-1α is stablized and translocated to the nucleus, where it dimerizes with the constitutively expressed HIF-1β. This complex binds to target genes through HRE sites with transcriptional co-activators such as p300/CBP. Targeted genes are then expressed

with HIF-α protein stabilization in hypoxic conditions being dependent on ROS (a suggested mechanism being decreased availability of Fe^{++} to the hydoxylases due to enhanced H_2O_2 levels). Others provide substantial evidence that PHD inhibition mimicks the hypoxic response (for details see [68, 150]).

Whatever the exact mechanism, the stabilized HIF-α subunits translocate to the nucleus where they dimerize with a HIF-β subunit, bind to hypoxia-responsive elements (HRE) mostly within promoters of specific genes, and, in the case of HIF-1α and HIF-2α, recruit transcriptional co-factors such as CBP/p300 and p160/steroid receptor coactivator-1 to the C-TAD [67, 151–155]. It is notable that, under normoxic conditions, hydroxylation of a conserved asparagine in the C-TAD of HIF-1α and HIF-2α subunits (Asn803 in HIF-1α and Asn851 in HIF-2α) by factor inhibiting HIF (FIH), an asparaginyl hydroxylase, which, like the other hydroxylases, is an Fe^{2+} dependent and 2-OG-dependent dioxygenase, provides a further level of control. This asparagine hydroxylation abolishes HIF-1α and HIF-2α interactions with transcriptional co-activators, CBP/p300, further inhibiting their function in normoxic conditions. Hypoxia, iron chelators, and transition metals like cobalt inactivate FIH and stimulate the transcription of HIF-1α and -2α target genes [156–158]. Furthermore, hypoxia and HIFs activate histone modifying proteins (HDACs) and hence chromatin accessibility (reviewed in [1]).

The result of HIF-1α and HIF-2α stabilization in hypoxia is the expression of a plethora of genes that help the cell adapt to its hypoxic environment by regulating the self-renewal, survival, motility and metabolism of the cell, the integrity of basement membranes and vasomotor control, and promoting such processes as hematopoiesis and angiogenesis (reviewed in [3, 4, 6, 89, 95, 96]).

In our laboratory, we have examined the transcriptional profiles of genes expressed in human cord blood CD133$^+$ HSC/HPC and their associated bone marrow stromal niche cells after short exposures to acute hypoxia (1.5% O_2) using high-density Affymetrix oligo arrays [3, 89]. A total of 183 genes were differentially regulated by hypoxia in CD133$^+$ cells, with almost half the genes being involved in metabolism, cell proliferation/survival, transcription, signal transduction, and transport. This compared to 45 genes that were differentially regulated by short exposures to hypoxia in bone marrow mesenchymal stem cell-derived stromal cells, with more than half being assigned to signal transduction, cell proliferation/survival, metabolism, transcription, and nucleic acid metabolism. Only a small proportion of the hypoxia-responsive genes were shared between the two cell populations suggesting that the molecular response to low oxygen levels is largely cell type specific. However, although HIF-1α was found to be upregulated in both cell types, the regulation in hypoxia of these known and novel hypoxia-regulated genes by HIF-1α or HIF-2α was not defined. Other array profiling and HIF-α knock-out/knock-in gene strategies, although often carried out in cell lines, indicate that HIF-1α and HIF-2α can activate the transcription of common and unique gene targets (see [5, 159] and references therein). For example, HIF-1α is reported to be involved more heavily with the glycolytic metabolic response and metabolic adaptation to hypoxia although it also up-regulates molecules involved in cell migration and/or apoptotic pathways such CXCR4, BNIP3 and BNIP3L [113, 115, 152, 153, 160], while HIF-2α has been shown to preferentially target such genes as VEGF, VEGFR2, TGF-α, lysyl oxidase, cyclin D1, CITED2, and Oct-4, some of which are involved in cell proliferation, maintaining an undifferentiated state and/or tumorigenesis (reviewed in [5, 152, 161]; [113, 162–165]). This will of course depend on the expression profile of these HIF-α subunits within particular cells. In vitro mES cells lacking HIF-1α show decreased expression of known targets genes such glucose transporters, metabolic enzymes, and VEGF. In other stem/progenitor cells, HIF-1α is, for example, known to regulate Sox9 expression and hence is associated with chondrogenesis from MSC, [65] and also to activate expression of the ABC transporter ABCG2/BCRP-1 [166] which is expressed by primitive HSC and confers a survival advantage on these cells (see [76] and references therein).

Not surprisingly and as discussed earlier, cell cycle regulatory proteins such as the cyclins, cyclin-dependent kinases, and their inhibitors (CKIs) have also been implicated in responses to hypoxia, with many such as cyclin D1 being targets of HIF themselves [164, 167].

Knockout of the CKI family members has been a useful strategy to expand HSC numbers in vitro [168 , 169].

Interestingly, c-Myc, which has been shown to play a role in determining stem cell fate in the HSC osteoblastic niche [58] is repressed by HIF-1α [170], which then induces cell cycle arrest. As suggested earlier, such a mechanism once activated might encourage quiescence of HSC in hypoxia.

A host of HIF regulatory mechanisms other than those described above exists (see examples in Table 1) (reviewed in [1, 2, 4–6, 68, 70, 95, 96, 134, 154, 171]).

Additional molecules such as Trx-1 and Ref-1 promote the interaction of HIF-1α with CBP/p300 through redox-dependent processes (reviewed in [4, 96]; [172]). Downstream mediators of the cellular response to oxidative stress that can regulate HIF-1α (e.g., the FoxO4 transcription factor) [79] have been described elsewhere ([173]; reviewed in [174]).

HIF-1α stability or half-life also depends on HSP90 heat shock protein binding [175], or on the interaction of RACK1 with HIF-1α and then Elongin C, which enhances HIF-1α proteosomal degradation independently of VHL, PHD2, and O_2 (reviewed in [2]; [176]). In addition, other oxygen-independent pathways exist. These include the activation of the phosphatidylinositol 3-kinase (PI3K), protein kinase B (AKT), and mitogen-activated protein kinase (MAPK) pathways, which can result in increased HIF-1α protein synthesis [96, 177–179].

Additional negative feedback loops also regulate HIF-α subunit function. Under hypoxic conditions (6% O_2 for 6 h), HIF-3α2 or IPAS transcription can be up-regulated by interaction of HIF-1α with hypoxia-response elements (HRE) within the IPAS promoter and potentially augment IPAS splicing [98]. The subsequent interaction of IPAS with the amino-terminal region of HIF-α subunits prevents their DNA binding, thus acting as a dominant negative inhibitor of HIF-α activity [98, 109 , 180].

Table 1 Examples of mechanisms/molecules regulating HIF-α stability and/or transcriptional activity

- Oxygen/hypoxia/reactive oxygen species
- Propyl and asparaginyl hydroxylases (e.g., PHD1, 2, and 3)
- Regulators of function/stability of PHDs (e.g., OS9, Siah 1a and 2 E3 ubiquitin ligases, SSAT2 acetyl transferase) and of apparaginyl hydroxylase/FIH (e.g., desferrioxamine, Co^{++}, Ni^{++}, Mn^{++}, PKCζ, Cut-like homeodomain protein)
- Non-heme Fe^{++}
- Ascorbate
- Redox processes
- Intermediates of the TCA cycle
- Signaling pathways (e.g., mTOR, PI3K, AKT, p42/44 MAPK, Ras/Raf/ERK1/2, diacyl glycerol kinases, nitric oxide, carbon monoxide)
- Growth factors (e.g., EGF, insulin, IGF-1 and 2, angiotensin II, thrombin, PDGF, TGF-β1, IL-1β, TGF-α, FSH, androgens, HGF, TNF-α)
- Oncogene and tumor suppressor gene pathways (e.g., mutations involving VHL, SDHD, FH, TSC1 and 2, PTEN, p53, Ha-Ras, v-Src, c-Myc, bcr-abl, PML, AML-ETO, p14^{ARF})
- Other interacting molecules/pathways (e.g., HSP90, RACK1, SUMOylation, VHL deubiquitinating enzyme 2, HIF-3α2, Runx-1, CITED2/CITED4, GSK3β, FoxO3, FoxO4, p300)

Abbreviations: PHD, propyl hydroxylase domain enzymes; VHL, von Hipple-Lindau tumor suppressor gene; SHDH, succinate dehydrogenase deficiency; FH, fumarate hydratase; TSC, tuberous sclerosis tumor suppressor; PTEN, phosphate and tensin homolog; Ha-Ras, Harvey Ras; PML, promyelocytic leukemia tumor suppressor.

16 Is There Cross-Talk Between HIFS and novel Signaling Pathways in Stem Cells?

With the advent of genomics and various proteomic approaches, the molecular networks regulating stem cell survival, self-renewal, differentiation, and migration are now receiving a great deal of attention [181–183].

First described using massive parallel array technology, which could assess the expression of thousands of genes in parallel, this approach was used effectively to identify stem cell–specific genes expressed between various embryonic and somatic stem cells [184–186] and those expressed at various stages of human ES cell [187–189] and HSC activation and differentiation [190–192].

Novel approaches to understand this vast array of data include use of combined expression data sets, information from transcription factors inhibition studies, binding data from chromatin immunoprecipitation, and genome sequence data in a systematic global approach to identify novel core transcription factor networks in ES cells [181, 193, 194] and HSC/HPC [182, 183, 195]. These approaches tend to confirm gene regulatory networks that are known to exist (i.e., the Oct4-Nanog-Sox2 core regulatory network) in ES cells that had been described previously in separate genetic and biochemical studies (reviewed in [196] and references therein), but are also able to identify a host of other novel core regulators [181].

Ultimately however, because of the nature of these bioinformatics based approaches and the many thousands of genes identified with the many interactions which still need to be verified experimentally, it is still the classical developmental pathways associated with pluripotency, self-renewal, and proliferation that are most well understood. A selection of these is explored below in the context of hypoxia and HIF-α signaling.

17 HIFS and Genes Regulating Pluripotency

Recent studies have demonstrated that 3–4 genes, Oct-3/4, Sox-2, and Klf-4 with or without c-Myc, when ectopically overexpressed, can reprogram murine and human fibroblasts to a pluripotent-like state [197–200].

Interestingly, Oct-4 and c-Myc have been shown to be directly regulated by HIF-α, however, the mechanisms of action are quite different. Oct-4 is transcriptionally up-regulated by HIF-2α in a knock-in model into the HIF-1α locus both in HEK293T cells and mES cells [127], while HIF-1α and HIF-2α differentially regulate the activity of c-Myc in a direct association with c-Myc or its partners (reviewed in [5]; [97]). Although Klf-4 and Sox2 are the other components of this reprogramming machinery [181, 197], neither has yet been shown to be regulated by hypoxia and/or HIF-α signaling. Nanog, another central regulator of pluripotency [201 and 202] and suggested to be a permissive factor in reprogramming [197, 203], has also not yet been implicated in a HIF-α signaling pathway.

18 HIFS and Genes Regulating Self-Renewal or Proliferation

A host of other intrinsic factors commonly implicated in the maintenance or self-renewal of ES cells or HSC/HPC is now known. These include such molecules as Notch, Wnt/Catenin, BMP/Smad, and STAT3. Here, we explore whether potential novel interactions exist between HIFs and these stem cell regulatory molecules/pathways (see Fig. 3) and review their relevance to ES cells and/or HSC/HPC self-renewal.

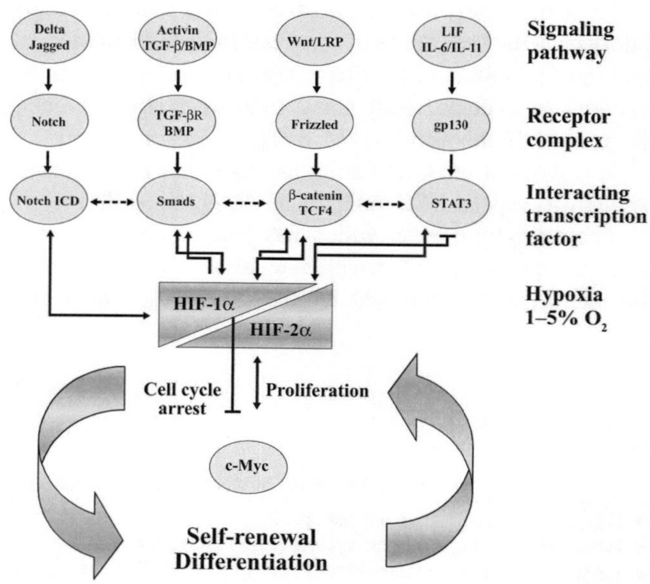

Fig. 3 Cross-talk of signaling pathways with HIF-α in stem cells and cancer. Pathways known to regulate cancer and stem cell maintenance and possible cross-talk with HIF-α are shown. These include Delta/Notch, TGF/Smad, Wnt/Catenin, and LIF/STAT3 signaling cascades. In most instances, reports of these developmentally regulated pathways in cross-talk with the HIF-α molecule are restricted to tumor cells. However, the Notch signaling cascade represents the first direct association of such a pathway with the HIF-α axis to be observed within a stem cell. Due to the conserved nature of signaling across cell types and organisms, these interactions with HIF-α may also extend to various stem cell compartments. Cross-talk with HIF-α signaling includes direct association of interacting transcription factor with HIF-α ($\leftrightarrow$), indirect transcriptional regulation by HIFs ($\rightarrow$), possible cross-talk suggested in other systems (...) and negative regulation, both direct and indirect ($\dashv$). See Section 6.6 for details

19 HIF-1α and targeting of Notch-Regulated Genes

It now appears that hypoxia can promote the undifferentiated cell state in neural stem/progenitor cell populations in a Notch-dependent manner [204].

Notch signaling components include the Notch receptors (Notch1-4) and Notch ligands (Delta-like 1, 3, 4 and Jagged 1 and 2). Binding of the these partners facilitates the cleavage and release of the intracellular domain of Notch (NICD), its translocation to the nucleus and its association with such binding partners as CSL, ICN, and p300. The Notch intracellular domain also interacts with HIF-1α, which is recruited with other co-activators such as p300 to Notch-responsive promoters (reviewed in [205–207]). The various Notch signaling molecules are known to be expressed in ES cells [208], but they appear inactive [209]. Instead, Notch and interacting molecules may play a role in more committed progenitors/restricted stem cells (reviewed in [205]). Examples include lymphopoiesis or repopulating HSCs at least in in vitro analyses (reviewed in [210]; [211–220]). Although yet to be demonstrated, given the potential roles of Notch and hypoxia in regulating HSC/HPC fate in vitro, it seems likely that a similar association of the Notch signaling complex and HIF-1α may be found at least when expanding this stem/progenitor cell compartment ex vivo.

20 Wnt and beta Catenin

Wnt/β-catenin is known to mediate many aspects of mammalian development [221]. Signaling is mediated by a large family of Wnt signaling ligands binding to their Frizzled receptors and the LRP 5/6 receptor to form heterodimeric receptor complexes. These signal to activated dishevelled (Dsh), which then sequesters components of the destruction complex (APC, Axin, and GSK-3β) to halt degradation of β-catenin [222]. β-catenin is then free to translocate to the nucleus, where it associates with TCF4/LEF to mediate target gene expression. Wnt signaling has been shown to have a regulatory role in nearly all adult stem cells, including HSC (reviewed in [210, 223]). Wnt5a was initially shown to promote murine fetal hematopoiesis in vitro [224], and, subsequently, purified Wnt3a was shown to mediate expansion of adult murine HSCs in vitro [225]. Wnt5a expression has also been demonstrated in human CD34^{+}Lin^{-} cells [226] and many other Wnt family members have since been shown to modulate HSC/HPC expansion in vitro [227]. Manipulation of Wnt signaling by overexpression of β-catenin can promote significant expansion of murine HSCs in vitro, up-regulating HoxB4 and Notch-1 expression in the process [228]. However, the role for β-catenin in murine HSC expansion in vivo has been questioned in recent studies, since β-catenin has been conditionally expressed or deleted without having an effect on hematopoiesis [229 , 230].

As part of the destruction complex, axin and GSK-3β mediate the phosphorylation, ubiqutination, and degradation of β-catenin in a basal state. Inhibitors of GSK3β gene have been shown to promote human and mouse ES cell maintenance [231] and hematopoietic repopulation in transplanted mice [232] by modulating gene targets not only of Wnt, but also of Notch and Hedgehog pathways. It may be that manipulation of β-catenin expression also affects sequestration of components from the destruction complex to bring about effects on ES and HSC self-renewal, effects that are subtle and context specific.

In relation to HIF-α signaling, lack of pVHL in TCF4 null mice (with pVHL being a downstream transcriptional target of TCF4) promotes stabilization and accumulation of nuclear HIF-1α [233] and suggests indirect regulation of HIF activity by the β-catenin pathway. We observe in our own studies [89] that TCF4 is up-regulated during hypoxia in umbilical cord CD133^{+} cells, while recent studies by Dao et al. [91] also show that hypoxia can induce inhibition of GSK-3β and β-catenin translocation to the nucleus in umbilical cord CD34^{+} HPC, which further suggests a degree of co-regulation by these pathways. Only recently has a direct association been demonstrated for β-catenin and HIF-1α, where HIF-1α sequesters β-catenin to the promoter region of HIF-1α target genes at the expense of TCF4 binding [234].

In the TCF4 null mouse described above, this may also indicate a second level of regulation for HIF-1α, that is, stabilization through lack of the pVHL factor, but also increased β-catenin binding and translocation to HIF-1α target sites in the nucleus. So, where once there appeared to be indirect transcriptional effects, there are now direct interactions between HIF and components of Wnt signaling. Co-operation of Wnt and Notch signaling pathways with each other (reviewed in [210]) and with HIF-1α to regulate HSC fate has been proposed, and fine-tuning of these regulatory networks may be important for enhancing HSC expansion in vitro.

21 The TGF-beta (β) Superfamily

The TGF-β superfamily includes bone morphogenic proteins (BMPs), activins and TGF-β itself. Bone morphogenetic proteins (BMPs) have been shown to contribute to many facets of mammalian development, including antero-posterior axis formation (reviewed in [235]), maintenance of mES cells [236–238], and hematopoiesis [238]. TGF-β deficiency in vivo does not appear to affect proliferation and differentiation of HSCs. However, TGF-β can inhibit hematopoietic progenitor proliferation in vitro and is a strong inducer of quiescience [239–242]. While TGF-β has been shown to have inhibitory effects on HSC expansion in vitro, BMP-4 has differential effects on human HSC depending on dose, evidently maintaining but not expanding human HSC/HPC in vitro (reviewed in [210]; [243]). BMP has been identified as the critical serum derived factor that maintains mES in a state of self-renewal [236], although it is replaced by a requirement for Activin (and FGF2) in hES cells [244] and murine epiblast stem cells [245 , 246].

Signals from this diverse array of growth factors and morphogens are all transduced by the Smad family of transcription factors ([247]; reviewed in [210, 248]). The effects of precise events of Smad signaling and transduction through the Smad family upon hematopoiesis from ES cells and HSC self-renewal are complex and context dependent, with the role of Smads 2, 3, 4, 5, and 7 in hematopoiesis being reviewed recently elsewhere [210]. Of note are the following pertinent experimental observations. The BMP signaling mediator, Smad5 has been found to play a role in murine embryonic, but not adult, hematopoiesis [249]. While overexpression of inhibitory Smad7 was found to promote self-renewal of murine HSCs in vivo [250] and to improve myeloid engraftment at the expense of lymphoid engraftment of human HSC/HPC in SCID mice [251], deletion of Smad4 (predicted to work similarly to Smad7) resulted in a reduced ability of murine HSCs for primary and secondary repopulation of hematopoiesis (reviewed in [210]; [252]). Blank et al. [210] have suggested that Smad4 may also participate in Wnt and Notch signaling pathways or may interact with HoxA9 regulating its function and HSC proliferation or self-renewal.

Cross-talk between hypoxia, HIF-1α, and TGF-β/Smad signaling has in fact been demonstrated in relation to CXCL12 induction and subsequent HSC/HPC induced homing towards this chemokine [253]. There also seems to be an association between components of TGF-β signaling and HIF-1α in other cell types. TGF-β in hypoxia has been shown to promote VEGF, EPO, and endoglin expression in endothelial and hepatoma cells [254–257] and while HIF-1α was shown to be stabilized by TGF-β–mediated inhibition of PHD2 [258], a direct interaction between Smad3, HIF-1α, and Sp1 (with HNF4 and CBP/p300) has also been demonstrated [257].

Clearly, signaling pathways are conserved between cells, and it is likely that this association may hold true in HSCs and other stem cells.

22 HIF-1α and STAT3 Cross-Talk

In mES cells, LIF which signals through the gp130 receptor to activate STAT3 (signal transducer and activator of transcription) can maintain mES cells in the absence of feeder layers (reviewed in [259]), while constitutively active STAT3 can maintain mES cells in the absence of LIF [260].

Interestingly in mES cells, HIF-1α has been shown to inhibit mES self-renewal by negative regulation of the LIF/STAT3 pathway [261].

This negative effect of hypoxia on mES maintenance is not evident in human cells [30, 262], but, instead, hES cells are maintained by Activin and FGF2 ([244]; reviewed in [263]), with at least the latter gene known to be regulated by HIF-1α [264].

A further level of cross-talk however may exist between these pathways since LIF/STAT3 signaling components are expressed in hES upon differentiation, and hence the effects of hypoxia and cross-talk with HIF-1α could become more critical in these differentiating tissues. LIF is a member of the cytokine family that includes IL-6 and IL-11. These cytokines share a common gp130 β receptor, and thus signal through STAT3, the major downstream mediator of gp130 signaling. These cytokines promote HSC/HPC expansion or differentiation to varying degrees (reviewed in [210]; [265]). Although IL-11 deficency does not affect murine hematopoiesis in vivo [266], deregulated STAT3 signaling can affect HSC self-renewal, with overexpression of constitutively active transcripts improving HSC self-renewal and repopulation potential in irradiated mice [267], after which they become latent.

Recently STAT3, which is up-regulated in many carcinomas, has been shown to regulate HIF-1α–mediated transcription [268 , 269] with an association being reported between STAT3, HIF-1α, p300/CBP, and Ref-1/APE [268] in carcinoma cells. This regulatory network is therefore known to exist. This physical interaction of STAT3 with HIF-1α is also responsible for increased stability of HIF-1α during hypoxia [269], and is a further example of mechanisms involved in HIF-1α signaling that augment HIF-1α signaling beyond that of the classical pVHL degradation pathway.

Further regulatory molecules and pathways that affect HSC self-renewal and survival have been identified. As well as Notch, Wnt, and the TGF-β family signaling pathways (reviewed in [210]), HoxB4 [270], and FGFs [271], the angiopoietin-like [272] and the nephroblastoma overexpressed (NOV) [273] proteins are of particular note. Whether these are regulated by hypoxia or HIFs is not known, although the studies described above would suggest that there is likely to be cross-talk between the HIFs and other regulatory pathways in stem cells that requires further exploration.

23 Conclusions

Oxygen is essential to human life. Here, we have reviewed the current knowledge available on oxygen tensions in specific cells and tissues in relation to organogensis and hematopoiesis. We have further shown that stem/progenitor cells involved in these processes can experience low oxygen environments in the stem/progenitor cell niches they occupy, and that this can finely tune their fate decisions. Only recently has sufficient information become available to define genes regulated by hypoxia in stem/progenitor cells, yet the pathways remain to be fully defined. It will be important in the future to understand the interacting pathways that connect hypoxia and HIFs to other stem cell signaling networks and to use this information in promoting stem cell expansion, engraftment and differentiation.

Acknowledgments Supported by NHS Blood and Transplant, the NHS R&D Directorate, the Leukaemia Research Fund, the BBSRC, the British Heart Foundation, the Medical Research Council, NIHR, an ECMC award, and the European Union Thercord and Cascade Programs.

We would like to thank Mrs. Brenda Cooley for assistance with the references.

References

1. Okazaki K, Maltepe E. Oxygen, epigenetics and stem cell fate. Regen Med. 2006;1:71–83.
2. Semenza GL. Life with oxygen. Science. 2007;318:62–4.
3. Watt SM, Smythe J, Fox A, et al. Blueprint for the response of blood and bone marrow derived stem cells and their progeny to hypoxia. In: Habib N et al. editors. Stem cell repair and regeneration. London, UK: Imperial College Press, 2007. pp. 61–84.
4. Patiar S, Harris AL. Role of hypoxia-inducible factor-1alpha as a cancer therapy target. Endocr Relat Cancer. 2006;13 Suppl 1: S61–75.
5. Gordan JD, Simon MC. Hypoxia-inducible factors: central regulators of the tumor phenotype. Curr Opin Genet Dev. 2007;17:71–7.
6. Keith B, Simon MC. Hypoxia-inducible factors, stem cells, and cancer. Cell. 2007;129:465–72.
7. Walmsley SR, McGovern NN, Whyte MK, et al. The HIF/VHL pathway: from oxygen sensing to innate immunity. Am J Respir Cell Mol Biol. 2008;38:251–5.
8. Webster WS, Abela D. The effect of hypoxia in development. Birth Defects Res C Embryo Today. 2007;81:215–28.
9. Mitchell JA, Yochim JM. Intrauterine oxygen tension during the estrous cycle in the rat: its relation to uterine respiration and vascular activity. Endocrinology. 1968;83:701–5.
10. Yochim JM, Mitchell JA. Intrauterine oxygen tension in the rat during progestation: its possible relation to carbohydrate metabolism and the regulation of nidation. Endocrinology. 1968;83:706–13.
11. Rodesch F, Simon P, Donner C, et al. Oxygen measurements in endometrial and trophoblastic tissues during early pregnancy. Obstet Gynecol. 1992;80:283–5.
12. Fischer B, Bavister BD. Oxygen tension in the oviduct and uterus of rhesus monkeys, hamsters and rabbits. J Reprod Fertil. 1993;99:673–9.
13. Kaufman DL, Mitchell JA. Intrauterine oxygen tension during the oestrous cycle in the hamster: patterns of change. Comp Biochem Physiol Comp Physiol. 1994;107:673–8.
14. Jaffe R, Jauniaux E, Hustin J, et al. Maternal circulation in the first-trimester human placenta-myth or reality? Am J Obstet Gynecol. 1997;176:695–705.
15. Van Blerkom J, Antczak M, Schrader R, et al. The developmental potential of the human oocyte is related to the dissolved oxygen content of follicular fluid: association with vascular endothelial growth factor levels and perifollicular blood flow characteristics. Hum Reprod. 1997;12:1047–55.
16. Burton GJ, Jauniaux E, Watson AL, et al. Maternal arterial connections to the placental intervillous space during the first trimester of human pregnancy: the Boyd collection revisited. Am J Obstet Gynecol. 1999;181:718–24.
17. Rossant J. Stem cells from the mammalian blastocyst. Stem Cells. 2001;19:477–82.
18. James JL, Stone PR, Chamley LW, et al. The regulation of trophoblast differentiation by oxygen in the first trimester of pregnancy. Hum Reprod Update. 2006;12:137–44.
19. Torres-Padilla ME, Parfitt DE, Kouzarides T, et al. Histone arginine methylation regulates pluripotency in the early mouse embryo. Nature. 2007;445:214–8.
20. Piotrowska K, Zernicka-Goetz M. Role for sperm in spatial patterning of the early mouse embryo. Nature. 2001;409:517–21.
21. Plusa B, Hadjantonakis AK, Gray D, et al. The first cleavage of the mouse zygote predicts the blastocyst axis. Nature. 2005;434: 391–5.
22. Morriss GM, New DA. Effect of oxygen concentration on morphogenesis of cranial neural folds and neural crest in cultured rat embryos. J Embryol Exp Morphol. 1979;54:17–35.
23. Pabon JE, Findley WE, Gibbons WE, et al. The toxic effect of short exposures to the atmospheric oxygen concentration on early mouse embryonic development. Fertil Steril. 1989;51:896–900.
24. Thompson JG, Simpson AC, Pugh PA, et al. Effect of oxygen concentration on in-vitro development of preimplantation sheep and cattle embryos. J Reprod Fertil. 1990;89:573–8.
25. Umaoka Y, Noda Y, Narimoto K, et al. Effects of oxygen toxicity on early development of mouse embryos. Mol Reprod Dev. 1992;31:28–33.
26. Li J, Foote RH. Culture of rabbit zygotes into blastocysts in protein-free medium with one to twenty per cent oxygen. J Reprod Fertil. 1993;98:163–7.
27. Eppig JJ, Wigglesworth K. Factors affecting the developmental competence of mouse oocytes grown in vitro: oxygen concentration. Mol Reprod Dev. 1995;42:447–56.
28. Bernardi ML, Flechon JE, Delouis C. Influence of culture system and oxygen tension on the development of ovine zygotes matured and fertilized in vitro. J Reprod Fertil. 1996;106:161–7.
29. Catt JW, Henman M. Toxic effects of oxygen on human embryo development. Hum Reprod. 2000;15 Suppl 2:199–206.
30. Ezashi T, Das P, Roberts RM. Low O2 tensions and the prevention of differentiation of hES cells. Proc Natl Acad Sci U S A. 2005;102:4783–8.
31. Michalska AE, Pegah J, Tellis I, et al. Effect of neurotrophins and low oxygen on cloning efficiency of human embryonic stem cells. 5th ISSCR Annual Meeting, Cairns, Australia. 2007;Abstract 292:142.
32. Draper JS, Smith K, Gokhale P, et al. Recurrent gain of chromosomes 17q and 12 in cultured human embryonic stem cells. Nat Biotechnol. 2004;22:53–4.
33. Jauniaux E, Watson A, Burton G. Evaluation of respiratory gases and acid-base gradients in human fetal fluids and uteroplacental tissue between 7 and 16 weeks gestation. Am J Obstet Gynecol. 2001;184:998–1003.
34. Aplin JD, Kimber SJ. Trophoblast-uterine interactions at implantation. Reprod Biol Endocrinol. 2004;2:48.
35. Samokhvalov IM, Samokhvalova NI, Nishikawa S. Cell tracing shows the contribution of the yolk sac to adult haematopoiesis. Nature. 2007;446:1056–61.
36. Tavian M, Peault B. Embryonic development of the human hematopoietic system. Int J Dev Biol. 2005;49:243–50.

37. Cumano A, Godin I. Ontogeny of the hematopoietic system. Annu Rev Immunol. 2007;25:745–85.
38. Bloom W, Bartelmez GW. Hematopoiesis in young human embryos. Am J Anat. 1940;67:21–53.
39. Fukuda T. Undifferentiated mononuclear cell in human embryonic liver; presumptive hematopoietic stem cell. Virchows Arch B Cell Pathol. 1973;14:31–4.
40. Tavassoli M, Yoffey JM. Bone marrow structure and function. New York: Alan R Liss Press Inc.; 1983.
41. Moore MA, Metcalf D. Ontogeny of the haemopoietic system: yolk sac origin of in vivo and in vitro colony forming cells in the developing mouse embryo. Br J Haematol. 1970;18:279–96.
42. Zambidis ET, Sinka L, Tavian M, et al. Emergence of human angiohematopoietic cells in normal development and from cultured embryonic stem cells. Ann N Y Acad Sci. 2007;1106:223–32.
43. Kumaravelu P, Hook L, Morrison AM, et al. Quantitative developmental anatomy of definitive haematopoietic stem cells/long-term repopulating units (HSC/RUs): role of the aorta-gonad-mesonephros (AGM) region and the yolk sac in colonisation of the mouse embryonic liver. Development. 2002;129:4891–9.
44. Zeigler BM, Sugiyama D, Chen M, et al. The allantois and chorion, when isolated before circulation or chorio-allantoic fusion, have hematopoietic potential. Development. 2006;133:4183–92.
45. Gekas C, Dieterlen-Livre F, Orkin SH, et al. The placenta is a niche for hematopoietic stem cells. Dev Cell. 2005;8:365–75.
46. Yokota T, Huang J, Tavian M, et al. Tracing the first waves of lymphopoiesis in mice. Development. 2006;133:2041–51.
47. Charbord P, Tavian M, Humeau L, et al. Early ontogeny of the human marrow from long bones: an immunohistochemical study of hematopoiesis and its microenvironment. Blood. 1996;87:4109–19.
48. Guillot PV, Donoghue O, Kurata H, et al. Fetal stem cells: betwixt and between. Semin Reprod Med. 2006;24:340–7.
49. Campagnoli C, Roberts IA, Kumar S, et al. Identification of mesenchymal stem/progenitor cells in human first-trimester fetal blood, liver, and bone marrow. Blood. 2001;98:2396–402.
50. In't Anker PS, Scherjon SA, Kleiburg-van der Keur C, et al. Amniotic fluid as a novel source of mesenchymal stem cells for therapeutic transplantation. Blood. 2003;102:1548–9.
51. De Coppi P, Bartsch G, Siddiqui MM, et al. Isolation of amniotic stem cell lines with potential for therapy. Nat Biotechnol. 2007;25:100–6.
52. Towler DA. Vascular biology and bone formation: hints from HIF. J Clin Invest. 2007;117:1477–80.
53. Watt SM, Forde SP. The central role of the chemokine receptor, CXCR4, in haemopoietic stem cell transplantation. Will CXCR4 contagonists contribute to the treatment of blood disorders. Vox Sang. 2008;94:18–32.
54. Trentin JJ. Hemopoietic microenvironments. Transplant Proc. 1978;10:77–82.
55. Bizzozero GN. Sulla funzione ematopoietici del midollo delle oss. R C R 1st Lomb Sci Lett. 1868;2:815–8.
56. Bowie MB, McKnight KD, Kent DG, et al. Hematopoietic stem cells proliferate until after birth and show a reversible phase-specific engraftment defect. J Clin Invest. 2006;116:2808–16.
57. Kikuchi K, Kondo M. Developmental switch of mouse hematopoietic stem cells from fetal to adult type occurs in bone marrow after birth. Proc Natl Acad Sci U S A. 2006;103:17852–7.
58. Wilson A, Trumpp A. Bone-marrow haematopoietic-stem-cell niches. Nat Rev Immunol. 2006;6:93–106.
59. Scadden DT. The stem cell niche in health and leukemic disease. Best Pract Res Clin Haematol. 2007;20:19–27.
60. Maloney MA, Patt HM. Origin in repopulating cells after localized bone marrow depletion. Science. 1968;165:71–3.
61. Kiel MJ, Morrison SJ. Maintaining hematopoietic stem cells in the vascular niche. Immunity. 2006;25:862–4.
62. Haylock DN, Williams B, Johnston HM, et al. Hemopoietic stem cells with higher hemopoietic potential reside at the bone marrow endosteum. Stem Cells. 2007;25:1062–69.
63. Sugiyama T, Kohara H, Noda M, et al. Maintenance of the hematopoietic stem cell pool by CXCL12-CXCR4 chemokine signaling in bone marrow stromal cell niches. Immunity. 2006;25:977–88.
64. Kokovay E, Temple S. Taking neural crest stem cells to new heights. Cell. 2007;131:234–6.
65. Amarilio R, Viukov SV, Sharir A, et al. HIF1-α regulation of Sox9 is necessary to maintain differentiation of hypoxic prechondrogenic cells during early skeletogenesis. Development. 2007;134:3917–28.
66. Chow DC, Wenning LA, Miller WM, et al. Modeling pO(2) distributions in the bone marrow hematopoietic compartment. II. Modified Kroghian models. Biophys J. 2001;81:685–96.
67. Harris AL. Hypoxia-a key regulatory factor in tumour growth. Nat Rev Cancer. 2002;2:38–47.
68. Schofield CJ, Ratcliffe PJ. Signalling hypoxia by HIF hydroxylases. Biochem Biophys Res Commun. 2005;338:617–26.
69. Semenza GL. Regulation of physiological responses to continuous and intermittent hypoxia by hypoxia-inducible factor 1. Exp Physiol. 2006;91:803–6.
70. Semenza GL. Hypoxia-inducible factor 1 (HIF-1) pathway. Sci STKE. 2007;2007(407):cm8.
71. Levesque JP, Winkler IG, Hendy J, et al. Hematopoietic progenitor cell mobilization results in hypoxia with increased hypoxia-inducible transcription factor-1 alpha and vascular endothelial growth factor A in bone marrow. Stem Cells. 2007;25:1954–65.
72. Kopp HG, Avecilla ST, Hooper AT, et al. The bone marrow vascular niche: home of HSC differentiation and mobilization. Physiology (Bethesda). 2005;20:349–56.
73. Thomlinson, RH, Gray LH. The histological structure of some human lung cancers and the possible implications for radiotherapy. Br J Cancer. 1955;9:539–49.
74. Harrison JS, Rameshwar P, Chang V, et al. Oxygen saturation in the bone marrow of healthy volunteers. Blood. 2002;99:394.
75. Skouby AR. Haematologic adaptation in patients with chronic bronchitis and pulmonary insufficiency. Acta Med Scand. 1976;199:185–90.
76. Parmar K, Mauch P, Vergilio JA, et al. Distribution of hematopoietic stem cells in the bone marrow according to regional hypoxia. Proc Natl Acad Sci U S A. 2007;104:5431–6.
77. Jang YY, Sharkis SJ. A low level of reactive oxygen species selects for primitive hematopoietic stem cells that may reside in the low-oxygenic niche. Blood. 2007;110:3056–63.
78. Ito K, Hirao A, Arai F, et al. Regulation of oxidative stress by ATM is required for self-renewal of haematopoietic stem cells. Nature. 2004;431:997–1002.
79. Tothova Z, Kollipara R, Huntly BJ, et al. FoxOs are critical mediators of hematopoietic stem cell resistance to physiologic oxidative stress. Cell. 2007;128:325–39.
80. Miyamoto K, Araki KY, Naka K, et al. Foxo3a is essential for maintenance of the hematopoietic stem cell pool. Cell Stem Cell. 2007;1:101–12.
81. Winkler IG, Levesque JP. Mechanisms of hematopoietic stem cell mobilization: when innate immunity assails the cells that make blood and bone. Exp Hematol. 2006;34:996–1009.
82. Rochefort GY, Delorme B, Lopez A, et al. Multipotential mesenchymal stem cells are mobilized into peripheral blood by hypoxia. Stem Cells. 2006;24:2202–8.
83. Ceradini DJ, Kulkarni AR, Callaghan MJ, et al. Progenitor cell trafficking is regulated by hypoxic gradients through HIF-1 induction of SDF-1. Nat Med. 2004;10:858–64.

84. Fox A, Smythe J, Fisher N, et al. Mobilization of endothelial progenitor cells into the circulation in burned patients. Br J Surg. 2008;95:244–51.
85. Tepper OM, Capla JM, Galiano RD, et al. Adult vasculogenesis occurs through in situ recruitment, proliferation, and tubulization of circulating bone marrow-derived cells. Blood. 2005;105: 1068–77.
86. Cipolleschi MG, Dello, Sbarba P, Olivotto M. The role of hypoxia in the maintenance of hematopoietic stem cells. Blood. 1993;82:2031–7.
87. Ivanovic Z, Hermitte F, Brunet de la, Grange, P, et al. Simultaneous maintenance of human cord blood SCID-repopulating cells and expansion of committed progenitors at low O2 concentration (3%). Stem Cells. 2004;22:716–24.
88. Danet GH, Pan Y, Luongo JL, et al. Expansion of human SCID-repopulating cells under hypoxic conditions. J Clin Invest. 2003;112:126–35.
89. Martin-Rendon E, Hale SJ, Ryan D, et al. Transcriptional profiling of human cord blood CD133^{+} and cultured bone marrow mesenchymal stem cells in response to hypoxia. Stem Cells. 2007;25:1003–12.
90. Hermitte F, Brunet de la Grange P, Belloc F, et al. Very low O_2 concentration (0.1%) favors G_0return of dividing CD34^{+} cells. Stem Cells. 2006;24:65–73.
91. Dao MA, Creer MH, Nolta JA, et al. Biology of umbilical cord blood progenitors in bone marrow niches. Blood. 2007;110: 74–81.
92. Annabi B, Lee YT, Turcotte S, et al. Hypoxia promotes murine bone-marrow-derived stromal cell migration and tube formation. Stem Cells. 2003;21:337–47.
93. Fehrer C, Brunauer R, Laschober G, et al. Reduced oxygen tension attenuates differentiation capacity of human mesenchymal stem cells and prolongs their lifespan. Aging Cell. 2007;6: 745–57.
94. Hung SC, Pochampally RR, Hsu SC, et al. Short-term exposure of multipotent stromal cells to low oxygen increases their expression of CX3CR1 and CXCR4 and their engraftment in vivo. PLoS ONE. 2007;2:e416.
95. Mole DR, Ratcliffe PJ. Cellular oxygen sensing in health and disease. Pediatr Nephrol. 2008;23:681–94.
96. Bardos JI, Ashcroft M. Negative and positive regulation of HIF-1: a complex network. Biochim Biophys Acta. 2005;1755: 107–20.
97. Gordan JD, Bertout JA, Hu CJ, et al. HIF-2alpha promotes hypoxic cell proliferation by enhancing c-myc transcriptional activity. Cancer Cell. 2007;11:335–47.
98. Makino Y, Uenishi R, Okamoto K, et al. Transcriptional up-regulation of inhibitory PAS domain protein gene expression by hypoxia-inducible factor 1 (HIF-1): a negative feedback regulatory circuit in HIF-1-mediated signaling in hypoxic cells. J Biol Chem. 2007;282:14073–82.
99. Maynard MA, Qi H, Chung J, et al. Multiple splice variants of the human HIF-3 alpha locus are targets of the von Hippel-Lindau E3 ubiquitin ligase complex. J Biol Chem. 2003;278:11032–40.
100. Jiang BH, Rue E, Wang GL, et al. Dimerization, DNA binding, and transactivation properties of hypoxia-inducible factor 1. J Biol Chem. 1996;271:17771–8.
101. Huang LE, Gu J, Schau M, et al. Regulation of hypoxia-inducible factor 1alpha is mediated by an O2-dependent degradation domain via the ubiquitin-proteasome pathway. Proc Natl Acad Sci U S A. 1998;95:7987–92.
102. Jiang BH, Zheng JZ, Leung SW, et al. Transactivation and inhibitory domains of hypoxia-inducible factor 1alpha. Modulation of transcriptional activity by oxygen tension. J Biol Chem. 1997;272:19253–60.
103. Jain S, Maltepe E, Lu MM, et al. Expression of ARNT, ARNT2, HIF1 alpha, HIF2 alpha and Ah receptor mRNAs in the developing mouse. Mech Dev. 1998;73:117–23.
104. Takahashi T, Sugishita Y, Nojiri T, et al. Cloning of hypoxia-inducible factor 1alpha cDNA from chick embryonic ventricular myocytes. Biochem Biophys Res Commun. 2001;281:1057–62.
105. Tian H, McKnight SL, Russell DW. Endothelial PAS domain protein 1 (EPAS1), a transcription factor selectively expressed in endothelial cells. Genes Dev. 1997;11:72–82.
106. Compernolle V, Brusselmans K, Acker T, et al. Loss of HIF-2alpha and inhibition of VEGF impair fetal lung maturation, whereas treatment with VEGF prevents fatal respiratory distress in premature mice. Nat Med. 2002;8:702–10.
107. Wiesener MS, Jrgensen JS, Rosenberger C, et al. Widespread hypoxia-inducible expression of HIF-2alpha in distinct cell populations of different organs. FASEB J. 2003;17:271–3.
108. Makino Y, Cao R, Svensson K, et al. Inhibitory PAS domain protein is a negative regulator of hypoxia-inducible gene expression. Nature. 2001;414:550–4.
109. Makino Y, Kanopka A, Wilson WJ, et al. Inhibitory PAS domain protein (IPAS) is a hypoxia-inducible splicing variant of the hypoxia-inducible factor-3alpha locus. J Biol Chem. 2002;277:32405–8.
110. Talks KL, Turley H, Gatter KC, et al. The expression and distribution of the hypoxia-inducible factors HIF-1alpha and HIF-2alpha in normal human tissues, cancers, and tumor-associated macrophages. Am J Pathol. 2000;157:411–21.
111. Stroka DM, Burkhardt T, Desbaillets I, et al. HIF-1 is expressed in normoxic tissue and displays an organ-specific regulation under systemic hypoxia. FASEB J. 2001;15:2445–53.
112. Ramirez-Bergeron DL, Runge A, Dahl KD, et al. Hypoxia affects mesoderm and enhances hemangioblast specification during early development. Development. 2004;131:4623–34.
113. Hu CJ, Iyer S, Sataur A, et al. Differential regulation of the transcriptional activities of hypoxia-inducible factor 1 alpha (HIF-1alpha) and HIF-2alpha in stem cells. Mol Cell Biol. 2006;26:3514–26.
114. Ryan HE, Lo J, Johnson RS. HIF-1 alpha is required for solid tumor formation and embryonic vascularization. EMBO J. 1998;17:3005–15.
115. Iyer NV, Kotch LE, Agani F, et al. Cellular and developmental control of O2 homeostasis by hypoxia-inducible factor 1 alpha. Genes Dev. 1998;12:149–62.
116. Cowden Dahl KD, Fryer BH, Mack FA, et al. Hypoxia-inducible factors 1alpha and 2alpha regulate trophoblast differentiation. Mol Cell Biol. 2005;25:10479–91.
117. Scortegagna M, Morris MA, Oktay Y, et al. The HIF family member EPAS1/HIF-2alpha is required for normal hematopoiesis in mice. Blood. 2003;102:1634–40.
118. Scortegagna M, Ding K, Zhang Q, et al. HIF-2alpha regulates murine hematopoietic development in an erythropoietin-dependent manner. Blood 2005;105:3133–40.
119. Gruber M, Hu CJ, Johnson RS, et al. Acute postnatal ablation of Hif-2alpha results in anemia. Proc Natl Acad Sci U S A. 2007;104:2301–6.
120. Yoon D, Pastore YD, Divoky V, et al. Hypoxia-inducible factor-1 deficiency results in dysregulated erythropoiesis signaling and iron homeostasis in mouse development. J Biol Chem. 2006;281:25703–11.
121. Wang Y, Wan C, Deng L, et al. The hypoxia-inducible factor alpha pathway couples angiogenesis to osteogenesis during skeletal development. J Clin Invest. 2007;117:1616–26.
122. Kojima H, Sitkovsky MV, Cascalho M. HIF-1 alpha deficiency perturbs T and B cell functions. Curr Pharm Des. 2003;9: 1827–32.

123. Cramer T, Yamanishi Y, Clausen BE, et al. HIF-1alpha is essential for myeloid cell-mediated inflammation. Cell. 2003;112:645–57.
124. Kong T, Eltzschig HK, Karhausen J, et al. Leukocyte adhesion during hypoxia is mediated by HIF-1-dependent induction of beta2 integrin gene expression. Proc Natl Acad Sci U S A. 2004;101:10440–5.
125. Walmsley SR, Cadwallader KA, Chilvers ER. The role of HIF-1alpha in myeloid cell inflammation. Trends Immunol. 2005;26:434–9.
126. Covello KL, Simon MC, Keith B. Targeted replacement of hypoxia-inducible factor-1alpha by a hypoxia-inducible factor-2alpha knock-in allele promotes tumor growth. Cancer Res. 2005;65 2277–86.
127. Covello KL, Kehler J, Yu H, et al. HIF-2alpha regulates Oct-4: effects of hypoxia on stem cell function, embryonic development, and tumor growth. Genes Dev. 2006;20:557–70.
128. Dolwick KM, Swanson HI, Bradfield CA. In vitro analysis of Ah receptor domains involved in ligand-activated DNA recognition. Proc Natl Acad Sci U S A. 1993;90:8566–70.
129. Moffett P, Reece M, Pelletier J. The murine Sim-2 gene product inhibits transcription by active repression and functional interference. Mol Cell Biol. 1997;17:4933–47.
130. Maltepe E, Schmidt JV, Baunoch D, et al. Abnormal angiogenesis and responses to glucose and oxygen deprivation in mice lacking the protein ARNT. Nature. 1997;386:403–7.
131. Kozak KR, Abbott B, Hankinson O. ARNT-deficient mice and placental differentiation. Dev Biol. 1997;191:297–305.
132. Adelman DM, Maltepe E, Simon MC. Multilineage embryonic hematopoiesis requires hypoxic ARNT activity. Genes Dev. 1999;13:2478–83.
133. Adelman DM, Gertsenstein M, Nagy A, et al. Placental cell fates are regulated in vivo by HIF-mediated hypoxia responses. Genes Dev. 2000;14:3191–203.
134. Masson N, Ratcliffe PJ. HIF prolyl and asparaginyl hydroxylases in the biological response to intracellular O(2) levels. J Cell Sci. 2003;116:3041–9.
135. Masson N, Willam C, Maxwell PH, et al. Independent function of two destruction domains in hypoxia-inducible factor-alpha chains activated by prolyl hydroxylation. EMBO J. 2001;20:5197–206.
136. Jaakkola P, Mole DR, Tian YM, et al. Targeting of HIF-alpha to the von Hippel-Lindau ubiquitylation complex by O2-regulated prolyl hydroxylation. Science. 2001;292:468–72.
137. Ohh M, Takagi Y, Aso T, et al. Synthetic peptides define critical contacts between elongin C, elongin B, and the von Hippel-Lindau protein. J Clin Invest. 1999;104:1583–91.
138. Berra E, Benizri E, Ginouvs A, et al. HIF prolyl-hydroxylase 2 is the key oxygen sensor setting low steady-state levels of HIF-1alpha in normoxia. EMBO J. 2003;22:4082–90.
139. Metzen E, Berchner-Pfannschmidt U, Stengel P, et al. Intracellular localisation of human HIF-1 alpha hydroxylases: implications for oxygen sensing. J Cell Sci. 2003;116:1319–26.
140. Stiehl DP, Wirthner R, Koeditz J, et al. Increased prolyl 4-hydroxylase domain proteins compensate for decreased oxygen levels. Evidence for an autoregulatory oxygen-sensing system. J Biol Chem. 2006;281:23482–91.
141. Nakayama K, Frew IJ, Hagensen M, et al. Siah2 regulates stability of prolyl-hydroxylases, controls HIF1alpha abundance, and modulates physiological responses to hypoxia. Cell. 2004;117:941–52.
142. Appelhoff RJ, Tian YM, Raval RR, et al. Differential function of the prolyl hydroxylases PHD1, PHD2, and PHD3 in the regulation of hypoxia-inducible factor. J Biol Chem. 2004;279:38458–65.
143. Takeda K, Ho VC, Takeda H, et al. Placental but not heart defects are associated with elevated hypoxia-inducible factor alpha levels in mice lacking prolyl hydroxylase domain protein 2. Mol Cell Biol. 2006;26:8336–46.
144. Baek JH, Mahon PC, Oh J, et al. OS-9 interacts with hypoxia-inducible factor 1alpha and prolyl hydroxylases to promote oxygen-dependent degradation of HIF-1alpha. Mol Cell. 2005;17:503–12.
145. Jeong JW, Bae MK, Ahn MY, et al. Regulation and destabilization of HIF-1alpha by ARD1-mediated acetylation. Cell. 2002;111:709–20.
146. Bilton R, Mazure N, Trottier E, et al. Arrest-defective-1 protein, an acetyltransferase, does not alter stability of hypoxia-inducible factor (HIF)-1alpha and is not induced by hypoxia or HIF. J Biol Chem. 2005;280:31132–40.
147. Baek JH, Liu YV, McDonald KR, et al. Spermidine/spermine-N1-acetyltransferase 2 is an essential component of the ubiquitin ligase complex that regulates hypoxia-inducible factor 1alpha. J Biol Chem. 2007;282:23572–80.
148. Li Z, Wang D, Messing EM, et al. VHL protein-interacting deubiquitinating enzyme 2 deubiquitinates and stabilizes HIF-1alpha. EMBO Rep. 2005;6:373–8.
149. Bell EL, Chandel NS. Mitochondrial oxygen sensing: regulation of hypoxia-inducible factor by mitochondrial generated reactive oxygen species. Essays Biochem. 2007;43:17–27.
150. McDonough MA, Li V, Flashman E, et al. Cellular oxygen sensing: Crystal structure of hypoxia-inducible factor prolyl hydroxylase (PHD2). Proc Natl Acad Sci U S A. 2006;103:9814–9.
151. Pouyssgur J, Mechta-Grigoriou F. Redox regulation of the hypoxia-inducible factor. Biol Chem. 2006;387:1337–46.
152. Ratcliffe PJ. HIF-1 and HIF-2: working alone or together in hypoxia? J Clin Invest. 2007;117:862–5.
153. Wang V, Davis DA, Haque M, et al. Differential gene up-regulation by hypoxia-inducible factor-1alpha and hypoxia-inducible factor-2alpha in HEK293T cells. Cancer Res. 2005;65: 3299–306.
154. Sharp FR, Bernaudin M. HIF1 and oxygen sensing in the brain. Nat Rev Neurosci. 2004;5:437–48.
155. Semenza GL. Expression of hypoxia-inducible factor 1: mechanisms and consequences. Biochem Pharmacol. 2000;59:47–53.
156. Lando D, Peet DJ, Whelan DA, et al. Asparagine hydroxylation of the HIF transactivation domain a hypoxic switch. Science. 2002;295:858–61.
157. Hewitson KS, McNeill LA, Riordan MV, et al. Hypoxia-inducible factor (HIF) asparagine hydroxylase is identical to factor inhibiting HIF (FIH) and is related to the cupin structural family. J Biol Chem. 2002;277:26351–5.
158. Sang N, Fang J, Srinivas V, et al. Carboxyl-terminal transactivation activity of hypoxia-inducible factor 1 alpha is governed by a von Hippel-Lindau protein-independent, hydroxylation-regulated association with p300/CBP. Mol Cell Biol. 2002;22:2984–92.
159. Raval RR, Lau KW, Tran MG, et al. Contrasting properties of hypoxia-inducible factor 1 (HIF-1) and HIF-2 in von Hippel-Lindau-associated renal cell carcinoma. Mol Cell Biol. 2005;25:5675–86.
160. Lum JJ, Bui T, Gruber M, et al. The transcription factor HIF-1alpha plays a critical role in the growth factor-dependent regulation of both aerobic and anaerobic glycolysis. Genes Dev. 2007;21:1037–49.
161. Lfstedt T, Fredlund E, Holmquist-Mengelbier L, et al. Hypoxia inducible factor-2alpha in cancer. Cell Cycle. 2007;6:919–26.
162. Gunaratnam L, Morley M, Franovic A, et al. Hypoxia inducible factor activates the transforming growth factor-alpha/epidermal growth factor receptor growth stimulatory pathway in VHL(-/-) renal cell carcinoma cells. J Biol Chem. 2003;278:44966–74.
163. Hu CJ, Wang LY, Chodosh LA, et al. Differential roles of hypoxia-inducible factor 1alpha (HIF-1alpha) and HIF-2alpha in hypoxic gene regulation. Mol Cell Biol. 2003;23:9361–74.
164. Wykoff CC, Sotiriou C, Cockman ME, et al. Gene array of VHL mutation and hypoxia shows novel hypoxia-induced genes and

that cyclin D1 is a VHL target gene. Br J Cancer. 2004;90: 1235–43.
165. Elvidge GP, Glenny L, Appelhoff RJ, et al. Concordant regulation of gene expression by hypoxia and 2-oxoglutarate-dependent dioxygenase inhibition: the role of HIF-1alpha, HIF-2alpha, and other pathways. J Biol Chem. 2006;281:15215–26.
166. Comerford KM, Wallace TJ, Karhausen J, et al. Hypoxia-inducible factor-1-dependent regulation of the multidrug resistance (MDR1) gene. Cancer Res. 2002;62:3387–94.
167. Baba M, Hirai S, Yamada-Okabe H, et al. Loss of von Hippel-Lindau protein causes cell density dependent deregulation of CyclinD1 expression through hypoxia-inducible factor. Oncogene. 2003;22:2728–38.
168. Yuan Y, Shen H, Franklin DS, et al. In vivo self-renewing divisions of haematopoietic stem cells are increased in the absence of the early G1-phase inhibitor, p18INK4C. Nat Cell Biol. 2004;6:436–42.
169. Cheng T, Rodrigues N, Dombkowski D, et al. Stem cell repopulation efficiency but not pool size is governed by p27(kip1). Nat Med. 2000;6:1235–40.
170. Koshiji M, Kageyama Y, Pete EA, et al. HIF-1alpha induces cell cycle arrest by functionally counteracting Myc. EMBO J. 2004;23:1949–56.
171. Ulrich HD. SUMO teams up with ubiquitin to manage hypoxia. Cell. 2007;131:446–7.
172. Welsh SJ, Bellamy WT, Briehl MM, et al. The redox protein thioredoxin-1 (Trx-1) increases hypoxia-inducible factor 1 alpha protein expression: Trx-1 overexpression results in increased vascular endothelial growth factor production and enhanced tumor angiogenesis. Cancer Res. 2002;62:5089–95.
173. Bakker WJ, Harris IS, Mak TW. FOXO3a is activated in response to hypoxic stress and inhibits HIF1-induced apoptosis via regulation of CITED2. Mol Cell. 2007;28:941–53.
174. Arden KC. FoxOs in tumor suppression and stem cell maintenance. Cell. 2007;128:235–7.
175. Minet E, Mottet D, Michel G, et al. Hypoxia-induced activation of HIF-1: role of HIF-1alpha-Hsp90 interaction. FEBS Lett. 1999;460:251–6.
176. Liu YV, Semenza GL. RACK1 vs. HSP90: competition for HIF-1 alpha degradation vs. stabilization. Cell Cycle. 2007;6:656–9.
177. Richard DE, Berra E, Gothi E, et al. p42/p44 mitogen-activated protein kinases phosphorylate hypoxia-inducible factor 1alpha (HIF-1alpha) and enhance the transcriptional activity of HIF-1. J Biol Chem. 1999;274:32631–7.
178. Jin HO, An S, Lee HC, et al. Hypoxic condition- and high cell density-induced expression of Redd1 is regulated by activation of hypoxia-inducible factor-1alpha and Sp1 through the phosphatidylinositol 3-kinase/Akt signaling pathway. Cell Signal. 2007;19:1393–403.
179. Sutton KM, Hayat S, Chau NM, et al. Selective inhibition of MEK1/2 reveals a differential requirement for ERK1/2 signalling in the regulation of HIF-1 in response to hypoxia and IGF-1. Oncogene. 2007;26:3920–29.
180. Hara S, Hamada J, Kobayashi C, et al. Expression and characterization of hypoxia-inducible factor (HIF)-3alpha in human kidney: suppression of HIF-mediated gene expression by HIF-3alpha. Biochem Biophys Res Commun. 2001;287:808–13.
181. Zhou Q, Chipperfield H, Melton DA, et al. A gene regulatory network in mouse embryonic stem cells. Proc Natl Acad Sci U S A. 2007;104:16438–43.
182. Swiers G, Patient R, Loose M. Genetic regulatory networks programming hematopoietic stem cells and erythroid lineage specification. Dev Biol. 2006;294:525–40.
183. Soneji S, Huang S, Loose M, et al. Inference, validation, and dynamic modeling of transcription networks in multipotent hematopoietic cells. Ann N Y Acad Sci. 2007;1106:30–40.
184. Ivanova NB, Dimos JT, Schaniel C, et al. A stem cell molecular signature. Science. 2002;298:601–4.
185. Ramalho-Santos M, Yoon S, Matsuzaki Y, et al. 'Stemness': transcriptional profiling of embryonic and adult stem cells. Science. 2002;298:597–600.
186. Fortunel NO, Otu HH, Ng HH, et al. Comment on 'Stemness: transcriptional profiling of embryonic and adult stem cells' and 'a stem cell molecular signature'. Science. 2003;302:393.
187. Sperger JM, Chen X, Draper JS, et al. Gene expression patterns in human embryonic stem cells and human pluripotent germ cell tumors. Proc Natl Acad Sci U S A. 2003;100:13350–5.
188. Ivanova N, Dobrin R, Lu R, et al. Dissecting self-renewal in stem cells with RNA interference. Nature. 2006;442:533–8.
189. Adewumi O, Aflatoonian B, hrlund-Richter L, et al. Characterization of human embryonic stem cell lines by the International Stem Cell Initiative. Nat Biotechnol. 2007;25:803–16.
190. Venezia TA, Merchant AA, Ramos CA, et al. Molecular signatures of proliferation and quiescence in hematopoietic stem cells. PLoS Biol. 2004;2:e301.
191. Bruno L, Hoffmann R, McBlane F, et al. Molecular signatures of self-renewal, differentiation, and lineage choice in multipotential hemopoietic progenitor cells in vitro. Mol Cell Biol. 2004;24:741–56.
192. Passegue E, Wagers AJ, Giuriato S, et al. Global analysis of proliferation and cell cycle gene expression in the regulation of hematopoietic stem and progenitor cell fates. J Exp Med. 2005;202:1599–611.
193. Boyer LA, Lee TI, Cole MF, et al. Core transcriptional regulatory circuitry in human embryonic stem cells. Cell. 2005;122:947–56.
194. Loh YH, Wu Q, Chew JL, et al. The Oct4 and Nanog transcription network regulates pluripotency in mouse embryonic stem cells. Nat Genet. 2006;38:431–40.
195. Donaldson IJ, Chapman M, Kinston S, et al. Genome-wide identification of cis-regulatory sequences controlling blood and endothelial development. Hum Mol Genet. 2005;14:595–601.
196. Chambers I. The molecular basis of pluripotency in mouse embryonic stem cells. Cloning Stem Cells. 2004;6:386–91.
197. Nakagawa M, Koyanagi M, Tanabe K, et al. Generation of induced pluripotent stem cells without Myc from mouse and human fibroblasts. Nat Biotechnol. 2008;26:101–6.
198. Wernig M, Meissner A, Foreman R, et al. In vitro reprogramming of fibroblasts into a pluripotent ES-cell-like state. Nature. 2007;448:318–24.
199. Takahashi K, Tanabe K, Ohnuki M, et al. Induction of pluripotent stem cells from adult human fibroblasts by defined factors. Cell. 2007;13:861–72.
200. Maherali N, Sridharan R, Xie W, et al. Directly reprogrammed fibroblasts show global epigenetic remodelling and widespread tissue contribution. Cell Stem Cell. 2007;1:55–70.
201. Chambers I, Colby D, Robertson M, et al. Functional expression cloning of Nanog, a pluripotency sustaining factor in embryonic stem cells. Cell. 2003;113:643–55.
202. Mitsui K, Tokuzawa Y, Itoh H, et al. The homeoprotein Nanog is required for maintenance of pluripotency in mouse epiblast and ES cells. Cell. 2003;113:631–42.
203. Pan G, Thomson JA. Nanog and transcriptional networks in embryonic stem cell pluripotency. Cell Res. 2007;17:42–9.
204. Gustafsson MV, Zheng X, Pereira T, et al. Hypoxia requires Notch signaling to maintain the undifferentiated cell state. Dev Cell. 2005;9:617–28.
205. Chiba S. Notch signaling in stem cell systems. Stem Cells. 2006;24:2437–47.
206. Roy M, Pear WS, Aster JC. The multifaceted role of Notch in cancer. Curr Opin Genet Dev. 2007;17:52–9.
207. Fiza UM, Arias AM. Cell and molecular biology of Notch. J Endocrinol. 2007;194:459–74.

208. Walsh J, Andrews PW. Expression of Wnt and Notch pathway genes in a pluripotent human embryonal carcinoma cell line and embryonic stem cell. APMIS 2003;111:197–210.
209. Noggle SA, Weiler D, Condie BG. Notch signaling is inactive but inducible in human embryonic stem cells. Stem Cells. 2006;24:1646–53.
210. Blank U, Karlsson G, Karlsson S. Signaling pathways governing stem cell fate. Blood. 2008;111:492–503.
211. Radtke F, Wilson A, Mancini SJ, MacDonald HR. Notch regulation of lymphocyte development and function. Nat Immunol. 2004;5:247–53.
212. Duncan AW, Rattis FM, DiMascio LN, et al. Integration of Notch and Wnt signaling in hematopoietic stem cell maintenance. Nat Immunol. 2005;6:314–22.
213. Karanu FN, Murdoch B, Gallacher L, et al. The Notch ligand Jagged-1 represents a novel growth factor of human hematopoietic stem cells. J Exp Med. 2000;192:1365–72.
214. Karanu FN, Murdoch B, Miyabayashi T, et al. Human homologues of Delta-1 and Delta-4 function as mitogenic regulators of primitive human hematopoietic cells. Blood. 2001;97:1960–7.
215. Karanu FN, Yuefei L, Gallacher L, et al. Differential response of primitive human $CD34^-$ and $CD34^+$ hematopoietic cells to the Notch ligand Jagged-1. Leukemia. 2003;17:1366–74.
216. Stier S, Cheng T, Dombkowski D, et al. Notch1 activation increases hematopoietic stem cell self-renewal in vivo and favors lymphoid over myeloid lineage outcome. Blood. 2002;99: 2369–78.
217. Washburn T, Schweighoffer E, Gridley T, et al. Notch activity influences the alphabeta versus gammadelta T cell lineage decision. Cell. 1997;88:833–43.
218. Robey E, Chang D, Itano A, et al. An activated form of Notch influences the choice between CD4 and CD8 T cell lineages. Cell. 1996;87:483–92.
219. Ellisen LW, Bird J, West DC, et al. TAN-1, the human homolog of the Drosophila Notch gene, is broken by chromosomal translocations in T lymphoblastic neoplasms. Cell. 1991;66:649–61.
220. Mancini SJ, Mantei N, Dumortier A, et al. Jagged1-dependent Notch signaling is dispensable for hematopoietic stem cell self-renewal and differentiation. Blood. 2005;105:2340–2.
221. Cadigan KM, Nusse R. Wnt signaling: a common theme in animal development. Genes Dev. 1997;11:3286–305.
222. Willert K, Jones KA. Wnt signaling: is the party in the nucleus? Genes Dev. 2006;20:1394–404.
223. Reya T, Clevers H. Wnt signalling in stem cells and cancer. Nature. 2005;434:843–50.
224. Austin TW, Solar GP, Ziegler FC, et al. A role for the Wnt gene family in hematopoiesis: expansion of multilineage progenitor cells. Blood. 1997;89:3624–35.
225. Willert K, Brown JD, Danenberg E, et al. Wnt proteins are lipid-modified and can act as stem cell growth factors. Nature. 2003;423:448–52.
226. Van Den Berg DJ, Sharma AK, Bruno E, Hoffman R. Role of members of the Wnt gene family in human hematopoiesis. Blood. 1998;92:3189–202.
227. Nikolova T, Wu M, Brumbarov K, et al. WNT-conditioned media differentially affect the proliferation and differentiation of cord blood-derived $CD133^+$ cells in vitro. Differentiation. 2007; 75:100–11.
228. Reya T, Duncan AW, Ailles L, et al. A role for Wnt signalling in self-renewal of haematopoietic stem cells. Nature. 2003;423: 409–14.
229. Kirstetter P, Anderson K, Porse BT, et al. Activation of the canonical Wnt pathway leads to loss of hematopoietic stem cell repopulation and multilineage differentiation block. Nat Immunol. 2006;7:1048–56.
230. Cobas M, Wilson A, Ernst B, et al. Beta-catenin is dispensable for hematopoiesis and lymphopoiesis. J Exp Med. 2004;199:221–9.
231. Sato N, Meijer L, Skaltsounis L, et al. Maintenance of pluripotency in human and mouse embryonic stem cells through activation of Wnt signaling by a pharmacological GSK-3-specific inhibitor. Nat Med. 2004;10:55–63.
232. Trowbridge JJ, Xenocostas A, Moon RT, et al. Glycogen synthase kinase-3 is an in vivo regulator of hematopoietic stem cell repopulation. Nat Med. 2006;12:89–98.
233. Giles RH, Lolkema MP, Snijckers CM, et al. Interplay between VHL/HIF1alpha and Wnt/beta-catenin pathways during colorectal tumorigenesis. Oncogene. 2006;25:3065–70.
234. Kaidi A, Williams AC, Paraskeva C. Interaction between beta-catenin and HIF-1 promotes cellular adaptation to hypoxia. Nat Cell Biol. 2007;9:210–7.
235. Hogan BL. Bone morphogenetic proteins in development. Curr Opin Genet Dev. 1996;6:432–8.
236. Ying QL, Nichols J, Chambers I, et al. BMP induction of Id proteins suppresses differentiation and sustains embryonic stem cell self-renewal in collaboration with STAT3. Cell. 2003;115:281–92.
237. Winnier G, Blessing M, Labosky PA, et al. Bone morphogenetic protein-4 is required for mesoderm formation and patterning in the mouse. Genes Dev. 1995;9:2105–16.
238. Johansson BM, Wiles MV. Evidence for involvement of activin A and bone morphogenetic protein 4 in mammalian mesoderm and hematopoietic development. Mol Cell Biol. 1995;15:141–51.
239. Keller JR, Mantel C, Sing GK, et al. Transforming growth factor beta 1 selectively regulates early murine hematopoietic progenitors and inhibits the growth of IL-3-dependent myeloid leukemia cell lines. J Exp Med. 1988;168:737–50.
240. Hatzfeld J, Li ML, Brown EL, et al. Release of early human hematopoietic progenitors from quiescence by antisense transforming growth factor beta 1 or Rb oligonucleotides. J Exp Med. 1991;174:925–9.
241. Bruno E, Horrigan SK, Van Den BD, et al. The Smad5 gene is involved in the intracellular signaling pathways that mediate the inhibitory effects of transforming growth factor-beta on human hematopoiesis. Blood. 1998;91:1917–23.
242. Fortunel N, Batard P, Hatzfeld A, et al. High proliferative potential-quiescent cells: a working model to study primitive quiescent hematopoietic cells. J Cell Sci. 1998;111:1867–75.
243. Bhatia M, Bonnet D, Wu D, et al. Bone morphogenetic proteins regulate the developmental program of human hematopoietic stem cells. J Exp Med. 1999;189:1139–48.
244. Ludwig TE, Bergendahl V, Levenstein ME, et al. Feeder-independent culture of human embryonic stem cells. Nat Methods. 2006;3:637–46.
245. Brons IG, Smithers LE, Trotter MW, et al. Derivation of pluripotent epiblast stem cells from mammalian embryos. Nature. 2007;448:191–5.
246. Tesar PJ, Chenoweth JG, Brook FA, et al. New cell lines from mouse epiblast share defining features with human embryonic stem cells. Nature. 2007;448:196–9.
247. Heldin CH, Miyazono K, ten Dijke P, et al. TGF-beta signalling from cell membrane to nucleus through SMAD proteins. Nature. 1997;390:465–71.
248. Larsson J, Karlsson S. The role of Smad signaling in hematopoiesis. Oncogene. 2005;24:5676–92.
249. Singbrant S, Moody JL, Blank U, et al. Smad5 is dispensable for adult murine hematopoiesis. Blood. 2006;108:3707–12.
250. Blank U, Karlsson G, Moody JL, et al. Smad7 promotes self-renewal of hematopoietic stem cells. Blood. 2006;108:4246–54.
251. Chadwick K, Shojaei F, Gallacher L, Bhatia M. Smad7 alters cell fate decisions of human hematopoietic repopulating cells. Blood. 2005;105:1905–15.

252. Karlsson G, Blank U, Moody JL, et al. Smad4 is critical for self-renewal of hematopoietic stem cells. J Exp Med. 2007;204:467–74.
253. Tabatabai G, Frank B, Moehle R, et al. Irradiation and hypoxia promote homing of haematopoietic progenitor cells towards gliomas by TGF-beta-dependent HIF-1alpha-mediated induction of CXCL12. Brain. 2006;129:2426–35.
254. Akman HO, Zhang H, Siddiqui MA, et al. Response to hypoxia involves transforming growth factor-beta2 and Smad proteins in human endothelial cells. Blood. 2001;98:3324–31.
255. Sanchez-Elsner T, Botella LM, Velasco B, et al. Synergistic cooperation between hypoxia and transforming growth factor-beta pathways on human vascular endothelial growth factor gene expression. J Biol Chem. 2001;276:38527–35.
256. Sanchez-Elsner T, Botella LM, Velasco B, et al. Endoglin expression is regulated by transcriptional cooperation between the hypoxia and transforming growth factor-beta pathways. J Biol Chem. 2002;277:43799–808.
257. Sanchez-Elsner T, Ramirez JR, Sanz-Rodriguez F, et al. A cross-talk between hypoxia and TGF-beta orchestrates erythropoietin gene regulation through SP1 and Smads. J Mol Biol. 2004;336:9–24.
258. McMahon S, Charbonneau M, Grandmont S, et al. Transforming growth factor beta1 induces hypoxia-inducible factor-1 stabilization through selective inhibition of PHD2 expression. J Biol Chem. 2006;281:24171–81.
259. Smith AG. Mouse embryo stem cells: their identification, propagation and manipulation. Semin Cell Biol. 1992;3:385–99.
260. Matsuda T, Nakamura T, Nakao K, et al. STAT3 activation is sufficient to maintain an undifferentiated state of mouse embryonic stem cells. EMBO J. 1999;18:4261–9.
261. Jeong CH, Lee HJ, Cha JH, et al. Hypoxia-inducible factor-1 alpha inhibits self-renewal of mouse embryonic stem cells in vitro via negative regulation of the leukemia inhibitory factor-STAT3 pathway. J Biol Chem. 2007;282:13672–9.
262. Dahron L, Opitz SL, Zaehres H, et al. LIF/STAT3 signaling fails to maintain self-renewal of human embryonic stem cells. Stem Cells. 2004;22:770–8.
263. Lovell-Badge R. Many ways to pluripotency. Nat Biotechnol. 2007;25:1114–6.
264. Black SM, Devol JM, Wedgwood S. Regulation of fibroblast growth factor-2 expression in pulmonary arterial smooth muscle cells involves increased reactive oxygen species generation. Am J Physiol Cell Physiol. 2008;294:C345–54.
265. Rappold I, Watt SM, Kusadasi N, Rose-John S, Hatzfeld J, Ploemacher RE. Gp130-signaling synergizes with FL and TPO for the long-term expansion of cord blood progenitors. Leukemia. 1999;13:2036–48.
266. Nandurkar HH, Robb L, Tarlinton D, Barnett L, Köntgen F, Begley CG. Adult mice with targeted mutation of the interleukin-11 receptor (IL11Ra) display normal hematopoiesis. Blood. 1997;90:2148–59.
267. Chung YJ, Park BB, Kang YJ, et al. Unique effects of Stat3 on the early phase of hematopoietic stem cell regeneration. Blood. 2006;108:1208–15.
268. Gray MJ, Zhang J, Ellis LM, et al. HIF-1alpha, STAT3, CBP/p300 and Ref-1/APE are components of a transcriptional complex that regulates Src-dependent hypoxia-induced expression of VEGF in pancreatic and prostate carcinomas. Oncogene. 2005;24:3110–20.
269. Jung JE, Lee HG, Cho IH, et al. STAT3 is a potential modulator of HIF-1-mediated VEGF expression in human renal carcinoma cells. FASEB J. 2005;19:1296–98.
270. Zhang XB, Schwartz JL, Humphries RK, Kiem HP. Effects of HOXB4 overexpression on ex vivo expansion and immortalization of hematopoietic cells from different species. Stem Cells. 2007;25:2074–81.
271. Yeoh JS, de Haan G. Fibroblast growth factors as regulators of stem cell self-renewal and aging. Mech Ageing Dev. 2007;128:17–24.
272. Zhang CC, Kaba M, Ge G, et al. Angiopoietin-like proteins stimulate ex vivo expansion of hematopoietic stem cells. Nat Med. 2006;12:240–5.
273. Gupta R, Hong D, Iborra F, Sarno S, Enver T. NOV (CCN3) functions as a regulator of human hematopoietic stem or progenitor cells. Science. 2007;316:590–3.

Part III
Epigenetic Mechanisms in Stem Cells

Stem Cell Epigenetics

Joyce E. Ohm and Stephen B. Baylin

Abstract Epigenetics is the study of heritable changes in gene expression that occur independently of alterations in the primary DNA sequence. Normal development requires a carefully orchestrated epigenetic program on both a global and gene-specific/tissue-specific level. A tightly regulated program of active and repressive histone modifications working in close concert with the methylation of CpG dinucleotides in DNA are responsible for this epigenetic reprogramming, which includes the permanent silencing of genes required for stem/progenitor cell maintenance and the variable regulation of tissue/cell specific activation of genes required for successful differentiation of alternate cell lineages. In this chapter, we will discuss the normal stem/progenitor cell epigenetic remodeling proteins, including polycomb group (PcG) proteins, and the histone modifications that help modulate gene expression during development. We will highlight the essential role of DNA methylation in gene silencing programs during normal development, including both maternal and paternal imprinting and x-chromosome inactivation, and the permanent silencing of pluripotency-associated genes with terminal differentiation. We will conclude with a more extensive discussion of how the regulatory networks controlling stem cell epigenetics may go awry in cells that give rise to tumors and during tumor progression. The large numbers of epigenetically silenced genes that may be present in any given tumor, and the clustering of silenced genes within single cell pathways, begs the question of whether gene silencing is a series of random events resulting in an enhanced survival of a premalignant clone, or whether silencing is the result of a directed, instructive program for silencing initiation reflective of the cells of origin for tumors. In this regard, the current review stresses the latter hypothesis and the important possibility that the program is linked, at least for silencing of some cancer genes, to the epigenetic control of stem/precursor cell gene expression patterns.

Keywords Stem cells · Epigenetics · Chromatin · DNA methylation · Polycomb · Cancer

1 Introduction

Epigenetics is the study of heritable changes in gene expression that occur independently of alterations in the primary DNA sequence. Gene expression changes occurring during development are, by their nature, epigenetic. The precision required for the generation, from a single fertilized egg, of upwards of ten trillion cells in a human body that are capable of representing all of the various biology and functional components of differentiated organs, requires a masterfully coordinated, multifaceted program. Assuming that within an individual the genetic code is faithfully replicated and remains uniform from cell to cell, then epigenetics is left to orchestrate the activation and inactivation, or lack of activation, of genes required for successful development of alternate cell lineages during development. Epigenetics is also responsible for the conversion of stem/progenitor cell genes into terminally differentiated cells during adult cell renewal. In this chapter, we will briefly address differences in regulation of the proteins and histone modifications that help modulate gene expression during development and the role of DNA methylation during normal development. We will, then, more extensively discuss how the regulatory networks controlling stem cell epigenetics may go awry in cells that give rise to tumors and during tumor progression.

2 Epigenetic Modifications

In order to fully understand the complexity of epigenetic regulation and the role of these processes in normal stem/progenitor cell differentiation, we must first begin with

S.B. Baylin (✉)
Cancer Biology Division, The Sidney Kimmel Comprehensive Cancer Center, The Johns Hopkins University Medical Institutions; 1650 Orleans Street, Suite 541, Baltimore, MD, 21231
e-mail: sbaylin@jhmi.edu

V.K. Rajasekhar, M.C. Vemuri (eds.), *Regulatory Networks in Stem Cells*, Stem Cell Biology and Regenerative Medicine,
DOI 10.1007/978-1-60327-227-8_19, © Humana Press, a part of Springer Science+Business Media, LLC 2009

an understanding of the tools that a cell uses to control the chromatin structure and how DNA is packaged for function – this packaging constitutes the groundwork of epigenetics. Within a cell, DNA is packaged into chromatin by wrapping 147 bp of DNA around nucleosome complexes composed of pairs of four different histone proteins: H2A, H2B, H3, and H4 (Fig. 1a). The resulting octamer and its surrounding DNA, or the nucleosome, is the fundamental unit of chromatin and plays a key role in gene expression [1, 2]. Chromatin can be divided into two main subtypes: first there is euchromatin; in which nucleosomes are more variably spaced and linearly arranged, and which is more permissive for gene transcription. The second chromatin subtype is heterochromatin; which is densely packed in terms of

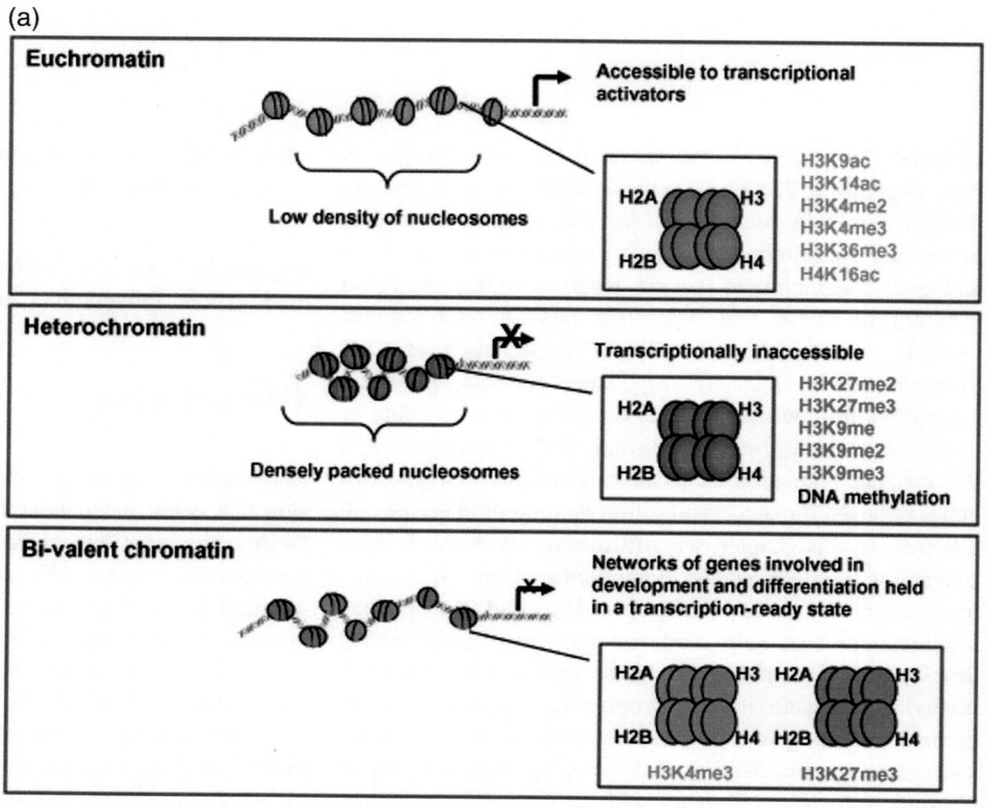

(b)

Histone Modification	Histone acetyltransferases	Histone Deacetylase	Lysine Methyltransferase	Lysine Demethylase
H3K4			MLL1 MLL2 MLL3 MLL4 MLL5 SET1A SET1B ASH1 Sc/SpSET1	LSD1/BHC110
H3K9	PCAF/GCN5	SirT1	SUV39H1 SUV39H2 G9a ESET/SETDB1 EuHMTase/GLP CLL8 SpClr4	JHDM2a JHDM2b JMJD2A/JHDM3a JMJD2B JMJD2C/GASC1 JMJD2D
H3K14	CBP/P300 PCAF/GCN5 TIP60 ScSAS3			
H3K27			EZH2 (PcG)	
H3K36			SET2 (Sc/Sp SET2) NSD1 SYMD2	JHDM1a JHDM1b JMJD2A/JHDM3a JMJD2C/GASC1
H4K16	TIP60 ScSAS2(SpMST2)	SirT1 SirT2 (ScSir2)		

Fig. 1 Chromatin structure and histone modifications. (**a**) Euchromatin (*top panel*) is associated with actively transcribed genes (*black arrow*). H3 is preferentially modified with active chromatin marks, and histones bearing these active marks (*light grey ovals*) are widely dispersed providing an open chromatin setting favorable for active transcription. Conversely, heterochromatin (*middle panel*) is characteristic of silent gene promoter (*black arrow with* "x"), and is associated with histones arranged in a more compacted nucleosomal configuration typical for transcriptionally repressed promoters. On these histones (*dark grey ovals*), H3 is preferentially modified with several repressive methylation marks, including, in some cases, DNA methylation. Networks of genes involved in development and differentiation are sometimes held in a transcription-ready state in stem cells called bivalent chromatin (*lower panel*). In this intermediate state, the histones contain the transcriptional activating mark H3K4me3 and the repressive modification H3K27me3. (**b**) Known histone acetyltransferases, histone deacetylases, lysine methyltransferases, and lysine de-methylases are shown for histone modifications to histone H3, lysine 4, 9, 14, 27, and 36, and histone H4, lysine 16 [5, 6]

both nucleosome spacing along the DNA and higher order arrangement of the nucleosomes, and thus restricted for the access of transcription factors, facilitating transcriptional repression and gene silencing [1, 2] (Fig. 1a).

The structure of chromatin and the associated gene transcription is controlled by covalent modifications to the tails of the histone proteins resulting in what has been termed a "histone code" that regulates chromatin-DNA interactions. The histone code is much more complex than originally proposed [3, 4], and to date, at least eight different classes of histone modifications have been characterized (reviewed in [5]). These include acetylation, lysine and arginine methylation, phosphorylation, ubiquitylation, sumoylation, ADP ribosylation, deimination, and proline isomerization. For the purpose of this chapter, we will focus upon a subset of these modifications that have been closely linked to the activation or repression of genes. Key marks for active transcription include the acetylation of H3 lysine 9 and 14 (H3K9ac, H3K14ac) and H4K16 (H4K16ac), the di- and tri-methylation of H3 Lysine 4 (H3K4me2, H3K4me3), and the tri-methylation of H3 lysine 36 (H3K36me3). Chromatin modifications linked to transcriptional repression include di-and tri-methylated H3 lysine 27 (H3K27me2, H3K27me3), and mono-, di-, and tri-methylation of H3 lysine 9 (H3K9me, H3K9me2, H3K9me3) (Fig. 1a). Many of the enzymes responsible for these histone modifications have

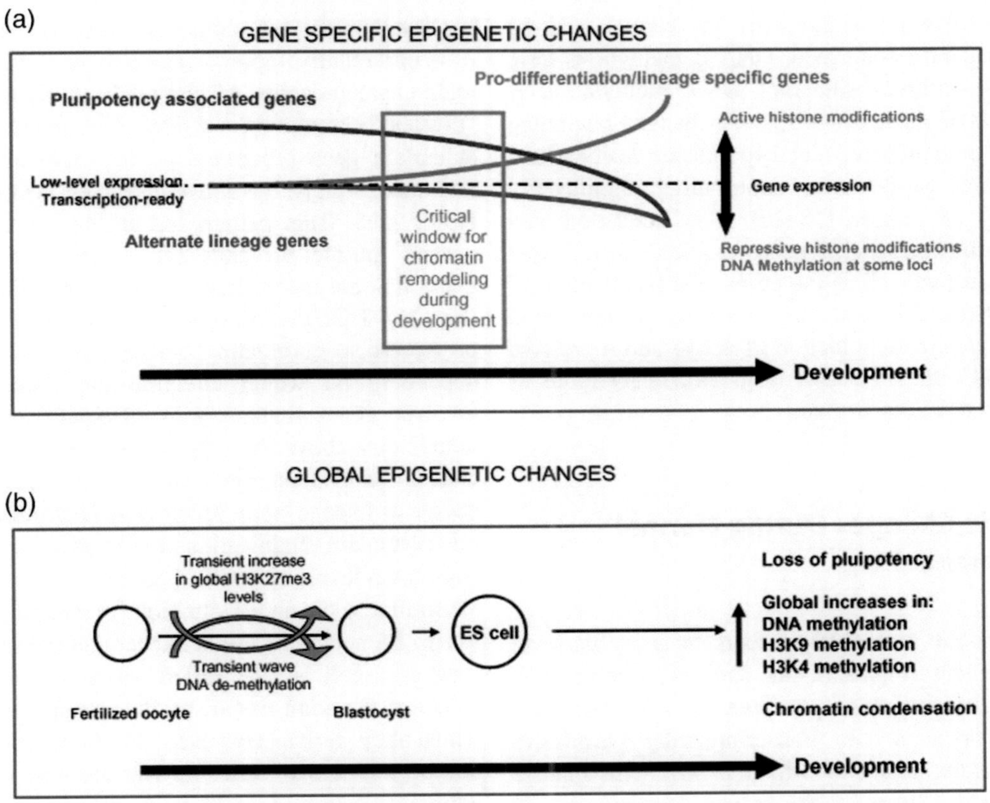

Fig. 2 Gene specific and global epigenetic changes during development. (**a**) A model of normal activation and gene silencing of different groups of developmental genes on a local, gene specific level, highlighting a critical window for chromatin remodeling during development. In stem and progenitor cells, important developmental and lineage specific genes are held in a bivalent, transcription-ready, low expression state (*dotted black line*). With differentiation, and in a specific lineage, some pro-differentiation and lineage specific genes must be activated, while those genes required for alternate lineages must be permanently silenced. The divergent expression status of these two groups of genes is controlled at least in part by the enrichment of active and repressive histone modifications at their promoters. Pluripotency-associated genes must be permanently silenced in terminally differentiated cells, and this is accomplished by the combination of enrichment for repressive chromatin modifications and for at least some genes, DNA methylation. The complex remodeling of these different classes of genes results in a critical window for chromatin remodeling that may leave genes vulnerable to inappropriate silencing or activation. (**b**) Histone modifications and DNA methylation are also modulated on a global level during development. In the pre-blastocyst phase, the key long-term gene transcription repression mark, H3K27me3, mediated by the Polycomb group (PcG) of transcriptional repressors, and DNA methylation undergo inversely correlated waves of enrichrichment. With terminal differentiation in normal cells, we see a loss of pluripotency, chromatin condensation, and global increases in DNA methylation and H3K4 and H3K9 methylation compared to ES cells

also been identified. Known acetyltransferases, deacetylases, lysine methyltransferases, and lysine demethylases for these modifications are shown in Fig. 1b [5, 6].

Another important epigenetic modification that is closely associated with gene silencing is DNA methylation. In this direct modification of DNA, in mammals, the cytosines are subject, in the context of CpG dinucleotides, to the addition of methyl groups by DNA methyltransferases (DNMT). These methyltransferases include DNMT1, DNMT3a, DNMT3b, and DNMT3L [7–9]. With the exception of the transient wave of global demethylation that occurs during embryonic development, DNA methylation is considered, relative to the histone modifications as will be discussed, to be a more permanent, heritable mark. While distinct experimental evidence exists to suggest that active and direct DNA demethylation can take place [10–12], the specific mechanisms for this, outside of base excision corrections by DNA glycosylases [11] are still not known, and no known, independently confirmed, DNA demethylases have been directly identified. Abberant DNA methylation is also a hallmark of most cancers, with changes beginning in the earliest premalignant and hyperplastic lesions, and throughout tumor progression, contributing significantly to the biology of cancer [13–16]. The activation and inactivation of lineage-specific genes required for different tissues and organs, as well as the permanent inactivation of pluripotency-associated genes in terminally differentiated cells, uses a complex combination of active and repressive histone and DNA modifications (Fig. 2a), many more of which are likely to be discovered.

3 Epigenetic Changes During Normal Development

Extensive chromatin remodeling occurs on a global level during development (Fig. 2b). As embryonic stem (ES) cells differentiate, they lose pluripotency. ES cells have a remarkably open, active, transcriptionally permissive chromatin structure, much of which is lost with lineage commitment [17]. There is an increase in regions of condensed heterochromatin and an increase in the global levels of the accompanying repressive histone modifications associated with differentiation [17, 18]. Structural chromatin proteins also become more stably associated with chromatin in differentiated cells (reviewed in [17]). The extent that adult tissue-specific stem/progenitor cells retain this global permissive chromatin status is likely tied to the extent with which they retain pluripotency. Enrichment of individual histone modifications also varies on a global level during different stages of development. For example, the methylation of H3K9 is increased with differentiation, which likely corresponds to the enhanced condensation of chromatin, and H3K4me is also enhanced globally in the postblastocyst phase as developmental and tissue specific genes are activated [18]. Additionally, the key long-term gene transcription repression mark, H3K27me3, mediated by the Polycomb group (PcG) of transcriptional repressors (Fig. 3), and DNA methylation undergo inversely correlated waves of enrichment on a global level, with significant decreases in DNA methylation in both the zygote and the embryonic germ cells (Fig. 2b) [18]. Several seminal studies of the epigenetic profiles of PcG genes in human ES and embryonic fibroblasts (EF) have been recently published. These studies utilized genome tiling array approaches in order to show that PcG proteins and/or the H3K27methylation mark are enriched at the promoters of between ~8% (ES cells) [19] and ~14% (EF cells) [20] of the annotated genome, and identify PcG enrichment as a key regulator of genes essential for development, morphogenesis, and organogenesis.

A particularly important component of the above genome-wide characterization of chromatin involves the concept of "bivalent" chromatin consisting of the simultaneous presence at certain gene promoters of an active mark, H3K4me3, and, interestingly, the above-discussed PcG mediated mark, H3K27me3. This pattern has recently been described in normal murine ES cells for a subset of developmental genes that are maintained in a low expression state [21–23] (Fig. 1a). This bivalent state is resolved to a primarily active or repressive chromatin conformation with differentiation depending on which direction the transcription of the involved genes changes with differentiation cues [21]. As noted in the above study by Bernstein et al. [21] defining the concept of bivalent chromatin among embryonic cells, the degree of lineage commitment may determine the balance of this chromatin modification at gene promoters and the basal expression levels of involved genes.

Similar to the above situation for commitment steps relative to ES cells, one might expect that the global chromatin state of a cell is also varied between adult differentiated cells and the adult tissue specific stem/precursor cells from which they derive. However, this is an essential question that begs extensive research over the next few years. Most adult stem/precursor cells have already significantly committed to a specific lineage, and likely have programmed, through epigenetic packaging of the genome, a considerable portion of their gene expression patterns. We might still expect, however, that the extent to which adult tissue-derived stem/precursor cells retain the ability to differentiate into various differentiated lineages is reflective of their global chromatin structure and their ability to modify it. At least some adult stem cell populations retain the bivalent chromatin seen in ES cells for a subset of genes. Thus, a significant number of genes have been shown to retain this bivalent chromatin structure in mouse neural progenitor cells (NPC) and

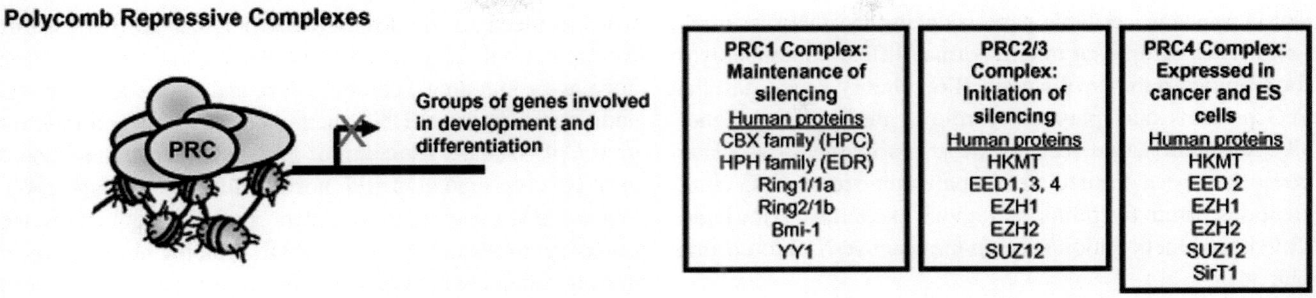

Fig. 3 The polycomb repressive complexes in humans.The human proteins known to be associated with polycomb repressive complex (PRC) 1, 2/3, and 4 are shown here [69, 72]. The PRC1 complex is thought to be responsible for the long-term maintenance of silencing by polycomb [72, 90]. PRC2/3 is thought to be responsible for the initiation of silencing, and both PRC 2/3 and PRC 4 contain the histone lysine methyltransferase EZH2, the protein responsible for H3K27 methylation [69]. PRC4 is similar to PRC 2/3, but contains the EED 2 isoform, which has only been observed in ES cells and cancer cells and the NAD+ dependent histone deacetylase SirT1 [69], which is responsible for deacetylation of H4K16 and H3K9 and regulates lifespan in multiple organisms [6]

mouse embryonic fibroblasts (EF) [23]. While there is significant overlap, the subset of gene promoters with bivalent chromatin in NPC and EF cells are distinct from each other and from the mouse ES cells [23]. The advent of ChIP-on-ChIP and ChIP-sequencing technologies has opened huge new opportunities for the genome-wide study of histone modifications. These assays consist of immunoprecipitation with antibodies to various histone modifications or transcriptional activators and repressors, followed by microarray analysis or genomic sequencing [20–26]. ChIP-sequencing, in particular, because of its ability to handle relatively small numbers of cells and cover the genome with great density [23, 25], will hopefully provide a platform to investigate the global chromatin signatures of various cell populations, including rare adult stem cells, adult precursor cells, and multiple types of differentiated cells which constitute an adult tissue.

Just as normal stem/progenitor cells are significantly remodeling chromatin during differentiation, it is also important to note that these cells also use DNA methylation, to collaborate with chromatin configuration, to stabilize key gene expression patterns that emerge during normal development and adult tissue cell turnover. For example, localized DNA methylation changes occur at key cytokine target genes during embryonic and adult differentiation and maturation of lymphocytes [27]. In global genomic packaging, for all types of cells, DNA methylation may be a key component of repressive chromatin which functions to give long-term silencing of transposons, stabilization of silenced genes in the processes of imprinting and x-inactivation, as well as a mechanism to permanently silence important pluipotency-associated genes [2]. Mammalian females use x-chromosome inactivation to equalize the imbalance of X-chromosome gene expression created by females having two X chromosomes in contrast to the male XY. Random X inactivation is initiated in the epiblast, or future embryo [28], and utilizes both histone and DNA methylation associated with gene silencing. The inactivated X chromosome (Xi) acquires both H3K27me3 and H3K9me2, while the activating H3K4me3 histone modification is lost [29, 30]. Dense regions of CpG dinucleotides, termed CpG islands, and discussed in great detail later below, along the entire Xi chromosome are also DNA hypermethylated in the embryo [31]. This multifaceted collaboration of histone modifications, nucleosome remodeling, and DNA methylation provides an elegant control system to produce heritable patterns of gene expression to guarantee the complex functions of mammalian organisms.

Both sexes use genomic imprinting to control the expression of approximately 100 imprinted genes and allowing monoallelic expression from either the maternal or paternal allele, and most imprinted genes regulate placental and fetal growth [2]. There are many similarities between X-chromosome inactivation and genomic imprinting [32] and both H3K27me3 and H3K9me2 histone methylation and DNA methylation are involved in the normal silencing of imprinted genes. For detailed reviews of X-inactivation and genomic imprinting see [2, 32].

DNA methylation also plays a key role in the permanent silencing of pluripotency-associated genes during normal development (Fig. 2a). hTERT, Oct-4, and Nanog are all strongly expressed in stem/progenitor cells and have been shown to accumulate DNA methylation in their promoter regions associated with gene silencing during differentiation of teratocarcinoma and ES cells [33–38]. In the case of hTERT, this methylation is associated with an increase in global de novo methylation and increases in both Dnmt1 and Dnmt3b activity midway through the differentiation process [33]. Silencing of Oct-4 in postmitotic neurons has been shown to be associated with loss of the active H3K4me2 mark, an increase in the polycomb regulated H3K27me3 modification, and CpG DNA methylation [39]. Together, these data suggest that permanent, heritable gene silencing via DNA methylation plays a significant role in normal differentiation, at least for a limited number of stem cell regulatory genes, and that PcG proteins may play a key role in this process. Generally, the methylation seen for these pluripotency-associated genes has been limited to a small number of CpG dinucleotides within the promoter regions in contrast to the dense CpG island methylation seen on the inactive X chromosome [33, 34].

4 Cancer Stem Cells

Many groups have suggested that stem cells or early progenitor cells may be the precursors from which cancer cells are derived [40–42]. All of these studies cite the fact that tumors harbor stem/precursor cell population markers, have heterogeneous cell populations among which only cell subsets have tumor regeneration capacity, and are driven by the up-regulation of many pathways important to maintenance of normal stem/precursor cells including PcG proteins, and the Wnt, Shh, and Notch pathways [41]. The cancer stem cell hypothesis has also been used to explain cancer recurrence. Drugs that target the rapidly proliferating bulk tumor population may leave the relatively quiescent cancer stem cells unphased and allow for tumor regrowth [41]. Similarly, both normal and malignant stem cells are known to be comparatively resistant to chemotherapy [43].

Despite the evidence for a potential stem cell of origin for cancer, this concept remains controversial, and there is no evidence to rule out the possibility that cancer is derived from terminally differentiated cells that have acquired these characteristics of stem/progenitor cells by de-differentiation [44]. For the purposes of this chapter, we assume that both the cancer stem cell hypothesis and the de-differentiation hypothesis may be true for different cancers and in different patients. The vast heterogeneity seen between different tumors may be reflective of the state of differentiation of their cell of origin, and tumors may arise from cells at all points along a differentiation pathway (Fig. 4).

5 Epigenetic Silencing of Tumor Suppressor Genes in Cancer – Links to Stem/Precursor Cell Chromatin

Data are now emerging showing a striking link between the stem cell epigenetic regulatory pathways discussed in the preceding section and the aberrant silencing of tumor suppressor genes in cancer. There is now tantalizing evidence that the mechanisms discussed earlier that are responsible for the normal silencing of stem cell regulatory genes have gone awry in a tumor cell of origin, and have stabilized, via addition of abnormal DNA methylation to gene promoter region CpG islands, silencing of groups of genes that should have been up-regulated with normal differentiation [45–47]. While we are beginning to understand more completely the role of repressive chromatin and DNA methylation as it pertains to tumor suppressor gene silencing in cancer, questions remain as to the origin of these marks and their potential relationship to the formation of heterochromatin and the silencing of stem/progenitor cell genes and alternative lineage genes during normal differentiation. We hypothesize, from recent of our data [45], that failure of the stem cell epigenetic regulatory networks likely contributes to the diseased state in cancer – specifically, hypermethylation of developmental regulator genes and tumor suppressor genes. This concept and the holistic concept of abnormal epigenetic silencing of key cancer genes in association with abnormal promoter region DNA methylation are discussed below.

Increased epigenetic silencing of tumor suppressor genes may facilitate abnormal survival of stem/progenitor cells under chronic stress conditions, such as chronic wound healing and inflammation [48, 49] (Fig. 4). Epigenetic gene silencing, and associated promoter CpG island DNA hypermethylation, is an alternative mechanism to mutations by which tumor suppressor genes may be inactivated within a cancer cell [13–16, 50–52]. These epigenetic changes are prevalent in all types of cancer, and their appearance may precede genetic changes in premalignant cells and foster the accumulation of additional genetic and epigenetic hits [49]. These epigenetically modified genes constitute important categories of tumor suppressor genes including cell cycle regulators, pro-differentiation factors, and anti-apoptotic genes [15], and many of these genes are known to play a role in normal development [53–55]. The dynamic, coordinated regulation of these genes during normal development suggests that a normal process gone awry might be responsible for instructively silencing these genes in cancer.

A panel of tumor suppressor and candidate tumor suppressor genes identified by our group and others to be frequently DNA hypermethylated and silenced in adult cancers [45] yields a large list of genes that are virtually all tied to normal differentiation and development, and likely to normally undergo remodeling of their chromatin in stem/progenitor cells undergoing differentiation. Genes have been identified from a mix of cancer cell lines and primary tumor samples, and encompass a wide range of tissue types including brain, breast, colon, esophageal, gastric, head and neck, hematopoetic, kidney/renal, liver, lung, ovarian, pancreatic, prostate, and uterine cancer (Fig. 5). This list is not exhaustive for genes known to be hypermethylated in cancer, but instead is a compilation of genes that are particularly sensitive to

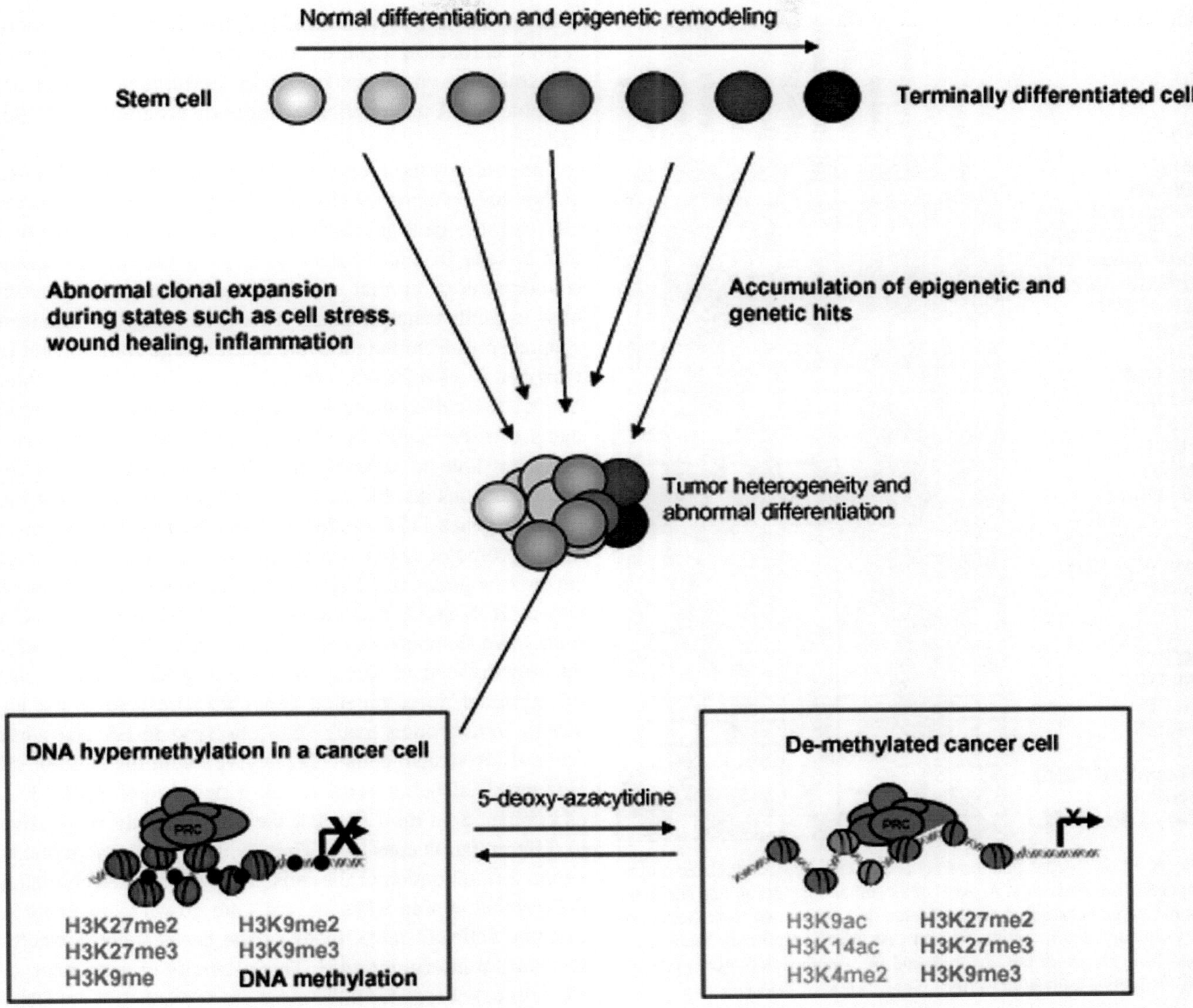

Fig. 4 Cancer stem cells, epigenetic gene silencing events, and tumorigenesis. For at least some pluripotency-associated and alternate lineage genes, gene silencing is a normal event in stem/progenitor cells as adult cell renewal takes place. Silencing is mediated by repressive chromatin modifications, which are malleable such that the genes might be activated for transcription as required by cell maturation. This homeostasis allows stem/progenitor cells to move properly along the differentiation pathway for a given tissue or cell system (*top panel*). During chronic cell injury, stress, wound healing or inflammation, the pressure for adaptive cell renewal draws from the stem/progenitor cell pool for ongoing repair. This pressure might allow for abnormal DNA methylation to be recruited to promoters of normally silenced genes by the chromatin that is in place (*lower left*, DNA hypermethylation, *black circles*). The result is a tight heritable silencing that cannot be easily reactivated as a cell clone expands, thereby channeling the cells towards the abnormal expansion, at the expense of differentiation. This entire scenario leads to benign precursor tumors that are at risk for progression. This progression would be fostered by subsequent genetic or epigenetic events. Chemically depleting DNA methylation using 5-aza-deoxy-cytidine can induce gene expression (*lower right*, demethylated cancer cell), but the gene does not return to the fully euchromatic state seen at a normally transcribed gene [67]. In this intermediate state, the histones contain the transcriptional activating marks H3K4me2 and H3K9/14ac and HP1a is not present. However, several repressive modifications remain and apparently need not change for significant transcription to occur. Histones retaining this mixture of activating and repressive chromatin remain in a compacted, closed, nucleosomal arrangement similar to that observed for normal bivalent chromatin for these same genes in a stem/progenitor cell

methylation, as is represented by their frequent targeting for this epigenetic change. All of the genes were found to be hypermethylated in multiple tumor types, and hypermethylation of several of these genes (CDH1, CDKN2A, and RASSF1) has been observed in all of these tumor types. The high frequency of targeting of genes involved in normal development, including genes such as CDKN2a (*p16*) that can regulate stem cell number and cell cycle functions

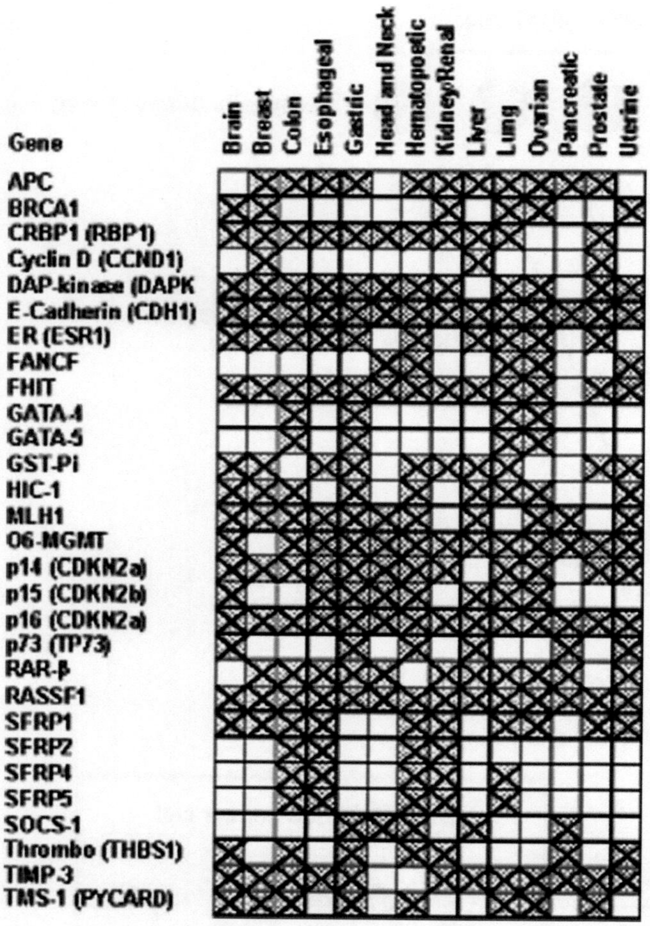

Fig. 5 Known genes that are frequently DNA hypermethylated and silenced in adult cancers. A panel of 29 tumor suppressor, and candidate tumor suppressor genes were selected that are known to be frequently hypermethylated in various cancer cell lines and primary tumor samples from review of the literature and from studies in our own labs (all utilized refs are given in [45])

[56–61], *GATA* transcription factors that regulate differentiation [62], morphogenesis control genes such as *CDH1*, and anti-apoptotic genes such as *DAP-kinase* [15] suggests that the underlying epigenetic controls responsible for controlling expression of these genes in normal stem and progenitor cells may provide an explanation for the targeting of at least some genes for DNA hypermethylation and gene silencing during tumor initiation and progression.

Multiple groups [45–47], have recently compiled evidence that transcriptionally repressive chromatin modifications associated with the promoters of these silenced genes in cancer cells resemble the chromatin modifications inherent to those same genes in normal embryonic stem cells (ES). In this embryonic setting, as mentioned earlier, these modifications may normally help prevent stem/precursor cells from conversion to committed cells until programmed to do so. We have hypothesized [45] that these "transcription-ready" chromatin patterns may render groups of genes vulnerable to errors that result in recruitment of aberrant promoter DNA methylation during early progression of adult cancers.

Interestingly, both normal embryonic stem cells (ES) and embryonal carcinoma cells (EC; Tera-1 and Tera-2) generally lack the gene promoter CpG island DNA methylation that is seen in adult cancer cells [45]. EC are malignant counterparts of normal embryonic germ cells, but in contrast to adult cancers, retain a marked ability to manifest spontaneous differentiation and multilineage commitment *in vitro* and *in vivo* [63–66], though not to the degree of their normal ES cell counterparts. Both ES and EC cells hold many of these genes in a "transcription-ready" state, leaving them able to respond to differentiation cues, and retaining a considerable degree of plasticity for expression of these genes [45]. Analysis of the histone modifications at the promoter regions of many of these important tumor suppressor genes in ES and EC cells reveals a mixture of active (H3K4me2) and repressive (H3K27me3) chromatin marks and demonstrates that the genes that get frequently hypermethylated in cancer are generally within the cluster of important developmental genes that are held in the bivalent, transcription ready state observed in ES and adult tissue-derived stem cells (Fig. 2). Even with the addition of H3K9me3 silencing mark to the genes studied in the EC, as compared to the ES cells, the genes remain responsive to differentiation cues [45]. Thus, with retinoic acid induced neural differentiation of the cells, or with increased epithelial differentiation seen when EC cells are grown as explants in immune deficient mice, many of the genes show distinctly increased transcription [45]. This plasticity of expression in EC cells is not seen for these same genes when they are DNA methylated and deeply, heritably repressed in adult cancers and difficult to reactivate unless the DNA methylation is removed [67].

While the silencing of these genes may play an essential role in tumor initiation or progression, the mechanisms underlying the specific targeting of these genes for DNA hypermethylation remains to be determined. The large numbers of epigenetically silenced genes that may be present in any given tumor, and the clustering of silenced genes within single cell pathways [68], begs the question of whether gene silencing is a series of random events resulting in an enhanced survival of a premalignant clone, or whether silencing is the result of a directed, instructive program for silencing initiation reflective of the cell of origin for the tumors. It in an intriguing possibility that the epigenetic program is linked, at least for silencing of some cancer genes, to the epigenetic control of stem/precursor cell gene expression patterns.

6 Polycomb Group of Transcriptional Repressors

High steady state levels of PcG complexes are especially present in stem and progenitor cells, as well as in tumor cells that have similar properties [69–71]. Recent research largely implicates the polycomb group (PcG) of transcriptional repressors as key players in DNA hypermethylation of tumor suppressor genes in cancer, primarily through the enrichment of the repressive H3K27me3 mark [45–47]. This histone modification is catalyzed by EZH2, a key component of the Polycomb group (PcG) complexes which maintain long-term gene silencing from lower organisms to man and are essential for the normal state of stem/progenitor cells and their commitment to various tissue types [69, 72–74]. Recent studies have linked EZH2 to promoter recruitment of the enzymes that catalyze DNA methylation, the DNA methyltransferases (DNMT's), and have suggested a role for this protein during the induction and targeting of DNA methylation [75]. Interestingly, for many genes frequently hypermethylated in cancer, both normal embryonic stem (ES) cells and embryonal carcinomas (EC) demonstrate an enrichment of key PcG proteins including SUZ12, EZH2, and SirT1, plus the H3K27me3 mark to their promoters [45]. When compared to the above mentioned genome-wide studies of PcG enrichment at human gene promoters in ES and embryonic fibroblasts (EF), approximately 68% of the genes on our panel of genes frequently DNA hypermethylated in adult cancers were also associated with PcG in either the PcG targing studies of ES and/or EF cells [45]. These data were striking, particularly considering, as noted previously above, that for each of the candidate genes we have studied in depth in adult cancer cells, DNA hypermethylation is accompanied by a low, but measurable enrichment of the H3K27 repressive mark [67]. Similar results were seen using 23 genes newly identified as hypermethylated in HCT-116 colon cancer cells [76] for which 56.5% were identified as PcG targets in either ES or EF cells, and is in agreement with data published simultaneously by two additional groups. Thus, Widschwendter et al. [46] show that PRC targets are up to 12-fold more likely to have cancer-specific DNA hypermethylation than nontargets when comparing 177 genes identified as hypermethylated in primary human colorectal tumors. Additionally, Schlesinger et al. [47] demonstrate that of 24 genes identified as specifically DNA hypermethylated in the colon cancer line Caco-2, all also demonstrate concurrent enrichment of the H3K27me3 modification, which is generally lacking in unmethylated controls, and >60% of these genes are PRC regulated in the above studies of ES and EF cells.

Interestingly, for the genes from our original list that are identified as PcG targets, approximately 50% were listed as PcG targets in both ES and EF cells. However, the remaining 50% were listed as unique PcG targets in either ES or EF cells. Those that were PcG targets in ES cells were generally expressed at low levels in EC and up-regulated with differentiation, and those for EF cells were generally expressed at higher levels in EC cells and down-regulated with differentiation [45]. This could explain how different patterns of hypermethylated genes in adult cancers might, then, be reflective of their chromatin status in a cell of origin.

These above links between cancer genes that become DNA hypermethylated and abnormally silenced and Polycomb group genes in stem/precursor cells provides key clues to the actual mechanisms leading to abnormal recruitment of the DNA methylation. PcG marking of genes show a dynamic and fascinating regulation during differentiation. Both the global and promoter specific levels of PcG complex related proteins, including SUZ12, EZH2, and SirT1, fall with *in vitro* differentiation for several genes that are frequently hypermethylated in cancer, such as GATA4 and CDKN2a (p16) [45]. Additionally, several PcG proteins, including Bmi1, Suz12, and Sfmbt show a transient increase in expression at various points during the differentiation process, followed by a lowering of expression as cells enter a more differentiated state [45]. Such data support the extensive work of others in discerning a role for this family during normal differentiation [69]. While low-level enrichment of H3K27me3 mark appears to be nearly ubiquitous at the small subset of DNA hypermethylated cancer gene promoters we have studied to date [67], direct global comparisons of PcG regulation between cancer cells and their proposed stem or progenitor cell of origin should help clarify and measure the full extent of the link between PcG regulation and DNA hypermethylation.

If PcG regulation in stem/progenitor cells leaves developmental genes vulnerable to DNA hypermethylation in a cancer cell of origin, what other repressive marks are responsible for the transition of a gene promoter from a transient "transcription-ready" state to one of heritable, permanent gene silencing and recruitment of DNA hypermethylation in a premalignant cell? While PcG complexes and the H3K27me marks have been associated with recruitment of DNA methylation [67, 75], additional studies suggest that these PcG constituents are not required to maintain such methylation [77]. While EC cells retain the bivalent marks found in ES cells, they show additional enrichment of two key repressive marks: H3K9me3, which is characteristic of silenced transcription in pericentromeric regions [78], and to a lower and more variable extent, H3K9me2 [45]. Both of these H3K9me marks are characteristic of DNA hypermethylated genes in adult cancers [67, 79–84]. In both Neurospora and Arabidopsis, mutations in histone methyltransferases that catalyze H3K9 methylation cause significant loss of genomic DNA methylation [85–89]. Interestingly, in the EC cells, global levels of both of the H3K9me repressive marks are increased considerably as compared to ES cells [45], suggesting a permissive background for the promoter changes in the neoplastic cells and/or a more differentiated cell of origin. In EC cells the fully methylated RASSF1 gene and the minimally methylated sFRP5 gene both demonstrate an increased presence of the H3K9me2 mark at their promoters as compared to unmethylated genes, and overexpression of the PcG protein Bmi1 can cause progressive increases in promoter DNA methylation of the SFRP5 gene in EC cells and a relative increase in H3K9me2 levels localized to the promoter of this gene [45]. In adult cancers, the repressive chromatin present for DNA hypermethylated genes is initially more enriched for H3K9me2 [67, 81] than is seen in the EC cells, and this mark is the only repressive mark that we have studied which is uniformly reduced when DNA hypermethylated genes are demethylated in adult cancer cells [67] (Fig. 4). Studies in colon and breast cancer cells indicate that enrichment of multiple components of transcriptionally repressive chromatin is characteristic of cancer gene promoters silenced in association with aberrant DNA methylation, including H3K27me3 and both H3K9me2 and H3K9me3 [67]. Perhaps most interesting have been findings that these silenced cancer gene promoters, when reactivated by DNA demethylating agents, do not return to a fully euchromatic chromatin state [67]. Rather, while active marks are restored, most repressive histone modification marks remain, including H3K27me3, which is generally increased. The only histone modification that we have found to be consistently altered with chemical demethylation of cancer cell lines is the H3K9me2 modification, and the resulting chromatin absent of DNA methylation is remarkably similar to the bivalent state observed in ES, EF, and EC cells (Fig. 4).

In terms of human cancer biology, our findings and those of others, suggest that a stem cell like promoter "ground state" for these genes may be indicative of the contribution of stem cell and/or progenitor cells to the derivation of adult cancers. These cells may be especially at risk for cancer initiation due to their continued expansion during states such as chronic wound healing and inflammation [48, 67]. Deregulation of this process in premalignant cells during a critical window for chromatin remodeling may result in the inappropriate silencing and DNA hypermethylation of polycomb-regulated tumor suppressor genes in an abnormal clone. Pro-differentiation/lineage specific genes that are similarly PcG regulated, but should have been activated with differentiation may be specifically targeted for silencing. Instead of gene activation, this transient repression and transcription-ready state in stem/progenitor cells may be converted to one of heritable silencing via DNA hypermethylation, and packaged into regions of dense heterochromatin in premalignant cells. The specific pro-differentiation, growth control properties of these genes enhances the likelihood, in these abnormal cell clones, of subsequent tumor initiation and progression (Fig. 4). Determining the localization, targeting and composition of this PcG repressive complex, and the association of this complex with known DNA and histone methyltransferases during differentiation will hopefully increase our understanding of the potential contribution of stem cell epigenetic regulatory networks to cancer initiation and abnormal silencing in a tumor cell of origin.

References

1. Horn PJ, Peterson CL. Molecular biology. Chromatin higher order folding-wrapping up transcription. Science. 2002;297:1824–7.
2. Kiefer JC. Epigenetics in development. Dev Dyn. 2007;236:1144–56.
3. Strahl BD, Allis CD. The language of covalent histone modifications. Nature. 2000;403:41–5.
4. Jenuwein T, Allis CD. Translating the histone code. Science. 2001;293:1074–80.
5. Kouzarides T. Chromatin modifications and their function. Cell. 2007;128:693–705.
6. Pruitt K, Zinn RL, Ohm JE, et al. Inhibition of SIRT1 reactivates silenced cancer genes without loss of promoter DNA hypermethylation. PLoS Genet. 2006;2:e40.
7. Tucker KL, Beard C, Dausmann J, et al. Germ-line passage is required for establishment of methylation and expression patterns of imprinted but not of nonimprinted genes. Genes Dev. 1996;10:1008–20.
8. Xie S, Wang Z, Okano M, et al. Cloning, expression and chromosome locations of the human DNMT3 gene family. Gene. 1999;236:87–95.
9. Bourc'his D, Xu GL, Lin CS, Bollman B, Bestor TH. Dnmt3L and the establishment of maternal genomic imprints. Science. 2001;294:2536–9.
10. Penterman J, Zilberman D, Huh JH, Ballinger T, Henikoff S, Fischer RL. DNA demethylation in the Arabidopsis genome. Proc Natl Acad Sci U S A. 2007;104:6752–7.
11. Agius F, Kapoor A, Zhu JK. Role of the Arabidopsis DNA glycosylase/lyase ROS1 in active DNA demethylation. Proc Natl Acad Sci U S A. 2006;103:11796–801.
12. Bruniquel D, Schwartz RH. Selective, stable demethylation of the interleukin-2 gene enhances transcription by an active process. Nat Immunol. 2003;4:235–40.
13. Jones PA, Laird PW. Cancer epigenetics comes of age. Nat Genet. 1999;21:163–167.
14. Feinberg AP, Tycko B. The history of cancer epigenetics. Nat Rev Cancer. 2004;4:143–53.
15. Herman JG, Baylin SB. Gene silencing in cancer in association with promoter hypermethylation. N Engl J Med. 2003;349:2042–54.
16. Jones PA, Baylin SB. The fundamental role of epigenetic events in cancer. Nat Rev Genet. 2002;3:415–28.
17. Meshorer E, Misteli T. Chromatin in pluripotent embryonic stem cells and differentiation. Nat Rev Mol Cell Biol. 2006;7:540–6.
18. Reik W. Stability and flexibility of epigenetic gene regulation in mammalian development. Nature. 2007;447:425–32.

19. Lee TI, Jenner RG, Boyer LA, et al. Control of developmental regulators by polycomb in human embryonic stem cells. Cell. 2006;125:301–13.
20. Bracken AP, Dietrich N, Pasini D, Hansen KH, Helin K. Genome-wide mapping of Polycomb target genes unravels their roles in cell fate transitions. Genes Dev. 2006;20:1123–1136.
21. Bernstein BE, Mikkelsen TS, Xie X, et al. A bivalent chromatin structure marks key developmental genes in embryonic stem cells. Cell. 2006;125:315–226.
22. Azuara V, Perry P, Sauer S, et al. Chromatin signatures of pluripotent cell lines. Nat Cell Biol. 2006;8:532–8.
23. Mikkelsen TS, Ku M, Jaffe DB, et al. Genome-wide maps of chromatin state in pluripotent and lineage-committed cells. Nature. 2007;448:553–60.
24. Ren B, Robert F, Wyrick JJ, et al. Genome-wide location and function of DNA binding proteins. Science. 2000;290:2306–9.
25. Barski A, Cuddapah S, Cui K, et al. High-resolution profiling of histone methylations in the human genome. Cell. 2007;129: 823–37.
26. Lee TI, Jenner RG, Boyer LA, et al. Control of developmental regulators by Polycomb in human embryonic stem cells. Cell. 2006;125:301–313.
27. Sakashita K, Koike K, Kinoshita T, et al. Dynamic DNA methylation change in the CpG island region of p15 during human myeloid development. J Clin Invest. 2001;108:1195–204.
28. Allegrucci C, Thurston A, Lucas E, Young L. Epigenetics and the germline. Reproduction. 2005;129:137–49.
29. Plath K, Fang J, Mlynarczyk-Evans SK, et al. Role of histone H3 lysine 27 methylation in X inactivation. Science. 2003;300:131–5.
30. Valley CM, Pertz LM, Balakumaran BS, Willard HF. Chromosome-wide, allele-specific analysis of the histone code on the human X chromosome. Hum Mol Genet. 2006;15:2335–47.
31. Kratzer PG, Chapman VM, Lambert H, Evans RE, Liskay RM. Differences in the DNA of the inactive X chromosomes of fetal and extraembryonic tissues of mice. Cell. 1983;33:37–42.
32. Reik W, Lewis A. Co-evolution of X-chromosome inactivation and imprinting in mammals. Nat Rev Genet. 2005;6:403–10.
33. Lopatina NG, Poole JC, Saldanha SN, et al. Control mechanisms in the regulation of telomerase reverse transcriptase expression in differentiating human teratocarcinoma cells. Biochem Biophys Res Commun. 2003;306:650–9.
34. Hattori N, Nishino K, Ko YG, et al. Epigenetic control of mouse Oct-4 gene expression in embryonic stem cells and trophoblast stem cells. J Biol Chem. 2004;279:17063–9.
35. Deb-Rinker P, Ly D, Jezierski A, Sikorska M, Walker PR. Sequential DNA methylation of the Nanog and Oct-4 upstream regions in human NT2 cells during neuronal differentiation. J Biol Chem. 2005;280:6257–60.
36. Feldman N, Gerson A, Fang J, et al. G9a-mediated irreversible epigenetic inactivation of Oct-3/4 during early embryogenesis. Nat Cell Biol. 2006;8:188–94.
37. Hattori N, Imao Y, Nishino K, et al. Epigenetic regulation of Nanog gene in embryonic stem and trophoblast stem cells. Genes Cells. 2007;12:387–96.
38. Yeo S, Jeong S, Kim J, Han JS, Han YM, Kang YK. Characterization of DNA methylation change in stem cell marker genes during differentiation of human embryonic stem cells. Biochem Biophys Res Commun. 2007;359:536–42.
39. Aoto T, Saitoh N, Ichimura T, Niwa H, Nakao M. Nuclear and chromatin reorganization in the MHC-Oct3/4 locus at developmental phases of embryonic stem cell differentiation. Dev Biol. 2006;298:354–67.
40. Al-Hajj M, Wicha MS, Benito-Hernandez A, Morrison SJ, Clarke MF. Prospective identification of tumorigenic breast cancer cells. Proc Natl Acad Sci U S A. 2003;100:3983–8.
41. Reya T, Morrison SJ, Clarke MF, Weissman IL. Stem cells, cancer, and cancer stem cells. Nature. 2001;414:105–11.
42. Clarke MF, Fuller M. Stem cells and cancer: two faces of eve. Cell. 2006;124:1111–5.
43. Harrison DE, Lerner CP. Most primitive hematopoietic stem cells are stimulated to cycle rapidly after treatment with 5-fluorouracil. Blood. 1991;78:1237–40.
44. Rapp UR, Ceteci F, Schreck R. Oncogene-induced plasticity and cancer stem cells. Cell Cycle. 2007;7.
45. Ohm JE, McGarvey KM, Yu X, et al. A stem cell-like chromatin pattern may predispose tumor suppressor genes to DNA hypermethylation and heritable silencing. Nat Genet. 2007;39: 237–42.
46. Widschwendter M, Fiegl H, Egle D, et al. Epigenetic stem cell signature in cancer. Nat Genet. 2007;39:157–8.
47. Schlesinger Y, Straussman R, Keshet I, et al. Polycomb-mediated methylation on Lys27 of histone H3 pre-marks genes for de novo methylation in cancer. Nat Genet. 2007;39:232–6.
48. Beachy PA, Karhadkar SS, Berman DM. Tissue repair and stem cell renewal in carcinogenesis. Nature. 2004;432:324–31.
49. Baylin SB, Ohm JE. Epigenetic gene silencing in cancer – a mechanism for early oncogenic pathway addiction? Nat Rev Cancer. 2006;6:107–16.
50. Hahn WC, Counter CM, Lundberg AS, Beijersbergen RL, Brooks MW, Weinberg RA. Creation of human tumour cells with defined genetic elements. Nature. 1999;400:464–8.
51. Aaltonen LA, Peltomaki P, Leach FS, et al. Clues to the pathogenesis of familial colorectal cancer. Science. 1993;260:812–6.
52. Kinzler KW, Vogelstein B. Cancer-susceptibility genes. Gatekeepers and caretakers. Nature. 1997;386:761–763.
53. Gregorieff A, Clevers H. Wnt signaling in the intestinal epithelium: from endoderm to cancer. Genes Dev. 2005;19:877–90.
54. Furukawa Y. Cell cycle control genes and hematopoietic cell differentiation. Leuk Lymphoma. 2002;43:225–31.
55. Burch JB. Regulation of GATA gene expression during vertebrate development. Semin Cell Dev Biol. 2005;16:71–81.
56. Park IK, Qian D, Kiel M, et al. Bmi-1 is required for maintenance of adult self-renewing haematopoietic stem cells. Nature. 2003;423:302–5.
57. Sharpless NE, Alson S, Chan S, Silver DP, Castrillon DH, DePinho RA. p16(INK4a) and p53 deficiency cooperate in tumorigenesis. Cancer Res. 2002;62:2761–5.
58. Collado M, Blasco MA, Serrano M. Cellular senescence in cancer and aging. Cell. 2007;130:223–33.
59. Molofsky AV, Slutsky SG, Joseph NM, et al. Increasing p16INK4a expression decreases forebrain progenitors and neurogenesis during ageing. Nature. 2006;443:448–52.
60. Janzen V, Forkert R, Fleming HE, et al. Stem-cell ageing modified by the cyclin-dependent kinase inhibitor p16INK4a. Nature. 2006;443:421–6.
61. Krishnamurthy J, Ramsey MR, Ligon KL, et al. p16INK4a induces an age-dependent decline in islet regenerative potential. Nature. 2006;443:453–7.
62. Laverriere AC, MacNeill C, Mueller C, Poelmann RE, Burch JB, Evans T. GATA-4/5/6, a subfamily of three transcription factors transcribed in developing heart and gut. J Biol Chem. 1994;269:23177–84.
63. Andrews PW. Human teratocarcinomas. Biochim Biophys Acta. 1988;948:17–36.
64. Andrews PW. Retinoic acid induces neuronal differentiation of a cloned human embryonal carcinoma cell line in vitro. Dev Biol. 1984;103:285–93.
65. Mintz B, Illmensee K. Normal genetically mosaic mice produced from malignant teratocarcinoma cells. Proc Natl Acad Sci U S A. 1975;72:3585–9.

66. Palmiter RD, Chen HY, Brinster RL. Differential regulation of metallothionein-thymidine kinase fusion genes in transgenic mice and their offspring. Cell. 1982;29:701–10.
67. McGarvey KM, Fahrner JA, Greene E, Martens J, Jenuwein T, Baylin SB. Silenced tumor suppressor genes reactivated by DNA demethylation do not return to a fully euchromatic chromatin state. Cancer Res. 2006;66:3541–3549.
68. Jones PA, Baylin SB. The epigenomics of cancer. Cell. 2007;128:683–92.
69. Kuzmichev A, Margueron R, Vaquero A, et al. Composition and histone substrates of polycomb repressive group complexes change during cellular differentiation. PNAS. 2005;102: 1859–1864.
70. Kleer CG, Cao Q, Varambally S, et al. EZH2 is a marker of aggressive breast cancer and promotes neoplastic transformation of breast epithelial cells. Proc Natl Acad Sci U S A. 2003;100: 11606–11.
71. Kirmizis A, Bartley SM, Farnham PJ. Identification of the polycomb group protein SU(Z)12 as a potential molecular target for human cancer therapy. Mol Cancer Ther. 2003;2:113–21.
72. Lund AavL, M. Polycomb complexes and silencing mechanisms. Curr Opin Genet Dev 2004;16:1–8.
73. Valk-Lingbeek ME, Bruggeman SW, van Lohuizen M. Stem cells and cancer;the polycomb connection. Cell. 2004;118:409–18.
74. Otte AP, Kwaks TH. Gene repression by Polycomb group protein complexes: a distinct complex for every occasion? Curr Opin Genet Dev. 2003;13:448–54.
75. Vire E, Brenner C, Deplus R, et al. The Polycomb group protein EZH2 directly controls DNA methylation. Nature. 2006;439: 871–4.
76. Schuebel KE, Chen W, Cope L, et al. Comparing the DNA hypermethylome with gene mutations in human colorectal cancer. PLoS Genet. 2007;3:e157.
77. McGarvey KM, Greene E, Fahrner JA, Jenuwein T, Baylin SB. DNA methylation and complete transcriptional silencing of cancer genes persist after depletion of EZH2. Cancer Res. 2007;67: 5097–102.
78. Schotta G, Lachner M, Sarma K, et al. A silencing pathway to induce H3-K9 and H4-K20 trimethylation at constitutive heterochromatin. Genes Dev. 2004;18:1251–62.
79. Lachner M, O'Sullivan RJ, Jenuwein T. An epigenetic road map for histone lysine methylation. J Cell Sci. 2003;116:2117–24.
80. Nguyen CT, Gonzales FA, Jones PA. Altered chromatin structure associated with methylation-induced gene silencing in cancer cells: correlation of accessibility, methylation, MeCP2 binding and acetylation. Nucleic Acids Res. 2001;29:4598–606.
81. Fahrner JA, Eguchi S, Herman JG, Baylin SB. Dependence of histone modifications and gene expression on DNA hypermethylation in cancer. Cancer Res. 2002;62:7213–8.
82. Kouzarides T. Histone methylation in transcriptional control. Curr Opin Genet Dev. 2002;12:198–209.
83. Briggs SD, Xiao T, Sun ZW, et al. Gene silencing: trans-histone regulatory pathway in chromatin. Nature. 2002;418:498.
84. Fischle W, Wang Y, Allis CD. Histone and chromatin cross-talk. Curr Opin Cell Biol. 2003;15:172–83.
85. Tamaru H, Selker EU. A histone H3 methyltransferase controls DNA methylation in Neurospora crassa. Nature. 2001;414: 277–83.
86. Tamaru H, Zhang X, McMillen D, et al. Trimethylated lysine 9 of histone H3 is a mark for DNA methylation in Neurospora crassa. Nat Genet. 2003;34:75–9.
87. Johnson L, Cao X, Jacobsen S. Interplay between two epigenetic marks. DNA methylation and histone H3 lysine 9 methylation. Curr Biol. 2002;12:1360–7.
88. Malagnac F, Bartee L, Bender J. An Arabidopsis SET domain protein required for maintenance but not establishment of DNA methylation. EMBO J. 2002;21:6842–52.
89. Jackson JP, Johnson L, Jasencakova Z, et al. Dimethylation of histone H3 lysine 9 is a critical mark for DNA methylation and gene silencing in Arabidopsis thaliana. Chromosoma. 2004;112: 308–15.
90. Schuettengruber B, Chourrout D, Vervoort M, Leblanc B, Cavalli G. Genome regulation by polycomb and trithorax proteins. Cell. 2007;128:735–45.

Epigenetic Signature of Embryonal Stem Cells: A DNA Methylation Perspective

Monther Abu-Remaileh and Yehudit Bergman

Abstract Specific epigenetic features underpin the pluripotency of ES cells. ES cells have a unique DNA methylation signature, express high levels of DNA de novo methyltransferases, and, unlike somatic cells, are capable of methylating exogenously introduced DNA. At the same time, ES cells protect specific CpG dinucleotides from undergoing de novo methylation, a process that clearly mimics what occurs in the normal embryo. ES cells protect CpG island– as well as non-CpG island–promoters that direct expression of genes involved in stem cell identity from de novo methylation. These promoters are apparently protected by virtue of inherent common sequence elements through binding of transcription machinery related factors. These mechanisms are critical for setting up the correct genome methylation pattern, which is mostly stable in somatic cells. Genes that belong to a self-organizing network of transcription that prevents differentiation and promote proliferation and pluripotency, such as Oct-3/4 and Nanog, are silenced during differentiation by histone modification as well as by DNA methylation. Indeed, genetic experiments have supported the notion that histone modification directs DNA methylation, which represents a second-line epigenetic change, the role of which is to permanently silence gene expression, thereby preventing reprogramming.

Keywords ES cells · Epigenetics · DNA methylation · CpG islands · Pluripotency · Reprogramming

1 Introduction

Normal development appears to take place through a unidirectional process characterized by a step-wise decrease in developmental potential. Changes in gene expression during development are accompanied or caused by epigenetic regulation. Epigenetic mechanisms are defined as a heritable code, other than the genomic sequence, that regulate and maintain gene expression patterns through DNA replication. The epigenetic information encompasses methylation of DNA at CpG sequences, modification of histone tails, variant histones, nuclear localization, replication timing, and the presence of non-nucleosomal chromatin associated proteins. DNA methylation occurs almost exclusively in a CpG dinucleotide context by adding a methyl group on the cytosine nucleotide [1]. On the other hand, the core histones that make up the nucleosome are subject to more than 100 modifications including acetylation, methylation, phosphorylation, and ubiquitination [2]. Histone modifications are known to affect the chromatin structure and thus regulate gene expression at the transcriptional level [3]. For example, acetylation of histone 3 and histone 4 (H3 and H4) [4] or di- or trimethylation (me) of lysine 4 of H3 (H3K4me) [5] are associated with active transcription. In contrast, methylation of lysine 9 of H3 (H3K9me) [6] and methylation of lysine 27 of H3 (H3K27me) [7, 8] mark repressed genes.

DNA methylation regulates a number of biological processes, including genomic imprinting, X chromosome inactivation, silencing of tumor suppressor genes, and repression of retroviral elements [1, 9]. Loss of methylation in mice results in severe developmental defects and early embryonic lethality [10–12]. Moreover, aberrant methylation patterns are thought to be involved in tumorigenesis [13–15] causing genomic instability, abnormal imprinting, and deregulated expression of oncogenes or tumor suppressor genes.

DNA methylation has been divided into two functionally distinct groups: de novo and maintenance methylation. De novo methylation is catalyzed by DNA methylatransferase 3a and 3b (DNMT3a and DNMT3b) and is important for the establishment of methylation patterns in early embryos, during development, and during carcinogenesis [11]. In order to maintain these methylation patterns set by de novo methylation, DNMT1 conducts maintenance methylation [16, 17]. It shows high affinity for hemimethylated

Y. Bergman (✉)
The Hubert Humphrey Center for Experimental Medicine and Cancer Research, Hebrew University Medical School, Ein Kerem, Jerusalem 91120, Israel
e-mail: yehudit.bergman@huji.ac.il

V.K. Rajasekhar, M.C. Vemuri (eds.), *Regulatory Networks in Stem Cells*, Stem Cell Biology and Regenerative Medicine,
DOI 10.1007/978-1-60327-227-8_20, © Humana Press, a part of Springer Science+Business Media, LLC 2009

substrates and is localized to the replication fork during cellular division via association with PCNA. More recently, it was shown that the protein UHRF1 (ubiquitin-like containing PHD and RING finger domains 1) colocalizes with DNMT1 throughout S phase. It appears to tether DNMT1 to chromatin through direct interaction with DNMT1. In addition, UHRF1 shows strong preferential binding to hemimethylated CG sites, the physiological substrate of DNMT1. Thus, UHRF1 may serve as a cofactor that helps in recruitment of DNMT1 to hemimethylated DNA to facilitate faithful maintenance of DNA methylation [18].

DNA methylation patterns undergo genome-wide alterations that occur immediately after fertilization and during early-preimplantation development. There is extensive and rapid genome-wide demethylation of the paternal genome [19, 20]. This demethylation occurs after the removal of protamines (basic proteins that are associated with DNA in the sperm) and the acquisition of histones during the long G1 phase, prior to DNA replication [21]. Interestingly, the maternal genome escapes this process and it seems that PGC7/Stella functions as a protector against the global demethylation, since in PGC7/Stella-deficient oocytes, the maternal genome is massively demethylated [22]. The molecular mechanism that underlies this active demethylation (i.e., replication independent) in the zygote is still unknown. However, there are several candidates that have been indicated as possibly being involved, such as, AID [23] and Gadd45 [24]. The demethylation of the paternal genome is followed by total genome demethylation, reaching an overall low methylation levels at the blastocyst stage [21]. Following implantation, the ICM of the blastocyst starts to reacquire methylation marks at CpG dinucleotides. In contrast, CpG islands that are defined as being longer than 500 bp and having a GC content greater than 55% and an observed CpG/expected CpG ratio of 0.65, are mostly protected from this wave of de novo methylation (CpG islands in inactivated genes on the X chromosome do however become methylated). Thus, in somatic cells between 70% and 90% of CpG dinucleotides in the genome are methylated, whereas most CpG islands are unmodified [13].

2 ES Cells

Embryonic stem (ES) cells, which are derived from the inner cell mass (ICM) of mammalian blastocysts, are characterized by their unlimited potential for self-renewal and their capacity to differentiate into all kinds of somatic cell types in vitro and in vivo [25, 26]. More recently, pluripotent stem cells have been derived also from spermatogonial stem cells [27]. This suggests that the network of transcription factors and epigenetic regulators capable of supporting pluripotency may be maintained during germ cell development. The ES cells exhibit a pluripotent state in vitro that may correspond to, but is not identical to, the transient pluripotent phenotype of the primitive ectoderm cells in the ICM, that are thought to progress rapidly to form the epiblast cells [28]. Thus, ES epigenetic state most probably is not identical to the one of the primitive ectoderm cells. With this reservation, ES cells are still the best model system for the unique epigenome of the ICM cells during the early stages of embryogenesis when plasticity of the genome is at its best. For this reason, in this review we will confine our discussion to the recent advances in studying the ES cell epigenome, while concentrating on the DNA methylation in order to understand its vital role both in self-renewal and in pluripotency of embryonic stem cells.

2.1 A Transcription Factor Network Controls ES Cells Identity

In order to elucidate the vital role of the ES cell epigenome, including the pluripotency-specific DNA methylation pattern, it is noteworthy to review the recent advances in studying the molecular transcriptional networks that maintain ES cell identity. Oct-3/4, Nanog and Sox2 transcription factors have been identified as crucial regulators of pluripotency. Oct-3/4 is the earliest expressed transcription factor that is known to be crucial in murine pre-implantation development [29]. The Oct-3/4 gene is a member of the POU family of transcription factors; it is expressed in ES cells and in embryonic carcinoma (EC) cells [29, 30]. Oct-3/4 is essential for the pluripotent identity of the founder cell population in the ICM [31]. Furthermore, it was shown that a critical amount of Oct-3/4 is required to sustain stem cell self-renewal, and any up- or down-regulation induces divergent developmental programs [32]. Nanog is an NK-2 class homeobox transcription factor that is expressed throughout the pluripotent cells of the ICM but is down-regulated in extraembryonic lineages and in pluripotent cells of the peri-implantation embryo [33]. Forced expression of Nanog, unlike that of Oct-3/4, is sufficient to maintain the pluripotent state in mouse ES cells in the absence of LIF (leukemia inhibitory factor) [33, 34], and in human ES cells results in cells that can be propagated in the absence of feeder cells and associated signaling [35]. Nanog-null embryos fail to maintain the pluripotent lineage and arrest at peri-implantation. The third transcription factor, Sox2, is a member of high mobility group (HMG) protein family that is expressed in the ICM, early primitive ectoderm, anterior primitive ectoderm, germ cells, and multipotent extra embryonic ectoderm cells [36, 37]. Although Sox2 expression is not restricted to pluripotent cells it plays an important role in the maintenance of pluripotency and lineage specification. Sox2-null embryos arrest at a similar time to Oct-3/4- and Nanog-null embryos, that is, around the time of implantation. Knockdown of Sox2 in mouse ES cells induces differentiation into multiple lineages, including trophectoderm [38], supporting its role in the maintenance of pluripotency [36].

These three transcription factors, which determine early cell fate decisions and regulate ES cell pluripotency, bind to their own promoters to form an interconnected self-organizing network [39]. Recent studies have enabled the construction of transcriptional regulatory networks in ES cells that provide a foundation for understanding how these factors control pluripotency and influence subsequent differentiation events. Using RNA interference technology, microarray analysis, and genome-wide chromatin immunoprecipitation experiments, numerous target genes bound by Oct-3/4, Nanog, and Sox2 have been identified. These factors appear to form a tight transcriptional regulatory circuit that maintains mouse and human ES cells in a pluripotent state [38–41]. Interestingly, among Oct-3/4 bound genes half are also bound by Sox2, and more than 90% of the promoter regions occupied by Oct-3/4 and Sox2 are bound by Nanog, thus forming an interconnected autoregulating loop that maintains ES cell identity.

The growing understanding of the transcriptional regulation in ES cells has led to the discovery of the factors that can turn terminally differentiated cells into ES cell-like cells. Co-introduction of four transgenes encoding the transcription factors Oct-3/4, Sox2, c-Myc, and Klf4 into somatic cells, such as embryonic and adult tail-tip fibroblasts, resulted in the generation of induced pluripotent stem (iPS) cells, which gave rise to chimeric embryos following their injection into mouse blastocysts (Fig. 1) [42]. Several studies published this year not only reproduced but also extended these findings by demonstrating the pluripotency and differentiation potential of mouse iPS cells in rigorous developmental assays [43, 44]. Both Klf4 and c-Myc are oncogenes [45, 46], which play a role in reducing ES differentiation [47] and promoting ES self-renewal [48], respectively. It is thought that these four factors establish pluripotency in somatic cells

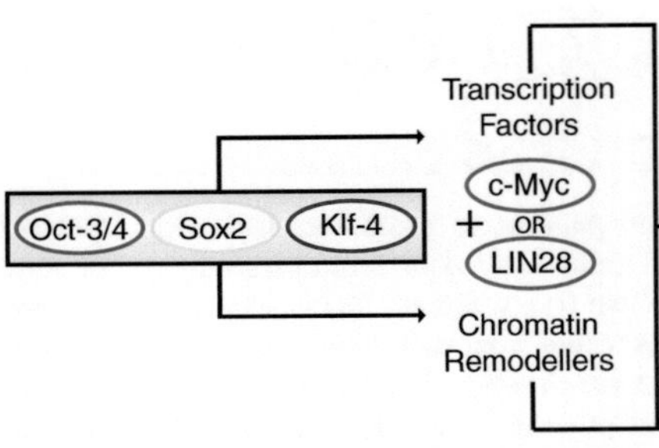

Fig. 1 Transcriptional networks that control reprogramming. A set of four transcription factors apparently regulates transcription factors and chromatin remodelers that in turn induce reprogramming

as follows. First, c-Myc promotes DNA replication, thus relaxing the chromatin structure, which in turn allows Oct-3/4 to access its target genes. Sox2 and Klf4 also co-operate with Oct-3/4 to activate target genes that encode transcription factors that establish the pluripotent transcription factor network (such as Nanog). Interestingly, these transcription factors activate epigenetic regulators, including several histone demethylases that were shown to be positively regulated by Oct-3/4 [49], and that are involved in establishing the pluripotent epigenome. The induction of pluripotency was recently achieved in human cells as well. Human fibroblasts were reprogrammed into human iPS [50, 51], using either the same four factors described above, or a new "mix" containing Oct-3/4, Sox2, Klf4, and LIN28, the latter being an RNA-binding protein that is highly expressed in human ES cells but down-regulated during ES cell differentiation [52].

2.2 Pluripotency and Chromatin Structure

Specific epigenetic features underpin the pluripotency of ES cells. Recent studies have demonstrated that ES cell chromatin is in a highly dynamic state with an apparently transient association of chromatin structural proteins, which is reflected in the relatively decondensed chromatin of ES cells [53]. This dynamic exchange of chromatin proteins, including histones within intact chromatin in ES cells, is not a function of replication, but rather a potentially ES-unique mechanism whereby histone modifications might be dynamically deposited on and off the chromatin at development control genes. Moreover, there is a general abundance of transcriptionally active chromatin marks such as trimethylation of lysine 4 of histone H3 (H3K4me3) and acetylation of histone H4 (H4Ac) [54]. Genome-wide analyses of histone modifications have suggested that, in ES cells, genes encoding regulators of early development are associated with bivalent chromatin domains (Fig. 2) [54, 55]. These regions have the repressive mark of trimethylation of lysine 27 on histone H3 (H3K27me3) as well as the H3K4me3 mark that is associated with active genes. The presence of these opposing marks suggests that these genes are poised to be released from repression as soon as ES cells will be induced to differentiate. Interestingly, many of the regions that are bivalently marked are also bound by the key transcription factors associated with pluripotency; Oct-3/4, Nanog and Sox2 (see above). However, more recent studies indicate the existence of bivalent domains in differentiated cells, which may indicate that they play different functions in these cells [56, 57]. Alternatively, one might suggest that the bivalent chromatin domains in ES cells may harbor an additional mark, unique to ES cells, turning them into trivalent domains (Fig. 2).

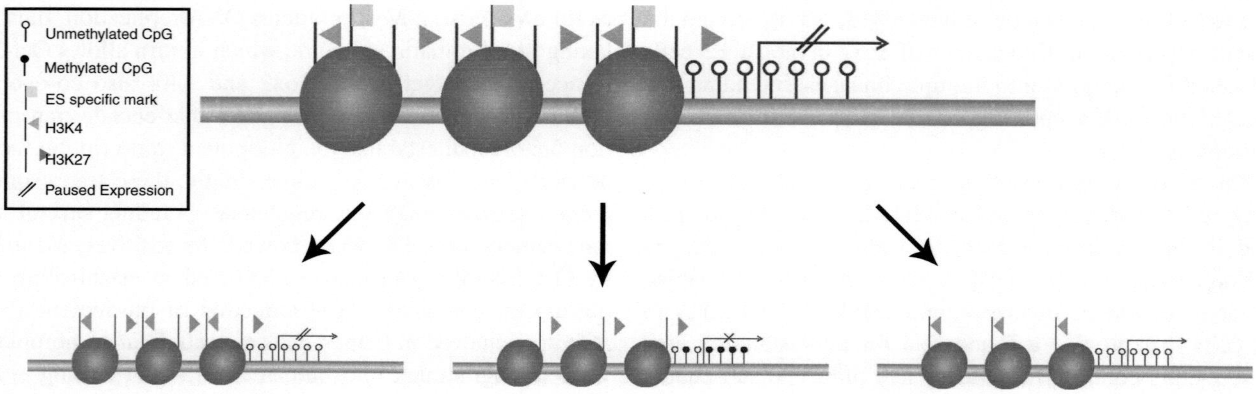

Fig. 2 A model for ES cell specific chromatin marking. In ES cells genes encoding regulators of early development are associated with trivalent chromatin domains (H3K27me3, H3K4me3, and an unidentified ES-specific mark). Upon differentiation, active genes are marked with H3K4me3, inactive genes with H3K27me3 and CpG methylation, and poised genes with both H3K4me3 and H3K27me3

3 DNA Methylation Pattern in ES Cells

ES cells have several unique features that are required to maintain their self-renewal and pluripotent potential. Unlike somatic cells, ES cells are capable of de novo methylating exogenously introduced DNA while at the same time protecting CpG islands sequences. This process clearly mimics what occurs in the normal embryo [58]. A more recent study shows that ES cells have a unique DNA methylation signature when compared to either differentiated somatic cells or cancer cells, which may indicate the importance of the methylation profile in determining ES cell character [59].

3.1 De Novo Methylation in ES Cells

The establishment of DNA methylation patterns requires de novo methylation that occurs mainly during early development and gametogenesis in mice. De novo methylation activity is detected predominantly in early embryos, embryonic carcinoma (EC) cells and ES cells [60–62]. ES cells express, in addition to DNMT1, high levels of the de novo methylases DNMT3a, DNMT3a2 (a shorter isoform of DNMT3a that is the predominant form in ES cells), DNMT3b1, and DNMT3b6 (two isoforms of DNMT3b) (Fig. 3A). Accordingly, ES cells show much higher rates of de novo methylation than do differentiated somatic cells (Fig. 4). Deletion of DNMT3a and DNMT3b results in global hypomethylation of genomic DNA and partial resistance to differentiation of mouse ES cells [63, 64]. These findings indicate that DNA methylation plays a pivotal role in gene regulation during differentiation and development. This de novo methylation activity makes stem cells a preferred system to study de novo methylation mechanisms.

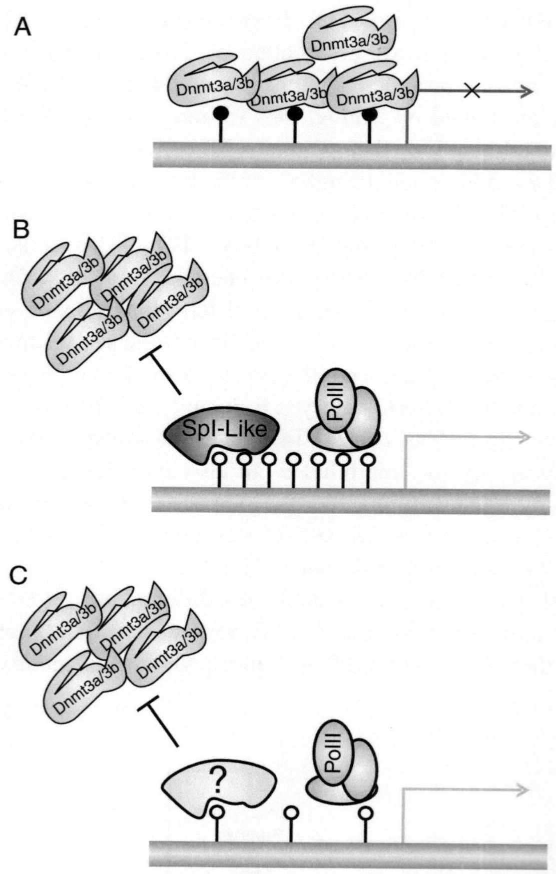

Fig. 3 DNA methylation patterns of promoters harboring CpG and non-CpG islands. In ES cells, CpG poor promoters undergo de novo methylation and are inactive (**A**), whereas CpG islands containing promoters are active and protected from undergoing de novo methylation by virtue of binding factors such as Sp1-like and pre-initiation transcriptional complex (**B**). Non-CpG island promoters that should be active are also protected in ES cells by *trans*-acting factors (**C**)

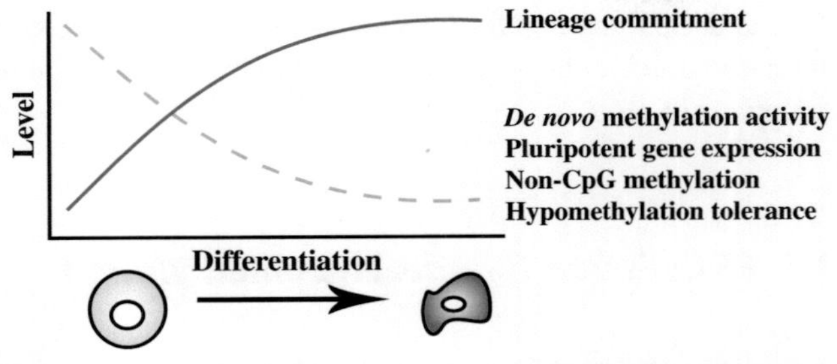

Fig. 4 Changes occurring as a function of ES cell differentiation (see text)

Recently, a novel mechanism for de novo methylation of imprinted genes during gametogenesis has been identified. Dnmt3L, an enzymatically inactive member of Dnmt3 family, was shown to be required for the establishment of maternal methylation imprints in mouse [65] and for the de novo methylation of dispersed repeated sequence in the male genome [66]. Moreover, it was shown that Dnmt3L physically interacts with DNMT3a and DNMT3b [67]. Co-crystallization experiments have shown that Dnmt3L recognizes histone H3 tails that are unmethylated at lysine 4 and induces de novo DNA methylation by recruitment or activation of DNMT3a2, a germline-specific isoform of DNMT3a [68]. Thus, DnmtL has dual functions of binding the unmethylated histone H3 tail and activating the DNA methyltransferase. Hence, H3K4 methylation could protect unmethylated sequences from DNA methylation by the DNMT3a-Dnmt3L complex. Furthermore, a recently published study has shown that DNMT3a binds to Dntmt3L in a conformation that allows the methylation of CpG sites at distances of 8–10 nucleotides [69]. This periodic pattern can be seen in regions other than imprinting control regions suggesting that this novel code is a general target site for de novo methylation by DNMT3a-Dnmt3L complex in other cell systems. ES cells are very good candidates for such methylation model, given that these cells express both DNMT3a2 and Dnmt3L at high levels [70, 71]. Thus, Dnmt3L could convert patterns of histone H3 methylation, which are unknown to be transmitted by mitotic inheritance, into patterns of DNA methylation that mediate the heritable transcriptional silencing of the affected sequences.

Although the enzymes responsible for methylation patterns have been identified, the precise molecular mechanisms, including cofactors that lead to recruitment and efficient targeting of the enzymatic machinery to its appropriate sites, are not all known. In addition to the Dnmt3L, Lsh is another cofactor that affects the activity of the Dnmts. It belongs to the SNF2 family of proteins that participate in chromatin remodeling and specifically interacts with DNMT3a and DNMT3b, but not DNMT1, in ES cells [72]. Interestingly, Lsh-deficient ES cells (and not 3T3 cells) show genome-wide CpG hypomethylation, supporting its role in de novo methylation [72].

3.2 CpG Islands Are Demethylated in ES Cells

CpG islands are often but not always found in promoter regions, and about 40% of genes contain CpG islands that are situated at the end of the 5′ region (promoter, untranslated region, and exon 1) [13]. The rest of the genome, such as the intergenic and the intronic regions, is considered to be CpG poor. In normal cells, CpG poor regions are usually methylated whereas CpG islands are generally hypomethylated, with a few exceptions including the inactive X chromosome. During the development of cancer, many CpG islands undergo hypermethylation while the CpG poor regions become hypomethylated. This alteration in DNA methylation pattern leads to changes in chromatin structure causing the silencing of tumor suppressor genes and instability of the genome [13]. This change in methylation pattern during cancer is similar to the pattern observed on the inactive X chromosome.

In ES cells it is thought that CpG islands are protected from methylation by virtue of inherent common sequence elements (Fig. 3B). One mechanism by which the islands may be protected is by *trans* acting factors that bind specific *cis* regulatory elements. Indeed, *cis* regulatory sequences, such as Sp1-like binding sites were shown to play a key role in protecting CpG islands from de novo methylation [62]. The loss of these elements induces the methylation of exogenously introduced CpG islands in ES cells [58]. Interestingly, ES cells have a unique ability to demethylate CpG islands that are in vitro methylated [58]. In terminally differentiated fibroblasts, however, these CpG islands are not demethylated [73]. This may indicate that the demethylation of CpG islands could be restricted to ES cells.

The full spectra of the molecular mechanisms that are involved in CpG island recognition are still not known. Most probably, binding of the transcriptional pre-initiation complex is one of them. These mechanisms are critical for setting up the correct genome methylation pattern since those that have been acquired during development are mostly stable in somatic cells and, in turn, are involved in regulating gene expression.

3.3 ES Cells Can Protect Other CpGs from De Novo Methylation

The ability of ES cells to protect CpGs from undergoing de novo methylation in the presence of a predominant de novo methylation process [11, 70] is not limited to CpG islands. This phenomenon is of great importance in maintaining pluripotency and self-renewal of ES cells. Oct-3/4 is the earliest expressed transcription factor that is known to be crucial in murine pre-implantation development [29]. The promoter region of Oct-3/4, which is not a bona fide CpG island, is hypomethylated in the blastula stage and in cultured ES cells but undergoes de novo methylation during subsequent stages of embryogenesis and in differentiated ES cells [74, 75]. Moreover, the Oct-3/4 enhancer element was shown to play a role in protecting the Oct-3/4 from de novo methylation, both in ES cells as well as during early embryogenesis. Additional studies have also identified this phenomenon in other genes that are required for the maintenance of the stemness identity, such as Nanog and Fgf4 [76, 77]. These studies show that the ability of ES cells to maintain a constitutive expression of a subset of genes required for stem cell identity is governed, in part, by an active mechanism that recognizes the regulatory elements in these genes and protects the CpG sites from undergoing de novo methylation (Fig. 3C).

It seems that the activity of some components that play a role in this mechanism is either lost or inhibited during ES cells differentiation and embryonic development, which allows these regions to become methylated and thus irreversibly repressed. Interestingly, a recent study showed that this protection is not limited to CpG sites located in promoter regions of active genes since unmethylated windows of CpG dinucleotides were found to mark enhancers of some tissue-specific genes in ES cells. The unmethylated windows expand in cells that express the gene and contract, disappear, or remain unchanged in nonexpressing tissues. Moreover, ES cells were capable of demethylating premethylated constructs containing these enhancers when these were integrated into the ES cell genome and the unmethylated windows readily appeared. This demethylation was not achieved in somatic cells, even in those that express these genes [78]. Thus, an enhancer capable of promoting the loss of DNA methylation in ES cells cannot promote the loss of methylation in cells that express the corresponding endogenous genes. This unique property of ES cells calls for further investigations.

3.4 Non-CpG Methylation Is Prevalent in ES Cells

Non-CpG methylation has been extensively analyzed in plants and fungi and might add an additional layer of complexity to the epigenetic code. For example, it has been shown that non-CpG methylation by the Arabidopsis CHROMOMETHYLASE3 plays a specific role in the silencing of retroviral gene sequences [79]. Nevertheless, the existence of non-CpG methylation in mammalian DNA has been a contentious issue since it was thought that mammalian methylation occurs predominantly (or even exclusively) at CpG dinucleotides [80]. Recent advances in the technology used to detect DNA methylation in mammalian genome have allowed the researchers to detect CpA and CpT methylation. Cytosine methylation levels in these two dinucleotides was much higher in ES cells than in somatic cells (Fig. 4) [11, 81].

3.5 ES Cells Can Tolerate Hypomethylation

Reduction of DNMT1 activity in mice has been shown to induce tumors, chromosomal instability [82, 83], and global hypomethylation in male germ cells that results in meiotic abnormality [66]. Inactivation of both DNMT3a and DNMT3b abolishes de novo methylation in mouse embryos [11]. ES cells lacking DNMT1 or/and DNMT3a/3b are able to survive and maintain their self-renewal capacity [63, 64, 84], however, these cells lose their pluripotency and their differentiation potential was severely blocked. A similar phenomenon was also observed when ES cells were deprived of CpG binding protein [85]. This situation is in sharp contrast to DNMT3a/DNMT3b–deficient hematopoietic stem cells (adult stem cells), which progressively lose their replication potential, but not differentiation potential [86]. The ability of undifferentiated ES cells to retain chromosomal stability without CpG methylation may indicate that ES cells are able to maintain stable heterochromatin and intact chromosomes by an epigenetic mechanism that is independent of CpG methylation. This idea is in agreement with previous studies that have shown the existence of specific features of chromatin structure in ES cells [53].

4 Genetic and Epigenetic Reprogramming

It is now well established that somatic nuclei can be reprogrammed into a pluripotent state by several experimental manipulations, such as somatic nuclear transfer into oocyte cytoplasm (SCNT) [87], fusion with ES cells [88], and ectopic expression of selected transcription factors [42]. The low efficacy of the reprogramming process indicates that epigenetic modification must occur to allow this process to proceed.

As development and differentiation proceed, differentiated cells accumulate epigenetic marks that differ from those of pluripotent cells. Epigenetic marks can be of short-term flexibility, which can be removed independently of or within very few cell divisions, and of long-term stability and heritability. During the early stages of development, genes that are required later in development are transiently held in a repressed state of histone modifications, which are highly flexible and easily reversed when the expression of these genes is needed. Interestingly, genes that are crucial for pluripotency are silenced during differentiation by histone modifications as well as by DNA methylation. For example, expression of Oct-3/4 and Nanog, two genes that encode pluripotency-sustaining transcription factors, are silenced both by histone modifications and by DNA methylation in differentiated ES and somatic tissues [74–76].

Repressed chromatin is characterized by methylation of DNA at CpG dinucleotides and by chromatin modifications. Among the various histone modifications, lysine methylation stands out due to its multivalency, relative stability, and potential cross-talk with effector proteins. Di- and trimethylation of histone H3K9 create binding sites for the chromodomain-containing proteins of the HP1 family [89]. There are several euchromatic histone methyltransferases, such as Suv39h [90], G9a [91], the closely related GLP/Eμ-HMTase [92], and Eset/Setdb1 [93, 94], that can repress gene activity by inducing local H3K9me2 and H3K9me3 modifications at target promoters. Suv39h1 and 2 play dominant roles in pericentromeric heterochromatin formation, whereas G9a is involved in methylating nonheterochromatic loci that are involved with transcriptionally active genes [95, 96].

For many years it has been suspected that the chromatin structure could recruit DNA methylatransferases to specific loci. Indeed, links between histone methylation and DNA methylation have been emerging [97, 98]. Genetic studies have suggested that Suv39h is specifically required for DNMT3b-dependent DNA methylation at pericentric repeats [99]. In Arabidopsis, mutations in the histone H3K9 methyltransferase caused significant loss of genomic DNA methylation [100]. It was recently shown that Ezh2, a histone H3 lysine 27–specific methylase that brings about heterochromatinization through the binding of a chromodomain protein, is also capable of causing local de novo DNA methylation by directly recruiting Dnmts [97]. Tethering of the methylases to the DNA is most probably directed by specific factors [101]. It should be noted that several studies have demonstrated that HP1 itself, a protein that specifically binds to methylated lysine 9 on histone H3, can recruit Dnmts [98, 102, 103], and this may provide an auxiliary mechanism for targeting and maintaining DNA methylation at heterochromatin regions in animal cells. Only in the case of Neurospora, however, has it been shown that an HP1-like molecule is actually required for local de novo methylation [98].

Genetic disruption of G9a in mouse ES cells resulted in significant disruption of histone H3K9me2 with concomitant redistribution of HP1γ from euchromatin to pericentromeric heterochromatin [91, 104]. This deficiency causes DNA demethylation at 1.6% of the total Not1 sites analyzed in ES cells, indicating that G9a site-selectively contributes to DNA methylation in the genome [105]. Furthermore, it was shown that histone methylation by G9a increases recruitment of HP1, which in turns interacts with DNMT1, resulting in a further increase in DNA methylation [103]. G9a was also shown to both directly bind DNMT1 and colocalize with it in the nucleus during replication. In this case, DNMT1 is the primary loading factor since siRNA knockdown of DNMT1 impairs G9a loading as well as histone H3 methylation and de novo DNA methylation [103]. Furthermore, G9a was shown to be required for DNA methylation at the imprinted gene Snrpn, which is located in euchromatin [106] as well as for the Oct-3/4 transcription factor [75].

It seems that G9a plays multiple roles in Oct-3/4 heterochromatinization. First, it is involved in histone deacetylation at the Oct-3/4 promoter region. Once histone H3K9 is deacetylated, G9a instigates di- and tri-methylation of lysine 9. In this regard, it should be noted that G9a actually forms a stoichiometric heterodimeric complex with another methylase, GLP/Eμ-Hmtase1, and both probably function cooperatively to mediate this K9 modification [104]. Then, DNA methylation is catalyzed by DNMT3a and DNMT3b, which are recruited to the Oct-3/4 promoter through the interaction with G9a [107]. Genetic experiments have supported the notion that histone methylation directs DNA methylation, since when DNA methylation is inhibited there is no effect on histone methylation of the Oct-3/4 promoter [75]. The same is true for repeat sequences or CpG islands [108]. Thus, DNA methylation represents a second-line epigenetic change, the role of which is to permanently silence the Oct-3/4 expression.

Embryonic stem cells that have been induced to differentiate in culture are not capable of recovering totipotency even when replated in an environment conducive to the undifferentiated state. It has been shown that G9a plays a critical role in this restriction process, since, in the absence of this gene, ES cells actually retain the ability to undergo reprogramming even after extensive retinoic acid (RA)-induced differentiation. At the molecular level, G9a operates by bringing about irreversible epigenetic silencing of early embryonic genes such as Oct-3/4, through its inherent ability to carry out H3K9me3 mediated heterochromatinization as well as de novo methylation at the promoter region [107]. Taken together, these results indicate that direct interactions between histone and DNA methylases may represent a general mechanism to inhibit gene expression, thereby preventing reprogramming.

Clearly, the interaction between the transcriptional network and epigenetic mechanisms are in the heart of generating pluripotent cells. Recent studies described above indeed support the notion that the combination of a set of transcription factors as well as epigenetic regulators (such as Jmjd1a and Jmjd2c histone demethylases [49]) are needed for a more complete reprogramming of somatic cells to a pluripotent state. These transcription factors are part of a self-organizing network of transcription factors that prevents differentiation and promote proliferation and pluripotency [28].

ES cells are considered to be an attractive model for studying the molecular systems regulating many biological processes. The ability to induce ES cell differentiation in vitro has been used to elucidate several biological roles of epigenetic modifications during development. DNA methylation is among the marks that have been shown to play a vital role in maintaining ES cell identity. Furthermore, DNA methylation has been shown to be required for irreversible differentiation and thus for normal development. Nevertheless, DNA methylation patterns in ES cells are poorly investigated and little is known about the molecular mechanisms that regulate these patterns. The ability of ES cells to protect certain genomic regions, whether expressed or not, from de novo methylation while harboring a predominant de novo methylation machinery calls for further investigation. The relationship between histone modifications and DNA methylation should also be studied, both in the pluripotent state and during differentiation. These studies may facilitate the understanding of the ES cells epigenome, which may have a great influence on the ongoing research that aims at producing ethical ES cells suitable to be used in regenerative medicine.

Acknowledgments We thank C. Rosenbluh for stimulating discussions and constructive reading of the manuscript. This work was supported by grants from the Israel Academy of Science (Y.B.), Philip Morris USA Inc. and Philip Morris International (Y.B.), the National Institutes of Health (Y.B.) and the Israel Cancer Research Fund (Y.B.).

References

1. Bird A. DNA methylation patterns and epigenetic memory. Gene Dev. 2002;16(1):6–21.
2. Kouzarides T. Chromatin modifications and their function. Cell. 2007;128(4):693–705.
3. Narlikar GJ, Fan HY, Kingston RE. Cooperation between complexes that regulate chromatin structure and transcription. Cell. 2002;108(4):475–87.
4. Struhl K. Histone acetylation and transcriptional regulatory mechanisms. Genes Dev. 1998;12(5):599–606.
5. Santos-Rosa H, Schneider R, Bannister AJ, et al. Active genes are tri-methylated at K4 of histone H3. Nature. 2002;419(6905): 407–11.
6. Nakayama J, Rice JC, Strahl BD, Allis CD, Grewal SI. Role of histone H3 lysine 9 methylation in epigenetic control of heterochromatin assembly. Science. 2001;292(5514):110–3.
7. Plath K, Fang J, Mlynarczyk-Evans SK, et al. Role of histone H3 lysine 27 methylation in X inactivation. Science . 2003;300(5616):131–5.
8. Koyanagi M, Baguet A, Martens J, Margueron R, Jenuwein T, Bix M. EZH2 and histone 3 trimethyl lysine 27 associated with Il4 and Il13 gene silencing in Th1 cells. J Biol Chem. 2005;280(36):31470–7.
9. Li E. Chromatin modification and epigenetic reprogramming in mammalian development. Nat Rev. 2002;3(9):662–73.
10. Li E, Bestor TH, Jaenisch R. Targeted mutation of the DNA methyltransferase gene results in embryonic lethality. Cell. 1992;69(6):915–26.
11. Okano M, Bell DW, Haber DA, Li E. DNA methyltransferases Dnmt3a and Dnmt3b are essential for de novo methylation and mammalian development. Cell. 1999;99(3):247–57.
12. Dennis K, Fan T, Geiman T, Yan Q, Muegge K. Lsh, a member of the SNF2 family, is required for genome-wide methylation. Genes Dev. 2001;15(22):2940–4.
13. Jones PA, Baylin SB. The fundamental role of epigenetic events in cancer. Nat Rev. 2002;3(6):415–28.
14. Chen W, Cooper TK, Zahnow CA, et al. Epigenetic and genetic loss of Hic1 function accentuates the role of p53 in tumorigenesis. Cancer Cell. 2004;6(4):387–98.
15. Yu L, Liu C, Vandeusen J, et al. Global assessment of promoter methylation in a mouse model of cancer identifies ID4 as a putative tumor-suppressor gene in human leukemia. Nat Genet. 2005;37(3):265–74.
16. Leonhardt H, Page AW, Weier HU, Bestor TH. A targeting sequence directs DNA methyltransferase to sites of DNA replication in mammalian nuclei. Cell. 1992;71(5):865–73.
17. Liu Y, Oakeley EJ, Sun L, Jost JP. Multiple domains are involved in the targeting of the mouse DNA methyltransferase to the DNA replication foci. Nucleic Acids Res. 1998;26(4): 1038–45.
18. Bostick M, Kim JK, Esteve PO, Clark A, Pradhan S, Jacobsen SE. UHRF1 plays a role in maintaining DNA methylation in mammalian cells. Science. 2007;317(5845):1760–4.
19. Mayer W, Niveleau A, Walter J, Fundele R, Haaf T. Demethylation of the zygotic paternal genome. Nature. 2000; 403(6769):501–2.
20. Oswald J, Engemann S, Lane N, et al. Active demethylation of the paternal genome in the mouse zygote. Curr Biol. 2000;10(8): 475–8.
21. Reik W, Dean W, Walter J. Epigenetic reprogramming in mammalian development. Science. 2001;293(5532):1089–93.
22. Nakamura T, Arai Y, Umehara H, et al. PGC7/Stella protects against DNA demethylation in early embryogenesis. Nat Cell Biol. 2007;9(1):64–71.
23. Morgan HD, Dean W, Coker HA, Reik W, Petersen-Mahrt SK. Activation-induced cytidine deaminase deaminates 5-methylcytosine in DNA and is expressed in pluripotent tissues: implications for epigenetic reprogramming. J Biol Chem. 2004;279(50):52353–60.
24. Barreto G, Schafer A, Marhold J, et al. Gadd45a promotes epigenetic gene activation by repair-mediated DNA demethylation. Nature. 2007;445(7128):671–5.
25. Evans MJ, Kaufman MH. Establishment in culture of pluripotential cells from mouse embryos. Nature. 1981;292(5819): 154–6.
26. Bradley A, Evans M, Kaufman MH, Robertson E. Formation of germ-line chimaeras from embryo-derived teratocarcinoma cell lines. Nature. 1984;309(5965):255–6.
27. Kanatsu-Shinohara M, Toyokuni S, Shinohara T. Genetic selection of mouse male germline stem cells in vitro: offspring from single stem cells. Biol Reprod. 2005;72(1):236–40.
28. Surani MA, Hayashi K, Hajkova P. Genetic and epigenetic regulators of pluripotency. Cell. 2007;128(4):747–62.
29. Scholer HR. Octamania: the POU factors in murine development. Trends Genet. 1991;7(10):323–9.
30. Herr W, Cleary MA. The POU domain: versatility in transcriptional regulation by a flexible two-in-one DNA-binding domain. Genes Dev. 1995;9(14):1679–93.
31. Nichols J, Zevnik B, Anastassiadis K, et al. Formation of pluripotent stem cells in the mammalian embryo depends on the POU transcription factor Oct4. Cell. 1998;95(3):379–91.
32. Niwa H, Miyazaki J, Smith AG. Quantitative expression of Oct-3/4 defines differentiation, dedifferentiation or self-renewal of ES cells. Nat Genet. 2000;24(4):372–6.

33. Chambers I, Colby D, Robertson M, et al. Functional expression cloning of Nanog, a pluripotency sustaining factor in embryonic stem cells. Cell. 2003;113(5):643–55.
34. Mitsui K, Tokuzawa Y, Itoh H, et al. The homeoprotein Nanog is required for maintenance of pluripotency in mouse epiblast and ES cells. Cell. 2003;113(5):631–42.
35. Darr H, Mayshar Y, Benvenisty N. Overexpression of NANOG in human ES cells enables feeder-free growth while inducing primitive ectoderm features. Development. 2006;133(6): 1193–201.
36. Avilion AA, Nicolis SK, Pevny LH, Perez L, Vivian N, Lovell-Badge R. Multipotent cell lineages in early mouse development depend on SOX2 function. Genes Dev. 2003;17(1):126–40.
37. Wood HB, Episkopou V. Comparative expression of the mouse Sox1, Sox2 and Sox3 genes from pre-gastrulation to early somite stages. Mech Dev. 1999;86(1–2):197–201.
38. Ivanova N, Dobrin R, Lu R, et al. Dissecting self-renewal in stem cells with RNA interference. Nature. 2006;442(7102): 533–8.
39. Boyer LA, Lee TI, Cole MF, et al. Core transcriptional regulatory circuitry in human embryonic stem cells. Cell. 2005;122(6): 947–56.
40. Boyer LA, Mathur D, Jaenisch R. Molecular control of pluripotency. Curr Opin Genet Dev. 2006;16(5):455–62.
41. Loh YH, Wu Q, Chew JL, et al. The Oct4 and Nanog transcription network regulates pluripotency in mouse embryonic stem cells. Nat Genet. 2006;38(4):431–40.
42. Takahashi K, Yamanaka S. Induction of pluripotent stem cells from mouse embryonic and adult fibroblast cultures by defined factors. Cell. 2006;126(4):663–76.
43. Okita K, Ichisaka T, Yamanaka S. Generation of germline-competent induced pluripotent stem cells. Nature. 2007; 448(7151):313–7.
44. Wernig M, Meissner A, Foreman R, et al. In vitro reprogramming of fibroblasts into a pluripotent ES-cell-like state. Nature. 2007;448(7151):318–24.
45. Rowland BD, Peeper DS. KLF4, p21 and context-dependent opposing forces in cancer. Nat Rev Cancer. 2006;6(1):11–23.
46. Knoepfler PS. Myc goes global: new tricks for an old oncogene. Cancer Res. 2007;67(11):5061–3.
47. Li Y, McClintick J, Zhong L, Edenberg HJ, Yoder MC, Chan RJ. Murine embryonic stem cell differentiation is promoted by SOCS-3 and inhibited by the zinc finger transcription factor Klf4. Blood. 2005;105(2):635–7.
48. Cartwright P, McLean C, Sheppard A, Rivett D, Jones K, Dalton S. LIF/STAT3 controls ES cell self-renewal and pluripotency by a Myc-dependent mechanism. Development. 2005;132(5): 885–96.
49. Loh YH, Zhang W, Chen X, George J, Ng HH. Jmjd1a and Jmjd2c histone H3 Lys 9 demethylases regulate self-renewal in embryonic stem cells. Genes Dev. 2007;21(20):2545–57.
50. Takahashi K, Tanabe K, Ohnuki M, et al. Induction of pluripotent stem cells from adult human fibroblasts by defined factors. Cell. 2007;131(5):861–72.
51. Yu J, Vodyanik MA, Smuga-Otto K, et al. Induced pluripotent stem cell lines derived from human somatic cells. Science. 2007;318(5858):1917–20
52. Richards M, Tan SP, Tan JH, Chan WK, Bongso A. The transcriptome profile of human embryonic stem cells as defined by SAGE. Stem Cells. 2004;22(1):51–64.
53. Meshorer E, Yellajoshula D, George E, Scambler PJ, Brown DT, Misteli T. Hyperdynamic plasticity of chromatin proteins in pluripotent embryonic stem cells. Developmental Cell. 2006;10(1):105–16.
54. Azuara V, Perry P, Sauer S, et al. Chromatin signatures of pluripotent cell lines. Nat Cell Biol. 2006;8(5):532–8.
55. Bernstein BE, Mikkelsen TS, Xie X, et al. A bivalent chromatin structure marks key developmental genes in embryonic stem cells. Cell. 2006;125(2):315–26.
56. Pan G, Tian S, Nie J, et al. Whole-genome analysis of histone H3 lysine 4 and lysine 27 methylation in human embryonic stem cells. Cell Stem Cell. 2007;1(3): :299–312.
57. Zhao XD, Han X, Chew JL, et al. Whole-genome mapping of histone h3 lys4 and 27 trimethylations reveals distinct genomic compartments in human embryonic stem cells. Cell Stem Cell. 2007;1(3):286–98.
58. Frank D, Keshet I, Shani M, Levine A, Razin A, Cedar H. Demethylation of CpG islands in embryonic cells. Nature. 1991;351(6323):239–41.
59. Bibikova M, Chudin E, Wu B, et al. Human embryonic stem cells have a unique epigenetic signature. Genome Res. 2006;16(9):1075–83.
60. Jahner D, Stuhlmann H, Stewart CL, et al. De novo methylation and expression of retroviral genomes during mouse embryogenesis. Nature. 1982;298(5875):623–8.
61. Stewart CL, Stuhlmann H, Jahner D, Jaenisch R. De novo methylation, expression, and infectivity of retroviral genomes introduced into embryonal carcinoma cells. Proc Natl Acad Sci U S A. 1982;79(13):4098–102.
62. Brandeis M, Frank D, Keshet I, et al. Sp1 elements protect a CpG island from de novo methylation. Nature. 1994;371(6496):435–8.
63. Jackson M, Krassowska A, Gilbert N, et al. Severe global DNA hypomethylation blocks differentiation and induces histone hyperacetylation in embryonic stem cells. Mol Cell Biol. 2004;24(20):8862–71.
64. Tsumura A, Hayakawa T, Kumaki Y, et al. Maintenance of self-renewal ability of mouse embryonic stem cells in the absence of DNA methyltransferases Dnmt1, Dnmt3a and Dnmt3b. Genes Cells. 2006;11(7):805–14.
65. Bourc'his D, Xu GL, Lin CS, Bollman B, Bestor TH. Dnmt3L and the establishment of maternal genomic imprints. Science. 2001;294(5551):2536–9.
66. Bourc'his D, Bestor TH. Meiotic catastrophe and retrotransposon reactivation in male germ cells lacking Dnmt3L. Nature. 2004;431(7004):96–9.
67. Hata K, Okano M, Lei H, Li E. Dnmt3L cooperates with the Dnmt3 family of de novo DNA methyltransferases to establish maternal imprints in mice. Development. 2002;129(8):1983–93.
68. Ooi SK, Qiu C, Bernstein E, et al. DNMT3L connects unmethylated lysine 4 of histone H3 to de novo methylation of DNA. Nature. 2007;448(7154):714–7.
69. Jia D, Jurkowska RZ, Zhang X, Jeltsch A, Cheng X. Structure of Dnmt3a bound to Dnmt3L suggests a model for de novo DNA methylation. Nature. 2007;449(7159):248–51.
70. Chen T, Ueda Y, Xie S, Li E. A novel Dnmt3a isoform produced from an alternative promoter localizes to euchromatin and its expression correlates with active de novo methylation. J Biol Chem. 2002;277(41):38746–54.
71. Kaneda M, Okano M, Hata K, et al. Essential role for de novo DNA methyltransferase Dnmt3a in paternal and maternal imprinting. Nature. 2004;429(6994):900–3.
72. Zhu H, Geiman TM, Xi S, et al. Lsh is involved in de novo methylation of DNA. EMBO J. 2006;25(2):335–45.
73. Stein R, Razin A, Cedar H. In vitro methylation of the hamster adenine phosphoribosyltransferase gene inhibits its expression in mouse L cells. Proc Natl Acad Sci U S A. 1982;79(11):3418–22.
74. Gidekel S, Bergman Y. A unique developmental pattern of Oct-3/4 DNA methylation is controlled by a cis-demodification element. J Biol Chem. 2002;277(37):34521–30.
75. Feldman N, Gerson A, Fang J, et al. G9a-mediated irreversible epigenetic inactivation of Oct-3/4 during early embryogenesis. Nat Cell Biol. 2006;8(2):188–94.

76. Hattori N, Imao Y, Nishino K, et al. Epigenetic regulation of Nanog gene in embryonic stem and trophoblast stem cells. Genes Cells. 2007;12(3):387–96.
77. Imamura M, Miura K, Iwabuchi K, et al. Transcriptional repression and DNA hypermethylation of a small set of ES cell marker genes in male germline stem cells. BMC Dev Biol 2006;6:34.
78. Xu J, Pope SD, Jazirehi AR, et al. Pioneer factor interactions and unmethylated CpG dinucleotides mark silent tissue-specific enhancers in embryonic stem cells. Proc Nat Acad Sci U S A. 2007;104(30):12377–82.
79. Lindroth AM, Cao X, Jackson JP, et al. Requirement of CHROMOMETHYLASE3 for maintenance of CpXpG methylation. Science. 2001;292(5524):2077–80.
80. Gruenbaum Y, Stein R, Cedar H, Razin A. Methylation of CpG sequences in eukaryotic DNA. FEBS Lett. 1981;124(1):67–71.
81. Ramsahoye BH, Biniszkiewicz D, Lyko F, Clark V, Bird AP, Jaenisch R. Non-CpG methylation is prevalent in embryonic stem cells and may be mediated by DNA methyltransferase 3a. Proc Natl Acad Sci U S A. 2000;97(10):5237–42.
82. Eden A, Gaudet F, Waghmare A, Jaenisch R. Chromosomal instability and tumors promoted by DNA hypomethylation. Science. 2003;300(5618):455.
83. Gaudet F, Hodgson JG, Eden A, et al. Induction of tumors in mice by genomic hypomethylation. Science. 2003;300(5618):489–92.
84. Lei H, Oh SP, Okano M, et al. De novo DNA cytosine methyltransferase activities in mouse embryonic stem cells. Development. 1996;122(10):3195–205.
85. Carlone DL, Lee JH, Young SR, et al. Reduced genomic cytosine methylation and defective cellular differentiation in embryonic stem cells lacking CpG binding protein. Mol Cell Biol. 2005;25(12):4881–91.
86. Tadokoro Y, Ema H, Okano M, Li E, Nakauchi H. De novo DNA methyltransferase is essential for self-renewal, but not for differentiation, in hematopoietic stem cells. J Exp Med. 2007;204(4):715–22.
87. Gurdon JB. From nuclear transfer to nuclear reprogramming: the reversal of cell differentiation. Annu Rev Cell Dev Biol. 2006;22:1–22.
88. Tada M, Takahama Y, Abe K, Nakatsuji N, Tada T. Nuclear reprogramming of somatic cells by in vitro hybridization with ES cells. Curr Biol. 2001;11(19):1553–8.
89. Bannister AJ, Zegerman P, Partridge JF, et al. Selective recognition of methylated lysine 9 on histone H3 by the HP1 chromo domain. Nature. 2001;410(6824):120–4.
90. Rea S, Eisenhaber F, O'Carroll D, et al. Regulation of chromatin structure by site-specific histone H3 methyltransferases. Nature. 2000;406(6796):593–9.
91. Tachibana M, Sugimoto K, Nozaki M, et al. G9a histone methyltransferase plays a dominant role in euchromatic histone H3 lysine 9 methylation and is essential for early embryogenesis. Genes Dev. 2002;16(14):1779–91.
92. Ogawa H, Ishiguro K, Gaubatz S, Livingston DM, Nakatani Y. A complex with chromatin modifiers that occupies E2F- and Myc-responsive genes in G0 cells. Science. 2002;296(5570):1132–6.
93. Yang L, Xia L, Wu DY, et al. Molecular cloning of ESET, a novel histone H3-specific methyltransferase that interacts with ERG transcription factor. Oncogene. 2002;21(1):148–52.
94. Schultz DC, Ayyanathan K, Negorev D, Maul GG, Rauscher FJ, 3rd. SETDB1: a novel KAP-1-associated histone H3, lysine 9-specific methyltransferase that contributes to HP1-mediated silencing of euchromatic genes by KRAB zinc-finger proteins. Genes Dev. 2002;16(8):919–32.
95. Peters AH, Kubicek S, Mechtler K, et al. Partitioning and plasticity of repressive histone methylation states in mammalian chromatin. Mol Cell. 2003;12(6):1577–89.
96. Rice JC, Briggs SD, Ueberheide B, et al. Histone methyltransferases direct different degrees of methylation to define distinct chromatin domains. Mol Cell. 2003;12(6):1591–8.
97. Vire E, Brenner C, Deplus R, et al. The Polycomb group protein EZH2 directly controls DNA methylation. Nature. 2006;439(7078):871–4.
98. Freitag M, Hickey PC, Khlafallah TK, Read ND, Selker EU. HP1 is essential for DNA methylation in neurospora. Mol Cell. 2004;13(3):427–34.
99. Lehnertz B, Ueda Y, Derijck AA, et al. Suv39h-mediated histone H3 lysine 9 methylation directs DNA methylation to major satellite repeats at pericentric heterochromatin. Curr Biol. 2003;13(14):1192–200.
100. Jackson JP, Lindroth AM, Cao X, Jacobsen SE. Control of CpNpG DNA methylation by the KRYPTONITE histone H3 methyltransferase. Nature. 2002;416(6880):556–60.
101. Vassen L, Fiolka K, Moroy T. Gfi1b alters histone methylation at target gene promoters and sites of gamma-satellite containing heterochromatin. EMBO J. 2006;25(11):2409–19.
102. Fuks F, Hurd PJ, Deplus R, Kouzarides T. The DNA methyltransferases associate with HP1 and the SUV39H1 histone methyltransferase. Nucleic Acids Res. 2003;31(9):2305–12.
103. Smallwood A, Esteve PO, Pradhan S, Carey M. Functional cooperation between HP1 and DNMT1 mediates gene silencing. Genes Dev. 2007;21(10):1169–78.
104. Tachibana M, Ueda J, Fukuda M, et al. Histone methyltransferases G9a and GLP form heteromeric complexes and are both crucial for methylation of euchromatin at H3-K9. Genes Dev. 2005;19(7):815–26.
105. Ikegami K, Iwatani M, Suzuki M, et al. Genome-wide and locus-specific DNA hypomethylation in G9a deficient mouse embryonic stem cells. Genes Cells. 2007;12(1):1–11.
106. Xin Z, Tachibana M, Guggiari M, Heard E, Shinkai Y, Wagstaff J. Role of histone methyltransferase G9a in CpG methylation of the Prader-Willi syndrome imprinting center. J Biol Chem. 2003;278(17):14996–5000.
107. Epsztejn-Litman S, Feldman N, Abu-Remaileh M, et al. De novo DNA methylation promoted by G9a prevents reprogramming of embryonically silenced genes. Nat Struct Mol Biol. In press.
108. McGarvey KM, Fahrner JA, Greene E, Martens J, Jenuwein T, Baylin SB. Silenced tumor suppressor genes reactivated by DNA demethylation do not return to a fully euchromatic chromatin state. Cancer Res. 2006;66(7):3541–9.

Epigenetic Basis for Differentiation Plasticity in Stem Cells

Philippe Collas, Sanna Timoskainen and Agate Noer

Abstract Stem cells possess the remarkable property of being able to self-renew and give rise to at least one more differentiated cell type. Embryonic stem cells have the ability to differentiate into all cell types of the body and have unlimited self-renewal potential. Somatic stem cells are found in many adult tissues. They have an extensive but finite lifespan and can differentiate into a more restricted range of cell types. Increasing evidence suggests that the multilineage differentiation ability of stem cells is brought about by the potential for expression of developmentally regulated transcription factors and of lineage-specification genes. Potential for gene expression is largely controlled by epigenetic modifications of DNA (DNA methylation) and chromatin (such as post-translational histone modifications) on regulatory regions. These modifications modulate chromatin organization not only on specific genes but also at the level of the whole nucleus. They can also influence the timing of DNA replication. This chapter highlights how epigenetic mechanisms that poise genes for transcription in undifferentiated stem cells are being uncovered through, notably, genome-wide mapping of DNA methylation, histone modifications, and transcription factor binding. Epigenetic marks on developmentally regulated and lineage-specifying genes in undifferentiated stem cells seem to define a pluripotent state.

Keywords Chromatin · Differentiation · DNA methylation · Embryonic stem cell · Epigenetics · Mesenchymal stem cell

1 Introduction: Embryonic Stem Cells, Mesenchymal Stem Cells, and Epigenetics

Stem cells are defined by their ability to self-renew and to give rise to at least one more differentiated cell type. Embryonic stem cells (ESCs), in vitro derivatives of the inner cell mass of blastocysts, retain the ability of the inner cell mass to differentiate into all cell types of the body and acquire unlimited self-renewal potential. For these reasons, human ESCs (hESCs) have received considerable attention since their derivation 10 years ago [1] due to their perceived use in regenerative medicine. Multiple extracellular factors are required for the establishment and maintenance of pluripotency in ESCs and these have been reviewed elsewhere recently [2, 3]. The multilineage differentiation ability of ESCs is defined by the potential for expression of lineage-specification genes. This chapter analyzes epigenetic mechanisms by which these genes are poised for transcription.

Stem cells have in recent years also been identified in many adult organs. Stromal stem cells, present in a variety of mesenchymal tissues, are also being scrutinized due to their potential use in autologous cell replacement therapy [4, 5]. In contrast to ESCs, mesenchymal stem cells (MSCs) seem to be restricted to forming preferentially mesodermal cell types such as adipocytes, myocytes, osteocytes, and chondrocytes. However, rare subsets of MSCs identified in bone marrow seem to have the ability to form cell types of all three germ layers and have challenged the limited differentiation potential of somatic stem cells [6]. An abundant source of MSCs is adipose tissue-derived stem cells (ASCs) isolated from liposuction material [7, 8]. Like bone marrow-derived MSCs, ASCs can differentiate into mesodermal cell types; however, recent findings suggest a limited differentiation ability even within mesodermal lineages [9, 10]. So, although MSCs retain the ability of express various lineage-specific genes upon differentiation, this potential is more restricted than in ESCs.

The potential for gene expression in stem cells is regulated by epigenetic processes that confer a specific chromatin configuration on gene regulatory regions and on coding sequences. Epigenetic mechanisms refer to heritable modifications on DNA and chromatin that do not affect DNA sequence. The best-characterized epigenetic modification is cytosine methylation on DNA, which is in general associated with gene silencing. Epigenetic modifications of chromatin regroup post-translational alteration of histones including

P. Collas (✉)
Institute of Basic Medical Sciences, Department of Biochemistry, Faculty of Medicine, University of Oslo, PO Box 1112, Blindern, 0317 Oslo, Norway
e-mail: philippe.collas@medisin.uio.no

V.K. Rajasekhar, M.C. Vemuri (eds.), *Regulatory Networks in Stem Cells*, Stem Cell Biology and Regenerative Medicine, DOI 10.1007/978-1-60327-227-8_21,

phosphorylation, acetylation, methylation, ubiquitination, and SUMOylation, in combination with the dynamic replacement of core histone by histone variants, such as the deposition of histone H3.3 on transcriptionally active promoters [11, 12]. In addition to epigenetic modifications, the positioning of transcriptional activators, transcriptional repressors, other chromatin remodeling enzymes, and small interfering RNAs on target genes also regulate gene expression. This chapter highlights our current view of the epigenetic landscape of ESCs and MSCs, and how the epigenetic landscape is likely to provide a molecular basis for gene activation and multilineage differentiation potential.

2 DNA Methylation Patterns in Embryonic Stem Cells

2.1 DNA Methylation and Gene Expression

DNA methylation consists in the addition of a methyl group to the 5 position of a cytosine in a cytosine-phosphate-guanine (CpG) dinucleotide (Fig. 1A). CpG methylation is symmetrical (it occurs on both DNA strands) and targets isolated CpGs, clustered CpGs, or even clustered CpGs within a CpG island. A CpG island is defined as a sequence in which the observed/expected C frequency is greater than 0.6 with a GC dinucleotide content greater than 50%. According to Gardiner-Garden and Frommer [13], the expected number of CpG dimers in a given 200 bp window is calculated as the number of C's in the window multiplied by the number of Gs in the window, divided by window length. This 200 bp window is moving across the sequence of interest at 1 bp intervals. CpG islands are often found in the 5′ regulatory regions of vertebrate housekeeping genes. CpG islands are often protected from methylation, enabling constitutive expression of these genes. CpG islands in the promoter of tumor suppressor genes, for instance, are unmethylated in normal cells, whereas a hallmark of cancer is de novo methylation of these CpG islands, resulting in repression of tumor suppressor genes and triggering of an uncontrolled cell cycle. DNA methylation of tumor suppressor genes constitutes the basis of a number of anticancer therapies relying on the inhibition of DNA methyl transferases [14].

CpG methylation is catalyzed by DNA methyltransferases (DNMTs). The maintenance DNA methyltransferase DNMT1 specifically recognizes hemi-methylated DNA after replication and methylates the daughter strand, ensuring fidelity in the methylation profile after replication [15]. In contrast to DNMT1, DNMT3a and DNMT3b are implicated in de novo DNA methylation that takes place during embryonic development and cell differentiation [16] as a means of shutting down genes whose activity is no longer required as cells differentiate (e.g., that of pluripotency-associated genes). The fourth DNMT, DNMT2, has to date no clear ascribed function in DNA methylation [17–21], but has been shown to have cytoplasmic transfer RNA methyltransferase activity [22, 23].

DNA methylation is a hallmark of long-term gene silencing (Fig. 1B). The methyl groups create target sites for methyl-binding proteins that induce transcriptional repression by recruiting co-repressors such as histone deacetylases [24]. So DNA methylation largely contributes to gene silencing [25, 26] and as such it is essential for development [27–30], X chromosome inactivation [31], and genomic imprinting [32–35]. The relationship between DNA methylation and gene expression is complex [36] and recent evidence based on genome-wide CpG methylation profiling highlights promoter CpG content as a component of this complexity

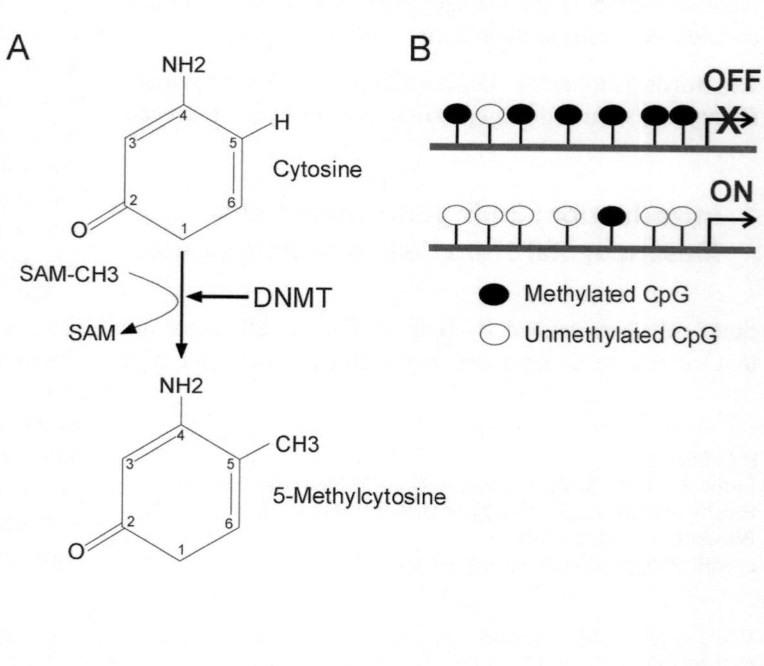

Fig. 1 Principles of DNA methylation. (**A**) Mechanism of DNA methylation. (**B**) Textbook view of the relationship between DNA methylation and gene expression

[37] (see below). In vitro differentiation of ESCs and embryonal carcinoma (EC) cells also correlates with changes in DNA methylation notably on the promoter of developmentally regulated genes expressed in pluripotent ESCs such as the transcription factors *OCT4* and *NANOG* [38–40]. However, to date, only sporadic indications of CpG methylation changes have been reported during differentiation of MSCs or precursor cells [9, 10, 41, 42].

2.2 CpG Methylation Patterns in Murine ES Cells

Limited evidence suggests that the DNA methylation signature of ESCs is distinct from that of differentiated somatic cells; however, whether this reflects differences in gene expression or the true pluripotent nature of ESCs is unclear. Restriction enzyme digestion-mediated analyses of global DNA methylation show that mouse ESC genomes are less methylated than those of differentiated somatic cells [43, 44]. Notably, XX mouse ESCs are hypomethylated relative to XY ESCs. Hypomethylation affects both repetitive and unique sequences, including differentially methylated regions that regulate expression of paternally imprinted loci [44]. Increased hypomethylation of XX ESCs has been attributed to the presence of two active X chromosomes (active X is hypomethylated relative to inactive X) and to reduced levels of DNMT3a and 3b. However, in DNMT-deficient [$Dnmt3a^{-/-}$ $Dnmt3b^{-/-}$] mouse ESCs, only 0.6% of CpGs are demethylated [43] so the extent to which DNMT3a and 3b contribute to global DNA methylation in mouse ESCs remains uncertain. DNMT1 deficiency, in contrast, reduces global methylation levels from 65% to 20%, a condition that blocks differentiation potential [43]. Unfortunately, no indication currently exists on the methylation status of regulatory regions of lineage-specific genes in mouse ESCs, which could account for their potential for expression upon differentiation.

2.3 The State of DNA Methylation in Human ES Cells

DNA methylation analyses of hESCs have been promoted by in vitro fertilization data on the unexpectedly high incidence of imprinting and other epigenetic abnormalities in embryos [45], suggesting that hESCs may also display variation in their epigenetic makeup. A restriction analysis-based methylation profiling of over 1500 CpG sites from over 370 genes in 14 hESC lines [46] revealed an average of 35% methylation, a value substantially lower than that reported for mouse ES cells [44]. hESC methylation profiles were segregated from those of normal and cancer cell lines, normal tissue, and somatic stem cells, reflecting an epigenetic distance between hESCs and other cell types [46]. Interestingly, less than 50 CpGs within 40 genes contributed to this difference. Another 25 CpG sites from over 20 genes distinguished hESCs from normal differentiated cells and somatic stem cells; these sites were found to represent markers of developmental potential [46]. Other genes differentially methylated in hESCs relative to somatic cells are markers of pluripotency such as *OCT4* and *NANOG*, which are unmethylated in undifferentiated hESCs [47], while being partially methylated in human MSCs in which they are not expressed (ST and PC, unpublished data). Thus, on the basis of these analyses, it appears that the methylation pattern of a relatively small number of developmentally controlled genes may constitute an epigenetic mark unique to hESCs.

The need for large-scale expansion of hESCs for any therapeutic use raises the question of epigenetic stability of hESCs in long-term culture. The consensus from published reports is that extended culture of hESCs alters DNA methylation. Restriction landmark genome scanning analysis of ~2000 loci has identified epigenetic variations between hESC lines at loci important for differentiation [48]. Most changes occur shortly after hESC derivation and are heritable, whereas some alterations are maintained even after in vitro differentiation. This study is supported by a similar methylation drift at a small number of promoters examined in late passage cultures of other hESC lines [46, 49]. In contrast, however, stable methylation patterns have been reported by bisulfite genomic sequencing in a small number of imprinted loci in four different hESC lines [50]. So epigenetic variation occurs during extended culture of hESCs, but the timing and degree of this epigenetic drift are likely to be cell line dependent.

An odd feature of DNA methylation changes reported in hESCs by Allegrucci and colleagues [48] is heritability upon long-term expansion, suggesting that long-term culture may elicit a reprogramming of the hESC epigenome. In contrast, as discussed below, we found that human ASCs undergo stochastic methylation changes upon culture [9, 10, 51]. Thus, the hypothesis of programmed CpG methylation changes during culture may not necessarily hold for cell types other than hESCs. Several reports on random methylation events in human cell cultures support this view [52–54].

A picture currently missing from the ESC epigenetics is a high-resolution genome-wide DNA methylation profiling across regulatory and coding regions. Methyl-DNA immunoprecipitation (MeDIP) assays coupled to genomic array hybridization are particularly well suited for whole-genome and promoter investigations [37, 55]. Such data can be superimposed onto transcription factor binding [37, 55] and histone modification maps to elaborate a multilayered epigenetic profile characteristic of pluripotent cells. We welcome new data on the DNA methylation landscape of ESCs.

2.4 Unexpected DNA Methylation Events During ES Cell Differentiation?

Whether unscheduled CpG methylation occurs upon in vitro differentiation of hESCs remains to be established but appears as a possibility. Analysis of over 4600 CpG islands revealed that 1.4% undergo unexpected hypermethylation upon neurogenic differentiation of hESCs, in regulatory regions of genes involved in metabolism, signal transduction and differentiation [56]. Although distinct from tumor suppressor CpG island methylation, this hypermethylation leads to the down-regulation of the affected genes, and as such has been suggested to have implications in the development of metabolic diseases [56]. Thus, the risk of aberrant CpG island methylation upon hESC differentiation should be considered when optimizing differentiation protocols, in particular if they are going to be used in therapeutic applications.

3 Bivalent Histone Marks on Developmentally Regulated Genes in ESCs May Poise Promoters for Activation

3.1 Post-translational Histone Modifications

The eukaryotic genome is packaged by interactions of DNA with proteins into chromatin. The core element of chromatin is the nucleosome, which consists of 147 base pairs of DNA wrapped around two subunits of each of histone H2A, H2B, H3, and H4. Nucleosomes are spaced by the linker histone H1. The amino-terminal tails of histones are post-translationally modified to confer physical properties that affect their interactions with DNA on gene regulatory sequences. Histone modifications not only influence chromatin packaging but are also "read" by adaptor molecules, chromatin-modifying enzymes, transcription factors, and transcriptional repressors, and thereby contribute to the regulation of transcription [57, 58, 59]. Epigenetic histone modifications have been best characterized so far for H3 and H4 and include combinatorial phosphorylation, ubiquitination, SUMOylation, acetylation, and methylation (Fig. 2). In particular, di- and trimethylation of H3 lysine 9 (H3K9m2/m3) and trimethylation of H3K27 (H3K27m3) elicit the formation of repressive heterochromatin through the recruitment of heterochromatin protein 1 (HP1) [59, 60] and polycomb group (PcG) proteins, respectively [61, 62, 63]. However, whereas H3K9m3 marks constitutive heterochromatin [64], H3K27m3 characterizes facultative heterochromatin, or chromatin domains harboring transcriptionally repressed genes that can be activated upon stimulation [65, 66]. In contrast, acetylation of histone tails loosens their interaction with DNA and creates a chromatin conformation suitable for targeting of transcriptional activators. Thus, acetylation on H3K9 (H3K9ac) and H4K16 (H4K16ac), together with di- or trimethylation of H3K4 (H3K4m2/3), are exclusively found in euchromatin, often in association with transcriptionally active genes [67, 68, 69]. In addition to altering histone-DNA interactions, H3K4m3 and H3K9ac mediate the recruitment and tethering of transcriptional activators [70, 71]. Mapping of the positioning of histone modifications throughout the genome or on given promoters has been enabled by chromatin immunoprecipitation (ChIP) assays, whereby a specific histone modification is immunoprecipitated and associated DNA sequences are identified by polymerase chain reaction (PCR) or by labeling and hybridization onto genomic arrays [72].

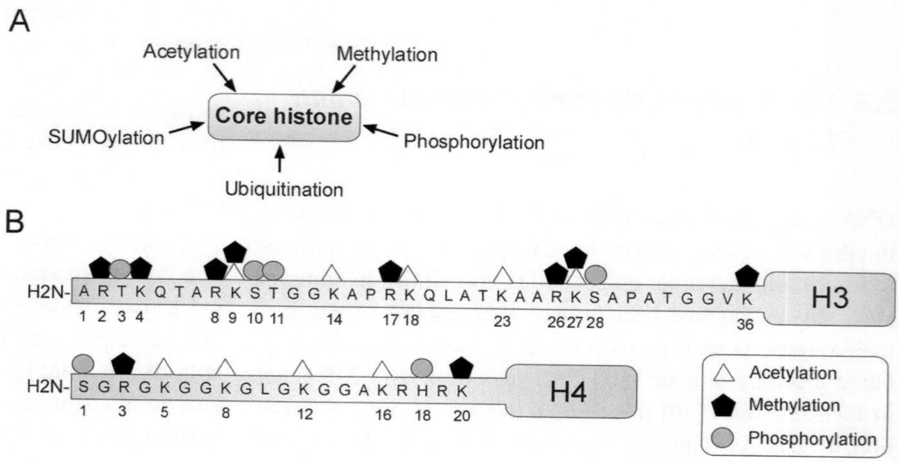

Fig. 2 Post-translational histone modifications. (**A**) Core histones can be methylated, acetylated, phosphorylated, ubiquitinated, or SUMOylated to modulate gene expression. (**B**) Site and nature of known post-translational modifications on the amino-terminal tails of the core histones H3 and H4

3.2 Mapping of Histone Modifications in the ES Cell Genome

Recent mapping of histone modifications has shown that lineage-specific genes, which are either silent or active in differentiated somatic cells, are in a potentially active state in pluripotent ESCs. Genome-wide and locus-specific ChIP analyses reveal that repressed but potentially active promoters are associated with so-called "bivalent" histone modifications characterized by H3K4m3, a mark of active genes, and H3K27m3, which associates with inactive genes [65, 66] (Fig. 3A). Azuara et al. [65] have shown that several transcription factors essential for lineage specification are not expressed in mouse ESCs but are marked on their promoter by H3K4m3, H3K27m3, and H3K9ac. Unscheduled expression of these genes is induced in ESCs deficient for embryonic ectoderm development protein, a component of the polycomb repressor complex PRC2 (see below), which harbors H3K27 methyltransferase activity [73], demonstrating the essential role of trimethylation of H3K27 in maintaining a transcriptional brake in a context of transcriptionally permissive chromatin. At the genome-wide level, these "bivalent domains" consist of large regions of H3K27 trimethylation embedding smaller areas of H3K4 trimethylation [66]. Consistent with the data of Azuara et al. [65], these domains include transcription factor encoding genes that are repressed or expressed at low levels. Intriguingly, the correlation between histone methylation marks and genomic sequence in ESCs raises the hypothesis that DNA sequence may prime the epigenetic landscape in pluripotent cells [66]. Nevertheless, not all lineage-control genes in ESCs are associated with bivalent histone modifications; rather, they are marked by H3K4m3 only or do not display H3K4m3 or H3K27m3 [66]. The critical role of these genes in lineage determination suggests that they are also in a transcriptionally poised state and await, through yet unknown epigenetic mechanisms, permission for transcription.

3.3 Hyperdynamic Chromatin in ES Cells

A dynamic reorganization of chromatin domains is essential for setting up heritable transcriptional programs in the context of differentiation [74]. Many structural chromatin proteins such as heterochromatin protein 1 (HP1) and histones have been shown to bind more loosely to chromatin of ESCs than differentiated or somatic cells [75]. These proteins are also hyperdynamic in ESCs relative to differentiated cells. Fluorescence recovery after photobleaching studies have shown that all three isoforms of HP1 fused with green fluorescence protein exchange faster in heterochromatic foci of undifferentiated mouse ESCs than after differentiation [75]. Likewise, exchange rates of fluorescently tagged histones H1, H2B, and H3 are significantly higher in pluripotent ESCs than in differentiated counterparts. These studies unravel the existence of a greater fraction of loosely bound HP1 as well as core and linker histones in ESCs. The hyperdynamic nature of chromatin-associated proteins in pluripotent ESCs reflects some plasticity in chromatin organization and thereby provides a basis for pluripotency. The concept of hyperdynamic chromatin in ESCs is line with an attractive yet highly speculative "histone modification pulsing" model whereby developmentally regulated genes would be marked by transient histone modifications in pluripotent cells to enable the appropriate response upon differentiation [76].

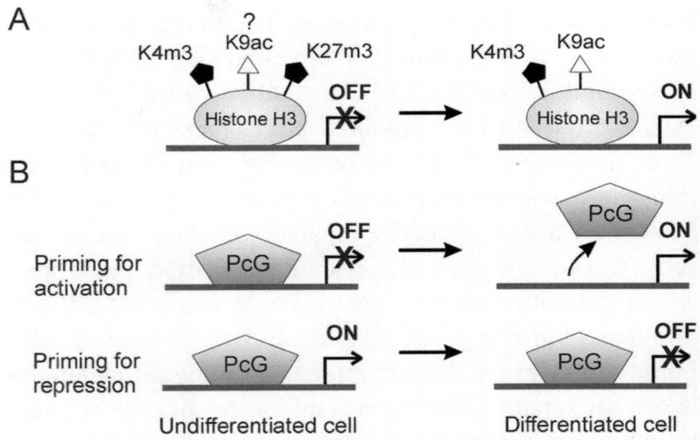

Fig. 3 Control of lineage-specific gene expression by histone H3K27 methylation and PcGs. (**A**) In undifferentiated cells, repressed lineage-specific genes are marked by trimethylation of K4 and K27 (the bivalent marks) and acetylation of H3K9. These marks are believed to primed genes for activation. Upon differentiation, demethylation of H3K27 results in transcriptional activation of the gene. (**B**) In undifferentiated cells, repressed lineage-specific genes can be either primed for activation by occupancy of PcGs on the promoter; differentiation coincides with removal of the PcG complex and activation of the gene. However, genes expressed in undifferentiated cells can also be primed for transcriptional repression by PcG complexes on the promoter

4 Polycomb Group Proteins Impose a Transcriptional Brake on Lineage-Priming Genes

PcGs are transcriptional repressors [77, 78] found within two distinct and conserved PRCs (PRC1 and PCR2) working cooperatively [79]. Involvement of PRCs in pluripotency has been suggested by the requirement of PcG proteins for the patterning of gene expression during development, for establishing pluripotent ESCs, and for maintaining somatic stem cell cultures (reviewed in [80]).

In undifferentiated ESCs, PcGs preferentially occupy genes that are activated upon differentiation, consistent with the view that these genes are poised for transcription [81–83] (Fig. 3B). Histone methyltransferase activity of Eed and enhancer of zeste homologue 2 (Ezh2; another PRC2 component) is responsible for trimethylation of H3K27 on these target genes [61, 62]. In addition, trimethylation of H3K4 is mediated by Trithorax group (Trx) proteins [78]. Thus, the known interplay between PcG and Trx proteins is also likely to establish bivalent domains of histone modifications on developmentally regulated genes in pluripotent cells. PcGs, however, are also dynamic and not always associated with repressed genes. For genes activated upon differentiation, PcGs are displaced from promoters [83]. Furthermore, genes that are repressed during differentiation have also paradoxically been found to be already occupied by PcG proteins in undifferentiated cells, while in a state of activity. These findings suggest that PRCs constitute a "pre-programmed memory system" established during embryogenesis [83]. This program would mark certain genes for transcriptional repression upon differentiation, while other genes would be primed for activation (Fig. 3B). It will be interesting to determine whether genes poised for transcriptional activation or repression by PcG proteins are marked by distinct histone modifications (e.g., different levels of the active H3K9ac mark) or by a specific CpG methylation status. An increasing body of evidence, therefore, suggests that unique combinations of CpG methylation, histone modifications, PcG occupancy, and nucleosome positioning [84–87] on developmentally regulated gene promoters, in a context of hyperdynamic chromatin, define a pluripotent genomic organization in ESCs.

5 The Epigenetics of Mesenchymal Stem Cells

The epigenetic landscape of somatic (adult) stem cells remains to date largely unraveled. This section highlights recent published and unpublished findings on the relationship between DNA methylation of lineage-specification genes, gene expression, and potential for cell differentiation in MSCs. Focus is on adipose tissue-derived stem cells (ASCs) on which most epigenetic studies reported up to now have been conducted. The emerging concept is that a CpG methylation pattern preprograms ASCs for adipogenic differentiation preferentially over other differentiation pathways.

5.1 DNA Methylation on Promoters of Lineage-Specification Genes: Programming Stem Cells for Differentiation Potential?

The adipose tissue constitutes a rich source of MSCs [7, 8, 88, 89]. ASCs with a $CD34^{+}CD105^{+}CD45^{-}CD31^{-}$ phenotype have been isolated by negative selection against CD45 and CD31 with high purity (~99%) from the stromal vascular fraction of human liposuction material [7]. Notably, cultured ASCs display a gene expression profile and surface antigen phenotype similar to bone marrow-derived MSCs [7, 90–92], suggesting a common mesodermal ancestor. ASCs exhibit primarily mesodermal differentiation abilities in vitro and can promote neuronal functions, osteogenic repair, and reconstitution of the immune system in vivo [41, 88]. ASCs can also differentiate toward the endothelial cell lineage in vitro and contribute to revascularization of ischemic tissue; nonetheless, whether their contribution is direct or indirect remains debated [9, 93]. Transcriptional profiling of freshly isolated, uncultured ASCs reveals expression of genes extending across the three germ layers, suggestive of a differentiation potential toward nonmesodermal lineages [7]. However, whether ASCs form functional tissues of these lineages in vivo remains under debate.

Recent studies have started to unveil the DNA methylation profile of tissue-specific genes in human ASCs (Fig. 4). Bisulfite genomic sequencing analysis of four adipogenic specification promoters (namely, leptin [*LEP*], peroxisome proliferator activated receptor gamma 2 [*PPARG2*], fatty acid-binding protein 4 [*FABP4*], and lipoprotein lipase [*LPL*]) reveals several DNA methylation features in freshly isolated ASCs [10]. Firstly, these promoters are globally hypomethylated, with a mere 5–30% of CpGs being methylated. Secondly, CpG methylation profiles are mosaic between ASC donors and within donors. This mosaicism is consistent with that observed in stem cells from intestinal

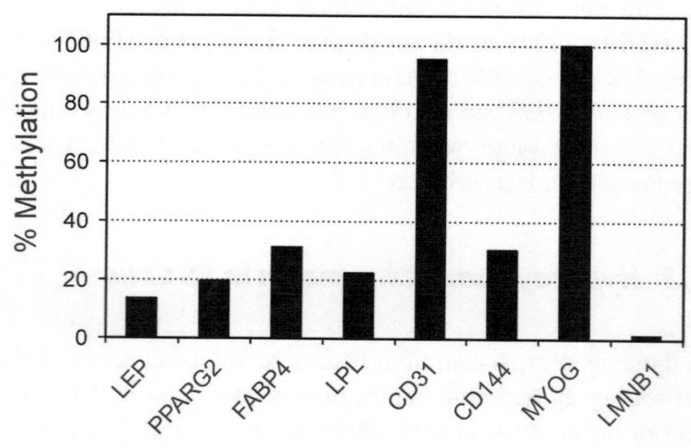

Fig. 4 Extent of CpG methylation in the promoter region of lineage-specific and housekeeping genes in undifferentiated ASCs. Genes indicative of the adipogenic lineage (*LEP*, *PPARG2*, *FABP4*, *LPL*), endothelial cell lineage (*CD331*, *CD144*), and myogenic lineage (*MYOG*) are represented. Lamin B1 (*LMMB1*) is a constitutively expressed gene. The mean percentage of methylation across indicated promoter regions is shown. Note the greater percentage of methylation in *CD31* and *MYOG* relative to adipogenic promoters ($p < 0.001$; t-tests). The *CD144* promoter appears relatively hypomethylated due to unmethylation of the 5′ half of the region examined, while the 3′ half is fully methylated in undifferentiated ASCs [9]. The *LMNB1* promoter is unmethylated

crypts [94–96]. Mosaicism is believed to result from stochastic methylation that accumulates independently in different cells as a result of exposure to environmental, aging, and health factors [14, 96–99], together with a propensity of certain CpGs to be hypermethylated [53, 100]. Indeed, it is clear that each gene contains cytosines more susceptible to methylation than others [10].

It appears that, in contrast to adipogenic promoters, nonadipogenic promoters, such as myogenic or endothelial cell regulatory regions, display significantly more CpG methylation [9, 10] (Fig. 4). The myogenic promoter myogenin (*MYOG*) is completely methylated in freshly isolated ASCs. *MYOG* is also methylated in endothelial cells as expected from this cell type (our unpublished data). In addition, regulatory regions of the *CD31* (also called platelet endothelial cell adhesion molecule-1 or *PECAM1*) and *CD144* (also called vascular endothelium cadherin or *CDH5*) genes are also extensively methylated in ASCs but not in endothelial progenitor or differentiated cells [9]. Housekeeping genes such as *GAPDH* and *LMNB1* are unmethylated, as expected from their constitutive expression. Current results, therefore, illustrate the hypomethylation of adipogenic genes in freshly isolated ASCs. Nonadipogenic lineage-specific promoters seem to be, in contrast, more methylated. This raises the view of an epigenetic programming of ASCs for adipogenic differentiation by a DNA methylation pattern at critical promoters.

Long-term culture of human ASCs does not significantly alter methylation states. Few CpGs in the *LEP*, *FABP4*, and *LPL* promoters become methylated upon culture of ASCs while even fewer are demethylated; however, the significance of these methylation changes remains uncertain. Indeed, increased mosaicism in CpG methylation is detected between cell clones relative to that detected between individual ASC donors [10], but culture to senescence does not enhance mosaicism [51]. In contrast to a previous report on CpG methylation in hESCs [48], we have no evidence of heritable methylation changes in cultured ASCs, suggesting randomness of (de)methylation events. In addition to presumed defects in DNMT1 function, different cells in the initial ASC population display mosaic CpG methylation. Moreover, asymmetric cell division, a characteristic of pluripotent stem cells, is expected to generate a different epigenetic pattern in each daughter cell within a clonal population. Studies available to date, therefore, argue that hypomethylation of adipogenic promoters, in contrast to other lineage-specific promoters, constitutes an epigenetic signature of human ASCs. A working hypothesis, then, is that MSCs are preprogrammed by DNA methylation of lineage-specific genes to preferentially differentiate into the cell type(s) of the tissues in which they reside.

Hypomethylation of adipogenic promoters in undifferentiated ASCs raises the issue of how DNA methylation correlates with transcription. All genes examined in the above study are expressed (at low level) in freshly isolated ASCs, and a fraction of these genes become inactivated upon culture despite the maintenance of a hypomethylated state [10, 51]. Conversely, DNA methylation does not preclude expression of a gene, as exemplified by transcription of the methylated *CD31* and *CD144* loci in ASCs [9] (see also below). So gene expression in ASCs does not correlate with a specific methylation pattern in any of the genes examined to date, an observation not restricted to pluripotent cells [36, 101].

5.2 What Is the Relationship Between Promoter DNA Methylation and Transcriptional Activity?

An elegant genome-wide DNA methylation profiling in several somatic cell types and in sperm shows that the relationship between promoter DNA methylation and promoter activity depends on the CpG content of the promoter [37]. Promoters with low CpG content show no correlation between activity (determined by RNA polymerase II occupancy) and abundance of methylated CpGs; therefore, transcriptionally active low CpG promoters (LCPs) are not necessarily un- or hypomethylated [37]. It seems in fact that most low CpG promoters are methylated regardless of their activity status. On the contrary, the activity of intermediate CpG content promoters (ICPs) and high CpG content promoters (HCPs) is inversely correlated to the extent of methylation [37]. In these categories, the proportion of transcriptionally active promoters decreases with increasing DNA methylation, arguing that methylation of ICPs and HCPs is incompatible with transcription. Further analysis, however, shows that inactive ICPs and HCPs differ in their DNA methylation status: most inactive HCPs are unmethylated, whereas a high proportion of inactive ICPs are methylated. So, collectively, the work of Weber and colleagues [37] argues that inactive HCPs globally remain unmethylated, inactive ICPs are often methylated, whereas LCPs are frequently methylated irrespective of their activation status. A genome-wide analysis of CpG methylation profiling in different MSC populations, not examined in the Weber study, will be welcome to assess the relationship between CpG content, methylation state and transcriptional status in these cells.

5.3 What Is Known on Histone Modifications Associated with Differentiation-Regulated Genes?

The nature of histone modifications marking promoters regulated by differentiation in MSCs remains at present largely unknown. Analyses have up to now been limited to normal differentiated cultured cells, cancer cell lines, and mouse ESCs. The availability of ChIP assays suitable for chromatin from small cell numbers [39, 102], however, opens avenues for investigating limiting cell samples such as embryonic cells [102]. Preliminary observations from our laboratory point to the presence of the activating H3K4m3 mark (together with acetylated H3K9) and of the repressive H3K27m3 modification on lineage-specific promoters of undifferentiated ASCs (AN and PC, manuscript submitted) (Fig. 5, MSCs). So together with the hypomethylated state of these promoters [10], these presumably bivalent histone marks (co-occupancy on the same nucleosome needs to be demonstrated) reinforce the view of adipogenic promoters preprogrammed for activation upon adipogenic stimulation. In contrast, the relatively hypermethylated state of the *MYOG* locus, despite the presence of trimethylated H3K4 and H3K27, seems to "lock" the gene into a repressive state. Upon adipogenic differentiation, adipogenic gene activation is accompanied by a demethylation of H3K27, while, interestingly, the *MYOG* promoter remains trimethylated on K27 (AN and PC, manuscript submitted). H3K27 demethylation may result from PRC2 removal from the promoter or active demethylation of H3K27 (Fig. 5). Inactivation of a promoter upon lineage-specific differentiation would, conversely, lead to deacetylation and trimethylation of H3K9 and maintenance of trimethylated H3K27. These hypotheses remain to be tested.

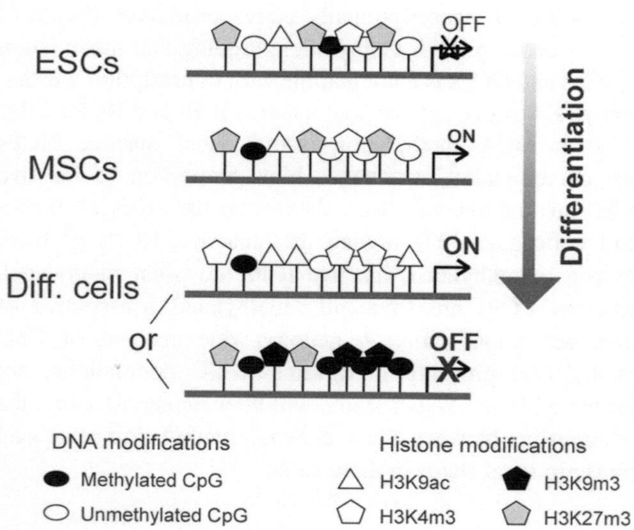

Fig. 5 Epigenetic landscape of genes associated with lineage specification as function of differentiation. ESCs, undifferentiated embryonic stem cells; MSCs, undifferentiated mesenchymal stem cells; Diff. cells, differentiated somatic cells. Two scenarios are presented for lineage-specific genes in differentiated cells, depending on whether the gene is activated or upregulated (ON), or turned off (OFF). Note that relationship between promoter DNA methylation and promoter activity depends on CpG content of the promoter

6 Linking DNA Methylation and Histone Modifications to Replication Timing

DNA methylation has long been implicated in the organization of the nuclear compartment, particularly in regions of constitutive heterochromatin (see [103] for an overview of the evidence). A recent study shed light on the nature of the relationship between global DNA methylation levels and chromatin organization [103]. Indeed, *Dnmt3a*$^{-/-}$ *Dnmt3b*$^{-/-}$ mouse ESCs lacking DNA methylation have been shown to exhibit enhanced clustering of pericentric heterochromatin and major changes in chromatin structure [103]. Levels of H3K9m2 are reduced (H3K9m3 remains surprisingly unaltered) while levels of acetylated H3K9, H4K5, and H4K16 increase, both globally and on major satellite repeats, suggesting a reorganization of heterochromatin in these cells. Mobility of the linker histones H1 and H5 is also reduced. In contrast, absence of DNA methylation does not seem to affect compaction of bulk and heterochromatin, on the basis of nuclease digestion, nucleosome spacing, and chromatin fractionation [103]. Genes reactivated by elimination of DNMT1 in mouse ESCs become enriched in acetylated H3K9 and H3K14, H4ac, and H3K4m3, while those not reactivated by removal of DNA methylation show no hyperacetylation [104]. Thus, some methylated genes in ESCs are subject to additional repressive mechanisms affecting histone H3 acetylation. These studies illustrate how DNA methylation affects global chromatin packaging and subsequently, organization of the nucleus, but in a manner that does not involve chromatin compaction. Despite these global changes, however, different classes of genes respond differently to the absence of DNA methylation.

Timing of DNA replication has been shown to be influenced by the state of chromatin (active vs. inactive), albeit not always by transcription per se [65, 105]. Replication timing has been introduced as an additional epigenetic component [106], although whether it qualifies as an "epigenetic" component on the basis of the definition of epigenetics remains questionable (replication timing is not a modification of DNA or chromatin). Interestingly, in mouse ESCs, a number of genes not necessarily expressed but which may be important later during differentiation have been shown to replicate early in S phase [65]. Genes that are not needed, however, replicate later in S phase. Indeed, genes encoding key neuronal-specific transcription factors replicate early in undifferentiated ESCs, but late in hematopoietic stem cells in which these genes are not required [65]. Therefore, lineage-specification genes are able to undergo modifications in chromatin organization and switch from early to late replication timing in the course of differentiation.

Early replication timing has been linked to enriched histone acetylation [107, 108], but how replication timing functionally relates to DNA methylation remains to be explored. Recent evidence indicates that genes whose expression is dependent on DNA demethylation in ESCs consistently replicate early in S phase, while half of those genes not reactivated by DNA demethylation replicate late [104]. Nonetheless, the overall replication timing pattern does not seem to be dependent on CpG methylation [109] and methylation is not necessarily affected by replication timing profile, suggesting that replication timing and DNA methylation are independently established [104].

7 Perspectives

Genome-wide technologies have provided a wealth of information on mechanisms regulating gene expression in the context of development, cell differentiation, and disease. These studies have also started to unravel the epigenetic landscape of ESCs and somatic stem cells, providing a molecular frame for the pluripotent state. Such approaches have, in our opinion, been welcome because defining pluripotency simply on the basis of gene expression in ESCs has been deceptive [110].

7.1 What Else Do We Need to Know?

Multiple aspects of stem cell function remain to be investigated. We are looking the tip of the iceberg in the epigenetic landscape of stem cells. Mapping of DNA methylation marks, of novel histone modifications, and of novel transcriptional regulators [111], together with improved bioinformatics tools, will enhance the resolution of the current stem cell epigenetic map.

A totally unexplored area is in vivo epigenetics [9, 10]. The fate of ESCs after transplantation into animal models is being studied, but the extent of contribution of MSCs to various tissues remains debated. Our analyses of DNA methylation in ASCs after in vitro differentiation suggest that the cells retain an undifferentiated ASC epigenetic program despite phenotypic changes [41]. In the event MSCs do directly contribute to host tissue in vivo, a hypothesis is that the target tissue provides a beneficial environment for stem cell function. Intuitively, the in vivo environment may be more conducive to epigenetic commitment of MSCs than the culture flask.

Broader application of imaging techniques to chromatin dynamics, gene expression and epigenetics [74, 75, 112–114] will undoubtedly contribute to our understanding of genome organization in stem cells. Ultimately, compilation of nucleus-wide four-dimensional imaging data and genome-wide biochemical and genetic data will provide an integrated representation of functional genome organization in stem cells.

7.2 Reprogramming Somatic Stem Cells to Pluripotency?

The restricted differentiation potential of MSCs currently limits their application to regenerative medicine. Qualities of the ideal stem cell in a clinical setting are expected to be extensive ability to be expanded in culture without genetic and epigenetic abnormalities, ability to form functional cell types in vitro and in vivo, and immuno-compatibility with the patient. Patient-derived somatic stem cells fulfill the latter requirement, however, they currently do not meet the first two. Attempts to alleviate limited differentiation potential of MSCs aim at enhancing differentiation plasticity through a nuclear reprogramming process. Current strategies for reprogramming somatic cells to pluripotency include nuclear transplantation [115–117], fusion with ESCs [118–120], treatment with extracts from eggs [121], ESCs, or other pluripotent cells [40], and retroviral transduction of pluripotency-associated factors [122]. These approaches have been reviewed elsewhere [123, 124]. Notably, recent attempts at reprogramming somatic stem cells have been reported, and results suggest that the reprogramming efficiency by nuclear transfer of progenitor cells compared to terminally differentiated cells is not improved [125]. Clearly, more efforts are needed to determine whether somatic stem cells will some day be safely reprogrammed to a pluripotent state to enable their use in autologous cell therapy.

Acknowledgments Work by the authors is supported by the FUGE, STORFORSK, YFF, and STAMCELLER programs of the Research Council of Norway and by the Norwegian Cancer Society (PC).

References

1. Thomson JA, Itskovitz-Eldor J, Shapiro SS, Waknitz MA, Swiergiel JJ, Marshall VS, et al. Embryonic stem cell lines derived from human blastocysts. Science. 1998;282:1145–7.
2. Boiani M, Scholer HR. Regulatory networks in embryo-derived pluripotent stem cells. Nat Rev Mol Cell Biol. 2005;6:872–84.
3. Hoffman LM, Carpenter MK. Characterization and culture of human embryonic stem cells. Nat Biotechnol. 2005;23:699–708.
4. Jahagirdar BN, Verfaillie CM. Multipotent adult progenitor cell and stem cell plasticity. Stem Cell Rev. 2005;1:53–9.
5. Verfaillie C. Stem cell plasticity. Hematology. 2005;10 Suppl 1:293–6.
6. Jiang Y, Jahagirdar BN, Reinhardt RL, Schwartz RE, Keene CD, Ortiz-Gonzalez XR, et al. Pluripotency of mesenchymal stem cells derived from adult marrow. Nature. 2002;418:41–9.
7. Boquest AC, Shahdadfar A, Fronsdal K, Sigurjonsson O, Tunheim SH, Collas P, et al. Isolation and transcription profiling of purified uncultured human stromal stem cells: alteration of gene expression after in vitro cell culture. Mol Biol Cell. 2005;16:1131–41.
8. Zuk PA, Zhu M, Mizuno H, Huang J, Futrell JW, Katz AJ, et al. Multilineage cells from human adipose tissue: implications for cell-based therapies. Tissue Eng. 2001;7:211–28.
9. Boquest AC, Noer A, Sorensen AL, Vekterud K, Collas P. CpG methylation profiles of endothelial cell-specific gene promoter

regions in adipose tissue stem cells suggest limited differentiation potential toward the endothelial cell lineage. Stem Cells. 2007;25:852–61.
10. Noer A, Sørensen AL, Boquest AC, Collas P. Stable CpG hypomethylation of adipogenic promoters in freshly isolated, cultured and differentiated mesenchymal stem cells from adipose tissue. Mol Biol Cell. 2006;17:3543–56.
11. Mito Y, Henikoff JG, Henikoff S. Genome-scale profiling of histone H3.3 replacement patterns. Nat Genet. 2005;37:1090–7.
12. Mito Y, Henikoff JG, Henikoff S. Histone replacement marks the boundaries of cis-regulatory domains. Science. 2007;315:1408–11.
13. Gardiner-Garden M, Frommer M. CpG islands in vertebrate genomes. J Mol Biol. 1987;196:261–82.
14. Laird PW. Cancer epigenetics. Hum Mol Genet 2005;14:R65–76.
15. Jaenisch R, Bird A. Epigenetic regulation of gene expression: how the genome integrates intrinsic and environmental signals. Nat Genet. 2003;33 Suppl:245–54.
16. Turek-Plewa J, Jagodzinski PP. The role of mammalian DNA methyltransferases in the regulation of gene expression. Cell Mol Biol Lett 2005;10:631–47.
17. Jeltsch A, Nellen W, Lyko F. Two substrates are better than one: dual specificities for Dnmt2 methyltransferases. Trends Biochem Sci. 2006;31:306–8.
18. Kunert N, Marhold J, Stanke J, Stach D, Lyko F. A Dnmt2-like protein mediates DNA methylation in Drosophila. Development. 2003;130:5083–90.
19. Tang LY, Reddy MN, Rasheva V, Lee TL, Lin MJ, Hung MS, et al. The eukaryotic DNMT2 genes encode a new class of cytosine-5 DNA methyltransferases. J Biol Chem. 2003;278:33613–6.
20. Hermann A, Schmitt S, Jeltsch A. The human Dnmt2 has residual DNA-(cytosine-C5) methyltransferase activity. J Biol Chem. 2003;278:31717–21.
21. Liu K, Wang YF, Cantemir C, Muller MT. Endogenous assays of DNA methyltransferases: evidence for differential activities of DNMT1, DNMT2, and DNMT3 in mammalian cells in vivo. Mol Cell Biol 2003;23:2709–19.
22. Rai K, Chidester S, Zavala CV, Manos EJ, James SR, Karpf AR, et al. Dnmt2 functions in the cytoplasm to promote liver, brain, and retina development in zebrafish. Genes Dev. 2007;21:261–6.
23. Goll MG, Kirpekar F, Maggert KA, Yoder JA, Hsieh CL, Zhang X, et al. Methylation of tRNAAsp by the DNA methyltransferase homolog Dnmt2. Science. 2006; 311:395–8.
24. Nan X, Ng HH, Johnson CA, Laherty CD, Turner BM, Eisenman RN, et al. Transcriptional repression by the methyl-CpG-binding protein MeCP2 involves a histone deacetylase complex. Nature. 1998;393:386–9.
25. Hoffman AR, Hu JF. Directing DNA methylation to inhibit gene expression. Cell Mol Neurobiol. 2006;26:425–38.
26. Klose RJ, Bird AP. Genomic DNA methylation: the mark and its mediators. Trends Biochem Sci. 2006;31:89–97.
27. Morgan HD, Santos F, Green K, Dean W, Reik W. Epigenetic reprogramming in mammals. Hum Mol Genet. 2005;14:R47–58.
28. Young LE, Beaujean N. DNA methylation in the preimplantation embryo: the differing stories of the mouse and sheep. Anim Reprod Sci. 2004;82–83:61–78.
29. Mann JR. Imprinting in the germ line. Stem Cells. 2001;19:287–94.
30. Razin A, Shemer R. DNA methylation in early development. Hum Mol Genet. 1995;4:1751–5.
31. Hellman A, Chess A. Gene body-specific methylation on the active X chromosome. Science. 2007;315:1141–3.
32. Tremblay KD, Saam JR, Ingram RS, Tilghman SM, Bartolomei MS. A paternal-specific methylation imprint marks the alleles of the mouse H19 gene. Nat Genet. 1995;9:407–13.
33. Reik W, Howlett SK, Surani MA. Imprinting by DNA methylation: from transgenes to endogenous gene sequences. Development. 1990;1:99–106.
34. Sapienza C, Peterson AC, Rossant J, Balling R. Degree of methylation of transgenes is dependent on gamete of origin. Nature. 1987;328:251–4.
35. Reik W, Collick A, Norris ML, Barton SC, Surani MA. Genomic imprinting determines methylation of parental alleles in transgenic mice. Nature. 1987;328:248–51.
36. Jones PA, Takai D. The role of DNA methylation in mammalian epigenetics. Science. 2001;293:1068–70.
37. Weber M, Hellmann I, Stadler MB, Ramos L, Paabo S, Rebhan M, Schubeler D. Distribution, silencing potential and evolutionary impact of promoter DNA methylation in the human genome. Nat Genet. 2007;39:457–66.
38. Deb-Rinker P, Ly D, Jezierski A, Sikorska M, Walker PR. Sequential DNA methylation of the Nanog and Oct-4 upstream regions in human NT2 cells during neuronal differentiation. J Biol Chem. 2005;280:6257–60.
39. Dahl JA, Collas P. Q^2ChIP, a quick and quantitative chromatin immunoprecipitation assay unravels epigenetic dynamics of developmentally regulated genes in human carcinoma cells. Stem Cells. 2007;25:1037–46.
40. Freberg CT, Dahl JA, Timoskainen S, Collas P. Epigenetic reprogramming of OCT4 and NANOG regulatory regions by Embryonal Carcinoma cell extract. Mol Biol Cell. 2007;18:1543–53.
41. Boquest AC, Noer A, Collas P. Epigenetic programming of mesenchymal stem cells from human adipose tissue. Stem Cell Rev. 2006;2:319–29.
42. Brero A, Easwaran HP, Nowak D, Grunewald I, Cremer T, Leonhardt H, et al. Methyl CpG-binding proteins induce large-scale chromatin reorganization during terminal differentiation. J Cell Biol. 2005;169:733–43.
43. Jackson M, Krassowska A, Gilbert N, Chevassut T, Forrester L, Ansell J, et al. Severe global DNA hypomethylation blocks differentiation and induces histone hyperacetylation in embryonic stem cells. Mol Cell Biol. 2004;24:8862–71.
44. Zvetkova I, Apedaile A, Ramsahoye B, Mermoud JE, Crompton LA, John R, et al. Global hypomethylation of the genome in XX embryonic stem cells. Nat Genet. 2005;37:1274–9.
45. Jacob S, Moley KH. Gametes and embryo epigenetic reprogramming affect developmental outcome: implication for assisted reproductive technologies. Pediatr Res. 2005;58:437–46.
46. Bibikova M, Chudin E, Wu B, Zhou L, Garcia EW, Liu Y, et al. Human embryonic stem cells have a unique epigenetic signature. Genome Res. 2006;16:1075–83.
47. Lagarkova MA, Volchkov PY, Lyakisheva AV, Philonenko ES, Kiselev SL. Diverse epigenetic profile of novel human embryonic stem cell lines. Cell Cycle. 2006;5:416–20.
48. Allegrucci C, Wu YZ, Thurston A, Denning CN, Priddle H, Mummery CL, et al. Restriction landmark genome scanning identifies culture-induced DNA methylation instability in the human embryonic stem cell epigenome. Hum Mol Genet. 2007;16:1253–68.
49. Maitra A, Arking DE, Shivapurkar N, Ikeda M, Stastny V, Kassauei K, et al. Genomic alterations in cultured human embryonic stem cells. Nat Genet. 2005;37:1099–103.
50. Rugg-Gunn PJ, Ferguson-Smith AC, Pedersen RA. Epigenetic status of human embryonic stem cells. Nat Genet. 2005;37:585–7.
51. Noer A, Boquest AC, Collas P. Dynamics of adipogenic promoter DNA methylation during clonal culture of human adipose stem cells to senescence. BMC Cell Biol. 2007;8:18–29.
52. Graff JR, Gabrielson E, Fujii H, Baylin SB, Herman JG. Methylation patterns of the E-cadherin 5'CpG island are unstable

and reflect the dynamic, heterogeneous loss of E-cadherin expression during metastatic progression. J Biol Chem. 2000;275: 2727–32.
53. Silva AJ, Ward K, White R. Mosaic methylation in clonal tissue. Dev Biol. 1993;156:391–8.
54. Zhu X, Deng C, Kuick R, Yung R, Lamb B, Neel JV, et al. Analysis of human peripheral blood T cells and single-cell-derived T cell clones uncovers extensive clonal CpG island methylation heterogeneity throughout the genome. Proc Natl Acad Sci U S A. 1999;96:8058–63.
55. Weber M, Davies JJ, Wittig D, Oakeley EJ, Haase M, Lam WL, et al. Chromosome-wide and promoter-specific analyses identify sites of differential DNA methylation in normal and transformed human cells. Nat Genet. 2005;37:853–62.
56. Shen Y, Chow J, Wang Z, Fan G. Abnormal CpG island methylation occurs during in vitro differentiation of human embryonic stem cells. Hum Mol Genet. 2006;15:2623–35.
57. Jenuwein T, Allis CD. Translating the histone code. Science. 2001;293:1074–80.
58. Mellor J. It takes a PHD to read the histone code. Cell. 2006;126:22–4.
59. Kouzarides T. Chromatin modifications and their function. Cell. 2007;128:693–705.
60. Lachner M, O'Carroll D, Rea S, Mechtler K, Jenuwein T. Methylation of histone H3 lysine 9 creates a binding site for HP1 proteins. Nature. 2001;410:116–20.
61. Cao R, Wang L, Wang H, Xia L, Erdjument-Bromage H, Tempst P, et al. Role of histone H3 lysine 27 methylation in Polycomb-group silencing. Science. 2002;298:1039–43.
62. Cao R, Zhang Y. The functions of E(Z)/EZH2-mediated methylation of lysine 27 in histone H3. Curr Opin Genet Dev. 2004;14:155–64.
63. Pasini D, Bracken AP, Jensen MR, Lazzerini DE, Helin K. Suz12 is essential for mouse development and for EZH2 histone methyltransferase activity. EMBO J. 2004;23:4061–71.
64. Lachner M, Jenuwein T. The many faces of histone lysine methylation. Curr Opin Cell Biol. 2002;14:286–98.
65. Azuara V, Perry P, Sauer S, Spivakov M, Jorgensen HF, John RM, et al. Chromatin signatures of pluripotent cell lines. Nat Cell Biol. 2006;8:532–8.
66. Bernstein BE, Mikkelsen TS, Xie X, Kamal M, Huebert DJ, Cuff J, et al. A bivalent chromatin structure marks key developmental genes in embryonic stem cells. Cell. 2006;125:315–26.
67. Struhl K. Histone acetylation and transcriptional regulatory mechanisms. Genes Dev. 1998;12:599–606.
68. Santos-Rosa H, Schneider R, Bannister AJ, Sherriff J, Bernstein BE, Emre NC, et al. Active genes are tri-methylated at K4 of histone H3. Nature. 2002;419:407–11.
69. Schubeler D, MacAlpine DM, Scalzo D, Wirbelauer C, Kooperberg C, van LF, et al. The histone modification pattern of active genes revealed through genome-wide chromatin analysis of a higher eukaryote. Genes Dev. 2004;18:1263–71.
70. Kingston RE, Narlikar GJ. ATP-dependent remodeling and acetylation as regulators of chromatin fluidity. Genes Dev. 1999;13:2339–52.
71. Pray-Grant MG, Daniel JA, Schieltz D, Yates JR, III, Grant PA. Chd1 chromodomain links histone H3 methylation with SAGA- and SLIK-dependent acetylation. Nature. 2005;433:434–8.
72. Collas P, Dahl JA. Chop it, ChIP it, check it: the current status of chromatin immunoprecipitation. Front Biosci. 2008;13:929–43.
73. Montgomery ND, Yee D, Chen A, Kalantry S, Chamberlain SJ, Otte AP, et al. The murine polycomb group protein Eed is required for global histone H3 lysine-27 methylation. Curr Biol. 2005;15:942–7.
74. Meshorer E, Misteli T. Chromatin in pluripotent embryonic stem cells and differentiation. Nat Rev Mol Cell Biol. 2006;7:540–6.
75. Meshorer E, Yellajoshula D, George E, Scambler PJ, Brown DT, Misteli T. Hyperdynamic plasticity of chromatin proteins in pluripotent embryonic stem cells. Dev Cell. 2006;10:105–16.
76. Gan Q, Yoshida T, McDonald OG, Owens GK. Concise review: epigenetic mechanisms contribute to pluripotency and cell lineage determination of embryonic stem cells. Stem Cells. 2007;25:2–9.
77. Kennison JA. The Polycomb and trithorax group proteins of Drosophila: trans-regulators of homeotic gene function. Annu Rev Genet. 1995;29:289–303.
78. Ringrose L, Paro R. Polycomb/Trithorax response elements and epigenetic memory of cell identity. Development. 2007;134: 223–32.
79. Otte AP, Kwaks TH. Gene repression by Polycomb group protein complexes: a distinct complex for every occasion? Curr Opin Genet Dev. 2003;13:448–54.
80. Boyer LA, Mathur D, Jaenisch R. Molecular control of pluripotency. Curr Opin Genet Dev. 2006;16:455–62.
81. Boyer LA, Plath K, Zeitlinger J, Brambrink T, Medeiros LA, Lee TI, et al. Polycomb complexes repress developmental regulators in murine embryonic stem cells. Nature. 2006;441:349–53.
82. Lee TI, Jenner RG, Boyer LA, Guenther MG, Levine SS, Kumar RM, et al. Control of developmental regulators by Polycomb in human embryonic stem cells. Cell. 2006;125:301–13.
83. Bracken AP, Dietrich N, Pasini D, Hansen KH, Helin K. Genome-wide mapping of Polycomb target genes unravels their roles in cell fate transitions. Genes Dev. 2006;20:1123–36.
84. Pusarla RH, Vinayachandran V, Bhargava P. Nucleosome positioning in relation to nucleosome spacing and DNA sequence-specific binding of a protein. FEBS J. 2007;274:2396–410.
85. Ozsolak F, Song JS, Liu XS, Fisher DE. High-throughput mapping of the chromatin structure of human promoters. Nat Biotechnol. 2007;25:244–8.
86. Segal E, Fondufe-Mittendorf Y, Chen L, Thastrom A, Field Y, Moore IK, et al. A genomic code for nucleosome positioning. Nature. 2006;442:772–8.
87. Saha A, Wittmeyer J, Cairns BR. Chromatin remodelling: the industrial revolution of DNA around histones. Nat Rev Mol Cell Biol. 2006;7:437–47.
88. Fraser JK, Wulur I, Alfonso Z, Hedrick MH. Fat tissue: an underappreciated source of stem cells for biotechnology. Trends Biotechnol. 2006;24:150–4.
89. Zuk PA, Zhu M, Ashjian P, De Ugarte DA, Huang JI, Mizuno H, et al. Human adipose tissue is a source of multipotent stem cells. Mol Biol Cell. 2002;13:4279–95.
90. Katz AJ, Tholpady A, Tholpady SS, Shang H, Ogle RC. Cell surface and transcriptional characterization of human adipose-derived adherent stromal (hADAS) cells. Stem Cells. 2005;23:412–23.
91. Kern S, Eichler H, Stoeve J, Kluter H, Bieback K. Comparative analysis of mesenchymal stem cells from bone marrow, umbilical cord blood or adipose tissue. Stem Cells. 2006;24: 1294–301.
92. Urs S, Smith C, Campbell B, Saxton AM, Taylor J, Zhang B, et al. Gene expression profiling in human preadipocytes and adipocytes by microarray analysis. J Nutr. 2004;134:762–70.
93. Cousin B, Andre M, Arnaud E, Penicaud L, Casteilla L. Reconstitution of lethally irradiated mice by cells isolated from adipose tissue. Biochem Biophys Res Commun. 2003;301: 1016–22.
94. Kim JY, Beart RW, Shibata D. Stability of colon stem cell methylation after neo-adjuvant therapy in a patient with attenuated familial adenomatous polyposis. BMC Gastroenterol. 2005;5: 19–25.
95. Kim JY, Tavare S, Shibata D. Counting human somatic cell replications: methylation mirrors endometrial stem cell divisions. Proc Natl Acad Sci U S A. 2005;102:17739–44.

96. Yatabe Y, Tavare S, Shibata D. Investigating stem cells in human colon by using methylation patterns. Proc Natl Acad Sci U S A. 2001;98:10839–44.
97. Esteller M. Aberrant DNA methylation as a cancer-inducing mechanism. Annu Rev Pharmacol Toxicol. 2005;45:629–56.
98. Hoffman LM, Carpenter MK. Human embryonic stem cell stability. Stem Cell Rev. 2005;1:139–44.
99. Ushijima T, Okochi-Takada E. Aberrant methylations in cancer cells: where do they come from? Cancer Sci. 2005;96:206–11.
100. Pfeifer GP, Steigerwald SD, Hansen RS, Gartler SM, Riggs AD. Polymerase chain reaction-aided genomic sequencing of an X chromosome-linked CpG island: methylation patterns suggest clonal inheritance, CpG site autonomy, and an explanation of activity state stability. Proc Natl Acad Sci U S A. 1990;87:8252–56.
101. Kaneko KJ, Rein T, Guo ZS, Latham K, Depamphilis ML. DNA methylation may restrict but does not determine differential gene expression at the Sgy/Tead2 locus during mouse development. Mol Cell Biol. 2004;24:1968–82.
102. O'Neill LP, Vermilyea MD, Turner BM. Epigenetic characterization of the early embryo with a chromatin immunoprecipitation protocol applicable to small cell populations. Nat Genet. 2006;38:835–41.
103. Gilbert N, Thomson I, Boyle S, Allan J, Ramsahoye B, Bickmore WA. DNA methylation affects nuclear organization, histone modifications, and linker histone binding but not chromatin compaction. J Cell Biol. 2007;177:401–11.
104. Lande-Diner L, Zhang J, Ben-Porath I, Amariglio N, Keshet I, Hecht M, et al. Role of DNA methylation in stable gene repression. J Biol Chem. 2007;282:12194–200.
105. Azuara V, Brown KE, Williams RR, Webb N, Dillon N, Festenstein R, et al. Heritable gene silencing in lymphocytes delays chromatid resolution without affecting the timing of DNA replication. Nat Cell Biol. 2003;5:668–74.
106. Spivakov M, Fisher AG. Epigenetic signatures of stem-cell identity. Nat Rev Genet. 2007;8:263–71.
107. Schubeler D, Scalzo D, Kooperberg C, van SB, Delrow J, Groudine M. Genome-wide DNA replication profile for Drosophila melanogaster: a link between transcription and replication timing. Nat Genet. 2002;32:438–42.
108. Vogelauer M, Rubbi L, Lucas I, Brewer BJ, Grunstein M. Histone acetylation regulates the time of replication origin firing. Mol Cell. 2002;10:1223–33.
109. Gribnau J, Hochedlinger K, Hata K, Li E, Jaenisch R. Asynchronous replication timing of imprinted loci is independent of DNA methylation, but consistent with differential subnuclear localization. Genes Dev. 2003;17:759–73.
110. Fortunel NO, Otu HH, Ng HH, Chen J, Mu X, Chevassut T, et al. Comment on "'Stemness': transcriptional profiling of embryonic and adult stem cells" and "a stem cell molecular signature". Science. 2003;302:393.
111. Sugiyama T, Cam HP, Sugiyama R, Noma K, Zofall M, Kobayashi R, et al. SHREC, an effector complex for heterochromatic transcriptional silencing. Cell. 2007;128:491–504.
112. Heard E, Bickmore W. The ins and outs of gene regulation and chromosome territory organisation. Curr Opin Cell Biol. 2007;19:311–6.
113. Morey C, Da Silva NR, Perry P, Bickmore WA. Nuclear reorganisation and chromatin decondensation are conserved, but distinct, mechanisms linked to Hox gene activation. Development. 2007;134:909–19.
114. Gilbert N, Bickmore WA. The relationship between higher-order chromatin structure and transcription. Biochem Soc Symp. 2006;73:59–66.
115. Gurdon JB, Byrne JA. The first half-century of nuclear transplantation. Proc Natl Acad Sci U S A. 2003;100:8048–52.
116. Wilmut I, Beaujean N, De Sousa PA, Dinnyes A, King TJ, Paterson LA, et al. Somatic cell nuclear transfer. Nature. 2002;419:583–6.
117. Yang X, Smith SL, Tian XC, Lewin HA, Renard JP, Wakayama T. Nuclear reprogramming of cloned embryos and its implications for therapeutic cloning. Nat Genet. 2007;39: 295–302.
118. Kimura H, Tada M, Nakatsuji N, Tada T. Histone code modifications on pluripotential nuclei of reprogrammed somatic cells. Mol Cell Biol. 2004;24:5710–20.
119. Ying QL, Nichols J, Evans EP, Smith AG. Changing potency by spontaneous fusion. Nature. 2002;416:545–8.
120. Cowan CA, Atienza J, Melton DA, Eggan K. Nuclear reprogramming of somatic cells after fusion with human embryonic stem cells. Science .2005;309:1369–73.
121. Miyamoto K, Furusawa T, Ohnuki M, Goel S, Tokunaga T, Minami N, et al. Reprogramming events of mammalian somatic cells induced by Xenopus laevis egg extracts. Mol Reprod Dev. 2007;74:1268–77.
122. Takahashi K, Yamanaka S. Induction of pluripotent stem cells from mouse embryonic and adult fibroblast cultures by defined factors. Cell. 2006;126:663–76.
123. Hochedlinger K, Jaenisch R. Nuclear reprogramming and pluripotency. Nature. 2006;441:1061–7.
124. Collas P, Gammelsaeter R. Novel approaches to epigenetic reprogramming of somatic cells. Cloning Stem Cells. 2007;9: 26–32.
125. Sung LY, Gao S, Shen H, Yu H, Song Y, Smith SL, et al. Differentiated cells are more efficient than adult stem cells for cloning by somatic cell nuclear transfer. Nat Genet. 2006;38: 1323–8.

Role of DNA Methylation and Epigenetics in Stem Cells

Bhaskar Thyagarajan and Mahendra Rao

Abstract In recent years, great strides have been made in our understanding of the biology of human embryonic stem cells and their ability to differentiate into multiple lineages. Although it has always been obvious that the differentiation of stem cells does not come about due to changes to the primary sequence of the genome, only now are we beginning to understand the mechanisms involved in this process. Chromatin modification has been shown to control the expression of key genes involved in the progression of stem cells into their differentiated progeny. In this chapter, some of the key regulatory mechanisms involved in epigenetic modification of the genome are discussed.

Keywords Epigenetic regulation · DNA methylation · miRNA · Histone acetylation · X-chromosome inactivation

1 Introduction

Transcriptional regulation plays a central role in defining the state of a cell. Activating and inhibiting signals establish a dynamic cascade of coordinated gene expression changes in response to extrinsic signals and intrinsic programming. Integration of these instructions occurs in the nucleus through combinations of signal-activated and tissue-restricted transcription factors (TFs) binding to and controlling related enhancers or *cis*-regulatory modules (CRMs) of co-expressed genes. Additional regulation is provided by previously unappreciated epigenetic mechanisms such as histone modulation and CpG island methylation and by small untranslated RNA molecules such as microRNA (miRNA). The set of individual components controlling a particular biological process and the various interactions among them define a regulatory network and the sum of interacting regulatory networks defines the state (transcriptome) of the cell.

B. Thyagarajan (✉)
Invitrogen Corporation, 5781 Van Allen Way, Carlsbad, CA 92008
e-mail: bhaskar.thyagarajan@invitrogen.com

The ability to obtain pure populations of cells by either growing cells in feeder-free conditions, to harvest feeder cells away from ES cells, and to harvest RNA from single cells or small amounts of tissue, enhancements in library construction, and the improving quality of genomic information allowing short reads to be unambiguously mapped have all contributed to our ability to collect detailed information on gene expression in ESC. Equally important, the ability to access the detailed information developed by IVF clinics and to map the information to a well-annotated genomic database has allowed these large-scale techniques to be applied to human cells. These and other technical advances have allowed large-scale genomic analysis to be performed by a variety of techniques and the publication of several recent reports [1–7].

In recent years it has become clear that, in addition to regulation of gene expression by transcription and post-transcriptional methods, other regulatory processes may be equally important. Genome modifications resulting from epigenetic changes are believed to play a critical role in the development and/or progression of stem cells. Experimental evidence suggests that methylation and epigenetic changes could also be critical determinants of cellular senescence and aging [8–11]. Here, we review the current evidence and discuss how imbalances in chromatin remodeling might regulate stem cell self-renewal, proliferation rate, and differentiation.

2 Epigenetic Regulation

DNA methylation, along with histone modification (acetylation, methylation, phosphorylation, ubiquitination, etc.) and miRNA regulation are considered to be mechanisms of epigenetic regulation. Epigenetic modulation changes the expression of a gene without making any changes to the DNA. Such epigenetic differences allow two cells within the same organism that contain the same genetic complement of DNA to express a unique subset of genes and to differentiate.

V.K. Rajasekhar, M.C. Vemuri (eds.), *Regulatory Networks in Stem Cells*, Stem Cell Biology and Regenerative Medicine,
DOI 10.1007/978-1-60327-227-8_22, © Humana Press, a part of Springer Science+Business Media, LLC 2009

Chromatin remodeling is an epigenetic phenomenon that affects gene transcription within a cell and, thereby, its phenotype. In the nucleus, DNA is packaged into chromatin, consisting of strings of nucleosomes, each containing 147 base pairs of genomic DNA wrapped twice around a highly conserved histone complex. Nucleosomes are then further condensed and organized into increasingly highly ordered chromatin structures. The most open chromatin domains are known as euchromatin. Euchromatin domains are generally transcriptionally active. On the other hand, heterochromatin domains are generally inaccessible to DNA binding factors and, therefore, transcriptionally silent. Chromatin must be remodeled to allow transcription factors and RNA polymerase to interact with the promoter region of a gene. Histone modifying enzymes are known to remodel and regulate chromatin structure through acetylation, methylation, and phosphorylation of the histone proteins [12].

In addition to CpG methylation, histone modification can also play a role in transcriptional regulation. Actively transcribed regions of the genome generally have low levels of methylated DNA and high levels of acetylated histones, whereas silent regions of the genome generally have high levels of methylated DNA and low levels of acetylated histones [10, 13–16]. It has been proposed that histone modifications form a "histone code" that is read by other proteins in order to regulate transcription of a genomic locus [17].

Recent studies also implicate noncoding RNA molecules, especially microRNA, in transcriptional regulation [18], and it has been shown that these molecules play a very important role in maintenance, growth, and differentiation of stem cells. Until recently, the analysis of stem cells and their lineages has largely focused on transcriptional regulation. Emerging evidence suggests that epigenetic control and post-translational regulation, including that by small noncoding RNAs, are essential components of stem cell biology [19]. Perhaps the most well-known examples of epigenetic regulation are CpG island methylation, X-chromosome inactivation, and imprinting. These phenomena are described in the following sections.

3 DNA Methylation in Stem Cells

Methylation of cytosine is the only known endogenous modification of DNA in mammals and occurs by the enzymatic addition of a methyl group to the carbon-5 position of cytosine [20]. This methyl group at the fifth position of the cytosine pyrimidine ring, which is present in about 80% of CpG-dinucleotides in the human genome, can be of major functional significance [21]. DNA methyltransferase enzymes are known to be involved in adding the methyl group to the cytosine that resides in the CpG islands in a promoter [20]. It is believed that these methyl groups can attract and bind gene-silencing proteins. Although the exact mechanism by which methylation of promoter regions leads to silencing is unknown, it is believed that these methyl groups can attract and bind gene-silencing proteins [22, 23].

As our understanding of the complex interactions that regulate cell and tissue specific gene expression have grown, so have techniques evolved to allow one to analyze these aspects of cell regulation. Some of these methods are described below.

The Genome Institute of Singapore has developed a platform called 5′ and 3′ LongSAGE (Serial Analysis of Gene Expression), which allows for rapid, detailed analysis of the genome by precisely defining the boundaries of every gene with chemical tags [24]. This technique has been used to generate new information about transcripts of various kinds, including anti-sense and miRNA [24–29]. This information will aid in identification of transcription initiation sites, as well as polyA sites. This in turn will allow easier identification of promoter regions in the genome.

Illumina, Inc. (San Diego, CA) has developed an array-based method for analyzing the methylation status of 1536 CpG sites selected from the 5′-regulatory region of 371 genes [30]. These include known imprinted genes (e.g., the gene cluster localized to the BWS imprinted region), tumor suppressor genes and oncogenes, cell cycle checkpoint genes, genes regulated by various signaling pathways and/or responsible for altered cell growth, differentiation and apoptosis, genes for DNA damage repair and oxidative metabolism, and genes previously reported to be differentially methylated. These assays can be carried out using as little as 200 ng of bisulfite-treated genomic DNA [30]. This assay was used to obtain a quantitative measure of the methylation level at each CpG site in human embryonic stem (hES) cells [31].

Equally important have been developing bioinformatics and software to be able to obtain a global profile of expression and correlate information from these different methodologies to enable efficient data mining. In the subsequent sections we highlight some of the emerging results.

Three DNA methyl transferase enzymes, *DNMT1, DNMT3A,* and *DNMT3B*, have been identified in mammalian cells [32, 33]. Deletion of any of these genes in mice is lethal [34, 35]. Mouse embryos carrying homozygous deletions for *Dnmt1* and *Dnmt3b* die before birth, but *Dnmt3a* leads to death around 3 weeks after birth [34, 35]. Mice that are heterozygous mutants for any of these genes are normal and fertile [34, 35]. In another embryonic stem (ES) study, Hsieh et al. [36] demonstrated that Dnmt3a and Dnmt3b have some overlapping target sites. *DNMT3B* participates in the methylation of centromeric minor satellite repeats while *DNMT3A* is not involved [35]. Interestingly, expression of *DNMT3A* and *DNMT3B* is down-regulated in adult somatic tissues compared to undifferentiated ES cells [33].

De novo methyl-transferase activity is necessary for early methylation of DNA, and this activity is mostly present during early embryo development [37]. Other than DNA methyl-transferases, there are several methyl-CpG-binding proteins (MBDs) involved in DNA methylation process. Most of these are proteins involved in transcriptional regulation controlled by CpG methylation pattern. One of these MBDs, MBD4, can remove thymine or uracil from a G:T or G:U mismatch at CpG sites [38].

Fan and colleagues [31] have examined multiple ESC lines in a comprehensive DNA methylation profiling approach to understand the characteristics and capabilities of hES cells and their differentiated derivatives. They measured ~1500 CpG sites chosen from 371 genes in 15 hES cell lines [39–44], 4 adult stem cell lines, 1 pluripotent embryonic carcinoma clone (NTERA 2) [45], and 1 fibroblast-like cell line derived from an ES cell. They showed by cluster analysis, based on DNA methylation profiles, that all hES cell lines group together, regardless of their laboratory of origin, and that one can separate undifferentiated cells from their differentiated products, from normal lymphoblastoid cell lines and from cancer cell lines. The authors also monitored epigenetic stability through multiple passages of nine hES cells and showed that the degree of overall change in methylation was proportional to the number of passages separating compared preparations. In other words, the greater the passage number of a particular cell line, the more changes observed. However, the authors were unable to identify specific sets of genes that were altered during time in culture, and concluded that these methylation changes are small compared to the differences between cell types.

It is known that female germ cells are less methylated compared to their male counterparts. Gamete methylation pattern is erased (demethylated) near the eighth day of blastocyst formation [46, 47]. Interestingly, non-CpG island methylation is also been reported in several occasions [48, 49]. Ramsahoye et al. [50] have used a dual-labeling nearest neighbor technique and bisulfite genomic sequencing methods to investigate the nearest neighbors of 5-methylcytosine residues in mammalian DNA. They found that embryonic stem cells, but not somatic tissues, have significant cytosine-5 methylation at CpA and, to a lesser extent, at CpT.

4 Imprinting

Gene imprinting is another well-studied phenomenon in DNA methylation research. Genes get imprinted (silenced), mainly because of the methylation changes in the promoters. Imprinted genes are a unique group of genes that are important for fetal growth and development, especially in the placenta, as well as for postnatal behavior and cognition. The expression of imprinted genes does not follow a Mendelian pattern of inheritance but instead depends on the parent of origin to dictate its expression [51–54]. Regulatory regions of such genes are typically methylated in the silent allele and are exempt from the large-scale, genome-wide demethylation that occurs during pre-implantation development. Imprinted genes are particularly sensitive to environmental changes [55]. Recent reports from IVF clinics have suggested an unexpectedly high occurrence of imprinting and other epigenetic abnormalities in early-stage human embryos [56], raising the possibility that cultured embryonic stem cells may vary considerably in their epigenetic status. These changes may influence functional differences in differentiation ability [31]. It is known that abnormal imprinting can lead to a variety of inherited syndromes in humans (Table 1) [57]. It has also been shown that abnormal imprinting has been associated with a number of tumors (Table 1) [58]. A proper understanding of imprinting and the regulatory mechanisms involved in the process will be helpful in developing strategies to counter these diseases.

Table 1 Diseases arising from improper imprinting

Disease	Imprinted region
Neuroblastoma[b]	1p36
Transient Neonatal Diabetes[a]	6q24
Beckwith-Wiedemann Syndrome[a]	11p15.5
Rhabdomyosarcoma[b]	11p15.5
Silver-Russell Syndrome[a]	11p15.5
Wilm's Tumor[b]	11p15.5
Glomus Tumors[b]	11q13, 11q22.3-q23.3
Maternal Uniparental Disomy Syndrome[a]	14q32
Paternal Uniparental Disomy Syndrome[a]	14q32
Angelman Syndrome[a]	15q11-q13
Prader-Willu Syndrome[a]	15q11-q13
Pseudohypoparathyroidism Type 1b[a]	20q12-q13

[a] Reference [57].
[b] Reference [58].

5 X-Chromosome Inactivation

Carpenter et al. [59] have studied X-chromosome inactivation in human embryonic stem cells, and have reported that individual hES lines exhibit different patterns of X-inactivation. While the male hES cells tested in this study uniformly exhibit X-inactivation, female lines are not consistent. The female ES lines H9 and CyT25 show X-inactivation, whereas the line H7 does not. Although the expression of XIST, the RNA involved in marking the chromosome for inactivation, can be rescued by treatment with a demethylating agent, this has no effect on X-inactivation. The lack of X-inactivation does not influence the expression of commonly used stem cell markers. These data show that human ES cells do undergo X chromosome inactivation, but also that they show varying patterns of inactivation. Further studies need to be carried out to elucidate the mechanisms and the time course of this process in hES cells.

Recently, Shiota et al. [60] investigated the genome-wide DNA methylation pattern of CpG islands by restriction landmark genomic scanning in mouse stem cells before and after differentiation, in sperm as well as in somatic tissues. Results indicated that CpG islands were numerous and widespread in all cells tested. Their location was dependent on tissue type, cell lineage, and stage of differentiation. Taken together with the results described by Hoffman et al. [59], these data indicate that genomic loci with altered methylation status seem to be more common than has hitherto been realized.

6 Histone Modification

Several efforts have been undertaken to assess histone modifications in stem cells. Perhaps the most common technique used is Chromatin immunoprecipitation (ChIP) and its modifications (reviewed in [61]). ChIP is a widely used technique to identify and quantify factors that bind to DNA during chromatin modification. ChIP defines the genomic distribution of proteins and their modifications but is limited by the cell numbers. Recently, O'Neill et al. [62] describe a protocol that uses carrier chromatin and PCR, "carrier" ChIP (CChIP), to permit analysis of as few as 100 mouse ES cells. They assayed histone modifications at key regulator genes (such as Nanog, Pou5f1 [also known as Oct4] and Cdx2) by CChIP in mouse embryonic stem (ES) cells, in inner cell mass (ICM), and trophectoderm of cultured blastocysts. Activating and silencing modifications (H4 acetylation and H3K9 methylation) mark active and silent promoters as predicted, and they find close correlation between values derived from CChIP and conventional ChIP. Studies on genes silenced in both ICM and ES cells (Cdx2, Cfc1, Hhex, and Nkx2-2) show that the intensity of silencing marks is relatively diminished in ES cells, indicating a possible relaxation of some components of silencing on adaptation to culture. Meshorer et al. [63] have shown that structural chromatin proteins and histones are bound much more loosely to chromatin of ES cells compared to differentiated or somatic cells. This suggests that the chromatin in ESC is in a much looser configuration when compared to differentiated cells. Other data have shown that promoter regions of genes repressed in ES cells are associated with bivalent histone modifications [64, 65]. These are mostly genes that encode transcription factors that are turned on upon differentiation of the cells. The presence of both H3K4m3, which is associated with active genes, and H3K27m3, which is associated with inactive genes, has been described at these loci. There are large regions of H3K27 trimethylation, with smaller islands of H3K4 trimethylation [65]. ChIP techniques combined with single molecule sequencing have been use to elucidate genome-wide chromatin maps of ES cells and their differentiated progeny [66], and they confirm the association of H3K4m3 and H3K27m3 association with transcriptionally active and inactive regions respectively. In addition, this study shows a strong correlation between lysine 36 trimethylation and coding and noncoding transcripts, which will facilitate gene annotation. The authors also show that trimethylation of lysine 4 and lysine 9 serves as a marker for imprinting control regions.

These data suggest that these genes are in a state of readiness, and when the appropriate signals are received, they commit to a lineage. Several studies demonstrate the involvement of the Polycomb group of proteins (PcGs) in this process [67–80]. These proteins are part of two distinct Polycomb Repressor Complexes (PRCs), PRC1 and PRC2, and are believed to play a role in gene expression at different stages of development [78, 81]. It is believed that these complexes bind to the marked regions of the genome, and repress the transcription of those genes. Once appropriate signals are delivered, the repression is removed, and the cells go through the process of differentiation. More recently, it has been shown that the histone modification in embryonic stem cells is controlled by STAT3 and Oct4 through the induction of Eed [82]. The data generated from these studies strongly suggest that the pluripotency of stem cells is maintained by epigenetic mechanisms.

7 Noncoding RNA in Stem Cells

There is emerging evidence that miRNA can play a role in regulation of transcription. Many different classes of small non-coding RNAs are present in human cells [83]. The term noncoding RNA (ncRNA) is commonly employed for RNA that does not encode a protein, but this does not mean that such RNA molecules do not contain information nor have function. Although it has been generally assumed that most genetic information is transacted by proteins, recent evidence suggests that the majority of the genomes of mammals and other complex organisms is in fact transcribed into ncRNA, many of which are alternatively spliced and/or processed into smaller products. These ncRNAs include microRNAs and snoRNAs (many, if not most, of which remain to be identified), as well as likely other classes of yet-to-be-discovered small regulatory RNAs, and tens of thousands of longer transcripts (including complex patterns of interlacing and overlapping sense and antisense transcripts), many of which have unknown function. These RNAs (including those derived from introns) appear to comprise a hidden layer of internal signals that control various levels of gene expression in physiology and development, including chromatin architecture/epigenetic memory, transcription, RNA splicing, editing, translation, and turnover. RNA regulatory networks may determine most of our complex characteristics, play a significant role in disease, and constitute an unexplored world of genetic variation both within and between species [83].

Small double-stranded modulatory RNAs have been proposed to regulate the generation of neurons from adult neural stem cells by binding to REST [84], although the mechanism by which this occurs remains unclear. Another recently identified species of small noncoding RNAs are microRNAs (miRNAs), which are likely key post-transcriptional players in stem cells and their differentiated progeny. It has been shown that embryonic stem cells express a unique set of miRNAs [85, 86, 87]. The species identified in these studies include miR-302b*, miR-302b, miR-302c*, miR-302c, miR-302a*, miR-302d, miR-367, miR-200c, miR-368, miR-154*, miR-371, miR-372, miR-373*, and miR-373. Recent global miRNA expression profiling data indicate that several differentially expressed miRNA clusters can be seen in different stages of differentiation in mouse (Perera and Lim, personal communication), and human stem cells [88]. It is generally believed that miRNA affect protein levels by either suppressing translation or by inducing degradation of mRNA [83]. Identification of global differences between pluripotent and differentiated cells allows one to then focus on the role of individual miRNAs. It has been shown that miRNA-1-2 is involved in cardiogenesis [89], and deletion of this miRNA disrupts numerous cardiac functions. It has also been shown that miR-134 can induce differentiation into ectodermal lineage by attenuating the expression of Nanog and LRH1 [90]. Other studies have shown a role for miRNA in neuronal gene expression [91], brain morphogenesis [92], muscle differentiation [93], and stem cell division [94]. A more recent study has estimated that each ES cell has as many as 110,000 miRNA molecules [95], with the majority of them originating from six distinct loci. Four of these loci have been implicated in cell cycle control and oncogenesis.

These findings should have broad implications for the use of embryonic stem cells and cells derived from them therapeutically, and also for the cloning of animals by the transfer of somatic cell nuclei.

8 Conclusion

The field of DNA methylation has grown dramatically and become one of the most dynamic and rapidly developing branches of molecular biology. By comparing DNA methylation patterns of individual hES cell lines with their respective gene expression patterns and other genome characteristics, we hope to contribute to an understanding of the nature of pluripotence in human embryonic stem cells.

A general schematic showing the epigenetic factors affecting protein levels in a cell is shown in Fig. 1. The ability to access the epigenomic information for a large number of genes or the entire genome [96–99] should greatly facilitate the understanding of the nature of pluripotence in embryonic stem cells. It appears likely that the epigenomic profile of a

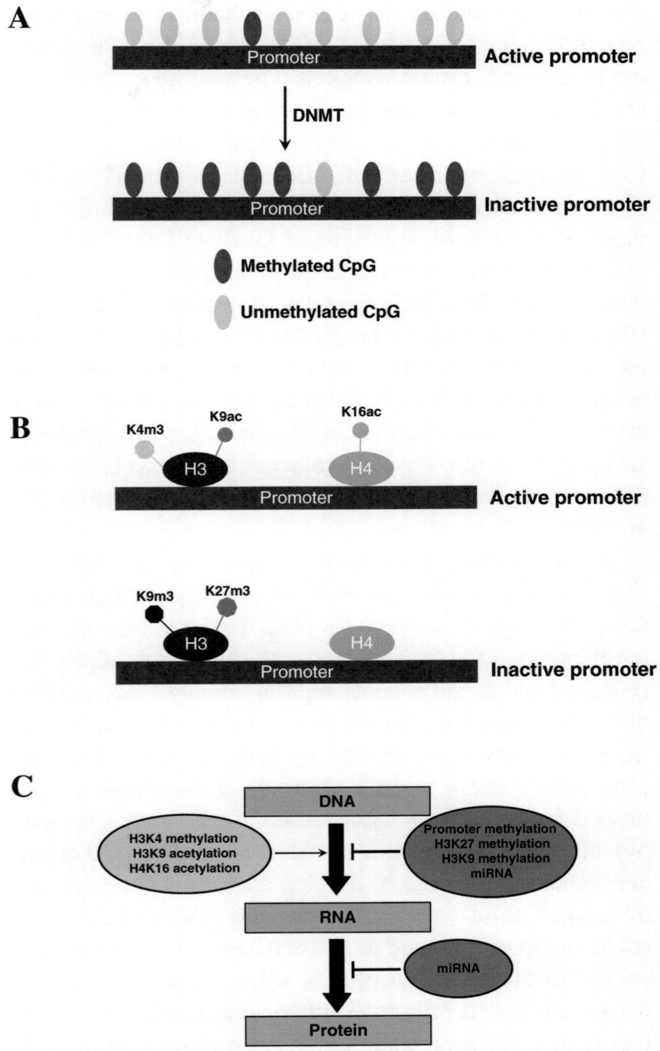

Fig. 1 Epigenetic mechanisms regulating protein levels. (**A**) CpG methylation at the promoter regions can influence activity of the promoter. Highly methylated areas are associated with inactive promoters, and vice-versa. (**B**) Histone modification in the promoter regions affects transcriptional activity of the promoter. Tri-methylation of H3K9 and H3K27 is associated with inactive promoters. Tri-methylation of H3K4 and acetylation of H3K9 and H4K16 are associated with active promoter regions. (**C**) Various factors influencing protein levels are presented in schematic form. DNA methylation, histone methylation, histone acetylation, and microRNA influence protein levels in a cell by affecting transcription as well as translation

cell type is unique and can be used to characterize and differentiate between cell types [18, 100]. This ability to determine the epigenomic profile of a cell type will be of invaluable assistance in evaluating novel technologies. Large efforts have been directed to the development of new culture systems for the maintenance of hES cells *in vitro* [101–106]. These studies potentially represent important technical advances for the field. But without knowing how much particular cell lines differentiate under a given set of growth conditions, or how the properties of a given cell line vary when grown in different laboratories, it is difficult to assess the utility of any new protocol [107]. Comparing the epigenomic profile of cells grown under different conditions would provide a simple solution. Studying the epigenomic profile of stem cells will also have a significant impact on the studies of human epigenetic disorders and assisted reproduction [56, 108, 109].

Recent reports have demonstrated the derivation of pluripotent stem cells from both mouse and human terminally differentiated cells [110–114]. This is a very exciting development in the field, and raises the possibility of deriving cells from individual patients, reprogramming them, and then differentiating them into the required tissue without risk of immune rejection and eliminating potential ethical concerns about embryonic stem cells. Currently, this has been accomplished by viral transduction of adult cells with a cocktail of factors. This approach is not ideal for therapy, and efforts are currently underway to develop alternative means. Since differentiation of embryonic stem cells is accomplished by changing the epigenomic profile of the cell, it is reasonable to assume that reversing the process will allow us to effectively derive pluripotent cells from differentiated cells. It is therefore critical to understand the various processes involved in maintaining and changing the epigenomic profile of cells.

References

1. Bhattacharya B, Miura T, Brandenberger R, et al. Gene expression in human embryonic stem cell lines: unique molecular signature. Blood. 2004;103:2956–64.
2. Brandenberger R, Khrebtukova I, Thies RS, et al. MPSS profiling of human embryonic stem cells. BMC Dev Biol. 2004;4:10.
3. Brandenberger R, Wei H, Zhang S, et al. Transcriptome characterization elucidates signaling networks that control human ES cell growth and differentiation. Nat Biotechnol. 2004;22:707–16.
4. Ginis I, Luo Y, Miura T, et al. Differences between human and mouse embryonic stem cells. Dev Biol. 2004; 269:360–80.
5. Sato N, Sanjuan IM, Heke M, et al. Molecular signature of human embryonic stem cells and its comparison with the mouse. Dev Biol. 2003;260:404–13.
6. Sperger JM, Chen X, Draper JS, et al. Gene expression patterns in human embryonic stem cells and human pluripotent germ cell tumors. Proc Natl Acad Sci U S A . 2003;100:13350–5.
7. Abeyta MJ, Clark AT, Rodriguez RT, et al. Unique gene expression signatures of independently-derived human embryonic stem cell lines. Hum Mol Genet. 2004;13:601–8.
8. Fraga MF, Agrelo R, Esteller M. Cross-talk between aging and cancer: the epigenetic language. Ann N Y Acad Sci. 2007;1100:60–74.
9. Lu Q, Qiu X, Hu N, et al. Epigenetics, disease, and therapeutic interventions. Ageing Res Rev. 2006;5:449–67.
10. Rodenhiser D, Mann M. Epigenetics and human disease: translating basic biology into clinical applications. CMAJ 2006;174: 341–8.
11. Moss TJ, Wallrath LL. Connections between epigenetic gene silencing and human disease. Mutat Res. 2007;618:163–74.
12. Grewal SI, Moazed D. Heterochromatin and epigenetic control of gene expression. Science. 2003;301:798–802.
13. Elgin SC, Grewal SI. Heterochromatin: silence is golden. Curr Biol. 2003;13:R895–8.
14. Espino PS, Drobic B, Dunn KL, et al. Histone modifications as a platform for cancer therapy. J Cell Biochem. 2005;94:1088–102.
15. Hendrich B, Guy J, Ramsahoye B, et al. Closely related proteins MBD2 and MBD3 play distinctive but interacting roles in mouse development. Genes Dev. 2001;15:710–23.
16. Peterson CL, Laniel MA. Histones and histone modifications. Curr Biol. 2004;14:R546–51.
17. Strahl BD, Allis CD. The language of covalent histone modifications. Nature. 2000;403:41–5.
18. Zhan M. Genomic studies to explore self-renewal and differentiation properties of embryonic stem cells. Front Biosci. 2008;13:276–83.
19. Cheng LC, Tavazoie M, Doetsch F. Stem cells: from epigenetics to microRNAs. Neuron. 2005;46:363–7.
20. Chiang PK, Gordon RK, Tal J, et al. S-Adenosylmethionine and methylation. FASEB J. 1996;10:471–80.
21. Bird AP. CpG-rich islands and the function of DNA methylation. Nature. 1986;321:209–13.
22. Wade PA. Methyl CpG-binding proteins and transcriptional repression. Bioessays. 2001;23:1131–7.
23. Wade PA. SWItching off methylated DNA. Nat Genet. 2005;37:212–3.
24. Wei CL, Ng P, Chiu KP, et al. 5' Long serial analysis of gene expression (LongSAGE) and 3' LongSAGE for transcriptome characterization and genome annotation. Proc Natl Acad Sci U S A. 2004;101:11701–6.
25. Ge X, Wu Q, Jung YC, et al. A large quantity of novel human antisense transcripts detected by LongSAGE. Bioinformatics. 2006;22:2475–9.
26. Ge X, Wu Q, Wang SM. SAGE detects microRNA precursors. BMC Genomics. 2006;7:285.
27. Nielsen KL, Hogh AL, Emmersen J. DeepSAGE-digital transcriptomics with high sensitivity, simple experimental protocol and multiplexing of samples. Nucleic Acids Res. 2006;34:e133.
28. Wahl MB, Heinzmann U, Imai K. LongSAGE analysis significantly improves genome annotation: identifications of novel genes and alternative transcripts in the mouse. Bioinformatics. 2005;21:1393–400.
29. Wahl MB, Heinzmann U, Imai K. LongSAGE analysis revealed the presence of a large number of novel antisense genes in the mouse genome. Bioinformatics. 2005;21:1389–92.
30. Bibikova M, Lin Z, Zhou L, et al. High-throughput DNA methylation profiling using universal bead arrays. Genome Res. 2006;16:383–393.
31. Bibikova M, Chudin E, Wu B, et al. Human embryonic stem cells have a unique epigenetic signature. Genome Res. 2006;16: 1075–83.
32. Bestor T, Laudano A, Mattaliano R, et al. Cloning and sequencing of a cDNA encoding DNA methyltransferase of mouse cells. The carboxyl-terminal domain of the mammalian enzymes is related to bacterial restriction methyltransferases. J Mol Biol. 1988;203:971–83.

33. Okano M, Xie S, Li E. Cloning and characterization of a family of novel mammalian DNA (cytosine-5) methyltransferases. Nat Genet. 1998;19:219–20.
34. Li E, Bestor TH, Jaenisch R. Targeted mutation of the DNA methyltransferase gene results in embryonic lethality. Cell. 1992;69:915–26.
35. Okano M, Bell DW, Haber DA, et al. DNA methyltransferases Dnmt3a and Dnmt3b are essential for de novo methylation and mammalian development. Cell. 1999;99:247–57.
36. Hsieh CL. In vivo activity of murine de novo methyltransferases, Dnmt3a and Dnmt3b. Mol Cell Biol. 1999;19:8211–8.
37. Jahner D, Stuhlmann H, Stewart CL, et al. De novo methylation and expression of retroviral genomes during mouse embryogenesis. Nature. 1982;298:623–8.
38. Hendrich B, Hardeland U, Ng HH, et al. The thymine glycosylase MBD4 can bind to the product of deamination at methylated CpG sites. Nature. 1999;401:301–4.
39. Amit M, Itskovitz-Eldor J. Derivation and spontaneous differentiation of human embryonic stem cells. J Anat. 2002;200: 225–32.
40. Brimble SN, Zeng X, Weiler DA, et al. Karyotypic stability, genotyping, differentiation, feeder-free maintenance, and gene expression sampling in three human embryonic stem cell lines derived prior to August 9, 2001. Stem Cells Dev. 2004;13: 585–97.
41. Heins N, Englund MC, Sjoblom C, et al. Derivation, characterization, and differentiation of human embryonic stem cells. Stem Cells. 2004;22:367–76.
42. Reubinoff BE, Pera MF, Fong CY, et al. Embryonic stem cell lines from human blastocysts: somatic differentiation in vitro. Nat Biotechnol. 2000;18:399–404.
43. Thomson JA, Itskovitz-Eldor J, Shapiro SS, et al. Embryonic stem cell lines derived from human blastocysts. Science. 1998;282:1145–7.
44. Maitra A, Arking DE, Shivapurkar N, et al. Genomic alterations in cultured human embryonic stem cells. Nat Genet. 2005;37:1099–103
45. Andrews PW, Damjanov I, Simon D, et al. Pluripotent embryonal carcinoma clones derived from the human teratocarcinoma cell line Tera-2. Differentiation in vivo and in vitro. Lab Invest. 1984;50:147–62.
46. Monk M, Boubelik M, Lehnert S. Temporal and regional changes in DNA methylation in the embryonic, extraembryonic and germ cell lineages during mouse embryo development. Development. 1987;99:371–82.
47. Kafri T, Ariel M, Brandeis M, et al. Developmental pattern of gene-specific DNA methylation in the mouse embryo and germ line. Genes Dev. 1992;6:705–14.
48. Woodcock DM, Crowther PJ, Diver WP. The majority of methylated deoxycytidines in human DNA are not in the CpG dinucleotide. Biochem Biophys Res Commun. 1987;145: 888–94.
49. Toth M, Muller U, Doerfler W. Establishment of de novo DNA methylation patterns. Transcription factor binding and deoxycytidine methylation at CpG and non-CpG sequences in an integrated adenovirus promoter. J Mol Biol. 1990;214:673–83.
50. Ramsahoye BH, Biniszkiewicz D, Lyko F, et al. Non-CpG methylation is prevalent in embryonic stem cells and may be mediated by DNA methyltransferase 3a. Proc Natl Acad Sci U S A. 2000;97:5237–42.
51. Constancia M, Hemberger M, Hughes J, et al. Placental-specific IGF-II is a major modulator of placental and fetal growth. Nature. 2002;417:945–8.
52. Frank D, Fortino W, Clark L, et al. Placental overgrowth in mice lacking the imprinted gene Ipl. Proc Natl Acad Sci U S A. 2002;99:7490–5.
53. Lin SP, Youngson N, Takada S, et al. Asymmetric regulation of imprinting on the maternal and paternal chromosomes at the Dlk1-Gtl2 imprinted cluster on mouse chromosome 12. Nat Genet. 2003;35:97–102.
54. Ferguson-Smith AC, Surani MA. Imprinting and the epigenetic asymmetry between parental genomes. Science. 2001;293: 1086–9.
55. Armstrong L, Lako M, Dean W, et al. Epigenetic modification is central to genome reprogramming in somatic cell nuclear transfer. Stem Cells. 2006;24:805–14.
56. Jacob S, Moley KH. Gametes and embryo epigenetic reprogramming affect developmental outcome: implication for assisted reproductive technologies. Pediatr Res . 2005;58:437–46.
57. Temple IK. Imprinting in human disease with special reference to transient neonatal diabetes and Beckwith-Wiedemann syndrome. Endocr Dev. 2007;12:113–23.
58. Santos-Reboucas CB, Pimentel MM. Implication of abnormal epigenetic patterns for human diseases. Eur J Hum Genet. 2007;15:10–7.
59. Hoffman LM, Hall L, Batten JL, et al. X-inactivation status varies in human embryonic stem cell lines. Stem Cells. 2005;23: 1468–78.
60. Shiota K, Kogo Y, Ohgane J, et al. Epigenetic marks by DNA methylation specific to stem, germ and somatic cells in mice. Genes Cells. 2002;7:961–9.
61. Hanlon SE, Lieb JD. Progress and challenges in profiling the dynamics of chromatin and transcription factor binding with DNA microarrays. Curr Opin Genet Dev. 2004;14:697–705.
62. O'Neill LP, Vermilyea MD, Turner BM. Epigenetic characterization of the early embryo with a chromatin immunoprecipitation protocol applicable to small cell populations. Nat Genet. 2006;38:835–41.
63. Meshorer E, Yellajoshula D, George E, et al. Hyperdynamic plasticity of chromatin proteins in pluripotent embryonic stem cells. Dev Cell. 2006;10:105–16.
64. Azuara V, Perry P, Sauer S, et al. Chromatin signatures of pluripotent cell lines. Nat Cell Biol. 2006;8:532–8.
65. Bernstein BE, Mikkelsen TS, Xie X, et al. A bivalent chromatin structure marks key developmental genes in embryonic stem cells. Cell. 2006;125:315–26.
66. Mikkelsen TS, Ku M, Jaffe DB, et al. Genome-wide maps of chromatin state in pluripotent and lineage-committed cells. Nature. 2007;448:553–60.
67. Boyer LA, Lee TI, Cole MF, et al. Core transcriptional regulatory circuitry in human embryonic stem cells. Cell. 2005;122:947–56.
68. Guenther MG, Levine SS, Boyer LA, et al. A chromatin landmark and transcription initiation at most promoters in human cells. Cell. 2007;130:77–88.
69. Lee TI, Jenner RG, Boyer LA, et al. Control of developmental regulators by Polycomb in human embryonic stem cells. Cell. 2006;125:301–13.
70. Boyer LA, Plath K, Zeitlinger J, et al. Polycomb complexes repress developmental regulators in murine embryonic stem cells. Nature. 2006;441:349–53.
71. Bracken AP, Dietrich N, Pasini D, et al. Genome-wide mapping of Polycomb target genes unravels their roles in cell fate transitions. Genes Dev. 2006;20:1123–36.
72. Buszczak M, Spradling AC. Searching chromatin for stem cell identity. Cell. 2006;125:233–6.
73. de la Cruz CC, Kirmizis A, Simon MD, et al. The polycomb group protein SUZ12 regulates histone H3 lysine 9 methylation and HP1 alpha distribution. Chromosome Res. 2007;15:299–314.
74. Jorgensen HF, Giadrossi S, Casanova M, et al. Stem cells primed for action: polycomb repressive complexes restrain the expression of lineage-specific regulators in embryonic stem cells. Cell Cycle. 2006;5:1411–4.

75. Lee ER, Murdoch FE, Fritsch MK. High histone acetylation and decreased polycomb repressive complex 2 member levels regulate gene specific transcriptional changes during early embryonic stem cell differentiation induced by retinoic acid. Stem Cells. 2007;25:2191–9.
76. Pasini D, Bracken AP, Hansen JB, et al. The polycomb group protein Suz12 is required for embryonic stem cell differentiation. Mol Cell Biol. 2007;27:3769–79.
77. Rajasekhar VK, Begemann M. Concise review: roles of polycomb group proteins in development and disease: a stem cell perspective. Stem Cells. 2007;25:2498–510.
78. Spivakov M, Fisher AG. Epigenetic signatures of stem-cell identity. Nat Rev Genet. 2007;8:263–71.
79. Tanay A, O'Donnell AH, Damelin M, et al. Hyperconserved CpG domains underlie Polycomb-binding sites. Proc Natl Acad Sci U S A. 2007;104:5521–6.
80. Rao M. Conserved and divergent paths that regulate self-renewal in mouse and human embryonic stem cells. Dev Biol. 2004;275:269–86.
81. Collas P, Noer A, Timoskainen S. Programming the genome in embryonic and somatic stem cells. J Cell Mol Med. 2007;11: 602–20.
82. Ura H, Usuda M, Kinoshita K, et al. STAT3 and Oct-3/4 control histone modification through induction of Eed in embryonic stem cells. J Biol Chem. 2008;283:9713–23
83. Mattick JS, Makunin IV. Non-coding RNA. Hum Mol Genet. 2006;15 Spec No 1:R17–29.
84. Kuwabara T, Hsieh J, Nakashima K, et al. A small modulatory dsRNA specifies the fate of adult neural stem cells. Cell. 2004;116:779–93.
85. Houbaviy HB, Murray MF, Sharp PA. Embryonic stem cell-specific MicroRNAs. Dev Cell. 2003;5:351–8.
86. Suh MR, Lee Y, Kim JY, et al. Human embryonic stem cells express a unique set of microRNAs. Dev Biol. 2004;270:488–98.
87. Josephson R, Ording CJ, Liu Y, et al. Qualification of embryonal carcinoma 2102Ep as a reference for human embryonic stem cell research. Stem Cells. 2007;25:437–46.
88. Lakshmipathy U, Love B, Goff LA, et al. MicroRNA expression pattern of undifferentiated and differentiated human embryonic stem cells. Stem Cells Dev. 2007;16:1003–16.
89. Zhao Y, Ransom JF, Li A, et al. Dysregulation of cardiogenesis, cardiac conduction, and cell cycle in mice lacking miRNA-1-2. Cell. 2007;129:303–17.
90. Tay YM, Tam WL, Ang YS, et al. MicroRNA-134 modulates the differentiation of mouse embryonic stem cells, where it causes post-transcriptional attenuation of Nanog and LRH1. Stem Cells. 2008;26:17–29.
91. Klein ME, Impey S, Goodman RH. Role reversal: the regulation of neuronal gene expression by microRNAs. Curr Opin Neurobiol. 2005;15:507–13.
92. Giraldez AJ, Cinalli RM, Glasner ME, et al. MicroRNAs regulate brain morphogenesis in zebrafish. Science. 2005;308:833–8.
93. Naguibneva I, Ameyar-Zazoua M, Polesskaya A, et al. The microRNA miR-181 targets the homeobox protein Hox-A11 during mammalian myoblast differentiation. Nat Cell Biol. 2006;8: 278–84.
94. Hatfield SD, Shcherbata HR, Fischer KA, et al. Stem cell division is regulated by the microRNA pathway. Nature. 2005;435:974–8.
95. Calabrese JM, Seila AC, Yeo GW, et al. RNA sequence analysis defines Dicer's role in mouse embryonic stem cells. Proc Natl Acad Sci U S A. 2007;104:18097–102.
96. Heisler LE, Torti D, Boutros PC, et al. CpG Island microarray probe sequences derived from a physical library are representative of CpG Islands annotated on the human genome. Nucleic Acids Res. 2005;33:2952–61.
97. Rakyan VK, Hildmann T, Novik KL, et al. DNA methylation profiling of the human major histocompatibility complex: a pilot study for the human epigenome project. PLoS Biol. 2004;2:e405.
98. Weber M, Davies JJ, Wittig D, et al. Chromosome-wide and promoter-specific analyses identify sites of differential DNA methylation in normal and transformed human cells. Nat Genet. 2005;37:853–62.
99. Murrell A, Rakyan VK, Beck S. From genome to epigenome. Hum Mol Genet. 2005;14 Spec No 1:R3–10.
100. Pal R, Mandal A, Rao HS, et al. A panel of tests to standardize the characterization of human embryonic stem cells. Regenerative Med. 2007;2:179–92.
101. Yao S, Chen S, Clark J, et al. Long-term self-renewal and directed differentiation of human embryonic stem cells in chemically defined conditions. Proc Natl Acad Sci U S A. 2006;103:6907–12.
102. Lei T, Jacob S, Ajil-Zaraa I, et al. Xeno-free derivation and culture of human embryonic stem cells: current status, problems and challenges. Cell Res. 2007;17:682–8.
103. Skottman H, Hovatta O. Culture conditions for human embryonic stem cells. Reproduction. 2006;132:691–8.
104. Dang SM, Zandstra PW. Scalable production of embryonic stem cell-derived cells. Methods Mol Biol. 2005;290:353–64.
105. Rao BM, Zandstra PW. Culture development for human embryonic stem cell propagation: molecular aspects and challenges. Curr Opin Biotechnol. 2005;16:568–76.
106. Mallon BS, Park KY, Chen KG, et al. Toward xeno-free culture of human embryonic stem cells. Int J Biochem Cell Biol. 2006;38:1063–75.
107. Andrews PW, Benvenisty N, McKay R, et al. The International Stem Cell Initiative: toward benchmarks for human embryonic stem cell research. Nat Biotechnol. 2005;23:795–7.
108. Allegrucci C, Denning C, Priddle H, et al. Stem-cell consequences of embryo epigenetic defects. Lancet. 2004;364:206–8.
109. Cowan CA, Atienza J, Melton DA, et al. Nuclear reprogramming of somatic cells after fusion with human embryonic stem cells. Science. 2005;309:1369–73.
110. Nakagawa M, Koyanagi M, Tanabe K, et al. Generation of induced pluripotent stem cells without Myc from mouse and human fibroblasts. Nat Biotechnol. 2008;26:101–6.
111. Okita K, Ichisaka T, Yamanaka S. Generation of germline-competent induced pluripotent stem cells. Nature. 2007;448: 313–17.
112. Takahashi K, Tanabe K, Ohnuki M, et al. Induction of pluripotent stem cells from adult human fibroblasts by defined factors. Cell. 2007;131:861–72.
113. Takahashi K, Yamanaka S. Induction of pluripotent stem cells from mouse embryonic and adult fibroblast cultures by defined factors. Cell. 2006;126:663–76.
114. Yu J, Vodyanik MA, Smuga-Otto K, et al. Induced pluripotent stem cell lines derived from human somatic cells. Science. 2007;318:1917–20.

DNA Methylation and the Epigenetic Program in Stem Cells

Laurie Jackson-Grusby

Abstract Epigenetic regulation of gene expression, which refers to stable and heritable changes in gene expression potential, is essential for normal embryonic development and cellular differentiation. Epigenetic mechanisms provide a memory of developmental history of a cell, and are also responsive to environmental inputs. These extracellular cues serve to direct programs of gene expression that restrict developmental potency as the organism proceeds from a totipotent single cell to a fully mature state. Mounting experimental evidence suggests that epigenetic modifications, either DNA methylation or post-translational modification of histones, are functionally required for establishing and maintaining heritable states of gene expression. Importantly, these codes also serve to identify states of developmental potency, and as such may be useful diagnostic and prognostic markers for diseases involving abnormal tissue homeostasis. There is an emerging view that epigenetic alterations play an important role in a wide range of multifactorial disorders. The importance of epigenetic control in developmental regulation of stem cells supports the notion that similar mechanisms regulate adult tissue homeostasis at the level of adult stem cells. Altered cellular plasticity may be a general mechanism through which aberrant epigenetic programs exert their effects. Loss of epigenetic control can lead to reduced self-renewal and accelerated aging of stem cells. Conversely, enhanced self-renewal through epimutation in tissue stem cells, or reacquisition of stem cell expression states by faulty reprogramming mechanisms are thought to represent early events in cancer. The inherent reversibility of these epigenetic states provides a potential therapeutic opportunity to reset the balance of tissue homeostasis through these pathways.

Keywords Epigenetics · Stem cells · DNA methylation · Chromatin · Reprogramming

L. Jackson-Grusby (✉)
Children's Hospital Boston, Harvard Stem Cell Institute, Harvard Medical School, Boston, MA 02115
e-mail: laurie.jackson-grusby@childrens.harvard.edu

1 Control of the DNA Methylation Cycle

Epigenetic regulation of the genome depends on DNA methylation [1, 2], the post-replication modification of cytosine bases in the target sequence CpG. Riggs [3] and Holliday [4] proposed over three decades ago that DNA methylation might serve to stably maintain states of gene expression as chromosomes replicate during cell division. Genomic methylation patterns are propagated during S phase of the cell cycle by the action of the maintenance DNA methyltransferase Dnmt1, an enzyme associated with replication machinery [5] (Fig. 1). Dnmt1 preferentially associates with hemi-methylated DNA via its association with the SET-RING finger-associated protein Np95 [6, 7], and methylates newly synthesized CpG sequences opposite methylated CpGs on the template DNA strand. Mechanistically, DNA methylation acts in concert with post-synthetic modifications of histones to silence transcription by establishing and maintaining a repressive chromatin state [8, 9]. Dnmt1 serves to consolidate the silenced gene state by recruiting histone deacetylases (HDACs), which further mediate chromatin compaction and gene silencing [1, 2].

Roughly 80% of CpGs in the genome are methylated, however, regions enriched in CpG, known as CpG islands, are devoid of methylation when associated with gene promoters [10]. The establishment of de novo methylation patterns requires a family of enzymes encoded by two genes,

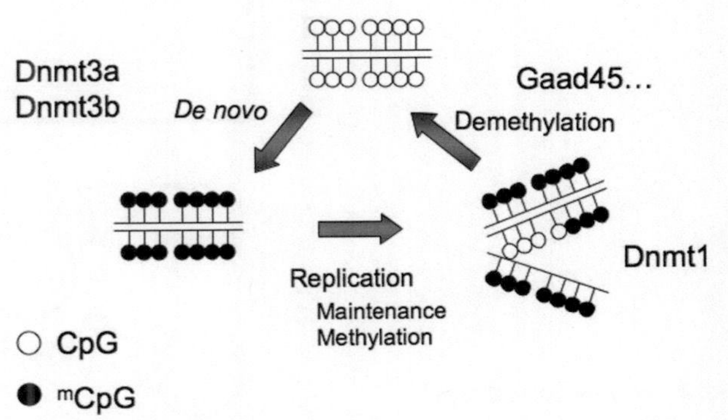

Fig. 1 The DNA methylation cycle

V.K. Rajasekhar, M.C. Vemuri (eds.), *Regulatory Networks in Stem Cells*, Stem Cell Biology and Regenerative Medicine,
DOI 10.1007/978-1-60327-227-8_23, © Humana Press, a part of Springer Science+Business Media, LLC 2009

Dnmt3a and Dnmt3b [11]. The balance of isoforms for the de novo enzymes may serve to regulate the processivity and specificity of de novo methylation. For example, selective induction of the Dnmt3b2 isoform in the intestine causes chronic overmethylation of the H19/Igf2 imprinted domain, which leads to elevated intestinal tumor incidence in mice[12]. The implications of the role this process plays in normal and malignant stem cells will be discussed below. Uncovering the mechanisms regulating sequence specificity for de novo methylation under physiologic or pathophysiologic levels of Dnmt3a or Dnmt3 expression, which are largely unknown at present, remains an exciting and open area of current investigation.

An emerging new area of research is the elucidation of the mechanism of regulated DNA demethylation of specific sequences. DNA demethylation was initially thought to be predominantly passive, through DNA replication with inhibition of the maintenance methylation [13], and this may be the mechanism of therapeutic action of compounds that reactivate epigenetically silenced genes. Evidence for replication-independent active demethylation was observed in interleukin-2 gene regulation in T lymphocytes [14]. Induced changes in promoter methylation at this locus provide an epigenetic memory leading to long-term rapid inducibility of IL-2 in induced or reprogrammed T cells [15]. Pharmacologic inhibitors of HDACs can induce DNA demethylation in terminally differentiated GABAergic interneurons in mice [16], however, the cross-talk mechanisms between histone acetylation and DNA demethylation have not been resolved to determine whether this is a direct or an indirect effect. Niehrs, Lyko, and colleagues [17] have provided a potential molecular mechanism for DNA demethylation with the demonstration that the Gadd45a-dependent repair pathway mediates demethylation and epigenetic gene activation. Understanding how the Gadd45a pathway, and potentially yet-to-be-discovered pathways, select gene targets for epigenetic reprogramming through sequence-specific DNA demethylation, and the biologic consequences of these pathways will undoubtedly provide new insights into cellular plasticity in the coming years.

2 DNA Methylation, Lineage Restriction, and Developmental Potency

An extension of the Riggs-Holiday hypothesis on DNA methylation-dependent heritable gene silencing is a corollary notion that DNA methylation state marks or even defines the state of cellular differentiation. Consistent with this idea, gene expression and DNA methylation status are inversely associated for many genes and developmentally controlled with cell-type specificity as defined by molecular and genetic studies of mouse embryogenesis. For example, the pluripotency gene Oct 4 is expressed in preimplantation embryos and in the germline, and its promoter is methylated and silenced in differentiated cells [18, 19]. The importance of DNA methylation in determining cellular plasticity is further evidenced by an intriguing inverse relationship with global levels of DNA methylation and the developmental potency of a cell [20] (Fig. 2). Following fertilization

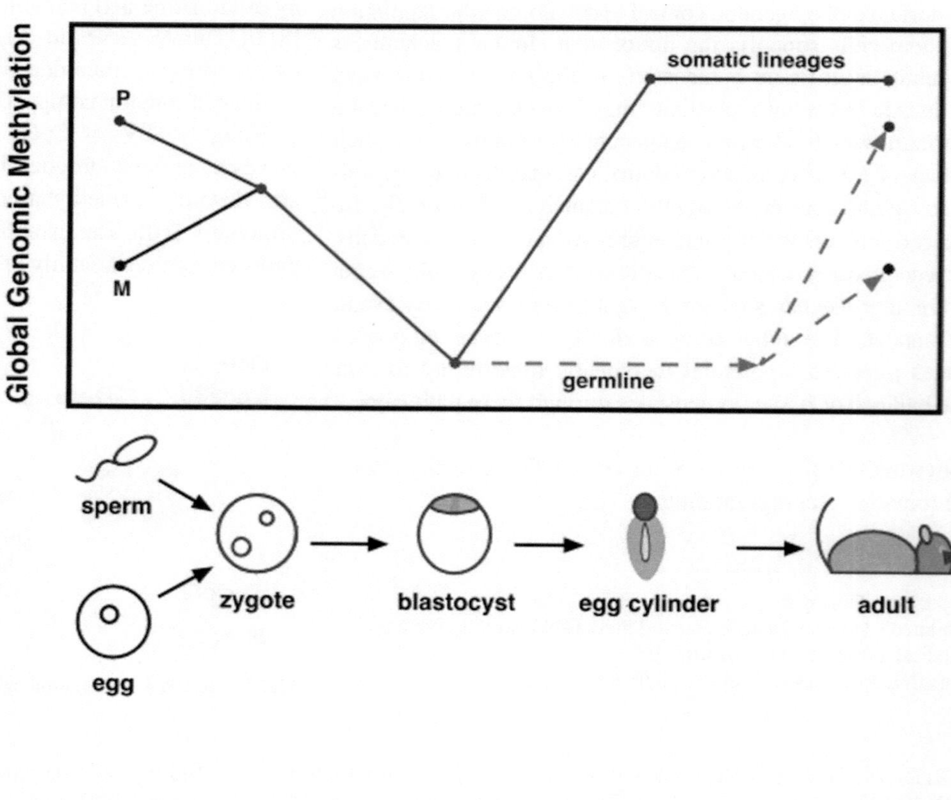

Fig. 2 Dynamic regulation of DNA methylation during development

and during preimplantation development, maternal and paternal haploid genomes undergo global passive and active demethylation, respectively. This demethylase activity within the egg cytoplasm may be one of the key mechanisms that allows a somatic cell nucleus to be reprogrammed to developmental totipotency [21, 22]. As an aid to the biochemical characterization of this process, reprogramming activity is not constrained to the relatively scarce egg cytoplasm, but is also present in cleavage stage embryos, embryonic stem (ES) cells, embryonic germ (EG) cells, and stem cells derived by genetically induced pluripotency (iPS) [23–25]. Whether induction of this demethylation activity is a direct consequence of overexpression of one of the reprogramming factors remains to be determined.

Several lines of evidence support the essential role of DNA methylation in cell type-specific gene regulation and cell fate control. First, following implantation the de novo methyltransferases Dnmt3a and 3b catalyze cell type-specific methylation patterns as tissues differentiate [11, 26], and these methylation states are stably propagated by Dnmt1 as cells divide [5]. Indeed, the first key steps in differentiation require the silencing of the pluripotency genes oct4 and nanog, which depend on synergistic activity from both Dnmt3a and Dnmt3b. Lineage restriction during development may be exquisitely controlled by such gene silencing events (Fig. 3). Evidence from hematopoietic stem cells (HSCs) has shown that lineage directing genes that are actively transcribed are accompanied by unmethylated promoter sequences as expected, yet this methylation pattern marks poised transcription already established in ES cells through the binding of sequence specific transcription factors. The implication is that loss of the activator in this case accompanies gene silencing, promoter methylation, and lineage restriction during differentiation [27, 28].

Sequence specificity for developmentally controlled de novo methylation reactions are not well understood, and may depend on diverse mechanisms depending on the genomic context. A recent convergence of fields of research linking small RNA biology and chromatin biology has led to a rapidly emerging molecular pathway that directs epigenetic silencing [29–31]. RNA interference has been revealed over the past 10 years as a mechanism of post-transcriptional gene silencing, however, the RNAi pathway has also been shown to effect changes in histone methylation and DNA methylation that lead to long-term transcriptional gene silencing [32, 33]. A second mechanism of RNA-dependent silencing in stem cells is mediated by regulated antisense transcription, exemplified by RNA-directed DNA methylation at the Xist locus during X chromosome inactivation in female ES cells [34], and silencing of a human p15 tumor suppressor transgene in mouse ES cells.

Maintenance of silencing among differentiated cell types from mouse postgastrulation embryos, embryonic fibroblasts, developing T cells, or neural progenitors requires the maintenance DNA methyltransferase Dnmt1 as these somatic cell types undergo p53-dependent apoptosis in the absence of Dnmt1 [35, 36]. Dnmt1 loss in somatic cells is associated with widespread ectopic gene expression in both mouse and Xenopus. Developmental genomic methylation patterns, and genetic analysis of the Dnmts, suggest that DNA methylation may be dispensible in multipotent stem cells, despite its requirement in differentiating cell types. Pluripotent ES and EG cells, which are derived from embryonic cell types that are naturally hypomethylated, do not require any of the DNA methyltransferases to proliferate in an undifferentiated state [35, 37, 38]. By contrast, deficiency for Dnmt1 or Dnmt3a and Dnmt3b disrupts normal differentiation [11]. Together, these results demonstrate that DNA methylation is dispensible for embryonic stem cell renewal, but is required for transcriptional silencing and viability of differentiated cells. Evidence in HSCs suggests that self-renewal of adult tissue stem cells does depend on de novo Dnmt3 activity, however, yet paradoxically, differentiation of these cells is intact [39]. The extension of this observation into other adult tissue-derived stem cells is needed to reveal the generality of this mechanism.

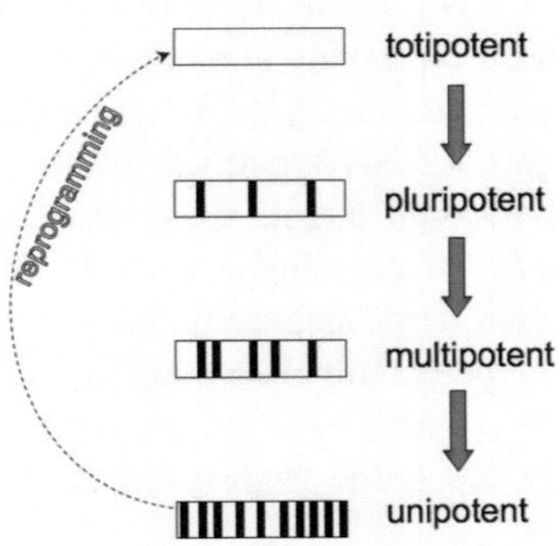

Fig. 3 Lineage restriction and reprogramming to developmental totipotency

3 Homeostatic Regulation of Genomic Imprinting

Genomic elements that direct parent-of-origin specific expression, termed imprinted regions, are an important subclass of epigenetically regulated sequences in the genome [1, 2]. Rare human syndromes with biased parental inheritance demonstrate the contributions of imprinted genes to human development, cancer, metabolism, and neurological function. Understanding the sets of molecular events leading to normal and aberrant regulation at these sequences has far-reaching impact to ensure the proper developmental programming of stem cells for therapeutic benefit. Additionally, the imprinted genes Igf2 and Dlk1 are known to directly control stem cell homeostasis in vivo, and this may be a more general attribute of imprinted genes that warrants further investigation [40].

A prominent feature of genomic parental imprinting is the reversibility of the mark that discriminates the maternal and paternal alleles. For example, a maternally inherited imprint is transmitted from female offspring, but is erased and replaced with paternal marks from male offspring [41]. In both cases these labile imprinting marks are removed during development of prospective germ cells, and the differential marking of alleles to be inherited from the maternal or paternal source is established during gametogenesis.

Sites of DNA methylation in imprinting control regions have been identified along with histone modifications that differentially mark the prospective status of many imprinted genes as having derived from either oocyte or sperm [42]. Establishment of these DNA methylation differences is dependent upon the de novo methyltransferase Dnmt3a [43], which acts in concert with an homologous protein Dnmt3L, which itself does not have resident methyltransferase activity [44, 45]. In preimplantation embryos, the maternally and paternally inherited chromosomes are differentially reprogrammed to direct early development [46]. The paternal genome is rapidly and actively demethylated [47], whereas the maternal genome is demethylated passively. Imprinting marks are retained during this reprogramming, and the mechanism for stabilizing maternal imprints is dependent upon an oocyte-specific form of the maintenance methyltransferase Dnmt1 [48].

The maintenance of most, if not all, imprints beyond postimplantation development is dependent upon the somatic form of Dnmt1 [42]. Based on the observation by Tucker et al. [38] that Dnmt1-deficient cells exhibit loss of imprinting (LOI) after insertional reactivation of Dnmt1, albeit with hypomorphic Dnmt1 levels, we created a new model for LOI using a dual-recombinase gene reactivation strategy for Dnmt1 that caused a transient demethylation in ES cells [49]. Analysis of chimeric animals derived from LOI ES cells showed cancer predisposition with an entirely epigenetic origin. The molecular basis for the elevated cancer risk has not been completely dissected, however, we observed that LOI caused fibroblasts to spontaneously immortalize. Feinberg and colleagues [50] have demonstrated that loss of IGF2 imprinting is a risk factor for human cancer, and forced Igf2 LOI in mice elevates adenoma formation in predisposed animals with concomitant expansion of the progenitor domain in the intestinal crypts. Recent analysis of imprint stability in human ES cells suggests that disruptions to monoallelic regulation is common among stem cells, however, the precise sequence susceptibilities and functional outcomes may be species specific [51]. The broader involvement of LOI in developmental syndromes with complex human pathologies suggests further analysis of imprinted loci will provide new connections between the developmental consequences and long-term pathophysiologic effects of LOI in vivo.

4 Aberrant DNA Methylation and Cancer Stem Cells

The stem cell and progenitor cell model for the origins of human cancers has emerged from several lines of experimental evidence [52], including the identification of stem cell markers that prospectively identify tumor-initiating cells, and the molecular characterization of mutational events that transform normal progenitor cells into cancer stem cells [53]. Epigenetic changes may be more prevalent than mutational events in these early stages of cancer initiation, causing both cell intrinsic and extrinsic alterations in the regulation of stem cell quiescence that shift the balance to enhance stem cell self-renewal [54]. A molecular paradigm for the acquisition of and selection for stable epigenetic gene silencing has emerged from the analysis of epigenetic control in normal embryonic stem cells. Using genome-wide profiling of promoter occupancy using chromatin-immunoprecipitation methods, the genomic targets of the polycomb repressive complexes were revealed to be transcriptional regulators that are silenced in ES cells but poised for activation upon differentiation [55, 56]. The chromatin marks that identify these target genes are termed bivalent domains because they contain both active (H3K4) and repressive (H3K27) chromatin modifications (Fig. 4) [57, 58]. The bivalent domain serves as a transient repression mark that can resolve into an activated response or a more stably silenced chromatin state depending on the direction of differentiation.

Two groups identified a connection between the bivalent chromatin status in stem cells, and the increased propensity for a subset of these same target genes to become stably silenced by DNA hypermethylation in tumors [59, 60]. These studies suggest that epigenetic alterations mark genes with known silencing in stem cells, providing evidence for the epigenetic progenitor model of cancer. In one study, overexpression of the PRC1 component Bmi1 in TERA-2 embryonic carcinoma cells could drive the acquisition of aberrant DNA hypermethylation at the SFRP5 locus. The working model holds that normal stem cells utilize a reversible silencing mechanism via the polycomb pathway. A shift in the balance of PRC1 components or aberrant cross-talk between PRC2 and Dnmt enzymes may facilitate early aberrant states of DNA methylation that become consolidated into long-term repression complexes coupled to complete and stable promoter methylation. In these first analyses, additional targets of DNA hypermethylation beyond the already described polycomb targets were reported, suggesting parallel pathways of silencing or yet to be described polycomb targets in adult tissue stem cells. As multiple targets in these pathways are potentially druggable [61], the reversibility of the silencing, especially in the incipient stages before stable silencing ensues, holds promise that adverse consequences of aberrant

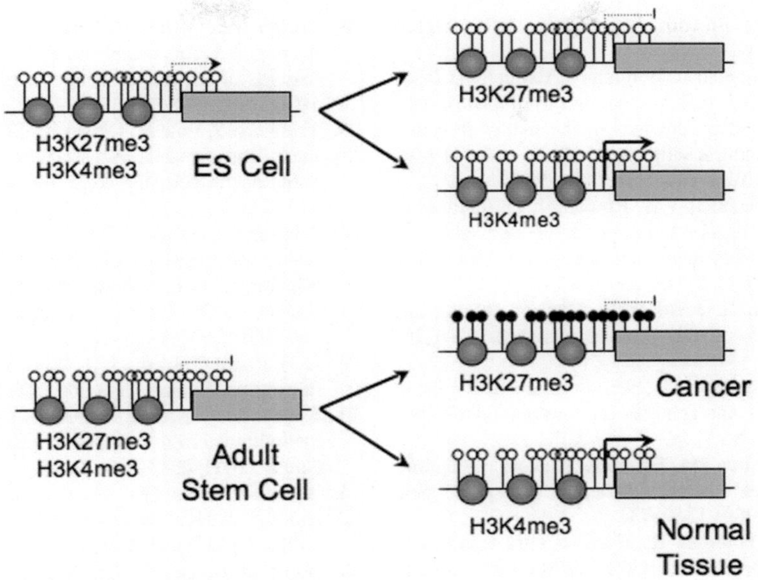

Fig. 4 Paradigms of gene silencing in normal and cancer stem cells

gene silencing in cancer stem cells may be an achievable long term goal.

5 Future Perspectives

The rapid pace of discovery of the role of epigenetics in stem cells makes prognostication particularly daunting. Undoubtedly, the era of genome-wide deep sequencing technologies that is upon us will reveal complex pathways of interconnection between patterns of genomic DNA methylation and associated histone modifications [62–64]. Together with proteomic studies that are aimed at revealing regulatory nodes in stem cells, within and outside the nucleus, we can anticipate a systems-level description of epigenetic regulation in stem cells to emerge quickly [65]. The challenges that remain will include connecting these global snapshots to the functional properties of stem cells, namely self-renewal and multi-lineage differentiation, as well as the drilled-down view of regulation within a single cell. Pushing the limits of our present technologies to single-cell resolution will be imperative to validate and extend the models developed from population studies to the ascribed functional properties of stem cells that are by definition possessed by a singular cell.

References

1. Feinberg AP. Phenotypic plasticity and the epigenetics of human disease. Nature. 2007;447(7143):433–40.
2. Reik W. Stability and flexibility of epigenetic gene regulation in mammalian development. Nature. 2007;447(7143):425–32.
3. Riggs AD. X inactivation, differentiation, and DNA methylation. Cytogenet Cell Genet. 1975;14(1):9–25.
4. Holliday R. Molecular aspects of genetic exchange and gene conversion. Genetics. 1974;78(1):273–87.
5. Leonhardt H, Page AW, Weier HU, Bestor TH. A targeting sequence directs DNA methyltransferase to sites of DNA replication in mammalian nuclei. Cell .1992;71(5):865–73.
6. Sharif J, Muto M, Takebayashi S, et al. The SRA protein Np95 mediates epigenetic inheritance by recruiting Dnmt1 to methylated DNA. Nature. 2007;450(7171):908–12.
7. Bostick M, Kim JK, Esteve PO, Clark A, Pradhan S, Jacobsen SE. UHRF1 plays a role in maintaining DNA methylation in mammalian cells. Science. 2007;317(5845):1760–4.
8. Del Valle L, Enam S, Lassak A, et al. Insulin-like growth factor I receptor activity in human medulloblastomas. Clin Cancer Res. 2002;8(6):1822–30.
9. Reik W, Dean W. DNA methylation and mammalian epigenetics. Electrophoresis. 2001;22(14):2838–43.
10. Jones PA. The DNA methylation paradox. Trends Genet. 1999;15(1):34–7.
11. Okano M, Bell DW, Haber DA, Li E. DNA methyltransferases Dnmt3a and Dnmt3b are essential for de novo methylation and mammalian development. Cell. 1999;99(3):247–57.
12. Linhart HG, Lin H, Yamada Y, et al. Dnmt3b promotes tumorigenesis in vivo by gene-specific de novo methylation and transcriptional silencing. Genes Dev. 2007;21(23):3110–22.
13. Riggs AD. DNA methylation and cell memory. Cell Biophys. 1989;15(1-2):1–13.
14. Bruniquel D, Schwartz RH. Selective, stable demethylation of the interleukin-2 gene enhances transcription by an active process. Nat Immunol. 2003;4(3):235–40.
15. Murayama A, Sakura K, Nakama M, et al. A specific CpG site demethylation in the human interleukin 2 gene promoter is an epigenetic memory. EMBO J. 2006;25(5):1081–92.
16. Dong E, Guidotti A, Grayson DR, Costa E. Histone hyperacetylation induces demethylation of reelin and 67-kDa glutamic acid decarboxylase promoters. Proc Natl Acad Sci U S A. 2007;104(11):4676–81.
17. Barreto G, Schafer A, Marhold J, et al. Gadd45a promotes epigenetic gene activation by repair-mediated DNA demethylation. Nature. 2007;445(7128):671–5.
18. Hattori N, Nishino K, Ko YG, Ohgane J, Tanaka S, Shiota K. Epigenetic control of mouse Oct-4 gene expression in embryonic

stem cells and trophoblast stem cells. J Biol Chem. 2004;279(17):17063–9.
19. Ben-Shushan E, Pikarsky E, Klar A, Bergman Y. Extinction of Oct-3/4 gene expression in embryonal carcinoma x fibroblast somatic cell hybrids is accompanied by changes in the methylation status, chromatin structure, and transcriptional activity of the Oct-3/4 upstream region. Mol Cell Biol. 1993;13(2):891–901.
20. Sanford JP, Clark HJ, Chapman VM, Rossant J. Differences in DNA methylation during oogenesis and spermatogenesis and their persistence during early embryogenesis in the mouse. Genes Dev. 1987;1(10):1039–46.
21. Rideout WM, 3rd, Eggan K, Jaenisch R. Nuclear cloning and epigenetic reprogramming of the genome. Science. 2001;293(5532):1093–8.
22. Jeanisch R, Eggan K, Humpherys D, Rideout W, Hochedlinger K. Nuclear cloning, stem cells, and genomic reprogramming. Cloning Stem Cells. 2002;4(4):389–96.
23. Cowan CA, Atienza J, Melton DA, Eggan K. Nuclear reprogramming of somatic cells after fusion with human embryonic stem cells. Science. 2005;309(5739):1369–73.
24. Wernig M, Meissner A, Foreman R, et al. In vitro reprogramming of fibroblasts into a pluripotent ES-cell-like state. Nature. 2007;448(7151):318–24.
25. Okita K, Ichisaka T, Yamanaka S. Generation of germline-competent induced pluripotent stem cells. Nature. 2007;448(7151):313–7.
26. Kato Y, Kaneda M, Hata K, et al. Role of the Dnmt3 family in de novo methylation of imprinted and repetitive sequences during male germ cell development in the mouse. Hum Mol Genet. 2007;16(19):2272–80.
27. Attema JL, Papathanasiou P, Forsberg EC, Xu J, Smale ST, Weissman IL. Epigenetic characterization of hematopoietic stem cell differentiation using miniChIP and bisulfite sequencing analysis. Proc Natl Acad Sci U S A. 2007;104(30):12371–6.
28. Xu J, Pope SD, Jazirehi AR, et al. Pioneer factor interactions and unmethylated CpG dinucleotides mark silent tissue-specific enhancers in embryonic stem cells. Proc Natl Acad Sci U S A. 2007;104(30):12377–82.
29. Morris KV, Chan SW, Jacobsen SE, Looney DJ. Small interfering RNA-induced transcriptional gene silencing in human cells. Science. 2004;305(5688):1289–92.
30. Kawasaki H, Taira K. Induction of DNA methylation and gene silencing by short interfering RNAs in human cells. Nature. 2004;431(7005):211–7.
31. Bartel DP, Chen CZ. Micromanagers of gene expression: the potentially widespread influence of metazoan microRNAs. Nat Rev Genet. 2004;5(5):396–400.
32. Morris KV. siRNA-mediated transcriptional gene silencing: the potential mechanism and a possible role in the histone code. Cell Mol Life Sci. 2005;62(24):3057–66.
33. Kawasaki H, Taira K, Morris KV. siRNA induced transcriptional gene silencing in mammalian cells. Cell Cycle. 2005;4(3):442–8.
34. Sun BK, Deaton AM, Lee JT. A transient heterochromatic state in Xist preempts X inactivation choice without RNA stabilization. Mol Cell. 2006;21(5):617–28.
35. Jackson-Grusby L, Beard C, Possemato R, et al. Loss of genomic methylation causes p53-dependent apoptosis and epigenetic deregulation. Nat Genet. 2001;27(1):31–9.
36. Lee PP, Fitzpatrick DR, Beard C, et al. A critical role for Dnmt1 and DNA methylation in T cell development, function, and survival. Immunity. 2001;15(5):763–74.
37. Li E, Bestor TH, Jaenisch R. Targeted mutation of the DNA methyltransferase gene results in embryonic lethality. Cell. 1992;69(6):915–26.
38. Tucker KL, Beard C, Dausmann J, et al. Germ-line passage is required for establishment of methylation and expression patterns of imprinted but not of nonimprinted genes. Genes Dev. 1996;10(8):1008–20.
39. Tadokoro Y, Ema H, Okano M, Li E, Nakauchi H. De novo DNA methyltransferase is essential for self-renewal, but not for differentiation, in hematopoietic stem cells. J Exp Med. 2007;204(4):715–22.
40. Charlier C, Segers K, Karim L, et al. The callipyge mutation enhances the expression of coregulated imprinted genes in cis without affecting their imprinting status. Nat Genet. 2001;27(4):367–9.
41. Barlow DP. Gametic imprinting in mammals. Science. 1995;270(5242):1610–3.
42. Li E, Beard C, Jaenisch R. Role for DNA methylation in genomic imprinting. Nature. 1993;366(6453):362–5.
43. Kaneda M, Okano M, Hata K, et al. Essential role for de novo DNA methyltransferase Dnmt3a in paternal and maternal imprinting. Nature. 2004;429(6994):900–3.
44. Bourchis D, Xu GL, Lin CS, Bollman B, Bestor TH. Dnmt3L and the establishment of maternal genomic imprints. Science. 2001;294(5551):2536–9.
45. Bourchis D, Bestor TH. Meiotic catastrophe and retrotransposon reactivation in male germ cells lacking Dnmt3L. Nature. 2004;431(7004):96–9.
46. Okano M, Li E. Genetic analyses of DNA methyltransferase genes in mouse model system. J Nutr. 2002;132(8 Suppl):2462S–5S.
47. Oswald J, Engemann S, Lane N, et al. Active demethylation of the paternal genome in the mouse zygote. Curr Biol. 2000;10(8):475–8.
48. Howell CY, Bestor TH, Ding F, et al. Genomic imprinting disrupted by a maternal effect mutation in the Dnmt1 gene. Cell. 2001;104(6):829–38.
49. Holm TM, Jackson-Grusby L, Brambrink T, Yamada Y, Rideout WM, 3rd, Jaenisch R. Global loss of imprinting leads to widespread tumorigenesis in adult mice. Cancer Cell. 2005;8(4):275–85.
50. Cui H, Niemitz EL, Ravenel JD, et al. Loss of imprinting of insulin-like growth factor-II in Wilms tumor commonly involves altered methylation but not mutations of CTCF or its binding site. Cancer Res. 2001;61(13):4947–50.
51. Rugg-Gunn PJ, Ferguson-Smith AC, Pedersen RA. Status of genomic imprinting in human embryonic stem cells as revealed by a large cohort of independently derived and maintained lines. Hum Mol Genet. 2007;16 Spec No. 2:R243–51.
52. Feinberg AP, Ohlsson R, Henikoff S. The epigenetic progenitor origin of human cancer. Nat Rev Genet. 2006;7(1):21–33.
53. Krivtsov AV, Armstrong SA. MLL translocations, histone modifications and leukaemia stem-cell development. Nat Rev Cancer. 2007;7(11):823–33.
54. Ting AH, McGarvey KM, Baylin SB. The cancer epigenome-components and functional correlates. Genes Dev. 2006;20(23):3215–31.
55. Bracken AP, Dietrich N, Pasini D, Hansen KH, Helin K. Genome-wide mapping of Polycomb target genes unravels their roles in cell fate transitions. Genes Dev. 2006;20(9):1123–36.
56. Boyer LA, Plath K, Zeitlinger J, et al. Polycomb complexes repress developmental regulators in murine embryonic stem cells. Nature. 2006;441(7091):349–53.
57. Bernstein BE, Mikkelsen TS, Xie X, et al. A bivalent chromatin structure marks key developmental genes in embryonic stem cells. Cell. 2006;125(2):315–26.
58. Azuara V, Perry P, Sauer S, et al. Chromatin signatures of pluripotent cell lines. Nat Cell Biol. 2006;8(5):532–8.
59. Widschwendter M, Fiegl H, Egle D, et al. Epigenetic stem cell signature in cancer. Nat Genet. 2007;39(2):157–8.

60. Ohm JE, McGarvey KM, Yu X, et al. A stem cell-like chromatin pattern may predispose tumor suppressor genes to DNA hypermethylation and heritable silencing. Nat Genet. 2007;39(2): 237–42.
61. de Vos D. Epigenetic drugs: a longstanding story. Semin Oncol. 2005;32(5):437–42.
62. Barski A, Cuddapah S, Cui K, et al. High-resolution profiling of histone methylations in the human genome. Cell. 2007;129(4): 823–37.
63. Mikkelsen TS, Ku M, Jaffe DB, et al. Genome-wide maps of chromatin state in pluripotent and lineage-committed cells. Nature. 2007;448(7153):553–60.
64. Cokus SJ, Feng S, Zhang X, et al. Shotgun bisulphate sequencing of the *Arabidopsis* genome reveals DNA methylation patterning. Nature. 2008;452(7184):215–9.
65. Wang J, Rao S, Chu J, et al. A protein interaction network for pluripotency of embryonic stem cells. Nature. 2006;444(7117): 364–8.

Polycomb Group Protein Homeostasis in Stem Cell Identity – A Hypothetical Appraisal

Vinagolu K. Rajasekhar

Abstract Embryonic stem cells are usually characterized by self-renewal and pluripotency as defining properties. On the other hand, adult stem cells are widely believed to be associated with an orderly quiescence and requisite multipotency. Selective epigenetic control at the level of reversible histone and DNA modifications have recently been realized to have a fundamental role in specifying and maintaining a particular identity of stem cells. Recent progress in deriving the inducible pluripotent cells from terminally differentiated cells basically relies on the successful reversal of the developmental specification and associated epigenetic restrictions. All of the epigenetic changes that take place on the DNA without altering its sequence appear to occur by the sequential and stepwise biochemical reactions mediated by a definite set of highly conserved multi-protein complexes called polycomb group proteins (PcG proteins). Although much is known about the involvement and the potential mechanism of the PcG proteins in maintaining cellular identities, how exactly the PcG protein levels are regulated during stem cell maintenance remains unknown. Moreover, during an aberrant differentiation such as that occur in various cancers, the levels of PcG proteins are differentially affected, resulting in an impairment of PcG protein homeostasis. In this chapter, various possibilities are discussed as to how the PcG protein homeostasis may be interpreted in the light of stem cell identity. Potential feedback regulatory loops that control the accumulation of the PcG proteins are hypothesized to play an important role in stem cell maintenance. Autoregulatory controls have long been realized as some of the many means of regulating a spectrum of multi-protein complexes in cell biology. A dysregulation in such feedback control of PcG proteins may result in an escape from regulated quiescence and self-renewal that may in turn lead to a loss of stem cell identity and the onset of an aberrant differentiation leading to diseases such as cancers.

V.K. Rajasekhar (✉)
Memorial Sloan-Kettering Cancer Center, New York, NY 10021, USA
e-mail: vinagolr@mskcc.org.

Keywords Embryonic stem cells · Cancer stem cells · Polycomb feedback loops · Polycomb proteins in stem cells · Pluripotency and transcription · Regulation of polycomb proteins · Stem cell epigenetics · Translational control in stem cells

1 Introduction

Self-renewal potential and pluripotency/multipotency are the most important hallmarks of stem cells, while the driving mechanisms behind these stem cell processes remain elusive. For about the first eight cell divisions (that are symmetric), preimplantation embryos contain totipotency [1]; each cell has the potential to develop into a complete organism. These cells form the embryonic inner cell mass of mammalian blastocysts, which become the source for pluripotent embryonic stem cells in cultures. However, the cells soon undergo asymmetrical cell divisions, resulting in a parent-like daughter cell with pluripotency and a distinct daughter cell with differentiation potential. It has long been known that cell fate is initiated even before the implantation of an embryo. On the other hand, by unknown mechanisms, small numbers of stem cells are sorted out during embryo development and maintained throughout the differentiation of adult tissues. That stem cells in adult organisms are indeed the remnants of embryonic stem cells has not yet been proved. Regardless of their origin, however, adult stem cells appear to be maintained in strict homeostasis in many multicellular organisms. Thus, adult stem cells in the human body can remain undifferentiated spatiotemporally and can self-renew, but they also proliferate under specialized/requisite conditions. Unlike embryonic stem cells with pluripotency, adult stem cells are largely multipotent, so each cell can differentiate into various limited cell types of the parent organ. Moreover, adult stem cells are endowed with limited proliferative capacity.

Epigenetic control of required gene expression patterns in the maintenance of stem cell identity has been identified as an integral part of metazoan development [2–4]. Recently, this has been intensely studied in an attempt to understand

V.K. Rajasekhar, M.C. Vemuri (eds.), *Regulatory Networks in Stem Cells*, Stem Cell Biology and Regenerative Medicine,
DOI 10.1007/978-1-60327-227-8_24, © Humana Press, a part of Springer Science+Business Media, LLC 2009

the molecular mechanisms and functional circuitry during normal development, aging, and disease. While maintaining the cellular phenotype, epigenetic regulation at the molecular level includes a reversible but unique gene-expression profile that is modified during cellular differentiation [5–7]. Polycomb group proteins (PcG) are well-conserved, novel, transcriptional repressors maintaining cellular homeostasis by functioning as multimeric complexes in the regulation of DNA accessibility in eukaryotes, which is necessary for DNA repair, DNA replication, and gene transcription [7, 8]. This phenomenon whereby cellular phenotype is determined in the absence of genotype changes is referred to as "epigenetic control." Essentially, this control includes selective transcriptional repression by PcG proteins that is accomplished at the level of nucleosomes through PcG-mediated post-translational modification of amino acids in the histones [9]. Such modifications affect chromatin access of the proteins associated with promoter activity and gene transcription. While methylations represent the predominant modifications in histones, monomethylations of at least histone3 lysine9 (H3K9), H3K27, and H3K79 are associated with gene activation and their trimethylations with repression [4, 10, 11]. These cellular manifestations appear to have originated in their cognate stem cells (see Fig. 2 of the Ref number 13, which is also reproduced as a bottom inset in the front cover page after permission from the Alpha Med Press).

Importantly, the individual PcG proteins exert their biological function on transcriptional repression in a strictly concerted and sequential manner [12–14]. At first, a set of multiple PcG proteins forming a polycomb repressive complex-2 (PRC2) initiates transcriptional repression. A different set of PcG proteins grouped as the polycomb repressive complex-1 (PRC1) maintains the transcriptional repression processes. Human PRC1 comprises a number of subunits: BMI1/ MEL18 (vertebrate ortholog of Posterior Sex Combs), RING1A/RING1B/RNF2 (Ring Finger Protein), hPC 1-3 (Polycomb), hPH1-3 (Polyhomeotic), and YY1 (Pleiohomeotic), among others [15]. On other hand, the PRC2 includes EZH2 (Enhancer of Zeste-2), SUZ12 (Suppressor of Zeste 12), and EED (Embryonic Ectoderm Development) as core components [14]. It is now accepted that many epigenetic modifiers such as histone-modifying enzymes, DNA methyltransferases, small RNAs, and their regulatory proteins cooperate with the PRCs during cell fate determination, maintenance, and cellular differentiation. It is also known that these PRCs may have many targets for methylation, phosphorylation, and some other unknown post-translational modifications, depending on the cellular contexts [16]. This suggests a possibility that additional novel accessory factors may affect a cell in a cellular context-dependent manner. This polycomb-mediated regulation of gene expression has been explored in stem cells, especially during their maintenance and differentiation. It is particularly fascinating to note that histone-dependent epigenetic memories could be maintained over 24 cell divisions even without transcription and independent of promoter DNA methylations. Also, an extended role of the PcG proteins during development and disease has recently drawn interest among researchers [13]. However, the molecular basis for regulation of PcG proteins remains unknown. This chapter evaluates the published data on possible regulation of these multiprotein complexes. A preliminary, testable consensus on this subject is proposed for future studies.

Current studies indicate that PcG levels are regulated in a cell-type-specific manner throughout organism development [16–18]. Such regulation is accomplished through control of both transcript and protein levels [18]. Therefore, control of PcG levels may be necessary in the normal development of metazoans. Data from an aberrant expression of the PcG mRNAs and their proteins resulting in embryonic defects and tumorigenesis [19, 20] support the contention. This leads to the next obvious query regarding the molecular mechanism behind such a fine-tuning between the levels of PcG transcripts and their protein products. Because of the innate self-renewal and differentiation properties of stem cells, the regulation aspect of PcG protein levels is of particular interest in the context of stem cell biology.

2 PcG Protein Homeostasis in Relation to Stem Cell Identity

Analyses of gene expression data and *cis*-regulatory genetic elements identified many genomic regulatory networks and the associated multicomponent signaling proteins in the development of multicellular organisms [21]. The regulation of multiprotein complexes using transcriptional and post-transcriptional feedback controls has been an important and evolutionarily conserved mechanism to regulate complex biochemical functions in cell biology [21–27]. It is well known that the function of macromolecular complexes can be disrupted by loss of one subunit, leading to a dysfunction and/or destabilization of the entire multisubunit complex. It is not known whether the individual components of PRCs are also similarly regulated. PRC2 protein levels in mouse 8.5dpc embryos and HeLa cells have been associated with the presence of the other components of the complex [28]. For example, the siRNA-mediated knockdown of Suz12 results in the proteasome-inhibitor-sensitive destabilization of only the Ezh2 protein level, but not the Eed level. However, the transcript levels of both Ezh2 and Eed remain comparable to that of control sets [28]. This is in contrast to the role of multiprotein complexes in gene repression, as mRNA levels are likely to be up-regulated upon functional loss of the other components. Furthermore, it is also not known whether any post-transcriptional effects such as altered translational efficiency, etc., may have been associated under the same experimental conditions. On the other hand, depletion of any one of the individual PRC2 members, such as EZH2, EED, and SUZ12, by siRNA-mediated knockdown in human embryonic diploid fibroblasts adversely affects protein levels and mRNA levels of the remaining components [29]. Thus, cell-type-specific regulation taking place even in the control of PRC accumulation, may have evolved as the yet uncharacterized mechanism of PcG protein homeostasis.

The above findings also indicate that the biosynthesis of PcG proteins may be controlled by different layers of feedback regulatory loops from the PRC components. It has been long recognized that autoregulatory feedback controls play a part in regulating a spectrum of functional multi-protein complexes in the cells. The following hypothesis is based on an appraisal of published data on the putative homeostasis of PcG proteins from various laboratories, although more experimental data are needed in order to substantiate the potential possibilities raised below. The following two types of feedback loops may play a role in regulating PRC2 components. These putative feedback loops may occur at the transcriptional and the posttranscriptional levels of gene expression. Thus, PRC2 may facilitate positive feedback regulation at the level of gene transcription of the other PRC2 components (Fig. 1A), and uncomplexed individual PRC2 components could negatively autoregulate their accumulation at the translational level in the absence of functional PRC2 (Fig. 1B). Each component of PRC2 co-immunoprecipitating the rest of the other components of PRC2 [18, 28, 30] is consistent with the possible existence of a functionally interacting protein complex. This model is also consistent with earlier observations where the mRNA levels of PRC2 components were found to be coordinately induced by E2F transcription factors and inhibited by pRB in WI38 human diploid fibroblasts [31]. Moreover, the PcG proteins bind to the Ph locus in Drosophila [29, 32, 33], and often occupy the promoter regions of the PcG genes in higher organisms [29].

Alternatively, it is possible that following the siRNA-mediated depletion of SUZ12, freely available EZH2 protein [29] may exert a feedback inhibition at the translational level (Fig. 1B). Therefore, it is important to consider that when one of the PRC2 components is completely inhibited by the siRNA knockdown, the protein levels of other PRC2 components are differentially affected compared to their controls [29]. It remains unknown whether other mRNAs of the PRC2 components may also exert similar function as portrayed in the proposed model. This is expected to affect the stoichiometry of PRC2 components and thereby impede the formation of a full and functional PRC2. This situation may be responsible for the suggested positive feedback control in transcription of the PRC2 components.

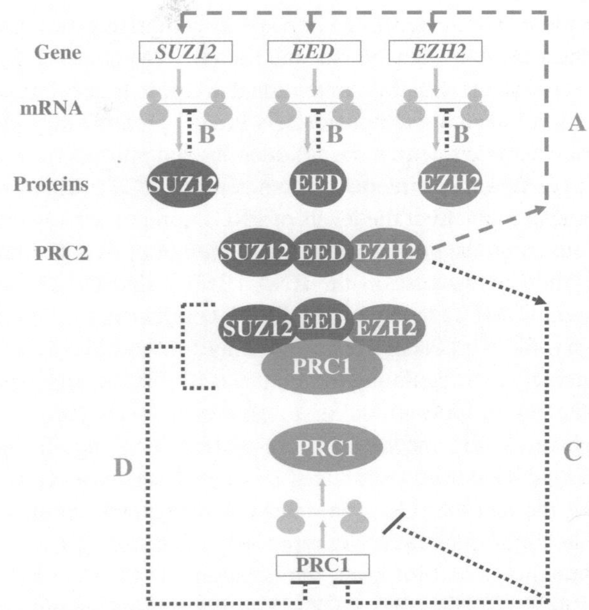

Fig. 1 Hypothetical regulatory feedback loops in the PcG protein homeostasis in stem cells. PRC1 represents multiprotein complexes. SUZ12, EED, and EZH2, mark the components of PRC2. Regulators are represented by colored objects: gene promoters by gray rectangles and functional mRNAs as ribosome-loaded objects. Positive feedback controls are shown by (green) dashed lines (**A**). Negative feedback controls appear in (red) dotted lines (**B**, **C**, and **D**). Steps in genes encoding regulators are denoted by solid (gray) arrows (*see* also Color Insert)

The suggested types of interacting feedback autoregulatory loops in the accumulation of PcG proteins may be comparable to other instances where uncomplexed free and individual ribosomal subunit proteins bind to the cognate mRNAs and inhibit their translation [34, 35]. The assembly of all the subunits into a functional ribosome fails to exert an autoregulation at the translational level. Just as the concentration of free/uncomplexed ribosomal proteins depends on levels of free ribosomal RNA during the assembly of the ribosome [34, 36], the levels of individual PRC2 components may also depend on the formation of a functional PRC2 complex. Thus, a possibility that the transcription of PRC2 genes may be regulated by the PRC2 complex similarly to the way the transcription of ribosomal RNA genes is regulated by ribosomal proteins becomes a subject for further study. The molecular mechanism of translational control that triggers and regulates the dynamic PcG protein homeostasis may turn out to be comparable to other findings. For example, upon sensing the activated Ras and Akt signaling, previously untranslated sets of mRNAs are quickly recruited into polysomes or the previously translating mRNAs are released from the polysomal machinery in the brain progenitor cells [37]. The observation that a large part of the active transcriptome in stem cells appears to comprise genes associated with RNA metabolism and protein biosynthesis [38] is also consistent with this contention that a post-transcriptional and/or a translational homeostasis in PcG proteins may play a considerable role in stem cell identity and pluripotency.

In addition, it is interesting to note that PRC2 components also appear to affect the levels of PRC1 components. For example, upon the siRNA-mediated depletion of SUZ12, there is a marked increase in the RNA (BMI 1 and CBX8) and protein (BMI 1) levels of the PRC1 components in human embryonic fibroblasts [29]. This may be viewed as PRC2 negatively autoregulating the synthesis of PRC1 components at the transcriptional and/or translational levels (Fig. 1C). Such regulatory mechanisms are of biological significance and may have been maintained by natural selection, considering the fact that PRC2 initiates transcriptional repression, while PRC1 maintains the repressive conditions from flies to humans [12]. Moreover, the depletion of PRC1 (BMI 1) or PRC2 (EZH2, EED, SUZ12) components augments the gene expression of other components measured at the level of transcript abundance [29]. Therefore, it appears reasonable to assume that the hierarchical PcG complex (PRC1 and PRC2 together) may fine-tune the overall functional competence of PcG proteins by negatively autoregulating the transcription of all PRC components (Fig. 1D).

Although, direct experimental proof for these possibilities is yet to be obtained, the tentative hypothesis evoked here accommodates the functional significance of PcG protein homeostasis in relation to the maintenance of stem cell pluripotency. Binding of PRC1 (CBX8) and PRC2 (SUZ12) components to a large number of PcG promoters may then be considered as preliminary evidence in support of this argument [29]. Similarly, genes bound by SUZ12 in human ES cells include CBX8 and BMI 1, and these genes are also upregulated in SUZ12-deficient cells derived from homozygous mutant mouse blastocysts [38]. Furthermore, PRC1 (Phc1 and Rnf2) and PRC2 (Suz12 and Eed) components occupy the Cbx4 and Cbx8 genes in the mouse ES cells [39]. But in the Eed-deficient mouse ES cells, where decreased levels of Ezh2 are also observed, the relative transcript accumulation of other components of PRC2 and PRC1, such as Suz12 and Rnf2, respectively, are only modestly affected [39]. Furthermore, human genome-wide interrogation reveals a relatively considerable amount of binding for CBX8 (PRC1) in embryonic fibroblasts [29] versus SUZ12 (PRC2) in ES cells [38]. Even during hematopoiesis, distinct PcG complexes are expressed and regulated in a nonoverlapping fashion in resting versus mature follicular B cells of the human germinal center [40]. Thus, it appears that understanding the underlying mechanisms of PcG protein homeostasis may provide molecular insights into the manifestation of definite stem cell identities.

3 Perspective

The fact that the embryonic stem cells lacking Eed [41] and Suz12 [42] lose pluripotency, and that deletion of Bmi1 results in loss of adult stem cells [43–46], highlights the importance of PcG protein homeostasis in stem cell identity. The hypothesis postulated in this chapter considers that the feedback autoregulatory mechanism regulates stem cell–specific PcG protein homeostasis, an aberration of which results in loss of stem cell identity. This hypothesis takes into account various correlative and circumstantial data obtained during biochemical studies affecting subunits of the PRC members. More direct experimental evidence substantiating this hypothesis is not yet available. However, recent evidence shows that the tumor microenvironment regulates cancer stem cells [47]. Therefore, it is expected that the identities of normal adult stem cells could also be influenced by another tier of control, such as their stem cell niches. In the latter scenario, it becomes important to understand whether the various organ tissue stem cells are similarly or differentially affected by niche-dependent signals, and, if so, whether this, in turn, affects the characteristics of PcG protein homeostasis in stem cells. Moreover, the translation of mRNAs encoding secretory/membrane proteins versus cytosolic proteins is compartmentalized in mammalian cells [48]. Different cell context/niche-dependent signaling pathways may also transduce selective effects on polysome recruitment of cellular mRNAs [49]. Therefore, it could be possible that cell-type-specific translation of various mRNAs, including that encoding for PcG proteins and thereby the altered PcG protein homeostasis, may influence the overall identity of a particular stem cell type, depending on its cellular context. Corroborating this contention, differential expression of PcG proteins was identified in various cancers, and the data were interpreted to reflect the situation within a subset of stem-like cancer-inducing tumor cells [50, 13, 51].

The above findings suggest that cell-type-specific PcG protein homeostasis may exist and may be widespread in metazoan. Dynamic autoregulation of PcG genes has been found in other systems such as plants [52, 53]. For example, Arabidopsis PcG protein called Medea, that has homology to Drosophila PRC2 subunit, the Enhancer of Zeste and H3K27 methylating activity, was recently discovered to be autoregulated [54, 52, 55, 22]. Thus, the above-proposed autoregulatory feedback mechanism governing the PcG protein homeostasis appears to be a broadly evolved biological phenomenon. Fine-tuning the stoichiometry of PcG protein complexes such that the specificity of particular stem cell identity is maintained could be the selection pressure behind the evolutionarily conserved and developmentally associated biological processes in multicellular organisms.

It is interesting to note that in the case of induced pluripotent cell systems [56], the levels of some of the pluripotency-inducing factors also appear to be regulated in a noncoordinated fashion between their gene transcription and protein accumulation. For example, induction of Nanog is evident at the mRNA level, but protein levels of the Nanog as well as other pluripotency-related transcription factors such as Oct4 and Sox2 contrast with the induction of their mRNA levels. This observation is especially interesting in view of the fact that the Nanog itself participates in a finely tuned autoregulatory transcription factor feedback network of Nanog, Oct4, and Sox2 in regulating stem cell identity and pluripotency [57–59]. The PcG proteins are suggested to associate with some of these transcription factors in chromatin binding during the maintenance of stem cell pluripotency [39]. Therefore, the expression levels of such pluripotency-related transcription factors that act in the same axis of PcG protein function are expected to be similarly feedback-regulated in a teleological sense. Future investigations are needed to determine whether common pathways exist between cross-regulatory mechanisms controlling feedback autoregulatory loops among PcG proteins and other pluripotency-related transcription factors in stem cells. The far-reaching implications of such studies are not only directly related to the molecular understanding of human development, but could also facilitate therapeutic interventions to human diseases including cancers.

Acknowledgments I wish to thank the Byrne Award for partial support. Thanks also to Birgit Baur and Jackie Arenz for help in the initial preparation of this manuscript, and to Carol Pearce, writer/editor with the MSKCC Department of Medicine, for editorial review. I am indebted to scientific encouragement from Drs. Sudhir Sopory, Hans Mohr, Howard I. Scher Lorenz Studer, and Nahum Sonenberg. Thanks are also due to Dr. David Thaler for helpful discussions on feedback control mechanisms in cell biology.

References

1. Torres-Padilla ME, Parfitt DE, Kouzarides T, Zernicka-Goetz M. Histone arginine methylation regulates pluripotency in the early mouse embryo. Nature. 2007;445:214–8.
2. Boiani M, Scholer HR. Regulatory networks in embryo-derived pluripotent stem cells. Nat Rev Mol Cell Biol. 2005;6:872–84.
3. Sasaki H, Matsui Y. Epigenetic events in mammalian germ-cell development: reprogramming and beyond. Nat Rev Genet. 2008;9:129–40.
4. Jaenisch R, Young R. Stem cells, the molecular circuitry of pluripotency and nuclear reprogramming. Cell. 2008;132:567–82.
5. Bernstein BE, Meissner A, Lander ES. The mammalian epigenome. Cell. 2007;128:669–81.
6. Surani MA, Hayashi K, Hajkova P. Genetic and epigenetic regulators of pluripotency. Cell. 2007;128:747–62.
7. Whitcomb SJ, Basu A, Allis CD, Bernstein E. Polycomb group proteins: an evolutionary perspective. Trends Genet. 2007;23: 494–502.
8. Quina AS, Buschbeck M, Di Croce L. Chromatin structure and epigenetics. Biochem Pharmacol. 2006;72:1563–9.
9. Barrera LO, Ren B. The transcriptional regulatory code of eukaryotic cells – insights from genome-wide analysis of chromatin organization and transcription factor binding. Curr Opin Cell Biol. 2006;18(3):291–8.
10. Barski A, Cuddapah S, Cui K, et al. High-resolution profiling of histone methylations in the human genome. Cell. 2007;129: 823–37.
11. Guenther MG, Levine SS, Boyer LA, Jaenisch R, Young RA. A chromatin landmark and transcription initiation at most promoters in human cells. Cell. 2007;130:77–88.
12. Cavalli G. Chromatin and epigenetics in development: blending cellular memory with cell fate plasticity. Development. 2006;133:2089–94.
13. Rajasekhar VK, Begemann M. Concise review: roles of polycomb group proteins in development and disease: a stem cell perspective. Stem Cells. 2007;25:2498–510.
14. Ringrose L, Paro R. Epigenetic regulation of cellular memory by the Polycomb and Trithorax group proteins. Annu Rev Genet. 2004;38:413–43.
15. Levine SS, King IF, Kingston RE. Division of labor in polycomb group repression. Trends Biochem Sci. 2004;29:478–85.
16. Kuzmichev A, Jenuwein T, Tempst P, Reinberg D. Different EZH2-containing complexes target methylation of histone H1 or nucleosomal histone H3. Mol Cell. 2004;14:183–93.
17. Kuzmichev A, Margueron R, Vaquero A, et al. Composition and histone substrates of polycomb repressive group complexes change during cellular differentiation. Proc Natl Acad Sci U S A. 2005;102:1859–64.
18. Pasini D, Bracken AP, Helin K. Polycomb group proteins in cell cycle progression and cancer. Cell Cycle. 2004;3:396–400.
19. Allegrucci C, Denning C, Priddle H, Young L. Stem-cell consequences of embryo epigenetic defects. Lancet. 2004;364: 206–8.
20. Lotem J, Sachs L. Epigenetics and the plasticity of differentiation in normal and cancer stem cells. Oncogene. 2006;25(59): 7663–72.
21. Stathopoulos A, Levine M. Genomic regulatory networks and animal development. Dev Cell. 2005;9:449–62.
22. Jullien PE, Katz A, Oliva M, Ohad N, Berger F. Polycomb group complexes self-regulate imprinting of the Polycomb group gene MEDEA in Arabidopsis. Curr Biol. 2006;16:486–92.
23. Yang Q, Inoki K, Kim E, Guan KL. TSC1/TSC2 and Rheb have different effects on TORC1 and TORC2 activity. Proc Natl Acad Sci U S A. 2006;103:6811–6.
24. Harris SL, Levine AJ. The p53 pathway: positive and negative feedback loops. Oncogene. 2005;24:2899–908.
25. Dai MS, Zeng SX, Jin Y, Sun XX, David L, Lu H. Ribosomal protein L23 activates p53 by inhibiting MDM2 function in response to ribosomal perturbation but not to translation inhibition. Mol Cell Biol. 2004;24:7654–68.
26. Guillier M, Allemand F, Raibaud S, Dardel F, Springer M, Chiaruttini C. Translational feedback regulation of the gene for L35 in Escherichia coli requires binding of ribosomal protein L20 to two sites in its leader mRNA: a possible case of ribosomal RNA-messenger RNA molecular mimicry. RNA. 2002;8: 878–89.
27. O'Geen H, Squazzo SL, Iyengar S, et al. Genome-wide analysis of KAP1 binding suggests autoregulation of KRAB-ZNFs. PLoS Genet. 2007;3:e89.
28. Pasini D, Bracken AP, Jensen MR, Lazzerini Denchi E, Helin K. Suz12 is essential for mouse development and for EZH2 histone methyltransferase activity. EMBO J. 2004;23:4061–71.
29. Bracken AP, Dietrich N, Pasini D, Hansen KH, Helin K. Genome-wide mapping of Polycomb target genes unravels their roles in cell fate transitions. Genes Dev. 2006;20:1123–36.

30. Cao R, Tsukada Y, Zhang Y. Role of Bmi-1 and Ring1A in H2A ubiquitylation and Hox gene silencing. Mol Cell. 2005;20:845–54.
31. Bracken AP, Pasini D, Capra M, Prosperini E, Colli E, Helin K. EZH2 is downstream of the pRB-E2F pathway, essential for proliferation and amplified in cancer. EMBO J. 2003;22:5323–35.
32. Bloyer S, Cavalli G, Brock HW, Dura JM. Identification and characterization of polyhomeotic PREs and TREs. Dev Biol. 2003;261:426–42.
33. Fauvarque MO, Zuber V, Dura JM. Regulation of polyhomeotic transcription may involve local changes in chromatin activity in Drosophila. Mech Dev. 1995;52:343–55.
34. Zengel JM, Jerauld A, Walker A, Wahl MC, Lindahl L. The extended loops of ribosomal proteins L4 and L22 are not required for ribosome assembly or L4-mediated autogenous control. RNA. 2003;9:1188–97.
35. Zengel JM, Lindahl L. Diverse mechanisms for regulating ribosomal protein synthesis in Escherichia coli. Prog Nucleic Acid Res Mol Biol. 1994;47:331–70.
36. Dennis PP, Ehrenberg M, Bremer H. Control of rRNA synthesis in Escherichia coli: a systems biology approach. Microbiol Mol Biol Rev. 2004;68:639–68.
37. Rajasekhar VK, Viale A, Socci ND, Wiedmann M, Hu X, Holland EC. Oncogenic Ras and Akt signaling contribute to glioblastoma formation by differential recruitment of existing mRNAs to polysomes. Mol Cell. 2003;12:889–901.
38. Lee TI, Jenner RG, Boyer LA, et al. Control of developmental regulators by Polycomb in human embryonic stem cells. Cell. 2006;125:301–13.
39. Boyer LA, Plath K, Zeitlinger J, et al. Polycomb complexes repress developmental regulators in murine embryonic stem cells. Nature. 2006;441(7091):349–53.
40. van Galen JC, Dukers DF, Giroth C, et al. Distinct expression patterns of polycomb oncoproteins and their binding partners during the germinal center reaction. Eur J Immunol. 2004;34:1870–81.
41. Faust C, Schumacher A, Holdener B, Magnuson T. The eed mutation disrupts anterior mesoderm production in mice. Development. 1995;121:273–85.
42. Pasini D, Bracken AP, Hansen JB, Capillo M, Helin K. The Polycomb Group protein Suz12 is required for embryonic stem cell differentiation. Mol Cell Biol. 2007;27(10):3769–79.
43. Kim JY, Sawada A, Tokimasa S, et al. Defective long-term repopulating ability in hematopoietic stem cells lacking the Polycomb-group gene rae28. Eur J Haematol. 2004;73:75–84.
44. Park IK, Qian D, Kiel M, et al. Bmi-1 is required for maintenance of adult self-renewing haematopoietic stem cells. Nature. 2003;423:302–5.
45. Molofsky AV, Pardal R, Iwashita T, Park IK, Clarke MF, Morrison SJ. Bmi-1 dependence distinguishes neural stem cell self-renewal from progenitor proliferation. Nature. 2003;425:962–7.
46. Liu S, Dontu G, Mantle ID, et al. Hedgehog signaling and Bmi-1 regulate self-renewal of normal and malignant human mammary stem cells. Cancer Res. 2006;66:6063–71.
47. Rajasekhar VK, Dalerba P, Passegue E, Lagasse E, Najbauer J. The 5th International Society for Stem Cell Research (ISSCR) Annual Meeting, June. 2007. Stem Cells. 2008;26:292–8.
48. Stephens SB, Nicchitta CV. Divergent regulation of protein synthesis in the cytosol and endoplasmic reticulum compartments of mammalian cells. Mol Biol Cell. 2008;19:623–32.
49. Rajasekhar VK, Holland EC. Postgenomic global analysis of translational control induced by oncogenic signaling. Oncogene. 2004;23:3248–64.
50. Pietersen AM, van Lohuizen M. Stem cell regulation by polycomb repressors: postponing commitment. Curr Opin Cell Biol. 2008;20(2):201–7.
51. Sparmann A, van Lohuizen M. Polycomb silencers control cell fate, development and cancer. Nat Rev Cancer. 2006;6:846–56.
52. Baroux C, Gagliardini V, Page DR, Grossniklaus U. Dynamic regulatory interactions of Polycomb group genes: MEDEA autoregulation is required for imprinted gene expression in Arabidopsis. Genes Dev. 2006;20:1081–6.
53. Putterill J, Laurie R, Macknight R. It's time to flower: the genetic control of flowering time. Bioessays. 2004;26:363–73.
54. Baubec T, Mittelsten Scheid O. Medea in full self-control. Trends Plant Sci. 2006;11:469–71.
55. Gehring M, Huh JH, Hsieh TF, et al. DEMETER DNA glycosylase establishes MEDEA polycomb gene self-imprinting by allele-specific demethylation. Cell. 2006;124:495–506.
56. Takahashi K, Yamanaka S. Induction of pluripotent stem cells from mouse embryonic and adult fibroblast cultures by defined factors. Cell. 2006;126:663–76.
57. Boyer LA, Lee TI, Cole MF, et al. Core transcriptional regulatory circuitry in human embryonic stem cells. Cell. 2005;122:947–56.
58. Boyer LA, Mathur D, Jaenisch R. Molecular control of pluripotency. Curr Opin Genet Dev. 2006;6(5):455–62.
59. Niwa H. How is pluripotency determined and maintained? Development. 2007;134:635–46.

Part IV
Signaling and Regulation in Select Stem Cell Types

Signaling Pathways in Embryonic Stem Cells

D. Reynolds, Ludovic Vallier, Zhenzhi Chng and Roger Pedersen

Abstract This chapter covers the different signaling pathways affecting embryonic stem (ES) cells. We discuss those signals governing maintenance of pluripotency, survival and proliferation, and induction of differentiation. We also cover the differences in signaling responsiveness and requirements between mouse and human ES cells, and discuss the developmental context of pluripotent tissue, which gives useful insights for the derivation and culture of ES cells as well as for directed differentiation to adult tissues.

Keywords Signaling · Human · Mouse · Embryonic stem cells · Development · Pluripotency · Differentiation

1 Introduction

Embryonic stem cells are a unique and fascinating cell type for their property of pluripotency, the ability to differentiate to form most adult tissues. Their designation not only denotes their origin, but also their potential: while a skin stem cell can generate any other part of the skin, embryonic stem (ES) cells can generate any part of the entire body.

Furthermore, unlike most primary cell lines, ES cells also display apparent immortality, in that they can be cultured indefinitely without signs of senescence, a property otherwise seen only in abnormal or oncogene-immortalized cell lines. Unlike such immortalized lines, ES cells exist in a metastable state: both their survival and growth, and the decision between proliferation and differentiation, are controlled by multiple chemical stimuli from their environment (Fig. 1). Consequently, ES cells are readily triggered to differentiate, even if this is not the intended outcome.

D. Reynolds (✉)
University of Cambridge Department of Surgery Laboratory for Regenerative Medicine West Forvie Building Forvie Site, Robinson Way Cambridge, CB2 0SZ
e-mail: d.reynolds.00@cantab.net

To practically culture ES cells, and to make use of them for experimental and therapeutic purposes, we must therefore understand what stimuli they respond to, as well as understand the molecular features of pluripotency itself.

Early successes with the derivation and culture of embryonic stem cells were achieved by growth in complex, serum-containing media and co-culture with other cell types, giving a diverse mixture of different signals and nutrients.

Building on an increasing understanding of developmental biology and cell signaling at the molecular level, recent advances have narrowed down which of these signals are in fact required to regulate the different aspects of ES cell behavior.

This knowledge has given us simpler and more reliable means of ES cell culture, shed light on interesting developmental differences between human and mouse, and paved the way to the directed in vitro differentiation of specialized adult tissues.

This chapter covers the developmental context of mouse and human embryonic stem cells, the signaling pathways that influence them, and how these signals interact, causing either proliferation and maintenance of the pluripotent state, or induction of differentiation towards the different embryonic and extra-embryonic cell types they are capable of forming.

2 Embryonic Stem Cells and Early Development

An understanding of the signals controlling embryonic cell growth and differentiation can be gained by examining such mechanisms in the embryos from which they are derived. A half-century of genetic studies has yielded many insights into the signals controlling early embryonic development, and these signals appear to be conserved between the vertebrate classes. ES cells, however, appear to be a largely mammalian phenomenon.

Pluripotent tissues exist in every embryo, from the formation of the zygote at fertilization until the appearance of irreversibly determined tissues.

V.K. Rajasekhar, M.C. Vemuri (eds.), *Regulatory Networks in Stem Cells*, Stem Cell Biology and Regenerative Medicine,
DOI 10.1007/978-1-60327-227-8_25,

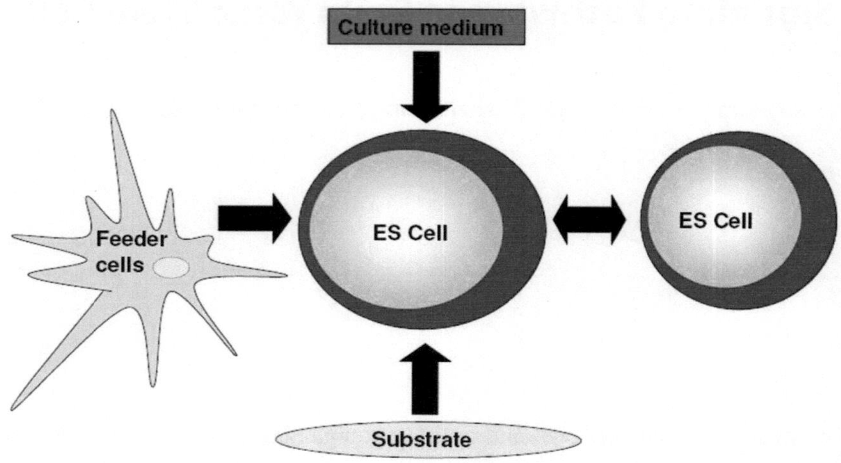

Fig. 1 Sources of factors influencing stem cells include dissolved signaling molecules contained in the culture medium, and interactions with the substrate they attach to. Co-cultured feeder cells and neighboring ES cells also produce dissolved signaling molecules and short-range paracrine factors, as well as interactions through cell-cell contact

In most organisms, pluripotent tissues are transient: in fish and amphibians, both the supporting structures and the nutrition needed to complete development are present *ab initio*, and following fertilization, morphogenesis and differentiation proceed at a rapid pace. In placental mammals, in contrast, the structures that will ultimately form the adult organism do not develop significantly until after the embryo has successfully implanted in the uterus. For implantation to occur, the embryo must first generate successive layers of extra-embryonic tissues, which will contribute to no part of the new organism but form the placenta and amniotic membranes needed to support fetal growth in utero.[1]

During the period of extra-embryonic tissue formation, some pluripotent cells are sequestered from any inductive influences that would result in their precocious differentiation and thereby thwart fetal development.

These undifferentiated, pluripotent cells are found in the inner cell mass of the blastocyst and the epiblast layer of the postimplantation embryo.

In summary, while in most organisms pluripotency is a fleeting phenomenon, the early placental embryo must maintain a mass of pluripotent stem cells for a prolonged period during its peri-implantation development. Thus, in placental mammals (and other amniotic embryos), maintenance of the undifferentiated state can be regarded as a cell fate decision akin to the decisions taken during later specification of the germ layers. By replicating the conditions required in vivo to induce this undifferentiated state, and providing a suitable substrate and source of nutrients, a population of pluripotent cells can be maintained indefinitely and expanded in vitro as embryonic stem cells. Figures 2 and 3 present a comparison of early development in amphibians and mammals.

3 Signaling Pathways Governing Pluripotency

3.1 Signaling Pathways Regulating the Pluripotent State of Mouse Embryonic Stem Cells

3.1.1 Early Derivations of Mouse ES Cells

The field of embryonic stem cell research arose in the 1980s as an offshoot of the study of mouse embryology. Early work with aggregation of morulae [1] and transfer of inner-cell mass cells between different blastulae [2] led to the formation of chimeric mice, demonstrating that cells at this stage of development retained the property of pluripotency. Further experiments with transplanting even quite advanced embryos to within the body cavity led to the discovery of teratocarcinomas, disorganized growths containing a variety of differentiated tissues. These growths were found to be sustained by a population of stem cells, and these embryonic carcinoma cells could be reintroduced to the inner cell mass and would contribute to a chimeric mouse [3]. The existence of such long-lived growths of pluripotent tissue suggested that cells from the inner cell mass might be stably cultured in vitro in the pluripotent state.

The conditions initially used to derive embryonic stem cells included fetal calf serum – a rich cocktail of nutrients and growth factors – and co-culture with mitotically arrested "feeder" cells, to provide further growth factors and extracellular matrix [4, 5]. Thus, it was not clear what specific compounds of the culture milieu were responsible for maintenance of pluripotency. The mechanisms of self-renewal

[1] An intermediate situation is seen in egg-laying amniotes. The reptile or bird embryo does not implant or generate a placenta but does give rise to the extra-embryonic tissue forming the yolk sac – found in vestigial form in placental mammals – and other external structures. Interestingly, embryonic stem cells have been derived from the chicken.

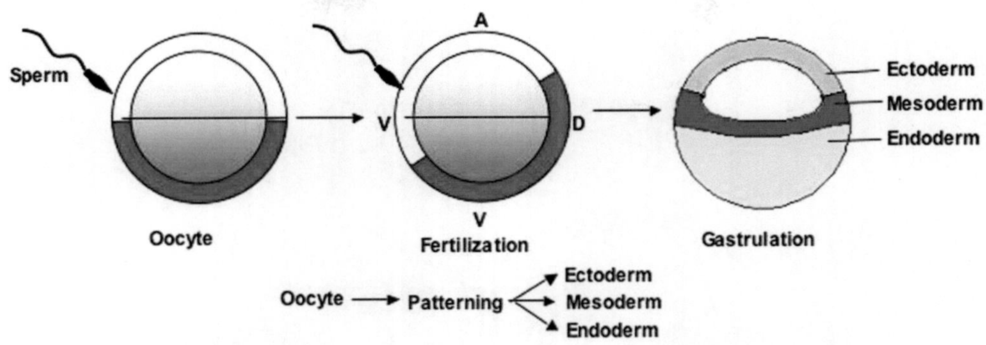

Fig. 2 Early development of the frog embryo. In amphibian development, the body axes (animal/vegetal, dorsal/ventral) are patterned by the event of fertilization. Development proceeds immediately afterwards, and the entire zygote contributes to the embryo (*see* also Color Insert)

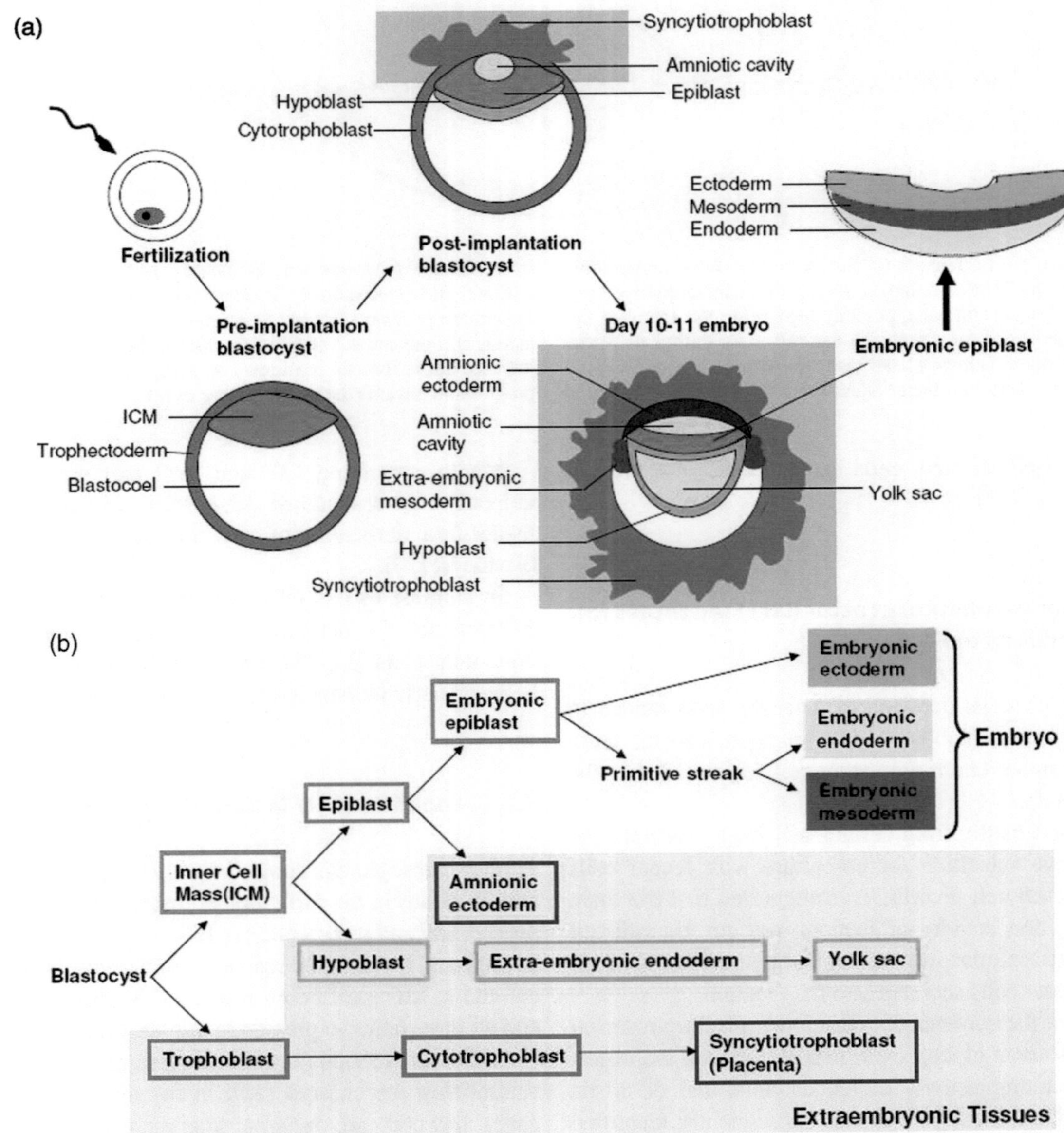

Fig. 3 (a) Early development of the human embryo. (b) Cell fate decisions in early human development. In early development in placental mammals, several layers of extra-embryonic tissues must be formed before, during, and after implantation in the uterine lining. Patterning of the embryo only begins after these structures are formed, and a population of pluripotent cells is maintained until this point. The need to generate extra-embryonic tissues may give amniotic organisms in general, and placental mammals in particular, developmental processes facilitating the derivation of embryonic stem cells. The possibility of long-term culture of pluripotent cells from species without extra-embryonic tissue cannot, however, be ruled out (*see* also Color Insert)

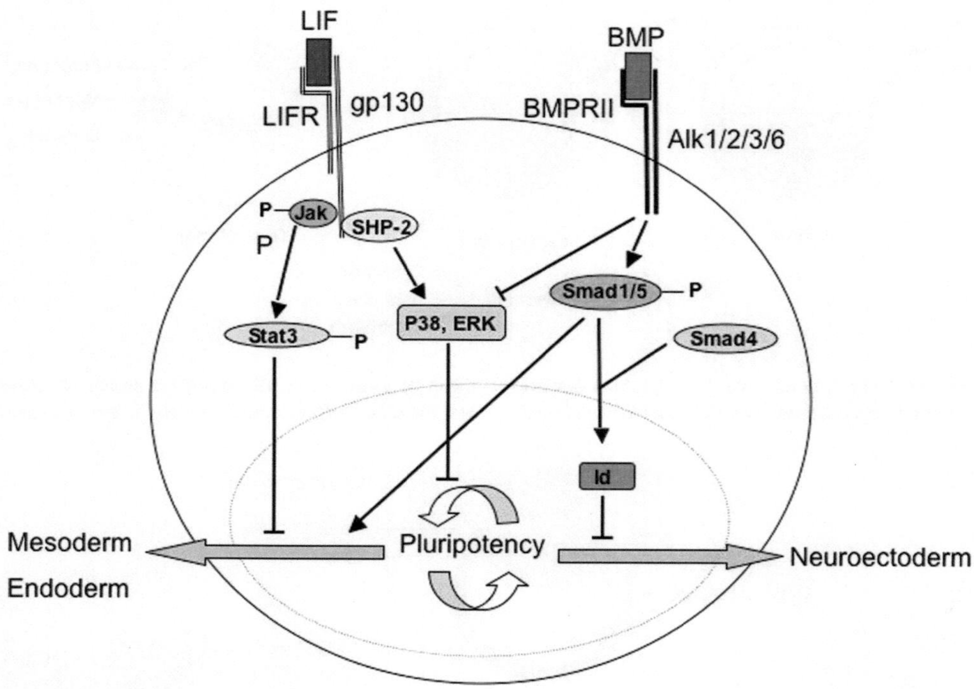

Fig. 4 Summary of maintenance of pluripotency in mouse embryonic stem cells. LIF and BMP signaling co-operate to maintain pluripotency of mouse embryonic stem cells, possibly replicating the signaling environment experienced by a mouse inner cell mass during diapause. Although both these pathways have many downstream targets, the activation of the transcription factor STAT3 and the expression of the Id transcriptional repressor proteins appear to be responsible for the effects LIF and BMP, respectively, in maintaining the undifferentiated state. How these pathways interact to induce mESCs to remain undifferentiated is still the subject of research, both to find how they interact with the core transcriptional machinery of pluripotency and how they block progression towards the various forms of differentiated tissue

in mouse embryonic stem cells became clear, however, in subsequent studies (Fig. 4).

3.1.2 Leukemia Inhibitory Factor (LIF) Substitutes for the Activity of Feeder Cells

Finding that ES cells deprived of a feeder layer tended to differentiate, researchers began to investigate how the feeders might promote embryonic stem cell self-renewal in the pluripotent state.

Conditioned media from certain cell lines was found to be an effective substitute for co-culture with feeder cells in promoting ES cell growth. This suggested that the main antidifferentiation activity of feeders was not via cell-cell contact or extracellular matrix, but through some diffusible signal the feeder cells secreted into the medium.

Analysis of the contents of conditioned medium revealed a particular soluble glycoprotein that reproduced the differentiation inhibiting activity of feeder cells, and of media conditioned by them [6]. This was subsequently identified as the cytokine leukemia inhibitory factor (LIF) [7, 8], previously known as an inducer of macrophage differentiation from cultured murine leukemic cells [9].

Media containing LIF and fetal calf serum supported not only maintenance of pre-existing mouse ES cells but feeder-free derivation of new mouse ES lines from the blastocyst [10].

In contrast to this effect on mouse ES cells, LIF is ineffective for the derivation and culture of human embryonic stem cells [11–13], suggesting significant differences between early human and mouse development.

3.1.3 Function of LIF in Early Development

Further investigations showed that LIF signaling has important functions in the early mouse embryo. The blastocyst expresses LIF and its receptors (LIFR and glycoprotein 130) in a reciprocal fashion, the receptors being present on the inner cell mass, LIF itself being expressed by the trophectoderm, and all three being expressed in the uterine lining [14].

Although Xenopus studies suggest a role for gp130 in ventralizing the embryo [15], in the mouse none of these genes however are *required* for early development: null mutants for all three genes will successfully implant. While LIFR and GP130 mutants display later-stage embryonic lethality, homozygote LIF mutants can grow to adulthood, although the females are infertile with their decidua failing to support implantation. There is some evidence that this function of LIF is conserved in other species, including humans [16, 17].

In seeking to explain how LIF signaling could be essential for embryonic stem cell renewal but not embryonic development per se, Smith and colleagues investigated the role of the LIF pathway in blastocyst diapause, also known as delayed implantation, a state of developmental stasis found in mice during which a fertilized embryo in a lactating mother is hormonally induced to temporarily halt its development before implantation.

Indeed, an early requirement for LIF signaling is seen only in embryos induced to enter diapause. The inner cell mass of mouse blastocysts lacking LIF receptors is incapable of surviving diapause and continuing development, with embryos with mutations in the LIF pathway suffering increased cell death in the inner cell mass [18]. This implies that mouse ES cells are developmentally equivalent to the inner cell mass of the blastocyst, which is plausible given their ability to reintegrate there and form chimeric embryos. This also explains why the techniques used for mouse ES culture are not transferable to human ES cells, as the human blastocyst cannot enter diapause and lacks an equivalent response.

3.1.4 LIF Promotes Self-Renewal by Activation of the Transcription Factor STAT3

The LIF receptor complex (LIFR recruiting LIF to GP130) has multiple subcellular effects, being able to activate several different signaling cascades.

These including the Janus-activating kinase (JAK)/STAT3 pathway, SHP-2, and the mitogen activated protein kinases extracellular-regulated kinase (ERK) 1 and 2. Of these, only STAT3 appears involved in maintaining the pluripotent state, with ERK acting instead to promote differentiation.

STAT3 is a transcription factor that enters the nucleus when phosphorylated. Interference with STAT3 signaling by expression of a dominant negative protein causes abrupt differentiation in mouse ES cells [19], while a STAT3-ERT2 fusion protein (which is activated by tamoxifen treatment, thus bypassing the protein's normal means of control) substitutes for LIF activation in mESC maintenance [15]. In contrast, receptor mutations preventing activation of SHP-2 and ERK do not cause differentiation, and in fact increase the tendency of mouse ES cells to remain undifferentiated and increase the effectiveness of LIF in maintaining the undifferentiated state [20]. Furthermore, chemical inhibition of ERK improves the success rate of mouse ES cell derivation [21]. Interestingly, ERK is also activated by insulin, which is present at a high concentration in cell culture media in order to promote cell growth, and by FGF signaling, which causes neural differentiation of mouse ES cells. Thus, LIF signaling seems to generate opposing subcellular effects, some promoting self-renewal, others differentiation.

Without the presence of serum, however, LIF does not prevent differentiation, with the cells steadily heading towards a neural fate. This suggests that it must interact with other signals present in serum in order to effectively maintain pluripotency [22].

3.1.5 BMP Substitutes for Serum in the Maintenance of Pluripotency

While LIF activity is sufficient to maintain pluripotency in mouse ES cells in place of feeders, it does not substitute for serum, without which neural differentiation begins to occur. This process can be prevented by activation of BMP signaling by BMP2 or BMP4, allowing long-term culture of mouse ES cells in chemically defined conditions [22].

This finding is surprising, as BMP signaling is a well-known inducer of differentiation, having a conserved role in the patterning of mesoderm and endoderm during embryogenesis, and similarly inducing mesoderm differentiation from mouse ES cells in the absence of LIF [23]. In vivo, BMP is required post-implantation for the mouse embryo to progress beyond the egg cylinder stage, and it is also involved in the specification of germ cells [24]. Thus, factors downstream of LIF and BMP must interact in some way to allow the maintenance of pluripotency.

3.1.6 Id Proteins Are Expressed Downstream of BMP and Block Neuroectodermal Differentiation

The canonical mechanism of BMP signaling is via the phosphorylation of Smad1 and Smad5 by the receptor complex of BMPR-I and BMPR-II. These R (receptor) Smads bind to the C (common) Smad4 and are translocated to the nucleus, where the complex then binds to transcription factors targeting specific promoter sequences, thereby driving expression of BMP target genes.

One set of genes transcribed downstream of BMP signaling encode the Id (Inhibitor of Differentiation) proteins, named for their previously characterized role in preventing terminal differentiation in muscle [25].

These are DNA binding helix-loop-helix proteins that lack the domains for transcriptional activation, and bind to promoter sites, effectively blocking transcription. Id genes 1, 2, and 3 were identified as BMP targets in mouse ES cells (even in the absence of LIF), as well as being co-localized with areas of BMP activity during later development [26]. Constitutive overexpression of Id 1 allows mouse ES cells to grow independently of BMP or serum without differentiating to a neural fate, implying Id to be a key effector of BMP in maintenance of pluripotency [22]. Consistent with observations in the embryo, Id does not block differentiation towards mesoderm and endoderm.

In summary, while Id expression is LIF independent and Id is able to substitute for BMP in maintenance of pluripotency, maintenance of pluripotency in embryonic stem cells expressing Id1 still depends on the presence of LIF.

As Id proteins do not themselves prevent mesendodermal differentiation (which is promoted by the independent action of BMP), some additional mechanism must be present by which BMP signaling is modulated by the presence of LIF. One possible way this could occur is via binding of Smad1 to the transcription factor STAT3 – the downstream effector of LIF signaling required for maintenance of pluripotency [22], or to the core pluripotency factor Nanog [27]. This binding could negatively regulate or redirect both pathways, and may reduce the effect of BMP signaling to some degree, so that only the more BMP-sensitive downstream genes are actually expressed. Such a suppression of BMP could provide a mechanism by which some BMP regulated genes – for instance, Id genes – might be expressed in ES cells, but not the full range of BMP targets, which would otherwise drive differentiation. It is also possible that the Smad1-STAT3 complex has its own transcriptional targets that would not be activated by either transcription factor alone.

Another potential interaction between BMP and LIF signaling is in BMP-driven inhibition of other pathways antagonistic to pluripotency, the ERK and MAPK cascades [28]. This could reduce the differentiation effects of LIF and block neuralizing influences from FGF, and thus could provide an explanation of the fact that ES like cells can be derived from Smad4 null blastocysts, which would be unable to transduce BMP signaling through the Smad1 mediated regulation of Id1 expression [29].

3.2 Signaling Pathways Regulating the Pluripotent State in Human Embryonic Stem Cells and Mouse Epiblast Stem Cells

Human embryonic stem cells are pluripotent stem cells derived from embryos at the blastocyst stage [30]. Their embryonic origin confers upon them the capacity to proliferate indefinitely in vitro while maintaining the property to differentiate into the progenitors of the three germ layers: ectoderm, mesoderm, and endoderm.

Despite their common origin and similar properties of self-renewal and pluripotency, mESCs and hESCs differ in several aspects. hESCs form flat colonies and their clonality (i.e., their capacity to be propagated after single cell dissociation) is very limited [31]. In addition, hESCs can differentiate into extra-embryonic tissues including trophectoderm [32], whereas mouse ESCs cannot form either trophectoderm or primitive endoderm [33]. Most importantly, whereas mouse ESCs rely on LIF and BMP4 to maintain their pluripotent status [22], the LIF signaling pathway is not effective in hESCs [12, 13] and BMP4 induces differentiation of hESCs into trophectoderm [32]. hESCs rely instead on Activin and FGF to maintain their pluripotent status [34, 35, 36] and this property is shared by pluripotent cells derived from the epiblast of post-implantation mouse embryos (EpiSCs) [37, 38]. In the following section we describe in detail each of these signaling pathways and their function in maintaining self-renewal and pluripotency of hESCs and EpiSCs in vitro, and in vivo in the mouse embryo.

3.2.1 Activin/Nodal/TGFβ Signaling Pathway

The transforming growth factor (TGF)β family includes almost 30 members, which are subdivided into the TGFβs themselves, the activins, the inhibins, Nodal, myostatin, the bone morphogenetic proteins (BMP), growth/differentiation factors (GDF), and the anti-Müllerian hormone. Expression of these growth factors can detected in most embryonic tissues and their adult derivatives. In agreement with their ubiquitous expression, TGFβ signaling pathways have pleiotropic functions, which include the control of proliferation, differentiation, apoptosis, and cell adhesion [39]. The TGFβ proteins act by binding heteromeric complexes between type I and type II receptors, which, once activated, phosphorylate the Smads proteins (Fig. 5). This phosphorylation permits their interaction with Smad4 and the resulting complexes can then shuttle into the nucleus to control a large number of activators or repressors of transcription [40].

Nodal and Activin represent two distinct growth factors that nevertheless activate the same type I receptors (Alk4 and Alk7) and the same type II receptor (ActRIIB) thereby activating the Smad2/3 signaling pathway. TGFβ 1 itself generally uses different receptors (Alk5, TβRII), which also activate the same Smad2/3 pathway. Despite these common downstream effectors, TGFβ, Activin, and Nodal can control different biological events in vitro and in vivo. Indeed, the activity of these growth factors is tightly controlled by numerous extracellular co-activators and co-inhibitors. Cripto/Criptic are strictly necessary for Nodal to activate the Alk4/ActRIIb receptors, whereas Lefty and Cerberus are capable of blocking Nodal access to the same receptors [41]. Activin is inhibited by follistatin, and Activin alone is sufficient for receptor activation without cofactors. Importantly, Cripto contains consensus sequences for glycosyl-phosphatidylinositol (GPI) attachment to the

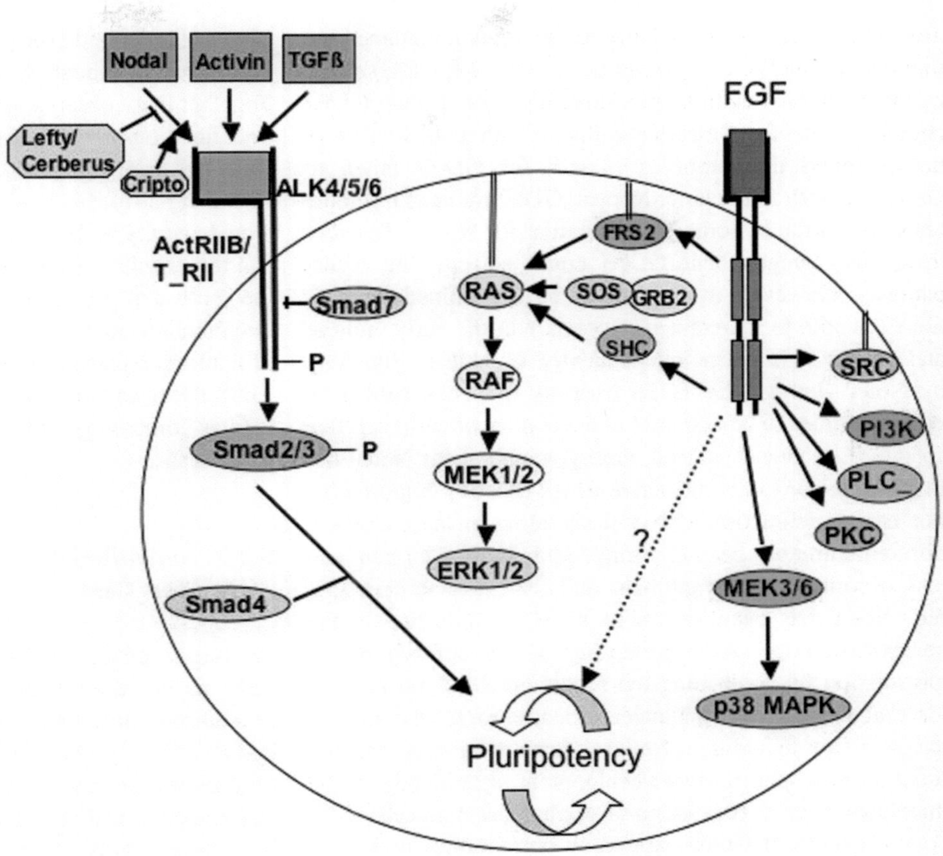

Fig. 5 Summary of maintenance of pluripotency in human embryonic stem cells and mouse epiblast stem cells. Nodal/Activin signaling co-operates with FGF to maintain pluripotency and self-renewal both in human ES cells and in mouse EpiSCs, the pluripotent stem cells derived from the late mouse epiblast. This parallels the requirement for these factors by the late epiblast stage in vivo. Research is ongoing into the mechanisms by which these pathways interact with the core transcriptional machinery of pluripotency

cell membrane, which interacts with the MAPK/AKT signaling pathways [42] suggesting that this extracellular modulator can also link Nodal to other signaling pathways. Altogether, these observations lead to the conclusion that TGFβ signaling is a complex pathway, which complicates the analysis of the requirement for and consequences of its activity.

The diversity of mechanisms controlled by TGFβ is also a direct consequence of the large number of proteins able to bind the Smad proteins. Indeed, more than 50 Smad partners have been characterized, including transcription factors (such as foxH1, Runx1 GATA3), repressors of transcription (E2F4, HoxC9), calcium binding protein (such as calmodulin), and intracellular trafficking signals (importin) [42].

Consequently, the level of expression of each of these factors may define the function of TGFβ signaling in a specific cell type, and may explain how TGFβ signaling can induces opposite effects on different cells (i.e., increase proliferation or cause growth arrest in cancer cell lines, or maintain pluripotency or drive mesendoderm differentiation in pluripotent cells). In the following section, we analyze the function of Activin/Nodal/TGFβ signaling in maintaining the pluripotent status of hESCs and the epiblast of mammalian embryos at post-implantation stages.

3.2.2 Function of Activin/Nodal/TGFβ Signaling During Early Mouse Development

TGFβ signaling components have distinct developmental roles, despite their apparently similar signaling effects. Genetic studies in the mouse have shown that embryos mutant for TGFβ1 or for the receptor TβRII died at mid-gestation, whereas those mutant for TGFβ2 or TGFβ3 died during perinatal development [43]. Consequently, these TGFβ components are not necessary for maintenance of pluripotency or in the control of early cell fate specification, which occur at earlier stages of development. Similar studies have shown that Activin A–deficient mice develop to term but die within 24 h of birth [44], lacking whiskers and lower incisors and with defects in their secondary palates, including cleft palate. Therefore, expression of Activin A is not essential for mesoderm formation in mice (in contrast with observations in lower vertebrates). Only BMPs, GDFs, and Nodal appear to be necessary for early mouse development [45–47]. BMP signaling is required for germ cell specification, proliferation of the epiblast, and differentiation of extra-embryonic tissues, including the visceral endoderm. The function of GDF in early development is less clear. In one recent study, GDF3 appeared to act as a BMP inhibitor and thus maintained the pluripotent state of embryonic stem cells [48]. In a second study, gain of function experiments showed that GDF3 could activate the Smad2/3 pathway, and genetic studies demonstrated that embryos mutant for GDF3 failed to gastrulate [49]. In the latter respect, GDF3 mutants resemble embryos with a hypomorphic mutation for Nodal. Together these data suggest that GDF3 could activate the Nodal pathway. However, it remains to be determined whether GDF3 is able to maintain pluripotency in the early embryo and in ESCs in vitro independently of Nodal. Signaling by Nodal through the TGFβ pathway has been shown by genetic studies to be essential in the control of early cell fate specification events. Indeed, embryos mutant for Nodal fail to gastrulate or form a primitive streak [50, 51]. Importantly, this essential function of Nodal signaling in mesendoderm differentiation has been confirmed and clarified by gain and loss of function in amphibian and fish development [52, 53]. Nodal has been shown to be also implicated in the anterior-posterior (A-P) patterning of the embryo before gastrulation by controlling the specification of the anterior visceral endoderm [54], which is essential for head and trunk development. In addition, Nodal appears to have an essential function in trophectoderm development, specifically in the maintenance of its population of trophoblast stem cells [55].

The pattern of Nodal expression corroborates its known roles. Nodal expression in mouse embryos is first seen throughout the late epiblast layer just after implantation [51, 56], and at the early streak stage, Nodal expression is disappears from the anterior part of the embryo although it remains strongly expressed in the posterior part of the embryo where the primitive streak has formed. Nascent mesoderm quickly loses Nodal expression. This dynamic pattern of expression could indicate a function for Nodal before and during gastrulation. This hypothesis has been confirmed by a detailed analysis of the tissues generated in Nodal mutants. In the absence of Nodal, expression of posterior markers of differentiation including Wnt3, FGF8, and Brachyury (T) cannot be detected in the epiblast [57], confirming an essential function for Nodal in mesendoderm specification. However these absences cannot be explained by a general inhibition of differentiation because the epiblast cell population in Nodal mutant is substantially decreased and the pluripotent marker OCT4 is not expressed [58, 57]. These observations suggest that the pregastrulation arrest of Nodal-deficient embryos could instead reflect an impaired pluripotency.

This hypothesis has recently been strongly reinforced by studies showing that Nodal is necessary to maintain the expression of pluripotent markers in the epiblast before gastrulation. Most notably, absence of Nodal results in ectopic expression of neuroectoderm markers [59, 60]. In addition, two independent studies have recently shown that pluripotent stem cells derived from the post-implantation late epiblast layer are dependent on Activin/Nodal signaling [37, 38]. Together these observations suggest that Nodal may have distinct functions in the mouse embryo proper at the late epiblast stage and during gastrulation. Its initial function would be to maintain pluripotency of the epiblast prior gastrulation. Then Nodal activity in the posterior part of the embryo would control the primitive streak formation. However, understanding the molecular mechanisms by which Nodal achieves these two functions in vivo remains a challenge due to the technical difficulties to perform large-scale studies on mouse embryos. Thus, the recent availability of mouse epiblast stem cells as well as continuing studies on hESCs could help resolve this major issue.

3.2.3 Nodal/Activin Signaling in Mouse Embryonic Stem Cells

Although Activin itself does not seem to be required during early mouse development, recombinant Activin increases mesodermal differentiation of mouse ES cells as embryoid bodies [23, 61], and TGF-β signaling is required for mES proliferation in culture [62]. Activin may substitute for the physiological activity of Nodal, which plays multiple roles in mouse embryogenesis. Although recombinant Nodal does not need to be added to mES cell medium to maintain pluripotency, Nodal seems to be produced by mESCs themselves as an autocrine signal. Blockade of Smad 2/3 phosphorylation with SB-431542 greatly decreases the proliferation of mouse ES cells, although in contrast to human ES or mouse EpiSCs this does not cause differentiation [62].

3.2.4 Functions of the Activin/Nodal/TGFβ Signaling Pathway in Human Embryonic Stem Cells and in Mouse Epiblast Stem Cells

There is now strong evidence that TGFβ signaling is involved in the control of hESC pluripotency. Members of the TGFβ family are less efficient in driving differentiation of hESCs than in maintaining pluripotency [63]. Feeder-free culture conditions invariably involve a source of TGFβ [64, 65] and, moreover, a large number of gene expression profiling experiments indicate a function for TGFβ signaling in hESCs [66–70]. In addition, constitutive expression of Nodal in hESCs is sufficient to block differentiation of hESCs grown as EBs in a chemically defined medium [34]. Finally inhibition of Nodal by overexpression of Lefty increases neuroectoderm differentiation, and blockade of TGFβ signaling by the pharmacological inhibitor SB4315452 systematically induces differentiation of hESCs regardless of the culture conditions used [35, 71]. Together, these observations indicate that Activin/Nodal/TGFβ signaling maintains pluripotency of hESCs. Interestingly, high doses of Activin in combination with serum induce differentiation of hESCs into mesoderm and endoderm [72], confirming that Activin/Nodal/TGFβ signaling also has an essential function in human differentiation. However, high doses of Activin are not sufficient to drive differentiation of hESCs grown in a chemically defined medium devoid of serum [36]. This observation suggests that co-factors are necessary for Activin signaling to induce differentiation and thus that its function in pluripotency is more complex. Studies at the molecular will be needed to understand the mechanisms by which Activin/Nodal signaling can both maintain pluripotency in hESCs and induce mesoderm and endoderm differentiation.

Clarification of the roles of specific components of the TGFβ signaling pathway (i.e., Smads and their binding partners) in hESCs and during mesoderm and endoderm differentiation, and identification of the target genes controlled by Smad2/3 transcriptional complexes will represent key steps to reveal the nature of the mechanisms involved in this dual function.

The recent derivation of EpiSCs might also offer new possibilities to address this question [37, 38]. This novel type of embryonic stem cell shares many features with hESCs including, their low clonality, a low capacity to colonize embryos at the blastocyst stage, the capacity to differentiate into trophectoderm when grown in the presence of BMP4, and a strict dependency on Activin/Nodal/TGFβ to maintain their pluripotent status. Importantly, microarray studies have shown that the gene expression profile of EpiSCs resembles that of their in vivo counterpart (the late epiblast layer), suggesting that EpiSCs maintain certain native characteristics during the process of derivation. Moreover, the target genes of OCT4 more closely resemble those of hESCs. Taken together, these observations suggest that hESCs are more closely related to pluripotent cells of post-implantation stages rather than the ICM.

However, it is important to underline existing differences between EpiSCs and hESCs. Contrary to EpiSCs, hESCs express the ICM marker Rex-1 and they display alkaline phophatase activity, while they do not express the ectoderm marker FGF5 [73]. Importantly, hESCs are derived by culturing embryos from the blastocyst stage, which implies that the cultured human ICM cells progress in vitro towards an epiblast stage during the process of derivation. Such in vitro progression could have consequences for the properties of hESCs, especially if one considers that the epiblast stage lasts several days in human development against one day in the mouse embryo. Therefore, distinct epiblast states could exist in human development, which could be translated into different states of pluripotency, thereby explaining the substantial variation in differentiation potential observed between different hESC lines. This hypothesis reinforces the necessity of developing robust method of hESC derivation, since this initial step might define the stage at which the development of human ICM cells is immortalized in vitro by the derivation process. Comparison of the epigenetic status of hESCs, EpiSCs, and also mESCs will provide valuable information concerning their embryonic identity since major epigenetic changes occur after implantation in mammalian development. Finally generation of EpiSCs in other mammalian species including non-human primate and domestic species will also help to resolve these questions.

3.2.5 FGF Signaling Pathway

The fibroblast growth factor (FGF) family includes 22 members (FGF1–22) that signal through four distinct receptors (FGF R1–R4). The genes encoding these receptors can produce numerous isoforms through alternative splicing. Interaction between FGFs and their receptors is promiscuous, as one receptor can be activated by a large number of FGF ligands. Importantly, stable interaction between FGFs and their receptors requires heparin or heparin sulphate.Receptor binding of FGFs activates an intrinsic receptor tyrosine kinase activity, which causes phosphorylation of multiple tyrosines on the receptor. These sites then recruit diverse proteins containing the SH2 (src homology-2) or PTB (phosphotyrosine binding) domains, which allows signaling complexes to be recruited and assembled. This results in a cascade of other phosphorylation events, through which FGFs are able to activate multiple signaling pathways, including the RAS-MAP kinase pathway (which encompass the ERK1/2, p38 and JNK kinases), the PI3 kinase-AKT pathway, and the PLCγ pathway [74].

Importantly, FGF signaling can also interact with other pathways, including the TGFβ pathway through phosphorylation of the Smad proteins, and the WNT pathway through the control of GSK3B phosphorylation by PI3 kinase and through the control of the expression of E-cadherin by the WNT-dependent transcription factor Snail [74]. This crosstalk between signaling pathways and the large number of downstream effectors controlled by FGFs partially account for the diversity of biological process they control, including proliferation, differentiation, or apoptosis during embryonic development and in adult tissues. This complexity also increases the difficulty of achieving a definitive understanding of the roles of FGF signaling in pluripotency. Here we propose to analyze the possible function of FGF in hESCs and in mEpiSCs.

3.2.6 Function of FGF Signaling During Early Mouse Development

A large number of FGF genes and their receptors have been knocked out in the mouse using gene targeting in embryonic stem cells [75]. Among FGF ligands, only absence of FGF4 and FGF8 alters early development [76, 77]. Embryos mutant for FGF4 die shortly after implantation due to abnormal trophectoderm differentiation and too slow ICM proliferation, while embryos mutant for FGF8 undergo abnormal gastrulation with reduced mesoderm differentiation. FGFR1 and FGFR2 are also necessary for early development [78–80], while absence of FGFR3 results in skeletal defects during later development [81, 82]. Interestingly, embryo mutants for FGF1R failed to gastrulate due to a failure of cell migration through the primitive streak. Absence of FGFR2 blocks early postimplantation development between implantation and the formation of the egg cylinder [80], confirming an essential role for FGF signaling in the early events controlling blastocyst formation. However, chimera experiments have shown that FGF1R mutant mouse ES cells fail to migrate properly through the anterior part of the primitive streak; they can colonize the posterior part of the streak, but then they form ectopic neural tube [83]. Consequently, FGF signaling is essential not only for mesoderm germ layer specification but also for migration and adhesion of the mesoderm and endoderm cells during gastrulation. These observations underscore the multiplicity of mechanisms controlled by FGF signaling and thus the difficulty to establish a clear function for FGF signaling in early development. Notwithstanding these difficulties, the genetic studies provide only limited evidence for an essential function for FGF signaling in the regulation of pluripotency, as only FGF4 and FGFR2 deficiency results in arrest before gastrulation. Moreover, the existing mutations do not provide detailed information concerning the molecular mechanisms by which FGF signaling affects germ layer specification. Consequently, more detailed studies of FGF4 and FGF8 and FGFR1 and FGFR2 mutants are required to define more precisely the function of FGF signaling in the early embryo, including analysis of their effect on expression of pluripotency markers such as Nanog, OCT4, and Sox2.

3.2.7 FGF Signaling and Mouse Embryonic Stem Cells

Mouse ES cells are sensitive to fibroblast growth factor (FGF) signaling and indeed express FGF4, though, in contrast to human ES cells and mouse EpiSCs, FGF has no pluripotency-maintaining effect on mouse ES cells [26].

In early development, FGF signaling is required for cell proliferation after implantation, but not for formation of the mouse blastocyst and ICM. While mutant embryos lacking FGF receptor 1 die before or during gastrulation, it is possible to grow FGFR-null mouse ES cells [79]. This contrasts with human ES cells, which require FGF for cell proliferation, but is unsurprising given the ICM-like properties of mouse ES cells.

Although FGF has many downstream effects in common with LIF, activating the same signaling cascades through PI3 kinase and the ERK and MAPK pathways, FGF does not appear activate of STAT3 in mESCs, the element of LIF signaling active in maintenance of mESC pluripotency. Autocrine FGF signaling by FGF4 is active in neural differentiation from mESCs, as well as modulating the BMP signal in differentiation to non-neural fates [84].

3.2.8 Functions of FGF Signaling Pathway in Self-Renewal of Human Embryonic Stem Cells and in Mouse Epiblast Stem Cells

Most of the culture media that have been used to cultivate hESCs are supplemented with FGF2, which is essential for hESC survival and growth [31]. High doses of FGF2 appear to be sufficient for maintaining pluripotency [85] of hESCs grown in commercial serum replacer and on Matrigel. FGF2 is also required in combination with Activin to propagate mouse EpiSCs in vitro without serum components or feeder layers in simplified [37], chemically defined culture medium (CDM) [23]. However, FGF is not sufficient for maintenance of pluripotency when hESCs are grown in CDM without Matrigel, and more recent studies have demonstrated that FGF is not able to maintain the pluripotency of hESCs without Activin [36]. Nevertheless, FGF blocks differentiation induced by the BMP4-like activity contained in media containing serum components, and thus its action on pluripotency appears to interact with other growth factors that may be present in complex culture environments [86]. Importantly, high doses of Activin are sufficient to maintain the pluripotent status of hESCs grown for several passages in a chemically defined medium, even in the absence of FGF signaling [36]. Thus, the combination of Activin and FGF is only required for long-term maintenance of hESCs, suggesting that the function of FGF lies more in sustaining self renewal of embryonic stem cells in vitro (e.g., by promoting their proliferation and adhesion) rather than by maintaining their capacity for differentiation. Interestingly, the addition of FGF to cultures of rat EpiSCs induced their differentiation, revealing that rat EpiSCs require only Activin to maintain their pluripotency [37]. This finding indicates that the role of FGF in self-renewal of pluripotent stem cells is species dependent. Taken together, these observations show that the precise functions of FGF in hESCs and EpiSCs remain unclear and that further studies at the molecular level are required to define the consequences of FGF signaling in pluripotent cells. However, the complexity of FGF signaling renders these studies difficult. The inefficiency of homologous recombination in hESCs compounds this difficulty and forces researchers to rely on chemical inhibitors of downstream pathways of FGF signaling to acquire new insights. The potential for harnessing mouse functional genomics through the use of EpiSCs offers particular promise for addressing this problem.

4 Other Signaling Pathways Influencing Embryonic Stem Cells

In addition to responding to the signals described above which control maintenance of pluripotency, embryonic stem cells are sensitive to numerous other factors, which influence proliferation, survival, and differentiation.

4.1 Wnt Signaling

The Wnt family of signaling proteins is conserved throughout animal evolution and have many roles during development and tissue homeostasis [87]. Canonical WNT signaling acts by blocking degradation of β-catenin downstream of glycogen synthase kinase, (GSK) 3. β-Catenin drives transcription via TCF-LEF proteins as well as being involved in cell-cell adhesion via the cadherins. This gives Wnt signaling both a transcriptional and potentially morphogenetic effect. The involvement of GSK3 also links Wnt signaling to metabolic processes downstream of insulin and P13 kinase.

WNT signaling is involved in embryogenesis during gastrulation and patterning of mesoderm, and plays a part in regulating various stem cell niches in adulthood.

Both mouse and human embryonic stem cells are sensitive to WNT signaling, and it appears to act as a paracrine signal, influencing pluripotency and self-renewal. In mouse ES cells, this appears to be through increased expression of STAT3, thus increasing the pluripotency maintaining effect of LIF [88, 89]. It is also possible that both pathways cause expression of c-Myc, which promotes both proliferation and telomerase activity [90].

Some research has also indicated an activity for WNT signaling in self-renewal of human embryonic stem cells [91]. WNT signaling alone appears insufficient to maintain pluripotency, Its role seems to increase the rate of cell proliferation, both in the undifferentiated state (in cells exposed to other pluripotency-maintaining factors) and during differentiation [92]. This may parallel its role in adult stem cell niches.

Noncanonical WNT signaling via calcium levels could also play a role in human ES cells, which appear to show a strong calcium response when exposed to serum [93], although the significance of this is not clearly understood.

4.2 Insulin-Like Growth Factors

Most culture medium used for the growth of mouse and human ES cells contains high concentrations of insulin. Aside from encouraging uptake of glucose, activation of the insulin receptor and IGF1 receptor seems to directly encourage ES cell self-renewal, an effect that seems to interact with FGF and EGF signaling [94, 95]. Such work towards understanding which receptors are activated in cultured of ES cells is valuable to improving methods of cell culture, as using the correct signals (rather than close analogues) may reduce the potential to induce epigenetic changes over time due to selection of cells culture-adapted to grow in suboptimal conditions.

4.3 Factors Improving ES Cell Survival

Human ES and mouse epiblast stem cells appear to have poorer clonality than mouse ES cells – in other words, a poor survival rate if separated into single cells, even if grown in otherwise optimal conditions for maintenance of pluripotency. This may imply a direct need for cell–cell contact [96], and suggests that they require some additional forms of signaling to encourage growth and prevent apoptosis.

Screening of mitogenic agents present in serum led to the finding that a combination of sphingosine-1-phosphate (S1P) and platelet-derived growth factor (PDGF) could substitute for the presence of serum in feeder- or Matrigel-based culture of human ES cells [97]. These factors are known to drive proliferation of a number of cell types, and in human ES cells appear to have an anti-apoptotic effect [93].

Another molecule improving ES cell survival is pleotrophin, a paracrine factor secreted by feeder cells [98]. This seems to improve the success rate of clonal propagation of human ES cells. Inhibition of Rho-assisted kinase (ROCK) also appears to significantly reduce cell death in dissociated hESCs, indicating something of the mechanism behind the observed apoptosis and offering a valuable experimental tool [99, 100].

5 Conclusions and Future Directions

A significant body of knowledge has now been amassed concerning what signaling pathways are required to maintain pluripotency and self-renewal in both mouse and human embryonic stem cells. This has allowed researchers to culture these cells in better-defined conditions, improving the reproducibility of results and the prospects for the range of basic and clinical applications that ESCs promise. Often, several different conditions appear to give the same outcome, which may reflect the inter-related nature of many of the pathways coupled with the wide variety of growth factors secreted by the embryonic stem cells themselves. A large dose of one factor may stimulate production of other factors, or drive activation of other related pathways.

The differences in growth factors responsible for pluripotency between mouse and human ESCs were initially surprising. There was no obvious explanation for how two cell types with apparently similar developmental potency and transcriptional machinery could require such different signals to maintain them in culture. This now appears to be resolved by the knowledge that the different signals reflect stage-specific differences in the development of pluripotent cells of the embryo itself and the stem cells that can be derived from them.

The embryonic mouse employs LIF/STAT3 signaling to maintain the inner cell mass during the developmental stasis of diapause, a specific adaptation to their mode of reproduction not found in humans. The late human blastocyst only relies on LIF signaling for implantation, and then only as a maternal factor. This seems to explain why LIF can maintain mouse embryonic stem cells in culture, but has no effect on human ES cells. The conditions required for human ES cell culture, Nodal/Activin and FGF, seem to replicate the signals required to maintain pluripotency and proliferation in the mouse epiblast. The fact that these same conditions do indeed support in vitro culture of mouse epiblast stem cells, which closely resemble human ESCs in their behavior and responsiveness, suggests that this mechanism is conserved between species. This explanation of the differences and the finding of apparent commonality between mouse and human pluripotent tissues opens up many possibilities.

Early researchers noted how early vertebrate embryos were similar in appearance, only diverging in the later stages of development diverge to produce the adult organisms of different species. It now seems that certain molecular aspects of the initial steps are also widely divergent even between different mammals, reflecting subtly different modes of reproduction.

This understanding, and the likely conservation of TGFβ signaling in maintenance of pluripotency at the epiblast stage (as in EpiSCs and hESCs) may aid the derivation of embryonic stem cells from previously intractable mammalian species.

The fact that mouse and human pluripotent tissues appear to progress through a similar molecular state during embryogenesis suggests that the signals required for patterning and morphogenesis are also likely to be conserved. This is good news for the field, as it seems likely that our increasing knowledge of mouse development and acquisition of massive genetic resources can be help us towards understanding how to achieve directed differentiation of useful adult tissues from human pluripotent cells.

Many basic questions still remain to be answered, and solving them will give us an improved understanding of how to sustain pluripotency and to achieve differentiation in chemically defined conditions. It will also deepen our understanding of our own early development.

We currently have a good understanding of the signaling requirements of embryonic stem cells, and also of the transcription factors that form the core transcriptional circuit regulating pluripotency. There are also data on the epigenetic changes that occur during differentiation. The major remaining task is to define exactly how activation of the growth factor signaling pathways active in pluripotent cells maintains the pluripotent state.

Knowing the identities of core transcription factors has already enabled researchers to reprogram adult tissues to an ES like state, however such methods have a low success rate and may not be acceptable for medical use in their current form. A clearer understanding of how pluripotency mechanisms act might allow novel approaches to achieve reprogramming of mature cells to a pluripotent state.

Understanding the differences between mouse ES cells and mouse epiblast stem cells will be of great interest as these are cells of the same species in which pluripotency is maintained downstream of different signal pathways. These two cell types could show what changes occur between blastocyst and epiblast. Understanding these changes might indicate means to reverse them, potentially unlocking greater plasticity for human ES cells, or allowing production of germline chimaeras from other species' stem cells derived from the epiblast. The principles controlling fate decisions in embryonic stem cells may also be involved in later developmental processes, and in the activity of adult stem cells.

Finally, a deeper understanding of the core transcriptional circuits for the differentiated progeny of stem cells may enable shortcuts to producing intermediate or mature cells using approaches akin to reprogramming. Future developments in this field are therefore likely to be of great interest both in pure science and for their medical applications.

References

1. Tarkowski AK. Mouse chimaeras developed from fused eggs. Nature. 1961;190:857–60.
2. Gardner RL. Mouse chimeras obtained by the injection of cells into the blastocyst. Nature. 1968;220(5167):596–7.
3. Martin GR. Teratocarcinomas and mammalian embryogenesis. Science. 1980;209(4458):768–76.
4. Evans MJ, Kaufman MH. Establishment in culture of pluripotential cells from mouse embryos. Nature. 1981;292(5819):154–6.
5. Martin GR. Isolation of a pluripotent cell line from early mouse embryos cultured in medium conditioned by teratocarcinoma stem cells. Proc Natl Acad Sci U S A. 1981;78(12):7634–8.

6. Smith AG, Hooper ML. Buffalo rat liver cells produce a diffusible activity which inhibits the differentiation of murine embryonal carcinoma and embryonic stem cells. Dev Biol. 1987;121(1): 1–9.
7. Smith AG, Heath JK, Donaldson DD, et al. Inhibition of pluripotential embryonic stem cell differentiation by purified polypeptides. Nature. 1988;336(6200):688–90.
8. Williams RL, Hilton DJ, Pease S, et al. Myeloid leukaemia inhibitory factor maintains the developmental potential of embryonic stem cells. Nature. 1988;336(6200):684–7.
9. Gearing DP, Gough NM, King JA, et al. Molecular cloning and expression of cDNA encoding a murine myeloid leukaemia inhibitory factor (LIF). EMBO J. 1987;6(13):3995–4002.
10. Nichols J, Evans EP, Smith AG. Establishment of germ-line-competent embryonic stem (ES) cells using differentiation inhibiting activity. Development. 1990;110(4):1341–8.
11. Bongso A, Fong CY, Ng SC, Ratnam S. Isolation and culture of inner cell mass cells from human blastocysts. Hum Reprod. 1994;9(11):2110–7.
12. Daheron L, Opitz SL, Zaehres H, et al. LIF/STAT3 signaling fails to maintain self-renewal of human embryonic stem cells. Stem Cells. 2004;22(5):770–8.
13. Humphrey RK, Beattie GM, Lopez AD, et al. Maintenance of pluripotency in human embryonic stem cells is STAT3 independent. Stem Cells. 2004;22(4):522–30.
14. Nichols J, Davidson D, Taga T, Yoshida K, Chambers I, Smith A. Complementary tissue-specific expression of LIF and LIF-receptor mRNAs in early mouse embryogenesis. Mech Dev. 1996;57(2):123–31.
15. Nishinakamura R, Matsumoto Y, Matsuda T, et al. Activation of Stat3 by cytokine receptor gp130 ventralizes Xenopus embryos independent of BMP-4. Dev Biol. 1999;216(2):481–90.
16. Lass A, Weiser W, Munafo A, Loumaye E. Leukemia inhibitory factor in human reproduction. Fertil Steril. 2001;76(6):1091–6.
17. Aghajanova L. Leukemia inhibitory factor and human embryo implantation. Ann N Y Acad Sci. 2004;1034:176–83.
18. Nichols J, Chambers I, Taga T, Smith A. Physiological rationale for responsiveness of mouse embryonic stem cells to gp130 cytokines. Development. 2001;128(12):2333–9.
19. Niwa H, Burdon T, Chambers I, Smith A. Self-renewal of pluripotent embryonic stem cells is mediated via activation of STAT3. Genes Dev. 1998;12(13):2048–60.
20. Burdon T, Stracey C, Chambers I, Nichols J, Smith A. Suppression of SHP-2 and ERK signalling promotes self-renewal of mouse embryonic stem cells. Dev Biol. 1999;210(1):30–43.
21. Buehr M, Smith A. Genesis of embryonic stem cells. Philos Trans R Soc Lond B Biol Sci. 2003;358(1436):1397–402; discussion 402.
22. Ying QL, Nichols J, Chambers I, Smith A. BMP induction of Id proteins suppresses differentiation and sustains embryonic stem cell self-renewal in collaboration with STAT3. Cell. 2003;115(3):281–92.
23. Johansson BM, Wiles MV. Evidence for involvement of activin A and bone morphogenetic protein 4 in mammalian mesoderm and hematopoietic development. Mol Cell Biol. 1995;15(1): 141–51.
24. Ying Y, Qi X, Zhao GQ. Induction of primordial germ cells from murine epiblasts by synergistic action of BMP4 and BMP8B signaling pathways. Proc Natl Acad Sci U S A. 2001;98(14): 7858–62.
25. Benezra R, Davis RL, Lockshon D, Turner DL, Weintraub H. The protein Id: a negative regulator of helix-loop-helix DNA binding proteins. Cell. 1990;61(1):49–59.
26. Hollnagel A, Oehlmann V, Heymer J, Ruther U, Nordheim A. Id genes are direct targets of bone morphogenetic protein induction in embryonic stem cells. J Biol Chem. 1999;274(28):19838–45.
27. Suzuki A, Raya A, Kawakami Y, et al. Nanog binds to Smad1 and blocks bone morphogenetic protein-induced differentiation of embryonic stem cells. Proc Natl Acad Sci U S A. 2006;103(27):10294–9.
28. Qi X, Li TG, Hao J, et al. BMP4 supports self-renewal of embryonic stem cells by inhibiting mitogen-activated protein kinase pathways. Proc Natl Acad Sci U S A . 2004;101(16):6027–32.
29. Sirard C, de la Pompa JL, Elia A, et al. The tumor suppressor gene Smad4/Dpc4 is required for gastrulation and later for anterior development of the mouse embryo. Genes Dev. 1998;12(1):107–19.
30. Thomson JA, Itskovitz-Eldor J, Shapiro SS, et al. Embryonic stem cell lines derived from human blastocysts. Science. 1998;282(5391):1145–7.
31. Amit M, Carpenter MK, Inokuma MS, et al. Clonally derived human embryonic stem cell lines maintain pluripotency and proliferative potential for prolonged periods of culture. Dev Biol. 2000;227(2):271–8.
32. Xu RH, Chen X, Li DS, et al. BMP4 initiates human embryonic stem cell differentiation to trophoblast. Nat Biotechnol. 2002;20(12):1261–4.
33. Beddington RS, Robertson EJ. An assessment of the developmental potential of embryonic stem cells in the midgestation mouse embryo. Development. 1989;105(4):733–7.
34. Vallier L, Reynolds D, Pedersen RA. Nodal inhibits differentiation of human embryonic stem cells along the neuroectodermal default pathway. Dev Biol. 2004;275(2):403–21.
35. James D, Levine AJ, Besser D, Hemmati-Brivanlou A. TGF {beta}/activin/nodal signaling is necessary for the maintenance of pluripotency in human embryonic stem cells. Development. 2005;132(6):1273–82.
36. Vallier L, Alexander M, Pedersen RA. Activin/Nodal and FGF pathways cooperate to maintain pluripotency of human embryonic stem cells. J Cell Sci. 2005;118(Pt 19):4495–509.
37. Brons IG, Smithers LE, Trotter MW, et al. Derivation of pluripotent epiblast stem cells from mammalian embryos. Nature. 2007;448(7150):191–5.
38. Tesar PJ, Chenoweth JG, Brook FA, et al. New cell lines from mouse epiblast share defining features with human embryonic stem cells. Nature. 2007;448(7150):196–9.
39. Massague J, Seoane J, Wotton D. Smad transcription factors. Genes Dev. 2005;19(23):2783–810.
40. Miyazawa K, Shinozaki M, Hara T, Furuya T, Miyazono K. Two major Smad pathways in TGF-beta superfamily signalling. Genes Cells. 2002;7(12):1191–204.
41. Schier AF. Nodal signaling in vertebrate development. Annu Rev Cell Dev Biol. 2003;19:589–621.
42. Strizzi L, Bianco C, Normanno N, Salomon D. Cripto-1: a multifunctional modulator during embryogenesis and oncogenesis. Oncogene. 2005;24(37):5731–41.
43. Goumans MJ, Mummery C. Functional analysis of the TGFbeta receptor/Smad pathway through gene ablation in mice. Int J Dev Biol. 2000;44(3):253–65.
44. Matzuk MM, Kumar TR, Vassalli A, et al. Functional analysis of activins during mammalian development. Nature. 1995;374(6520):354–6.
45. Winnier G, Blessing M, Labosky PA, Hogan BL. Bone morphogenetic protein-4 is required for mesoderm formation and patterning in the mouse. Genes Dev. 1995;9(17):2105–16.
46. Mishina Y, Suzuki A, Ueno N, Behringer RR. Bmpr encodes a type I bone morphogenetic protein receptor that is essential for gastrulation during mouse embryogenesis. Genes Dev. 1995;9(24):3027–37.
47. Lawson KA, Dunn NR, Roelen BA, et al. Bmp4 is required for the generation of primordial germ cells in the mouse embryo. Genes Dev. 1999;13(4):424–36.

48. Levine AJ, Brivanlou AH. GDF3, a BMP inhibitor, regulates cell fate in stem cells and early embryos. Development. 2006;133(2): 209–16.
49. Chen C, Ware SM, Sato A, et al. The Vg1-related protein Gdf3 acts in a Nodal signaling pathway in the pre-gastrulation mouse embryo. Development. 2006;133(2):319–29.
50. Conlon FL, Barth KS, Robertson EJ. A novel retrovirally induced embryonic lethal mutation in the mouse: assessment of the developmental fate of embryonic stem cells homozygous for the 413.d proviral integration. Development. 1991;111(4):969–81.
51. Conlon FL, Lyons KM, Takaesu N, et al. A primary requirement for nodal in the formation and maintenance of the primitive streak in the mouse. Development. 1994;120(7):1919–28.
52. Jones CM, Kuehn MR, Hogan BL, Smith JC, Wright CV. Nodal-related signals induce axial mesoderm and dorsalize mesoderm during gastrulation. Development. 1995;121(11):3651–62.
53. Rebagliati MR, Toyama R, Fricke C, Haffter P, Dawid IB. Zebrafish nodal-related genes are implicated in axial patterning and establishing left-right asymmetry. Dev Biol. 1998;199(2):261–72.
54. Robertson EJ, Norris DP, Brennan J, Bikoff EK. Control of early anterior-posterior patterning in the mouse embryo by TGF-beta signalling. Philos Trans R Soc Lond B Biol Sci. 2003;358(1436):1351–7; discussion 7.
55. Guzman-Ayala M, Ben-Haim N, Beck S, Constam DB. Nodal protein processing and fibroblast growth factor 4 synergize to maintain a trophoblast stem cell microenvironment. Proc Natl Acad Sci U S A. 2004;101(44):15656–60.
56. Varlet I, Collignon J, Robertson EJ. nodal expression in the primitive endoderm is required for specification of the anterior axis during mouse gastrulation. Development. 1997;124(5):1033–44.
57. Brennan J, Lu CC, Norris DP, Rodriguez TA, Beddington RS, Robertson EJ. Nodal signalling in the epiblast patterns the early mouse embryo. Nature. 2001;411(6840):965–9.
58. Varlet I, Collignon J, Norris DP, Robertson EJ. Nodal signaling and axis formation in the mouse. Cold Spring Harb Symp Quant Biol. 1997;62:105–13.
59. Mesnard D, Guzman-Ayala M, Constam DB. Nodal specifies embryonic visceral endoderm and sustains pluripotent cells in the epiblast before overt axial patterning. Development. 2006;133(13):2497–505.
60. Camus A, Perea-Gomez A, Moreau A, Collignon J. Absence of Nodal signaling promotes precocious neural differentiation in the mouse embryo. Dev Biol. 2006;295(2):743–55.
61. Wiles MV, Johansson BM. Embryonic stem cell development in a chemically defined medium. Exp Cell Res. 1999;247(1):241–8.
62. Ogawa K, Saito A, Matsui H, et al. Activin-Nodal signaling is involved in propagation of mouse embryonic stem cells. J Cell Sci. 2007;120(Pt 1):55–65.
63. Schuldiner M, Yanuka O, Itskovitz-Eldor J, Melton DA, Benvenisty N. Effects of eight growth factors on the differentiation of cells derived from human embryonic stem cells. Proc Natl Acad Sci U S A. 2000;97(21):11307–12.
64. Amit M, Shariki C, Margulets V, Itskovitz-Eldor J. Feeder layer- and serum-free culture of human embryonic stem cells. Biol Reprod. 2004;70(3):837–45.
65. Xu C, Inokuma MS, Denham J, et al. Feeder-free growth of undifferentiated human embryonic stem cells. Nat Biotechnol. 2001;19(10):971–4.
66. Calhoun JD, Rao RR, Warrenfeltz S, et al. Transcriptional profiling of initial differentiation events in human embryonic stem cells. Biochem Biophys Res Commun. 2004;323(2):453–64.
67. Rosler ES, Fisk GJ, Ares X, et al. Long-term culture of human embryonic stem cells in feeder-free conditions. Dev Dyn. 2004;229(2):259–74.
68. Miura T, Luo Y, Khrebtukova I, et al. Monitoring early differentiation events in human embryonic stem cells by massively parallel signature sequencing and expressed sequence tag scan. Stem Cells Dev. 2004;13(6):694–715.
69. Brandenberger R, Wei H, Zhang S, et al. Transcriptome characterization elucidates signaling networks that control human ES cell growth and differentiation. Nat Biotechnol. 2004;22(6): 707–16.
70. Zeng X, Miura T, Luo Y, et al. Properties of pluripotent human embryonic stem cells BG01 and BG02. Stem Cells. 2004;22(3):292–312.
71. Smith JR, Vallier L, Lupo G, Alexander M, Harris WA, Pedersen RA. Inhibition of Activin/Nodal signaling promotes specification of human embryonic stem cells into neuroectoderm. Dev Biol. 2008;313(1):107–17.
72. D'Amour KA, Agulnick AD, Eliazer S, Kelly OG, Kroon E, Baetge EE. Efficient differentiation of human embryonic stem cells to definitive endoderm. Nat Biotechnol. 2005;23(12): 1534–41.
73. Ginis I, Luo Y, Miura T, et al. Differences between human and mouse embryonic stem cells. Dev Biol. 2004;269(2):360–80.
74. Dailey L, Ambrosetti D, Mansukhani A, Basilico C. Mechanisms underlying differential responses to FGF signaling. Cytokine Growth Factor Rev. 2005;16(2):233–47.
75. Rossant J, Ciruna B, Partanen J. FGF signaling in mouse gastrulation and anteroposterior patterning. Cold Spring Harb Symp Quant Biol. 1997;62:127–33.
76. Feldman B, Poueymirou W, Papaioannou VE, DeChiara TM, Goldfarb M. Requirement of FGF-4 for postimplantation mouse development. Science. 1995;267(5195):246–9.
77. Sun X, Meyers EN, Lewandoski M, Martin GR. Targeted disruption of Fgf8 causes failure of cell migration in the gastrulating mouse embryo. Genes Dev. 1999;13(14):1834–46.
78. Yamaguchi TP, Harpal K, Henkemeyer M, Rossant J. fgfr-1 is required for embryonic growth and mesodermal patterning during mouse gastrulation. Genes Dev. 1994;8(24):3032–44.
79. Deng CX, Wynshaw-Boris A, Shen MM, Daugherty C, Ornitz DM, Leder P. Murine FGFR-1 is required for early postimplantation growth and axial organization. Genes Dev. 1994;8(24):3045–57.
80. Arman E, Haffner-Krausz R, Chen Y, Heath JK, Lonai P. Targeted disruption of fibroblast growth factor (FGF) receptor 2 suggests a role for FGF signaling in pregastrulation mammalian development. Proc Natl Acad Sci U S A. 1998;95(9):5082–7.
81. Deng C, Wynshaw-Boris A, Zhou F, Kuo A, Leder P. Fibroblast growth factor receptor 3 is a negative regulator of bone growth. Cell. 1996;84(6):911–21.
82. Weinstein M, Xu X, Ohyama K, Deng CX. FGFR-3 and FGFR-4 function cooperatively to direct alveogenesis in the murine lung. Development. 1998;125(18):3615–23.
83. Ciruna BG, Schwartz L, Harpal K, Yamaguchi TP, Rossant J. Chimeric analysis of fibroblast growth factor receptor-1 (Fgfr1) function: a role for FGFR1 in morphogenetic movement through the primitive streak. Development. 1997;124(14): 2829–41.
84. Kunath T, Saba-El-Leil MK, Almousailleakh M, Wray J, Meloche S, Smith A. FGF stimulation of the Erk1/2 signalling cascade triggers transition of pluripotent embryonic stem cells from self-renewal to lineage commitment. Development. 2007;134(16):2895–902.
85. Xu C, Rosler E, Jiang J, et al. Basic fibroblast growth factor supports undifferentiated human embryonic stem cell growth without conditioned medium. Stem Cells. 2005;23(3):315–23.
86. Xu RH, Peck RM, Li DS, Feng X, Ludwig T, Thomson JA. Basic FGF and suppression of BMP signaling sustain undifferentiated proliferation of human ES cells. Nat Methods. 2005;2(3):185–90.
87. Katoh M, Katoh M. WNT signaling pathway and stem cell signaling network. Clin Cancer Res. 2007;13(14):4042–5.

88. Hao J, Li TG, Qi X, Zhao DF, Zhao GQ. WNT/beta-catenin pathway up-regulates Stat3 and converges on LIF to prevent differentiation of mouse embryonic stem cells. Dev Biol. 2006;290(1):81–91.
89. Ogawa K, Nishinakamura R, Iwamatsu Y, Shimosato D, Niwa H. Synergistic action of Wnt and LIF in maintaining pluripotency of mouse ES cells. Biochem Biophys Res Commun. 2006;343(1):159–66.
90. Kristensen DM, Kalisz M, Nielsen JH. Cytokine signalling in embryonic stem cells. APMIS. 2005;113(11–12):756–72.
91. Sato N, Meijer L, Skaltsounis L, Greengard P, Brivanlou AH. Maintenance of pluripotency in human and mouse embryonic stem cells through activation of Wnt signaling by a pharmacological GSK-3-specific inhibitor. Nat Med. 2004;10(1): 55–63.
92. Dravid G, Ye Z, Hammond H, et al. Defining the role of Wnt/beta-catenin signaling in the survival, proliferation, and self-renewal of human embryonic stem cells. Stem Cells. 2005;23(10): 1489–501.
93. Wong RC, Tellis I, Jamshidi P, Pera M, Pebay A. Anti-apoptotic effect of sphingosine-1-phosphate and platelet-derived growth factor in human embryonic stem cells. Stem Cells Dev. 2007;16(6):989–1001.
94. Bendall SC, Stewart MH, Menendez P, et al. IGF and FGF cooperatively establish the regulatory stem cell niche of pluripotent human cells in vitro. Nature. 2007;448(7157):1015–21.
95. Wang L, Schulz TC, Sherrer ES, et al. Self-renewal of human embryonic stem cells requires insulin-like growth factor-1 receptor and ERBB2 receptor signaling. Blood. 2007;110(12):4111–9.
96. Wong RC, Dottori M, Koh KL, Nguyen LT, Pera MF, Pebay A. Gap junctions modulate apoptosis and colony growth of human embryonic stem cells maintained in a serum-free system. Biochem Biophys Res Commun. 2006;344(1):181–8.
97. Pebay A, Wong RC, Pitson SM, et al. Essential roles of sphingosine-1-phosphate and platelet-derived growth factor in the maintenance of human embryonic stem cells. Stem Cells. 2005;23(10):1541–8.
98. Soh BS, Song CM, Vallier L, et al. Pleiotrophin enhances clonal growth and long-term expansion of human embryonic stem cells. Stem Cells. 2007;25(12):3029–37.
99. Watanabe K, Ueno M, Kamiya D, et al. A ROCK inhibitor permits survival of dissociated human embryonic stem cells. Nat Biotechnol. 2007;25(6):681–6.
100. Peerani R, Rao BM, Bauwens C, et al. Niche-mediated control of human embryonic stem cell self-renewal and differentiation. EMBO J. 2007;26(22):4744–55.

Regulation of Stem Cell Systems by PI3K/Akt Signaling

Tohru Kimura and Toru Nakano

Abstract Stem cells can replenish their own population while supplying the cells necessary to maintain tissue homeostasis. Pluripotent stem cells, which have broader developmental potency than tissue stem cells, are derived from the same source in mice and humans. In this review, we summarize the functions of phosphoinositide-3 kinase (PI3K) and its downstream serine/threonine kinase Akt in a variety of stem cell systems. Primordial germ cells (PGCs), which are embryonic germ cell precursors, are unique in that they acquire pluripotency under cultural and pathological conditions. PGCs lacking *Pten*, which encodes a phosphatase that antagonizes PI3K signaling, give rise to early-onset testicular teratomas in vivo and augment the derivation of pluripotent embryonic germ (EG) cells in vitro. Transient activation of Akt sufficiently recapitulates the effects of *Pten* deficiency on EG cell derivation. Enhanced EG cell derivation is brought about by the Akt-mediated inhibition of the tumor suppressor p53. In embryonic stem (ES) cells, PI3K/Akt signaling plays a pivotal role in maintaining pluripotency in part via transcriptional activation of the pluripotent transcription factor Nanog. In turn, the expression of Tcl1, a cofactor of Akt, is activated by pluripotent transcription factors, including Oct-3/4. Therefore, PI3K/Akt signaling and the transcription factor network constitute the positive feedback circuitry necessary to maintain pluripotency in ES cells. In tissue stem cells, such as hair follicular, intestinal, and hematopoietic stem cells, PI3K/Akt signaling activates quiescent stem cells, leading to the generation of committed progenitors and cancer stem cells. These findings underscore the idea that PI3K/Akt signaling regulates "stemness" in many stem cell systems.

Keywords PI3K · Akt · Pluripotency · Dedifferentiation · Primordial germ cells · EG cells · ES cells · Tissue stem cells · Cancer stem cells

1 PI3K/Akt Signaling

Phosphoinositide-3 kinase (PI3K)/Akt signaling is implicated in the regulation of diverse cellular functions, including cell proliferation, growth, survival, migration, metabolism, angiogenesis, and tumorigenesis [1, 2]. Several growth factors, cell adhesion molecules, and chemokines activate PI3K, leading to the production of phosphatidylinositol (3,4,5)-triphosphate (PIP3) from phosphatidylinositol (4,5)-bisphosphate (PIP2) (Fig. 1). PIP3 then transmits the signal through such downstream effectors as the serine/threonine kinase Akt and the GTPases Rac and Cdc42. The strength and duration of the PI3K/Akt signal is balanced by the tumor suppressor PTEN (phosphatase and tensin homolog deleted on chromosome 10), which catalyzes the dephosphorylation of PIP3 to PIP2 [3]. Akt exerts its physiological and pathological effects through the phosphorylation of various target proteins, such as those depicted in Fig. 1. The function and regulation of these substrates are described in detail in the following sections.

In this review, we summarize the roles of PI3K/Akt signaling in three stem cell systems. First, we focus on the dedifferentiation of germline-committed cells into pluripotent stem cells, as this system provides a unique opportunity to study pluripotency and reprogramming. Recent experimental evidence has shown that PI3K/Akt signaling promotes this dedifferentiation process. We then discuss the function and regulation of the PI3K/Akt signal in another pluripotent stem cell line, embryonic stem (ES) cells. Finally, we summarize recent evidence demonstrating that PI3K/Akt signaling regulates the balance between activation and quiescence in tissue-specific stem cells.

T. Kimura (✉)
Department of Pathology, Graduate School of Medicine, Osaka University, 2-2 Yamada-oka, Suita, Osaka, Japan 565-0871.
e-mail: tkimura@patho.med.osaka-u.ac.jp

V.K. Rajasekhar, M.C. Vemuri (eds.), *Regulatory Networks in Stem Cells*, Stem Cell Biology and Regenerative Medicine,
DOI 10.1007/978-1-60327-227-8_26,

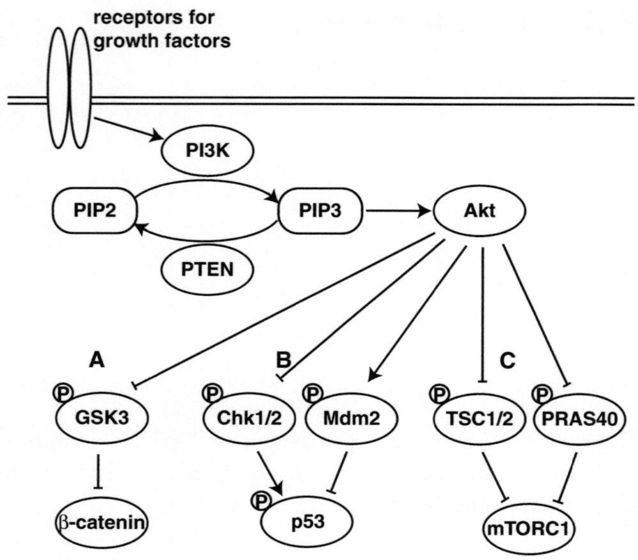

Fig. 1 PI3K/Akt signaling The activation of PI3K by growth factor receptors generates PIP3, which activates the serine/threonine kinase Akt. PI3K/Akt signaling is counteracted by the phosphatase PTEN. Akt phosphorylates several target proteins, but only the substrates discussed in the text are illustrated in the figure. (**A**) The kinase activity of GSK3 is inhibited by Akt-mediated phosphorylation. Upon Akt activation, β-catenin, a transcriptional cofactor degraded via GSK3 phosphorylation, translocates to the nucleus where it controls target genes implicated in stem cell regulation. In addition, Akt directly phosphorylates β-catenin, thereby facilitating its nuclear accumulation. This pathway is involved in crypt formation and the incidence of colorectal cancer. (**B**) p53 activity is suppressed by Akt signaling in two ways. First, Akt inhibits the checkpoint kinase Chl1/2, which phosphorylates p53 for its activation. Second, Akt promotes the stability and nuclear import of Mdm2, which degrades p53 via a ubiquitin-related pathway. The PI3K/Akt-mediated inhibition of p53, but not of GSK3, contributes to the dedifferentiation of PGCs into pluripotent stem cells. (**C**) The kinase activity of mTORC1, which activates translation and ribosome biogenesis, is suppressed by TSC1/2 and PRAS40. Akt activates the growth-promoting mTORC1 pathway, which is critical for the maintenance of leukemia stem cells, by inactivating TSC1/2 and PRAS40

2 PI3K/Akt Signaling and Germ Cell Dedifferentiation

2.1 The "Dedifferentiation" of PGCs into Pluripotent Stem Cells

Founder germ cells, or primordial germ cells (PGCs), are derived from pluripotent epiblasts at embryonic day 7.5 (E7.5) in mice [4–6]. PGCs migrate from the extraembryonic region through the hindgut and dorsal mesentery to the genital ridges at E11.5, when they undergo a series of sexual differentiation steps. In males, the PGCs enter mitotic arrest at E13.5. The spermatogonial stem cells, which resume mitosis shortly after birth, begin to produce sperm via spermatogenesis at puberty. In females, the PGCs begin meiosis at E13.5; however, after birth, the oogonia become quiescent at prophase I. Oocytes are then selected to mature and ovulate after puberty.

PGCs ultimately differentiate into oocytes or sperm during normal development, and they do not contribute to chimeric mice when injected into blastocysts [7]. These facts suggest that PGCs are already restricted to germline development and do not exhibit pluripotency under physiological conditions. Nonetheless, PGCs acquire pluripotency as follows [8]. First, PGCs are the originators of testicular teratomas, tumors containing various differentiated tissues and undifferentiated embryonal carcinoma (EC) cells [9, 10]. Second, pluripotent stem cells, known as embryonic germ (EG) cells, can be established from PGCs cultured in the presence of leukemia inhibitory factor (LIF), stem cell factor (SCF), and basic fibroblast growth factor (bFGF) [11, 12]. EG cells are able to differentiate into all three germ layers and into germ cells in chimeras after introduction into blastocysts; thus, their developmental potency is essentially the same as that of ES cells [13, 14]. Therefore, PGCs are unique from the perspective of stem cell biology in that they "dedifferentiate" from germ lineage cells into pluripotent stem cells under pathological and experimental conditions.

2.2 PI3K/Akt Signaling and the Dedifferentiation of PGCs

2.2.1 Effects of *Pten* Deficiency

The critical role of PI3K signaling in the dedifferentiation of PGCs was first demonstrated in conditional *Pten*-deficient mice [15]. Transgenic mice expressing Cre recombinase from the *TNAP* (tissue nonspecific alkaline phosphatase) locus were used to generate PGC-specific *Pten*-deficient mice. The expression of a germ cell–specific marker (Mvh, mouse vasa homolog) was up-regulated in the PGCs of the *Pten*-deficient embryos at E13.5, as in their control littermates, suggesting that germ cell commitment normally takes place in the absence of *Pten*. However, while the control male PGCs entered mitotic quiescence at E13.5, many PGCs in the mutant males continued to proliferate after E14.5. Simultaneously, most of the PGCs were lost by apoptosis, leading to germ cell deficiency in the newborn mutant males. However, a few germ cells survived to produce teratomatous foci at E16.5 and fully developed testicular teratomas at birth. Mvh expression was down-regulated and Akt phosphorylation was elevated in the teratomas. Furthermore, EG cell formation was enhanced in the mutants, irrespective of gender, when E11.5 PGCs were cultured with LIF, SCF, and bFGF. These observations

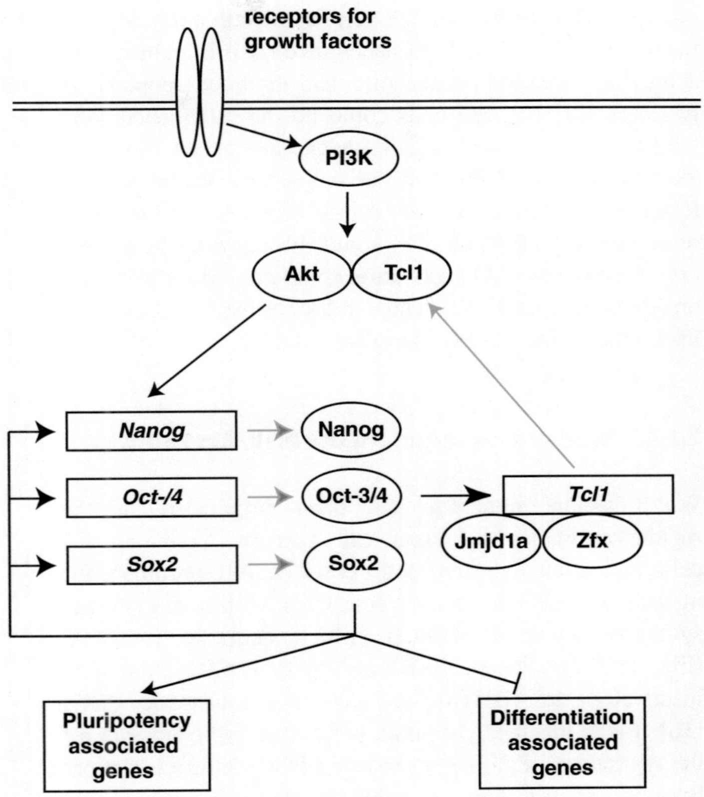

Fig. 2 Interplay of PI3K/Akt signaling and the pluripotent transcriptional network The pluripotent transcription factors Nanog, Oct-3/4, and Sox2 maintain their own expression. In addition, three factors activate other pluripotent-associated genes and repress differentiation-associated genes in ES cells. PI3K/Akt signaling and its cofactor Tcl1 activate *Nanog* expression, presumably via inhibition of GSK3. Oct-3/4 activates *Tcl1* expression together with other pluripotent transcriptional regulators such as Zfx and Jmjd1a, thereby forming a positive feedback loop that maintains pluripotency

demonstrate that suppression of PI3K signaling by Pten is prerequisite for male germ cell development and for the prevention of PGC dedifferentiation. Since strain 129 mice develop spontaneous testicular teratomas [10], note that the *Pten*-deficient mice were analyzed in a mixed background and not in a pure 129 background.

2.2.2 Effects of Akt Activation

Transgenic mice expressing an Akt–Mer fusion protein were used to show that Akt is a critical downstream effector of PI3K with respect to PGC dedifferentiation [16]. The Akt–Mer fusion protein was composed of a constitutively active version of Akt and modified estrogen receptor (Mer). While Akt–Mer is in an enzymatically inactive form in the absence of the Mer ligand 4-hydroxytamoxifen (4OHT), it is rapidly activated in the presence of 4OHT. Thus, Akt–Mer transgenic mice provide a valuable system with which to examine the effects of transient Akt activation in vitro and in vivo.

When PGCs from the Akt–Mer transgenic embryos were cultured with LIF, SCF, and bFGF, EG cell formation was augmented by the addition of 4OHT. The efficiency of EG cell formation was similar to that in the *Pten*-deficient PGCs. The enhancement of EG cell derivation by 4OHT was observed both in E8.5 migratory PGCs and E11.5 gonadal PGCs. However, inhibition of PI3K abolished EG cell derivation, although the inhibitory effects were completely rescued by Akt activation, showing that activation of PI3K/Akt signaling is required for EG cell derivation.

bFGF is essential for EG cell derivation, since EG cells were never established from wild-type PGCs cultured with LIF and SCF but without bFGF. Furthermore, bFGF, but neither LIF nor SCF, efficiently induced Akt phosphorylation in the cultured PGCs. However, in the 4OHT-treated transgenic PGCs, EG cells could be derived in the absence of bFGF as efficiently as in wild-type PGCs cultured with bFGF, which demonstrates that Akt mediates the effects of bFGF. Transient Akt activation for 24 h after the start of the culture period was sufficient to produce EG cells, indicating that events critical to the acquisition of pluripotency occurred at an early stage of culture.

2.2.3 Mitotic Activity and the Acquisition of Pluripotency in PGCs

The ability to acquire pluripotency is well correlated with the mitotic activity of PGCs. Male and female germ cells enter mitotic arrest and meiosis, respectively, at E13.5; however, prolonged proliferation was observed in PGC-specific *Pten*-deficient mice after E14.5 [15]. Continued proliferation has also been reported in 129-*Ter/Ter* mice that develop testicular teratomas with high penetrance [17]. In culture, Akt activation enhanced proliferation and inhibited apoptosis in E11.5 PGCs [16]. EG cells could not be established from wild-type germ cells after E14.5, regardless of sex. Similarly, although enhanced EG cell derivation was observed up to E14.5 and E13.5 in Akt–Mer transgenic males and females, respectively, no EG cell lines could subsequently be derived [16]. Consistently, Akt activation could not induce germ cell proliferation after E15.5. Thus, mitotic activation appears to be a critical factor in PGC dedifferentiation.

2.2.4 Downstream Targets of the PI3K/Akt Pathway

Wnt/β-catenin signaling has been implicated in the regulation of various stem cell systems, including ES cells and epithelial stem cells [18–21]. Akt promotes the nuclear accumulation of β-catenin via inhibition of glycogen synthase kinase 3 (GSK3), which degrades β-catenin (Fig. 1) [22]. Indeed, GSK3 is hyperphosphorylated (i.e., inactivated) in 4OHT-treated Akt–Mer transgenic PGCs [16]. Taken together, β-catenin is an attractive candidate for the regulation of PI3K/Akt-induced PGC dedifferentiation; however, several lines of evidence show that this is not the case. First, nuclear localization of β-catenin was not observed in *Pten*-deficient PGCs [23]. Second, PGC-specific expression of nuclear-localized β-catenin did not lead to testicular teratoma, but instead caused a delay in cell cycle progression [23]. Third, treatment of PGCs with the GSK3 inhibitor 6-bromoindirubin-3′-oxime (BIO) did not augment EG cell derivation in cells cultured with LIF, SCF, and bFGF [16].

Another candidate downstream target of Akt signaling is the tumor suppressor p53 (Fig. 1). The activity and nuclear localization of p53 is suppressed in ES cells [24], and *Nanog* expression is suppressed by p53 during ES cell differentiation [25]. The derivation of pluripotent stem cells from neonatal spermatogonial stem cells was promoted by *p53* deficiency [26]. In cultured Akt–Mer transgenic PGCs, Akt activation appears to suppress p53 function because 4OHT treatment enhanced the stability and nuclear localization of Mdm2, which is a ubiquitin ligase for p53, and it suppressed the phosphorylation of p53, which is required for its activation [16]. Most intriguingly, *p53* deficiency recapitulated the effects of Akt activation on EG cell derivation; that is, when cultured with the aforementioned three growth factors, the efficiency of derivation was enhanced in *p53*-deficient mice, and EG cell lines could be established from the mice in the absence of bFGF [16]. Thus, the PI3K/Akt-mediated inhibition of p53 contributes to the dedifferentiation of PGCs in culture. However, other downstream targets of Akt likely exist, particularly in vivo, considering that unlike *Pten*-deficient mice, *p53*-deficient mice with a mixed background showed normal PGC development and did not develop testicular teratomas at birth [27].

2.3 The "Dedifferentiation" of Spermatogonial Stem Cells into Pluripotent Stem Cells

Spermatogonial stem cells resume mitosis after birth. Spermatogonial stem cell lines, known as germline stem (GS) cells, can be established from neonatal mice and cultured for long periods in the presence of glial cell line–derived neurotrophic factor (GDNF) [28]. GS cell lines retain a unipotent differentiation ability in vitro because they differentiate into functional sperm but do not develop teratomas after testicular implantation. However, pluripotent cells, or multipotent GS (mGS) cells, occasionally emerge early on during the culture period [26]. These cells can be propagated in the presence of LIF, but not GDNF, and they are able to differentiate into three germ layers and germ cells after introduction into blastocysts. Similar pluripotent cell lines can also be derived from cultured adult mouse spermatogonial stem cells [29].

GDNF is essential for the self-renewal of GS cells in vitro and of spermatogonia in vivo [28, 30]. Inhibition of PI3K prevented GS cell proliferation in the presence of GDNF [31]. GS cells carrying an *Akt–Mer* expression plasmid exhibited self-renewal in the absence of GDNF as long as Akt signaling was activated by the addition of 4OHT. Therefore, the self-renewing proliferation of GS cells depends on PI3K/Akt signaling; however, the derivation of mGS cells from Akt–Mer-expressing GS cells was not promoted by 4OHT [31], indicating that Akt provokes distinct responses in germ cells depending on their developmental stage.

3 PI3K/Akt Signaling in ES Cells

3.1 Mechanisms Regulating ES Cell Pluripotency

ES cells are pluripotent cells derived from the inner cell mass of blastocysts [32]. ES cell pluripotency is regulated by extrinsic signaling pathways and an intrinsic pluripotent transcription factor network [33]. The pluripotency of mouse ES cells, but not of primate ES cells, can be maintained under a feeder-free condition by LIF, which induces the nuclear translocation of signal transducers and activators of transcription 3 (STAT3) [34–36]. The self-renewing proliferation of mouse ES cells is also supported by

Wnt/β-catenin, bone morphogenetic protein (BMP), and Activin/Nodal signaling [21, 37–40]. In comparison, human ES cells can be propagated by insulin-like growth factor II (IGF-II) without feeder layers [41]. Several genes known as pluripotency-associated or differentiation-associated genes are activated or suppressed, respectively, by the transcription factors Oct-3/4, Sox2, and Nanog (Fig. 2) [42–44].

3.2 Maintenance of ES Cell Pluripotency by PI3K/Akt Signaling

Accumulating evidence indicates that PI3K/Akt signaling plays critical roles in both the promotion of self-renewing proliferation and the maintenance of pluripotency. *Pten*-deficient ES cells showed enhanced proliferation in vitro and formed large tumorigenic teratomas composed of undifferentiated cells after implantation into immunodeficient mice [45]; however, these phenotypes were reverted by the additional deletion of *Akt-1*, indicating that Akt-1 mediates the effects of *Pten* deficiency [46]. Furthermore, the small GTPase ERas, which activates PI3K/Akt signaling, was required for the proliferative and tumorigenic activites of the ES cells [47].

PI3K/Akt signaling is activated by LIF in mouse ES cells [48]. Inhibition of PI3K by pharmacological means, or by the use of a dominant-negative version of PI3K, reduced mouse ES cell self-renewal in the presence of LIF, showing that PI3K is required for self-renewing cellular proliferation [48]. Furthermore, mouse ES cells expressing a constitutively active form of Akt-1 maintained an undifferentiated phenotype in the absence of LIF [49]. The effects of Akt activation were reversible because LIF dependence and the pluripotent differentiation activity of the cells were restored by deletion of the *Akt-1* construct. In addition, cynomolgous monkey ES cells carrying an *Akt–Mer* expression cassette self-renewed without feeder cells in the presence of 4OHT, but not in the absence of 4OHT [49]. After the withdrawal of 4OHT, the ES cells differentiated into three germ layers in embryoid bodies. Therefore, Akt activation is sufficient for the maintenance of ES cell pluripotency in both mice and primates.

The mechanism whereby PI3K/Akt signaling supports ES cell pluripotency remains to be elucidated. Considering that this signal regulates both the proliferation and pluripotency of ES cells, mitotic activation may be a key event in the PI3K/Akt-mediated self-renewal of ES cells, as discussed above in Section 2.2.3. One candidate downstream target of the PI3K/Akt pathway is the pluripotent transcription factor Nanog. *Nanog* expression was down-regulated by PI3K inhibition [50, 51]; however, its expression was restored by the inhibition of GSK3 with BIO [50], raising the possibility that Akt-mediated GSK3 inhibition may play a role in *Nanog* expression.

3.3 T Cell Leukemia 1 (Tcl1) Is a Cofactor of Akt

Tcl1 enhances Akt kinase activity and mediates its nuclear translocation [52]. *Tcl1* is highly expressed in ES cells and its expression level is proportional to the level of Akt phosphorylation [53]. The knockdown of *Tcl1* in mouse ES cells resulted in differentiation or reduced proliferation in the presence of LIF [53, 54]. Because the phenotype was restored by expression of a constitutively active form of Akt [53], Tcl1 likely functions via Akt in self-renewing proliferation.

Genome-wide chromatin-immunoprecipitation analysis identified *Tcl1* as a target of Oct-3/4 [53]. Since the conditional removal of Oct-3/4 led to reduced *Tcl1* expression, *Tcl1* transcription is positively regulated by Oct-3/4 in ES cells. Inactivation of the zinc finger transcription factor Zfx impaired the self-renewal of mouse ES cells and reduced the transcription of several self-renewal regulators, including *Tcl1* [55]. Furthermore, deletion of Jmjd1a, which demethylates silencing-associated histone H3 lysine 9 dimethylation (H3K9me2), caused ES cell differentiation [56]. *Tcl1* transcription decreased in ES cells lacking Jmjd1a, with accumulation of H3K9me2 at its promoter region. Thus, *Tcl1* expression is tightly regulated in ES cells by pluripotent transcriptional regulators such as Oct-3/4, Zfx, and Jmjd1a. Taken together with the observation that *Nanog* is under the control of PI3K/Akt signaling, this pluripotent transcriptional network and PI3K/Akt signaling constitute a positive feedback circuit in ES cells (Fig. 2).

4 PI3K/Akt Signaling in Tissue Stem Cells

4.1 Tissue Stem Cells and Cancer Stem Cells

Unlike the culture-adapted stem cells discussed above, tissue stem cells are generally located within niches and exist in a mitotically quiescent state. Upon an appropriate stimulus or tissue damage, these quiescent cells begin to proliferate and produce progenitor cells with high mitotic activity. The progenitor cells gradually lose their multipotent differentiation capacities and eventually produce terminally differentiated cells.

Disorders in the self-renewing ability of stem cells and progenitors could theoretically give rise to dysfunctionally replicating cells, namely, cancer stem cells or cancer initiating cells [57]. In fact, stem cell-like cells have been detected in a range of human cancers; some cancer-initiating cells are likely derived from stem cells while others acquire stem cell features (e.g., self-renewal). Mutations that activate PI3K/Akt signaling have been identified in several human cancers [3], and recent studies have suggested that aberrant activation of PI3K/Akt signaling promotes the production and maintenance of cancer stem cells.

4.2 Hair Follicular and Epidermal Stem Cells

The epidermis of the skin consists of a multilayered epithelium, called the interfollicular epidermis (IFE), and associated appendages, including hair follicles, sebaceous glands, and sweat glands [58, 59]. Epidermal and follicular stem cells reside at the basal layer of the IFE and in the bulge of each hair follicle, respectively. In the IFE, progenitors produced from epidermal stem cells differentiate sequentially upward to generate the epithelial layers. Hair follicles undergo cyclic phases of rest (telogen), growth (anagen), and regression (catagen). The initiation of a new anagen cycle, that is, new hair growth, depends on the proliferation of quiescent follicular stem cells. These activated stem cells in turn generate highly proliferative progenitor cells, which give rise to all of the cell types found in new hair follicles.

A critical role for PI3K/Akt signaling in postnatal epidermal development has been proposed based on analyses of mutant mice. The deletion of *Akt-1* resulted in thin epidermal layers and retarded hair follicle development [60]. A more severe reduction in the epidermal layers and hair follicles was reported in newborn *Akt-1*$^{-/-}$ *Akt-2*$^{-/-}$ and *Akt-1*$^{-/-}$ *Akt-3*$^{+/-}$ mutant mice [61, 62]. Conversely, the keratinocyte-specific deletion of *Pten* caused epidermal hyperplasia with subsequent tumorigenesis and accelerated hair morphogenesis [63, 64]. These findings raise the question of whether PI3K/Akt signaling regulates epidermal and hair follicular stem cell systems.

Akt–Mer transgenic mice were utilized to address this question [65]. In mice, hair follicles synchronously enter quiescent telogen at 6 weeks of age and remain in a quiescent state for at least 3 weeks. Consistently, when wild-type mice were shaved at 6 weeks of age and treated topically with 4OHT for 3 weeks, no hair growth occurred. However, Akt–Mer transgenic mice had prominent new hair growth at 3 weeks after treatment with 4OHT, but not with vehicle ethanol. Consistent with this result, histological analysis showed that the telogenic follicles of the transgenic mice entered into anagen. In addition, many proliferating epidermal cells, presumably progenitors, were detectable at the basal layers of the IFE and in the hair follicles of the mice following 4OHT treatment. Flow cytometric and holoclone analyses revealed that the number of CD34-positive, α6-integrin-negative progenitor cells increased in the transgenic mice, clearly demonstrating that Akt signaling activates quiescent stem cells and produces committed progenitors in epidermal and hair follicular stem cells.

4.3 Hematopoietic Stem Cells

The impact of PI3K/Akt activation on tissue stem cell quiescence has been clearly demonstrated in hematopoiesis. Hematopoietic stem cells (HSCs) reside in osteoblastic and endothelial niches within the bone narrow, and they differentiate into more than 10 types of blood cells [66]. During hematopoietic differentiation, HSCs progressively give rise to progenitor cells with more restricted and distinct developmental potencies, which in turn generate mature blood cells [67].

The deletion of *Pten* during hematopoiesis had three effects on the homeostasis of the stem cell pool [68, 69]. First, *Pten* inactivation transiently increased but subsequently exhausted the supply of HSCs, indicating HSC hyperactivation. Second, the *Pten*-deficient HSCs showed a skewed differentiation into myeloid and T-lymphoid lineages. Third, the *Pten*-deficient mice progressed to leukemia, including acute myeloblastic and acute lymphoblastic leukemia (AML and ALL). The leukemic cells contained transplantable leukemia-initiating cells (i.e., leukemia stem cells), which were enriched in the populations expressing HSC markers. Pharmacological inhibition of mTOR complex 1 (mTORC1 or mTOR-raptor complex), which promotes cell growth via activation of translation and ribosome biogenesis (Fig. 1), not only rescued HSC exhaustion but also inhibited the incidence of leukemia and self-renewing proliferation of the cancer stem cells [69]. This finding is relevant particularly in terms of cancer therapy because anti-mTORC1 therapy in mice successfully depleted the cancer stem cells without damaging the normal stem cells.

4.4 Epithelial Stem Cells

In addition to the epidermis and hair follicles, stem cell populations exist in the epithelial tissues of other organs, including the intestine, lung, mammary glands, and prostate. Intestinal polyposis, a precancerous neoplasia, results from an aberrant increase in the number of crypts, which contain intestinal stem cells (ISCs). Inactivation of *Pten* in the intestinal epithelium brought about de novo crypt formation due to excess ISCs and hamartomatous intestinal polyps [70]. The effect, which was mediated by the nuclear accumulation of β-catenin, was induced by direct phosphorylation by Akt. Therefore, the nuclear targeting of β-catenin, which is promoted by both canonical Wnt and PI3K/Akt signaling, promotes the cell cycle entry and proliferation of ISCs.

The prostate-specific deletion of *Pten* led to metastatic prostate cancer with prostatic intraepithelial neoplasia (PIN) [71]. Akt mediates the onset of PIN because prostate-restricted expression of activated Akt induced PIN [72] and

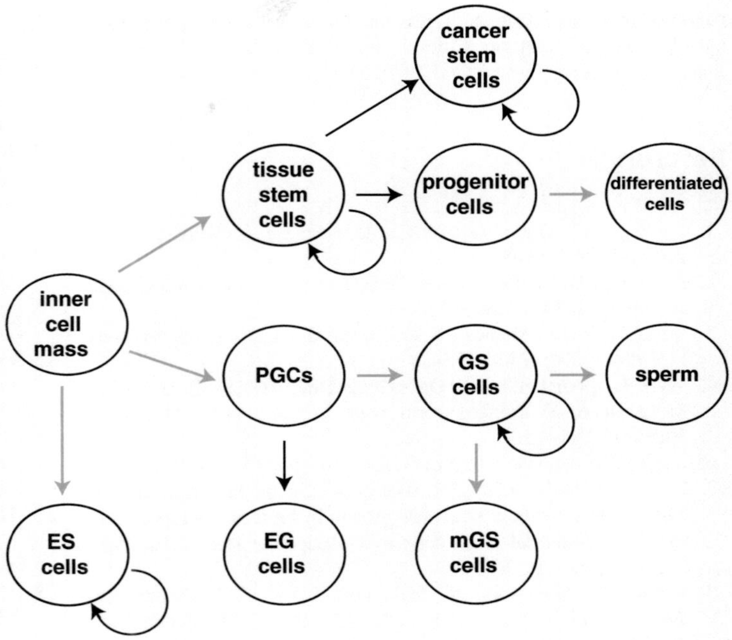

Fig. 3 Regulation of "stemness" by PI3K/Akt signaling PI3K/Akt signaling regulates two pluripotent stem cell systems: the dedifferentiation of PGCs into EG cells and the maintenance of ES cell pluripotency. In tissue stem cells, this signaling promotes the self-renewal of spermatogonial stem cells and activates quiescent somatic stem cells, leading to the production of committed progenitor cells and cancer stem cells. The processes promoted by PI3K/Akt signaling are indicated by black arrows

haplodeficiency of *Akt-1* in *Pten*$^{+/-}$ mice attenuated PIN [73]. Reconstitution of prostatic tissues by cell transplantation indicated a high proportion of stem/progenitor cells among the Sca-1 (stem cell antigen-1)-positive cells [74]. The number of Sca-1-positive cells increased in the prostate glands of the *Pten*-deficient mice [75], and Akt activation in the epithelial cells of the glands was sufficient to initiate tumorigenesis [74].

Similarly, bronchioalveolar epithelium-specific *Pten*-deficient mice developed lung adenocarcinoma with an increased number of bronchioalveolar stem cells (BASCs) [76], while mammary cells lacking *Pten* displayed precocious lobulo-alveolar development, excess ductal branching, and tumor development [77]. However, whether the deletion of*Pten* affects mammary stem cells (MaSCs) is unknown. Taken together, these data demonstrate that the quiescence and activation of tissue stem cells are regulated by the balanced action of PI3K/Akt signaling and *Pten*, and disruption of this balance causes aberrant tissue stem cell activation, leading to the exhaustion of normal stem cells and generation of cancer stem cells.

5 Future Perspectives

Recent findings have revealed that PI3K/Akt signaling regulates "stemness" in various stem cell systems (Fig. 3). In pluripotent stem cell systems, PI3K/Akt signaling supports pluripotency in the dedifferentiation of PGCs and maintenance of ES cells. In tissue stem cells, PI3K/Akt signaling activates quiescent stem cells, leading to the production of committed progenitors and cancer stem cells. In these stem cell systems, the PI3K/Akt signal appears to exert its effects at least in part through mitotic activation.

Still, several critical questions remain unanswered. First, how does PI3K/Akt signaling reprogram germline-committed cells to become pluripotent cells? As *p53* deficiency promotes the dedifferentiation process, the transcriptional network regulated by p53 should be investigated. The knowledge obtained through studies of ES cell pluripotency can be extrapolated to this process. In addition, it would be interesting to examine the relationship between PI3K/Akt signaling and deadend-1 (Dnd1), whose mutation (*Ter*) produced a phenotype similar to that of PGC-specific *Pten*-deficient mice [78, 79]. Second, what are the targets of PI3K/Akt signaling in each stem cell system? Analyses of such Akt downstream targets as mTORC1 and forkhead transcription factors (FOXOs) will provide valuable information regarding how PI3K/Akt regulates the "stemness" of each stem cell. Third, which growth factors and adhesion molecules activate or inhibit PI3K/Ak signaling in vivo? Fourth, does cross-talk exist between PI3K/Akt and other signaling pathways? In this regard, BMP signaling appears to inhibit PI3K/Akt signaling in epidermal stem cells because the skin-specific deletion of *Bmpr1* (BMP receptor 1A) produced a phenotype similar to that of *Pten*-deficient and 4OHT-treated Akt–Mer transgenic mice [80]. Further investigation of the molecular mechanisms whereby PI3K/Akt signaling regulates "stemness" will provide a valuable platform from which to advance not only stem cell biology but also the future of regenerative medicine.

Acknowledgments This study was supported in part by grants from the Ministry of Education, Culture, Sports, Science and Technology, Japan, and the 21st Century COE "CICET."

References

1. Cantley LC. The phosphoinositide 3-kinase pathway. Science. 2002;296(5573):1655–7.
2. Manning BD, Cantley LC. AKT/PKB signaling: navigating downstream. Cell. 2007;129(7):1261–74.
3. Stiles B, Groszer M, Wang S, Jiao J, Wu H. PTENless means more. Dev Biol. 2004;273(2):175–84.
4. Wylie C. Germ cells. Curr Opin Genet Dev. 2000;10(4):410–3.
5. McLaren A. Primordial germ cells in the mouse. Dev Biol. 2003;262(1):1–15.
6. Rugh R. The mouse. Oxford: Oxford University Press; 1968.
7. Durcova-Hills G, Adams IR, Barton SC, Surani MA, McLaren A. The role of exogenous fibroblast growth factor-2 on the reprogramming of primordial germ cells into pluripotent stem cells. Stem Cells. 2006;24(6):1441–9.
8. Kimura T, Murayama K, Nakamura T, et al. Testicular teratomas: back to pluripotent stem cells. Prog Mol Subcell Biol. 2005;40:133–50.
9. Stevens LC. Origin of testicular teratomas from primordial germ cells in mice. J Natl Cancer Inst. 1967;38(4):549–52.
10. Stevens LC. Teratocarcinogenesis and spontaneous parthenogenesis in mice. Results Probl Cell Differ. 1980;11:265–74.
11. Matsui Y, Zsebo K, Hogan BL. Derivation of pluripotential embryonic stem cells from murine primordial germ cells in culture. Cell. 1992;70(5):841–7.
12. Resnick JL, Bixler LS, Cheng L, Donovan PJ. Long-term proliferation of mouse primordial germ cells in culture. Nature. 1992;359(6395):550–1.
13. Labosky PA, Barlow DP, Hogan BL. Mouse embryonic germ (EG) cell lines: transmission through the germline and differences in the methylation imprint of insulin-like growth factor 2 receptor (Igf2r) gene compared with embryonic stem (ES) cell lines. Development. 1994;120(11):3197–204.
14. Stewart CL, Gadi I, Bhatt H. Stem cells from primordial germ cells can reenter the germ line. Dev Biol. 1994;161(2):626–8.
15. Kimura T, Suzuki A, Fujita Y, et al. Conditional loss of PTEN leads to testicular teratoma and enhances embryonic germ cell production. Development. 2003;130(8):1691–700.
16. Kimura T, Tomooka M, Yamano N, et al. Akt signaling promotes derivation of embryonic germ cells from primordial germ cells. Development. 2008;135(5):869–79.
17. Noguchi T, Stevens LC. Primordial germ cell proliferation in fetal testes in mouse strains with high and low incidences of congenital testicular teratomas. J Natl Cancer Inst. 1982;69(4): 907–13.
18. Lo Celso C, Prowse DM, Watt FM. Transient activation of beta-catenin signalling in adult mouse epidermis is sufficient to induce new hair follicles but continuous activation is required to maintain hair follicle tumours. Development. 2004;131(8): 1787–99.
19. Van Mater D, Kolligs FT, Dlugosz AA, Fearon ER. Transient activation of beta-catenin signaling in cutaneous keratinocytes is sufficient to trigger the active growth phase of the hair cycle in mice. Genes Dev. 2003;17(10):1219–24.
20. van de Wetering M, Sancho E, Verweij C, et al. The beta-catenin/TCF-4 complex imposes a crypt progenitor phenotype on colorectal cancer cells. Cell. 2002;111(2):241–50.
21. Sato N, Meijer L, Skaltsounis L, Greengard P, Brivanlou AH. Maintenance of pluripotency in human and mouse embryonic stem cells through activation of Wnt signaling by a pharmacological GSK-3-specific inhibitor. Nat Med. 2004;10(1):55–63.
22. Moon RT, Kohn AD, De Ferrari GV, Kaykas A. WNT and beta-catenin signalling: diseases and therapies. Nat Rev Genet. 2004;5(9):691–701.
23. Kimura T, Nakamura T, Murayama K, et al. The stabilization of β-catenin leads to impaired primordial germ cell development via aberrant cell cycle progression. Dev Biol. 2006;300: 545–53.
24. Aladjem MI, Spike BT, Rodewald LW, et al. ES cells do not activate p53-dependent stress responses and undergo p53-independent apoptosis in response to DNA damage. Curr Biol. 1998;8(3): 145–55.
25. Lin T, Chao C, Saito S, et al. p53 induces differentiation of mouse embryonic stem cells by suppressing Nanog expression. Nat Cell Biol. 2005;7(2):165–71.
26. Kanatsu-Shinohara M, Inoue K, Lee J, et al. Generation of pluripotent stem cells from neonatal mouse testis. Cell. 2004;119(7): 1001–12.
27. Harvey M, McArthur MJ, Montgomery CA, Jr., Bradley A, Donehower LA. Genetic background alters the spectrum of tumors that develop in p53-deficient mice. Faseb J. 1993;7(10):938–43.
28. Kanatsu-Shinohara M, Ogonuki N, Inoue K, et al. Long-term proliferation in culture and germline transmission of mouse male germline stem cells. Biol Reprod. 2003;69(2):612–6.
29. Seandel M, James D, Shmelkov SV, et al. Generation of functional multipotent adult stem cells from GPR125+ germline progenitors. Nature. 2007;449(7160):346–50.
30. Meng X, Lindahl M, Hyvonen ME, et al. Regulation of cell fate decision of undifferentiated spermatogonia by GDNF. Science. 2000;287(5457):1489–93.
31. Lee J, Kanatsu-Shinohara M, Inoue K, et al. Akt mediates self-renewal division of mouse spermatogonial stem cells. Development. 2007;134(10):1853–9.
32. Chambers I, Smith A. Self-renewal of teratocarcinoma and embryonic stem cells. Oncogene. 2004;23(43):7150–60.
33. Niwa H. How is pluripotency determined and maintained? Development. 2007;134(4):635–46.
34. Daheron L, Opitz SL, Zaehres H, et al. LIF/STAT3 signaling fails to maintain self-renewal of human embryonic stem cells. Stem Cells. 2004;22(5):770–8.
35. Humphrey RK, Beattie GM, Lopez AD, et al. Maintenance of pluripotency in human embryonic stem cells is STAT3 independent. Stem Cells. 2004;22(4):522–30.
36. Matsuda T, Nakamura T, Nakao K, et al. STAT3 activation is sufficient to maintain an undifferentiated state of mouse embryonic stem cells. EMBO J. 1999;18(15):4261–9.
37. Kielman MF, Rindapaa M, Gaspar C, et al. Apc modulates embryonic stem-cell differentiation by controlling the dosage of beta-catenin signaling. Nat Genet. 2002;32(4):594–605.
38. Qi X, Li TG, Hao J, et al. BMP4 supports self-renewal of embryonic stem cells by inhibiting mitogen-activated protein kinase pathways. Proc Natl Acad Sci U S A. 2004;101(16):6027–32.
39. Ying QL, Nichols J, Chambers I, Smith A. BMP induction of Id proteins suppresses differentiation and sustains embryonic stem cell self-renewal in collaboration with STAT3. Cell. 2003;115(3):281–92.
40. Ogawa K, Saito A, Matsui H, et al. Activin-Nodal signaling is involved in propagation of mouse embryonic stem cells. J Cell Sci. 2007;120(Pt 1):55–65.
41. Bendall SC, Stewart MH, Menendez P, et al. IGF and FGF cooperatively establish the regulatory stem cell niche of pluripotent human cells in vitro. Nature. 2007;448(7157):1015–21.
42. Boyer LA, Lee TI, Cole MF, et al. Core transcriptional regulatory circuitry in human embryonic stem cells. Cell. 2005;122(6): 947–56.

43. Loh YH, Wu Q, Chew JL, et al. The Oct4 and Nanog transcription network regulates pluripotency in mouse embryonic stem cells. Nat Genet. 2006;38(4):431–40.
44. Masui S, Nakatake Y, Toyooka Y, et al. Pluripotency governed by Sox2 via regulation of Oct3/4 expression in mouse embryonic stem cells. Nat Cell Biol. 2007;9(6):625–35.
45. Sun H, Lesche R, Li DM, et al. PTEN modulates cell cycle progression and cell survival by regulating phosphatidylinositol 3,4,5,-trisphosphate and Akt/protein kinase B signaling pathway. Proc Natl Acad Sci U S A. 1999;96(11):6199–204.
46. Stiles B, Gilman V, Khanzenzon N, et al. Essential role of AKT-1/protein kinase B alpha in PTEN-controlled tumorigenesis. Mol Cell Biol. 2002;22(11):3842–51.
47. Takahashi K, Mitsui K, Yamanaka S. Role of ERas in promoting tumour-like properties in mouse embryonic stem cells. Nature. 2003;423(6939):541–5.
48. Paling NR, Wheadon H, Bone HK, Welham MJ. Regulation of embryonic stem cell self-renewal by phosphoinositide 3-kinase-dependent signaling. J Biol Chem. 2004;279(46): 48063–70.
49. Watanabe S, Umehara H, Murayama K, Okabe M, Kimura T, Nakano T. Activation of Akt signaling is sufficient to maintain pluripotency in mouse and primate embryonic stem cells. Oncogene. 2006;25(19):2697–707.
50. Storm M, Bone H, Beck C, et al. Regulation of Nanog expression by phosphoinositide 3-kinase-dependent signaling in murine embryonic stem cells. J Biol Chem. 2007;282(9): 6265–73.
51. Takahashi K, Murakami M, Yamanaka S. Role of the phosphoinositide 3-kinase pathway in mouse embryonic stem (ES) cells. Biochem Soc Trans. 2005;33(Pt 6):1522–5.
52. Pekarsky Y, Koval A, Hallas C, et al. Tcl1 enhances Akt kinase activity and mediates its nuclear translocation. Proc Natl Acad Sci U S A. 2000;97(7):3028–33.
53. Matoba R, Niwa H, Masui S, et al. Dissecting Oct3/4-regulated gene networks in embryonic stem cells by expression profiling. PLoS ONE. 2006;1:e26.
54. Ivanova N, Dobrin R, Lu R, et al. Dissecting self-renewal in stem cells with RNA interference. Nature. 2006;442:533–8.
55. Galan-Caridad JM, Harel S, Arenzana TL, et al. Zfx controls the self-renewal of embryonic and hematopoietic stem cells. Cell. 2007;129(2):345–57.
56. Loh YH, Zhang W, Chen X, George J, Ng HH. Jmjd1a and Jmjd2c histone H3 Lys 9 demethylases regulate self-renewal in embryonic stem cells. Genes Dev. 2007;21(20):2545–57.
57. Reya T, Morrison SJ, Clarke MF, Weissman IL. Stem cells, cancer, and cancer stem cells. Nature. 2001;414(6859):105–11.
58. Watt FM. Stem cell fate and patterning in mammalian epidermis. Curr Opin Genet Dev. 2001;11(4):410–7.
59. Fuchs E, Tumbar T, Guasch G. Socializing with the neighbors: stem cells and their niche. Cell. 2004;116(6):769–78.
60. Di-Poi N, Ng CY, Tan NS, et al. Epithelium-mesenchyme interactions control the activity of peroxisome proliferator-activated receptor beta/delta during hair follicle development. Mol Cell Biol. 2005;25(5):1696–712.
61. Yang ZZ, Tschopp O, Di-Poi N, et al. Dosage-dependent effects of Akt1/protein kinase Balpha (PKBalpha) and Akt3/PKBgamma on thymus, skin, and cardiovascular and nervous system development in mice. Mol Cell Biol. 2005;25(23): 10407–18.
62. Peng XD, Xu PZ, Chen ML, et al. Dwarfism, impaired skin development, skeletal muscle atrophy, delayed bone development, and impeded adipogenesis in mice lacking Akt1 and Akt2. Genes Dev. 2003;17(11):1352–65.
63. Suzuki A, Itami S, Ohishi M, et al. Keratinocyte-specific Pten deficiency results in epidermal hyperplasia, accelerated hair follicle morphogenesis and tumor formation. Cancer Res. 2003;63(3): 674–81.
64. Backman SA, Ghazarian D, So K, et al. Early onset of neoplasia in the prostate and skin of mice with tissue-specific deletion of Pten. Proc Natl Acad Sci U S A. 2004;101(6):1725–30.
65. Murayama K, Kimura T, Tarutani M, et al. Akt activation induces epidermal hyperplasia and proliferation of epidermal progenitors. Oncogene. 2007;26(33):4882–8.
66. Martinez-Agosto JA, Mikkola HK, Hartenstein V, Banerjee U. The hematopoietic stem cell and its niche: a comparative view. Genes Dev. 2007;21(23):3044–60.
67. Iwasaki H, Akashi K. Hematopoietic developmental pathways: on cellular basis. Oncogene. 2007;26(47):6687–96.
68. Zhang J, Grindley JC, Yin T, et al. PTEN maintains haematopoietic stem cells and acts in lineage choice and leukaemia prevention. Nature. 2006;441(7092):518–22.
69. Yilmaz OH, Valdez R, Theisen BK, et al. Pten dependence distinguishes haematopoietic stem cells from leukaemia-initiating cells. Nature. 2006;441(7092):475–82.
70. He XC, Yin T, Grindley JC, et al. PTEN-deficient intestinal stem cells initiate intestinal polyposis. Nat Genet. 2007;39(2):189–98.
71. Wang S, Gao J, Lei Q, et al. Prostate-specific deletion of the murine Pten tumor suppressor gene leads to metastatic prostate cancer. Cancer Cell. 2003;4(3):209–21.
72. Majumder PK, Yeh JJ, George DJ, et al. Prostate intraepithelial neoplasia induced by prostate restricted Akt activation: the MPAKT model. Proc Natl Acad Sci U S A. 2003;100(13):7841–6.
73. Chen ML, Xu PZ, Peng XD, et al. The deficiency of Akt1 is sufficient to suppress tumor development in Pten+/- mice. Genes Dev. 2006;20(12):1569–74.
74. Xin L, Lawson DA, Witte ON. The Sca-1 cell surface marker enriches for a prostate-regenerating cell subpopulation that can initiate prostate tumorigenesis. Proc Natl Acad Sci U S A. 2005;102(19):6942–7.
75. Wang S, Garcia AJ, Wu M, Lawson DA, Witte ON, Wu H. Pten deletion leads to the expansion of a prostatic stem/progenitor cell subpopulation and tumor initiation. Proc Natl Acad Sci U S A. 2006;103(5):1480–5.
76. Yanagi S, Kishimoto H, Kawahara K, et al. Pten controls lung morphogenesis, bronchioalveolar stem cells, and onset of lung adenocarcinomas in mice. J Clin Invest. 2007;117(10):2929–40.
77. Li G, Robinson GW, Lesche R, et al. Conditional loss of PTEN leads to precocious development and neoplasia in the mammary gland. Development. 2002;129(17):4159–70.
78. Youngren KK, Coveney D, Peng X, et al. The Ter mutation in the dead end gene causes germ cell loss and testicular germ cell tumours. Nature. 2005;435(7040):360–4.
79. Kedde M, Strasser MJ, Boldajipour B, et al. RNA-binding protein Dnd1 inhibits microRNA access to target mRNA. Cell. 2007;131(7):1273–86.
80. Kobielak K, Stokes N, de la Cruz J, Polak L, Fuchs E. Loss of a quiescent niche but not follicle stem cells in the absence of bone morphogenetic protein signaling. Proc Natl Acad Sci U S A. 2007;104(24):10063–8.

Endothelial Ontogeny During Embryogenesis: Role of Cytokine Signaling Pathways

Daylon James, Marco Seandel and Shahin Rafii

Abstract Efficient distribution of oxygen and nutrients are fundamental and universal requirements of tissue homeostasis. In adults, angiogenesis from existing vessels occurs during routine tissue repair and is a prerequisite for the advance of tumor growth. Indeed, proliferation of any solid tissue beyond microscopic scales, in vivo or in vitro, is inherently dependent on vascularization. Endothelial ontogeny is hugely relevant to clinical science, as millions of patients each year succumb to vascular diseases resulting in mortality and morbidity due to tissue ischemia. Medical interventions are only partially effective in revascularizing ischemic tissues; moreover, recent clinical trials to induce revascularization using adult marrow-derived stem cells have yielded only minor benefit. Human embryonic stem cells (ESCs) have the potential to provide an unlimited source of vascular tissue for cell-based therapies, while also providing insight into the molecular events underlying human vascular development. In order for human ESCs to fulfill this potential, however, it will first be necessary to define the developmental events that influence endothelial differentiation and blood vessel formation. The molecular pathways involved in developmental vasculogenesis in mouse have been well studied, yet many elements of this process are obscure, and the correlation of these studies to human biology remains to be seen. Below, we review the current understanding of mammalian vascular development gained from studies of mouse genetic models and ESCs, focusing on the cytokine-mediated signaling pathways that govern this process.

Keywords Endothelial stem cells · Growth factor signaling · Angiogenesis · Vascular diseases · VEGF · TGF · BMP · Hedgehog · Notch · PDGF

S. Rafii (✉)
Howard Hughes Medical Institute, Ansary Center for Stem Cell Therapeutics, Department of Genetic Medicine, Division of Hematology-Oncology, Weill Cornell Medical College, 1300 York Avenue, Room A-863, New York, NY 10021
e-mail: srafii@med.cornell.edu

1 Introduction

Endothelial cells (EC), the essential components of blood vessels, first emerge in the yolk sac from the hemangioblast, a mesoderm-derived progenitor that gives rise to both hematopoietic and endothelial cell lineages [1]. The primary role of ECs is the formation of vessels to circulate oxygen and nutrients: first in the yolk sac prior to integration with maternal circulation; later in establishing embryonic and fetal vasculature; and throughout adulthood during the maintenance and repair of ischemic and/or damaged tissues. Yet ECs also provide molecular cues essential to embryonic organogenesis [2, 3], and in adults, endothelial precursor cells are recruited from circulation to support tumor neoangiogenesis [4]. This diversity of roles in embryonic, adult, and pathologic contexts has invited considerable scrutiny to the biology of endothelium in recent years. Many efforts have utilized in vivo platforms (i.e., genetic mutants, wound healing, and tumor models) to shed light on the molecular factors that direct embryonic and adult vasculogenesis/angiogenesis. However, data generated from these studies can be difficult to interpret, as the generation and assembly of vascular tissues are complex multifactorial processes, requiring collaboration between multiple cell types and signaling pathways. In vitro experiments using embryonic stem cells (ESCs) have proven to be a valuable complement to in vivo studies of vascular development. ESCs are a population of cells that have the unique ability to give rise to all the specialized cell types of the adult while also renewing themselves indefinitely in culture. Indeed, the differentiation of specialized tissues in cultured ESCs loosely recapitulates the temporal emergence of their correlates during embryogenesis [5]. In vitro, studies of vascular processes using ESCs can be advantageous because they allow the isolation and discrete assessment of molecular and environmental variables that regulate the many facets of vessel formation. Herein, we review experiments performed in mouse that have revealed some of the signaling pathways that are fundamental to vascular ontogeny and discuss their respective roles in that process.

V.K. Rajasekhar, M.C. Vemuri (eds.), *Regulatory Networks in Stem Cells*, Stem Cell Biology and Regenerative Medicine,
DOI 10.1007/978-1-60327-227-8_27, © Humana Press, a part of Springer Science+Business Media, LLC 2009

2 Endothelial Specification and Proliferation

During embryogenesis, an initial population of cells within the embryo, the inner cell mass, expands and becomes differentiated to the myriad cell types that will make up all of the specialized tissues present in the adult. Similarly, the establishment of the vascular system first requires the specification of vascular progenitors that subsequently expand to give rise to the diverse array of endothelial subtypes that make up the vascular tree. As a whole, vascular development is a complex process, demanding a coordinated sequence of cell fate decisions and intercellular communication. Yet the early events that dictate endothelial specification and proliferation have been well studied, and a few molecular pathways have emerged as fundamental players in this process.

2.1 VEGF Signaling

Vascular endothelial growth factors (VEGFs) are highly homologous proteins of the platelet-derived growth factor family, which act as endothelial-specific mitogens in vitro, and modulate both physiological and pathological vascular processes in vivo [6]. Of the five members of the VEGF family (VEGF-A, -B, -C, -D, and placental growth factor [PlGF]), VEGF-A has been the best characterized and has demonstrated a role in endothelial specification, proliferation, and migration during embryonic vascular development [7]. Alternative splicing of the VEGF-A gene results in five distinct isoforms, VEGF-A_{115}, -A_{120}, -A_{144}, -A_{164}, and -A_{188} [8]. These splice variants vary in their ability to bind heparin sulfate proteoglycans in the extracellular matrix, thereby affecting their tissue permeability. Smaller isoforms lack two heparin-binding domains, have low affinity for ECM and thus diffuse more deeply into tissues than larger isoforms.

Faithful VEGF-A expression is absolutely essential to the formation of embryonic vasculature. Null mutations of VEGF-A in mice are embryonic lethal between ED8.5 and 9.5 and fail to form blood islands. In fact, inactivation of a single allele of the VEGF-A gene results in embryonic lethality at mid-gestation with defects in blood island formation and endothelial differentiation [9, 10]. Specifically, while primitive blood and endothelial cells are formed, they are reduced in number and are not properly organized to form the vascular system. The lethal phenotype of heterozygous mutants indicates that an appropriate level of VEGF-A signaling is crucial to vascular development. Indeed, the levels of VEGF-A available to ECs can induce either proliferation (high) or migration (low) during vascular morphogenesis [11, 12]. This ability to elicit markedly different cellular responses at different doses is what defines morphogens, and is a fundamental requirement for growth and pattern formation in both embryonic and adult tissues.

The level of VEGF signal integration is regulated by multiple mechanisms. Apart from alternative splicing that varies the permeability of VEGF-A, there is also regulation at the receptor level. VEGF-A stimulates ECs mainly through binding to two tyrosine receptor kinases, VEGFR2 (Flk1, KDR), and VEGFR1 (Flt1) [6]. The affinity of VEGF-A for VEGFR1 is 10-fold higher than it is for VEGFR2 [13], though in spite of the presence of an intracellular kinase domain on VEGFR1, it is thought that this receptor may act to repress the amount of VEGF signal integration within a cell. Null mutations of either VEGFR1 [14, 15] or VEGFR2 [16] are embryonic lethal (VEGFR2 pheno-copies VEGF-A null mutants), but the phenotype of VEGFR1 mutants is vascular disorganization due to uncontrolled endothelial progenitor cell proliferation; and mice in which the intracellular kinase domain alone of VEGFR1 is genetically inactivated show normal development and angiogenesis [13]. Taken together, these findings suggest that VEGFR1 acts to attenuate VEGF signal integration by sequestering VEGF-A ligand away from VEGFR2. During embryonic development, VEGFR2 is first expressed on endothelial precursors (angioblasts), where it functions to induce endothelial differentiation and proliferation [17]. VEGFR1 expression comes at a later timepoint, when it may serve as a “sink” to reduce the pool of soluble VEGF-A, thereby altering EC fate from a proliferative state to a migratory/morphogenetic state.

VEGFs have also been shown to be a key cytokine regulating the function of adult vascular progenitors [18]. Circulating endothelial progenitors express VEGFR2 [19], and endothelial colony forming units are increased with systemic exposure to VEGF [20]. This effect is abrogated by treatment with anti-VEGF therapy. Concurrently, VEGF was found to increase incorporation of endothelial progenitors into the vasculature of newborn animals [21]. Subsequently, it was found that the VEGF-mediated effect required activity of MMP-9 to release kit-ligand in the bone marrow [22]. The functional significance of such VEGF-dependent endothelial progenitors is illustrated by the impaired tumor growth observed in the absence of normal progenitor mobilization [23, 24].

2.2 The TGFβ Superfamily

In addition to VEGF, there are numerous other signaling pathways involved in vascular development. The earliest developmental signals, required for specification/commitment of endothelial cells and their progenitors, are in the transforming growth factor beta (TGFβ) pathway. The TGFβ superfamily includes more than 40 ligands in the human genome and has diverse functions in both embryonic development and adult tissue homeostasis [25]. The pathway uses a simple mechanism to signal to the nucleus, whereby ligand binding of results in heterodimerization type I and type II receptors, leading to phosphorylation of effector SMADs, which translocate to the nucleus to modulate gene expression. The TGFβ superfamily is divided into two main branches, which signal through distinct type I receptors and effector Smads: bone morphogenetic proteins (BMPs), which employ the type I receptors ALK1, 2, 3, and 6 along with Smads 1 and 5; and TGFβ/Activin, which employ ALK4, 5 and 7 and Smads 2 and 3. Both branches of TGFβ signaling play roles in vascular specification and remodeling at various points in EC developmental ontogeny. Specifically, this signaling axis mediates the balance of endothelial cell specification, proliferation and organization during embryogenesis.

2.2.1 BMP Signaling

Bone morphogenetic proteins (BMPs) were initially isolated for their potential to induce ectopic bone and cartilage in rats [26], but this family of ligands has since been shown to regulate several cellular/developmental processes including growth, differentiation, survival, and patterning. Experiments across multiple model organisms have implicated a role for BMPs in dorso-ventral axis formation, specifically in assigning ventral identity. In *Xenopus laevis*, high BMP activity has been noted in the ventral side of gastrula stage embryos, a region fated to give rise to blood and vascular cells [27]. And in human embryos, BMPs are also highly expressed ventrally, in mesoderm surrounding the dorsal aorta and containing the AGM [28]. In mice, multiple targeted deletion models within the BMP pathway (BMP4, Alk2/3, Smad1/5) result in an early embryonic lethal phenotype [29–33] – BMP4 null mutations are embryonic lethal around ED7.5, displaying defects in mesoderm formation and patterning, but embryos can survive as late as ED9.5 and show vascular abnormalities at this stage with reduced blood islands. Furthermore, mutations of the BMP family effector molecules, Smad1 and Smad5, result in a similar phenotype of impaired yolk sac circulation, though Smad5 mutant embryos display an over-proliferation of primitive blood cells at ED8.5 and the absence of blood cells at ED9.5. Collectively, these data suggest that BMP signaling is necessary for proper vascular development in vivo, but the fundamental requirement for BMPs in the specification of ventral mesoderm obscures the precise role of these molecules during vascular development.

Many studies have assessed the effect of BMPs on endothelial differentiation and patterning using in vitro cell culture. These experiments have the benefits of precisely defining the temporal windows in which BMPs have their effect while also allowing direct observation of EC morphogenesis by microscopy. In mouse ESC models of vasculogenesis, it has been shown that BMP signaling plays a dual role [34]. First, BMP4 is necessary for mesoderm induction around two days of differentiation (correlates to ~ED5.5), then, it stimulates the expression of VEGFR2 (Flk1) beginning after 3 days (~ED6.5). And in cultured human umbilical vein endothelial cells (HUVEC), overexpressing components of the BMP pathway results in increased capillary tube formation, a product of EC activation, migration, and organization [35]. These in vitro results are consistent with the phenotypes observed in genetic models of BMP deficiency, and in combination with these studies, suggests a context specific role for BMP signaling in vascular development: first, BMPs play a role in the induction of ventral mesoderm; and second, BMPs induce differentiation of hematopoietic and vascular progenitor cells.

BMPs have also been shown to partake in vascular processes in the context of adult tissues. Endothelial progenitor cells (EPC) express the BMP receptor, BMPR2, and BMP2 has been shown to protect them from apoptosis [36]. In contrast, another study found that BMP2 had no effect on EPC viability, but did induce migration [37]. In transgenic mice for which the expression of BMPR2 is silenced by RNA interference, the blood vessels of adults are dilated and lack integrity, resulting in shortened lifespan [38]. These effects are attributed to dysfunctions in apoptosis and endothelial cell migration. Taken together, the results from mouse null mutants, ESCs, and physiologic studies of BMP dysfunction in adult mice suggest a biphasic function of BMP signaling in developmental vasculogenesis, with a sustained role in maintenance of vascular integrity in adults:

2.2.2 TGFβ Signaling

In addition to the BMP family, there is an alternate branch of TGFβ superfamily signaling, which includes the TGFβ, Activin and Nodal ligands, and which is also implicated in a diverse array of developmental, physiological and pathological processes [25]. Similar to BMP, studies of TGFβ in multiple animal models have demonstrated a fundamental requirement for this pathway in mesoderm specification [39]. Yet, mouse null mutants of multiple TGFβ signaling components show severe cardiovascular defects resulting in embryonic lethality [40–42]. Specifically, TGFβ signals, in concert with BMPs, are thought to mediate the balance between endothelial proliferation/activation and maturation. TGFβs have been shown to signal in ECs via two distinct receptor complexes [43, 44]; canonical signaling, via the type I receptor, ALK-5 and downstream phosphorylation of Smad2 and Smad3, results in the inhibition of proliferation and sheet formation. In endothelial cells deficient for

ALK-5, TGFβ signaling was largely abrogated and this effect was enough to produce embryonic lethality in homozygous animals, attributed to vascular defects downstream of abnormal extracellular matrix production and migration in endothelial precursors [42]. Yet ECs also express ALK-1, the type I receptor that mediates BMP signal integration via Smad1 and Smad5. Moreover, ECs also express endoglin, a type I membrane glycoprotein which is part of the TGFβ receptor complex [45]. This reflects a unique capacity for ECs to respond to TGFβs via a noncanonical pathway, comprised of TGFβRII, ALK-1 and endoglin resulting in phosphorylation of the canonical BMP effectors, Smad1 and Smad5. The exact role of this signaling in balancing EC activation state remains unclear.

During mouse embryogenesis, dominant negative mutation of the type II receptor TBFβRII results in failed development at the two-cell stage [46], but homozygous null mutation results in lethality at around ED10.5 [47]. The vascular phenotype of these embryos is characterized by impaired vessel integrity, with a defect in terminal differentiation of ECs. Endoglin null mice are lethal at around ED11.5, with disorganized vasculature and failure to form mature blood vessels [45]. While overt disruption of vessel organization is evident at ED9.5–10.5, vascular smooth muscle cells show impaired development by ED8.5, and at ED9.5, they fail to surround major vessels. ALK-1 deficient embryos show a similar phenotype, dying around ED11.5, and exhibiting impaired vascular development [48]. Specifically, at ED9.5 the yolk sac lacks mature vessels and the embryo proper exhibits excessive fusion of capillary networks, with abnormally dilated vessels in both tissues. ALK-1 signaling seems mainly to be mediated by Smad5, as mouse null mutants of this gene die around ED10.5–11.5 due to similar circulatory defects [33]. At ED9.0, Smad5–/– embryos lack well-organized vasculature in the yolk sac and embryo proper, and vessels are enlarged and surrounded by a decreased number of vascular smooth muscle cells. These studies point toward a requirement for non-canonical, ALK-1 mediated TGFβ signaling in the maturation and stabilization of endothelial tubes, yet it is unclear how ECs respond to this input at the cellular level. Owing to the fact that ALK-1 null embryos show increased expression of VEGF, it has been proposed that noncanonical TGFβ signaling in mouse embryogenesis has an inhibitory influence on EC proliferation. Yet in vitro, specific activation or inhibition of ALK-1 results in increased or reduced EC proliferation/migration, respectively. The opposite effects occur downstream of ALK-5 activation/inhibition [43, 44]. Taken together, these results suggest a dual role for the TGFβ family in EC regulation. Yet unlike VEGF, for which disparate cellular responses are elicited depending on the concentration of ligand, TGFβs induce contrary responses depending on the receptor complex that is bound.

2.3 Wnt Signaling

Wnt signals have diverse functions in embryonic and adult cells, including cell fate determination, cell movement, and establishment of tissue polarity. Within the canonical Wnt signaling pathway, signals are integrated via the Frizzled family of receptors and transmitted to the nucleus by the β-catenin cascade. Wnts are necessary in early embryogenesis for the specification of the primitive streak and mesoderm [39] and, as such, null mutations of Wnt signaling components obscure the role of this pathway in vascular development. Studies in ESCs have overcome this impediment to address the role of this pathway in vascular development. Recent work has demonstrated a role for Wnt signaling in endothelial specification by using global gene expression analysis [49]. The study isolated the Flk1+ and Flk1– populations from differentiating mESCs and compared their expression profiles using microarray. This revealed differential regulation of Wnt ligands and receptors between the two populations and furthermore, activation of the pathway resulted in an expansion of Flk1+ vascular progenitors while inhibition resulted in diminished Flk1+ progenitors and a reduction in mature capillary-like structures. More recently, a regimented analysis of vascular differentiation from mESCs revealed a stage-specific role for Wnt signaling that occurred in two distinct phases [50]. Early in differentiation, around the time of initial mesodermal specification, stimulation of the Wnt/b-Catenin pathway enhanced cardiomyocyte differentiation at the expense of vascular and hematopoietic lineages. Conversely, activation of the pathway at a later phase, following induction of vascular and hematopoietic progenitors, resulted in increased expression of vascular and hematopoietic genes. These studies suggest that Wnt signal transduction executes diverse functions at different stages of vascular development. First, Wnt signaling is required for initial specification of the primitive streak and mesoderm; second, mesodermal precursors (with potential to differentiate to cardiomyocyte, vascular, and hematopoietic cell types) are pushed toward the cardiac lineage by Wnts; and third, vascular and hematopoietic progenitors are enriched in response to Wnts.

3 Endothelial Stabilization, Polarity, and Migration

Specification and expansion of endothelial cells is obviously a prerequisite to blood vessel formation and homeostasis, yet diverse cellular processes and accessory cell types are also necessary to establish and maintain vessel integrity in the face of the constant stress of the circulatory system. In order for endothelium to generate functional vessels, it must first be segregated into the diverse array of EC subtypes present in the vasculature. Later, these cells must undergo maturation to form closed tubules and be stabilized by smooth muscle cells (SMCs) and pericytes. In larger order blood vessels (arterioles/venules, arteries/veins), these mural cells confer the elasticity necessary to accommodate the hydraulic pressure of the cardiovascular system. These processes are essential for the development of functional vasculature, and disruption of the pathways essential for EC morphogenesis and/or the interaction between ECs and SMCs/pericytes results in failure to form organized vascular networks.

3.1 Ephrins

The Eph superfamily of receptors is the largest subfamily of receptor tyrosine kinases, with at least 15 members identified in vertebrates [51]. These receptors are unique in that together with their membrane-bound ligands, ephrins, they mediate bidirectional signal transduction between the cells on which they are expressed. In vertebrates, Eph receptors and ephrin ligands play a fundamental role in developmental, physiological, and disease processes, including neurogenesis, angiongenesis, and oncogenesis.

During mammalian embryogenesis, Eph receptors and ephrin ligands play an important role in vascular development [52]. In particular, EphB4 and its cognate ligand, ephrinB2, are critical for the establishment of arterial versus venous identity [53–55]. Targeted mutation in mice of either ephrin2B or EphB4 results in virtually identical phenotypes, with embryonic lethality at ED10.5, attributed to defects in both arterial and venous remodeling and a failure of intercalation between arteries and veins. Early in development, the transmembrane ligand ephrinB2 is expressed on arterial endothelial cells, while its receptor EphB4 is present on venous endothelium, supporting a role for these signaling pathways in polarizing ECs into these two types of blood vessels. Yet these molecules seem to play a larger role, as they are not exclusively expressed in endothelial cells; as embryonic development proceeds, expression of ephrinB2 expands to arterial SMCs, perhaps mediating vessel organization and/or stabilization.

In vitro, signaling between ephrinB2 and EphB4 bearing cells is involved in many of the cellular processes underlying angiogenesis. Chimeric ephrinB2-Fc protein that is bound to the culture surface induces detachment of endothelial cells, indicating a role in EC adhesion. EphrinB2-Fc also induces cell migration in HUVEC via the PI3 Kinase pathway [56, 57]. In addition to adhesion and migration, ephrinB2 and EphB4-signaling is also involved in differentiation of ECs. In mESCs, the pace and frequency of differentiation to hemangioblast, blood, cardiomyocyte, and vascular cell types was impaired for EphB4–/– mice [58], perhaps reflecting an essential function in the response to mesoderm inducing signals. Taken together, these studies indicate that reciprocal signaling between ephrinB2 and EphB4 plays an important role in vascular development, especially in defining the boundary between and organization of arteries and veins.

3.2 Notch Signaling

The Notch signaling pathway plays diverse role in embryogenesis, mediating cell fate decisions by either maintaining primitive undifferentiated identities or inducing differentiation, depending on the developmental context. In mammals, the pathway is comprised of five ligands, called jagged1/2 and delta1/3/4, and four receptors, notch1–4 [59]. Similar to the Eph pathway components, Notch ligands and receptors are distinct in that they are both transmembrane and thus mediate *trans*-signaling events between neighboring homotypic and heterotypic cell types. Targeted deletion of either Notch ligands or receptors results in embryonic lethality with severe cardiovascular defects. Mutants null for notch1 exhibit intact vasculogenesis but have major defects in angiogenesis resulting in lethality by ED9.5 [60, 61]. Deletion of the ligands jagged1 [62] or delta4 [63, 64] results in lethality at ED10.5 and 9.5, respectively, with both exhibiting failures in vascular remodeling, These phenotypes confirm the relevance of Notch signaling to cardiovascular development, though the precise mechanism of its involvement with blood vessel formation remains obscure.

3.3 PDGFβ Signaling

Platelet-derived growth factors (PDGF) comprise a family of molecules that are expressed on a wide array of cell types, and disregulation of this pathway has been implicated in a variety of human diseases including atherosclerosis and cancers [65]. PDGF ligands are homo and/or heterodimers that form from the combination of two of four polypeptides, PDGFs A, B, C, and D (notated as PDGF-XX). The most well studied PDGFs are the A and B forms, and similarly to VEGFs, these have been shown to contain sequences that confer high affinity binding to extracellular matrix and regulate their biological activity and availability. Notwithstanding this similarity, as well as significant homology to VEGFs, PDGF family components are clearly distinct, and display a different expression pattern and function to components of the VEGF pathway, yet at least some PDGF signaling components function in establishment of the cardiovascular system. PDGF-BB is expressed by endothelial cells and its cognate receptor, PDGFβ, is expressed in vascular smooth muscle and pericyte cells [65]. For this reason, it is thought that this signaling axis functions to mediate interaction and/or recruitment between ECs and the SMCs/pericytes that stabilize them. Indeed, null mutation of PDGF-B results in reduced pericyte association with capillaries, resulting in vascular fragility, hemorrhage and ultimately, embryonic lethality [66]. And knocking out the receptor PDGF-β results in a similar phenotype though the vascular pathology results from a severe shortage of vascular mural cells [67]. These phenotypes, along with the restricted expression of PDGF ligands and receptors on ECs and mural cells, respectively, support a model of vasculogenesis in which secretion of PDGF ligands by endothelium results in recruitment and proliferation of vascular SMCs and/or pericytes. As stabilization of blood vessels is required to accommodate the hydraulic forces present in a closed circulatory system, the PDGF signaling axis is an essential component of functional vasculogenesis.

3.4 Hedgehog Signaling

The cellular interactions that result in vessel stabilization are also mediated by other signaling pathways, in addition to PDGF. The Hedgehog pathway is a family of secreted molecules that serve diverse roles during embryonic development and adult tissue homeostasis. These proteins are bound by the receptor Patched1, relieving Patched1-mediated repression of Smoothened, which in turn promotes downstream gene expression by way of the transcription factor Gli [68]. Genetic studies in mice have scrutinized the role of Hedgehog signaling in vascular development, specifically in yolk sac vasculogenesis/hematopoiesis [69, 70]. Mutation of Indian Hedgehog (Ihh) does not abrogate the differentiation of yolk sac ECs, and these cells are able to form a primitive capillary plexus, however, the vessels that form are smaller and collapsed, and 50% of Ihh–/– embryos die at mid-gestation. Smoothened null embryos die earlier, at around ED9.5, with a similar, though more severe, phenotype, showing a complete lack of remodeling of the primitive vascular plexus. The disparate severity of the vascular phenotype between Ihh–/– and Smoothened–/– embryos is believed to be due to a compensation for Ihh absence by another Hedgehog family ligand, Desert Hedgehog, which begins to be expressed in the mesoderm of the yolk sac around mid-gestation.

The presence of ECs, albeit ECs compromised in their ability to undergo proper vascular morphogenesis, in Ihh–/– [70] and Smoothened–/– [71] mice suggests that the vascular defect in these embryos is not due to failures in EC specification. However, studies in mESCs have shown that Hedgehog signaling is required for endothelial differentiation. Although response to the Hedgehog signal in both embryos and mESC-derived embryoid bodies is restricted to the mesothelial and smooth muscle cells, Ihh–/– and Smo–/– mESCs do not form blood islands or ECs upon differentiation in embryoid bodies [70]. The inability to form ECs from Hedgehog-deficient mESCs is unexpected, as it contrasts with the phenotype of Ihh–/– or Smo–/– embryos, in which ECs are formed yet lack the ability to undergo vascular morphogenesis. Yet vascular smooth muscle and mesothelial cells, and not ECs, are the ones that incorporate the Hedgehog signaling input. Combined with the phenotype of the vascular defect (collapsed vessels that do not support blood flow), these results suggest a role for Hedgehog signaling in vessel stabilization and/or organization.

3.5 Stromal Cell-Derived Factor-1

Stromal cell–derived factor-1 (SDF-1) is an angiogenic chemokine expressed in multiple tissues such as lung, liver, skin, and bone marrow that signals through the CXCR4 G-protein coupled receptor [72]. The SDF1-CXCR4 signaling axis is thought to regulate cell trafficking – both mobilization from different tissue compartments and cellular retention elsewhere. Indeed, a variety of stimuli can up-regulate the constitutive level of SDF-1 expression [72]. This has been demonstrated in many different contexts including ischemic tissue, the immune system and in metastasis of tumor cells. The importance of the SDF1-CXCR4 signaling axis in development was revealed in the mouse knockout of CXCR4, which displayed abnormal superior mesenteric and stomach vessels, despite grossly normal organogenesis [73]. One of the mechanisms by which SDF1-CXCR4 signaling drives angiogenesis is by mobilization of bone marrow derived endothelial progenitors into the circulation and ultimately to sites of neoangiogenesis [22]. Paradoxically, a small molecule CXCR4 antagonist (AMD3100) causes an acute rise in circulating endothelial progenitors, concurrent with increased SDF1 [74]. The functional significance of SDF1 secretion in neoangiogenesis was demonstrated by Orimo et al. [75], who found that tumor-associated fibroblasts secrete SDF1 thereby inducing tumor growth through recruitment of endothelial progenitor cells from the circulation. Similarly, SDF1 can stimulate recovery from experimental hindlimb ischemia in mice by recruiting EPCs [76]. Futhermore, when CXCR4 signaling is impaired, function of endothelial progenitors is significantly reduced, as shown for CXCR4-expressing EPCs derived from patients with coronary artery disease [77]. The above experiments support a critical function for SDF1 in the maintenance of the vascular system. While genetic ablation of CXCR4 does not result in a dramatic vascular phenotype during embryogenesis, in adults, the SDF1/CXCR4 signaling axis is essential for proper mobilization of endothelial cells to sites of neoangiogenesis.

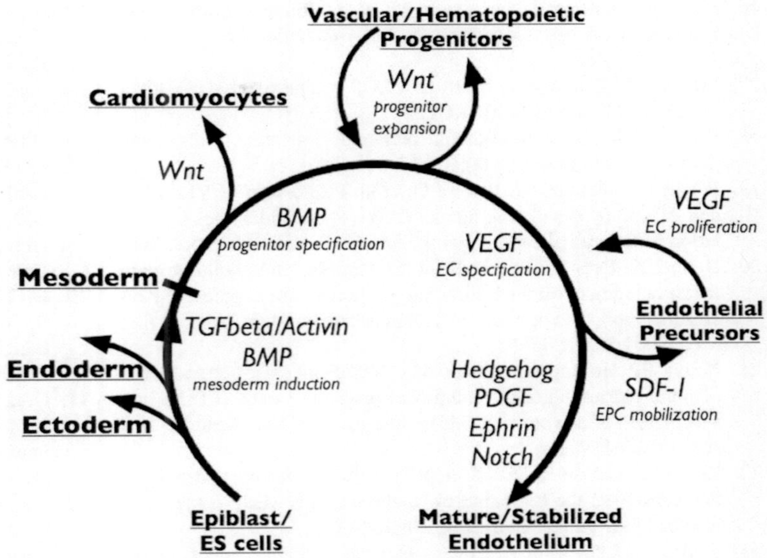

Fig. 1 During embryogenesis, blood vessels are formed by specification of vascular cells along a hierarchy of developmental stages.Mesoderm arises from pluripotent epiblast or ES cells and subsequently differentiates to multipotent vascular/hematopoietic progenitor cells. Endothelial derivatives of vascular/ hematopoietic progenitors undergo expansion and/or maturation to give rise to functional, stabilized blood vessels. Cell populations at each step in the differentiation process are underlined. Signal transduction pathways involved in the establishment of discreet cell fates are shown in italics

4 Conclusion

The formation and maintenance of a functional vascular system requires the cooperation of numerous cell types and the intersection of multiple signaling pathways (shown schematically in Figure 1). Assigning discreet roles to the molecular factors involved in this complex process is of critical importance because it promises to advance our understanding of angiogenesis in both developmental and pathologic contexts. Genetic approaches in mice have elucidated many of the pathways critical to proper vascular development in vivo, and ESC studies have extended this knowledge to define the cellular mechanisms that underlie vasculogenesis. This work has revealed a multifaceted process requiring the specification, proliferation, migration, and morphogenesis of ECs, all in the context of dynamic interactions, both between endothelial subtypes as well as with mural cells that are essential for vessel stabilization. As the picture of vascular specification and development has increased in resolution, tools have emerged that exploit vasculogenic processes for clinical treatments, including cancer. Furthermore, with human embryonic stem cells as a tool, current and future work can model human vascular tissue, in both healthy and diseased contexts, and generate virtually unlimited numbers of ECs for scientific and clinical study.

References

1. Baron MH. Molecular regulation of embryonic hematopoiesis and vascular development: a novel pathway. J Hematother Stem Cell Res. 2001;10:587–94.
2. Matsumoto K, Yoshitomi H, Rossant J, Zaret KS. Liver organogenesis promoted by endothelial cells prior to vascular function. Science. 2001;294:559–63.
3. Lammert E, Cleaver O, Melton D. Role of endothelial cells in early pancreas and liver development. Mech Dev. 2003;120:59–64.
4. Rafii S, Lyden D. Therapeutic stem and progenitor cell transplantation for organ vascularization and regeneration. Nat Med. 2003;9:702–12.
5. Smith AG. Embryo-derived stem cells: of mice and men. Annu Rev Cell Dev Biol. 2001;17:435–62.
6. Shibuya M, Claesson-Welsh L. Signal transduction by VEGF receptors in regulation of angiogenesis and lymphangiogenesis. Exp Cell Res. 2006;312:549–60.
7. Houck KA, Ferrara N, Winer J, Cachianes G, Li B, Leung DW. The vascular endothelial growth factor family: identification of a fourth molecular species and characterization of alternative splicing of RNA. Mol Endocrinol. 1991;5:1806–14.
8. Park JE, Keller GA, Ferrara N. The vascular endothelial growth factor (VEGF) isoforms: differential deposition into the subepithelial extracellular matrix and bioactivity of extracellular matrix-bound VEGF. Mol Biol Cell. 1993;4:1317–26.
9. Carmeliet P, Ferreira V, Breier G, et al. Abnormal blood vessel development and lethality in embryos lacking a single VEGF allele. Nature. 1996;380:435–9.
10. Ferrara N, Carver-Moore K, Chen H, et al. Heterozygous embryonic lethality induced by targeted inactivation of the VEGF gene. Nature. 1996;380:439–42.
11. Ruhrberg C, Gerhardt H, Golding M, et al. Spatially restricted patterning cues provided by heparin-binding VEGF-A control blood vessel branching morphogenesis. Genes Dev. 2002;16: 2684–98.
12. Gerhardt H, Golding M, Fruttiger M, et al. VEGF guides angiogenic sprouting utilizing endothelial tip cell filopodia. J Cell Biol. 2003;161:1163–77.
13. Hiratsuka S, Minowa O, Kuno J, Noda T, Shibuya M. Flt-1 lacking the tyrosine kinase domain is sufficient for normal development and angiogenesis in mice. Proc Natl Acad Sci U S A. 1998;95: 9349–54.
14. Kearney JB, Ambler CA, Monaco KA, Johnson N, Rapoport RG, Bautch VL. Vascular endothelial growth factor receptor Flt-1 negatively regulates developmental blood vessel formation by modulating endothelial cell division. Blood. 2002;99: 2397–407.
15. Fong GH, Zhang L, Bryce DM, Peng J. Increased hemangioblast commitment, not vascular disorganization, is the primary defect in flt-1 knock-out mice. Development. 1999;126:3015–25.

16. Shalaby F, Rossant J, Yamaguchi TP, et al. Failure of blood-island formation and vasculogenesis in Flk-1-deficient mice. Nature. 1995;376:62–6.
17. Matsumoto T, Claesson-Welsh L. VEGF receptor signal transduction. Sci STKE. 2001;2001:RE21.
18. Rafii S. Circulating endothelial precursors: mystery, reality, and promise. J Clin Invest. 2000;105:17–9.
19. Peichev M, Naiyer AJ, Pereira D, et al. Expression of VEGFR-2 and AC133 by circulating human CD34(+) cells identifies a population of functional endothelial precursors. Blood. 2000;95:952–8.
20. Hattori K, Dias S, Heissig B, et al. Vascular endothelial growth factor and angiopoietin-1 stimulate postnatal hematopoiesis by recruitment of vasculogenic and hematopoietic stem cells. J Exp Med. 2001;193:1005–14.
21. Young PP, Hofling AA, Sands MS. VEGF increases engraftment of bone marrow-derived endothelial progenitor cells (EPCs) into vasculature of newborn murine recipients. Proc Natl Acad Sci U S A. 2002;99:11951–6.
22. Heissig B, Hattori K, Dias S, et al. Recruitment of stem and progenitor cells from the bone marrow niche requires MMP-9 mediated release of kit-ligand. Cell. 2002;109:625–37.
23. Shaked Y, Ciarrocchi A, Franco M, et al. Therapy-induced acute recruitment of circulating endothelial progenitor cells to tumors. Science. 2006;313:1785–7.
24. Lyden D, Hattori K, Dias S, et al. Impaired recruitment of bone-marrow-derived endothelial and hematopoietic precursor cells blocks tumor angiogenesis and growth. Nat Med. 2001;7: 1194–201.
25. Massague J, Blain SW, Lo RS. TGFbeta signaling in growth control, cancer, and heritable disorders. Cell. 2000;103:295–309.
26. Urist MR. Bone: formation by autoinduction. Science. 1965;150:893–9.
27. De Robertis EM, Kuroda H. Dorsal-ventral patterning and neural induction in Xenopus embryos. Annu Rev Cell Dev Biol. 2004;20:285–308.
28. Marshall CJ, Kinnon C, Thrasher AJ. Polarized expression of bone morphogenetic protein-4 in the human aorta-gonad-mesonephros region. Blood. 2000;96:1591–3.
29. Winnier G, Blessing M, Labosky PA, Hogan BL. Bone morphogenetic protein-4 is required for mesoderm formation and patterning in the mouse. Genes Dev. 1995;9:2105–16.
30. Gu Z, Reynolds EM, Song J, et al. The type I serine/threonine kinase receptor ActRIA (ALK2) is required for gastrulation of the mouse embryo. Development. 1999;126:2551–61.
31. Mishina Y, Suzuki A, Ueno N, Behringer RR. Bmpr encodes a type I bone morphogenetic protein receptor that is essential for gastrulation during mouse embryogenesis. Genes Dev. 1995;9: 3027–37.
32. Tremblay KD, Dunn NR, Robertson EJ. Mouse embryos lacking Smad1 signals display defects in extra-embryonic tissues and germ cell formation. Development. 2001;128:3609–21.
33. Chang H, Huylebroeck D, Verschueren K, Guo Q, Matzuk MM, Zwijsen A. Smad5 knockout mice die at mid-gestation due to multiple embryonic and extraembryonic defects. Development. 1999;126:1631–42.
34. Park C, Afrikanova I, Chung YS, et al. A hierarchical order of factors in the generation of FLK1- and SCL-expressing hematopoietic and endothelial progenitors from embryonic stem cells. Development. 2004;131:2749–62.
35. Valdimarsdottir G, Goumans MJ, Rosendahl A, et al. Stimulation of Id1 expression by bone morphogenetic protein is sufficient and necessary for bone morphogenetic protein-induced activation of endothelial cells. Circulation. 2002;106:2263–70.
36. Teichert-Kuliszewska K, Kutryk MJ, Kuliszewski MA, et al. Bone morphogenetic protein receptor-2 signaling promotes pulmonary arterial endothelial cell survival: implications for loss-of-function mutations in the pathogenesis of pulmonary hypertension. Circ Res. 2006;98:209–17.
37. Raida M, Heymann AC, Gunther C, Niederwieser D. Role of bone morphogenetic protein 2 in the crosstalk between endothelial progenitor cells and mesenchymal stem cells. Int J Mol Med. 2006;18:735–9.
38. Liu D, Wang J, Kinzel B, et al. Dosage-dependent requirement of BMP type II receptor for maintenance of vascular integrity. Blood. 2007;110:1502–10.
39. Kimelman D. Mesoderm induction: from caps to chips. Nat Rev Genet. 2006;7:360–72.
40. Dickson MC, Martin JS, Cousins FM, Kulkarni AB, Karlsson S, Akhurst RJ. Defective haematopoiesis and vasculogenesis in transforming growth factor-beta 1 knock out mice. Development. 1995;121:1845–54.
41. Goumans MJ, Zwijsen A, van Rooijen MA, Huylebroeck D, Roelen BA, Mummery CL. Transforming growth factor-beta signalling in extraembryonic mesoderm is required for yolk sac vasculogenesis in mice. Development. 1999;126:3473–83.
42. Larsson J, Goumans MJ, Sjostrand LJ, et al. Abnormal angiogenesis but intact hematopoietic potential in TGF-beta type I receptor-deficient mice. EMBO J. 2001;20:1663–73.
43. Goumans MJ, Valdimarsdottir G, Itoh S, Rosendahl A, Sideras P, ten Dijke P. Balancing the activation state of the endothelium via two distinct TGF-beta type I receptors. EMBO J. 2002;21: 1743–53.
44. Lebrin F, Deckers M, Bertolino P, Ten Dijke P. TGF-beta receptor function in the endothelium. Cardiovasc Res. 2005;65:599–608.
45. Arthur HM, Ure J, Smith AJ, et al. Endoglin, an ancillary TGFbeta receptor, is required for extraembryonic angiogenesis and plays a key role in heart development. Dev Biol. 2000;217:42–53.
46. Goumans MJ, Mummery C. Functional analysis of the TGFbeta receptor/Smad pathway through gene ablation in mice. Int J Dev Biol. 2000;44:253–65.
47. Oshima M, Oshima H, Taketo MM. TGF-beta receptor type II deficiency results in defects of yolk sac hematopoiesis and vasculogenesis. Dev Biol. 1996;179:297–302.
48. Oh SP, Seki T, Goss KA, et al. Activin receptor-like kinase 1 modulates transforming growth factor-beta 1 signaling in the regulation of angiogenesis. Proc Natl Acad Sci U S A. 2000;97:2626–31.
49. Wang H, Charles PC, Wu Y, et al. Gene expression profile signatures indicate a role for Wnt signaling in endothelial commitment from embryonic stem cells. Circ Res. 2006;98:1331–9.
50. Naito AT, Shiojima I, Akazawa H, et al. Developmental stage-specific biphasic roles of Wnt/beta-catenin signaling in cardiomyogenesis and hematopoiesis. Proc Natl Acad Sci U S A. 2006;103:19812–7.
51. Gale NW, Yancopoulos GD. Ephrins and their receptors: a repulsive topic? Cell Tissue Res. 1997;290:227–41.
52. Zhang J, Hughes S. Role of the ephrin and Eph receptor tyrosine kinase families in angiogenesis and development of the cardiovascular system. J Pathol. 2006;208:453–61.
53. Gerety SS, Anderson DJ. Cardiovascular ephrinB2 function is essential for embryonic angiogenesis. Development. 2002;129: 1397–410.
54. Adams RH, Wilkinson GA, Weiss C, et al. Roles of ephrinB ligands and EphB receptors in cardiovascular development: demarcation of arterial/venous domains, vascular morphogenesis, and sprouting angiogenesis. Genes Dev. 1999;13:295–306.
55. Wang HU, Chen ZF, Anderson DJ. Molecular distinction and angiogenic interaction between embryonic arteries and veins revealed by ephrin-B2 and its receptor Eph-B4. Cell. 1998;93:741–53.
56. Maekawa H, Oike Y, Kanda S, et al. Ephrin-B2 induces migration of endothelial cells through the phosphatidylinositol-3 kinase pathway and promotes angiogenesis in adult vasculature. Arterioscler Thromb Vasc Biol. 2003;23:2008–14.

57. Fuller T, Korff T, Kilian A, Dandekar G, Augustin HG. Forward EphB4 signaling in endothelial cells controls cellular repulsion and segregation from ephrinB2 positive cells. J Cell Sci. 2003;116:2461–70.
58. Wang Z, Cohen K, Shao Y, Mole P, Dombkowski D, Scadden DT. Ephrin receptor, EphB4, regulates ES cell differentiation of primitive mammalian hemangioblasts, blood, cardiomyocytes, and blood vessels. Blood. 2004;103:100–9.
59. Artavanis-Tsakonas S, Rand MD, Lake RJ. Notch signaling: cell fate control and signal integration in development. Science. 1999;284:770–6.
60. Swiatek PJ, Lindsell CE, del Amo FF, Weinmaster G, Gridley T. Notch1 is essential for postimplantation development in mice. Genes Dev. 1994;8:707–19.
61. Krebs LT, Xue Y, Norton CR, et al. Notch signaling is essential for vascular morphogenesis in mice. Genes Dev. 2000;14:1343–52.
62. Xue Y, Gao X, Lindsell CE, et al. Embryonic lethality and vascular defects in mice lacking the Notch ligand Jagged1. Hum Mol Genet. 1999;8:723–30.
63. Duarte A, Hirashima M, Benedito R, et al. Dosage-sensitive requirement for mouse Dll4 in artery development. Genes Dev. 2004;18:2474–8.
64. Gale NW, Dominguez MG, Noguera I, et al. Haploinsufficiency of delta-like 4 ligand results in embryonic lethality due to major defects in arterial and vascular development. Proc Natl Acad Sci U S A. 2004;101:15949–54.
65. Betsholtz C, Karlsson L, Lindahl P. Developmental roles of platelet-derived growth factors. Bioessays. 2001;23:494–507.
66. Leveen P, Pekny M, Gebre-Medhin S, Swolin B, Larsson E, Betsholtz C. Mice deficient for PDGF B show renal, cardiovascular, and hematological abnormalities. Genes Dev. 1994;8:1875–87.
67. Soriano P. Abnormal kidney development and hematological disorders in PDGF beta-receptor mutant mice. Genes Dev. 1994;8:1888–96.
68. Varjosalo M, Taipale J. Hedgehog signaling. J Cell Sci. 2007;120:3–6.
69. Byrd N, Grabel L. Hedgehog signaling in murine vasculogenesis and angiogenesis. Trends Cardiovasc Med. 2004;14:308–13.
70. Byrd N, Becker S, Maye P, et al. Hedgehog is required for murine yolk sac angiogenesis. Development. 2002;129:361–72.
71. Zhang XM, Ramalho-Santos M, McMahon AP. Smoothened mutants reveal redundant roles for Shh and Ihh signaling including regulation of L/R symmetry by the mouse node. Cell. 2001;106:781–92.
72. Petit I, Karajannis MA, Vincent L, et al. The microtubule targeting agent, CA4P, regresses leukemic xenografts by disrupting interaction with vascular vells and mitochondrial-dependent cell death. Blood. 2008;111(4):1951–61.
73. Tachibana K, Hirota S, Iizasa H, et al. The chemokine receptor CXCR4 is essential for vascularization of the gastrointestinal tract. Nature. 1998;393:591–4.
74. Shepherd RM, Capoccia BJ, Devine SM, et al. Angiogenic cells can be rapidly mobilized and efficiently harvested from the blood following treatment with AMD3100. Blood.. 2006;108:3662–7.
75. Orimo A, Gupta PB, Sgroi DC, et al. Stromal fibroblasts present in invasive human breast carcinomas promote tumor growth and angiogenesis through elevated SDF-1/CXCL12 secretion. Cell. 2005;121:335–48.
76. Yamaguchi J, Kusano KF, Masuo O, et al. Stromal cell-derived factor-1 effects on ex vivo expanded endothelial progenitor cell recruitment for ischemic neovascularization. Circulation. 2003;107:1322–8.
77. Walter DH, Haendeler J, Reinhold J, et al. Impaired CXCR4 signaling contributes to the reduced neovascularization capacity of endothelial progenitor cells from patients with coronary artery disease. Circ Res. 2005;97:1142–51.

Signaling Networks in Mesenchymal Stem Cells

Vivek M. Tanavde, Lailing Liew, Jiahao Lim and Felicia Ng

Abstract Mechanisms governing the differentiation of mesenchymal stem cells (MSC) are poorly understood. Functional network analysis based on gene expression data is a powerful tool to study signaling mechanisms in cells. Such studies have yielded useful information in other stem cell systems like embryonic and hematopoietic stem cells. Studies dealing with single pathways or genes active in MSC form the building blocks for network analyses of MSC. This chapter describes how functional network analysis is useful in identifying signaling pathways that are active in MSC and how a core MSC network identifies known MSC pathways while predicting new pathways that are important in MSC differentiation.

Keywords Mesenchymal stem cells · MSC signaling · Network analysis of gene expression

1 Introduction

Mesenchymal stem cells (MSC) are stem cell populations that have the capacity to differentiate into cells of connective tissue lineages, including bone, fat, cartilage, and muscle [1, 2]. In addition, some reports show a wider differentiation potential that includes ectodermal and endodermal lineages [3, 4]. MSC also play a role in providing the stromal support system for hematopoietic stem cells (HSC) in the marrow [5]. MSC can be obtained in relatively large numbers from a variety of connective tissue sources throughout development and in the adult. These include adipose [6] and dermal tissue [7], synovial fluid [8, 9], deciduous teeth [10–12], cord blood [13, 14], amniotic fluid [15–17], and placenta [18–20]. However, the largest body of data available is on postpubertal stromal MSC [21, 22] from the bone marrow and these are often used as the prototypical MSC population. Adult stromal derived MSC have been shown to be easy to cultivate and expand [1], and they maintain pluripotency after prolonged culture and thus sufficient numbers of cells can be obtained for therapy both in autologous use and for allogenic transplants [23]. Although MSC are already being used therapeutically, little is known about the mechanisms that control their growth and differentiation. Understanding these mechanisms is essential for designing better media for culturing MSC as well as achieving the desired differentiated progeny in vivo or in vitro. Functional network analysis of gene expression data is a useful approach for unraveling the mechanisms involved in MSC growth and differentiation. Such network analyses draw upon approaches perfected for other stem cells as well as published literature on signaling pathways active in MSC. This chapter aims to elucidate how these approaches can be used to decipher MSC biology.

2 Network Analyses in Other Stem Cells

Analysis of signaling networks from other stem cells also serve as good models for devolving MSC signaling networks. Signaling networks have been developed for human ES cells based on EST analysis [24], signaling pathway kinetics [25] and Bayesian modeling signaling networks based on protein phosphorylation states [26]. Similarly in HSC, TGF-β modulates differentiation through regulation of KIT, IL-6 receptor, FLT3, and MPL receptors[27]. Genome-wide computational analysis of HSC has yielded the mechanism of how SCL, Smad-6, BMP-4, and Runx transcription factors interact with each other to control fate of HSC in early hematopoiesis [28]. Although there is no evidence of expression of SCL in MSC. Runx is down-regulated during MSC differentiation[29] and TGF-β signaling through Smad is important in osteogenic and chondrogenic differentiation of MSC [30] as well. It is possible that the same genes use similar signaling networks to control differentiation in different stem cells.

V.M. Tanavde (✉)
Genome & Gene Expression Data Analysis Group, Bioinformatics Institute, A* STAR Singapore, 30 Biopolis St., Matrix #07-01 Singapore 138671
e-mail: vivek@bii.a-star.edu.sg

V.K. Rajasekhar, M.C. Vemuri (eds.), *Regulatory Networks in Stem Cells*, Stem Cell Biology and Regenerative Medicine,
DOI 10.1007/978-1-60327-227-8_28, © Humana Press, a part of Springer Science+Business Media, LLC 2009

3 Cell Signaling in MSC

Most information about signaling pathways active in MSC is derived from studying the culture conditions that support MSC growth [31] or from signaling pathways implicated in other adult stem cells. For example, it is well established that β-FGF is necessary for MSC expansion [32, 33]. Similarly bone morphogenetic proteins (BMPs) have been shown to play a major role in MSC differentiation [34]. IL-6, which is a well-known differentiation cytokine in the hematopoietic system, induces osteogenic differentiation through gp130 signaling [35]. Wnt 3a promotes MSC proliferation [36], whereas Smad-3 is needed for MSC migration [37]. Epidermal growth factor receptor (EGFR) signaling through Her-1 has also been implicated in MSC proliferation and survival [38]. Jagged1 signaling through the Notch pathway promotes differentiation of MSC into cardiomyogenic cells [39]. These studies form the building blocks of functional network analysis. They provide experimentally verified information about the signaling molecules important in MSC differentiation. Many groups have also focused their efforts on signaling cascades controlling specific lineages of MSC differentiation. For example, the TGF-β family members play an important role in chondrogenic differentiation of MSC [30, 40–42], whereas Wnt signaling is a suppressor of osteogenic differentiation [36]. Attempts to elucidate signaling pathways in MSC differentiation have also focused on osteogenic and chondrogenic differentiation and the mechanisms controlling the "switch" between differentiation into these two lineages. Apart from TGF-β and Wnt pathways, signaling through the FGF receptor has been shown to have a mitogenic effect on undifferentiated MSC [32] as well as exert positive and negative effects on limb, frontonatal and mandibular chondrogenesis [43]. PDGF mainly from platelet extracts promotes MSC survival and expansion [44]. In addition, tethered EGF provides survival advantage to MSC [45], whereas IGF-1 regulates chondrogenic differentiation of periosteal MSC [46]. However, these studies mostly study the significance of individual signaling pathways in MSC differentiation. It is unclear how these pathways interact with other to regulate MSC differentiation. It is also unclear whether there are common pathways controlling MSC growth and if differentiation into tissues of mesodermal origin like bone and cartilage is governed by similar pathways as differentiation into nonmesodermal cells like neurons.

In addition to data from individual pathways and molecules, temporally ordered changes in global gene expression can be used to infer functional connectivity. This approach requires co-opting mathematical and computational tools from other disciplines such as engineering and physics.

4 Compilation of These Pathways into Networks

Reconstruction of biochemical networks is a complex task. The role of each protein in a signaling network is to communicate the signal from one node to the next. To accomplish this the protein has to be in a defined signaling "state." The state of a signaling molecule is characterized by covalent modifications of the native polypeptide, the ligands bound to the protein, its state of association with another protein and its cellular location. Interactions within and between functional states of molecules as well as transitions between functional states, provide the building blocks for reconstruction of a signaling network. Among the different approaches used for generating networks based on gene expression data, the clustering of co-expression profiles is most promising [47]. This approach allows the user to infer shared regulatory inputs and functional pathways. Analysis of large datasets of gene expression data (often collected from a number of clinical samples or at discrete intervals over a long period of time from few samples), can reveal casual connections between genes which may not be apparent in one to two experiments carried out on few samples. Such an approach is useful for reverse engineering of genetic networks from expression data and biological knowledge.

Insights into how complex signaling networks orchestrate MSC function can be derived from studies elucidating the role of multiple signaling pathways in MSC differentiation into specific lineages. For example, Wnt and Runx signaling is critical for differentiation of MSC to osteoblasts via networks involving TGF-β, FGF, hedgehog, and ephrin signaling [48]. This process may be regulated by phosphophoryn and also involves components of the MAP-kinase pathway [49]. Similarly, hypoxia also plays a role in chondrogenic differentiation of MSC through oxygen sensing networks [50]. Network analysis can be also be applied to understand the paracrine activity of MSC. For example, mechanically stimulated MSC secrete matrix metalloprotease 2, TGF-β, and FGF to stimulate angiogenesis in a paracrine manner [51]. Similarly a complex network of Activin, bone morphogenetic protein (BMP), FGF, and Follistatin regulates epithelial stem cell proliferation in teeth [52]. Since receptors of some of these cytokines are also expressed on the MSC surface it is possible that these cytokines also act in an autocrine manner on the MSC themselves.

Tsai et al. [53] has used such an approach to compare functional networks in MSC derived from different tissues.Using marrow-derived MSC as the standard they compared the transcriptomes of MSC derived from amniotic fluid, amniotic membrane, and umbilical cord blood to generate specific networks of genes active in these cells. They further classified these gene networks into functional groups. Using functional network analysis, they identified a core MSC network of genes expressed in MSC from all tissues. This network when subjected to functional network analysis suggests that TGF-β signaling is a common feature in all these cells, since they were able to identify components of the TGF-β pathway like Smurf-2 and Smad. Using functional analysis networks, they also identified the importance of the Wnt signaling pathway in MSC.

However, when looking at such data it must kept in mind that these studies provide a computational prediction of gene networks active in MSC. Whether these networks play a role in cell function remains to be proven. Experimental validation of these predictions is of paramount importance to prove the accuracy of such predictions.

The other approach to generating biological networks is based on published data of proteins or genes interacting with each other. There are a number of commercial platforms like Ingenuity Pathway analysis, Pathway Studio [54], and Metacore [55]. These platforms integrate gene expression data with published data on protein-protein interactions and generate networks of expressed genes. The genes in these lists are included based on their expression profile, whereas the relationships between genes (or nodes in a network) are based on published literature. One major disadvantage of these platforms, however, is they are not cell-type specific and therefore the results maybe erroneous if stem cells use different pathways compared to differentiated cells (or cell lines) on which these network relationships are based. Also the underlying databases used by these platforms seldom cover all the published information in every field. Therefore, there is chance that information published in a highly domain specific journal may not be represented in these databases. Thus, the experimentally proven relationship between these proteins may not be captured in any of these databases and therefore will not show up in the generated network. However, when combined with gene expression data and prior biological knowledge, these tools are useful in unraveling the relationships between specific genes of interest and other novel genes that may be implicated in stem cell functions.

We have used such a combination of gene expression analysis with networks based on published data for identifying signaling networks active in marrow MSC. Figure 1 shows the network generated using Ingenuity Pathway Analysis (IPA) from differentially expressed genes in undifferentiated MSC compared to osteoblasts, chondrocytes, and adipocytes derived from these MSC. Such network analysis of the approximately 7000 genes expressed by MSC is too complex and is not of much value to unravel MSC biology. We then looked at the differential expression of genes as MSC differentiate into osteoblasts, chondrocytes, or adipocytes and generated a signaling pathway network with the gene expression data overlaid on this network. Figure 2 shows networks for TGF-β and PDGF pathways with genes strongly expressed in undifferentiated MSC. This network yields useful information about the likely pathways active in MSC differentiation. Genes like Smurf-2, Smad-7, etc., are components of Activin signaling through TGF-β pathway. We combined such network analysis with biological insights from published literature and identified PDGF, TGF-β, and FGF signaling as critical for MSC survival and differentiation. The validity of our approach was tested in an experimental system where we inhibited these signaling networks using specific small molecule inhibitors of receptor tyrosine kinases.

The TGF-β pathway has been shown to be important in MSC differentiation into the osteogenic and chondrogenic lineages [59, 60]. Also in our study, Activin receptor was very strongly expressed on undifferentiated MSC. The Activin/Nodal pathway that signals through the TGF-β pathway co-operates with FGF signaling in maintaining the pluripotency of embryonic stem cells [61]. Therefore in this study, we investigated whether Activin signaling plays a similar role in MSC differentiation. Treatment with ALK-5 inhibitor SB431542 did not result in extensive cell death, but increased the population doubling time from 2 days to 4.5 days at 20 μM concentration. This observation further supports the fact that TGF-β signaling is an important event in MSC proliferation in a Smad-3 dependant manner [62]. TGF-β is also an inhibitor of adipogenesis via Smad 3 signaling [30]. Our results provide further evidence to this observation since blocking Smad 3 mediated TGF-β signaling resulted in enhanced adipogenic differentiation with complete inhibition of chondrogenic differentiation.

The PDGF signaling inhibitor Tyrphostin was toxic to MSC at 40 μM concentration. Inhibiting the PDGF receptor-α did not result in complete inhibition of adipogenic or chondrogenic differentiation in our perturbation experiments. Therefore, either PDGF signaling may not be critical for MSC differentiation into these lineages or there is some cross-talk between components of the PDGF pathway with other signaling pathways. Although PDGF signaling is important in MSC migration [56], there is little published evidence of the importance of PDGF in MSC differentiation, apart from inhibition of osteogenesis [57, 58]. Our results also support the hypothesis that, while PDGF signaling is important in MSC survival, it probably does not play an important role in differentiation.

FGF is an important growth factor for MSC expansion. Addition of β-FGF has been shown to increase the growth rate and life span of MSC from different species [32, 59]. Surprisingly the FGFR1 transcript was very weakly expressed in undifferentiated MSC and was not up-regulated during differentiation. FGFR1 is a key regulator of osteoblast maturation in osteogenesis [60, 61]. Therefore it is not surprising that inhibition of FGF signaling leads to

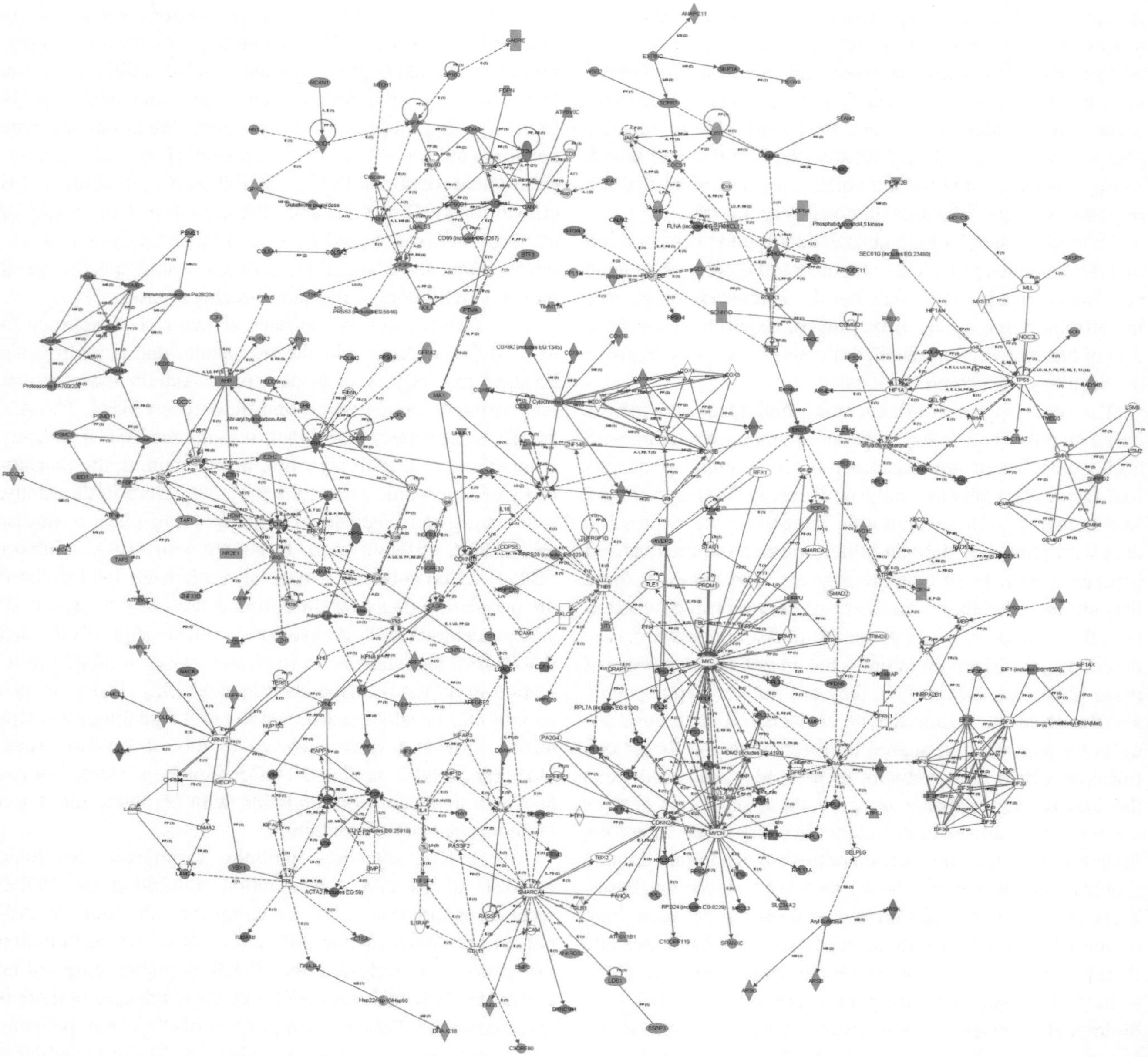

Fig. 1 Network of all expressed genes of MSC. This network was generated using the top and bottom 5th percentile of genes expressed by MSC in Ingenuity Pathway Analysis. The original network consisted of 50 subnetworks. For better visualization, a network comprising of only 10 subnetworks is represented here. The red boxes are genes that were up-regulated in MSC whereas the green boxes are genes that are down-regulated in MSC compared to MSC derived osteoblasts, chondrocytes, and adipocytes. It is clear that not much information can be obtained from such a network since it is too complex and represents too much data to be handled in an experimental system

an abrogation of osteogenic differentiation. FGF-1 has also been shown to play a role in chondrogenesis [59]. Human MSC treated with β-FGF gave rise to larger chondrogenic pellets with higher proteoglycan production [62]. Our results show that FGF signaling though not critical for chondrogenesis (since we did not observe a complete inhibition of chondrogenesis) is important and blocking this pathway results in reduced chondrogenic differentiation. FGFR signals through ERK and the ERK-MAPK pathway was significantly differentially expressed in all three lineages compared to MSC. FGF treatment of mesenchymal cells of the chick embryo wing bud elevated endogenous ERK phosphorylation in micromass cultures [43]. Therefore it is possible that the effect of FGF signaling on MSC differentiation is mediated through ERK. Further we are also able to show that these three pathways are sufficient for expanding MSC in serum free

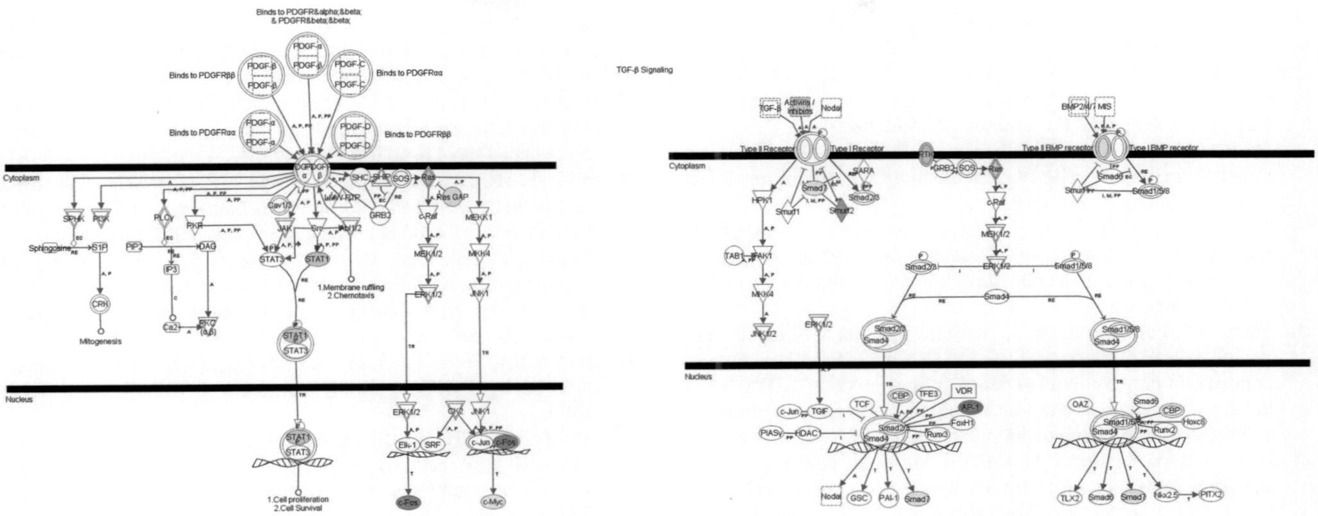

Fig. 2 Visualizing gene expression data as components of signaling pathways gives more information about cell function. Differentially expressed genes in MSC from the same experiment in Fig. 1 are shown as components of TGF-β and PDGF signaling pathways. Gene expression of undifferentiated MSC is compared to osteoblasts, chondrocytes, and adipocytes as in Fig. 1 . Components of the TGF-β pathway like Smurf and AP-1 are strongly expressed, as is FOS from the PDGF pathway. Such pathway maps generated over different time points can yield information about the changes of expression levels of their components over time. This information can then be used for predicting whether a pathway is active in MSC at a particular state of differentiation

media. It will be interesting to see whether we can achieve robust MSC expansion and targeted differentiation through activating one pathway while blocking another one.

5 Challenges for Network Analysis of MSC

The elucidation of signaling and regulatory networks in MSC is still a complex process. The biggest challenge is developing high throughput assays for measuring levels and phosphorylation of a large number of molecules in MSC. Integrated microarray and proteomics studies on MSC cultured from different donors and at different passages will help unravel some of these networks controlling MSC function. Tools like multiparametric flow cytometry to study phosphorylation of multiple proteins at the single-cell level can also help unravel signaling networks at the cellular level [63]. The advantage of such studies is that it is possible to identify signaling networks in phenotypically defined cells, leading to a better explanation of some of the differences observed in different laboratories studying MSC signaling. The fact that the MSC phenotype is still poorly defined poses further challenges to such studies. Another challenge is that some of these networks may be cell specific. So networks modeled for one cell type may not be applicable to other cell types. Also it is unclear whether MSC from different tissues use different mechanisms for differentiation. The observed difference in the potential of MSC to differentiate into various lineages definitely suggests that different mechanisms may be involved. However, since these mechanisms are poorly understood (particularly for non-marrow-derived MSC), there is no explicit data to prove or disprove this hypothesis. Also, since computational networks depend heavily on the accuracy of the input data and the parameters controlling MSC differentiation are poorly understood, in silico network analyses maybe error prone. Therefore, experimental validation of the models based on such networks is of paramount importance. However, such analyses coupled with good experimental data studying the effect of perturbation of these networks on MSC function can definitely be useful to predict and control the fate of differentiating MSC.

References

1. Beyer Nardi N, da Silva Meirelles L. Mesenchymal stem cells: isolation, in vitro expansion and characterization. Handb Exp Pharmacol. 2006:249–82.
2. Deans RJ, Moseley AB. Mesenchymal stem cells: biology and potential clinical uses. Exp Hematol. 2000; 28:875–84.
3. Caplan AI, Bruder SP. Mesenchymal stem cells: building blocks for molecular medicine in the 21st century. Trends Mol Med. 2001;7:259–64.
4. Baksh D, Song L, Tuan RS. Adult mesenchymal stem cells: characterization, differentiation, and application in cell and gene therapy. J Cell Mol Med. 2004;8:301–16.
5. Dazzi F, Ramasamy R, Glennie S, Jones SP, Roberts I. The role of mesenchymal stem cells in haemopoiesis. Blood Rev. 2006;20: 161–71.

6. Nakagami H, Morishita R, Maeda K, Kikuchi Y, Ogihara T, Kaneda Y. Adipose tissue-derived stromal cells as a novel option for regenerative cell therapy. J Atheroscler Thromb.2006;13:77–81.
7. Chunmeng S, Tianmin C. Skin: a promising reservoir for adult stem cell populations. Med Hypotheses. 2004;62:683–8.
8. De Bari C, Dell'Accio F, Tylzanowski P, Luyten FP. Multipotent mesenchymal stem cells from adult human synovial membrane. Arthritis Rheum. 2001;44:1928–42.
9. De Bari C, Dell'Accio F, Vandenabeele F, Vermeesch JR, Raymackers JM, Luyten FP. Skeletal muscle repair by adult human mesenchymal stem cells from synovial membrane. J Cell Biol. 2003;160:909–18.
10. Yamada Y, Fujimoto A, Ito A, Yoshimi R, Ueda M. Cluster analysis and gene expression profiles: a cDNA microarray system-based comparison between human dental pulp stem cells (hDPSCs) and human mesenchymal stem cells (hMSCs) for tissue engineering cell therapy. Biomaterials. 2006;27:3766–81.
11. Pierdomenico L, Bonsi L, Calvitti M, et al. Multipotent mesenchymal stem cells with immunosuppressive activity can be easily isolated from dental pulp. Transplantation. 2005;80:836–42.
12. Suzuki A, Raya A, Kawakami Y, et al. Nanog binds to Smad1 and blocks bone morphogenetic protein-induced differentiation of embryonic stem cells. Proc Natl Acad Sci U S A. 2006;103: 10294–9.
13. Bieback K, Kern S, Kluter H, Eichler H. Critical parameters for the isolation of mesenchymal stem cells from umbilical cord blood. Stem Cells. 2004;22:625–34.
14. Lee MW, Choi J, Yang MS, et al. Mesenchymal stem cells from cryopreserved human umbilical cord blood. Biochem Biophys Res Commun. 2004;320:273–8.
15. De Coppi P, Callegari A, Chiavegato A, et al. Amniotic fluid and bone marrow derived mesenchymal stem cells can be converted to smooth muscle cells in the cryo-injured rat bladder and prevent compensatory hypertrophy of surviving smooth muscle cells. J Urol. 2007;177:369–76.
16. In t Anker PS, Scherjon SA, Kleijburg-van der Keur C, et al. Amniotic fluid as a novel source of mesenchymal stem cells for therapeutic transplantation. Blood. 2003;102:1548–9.
17. Tsai MS, Lee JL, Chang YJ, Hwang SM. Isolation of human multipotent mesenchymal stem cells from second-trimester amniotic fluid using a novel two-stage culture protocol. Hum Reprod. 2004;19:1450–6.
18. Li CD, Zhang WY, Li HL, et al. Isolation and identification of a multilineage potential mesenchymal cell from human placenta. Placenta. 2005.
19. Igura K, Zhang X, Takahashi K, Mitsuru A, Yamaguchi S, Takashi TA. Isolation and characterization of mesenchymal progenitor cells from chorionic villi of human placenta. Cytotherapy 2004;6: 543–53.
20. Miao Z, Jin J, Chen L, et al. Isolation of mesenchymal stem cells from human placenta: comparison with human bone marrow mesenchymal stem cells. Cell Biol Int. 2006;30:681–7.
21. Keating A. Mesenchymal stromal cells. Curr Opin Hematol. 2006;13:419–25.
22. Seshi B, Kumar S, King D. Multilineage gene expression in human bone marrow stromal cells as evidenced by single-cell microarray analysis. Blood Cells Mol Dis. 2003;31:268–85.
23. Almeida-Porada G, Porada C, Zanjani ED. Plasticity of human stem cells in the fetal sheep model of human stem cell transplantation. Int J Hematol. 2004;79:1–6.
24. Brandenberger R, Wei H, Zhang S, et al. Transcriptome characterization elucidates signaling networks that control human ES cell growth and differentiation. Nat Biotechnol. 2004;22:707–16.
25. Mahdavi A, Davey RE, Bhola P, Yin T, Zandstra PW. Sensitivity analysis of intracellular signaling pathway kinetics predicts targets for stem cell fate control. PLoS Comput Biol. 2007;3:e130.
26. Woolf PJ, Prudhomme W, Daheron L, Daley GQ, Lauffenburger DA. Bayesian analysis of signaling networks governing embryonic stem cell fate decisions. Bioinformatics. 2005;21:741–53.
27. Fortunel NO, Hatzfeld JA, Monier MN, Hatzfeld A. Control of hematopoietic stem/progenitor cell fate by transforming growth factor-beta. Oncol Res. 2003;13:445–53.
28. Pimanda JE, Donaldson IJ, de Bruijn MF, et al. The SCL transcriptional network and BMP signaling pathway interact to regulate RUNX1 activity. Proc Natl Acad Sci U S A. 2007;104:840–5.
29. Kumar S, Mahendra G, Ponnazhagan S. Determination of osteoprogenitor-specific promoter activity in mouse mesenchymal stem cells by recombinant adeno-associated virus transduction. Biochim Biophys Acta. 2005;1731:95–103.
30. Roelen BA, Dijke P. Controlling mesenchymal stem cell differentiation by TGFBeta family members. J Orthop Sci 2003;8: 740–8.
31. Im GI, Jung NH, Tae SK. Chondrogenic differentiation of mesenchymal stem cells isolated from patients in late adulthood: the optimal conditions of growth factors. Tissue Eng. 2006;12: 527–36.
32. Benavente CA, Sierralta WD, Conget PA, Minguell JJ. Subcellular distribution and mitogenic effect of basic fibroblast growth factor in mesenchymal uncommitted stem cells. Growth Factors. 2003;21:87–94.
33. Cheng L, Hammond H, Ye Z, Zhan X, Dravid G. Human adult marrow cells support prolonged expansion of human embryonic stem cells in culture. Stem Cells. 2003;21:131–42.
34. Diefenderfer DL, Osyczka AM, Reilly GC, Leboy PS. BMP responsiveness in human mesenchymal stem cells. Connect Tissue Res. 2003;44 Suppl 1:305–11.
35. Erices A, Conget P, Rojas C, Minguell JJ. Gp130 activation by soluble interleukin-6 receptor/interleukin-6 enhances osteoblastic differentiation of human bone marrow-derived mesenchymal stem cells. Exp Cell Res. 2002;280:24–32.
36. Boland GM, Perkins G, Hall DJ, Tuan RS. Wnt 3a promotes proliferation and suppresses osteogenic differentiation of adult human mesenchymal stem cells. J Cell Biochem. 2004;93:1210–30.
37. Epperly MW, Franicola D, Zhang X, et al. Reduced irradiation pulmonary fibrosis and stromal cell migration in Smad3-/- marrow chimeric mice. In Vivo. 2006;20:573–82.
38. Krampera M, Pasini A, Rigo A, et al. HB-EGF/HER-1 signaling in bone marrow mesenchymal stem cells: inducing cell expansion and reversibly preventing multilineage differentiation. Blood. 2005; 106:59–66.
39. Li H, Yu B, Zhang Y, Pan Z, Xu W. Jagged1 protein enhances the differentiation of mesenchymal stem cells into cardiomyocytes. Biochem Biophys Res Commun. 2006;341:320–5.
40. Bachner D, Ahrens M, Schroder D, et al. Bmp-2 downstream targets in mesenchymal development identified by subtractive cloning from recombinant mesenchymal progenitors (C3H10T1/2). Dev Dyn. 1998;213:398–411.
41. Chen T, Bai H, Shao Y, et al. Stromal cell-derived factor-1/CXCR4 signaling modifies the capillary-like organization of human embryonic stem cell-derived endothelium in vitro. Stem Cells. 2007;25:392–401.
42. Goessler UR, Bugert P, Bieback K, et al. In-vitro analysis of the expression of TGFbeta -superfamily-members during chondrogenic differentiation of mesenchymal stem cells and chondrocytes during dedifferentiation in cell culture. Cell Mol Biol Lett. 2005;10: 345–62.
43. Bobick BE, Thornhill TM, Kulyk WM. Fibroblast growth factors 2, 4, and 8 exert both negative and positive effects on limb, frontonasal, and mandibular chondrogenesis via MEK-ERK activation. J Cell Physiol. 2007;211:233–43.
44. Doucet C, Ernou I, Zhang Y, et al. Platelet lysates promote mesenchymal stem cell expansion: a safety substitute for animal

serum in cell-based therapy applications. J Cell Physiol. 2005;205: 228–36.
45. Fan VH, Tamama K, Au A, et al. Tethered EGF provides a survival advantage to mesenchymal stem cells. Stem Cells. 2007;25: 1241–51.
46. Fukumoto T, Sperling JW, Sanyal A, et al. Combined effects of insulin-like growth factor-1 and transforming growth factor-beta1 on periosteal mesenchymal cells during chondrogenesis in vitro. Osteoarthritis Cartilage. 2003;11:55–64.
47. DHaeseleer P, Liang S, Somogyi R. Genetic network inference: from co-expression clustering to reverse engineering. Bioinformatics. 2000;16:707–26.
48. Huang W, Yang S, Shao J, Li YP. Signaling and transcriptional regulation in osteoblast commitment and differentiation. Front Biosci. 2007;12:3068–92.
49. Jadlowiec J, Koch H, Zhang X, Campbell PG, Seyedain M, Sfeir C. Phosphophoryn regulates the gene expression and differentiation of NIH3T3, MC3T3-E1, and human mesenchymal stem cells via the integrin/MAPK signaling pathway. J Biol Chem. 2004;279: 53323–30.
50. Martin-Rendon E, Hale SJ, Ryan D, et al. Transcriptional profiling of human cord blood CD133+ and cultured bone marrow mesenchymal stem cells in response to hypoxia. Stem Cells. 2007;25:1003–12.
51. Kasper G, Dankert N, Tuischer J, et al. Mesenchymal stem cells regulate angiogenesis according to their mechanical environment. Stem Cells. 2007;25:903–10.
52. Wang XP, Suomalainen M, Felszeghy S, et al. An integrated gene regulatory network controls stem cell proliferation in teeth. PLoS Biol. 2007;5:e159.
53. Tsai MS, Hwang SM, Chen KD, et al. Functional network analysis on the transcriptomes of mesenchymal stem cells derived from amniotic fluid, amniotic membrane, cord blood, and bone marrow. Stem Cells. 2007;25:2511–23.
54. Nikitin A, Egorov S, Daraselia N, Mazo I. Pathway studio-the analysis and navigation of molecular networks. Bioinformatics. 2003;19:2155–7.
55. Ekins S, Nikolsky Y, Bugrim A, Kirillov E, Nikolskaya T. Pathway mapping tools for analysis of high content data. Methods Mol Biol. 2007;356:319–50.
56. Tsutsumi S, Shimazu A, Miyazaki K, et al. Retention of multilineage differentiation potential of mesenchymal cells during proliferation in response to FGF. Biochem Biophys Res Commun. 2001;288:413–9.
57. Jackson RA, Nurcombe V, Cool SM. Coordinated fibroblast growth factor and heparan sulfate regulation of osteogenesis. Gene. 2006;379:79–91.
58. Jacob AL, Smith C, Partanen J, Ornitz DM. Fibroblast growth factor receptor 1 signaling in the osteo-chondrogenic cell lineage regulates sequential steps of osteoblast maturation. Dev Biol. 2006; 296:315–28.
59. Solchaga LA, Penick K, Porter JD, Goldberg VM, Caplan AI, Welter JF. FGF-2 enhances the mitotic and chondrogenic potentials of human adult bone marrow-derived mesenchymal stem cells. J Cell Physiol. 2005;203:398–409.
60. Ozaki Y, Nishimura M, Sekiya K, et al. Comprehensive analysis of chemotactic factors for bone marrow mesenchymal stem cells. Stem Cells Dev. 2007;16:119–29.
61. Kratchmarova I, Blagoev B, Haack-Sorensen M, Kassem M, Mann M. Mechanism of divergent growth factor effects in mesenchymal stem cell differentiation. Science. 2005;308:1472–7.
62. Gruber R, Karreth F, Kandler B, et al. Platelet-released supernatants increase migration and proliferation, and decrease osteogenic differentiation of bone marrow-derived mesenchymal progenitor cells under in vitro conditions. Platelets. 2004;15:29–35.
63. Irish JM, Hovland R, Krutzik PO, et al. Single cell profiling of potentiated phospho-protein networks in cancer cells. Cell. 2004;118:217–28.

Single-Cell Approaches to Dissect Cellular Signaling Networks

Weijia Wang and Julie Audet

Abstract Progress in understanding signal transduction, especially in rare and heterogeneous stem cell populations, is dependent on advances in single-cell assays. Newly developed techniques based on flow cytometry, capillary electrophoresis and live-cell imaging have enabled researchers to study kinase activity at the single-cell level. Since kinase activation is central to the regulation of virtually all cellular functions, single-cell kinase assays promise to help elucidating the molecular mechanisms controlling stem cell fate decisions.

Keywords Stem cells · Kinase · Single-cell analysis · Signaling network · Flow cytometry · Capillary electrophoresis · FRET

1 Introduction

Kinase activity is central to the regulation of cell fate decisions. Intracellular kinases propagate the signals intracellularly through cascades, or pathways, after a ligand binds to its receptor on the cell surface. It has become clear that the cellular outcome (e.g., proliferation, differentiation, survival, and migration) is shaped by the dynamic interactions between pathways[1]. So far, most studies have used traditional biochemical and molecular assays such as Western blotting to study the activation of signaling pathways following stimulation. These conventional biochemical assays have two major limitations when used to measure kinase activities. First, they have poor sensitivity and require pooling a large number of cells which masks individual cell-to-cell variations (Fig. 1a). In addition, primary adult stem cell populations are very scarce. Sorting is generally labor intensive, and can become very costly to obtain sufficient number of cells for traditional assays. Furthermore, concerns have been raised up regarding the cell purity and introduction of artefacts to the follow-up analysis. This implies a fundamental need to study stem cells at the single-cell level. Second, conventional assays cannot detect a bistable (all-or-none) activation of the kinase; such behavior has been described in *Xenopus* oocytes [2] (see Fig. 1b). Studies of such behavior have been hampered in mammalian cells, which is attributable to the incapability of conventional assays to study nonlinearity in signaling systems. Therefore, quantitative kinase assays enabling kinase activity measurements within single cells are required to study the activation profiles in heterogeneous populations.

During the past few years, the development of new single-cell techniques has rendered the study of kinase activity in living cells widespread. Here we will provide a brief overview of the state-of-the-art approaches and highlight the achievements, potential applications, advantages and drawbacks (summarized in Table 1) of each.

2 Flow Cytometry

The use of kinase antibodies that recognize specific phosphoresidues in conjunction with flow cytometry represents a powerful approach to measure kinase activation in single cells in a to high-throughput manner [3, 4] (Fig. 2a). The method was first used by Muller et al. [5] to study the steps involved in the T-cell activation cascade. Measuring phosphotyrosine levels by flow cytometry revealed, for the first time, a direct correlation between the activation of protein tyrosine kinase and the level of T-cell antigen receptor occupancy in individual T cells upon antigenic stimulation. Chow et al. [6] then to combined the intracellular kinase staining using more specific antibodies that recognized phosphorylated residues on MEK (MAPK [mitogen-activated protein kinase]/ERK [extracellular signal-regulated kinase] kinase)

J. Audet (✉)
Assistant Professor, Institute of Biomaterials and Biomedical Engineering, Terrence Donnelly Centre for Cellular and Biomolecular Research, University of Toronto, 164 College St., Room 407, Toronto, ON, Canada, M5S 3G9, Tel: 416-946-0209, Fax: 416-978-4317
e-mail: julie.audet@utoronto.ca

V.K. Rajasekhar, M.C. Vemuri (eds.), *Regulatory Networks in Stem Cells*, Stem Cell Biology and Regenerative Medicine,
DOI 10.1007/978-1-60327-227-8_29, © Humana Press, a part of Springer Science+Business Media, LLC 2009

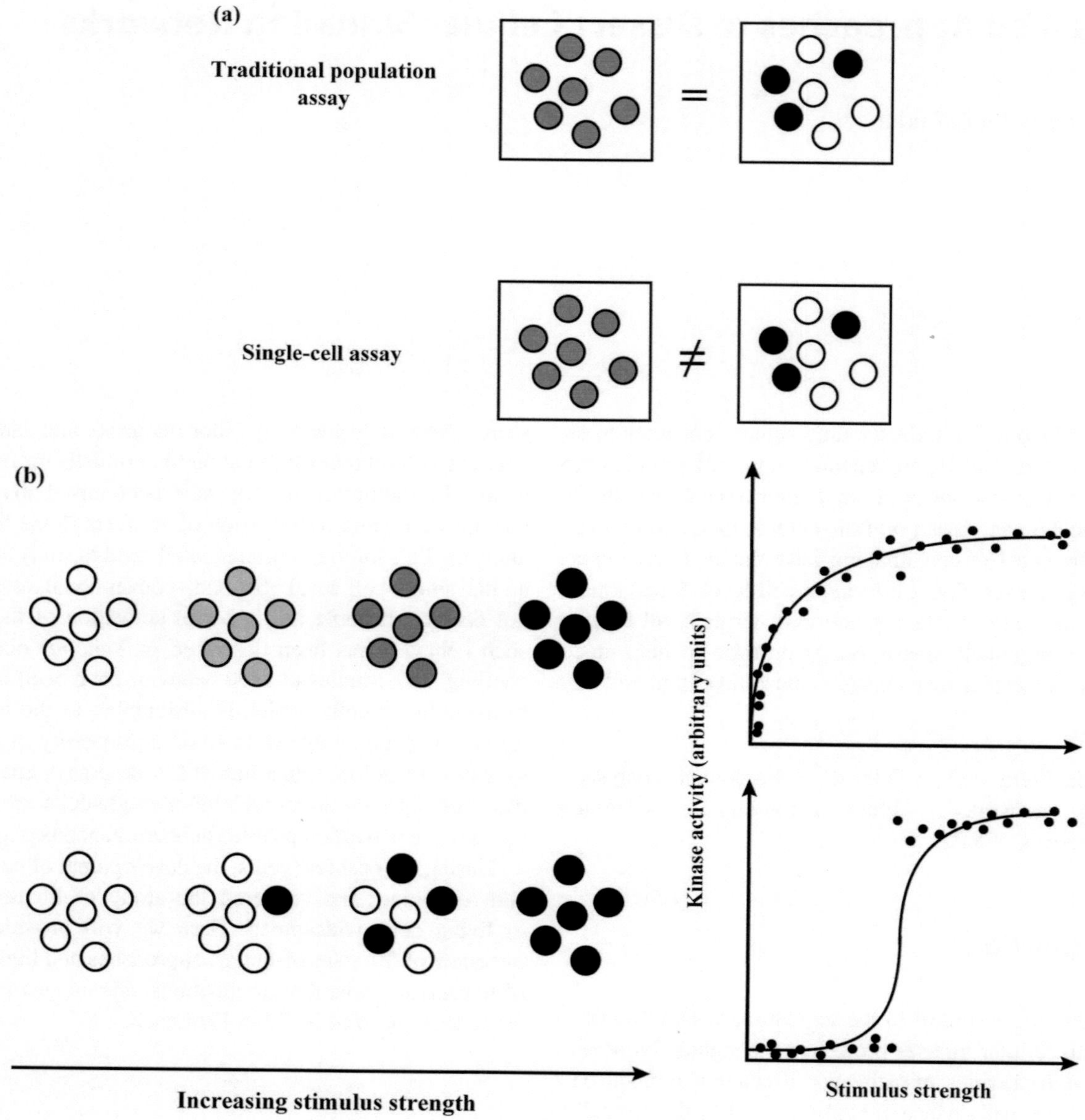

Fig. 1 (**a**) Kinase activity measured by traditional population assays vs. single-cell assays. For a heterogeneous cell population (presented in the right side box), individual cell responses are averaged out across the population when measured by conventional population assays. The heterogeneous population is "falsely" regarded as equal to a homogeneous population (presented in the left side box) with kinase activity level equivalent to the mean value of the heterogeneous population. In contrast, single-cell assays are capable of recognizing cell-to-cell differences. (**b**) Illustration of a graded and a bistable (all-or-none) response in cells. In the upper panel, the cells respond in a graded fashion with the increase in stimulus concentration, which is translated into the plot showing the activity of the kinase as a function of stimulus strength. Each dot in the plot represents a measurement from a single cell. The lower panel illustrates a switch-like (all-or-none) response in cells, i.e., in a single cell, either the kinase is switched on or off. Measurements of the population average cannot distinguish the populations represented in the two panels and are not capable of revealing some properties of signaling transduction such as bistability

Table 1 List of advantages and drawbacks of each single-cell approach reviewed

Flow cytometry	Capillary electrophoresis	Live-cell imaging
Advantages:		
• high throughput	• study multiple kinases and post-translational modifications at one time	• high spatiotemporal resolution
• measure multiple kinases at one time	• high resolution and sensitivity	• real-time measurements
• not requiring genetic manipulations	• minute consumption of reagents	
Drawbacks:		
• fixed-time points	• require introduction of reporter molecules	• limited number of kinases at one time
• lack of spatial resolution	• fixed-time points	• possible disruption of normal function of the target protein
• high dependence on the availability of phospho-specific antibodies	• lack of spatial resolution	• high demand on imaging instruments

and ERK with a cell surface marker (CD3) to monitor the MAPK activation in T-lymphocyte subset within peripheral blood cells. Treatment with two inhibitors that targeted different signaling molecules within the pathway resulted in dose-dependent but differential inhibition of MAPK activation. This suggests the potential use of flow cytometry for pharmacodynamic monitoring of signal transduction inhibitors in clinical trials. Multiparametric (13 colors) analysis was employed by Perez and Nolan [7], for the first time, to detect the activation of multiple (up to four) kinases simultaneously at a single-cell level. The study demonstrated the ability of multiparameter flow cytometry to identify functional cell subsets in a heterogeneous population based on kinase activation states. Such assay to measure the activity of multiple kinases within a single cell was further used to investigate signaling anomalies in cancer cells [8]. Six nodal phospho-proteins were selected and the changes in their phosphorylation level in response to five different stimuli were compared to their basal activity level. The results indicated that cancer cells remodelled their signaling networks differentially in response to extracellular cues, which can be used to categorize different cancer subtypes and correlate them with genetic mutations, pathological phenotypes, and differential responses to medical treatments. Analysis of alterations in cell signaling events by flow cytometry provides a useful way to identify biomarkers for cancer diagnosis, monitor disease progression and help to devise individualized therapeutic strategies. More comprehensive reviews on the use of flow cytometry to map cancer cell signaling networks and the potential of such approach in facilitating the development of targeted therapies for haematological cancer patients are available for readers who are particularly interested in this field [9, 10].

It is worth noting that the multiparametric measurement of flow cyotmetry is capable of resolving multiple kinase activation simultaneously, which makes it possible to study dynamic interactions between multiple pathways that are believed to govern cell behavior. It is also possible to reconstruct the signaling networks at a relatively large scale. Since this approach does not require to genetically engineer the cells with kinase activity reporters, it is easily applied to the study of primary cells. Nolan's group studied altered growth factor responses in phospho-protein signaling networks in peripheral blood cells from leukemia patients [8]. Our group has developed\ flow cytometric assays to detect and quantitate the activity of multiple kinases (e.g., ERK, STAT5) simultaneously within subsets of primary murine bone marrow–derived erythroid cells. These subsets are identified by labeling cell surface markers (such as c-Kit, CD-71 and Ter-119). Despite its ability to measure multiple parameters in a high-throughput manner, flow cytometry has several shortcomings that need to be mentioned here. The measurements are not in real time. However, activation kinetics at fixed time points can be obtained by varying the time interval between stimulation and fixation of the cells. In addition, semi-quantitative measurements can be obtained based on the median fluorescence intensity. Another caveat of the assay is that it highly depends on the availability of the phospho-specific antibodies and their specificity. The specificity of the antibody can be confirmed by using specific inhibitors that perturb the activity of the upstream kinase. However, it is important to bear in mind that the detection of one or two phosphorylated residue(s) on a kinase does not ensure its activation. The reasons are: (1) sometimes multiple phosphorylations or additional modifications are needed to confer activity; and (2) other proteins may interact with the

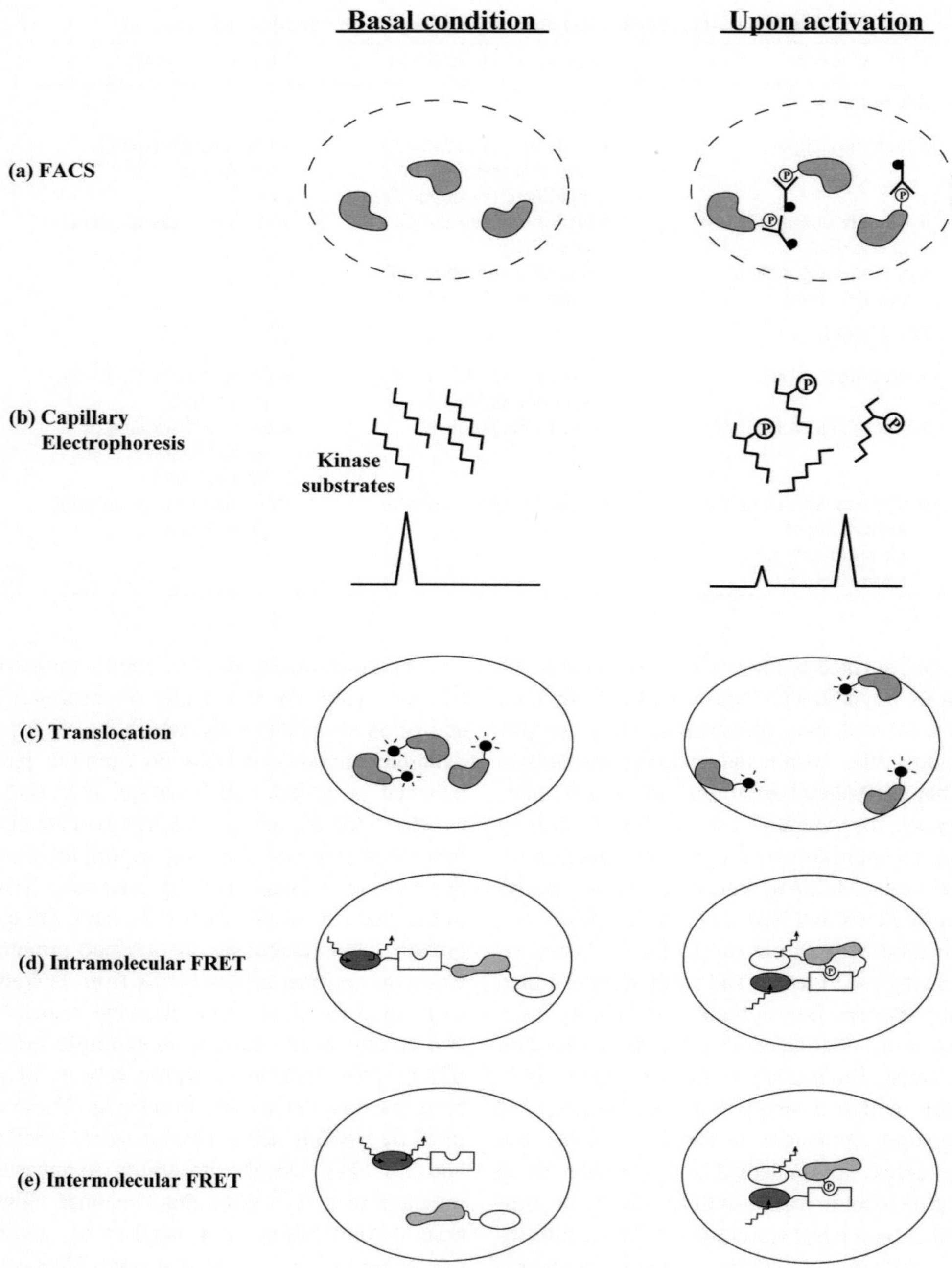

Fig. 2 Single-cell approaches to measure kinase activity. (**a**) Flow cytometry. It is capable of quantifying multiple kinase activities simultaneously by using phospho-specific antibodies. (**b**) Capillary electrophoresis. Fluorescently labelled kinase substrates are introduced into the cells. Upon cell stimulation, the cell is lysed and its contents are rapidly loaded into a capillary-based microanalytical separation device. The substrates and their products are separated and quantitated. (**c**) Translocation. Proteins tagged with GFP are genetically expressed in the cells, permitting visualization of the movement of the protein of interest upon stimulation. (**d**) Intramolecular FRET. FRET-based kinase biosensors employ a unimolecular design in which the kinase-specific substrate domain undergoes a conformational reorganization or the binding domain recognizes and binds to the activated state of the substrate domain to bring the FRET pair into close proximity to result in FRET. The designs may vary to meet the specific requirements of different applications. (**e**) Intermolecular FRET. Two GFP variants are attached to two different proteins to study protein-protein interactions. Derivation of a new approach involves using an antibody or a protein domain which recognizes and binds only to a specific conformation or post-translational modification of the protein of interest. The protein domain or antibody bears a dye that undergoes FRET when it is brought in close proximity to the GFP on the targeted protein

phosphorylated protein to inhibit its activity [11]. Last of all, when adherent cells are analysed, their detachment from their substrate might activate signaling [7].

3 Capillary Electrophoresis

With the advent of optical instrumentation which is capable of detecting minute amounts of biomolecules, capillary electrophoresis (CE) has emerged as an important platform for cellular biological studies [12]. CE combined with a laser-induced fluorescence (LIF) detection system has become a powerful tool for single-cell analysis. With such CE-LIF system, Allbritton and colleagues [13] have developed a strategy to measure kinase activity in single cells. The fluorescently labelled synthetic peptides that are substrates for a kinase of interest are loaded into the cells. Upon cellular stimulation, the cell is lysed and its contents are rapidly injected into a capillary-based microanalytical separation device. Cell lysis and injection of the cell contents in the separation column enable the termination of the biochemical reactions within milliseconds or less [14], and, subsequently, the separation and detection of the fluorescently labelled reporters (Fig. 2b). To date, only few reports presents measurements of kinase activities from a single cell using CE. Meredith et al. [13], for the first time, measured multiple kinases including protein kinase C (PKC), protein kinase A (PKA), calcium-calmodulin activated kinase II (CamKII), and cdc2 protein kinase (cdc2K) simultaneously. The activities of the kinases were correlated with the amount of phosphorylated peptides upon pharmacological and physiological stimulations. A detection limit of 10-20 mol enabled the system to detect substantially lower quantities of fluorophores. Therefore, reporters at concentrations lower than native substrates were loaded into the cell to eliminate artefacts due to competitive inhibition of the kinase by reporters. The same approach was used to quantitatively measure the activities of protein kinase B (PKB) and phosphatases. The results indicated that short peptides can serve as sensitive reporters of the activation state of native kinases within live cells, and cell-to-cell difference was also revealed [15].

The CE-based techniques have significant advantages, including high resolution, great sensitivity, minimum reagent consumption, and reduced analysis time. However, it also has shortcomings. One of the drawbacks is that real-time measurements cannot be performed. Different time points can be obtained but require multiple measurements on different, single cells. The peptides used as reporters must be loaded into the cells before the analysis. Several methods have been exploited, including physical or chemical means (microinjection, pinocytic loading) or making the reporters cell membrane permeant by conjugating them to a cell-penetrating peptide [16, 17]. In addition, the specificity and affinity of the peptides need to be carefully validated.

4 Live-Cell Imaging

Recent advances in fluorescent proteins, small molecule fluorophores, and imaging technology have provided new and powerful tools to investigate signal transduction at the level of single cells with a high spatiotemporal resolution. The introduction of the green fluorescent protein (GFP) isolated from the jellyfish *Aequorea* as the molecular biological tool in 1994 [18, 19] enabled tagging hundreds of proteins through genetic manipulations and allowed visualization of protein localization and movement in living cells. Further efforts have concentrated on overcoming the drawbacks of wild-type GFP, including dual peaked excitation spectra, poor photostability, and poor folding at 37°C. Researchers have modified GFP via both directed and random mutagenesis to produce a myriad of GFP derivatives with new colors, improved folding, enhanced intrinsic brightness, and/or altered pH sensitivity [20]. At the same time, with developments in highly automated live cell fluorescence microscopic techniques, real-time tracking of kinase dynamics in living cells has become available to delineate the mechanisms underlying the complex regulation of kinase activity.

4.1 Translocation

Some signaling molecules translocate from one compartment of the cell to another to reach their site of action upon cellular stimulation. Expressing a GFP fusion proteins tagged version of the signaling protein of interest enables monitoring and tracking the movement of the protein within the cell before, during, and after stimulation (Fig. 2c). Reversible translocation of GFP-tagged CaMKII from cytoskeletal sites to postsynaptic densities upon N-methyl-D-aspartate (NMDA) receptor stimulation was monitored in cultured hippocampal neurons. The study revealed that the dynamics of the translocation was regulated by two molecular events – calmodulin binding and autophosphorylation [21]. Hirose et al. [22] used the same approach to study translocation of GFP-tagged pleckstrin homology domain (PHD) from phospholipase C-δ1 in response to increased concentration of inositol-1,4,5-trisphosphate (IP_3). This is the first study that used fluorescent indicator to monitor IP_3 dynamics within single cells. It was shown that oscillations in IP_3 levels, indicated by translocation of GFP-PHD, were synchronized with Ca^{2+} oscillations, which shed lights on the mechanistic regulation of complex Ca^{2+} signaling.

Imaging GFP fusion proteins offers a simple, fast, and relatively reliable way to detect signaling protein activation if translocation is part of the activation mechanism. However, several caveats need to be kept in mind. First, protein translocation does not necessarily indicate protein activation. Tanimura et al. [23] studied the interplay of calcium, diacylglycerol, and phosphorylation in the regulation of PKCα. They found that some PKC inhibitors prevented PKC from dissociating from the plasma membrane, indicating that the location of the kinase does not always reflect its activity. Second, one must always be aware of the potential differences between the GFP fusion protein and the native protein in terms of the biological function and activity. Third, response kinetics may be limited by the time required for the engineered protein to be expressed, to mature, and to diffuse to the right locations, rather than the intrinsic mechanism. Fourth, the quantitative measurement of translocation depends on the expression level of the fusion protein and the imaging resolution. Overexpression is generally required to obtain adequate fluorescence intensity to reach detection threshold, and therefore it might not reflect the normal signaling context. Fifth, usually only one kinase is assayed at one time within the cells. The information regarding the interactions between signaling pathways is missing using such approach. A multicolor imaging system was used to track PKC and one of its downstream targets, tagged with GFP variants possessing different spectral characteristics[24]. The system can detect up to four colors, but the substantial spectral overlap among GFP mutants has most likely limited the number of kinases that can be measured at one time [24, 25].

4.2 FRET

The development of GFP-based fluorescence resonance energy transfer (FRET) technologies has significantly improved our ability to study signaling protein activation in live cells. FRET involves a radiationless energy transfer between two fluorophores: when the two fluorophores are brought in close proximity ($\leq$ 80 Å), excitation of the donor fluorophore results in emission from the acceptor. FRET is very sensitive regarding the distance between the two fluorophores; the efficiency of FRET falls off with the sixth power of the distance between the donor and the acceptor [26, 27]. In addition, FRET also depends on the spectral characteristics of the donor and the acceptor. The spectral overlap between the donor emission and the acceptor excitation should be sufficient, however, not excessive. Also, the brightness and photostability of the two fluorophores must be adequate. Within the known GFP mutants, the commonly used pair for FRET comprises the cyan and yellow mutants – CFP and YFP, which overcome the relatively poor brightness and photostability of the initial pair – blue fluorescent protein (BFP) and GFP [20]. Recent advancements in the development of new fluorescent proteins have provided us a variety of variants with new choice of colors and improved spectral properties [28, 29].

Intramolecular FRET involves attachment of two fluorophores on the same protein to detect conformational changes. Upon stimulation, the labelled protein undergoes changes in its tertiary structure, resulting in changes in the relative positions of the two fluorophores, and consequently, changes in spectra (Fig. 2d). However, only few studies have successfully used this method to monitor signaling protein activities, mainly because of the following two reasons: (1) Few naturally occurring conformational changes can alter the distance between the two attached dyes sufficiently to produce detectable FRET changes in vivo; (2) the cases in which attachment of two fluorophores at the appropriate positions resulting in an analog with fully intact biological function are scarce. Mochizuki et al. [30] extended the protein "transducer" approach (initially introduced by Miyawaki et al. [31] to report intracellular Ca^{2+}concentrations) and generated a domain biosensor to detect growth-factor induced Ras activation by genetically engineering a protein consisted of H-Ras, a Raf domain that only binds to activated Ras, and YFP and CFP. A similar biosensor for Rap1 was also made by replacing Ras with Rap1. These biosensors enabled not only the spatiotemporal imaging of the activation of Ras and Rap1, but also the relative quantitation of Ras and Rap1 activities attributed to the fixed ratio of the donor and the acceptor within cells. Besides the fixed donors to acceptors ratio, other advantages of such unimolecular design also include easy targeting, readily genetic encoding and less interference with endogenous interacting molecules [32]. The same approach, but with reporters containing different functional domains, has been used by researchers to gain information about dynamic regulation and spatial compartmentalization of kinase activation. (details in structural designs of different constructs are beyond the scope of this review but are covered in the paper of Ni et al. [33]). For instance, a reporter containing a phosphoamino acid binding domain and a kinase substrate domain was used to monitor PKA activities in living cells and to reveal the effects of substrate localization on PKA activation [34]. The same design has been utilized to develop reporters to study a range of serine/threonine and tyrosine kinases [34–44]. A different probe comprising the kinase of interest itself flanked by a FRET pair was used to directly visualize dynamic activation and translocation of PKB/Akt, CaMKII, ERK, and MAP kinase-activated protein kinase 2 (MK2) in single living cells by taking advantage of the fact that the kinase undergoes a conformational change upon phosphorylation[45–49]. One of the drawbacks of such FRET-based biosensors is that both termini of the kinase of interest need to be linked to the fluorophore; this approach may not be useful if the termini of the target protein are involved in the biological function.

Another alternative design which involved sandwiching a kinase substrate between a FRET pair was also utilized to monitor PKA, PKC, ERK, Smad1, and glycogen synthase kinase 3β (GSK3β) [50–54]. These FRET-based kinase biosensors are capable of measuring in real time and space and have provided high-resolution information of spatiotemporal regulation of kinases which is complementary to other approaches. In intermolecular FRET, the two fluorophores reside on different proteins that must be brought in close proximity to achieve FRET (Fig. 2e). Compared with intramolecular FRET, for intermolecular FRET it is relatively easier to obtain analogs that retain normal biological activity since one protein only needs to be labelled with one fluorophore. This FRET application has been used to study protein-protein interactions in signal transduction cascades where protein interactions are considered as an important indicator of signaling activity. Ruehr et al. [55] exploited the CFP-YFP pair to examine the binding of the type IIα isoform of the regulatory subunit of cAMP-dependent PKA to a peptide from an A-kinase anchoring protein in live CHO cells. Although the advantages of GFP are overriding, including the relative ease of expressing genetically tagged proteins and real-time measurements, even the best GFP mutants have considerably overlapping spectra which result in contamination of FRET signal by “bleed through” from direct excitation of the donor. Careful correction for this “bleed through” and proper controls are required to derive “accurate” FRET measurements. In addition, it was reported that dimerization of GFP itself can cause protein-protein interaction [56]. However, new mutants have been generated to diminish this tendency.

A new approach which involves combining dyes with GFPs for intermolecular FRET application in living cells has also been explored. Dyes are superior to GFPs in terms of their spectral properties, including wider selection of fluorescence wavelengths and more resistance to environmental changes. Fluorescently labelled antibodies or protein domains that bind exclusively to a specific state of the protein (e.g., phosphorylated, or other conformational changes) lead to FRET when they interact with their GFP tagged protein targets. Ng et al. [57] used fluorescently labelled phosphorylation site-specific antibodies to detect the activation of PKCα. FRET was generated when the antibody bound to the phosphorylated GFP-tagged PKCα; this made it possible to visualize PKCα activation in live cells. A similar approach was used to study auto-phosphorylation dynamics of the receptor tyrosine kinase epidermal growth factor receptor (EGFR) and ErbB1 with phosphotyrosine antibodies [58, 59]. These studies used fluorescence lifetime imaging microscopy (FLIM) to measure the lifetime of fluorescence emission from the donor and acceptor fluorophores, rather than FRET intensity, to simplify image-correction procedures as well as to reduce artefacts caused by interactions of the donor or acceptor with other molecules in the cell. However, FLIM still requires specialized equipment, which has limited its widespread usage [60]. Kraynov et al. [61] used a fluorescently labelled protein domain from p21-associated kinase (PAK) that only binds to activated Rac. The resulting FRET enabled the authors to monitor the dynamics of Rac activation in migrating fibroblasts over time and revealed a specific role of Rac in tail retraction. The caveats of these domain/antibody biosensors include: (1) restriction of access to some subcellular locations; (2) limited availability of target proteins due to structural hindrance; (3) competition between the biosensors and native ligands; and (4) steric blocking of the binding site, which can be readily identified by constitutively active GFP-tagged mutants. Some researchers have taken advantage of this aspect of those biosensors to study steric regulation of protein activities[60].

FRET and the FRET-based biosensors have provided us with a significant amount of information about spatial localization and temporal dynamics of protein activation in living cells. Despite the advantages of monitoring the activation events in real time and space and the convenience of genetic expression, the FRET-based techniques face new challenges. There remain many important signaling proteins that cannot be studied by current approaches. Some proteins are sterically not accessible to domain/antibody biosensors, or their functions are severely perturbed when they are conjugated to fluorophores. New methods are needed to overcome these hurdles. “Solvatochromic dyes” are designed to change fluorescence when the protein they are bound to, undergoes conformational changes or post-translational modifications. However, extensive work needs to be carried out before the dyes can be used for live-cell imaging [62]. There also have been new developments in site-specific protein labelling [63, 64] and in methodologies to load dye-labelled proteins into cells [65]. There are also limitations regarding the number of signaling proteins that can be measured at a time in FRET-based approaches.

5 Conclusions

Recent technological advances have enabled studies ranging from visualizing protein movement, monitoring spatiotemporal dynamics of protein activations, to simultaneously quantifying multiple kinase activitie within a single live cell. These single-cell kinase assays will provide substantial information on signaling dynamics and kinetics, and “cross-talk” or interactions between signaling pathways and will facilitate unraveling the mechanistic regulation of signaling networks. The emerging single-cell analytical technologies will contribute to developing a more comprehensive understanding of the activities of signaling proteins that regulate stem cell fate decisions.

References

1. Callard R, George AJ, Stark J. Cytokines, chaos and complexity. Immunity. 1999;11(5):507–13.
2. Ferrell Jr. JE, Machleder EM. The biochemical basis of an all-or-none cell fate switch in Xenopus oocytes. Science. 1998;280: 895–8.
3. Krutzik PO, Irish JM, Nolan GP, et al. Analysis of protein phosphorylation and cellular signaling events by flow cytometry: techniques and clinical applications. Clin Immunol. 2004;110: 206–21.
4. Krutzik PO, Nolan GP. Intracellular phospho-protein staining techniques for flow cytometry: monitoring single cell signaling events. Cytometry. 2003;55A:61–70.
5. Muller S, Demotz S, Bulliard C, et al. Kinetics and extent of protein tyrosine kinase activation in individual T cells upon antigenic stimulation. Immunology. 1999;97(2):287–93.
6. Chow S, Patel H, Hedley DW. Measurement of MAP kinase activation by flow cytometry using phospho-specific antibodies to MEK and ERK: potential for pharmacodynamic monitoring of signal transduction inhibitors. Cytometry. 2001;46:72–8.
7. Perez OD, Nolan GP. Simultaneous measurement of multiple active kinase states using polychromatic flow cytometry. Nat Biotechnol. 2002;20:155–62.
8. Irish JM, Hovland R, Krutzik PO, et al. Single cell profiling of potentiated phospho-protein networks in cancer cells. Cell. 2004;118:217–28.
9. Irish JM, Kotecha N, Nolan GP. Mapping normal and cancer cell signalling networks: towards single-cell proteomics. Nat Rev Cancer. 2006;6(2):146–55.
10. Nolan GP. Deeper insights into hematological oncology disorders via single-cell phospho-signaling analysis. Hematology. 2006: 123–7.
11. Sims CE, Allbritton NL. Single-cell kinase assays: opening a window onto cell behavior. Curr Opin Biotechnol. 2003;14(1):23–8.
12. Babu SC, Song EJ, Babar SM, et al. Capillary electrophoresis of signaling molecules. Electrophoresis. 2007;21(9):890–7.
13. Meredith GD, Sims CE, Soughayer JS, et al. Measurement of kinase activation in single mammalian cells. Nat Biotechnol. 2000;18:309–12.
14. Brown RB, Audet J. Sampling efficiency of a single-cell capillary electrophoresis system. Cytometry A. 2007;71(10):882–8.
15. Li H, Sims CE, Kaluzova M, et al. A quantitative single-cell assay for protein kinase B reveals important insights into the biochemical behavior of an intracellular substrate peptide. Biochemistry. 2004;43(6):1599–608.
16. Manceur A, Wu A, Audet J. Flow cytometric screening of cell-penetrating peptides for their uptake into embryonic and stem cells. Anal Biochem. 2007;364(1):51–9.
17. Soughayer JS, Wang Y, Li H, et al. Characterization of TAT-mediated transport of detachable kinase substrates. Biochemistry. 2004;43(26):8528–40.
18. Chalfie M, Tu Y, Euskirchen G, et al. Green fluorescent protein as a marker for gene expression. Science. 1994;263(5148):802–5.
19. Inouye S, Tsuji FI. Aequorea green fluorescent protein. Expression of the gene and fluorescence characteristics of the recombinant protein. FEBS Lett. 1994;341(2–3):277–80.
20. Tsien RY. The green fluorescent protein. Annu Rev Biochem. 1998;67:509–44.
21. Shen K, Meyer T. Dynamic control of CaMKII translocation and localization in hippocampal neurons by NMDA receptor stimulation. Science. 1999;284(5411):162–6.
22. Hirose K, Kadowaki S, Tanabe M, et al. Spatiotemporal dynamics of inositol 1,4,5-trisphosphate that underlies complex Ca2+ mobilization patterns. Science. 1999;284:1527–30.
23. Tanimura A, Nezu A, Morita T, et al. Interplay between calcium, diacylglycerol, and phosphorylation in the spatial and temporal regulation of PKCalpha-GFP. J Biol Chem. 2002;277: 29054–62.
24. Sawano A, Hama H, Saito N, et al. Multicolor imaging of Ca2+ and protein kinase C signals using novel epifluorescence microscopy. Biophys J. 2002;82:1076–85.
25. Pepperkok R, Squire A, Geley S, et al. Simultaneous detection of multiple green fluorescent proteins in live cells by fluorescence lifetime imaging microscopy. Curr Biol. 1999;9:269–72.
26. Lackowicz JR. Principles of fluorescence spectroscopy. New York: Plenum Press. 1986.
27. Stryer L, Haugland RP. Energy transfer: a spectroscopic ruler. Proc Natl Acad Sci U S A. 1967;58:719–26.
28. Shaner NC, Campbell RE, Steinbach PA, et al. Improved monomeric red, orange and yellow fluorescent proteins derived from Discosoma sp. red fluorescent protein. Nat Biotechnol. 2004;22(12):1567–72.
29. Shaner NC, Steinbach PA, Tsien RY. A guide to choosing fluorescent proteins. Nat Methods. 2005;2(12):905–9.
30. Mochizuki N, Yamashita S, Kurokawa K, et al. Spatio-temporal images of growth-factor-induced activation of Ras and Rap1. Nature. 2001;411:1065–8.
31. Miyawaki A, Llopis J, Heim R, et al. Fluorescent indicators for Ca2+ based on green fluorescent proteins and calmodulin. Nature. 1997;388:882–7.
32. Zhang J, Allen MD. FRET-based biosensors for protein kinases: illuminating the kinome. Mol BioSyst. 2007;3:759–65.
33. Ni Q, Titov DV, Zhang J. Analyzing protein kinase dynamics in living cells with FRET reporters. Methods. 2006;40:279–86.
34. Zhang J, Ma Y, Taylor SS, et al. Genetically encoded reporters of protein kinase A activity reveal impact of substrate tethering. Proc Natl Acad Sci U S A. 2001;98:14997–5002.
35. Wang Y, Botvinick EL, Zhao Y, et al. Visualizing the mechanical activation of Src. Nature. 2005;434:1040–5.
36. Ting AY, Kain KH, Klemke RL, et al. Genetically encoded fluorescent reporters of protein tyrosine kinase activities in living cells. Proc Natl Acad Sci U S A. 2001;98:15003–8.
37. Sato M, Ozawa T, Inukai K, et al. Fluorescent indicators for imaging protein phosphorylation in single living cells. Nat Biotechnol. 2002;20:287–94.
38. Violin JD, Zhang J, Tsien RY, et al. A genetically encoded fluorescent reporter reveals oscillatory phosphorylation by protein kinase C. J Cell Biol. 2003;161:899–909.
39. Sasaki K, Sato M, Umezawa Y. Fluorescent indicators for Akt/protein kinase B and dynamics of Akt activity visualized in living cells. J Biol Chem. 2003;278:30945–51.
40. Zhang J, Hupfeld CJ, S.Taylor S, et al. Insulin disrupts β-adrenergic signalling to protein kinase A in adipocytes. Nature. 2005;437:569–73.
41. Kunkel MT, Ni Q, Tsien RY, et al. Spatio-temporal dynamics of protein kinase B/Akt signaling revealed by a genetically encoded fluorescent reporter. J Biol Chem. 2005;280:5581–7.
42. Johnson SA, You Z, Hunter T. Monitoring ATM kinase activity in living cells. DNA Repair (Amst). 2007;6:1277–84.
43. Kunkel MT, Toker A, Tsien RY, et al. Calcium-dependent regulation of protein kinase D revealed by a genetically encoded kinase activity reporter. J Biol Chem. 2007;282:6733–42.
44. Sato M, Kawai Y, Umezawa Y. Genetically encoded fluorescent indicators to visualize protein phosphorylation by extracellular signal-regulated kinase in single living cells. Anal Chem. 2007;79:2570–5.
45. Calleja V, Ameer-Beg SM, Vojnovic B, et al. Monitoring conformational changes of proteins in cells by fluorescence lifetime imaging microscopy. Biochem J. 2003;372:33–40.

46. Calleja V, Alcor D, Laguerre M, et al. Intramolecular and intermolecular interactions of protein kinase B define its activation in vivo. PLoS Biol. 2007;5(4):e95.
47. Takao K, Okamoto K, Nakagawa T, et al. Visualization of synaptic Ca2+/calmodulin-dependent protein kinase II activity in living neurons. J Neurosci. 2005;25:3107–12.
48. Fujioka A, Terai K, Itoh RE, et al. Dynamics of the Ras/ERK MAPK cascade as monitored by fluorescent probes. J Biol Chem. 2006;281(13):8917–26.
49. Neininger A, Thielemann H, Gaestel M. FRET-based detection of different conformations of MK2. EMBO Rep. 2001;2:703–8.
50. Nagai Y, Miyazaki M, Aoki R, et al. A fluorescent indicator for visualizing cAMP-induced phosphorylation in vivo. Nat Biotechnol. 2000;18:313–6.
51. Schleifenbaum A, Stier G, Gasch A, et al. Genetically encoded FRET probe for PKC activity based on pleckstrin. J Am Chem Soc. 2004;126(38):11786–7.
52. Green HM, Alberola-Ila J. Development of ERK activity sensor, an in vitro, FRET-based sensor of extracellular regulated kinase activity. BMC Chem Biol. 2005;5:1.
53. Gromova KV, Friedrich M, Noskov A, et al. Visualizing Smad1/4 signaling response to bone morphogenetic protein-4 activation by FRET biosensors. Biochimica et Biophysica Acta. 2007;1773:1759–73.
54. Chun W, Johnson GVW. Activation of glycogen synthase kinase 3β promotes the intermolecular association of tau. The use of fluorescence resonance energy transfer microscopy. J Biol Chem. 2007;282:23410–7.
55. Ruehr ML, Zakhary DR, Damron DS, et al. Cyclic AMP-dependent protein kinase binding to A-kinase anchoring proteins in living cells by fluorescence resonance energy transfer of green fluorescent protein fusion proteins. J Biol Chem. 1999;274:33092–6.
56. Heikal AA, Hess ST, Baird GS, et al. Molecular spectroscopy and dynamics of intrinsically fluorescent proteins: coral red (dsRed) and yellow (Citrine). Proc Natl Acad Sci U S A. 2000;97:11996–2001.
57. Ng T, Squire A, Hansra G, et al. Imaging protein kinase Calpha activation in cells. Science. 1999;283(5410):2085–9.
58. Verveer PJ, Wouters FS, Reynolds AR, et al. Quantitative imaging of lateral ErbB1 receptor signal propagation in the plasma membrane. Science. 2000;290:1567–70.
59. Wouters FS, Bastiaens PI. Fluorescence lifetime imaging of receptor tyrosine kinase activity in cells. Curr Biol. 1999;9:1127–30.
60. Hahn k, Toutchkine A. Live-cell fluorescent biosensors for activated signaling proteins. Curr Opin Cell Biol. 2002;14:167–72.
61. Kraynov VS, chamberlain C, Bokoch GM, et al. Localized Rac activation dynamics visualized in living cells. Science. 2000;290:333–7.
62. Chamberlain C, Hahn KM. Watching proteins in the wild: fluorescence methods to study protein dynamics in living cells. Traffic. 2000;1:755–62.
63. Cornish VW, Hahn KM, Schultz PG. Site-specific protein modification using a ketone handle. J Am Chem Soc. 1996;118:8150–1.
64. Wang L, Brock A, Herberich B, et al. Expanding the genetic code of *Escherichia coli*. Science. 2001;292:498–500.
65. Schwarze SR, Dowdy SF. In vivo protein transduction: intracellular delivery of biologically active proteins, compounds and DNA. Trends Pharmcol Sci. 2000;21:45–8.

Hematopoietic Stem Cells

Malcolm A.S. Moore

Abstract Hematopoietic stem cells (HSCs) are the most well-characterized tissue-specific stem cell. Over 50 years of basic research and clinical application has provided insight into the molecular and cellular mechanisms of HSC biology. HSC undergo self-renewal by symmetric or asymmetric division, or differentiation to common myeloid progenitors and progressively more differentiated myeloid and lymphoid progeny. The chemokine SDF-1/ CXCL12 produced by marrow reticular cells plays a central role in regulating HSC migration and homing. HSC reside within a three-dimensional multicellular signaling unit or "niche" within the bone marrow. Within the niche HSC can be sustained in a dormant G0 state. Accumulating genetic and functional data indicate molecular cross-talk between HSCs and osteoblasts, endothelial cells, and perivascular reticular cells that compose the endosteal and vascular niches. This involves large number of molecules (cytokines, chemokines, integrins, morphogens, and their receptors). The cytokines c-Kit ligand and thrombopoietin in particular are critical for HSC maintenance and self-renewal. The morphogens (Notch ligands, Wnt, Hedgehog, TGFβ, and BMP) also have variable and overlapping roles in supporting HSC self-renewal. Many of these secreted signaling molecules bind to the extracellular matrix and do not diffuse far or are presented by niche cells in transmembrane form (e.g., Jagged-Notch, Kit Ligand-c-Kit, Angiopoietin1-Tie2). Increased understanding of the molecular basis of HSC regulation will ultimately lead to protocols for maintenance and expansion of HSCs in vitro for clinical use in cell and gene therapies.

Keywords Stem cell · Cytokine · Chemokine · Morphogen · Self-renewal · Asymmetric cell division · Integrins · Niche · Homeobox genes

M.A.S. Moore (✉)
Enid A. Haupt Professor of Cell Biology, Cell Biology Program, Memorial Sloan-Kettering Cancer Center, New York, NY 10065
e-mail: m-moore@ski.mskcc.org

1 Introduction

The nature and developmental potential of stem cells within the hematopoietic system has been debated for over a century. In 1896 Pappenheim used the term *stem cell* to describe a precursor cell capable of giving rise to both red and white blood cells, and subsequently Maximow, Dantschakoff, and Neumann began to use the term *stem cell* to refer to the common precursor of the blood system [1]. The early hematology community became divided into those who believed that there was a single type of stem cell generating all myeloid and lymphoid lineage cells (monophyletic hypothesis) and those that recognized a number of lineage-restricted stem cells (polyphyletic hypothesis). The prevailing view is a monophyletic one, with a pluripotent stem cell compartment and a hierarchy of progressively more lineage-restricted progenitor cells. The concept of a self-renewing, pluripotential hematopoietic stem cell (HSC) achieved experimental validation from the pioneering work of Till and McCulloch in the early 1960s using the spleen colony-forming assay (CFU-s) in irradiated mice [2, 3].

2 Functional Assays for HSC

2.1 *In Vitro Assays*

Clonogenic assays in semi-solid medium can detect at least some HSC provided appropriate early-acting cytokines are used. These colonies frequently contain a mixture of granulocytes, megakaryocytes, macrophages, and erythroid elements (CFU-GEMM). However, common myeloid progenitors and both short- and long-term repopulating HSC form such colonies so the clonogenic assay is not specific. A subtype of high proliferative potential colony-forming cells (HPP-CFC) that form colonies of $>10^5$ cells have been used as surrogate HSC assays [4]. More recent assays involve the development of cobblestone areas of phase-dark cells that develop beneath marrow stroma following 3–4 weeks (murine) or >5 weeks

V.K. Rajasekhar, M.C. Vemuri (eds.), *Regulatory Networks in Stem Cells*, Stem Cell Biology and Regenerative Medicine,
DOI 10.1007/978-1-60327-227-8_30, © Humana Press, a part of Springer Science+Business Media, LLC 2009

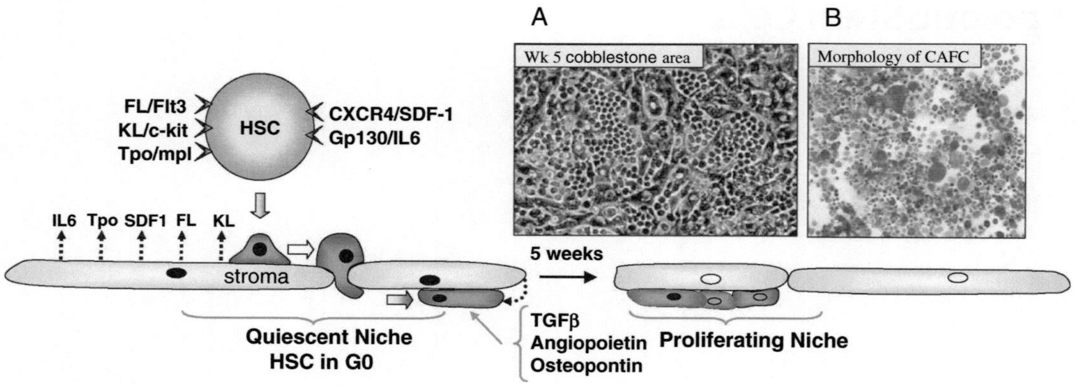

Fig. 1 Model illustrating the "late" cobblestone area-forming cell (CAFC) assay showing the interaction of the HSC with the bone marrow stromal niche using cell lines equivalent to CXCL12-abundant reticular cells of Sugiyama et al. [87]. SDF-1/ CXCL12 produced by the marrow stroma establishes a gradient that attracts CXCR4$^+$HSC that adhere via VLA-4-VCAM-1 mediated adhesion and then migrate beneath the stroma. Stromal membrane-associated cKit ligand (KL) and Flt3 ligand (FL) provide survival and proliferation signals to the HSC while negative regulatory influences from the stroma (TGFβ, angiopoietin. osteopontin) override proliferative signals and place the HSC in a G0 state that persists for some weeks. Escape from quiescence may be a stochastic process or driven by an alteration in the balance between proliferation stimulating and inhibiting factors. At this stage the HSC proliferates and differentiates, forming colonies of 5–100 phase dark cells by week 5 (A). At this stage each cobblestone area contains an average of 1–4 HSC and 5–20 progenitors (CFU-GM, BFU-E) as well as differentiating erythroid, megakaryocytic and granulocytic and monocytoid cells (B. hematoxylin and eosin stain)

(human) co-culture with bone marrow CD34$^+$ cells (Fig. 1) [5]. These "late" cobblestone area-forming cells (CAFC) have self-renewal and pluripotent differentiation potential and can be distinguished from committed progenitors forming "early" cobblestone areas at 1–2 weeks in mouse and 2–3 weeks in human assays. A modification of this assay involves quantitation of secondary progenitor erythroid and myeloid colony formation, for example, after 5 weeks of stromal or cytokine dependent culture (long-term culture-initiating cell assay [LTC-IC]). Quantitation of HSC can be achieved by undertaking the stromal co-culture assays under limiting dilution conditions. In a comparison of human HSC quantitation by LTC-IC assay versus in vivo NOD/SCID mouse limiting dilution engraftment (SRC assay) in vitro culture of cord blood cells on human marrow stroma led to to a decline of SRC six-fold while LTC-IC were maintained or increased over this time [6]. While the authors concluded that the two assays detected different cell populations, an equally plausible explanation is that initially quiescent HSC entered cell cycle in vitro and this has been shown to profoundly impair in vivo engraftment whereas cycle status does not influence HSC readout in vitro. These methods are reviewed in more detail by van Os et al. [7]

2.2 In Vivo Assays

The spleen colony-forming assay is no longer considered an HSC assay. About half of day 12 CFU-s are derived from the short-term repopulating HSC and multipotent progenitor (MPP) compartment and the other half are derived from the myegakaryocyte-erythroid progenitor (MEP) and common myeloid progenitor (CMP) populations [8, 21, 22], while the vast majority of day 8 CFU-s are derived from MEPs [8]. The in vivo assays for murine HSC involve limiting dilution, competitive (e.g., with addition of a genetically distinct marrow population) repopulation of irradiated recipients, with evaluation after 3–6 months. The immunodeficient SCID or NOD/SCID mouse supports human hematopoiesis following intravenous injection of $2 \times 10^4 - 2 \times 10^5$ CD34$^+$ cells and quantitation of engraftment by measure of human CD45$^+$ cells in the murine femoral bone marrow at 5 weeks and beyond. This can be used under limiting dilution conditions to quantify human HSC (SCID Repopulating Cells [SRC]) (reviewed in [9]). The SRC assay may overestimate long-term repopulating ability [10]. In a direct comparison of engraftment of virally transduced HSC into non-human primates and NOD/SCID mice, NOD/SCID repopulating clones were able to contribute to short-term repopulation in primates. However, no NOD/SCID repopulating clones contributed to primate hematopoiesis at 6 months or later [11]. Mazurier et al. [12] identified a new class of human HSC by direct intrafemoral injection in NOD/SCID mice that were CD34$^+$, CD38^{low}, and CD36$^-$. These "rapid-SRCs" rapidly generate high levels of human myeloid and erythroid regeneration within 2 weeks post-transplantation and subsequently enter the blood and colonize other marrow sites. Kimura et al. [13] also used intrafemoral injection to identify a very primitive long-term repopulating human HSC that did not express CD34, c-Kit, or Flt-3, however, this observation remains controversial [10].

2.3 Homing Efficiency of HSC

The absolute quantitation of HSC by in vivo assay requires that their "seeding" efficiency be known. The engraftment of HSCs into irradiated mice was thought to be an inefficient process. In early studies, seeding of CFU-s was determined by secondary transplantation, with 15–20% engrafted by 24h and the majority of these localizing within 3h [3]. The probability of any individual HSC to "seed" the bone marrow, when injected intravenously, has been calculated to be at best 10–20% [14, 15]. Advances in HSC enrichment technologies combined with strategies to test their engraftment efficiency by measuring competition of unpurified donor bone marrow cells with recipient cells in murine hosts or by tracking the engraftment of one highly purified stem cell injected per recipient showed that HSC engrafted with high efficiency (30–60%), or even absolute efficiency [16–20]. Hoechst dye effluxing mouse marrow side population (SP) HSC engrafted with 36% efficiency [20]. Matsuzaki et al. [18] used a combination of dye-efflux and cell-surface markers to purify a more homogenous HSC population and 96% of mice given transplants of such single Lin$^-$ CD34$^-$Kit$^+$Sca-1$^+$-SP cell exhibited long-term multilineage hematopoietic engraftment.

3 Phenotypic Characterization of HSC

3.1 Murine HSC

Spangrude et al. [21] identified mouse HSCs within a lineage negative (Lin$^-$) population of marrow cells expressing the pan-hematopoietic CD45 epitope, lacking markers for erythroid, myeloid T, B, and NK lineages and expressing Thy1^{lo}and Sca-1$^+$. Further selection for cytokine receptor c-Kit enriches for a population in which ~10% of cells have long-term reconstituting activity (LT-HSCs) [16, 22]. CD34$^+$ is expressed on LT-HSC of the murine fetus and neonate but decreases with age so that HSC of 10-week-old mice are CD34$^-$ [23, 24]. CD34 expression reflects the activation/kinetic state of HSC and expression is reversible. In the murine system, the majority of HSC are CD38$^+$ and there is a reciprocal relationship of CD34 and CD38 expression [23]. HSC enrichment is also obtained by selection for Rhodamine 123 or Hoechst 33288 dye exclusion, based on the ability of primitive hematopoietic cells to efflux certain small molecules by the ABC pump. The latter dye has been used to identify a FACS dye-excluding "side population." This SP fraction is highly enriched for HSC in the mouse marrow but not in human [25]. The surface markers for human subset analysis are in many cases different to those used in the mouse. In part this reflects clear species differences in location. For example Flt3 is a marker for human HSC but in mice is expressed on a common lymphoid progenitor [26]. Surface receptors of the SLAM family are differentially expressed among functionally distinct HSC and progenitor populations of the mouse. HSC are CD150$^+$ CD244$^-$ CD48$^-$, while multipotent progenitors are CD244$^+$ CD150$^-$ CD48$^-$ [27]. The combination of CD150 and CD48 represented only 0.0084% ± 0.0028% of whole bone marrow cells. Although CD41 is expressed by primitive HSCs, CD41 is down-regulated by HSCs during the transition to definitive hematopoiesis and most adult HSCs do not express CD41. Thirty-seven percent of CD150$^+$CD48$^-$cells were CD41$^-$and 45% of these gave long-term multilineage reconstitution [27]. The SLAM markers have not proven useful in characterization of human HSC. Weksberg et al. [28] reported that exclusive reliance on SLAM family markers to isolate HSCs neglected a substantial fraction of marrow HSC. Inclusion of canonical HSC markers (Sca-1, c-Kit, and lineage markers) in the SLAM scheme greatly augmented HSC purity. Furthermore, both CD150$^+$ and CD150$^-$cells were within the Hoechst dye SP population, and both populations can contribute to long-term multilineage reconstitution. The endothelial protein C receptor CD201 has recently been shown to explicitly identify murine (but not human) HSC with CD201-positive bone marrow Lin$^-$c-Kit$^+$Sca-1$^+$CD34$^-$ cells appearing more that 1000-fold enriched for HSC activity [29]. CD201 expression was found to correlate closely with both SP phenotype as well as the traditional Lin$^-$Kit$^+$Scal-1$^+$CD34$^-$ HSCs phenotype.

3.2 Human HSC

The surface markers for human subset analysis are in many cases different to those used in the mouse. In part this reflects clear species differences in location e.g. Flt3. Up-regulation of Flt3 expression on murine Lin$^-$Sca-1$^+$c-kit$^+$ HSC is accompanied by loss of self-renewal capacity but sustained lymphoid-restricted reconstitution potential [26]. CD34 is expressed on human HSC regardless of cycle status but is absent on quiescent murine HSC. Candidate human HSC with long-term (>5 weeks) engraftment of NOD/SCID mice (SCID-Repopulating Cells [SRCs]) are CD34 and CD133 positive regardless of cycle status [9]. However, 95–99% of CD34$^+$ cells and 70–75% of CD133$^+$ cells are progenitor populations. Since the latter are also CD38$^+$, the CD34$^+$, CD38$^-$phenotype has been most frequently used for human HSC enrichment [30]. CD38 is reversibly expressed on CD34$^+$ SRC between negative and low levels and corresponds to a change in the cell-cycle state. Thy-1 (CD90) is also expressed on human HSC but is also not an exclusive marker since it is also expressed on a subset of progenitors. Using NOD/SCID IL2Rγ null mice as few as 20,000 cord blood Lin$^-$ CD34$^+$ CD38$^-$ cells were shown to engraft [31]. The Rhodamine-123 dye effluxing fraction of the Lin$^-$ CD34$^+$ CD38$^-$ population of CB, contains SRC at a frequency of 1 in 30 cells [32]. The CD34$^+$CD38$^-$ population can be separated into three subpopulations: (1) CD90$^+$ CD45RA$^-$(30%) LT-HSC, with as few as 10 cells providing engraftment; (2) CD90$^-$ CD45RA$^+$ (32%), with little or no stem or progenitor activity; and (3) CD90$^-$, CD45RA$^-$(38%) multipotent progenitors [33]. Distinction between human LT-HSC and short-term reopopulating-HSC was provided by differential engraftment in NOD/SCID versus NOD/SCIDβ2 null mice. The former are only engrafted by LT-HSC whereas both populations engraft in the more NK deficient NOD/SCIDβ2null mice [9]. Telomerase activity and hTERT expression are up-regulated in the transition from more primitive SRCs to short-term SRCs lacking secondary SRC capacity [34]. The intracellular enzyme, aldehyde dehydrogenase (ALDH), protects BM progenitors from the cytotoxic effects of cyclophosphamide by deactivation of its metabolite, 4-hydroxycyclophosphamide [35]. SRCs can be enriched using a fluorescently labeled dye specific for ALDH activity with the 10% top ALDH$^+$ CD34$^+$, CD38$^-$ containing LT-HSC at a frequency of 1/360 [36].

4 Cell Cycle Regulation of HSC

In mammalian cells, entry into the cell cycle requires sequential activation of the cyclin-dependent kinases (CDK) 4/6 and CDK2, which are inhibited by the cell-cycle inhibitors (CKIs), comprising the INK4 proteins (p16INK4A, p15INK4B, p18INK4C, and p19INK4D) and the Cip/Kip proteins (p21Cip1/Waf1, p27kip1, and p57Kip2). Upon mitogenic stimulation, cyclin D is up-regulated and interacts with CDK4/6, resulting in Rb phosphorylation to initiate cell cycle progression. Although Cip/Kip proteins (such as p21) broadly inhibit CDK2 in late G1/S and possibly CDK1 in M phase, they are not capable of inhibiting CDK4/6 activity early in G1. In contrast, INK4 proteins (such as p18) are able to specifically compete with cyclin D to bind CDK4/6 in early G1. p16INK4A and p19ARF are downstream mediators of the Bmi-1 protein regulating HSC self-renewal [37]. p21Cip1/Waf1, a late G1-phase CKI, and p27kip1, have been shown to govern the pool size of HSC and progenitors respectively [37–39]. Enforced expression of the HOXB4 transcription factor and down-regulation of p21 (Cip1/Waf) can each independently increase murine HSC proliferation [38]. When p21 knockdown and HOXB4 overexpression were combined in HSC, long-term competitive repopulating cells expanded 100-fold in 5 days [40]. Absence of p21Cip1/Waf1 increased cell-cycle entry of HSC under homeostatic conditions, but caused premature exhaustion of HSC under conditions of stress [41]. In contrast, deletion of an early G1-phase CKI, p18, resulted in strikingly improved long-term engraftment, largely by increasing their self-renewal [41]. Therefore, different CKIs have highly distinct effects on the kinetics of HSCs, possibly because of their active position in the cell cycle. TGF-β1 has been implicated in the maintainence of HSC in a quiescent, or slowly cycling state (*vide infra*) and in human HSC this cell-cycle arrest is mediated by up-regulation of the CKI p57KIP2 [42].

4.1 Proliferative Status of the HSC Compartment

A subpopulation of HSC must enter cell cycle on a regular basis to generate transit amplifying progenitor cells sufficient to sustain steady-state production of mature blood cells. At any one time, approximately 5% of murine LT-HSC is in the S/G2/M phase of the cell cycle and another 20% are in G1 phase. Incorporation of the DNA label 5-bromo-2- deoxyuridine (BrdU) has been used to determine the rate at which different cohorts of HSC entered the cell cycle over time [43] with ~50% of LT-HSC incorporating BrdU by 6 days, >90% by 30 days, and 99% by 6 months. More recent studies on BrdU incorporation into CD150$^+$CD48$^-$CD41$^-$Lin$^-$Sca-1$^+$c-kit$^+$ cells (47% LT-HSC) showed 6.0% entered the cell cycle daily [44]. If a minority of HSCs were more deeply quiescent than most other HSCs, they should remain BrdU-negative, even after long periods of BrdU treatment. This was not observed since >99% of HSCs were labeled after 6 months and in pulse-chase studies 2% of HSCs were BrdU$^+$ after 120 days, demonstrating that the frequency of BrdU-retaining HSCs continues to decline over time rather than there being a deeply quiescent subset of HSCs that retains BrdU indefinitely. Nonetheless, data cannot exclude the possibility that a minority of HSCs divide more slowly. HSC in bone marrow (BM) cycle rapidly and expand their numbers in response to cytoreductive agents, such as cyclophosphamide (CY), and cytokines, such as G-CSF with virtually all HSC entering cell cycle. However, G-CSF/CY mobilized PB HSC are almost all in the G0 or G1 phase of the cell cycle. This has raised the question of whether, following G-CSF/CY treatment, a subset of noncycling HSC is selectively released from the bone marrow, or whether cycling HSC are mobilized after progression through M phase, but before the next S phase of the cell cycle. BrdU incorporation studies indicate the latter [45].

4.2 Negative Regulation of HSC Proliferation

While the extent and duration of HSC quiescence in the adult human bone marrow remains controversial, it is unequivocal that the majority of HSC are not in cell cycle at any one time. This noncycling state could be imposed by absence of positive stimulating factors or by presence of negative regulatory factors. The negative regulatory roles of Transforming Growth Factor-β (TGF-β), Angiopoietin, and Osteopontin are discussed in later sections in the context of quiescence of HSC in osteoblasts niches. Low concentrations of SDF-1 induce a G0/G1 transition and increase the survival of immature human $CD34^+$ cells, while high levels of this ligand cause human $CD34^+$ progenitor cell quiescence and retention in a nonmotile mode (reviewed in [46]). Cycling, normal $CD34^+CD38^+$ cells secrete low levels of SDF-1 in an autocrine manner, unlike the more primitive, quiescent human $CD34^+CD38^-$ cells. Importantly, cell cycle affects homing and retention of repopulating HSC. Tumor necrosis factor (TNF), if overexpressed, mediates bone marrow suppression and also negatively regulates HSC self-renewal [47, 48]. Murine $Lin^-Sca\text{-}1^+c\text{-}kit^+$ and human $CD34^+CD38^-$ HSC were capable of undergoing cell divisions in the presence of TNF but were severely compromised in their short- and long-term multilineage reconstituting ability. The ETS family transcription factor MEF (myeloid ELF-1-like factor, also known as ELF4) is reported to regulate quiescence of HSC [49]. MEF functions as a transcriptional activator of the IL-8, perforin, GM-CSF, and IL-3 genes in hematopoietic cells. MEF null HSC displayed increased residence in G0 with reduced BrdU incorporation in vivo and impaired cytokine proliferation in vitro. MEF null mice are consequently relatively resistant to the myelosupressive effects of chemotherapy or radiation. Lipid raft clustering is also reported to be a key event in the regulation of HSC dormancy [50]. Freshly isolated HSCs from the BM niche lack lipid raft clustering, exhibit repression of the Akt–Fox0 signaling pathway, and express abundant p57Kip2/Cdk1c cyclin-dependent kinase inhibitor. Lipid raft clustering induced by cytokines is essential for HSC re-entry into the cell cycle. Conversely, inhibition of lipid raft clustering caused sustained nuclear accumulation of Fox0 transcription factors and induced HSC hibernation ex vivo [50]. These data establish a critical role for lipid rafts in regulating the cell cycle, the survival, and the entry into apoptosis of HSCs and their hibernation. Tpo signaling through Mpl upregulates β-1 integrin and p57Kip2 [51]. p57 is highly expressed in mouse quiescent HSC and is very reduced in Mpl null mice. Exogenous Tpo transiently increased HSC quiesecence (12 h treatment protected HSC from 5-fluorouracil), but Tpo is subsequently important for increased HSC proliferation in vivo.

4.3 HSC Self-Renewal

A central question in cell biology is how two progeny cells adopt distinct fates? HSCs for example, must undergo asymmetric division to generate cells to sustain long-term hematopoiesis as well as to produce progeny cells of the distinct blood lineages (reviewed in [52, 53]). The concept of a self-renewing, pluripotential HSC achieved experimental validation in the early 1960s using CFU-s assays [2, 3]. Secondary passage of individually excised spleen colonies showed high variability in numbers of secondary colonies generated and this fitted a skewed (gamma) distribution. The probability that a single CFU-s upon division would generate a new CFU-s (self-renewal) was calculated as 0.6 while production of a differentiated progenitor cell was 0.4. The conclusion was that this process was random or stochastic. To sustain hematopoiesis and to maintain nearly constant numbers of HSCs it has been proposed that adult HSCs divide asymmetrically and this has been observed in culture [54]. Giebel et al. [55] showed that the vast majority of the most primitive, in vitro–detectable human hematopoietic cells give rise to daughter cells adopting different cell fates, either inheriting the developmental capacity of the mother cell or becoming more specified. Such models are generally classed as "instructive," in that systemic feedback signals or local environmental cues may regulate HSC decision making, or "stochastic," in which any given HSC is equally likely to exit the stem cell state at a given time, resulting in a constant pool size determined by the overall probability of self-renewal within the population [56]. It is important to note that asymmetric division cannot be the exclusive mode of HSC division, because expansion of the HSC pool is necessary and possible, for example, after its reduction by injury or irradiation. However, asymmetric division could be a central homeostatic mechanism controlling HSC self-renewal, possibly modulated by extrinsic signals under regenerative stress conditions. Much attention has revolved around the role played by hematopoietic growth factors in this process: do they play an instructive, that is, a deterministic role, or are they simply permissive or selective, that is, allow the survival and proliferation of independently committed cells? One model of HSC lineage specification posits that low-level multilineage gene activity establishes a ground state or level of noise from which regulatory networks can start to build and be amplified or diminished either through quasi-random or stochastic changes in the components (i.e., spontaneous transcription of an accessible activator or repressor) or through positive/negative reinforcement through extracellular signalling via stochastically expressed receptor molecules [56]. Such a model incorporates both cell-extrinsic and cell-intrinsic components. Experiments using mouse $CD34^-KLS$ cells have shown that culture with a cytokine combination (c-Kit L and IL3), which induce greater levels of differentiation, led to asymmetric divisions in 52% of cells whereas culture with cytokines (c-Kit L and Tpo), which tend to preserve undifferentiated cells, led to asymmetric division in only 17% of the cells [54]. Data suggest that cytokines that decrease differentiation do so by decreasing asymmetric division. Wu et al. [57] used transgenic Notch reporter mice, in which GFP fluorescence indicated the status of Notch signaling, and time-lapse microscopy was used to trace hematopoietic precursor division and define whether the pattern and rate of division changed in the context of different microenvironments. When GFP^+ HSC divided they did so through a combination of asymmetric and symmetric divisions. There is good evidence that HSCs can expand in vivo and be maintained in vitro in close contact to stroma cells. When placed on stroma that induce differentiation, HSC predominantly use asymmetric divisions or symmetric commitment divisions; in contrast, when they are placed on stroma that promote maintenance of HSC they proceed predominantly through symmetric renewal divisions. β-1-integrin-mediated interactions between stroma and HSC significantly increased the proportion of asymmetrically dividing cells and led to a substantial increase in HSC [58]. Although all these observations are in accordance with the model of asymmetric cell division in which HSC contain the potential to give birth to two intrinsically different daughter cells, it cannot be concluded that the observed differences are indeed the result of an asymmetric cell division. In principle, these differences could be established by postmitotic, extrinsic, decision processes. Beckmann et al. [59] identified proteins (CD53, CD62L/L-selectin, CD63/lamp-3, and CD71/transferrin receptor) in combination with CD34 and C133 that segregated differentially in about 20% of primitive human hematopoietic cells that divide in stroma-free cultures. The asymmetrically dividing $CD34^+$ $CD133^+$ cells obtaining more CD53 or CD62L or less CD63 are more primitive than their sister cells. Daughter cells inheriting more of the CD71 vesicular-like structures are more mature than their sister cells. This indicates that HSCs/HPCs have the capability to divide asymmetrically. There are at least two hypothetical mechanisms by which asymmetric cell division can be achieved: divisional asymmetry and environmental asymmetry. In divisional asymmetry, specific cell-fate determinants in the genome or cytoplasm (RNA or proteins) are distributed unequally during cell division. After cell division, only one daughter cell receives the determinants, thus retaining the HSC fate while the other daughter differentiates. In environmental asymmetry, one HSC produces two identical daughter cells initially; however, only one remains in the HSC niche and retains the stem cell identity, while the other enters a different environment favoring its differentiation.

The immortal strand hypothesis was proposed as a mechanism by which stem cells could avoid accumulating mutations that arise during DNA replication [60]. Whereas most cells segregate their chromosomes randomly it was argued that adult stem cells in steady-state tissues might retain older DNA strands during asymmetric self-renewing divisions, segregating newly synthesized strands to daughter cells fated to differentiate. In support of this hypothesis a high incidence of non-random template strand segregation and asymmetric fate determination has been reported in dividing muscle stem cells with daughter cells inheriting the older templates retained the more immature phenotype, whereas daughters inheriting the newer templates acquired a more differentiated phenotype [61, 62]. Kiel et al. [44] were unable to confirm this in HSC. Sequential administration of 5-chloro-2-deoxyuridine and 5-iodo-2-deoxyuridine indicated that all HSCs segregate their chromosomes randomly. Division of individual HSCs in culture revealed no asymmetric segregation of the label and no retention of older DNA strands during division.

There is increasing evidence that the most important HSC functions, self-renewal and differentiation, are epigenetically preprogrammed and therefore predictable. This contradicts older models of HSC behavior, which postulated a single type of HSC that can be continuously molded into different subtypes of HSC. McKenzie et al. [63] evaluated the repopulation and self-renewal of individual human SRCs, tracked on the basis of lentiviral integration sites, in serially transplanted immune-deficient mice. They demonstrated maintenance by self-renewing SRCs after an initial period of clonal instability, a result inconsistent with a clonal succession model. Wide variation was noted in proliferation kinetics and self-renewal among SRCs, as well as between SRC daughter cells that repopulated equivalently, suggesting that SRC fate is unpredictable before SRCs enter more rigid "downstream" developmental programs. Serial transplantation of clonally repopulating murine HSC shows that self-renewal is associated with distinct hematopoietic differentiation programs [64]. Daughter HSCs derived from individual clones were remarkably similar to each other in the extent and kinetics of repopulation, and showed equivalent contributions to the myeloid or lymphoid lineages. Lineage contribution could be followed because of the discovery of a new subset of HSCs that gave rise stably to skewed ratios of myeloid and lymphoid cells [64–66]. Mice transplanted with single $CD45^+$, Lin^-Rho^-, SP cells gave one of four types of LT-HSC termed α (myeloid-biased), β (balanced lympho-myeloid), γ (lymphoid biased but still multilineage at 4 months), and δ (lymphoid biased but no longer multi-lineage at 4 months) (Fig. 2) [66]. These data indicate that the HSC compartment consists of a limited number of functionally distinct HSC subsets, each with predictable behavior. This contradicts older models of HSC

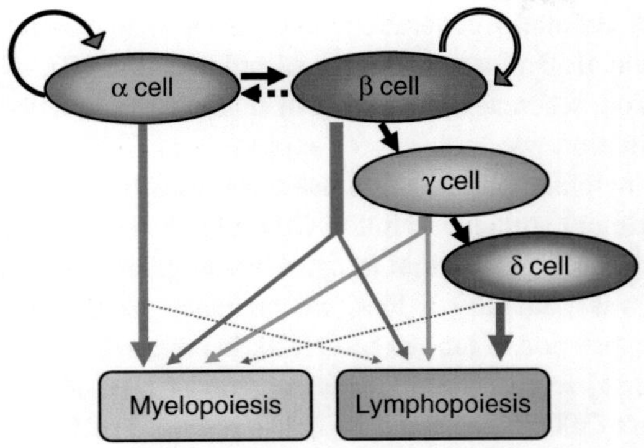

Fig. 2 Diagram of the relationship of different subtypes of long-term repopulating murine HSC. Single HSC transplantation and serial passage studies have identified four HSC subtype based on patterns of long-term (>16 weeks) generation of mature lympho-myeloid progeny [64–48]. α HSC are self-renewing but with a propensity for generating mature myeloid progeny that is sustained over multiple cycles of repopulation. It also has the potential to convert to HSC capable of providing substantial lymphoid repopulating activity (β HSC). β, γ, and δ HSC form a hierarchy with capacity to generate lymphoid and myeloid progenitors with myeloid potential diminishing progressively before multipotent repopulating capacity is lost. Modified after Dykstra et al. [66]

behavior, which postulated a single type of HSC that can be continuously molded into different subtypes of HSC.

5 Cytokine Stimulation of HSC Proliferation and Expansion

With the discovery of a number of hematopoietic growth factors (Interleukins, Colony Stimulating Factors) and their availability as recombinant proteins, in vitro culture systems were developed that supported extensive cell and progenitor expansion in the absence of stroma but in the presence of combinations of cytokines (reviewed in [69, 70]). Detailed studies of purified murine HSCs in recent years have shown that the receptors for thrombopoietin (Tpo) and c-Kit Ligand/Stem Cell Factor (KL/SCF), c-mpl, and c-kit, respectively, are both expressed on repopulating HSCs (reviewed in [69, 70]).

5.1 c-Kit

The proto-oncogene c-Kit was first identified as the cellular homologue of the oncogene v-Kit, and c-Kit was found to be allelic with the murine dominant white spotting (W) locus on chromosome 5, while its ligand, KL is is encoded by the murine Steel (Sl) locus on chromosome 10 (reviewed in [70, 71]). The several naturally occurring mutations in these two loci give rise to defects in proliferation, migration, and differentiation of HSC. KL is expressed as a glycosylated transmembrane protein and alternative splicing leads to two isoforms that differ in presence or absence of a particular proteolytic cleavage site [71]. The isoform containing the cleavage site undergoes proteolysis and become soluble upon release from the plasma membrane, whereas the isoform lacking the cleavage site remains cell associated. The two isoforms have different abilities to transmit signals. Stimulation with the soluble isoform leads to rapid and transient activation and autophosphorylation of c-Kit, as well as fast degradation, whereas stimulation with the membrane-associated isoform leads to a more sustained activation. Differences also exist in signaling downstream of c-Kit with the membrane-bound ligand inducing a more persistent activation of Erk1/2 and p38 MAPK as compared to the soluble ligand [71]. Czechowicz et al. [72] have recently found that HSCs, functionally displaced from their niches by in vivo treatment with anti-c-kit antibody, enhancing HSC engraftment after transplantation.

5.2 Thrombopoietin-Mpl

Recombinant Tpo itself was found to synergize with other proliferative cytokines to support HSC proliferation in vitro. Because Tpo, like other cytokines such as KL and IL-3, activates MAPK, AKT, and STAT pathways, such interactions might suggest overlapping, redundant functions [73]. However, the finding that Mpl mice have HSC deficiencies suggested that Tpo might have an important and unique role not shared by other cytokines [74]. Postnatal HSCs are dependent on Tpo for survival and maintenance unlike fetal HSC [51, 75]. Tpo–/– mice had normal numbers and function of HSC at birth but numbers begin to decline within a few weeks after birth and continue to do so throughout life, associated with reduced levels of p57Kip2 and p19Ink4D and independent of Bcl2 expression. Tpo and the cytokine signaling inhibitor adaptor molecule LNK, are opposing physiological regulators of HSC expansion [76]. Lnk negatively regulates self-renewal of HSCs by modifying Tpo-mediated signal transduction and Lnk-deficient LT-HSCs are hypersensitive to Tpo. Competitive repopulation revealed that long-term repopulating activity increases in Lnk-deficient HSCs and Lnk–/– HSCs continue to expand postnatally, up to 24-fold above normal by 6 mo of age [76]. A balance in positive and negative signals downstream from the Tpo signal plays a role in the regulation of the probability of self-renewal in HSCs and single-cell transplantation of paired HSC daughter cells indicated that a combination of KL and Tpo efficiently induces symmetrical self-renewal division in Lnk-deficient LT-HSCs [77]. Lnk–/– HSC expansion is dependent on Tpo, and 12-wk-old Lnk–/– Tpo–/– mice have 65-fold fewer LT-HSCs than Lnk–/– mice. Expansions of multiple myeloid, but not lymphoid, progenitors in Lnk–/– mice also proved Tpo-dependent.

5.3 Flt3

The class III receptor tyrosine kinase Flt3/Flk2 (CD135) is expressed on the majority of human $CD34^+$ cells including CFU-GM and CFU-GEMM and Flt3 Ligand (FL) responsive cells include candidate HSC that are $CD34^+CD38^-$, rhodamine 123^{dull}, and resistant to 4-hydroperoxycyclophosphamide (4-HC) [78]. Furthermore, in culture of single $CD34^+$ $CD38^-$cells with cKit-L, IL-3, IL-6, and G-CSF, addition of Flt3L increased two-fold the recruitment of these cells into cell cycle. Flt3L also regulated the function and expression of the β-integrins, VLA-4 and VLA-5, on primitive hematopietic cells that may be important for their homing and retention within the bone marrow microenvironment [79].

5.4 Cytokine Synergy in HSC Proliferation and Differentiation

In vitro functional assays have identified a number of cytokines that have distinct stimulatory effects on primitive hematopoietic cells, particularly when used in various combinations. These include FL, KL, Tpo, IL-1, IL-3, IL-6, IL-11, IL-12, G-CSF, and GM-CSF ([4, 75]; reviewed in [69, 70]). Using limiting dilution assays for competitive repopulating cells, modest degrees of HSC expansion have been reported in murine in vitro expansion systems using combinations of KL, FL, and IL-1. Optimal growth factor combinations are required to achieve in vitro HSC expansion and while KL and FL are sufficient to maintain survival and proliferation, retention of HSC function requires activation of additional pathway, for example, gp130, stimulated via IL-6 or IL-12 (reviewed in [80]). Haylock et al. [78] were first to report a CFC expansion of 66-fold in human $CD34^+$ cultures stimulated with IL-1, IL-3, IL-6, G-CSF, and GM-CSF. Subsequent studies showed the importance of specific cytokines for expansion of defined progenitor cell types. In a human $CD34^+$ expansion system, IL-6 plus soluble IL-6 receptor (to maximize IL-6 signaling where IL-6 receptor density could be limiting), together with FL, KL and Tpo resulted in a ~4-fold expansion of HSC as determined by NOD/SCID limiting dilution repopulation [79]. Both FL and IL-6/sIL-6 are important for expanding cord blood $CD34^+$-derived LTC-IC, while the addition of KL to either of these factors enhances generation of CFC [81]. In contrast, FL, KL, and IL-3 most efficiently stimulated BM LTC-IC proliferation, whereas the addition of IL-6/sIL-6R or Tpo to this combination was required for expansion of CFC. Regardless of the cytokine combination used, rigorous limiting dilution engraftment studies in NOD/SCID mice have, in the majority of studies, shown that in cytokine-supplemented cultures there was generally a loss, maintenance or modest (2–4-fold) expansion of human SRC [82]. In contrast, Gammaitoni et al. [83] reported a major expansion of LTC-IC and progenitors (3000-fold) and a 70-fold expansion of NOD/SCID engrafting HSC (SRC) in cultures of CB $CD34^+$ cells with FL, KL, Tpo, and IL-6. Under similar conditions but with $CD34^+$ cells from adult BM and mobilized peripheral blood (MPB), ex vivo expansion was shorter in duration and extent, with SRC expansion of only 6-fold by 3 weeks [83]. In a modified stromal-free culture system with cord blood-derived $CD34^+$ cells and a cocktail of four cytokines (KL, FL, Tpo, and IL-6), with $CD34^+$ re-isolation at monthly intervals, continuous expansion of HSC was observed over 10–20 weeks by in vitro CAFC and LTC-IC assays, and by NOD/SCID engraftment and secondary and tertiary passaging [83]. Despite extensive proliferation, the telomere length of cultured hematopoietic cells initially increased and it was only at late stages of culture that telomere shortening was detected. Telomere length stabilization correlated with high telomerase levels. In contrast cytokine stimulated adult $CD34^+$ cells from bone marrow or G-CSF mobilized blood showed $CD34^+$ and NOD/SCID engraftment expansions of 6-fold for only 3–4 weeks, with telomere shortening and low levels of telomerase activity. Human fetal liver-derived HSC with NOD/SCID repopulating capacity have been expanded (10–100-fold net expansion over 28 days of culture) with FL, Tpo, KL, and IL-6 and 8% human AB plasma [84]. The $CD133^+$ G0 cell subpopulation from cord blood is enriched for HSC and progenitors (CFC), with an LTC-IC incidence of 1 in 4.2 cells and a CFC incidence of 1 in 2.8 cells [85]. These cells could be expanded with cytokines in a serum-free, stromal-free culture system for up to 30 weeks resulting in a 100 million–fold amplification of progenitors. Quantitation of HSC expansion in vitro is complicated by divergence in results when in vitro HSC assays (CAFC, LTC-IC) are used as an end point, versus in vivo (SCR). While this has lead to considerable debate as to the validity of the assays, the issue is probably related to the efficiency of the respective systems. Clearly, the in vivo assay requires HSC survival and homing to the HSC niches of the mouse bone marrow and there is evidence that in vitro culture or cycling HSC may be less efficient in vivo while retaining potential to self renew and form CAFC/LTC-IC in vitro. In this context, Liu et al. [86] cultured CB $CD34^+$ cells for up to 5 days with a cocktail of cytokines, labeled the cells with 111-Indium and determined recovery at 48 h following intravenous injection into NOD/SCID mice. Cultured, cycling HSC and progenitors showed a reduction of marrow homing (from 11.3 to 5.4%) with reduced homing to spleen, liver and lung. Acute cytokine deprivation in the in vivo environment and Fas/CD95-mediated cell death may be responsible for reduced engraftment efficiency of cultured HSC.

6 HSC Homing and Extravasation

The chemokine SDF-1/CXC12 is expressed on bone marrow vascular endothelium, immature ostoblasts in the endosteal region and particularly the marrow reticular (CAR) cells [87], while its receptor, CXCR4, is expressed on HSC and progenitors (Fig. 3). The SDF-1/CXCR4 pathway plays a major role in regulating mobilization, migration and retention of HSC (reviewed in [46]). This chemokine pathway is essential for HSC seeding from the fetal liver to the bone marrow during development, however, CXCR4-deficient HSC can, with reduced efficiency, engraft adult irradiated mice (reviewed in [70, 88]). Elevation of plasma levels of SDF-1 occurs following intravenous injection of an adenovector expressing SDF-1 and the consequent reversal of the SDF-1gradient from blood to marrow leads to mobilization of HSC and progenitors [89]. Overexpression of CXCR4 on human CD34$^+$ cells by gene transfer increased their proliferation, migration, and NOD/SCID engraftment potential [46]. CXCR4 neutralization abolished human intravenous or intrafemoral CD34$^+$ engraftment in NOD/SCID mice, indicating the essential role of this receptor in BM seeding and colonization as well as homing [46]. HSC attachment and extravasation through the marrow endothelium involves the cooperation of adhesive interactions mediated by selectins and α4 integrins. The adhesion molecules P-selectin, E-selectin (CD62P and CD62E), and the α4-integrin ligand, vascular cell adhesion molecule-1 (VCAM-1/CD106), are constitutively expressed by marrow endothelium (reviewed in [90]). Extravasation begins with the rolling and endothelium tethering of circulating HSC and progenitors via selectin receptors. Firm adhesion and arrest requires both E-selectin–mediated and α-integrin–mediated interactions. HSC express two α4 integrins; α4β-1/VLA-4 and α4β7 (reviewed in [25]) (Fig. 3) with respective ligands VCAM-1/CD106 and the mucosal addressin cell adhesion molecule-1 expressed in marrow stroma. During this activation process, known as "inside-out signaling," integrin conformation is altered, enhancing the affinity and avidity to their ligands. α4β-1 Integrin is activated in vitro by SDF-1 [91] and a range of hematopoietic growth factors including Kit ligand, GM-CSF, and G-CSF, Tpo Flt3L, and HGF (reviewed in [25]). Once firmly adhered, a complex series of interactions between the HSCs and the endothelium leads to diapedesis of the HSCs between endothelial cells. This involves heterotypic adhesion molecules such as β2 integrins and homotypic cell adhesion molecules such as platelet-endothelial cell adhesion molecule-1 (CD31). HSC/HPC express adhesion receptors CD44 and Hyaluronan-mediated motility (RHAMM) that bind hyaluronan (HA), a glycosaminoglycan component of the extracellular matrix of the endothelial and endosteal BM microenvironments. Primitive hematopoietic cells synthesize and express HA, which was shown to be critical for the lodgment of transplanted HSCs within the endosteal versus the lineage-committed central marrow regions as demonstrated

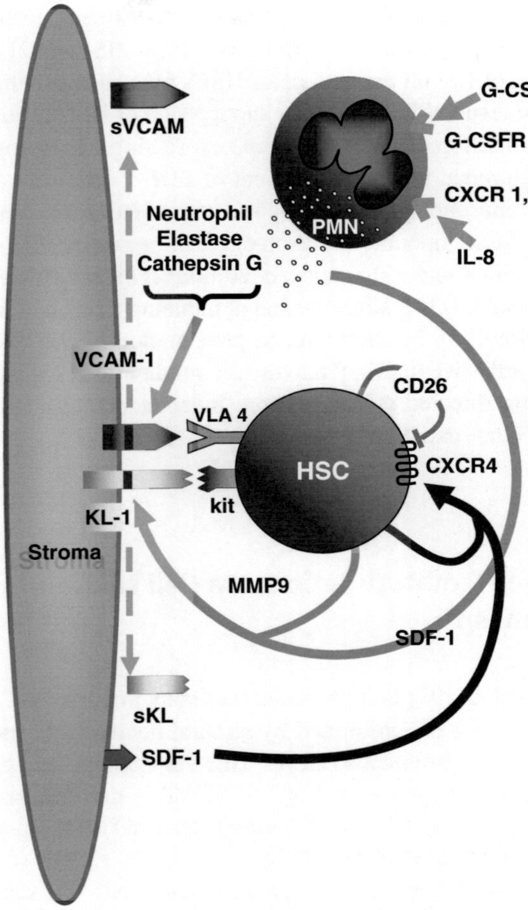

Fig. 3 HSC mobilization induced by G-CSF. HSC (CD34+) adherence to bone marrow stromal cells via VLA4 binding to VCAM-1 and cKit binding to transmembrane cKit ligand (KL-1). SDF-1 produced by the stroma induces matrix metalloproteinase-9 (MMP9) production by HSC and and this in turn cleaves the transmembrane KL-1, releasing soluble active ligand (sKL). The HSC mobilizing role of G-CSF is explained in part by increased numbers and activation of neutrophils within the bone marrow with secretion of a number of proteases (elastase, cathepsin G) that cleave VCAM-1 and "untether" HSC. Protease activity is also implicated in cleavage and inactivation of SDF-1 and CXCR4. HSC express the membrane bound ectopeptidase CD26 that removes dipeptides from the amino-terminus of proteins and acts as a negative regulator of the SDF-1-CXCR4 axis. Cycling, normal CD34+ cells secrete low levels of SDF-1 in an autocrine manner

by the significant alteration in spatial distribution resulting from the enzymatic removal of HA from HSCs [92]. The binding of HA on the surface of HSCs by a surrogate ligand in vitro resulted in a profound suppression of HSC proliferation and differentiation [91]. HSC expressing CD44 migrate on hyaluronan towards a gradient of SDF-1 acquiring a polarized morphology with CD44 concentrated at the leading edge of pseudopodia [93]. Proteoglycan-presented SDF-1 on endothelium under shear flow conditions is highly efficient at increasing $CD34^+$ adhesion and at inducing transendothelial migration [90]. Furthermore, co-presentation of chemokines via an adhesive matrix ("haptotactic gradient") is capable of inducing directed cell migration, independent of a soluble chemokine spatial gradient [28].

7 The Hematopoietic Stem Cell Niche Concept

Trentin et al. [94] first proposed that bone marrow HSC differentiation was determined by stromal hemopoietic inductive microenvironments (HIM). This view contrasted with a more stochastic view of HSC proliferation and differentiation proposed by Till and McCulloch, termed "hematopoiesis generated at random" (HER) [2, 3].

In reviewing this 36 years ago, I stated that "the stromal cells of hemopoietic organs must play a highly specialised role in hemopoiesis by creating special local conditions ('microenvironments') sustaining and directing the proliferation of immigrant hemopoietic stem cells" [95, p. 330] and concluded that "Hemopoietic cells contain all the information required for their further differentiation and specialisation; yet these processes require exogenous epigenetic stimuli. Both humoral factors and micro-environmental influences provide this exogenous stimulus for differentiation. The diversity of the products of such stimuli suggests that we are dealing with a chain of differentiative interactions leading to the production of micro-environmental induction mosaics. To develop such microenvironmental complexity multiple interactive stages must be present during the course of development" [95, p. 375]. To place this statement in the context of developments that have occurred in the intervening decades it should be realized that, apart from erythropoietin, no hematopoietic regulatory factors had been characterized and identification of colony stimulating factors required for in vitro hematopoietic colony formation (CFU-c assay) was just beginning. Furthermore, while the CFU-s assay provided a means of probing a primitive hematopoietic cell compartment, it was not recognized at this time that the assay detected multipotent and committed progenitors ("early" day 8 colonies), as well as a subpopulation of HSC distinct from more primitive pre-CFU-s ("late" day 12 colonies). It had also just been shown that CFU-c, as detected at that time, were committed myeloid progenitors (CFU-GM) distinct from the CFU-s [96].

7.1 The Endosteal HSC Niche

Lord et al. [97] first demonstrated that the highest concentrations of CFU-s exist near bone surfaces whereas the concentration of in vitro CFC increases to a peak value approximately 300 microns from the femoral axis with a low value at the bone surface. Thus, bone marrow cell populations conformed to a well-defined spatial organization corresponding to the chronologic relationships between marrow cells. Furthermore by tritiated thymidine "suicide " analysis they showed that CFU-s near the bone surface were proliferating at a *faster rate* than those more distant from the bone. This observation contrasted with later studies that emphasized that HSC in the endostel niche were preferentially quiescent [98, 99]. Gong et al. [100] showed that over one-half of the endosteal marrow cell population may be CFU-s. This physiologically limited microenvironment supporting HSC in vivo was first termed a niche by Schofield [101]. More recent analysis of the spatial distribution of HSC in the bone marrow has yielded conflicting data (Fig. 4). Nilsson et al. [92, 102] injected mice with CFSE fluorescence labeled HSC or progenitor enriched populations and after 15 h demonstrated that HSC enriched populations were significantly localized to the endosteal region, whereas mature terminally differentiated and lineage-committed cells selectively redistributed away from the endosteal region and were predominantly in the central marrow region. Haylock et al. [103] observed that >60% of CFSE-labelled hematopoietic cells and 33% of HSC identified by an LSK phenotype were retained at the bone surface following conventional recovery of bone marrow by femoral flushing. The latter could be recovered by grinding the bone and brief collagenase/diapase digestion. The bone endosteal associated LSK cells were modestly enriched for high proliferative potential (HPP)-CFC and had better short-term marrow homing ability and long term engraftment potential. Nilsson et al. [102] transplanted HSC enriched in the Rhodamine 123/Hoechst 33342^{dull} fraction into nonirradiated mice. Almost all donor cells detected identified by Y chromosome FISH up to 6 months post-transplant were within six cell diameters of the endosteal surface and were almost always single entities.

Direct visualization of engrafted murine GFP^+ HSC in candidate niches of the mouse bone has been obtained [104]. Under myeloablative conditions, GFP^+foci were first detected at the femoral epiphyses and some ribs and vertebra and thereafter spread to other bones. In nonmyeloablative

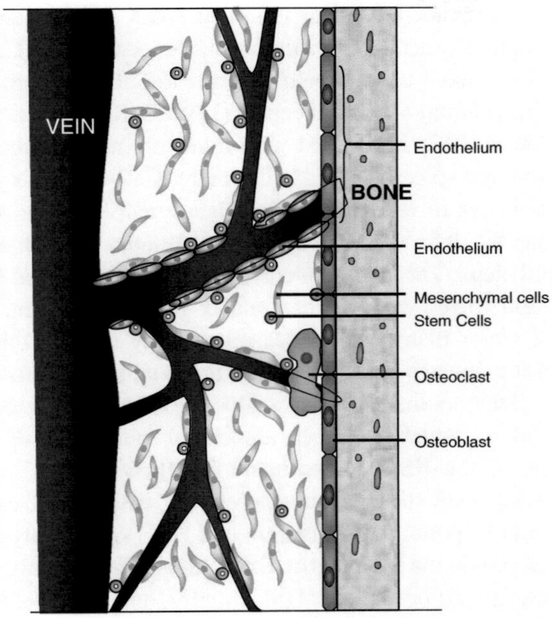

Fig. 4 Bone marrow HSC niches. Spindle-shaped osteoblasts at the endosteal bone surface serve as one form of niche to maintain HSC in a dormant form. Vasular niches contain both quiescent and proliferating HSC and promote differentiation and expansion along the megakaryocyte, erythroid and myeloid lineages. Reticular cells (CAR cells) express SDF-1/ CXCL12 and recruit HSC, providing a third potential niche. CAR cells also associate with osteoblasts and endothelium potentially providing a more complex three-dimensional microenvironment for HSC. The degree to which HSC reside adjacent to sinusoids in vascular niches marked by the presence of perivascular CAR cells, versus endosteum with or without adjacent CAR cells, and the significance of this with respect to quiescence, proliferation and differentiation remains controversial

transplants GFP^+ cells localized in femoral epiphyses and in some vertebra and ribs but remained quiescent for at least four months, presumably due to residence in quiescent niches. Primitive hematopoietic cells, including HSC, were retained in the endosteal niche, whereas lineage-committed cells localized to the central marrow regions [92]. Both human and murine HSCs express the glycosaminoglycan hyaluronan and the presence of this appears critical for their spatial distribution and endosteal localization following transplantion [92]. The transmembrane isoform of c-Kit Ligand (tm-KL) has been implicated in the adhesion of HSC to the extracellular matrix within the bone marrow microenvironment (reviewed in [105]). In Sl/Sld mice devoid of tm-KL the trans-membrane isoform does not appear to play a role in the homing of transplanted cells to the bone marrow, but is critical for the lodgment and detainment of HSC within their hemopoietic "niche" with a reduction of almost 30% within the endosteal marrow region by 15 h post-transplant [106]. The role of tm-KL was confirmed by analyzing the spatial distribution of HSC isolated using a neutralizing antibody to c-kit.

Suzuki et al. [107] found that the Gata2 promoter directed activity in all HSCs and these can be isolated efficiently from murine bone marrow by using Gata2-directed GFP fluorescence. The GFP^+ cells were Hoechst dye-excluding, non-cycling, $CD34^-Lin^-Sca\text{-}1^+c\text{-}Kit^+$ and 20% of were HSCs. Immunohistochemical analysis of fixed sections of the bone marrow employing an anti-GFP antibody further revealed that all cells that abundantly express GFP were in direct contact with osteoblasts (alkaline phosphatase-positive cells) and there was always only one brightly fluorescent GFP cell in a given niche with no evidence of GFP^+ clusters.

$BrdU^+$ long-term retaining HSCs are highly enriched at the bone surface compared to the center of marrow suggests that HSCs residing in the endosteal niche are more quiescent than the HSCs residing in the vascular niche [98]. This idea is reinforced by another study in which Angiopoietin-1 (Ang-1) was shown to be expressed mainly by osteoblasts [108]. Ang-1 signaling through its receptor Tie-2 expressed on HSC was reported to be important in sustaining the quiescence and long-term repopulating activity of HSCs in adult BM [99]. Immunofluorescence imaging in vivo has shown that $Tie2^+$ HSCs are localized to the bone surface in contact with Ang-1–expressing osteoblasts [99].

The osteoblast niche concept has been investigated in a number of studies using genetic models in mice in which osteoblast numbers are either increased or decreased and here again, somewhat contradictory results have been obtained. In a transgenic mouse model made to express receptors for parathyroid hormone and parathyroid hormone-related proteins under the control of the osteoblast specific, collagen-α-1 promoter, Calvi et al. [109] observed increased numbers of trabecular osteoblasts with a doubling of bone marrow $Sca\text{-}1^+c\text{-}Kit^+Lin^-$ cells, LTC-IC and long-LT- HSC. Marrow stroma derived from these mice contained more osteoblasts and was reported to be more efficient in supporting in vitro maintenance of HSCs than the stromal cells derived from normal mice. A model was proposed in which increased parathyroid hormone signaling in osteoblasts results in elevated levels of the Notch-1 ligand, Jagged-1 ultimately generating increased activation of Notch-1 signaling in HSCs and promoting HSC expansion. This expansion was abrogated by treatment with a gamma secretase inhibitor that blocked Notch activation.

Conditional knockout mice lacking the bone morphogenic protein receptor IA (BMPR-1A) in the BM stroma (including osteoblasts) also showed an increase in the number of both osteoblasts and long-term repopulating HSCs [98]. In the absence of ALK3 the number of spindle shaped osteoblastic cells expressing N-cadherin were increased and this was linked to a 2.4-fold increase in quiescent HSC and competitive repopulating HSC in the marrow. Cell-adhesion molecules expressed by osteoblasts, such as N-cadherin and β-1-integrin may be important for homing of HSC and their anchoring to the endosteal HSC niche [110]. HSCs were reported to reside in direct contact with osteoblasts via homophilic adhesion between N-cadherin-expressing HSCs and N-cadherin-expressing osteoblasts [98, 111]. N-cadherin expression by HSC may be regulated by c-Myc that has been reported to control the balance between HSC self-renewal and differentiation, possibly by regulating the interaction between HSC and their niches [111]. Conditional elimination of c-Myc activity in BM resulted in severe cytopenia and accumulation of self-renewing HSC in situ, with impaired differentiation. The c-Myc-deficient HSC appear trapped in stem cell niches, possibly due to up-regulation of N-cadherin and integrins. Enforced c-Myc expression in HSC repressed N-cadherin and integrins, leading to loss of self-renewal at the expense of differentiation. This concept has been challenged by Kiel et al. [44] who were unable to detect N-cadherin expression by CD150^{+}, CD48^{-} HSC either by immunostaining with anti-N-cadherin antibody or by analyzing β-galactosidase expression in N-cadherin lacZ gene trap mice. Furthermore, microarray analyses of highly purified HSCs did not detected N-cadherin [112].

Osteopetrotic (op/op) mice have an absolute CSF-1/M-CSF deficiency and impaired osteoblast development which in young mice leads to occlusion of the marrow cavity by excessive bone formation with a significant reduction in the space available for hematopoiesis [113]. At this time the spleen is a site of significant extramedullary hematopoiesis. However, older mice exhibit hematopoietic recovery and resolution of osteopetrosis, suggesting that the hematopoietic system has the capacity to use alternative mechanisms to compensate for the absence of an important multifunctional growth factor. Overexpression of the proto-oncogene c-fos in transgenic mice specifically affects bone, cartilage, and hematopoietic cell development [114]. Mice lacking the proto-oncogene c-fos develop severe osteopetrosis with deficiencies in bone remodeling and exhibit extramedullary hematopoiesis with altered B-cell development [114]. However HSC lacking Fos have full developmental potential and the defect in B-cell development is most likely due to the impaired bone marrow environment as a consequence of osteopetrosis [115].

A dramatic phenotype was observed in mice deficient in Cbfa-1, a transcription factor crucial for osteoblast progression. These animals do not develop osteoblasts, and they die at birth of respiratory failure. One of the major phenotypes in Cbfa-1 deficient mice is the complete absence of bone marrow, indicating that osteoblasts are required to initiate bone marrow hematopoiesis [116, 117]. In addition, analysis of embryonic hematopoiesis in these mutants showed normal hematopoietic development in liver and spleen until day E17.5. However, at day E18.5 both organs exhibit signs of excessive extramedullary hematopoiesis [118].

The candidate endosteal stem cell niche is characterized by a high extracellular calcium-ion concentration (Ca^{++}) and HSCs have been shown to express CaR, a G-protein-coupled calcium-sensing receptor [119]. CaR-deficient mice had fewer HSCs in the BM with relatively more in the circulation and spleen [119]. Furthermore, CaR–/– HSCs were less effective at repopulating irradiated recipients and were profoundly defective in localizing anatomically to the endosteal niche. This behavior correlated with diminished HSC adhesion to the extracellular matrix protein collagen, one of the major matrix molecules released by cells of the osteoblastic lineage and present at the endosteal surface of bone. Osteopontin is a matrix glycoprotein synthesized by osteoblasts, which has been associated with negative regulation of the HSC compartment [120]. Ostoblasts at the endosteal bone surface produce varying amounts of osteopontin in response to stimulation and HSC specifically bind to osteopontin via β-1integrin. Exogenous OPN potently suppresses the proliferation of HSC in vitro, and OPN-deficient mice have increased numbers of HSC with markedly enhanced cycling [102, 120]. Saturating levels of Tpo, IGF-1, c-Kit L, and FGF-1 produced 8-fold expansion of murine LT-HSC in 10 days. With addition of fetal liver–derived angiopoietin-like 2 and angiopoietin-like 3 net expansion increased to 24–30-fold [121].

7.2 The Endothelial Niche

The requirement for an endosteal niche for HSC maintenance has been questioned in a model of osteoblast depletion induced by ganciclovir treatment of transgenic mice expressing the herpesvirus thymidine kinase gene under the of control of a collagen promoter (Col2.3Delta-TK) [122, 123]. The mice lose lymphoid, including pre-pro-B and pro-B cells, erythroid, and myeloid progenitors in the bone marrow, preceding a decline in marrow HSCs, and an increase in extramedullary hematopoiesis in spleen and liver. There was no acute loss of HSC and the frequency of primitive hematopoietic precursors in bone marrow with a Lin^{-} c-kit^{+} Sca-1^{+} phenotype actually increased over time after osteoblast ablation. These results suggest that osteoblasts may not be obligate for HSC maintenance but are necessary for murine B-cell commitment and maturation via inductive signals that include VCAM-1-mediated adhesion to ostoblasts and locally secreted IL-7 and SDF-1 [123]. The endosteal niche concept has been challenged by data obtained from immunofluorescent imaging of CD150^{+} CD48^{-} Lin^{-}HSC (45% of cells of this phenotype give long-term HSC engraftment) [27]. Kiel et al. [27, 44] failed to demonstrate a preferential endosteal niche localization of these HSC with 60% localizing to sinusoidal perivascular areas in the bone marrow, while only 14% localized to the endosteum (Fig. 4). There was no evidence that HSCs that localize near sinusoids are different from the HSCs that localize near the endosteum. By BrDu retention nearly all $CD150^{+}CD48^{-}CD41^{-}$lineage cells are quiescent regardless of their location. Kiel et al. [44] also failed to detect a population of long-term quiescent HSC specifically associated with the endosteal surface and there was no evidence for a deeply quiescent subset of HSCs that goes months without dividing. Weksberg et al. [28] reported that exclusive reliance on SLAM family markers to isolate HSCs neglected a substantial fraction of marrow HSC and it is possible that these SP+, CD150– HSC may be preferentially associated with the endosteal surface. The significance of ostoblasts was also questioned by the data of Kiel et al. [124] using biglycan deficient mice. Biglycan is an extracellular matrix proteoglycan that is most prominently expressed by osteoblasts and chondrocytes. Biglycan deficiency leads to progressive osteoblast depletion with age and mice develop an osteoporosis-like phenotype, with less trabecular bone with fewer osteoblasts and osteoblast progenitors. Biglycan deficiency had no effect on hematopoiesis or on HSC frequency, absolute numbers or function irrespective of age, arguing against a role for osteoblast in HSC maintenance or proliferation [124]. However these studies do not exclude the possibility that a minority of HSC that localize to the endosteum may be regulated by osteoblasts and osteoclasts. HSCs may also be influenced at a distance (directly or indirectly) by extracellular factors that diffuse from the endosteum. Osteoblast cell lines secrete many (but not all) cytokines that promote the proliferation of hematopoietic cells in culture, and support the in vitro maintenance and expansion of HSCs [99, 102, 123–126].

HSCs are found in close contact with endothelial cells throughout development. Blood islands in the yolk sac can only develop in association with flk-1-positive vascular precursor cells [127] and CD34+ cells can be detected within the vessel wall of the aorta at embryonic day 35 [128] and later in perivascular locations of the fetal liver as well as in the adult bone marrow. Endothelial cells from the yolk sac are able to promote HSC/HPC proliferation in vitro [129]. Brain endothelial cells induce the concomitant expansion of purified human BM CD34+CD38– cells and SCID-repopulating cells [130]. Endothelial cells have been identified as a critical component of the neural stem cell niche [131]. Endothelial cells but not vascular smooth muscle cells release soluble factors that stimulate the self-renewal of neural stem cells, inhibit their differentiation, and enhance their neuron production. Human bone marrow microvascular endothelial cells support long-term proliferation and differentiation of human myeloid and megakaryocytic progenitors [132]. Long-term ex vivo expansion of human HSC was also reported in co-culture of mobilized peripheral blood CD34 cells on human endothelium transfected with adenovectors expressing thrombopoietin, c-kit ligand, and Flt-3 ligand [133]. Sipkins et al. [134] used dynamic in vivo confocal imaging to show that murine bone marrow contains unique anatomic regions defined by specialized endothelium. This vasculature expresses the E-selectin and SDF-1 in discrete, discontinuous areas that influence the homing HSC and lymphocytes as well as leukemia cells. Disruption of the interactions between SDF-1 and its receptor CXCR4 inhibits the homing to these vessels, suggesting that this molecularly distinct vasculature demarcates a microenvironment for HSC as well as early metastatic tumor spread in bone marrow. Chemokine-mediated interactions of megakaryocyte progenitors with sinusoidal bone marrow endothelial cells promotes Tpo-independent platelet production [135]. Megakaryocyte-active chemokines (SDF-1) and cytokines (FGF-4) restored thrombopoiesis in Tpo–/– and Mpl–/– mice. FGF-4 and SDF-1 enhanced vascular cell VCAM-1 and VLA-4-mediated localization of $CXCR4^{+}$ megakaryocyte progenitors to the vascular niche, promoting survival, maturation and platelet release. TPO supports progenitor cell expansion, whereas chemokine-mediated interaction of progenitors with the bone marrow vascular niche allows the progenitors to relocate to a microenvironment that is permissive and instructive for megakaryocyte maturation and thrombopoiesis [135].

7.3 The Role of Reticular Cells in Formation of the HSC Niche

Sugiyama et al. [87] identified a type of marrow reticular cell (CXC12-abundant reticular cell, CAR) that expresses high levels of the chemokine SDF-1/CXC12. CAR cells are present and interact with HSC in both the vascular and endosteal niches, which suggests a function linkage between these two niches (Fig. 4). This finding suggests that multiple stromal cell types are likely to be critical in the HSC niche since several cytokines implicated in HSC proliferation (e.g., c-Kit ligand, IL-6) are secreted by whole BM stromal cultures, but not in pure human osteoblast cultures. Thus HSC maintenance may require a complex multicellular niche.

7.4 Hypoxia and Reactive Oxygen Species (ROS)

Another peculiar property of the BM microenvironment is its hypoxic nature, a fact recently highlighted by Parmar et al. [136], who showed that HSCs are distributed predominantly at the lowest end of an oxygen gradient within the BM.

Using the hypoxia bioprobe pimonidazole, Levesque et al. [137] showed by confocal laser scanning microscopy that the endosteum at the bone-BM interface is hypoxic with constitutive expression of hypoxia-inducible transcription factor-1 (HIF-1) protein in steady-state mice. Interestingly, at the peak of HSC and progenitor cell mobilization induced by either G-CSF or cyclophosphamide, hypoxic areas expand through the central BM. A number of findings indicate that the level of oxidative stress influences HSC function. For example, mice deficient in the cell-cycle regulator *Atm* developed early onset BM failure, a phenotype accompanied by elevated levels of reactive oxygen species (ROS) in HSCs [67]. ROS activates the p38/MAPK pathway causing quiescent HSCs to cycle more frequently and eventually become exhausted [138]. Moreover, members of the FoxO subfamily of forkhead transcription factors have been shown to protect HSCs from oxidative stress by up-regulating genes involved in their detoxification. In light of these observations, it is conceivable that the hypoxic environment in which the HSCs reside may serve to protect them from oxygen radicals, ultimately keeping them quiescent. Furthermore, mesenchymal stem cells generate osteobasts efficiently in hypoxic conditions. FoxO3a, that acts downstream of the PTEN/PI3K/Akt pathway, is essential for HSC self-renewal and maintenance of the HSC pool [139]. While proliferation and differentiation of Fox03a null hematopoietic progenitors was normal, FoxO3a null HSCs can neither maintain quiescence nor support long-term reconstitution of hematopoiesis in a competitive transplantation assay. The mutant HSCs exhibit increased phosphorylation of p38MAPK, show a heightened sensitivity to cell cycle-specific myelotoxic injury, and lose self-renewal capacity during aging. Loss of FoxO3a was associated with decreased expression of p27 and p57, but not p21, in HSCs. Reduced expression of multiple negative regulators of the cell cycle, may account for the observed defect in the maintenance of HSC quiescence. The HSC compartment of FoxO3a null mice suffers from augmented levels of ROS as previously reported for mice deficient in Atm [67] and the defect could be rescued by administration of the antioxidant N-acetyl-L-cysteine (NAC). Tothova et al. [140] demonstrated that triple deletion of FoxO1, -3a, and -4 genes lead to marked decrease of $Lin^{-}Sca\text{-}1^{+}c\text{-}Kit^{+}$ short- and long-term repopulating HSC. Multi-FoxO-deficient bone marrow had defective long-term repopulating activity that correlated with increased cell cycling and apoptosis of HSC. There was also a marked context-dependent increase in ROS in FoxO-deficient HSC compared with wild-type HSC that correlated with changes in expression of genes that regulate ROS. FoxO proteins play essential roles in the response to physiologic oxidative stress and thereby mediate quiescence and enhanced survival in the HSC compartment, a function that is required for its long-term regenerative potential.

8 Morphogen Signaling Pathways Implicated in HSC Self-renewal

Hematopoietic growth factors and various morphogens activate a variety of signaling pathways implicated in general development and also HSC proliferation and differentiation (Fig. 5).

8.1 *The Pleiotropic Role in the TGF-β Family in Hematopoietic Regulation*

The pleiotropic effects of TGF-β already observed in various tissues are particularly well illustrated in the hematopoietic system. Depending on the degree of maturation of the hematopoietic cells, these factors are able to control either positively or negatively cell proliferation, differentiation, or apoptosis (reviewed in [141]). TGF-β family members bind two types of membrane serine/threonine kinases, the type I and type II receptors, forming a heteromeric receptor complex. The type II receptor then phosphorylates and activates the type I receptor, which in turn phosphorylates Smad transcription factors (Fig. 5) [142]. Of the eight Smad family members in human, five function as receptor substrates (RSmads). Smad2 and Smad3 do so as substrates of TGF-β, nodal and activin receptors, and Smads1, 5, and 8 as substrates of the receptors for bone morphogenetic proteins, myostatin and anti-muellerian hormone. Receptor-mediated phosphorylation triggers nuclear accumulation of RSmads and their binding to Smad4. Smad4, itself not a receptor substrate, is an essential partner of RSmads in transcriptional regulation of many genes. RSmad-Smad4 complexes bind a diverse group of DNA binding factors to achieve target gene selection and recruit transcriptional coactivators or corepressors for gene regulation [143]. Additionally, in a negative feedback mechanism, Smad6 and 7 inhibit TGF-β superfamily signaling by competing with R-Smads for Smad4 interaction and receptor binding and by targeting receptors for ubiquitination and degradation. Smad7 has been shown to inhibit both TGF-β/Activin signaling and BMP5 signaling by associating with activated type I receptors, thus preventing phosphorylation of RSmads [143]. Smad7 is also upregulated in response to TGF-β, Activin, and BMP, indicating its involvement in a negative feedback loop in response to stimulation by TGF-β and related ligands. These general mechanisms underly a large number of TGF-β gene responses controlling cell proliferation, organization, and fate. Blank et al. [144] reported that the self-renewal capacity of HSCs was promoted in vivo upon blocking of the entire Smad pathway by retroviral gene transfer of the inhibitory Smad7 to murine HSCs. HSCs overexpressing Smad7 had an unperturbed differentiation capacity as evidenced by normal contribution

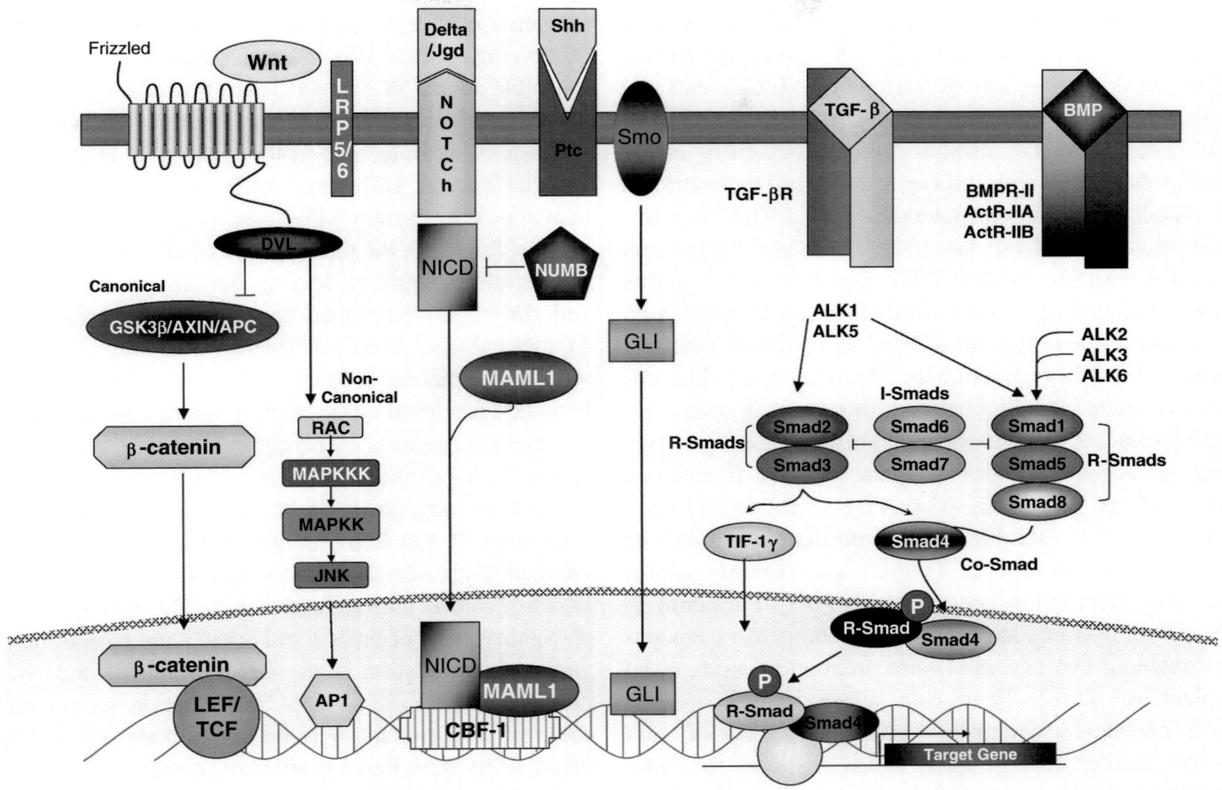

Fig. 5 Morphogen signaling pathways regulating HSC. When ligands of the Notch pathway, Delta 1-3 and Jagged 1-2 bind to the Notch receptors (Notch1-4), proteolytic events involving gamma-secretase lead to release and translocation of the intracellular domain of the receptor (NICD) to the nucleus. NICD form a complex with the transcription factor CSL and cofactors of the mastermind-like (MAML) family to activate transcription of the target genes. Numb is a ubiquitin ligases that is asymmetrically expressed in HSC progeny and degrades NICD, thus inhibiting Notch signaling. Wnt signaling is initiated when the ligands bind to Frizzled and LPR receptors at the cell surface. In the canonical pathway the receptor complex inhibits the phosphorylation of the downstream signal transducer β-catenin by glycogen synthase kinase (GSK3β) via DVL (mammalian homolog of disheveled). β-catenin accumulates and translocates to the nucleus where it interacts with the T-cell factor/Lymphoid enhancer factor (TCF) transcription factors to regulate gene expression. Non-canonical Wnt signaling can progress through DVL to the Rac/mitogen-activated protein (MAP) kinase pathway. TGF-β family members bind and signal through type I and type II serine/threonine receptors, both of which are necessary for signal transduction (ALK5/TGFβ type I receptor, ALKALK6/BMP type IB receptor, ALK3/BMP type IA, ALK2/Activin type I receptor, ALK4/Activin type IB receptor). Upon ligand binding and receptor activation, Smad proteins are activated through phosphorylation by type I receptors. There are three groups of Smads: receptor activated Smads (R-Smads), common-partner Smads (Co-Smads) and inhibitory Smads (I-Smads). Binding of BMPs to its receptor complex results in phosphorylation of regulatory Smad proteins transcription. TGF-β and activin signal via R-Smad2 and 3 while BMPs (BMP 1–7) signal through R-Smad1, 5, and 8. Phosphorylated R-Smads then form a complex with CoSmad4 and translocate to the nucleus to affect target gene transcription. The I-Smads, Smad6 and Smad7 function in a negative feedback loop to prevent activation of R-Smads. The ubiquitious nuclear protein Transcriptional Intermediary Factor 1gamma (TIF1γ) selectively binds receptor-phosphorylated Smad2/3 in competition with Smad4. Rapid and robust binding of TIF1γ to Smad2/3 occurs in human hematopoietic stem/progenitor cells, where TGFβ inhibits proliferation and stimulates erythroid differentiation. TIF1γ mediates the differentiation response while Smad4 mediates the antiproliferative response with Smad2/3 participating in both responses. Thus, Smad2/3-TIF1γ and Smad2/3-Smad4 function as complementary effector arms in the control of hematopoietic cell fate by the TGFβ/Smad pathway [149]. When bound by Sonic hedgehog (Shh) the Patched (Ptc) receptor's inhibition of Smoothened (Smo) is blocked allowing Smo to regulate transcription factors through Gli1 proteins. Modified after Ross and Li [240] and Blank et al. [241]

to both lymphoid and myeloid cell lineages, suggesting that the Smad pathway regulates self-renewal independently of differentiation. Low levels of TGF-β1 can modulate SDF-1 responsiveness of $CD34^+$ cells and thus may facilitate SDF-1-mediated retention and nurturing of HSC in bone marrow [145]. TGF-β has been implicated in the maintainence of HSC in a quiescent, or slowly cycling state and in human HSC this cell-cycle arrest is mediated by upregulation of the CKI p57KIP2 [42]. A variety of studies using antisense oligonucleotides or neutralizing antibodies against TGF-β ligand or TGF-β type II receptor have suggested that blockage of TGF-β signaling can provide a proliferative advantage to primitive hemopoietic progenitor cells that are otherwise quiescent and therefore difficult to be stimulated to proliferate [141]. Fortunel et al. [141] attributed the TGF-β affect in part to down-modulation of cell surface expression of tyrosine kinase receptors cKit, Flt3, IL6R, and the Tpo receptor Mpl. Transient blockage of autocrine TGF-β signaling in human HSC using a dominant negatively acting mutant of the TGF-β type II receptor enhanced their survival and overall proliferation potential [146]. This negative regulatory role of TGF-β has been challenged by Larsson et al. [147], who showed that TGFβ-R-null mice had normal in vivo hematopoiesis, a normal HSC cell cycle distribution and did not differ in long-term HSC repopulating potential compared to wild-type animals. Redundant mechanisms within the Smad signaling network or cross talk with other pathways relevant in the more complex and enduring in vivo setting may account for this discrepancy between in vitro and in vivo findings. While TGF-β1 or TGF-β3 showed a mouse strain-independent inhibition of the proliferation of murine HSC, the dose response of TGF-β2 was biphasic with a stimulatory effect at low concentrations [147]. Furthermore, adult TGFβ2^{+}/– mice have a defect in competitive repopulation potential that becomes more pronounced upon serial transplantation [148]. These data suggest that TGF-β2 is a novel, genetically determined, positive regulator of adult HSC that acts cell autonomously and is important for HSCs that have undergone replicative stress.

He et al. [149] provided evidence that hematopoiesis is controlled by a branching of the TGF-β pathway into the Smad4 branch and a distinct branch in which Transcriptional Intermediary Factor1γ (TIF1γ, also known as TRIM33, RFG7, PTC7, and Ectodermin) selectively binds receptor-phosphorylated Smad2/3 in competition with Smad4 (Fig. 5). In human HSC/progenitor cells, where TGF-β inhibits proliferation and stimulates erythroid differentiation, TIF1γ mediates the differentiation response while Smad4 mediates the antiproliferative response with Smad2/3 participating in both responses. Thus, Smad2/3-TIF1γ and Smad2/3-Smad4 function as complementary effector arms in the control of hematopoietic cell fate by the TGF-β/Smad pathway. Based on sequence similarity, chromosomal location, and exon boundaries, human TIF1γ is closely related to zebrafish moonshine (*mon*), which was identified as a gene required for blood formation [150]. Mutations in *mon* disrupt both embryonic and adult hematopoiesis. TIF1γ is required as a permissive cofactor for the erythroid lineage-specific control of hematopoietic gene expression and while early erythroid progenitors are formed in homozygous mutants, they fail to properly differentiate and instead undergo programmed cell death. siRNA depletion of TIF1γ or up-regulation of Smad4 in human CD34^{+} cells inhibited erythroid differentiation but produced HSC expansion in vitro as measured by week 5 CAFC and LTC-IC assay (Moore MAS, Dorn D, He W, Massague J, unpublished observation). The role of Smad4 in HSC function has remained elusive because of the early embryonic lethality of the conventional knockout. Karlsson et al. [151] used an inducible model of Smad4 deletion to show that systemic induction of Smad4 deletion was incompatible with survival because of anemia and histopathological changes in the colonic mucosa. Restriction of the Smad4 deletion to the hematopoietic system via transplantation demonstrated a role for Smad4 in the maintenance of HSC self-renewal and reconstituting capacity, leaving homing potential, viability, and differentiation intact. In an apparent balancing act, TGF-β–induced R-Smad–Smad4 complexes keep HSC quiescent, whereas R-Smad–TIF1γ complexes direct HSC to toward erythroid differentiation [149]. Absence of Smad4 tips the balance in favor of differentiation at the expense of self-renewing divisions [151].

RNA-binding protein with multiple splicing (RBPMS), a member of the RNA-binding protein family, physically interacts with Smad2, Smad3, and Smad4, and stimulates Smad-mediated transactivation possibly through enhanced phosphorylation of Smad2 and Smad3 and promotion of the nuclear accumulation of the Smad proteins [152]. While a direct role for RBPMS in HSC proliferation has not yet been demonstrated, it is likely to be involved since it was upregulated in transcriptional profiles of murine and human HSC [112, 153, 152–158].

8.2 Bone Morphogenic Protein (BMP)

BMPs are members of the TGF-β super-family of signalling molecules. These pleiotropic cytokine are important for embryonic tissue development and in regulating cell proliferation, differentiation, morphogenesis and apoptosis in multiple systems. During embryogenesis BMP-4 acts as an inducer of ventral mesoderm, tissue from which hematopoiesis originates [159]. Human HSC express the BMP-1 receptors activin-like kinase (ALK-3 and ALK-6) and their downstream transducers Smad-1, Smad-4, and Smad–5 (Fig. 5). Like TGF-β, high concentrations of BMPs (BMP-2, BMP-4, and BMP-7) inhibited HSC proliferation but maintained HSC long-term survival and repopulating potential whereas low concentrations of BMP-4 induced HSC proliferation and differentiation [160]. Stromal lines support HSC in vitro by mimicking the microenvironment necessary for HSC maintenance and expansion. They produce a variety of HSC stimulatory factors as well as proliferin-2 [161], and mKirre [162], Delta-like/preadipocyte factor-1(dlk) containing epidermal growth factor-like repeats that are related to those in the notch/delta/serrate family of proteins [163]. The ability of the murine fetal liver-derived stromal cell line AFT024 to maintain and expand HSC in long-term cultures has been well documented [164].

These stromal cells produce BMP-4 and this contributes significantly to expansion of co-cultured cord blood-derived HSC since neutralizing BMP4 monoclonal antibody reduced expansion [165]. The BMP4 gene has been reported to be a downstream target of both Sonic Hedgehog (Shh) and Wnt3a signaling however neither Shh nor Wnt3a were expressed by AFT024, suggesting that BMP4 is acting independently of these signalling pathways in this mode (Fig. 5) [164]. Early Hematopoietic Zinc Finger protein (EHZF), the human homologue to Evi3, is highly expressed in primitive human $CD34^+$ hematopoietic cells and declines rapidly during cytokine-driven differentiation [166]. EHZF and Evi3 share high homology (63.5%) with the co-transcription factor Oaz, implicated in B-lymphocyte differentiation, and in BMP signal transduction. In response to BMPs EHZF complexes SMADs1 and 4, binds to, and enhances the transcriptional activity of, a BMP2/4 responsive element and may thus play a role in the characteristic dose-related effects of BMP4 on early $CD34^+CD38^-Lin^-$ HSC [167].

8.3 Hedgehog Proteins

Three mammalian Hedgehog (Hh) proteins exist (Sonic, Indian, and Desert Hh), which are recognized and bound by the cell surface receptor Patched (Ptch). Binding of Hh to Ptch suspends the inhibition of its membrane-bound signaling partner Smoothened (Smo), which in turn initiates nuclear translocation and activation of the Gli family of transcription factors (Fig. 5). Shh, Ptch and Smo are expressed by human HSC [68]. In Ptch mutant mice the differentiation potential of the HSC and MPP is normal, although there was a slight decrease in their proliferative capacity [168]. In the absence of Ptch, the development of T- and B-lymphoid lineages is blocked at the level of the common lymphoid progenitor in the bone marrow. Consequently, the generation of peripheral T and B cells is abrogated but cells of the myeloid lineage develop normally. Shh-Ptch signaling appears acts as a master switch for proper lymphoid versus myeloid lineage commitment of HSC in the adult organism [168]. Soluble Shh enhanced cytokine-induced cord blood HSC expansion and Noggin, a specific inhibitor of BMP-4, inhibited this, indicating that Shh regulates HSC via mechanisms that are dependent on downstream BMP signals [68]. Downstream activation of the Shh signaling pathway induced cycling and expansion of HSC under homeostatic conditions and during acute regeneration [169]. However, this effect was at the expense of HSC function, because continued Shh activation during regeneration repressed expression of specific cell cycle regulators, leading to HSC exhaustion. In vivo treatment with cyclopamine, an inhibitor of the Hh pathway rescued these transcriptional and functional defects in HSCs. Indian hedgehog (Ihh)-expressing human stromal cells supported $CD34^+$ cells with markedly enhanced production of progenitor CFC [170]. The developmental protein Numb regulates the endocytic and ubiquitin-dependent processing of Notch1 and, consequently, the cell-fate decisions determined by Notch signaling (vide infra). HSS signaling through the hedgehog transcription factor Gli1 is also suppressed by Numb [171]. This effect involves the ubiquitin-regulated processing of Gli1, which is mediated by functional cooperation between Numb and the HECT domain E3 ubiquitin ligase Itch. There is the intriguing possibility that the activities of both Shh and Notch in stem cell self-renewal may be coordinated by a single signal.

8.4 Wnt

Wnt ligands are a family of secreted glycoproteins that can activate multiple signaling pathways by binding to members of the Frizzled family of receptors. A number of Wnt genes are expressed in bone marrow (Wnt2b, Wnt3a, Wnt5a, Wnt10b) and have been implicated in HSC self-renewal and differentiation (reviewed in [172, 173]). Gene profile analysis indicated both canonical and noncanonical Wnt pathway activation in HSC, including expression of multiple Wnt receptor genes (Fzd3, 4, 6, 7), as well as the tyrosine kinase Ryk, thought to be involved in noncanonical Wnt signal transduction. Stromal cells transduced with Wnt2b, Wnt5a, or Wnt10b showed enhanced ability to support primitive $CD34^+$ cells [174]. Wnt5a increased hematopoietic progenitor cell production in vitro [175] and when injected in vivo into immunodeficient mice enhanced engraftment by human $CD34^+$ cells [176]. HSC cultured with Wnt5a showed enhanced engraftment and multilineage repopulation compared to control cultured HSC [173]. This enhancement may be attributed to Wnt5a favoring persistence of HSC in the quiescent G0 phase, and G0 HSC engraft more efficiently than cycling cells [173]. Wnt3a palmitolylated protein has been purified and in one study it was shown to induce self-renewal of HSC [177]. The canonical Wnt pathway signals through β-catenin (Fig. 5) and Reya et al. [178] showed that overexpression of activated β-catenin expanded the pool of HSC in long-term culture and increased expression of HOXB4 and Notch1, both implicated in HSC self-renewal. Ectopic expression of axin or the frizzled ligand-binding domain, both inhibitors of the Wnt signaling pathway, inhibited HSC growth in vitro and reduced in vivo engraftment [178]. Notch signaling has been shown to be necessary for Wnt-mediated maintenance of undifferentiated HSC but not for their survival or entry into cell cycle [179]. Wnt signaling repressed glycogen synthase kinase-3β (GSK3β) leading to accumulation of β-catenin but also accumulation of intracellular fragments of Notch and activation of Notch target genes such as Hes-1. GSK-3 inhibitors modulate gene targets of Wnt, hedgehog and Notch pathways in HSC, without affecting mature cells and were shown to enhance in vivo hematopoietic repopulation by mouse or human HSC [180]. The inhibitor improved neutrophil and megakaryocyte recovery, mouse survival and long-term repopulation.

The essential role of β-catenin in hematopoiesis has been questioned in studies of mice with Cre-loxP-mediated inactivation of β-catenin since there was no impairment in HSC ability to self-renew and differentiate [181]. Scheller et al. [182] and Kirstetter et al. [183] used two different transgenic models in which HSC β-catenin was constitutively activated and in both studies canonical Wnt signaling *inhibited* HSC self-renewal and differentiation leading to bone marrow failure. Wnt ligands can also activate non-canonical signaling (Fig. 5) and in HSC exposed to either Wnt3a or Wnt5a, Wnt5a promoted β-catenin degradation via a noncanonical pathway and also inhibited Wnt3a-induced canonical signaling [173]. This discrepancy cannot be explained by a redundant and compensatory function of gamma-catenin, a close homolog of β-catenin that also associates with lymphoid enhancer factor/T cell factor transcription factors. Hematopoiesis, including thymopoiesis, was normal in the combined absence of beta- and gamma-catenin and HSC maintain long-term repopulation capacity and multilineage differentiation potential [174, 175]. Unexpectedly, ex vivo reporter gene assays show that Wnt signal transmission was maintained in double-deficient hematopoietic stem cells [184]. It is not yet possible to reconcile the divergent observations of various Wnts on HSC but is should be noted that the observations of Reya et al. [178] were made in bcl-2-expressing HSC cultured with a single cytokine (c-Kit ligand) and the results may not reflect what occurs in a more physiological system. Wnts are known to affect the development and function of the endosteal stem cell niche [173], thus both HSC intrinsic and extrinsic Wnt effects determine outcome and a balance between signaling by Wnt5a and Wnt3a may be necessary for normal hematopoiesis.

8.5 Notch

Among the proteins that have been postulated to be involved in HSC maintenance are the Notch receptors and their ligands. Mammals have four Notch receptors (Notch1–4) that bind five different ligands (Jagged-2, Delta-like-1–4). The ligands for Notch are transmembrane proteins expressed on adjacent cells and it is quite possible that signaling is bidirectional. A large Notch precursor protein is proteolytically cleaved to form the mature cell-surface receptor (Fig. 5). Ligand binding induces additional proteolytic events followed by translocation of the intracellular domain to the nucleus. There, Notch interacts with transcription factors such as RBPJ, activating transcription of basic helix-loop-helix genes such as HES-1. These in turn regulate expression of tissue-specific transcription factors that influence lineage commitment and other events. Notch is critical for the developmental specification of HSC fate and the subsequent homeostasis of HSC number in zebrafish [185]. In differential gene display, Notch-1 has consistently been shown to characterize the murine and human HSC [112, 153–158]. Osteoblasts expressing Jagged-1 were identified as being part of the HSC niche [109]. Notch-1 null mouse embryos showed defective hematopoietic (and vascular) development and were devoid of HSC. A number of reports have shown that Notch signaling mediated by both Delta and Jagged ligands expands the HSC compartment while blocking or delaying terminal myeloid differentiation [186–188]. Expression of activated Notch1 enhanced HSC self-renewal and a similar effect of differentiation inhibition and progenitor/HSC expansion was reported with activated Notch4 (Int3) [187]. The Delta-1 extracellular domain fused to the Fc portion of human IgG1, together with cytokines (KL, IL-6, IL-11, and Flt3L) inhibited myeloid differentiation and promoted several log increases in precursors capable of short-term lymphoid and myeloid repopulation [186, 187]. Addition of IL-7 promoted T-lymphocyte development whereas GM-CSF induced myeloid differentiation. Delaney et al. [188] showed that quantitative differences in Notch signaling determining HSC fate. In $CD34^+$ cord blood cultures, low densities of the Notch ligand Delta-1 enhanced in vitro generation of $CD34^+$ cells as well as CD14 and CD7 cells consistent with myeloid and lymphoid differentiation whereas higher concentrations induced apoptosis of $CD34^+$ cells but not CD7 T cell precursors. Soluble forms of human Delta-1 and Delta-4 proteins augment the proliferation of primitive human hematopoietic progenitors in vitro and expanding HSC with in vivo repopulating capacity [189]. Addition of a soluble form of human Jagged-1 to cultures of human $CD34^+$ cells had modest effects in augmenting cytokine-induced proliferation of progenitors but induced the survival and expansion of HSC capable of in vivo pluripotent repopulating capacity [190]. Stimulation of Notch and cytokine-induced signaling of c-Kit, Flt3 and Mpl had combinatorial effects in vitro resulting in a 10–20-fold expansion of murine HSC with long-term repopulating potential [191]. Inhibition of Notch signaling led to accelerated differentiation of HSC in vitro and depletion of HSC in vivo [179]. Wu et al. [57] used a Notch-GFP reporter mouse, in which GFP is highly expressed in populations enriched for HSCs and downregulated as these cells differentiate. When GFP^+ HSC are placed on stroma that induce differentiation they predominantly use asymmetric divisions or symmetric commitment divisions; in contrast, when they are placed on stroma that promote maintenance of immature cells (OP9), they proceed predominantly through symmetric renewal divisions. A number of studies have questioned the role of Notch activation on HSC proliferation. Inhibitory effects on the proliferation and survival of human hematopoietic CD34$^+$ cells have been reported. CD34$^+$ cells retrovirally transduced with the constitutively active human Notch-1 intracellular domain proliferated to a lesser extent in vitro than cells transduced with vector alone, concurrent with upregulation of p21 and induction of apoptosis [192]. This was accompanied by a reduction in absolute number of CD34$^+$ Thy$^+$Lin$^-$HSCs. HES-1, a major downstream effector of the Notch pathway is expressed at high levels in HSC-enriched CD34$^+$, CD38$^-$ subpopulations [179]. Transduction of HSC with HES-1 has produced disparate results. Shojaei et al. [193] observed enhanced in vivo reconstitution ability that correlated with increased cycling frequency. However, Chadwick et al. [192] reported no effect on HSC proliferation and conditional expression of HES-1 in murine and human HSC inhibited cell cycling in vitro and cell expansion in vivo with intact long-term reconstituting function. Activation of Notch signaling in hemangioblasts dramatically reduced their survival and proliferative capacity and lowered the levels of the hematopoietic stem cell markers CD34. Mancini et al. [194] reported that mice with simultaneous inactivation of Jagged-1 and Notch1 survived normally, even following chemotherapy-induced myelosuppression. However it is possible that other Notch receptors and/or ligands may substitute for Notch1 and/or Jagged. HSC regulation is complex, cell-context dependent and plagued by potential compensation systems and while the study excludes an essential role for Jagged-1 and Notch1 during hematopoieis, there are four Notch receptors and five ligands and the Notch pathway cross-talks with the Wnt pathway with more than nine frizzled receptors and 12 ligands [179].

8.5.1 Numb

The mammalian ortholog of the conserved Drosophila adaptor protein Numb (Nb) and its homolog Numblike (Nbl) modulate stem cell fate determination at least in part by antagonizing Notch signaling (Fig. 5). Using immunofluorescence, Wu et al. [57] found that Numb was either symmetrically or asymmetrically distributed in Notch-GFP reporter HSC undergoing division. While 56% showed equivalent distribution of Numb between the two incipient daughters, 44% of dividing cells displayed preferential localization of Numb to one incipient daughter at various stages of mitosis. Overexpression of Numb in HSC significantly reduced the frequency of cells responding to Notch (from 80% to 36%) Additionally, Notch reporter activity was progressively more repressed with increasing expression of Numb [57]. Cells carrying Numb had 4-fold fewer lineage-negative cells compared to control infected cells. These data suggest that the asymmetric segregation of Numb may functionally result in inhibition of Notch signaling and a more differentiated state. Surprisingly, simultaneous deletion of both Numb and Numblike in murine bone marrow precursors did not affect the ability of HSC to self-renew or to give rise to differentiated myeloid or lymphoid progeny, even under competitive conditions in mixed chimeras [195]. Furthermore, T cell fate specification and intrathymic T cell development were unaffected in the combined absence of Numb and Numblike. Collectively, data indicate that the Numb family of adaptor proteins is dispensable for hemopoiesis and lymphopoiesis in mice, despite their proposed role in neuronal stem cell development.

9 Molecular Pathways Regulating HSC Self-Renewal and Differentiation

There is increased but still incomplete understanding of the molecular pathways downstream of the cytokine and morphogen receptors that influence HSC self-renewal and differentiation. Progress has been made in identification of "stemness" genes and pathways necessary for maintenance of the self-renewing undifferentiated state and the regulatory pathways determining quiescence, proliferation, and lineage-restricted differentiation.

9.1 JAK/STAT Pathway

The JAKs are nontransmembrane protein tyrosine kinases and Jak1 and Jak2 can specifically phosphorylate signal transducer and activator of transcription STAT1 and STAT5, respectively. Binding of c-Kit ligand to a c-Kit receptor leads to activation of constitutively associated members of JAKs through cross-phosphorylation of JAKs on tyrosine residues. Activated JAKs, in turn, phosphorylate the c-KIT receptor, creating docking sites for specific signaling proteins, including STAT proteins (reviewed in [71]). The STAT1 protein is activated in response to c-Kit Ligand, suggesting that STAT1 is a critical component of the c-Kit signal transduction pathway. Furthermore c-kit directly phosphorylates STAT1 fusion proteins in in vitro kinase assays. Subsequently, the STATs form stable homodimers and heterodimers by interactions between the SH2 domain of one STAT protein and the phosphotyrosines present in the carboxyl terminal end of the kinase receptor. This leads to release of STAT proteins from the homodimer/heterodimer receptor complex, where they translocate to the nucleus and influence transcription of target genes by binding to specific regulatory sequences. Transduction of mouse HSC with constitutively activated STAT3 enhanced HSC self-renewal under stimulated, but not homeostatic conditions while a dominant negative form of STAT3 suppressed self-renewal [196]. The STAT5 pathway is strongly activated following ligand binding to the erythropoietin and IL-3 receptors, and weakly following FL binding to Flt3. However, constitutively activating mutants of Flt3, found in ~25% of human acute myeloid leukemia, are associated with strong activation of STAT5 (reviewed in [70]). A constitutively activated double mutant of STAT5a (STAT5a[1*6]) transduced into $CD34^+$ cells promoted enhanced HSC self-renewal as measured by CAFC assay and promoted enhanced erythroid differentiation relative to myeloid. This was causally linked to downregulation of C/EBPα, a transcription factor uniquely associated with granulocytic differentiation.

9.2 The Role of Polycomb Group (PcG) Proteins in HSC Self-Renewal

During development, PcG protein complexes are thought to maintain long-term and heritable gene silencing through local alterations of the chromatin structure. Two distinct Polycomb repressive complexes, PRCs 1 and 2, have been identified. Mammalian PRC1 contains Cbx, Mph, Ring, Bmi-1, and Mel18 and is thought to be important in the maintenance of gene repression. The second complex, PRC2, contains Ezh2, Eed, and Su(z)12 and is thought to be involved in initiation of gene repression [197]. PcG complexes are targeted to *cis*-regulatory Polycomb response elements (PREs) by DNA-binding transcription factors. The role of only a few PcG proteins in murine hematopoiesis has been established. Melo18 (Pcqf2) negatively regulates self-renewal of HSCs because its loss leads to an increase of HSCs in G0 and to enhanced HSC self-renewal [198]. Mph1/Rae28 (Phc1) mutant mice are embryonic lethal because HSC activity in these animals is not sufficient to maintain hematopoiesis during embryonic development [199]. Enhancer of zeste homolog 2 (Ezh2) is a PcG involved in histone methylation and deacetylation, stabilizating chromatin structure and preserving HSC potential after replicative stress [200]. Whereas normal HSCs are rapidly exhausted after serial transplantations, HSC overexpressing Ezh2 completely conserved long-term repopulating potential.

9.2.1 Bmi-1

Bmi-1has been implicated in HSC maintenance. Although Bmi-1–/– mice show normal development of embryonic hematopoiesis, adult Bmi-1–/– HSCs have a profound defect in self-renewal capacity. They cannot repopulate hematopoiesis long term, leading to progressive postnatal pancytopenia [37, 108, 201]. In Bmi-1- GFP knock-in mice, in which GFP was expressed under the endogenous transcriptional regulatory elements of the Bmi-1 gene, GFP was expressed at its highest levels in HSC and down-regulated upon HSC differentiation [202]. The frequency of long-term, repopulating HSCs was 1/16 in $Bmi\text{-}1^{high}c\text{-}kit^+Lin^-Sca\text{-}1^+$ bone marrow cells. Enforced expression of Bmi-1 in cord blood $CD34^+$ cells results in long-term maintenance and self-renewal of human HSC and progenitor cells. LTC-IC frequencies were increased and week 5 CAFCs from stromal cocultures could be serially replated to give rise to secondary CAFCs. Bmi-1-transduced cells could proliferate in stroma-free cytokine-dependent cultures for over 20 weeks and engrafted more efficiently in NOD-SCID mice with secondary NOD-SCID engraftment only achieved with cells overexpressing Bmi-1 [203]. Bmi-1 overexpression down-regulated expression of p16 and p19Arf, which are encoded by Ink4a, and enhanced HSC symmetrical division resulting in expansion of multipotent progenitors in vitro and enhanced HSC repopulation in vivo [108, 204]. Chagraoui et al. [205] identified E4F1, an inhibitor of cellular proliferation, as a novel Bmi1-interacting partner in hematopoietic cells. E4F1 genetically interacted to regulate hematopoietic cellular proliferation. Reduction of E4F1 levels through RNA interference-mediated knockdown rescued the clonogenic and repopulating ability of Bmi1–/– hematopoietic cells. HoxB4 requires Bmi-1 to execute its function as an HSC activator. Bmi-1 could be epistatic to HoxB4 or these two molecules may have some functional crosstalk in the regulation of HSC self-renewal and/or differentiation.

9.3 SCL/TAL-1 and Id1

E-proteins, encoded by E2A, HEB, and E2–2 genes, are a family of activators that may influence HSC maintenance by influencing transcription of p21cip1/waf1 and p16INK4a genes. E-protein function is controlled by Helix-loop-helix (HLH) inhibitors such as Id and the stem cell leukemia SCL/TAL-1 proteins, which recently have been suggested to play a role in HSC differentiation [206, 207]. These proteins can function as either transcriptional activators or repressors. Specification of HSCs from mesoderm during embryonic development requires SCL [206]. Forced expression of SCL strongly induces blood formation in embryos, indicating that this gene has a dominant role in commitment to hematopoiesis. In the adult hematopoietic system, expression of SCL is enriched in HSCs and multipotent progenitors, and in erythroid and megakaryocytic lineages, consistent with roles for this factor in adult hematopoiesis. Conditional gene targeting of the SCL gene in adult mice showed SCL to be dispensable for maintenance of the HSC properties, such as the long-term repopulating activity and multipotency, but not for the proper erythroid and megakaryocyte generation [208]. A transplantation defect of SCL-deleted cells was observed within 4 weeks of transplantation, indicating a defect in a multipotent progenitor or short-term repopulating HSCs [209]. Perry et al. [207] used a knock-in mouse model to evaluate Id1 expression on an individual-cell basis in LT- and ST-HSCs. This was done by inserting the GFP sequence into the Id1 locus such that its expression was driven by the Id1 promoter. Since mice homozygous for this insertion did not produce Id1 protein, it was also possible to evaluate how Id1 ablation affected the frequency of phenotypic HSCs in bone marrow, as well as the function of Id1–/– marrow in transplant assays. These data indicate a specific role for Id1 in modulating LT-HSC but not ST-HSC renewal and differentiation. Similar experiments with Id3–/– marrow indicate that Id3 does not contribute to LT-HSC activity.

9.4 Gfi-1

Gfi-1, a transcriptional repressor, is expressed in several compartments of the hematopoietic system and is a member of the same oncogenic complementation group as Bmi-1 The absence of Gfi-1 during embryonic development does not seem to alter the specification mechanisms of HSCs and mutant mice are born with a close to normal HSC pool size [210, 211]. Gfi-1 has a role in adult haematopoiesis since in Gfi-1 null mice T and B cells are produced in reduced numbers with impaired CLP and short-term repopulating HSC frequencies, a milder reduction in CMP numbers and a significant increase in GMPs. Gfi-1 null bone marrow cells could provide short-term and long-term reconstitution of hematopoiesis when transplanted into lethally irradiated recipients without competitive cells and in large excess. However, secondary transplantation revealed deficiencies in this ability that were not detected when WT bone marrow was serially transplanted, suggesting that Gfi-1 null stem cells are defective. Indeed, when measured against WT bone marrow competitor cells, a severe impairment of long-term repopulating activities of Gfi-1 null bone marrow cells became apparent. In chimeric mice whose cells had either hemizygous or homozygous disruption of Gfi-1, Hock et al. [210] demonstrated both populations contributed to the hematopoietic tissues of young (<3 weeks) mice; Gfi-1 null cells were no longer detectable in older mice (>2 months), except in tissues unrelated to the blood system. Similarly, hemoglobin isoform studies revealed a progressive decline in Gfi-1 null contribution to red blood cells, suggesting that Gfi-1 null cells can initiate but not sustain hematopoiesis. Serial transplantation experiments, clearly demonstrate the failure of Gfi-1 null HSCs to survive adoptive transfer (perhaps secondary to a failure to symmetrically divide and expand). Gfi-1 null mice show a surprisingly high proportion of actively cycling HSCs, suggesting that Gfi-1 restrains proliferation of HSCs and thereby regulates their self-renewal and long-term engraftment abilities [211].

9.5 Nov/CCN3

The matricellular protein Nephroblastoma Overexpressed (Nov/CCN3) has recently been shown to be essential for HSC functional integrity [212]. Nov expression is restricted to the $CD34^+$ compartments of cord blood, and its knockdown in these cells by lentivirus-mediated RNA interference abrogated HSC self-renewal, LTC-IC readout, and NOD/SCID engraftment. Expression of Bmi-1 and Gfi-1, genes previously implicated in HSC self-renewal, was similarly diminished but not the differentiation capacity of more mature progenitor cells in CFC assays. Conversely, forced expression of Nov and addition of recombinant Nov protein both enhanced HSC and/or progenitor activity in vitro and in vivo.

9.6 Cited2 and CREB-Binding Protein (CBP)

The transcriptional coactivators CBP and p300 are fate decision factors for HSCs [213]. Despite high protein homology and the performance of many overlapping functions, their roles are, at least in part, distinct during hematopoietic development. CBP is essential for HSC self-renewal, whereas p300 is not but instead plays a role in hematopoietic differentiation. GATA2, a critical regulator for the maintenance and expansion of HSC is essential for definitive hematopoiesis [214] and is stimulated by CBP. Cited2, a CBP/p300-dependent transcriptional modulator, binds directly with high affinity to the first cysteine-histidine–rich region of p300 and CBP. A role for Cited2 in hematopoiesis was shown by Chen et al. [215] in studies demonstrating that Cited2–/–fetal liver had greatly reduced HSC by phenotype and competitive repopulating capacity. Exactly how Cited2 acts is unknown but it is likely that some of its functions in hematopoiesis are partially mediated by CBP and p300. Microarray analysis showed decreased expression of Wnt5a and a panel of myeloid markers in Cited2–/–fetal livers, as well as decreased expression of Bmi-1, Notch, LEF-1, Mcl-1 and GATA-2 [215].

9.7 Hepatic Leukemia Factor (HLF)

HLF is closely related to zipper-containing transcription factors that have roles in developmental stage-specific gene expression and it has recently been shown to activate the LMO2 promoter [216] necessary for initiation of mammalian embryonic hematopoiesis. It has been identified as a candidate HSC regulator in gene profiling studies [112, 154, 193]. Mice transplanted with HLF trangenic HSCs revealed a dramatic increase in the level of BM reconstitution and number of HSC [193]. HLF transgenic cells had a similar cell cycling profile to vector-transduced cells, but had upregulated Bcl-2 and were less prone to apoptotic cell death.

9.8 The Hox Family of Hematopoietic Regulators

The Hox homeodomain (HD) proteins are DNA binding transcription factors that have long been recognized as key regulators of development and hematopoiesis. Hox genes are expressed at various stages during hematopoietic development. Lack-of-function studies using HoxB4, HoxA9, or HoxB3 null mice demonstrate that all these mutations compromise the repopulating ability of HSC. Magnusson et al. [217] investigated mice with a compound deficiency in HoxA9, HoxB3, and HoxB4 for evidence of synergy between these genes. While all three genes had overlapping functions in hematopoietic cells, none was absolutely essential for generation or maintenance of all major blood lineages.

9.8.1 HoxA9

HoxA9 plays an important role in normal hematopoiesis. Targeted disruption of HoxA9 in mice severely reduces the number of HSC and progenitor cells, while enforced expression of HoxA9 promotes proliferative expansion of HSC and progenitor cells and subsequently inhibits their differentiation (reviewed in [69, 70, 218, 219]). These data highlight the importance of precise control of HoxA9 protein levels during hematopoiesis. The pathways by which HoxA9 acts are largely unknown, but it has been suggested that HoxA9 positively regulates Pim1, an oncogenic kinase [220]. The hematological phenotypes of HoxA9- and Pim1-deficient animals are strikingly similar and HOXA9 protein binds to the Pim1 promoter and induces Pim1 mRNA and protein in hematopoietic cells [220]. Pim1 protein is diminished in HoxA9–/– cells, and HoxA9 and Pim1 mRNA levels track together in early hematopoietic compartments. Induction of Pim1 protein by HoxA9 increases the phosphorylation and inactivation of the pro-apoptotic BAD protein, a target of Pim1. HoxA9–/– cells show increased apoptosis and decreased proliferation, defects that are ameliorated by re-introduction of Pim1 [220]. Thus Pim1 appears to be a direct transcriptional target of HoxA9 and a mediator of its anti-apoptotic and pro-proliferative effects.

9.8.2 Hox10

HoxA10 is critical for normal development of the erythroid and megakaryocytic lineages and the proliferation induction of HSC by HoxA10 was dependent on its concentration [221]. High levels of HoxA10 had no effect on HSC proliferation and blocked erythroid and megakaryocyte development while intermediate levels of HoxA10 induced a 15-fold increase in HSC repopulating capacity. The HoxA10-mediated effects on hematopoietic cells were associated with altered expression of genes that govern stem cell self-renewal and lineage commitment, for example, HLF, Dickkopf-1 (Dkk-1), Gfi-1, and Gata-1 [221]. Interestingly, binding sites for HoxA10 were found in HLF, Dkk-1 and Gata-1, and Dkk-1 and Gfi-1 were transcriptionally activated by HoxA10. These findings reveal novel molecular pathways that act downstream of HoxA10 and identify HoxA10 as a master regulator of postnatal hematopoietic development.

9.8.3 HoxB4

HoxB4 plays critical role in promoting HSC self-renewal, and engraftment (reviewed in [218, 219]). Combinations of early acting cytokines increased HoxB4 promoter activity in primitive hematopoietic cells and Tpo acting via Mpl and p38MAPK increased HoxB4 expression 2–3-fold in primitive hematopoietic cells [222]. In Tpo–/– mice hematopoietic HoxB4 expression was 2–5-fold lower than in wild-type animals [222]. Wnt signaling in primitive hematopoietic cells also induced HoxB4 expression [178, 179]. HoxB4 null mice had reduced HSC and progenitor cells due to impaired proliferative capacity, but did not show perturbed lineage commitment. Retroviral-mediated ectopic expression of HoxB4 resulted in a rapid increase in proliferation of murine HSC both in vivo (1,000-fold increase in transduced HSC in a murine transplant model), and in vitro (40-fold expansion of murine HSC), with retention of lympho-myeloid repopulating potential and enhanced regenerative capability in mice (reviewed in [218, 219]). However, high levels of HoxB4 expression in human umbilical cord blood $CD34^+$ cells were recently reported to either increase proliferation of HSCs and inhibit differentiation [223], or direct the cells toward a myeloid differentiation program rather than increasing proliferation [224]. These studies suggest that in human hematopoietic progenitors, HOXB4 affects cell fate decisions (self-renewal, differentiation, or a differentiation block) in a proteinlevel-dependent manner. Therefore, like HoxA9, the relative abundance of HoxB4 requires precise regulation. Molecular mechanisms and target genes responsible for HoxB4-induced HSC expansion remain to be elucidated. Overexpression of HoxB4 point mutations lacking the capacity to bind DNA (HoxB4 [$N^5 1 \rightarrow A$]) failed to enhance proliferative activity of transduced BM populations, whereas mutants that blocked the capacity of Hox to cooperate with the transcription factor PBX in DNA-binding (HoxB4 [$W \rightarrow G$]) conferred a pronounced proliferative advantage in vitro and in vivo to transduced BM populations [225]. This mutant was comparable to wild-type HoxB4 in that its elevated level promoted a comparable degree of HSC expansion. This was distinct from results obtained by knocking down the expression of Pbx in HoxB4 overexpressing BM cells using lentivector transduction with a Pbx antisense construct. In this system the HSC were >20 times more competitive than HSC overexpressing HoxB4 with Pbx levels intact which were, in turn, 20–50 times more competitive than wild-type BM [226]. It appears that the likely explanation for these observations is that HoxB4 and Pbx genes act on distinct pathways in the HSC, the former promoting self-renewal and the latter inhibiting it.

Some transcription factors travel between cells because they contain protein domains that allow them to do so. This is the case for the HIV transcription factor TAT and for several homeoproteins such as Engrailed, HoxA5, HoxB4, HoxC8, and PAX6. Direct paracrine homeoprotein activity has not been considered, yet in theory it would represent a way for neighboring hematopoietic cells to exchange proliferative and differentiative signals. Bone marrow stromal cells (murine MS-5 stroma) have been lentivector transduced with a vector expressing HoxB4 linked to an immunoglobulin kappa chain leader sequence (signal peptide) that is cleaved during protein secretion. Co-culture of human CB $CD34^+$ cells on this stroma resulted in a 2–3-fold greater expansion of cells and progenitors, a 4-fold greater expansion of LTC-IC, and a 2.5-fold expansion of NOD/SCID repopulating cells relative to control over 5 weeks of culture [227]. A biologically active TAT-HoxB4 fusion protein has been produced and expressed as a recombinant protein that upon purification could be added to murine HSC cultures (reviewed in [219]). Since the half life of intracellular HOXB4 was $\sim$1 h, TAT-HoxB4 was added to cell culture every 3 h for 4 days together with cytokines (KL, IL-6, IL-3) resulting in a 5–6-fold expansion of HSC as measured by competitive repopulation. Ectopic overexpression of HoxB4 in murine bone marrow produced HSC that were $\sim$40 fold more competitive than un-transduced cells in mouse repopulation assay and by 3–5 months the HSC pool size of reconstituted mice equaled, but never exceeded, that of untreated control mice (reviewed in [219]). High-level ectopic expression of HOXB4 in human cord blood $CD34^+$ cells had a selective growth advantage in NOD/SCID mice but with substantial impairment in myelo-erythroid differentiation and B cell development [193]. In non-human primate competitive repopulating transplantation models, HoxB4 overexpressing $CD34^+$ cells had a 56-fold higher short term, and 5-fold higher long-term (6 months) engraftment than control cells [228]. Zhang et al. [229] transduced $CD34^+$ cells from nonhuman primates, dogs and humans with a HoxB4-expressing gammaretroviral vector. Compared to the control vector, HoxB4 overexpression resulted in a much larger increase in CFCs with dog cells (28-fold) compared to human peripheral blood, human cord blood, and baboon cells (2-, 4-, and 5-fold, respectively). Furthermore, HoxB4 overexpression resulted in immortalization with sustained growth ($>$ 2 months) of primitive hematopoietic cells from mice and dogs but not from monkeys and humans. This difference correlated with increased levels of retrovirally overexpressed HOXB4 in dog and mouse cells compared to human and nonhuman primate cells. These findings suggest that the growth promoting effects of HoxB4 are critically dependent on HoxB4 expression levels and that this can result in important species-specific differences in potency.

9.8.4 Cul4A Mediated Hox Ubiquitination

The abundance of a given cellular protein is regulated by the interplay between its biosynthesis and degradation. During normal hematopoietic development, HoxA9 is strongly expressed in the $CD34^+$ populations enriched in early myeloid progenitors, and is turned off when cells exit the $CD34^+$ compartment and undergo terminal differentiation. In conjunction with decreased biosynthesis, rapid turnover of HoxA9 would ensure low steady-state levels, which is necessary for proper execution of differentiation into myeloid lineages. HoxB4 is also expressed at high levels in HSC/progenitor compartments, and is down-regulated, but maintains low-level expression during differentiation. The studies of the biochemical mechanisms controlling the activities of HoxA9 and HoxB4 thus far have been focused primarily on transcriptional regulation. Little is known about how their cellular abundance is controlled at the posttranslational level. Identification of proteins involved in the removal of HoxA9 and HoxB4 will be necessary for understanding the elaborate regulatory circuitry governing hematopoiesis. Post-translational regulation of HOX protein levels has been linked to their ubiquitin-dependent proteolysis by the CUL-4A ubiquitin-ligase [230, 231].

9.8.5 Cdx4

Cdx4 belongs to the caudal family of homeobox genes which have been implicated in antero-posterior patterning of the axial skeleton and regulation of Hox gene expression. A Cdx4 mutation in Zebrafish causes severe anemia with complete absence of Runx-1 in blood cells [232]. Injection of mutants with HoxB7 and HoxA9 mRNA almost completely rescued Cdx4 mutants. Retroviral transduction of mouse ES-derived embryoid body cells with Cdx4 increased expression of HoxB4, HoxB3, HoxB8, and HoxA9 – all implicated in HSC or progenitor expansion – and Cdx4 was more potent than HoxB4 in stimulating hematopoiesis and CFU-GEMM production [232]. Using an embryonic stem cell line engineered with tetracycline-inducible Cdx4, Wang et al. [233] demonstrated that ectopic Cdx4 expression promotes hematopoietic mesoderm specification, increased hematopoietic progenitor formation, and, together with HoxB4, enhanced multilineage hematopoietic engraftment of lethally irradiated adult mice. Brief pulses of ectopic Cdx4 or HoxB4 expression are sufficient to enhance hematopoiesis during ESC differentiation, presumably by acting as developmental switches to activate posterior Hox genes [234].

9.9 NF-Ya

NF-Ya, the regulatory and DNA-binding subunit of the trimeric transcription factor NF-Y is preferentially expressed in HSC and its expression rapidly declines with differentiation [235, 236]. HSCs overexpressing NF-Ya are biased toward primitive hematopoiesis in vitro and show strikingly increased in vivo repopulating abilities after single or sequential BM transplantation. Overexpression of NF-Ya in primitive hematopoietic cells activates the HoxB4 promoter in cooperation with upstream/ubiquitous stimulating factor1 and 2 (USF-1/2) as well as Notch1, LEF-1, and telomerase RNA [235]. Thus, NF-Ya is considered to be a potent cellular regulator of HSC self-renewal [236].

9.10 Tel/Etv6

The Ets-related transcription factor Tel/Etv6, the product of a locus frequently involved in translocations in leukemia, is a selective regulator of HSC survival [237]. Following inactivation of Tel/Etv6, HSCs are lost in the adult bone marrow but their progeny are unaffected and transiently sustain blood formation. Absence of Tel/Etv6 after lineage commitment impaired maturation of megakaryocytes [237].

9.11 Zfx

Zfx is a zinc finger protein of the Zfy family, whose members are highly conserved in vertebrates. Zfx is expressed at elevated levels in several SC types. Conditional Zfx deletion abolished the maintenance of adult murine HSC but did not affect erythromyeloid progenitors or fetal HSC [238]. Zfx-deficient HSC interacted normally with their BM niche but showed increased apoptosis and SC-specific upregulation of stress inducible genes. The phenotype was very similar to that observed after the inducible deletion of Tel/Etv6 (243), FoxO1/3/4 [130, 140] or after germline deletion of Bmi-1 [37, 108, 201]. Similar to the latter two models, lymphopoiesis was severely impaired after the loss of Zfx. Thus, Zfx and possibly other HSC regulators appear to control both HSC self-renewal and lymphoid differentiation, but not erythromyeloid differentiation.

10 Conclusion

The current dogma recognizes that specific factors determine HSC fates, orchestrating HSC proliferative status, self-renewal (symmetric and asymmetric division) and commitment via successively more restricted progenitors. In this context "factors" may be classic hematopoietic growth factors and morphogens presented by specific interative niche cell components (osteoblasts, endothelial cells, reticular cells). HSC fate may be governed by (small) quantitative shifts in the relative activation of known signaling pathways [239]. Alterations in the relative levels of pathway activation may arise from dynamic shifts in the expression of signaling components (e.g., receptors, ligands). As extracellular conditions vary over small ranges, cells are induced to overcome signaling threshold barriers, and as a consequence, adopt new, stable cell fates. Functional genomics and other newly developing technologies will continue to extend our understanding of the development, self-renewal, and differentiation of HSCs. The use of knockout and ectopic overexpression models has revealed the role of a number of specific genes in HSC regulation. The often-conflicting data that has resulted indicates HSC regulation is a finely tuned system with tight transcriptional regulation and possibly more critical post translational regulation of protein levels [239–241]. The ability to maintain a self-renewing HSC pool throughout life, despite pressures for differentiation and proliferative exhaustion, further indicates that there are numerous built-in safety systems as reflected in the substantial redundancy of HSC signaling pathways. Ex vivo HSC culture systems can be manipulated quantititatively by varying the magnitude of signaling pathway activation influencing cell fate decisions; temporally, by defining windows of opportunity for stimulation; and, spatially, by fixed location-dependent signaling (niches). The picture emerging from accumulating genetic and functional data indicate that molecular cross-talk between HSCs and niche cells involves a large number of molecules (cytokines, chemokines, integrins, morphogens, and their receptors) [239–241]. Since many of these secreted signaling molecules bind to the extracellular matrix and do not diffuse far or are presented in transmembrane form (e.g., Jagged-Notch ligand, Kit Ligand-c-Kit, Ang1-Tie2), HSC must reside within a three-dimensional multicellular signaling unit. By analogy to the neuronal or immunological synapses the term HSC niche synapse has been proposed for this signaling unit [242]. Deeper insight into the molecular and cellular mechanisms that govern HSC self-renewal will ultimately lead to protocols for generation and expansion of HSCs in vitro for clinical use in cell and gene therapies.

References

1. Ramalho-Santos M, Willenbring H. On the origin of the term "stem cell". Cell Stem Cell. 2007;1:35–8.
2. Till J, McCulloch E. A direct measurement of the radiation sensitivity of normal mouse bone marrow cells. Radiat Res. 1961;14:213–22.
3. McCulloch EA, Till JE. Perspectives on the properties of stem cells. Nat Med. 2005;11:1026–8.
4. Muench MO, Firpo MT, Moore MAS. Bone marrow transplantation with Interleukin-1 plus kit-ligand ex vivo expanded bone marrow accelerates hematopoietic reconstitution in mice without the loss of stem cell lineage and proliferative potential. Blood. 1993;81:3463–73.
5. Jo D-Y, Rafii S, Hamada T, Moore MAS. Chemotaxis of primitive hematopoietic cells in response to stromal cell-derived factor-1. J Clin Invest. 2000;105:101–11.
6. Gan OI, Murdoch B, Larochelle A, Dick JE. Differential maintenance of primitive human SCID-repopulating cells, clonogenic progenitors, and long-term culture-initiating cells after incubation on human bone marrow stromal cells. Blood. 1997;90:641–50.
7. van Os R, Kamminga LM, de Haan G. Stem cell assays: something old, something new, something borrowed. Stem Cells. 2004;22:1181–90.
8. Na Nakorn T, Traver D, Weissman IL, Akashi K. Myeloerythroid-restricted progenitors are sufficient to confer radioprotection and provide the majority of day 8 CFU-S. J Clin Invest. 2002;109:1579–85.
9. Dick JE, Lapidot T. Biology of normal and acute myeloid leukemia stem cells. Int J Hematol. 2005;82:389–96.
10. Horn PA, Blasczyk R. Severe combined immunodeficiency-repopulating cell assay may overestimate long-term repopulation ability. Stem Cells. 2007;25:3271–2.
11. Schmidt M, Zickler P, Hoffmann G, et al. Polyclonal long-term repopulating stem cell clones in a primate model. Blood. 2002;100:2737–43.
12. Mazurier F, Doedens M, Gan OI, Dick JE. Rapid myeloerythroid repopulation after intrafemoral transplantation of NOD-SCID mice reveals a new class of human stem cells. Nat Med. 2003;9:959–63.
13. Kimura T, Asada R, Wang J, et al. Identification of long-term repopulating potential of human cord blood-derived CD34-flt3-severe combined immunodeficiency-repopulating cells by intrabone marrow injection. Stem Cells. 2007;25:1348–55.
14. van der Loo JC, Ploemacher RE. Marrow- and spleen-seeding efficiencies of all murine hematopoietic stem cell subsets are decreased by preincubation with hematopoietic growth factors. Blood. 1995;85:2598–606.
15. Lanzkron SM, Collector MI, Sharkis SJ. Hematopoietic stem cell tracking in vivo: a comparison of short-term and long-term repopulating cells. Blood. 1999;93:1916–21.
16. Osawa M, Hanada K, Hamada H, Nakauchi H. Long-term lymphohematopoietic reconstitution by a single CD34-low/negative hematopoietic stem cell. Science. 1996;273:242–5.
17. Benveniste P, Cantin C, Hyam D, Iscove NN. Hematopoietic stem cells engraft in mice with absolute efficiency. Nat Immunol. 2003;4:708–13.
18. Matsuzaki Y, Kinjo K, Mulligan RC, Okano H. Unexpectedly efficient homing capacity of purified murine hematopoietic stem cells. Immunity. 2004;20:87–93.
19. Uchida N, Dykstra B, Lyons KJ, Leung FY, Eaves CJ. Different in vivo repopulating activities of purified hematopoietic stem cells before and after being stimulated to divide in vitro with the same kinetics. Exp Hematol. 2003;31:1338–47.
20. Camargo FD, Chambers SM, Drew E, McNagny KM, Goodell MA. Hematopoietic stem cells do not engraft with absolute efficiencies. Blood. 2006;107:501–7.
21. Spangrude GJ, Heimfeld S, Weissman IL. Purification and characterization of mouse hematopoietic stem cells. Science. 1988;241:58–62.
22. Morrison SJ, Weissman IL. The long-term repopulating subset of hematopoietic stem cells is deterministic and isolatable by phenotype. Immunity. 1994;1:661–73.
23. Ogawa M. Changing phenotypes of hematopoietic stem cells. Exp Hematol. 2002;30:3–6.
24. Matsuoka S, Ebihara Y, Xu M, et al. CD34 expression on long-term repopulating hematopoietic stem cells changes during developmental stages. Blood. 2001;97:419–25.
25. Lin KK, Goodell MA. Purification of of hematopoietic stem cells using the side population. Methods Enzymol. 2006;420: 255–64.
26. Adolfsson J, Mansson R, Buza-Vidas N, et al. Identification of flt3(+) lympho-myeloid stem cells lacking erythro-megakaryocytic potential a revised road map for adult blood lineage commitment. Cell. 2005;121:295–306.
27. Kiel MJ, Yilmaz OH, Iwashita T, Yilmaz OH, Terhorst C, Morrison SJ. SLAM family receptors distinguish hematopoietic stem and progenitor cells and reveal endothelial niches for stem cells. Cell. 2005;121:1109–21.
28. Weksberg DC, Chambers SM, Boles NC, Goodell MA. CD150 negative Side Population cells represent a functionally distinct population of long-term hematopoietic stem cells. Blood. 2008;111:2444–51.
29. Balazs AB, Fabian AJ, Esmon CT, Mulligan RC. Endothelial protein C receptor (CD201) explicitly identifies hematopoietic stem cells in murine bone marrow. Blood. 2006;107:2317–21.
30. Ishikawa F, Livingston AG, Minamiguchi H, Wingard JR, Ogawa M. Human cord blood long-term engrafting cells are CD34+ CD38-. Leukemia. 2003;17:960–4.
31. Ishikawa F, Yasukawa M, Lyons B, et al. Development of functional human blood and immune systems in NOD/SCID/IL2 receptor chain(null) mice. Blood. 2005;106:1565–73.
32. McKenzie JL, Takenaka K, Gan OI, Doedens M, Dick JE. Low rhodamine 123 retention identifies long-term human hematopoietic stem cells within the Lin-CD34+CD38- population. Blood. 2007;109:543–5.
33. Yoshihara H, Arai F, Hosokawa K, Hagiwara H, Majeti R, Park CY, Weissman IL. Identification of a hierarchy of multipotent

hematopoietic progenitors in human cord blood. Cell Stem Cell. 2007;1:635–45.
34. Jarås M, Edqvist A, Rebetz J, Salford LG, Widegren B, Fan X. Human short-term repopulating cells have enhanced telomerase reverse transcriptase expression. Blood. 2006;108:1084–91.
35. Colvin M, Russo J, Hilton J, Dulik D, Fenselau C. Enzymatic mechanisms of resistance to alkylating agents in tumor cells and normal tissues. Adv Enzyme Regul. 1988;27:211–21.
36. Christ O, Lucke K, Imren S, Leung K, Hamilton M, Eaves A, Smith C, Eaves C, et al. Improved purification of hematopoietic stem cells based on their elevated aldehyde dehydrogenase activity. Haematologica. 2007;92:1165–72.
37. Park IK, Qian D, Kiel M, et al. Bmi-1 is required for maintenance of adult self-renewing haematopoietic stem cells. Nature. 2003;423:302–5.
38. Cheng T, Rodrigues N, Shen H, et al. Hematopoietic stem cell quiescence maintained by p21cip1/waf1. Science. 2000;287: 1804–8.
39. Cheng T, Scadden DT. Cell cycle entry of hematopoietic stem cells and progenitor cells controlled by distinct cyclin-dependent kinase inhibitors. Int J Hematol. 2002;75:460–5.
40. Miyake N, Brun AC, Magnusson M, Miyake K, Scadden DT, Karlsson S. HOXB4-induced self-renewal of hematopoietic stem cells is significantly enhanced by p21 deficiency. Stem Cells. 2006;24:653–61.
41. Yuan Y, Shen H, Franklin DS, Scadden DT, Cheng T. In vivo self-renewing divisions of haematopoietic stem cells are increased in the absence of the early G1-phase inhibitor, p18INK4C. Nat Cell Biol. 2004;6:436–42.
42. Scandura JM, Boccuni P, Massague J, Nimer SD. Transforming growth factor beta-induced cell cycle arrest of human hematopoietic cells requires p57KIP2 up-regulation. Proc Natl Acad Sci U S A. 2004;101:15231–6.
43. Cheshier SH, Morrison SJ, Liao X, Weissman IL. In vivo proliferation and cell cycle kinetics of long-term self-renewing hematopoietic stem cells. Proc Natl Acad Sci U S A. 1999;96:3120–5.
44. Kiel MJ, He S, Ashkenazi R, Gentry SN, Teta M, Kushner JA, Jackson TL, Morrison SJ. Haematopoietic stem cells do not asymmetrically segregate chromosomes or retain BrdU. Nature. 2007;449:238–42.
45. Wright DE, Cheshier SH, Wagers AJ, Randall TD, Christensen JL, Weissman IL. Cyclophosphamide/granulocyte colony-stimulating factor causes selective mobilization of bone marrow hematopoietic stem cells into the blood after M phase of the cell cycle. Blood. 2001;97:2278–85.
46. Lapidot T, Dar A, Kollet O. How do stem cells find their way home? Blood. 2005;106:1901–10.
47. Bryder D, Ramsfjell V, Dybedal I, et al. Self-renewal of multipotent long-term repopulating hematopoietic stem cells is negatively regulated by Fas and Tumor Necrosis Factor Receptor Activation. J Exp Med. 2001;194:941–52.
48. Dybedal I, Bryder D, Fossum A, Rusten LS, Jacobsen SEW. Tumor necrosis factor (TNF)–mediated activation of the p55 TNF receptor negatively regulates maintenance of cycling reconstituting human hematopoietic stem cells. Blood. 2001;98:1782–91.
49. Lacorazza HD, Yamada T, Liu Y, et al. The transcription factor MEF/ELF4 regulates the quiescence of primitive hematopoietic cells. Cancer Cell. 2006;9:175–87.
50. Yoshihara H, Arai F, Hosokawa K, et al. Thrombopoietin/MPL signaling regulates hematopoietic stem cell quiescence and interaction with the osteoblastic niche. Cell Stem Cell. 2007;1:685–97.
51. Yoshihara H, Arai F, Hosokawa K, Hagiwara T, et al. Thrombopoietin/MPL signaling regulates hematopoietic stem cell quiescence and interaction with the osteoblastic niche. Cell Stem Cell. 2007;1:685–97.
52. Ho AD, Wagner W. The beauty of asymmetry: asymmetric divisions and self-renewal in the haematopoietic system. Curr Opin Hematol. 2007;14:330–6.
53. Morrison SJ, Kimble J. Asymmetric and symmetric stem-cell divisions in development and cancer. Nature. 2006;441: 1068–74.
54. Takano H, Ema H, Sudo K, Nakauchi H. Asymmetric division and lineage commitment at the level of hematopoietic stem cells: inference from differentiation in daughter cell and granddaughter cell pairs. J Exp Med. 2004;199:295–302.
55. Giebel B, Zhang T, Beckmann J, et al. Primitive human hematopoietic cells give rise to differentially specified daughter cells upon their initial cell division. Blood. 2006;107:2146–52.
56. Enver T, Heyworth CM, Dexter TM. Do stem cells play dice? Blood. 1998;92:348–51.
57. Wu M, Kwon HY, Rattis F, Blum J, Zhao C, Ashkenazi R, Jackson TL, Gaiano N, Oliver T, Reya T. Imaging hematopoietic precursor division in real time. Cell Stem Cell. 2007;1:541–54.
58. Gottschling S, Saffrich R, Seckinger A, et al. Human mesenchymal stromal cells regulate initial self-renewing divisions of hematopoietic progenitor cells by a beta1-integrin-dependent mechanism. Stem Cells. 2007;25:798–806.
59. Beckmann J, Scheitza S, Wernet P, Fischer JC, Giebel B. Asymmetric cell division within the human hematopoietic stem and progenitor cell compartment: identification of asymmetrically segregating proteins. Blood. 2007;109:5494–501.
60. Cairns, J. Mutation selection and the natural history of cancer. Nature. 1975;255:197–200.
61. Shinin V, Gayraud-Morel B, Gomes D, Tajbakhsh S. Asymmetric division and cosegregation of template DNA strands in adult muscle satellite cells. Nature Cell Biol. 2006;8:677–87.
62. Conboy MJ, Karasov AO, Rando TA. High incidence of non-random template strand segregation and asymmetric fate determination in dividing stem cells and their progeny. PLoS Biol. 2007;5:e102.
63. McKenzie JL, Gan OI, Doedens M, Wang JC, Dick JE. Individual stem cells with highly variable proliferation and self-renewal properties comprise the human hematopoietic stem cell compartment. Nat Immunol. 2006;7:1225–33.
64. Sieburg HB, Cho RH, Dykstra B, Uchida N, Eaves CJ, Muller-Sieburg CE. The hematopoietic stem compartment consists of a limited number of discrete stem cell subsets. Blood. 2006;107:2311–16.
65. Dykstra B, Ramunas J, Kent D, et al. High-resolution video monitoring of hematopoietic stem cells cultured in single-cell arrays identifies new features of self-renewal. Proc Natl Acad Sci U S A. 2006 23;103:8185–90.
66. Dykstra B, Kent D, Michelle Bowie M, McCaffrey L, Hamilton M, Lyons K, Lee SJ, Brinkman R, Eaves C. Long-term propagation of distinct hematopoietic differentiation programs in vivo. Cell Stem Cell. 2007;1:218–29.
67. Ito K, Hirao A, Arai F, et al. Regulation of oxidative stress by ATM is required for self-renewal of haematopoietic stem cells. Nature. 2004;431:997–1002.
68. Bhardwaj G, Murdoch B, Wu D, et al. Sonic hedgehog induces the proliferation of primitive human hematopoietic cells via BMP regulation. Nat Immunol. 2001;2:172–80.
69. Moore MAS. Hematopoietic stem cells. In: Lanza R, Langer and Vacanti, editors. Principles of tissue engineering. 3rd ed. Burlington, MA: Elsevier Academic Press; 2007. p. 735–48.
70. Moore, MAS. Converging pathways in leukemogenesis and stem cell self-renewal. Exp Hematol. 2005;33:719–37.
71. Sharma S, Gurudutta GU, Satija NK, et al. Stem cell c-KIT and HOXB4 genes: critical roles and mechanisms in self-renewal, proliferation, and differentiation. Stem Cells Dev. 2006;15: 755–78.

72. Czechowicz A, Kraft D, Weissman IL, Bhattacharya D. Efficient transplantation via antibody-based clearance of hematopoietic stem cell niches. Science. 2007 23;318:1296–9.
73. Dorsch M, Fan PD, Danial NN, Rothman PB, Goff SP. The thrombopoietin receptor can mediate proliferation without activation of the Jak-STAT pathway. J Exp Med. 1997;186:1947–55.
74. Antonchuk J, Hyland CD, Hilton DJ, Alexander WS. Synergistic effects on erythropoiesis, thrombopoiesis, and stem cell competitiveness in mice deficient in thrombopoietin and steel factor receptors. Blood. 2004;104:1306–13.
75. Qian H, Buza-Vidas N, Hyland CD, et al. Critical role of thrombopoietin in maintaining adult quiescent hematopoietic stem cells. Cell Stem Cell. 2007;1:671–84.
76. Buza-Vidas N, Antonchuk J, Qian H, et al. Cytokines regulate postnatal hematopoietic stem cell expansion: opposing roles of thrombopoietin and LNK. Genes Dev. 2006;20:2018–23.
77. Seita J, Ema H, Ooehara J, et al. Lnk negatively regulates self-renewal of hematopoietic stem cells by modifying thrombopoietin-mediated signal transduction. Proc Natl Acad Sci U S A. 2007;104:2349–54.
78. Haylock DN, Horsfall MJ, Dowse TL, et al. Increased recruitment of hematopoietic progenitor cells underlies the ex vivo expansion potential of FLT3 ligand. Blood. 1997;90:2260–72.
79. Solanilla A, Grosset C, Duchez P, et al. Flt3-ligand induces adhesion of haematopoietic progenitor cells via a very late antigen (VLA)-4- and VLA-5-dependent mechanism. Br J Haematol. 2003;120:782–6.
80. Sauvageau G, Iscove NN, Humphries R. In vitro and in vivo expansion of hematopoietic stem cells. Oncogene. 2004;23: 7223–32.
81. Zandstra PW, Conneally E, Piret JM, Eaves CJ. Ontogeny-associated changes in the cytokine responses of primitive human haematopoietic cells. Br J Haematol. 1998;101:770–8.
82. Ueda T, Tsuji K, Yoshino H, Ebihara Y, et al. Expansion of human NOD/SCID-repopulating cells by stem cell factor, Flk2/Flt3 ligand, thrombopoietin, IL-6 and soluble IL-6 receptor. J Clin Invest. 2000;105:1013–21.
83. Gammaitoni L, Weisel KC, Gunetti M, et al. Elevated telomerase activity and minimal telomere loss in cord blood long-term cultures with extensive stem cell replication. Blood. 2004;103: 4440–8.
84. Rollini P, Kaiser S, Faes-van't Hull E, Kapp U, Leyvraz S. Long-term expansion of transplantable human fetal liver hematopoietic stem cells. Blood. 2004;103:1166–70.
85. Summers YJ, Heyworth CM, de Wynter EA, et al. CD34+ G0 cells from cord blood show a high incidence of long-term culture-initiating cells and a capacity for more than 100 million-fold amplification of colony-forming cells in vitro. Stem Cells. 2004;22:704–15.
86. Liu B, Buckley SM, Lewis ID, et al. Homing defect of cultured human hematopoietic cells in the NOD/SCID mouse is mediated by Fas/CD95. Exp Hematol. 2003;31:824–32.
87. Sugiyama T, Kohara H, Noda M, Nagasawa T. Maintenance of the hematopoietic stem cell pool by CXCL12–CXCR4 chemokine signaling in bone marrow stromal cell niches. Immunity. 2006;25:977–88.
88. Broxmeyer HE. Chemokines in hematopoiesis. Curr Opin Hematol. 2008;15:49–58.
89. Hattori K, Heissig B, Tashiro K, et al. Plasma elevation of stromal derived factor-1 induces mobilization of mature and immature hematopoietic progenitor and stem cells. Blood. 2001;97: 3354–60.
90. Kopp HG, Avecilla ST, Hooper AT, Rafii S. The bone marrow vascular niche: home of HSC differentiation and mobilization. Physiology. 2005;20:349–56.
91. Peled A, Kollet O, Ponomaryov T, et al. The chemokine SDF-1 activates the integrins LFA-1, VLA-4, and VLA-5 on immature human CD34D cells: role in transendothelial/stromal migration and engraftment of NOD/SCID mice. Blood. 2000;95:3289–96.
92. Nilsson SK, Haylock DN, Johnston HM, Occhiodoro T, Brown TJ, Simmons PJ. Hyaluronan is synthesized by primitive hemopoietic cells, participates in their lodgment at the endosteum following transplantation, and is involved in the regulation of their proliferation and differentiation in vitro. Blood. 2003;101: 856–62.
93. Avigdor A, Goichberg P, Shivtiel S, et al. CD44 and hyaluronoic acid cooperate with SDF-1 in the trafficking of human CD34+ stem/progenitor cells to bone marrow. Blood. 2004;103: 2981–9.
94. Trentin JJ. Determination of bone marrow stem cell differentiation by stromal hemopoietic inductive microenvironments (HIM). Am J Pathol. 1971;65:621–8.
95. Metcalf D, Moore MAS. Haemopoietic cells. Amsterdam: North-Holland Publishing Company; 1971.
96. Haskill JS, McNeill TA, Moore MAS. Density distribution analysis of in vivo and in vitro colony forming cell in bone marrow. J Cell Physiol. 1970;75:157–79.
97. Lord BI, Testa NG, Hendry JH. The relative spatial distributions of CFUs and CFUc in the normal mouse femur. Blood. 1975;46:65–72.
98. Zhang J, Niu C, Huang H, et al. Identification of the haematopoietic stem cell niche and control of the niche size. Nature. 2003;425:836–41.
99. Arai F, Hirao A, Ohmura M, et al. Tie2/Angiopoietin-1 signaling regulates hematopoietic stem cell quiescence in the bone marrow niche. Cell. 2004;118:149–61.
100. Gong JK. Endosteal marrow: a rich source of hematopoietic stem cells. Science. 1978;199:1443–5.
101. Schofield R. The relationship between the spleen colony-forming cell and the haemopoietic stem cell. Blood Cells. 1978;4:7–25.
102. Nilsson SK, Johnston HM, Whitty GA, et al. Osteopontin, a key component of the hematopoietic stem cell niche and regulator of primitive hematopoietic progenitor cells. Blood. 2005;106: 1232–9.
103. Haylock DN, Williams B, Johnston HM, et al. Hemopoietic stem cells with higher hemopoietic potential reside at the bone marrow endosteum. Stem Cells. 2007;25:1062–9.
104. Yoshimoto M, Shinohara T, Heike T, et al. Direct visualization of transplanted hematopoietic cell reconstitution in intact mouse organs indicates the presence of a niche. Exp Hematol. 2003;31:733–40.
105. Sharma S, Gurudutta GU, Satija NK, et al. Stem cell c-KIT and HOXB4 genes: critical roles and mechanisms in self-renewal, proliferation, and differentiation. Stem Cells Dev. 2006;15:755–78.
106. Driessen RL, Johnston HM, Nilsson SK. Membrane-bound stem cell factor is a key regulator in the initial lodgment of stem cells within the endosteal marrow region. Exp Hematol. 2003;31: 1284–91.
107. Suzuki N, Ohneda O, Minegishi N, et al. Combinatorial Gata2 and Sca1 expression defines hematopoietic stem cells in the bone marrow niche. Proc Natl Acad Sci U S A. 2006;103:2202–7.
108. van der Lugt NM, Domen J, Linders K, et al. Posterior transformation, neurological abnormalities, and severe hematopoietic defects in mice with a targeted deletion of the bmi-1 proto-oncogene. Genes Dev. 1994;8:757–69.
109. Calvi LM, Adams GB, Weibrecht KW, et al. Osteoblastic cells regulate the haematopoietic stem cell niche. Nature. 2003;425:841–6.
110. Potocnik AJ, Brakebusch C, Fassler R. Fetal and adult hematopoietic stem cells require beta1 integrin function for colonizing fetal liver, spleen, and bone marrow. Immunity. 2000;12:653–63.

111. Wilson A, Murphy MJ, Oskarsson T, et al. c-Myc controls the balance between hematopoietic stem cell self-renewal and differentiation. Genes Dev. 2004;18:2747–63.
112. Ivanova NB, Dimos JT, Schaniel C, et al. H. A stem cell molecular signature. Science. 2002;298:601–4.
113. Begg SK, Radley JM, Pollard JW, et al. Delayed hematopoietic development in osteopetrotic (op/op) mice. J Exp Med. 1993;177:237–42.
114. Wang ZQ, Ovitt C, Grigoriadis AE, et al. Bone and haematopoietic defects in mice lacking c-fos. Nature. 1992;360:741–5.
115. Okada S, Wang ZQ, Grigoriadis AE, Wagner EF, von Rüden T. Mice lacking c-fos have normal hematopoietic stem cells but exhibit altered B-cell differentiation due to an impaired bone marrow environment. Mol Cell Biol. 1994;14:382–90.
116. Otto F, Thornell AP, Crompton T, et al. Cbfa1, a candidate gene for cleidocranial dysplasia syndrome, is essential for osteoblast differentiation and bone development. Cell. 1997;89:765–71.
117. Komori T, Yagi H, Nomura S, et al. Targeted disruption of Cbfa1 results in a complete lack of bone formation owing to maturational arrest of osteoblasts. Cell. 1997;89:755–64.
118. Deguchi K, Yagi H, Inada M, et al. Excessive extramedullary hematopoiesis in Cbfa1-deficient mice with a congenital lack of bone marrow. Biochem Biophys Res Commun. 1999;255:352–9.
119. Adams GB, Chabner KT, Alley IR, et al. Stem cell engraftment at the endosteal niche is specified by the calcium-sensing receptor. Nature. 2006;439:599–603.
120. Stier S, Ko Y, Forkert R, et al. Osteopontin is a hematopoietic stem cell niche component that negatively regulates stem cell pool size. J Exp Med. 2005;201:1781–91.
121. Zhang CC, Kaba M, Ge G, et al. Angiopoietin-like proteins stimulate ex vivo expansion of hematopoietic stem cells. Nat Med. 2006;12:240–5.
122. Visnjic D, Kalajzic Z, Rowe DW, et al. Hematopoiesis is severely altered in mice with an induced osteoblast deficiency. Blood. 2004;103:3258–64.
123. Zhu J, Garrett R, Jung Y, et al. Osteoblasts support B-lymphocyte commitment and differentiation from hematopoietic stem cells. Blood. 2007;109:3706–12.
124. Kiel MJ, Radice GL, Morrison SJ. Lack of evidence that hematopoietic stem cells depend on N-cadherin-mediated adhesion to osteoblasts for their maintenance. Cell Stem Cell. 2007;1:204–17.
125. Taichman RS, Reilly MJ, Emerson SG. Human osteoblasts support human hematopoietic progenitor cells in vitro bone marrow cultures. Blood. 1996;87:518–24.
126. Stier S, Ko Y, Forkert R, et al. Osteopontin is a hematopoietic stem cell niche component that negatively regulates stem cell pool size. J Exp Med. 2005;201:1781–91.
127. Shalaby F, Rossant J, Yamaguchi TP, Gertsenstein M, Wu XF, Breitman ML, Schuh AC. Failure of blood-island formation and vasculogenesis in Flk-1-deficient mice. Nature. 1995;376:62–66.
128. Tavian M, Coulombel L, Luton D, et al. Aorta-associated CD34+ hematopoietic cells in the early human embryo. Blood. 1996;87:67–72.
129. Lu LS, Wang SJ, Auerbach R. In vitro and in vivo differentiation into B cells, T cells, and myeloid cells of primitive yolk sac hematopoietic precursor cells expanded > 100-fold by coculture with a clonal yolk sac endothelial cell line. Proc Natl Acad Sci U S A. 1996;93:14782–7.
130. Chute JP, Muramoto GG, Fung J, Oxford C. Soluble factors elaborated by human brain endothelial cells induce the concomitant expansion of purified human BM CD34+CD38- cells and SCID-repopulating cells. Blood. 2005;105:576–83.
131. Shen Q, Goderie SK, Jin L, et al. Endothelial cells stimulate self-renewal and expand neurogenesis of neural stem cells. Science. 2004;304:1338–40.
132. Rafii S, Shapiro F, Pettengell R, et al. Human bone marrow microvascular endothelial cells support long-term proliferation and differentiation of myeloid and megakaryocytic progenitors. Blood. 1995;86:3353–63.
133. Feugier P, Jo DY, Shieh JH, et al. Ex vivo expansion of stem and progenitor cells in co-culture of mobilized peripheral blood CD34cells on human endothelium transfected with adenovectors expressing thrombopoietin, c-kit ligand, and Flt-3 ligand. J Hematother Stem Cell Res. 2002;11:127–38.
134. Sipkins DA, Wei X, Wu JW, et al. In vivo imaging of specialized bone marrow endothelial microdomains for tumour engraftment. Nature. 2005;435:969–73.
135. Avecilla ST, Hattori K, Heissig B, et al. Chemokine-mediated interaction of hematopoietic progenitors with the bone marrow vascular niche is required for thrombopoiesis. Nat Med. 2004;10:64–71.
136. Parmar K, Mauch P, Vergilio JA, Sackstein R, Down JD. Distribution of hematopoietic stem cells in the bone marrow according to regional hypoxia. Proc Natl Acad Sci U S A. 2007;104: 5431–6.
137. Lévesque JP, Winkler IG, Hendy J, et al. Hematopoietic progenitor cell mobilization results in hypoxia with increased hypoxia-inducible transcription factor-1 alpha and vascular endothelial growth factor A in bone marrow. Stem Cells. 2007;25:1954–65.
138. Ito K, Hirao A, Arai F, et al. Reactive oxygen species act through p38 MAPK to limit the lifespan of hematopoietic stem cells. Nat Med. 2006;12:446–51.
139. Miyamoto K, Araki KY, Naka K, et al. Fox03a is essential for maintenance of the hematopoietic stem cell pool. Cell Stem Cell. 2007;1:101–12.
140. Tothova Z, Kollipara R, Huntly BJ, et al. FoxOs are critical mediators of hematopoietic stem cell resistance to physiologic oxidative stress. Cell. 2007;128:325–9.
141. Fortunel NO, Hatzfeld JA, Monier MN, Hatzfeld A. Control of hematopoietic stem/progenitor cell fate by transforming growth factor-beta. Oncol Res. 2003;13:445–53.
142. Shi Y, Massagué J. Mechanisms of TGF-beta signaling from cell membrane to the nucleus. Cell. 2003;113:685–700.
143. Souchelnytskyi S, Nakayama T, Nakao A, et al. Physical and functional interaction of murine and Xenopus Smad7 with bone morphogenetic protein receptors and transforming growth factor-beta receptors. J Biol Chem. 1998;273:25364–70.
144. Blank U, Karlsson G, Moody JL, et al. Smad7 promotes self-renewal of hematopoietic stem cells. Blood. 2006;108:4246–54.
145. Basu S, Broxmeyer HE. Transforming growth factor-{beta}1 modulates responses of CD34+ cord blood cells to stromal cell-derived factor-1/CXCL12. Blood. 2005;106:485–93.
146. Fan X, Valdimarsdottir G, Larsson J, et al. Transient disruption of autocrine TGF-beta signaling leads to enhanced survival and proliferation potential in single primitive human hemopoietic progenitor cells. J Immunol. 2002;168:755–62.
147. Larsson J, Blank U, Helgadottir H, et al. TGF-beta signaling-deficient hematopoietic stem cells have normal self-renewal and regenerative ability in vivo despite increased proliferative capacity in vitro. Blood. 2003;102:3129–35.
148. Langer JC, Henckaerts E, Orenstein J, Snoeck HW. Quantitative trait analysis reveals transforming growth factor-beta2 as a positive regulator of early hematopoietic progenitor and stem cell function. J Exp Med. 2004;199:5–14.
149. He W, Dorn DC, Erdjument-Bromage H, et al. Hematopoiesis controlled by distinct TIF1gamma and Smad4 branches of the TGFbeta pathway. Cell. 2006;125:929–41.
150. Ransom DG, Bahary N, Niss K, et al. The zebrafish moonshine gene encodes transcriptional intermediary factor 1gamma, an essential regulator of hematopoiesis, PLoS Biol. 2004;2: E237.

151. Karlsson G, Blank U, Moody JL, et al. Smad4 is critical for self-renewal of hematopoietic stem cells. J Exp Med. 2007;204: 467–74.
152. Sun Y, Ding L, Zhang H, et al. Potentiation of Smad-mediated transcriptional activation by the RNA-binding protein RBPMS. Nucleic Acids Res. 2006;34:6314–26.
153. Ramalho-Santos M, Yoon S, Matsuzaki Y, Mulligan RC, Melton DA. "Stemness": transcriptional profiling of embryonic and adult stem cells. Science. 2002;298:597–600.
154. Georgantas RW 3rd, Tanadve V, Malehorn M, et al. Microarray and serial analysis of gene expression analyses identify known and novel transcripts overexpressed in hematopoietic stem cells. Cancer Res. 2004;64:4434–41.
155. Wagner W, Ansorge A, Wirkner U, et al. Molecular evidence for stem cell function of the slow-dividing fraction among human hematopoietic progenitor cells by genome-wide analysis. Blood. 2004;104:675–86.
156. Ng YY, van Kessel B, Lokhorst HM, et al. Gene-expression profiling of CD34+ cells from various hematopoietic stem-cell sources reveals functional differences in stem-cell activity. J Leukoc Biol. 2004;75:314–23.
157. Toren A, Bielorai B, Jacob-Hirsch J, et al. CD133-positive hematopoietic stem cell "stemness" genes contain many genes mutated or abnormally expressed in leukemia. Stem Cells. 2005;23:1142–53.
158. Jaatinen T, Hemmoranta H, Hautaniemi S, et al. Global gene expression profile of human cord blood-derived CD133+ cells. Stem Cells. 2006;24:631–41.
159. Winnier G, Blessing M, Labosky PA, Hogan BL. Bone morphogenetic protein-4 is required for mesoderm formation and patterning in the mouse. Genes Dev. 1995;9:2105–16.
160. Bhatia M, Bonnet D, Wu D, et al. Bone morphogenetic proteins regulate the developmental program of human hematopoietic stem cells. J Exp Med. 1999;189:1139–48.
161. Choong ML, Tan AC, Luo B, Lodish HF. A novel role for proliferin-2 in the *ex vivo* expansion of hematopoietic stem cells. FEBS Lett. 2003;550:155–62.
162. Ueno H, Sakita-Ishikawa M, Morikawa Y, et al. A stromal cell-derived membrane protein that supports hematopoietic stem cells. Nat Immunol. 2003;4:457–63.
163. Moore KA, Pytowski B, Witte L, Hicklin D, Lemischka IR. Hematopoietic activity of a stromal cell transmembrane protein containing epidermal growth factor-like repeat motifs. Proc Natl Acad Sci U S A. 1997;94:4011–6.
164. Nolta JA, Thiemann FT, Arakawa-Hoyt J, et al. The AFT024 stromal cell line supports long-term ex vivo maintenance of engrafting multipotent human hematopoietic progenitors. Leukemia. 2002;16:352–61.
165. Hutton JF, Rozenkov V, Khor FS, D'Andrea RJ, Lewis ID. Bone morphogenetic protein 4 contributes to the maintenance of primitive cord blood hematopoietic progenitors in an ex vivo stroma-noncontact co-culture system. Stem Cells Dev. 2006;15:805–13.
166. Bond HM, Mesuraca M, Carbone E, et al. Early hematopoietic zinc finger protein (EHZF), the human homologue for Evi3 is highly expressed in primitive human hematopoietic cells. Blood. 2003;103:2062–70.
167. Bond HM, Mesuraca M, Amodio N, et al. Early hematopoietic zinc finger protein-zinc finger protein 521: A candidate regulator of diverse immature cells. Int J Biochem Cell Biol. 2008;40: 848–54.
168. Uhmann A, Dittmann K, Nitzki F, et al. The Hedgehog receptor Patched controls lymphoid lineage commitment. Blood. 2007;110:1814–23.
169. Trowbridge JJ, Scott MP, Bhatia M. Hedgehog modulates cell cycle regulators in stem cells to control hematopoietic regeneration. Proc Natl Acad Sci U S A. 2006;103:14134–9.
170. Kobune M, Ito Y, Kawano Y, et al. Indian hedgehog gene transfer augments hematopoietic support of human stromal cells including NOD/SCID-β2m-/- repopulating cells. Blood. 2004;104: 1002–9.
171. Di Marcotullio L, Ferretti E, Greco A, et al. Numb is a suppressor of Hedgehog signalling and targets Gli1 for Itch-dependent ubiquitination. Nat Cell Biol. 2006;8:1415–23.
172. Reya T, Cleavers H. Wnt signalling in stem cells and cancer. Nature. 2005;434:843–50.
173. Nemeth MJ, Bodine DM. Regulation of hematopoiesis and the hematopoietic stem cell niche by Wnt signaling pathways. Cell Res. 2007;17:746–58.
174. Van Den Berg DJ, Sharma AK, Bruno E, Hoffman R. Role of members of the Wnt gene family in human hematopoiesis. Blood. 1998;92:3189–202.
175. Austin TW, Solar GP, Ziegler FC, Liem L, Matthews W. A role for the Wnt gene family in hematopoiesis: expansion of multilineage progenitor cells. Blood. 1997;89:3624–35.
176. Murdoch B, Chadwick K, Martin M, et al. Wnt-5A augments repopulating capacity and primitive hematopoietic development of human blood stem cells in vivo. Proc Natl Acad Sci U S A. 2004;100:3422–7.
177. Willert K, Brown JD, Danenberg E, et al. Wnt proteins are lipid-modified and can act as stem cell growth factors. Nature. 2003;423:448–52.
178. Reya T, Duncan AW, Ailles L, et al. A role for Wnt signalling in self-renewal of haematopoietic stem cells. Nature. 2003;423: 409–14.
179. Duncan AW, Rattis FM, DiMascio LN, et al. Integration of Notch and Wnt signaling in hematopoietic stem cell maintenance. Nat Immunol. 2005;6:314–22.
180. Trowbridge JJ, Xenocostas A, Moon RT, Bhatia M. Glycogen synthase kinase-3 is an in vivo regulator of hematopoietic stem cell repopulation. Nat Med. 2006;12:89–98.
181. Cobas M, Wilson A, Ernst B, et al. Beta-catenin is dispensable for hematopoiesis and lymphopoiesis. J Exp Med. 2004;199:221–9.
182. Scheller M, Huelsken J, Rosenbauer F, et al. Hematopoietic stem cell and multilineage defects generated by constitutive beta-catenin activation. Nat Immunol. 2006;7:1037–47.
183. Kirstetter P, Anderson K, Porse BT, Jacobsen SE, Nerlov C. Activation of the canonical Wnt pathway leads to loss of hematopoietic stem cell repopulation and multilineage differentiation block. Nat Immunol. 2006;7:1048–56.
184. Koch U, Wilson A, Cobas M, Kemler R, Macdonald HR, Radtke F. Simultaneous loss of beta- and {gamma}-catenin does not perturb hematopoiesis or lymphopoiesis. Blood. 2008;111:160–4.
185. Burns CE, Traver D, Mayhall E, Shepard JL, Zon LI. Hematopoietic stem cell fate is established by the Notch-Runx pathway. Gens Dev. 2005;19:2331–42.
186. Varnum-Finney B, Brashem-Stein C, Bernstein ID. Combined effects of Notch signaling and cytokines induce a multiple log increase in precursors with lymphoid and myeloid reconstituting ability. Blood. 2003;101:1784–9.
187. Ye Q, Shieh J-H, Morrone G, Moore MAS. Expression of constitutively active Notch4 (Int-3) modulates myeloid proliferation and differentiation and promotes expansion of hematopoietic progenitors. Leukemia. 2004;18:777–87.
188. Delaney C, Varnum-Finney B, Aoyama K, Brashem-Stein C, Bernstein ID. Dose-dependent effects of the Notch ligand Delta1 on ex vivo differentiation and in vivo marrow repopulating ability of cord blood cells. Blood. 2005;106:2693–9.
189. Karanu FN, Murdoch B, Miyabayashi T, et al. Human homologues of Delta-1 and Delta-4 function as mitogenic regulators of primitive human hematopoietic cells. Blood. 2001;97: 1960–7.

190. Karanu FN, Murdoch B, Gallacher L, et al. The notch ligand jagged-1 represents a novel growth factor of human hematopoietic stem cells. J Exp Med. 2000;192:1365–72.
191. Kertesz Z, Vas V, Kiss J, et al. In vitro expansion of long-term repopulating hematopoietic stem cells in the presence of immobilized Jagged-1 and early acting cytokines. Cell Biol Int. 2006;30:401–5.
192. Chadwick N, Nostro MC, Baron M, et al. Notch signaling induces apoptosis in primary human CD34+ hematopoietic progenitor cells. Stem Cells. 2007;25:203–10.
193. Shojaei F, Trowbridge J, Gallacher L, et al. Hierarchical and ontogenic positions serve to define the molecular basis of human hematopoietic stem cell behavior. Dev Cell. 2005;8:651–63.
194. Mancini SJ, Mantei N, Dumortier A, et al. Jagged1-dependent Notch signaling is dispensable for hematopoietic stem cell self-renewal and differentiation. Blood. 2005;105:2340–2.
195. Wilson A, Ardiet DL, Saner C, et al. Normal hemopoiesis and lymphopoiesis in the combined absence of numb and numblike. J Immunol. 2007;178:6746–51.
196. Chung YJ, Park BB, Kang YJ, et al. Unique effects of STAT3 on the early phase of hematopoietic stem cell regeneration. Blood. 2006;108:1208–15.
197. Lund AH, van Lohuizen M. Polycomb complexes and silencing mechanisms. Curr Opin Cell Biol. 2004;16:239–46.
198. Kajiume T, Ninomiya Y, Ishihara H, Kanno R, Kanno M. Polycomb group gene mel-18 modulates the self-renewal activity and cell cycle status of hematopoietic stem cells. Exp Hematol. 2004;32:571–8.
199. Ohta H, Sawada A, Kim JY, et al.Polycomb group gene rae28 is required for sustaining activity of hematopoietic stem cells. J Exp Med. 2002;195:759–70.
200. Kamminga LM, Bystrykh LV, de Boer A, et al. The Polycomb group gene Ezh2 prevents hematopoietic stem cell exhaustion. Blood. 2006;107:2170–9.
201. Lessard J, Sauvageau G. Bmi-1 determines the proliferative capacity of normal and leukaemic stem cells. Nature. 2003;423:255–60.
202. Hosen N, Yamane T, Muijtjens M, et al. Bmi-1-green fluorescent protein-knock-in mice reveal the dynamic regulation of bmi-1 expression in normal and leukemic hematopoietic cells. Stem Cells. 2007;25:1635–44.
203. Rizo A, Dontje B, Vellenga E, de Haan G and Schuringa JJ. Long-term maintenance of human hematopoietic stem/progenitor cells by expression of BMI1. Blood. 2008;111:2621–30.
204. Iwama A, Oguro H, Negishi M, et al. Enhanced self-renewal of hematopoietic stem cells mediated by the polycomb gene product Bmi-1. Immunity. 2004;21:843–51.
205. Chagraoui J, Niessen SL, Lessard J, et al. E4F1: a novel candidate factor for mediating BMI1 function in primitive hematopoietic cells. Genes Dev. 2006;20:2110–20.
206. Shivdasani RA, Mayer EL, Orkin SH. Absence of blood formation in mice lacking the T cell leukemia oncoprotein tal-1/SCL. Nature. 1995;373:432–4.
207. Perry SS, Zhao Y, Nie L, et al. Id1, but not Id3, directs long-term repopulating hematopoietic stem-cell maintenance. Blood. 2007;110:2351–60.
208. Mikkola HK, Klintman J, Yang H, et al. Haematopoietic stem cells retain long-term repopulating activity and multipotency in the absence of stem-cell leukaemia SCL/tal-1 gene. Nature. 2003;421:547–51.
209. Curtis DJ, Hall MA, Van Stekelenburg LJ, et al. SCL is required for normal function of short-term repopulating hematopoietic stem cells. Blood. 2004;103:3342–8.
210. Hock H, Hamblen MJ, Rooke HM et al. Gfi-1 restricts proliferation and preserves functional integrity of haematopoietic stem cells. Nature. 2004;431:1002–7.
211. Zeng H, Y`ucel R, Kosan C, Klein-Hitpass L, M´oröy T. Transcription factor Gfi1 regulates self-renewal and engraftment of hematopoietic stem cells. EMBO J. 2004;23:4116–25.
212. Gupta R, Hong D, Iborra F, Sarno S, Enver T. NOV (CCN3) functions as a regulator of human hematopoietic stem or progenitor cells. Science. 2007;316:590–3.
213. Rebel VI, Kung AL, Tanner EA, Yang H, Bronson RT, Livingston DM. Distinct roles for CREB-binding protein and p300 in hematopoietic stem cell self-renewal. Proc Natl Acad Sci U S A. 2002;99:14789–94.
214. Tsai FY, Orkin SH. Transcription factor GATA-2 is required for proliferation/survival of early hematopoietic cells and mast cell formation, but not for erythroid and myeloid terminal differentiation. Blood. 1997;89:3636–43.
215. Chen Y, Haviernik P, Bunting KD, Yang YC. Cited2 is required for normal hematopoiesis in the murine fetal liver. Blood. 2007;110:2889–98.
216. Crable SC, Anderson KP. A PAR domain transcription factor is involved in the expression from a hematopoietic-specific promoter for the human LMO2 gene. Blood. 2003;101:4757–64.
217. Magnusson M, Brun AC, Lawrence HJ, Karlsson S. Hoxa9/hoxb3/hoxb4 compound null mice display severe hematopoietic defects. Exp Hematol. 2007;35:1421–8.
218. Moore MA, Chung KY, Plasilova M, et al. NUP98 dysregulation in myeloid leukemogenesis. Ann N Y Acad Sci. 2007;1106: 114–42.
219. Argiropoulos B, Humphries RK. Hox genes in hematopoiesis and leukemogenesis. Oncogene. 2007;26:6766–76.
220. Hu YL, Passegué E, Fong S, Largman C, Lawrence HJ. Evidence that the Pim1 kinase gene is a direct target of HOXA9. Blood. 2007;109:4732–8.
221. Magnusson M, Brun AC, Miyake N, et al. HOXA10 is a critical regulator for hematopoietic stem cells and erythroid/megakaryocyte development. Blood. 2007;109:3687–96.
222. Kirito K, Fox N, Kaushansky K. Thrombopoietin stimulates HoxB4 expression: An explanation for the favorable effects of Tpo on hematopoietic stem cells. Blood. 2003;102:3172–80.
223. Schiedlmeier B, Klump H, Will E, et al. High-level ectopic HOXB4 expression confers a profound in vivo competitive growth advantage on human cord blood CD34+ cells, but impairs lymphomyeloid differentiation. Blood. 2003;101:1759–68.
224. Brun AC, Fan X, Björnsson JM, Humphries RK, Karlsson S. Enforced adenoviral vector-mediated expression of HOXB4 in human umbilical cord blood CD34+ cells promotes myeloid differentiation but not proliferation. Mol Ther. 2003;8: 618–28.
225. Beslu N, Krosl J, Laurin M, et al. Molecular interactions involved in HOXB4-induced activation of HSC self-renewal. Blood. 2004;104:2307–14.
226. Krosl J, Beslu N, Mayotte N, Humphries RK, Sauvageau G. The competitive nature of HOXB4-transduced HSC is limited by PBX1: the generation of ultra-competitive stem cells retaining full differentiation potential. Immunity. 2003;18:561–71.
227. Amsellem S, Pflumio F, Bardinet D, et al. Ex vivo expansion of human hematopoietic stem cells by direct delivery of the HOXB4 homeoprotein. Nature Med. 2003;9:1423–7.
228. Zhang XB, Beard BC, Beebe K, et al. Differential effects of HOXB4 on nonhuman primate short- and long-term repopulating cells. PLoS Med. 2006;3:e173.
229. Zhang XB, Schwartz JL, Humphries RK, Kiem HP. Effects of HOXB4 overexpression on ex vivo expansion and immortalization of hematopoietic cells from different species. Stem Cells. 2007;25:2074–81.
230. Zhang Y, Morrone G, Zhang J, et al. CUL-4A stimulates ubiquitylation and degradation of the HOXA9 homeodomain protein. EMBO J. 2003;22:6057–67.

231. Chung KY, Morrone G, Schuringa JJ, et al. Enforced expression of NUP98-HOXA9 in human CD34+ cells provides a proliferative advantage and enhances stem cell self-renewal. Cancer Res. 2006;66:11781–91.
232. Davidson AJ, Ernst P, Wang Y, et al. *Cdx4* mutants fail to specify blood progenitors and can be rescued by multiple *hox* genes. Nature. 2003;425:300–6.
233. Wang Y, Yates F, Naveiras O, Ernst P, Daley GQ. Embryonic stem cell-derived hematopoietic stem cells. Proc Natl Acad Sci U S A. 2005;102:19081–6.
234. Lengerke C, McKinney-Freeman S, Naveiras O, et al. The cdx-hox pathway in hematopoietic stem cell formation from embryonic stem cells. Ann N Y Acad Sci. 2007;1106:97–208.
235. Zhu J, Giannola DM, Zhang, Y, Rivera AJ, Emerson SG. NF-Y cooperates with USF1/2 to induce the hematopoietic expression of HOXB4. Blood. 2003;102:2420–7.
236. Zhu J, Zhang Y, Joe GJ. Pompetti R, Emerson SG. NF-Ya activates multiple hematopoietic stem cell (HSC) regulatory genes and promotes HSC self-renewal. Proc Natl Acad Sci U S A. 2005;102:11728–33.
237. Hock H, Meade E, Medeiros S, et al. Tel/Etv6 is an essential and selective regulator of adult hematopoietic stem cell survival. Genes and Dev. 2004;18:2336–41.
238. Galan-Caridad JM, Harel S, Arenzana TL, et al. Zfx controls the self-renewal of embryonic and hematopoietic stem cells. Cell. 2007;129:356–7.
239. Davey RE, Zandstra PW. Signal processing underlying extrinsic control of stem cell fate. Curr Opin Hematol. 2004;11: 95–101.
240. Ross J, Li L. Recent advances in understanding extrinsic control of hematopoietic stem cell fate. Curr Opin Hematol. 2006;13:237–42.
241. Blank U, Karlsson G, Karlsson S. Signaling pathways governing stem cell fate. Blood. 2008;111:492–503.
242. Wilson A, Trumpp A. Bone-marrow haematopoietic-stem-cell niches. Nat Rev Immunol. 2006;6:93–106.

Renal Stem Cells and Kidney Regeneration

Takashi Yokoo, Akira Fukui, Kei Matsumoto and Tetsuya Kawamura

Abstract Significant advances have been made in stem cell research over the past decade. A number of non-hematopoietic sources of stem cells (or progenitor cells) have been identified including endothelial stem cells and neural stem cells. These discoveries have been a major step towards the potential regeneration of organs for clinical applications using stem cells. The worldwide shortage of donor kidneys means that this approach has garnered significant attention in the field of nephrology. Here, we review recent findings on renal stem cells and their possible therapeutic application for renal diseases.

Keywords Kidney regeneration · Mesenchymal stem cell · Renal failure · Xenobiology · Metanephros · Embryo

1 Introduction

Recent advances in stem cell research have brought the possibility of organ regeneration using somatic stem cells for clinical organ replacement one step closer to realization. However, anatomically complicated organs, such as the kidney and liver, have proven more refractory to stem cell–based regenerative techniques.

The kidney retains the potential to regenerate itself if the damage is not too severe and the kidney structure remains intact. Therefore, regenerative medicine for kidney diseases should aim to activate or support such potential. However, in cases of irreversible damage to the kidney, as can occur with long-term dialysis, the self-renewal function is lost totally. Thus, any application of regenerative medicine in chronic renal disease will require the establishment of an entire functional kidney de novo.

Acute renal failure (ARF) causes both apoptosis and necrosis of renal tubular epithelial cells. Over time, the injured tubules regenerate through cell proliferation, although the source of the cells that repopulate the injured nephron is not clear [1]. Tissue regeneration may result from the proliferation of surviving dedifferentiated cells, from renal stem cells that reside inside the kidney and migrate to the site of regeneration, or from bone marrow cells that gain access to the injured epithelium and differentiate into mature cells [2]. It is now believed that all solid organs possess both endogenous and exogenous stem cells. After tissue injury, intrinsic tissue stem cells replace damaged tissue as a first line of defense. If the pool of endogenous stem cells is exhausted, exogenous circulating stem cells are signaled to replenish the pool and participate in tissue repair as a backup rescue system [3]. Therefore, the aim of regenerative medicine for ARF is to identify renal stem cells that can be supplied exogenously when the kidney is injured and to identify key molecules that can "reprogram" quiescent tissue stem cells to participate in renal repair. These aims for the treatment of ARF are being addressed using established technologies in regenerative medicine that are currently used for other organs, such as the heart [4] and vessels [5].

Chronic renal failure (CRF) is the second type of renal failure to be considered in regenerative medicine. Most patients with CRF entering dialysis programs have type 2 diabetes, chronic glomerulonephritis, or hypertension [6]. The number of CRF patients requiring dialysis has increased markedly worldwide, mainly due to the significantly extended acceptance criteria for dialysis, which now include more elderly and diabetic patients, as well as those with other severe comorbidities [7]. In addition, long-term replacement therapy with hemo- or peritoneal-dialysis has markedly improved the prognosis for CRF patients. Current trends in maintenance dialysis population dynamics show an estimated annual worldwide cost of maintenance CRF therapy at close to US$75 billion. The size of this global maintenance dialysis population is expanding at a rate of

T. Yokoo (✉)
Project Team for Kidney Regeneration, Institute of DNA Medicine, Division of Nephrology and Hypertension, Department of Internal Medicine, The Jikei University School of Medicine, 3-25-8, Nishi-Shimbashi, Minato-ku, Tokyo, 105-8461 Japan,
e-mail: tyokoo@jikei.ac.jp

V.K. Rajasekhar, M.C. Vemuri (eds.), *Regulatory Networks in Stem Cells*, Stem Cell Biology and Regenerative Medicine,
DOI 10.1007/978-1-60327-227-8_31, © Humana Press, a part of Springer Science+Business Media, LLC 2009

7% per year [7]. If this current trend continues, the dialysis population will exceed 2 million patients by the year 2010 and the aggregate cost will be more than US$1 trillion over the coming decade [8]. This will render dialysis impractical in the near future as a therapeutic choice for CRF patients. Although kidneys may be transplanted successfully, the lack of suitable transplantable organs has prevented kidney transplantation from becoming a practical solution for most cases of CRF. Thus, there is a need for a new type of therapy through which patients with CRF may discontinue dialysis and, in this regard, kidney regeneration has considerable potential. However, the kidney is anatomically complicated and resident cells must communicate with each other to function. Therefore, a regenerated, whole, therapeutic kidney must contain fully organized and orchestrated cells able to fulfill their function. Thus, unlike for the treatment of ARF, regenerative medicine for CRF requires a novel approach to build a functional, whole kidney de novo. Current attempts to regenerate kidneys are based on the use of foreign human mesenchymal stem cells and our background knowledge of kidney organogenesis.

The present article reviews the challenges and recent advances in renal stem cell research and discusses its potential for clinical application in the treatment of kidney diseases.

2 Renal Stem Cells in Bone Marrow

It has been reported that, in male patients who receive kidney transplants from female donors, Y chromosome-positive tubular cells are observed in the kidneys [9] and approximately 1% of tubular cells are Y chromosome-positive after the kidneys recover from acute tubular necrosis [10]. Bone marrow stem cells can contribute to the formation of kidney cells, including mesangial cells [11, 12], tubular epithelial cells, and podocytes [13], which gives rise to the hypothesis that some renal stem cells are resident in and mobilized from the bone marrow. Therefore, many researchers initially tried to identify renal stem cells from extrarenal sources, within the bone marrow or circulation, and many experiments were performed using bone marrow transplantation of marked donor cells to trace their progeny. Lin et al. [14] were the first to report that bone marrow stem cells ($Rh^{low}Lin^{-}Sca\text{-}1^{+}c\text{-}kit^{+}$cells) isolated from male Rosa26 mice, which ubiquitously express the *LacZ* gene, may differentiate into renal proximal tubular cells and contribute to renal tubular regeneration when transplanted into female mice with renal ischemia/reperfusion injury. Kale et al. [15] subsequently reported the therapeutic potential of bone marrow stem cell infusion for the treatment of ARF. They induced ischemia/reperfusion injury in mice with and without bone marrow ablation to control for the contribution of bone marrow–derived cells. Renal damage in mice with bone marrow ablation was much worse than that observed in normal mice and this exacerbation was reversed by stem cell transfusion. That report provided the conceptual basis for the development of therapeutic strategies involving exogenous renal stem cells to enhance recovery from ARF. Since then, many papers have reported using different bone marrow fractions or different experimental models to investigate these potential therapies. The studies generally involve the transplantation of bone marrow cells marked with *LacZ*, enhanced green fluorescent protein (EGFP), or a genetic marker (Y chromosome) and their detection after the induction of renal damage. The progeny of these donor cells are detected using X-gal staining, fluorescence microscopy, or fluorescent in situ hybridization for the Y chromosome, respectively. Renal tubular cells bearing these markers were detected, indicating that some fraction of the transplanted bone marrow cells (i.e., bulk fraction, hematopoietic stem cells, and/or mesenchymal stroma) contributed to the renal repair following experimental ARF. The percentage of stem cells incorporated varies widely, but it is usually below 1% in a given organ; the magnitude of cells incorporated depends on the disease model studied [16]. This raises the possibility for the therapeutic infusion of stem cells fused with residential cells [17], with the effects mediated by immunomodulation [18] or paracrine mechanisms elicited through trophic mediators [19]. In fact, Kunter et al. [20] reported that marked acceleration of renal recovery following the infusion of low numbers of cells is related to paracrine growth factors, including vascular endothelial growth factor (VEGF) and transforming growth factor (TGF)-β1, and not to differentiation into resident renal cells or monocytes/macrophages. More recently, Duffield et al. [21] demonstrated that all the detection systems used in these earlier studies, although well established, were capable of producing false-positive results, which could overestimate the regenerative contribution of bone marrow-derived cells to renal repair.

3 Renal Stem Cells in the Adult Kidney

Currently, there are three ways in which tissue stem cells can be isolated, based on work with other solid organs. Most conventional methods use cell surface markers that can be expressed in tissue stem cells. CD133 expression was originally shown in hematopoietic stem and progenitor cells [22], but this marker is also expressed by stem cells of other tissues, such as vessels [23] and neurons [24], leading to speculation that CD133 is a universal cell surface marker of tissue stem cells. Recently, CD133-positive cells were identified in human adult kidney [25], in the interstitium of the renal cortex. These human CD133-positive kidney cells differentiated into renal tissue in vivo when injected into immunocompromised SCID mice. In addition, intravenously injected CD133-positive cells appeared to integrate with cells of kidney tubules that had been treated with glycerol to induce injury. These data implicate CD133-positive cells as markers of intrinsic renal stem cells [25].

Another proposed marker of renal stem cells is Nestin, which is a marker of multilineage stem cells expressed in neuroepithelial stem cells [26]. $Nestin^+$ cells are localized in large clusters within the papilla and, less prominently, in the glomeruli and juxtaglomerular arterioles in mice. Following ischemic insult, $Nestin^+$ cells migrate to cortices at a rate of 40 μm/30 min during the first 3 h [27], suggesting they are recruited to participate in the recovery from ischemia. Although it is not clear whether these $Nestin^+$ cells differentiate into mature resident cells or simply secrete renotrophic molecules at the site of injury, Nestin is another candidate marker to detect tissue stem cells in the adult kidney.

Dekel et al. [28] showed that nontubular $Sca\text{-}1^+Lin^-$ cells, which reside in the renal interstitial space, can be differentiated into myogenic, osteogenic, adipogenic, and neural linkages. After direct injection into the renal parenchyma, these cells may adopt a tubular phenotype following ischemic insult, suggesting that this population of cells may behave as tissue stem cells, contributing to the regeneration of injured kidneys.

Gene expression profiling of mesenchymal cells from embryonic kidney has been used to identify other potential cell surface markers of renal stem cells [29]. Challen et al. [29] found 21 genes that were selectively up-regulated in cells destined to differentiate into renal tissue. They highlighted CD24 and cadherin-11 as surface proteins that may be useful in the isolation of viable progenitor cells from the adult kidney. Cells expressing CD24 were incorporated into newly forming tubules, whereas cadherin-11 was expressed primarily on cells that formed the interstitium. Using CD133 and CD24, Sagrinati et al. [30] isolated multipotent progenitor cells with the ability to differentiate in vitro into proximal and distal tubules, osteogenic cells, adipocytes, and neuronal cells, as well as a subset of parietal epithelial cells (PEC) in Bowman's capsule of the adult human kidney. Intravenous injection of $CD24^+CD133^+$ PEC into SCID mice with glycerol-induced ARF may regenerate tubular structures in different portions of the nephron, thus reducing any associated morphological and functional damage in the kidney [30].

In a different approach that does not use cell surface markers, Kitamura et al. [31] attempted to establish renal progenitor cells using microdissection. Segments of nephron were cultured separately and, after simple limiting dilution, the cell line exhibiting the most potent growth was isolated. This cell line (rKS56) had the potential to differentiate into mature tubular cells in vitro, thus replacing injured tubules and improving renal function after implantation in vivo. It would be interesting to determine the cell surface marker profile of these cells with the aim of identifying a universal marker for renal stem cells.

Detection of side-population (SP) cells is another common method used to identify stem cells. This technique was first used to obtain an enriched population of hematopoietic cells from adult mouse bone marrow using Hoechst 33342 dye and fluorescence-activated cell sorting (FACS) [32], with cells negative for this staining deemed SP cells. This property of SP cells is due to the expression of efflux pumps belonging to the ATP-binding cassette superfamily of membrane transporters [33] and confers a survival advantage. Therefore, the SP phenotype can be used to purify a stem cell-rich fraction. Although the original study isolated a population of uncommitted hematopoietic stem cells, recent studies have shown that SP cells may populate other organs [34] while maintaining their potential as tissue stem cells [35]. To identify renal stem cells using this technique, Hishikawa et al. [36] isolated kidney SP cells from two congenital mouse models of renal failure and matched controls. Microarray analysis revealed the gene *Musculin/MyoR*, which is mainly expressed during muscle development [37], to be highly expressed in SP cells, suggesting that it may be used as a marker of renal stem cells. The musculin/MyoR-positive cells were localized in the interstitial space of the kidney and systemic injection of these SP cells demonstrated therapeutic potential in the cisplatin-induced ARF model. Although SP cells may, indeed, be renal stem cells, it is possible that the therapeutic effect observed was not a direct effect due to the differentiation of SP cells and their integration into injured tubule cells. Rather, an indirect paracrine effect on the growth of surrounding cells via hepatocyte growth factor (HGF), VEGF, and bone morphogenetic protein (BMP) 7 is more likely, because the number of redistributed SP cells after systemic injection was too small to account for a direct effect.

Another way to identify tissue stem cells is to use the DNA marker bromodeoxyuridine (BrdU), based on the assumption that tissue stem cells cycle very slowly and differentiate only as demanded by tissue turnover. This technique has been used to identify slow-cycling stem cells in other organs, including the skin [38], intestine [39], and lung [40]. However, these tissues all have a rapid cell turnover and it was thought that this approach may prove less successful in slow-turnover tissues, such as the kidney, because many resident renal cells such as podocytes divide slowly and therefore may be falsely identified as stem cells. Maeshima et al. [41] did report some success in identifying stem cells in renal tissue by injecting BrdU intraperitoneally into adult rats once a day for 7 days and inducing ischemia/reperfusion injury after 14 days. BrdU-positive cells were observed in the tubules, but not in the glomeruli and capillary vessels. Quantitative analysis showed a twofold increase in the number of

BrdU-positive cells after reperfusion, suggesting that most proliferating cells in the recovering kidney after renal ischemia were derived from BrdU-positive renal progenitor cells.

More recently, Oliver et al. [41] injected 3-day-old rats subcutaneously twice a day with BrdU for 3.5 days and then, after 2 months, localized BrdU-positive cells. During this stage of rat development, the kidney is still growing and resident cells are proliferating; thus, slow-cycling cells should be more easily distinguished from other cells. Surprisingly, there were numerous BrdU-positive cells in the renal papilla and only small numbers in the outer cortex, mid-cortex, and medulla, where they localized mainly within the interstitial area. The BrdU-positive cells were FACS sorted; they developed epithelial characteristics in vitro and, in vivo, were shown to migrate and incorporate into mature tubules. Following a transient episode of ischemia, the BrdU-positive cells quickly entered the cell cycle and disappeared from the papilla, implicating these cells in renal repair [42]. These results suggest that the renal papilla contains a population of adult kidney stem cells involved in kidney maintenance and repair, although the signals and access pathway involved remain unclear.

4 Renal Stem Cell Niche

Recent progress in stem cell biology has demonstrated that renal stem cells, with the capability to differentiate into mature renal cells, do exist in adult individual; however, the debate is ongoing regarding their niche. Suggestions include the interstitium of the cortex [25, 36] and papilla [42], tubules [27], and bone marrow [14, 15]. Recent evidence suggested that liver progenitor cells, called oval cells, which transdifferentiate into hepatocytes or biliary epithelial cells, may originate not only from the point where the terminal bile ducts meet the periportal hepatocytes, but also from the bone marrow [43]. It is equally likely that other stem cells, including those in the kidney, are not restricted to one place and may be supplied from different places depending on the severity, location, and duration of damage. However, resident tissue stem cells constitute only a small percentage of the total cellularity of an organ [44], suggesting that tissue-specific stem cells are not sufficient in number for therapeutic regeneration after tissue injury and must either be expanded in vivo or supplied on demand from the circulation. A recent study by Lin et al. [45] demonstrated definitive findings to cement the current consensus. These authors established chimera mice in which mature renal tubular epithelial cells and their progeny are permanently labeled with EGFP. Following ischemia/reperfusion injury in the mice, EGFP-positive cells showed incorporation of BrdU and expression of the dedifferentiation markers vimentin and Pax2. Furthermore, these cells began to express aquaporin (AQP)-3 at the basolateral membrane and ZO-1 in the intercellular junction, showing that dedifferentiated intrinsic tubular cells were redifferentiated into mature tubular cells. These data provided the direct evidence that regenerating tubule cells are derived from renal tubular epithelial cells. To address the relative contribution of intrinsic versus exogenous populations of cells to renal repair after ischemia/reperfusion injury, quantitative analyses of regenerating BrdU-positive cells and bone marrow-derived Y-positive cells revealed that 89% of the proliferating epithelial cells originated from the host cells and the remaining 11% originated from donor bone marrow cells [45]. In agreement with the earlier work of Duffield et al. [21], these data demonstrated that extrarenal bone marrow-derived cells can be incorporated into renal tubules after ischemic injury, but that intrarenal cells are the major source of tubular regeneration. Future studies need to provide a better understanding of what controls the contribution of renal stem cells in a given pathophysiological setting if they are to be applied therapeutically in the treatment of human disease.

5 Other Stem Cell Sources for Kidney Regeneration

Adult kidney stem cells are not necessarily the only possible source for kidney regeneration. This section reviews the possibility of using other sources for renal stem cells.

5.1 Embryonic Stem Cells

Embryonic stem (ES) cells are undifferentiated pluripotent stem cells isolated from the inner cell mass of blastocysts [46]. ES cells have the capacity to differentiate into several cell types of mesodermal, endodermal, and ectodermal lineage, depending on culture conditions, and are a potential source of cells for tissue regeneration. The application of ES cells for regenerative medicine has been approved in many disease models, including Parkinson's disease and diabetes [47, 48]. Since human ES cells have the capacity to differentiate into kidney structures when injected into immunosuppressed mice [49, 50], studies have focused on identifying the precise culture conditions that allow the differentiation of ES cells into renal cells in vitro. Schuldiner et al. showed that human ES cells cultured with eight growth factors, including HGF and activin A, differentiated into cells expressing WT-1 and renin [51]. More recently, it was reported that mouse ES cells stably transfected with Wnt4 (Wnt4-ES cells) differentiate into tubular-like structures expressing AQP-2 when cultured in the presence of HGF and activin A [52]. Using such in vitro techniques, it may eventually be possible to identify the key molecules [51] that determine the fate of ES cells, although it may be difficult to establish a whole, functional kidney in vitro for clinical use with this technology. An ex vivo culture system, in which ES cells (or ES-derived cells) were cultured in the developing metanephros, was investigated to determine the capacity of ES cells to differentiate into kidney cells integrated into the kidney structure. ROSA26 ES cells were stimulated with developmental signals in the microenvironment of a developing kidney following injection into a metanephros cultured in vitro. ES cell-derived, β-galactosidase-positive cells were identified in epithelial structures resembling tubules with an efficiency approaching 50% [53]. Based on these results, Kim and Dressler [54] attempted to identify the nephrogenic growth factors inducing the differentiation of ES cells into renal epithelial cells. When injected into a developing metanephros, ES cells treated with retinoic acid, activin A, and BMP7 contributed to tubular epithelia with near 100% efficiency [54]. Furthermore, Vignearu et al. [55] showed that ES cells expressing brachyury, which denotes mesoderm specification, may become a renal progenitor population in the presence of activin A. After injection into a developing metanephros, these cells may be incorporated into the blastimal cells of the nephrogenic zone. In addition, after single injection into developing live newborn mouse kidneys, these cells stably integrated into proximal tubules with normal morphology and polarization for 7 months without teratoma formation [55]. Taken together, these data highlight ES cells as a potential source of renal stem cells for regenerative therapy.

5.2 Induced Pluripotent Stem Cells

Major obstacles to using ES cells include: (i) the use of donated eggs; and (ii) the possible immune response to non-self cells. Therefore, an ideal cell source may be cells with all the properties of ES cells, but derived from an adult source, such as the skin. The first attempt to make such patient-specific stem cells used somatic cell nuclear transfer or cloning. This involved reprogramming DNA from an adult cell by transplanting it into the cytoplasmic environment of an unfertilized egg [56] or, more recently, of a newly fertilized egg [57]. ES cells derived through nuclear transfer have been generated in mice, albeit at fairly low rate. The first reported successful cloning in humans by Hwang et al. [58] was probably a parthenogenetic ES cell derived from a blastocyst; however, later claims by this group were found to be fraudulent. Theoretically, it should be possible to produce pluripotent stem cells from an adult human source and, indeed, non-human primate pluripotent stem cells have been produced from adult skin fibroblasts [59]. However, as yet there are no publications reporting successful development of a human ES cell line using the nuclear transfer technique. In this context, Takahashi and Yamanaka [60] reported that pluripotent ES-like cells can be produced from cultured somatic cells by retroviral transfer with *Ocr3/4*, *Sox2*, *c-Myc*, and *Klf4*, which are transcription factors associated with pluripotency. These cells were termed induced pluripotent stem (iPS) cells [60]. The rate-reprogrammed iPS cells can be selected by the reactivation of Fbx15 [61], Oct4 [61], or Nanog [62], all of which carry a drug-resistance marker inserted into the respective endogenous locus by homologous recombination or a transgene containing the Nanog promoter. Furthermore, iPS cells can be isolated on the basis of their morphology, resembling ES-like cells, without the use of transgenic donor cells [63]. The therapeutic potential of autologous iPS cells in a mouse model of a hereditary disease has been reported [64]. Recently, human iPS cells were successfully induced from adult skin fibroblasts using the same four factors [65]. iPS cells are epigenetically and biologically indistinguishable from normal ES cells and, therefore, may be another source of patient-specific renal stem cells for regenerative therapy.

5.3 Mesenchymal Stem Cells

Studies in mice monitoring the fate of mesenchymal stem cells (MSCs) or MSC-like populations after intravenous or intraperitoneal transplantation demonstrated the presence of donor cells in the bone marrow, spleen, bone, cartilage, and lung up to 5 months later [66, 67]. More recently, Liechty et al. [68] reported that human MSCs xenotransplanted intraperitoneally into sheep embryos are capable of engraftment in multiple tissues, including chondrocytes, adipocytes, myocytes, cardiomyocytes, bone marrow stromal cells, and thymic stroma. These findings suggest that MSCs have the capacity for site-specific differentiation into various tissue types. It has been reported that injection of MSCs may preserve renal function after the induction of various experimental models of renal failure [69]. However, Kunter et al. [70] reported recently that, following the intraglomerular injection of MSCs, these cells maldifferentiated into adipocytes in vivo, which offset the early beneficial effect of the MSCs in preserving damaged glomeruli and maintaining renal function. The glomerular microenvironment necessary for the differentiation of MSCs should be examined more precisely to enable cell therapy for ARF using MSCs.

In our experiments investigating generation of the neo-kidney, we used primary human (h) MSCs obtained from the bone marrow of healthy volunteers. hMSCs were shown recently to retain plasticity and the capacity to differentiate into several cells types depending on the microenvironment [71]. As discussed above, ES cells are ideal candidates as the source of cells for kidney regeneration [72]. Unlike ES cells, hMSCs injected into established metanephroi may not integrate into renal structures during organ culture, which is a distinct disadvantage in terms of the feasibility of forming functional renal structures. Our finding that hMSCs do not express WT1 or Pax2 (unpublished data) suggests that hMSCs do not possess a complete set of nephrogenic molecular features. An advantage of hMSCs over ES cells is the ease with which adult MSCs can be isolated from autologous bone marrow and used therapeutically without serious ethical issues or the requirement for immunosuppressants.

6 De Novo Establishment of a Whole Kidney from Stem Cells

Several groups have been working on building a kidney de novo, as a whole organ, as an absolute solution for kidney diseases.

Woolf et al. [73] reported that the metanephros may continue to grow if it is transplanted into the renal cortex of host mice. The developed transplant contains vascularized glomeruli and mature proximal tubules, and may have the capacity for glomerular filtration. Collecting duct-like structures appear to extend from the transplant towards the papilla of the host. Although there is no direct evidence that these collecting duct-like structures connect with the host's collecting system or that the transplant functions in a manner similar to native kidney, the results provide the rationale for the usefulness of the metanephros from early embryos as a potential source of transplantable regenerated kidney to address the shortage of organs for kidney transplantation. Potential problems with this system include questions as to the suitability of the renal capsule of dialysis patients as a transplant site, given the significant disruptions to this area, including to the vasculature, and the fact that space limitations beneath the renal capsule may hinder the growth of transplants. These concerns may be overcome by the system established by Rogers et al. [74], who also used the metanephros as a source of transplantable artificial kidney, but transplanted the graft into a host omentum, which is not confined by a tight organ capsule and has not been disturbed by dialysis. Metanephroi from rat, mouse, and pig were implanted into the omentum of the rat or mouse and the success of transplanting across xenogeneic barriers and the extent of differentiation into a functional nephron were evaluated. This experiment was based on previous studies showing minimal immunogenicity in tissues harvested at earlier gestational stages, including the metanephros [75]. In cases of allotransplantation (rat metanephros to rat omentum), transplants assumed a kidney-like shape in situ that was approximately one-third the diameter of the native kidney. Histologically, the transplants contained well-differentiated kidney structures. Importantly, this transplant technique can be performed without immunosuppression. With xenotransplantation, the pig metanephros grew and differentiated into renal tissue in the rat omentum, showing glomeruli, proximal tubules, and collecting ducts; however, immunosuppressants were required because, without these agents, the transplants disappeared soon after transplantation. Interestingly, the graft pig metanephros was slightly larger in volume (diameter and weight) than a normal rat kidney. Furthermore, the transplanted tissue produced urine and, surprisingly, after intact ureteroureterostomy with the ureter of the kidney that was removed, anephric rats started to void and showed a prolonged lifespan [76]. This success provides promise of a new and practical therapeutic strategy for CRF that establishes a functional renal unit by implanting xenometanephros together with immunosuppression.

In terms of a functional whole kidney, Chan et al. [77] reported the first attempt to establish a functional whole renal unit by developing a transplantable pronephros in *Xenopus*. *Xenopus* presumptive ectoderm, which becomes epidermis and neural tissue in normal development, contains pluripotent stem cells and can be differentiated into multilineage tissue cells under particular culture conditions [78]. Chan et al. [77] designed conditions for the induction of pronephric tubule-like structures from animal caps that involved a combination of activin and retinoic acid for only 3 h. This pronephros-like tissue was transplanted into bilaterally nephrectomized tadpoles to test for functional integrity as a pronephros. Bilateral pronephrectomy induces severe edema in tadpoles owing to the inability to excrete internal water, and tadpoles die within 9 days; transplantation of the pronephros-like unit at least partially corrected the edema and tadpoles survived for up to 1 month. To our knowledge, this study remains the only one to establish a transplantable functional whole kidney unit in vitro, although the pronephros structure formed is too primitive for any clinical application in humans.

To address the adverse effects of immunosuppressants, Lanza et al. [79] attempted to establish a self-kidney unit to eliminate the immune response problem and, therefore, the need for immunosuppression. To generate histocompatible kidney for artificial organ transplantation, Lanza et al. [79] used a nuclear transplantation technique, in which dermal fibroblasts isolated from adult cow were transferred into enucleated bovine oocytes and transferred nonsurgically into progestin-synchronized recipients. After 6–7 weeks, metanephroi were isolated from embryos, digested using collagenase, and expanded until the desired cell number was obtained by culture in vitro. These cells were then seeded on a specialized polymer tube, followed by implantation into the same cow from which the cells were cloned. Strikingly, this renal device seeded with cloned metanephric cells appeared to produce urine-like liquid, whereas those without cells or seeded with allogeneic cells did not. Histological analysis of the explant revealed a well-developed renal structure comprised of organized glomeruli-like, tubular-like, and vascular elements that were clearly distinct from each other but continuous within the structure. Therefore, these renal tissues appeared integrally connected in a unidirectional manner to the reservoirs, resulting in the excretion of urine into the collecting system. Although it is not clear how the cultured cells digested from the metanephros gained polarity and self-assembled into glomeruli and tubules, this technique successfully used nuclear transplantation for renal regeneration without the risks and long-term effects of immunosuppression.

Recently, Osafune et al. [80] reported an in vitro culture system in which a single Sall1 highly expressing cell from the metanephric mesenchyme forms a three-dimensional kidney structure consisting of glomeruli and renal tubules. This system is useful for examining mechanisms of renal progenitor differentiation, but also suggests the possibility of establishing a whole kidney from a single stem cell.

7 Establishment of Self-kidney from Autologous MSCs

7.1 Nephron Construction from hMSCs Using a Relay Culture System

The anatomical complexity of the kidney and the need for all resident cells to communicate with each other to produce urine means that an artificial kidney structure must include a glomerulus, tubules, interstitium, and vessels. It does not, however, need to be the same size as the native kidney, as long as the glomerular filtration rate exceeds 10 mL/min and the volume is at least 10% the volume of the native kidney. Ideally, the artificial kidney should also be maintained and grown with no, or only minimum, requirement for immunosuppression. An artificial kidney has several important functions in addition to the production of urine, including blood pressure control, maintaining the calcium–phosphorus balance, and the production of erythropoietin (EPO).

With these features in mind, we attempted to establish an ideal artificial kidney, addressing each of the issues one by one. First, we tried to reconstruct an organized and functional kidney structure using the developing heterozoic embryo as an "organ factory." During embryogenesis, a single fertilized cell develops into a whole body within 266 days in humans and 20 days in rodents. This neonate has every organ positioned correctly, indicating that a single fertilized ovum contains a blueprint from which the body, including the kidney, can be built. Therefore, we sought to "borrow" this programming of a developing embryo by applying the stem cells at the niche of organogenesis.

During development of the metanephros (the permanent kidney), the metanephric mesenchyme initially forms from the caudal portion of the nephrogenic cord [81] and secretes glial cell line-derived neurotrophic factor (GDNF), which induces the nearby Wolffian duct to produce a ureteric bud [82]. The metanephric mesenchyme consequently forms the glomerulus, proximal tubule, loop of Henle, and distal tubule, as well as the interstitium, as a result of reciprocal epithelial–mesenchymal induction between the ureteric bud and metanephric mesenchyme [83]. For this epithelial–mesenchymal induction to occur, GDNF must interact with its receptor, c-ret, which is expressed in the Wolffian duct [83]. We hypothesized that GDNF-expressing mesenchymal stem cells may differentiate into kidney structures if positioned at the budding site and stimulated by numerous factors spatially and temporally identical to those found in the developmental milieu.

To investigate this hypothesis, hMSCs were initially injected into the developing metanephros in vitro, although this was not sufficient to achieve kidney organogenesis or even integration of hMSCs into the developing rodent metanephros. No kidney structure was established, nor were any kidney specific genes expressed [84], suggesting that the hMSCs must be placed before the metanephros begins to develop in a specific, defined embryonic niche to allow their exposure to the repertoire of nephrogenic signals required to generate the organ. This can be best achieved by implanting hMSCs into the nephrogenic site of a developing embryo. However, once embryos are removed for cell implantation, they cannot be returned to the uterus for further development. Therefore, we established a culture system in combination with a whole embryo culture system, followed by metanephric organ culture. This "relay culture" allows the development of the metanephros from structures present before budding until the occurrence of complete organogenesis ex utero. In this system, embryos were isolated from the mother before budding and were grown in a culture bottle until the formation of a rudimentary kidney so that it could be further developed by organ culture in vitro [84]. Using this combination, rudimentary kidneys continued to grow in vitro, as assessed by the observation of fine tubulogenesis and ureteric bud branching, indicating that the metanephros can complete development ex utero even if the embryo is dissected prior to sprouting of the ureteric bud.

Based on these results, hMSCs were microinjected at the site of budding and subjected to relay culture. Before injection, the hMSCs were genetically engineered to express GDNF temporally using adenovirus and were also labeled with the *LacZ* gene and dioctadecyl-3,3,3′,3′-tetrametylindocarbocyanine (DiI). Soon after injection, the embryos, together with the placenta, were transferred to the incubator for whole embryos. After the relay culture, X-gal-positive cells were scattered throughout the rudimentary metanephros and were morphologically identical to tubular epithelial cells, interstitial cells, and glomerular epithelial cells [84]. In addition, reverse transcription–polymerase chain reaction revealed the expression of several podocyte- and tubule-specific genes [84]. These data demonstrated that using a xenobiotic developmental process for growing embryos allows endogenous hMSCs to undergo an epithelial conversion and be transformed into an orchestrated nephron consisting of glomerular epithelial cells (podocytes) and tubular epithelial cells that are linked. hMSCs can also differentiate into renal stroma after renal development [84].

7.2 Urine Production from the "Neo"-Kidney Using the Modified Relay Culture System

We then examined the next issue in the successful establishment of an artificial kidney de novo: urine production. This requires that the kidney formed has the vascular system of the recipient; therefore, the primary system must be modified to allow for vascular integration from the recipient to form a functional nephron. We used the findings of Rogers et al. [74] described above, whereby the metanephros can grow and differentiate into a functional renal unit with integration of recipients vessels if it is implanted into the omentum. To incorporate this modification into our relay culture system, we needed to know at how early a stage the metanephros could develop in the omentum. We transplanted metanephroi from different embryonic stages into the omentum and found after 2 weeks that only metanephroi from rat embryos older than embryonic day (E) 13.5 developed successfully. Therefore, the relay culture system was modified so that organ culture was terminated within 24 h, by which time the metanephros was allowed to develop sufficiently and the kidney primordia could be transplanted into the omentum (termed "modified relay culture system"). As a result, an hMSC-derived "neo"-kidney was generated that was equivalent to a human nephron [85].

To examine the origin of the vasculature in the neo-kidney, we generated *LacZ*-transgenic rats [86] as recipients so that donor- and recipient-derived tissues were distinguishable by X-gal assay. The usefulness of genetically marked transgenic (tg) rats in organogenesis has been confirmed previously, using GFP as a marker, in the rat [87]. Thus, we chose to use the *LacZ* tg rat, which expresses the marker gene ubiquitously, to determine vascular origin in our experiments [85]. Using the modified relay culture, several vessels from the omentum appeared to be integrated into the neo-kidney and X-gal staining showed that most of the peritubular capillaries were LacZ positive, suggesting they were of recipient origin. Furthermore, electron microscopic analysis revealed red blood cells in the glomerular vasculature. These data indicated that the vasculature of the neo-kidney in the omentum originated from the host and communicated with the host circulation, suggesting its viability to collect and filter the host blood to produce urine. To verify this, the neo-kidney was left in the omentum for 4 weeks to develop further. Surprisingly, the structure developed hydronephrosis, confirming the ability of the neo-kidney to produce urine: if the ureter were buried under the fat of the omentum, the urine would have no egress, resulting in hydronephrosis. Analysis of the liquid from the expanded ureter showed higher urea nitrogen and creatinine concentrations compared with the recipient sera that were comparable with concentrations in native urine [85]. Therefore, we concluded that the neo-kidney that developed in the omentum was capable of producing urine by filtration of the recipient's blood.

7.3 Acquirement of Renal Functions Other than Urine Production

Finally, we addressed the final goal for the development of an ideal regenerative kidney. The kidney plays an important local role in removing uremic toxins and excess fluid by producing urine and contributes to homeostasis through hematopoiesis, blood pressure control, and maintaining the calcium–phosphorus balance. Therefore, we investigated whether the neo-kidney produced by our system was biologically viable using mouse models of hereditary renal diseases, focusing on Fabry disease. Fabry disease is an X-linked lysosomal storage disease that is caused by a deficiency of the α-galactosidase A (α-gal A) enzyme. This leads to the abnormal accumulation of glycosphingolipid with terminal α-galactosyl residues (Gb3) in various organs, including the kidney, which causes CRF [88]. Renal involvement is characterized by Gb3 deposits mainly in podocytes and tubular epithelial cells, resulting in glomerulosclerosis, tubular atrophy, and interstitial fibrosis [88]. Fabry mice lacking α-gal A appear normal and there are no significant differences in renal histology by periodic acid-Schiff (PAS), Masson, and Sudan IV staining compared with wild-type mice (unpublished data). This is because the accumulation of abnormal lipid is very slow and the mice die before any manifestation of renal failure. We assessed the viability of the regenerated metanephros on the basis of α-gal A activity and clearance of abnormal Gb3 accumulated in Fabry mice. To this end, hMSCs were transfected with GDNF and injected into E9.5 Fabry mouse embryos, and then subjected to relay culture to regenerate the kidney. Compared with wild-type mice, basal levels of α-gal A bioactivity in the metanephros from the Fabry mouse was quite low, whereas metanephroi regenerated with hMSCs expressed significantly more α-gal A [84]. Gb3 started to accumulate within the ureteric bud and S-shaped bodies in metanephroi of Fabry mice and this accumulation was markedly reduced by replacement of the metanephroi with an hMSCs-derived nephron, which has α-gal A activity [84]. These data indicated that the regenerated neo-kidney is viable when it comes to maintaining the local environment. Furthermore, we confirmed that the neo-kidney could produce human proteins and participate in human homeostasis. For example, we examined the nucleotide sequences of 1α hydroxylase, parathyroid hormone (PTH) receptor-1, and erythropoietin from RNA extracted from the neo-kidney and identified human-specific products (unpublished data), suggesting that the established organ was integrated properly into the host endocrine system. Taken together, these data suggest that the neo-kidney developed in the omentum may be able to fulfill normal renal function in addition to urine production.

Production of EPO to maintain erythropoiesis is another important function of the kidney. EPO stimulates red blood cell production and is produced mainly in the kidneys. Although recombinant human EPO (rHuEPO) is widely administered to treat and mitigate renal anemia in CRF patients [89], improving the quality of life of these patients and reducing mortality and morbidity [90, 91], the cost of rHuEPO treatment (more than US$9000 per person per year) accounts for the highest annual drug sales worldwide [92]. There were three major findings regarding the hMSC-derived neo-kidney in rats: (i) human EPO is produced in rats harboring an organoid derived from autologous human bone marrow cells; (ii) human EPO production is stimulated by induction of anemia, suggesting that this system preserves the normal physiological regulation of EPO levels; and (iii) levels of EPO generated by the neo-kidney in response to anemia in rats in which native EPO was suppressed are sufficient to restore red cell recovery to a rate similar to that in control rats [93]. Taken together, these data suggest that the neo-kidney derived from hMSCs may be able to fulfill all renal functions, including urine production. This work could lead to a new generation of therapy modalities for CRF.

8 Conclusion

Therapies involving renal regeneration need first to discriminate between the two disease states of ARF and CRF. In the present article, we reviewed recent research in the field of regenerative medicine for kidney diseases and proposed possible therapeutic applications for these technologies in the treatment of ARF and CRF.

Recent progress in stem cell biology has demonstrated that renal stem cells that have the capacity to differentiate into mature renal cells exist in adults. Therefore, the current focus in this field is how to harness these cells for the treatment of ARF.

In contrast, establishment of a functional, whole kidney for the treatment for CRF has received little public attention owing to the challenging nature of this research. With the publication recently of the high Medicaid costs for dialysis-related diseases, there is renewed focus on developing innovative therapy for CRF that provides an alternative to dialysis. We would like to propose a putative scenario for the application of regenerative medicine to CRF therapy (Fig. 1).

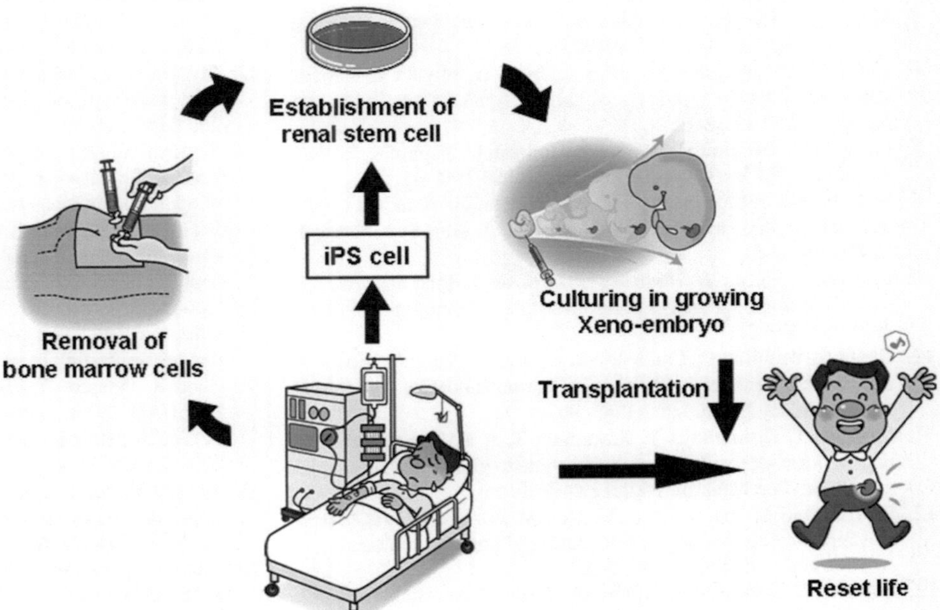

Fig. 1 Putative scenario for the application of stem cell biology to kidney regeneration. Renal stem cells derived from bone marrow cell or skin fibroblast of a dialysis patient are cultured in growing xeno-embryo for a given time to develop into kidney primordia, followed by autologous implantation into the omentum of the same patient. The kidney primordia eventually becomes a self-organ that performs mature renal function, relieving the patient from dialysis and completely resetting his/her body without renal disease

Bone marrow aspiration or skin biopsy is done from a patient with CRF and established renal stem cells are cultured in growing embryos for a given time to develop into kidney primordia, followed by autologous implantation into the omentum of the same patient. Kidney primordia eventually become self-organ that produces the patient's urine. The patient might relief from dialysis and completely reset his/her body without renal disease.

It should be noted that regenerative medicine for renal diseases is still in the developmental phase and a long way from being established; however, regenerative medicine provides significant hope for patients with renal disease who are dependent on dialysis. We believe that emerging knowledge of kidney stem cell biology and developmental biology will enable the development of new, regenerative therapeutic strategies for the treatment of renal diseases that aim to regain damaged components in the kidney or restore kidney function.

Acknowledgments This work was supported by grants from the Ministry of Education, Culture, Sports, Science and Technology of Japan, from the Kanae Foundation.

References

1. Thadhani R, Pascual M, Bonventre JV. Medical progress: acute renal failure. N Engl J Med. 1996;334:1448–60.
2. Krause D, Cantley LG. Bone marrow plasticity revisited: protection or differentiation in the kidney tubule? J Clin Invest. 2005;115:1705–8.
3. Korbling M, Estrov Z. Adult stem cells for tissue repair-a new therapeutic concept? N Engl J Med. 2003;349:570–82.
4. Orlic D, Hill JM, Arai AE. Stem cells for myocardial regeneration. Circ Res. 2002;91:1092–102.
5. Madeddu P. Therapeutic angiogenesis and vasculogenesis for tissue regeneration. Exp Physiol. 2005;90:315–26.
6. Kurokawa K, Nangaku M, Saito A, Inagi R, Miyata T. Current issues and future perspectives of chronic renal failure. J Am Soc Nephrol. 2002;13:S3–6.
7. Locatelli F, Vecchio LD, Possoni P, Monzoni C. Nephrology: main advances in the last 40 years. J Nephrol. 2006;19:6–11.
8. Lysaght MJ. Maintenance dialysis population dynamics: current trends and long-term implications. J Am Soc Nephrol. 2002;13:S37–40.
9. Paulsom R, Forbes SJ, Hodivala-Dilke K, et al. Bone marrow contributes to renal parenchymal turnover and regeneration. J Pathol. 2001;195:229–35.
10. Gupta S, Verfaillie C, Chmielewski D, Kim Y, Rosenberg ME. A role for extrarenal cells in the regeneration following acute renal failure. Kidney Int. 2002;62:1285–90.
11. Imasawa T, Utsunomiya Y, Kawamura T, et al. The potential of bone marrow-derived cells to differentiate to glomerular mesangial cells. J Am Soc Nephrol. 2001;12:1401–9.
12. Ito T, Suzuki A, Imai E, Okabe M, Hori M. Bone marrow is a reservoir repopulating mesangial cells during glomerular remodeling. J Am Soc Nephrol. 2001;12:2625–35.
13. Poulsom R, Alison MR, Forbes SJ, Wright NA. Adult stem cell plasticity. J Pathol. 2002;197:441–56.
14. Lin F, Cordes K, Li L, et al. Hematopoietic stem cells contribute to the regeneration of renal tubules after renal ischemia-reperfusion injury in mice. J Am Soc Nephrol. 2003;14:1188–99.
15. Kale SKA, Clark PR, Kashigarian M, Krause DS, Cantley LG. Bone marrow stem cells contribute to repair of the ischemically injured renal tubule. J Clin Invest. 2003;112:42–9.
16. Togel F, Weiss K, Yang Y, et al. Vasculotropic, paracrine actions of infused mesenchymal stem cells are important to the recovery from acute kidney injury. Am J Physiol Renal Physiol. 2007;292: F1626–35.
17. Wurmser AE, Gage FH. Stem cells: cell fusion causes confusion. Nature. 2002;416:485–7.
18. Aggarwal S, Pittenger MF. Human mesenchymal stem cells modulate allogeneic immune cell responses. Blood. 2005;105:1815–22.
19. Togel F, Hu Z, Weiss K, et al. Administrated mesenchymal stem cells protect against ischemic acute renal failure through differentiation-independent mechanisms. Am J Physiol Renal Physiol. 2005;289:F31–42.
20. Kunter U, Rong S, Djuric Z, et al. Transplanted mesenchymal stem cells accelerate glomerular healing in experimental glomerulonephritis. J Am Soc Nephrol. 2006;17: 2202–12.
21. Duffield JS, Park KM, Hsiao L-L, et al. Restoration of tubular epithelial cells during repair of the postischemic kidney occurs independently of bone marrow-derived stem cells J Clin Invest. 2005;115:1743–55.
22. Handgretinger R, Gordon PR, Leimig T, et al. Biology and plasticity of $CD133^{+}$ hematopoietic stem cells. Ann N Y Acad Sci. 2003;996:141–51.
23. Urbich C, Dimmeler S. Endothelial progenitor cells: characterization and role in vascular biology. Circ Res. 2004;95:343–53.
24. Uchida N, Buck DW, He D, et al. Direct isolation of human central nervous system stem cells. Proc Natl Acad Sci U S A. 2000;97:14720–5.
25. Bussolati B, Bruno S, Grange C, et al. Isolation of renal progenitor cells from adult human kidney. Am J Pathol. 2005;166:545–55.
26. Wiese C, Rolletschek A, Kania G, et al. Nestin expression – a property of multi-lineage progenitor cells? Cell Mol Life Sci. 2004;61:2510–22.
27. Patschan D, Michurina T, Shi HK, et al. Normal distribution and medullary-to-cortical shift of Nestin-expressing cells in acute renal ischemia. 2007;71:744–54.
28. Dekel B, Zangi L, Shezen E, et al. Isolation and characterization of nontubular $Sca\text{-}1^{+}Lin^{-}$ multipotent stem/progenitor cells from adult mouse kidney. J Am Soc Nephrol. 2006;17:3300–14.
29. Challen GA, Martinez G, Davis MJ, et al. Identifying the molecular phenotype of renal progenitor cells. J Am Soc Nephrol. 2004;15:2344–57.
30. Sagrinati C, Netti GS, Mazzinghi B, et al. Isolation and characterization of multipotent progenitor cells from the Bowman's capsule of adult human kidneys. J Am Soc Nephrol. 2006;17:2443–56.
31. Kitamura S, Yamasaki Y, Kinomura M, et al. Establishment and characterization of renal progenitor like cells from S3 segment of nephron in rat adult kidney. FASEB J. 2005; 19:1789–97.
32. Googell MA, Brose K, Paradis G, Conner AS, Mulligan RC. Isolation and functional properties of murine hematopoietic stem cells that are replicating in vivo. J Exp Med. 1996;183:1797–06.
33. Zhou S, Schuetz JD, Bunting KD, et al. The ABC transporter Bcrp1/ABCG2 is expressed in a wide variety of stem cells and is a molecular determinant of the side-population phenotype. Nat Med. 2001;7:1028–34.
34. Jackson K, Mi T, Goodwell M. Hematopoietic potential of stem cells isolated from murine skeletal muscle. Proc Natl Acad Sci U S A. 1999;96:14482–6.
35. Gussoni E, Soneoka Y, Strickland CD, et al. Dystrophin expression in the mdx mouse restored by stem cell transplantation. Nature. 1999;401:390–4.

36. Hishikawa K, Marumo T, Miura S, et al. Musculin/MyoR is expressed in kidney side population cells and can regulate their function. J Cell Biol. 2005;169:921–8.
37. Lu J, Webb R, Richardson JA, Olson EN. MyoR: a muscle-restricted basic helix-loop-helix transcription factor that antagonizes the action of MyoD. Proc Natl Acad Sci U S A. 1999;96: 552–7.
38. Cotsarelis G, Sun TT, Lavker RM. Label-retaining cells reside in the bulge area of pilosebaceous unit: Implication for follicular stem cells, hair cycle, and skin carcinogenesis. Cell. 1990;61:1329–37.
39. Bjerknes M, Cheng H. Clonal analysis of mouse intestinal epithelial progenitors. Gastroenterology. 1999;116:7–14.
40. Hong KU, Reynolds SD, Giangreco A, Hurley CM, Stripp BR. Clara cell secretory protein-expressing cells of the airway neuroepithelial body microenbironment include a label-retaining subset and are critical for epithelial renewal after progenitor cell depletion. Am J Respir Cell Mol Biol. 2001;24:671–81.
41. Maeshima A, Yamashita S, Mojima Y. Identification of renal progenitor-like tubular cells that participate in the regeneration processes of the kidney. J Am Soc Nephrol. 2003;14:3138–46.
42. Oliver JA, Maarouf O, Cheema FH, Martens TP, Al-Awqati Q. The renal papilla is a niche for adult kidney stem cells. J Clin Invest. 2004;114:795–804.
43. Fobes SJ, Poulsom R, Wright NA. Hepatic and renal differentiation from blood-borne stem cells. Gene Ther. 2002;9:625–30.
44. Alison MR, Poulsom R, Forbes S, Wright NA. An introduction to stem cells. J Pathol. 2002;197:419–23.
45. Lin F, Moran A, Igarashi P. Intrarenal cells, not bone marrow-derived cells, are the major source for regeneration in postischemic kidney. J Clin Invest. 2005;115:1756–65.
46. Martin GR: Isolation of a pluripotent cell line from early mouse embryos cultured in medium conditioned by teratocarcinoma stem cell. Proc Natl Acad Sci U S A. 1981;78:7634–38.
47. Bjorklund LM, Sanchez-Pernaute R, Chung S, et al. Embryonic stem cells develop into functional depaminergic neurons after transplantation in a Parkinson rat model. Proc Natl Acad Sci U S A. 2002;99:2344–9.
48. Blyszczuk P, Czyz J, Kania G, et al. Expression of Pax4 in embryonic stem cells promotes differentiation of nestin-positive progenitor and insulin-producing cells. Proc Natl Acad Sci U S A. 2003;100:998–1003.
49. Reubinoff BE, Pera MF, Fong CY, Trounson A, Bongso A. Embryonic stem cell lines from human blastcysts: Somatic differentiation in vitro. Nat Biotechnol. 2000;18:399–404.
50. Thomson JA, Etskovitz-Eldor J, Shapiro SS, et al. Embryonic stem cell lines from human blastcysts. Science. 1998;282:1145–7.
51. Schuldiner M, Yanuka O, Itskovitz-Ekdor J, Melton DA, Benvenisty N. Effects of eight growth factors on the differentiation of cells derived from human embryonic stem cells. Proc Nathl Acad Sci U S A. 2000;97:11307–12.
52. Kobayashi T, Tanaka H, Kuwana H, et al. Wnt4-transformed mouse embryonic stem cells differentiate into renal tubular cells. Biochem Biophys Res Commun. 2005;336:585–95.
53. Steenhard BM, Isom KS, Cazcarro P, et al. Integration of embryonic stem cells in metanephric kidney organ culture. J Am Soc Nephrol. 2005;16:1623–31.
54. Kim D, Dressler G: Nephrogenic factors promote differentiation of mouse embryonic stem cells into renal epithelia. J Am Soc Nephrol. 2005;16:3527–34.
55. Vignewu C, Polgar K, Striker G, et al. Mouse embryonic stem cell-derived embryoid bodies generate progenitors that integrate long term into renal proximal tubules in vivo. J Am Soc Nephrol. 2007;18:1709–20.
56. Wilmut I, Schnieke AE, McWhir J, Kind AJ, Cambell KH. Viable offspring derived from fetal and adult mammalian cells. Nature. 1997;385:810–3.
57. Egli D, Rosains J, Birkhoff G, Eggan K. Developmental reprogramming after chromosome transfer into mitotic mouse zygotes. Nature. 2007;447:679–85.
58. Hwang WS, Ryu YJ, Park JH, et al. Evidence of a pluripotent human embryonic stem cell line derived from a cloned blastocyst. Science. 2004;303:1669–74. (Retraction in Kennedy D. Science. 2006;311:335)
59. Byrne JA, Pedersen DA, Clepper LL, et al. Producing primate embryonic stem cells by somatic cell nuclear transfer. Nature. 2007;450:497–502.
60. Takahashi K, Yamanaka S. Induction of pluripotent stem cells from mouse embryonic and adult fibroblast cultures by defined factors. Cell. 2006;126:663–76.
61. Okita K, Ichisaka T, Yamanaka S. Generation of germline-competent induced pluripotent stem cells. Nature. 2007;448: 313–17.
62. Wernig M, Meissner A, Foreman R, et al. In vitro reprogramming of fibroblasts into a pluripotent ES-cell-like state. Nature. 2007;448:318–24.
63. Meissner A, Wernig M, Jaenisch R. Direct reprogramming of genetically unmodified fibroblasts into pluripotent stem cells. Nature Biotechnol. 2007;25:1177–81.
64. Hanna J, Wernig M, Markoulaki S, et al. Treatment of sick cell anemia mouse model with iPS cells generated from autologous skin. Science. 2007 21;318:1920–3.
65. Takahashi K, Tanabe K, Ohnuki M, et al. Induction of pluripotent stem cells from adult human fibroblasts by defined factors. Cell. 2007;131:1–12.
66. Pereira RF, Halford KW, O'Hara MD, et al. Cultured adherent cells from marrow can serve as long-lasting precursor cells for bone, cartilage, and lung in irradiated mice. Proc Natl Acad Sci U S A. 1995;92:4857–61.
67. Pereira RF, O'Hara MD, Laptev AV, et al. Marrow stromal cells as a source of progenitor cells for non-hematopoietic tissues in transgenic mice with a phenotype of osteogenesis imperfecta. Proc Natl Acad Sci U S A. 1998;95:1142–7.
68. Liechty KW, MacKenzie TC, Shaaban AF, et al. Human mesenchymal stem cells engraft and demonstrate site-specific differentiation after in utero transplantation in sheep. Nat Med. 2000;6: 1282–6.
69. Bussolati B, Camussi G. Stem cells in acute kidney injury. Contrib Nephrol. 2007;156:250–8.
70. Kunter U, Rong S, Boor P, et al. Mesenchymal stem cells prevent progressive experimental renal failure but maldifferentiate into glomerular adipocytes. J Am Soc Nephrol. 2007;18: 1754–64.
71. Prockop DJ: Marrow stromal cells as stem cells for nonhematopoietic tissues. Science. 1997;276:71–4.
72. Yamamoto M, Cui L, Johkura K, et al. Branching ducts similar to mesonephric ducts or ureteric buds in teratomas originating from mouse embryonic stem cells. Am J Physiol Renal Physiol. 2005;290:F52–60.
73. Woolf AS, Palmer SJ, Snow ML, Fine LG. Creation of functioning chimeric mammalian kidney. 1990;38:991–7.
74. Rogers S, Lowell JA, Hammerman NA, Hammerman MR. Transplantation of developing metanephroi into adult rats. Kidney Int. 1998;54:27–37.
75. Dekel B, Marcus H, Herzel BH, et al. In vivo modulation of the alogeneic immune response by human fetal kidneys: the role of cytokines, chemokines and cytolytic effecter molecules. Transplantation. 2000;69:1470–8.
76. Hammerman MR. Tissue engineering the kidney. Kidney Int. 2003;63:1195–204.
77. Chan T, Ariizumi T, Asashima M. A model system for organ engineering: transplantation of in vitro induced embryonic kidney. Naturwissenschaften. 1999;86:224–7.

78. Okabayashi K, Asashima M. Tissue generation from amphibian animal caps. Curr Opin Genet Dev. 2003;13:502–7.
79. Lanza RP, Chuug HY, Yoo JJ, et al. Generation of histocompatible tissues using nuclear transplantation. Nat Biotech. 2002;20:689–96.
80. Osafune K, Takasato M, Kispert A, Asashima M, Nishinakamura R. Identification of multipotent progenitors in the embryonic mouse kidney by a novel colony-forming assay. Development. 2005;133:151–61.
81. Saxen L. Organogenesis of the kidney. Cambridge, UK: Cambridge University Press; 1987.
82. Davies JA, Fisher CE. Genes and protein in renal development. Exp Nephrol. 2002;10:102–13.
83. Lipschuts JH. Molecular development of the kidney: a review of the results of gene disruption studies. Am J Kid Dis. 1998;31:383–97.
84. Yokoo T, Ohashi T, Shen J-S, et al. Human mesenchymal stem cells in rodent whole-embryo culture are reprogrammed to contribute to kidney tissue. Proc Natl Acad Sci U S A. 2005;102:3296–300.
85. Yokoo T, Fukui A, Ohashi T, et al. Xenobiotic kidney organogenesis from human mesenchymal stem cells using a growing rodent embryo. J Am Soc Nephrol. 2006;17:1026–34.
86. Inoue H, Osawa I, Murakami T, et al. Development of new inbred transgenic strains of rats with LacZ or GFP. Biochem Biophys Res Commun. 2005;329:288–95.
87. Sawada H, Sheng H-M, Hakamata Y, et al. Contribution of subcutaneous connective tissues to the epithelialization and cyst formation by the skin transplanted subcutaneously. Organogenesis. 2004;1:55–9.
88. Alroy J, Sabnis S, Kopp JB. Renal pathology in Fabry disease. J Am Soc Nephrol. 2002;13:S134–38.
89. Eschbach JW, Egrie JC, Downing MR, Browne JK, Adamson JW. Correction of the anemia of end-stage renal disease with recombinant human erythropoietin. Results of a combined phase I and II clinical trial. N Engl J Med. 1987;316:73–8.
90. Erslev AJ, Besarab A. Erythropoietin in the pathogenesis and treatment of the anemia of chronic renal failure. Kidney Int. 1997;51:622–30.
91. Tsakiris D. Morbidity and mortality reduction associated with the use of erythropoietin. Nephron. 2000;85(Suppl 1):S2–8.
92. Eckardt K-U, After 15 years of success – perspectives of erythropoietin therapy. Nephrol Dial Transpl. 2001;16:1745–49.
93. Yokoo T, Fukui A, Ohashi T, et al. Generation of transplantable erythropoietin-producer derived from human mesenchymal stem cells. Transplantation. 2008;85:1654–8.

The Endometrium: A Novel Source of Adult Stem/Progenitor Cells

Caroline E. Gargett and Kjiana E. Schwab

Abstract The human endometrium (lining of the uterus) is a dynamic remodeling tissue undergoing more than 400 cycles of regeneration, differentiation, and shedding during a woman's reproductive years. Endometrial regeneration also follows childbirth, almost complete resection and in post-menopausal women taking estrogen replacement therapy. In nonmenstruating species (e.g. rodents) there are cycles of endometrial growth and apoptosis, rather than physical shedding. The endometrium comprises epithelial-lined glands extending from the surface epithelium to the myometrium supported by extensive stroma. Remodeling of the endometrium is regulated by the co-ordinated and sequential actions of estrogen and progesterone in preparation for blastocyst implantation on a monthly basis in women and every 4–5 days in mice. Adult stem/progenitor cells are likely responsible for endometrial regeneration. Since there are no specific stem cell markers, initial studies using functional approaches identified candidate epithelial and stromal endometrial stem/progenitor cells as colony-forming cells/units (CFU) and side population (SP) cells. Recently, a subpopulation of human endometrial stromal cells with mesenchymal stem cell–like properties of CFU activity and multilineage differentiation have been isolated by their co-expression of CD146 and PDGF-receptor β. Candidate epithelial and stromal stem/progenitor cells have also been identified in mouse endometrium as rare label-retaining cells (LRC) in the luminal epithelium and as perivascular cells at the endometrial myometrial junction, respectively. While epithelial and most stromal LRC do not express estrogen receptor α (ERα), they rapidly proliferate on estrogen stimulation, most likely mediated by neighboring ERα-expressing niche cells. It is likely that these newly identified endometrial stem/progenitor cells may play key roles in the development of gynecological diseases associated with abnormal endometrial proliferation such as endometriosis and endometrial cancer. The endometrium may also provide a readily available source of mesenchymal stem-like cells for tissue engineering purposes with possible applications not only to urogynecology but also to heart disease, and soft tissue and bone repair.

Keywords Endometrium · Adult stem cells · Epithelial progenitor cells · Mesenchymal stem cells · Human · Mouse

The uterus is fundamental to the survival of the species for all live-bearing mammals. The endometrium or mucosal lining of the uterus plays a critical role in providing a nourishing nonhostile environment for the embryo to implant and undergo full development in readiness for life outside the uterus. However, in humans and old world primates if an embryo fails to implant at the appropriate time into the receptive endometrium and pregnancy does not ensue, a significant proportion of the endometrial lining is sloughed off as menstrual effluent and is regenerated in the next menstrual cycle. Thus, the human endometrium is a dynamic remodeling tissue undergoing more than 400 cycles of regeneration, differentiation, and shedding during a woman's reproductive years [1–3]. Each month, 4–7 mm of mucosal tissue grows within 4–10 days in the first half, or proliferative stage, of the menstrual cycle [3]. Endometrial regeneration also follows parturition, extensive resection and occurs in post-menopausal women taking estrogen replacement therapy [1, 4]. This level of new tissue growth occurring on a monthly basis is at least equivalent to the cellular turnover in other highly regenerative organs such as blood forming tissue of the bone marrow, epidermis, and intestinal epithelium. In these regenerative tissues, adult stem cells, responsible for provision of replacement cells to maintain tissue homeostasis have been identified. Recently, rare populations of epithelial stem/progenitor cells and mesenchymal stem-like cells have been identified in human endometrium using functional approaches [1, 5]. Thus, human endometrium presents a novel model of cyclical tissue regeneration for the study of adult stem cell dynamics.

C.E. Gargett (✉)
Centre for Women's Health Research, Monash University Department of Obstetrics and Gynaecology, Monash Medical Centre, 246 Clayton Road, Clayton, Victoria, 3168, Australia
e-mail: Caroline.Gargett@med.monash.edu.au

V.K. Rajasekhar, M.C. Vemuri (eds.), *Regulatory Networks in Stem Cells*, Stem Cell Biology and Regenerative Medicine,
DOI 10.1007/978-1-60327-227-8_32, © Humana Press, a part of Springer Science+Business Media, LLC 2009

1 Human Endometrium – Structure and Function

The human and primate uterus comprises the endometrial mucosal lining, a highly regenerative tissue situated on the thick muscular myometrium (Figs. 1, 2). The endometrial-myometrial junction is irregular with no submucosal tissue to separate endometrial glandular tissue from the underlying smooth muscle of the myometrium [7]. The endometrium and subendometrial myometrium originate from the Müllerian ducts during embryonic life, while the outer myometrial layer develops in fetal life and has a non-Müllerian origin [7, 8]. The endometrium is structurally and functionally divided into two major zones, the upper two thirds comprising the functionalis, which contains glands extending from the surface epithelium loosely held together by supportive stroma, and the lower basalis, consisting of the basal region of the glands, dense stroma, large vessels, and lymphoid aggregates (Fig. 2) [7, 9–11]. Endometrial glands and the luminal epithelium are lined by a single layer of tightly packed columnar epithelial cells producing a pseudostratified appearance, while the extensive stroma comprises stromal fibroblasts, vascular cells, and leukocytes.

2 The Menstrual Cycle: A Model of Cyclical Tissue Regeneration

The dynamic cycles of cellular proliferation and differentiation occurring during the menstrual cycle in human and primate endometrium are regulated by the co-ordinated and sequential actions of ovarian sex steroid hormones, estrogen and progesterone (Fig. 1). These hormones mediate their actions in endometrium via their cognate nuclear receptors, estrogen receptor α (ERα) and the progesterone receptor A isoform (PR-A), transcription factors that also show dynamic cell-specific changes of expression during the menstrual cycle as they up- or down-regulate their own and each others' expression in functionalis epithelium and stroma [12, 13]. Furthermore, ERα and PR-A activate many other target genes generating critical mediators regulating the cyclic growth and differentiation of the endometrial functionalis. In contrast, the basalis layer is relatively insensitive to sex steroid hormone actions and undergoes little proliferation or differentiation [10, 13–15]. It is the functionalis layer of human and primate endometrium that is shed at menstruation, while the basalis remains and functions as a germinal compartment for regenerating the new functionalis in the subsequent cycle. It is hypothesized that endometrial stem/progenitor cells responsible for regenerating the functionalis layer would reside in the basalis [16, 17].

The initial stage of endometrial repair involves migration of epithelial cells from protruding stumps of remnant basalis glands over the denuded surface within 48 h of shedding [18, 19] and does not appear to require estrogen. Early repair occurs while circulating estrogen levels are very low and when epithelial cells lack ERα expression [18, 19]. Similarly, in a mouse model of menstruation and endometrial restoration, complete regeneration of all cellular components occurred in the absence of estrogen [20].

As estrogen levels rise during the proliferative stage of the menstrual cycle, ERα and PR-A are induced in the epithelium and stroma and the functionalis grows rapidly with extensive proliferation of glandular epithelial cells and to a

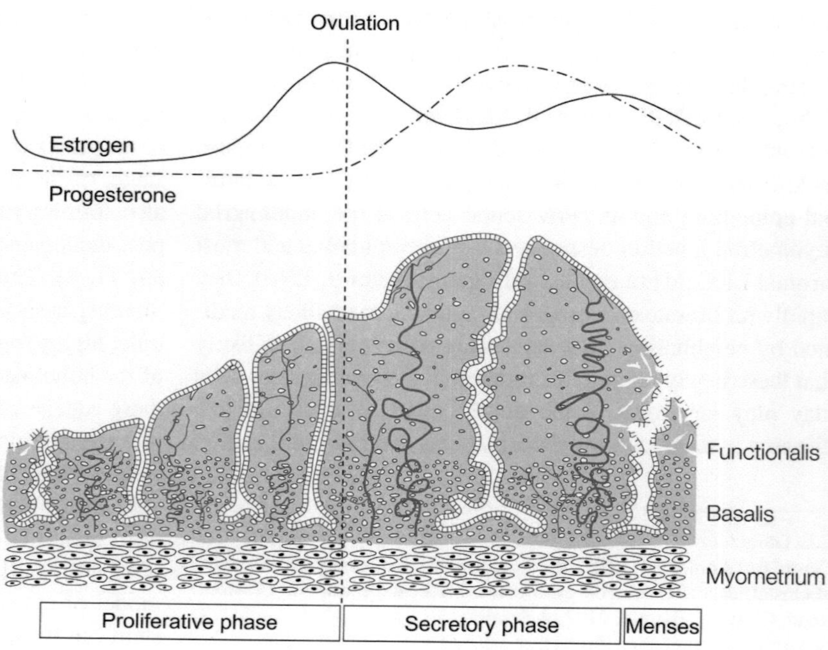

Fig. 1 Schematic of the human menstrual cycle illustrating the cycle of growth, functional differentiation and shedding of the functionalis layer in human endometrium regulated by circulating ovarian hormones. The basalis layer shows little change, is not shed, and is relatively insensitive to ovarian hormone action. The menstrual cycle is divided into three main phases: menses, proliferative (growth) phase, and secretory (functional differentiation) phase. Reprinted with permission from Elsevier [6]

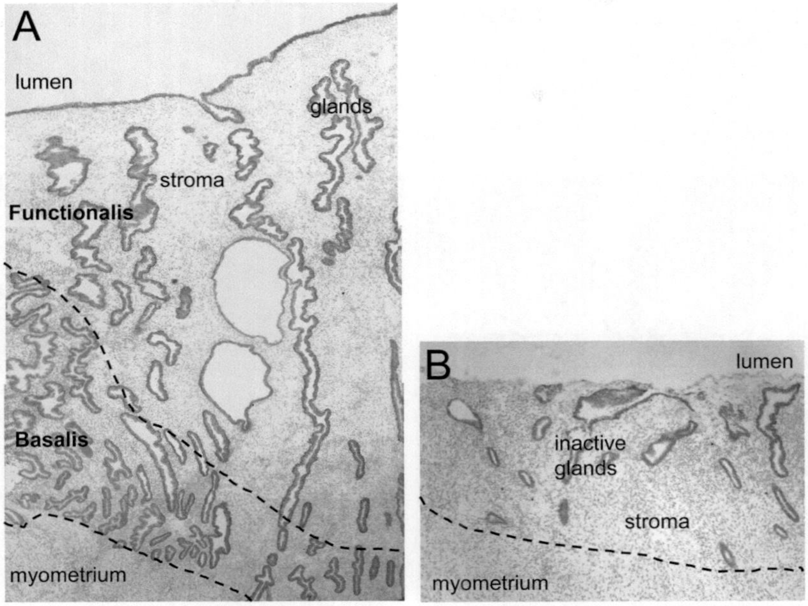

Fig. 2 Structure of human endometrium. (**A**) Full thickness active cycling human endometrium stained with epithelial marker EpCAM demonstrating the functionalis and basalis layers. The endometrial-myometrial junction is shown by the dotted line. Note the branching glands near the endometrial-myometrial junction and the direct apposition of endometrial glands on the myometrium. Reprinted with permission from Oxford University Press [1]. (**B**) Atrophic postmenopausal endometrium immumostained with EpCAM showing scanty inactive glands and stroma. Postmenopausal endometrium will regenerate into the active endometrium shown in (**A**) if estrogen replacement or tamoxifen, a partial agonist for endometrial estrogen receptors, is administered

lesser extent, stromal cells (Fig. 1) [14–16]. Following ovulation, proliferation gradually ceases and the estrogen-primed functionalis commences differentiation under the influence of progesterone (Fig. 1), which suppresses functionalis, but not basalis ERα and PR expression [2, 13]. In preparation for an implanting blastocyst the differentiating endometrial glands produce large quantities of glycogen and histotrophic secretory products [11, 13]. PR persists on stromal cells in the functionalis, which proliferate and differentiate into predecidual cells around spiral arterioles and beneath the luminal epithelium. Endometrial differentiation is also accompanied by dramatic changes in gene expression profiles [21, 22]. When implantation fails to occur, stromal decidualization becomes a terminal differentiation. The demise of the ovarian corpus luteum and subsequent fall in circulating estrogen and progesterone levels triggers menstruation, and the functionalis is shed (Fig. 1) [2, 23].

Thus the human endometrial functionalis shows dynamic changes in growth, differentiation, shedding, and regeneration each menstrual cycle. When menstrual cycles cease at menopause, the endometrium becomes very thin and atrophic, containing few glands in a basalis-like stroma (Fig. 2B) [24]. Similarly, women taking oral contraceptive pills (OCP) do not exhibit cyclic changes in circulating sex steroid hormones and their endometrium does not undergo cyclical growth, differentiation, and regression. Histologically OCP endometrium appears inactive with similar morphology to postmenopausal endometrium. However, when postmenopausal women take estrogen replacement therapy or women cease OCP medication, their endometrium regenerates, suggesting that resident stem/progenitor cells are capable of responding to appropriate hormonal cues to initiate tissue replacement.

3 Mouse Endometrium – Structure and Function

The murine uterus comprises two horns that have a similar histological organization to the human uterus. The luminal epithelium extends glands through the stroma to the myometrial layers of smooth muscle [25]. However, it differs since there is no defined functionalis and basalis layers, and mice do not menstruate. Instead mouse endometrium undergoes cycles of cellular proliferation and apoptosis during its 4–5 day estrus cycle (Fig. 3). In adult female mice, estrogen produced during proestrus stimulates proliferation of luminal and glandular epithelial cells in preparation for estrus when ovulation occurs [28]. During the subsequent progesterone dominant metestrus stage, epithelial cells differentiate to produce glycogen but full differentiation of the endometrial stroma requires a fertilized ovum [29].

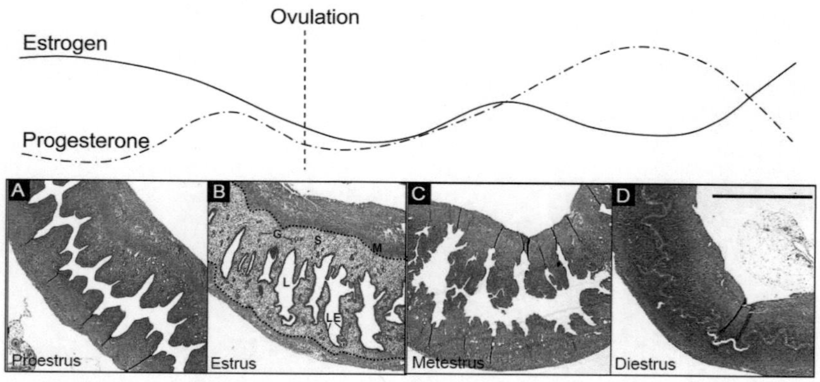

Fig. 3 Schematic of the murine estrus cycle illustrating changing circulating ovarian hormones and the related changes in the endometrium in longitudinal sections of the mouse uterus. Rapid growth of endometrial tissue occurs during proestrus under the influence of rising circulating estrogen and functional differentiation commences during metestrus, but is not complete until there is a fertilized ovum to initiate significant ovarian progesterone production and contact the endometrium to induce functional differentiation of the endometrium. Note that endometrial thickness and fluid accumulation is markedly increased at estrus. Ovarian hormone levels are redrawn from data presented in Walmer et al. [26] and histological images reprinted with permission from Bioscientifica [27]

Hence mice do not menstruate because the stroma does not terminally differentiate into decidual cells during cycles.

As in the human, estrogen also regulates endometrial cell survival, viability, and mitogenic effects via ERα in mouse endometrium [30]. Similarly, estrogen and progesterone co-ordinate the regulated expression of Esr1 (ERα) and PR-A to mediate proliferative and differentiating effects on mouse endometrial epithelium and stroma [31–33]. Tissue recombinant technology using Esr1-knockout and wild-type endometrium has also demonstrated the importance of stromal-epithelial interactions and shown that uterine stromal Esr1 mediates the mitogenic effects of estrogen on both human and murine endometrial epithelial cells [34, 35] by stimulating stromal production of growth factors that subsequently regulate epithelial proliferation and promote differentiation [36, 37].

Similar to women, female mice reach reproductive senescence due to ovarian failure. Female mice will undergo approximately 80 estrus cycles or produce 8–10 litters or a combination of these events during their reproductive life. Thus, an enormous turnover of mucosal tissue also occurs in mouse endometrium, despite the lack of menstruation.

4 The Estrus Cycle: Epithelial Remodeling and Models for Study of Growth and Differentiation of the Endometrium

The mouse is a well-established animal model for investigating endometrial function. Despite the lack of menstruation, ovariectomy, which removes the primary source of endogenous estrogen and progesterone, results in a thin atrophic endometrium with minimal glands, a substantially diminished luminal epithelium, and a dense stroma, which resembles basalis endometrium in humans. Furthermore, administration of estrogen to ovariectomized mice results in a rapid and substantial regeneration on the endometrium, with proliferation and restoration of the luminal epithelium, return of glands, and stromal proliferation with marked stromal edema [38–40], mimicking the proliferative stage of the human menstrual cycle [38, 41]. Endometrial differentiation during the human secretory stage of the menstrual cycle can be modeled in mice by examining endometrium within 5 days of mating at the earliest stages of pregnancy. Alternatively, sex steroid hormones can be manipulated in ovariectomized mice by giving a priming estrogen injection followed by repeated progesterone or progestin injections or subcutaneous slow release hormone containing pellets and a mechanical (decidual) stimulus [20]. A further advantage is that these mouse models of human endometrial regeneration can be examined using transgenic animals to enable dissection of molecular pathways regulating these processes, in particular the role of candidate endometrial stem/progenitor cells in these pathways. Another advantage of using mouse endometrium is that glandular development is a postnatal event [11, 42], which simplifies investigation of the role of endometrial stem/progenitor cells in endometrial growth during adenogenesis.

5 Evidence for Endometrial Stem/Progenitor Cells

The concept that basalis endometrium harbors stem/progenitor cells responsible for the remarkable regenerative capacity of endometrium was proposed many years ago [10, 43]. Indirect evidence for the existence of adult stem/progenitor cells in endometrium has accumulated over the

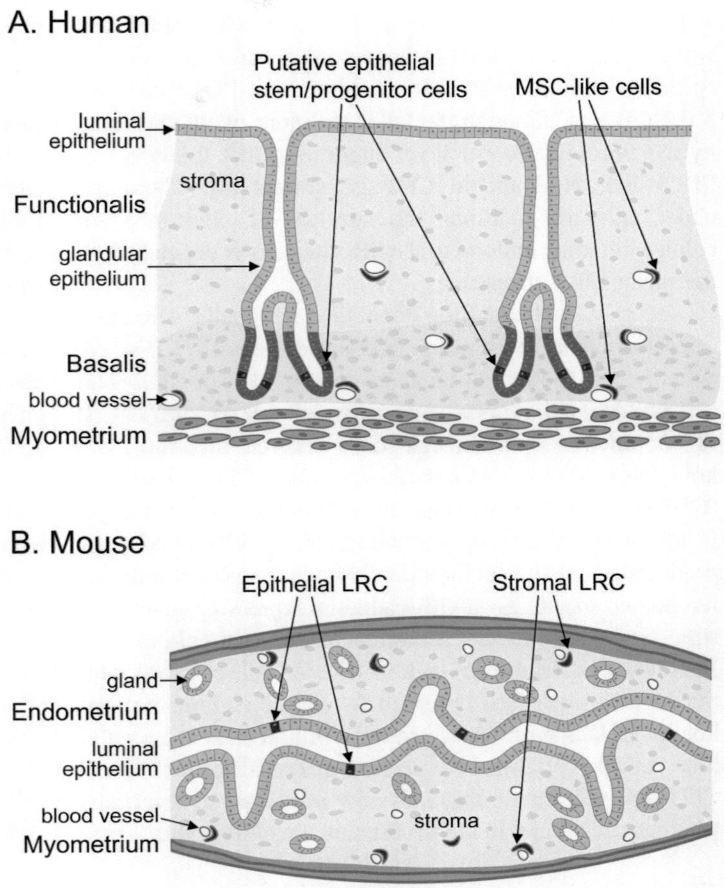

Fig. 4 Schematic showing the possible location of candidate endometrial stem/progenitor cells in human and mouse endometrium. (**A**) In human endometrium, it is predicted that epithelial stem/progenitors will be located in the basalis in the base of the glands. Recent data indicates that stromal stem/progenitors are located near blood vessels although it is not known if they reside in basalis and/or the functionalis. (**B**) In mouse endometrium, the location of epithelial and stromal label-retaining cells (LRC) (candidate stem/progenitor cells) that have the capacity to rapidly proliferate during estrogen-stimulated growth of regressed endometrium is shown in the luminal epithelium and mainly near blood vessels at the endometrial–myometrial junction, respectively. Reprinted with permission from Informa Healthcare [45]

intervening years [1]. Attempts to isolate, characterize, and locate endometrial stem/progenitor cells have recently been undertaken as experimental approaches to identify adult stem cells in other tissues have been developed [1].

The first published evidence for the existence of adult stem/progenitor cells in human endometrium identified rare, clonogenic epithelial and stromal cells, suggesting two types of adult stem/progenitor cell [44]. Rare epithelial and stromal colony-forming unit (CFU) cells were found in normal cycling and inactive perimenopausal endometrium, and in endometrium of women on oral contraceptives [46], suggesting that CFU may be responsible for regenerating cycling and atrophic endometrium.

6 Human Endometrial Epithelial Stem/Progenitor Cells

Accumulating evidence suggests that epithelial stem/progenitor cells exist in human endometrium. Cell cloning studies demonstrated that 0.2% of epithelial cells had CFU activity [44]. However, two types of CFU formed; large (0.09%) and small (0.14%), leading to the hypothesis that the large CFU were initiated by a stem/progenitor cell possibly located at the base of the glands in the basalis (Fig. 4A). Small CFU are possibly initiated by more differentiated transit amplifying cells, which are responsible for the extensive proliferation observed in the proliferative stage of the menstrual cycle and likely located in the functionalis layer [1, 4, 44]. Differential expression of epithelial markers was noted between large and small CFU. Small CFU expressed epithelial differentiation markers, cytokeratin, epithelial cell adhesion molecule (EpCAM) and α_6-integrin, but only the latter was expressed in cells of large CFU, which comprised small cells of high nuclear cytoplasmic ratio, suggesting an undifferentiated phenotype [44]. It is not known whether human endometrial epithelial stem/progenitor cells express ERα and/or PR-A. More definitive markers for these cells are required to investigate the role of ERα in human endometrial epithelial stem/progenitor cell function.

Several growth factors have been investigated for their potential roles in supporting endometrial CFU activity. Three growth factors supported both epithelial CFU activity in serum-free cultures: EGF, TGFα, and PDGF-BB [44, 46]. Fibroblast feeder layers were required for epithelial CFU activity in serum-free conditions, indicating the need for stromal-epithelial cell signaling and the importance of the stem cell niche for endometrial epithelial stem/progenitor cells. Four growth factors, IGF-1, LIF, SCF, and HGF, were weakly supportive of epithelial CFU, while bFGF was without effect. It is possible that colony-forming endometrial epithelial cells express EGF receptors and PDGF-receptor β (PDGF-Rβ), although the latter are more likely expressed by the fibroblast feeder layer suggesting that the effect of PDGF-BB on epithelial CFU is indirect [4]. Therefore, under certain conditions the survival of clonogenic or colony-initiating endometrial epithelial cells is dependent on the surrounding stroma.

Recently, epithelial Side Population (SP) cells were identified in human endometrial cell suspensions [47]. The SP phenotype is thought to be a universal marker of adult stem cell activity [48]. Fluorescence-activated cell sorting (FACS) of endometrial epithelial SP cells showed that most did not express mature endometrial epithelial (CD9) or stromal (CD13) markers suggesting an undifferentiated phenotype. In long-term culture, sorted endometrial epithelial SP cells proliferated slowly, but importantly these long-lived cells differentiated into CD9- and E-cadherin-expressing gland-like structures after a further 5 months in Matrigel culture [47]. This slow proliferation of endometrial SP cells contrasts with the rapid growth rate observed for single endometrial epithelial CFU that produced around a billion cells in a similar time period [49]. Since the percentages of SP cells [47] and CFU [44, 46] are similar, it is now important to determine whether cells found in the SP fraction are clonogenic.

7 Mouse Endometrial Epithelial Stem/Progenitor Cells

Label retaining cells (LRC) have been identified as candidate adult stem cells in vivo in mouse endometrium [40, 50, 51], exploiting the DNA synthesis label retention property of quiescent cells. A major advantage of this approach is that the location of $BrdU^+$ (LRC) candidate stem/progenitor cells and components of the surrounding stem cell niche is revealed when there are no specific adult stem cell markers, as is the case for mouse and human endometrium. Only one of the three studies demonstrated epithelial LRC [40]. During a 56-day chase following labeling of postnatal endometrium, the BrdU label diluted rapidly due to extensive proliferation of luminal epithelium as nascent glands developed during neonatal and prepubertal endometrial growth [40], and during subsequent estrus cycles. Epithelial LRC, comprising 3% of mouse endometrial epithelial cells were observed as separate cells in the luminal epithelium but not glands (Fig. 4B), suggesting that luminal epithelial stem/progenitor cells are responsible for the growth of glands during development and in the adult mouse. They may also have an important role in regenerating luminal epithelium, which undergoes substantial proliferation and apoptosis during the estrus cycle [27]. Likewise, the epithelial stem/progenitor population must be preserved in the diminished endometrial luminal epithelium following ovariectomy to effect regeneration of nascent glands and extensive luminal epithelium on exogenous estrogen administration.

Since estrogen and Esr1 have critical roles in the growth and regeneration of mouse endometrium, the expression of Esr1 on putative endometrial stem/progenitor cells was investigated in vivo in the LRC mouse model [40]. While Esr1 was expressed in the nuclei of mature epithelial cells, epithelial LRC lacked Esr1 expression, suggesting that estrogen stimulates proliferation of epithelial LRC through indirect mechanisms involving Esr1-expressing subepithelial stromal niche cells (Fig. 5A). Despite the lack of Esr1, endometrial epithelial LRC rapidly proliferate in response to estrogen, indicating the importance of stromal niche cells in transmitting proliferative signals to candidate stem/progenitor cells during estrogen-mediated cellular expansion (Fig. 5A) [1, 40].

The role of candidate endometrial stem/progenitor cells in mediating endometrial growth and regeneration was

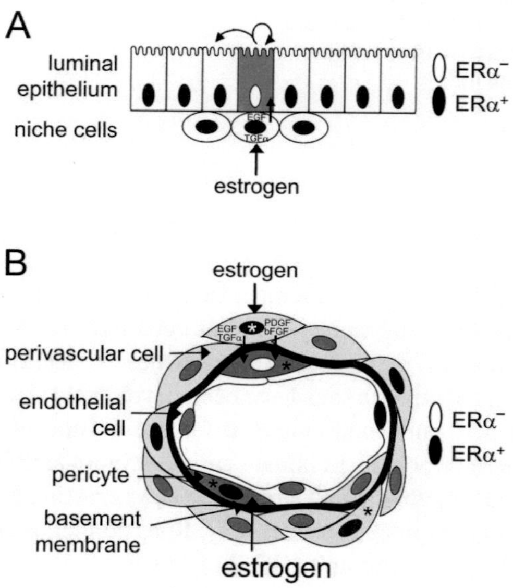

Fig. 5 Schematic of the putative endometrial epithelial and mesenchymal stem cell niche. (**A**) An $ER\alpha^-$ epithelial stem/progenitor cell (*shaded cell*) receives indirect proliferation signals from a surrounding ERα-expressing sub-epithelial stromal niche cell. Subsequent signaling between stromal niche cells and epithelial stem/progenitor cells may be mediated via estrogen-induced release of epidermal growth factor (EGF) and/or transforming growth factor-α (TGF-α) from the niche cell to interact with stem/progenitor cell EGF receptors. (**B**) Mesenchymal stem cell niche, indicating the possible pericyte and/or perivascular location of the mesenchymal stem cells (MSC)-like cells (*), some of which express ERα (black nuclei). Surrounding $ER\alpha^+$ perivascular cells or endothelial cells may act as niche cells to regulate MSC-like cell proliferation through production of PDGF-BB, EGF, TGF-α, or bFGF. Not all MSC-like cells respond to estrogen during endometrial regeneration. Endometrial MSC-like cells could be responsible for estrogen-induced growth of stromal tissue (perivascular MSC-like cells) and blood vessels (pericyte MSC-like cells). Reprinted with permission from Oxford University Press [1]

examined by comparing the kinetics of LRC proliferation during the final stages of endometrial development and maturation in prepubertal mice, and in adult cycling mice. A single dose of estrogen given to stimulate endometrial regeneration in these two groups of mice, whose endometria had been regressed by ovariectomy, showed that epithelial LRC responded differently in effecting endometrial growth in prepubertal and cycling mice [52]. It appears that $Esr1^-$ epithelial LRC drive endometrial growth in prepubertal mice that have never cycled. However, mature $Esr1^+$ epithelial cells have similar capacity to $Esr1^-$ epithelial LRC in initiating proliferative responses to regrow endometrial glands and surface epithelium in cycling mice.

8 Human Endometrial Stromal/Mesenchymal Stem-Like Cells

The first study to identify stromal stem/progenitor cell activity in human endometrium demonstrated that a small population (1.25%) of stromal cells possessed colony-forming ability [44]. Similar to epithelial CFU, two types of stromal CFU formed, with only 0.02% of stromal cells initiating large CFU, supporting the concept of a stromal cell hierarchy hypothesized to exist in human endometrium [1, 5]. Unlike epithelial CFU, both large and small stromal colonies expressed fibroblast markers, with some cells expressing α smooth muscle actin (αSMA), indicative of myofibroblast differentiation [44]. Endometrial stromal cell CFU activity was supported by bFGF as well as EGF, TGFα, and PDGF-BB, in serum-free cultures, the latter three also supporting epithelial CFU activity [44, 46]. The lack of bFGF support for epithelial clonogenicity suggests that there are two different stem/progenitor cells in human endometrium. The requirement for these growth factors by endometrial stem/progenitor cells in vivo has not yet been examined. The use of conditional knockout mice for these growth factors or their cognate receptors, or use of siRNA technologies would be important experimental approaches that would advance the understanding and functioning of endometrial stem/progenitor cells and their progeny.

Since PDGF-BB supports stromal CFU activity we reasoned that PDGF-Rβ would be expressed on clonogenic stromal cells, suggesting that this receptor would be useful for the prospective isolation of stromal stem/progenitor cells from human endometrium. Indeed, stromal stem/progenitor cells with mesenchymal stem cell (MSC)-like activity were isolated from human endometrium, using co-expression of CD146 and PDGF-Rβ, two perivascular cell markers [53]. The FACS-sorted $CD146^+PDGF\text{-}R\beta^+$ subpopulation of endometrial stromal cells were enriched 8-fold for CFU compared to unsorted stromal cells, and expressed typical MSC surface markers, CD44, CD73, CD90, and CD105 [53]. STRO-1, the classic marker used to prospectively isolate bone marrow MSC was not expressed by these cells or by clonogenic stromal CFU. This study also demonstrated multilineage differentiation of $CD146^+PDGF\text{-}R\beta^+$ cells into adipogenic, myogenic, chondrogenic, and osteoblastic lineages when cultured in appropriate induction media, the first demonstration of multipotent stem cells in human endometrium [53]. These data suggest that the $CD146^+PDGF\text{-}R\beta^+$ subpopulation of endometrial stromal cells contains MSC-like cells similar to MSC of bone marrow, fat and dental pulp [54–56]. Furthermore, confocal microscopy demonstrated that CD146 and PDGF-Rβ co-expressing cells were found in a perivascular location in functionalis and basalis blood vessels (Fig. 4A) [53]. Whether MSC-like cells in the basalis and/or functionalis are involved in regenerating endometrium awaits the identification of more specific markers. The demonstration of endometrial tissue reconstitution and differentiation in vivo would further support the existence of MSC in human endometrium.

Chondrogenic differentiation of 3% of a heterogeneous population of freshly isolated cultured endometrial stromal cells [57] suggests that this population is similar to that enriched with $CD146^+PDGF\text{-}R\beta^+$ cells, which all showed chondrocyte morphology and stained with Alcian blue, a standard histological stain for chondrocytes [53]. Endometrial stromal cells, but not myometrial, leiomyoma (smooth muscle cell tumor of myometrium), fallopian tube, or uterosacral ligament cells, had the capacity to differentiate into chondrocytes in vitro [57], suggesting that the endometrium is the only tissue of the female reproductive tract containing a significant population of multipotent stem cells. Alternatively, the resident stem cell populations of these tissues are not present in sufficient numbers for detection in this assay, since they are not as regenerative as endometrium.

The ability of some endometrial stromal cells to spontaneously differentiate into mesenchymal lineages has been noted in clinical samples. These observations have shown that human endometrium occasionally undergoes ossification, often after termination of pregnancy, and while the calcified tissue is not of fetal origin, it is usually associated with chronic inflammation and trauma [58], conditions known to promote incorporation of MSC into regenerating tissues [59]. In addition, tissues such as smooth muscle, bone, and cartilage can also be found in endometrium [60–62], suggesting an inappropriate differentiation of endometrial MSC-like cells. While these observational studies are retrospective and do not identify whether an endometrial MSC-like cell is responsible, they provide indirect evidence confirming our prospective studies demonstrating the presence of rare MSC-like cells in human endometrium [53].

Rare SP cells have been identified in human endometrial stromal cell suspensions, FACS sorted and examined for differentiation capacity in vitro [47]. These SP cells proliferated slowly in long-term culture similar to endometrial epithelial SP cells. Long-lived stromal SP cells differentiated into $CD13^+$ stromal-like clusters after 6 months in Matrigel culture [47]. Further studies are required to determine whether stromal SP cells form CFU and exhibit defining MSC properties such as multilineage differentiation.

More recently, a small population of SP cells (3%) was detected in freshly isolated human myometrial cells derived from the muscle layer of the uterus [63]. Myometrial SP cells were $CD34^+CD45^-$, indicating that they were not hemopoietic stem cells (HSC), and were quiescent with 98% in G_0 phase using Pyronine Y staining. Myometrial SP cells are relatively undifferentiated as they expressed lower levels of ERα, PR, and smooth muscle cell markers, calponin and smoothelin, than main population (MP) myometrial cells, and spontaneously differentiated into mature myometrial cells expressing αSMA and calponin in hypoxic conditions [63]. Some also underwent multilineage differentiation into osteogenic and adipogenic lineages when cultured in appropriate differentiation induction media. Myometrial SP cells expressed some bone marrow MSC surface markers including STRO-1, CD90, CD73, CD105, but not CD44 [63]. Human myometrial SP cells transplanted into the uterine horns of NOD/SCID/γ_c null (NOG) mice supplemented with estrogen incorporated into myometrium and co-expressed vimentin and αSMA. The functional capacity of human myometrial SP cells was further demonstrated by their expression of a pregnancy marker, oxytocin receptor mRNA, in pregnant but not non-pregnant NOG mice previously transplanted with SP cells [63]. This study supports the existence of myometrial stem cells with ability to produce mature myometrial cells in vitro and contribute to myometrial tissue in vivo, as well as exhibit some MSC properties. These interesting results raise questions about the source of the myometrial SP cells, and their relationship to endometrial MSC-like cells. Are they two distinct MSC populations or are they equivalent? From an ontological perspective, the inner layer of myometrial smooth muscle develops from Müllerian duct mesenchyme, the primordium of endometrial stroma, during fetal development, suggesting that myometrial SP cells and endometrial MSC-like cells may both be derived from Müllerian duct mesenchyme sometime during uterine development.

9 Mouse Endometrial Stromal/Mesenchymal Stem-Like Cells

Candidate stromal stem/progenitor cells have been identified in mouse endometrium as stromal LRC [40, 50, 51]. Our study identified 6% of stromal cells as LRC, with a third found beneath the luminal epithelium but not necessarily adjacent to epithelial LRC, while almost half were detected at the endometrial-myometrial junction (Fig. 4B) [40]. Likewise, another study identified a similar percentage of stromal LRC (9%) located near the endometrial-myometrial junction [50], correlating with the postulated basalis location in human endometrium [1]. Stromal LRC were further characterized for expression of various markers. Stromal LRC were $CD45^-$, indicating they were neither leukocytes nor derived from the bone marrow [40, 51] and a small population near the endometrial-myometrial junction expressed key stem cell markers, c-kit and Oct-4 [50]. In other studies, stem cell markers Sca-1 and c-kit were not expressed by stromal LRC [40, 51]. One third of stromal LRC expressed αSMA but not CD31, suggesting they are pericytes or vascular smooth muscle cells [40], confirming their perivascular location (Fig. 5B) identified for human MSC-like cells. Therefore, endometrial MSC-like cells occupy a perivascular location as do other MSC. Furthermore neural and HSC also have vascular niches [54, 64, 65]. Stromal LRC were functionally responsive to mitogenic signals associated with the estrus cycle and rapidly underwent cell division following estrogen or human chorionic gonadotropin administration [40, 51].

LRC have also been identified in the myometrium of 14 week BrdU pulse-chased mice and these colocalized with αSMA, Esr1, and β-catenin. They were not derived from bone marrow cells as they did not express hemopoietic cell markers, including CD45, CD34, or lineage markers [51]. Myometrial LRC were found on the periphery of the muscle bundles of the outer longitudinal muscle layer of the myometrium [51], which developmentally is derived from a non-Müllerian source [7, 8] and is likely to represent a different population of uterine stem/progenitor cells than endometrial MSC-like cells. Cells adjacent to myometrial LRC expressed Abcg2, the transporter responsible for the SP phenotype and SP cells were also detected in myometrial cell suspensions that appeared to be more differentiated transit amplifying cells [51]. This together with the human studies indicates that there may be a third stem/progenitor cell type in the uterus that may be either a derivative of the Müllerian mesenchyme and thus closely related to endometrial MSC-like cells, or may be from a separate lineage producing the outer myometrial muscle layer.

10 Origin of Endometrial Stem/Progenitor Cells

The endometrium offers a unique tissue in which to investigate the relative roles of resident stem/progenitor cells and transdifferentiated bone marrow-derived cells as the source of cells for its cyclical regeneration. Furthermore, substantial renewal of both epithelium and stroma occur on a regular basis throughout the reproductive years. This extensive stromal/vascular regeneration is not seen in other regenerating tissues.

11 Resident Fetal-Derived Stem/Progenitor Cells

The source of resident stem/progenitor cells in adult tissues, including human and mouse endometrium, is thought to be remnant fetal stem cells laid down as reserve cells during fetal life [1, 66]. The embryonic female reproductive tract originates from intermediate mesoderm which begins to form soon after gastrulation. As this embryonic tissue proliferates, it is thought that some cells undergo mesenchymal to epithelial transition to give rise to the coelomic epithelium that later invaginates to form the paramesonephric or Müllerian ducts [67]. These ducts comprise surface epithelium and underlying mesenchyme. Further growth and development of the endometrium occurs during fetal life in humans when the undifferentiated uterine surface epithelium invaginates into the underlying mesenchyme to form the nascent glands, and smooth muscle differentiation of the mesenchyme commences to form the inner myometrium [11]. In mice this stage occurs postnatally [11]. From a developmental perspective, it is more likely that endometrial epithelial and mesenchymal stem/progenitor cells are derived as separate lineages from the two distinct Müllerian cell types of the primordial uterus [1]. The different phenotypes, growth factor dependence, and frequency of clonogenic endometrial epithelial and stromal cells suggest that there are at least two endometrial progenitor cells. The alternate hypothesis is that an ultimate uterine stem cell has capacity to replace all endometrial and myometrial cells, including epithelial, stromal, vascular, and smooth muscle. To test this latter hypothesis, it is necessary to reconstitute the entire uterus from a single cell. A xeno-transplantation model of endometrial reconstitution using large numbers of unfractionated human endometrial cells transplanted into NOG mice produced organized endometrial and myometrial layers, thus providing the groundwork for testing this hypothesis [68].

12 Bone Marrow-Derived Circulating Stem Cells

It is well known that bone marrow stem cells, including HSC, MSC, and endothelial progenitors, circulate, albeit in low numbers. There is some clinical and scientific evidence that bone marrow cells home to sites of tissue damage to populate and incorporate into various organs, transdifferentiating into the cells of the new tissue in which they reside [69–71]. This plasticity of bone marrow derived cells is controversial as it generally appears to be a rare event that may result from cell fusion [72] or the paracrine action of growth factors released by transplanted MSC [73]. It has been advocated that rigorous assessment of engraftment and contribution to function of the new tissue must be demonstrated [72, 74]. Bearing these caveats in mind, there is accumulating evidence that bone marrow–derived cells may also be a potential source of stem cells that engraft the endometrium and may contribute to endometrial regeneration [75–77].

13 Human Bone Marrow-Derived Circulating Stem Cells

Observations of significant chimerism ranging from 0.2 to 52% in the endometrial glands and stroma of four women who had received single antigen HLA mismatched bone marrow transplants suggest that bone marrow stem cells contributed to endometrial regeneration in a setting of cellular turnover and inflammatory stimuli [77]. It appears that the level of chimerism increased with time elapsed since transplantation, although the number of cases was small. The high level of chimerism may also relate to the original degree of endometrial damage and loss of endogenous endometrial stem/progenitor cell populations resulting from pretransplant conditioning or ongoing graft versus host disease involving the endometrium [1]. However, the degree of endometrial chimerism observed is similar to that in other organs of bone marrow transplant recipients [71]. Donor-derived cells were found in focal areas of endometrial glands and stroma suggesting local proliferation of incorporated cells [77]. This observation, together with the similar percentage of donor derived epithelial and stromal cells in each patient, suggests that there may be a single endometrial stem cell responsible for production of both glands and stroma. While most gland profiles observed were exclusively of the donor or host type there was some chimerism within individual glands, suggesting that not all were monoclonal [77], consistent with the epigenetic error data observed in individual endometrial glands [78], but contrasting with gland monoclonality described for nontransplanted women [79]. Polyclonal glands containing only a few donor cells could result from fusion of bone marrow–derived cells with endometrial cells, although this possibility was excluded after examining cells for DNA content [77]. The endometrial epithelial and some stromal cells did not express CD45, suggesting transdifferentiation and distinguishing them from donor endometrial leukocytes. It is not known which bone marrow stem cells, or myeloid cells, contributed to endometrial regeneration in these women. Neither is it known whether bone marrow-derived cells regularly engraft the endometrium each menstrual cycle under normal physiological conditions or whether it occurs at the time of the bone marrow transplantation or subsequently on the resumption of endometrial cycling. It is known that large numbers of mature bone marrow–derived cells traverse the endometrium on a regular basis at precise times during the menstrual cycle [80]. Whether local tissue damage associated with menstruation is sufficient to attract bone marrow stem cells into the endometrium for permanent residence remains to be determined. Whether the bone marrow–derived stem cells incorporate into endometrial tissue in the basalis, functionalis, or both regions is also unknown. The studies required to answer some of these questions are not possible in humans and have been partly addressed by the use of transgenic reporter mice.

14 Mouse Bone Marrow–Derived Circulating Stem Cells

Further evidence for bone marrow stem cell contribution to endometrial repair comes from bone marrow transplant studies in lethally irradiated mice. In one study using a gender mismatched bone marrow transplant model, $<0.01\%$ of cytokeratin positive endometrial epithelial cells and $<0.1\%$ of stromal cells contained a Y chromosome 6 months after transplantation [76]. In these mice the possible contribution of resident endometrial stem/progenitor cells that may have accessed the circulation was excluded by hysterectomizing the recipients prior to transplant and by using male bone marrow donors. Bone marrow cell contribution to endometrial repair was minor in this model where estrus cycling did not occur due to ovarian damage from pretransplant bone marrow ablation [76]. It appears that engraftment of the endometrium seems more likely when there is cellular turnover associated with reproductive cycles or during repair after endometrial injury. A second study similarly demonstrated modest contribution of bone marrow cells to endometrial epithelium 10 months after bone marrow transplantation of irradiated mice with GFP-tagged cells [75], which assumed, but did not test for, resumption of estrus cycling in recipient mice. To further assess transdifferentiation of $CD45^+$ bone marrow cells to endometrial epithelium, a novel double transgenic CD45/Cre-Z/EG mouse was generated to track $CD45^+$ cells through expression of GFP [75]. Transdifferentiated GFP^+CD45^- endometrial epithelial cells increasingly contributed to the endometrial luminal epithelium ranging from 0% in 1week-old pups and 6week-old prepubertal mice, to 0.5% in 12 week-and 6% in 20week-old mice, which had undergone 8 and 22 estrus cycles, respectively [75]. Although there were insufficient animals per group for statistical analysis, this data suggests a small but increasing contribution of bone marrow–derived cells to the endometrial epithelium over time. Clonal expansion of transplant donor cells in the endometrium was apparent as GFP^+ cells were often found in clusters. These two studies in mice suggest that transdifferentiation of circulating bone marrow–derived cells makes a small contribution to regenerating endometrial epithelium of cycling mice but more work is required to precisely distinguish between HSC or more mature myelo-monocytic cells as the source of transdifferentiating cells. Whether bone marrow cells contribute to endometrial regeneration during the extensive endometrial epithelial growth and regeneration occurring during pregnancy and after parturition is yet to be determined, although circulating $CD45^+$ bone marrow cells were shown to contribute 82% of mouse uterine epithelium during pregnancy in a single pregnant double reporter *CD45/Cre-Z/EG* transgenic mouse [75]. Further studies are needed to examine the role of estrogen and progesterone in recruiting bone marrow cells to the endometrium, although progesterone may have a role during pregnancy.

15 Endometrium as a Model Tissue for Examining Adult Stem/Progenitor Cell Activity

The human endometrium is a dynamic remodeling tissue that undergoes cycles of regeneration, differentiation, and regression or shedding that presents an opportunity to investigate the role of stem/progenitor cells under dynamic, but also physiological conditions. While the type of studies that can be undertaken are limited in humans and the variation between subjects often masks significant differences, hysterectomy and curetting tissue is readily available as these are very common procedures in women. Tissues are also readily available from surgical procedures used to treat several common gynecological diseases that may be considered endometrial stem cell diseases, and these also present opportunities for investigation of the role of adult stem cells in their pathophysiology.

Menstrual blood is increasingly being examined as a source of sloughed stem/progenitor cells [81]. This source needs much further characterization and the stem cell status of isolated cells needs verification since menstrual blood not only contains fragments of shed endometrial functionalis, where adult stem cells are not expected to reside, but also peripheral blood that may contain small numbers of HSC, MSC, or endothelial progenitor cells. Furthermore, stem cells are rare cells in any tissue and it is difficult to understand a physiological system dispensing with these important cells so readily each month. Nevertheless menstrual blood is a potential source of adult stem cells that requires more investigation.

Mouse endometrium is also a dynamic remodeling tissue, although is not shed in such a dramatic manner as in humans. However, estrus cycles are easily mimicked and manipulated through the use of ovariectomy and hormonal replacement to induce cycles of endometrial regeneration and regression to examine adult stem/progenitor cell dynamics. Mouse genetics and the ability to conditionally express or ablate genes in many tissues provides a very attractive model for the study of endometrial and myometrial stem/progenitor cells. These have already been used to investigate the role of bone marrow derived cells and track their fate in endometrial tissues [51, 75, 76]. One of these models has also been used to investigate SP cells in mouse endometrium [51]. These and similar models will be invaluable to investigate the stem cell niche that regulates adult stem cell function as well as signaling pathways.

16 Endometrial Stem/Progenitor Cells and Endometriosis

Since adult stem cells regulate tissue homeostasis, it is expected that abnormal functioning of endometrial stem/progenitor cells and/or their surrounding niche cells may also be involved in the initiation and progression of gynecological diseases associated with abnormal endometrial proliferation, such as endometriosis and endometrial cancer [1, 82].

Endometriosis is characterized by the growth of ectopic endometrial tissue on pelvic organs and the peritoneum [83]. It is thought that retrograde menstruation, which occurs in most menstruating women, deposits menstrual debris into the peritoneal cavity. However, it is not known why only 6–10% of women develop endometriosis and its associated symptoms of inflammation, pain, and infertility. We postulate that in women who develop endometriosis, endometrial stem/progenitor cells are inappropriately shed during menstruation and reach the peritoneal cavity where they adhere and establish endometriotic implants [1, 84, 85]. This assertion is supported by the demonstration of clonogenic cells in a long-term culture derived from a sample of endometriotic tissue [86] and monoclonality of some endometriotic lesions [87, 88]. Bone marrow stem cells may also contribute to the progression of endometriosis lesion development as demonstrated recently in a mouse model [76]. Some forms of endometriosis may arise from remnant fetal Müllerian cells, which may behave like stem cells to establish ectopic growth of endometrial tissue. Clearly, the role of endometrial stem/progenitor cells or bone marrow stem cells in the development of endometriosis will require an extensive research effort in firstly identifying markers of the stem/progenitor cells. This has already commenced, following a recent report that LacZ-tagged bone marrow-derived cells engrafted ectopic endometrium in a murine model of endometriosis [76]. After 10 weeks, 0.1% of stromal and 0.04% of epithelial cells were of bone marrow origin, indicating that circulating bone marrow cells may contribute to the progression of endometriosis.

17 Endometrial Cancer Stem/Progenitor Cells

For many years it has been postulated that rare tumor cells (<1%) with stem cell properties are responsible for the growth and metastasis of tumors [89, 90]. Many features of carcinoma can be explained by the stem cell concept, including clonal origin and heterogeneity of tumors, the mesenchymal influence on cancer behavior, the local formation of precancerous lesions, and the plasticity of tumor cells [89]. Cancer stem cells differ from normal tissue stem cells in that their proliferation is no longer controlled by the neighboring cells of the stem cell niche [90, 91]. It is possible that cancer stem cells arise from adult stem cells that have accumulated multiple genetic and epigenetic changes over a period of time giving them selective proliferative advantage, allowing their clonal expansion and succession in the stem cell niche [89]. Thus, cancer stem cells are likely to be the key tumor cells involved in the initiation, progression, metastasis, and recurrence of tumors after treatment [92].

Cancer stem cells have been identified in leukemia [93], breast [94], and colon cancer [95–97] and glioblastoma [90], and emerging evidence suggests that a rare population of cancer stem cells may exist in endometrial adenocarcinoma [98], the commonest reproductive tract cancer in women and one of the few cancers that is increasing in incidence [99].

18 Potential of Endometrial Mesenchymal Stem-Like Cells in Tissue Engineering

There is great interest in the use of both embryonic and adult stem cells in tissue engineering applications for restoring function to aging or diseased tissues and organs. Medical advances have ensured increasing longevity and the aging population has many tissues in need of repair. The failure of artificial implants to last longer than 10–15 years and the problems associated with nondegradable synthetic materials make cell-based therapies for tissue replacement attractive [100]. A combination of temporary biological scaffold materials to provide initial support and stem cells to promote appropriate tissue genesis and regeneration of functional tissue is now the focus of current research [100]. These approaches could be adapted for tissue engineering support of the female reproductive tract [82]. Pelvic floor prolapse is a major problem resulting in 10% of women requiring surgery with approximately 30% of these requiring repeat surgery [101]. The use of artificial and biological scaffolds for pelvic floor prolapse surgery has improved outcomes to a limited degree. Thus, the use of tissue constructs comprising scaffolds and autologous endometrial mesenchymal stem/progenitor cells may provide a possible solution for treatment of pelvic floor prolapse in the future [101]. There is also the potential to use autologous endometrial MSC like cells for regenerative medicine applications on other soft tissues and cardiovascular disease for women.

19 Future Areas for Research and Conclusions

There is now sufficient evidence to conclude that rare populations of adult epithelial and stromal stem/progenitor cells exist in human and mouse endometrium. While some evidence has been obtained from in vivo studies in mice, identifying candidate stem/progenitor cells as LRC, all evidence collected to date for human endometrium has been from in vitro studies. It is most important that the capacity of these rare cells to reconstitute endometrial tissue in vivo using xenotransplantation approaches is undertaken. Such models have already been published and could be adapted to examine putative endometrial stem/progenitor populations [35, 68]. Extension of the LRC model to pregnancy and parturition would provide insight into the role and location of stem/progenitor cells during the greatest remodeling event in endometrium. There is also a pressing need to identify definitive markers for endometrial epithelial stem/progenitor cells and find markers that further purify the $CD146^{+}PDGF\text{-}R\beta^{+}$ endometrial MSC-like cells. Further characterization of the endometrial stem cell niches and the signaling pathways involved in the regulation of the resident stem/progenitor cells is also required. Investigation into the possible roles of developmental pathways involving bone morphogenetic protein, Hedgehog, Notch, and Wnt signaling in endometrial stem/progenitor cell self renewal and cell fate differentiation decisions would be a valuable starting point. More extensive studies examining how estrogen, progesterone, and the growth factors EGF, TGFα, PDGF, and bFGF interact with endometrial stem/progenitor cells and their niche cells would be useful. Such additional knowledge will assist the investigation into the role of endometrial stem/progenitor cells in gynecological disorders associated with abnormal endometrial proliferation and will not only increase our understanding of the pathophysiology of endometriosis and endometrial cancer, but it also has the potential to change the way these hormone-dependent diseases will be treated in the future. Insights from the study of endometrial stem/progenitor cells will also provide valuable information on the role of adult stem cells in maintaining tissue homeostasis in a dynamic remodeling system.

Acknowledgments The authors' work described in this review was supported by a grant from the National Health Medical and Research Council (NHMRC) of Australia to CEG (284344) and a Monash University Graduate Scholarship (to KES). C. Gargett is supported by an NHMRC RD Wright Career Development Award (465121).

References

1. Gargett CE. Uterine stem cells: What is the evidence? Hum Reprod Update. 2007;13:87–101.
2. Jabbour HN, Kelly RW, Fraser HM, Critchley HO. Endocrine regulation of menstruation. Endocr Rev. 2006;27:17–46.
3. McLennan CE, Rydell AH. Extent of endometrial shedding during normal menstruation. Obstet Gynecol. 1965;26:605–21.
4. Gargett CE. Identification and characterisation of human endometrial stem/progenitor cells. Aust N Z J Obstet Gynaecol. 2006;46:250–3.
5. Gargett CE, Chan RW, Schwab KE. Endometrial stem cells. Curr Opin Obstet Gynecol. 2007;19:377–83.
6. Gargett C, Chan R, Schwab K. Hormone and growth factor signaling in endometrial renewal: role of stem/progenitor cells. Mol Cell Endocrinol. 2008;288:22–9.
7. Uduwela AS, Perera MA, Aiqing L, Fraser IS. Endometrial-myometrial interface: relationship to adenomyosis and changes in pregnancy. Obstet Gynecol Surv. 2000;55:390–400.
8. Noe M, Kunz G, Herbertz M, Mall G, Leyendecker G. The cyclic pattern of the immunocytochemical expression of oestrogen and progesterone receptors in human myometrial and endometrial layers: characterization of the endometrial-subendometrial unit. Hum Reprod. 1999;14:190–7.
9. Okulicz WC, Ace CI, Scarrell R. Zonal changes in proliferation in the rhesus endometrium during the late secretory phase and menses. Proc Soc Exp Biol Med. 1997;214:132–8.
10. Padykula HA. Regeneration in the primate uterus: the role of stem cells. Ann N Y Acad Sci. 1991;622:47–56.
11. Spencer TE, Hayashi K, Hu J, Carpenter KD. Comparative developmental biology of the mammalian uterus. Curr Top Dev Biol. 2005;68:85–122.
12. Conneely OM, Mulac-Jericevic B, Lydon JP, De Mayo FJ. Reproductive functions of the progesterone receptor isoforms: lessons from knock-out mice. Mol Cell Endocrinol. 2001;179:97–103.
13. Slayden OD, Brenner RM. Hormonal regulation and localization of estrogen, progestin and androgen receptors in the endometrium of nonhuman primates: effects of progesterone receptor antagonists. Arch Histol Cytol. 2004;67:393–409.
14. Brenner RM, Slayden OD, Rodgers WH, Critchley HO, Carroll R, Nie XJ, et al. Immunocytochemical assessment of mitotic activity with an antibody to phosphorylated histone H3 in the macaque and human endometrium. Hum Reprod. 2003;18:1185–93.
15. Ferenczy A, Bertrand G, Gelfand MM. Proliferation kinetics of human endometrium during the normal menstrual cycle. Am J Obstet Gynecol. 1979;133:859–67.
16. Padykula HA, Coles LG, Okulicz WC, Rapaport SI, McCracken JA, King NW, Jr., et al. The basalis of the primate endometrium: a bifunctional germinal compartment. Biol Reprod. 1989;40: 681–90.

17. Gargett CE. Stem cells in gynaecology. Aust N Z J Obstet Gynaecol. 2004;44:380–6.
18. Ludwig H, Metzger H, Frauli M. Endometrium: tissue remodelling and regeneration. In: D'Arcangues C, Fraser IS, Newton JR, Odlind V, editors. Contraception and mechanisms of endometrial bleeding. Melbourne: Cambridge University Press; 1990. pp. 441–66.
19. Okulicz WC, Scarrell R. Estrogen receptor alpha and progesterone receptor in the rhesus endometrium during the late secretory phase and menses. Proc Soc Exp Biol Med. 1998;218: 316–21.
20. Kaitu'u-Lino TJ, Morison NB, Salamonsen LA. Estrogen is not essential for full endometrial restoration after breakdown: lessons from a mouse model. Endocrinology. 2007;148:5105–11.
21. Ponnampalam AP, Weston GC, Trajstman AC, Susil B, Rogers PA. Molecular classification of human endometrial cycle stages by transcriptional profiling. Mol Hum Reprod. 2004;10: 879–93.
22. Talbi S, Hamilton AE, Vo KC, Tulac S, Overgaard MT, Dosiou C, et al. Molecular phenotyping of human endometrium distinguishes menstrual cycle phases and underlying biological processes in normo-ovulatory women. Endocrinology. 2006;147:1097–121.
23. Salamonsen LA. Tissue injury and repair in the female human reproductive tract. Reproduction. 2003;125:301–11.
24. Feeley KM, Wells M. Hormone replacement therapy and the endometrium. J Clin Pathol. 2001;54:435–40.
25. Hu J, Gray CA, Spencer TE. Gene expression profiling of neonatal mouse uterine development. Biol Reprod. 2004;70:1870–6.
26. Walmer DK, Wrona MA, Hughes CL, Nelson KG. Lactoferrin expression in the mouse reproductive tract during the natural estrous cycle: correlation with circulating estradiol and progesterone. Endocrinology. 1992;131:1458–66.
27. Wood GA, Fata JE, Watson KL, Khokha R. Circulating hormones and estrous stage predict cellular and stromal remodeling in murine uterus. Reproduction. 2007;133:1035–44.
28. Finn CA, Martin L. The cellular response of the uterus of the aged mouse to oestrogen and progesterone. J Reprod Fertil. 1969;20:545–7.
29. Allen E. The oestrous cycle in the mouse. Am J Anat. 1922;30:297–371.
30. Couse JF, Korach KS. Estrogen receptor null mice: what have we learned and where will they lead us? Endocr Rev. 1999;20: 358–417.
31. Lydon JP, DeMayo FJ, Funk CR, Mani SK, Hughes AR, Montgomery CA, Jret al. Mice lacking progesterone receptor exhibit pleiotropic reproductive abnormalities. Genes Dev. 1995;9: 2266–78.
32. Kurita T, Young P, Brody JR, Lydon JP, O'Malley BW, Cunha GR. Stromal progesterone receptors mediate the inhibitory effects of progesterone on estrogen-induced uterine epithelial cell deoxyribonucleic acid synthesis. Endocrinology. 1998;139: 4708–4713.
33. Couse JF, Curtis SW, Washburn TF, Lindzey J, Golding TS, Lubahn DB, et al. Analysis of transcription and estrogen insensitivity in the female mouse after targeted disruption of the estrogen receptor gene. Mol Endocrinol. 1995;9:1441–54.
34. Cooke PS, Buchanan DL, Young P, Setiawan T, Brody J, Korach KS, et al. Stromal estrogen receptors mediate mitogenic effects of estradiol on uterine epithelium. Proc Natl Acad Sci U S A. 1997;94:6535–40.
35. Kurita T, Medina R, Schabel AB, Young P, Gama P, Parekh TV, et al. The activation function-1 domain of estrogen receptor alpha in uterine stromal cells is required for mouse but not human uterine epithelial response to estrogen. Differentiation. 2005;73:313–22.
36. Nelson KG, Takahashi T, Bossert NL, Walmer DK, McLachlan JA. Epidermal growth factor replaces estrogen in the stimulation of female genital-tract growth and differentiation. Proc Natl Acad Sci U S A. 1991;88:21–5.
37. Sato T, Wang G, Hardy MP, Kurita T, Cunha GR, Cooke PS. Role of systemic and local IGF-I in the effects of estrogen on growth and epithelial proliferation of mouse uterus. Endocrinology. 2002;143:2673–9.
38. Martin L, Finn CA. Interactions of oestradiol and progestins in the mouse uterus. J Endocrinol. 1970;48:109–15.
39. Martin L, Finn CA, Trinder G. Hypertrophy and hyperplasia in the mouse uterus after oestrogen treatment: an autoradiographic study. J Endocrinol. 1973;56:133–144.
40. Chan RW, Gargett CE. Identification of label-retaining cells in mouse endometrium. Stem Cells. 2006;24:1529–1538.
41. Martin L, Finn CA, Trinder G. Hypertrophy and hyperplasia in the mouse uterus after oestrogen treatment: an autoradiographic study. J Endocrinol. 1973;56:133–144.
42. Brody JR, Cunha GR. Histologic, morphometric, and immunocytochemical analysis of myometrial development in rats and mice: I. Normal development. Am J Anat. 1989;186:1–20.
43. Prianishnikov VA. On the concept of stem cell and a model of functional-morphological structure of the endometrium. Contraception 1978;18:213–223.
44. Chan RW, Schwab KE, Gargett CE. Clonogenicity of human endometrial epithelial and stromal cells. Biol Reprod. 2004;70:1738–50.
45. Gargett CE. Endometrial stem cells. In: Aplin JR, Fazleabas AT, Glasser SR, Giudice LC, editors. The endometrium: molecular, cellular and clinical perspectives. 2nd ed. London: Taylor and Francis Medical Publishing; 2007. pp. 135–153.
46. Schwab KE, Chan RW, Gargett CE. Putative stem cell activity of human endometrial epithelial and stromal cells during the menstrual cycle. Fertil Steril. 2005;84 Suppl 2:1124–30.
47. Kato K, Yoshimoto M, Kato K, Adachi S, Yamayoshi A, Arima T, et al. Characterization of side-population cells in human normal endometrium. Hum Reprod. 2007;22:1214–23.
48. Smalley MJ, Clarke RB. The mammary gland "side population": a putative stem/progenitor cell marker? J Mammary Gland Biol Neoplasia. 2005;10:37–47.
49. Gargett CE, Zillwood R, Schwab KE. Characterising the stem cell activity of human endometrial cells. Hum Reprod. 2005;20(Suppl 1):i95.
50. Cervello I, Martinez-Conejero JA, Horcajadas JA, Pellicer A, Simon C. Identification, characterization and co-localization of label-retaining cell population in mouse endometrium with typical undifferentiated markers. Hum Reprod. 2007;22:45–51.
51. Szotek PP, Chang HL, Zhang L, Preffer F, Dombkowski D, Donahoe PK, et al. Adult mouse myometrial label-retaining cells divide in response to gonadotropin stimulation. Stem Cells. 2007;25:1317–25.
52. Gargett CE, Chan RWS. Role of stem cells in estrogen induced endometrial regeneration. Reprod Sci. 2007;14(1 Suppl): 214A.
53. Schwab KE, Gargett CE. Co-expression of two perivascular cell markers isolates mesenchymal stem-like cells from human endometrium. Hum Reprod. 2007;22:2903–11.
54. Shi S, Gronthos S. Perivascular niche of postnatal mesenchymal stem cells in human bone marrow and dental pulp. J Bone Miner Res. 2003;18:696–704.
55. Zuk PA, Zhu M, Ashjian P, De Ugarte DA, Huang JI, Mizuno H, et al. Human adipose tissue is a source of multipotent stem cells. Mol Biol Cell. 2002;13:4279–95.
56. Gronthos S, Brahim J, Li W, Fisher LW, Cherman N, Boyde A, et al. Stem cell properties of human dental pulp stem cells. J Dent Res. 2002;81:531–5.

57. Wolff EF, Wolff AB, Hongling D, Taylor HS. Demonstration of multipotent stem cells in the adult human endometrium by in vitro chondrogenesis. Reprod Sci. 2007;14:524–33.
58. Biervliet FP, Maguiness SD, Robinson J, Killick SR. A successful cycle of IVF-ET after treatment of endometrial ossification; case report and review. J Obstet Gynaecol. 2004;24:472–3.
59. van Os R, Kamminga LM, de Haan G. Stem cell assays: something old, something new, something borrowed. Stem Cells. 2004;22:1181–90.
60. Bird CC, Willis RA. The production of smooth muscle by the endometrial stroma of the adult human uterus. J Path Bact. 1965;90:75–81.
61. Roth E, Taylor HB. Heterotopic cartilage in the uterus. Obstet Gynecol. 1966;27:838–44.
62. Mazur MT, Kraus FT. Histogenesis of morphologic variations in tumors of the uterine wall. Am J Surg Pathol. 1980;4:59–74.
63. Ono M, Maruyama T, Masuda H, Kajitani T, Nagashima T, Arase T, et al. Side population in human uterine myometrium displays phenotypic and functional characteristics of myometrial stem cells. Proc Natl Acad Sci U S A. 2007;104:18700–5.
64. Kiel MJ, Yilmaz OH, Iwashita T, Yilmaz OH, Terhorst C, Morrison SJ. SLAM family receptors distinguish hematopoietic stem and progenitor cells and reveal endothelial niches for stem cells. Cell. 2005;121:1109–21.
65. Palmer TD, Willhoite AR, Gage FH. Vascular niche for adult hippocampal neurogenesis. J Comp Neurol. 2000;425:479–94.
66. Snyder EY, Loring JF. A role for stem cell biology in the physiological and pathological aspects of aging. J Am Geriatr Soc. 2005;53:S287–91.
67. Kobayashi A, Behringer RR. Developmental genetics of the female reproductive tract in mammals. Nat Rev Genet. 2003;4: 969–80.
68. Masuda H, Maruyama T, Hiratsu E, Yamane J, Iwanami A, Nagashima T, et al. Noninvasive and real-time assessment of reconstructed functional human endometrium in NOD/SCID/gamma c(null) immunodeficient mice. Proc Natl Acad Sci U S A. 2007;104:1925–30.
69. Blau HM, Brazelton TR, Weimann JM. The evolving concept of a stem cell: entity or function? Cell. 2001;105:829–41.
70. Raff M. Adult stem cell plasticity: fact or artifact? Annu Rev Cell Dev Biol. 2003;19:1–22.
71. Korbling M, Estrov Z. Adult stem cells for tissue repair – a new therapeutic concept? N Engl J Med. 2003;349:570–582.
72. Wagers AJ, Weissman IL. Plasticity of adult stem cells. Cell. 2004;116:639–48.
73. Kinnaird T, Stabile E, Burnett MS, Shou M, Lee CW, Barr S, et al. Local delivery of marrow-derived stromal cells augments collateral perfusion through paracrine mechanisms. Circulation. 2004;109:1543–9.
74. Pauwelyn KA, Verfaillie CM. Transplantation of undifferentiated, bone marrow-derived stem cells. Curr Top Dev Biol. 2006;74:201–51.
75. Bratincsak A, Brownstein MJ, Cassiani-Ingoni R, Pastorino S, Szalayova I, Toth ZE, et al. CD45-positive blood cells give rise to uterine epithelial cells in mice. Stem Cells. 2007;25:2820–6.
76. Du H, Taylor HS. Contribution of bone marrow-derived stem cells to endometrium and endometriosis. Stem Cells. 2007;25:2082–6.
77. Taylor HS. Endometrial cells derived from donor stem cells in bone marrow transplant recipients. JAMA. 2004;292:81–5.
78. Kim JY, Tavare S, Shibata D. Counting human somatic cell replications: methylation mirrors endometrial stem cell divisions. Proc Natl Acad Sci U S A. 2005;102:17739–44.
79. Tanaka M, Kyo S, Kanaya T, Yatabe N, Nakamura M, Maida Y, et al. Evidence of the monoclonal composition of human endometrial epithelial glands and mosaic pattern of clonal distribution in luminal epithelium. Am J Pathol. 2003;163:295–301.
80. Salamonsen LA, Lathbury LJ. Endometrial leukocytes and menstruation. Hum Reprod Update. 2000;6:16–27.
81. Cui CH, Uyama T, Miyado K, Terai M, Kyo S, Kiyono T, et al. Menstrual blood-derived cells confer human dystrophin expression in the murine model of Duchenne muscular dystrophy via cell fusion and myogenic transdifferentiation. Mol Biol Cell. 2007;18:1586–94.
82. Gargett CE, Chan RW. Endometrial stem/progenitor cells and proliferative disorders of the endometrium. Minerva Ginecol. 2006;58:511–26.
83. Giudice LC, Kao LC. Endometriosis. Lancet. 2004;364:1789–99.
84. Leyendecker G, Herbertz M, Kunz G, Mall G. Endometriosis results from the dislocation of basal endometrium. Hum Reprod. 2002;17:2725–36.
85. Starzinski-Powitz A, Zeitvogel A, Schreiner A, Baumann R. In search of pathogenic mechanisms in endometriosis: the challenge for molecular cell biology. Curr Mol Med. 2001;1:655–64.
86. Tanaka T, Nakajima S, Umesaki N. Cellular heterogeneity in long-term surviving cells isolated from eutopic endometrial, ovarian endometrioma and adenomyosis tissues. Oncol Rep. 2003;10:1155–60.
87. Jimbo H, Hitomi Y, Yoshikawa H, Yano T, Momoeda M, Sakamoto A, et al. Evidence for monoclonal expansion of epithelial cells in ovarian endometrial cysts. Am J Pathol. 1997;150:1173–8.
88. Wu Y, Basir Z, Kajdacsy-Balla A, Strawn E, Macias V, Montgomery K, et al. Resolution of clonal origins for endometriotic lesions using laser capture microdissection and the human androgen receptor (HUMARA) assay. Fertil Steril. 2003;79 Suppl 1:710–7.
89. Miller SJ, Lavker RM, Sun TT. Interpreting epithelial cancer biology in the context of stem cells: tumor properties and therapeutic implications. Biochim Biophys Acta. 2005;1756:25–52.
90. Pardal R, Clarke MF, Morrison SJ. Applying the principles of stem-cell biology to cancer. Nat Rev Cancer. 2003;3:895–902.
91. Polyak K, Hahn WC. Roots and stems: stem cells in cancer. Nat Med. 2006;12:296–300.
92. Dalerba P, Cho RW, Clarke MF. Cancer stem cells: models and concepts. Annu Rev Med. 2007;58:267–284.
93. Lapidot T, Sirard C, Vormoor J, Murdoch B, Hoang T, Caceres-Cortes J, et al. A cell initiating human acute myeloid leukaemia after transplantation into SCID mice. Nature. 1994;367:645–8
94. Al-Hajj M, Wicha MS, Benito-Hernandez A, Morrison SJ, Clarke MF. Prospective identification of tumorigenic breast cancer cells. Proc Natl Acad Sci U S A. 2003;100:3983–8.
95. O'Brien CA, Pollett A, Gallinger S, Dick JE. A human colon cancer cell capable of initiating tumour growth in immunodeficient mice. Nature. 2007;445:106–10.
96. Dalerba P, Dylla SJ, Park IK, Liu R, Wang X, Cho RW, et al. Phenotypic characterization of human colorectal cancer stem cells. Proc Natl Acad Sci U S A. 2007;104:10158–63.
97. Ricci-Vitiani L, Lombardi DG, Pilozzi E, Biffoni M, Todaro M, Peschle C, et al. Identification and expansion of human colon-cancer-initiating cells. Nature. 2007;445:111–5.
98. Hubbard S, Chan RWS, Gargett CE. Preliminary evidence for progenitor/stem cells in human endometrial adenocarcinoma. In: 5th International Society for Stem Cell Research Annual Meeting; 2007 17–20 June; Cairns, Australia; 2007. p. 85.
99. Amant F, Moerman P, Neven P, Timmerman D, Van Limbergen E, Vergote I. Endometrial cancer. Lancet. 2005;366:491–505.
100. Rahaman MN, Mao JJ. Stem cell-based composite tissue constructs for regenerative medicine. Biotechnol Bioeng. 2005;91:261–84.
101. DeLancey JO. The hidden epidemic of pelvic floor dysfunction: achievable goals for improved prevention and treatment. Am J Obstet Gynecol. 2005;192:1488–95.

Epithelial Stem Cells and the Development of the Thymus, Parathyroid, and Skin

Chew-Li Soh, Joanna M.C. Lim, Richard L. Boyd and Ann P. Chidgey

Abstract It is evident that epithelial tissues are able to self-renew not only during normal homeostasis but also following damage. How this occurs has major implications for the development of therapies for degenerative conditions. While not fully understood, tissue self-renewal through the activation of endogenous stem cells is a likely possibility, however, regeneration through replication of pre-existing transit amplifying or mature cells may also occur. The current consensus is that stem cells reside in tissue-specific niches that are important in maintaining the stem cell state and exposing the cells to various differentiation signals. A comparative analysis of these niches from diverse tissue types has revealed similarities in signaling pathways for the maintenance, lineage determination, and differentiation of epithelial stem cells, but also has shown discrete differences. This review examines these issues, with a particular focus on the development of thymic epithelial cells. Understanding the nature of epithelial stem cell self-renewal and differentiation and the niches in which they reside will have important future clinical implications in regenerative medicine.

Keywords Epithelium · Epithelial stem cells · Niche · Thymus · Skin · Regenerative medicine

1 Introduction

1.1 The Epithelium

Epithelia are laterally coherent sheets of cells that cover the surfaces (e.g., epidermis of the skin), line the open cavities (e.g., gastrointestinal tract), or fashion the glands (e.g., thymus) of the body. Whereas most epithelial tissues are arranged in two-dimensional sheets with an underlying basement membrane, the thymus is unusual because the various epithelial subsets form tightly interconnected three-dimensional matrices without a clearly defined base. Possessing distinct apical-basolateral polarity, epithelia often function as an interface between different environments, and in doing so accomplish diverse functions including mechanical and pathogenic protection, nutrient absorption, waste filtration and excretion, hormonal secretion, and sensory reception. The epithelial apical surface encounters either the exterior surroundings or the luminal space; the basal domain mediates interaction between the cell and underlying extracellular matrix (ECM), while the lateral domain is involved in cell-cell adhesion and thus the maintenance of structural integrity [1]. Epithelial cells are customarily classified by a combination of three histological criteria: layering (simple, stratified), shape (squamous, cuboidal, columnar), and specialization (mucosal, keratinized, ciliated). Although their classification may be basic, their regulation is disparately rather complex. Epithelial morphogenesis requires the coordination and integration of an array of transcription factors and extracellular signals to achieve an appropriate biological outcome, whether that may be cell proliferation, differentiation, growth, or death. Most epithelia are able to self-renew by mitosis and exercise this property continuously to maintain tissue homeostasis; this property may be coupled to de novo generation of appropriate epithelia from resident stem cells at the specific site [2]. Given the potential clinical implications for regenerative medicine and disease treatment, a strong scientific focus now lies in the identification and characterization of epithelial stem cell populations and their relevance to induction, maintenance, and regeneration of epithelial-based tissues and organs.

1.2 Embryonic Origins of Epithelium

During embryogenesis, epithelialization commences following three to four cleavages of the zygote [3]. Shortly after, gastrulation occurs whereby a complex ensemble of cellular

Chew-Li Soh, Joanna M. C. Lim – these authors contributed equally

A.P. Chidgey (✉)
Monash Immunology and Stem Cell Laboratories (MISCL), Level 3, Building 75, Monash University, Wellington Road, Clayton, Victoria 3800, Australia
e-mail: ann.chidgey@med.monash.edu.au

V.K. Rajasekhar, M.C. Vemuri (eds.), *Regulatory Networks in Stem Cells*, Stem Cell Biology and Regenerative Medicine,
DOI 10.1007/978-1-60327-227-8_33, © Humana Press, a part of Springer Science+Business Media, LLC 2009

movements in the blastula result in the formation of a three-layered body plan, wherein primordial tissues are appropriately allocated to begin organogenesis. In mice and humans, gastrulation occurs between days 6.5–9 and days 13–16 of gestation, respectively [4, 5]. Initiation of gastrulation is marked by the emergence of the primitive streak and concludes with the formation of the three primary germ layers: the outer ectoderm, the middle mesoderm, and the inner definitive endoderm. Epiblast cells that ingress through the primitive streak form the mesoderm and definitive endoderm, whereas nonmigratory cells constitute the embryonic ectoderm [6]. Epithelia may be derived from ectoderm, mesoderm, or endoderm, and as organogenesis ensues, tissue-specific morphogenetic cues trigger the activation of signaling pathways, epithelial-mesenchymal transitions, and interactions that result in the structural and functional adaptation of diverse epithelia.

1.3 Embryonic and Adult Stem Cells

The appeal of stem cells for therapeutic application lies in their two classically defining characteristics: their unique ability to replenish themselves through self-renewal via cell division while maintaining the undifferentiated state, and their capacity to differentiate into different types of mature cells. These properties play a fundamental role in tissue organogenesis during embryonic development and tissue regeneration in the adult [7]. Embryonic stem cells are derived from the inner cell mass of the blastocyst [8, 9] and are functionally pluripotent; they possess the ability to give rise to all derivatives of the three embryonic germ layers and it is this theoretically unrivaled differentiation range that has prompted intensive research towards directing and exploiting their differentiation capabilities. As development proceeds and necessity for organogenesis arises, the embryo generates germline stem cells for reproduction and somatic stem cells for tissue formation. Although functionally diverse, both cell types retain the essential feature of indefinite self-renewal, but in line with the need for more specialized organs, they are progressively restricted in development – either being multipotent (limited to related lineages) or unipotent (single lineage cells) for certain tissues [7]. After birth, the surviving adult stem cells, which include both germline and somatic stem cells, are thoroughly dispersed among the various tissues, representing minor populations that occupy special niche microenvironments from where they serve their major role in cellular homeostasis [10].

The focus of this review will be on epithelial stem cells and the importance of the stem cell niche in providing the appropriate signals for their maintenance, activation, lineage commitment, and differentiation. Common epithelial properties and tissue-specific differences will be discussed with an emphasis on the development of the skin and thymus – two functional units with important applications in regenerative medicine.

2 The Epithelial Stem Cell Niche

Historically the term stem cell "niche" has been used simply to describe stem cell location [11], in the sense that it is a specific location in a tissue where stem cells can reside and self-renew for an indefinite period of time, producing, as necessary, differentiated progeny. However, it is now essential to include a description of the cellular components of the stem cell microenvironment, including the neighboring resident cells and the rich milieu of ECM they secrete, as well as the signals emanating from them in a definition of what constitutes an actual niche [7, 12]. Because the stem cell intimately depends on its surrounding environment for maintaining its unique properties, proper niche development is crucial [13] and is maintained by a constant dialogue between the stem cells and the surrounding niche cells [14]. These niches provide the protective microenvironment for stem cell survival and protection of their genetic profile. They minimize undesired differentiation cues and apoptotic stimuli that may challenge and exhaust reserves and prevent overproduction or abnormal proliferation of stem cells that may lead to dysregulation and cancer [15, 16].

2.1 Maintenance of Homeostasis

The capacity to divide symmetrically to generate two committed daughters, such that both enter differentiation pathways leading to a loss of self-renewal characteristics, is a feature of most somatic cells and is how organogenesis ensues (Fig. 1A). However, an added feature of stem cells is their ability to undergo asymmetric cell divisions, yielding one committed progenitor daughter and one stem cell daughter (Fig. 1B). In the niches, regulating the balance between symmetric and asymmetric stem cell divisions is critical in meeting the demand for differentiated cells within the surrounding tissue, and in maintaining homeostatic balance [17]. The committed daughter cells are typically referred to as the transit amplifying (TA) population; they expand cell numbers, proliferating more frequently than the anchored stem cell and give rise to descendents that travel down one of several terminal differentiation fates [13]. The presence of TA cells means that the tissue can maintain a high output of differentiated cells from a smaller number of stem cells [2]. The mechanism by which asymmetric stem cell

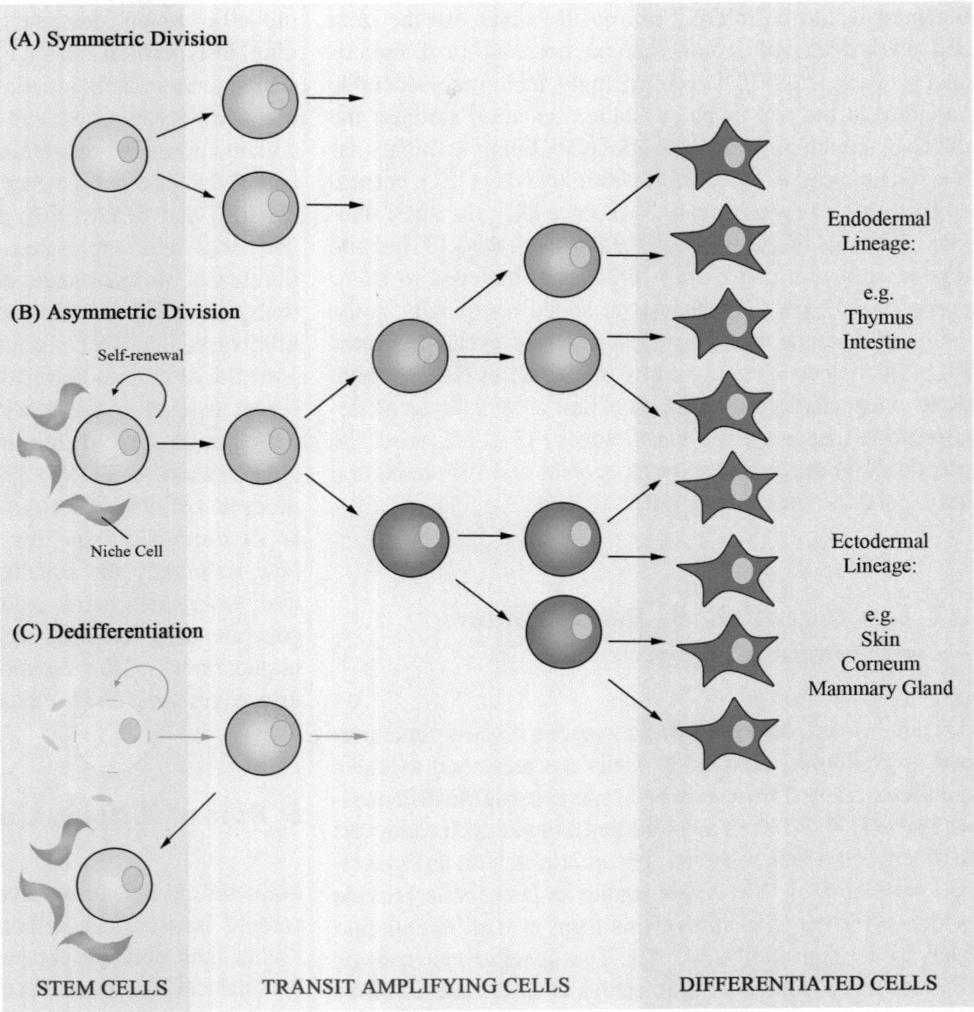

Fig. 1 Maintenance of tissue homeostasis. Examples of patterns of stem cell division and differentiation. **(A)** Symmetric Division. Each symmetric division generates two committed daughter cells that enter differentiation pathways. **(B)** Asymmetric Division. Each asymmetric division generates one amplifying committed progenitor daughter cell for subsequent division to produce functional terminally differentiated cells and one stem cell daughter for maintenance via self-renewal. The rates of self-renewal and differentiation are balanced to maintain tissue homeostasis. **(C)** Dedifferentiation. If a stem cell daughter is lost, a committed daughter cell is thought to be capable of reverting to a stem cell if restored to an appropriate niche (*see* also Color Insert)

division is achieved is believed to be heavily dependent on the orientation of the mitotic spindle. Contact of stem cells with a basal lamina allows the generation of cell polarity and differentiation-specific cell determinants. If the strongest spindle-polarizing signal emanates from the basal lamina pole, the cell determinants will be asymmetrically partitioned following mitosis. In contrast, if the strongest spindle-polarizing signal localizes at sites of cell–cell contact, the determinants are partitioned equivalently, leaving two identical daughters. Thus, the overriding key in deciding between asymmetric or symmetric division is whether the orientation of the mitotic spindle is parallel or perpendicular to the major cell polarity axis that aligns the cell fate determinants [12, 18–20]. The polarity is initiated through stimuli that prompt reorganization of the actin cytoskeleton, often making integrins (which regulate cell–matrix adhesion) and cadherins (which regulate cell–cell interaction) central to the process [1]. This type of stem cell–orientated division and hierarchical amplifying behavior has been well demonstrated in invertebrate model systems [21, 22], and appears to be typical of the renewal of most mammalian epithelial tissues, including the intestinal crypts [23], cornea [24], mammary glands [25], and skin [26, 27], though inevitably with varying degrees of structural complexity.

Consequently, although cell diversification is primarily complete at or shortly after birth, organisms set aside lifelong reservoirs of adult stems cells to maintain tissue homeostasis, replenishing cells as they die through natural apoptosis or by injury [12]. Evidently, to sustain this function throughout the lifespan of an organism, it is imperative to maintain a delicate balance between self-renewal and differentiation [7]. The pattern of symmetric division indicated above, in which more committed progeny arise, would be expected to occur under conditions where there is a loss of suitable niches or cues for stem cell maintenance or additional stimuli as a consequence of tissue damage promoting more rapid replenishment [17]. Asymmetric division, in contrast, would be maintained in a balanced homeostatic steady state. If stem cells are required for recruitment to newly formed niches or following tissue loss by wounding, the self-maintenance potential would be initially promoted to increase the stem cell pool size, with a concomitant decrease in differentiated progeny [28–30]. Not surprisingly, there is considerable variation in the rate of homeostatic regulation amongst the different epithelia. Intestinal epithelial lining is highly labile, completely self-renewing within five days [31]; corneal epithelium self-renews every 3–10 days [32]; the interfollicular epidermis of the skin takes about 8–10 days [33]; while thymic epithelial cell (TEC) turnover is believed to occur every 10–14 days [34–36]. As an added complexity, some epithelia undergo tissue homeostasis by a cyclical process [31]. This is best exemplified by hair follicles (HFs), which cycle constantly through periods of hair growth (anagen), degeneration (catagen), and rest (telogen) [37]. Likewise, the mammary gland cycles through growth and differentiation during and after pregnancy [38].

2.2 Dedifferentiation – Reacquisition of Stemness?

Asymmetric cell division is an irreversible process under normal physiological conditions. Cells are presumed to move unidirectionally from a stem cell state towards the differentiated state [17]. Whether a committed daughter cell can revert to a stem cell if restored to its appropriate niche is an interesting consideration. Two recent studies in *Drosophila* provide evidence for the existence of this form of dedifferentiation back to a "stem cell" [39, 40]. The specific mechanisms in place to stabilize the initial stages of differentiation may be reversed or overridden. In vertebrate systems, evidence for a similar pattern has arisen from radiation-induced stem cell apoptosis in the gut [41, 42]. Furthermore, cells later in the hierarchy have been shown to return to a stem cell state if transfected with viral oncogenes [43] or exposed to embryonic mesoderm [44]. A recent study in the hematopoietic system indicated that constitutive β-catenin expression in committed progenitors resulted in a reacquisition of some stem cell properties [45]. Moreover, it has been shown that when stem cells in the male testis are lost due to injury, the remaining progenitors, or progenitor cells acquired by transplantation, have the potential to acquire stem cell functions to repopulate the niche [46]. In short, such studies suggest that stem cell function is not strictly cell autonomous, and that there is a potential for some cells to "reattain stemness" if a suitable niche is intact (Fig. 1C). Consequently, if such mechanisms can be further defined, TA cells – which vastly outnumber true stem cells in vivo – may represent a valuable source of replenishment for normal or therapeutic tissue repair.

Indeed, the ability for human somatic cells to be reprogrammed has recently been reported by several studies [47–49], which describe methods to transform human fibroblasts into an induced pluripotent stem (iPS) cell state such that their differentiation potential and properties mimic that of embryonic stem cells. Employing a remarkably basic combination of four defined transcription factors – *OCT4* and *SOX2*, aided by either *KLF4* and *C-MYC* [47, 48] or *NANOG* and *LIN28* [49], the researchers demonstrate that the exogenous expression of these genes can reprogram fibroblasts differentiated from genetically modified cell lines, in addition to normal fetal, neonatal, and adult fibroblasts, into iPS cells, recapitulating the dedifferentiation potential of murine fibroblasts [50]. These reports follow the recent revelation that nonhuman primate embryonic stem cells can be derived from adult fibroblasts via somatic cell nuclear transfer (SCNT) [51]. Whether the reprogramming processes of iPS cells and SCNT-derived cells are equivalent is an important issue that needs to be addressed, as is how accurately the function of iPS cells reflects that of true embryonic stem cells. Nevertheless, the ability to generate stem cells from adult somatic cells has significant implications in the creation of patient-specific cells for disease research and eventual therapeutic application.

3 Role of Niche in Lineage Commitment

Maintaining a balance between stem cell quiescence and activity is a critical hallmark of a functional niche [16]. Rather than acting independently, it is the combination of the intrinsic characteristics of stem cells and their associated microenvironment that shapes their properties and defines their potential [12]. Molecular cross-talk between the epithelium and its adjacent mesenchyme, as well as key signaling pathways, are patterned to occur at the appropriate time and location, thereby providing the mechanisms by which the niche controls stem cell lineage commitment.

3.1 Epithelial-Mesenchymal Interactions Influence Epithelial Cell Fate

A feature of the epithelium is its close interaction with the underlying mesenchyme – a composition of nonpolarized, loosely associated connective tissue set in a gelatinous ECM [52]. The mesenchyme is known to directly influence the development of epithelial cells in embryonic rudiments such as the limb bud, where bone morphogenetic proteins (Bmps) and fibroblast growth factors (Fgfs) have been identified as molecular mediators of mesenchymal function [53, 54]. De novo HF formation and growth in skin is initiated when specialized mesenchymal dermal papilla cells transmit cues to multipotent epithelial stem cells in the hair bulge [55]. Mesenchymal cells, via their production of certain growth factors, are also required to support the proliferation of thymic epithelial progenitor cells, leading to expansion of the embryonic thymus [53]. Similarly, fibroblasts in the pericryptal mesenchyme of the intestine are thought to intimately influence the differentiating epithelium via growth factor secretion [56, 57]. Intriguingly, tissue recombination experiments have demonstrated that epithelia can be induced to adopt the cell fate of the mesenchymal tissue with which it has been incubated. For instance, mammary phenotypic expression has been induced in epidermal cells via culture with mammary mesenchyme [58]. The embryonic dermis can cause adult central corneal epithelium to form skin epidermis or to differentiate into pilosebaceous units or sweat glands [59]. Moreover, epithelial lung buds can be directed to form gastric glands in the presence of stomach mesenchyme or intestinal villi by intestinal mesenchyme and hepatic cords by liver mesenchyme [52]. Furthermore, signaling mechanisms are well conserved throughout species as shown when human embryonic stem cells were induced to differentiate into prostate epithelium using rodent urogenital sinus and seminal vesicle mesenchyme, respectively [60]. The immature and mature prostate cells induced were morphologically and functionally identical to human fetal and adult prostate. Collectively, these studies suggest that not only do mesenchymal factors permit epithelial growth and morphogenesis, they may also impart instructive signals that determine the further differentiation program of the epithelia. This is closely linked with the notion that epithelial progenitor populations can be reprogrammed.

3.2 Signaling Pathways Controlling Epithelial Stem Cell Function

Despite morphological and functional differences among epithelia, common signaling pathways appear to control epithelial stem cell maintenance, activation, lineage determination, and subsequent differentiation [31]. Although each epithelial niche possesses unique features to facilitate a specialized role, they likely share many universal aspects of regulation [13]. Some of these signaling pathways may act as direct regulators of the stem cell or their descendent transit amplifiers by modifying the proliferative capacity. Alternatively, others may function indirectly at the level of the stem cell niche. The necessity for precise control of stem cell fate and function is reflected by the fact that malignancy may result from fundamental disturbances in the homeostatic balance associated with normal stem cell self-renewal [17]. One of the key features of cancer cells is their unlimited capacity to divide, reminiscent of the proliferative potential of stem cells [61]. The niche minimizes tumorigenesis by controlling stem cell division and arrest. Therefore, any mutation that guides stem cells towards escape from niche regulation may facilitate uncontrolled proliferation. It thus follows that one of the primary differences between normal stem cells and cancer stem cells is that the latter may no longer be responsive to niche signaling [7].

Several signaling pathways have emerged as key regulators of stem cells, with members of the Wingless (Wnt), Bmp, Notch, and Hedgehog (Hh) families being implicated in the maintenance, self-renewal, or differentiation of various stem cell populations.

3.2.1 Wingless (Wnt)

The canonical Wnt signaling cascade orchestrates development and cellular homeostasis in a diverse range of morphogenesis programs [62]. In the context of stem cells, the Wnt signaling pathway, and its downstream transcription factor Lymphoid enhancer factor-1/T-cell factor (*Lef/Tcf*) are intimately associated with maintenance of the niche [13]. The secreted Wnt product serves as a ligand for the serpentine receptor, Frizzled [63]. Activation of the signaling cascade results in the inhibition of Glycogen synthase kinase-3β (Gsk3β), which in the absence of Wnt, normally maintains the cytoplasmic pool of β-catenin by phosphorylating β-catenin for targeted ubiquitin-mediated degradation via the adenomatous polyposis coli (APC)/axin/GSK3b complex. Otherwise, β-catenin is most commonly associated with cadherins at sites of intercellular adhesion junctions [64]. However in the presence of Wnt, the cytoplasmic pool of unphosphorylated β-catenin accumulates, allowing β-catenin to translocate to the nucleus to influence gene expression through the Lef/Tcf transcriptional complex [65, 66].

3.2.2 Bone Morphogenetic Protein (Bmp)

A variety of stem cell types utilize Bmp signals in a multitude of ways to define their fates, whether that may be maintaining pluripotency in embryonic stem cells, or regulating proliferation and differentiation in other niches [67]. Bmp belongs to the larger Transforming growth factor-β (Tgfβ) superfamily, which convey signals through type I and type II transmembrane serine-threonine kinase receptors to intracellular mediators known as Smad proteins [68]. In a simplified linear transduction model, Bmp ligand binding to the extracellular domain of type II receptors induces a conformational change that enables recruitment, phosphorylation and activation of type I receptors, which in turn phosphorylates and activates Smad 1, 5, or 8. The receptor-activated Smads then form a complex with Smad4, which undergoes nuclear translocation to stimulate gene transcription [67, 69, 70].

3.2.3 Notch

Notch signaling interacts intricately with other key signaling cascades in delineating cell fates and maintaining stem cell niches. Historically named after the *Drosophila* "notched"-winged mutant [71], the highly conserved *Notch* genes encode four receptors (Notch 1–4) [72]. Notch receptors and their corresponding five mammalian ligands, three Delta-like (1/3/4) and two Jagged (1/2), are single transmembrane proteins that mediate direct cell–cell contacts [73]. Notch ligand binding initiates a series of proteolytic cleavages of the receptor transmembrane subunits (S2, S3, and S4) by metalloproteinases and γ-secretases [74]. This results in the translocation of the Notch intracellular domain (NICD) into the nucleus where it associates with DNA-binding proteins such as RBP-J, switching them into transcriptional activators [31, 73]. Notch targets include the hairy-related basic helix-loop-helix transcription factors, *Hes* and *Hey* [75].

3.2.4 Hedgehog (Hh)

Hedgehog (Hh) signaling modulates developmental cascades and stem cell behavior, exerting differential effects on epithelial stem cell populations [31]. Hh proteins are potent mitogens and have been postulated to act downstream of Wnt in some situations [76, 77]. Sonic (Shh), Indian (Ihh), and Desert (Dhh) hedgehogs collectively make up the mammalian Hh family of secreted proteins, of which Shh is the most extensively studied [78]. During synthesis, Hh proteins undergo autocatalytic processing and are secreted by the Dispatched proteins [79]. Secreted Hhs then elicit cellular responses by binding to Patched-1 receptors, preventing Patched-1-mediated repression of Smoothened transmembrane proteins [78]. Smoothened proteins accumulate, allowing the activation of Gli transcription factors, which in turn regulate the expression of several target genes [79].

4 Adult Epithelial Stem Cell Niches

Epithelial stem cells reside in tissues of wide ranging architectural and functional types. Here we will discuss three examples of such diverse functional units – intestine, corneum, and mammary gland – in the context of the niches in which their epithelial stem cells reside, with particular emphasis on the role of epithelial-mesenchymal interactions and the integration of key signaling pathways in directing epithelial cell fate.

4.1 Intestinal Niche

The intestinal epithelium can be divided into two regions: a functional villus that accommodates the differentiated progeny and a proliferative crypt region (crypts of Lieberkühn) that represents the epithelial stem cell niche [13]. Multipotent epithelial stem cells reside in the crypts and give rise to the four principal epithelial lineages: absorptive enterocytes, mucin-secreting goblet cells, peptide hormone-producing enteroendocrine cells, and microbicidal Paneth cells [80, 81]. The first three of these lineages differentiate as they migrate up the crypt to the villus tip, after which they undergo apoptosis or are exfoliated into the lumen [13, 31]. The intestinal stem cells regenerate the entire villus every 3–5 days [82], and are generally proposed to be located at the fourth or fifth position from the base of the crypt [83–88]. The Paneth cells differentiate as they migrate downwards to reside at the crypt base. They are longer-lived cells, surviving 18–23 days before being phagocytosed by surrounding cells [89]. Although the location has been revealed, there is still uncertainty over the actual number of stem cells in each crypt owing to a current lack of markers that would permit their isolation from their putative crypt position [90]. At present there is debate over whether a single multipotent stem cell is selected to populate each adult niche [87, 91], or whether there are in fact four to six stem cells per crypt or even more [16, 92, 93].

The crypt epithelium contacts an underlying mesenchyme comprised of ECM, blood vessels, enteric neurons and fibroblasts [13], and cross-talk between the epithelium and mesenchyme is essential to support niche development. For instance, the pericryptal fibroblasts secrete hepatocyte growth factor (Hgf), Fgf7 (also known as keratinocyte growth factor) and Tgfβ, which have cognate receptors located on the adjacent epithelium [56]. Endothelial cells in the mesenchyme have been proposed to provide intestinal stem cells with survival signals such as Fgf [94]. Platelet-derived growth factor (Pdgf) may be another such factor, given that mice deficient for *Pdgfα* or its receptor (*Pdgfrα*) develop abnormal intestinal epithelium and depletion of the pericryptal mesenchyme [95]. Other clues to the regulation of the intestinal niche arise from analysis of key signaling pathways.

Wnt signaling is essential for maintaining the proliferative status of the intestinal stem cell niche. Mice lacking the Wnt-responsive transcription factor, *Tcf4*, display a neonatal intestinal epithelium composed entirely of inappropriately differentiated villus cells with severely compromised proliferation [96]. In homozygous Dickkopf-1 (*Dkk1*) transgenic intestines, overexpression of *Dkk1* – a soluble Wnt ligand inhibitor – also resulted in the suppression of proliferation [97]. Wnt is therefore critical for either maintaining proliferation within the stem cell niche or inhibiting the differentiation of the transit amplifying component. Wnt also functions in directing cell migration within the niche. Among the significant Wnt targets are the ephrin tyrosine kinase receptors, EphB2 and EphB3 [98], which are expressed in the intestinal crypts [99, 100]. Compound mutations of *EphB2* and *EphB3* result in the mispositioning of differentiated cells along the crypt-villus axis, suggesting that ephrins and their receptors generate complementary gradients to establish a boundary between the proliferating and differentiating compartments of the intestinal niche.

The Bmp signaling pathway functions as a negative regulator of intestinal stem cell proliferation, thereby completing a complementary axis with the Wnt cascade, via convergence on the control of nuclear β-catenin [84]. Bmp has been shown to repress nuclear β-catenin accumulation thereby suppressing the proliferative effects of Wnt signaling [84]. *Bmp4* is expressed in the intravillus mesenchyme, adjacent to the epithelial stem cells. Conditional inactivation of its receptor, *Bmpr1a*, in mice crypt cells caused an expansion of the intestinal stem and progenitor cell populations, leading to intestinal polyposis [84]. Similarly, intestinal overexpression of the Bmp inhibitor, *Noggin*, resulted in de novo ectopic crypt formation, which predisposed to intraepithelial neoplasia [101]. Ablation of *Smad4*, a Bmp transcriptional co-activator, also led to aberrant crypt formation [102], overall indicating a key role for Bmp signaling in directing crypt cell fate.

Notch signaling acts on intestinal stem cells to maintain proliferation of the absorptive enterocytes, while restricting the fate of the three secretory cell lineages – goblet, enteroendocrine, and Paneth cells. Notch activity leads to expression of *Hes1* and subsequent repression of *Math1*, prompting enterocyte differentiation. In the absence of Notch, *Math1* expression is permitted, resulting in differentiation of the secretory progenitors [103–106]. In support of this, a gain-of-function mutation in *Notch* – through constitutive expression of the NICD transcriptional co-activator – led to enhanced enterocyte progenitor proliferation in the crypt [107], whereas crypt cell proliferation was grossly impaired and goblet cell population enhanced following treatment with γ-secretase inhibitors or RBP-J transcriptional ablation [108].

Epithelial Hh signals have an impact on overall intestinal morphogenesis, and are required for patterning of the intestinal crypt-villus axis [109]. In an analysis of the effect of Hh signals on gastrointestinal development, it was shown that *Shh* mutants display intestinal transformation of the stomach epithelium, whereas *Ihh* mutants show reduced epithelial stem cell proliferation, as reflected by a substantial reduction in villus size [110]. Paradoxically, mice receiving anti-Hh monoclonal antibodies revealed a hyperproliferation of intestinal crypt epithelial cells and villus disorganization, however, cell migration, survival, and lineage commitment were apparently normal. More recently, Ihh has been identified as an antagonist of Wnt signaling in intestinal epithelial cells and may have a direct impact on enterocyte differentiation [111]. These contrasting reports may reflect a critical temporal requirement for Hh signaling during intestinal morphogenesis that varies between the embryo and adult.

4.2 Corneal (Limbal) Niche

The surface of the eye is comprised of the corneal and conjunctival epithelia, separated by a corneoscleral limbus [112]. Corneal epithelium is transparent, stratified, squamous, and avascular, and its integrity is maintained by a continuous cellular renewal every 3–10 days, ensuring visual clarity [32, 112]. The external corneal cells which are constantly being discarded into the lachrymal fluid due to homeostasis and mechanical abrasions, are replaced by differentiating peripheral corneal TA cells which have migrated centripetally from the limbus [113, 114]. These corneal TAs are derived from multipotent limbal epithelial stem cells (LESCs), which reside in clusters of 5-10 cells, interspersed with early limbal and corneal TA cells at the base of the "Palisades of Vogt," a vascularized, innervated, and undulating structural component of the limbal basal layer [115–117]. A recent study narrowed the specific niche location to within interpalisade extensions known as limbal epithelial crypts [118]. Due to the lack of appropriate and specific consensus markers, the LESCs isolated to date represent a heterogeneous population [113, 119]. Nevertheless, cells within this population display common epithelial stem cell properties, such as slow-cycling (label-retention), high clonogenicity, and proliferative capacity in vitro and in vivo [120, 121].

LESCs are interspersed with TA cells, neural crest (NC)-derived mesenchymal cells (e.g., melanocytes), and Langherhan's antigen presenting cells, and sit atop a basement membrane that overlies an ECM-rich stroma filled with mesenchymal cells and blood vessels [121]. Melanocytes (melanin-producing cells) deposit melanin granules into limbal basal epithelial cells, possibly conferring protection (e.g., against ultraviolet-related damage) in a similar manner as that provided to HF bulge stem cells [122, 123]. Tissue recombination experiments have shown that the limbal stroma and basal ECM components are important in maintaining the stemness of LESCs, while corneal stroma promotes limbal epithelial cell differentiation [112, 121, 124]. An interesting aspect to this inductive potency is the ability of rabbit limbal stroma to induce the dedifferentiation of corneal TA cells into basal-like cells with the concomitant loss of differentiation markers keratin-3 (K3) and connexin-43 [124]. LESCs also demonstrate the ability to transdifferentiate into fibroblasts [125], with inductive factors retained in the limbal stroma thought to act at the crossroads of corneal-versus-fibroblast fates in LESCs [125]. Moreover, limbal stroma may provide cytokines and growth factors required for limbal epithelial maintenance, proliferation and/or differentiation [126]. For example, the epithelial mitogen Fgf7 is differentially produced by limbal and corneal fibroblasts [126], and may induce the proliferation and differentiation of corneal TAs [127]. Furthermore, neurotrophins produced by limbal stromal cells may have roles in the regulation of LESCs [128, 129].

A range of *WNTs* and *WNT antagonists* are expressed in a reciprocal spatiotemporal manner during avian eye development, in the surface ocular ectoderm, corneal epithelium, and corneal endothelium [130, 131]. Wnt7a may have a role in corneal re-epithelialization post-injury, due to its elevated expression in wounded rat cornea and the ability of exogenous Wnt7a to induce proliferation of damaged corneal epithelial cells in vitro [132]. Moreover, Wnt ligands may be one of the cutaneous dermal-derived factors inducing the initial dedifferentiation and subsequent transdifferentiation of corneal epithelial cells towards an epidermal fate [133]. Conversely, corneal specification and differentiation in the external ocular epithelium may be regulated via the inhibition of Wnt signaling by Dkk2 [134].

The Bmp signaling pathway may have roles in corneal epithelial-stromal cross-talk and regulation, due to the presence of BMP ligands and receptors on human corneal epithelium and stroma [135]. Haploinsufficiency of *Bmp4* in mice results in a range of peripheral corneal abnormalities including corneal thinning, haziness, scarring, and vascularization, indicating a key role for Bmps in proper corneal development [136].

Notch1 display similar roles in the epidermis and limbus, possibly in the maintenance of stem cell niches, as evidenced by the extensive keratinization and hyperproliferation seen with its absence in the murine adult corneal epithelium [137]. The high expression of *Notch1* in putative rat LESCs, isolated by the Hoescht dye efflux method [138], and in the human limbal Palisades of Vogts [139] supports the notion of Notch signaling in LESC regulation. Conversely, Notch signaling may have an opposing role in the corneal epithelium whereby human corneal epithelial cell differentiation, indicated by elevated K3 expression, was observed upon pathway activation by the ligand, JAGGED-1 [140].

Shh and Patched-1 proteins are both expressed in rat adult limbal epithelial cells, while only low levels of Patched-1 (protein and transcripts) and *Dhh* transcripts are expressed in the rodent corneal epithelium [141, 142]. The low expression of Hh proteins and their receptors in corneal epithelium suggests a minor role in homeostatic maintenance. Supplementation of exogenous Shh to an in vitro rodent corneal wounding system caused proliferation of corneal epithelial cells, however in vivo suppressive mechanisms may be in place due to the unaltered proliferation rates of injured Shh-expressing corneal and limbal epithelium [141].

4.3 Mammary Niche

Mammary epithelia form the branching structures that invade specialized fat pads, and encompass three main cell types: myoepithelial (basal layer of ducts and alveoli), ductal, and alveolar (milk production) [143, 144]. Ductal and alveolar progenitors give rise to two other cell types: basal and luminal [38]. During stages of adolescence, pregnancy, lactation, and involution, the ductal and alveolar epithelia undergo cyclical growth and remodeling according to hormonal and stromal cues [145]. Extensive proliferation and multi-lineage differentiation of cap cells residing in transient club-shaped ductal terminal end buds suggests the presence of a stem cell niche [38, 146, 147]. However, the transient presence of these cap cells and terminal end buds, and the fact that nonterminal end bud mammary cells possess regenerative abilities raises the possibility of other niche locations [148]. These alternative niches may be occupied by putative stem and primary precursor cells, morphologically referred to in rodents as "small light cells," which have a scattered distribution throughout the mammary basement membrane [149, 150].

Putative stem cell/progenitor populations were able to reconstitute a complete mammary gland or either ductal or alveolar structures upon transplantation into de-epithelialized fat pads (with intact mesenchyme and vasculature) [151–154]. These in vivo studies demonstrated the existence of a hierarchical organization of mammary stem cells and lineage-restricted TA cells. Moreover, these cells retained the ability to self-renew following serial transplantation [155, 156]. Long-term multipotency of mouse and human mammary putative stem cells in vitro was also observed during the formation of mammospheres, spherical colonies of undifferentiated cells, in the presence of required growth factors [157, 158].

Interestingly, the ability of ductal branch generation by donor transplants is dependent on cues from the mammary fat pad and not the epithelium, as indicated by the infrequent chimerism observed in epithelial co-transplants [156]. This suggests an essential importance for mammary stroma in mammary gland reconstitution. Indeed, during embryogenesis and postnatal development, mesenchymal cells of the mammary mesenchyme and fat pad, although distinct, orchestrate various stages and aspects of mammary gland development [159]. However, only the mammary mesenchyme contains unique instructive abilities in mammary specification of embryonic epithelial cells, while the supportive role of the mammary fat pad may be replaced by adipose tissue from other organs [159].

An extensive array of Wnt ligands are expressed and required at different stages of mammary gland development during embryogenesis, puberty, pregnancy, and lactation [144, 160]. For instance, mammary glands were absent in mouse models in which Wnt/β-catenin signaling was inhibited, either by removal of the downstream transcription factor *Lef1* [161] or ectopic expression of the Wnt inhibitor *Dkk* [162], suggesting a role for Wnt signaling in the maintenance and survival of mammary stem/progenitor populations.

A requirement for β-catenin signaling in progenitor survival was demonstrated by the elevated apoptosis of mammary progenitors and resulting defects in development and pregnancy-related proliferation observed in its absence in mice [163]. Maintenance of stem/progenitor cell populations by Wnt signaling was further apparent when increasing concentrations of β-catenin and Wnt ligands caused a proportionate increase in mammary stem and/or progenitor cell numbers [164–166].

Notch signaling may have roles in maintaining self-renewal and inhibiting differentiation of mammary stem/progenitor cells. Overexpression of *Notch4* in mice caused improper in vivo differentiation profiles, sometimes resulting in the formation of mammary neoplasms [167–169]. Similarly, activation of NOTCH signaling in human mammary epithelial cultures caused a 10-fold elevation in in vitro mammosphere formation, results of which were reversed upon addition of NOTCH-inhibiting antibodies [170]. These effects were only observed in mammary stem/progenitor populations [170] supporting the notion of NOTCH signaling in mammary stem/progenitor stemness and lineage differentiation.

The Hh signaling pathway also serves an important role throughout mammary gland development via its regulation of epithelial-mesenchymal interactions [171]. Conditional removal of *Patched-1* and *Gli2* in mice resulted in distinct aberrations of ductal size and development [172, 173]. Moreover, Hh signaling may play an important part in stem cell self-renewal and progenitor cell proliferation of normal and cancerous mammary tissues. The SHH signaling receptor *PATCHED-1*, and mediators *GLI1* and *GLI2* were highly expressed in human mammospheres derived from stem/progenitor cells and exogenous SHH supplementation increased both the number and size of mammospheres [174]. In human breast neoplastic tissues, higher levels of SHH, PATCHED-1 and GLI1 were observed in comparison to non-neoplastic breast tissues [175]. Mammosphere formation of normal mammary stem/progenitor cells and proliferation of breast cancer cell lines were inhibited by cyclopamine-blockade of SHH signaling, further confirming the involvement of SHH signaling in these processes [174, 175].

5 Epithelial Functional Units: The Thymus and Skin

Many molecular and cellular characteristics are shared amongst epithelial tissues of the body. The thymic epithelium in particular has aspectual commonality to epithelia of multiple organs, especially those sharing similar endodermal embryonic origins (e.g., lung). Surprisingly the thymus also shares many common features with the skin, despite originating from different germ layers, although there is ectodermal input into the full development of the thymus.

Thymic epithelial cysts, present to a small degree in the normal thymus, but to a large degree in the epithelial rudiment of forkhead-box transcription factor n1 (*Foxn1*)-null (nude) mice, display mixtures of ciliated epithelium and cells that resemble mucus-secreting respiratory goblet cells, which may be a default pathway for thymic epithelial progenitor cells [176]. This cystic epithelium in the normal thymus may represent progeny that failed to receive the appropriate environmental cues that induce the required levels of *Foxn1* for differentiation into the thymic epithelial lineage and supports the notion that specialized tissue-specific properties may only reflect the expression of unique combinations of regulatory and structural genes that are widely expressed among different epithelia [176].

Thymus and skin epithelia express similar cell surface and intracellular antigens [177, 178], keratins [179], calcineurin [180], and *p63* [181] and produce thymulin hormone [182]. Moreover, the developmental program undertaken by medullary TECs to Hassall's corpuscles is comparable to that by skin epidermal basal cells to cornified cells [178]. The putative TEC progenitor marker MTS24 [183, 184] also marks a putative epidermal intermediate progenitor cell population residing in the hair follicular region between the bulge and sebaceous gland [185–187]. In addition, the nude phenotype incorporates both thymic and skin abnormalities resulting from defective differentiation of immature TECs and abnormal keratinization of the hair shaft and epidermis, respectively [188–190]. Although skin is of ectodermal origin, one of the more striking dual-organ epithelial similarities is the ability of human epidermal keratinocytes and fibroblasts to support T-cell formation from hematopoietic stem cells (HSC), albeit at a very low efficiency in an in vitro three-dimensional culture system [191]. This finding added a possible functional property to the ever-growing list of phenotypic thymus-skin similarities.

5.1 Skin Organogenesis

The skin is an impermeable barrier encasing an organism, which possesses vital protective (mechanical, chemical, immununological), thermoregulatory, and sensory functions. It consists of the interfollicular epidermis (IFE), dermis, and appendages. The ectoderm-derived epidermis is a stratified keratinized epithelium above a mesoderm- and NC-derived dermis [192]. The dermis consists of a connective network of ECM components (e.g., collagen), inductive fibroblasts, nerve receptors, lymph, and blood vessels [192]. The cutaneous appendages include the HFs, sebaceous glands, eccrine and apocrine sweat glands, nails, teeth, and mammary glands [192–194].

5.1.1 Epidermis

The epidermis originates from neuroectodermal cells committed to an epidermal fate by Wnt signals at E8.5 [195]. Wnt signaling inhibits the reactivity of neuroectodermal cells to Fgfs and enables the induction of Bmp signaling [196]. Basal cells proliferate and undergo stratification commitment and initiation at E9.5 and E10.5, respectively, as evident by the expression of the keratin markers K5, K14, and K15 [197–199]. Basal keratinocytes migrate upwards and give rise to the endothelial-like periderm, which confers transient protection and preserves the monolayer configuration of basal cells [200, 201]. At E13.5, asymmetric divisions of basal keratinocytes in a perpendicular mitotic plane form an intermediate suprabasal layer (embryo-specific proliferative cells which express K1 and K10), which proceeds to differentiate at E15.5 into a layer of spinous cells [27, 202]. Spinous cells then undergo further differentiation into granular cells, signified by K1 and K10 down-regulation, and expression of structural proteins loricin and filaggrin at E16.5 and E17.5, respectively. Cornification marks the conclusion of epidermal stratification, resulting in an impermeable barrier of dead, flattened, and anuclear cells (corneocytes or squames) in a cornified envelope at around E18.5, by which time the periderm has been shed [195, 203, 204].

Molecular mechanisms governing the initiation, stratification, and cornification of the epidermis has been comprehensively reviewed elsewhere [195, 198, 205, 206] and will thus only be briefly discussed below in the context of their respective signaling pathways and stem cell niche regulation.

5.1.2 Hair Follicle (HF)

The HF is a pivotal element of the pilosebaceous unit, which includes a pigmented hair shaft, sebaceous gland, and arrector pili muscle. Follicular morphogenesis occurs between E13 and E18.5, and is mediated by interactions between the epidermal epithelium, dermis, and NC-derived mesenchyme [207]. Central to appendage development are reciprocal epithelial-mesenchymal interactions and intercellular signaling proteins from the Wnt, Bmp, Notch, Hh, Fgf, and tumor necrosis factor (Tnf) families [207, 208].

There are several key stages of embryonic hair folliculogenesis [194, 198, 207, 209, 210]. The first stage involves the formation of hair placodes, which are systematically positioned small clusters of epidermal keratinocytes resulting from cues from the underlying mesenchyme. Reciprocal signals from the epithelial placode mediate the formation of a dermal condensate of fibroblasts. Continuous cross-talk and proliferation leads to the formation of a hair germ (or bud) and subsequently the hair peg, in which the invagination and enclosure of the dermal condensate by the epithelial placode, leads structurally to the dermal papilla. By E18.5, core matrix cells and sebaceous gland progenitors (in the upper HF) are present. Postnatal maturation until day 16 results in the formation of a complete HF, through proliferation and differentiation of matrix cells into eight concentric layers, which internally to externally incorporate the inner medulla, cortex, cuticle, three inner root sheath (IRS) layers, the companion layer, and the outer root sheath (ORS) layer, which is contiguous with the epidermal basal layer. The HF then undergoes degeneration (catagen), quiescence (telogen), and hair shaft shedding (exogen), followed by regeneration (anagen), which recapitulates embryonic follicogenesis. This cyclic regeneration throughout the life of the animal reconfirms the presence and qualities of the stem cells therein.

Hair shaft pigmentation occurs when cortical or medullary keratinocytes phagocytose melanin from donor melanocytes [211]. NC-derived melanoblasts (melanocyte progenitors) first migrate through the dermis, then epidermis, finally establishing themselves in the HF, where they mature into melanocytes when required [212]. Melanin in the epidermis and hair shaft serves as a protective mechanism against ultraviolet-mediated damages [213].

5.1.3 Adult Stem Cells in the Skin

Epithelial stem cell niches have been identified in the IFE and HF, with the latter also harboring stem cells and progenitors of NC-origin (e.g., melanoblasts) [192].

Stem Cells in the Interfollicular Epidermis

During homeostasis, external cornified cells of the IFE are continuously replenished by differentiating cells from deeper layers, a process with a turnover timeframe of ~8–10 days in mice [33] and ~14–75 days in humans [214]. In the mouse, one IFE stem cell has been proposed to exist at the core of each "epidermal proliferative unit," a hexagonal ensemble of ten basal cells (Fig. 2A) [215–217]. Human putative slow-cycling IFE stem cells exist in the basal layer [218], however, the location of their specific niche remains ambivalent [192, 219, 220]. In contrast to the mouse, it has been postulated that the human IFE stem cell niche resides in the tips of dermal papillae where specialized mesenchymal cells reside [221], tips or bases of shallow or deep rete ridges (epidermal undulations) [222–225], or both [218]. Other qualities of stemness such as quiescence, clonogenicity, and nondifferentiation have also been displayed for basal IFE stem cells [221–224, 226]. IFE stem cells normally possess restricted epidermal lineages, however during nonhomeostatic conditions (e.g., wounding and engraftments) IFE stem cells are

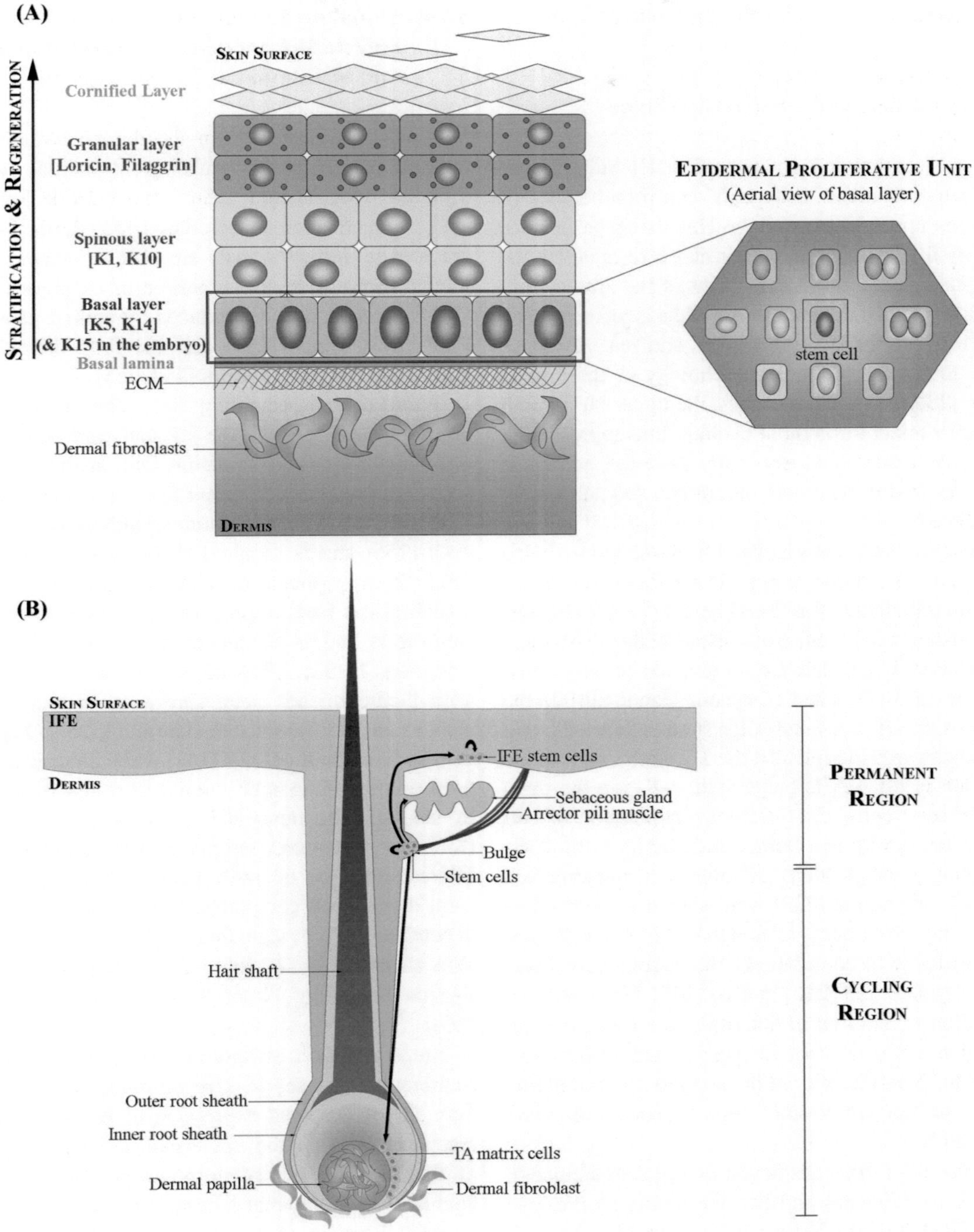

Fig. 2 Epidermal stem cell niches. **(A)** The IFE consists of four epithelial layers (basal, spinous, granular, and cornified) above a basal lamina and a lower dermis of inductive fibroblasts, ECM, nerve receptors (not shown), lymph and blood vessels (not shown). Cornified cells at the skin surface are constantly being shed and replaced by differentiating cells from the deeper layers. Epidermal homeostasis is maintained by IFE epithelial stem cells, whereby one putative stem cell is thought to reside at the center of each "epidermal proliferative unit" (a hexagonal structure of 10 basal cells). **(B)** The pilosebaceous unit consists of a hair shaft, sebaceous gland and arrector pili muscle. The hair follicle undergoes cyclical remodeling, whereby differentiating TA matrix cells interacting with the dermal papilla regenerate the follicular epithelial layers. Long-term epithelial stem cells reside in the bulge region, below the sebaceous gland and where the arrector pili muscle attaches. Upon activation, bulge cells migrate and repopulate TA matrix cells, sebaceous gland progenitors and IFE stem cells (transiently under non-homeostatic conditions). ECM, extracellular matrix; IFE, interfollicular epidermis; K, keratin; TA, transit amplifying (*see* also Color Insert)

also able to give rise to the epithelial components of the HF [227, 228].

Multipotent Stem Cells in the Hair Follicle Bulge

The anatomical location and structure of the HF bulge makes it an ideal stem cell niche. It appears as a protuberance of the ORS in the upper regions of the HF; its shape in mice is moulded by the lingering club hair and it rests below the sebaceous gland at the attachment point of the arrector pili muscle (Fig. 2B) [229]. In humans, the niche is located external, but in close proximity, to the bulge, and this difference may be due to the less defined morphology of the human bulge region [230, 231]. Nevertheless, the upper HF in both species remains intact throughout cycling, thus representing a niche that is permanent and physically sheltered, as well as generously enclosed in basement membrane and adequately vascularized [232].

Multipotent epithelial cells isolated from the mouse bulge region were able to remodel a complete HF and sebaceous gland, and re-epithelialize the IFE (Fig. 2B) [37, 233–235]. *Bona fide* self-renewing stem cells reside within the bulge, as single cells could be isolated, cultured, and serially transplanted to give rise to HFs and sebaceous glands [26]. In the regeneration of the HF, activated bulge stem cells are thought to migrate downwards to replenish the TA matrix cells at the base of the HF [198]. While bulge stem cells are the main source of HF stem cells, the transient repopulation of IFE stem cells occurs during non-homeostatic injury [233, 236–238]. Interestingly, the plasticity of bulge cells was stretched when Nestin$^+$ bulge cells [239] were shown to be capable of differentiating into a range of non-skin cell types in vitro [240] and were able to contribute to the endothelium [241] and to nerve regeneration [242] in vivo [243]. However, further clarification is required to determine the true extent of this plasticity in epithelial stem cells per se and resolve variables such as the heterogeneity of the starting cell population and differing locations of Nestin$^+$ cells between mouse and human HFs [244].

Bulge stem cells further displayed qualities of stemness. These cells were quiescent and had the ability to preserve labels (e.g., DNA analogs 5′-bromo-2′-deoxyuridine or tritiated thymidine) for at least 14 months in mice [245] and 4 months in humans [246]. Furthermore, the bulge may represent a long-term stem cell reservoir as more quiescent cells are located within the bulge than the IFE basal layer, however, this may also be due to the sparse distribution of IFE stem cells [247]. Despite the infrequent cycling of bulge cells, isolated bulge regions from mouse, rat, and human tissues displayed high proliferative capacity in clonogenic assays [248, 249, 250], conforming to undifferentiated stem/progenitor clones in culture, referred to as "holoclones" [251].

5.1.4 Signaling Pathways in Epidermal and Hair Follicle Morphogenesis, Stem Cell Activation and Maintenance

Mesenchymal signals from the dermal structures (i.e., fibroblast condensate and subsequent papilla) are critical in follicular specification, embryonic placode formation, and anagen reactivation in adulthood [55, 210]. As mentioned previously, dermal signals are able to convert corneal epithelium into epidermis inclusive of cutaneous appendages [59]. Inductive factors produced by dermal papillae include Fgf7 [252], Bmp4 [253], and the Bmp inhibitors Noggin, Ectodysplasin, and Gremlin [254, 255].

Wnt/β-catenin signaling has been implicated in skin morphogenesis of both the HF and epidermis, and in stem cell maintenance and activation. During embryogenesis, Wnt signaling specifies epidermal fates in neuroectodermal cells [196], dermal fates in dermomyotome-derived mesenchymal cells [256] and is required in HF placode initiation [162, 257]. In a postnatal-to-adulthood setting, Wnt/β-catenin signaling has been implicated in the maintenance of bulge stem cells and in follicogenesis. In the absence of Wnt signaling, Tcf3, a DNA binding protein that can associate with β-catenin, has been shown to maintain bulge stem cells in an undifferentiated state [258, 259]. *Lef1*-null mice display hair abnormalities [161] while increased or sustained β-catenin signaling in HF stem cells results in differentiation, in vivo and de novo HF generation, and pilomatricoma (follicular-associated neoplasm) development [260–262]. Interestingly, the de novo generation of pigmentless HFs from a stem cell compartment external to the HF bulge is dependent on Wnt signaling [227]. One of the downstream Wnt targets is Fgf4 [263], which may be one of the Fgf ligands regulating Notch signaling during follicogenesis [264].

Bmps and their antagonists have strictly choreographed spatiotemporal roles in skin morphogenesis and regeneration. The widespread expression of Bmp ligands and their receptors on both the epidermal and dermal components of the IFE and HF suggests its primary involvement in epithelial-mesenchymal interactions [265]. During embryogenesis, Bmp is required for the epidermal specification of neuroectodermal cells [266], while there is a temporal requirement for Bmp signaling during prenatal and postnatal follicogenesis. HF placode initiation requires Noggin-mediated Bmp antagonism, while subsequent cell delineation and differentiation of matrix cells needs active Bmp signaling [253, 267]. Conditional ablation of the Bmp receptor *Bmpr1a* in adult mice resulted in the expansion of activated bulge stem cells while overexpression caused premature terminal differentiation, indicating a critical window of Bmp signaling in maintaining the balance between quiescence and differentiation [268, 269].

During embryogenesis, Shh is required for epithelial-mesenchymal interactions in the developing hair germ and hair peg [270–272]. Shh is thought to act downstream of Wnt signaling [257], eliciting its action through the *Gli2* transcription factor [273]. In the adult HF, it is thought that Shh signaling mechanisms operate in TA matrix cells but not bulge stem cells [198]. In the IFE, Shh and Dhh have similar roles in regulating the proliferation of basal stem and progenitor cells [274], while Ihh influences sebocyte proliferation and differentiation and was up-regulated in mouse and human sebaceous tumors [275].

Notch signaling is required for the proper development of the intermediate suprabasal cells and its subsequent commitment and terminal differentiation into spinous cells [276–278]. *Notch1*-null mice display aberrant epidermal proliferation and differentiation [277]. Moreover, in the embryonic and postnatal HF, Notch is required for follicogenesis and homeostasis, as evidenced by *Notch1*-null embryos that display abnormal, thin, wavy, and short hair and post-neonates that lack hair altogether and are cystic [279, 280]. Thus, Notch signaling is required for lineage delineation of follicular stem and/or TA cells [281, 282]. In addition, Notch has pronounced roles in the sebaceous glands, given that in its absence, epidermal cysts form in place of the glands [279].

Microarray profiling of murine bulge stem cells [238, 249] and human IFE stem cells [283] displayed transcripts correlating to stemness, inhibitors of growth and differentiation, slow cell cycling, cell–cell and cell–ECM interactions. These studies highlighted the importance of the niche microenvironment in maintaining resident stem cells in an undifferentiated and quiescent state through constant interactions between the stem cells and their niche. Among the highly expressed genes in mouse bulge stem cells were genes encoding Wnt inhibitors and Bmps, indicating a suppression of follicular differentiation stimuli [37, 238, 249]. Moreover, the transcription factor *Lhx2*, expressed in murine follicular bulge stem cells, may act as a regulator promoting quiescence and inhibiting activation [284]. Another mechanism of retaining quiescence may be through inhibition of epidermal growth factor receptor (EGFR)-mediated proliferation by applying its antagonist, leucine-rich (Lrig1) repeats and immunoglobulin-like domains 1, in both mouse and human epidermal stem cell compartments [283]. Likewise, stem cell populations may be maintained by the Rho guanosine triphosphatase *Rac1* as conditional deletions of *Rac1* in embryonic and adult mice demonstrated cutaneous thinning and hair loss, possibly as a result of stem cell diminution [285, 286]. However, it is unclear whether the importance of *Rac1* is restricted to follicular stem cells [287, 288] or is also extendable to IFE stem cells [286]. Lastly, stem cell–ECM interaction is an important feature of the niche for retaining the stemness of resident cells. This may hold true for the IFE, in which detachment of keratinocytes from ECM or basement membrane may promote terminal differentiation programs [27, 289]. Cell–ECM cross-talk may also be a property of the HF bulge niche, as reflected by elevated expression of $\alpha6$ integrins in bulge stem cells [37, 224] and its ligand Tenascin-C in proximate ECM [238].

5.1.5 Skin in Regenerative Medicine

Human epidermal keratinocytes have been successfully cultured since 1975 [290], and this finding ultimately led to an improvement in the treatment of burns patients [291]. The more recent use of fibrin matrices has enabled the formation of epithelial sheets, significantly improving long-term skin engraftments [292, 293]. However, skin is a complex organ that brings together cells from two different embryological origins – ectoderm and mesoderm – integrated within a three-dimensional matrix. Thus, the ultimate use of stem cells in reconstructing such a complex tissue would rely on their ability to drive much of the regeneration process by the intrinsic bioengineering capacity of the cells in conjunction with the critically important role of the matrix in defining differentiation and three-dimensional organization [294]. To achieve this still requires a great deal of research and development, however, we are entering a new phase in regenerative medicine and biomedical engineering that promises to revolutionize the treatment of burns care, wound healing and other skin traumas.

5.2 Thymus Development

5.2.1 Thymus and Parathyroid Organogenesis

The thymus is a primary lymphoid organ that governs the maturation and tolerance induction of bone marrow-derived hematopoietic precursors into immunocompetent T-lymphocytes. The unique functions imparted by the thymus predominately reside within the epithelium, which forms the main constituent of the thymic stromal microenvironment. Based upon histological and functional grounds, the thymic stroma is divided into an outer cortex and inner medulla, each of which contain phenotypically distinct TEC subtypes that mediate a specialized aspect of thymocyte development [295]. To attain full functional competence, thymus organogenesis depends on interactions with all three embryonic germ layers: the endodermal epithelium [296], ectodermal NC-derived mesenchyme [53, 297], mesodermal hematopoietic precursors [298–300], and vasculature [301], and entails both morphological and molecular mechanisms of control. We will focus on the morphological events first, before a dissection of the transcription factors and signaling pathways involved.

The segmentation of the pharynx comprises the first step in thymus formation, and initiates a series of events that permit specification of epithelial cells towards a thymic fate [302]. The pharyngeal pouches are bilateral outpocketings of epithelial endoderm and are formed sequentially in a cranial-to-caudal order, apposing the pharyngeal clefts, which are invaginations of surface ectoderm [303, 304, 305]. Clustered homeobox (*Hox*) genes function in providing positional information along the pharynx, establishing a unique identity to each newly formed pharyngeal arch [306]. Third pouch axial identity, specified by *Hoxa3*, is essential for thymus positioning [307]. The initial stages of thymus organogenesis are closely associated with that of the parathyroid gland. Both organs are derived from a common primordium that arises from the proliferation of third pharyngeal pouch cells (Fig. 3). The first morphological indications of thymus commitment are seen by about day 10.5 of mouse embryogenesis (E10.5), although initiating signals may occur as early as E9.5 [308]. The common primordium appears to be spatially patterned into a prospective thymus domain ventrally, and a discrete parathyroid domain dorsally, governed by region-specific expression of *Foxn1* and *Gcm2*, respectively [309]. Each primordium is surrounded by a condensing mesenchymal capsule derived from the neural crest, which supports further outgrowth of the budding rudiment, and is believed to sustain the initial stages of TEC differentiation [53, 310]. At about E12.5, the shared primordia separate from the pharynx, begin resolution into distinct thymus and parathyroid organs and migrate towards the anterior chest cavity, assisted by signals from the surrounding mesenchyme [311]. Subsequently they reach their respective adult positions, with the thymus at the midline and the parathyroids at the lateral margins of the thyroid gland [312]. Hematopoietic T-cell precursors actually begin to seed the developing thymus anlage at E11.5, but the epithelial cells are unable to fully support T-cell maturation at this stage. As vascularization is yet to occur, the first colonizing hematopoietic cells leave the circulation and migrate through the perithymic mesenchyme in order to reach the thymic epithelium [313–315]. It is only upon further differentiation of the epithelial cells, from E12 onwards, into distinct stromal subtypes that the thymic epithelium acquires the capacity to completely support thymopoiesis [299, 316, 317]. This ability is believed to be aided by the unique three-dimensional meshwork organization of the thymic epithelium [184, 318].

Human thymus development exhibits comparable stages with the mouse, arising from the ventral elongation of the third pharyngeal pouch at the end of the fourth week of embryogenesis [34]. The endodermally generated tissue invades the underlying NC-derived mesenchyme, forming a lateral expansion from the pharynx that similarly compartmentalizes into ventral thymic and dorsal parathyroid domains [303]. By 7 weeks, the human thymic lobes begin migration to their definitive location in the mediastinum, undergoing fusion to form a typical bilobate gland by 8 weeks [34]. Later in the 8th week, the epithelial thymic rudiment is colonized by lymphoid precursors originating primarily from the fetal liver [178], although sustained thymopoiesis in postnatal life requires a continual influx of thymus-seeding progenitors from the bone marrow [319]. In week 10, mesodermal-derived fibrous connective tissue and vasculature invaginate the thymic rudiment effecting epithelial proliferation and differentiation, and thymic lobulation [178]. Differentiated lobules, subdivided into a lymphocyte-rich outer cortex and epithelial-rich medulla with more mature thymocytes, are evident by about the 12th week of human gestation, with the trabeculae demarcating each lobule and serving as a conduit for the vasculature and neural cells. Between weeks 14 and 16, mature lymphocytes exit the thymus to seed the periphery [178].

Despite a customary mediastinal location for the thymus, cervical thymic tissue has previously been documented in humans [320–322]. Clinically it presents as a cervical mass that may obstruct breathing or give rise to ectopic tumors. However the immunological significance of the human cervical thymus is currently unknown, and may not be functionally relevant owing to its diminutive size. A similar very small cervical thymus has recently been described in the mouse [323, 324]. This accessory mouse thymus is believed to possess the same general organization as the thoracic thymus, but in contrast to the situation in humans, has been shown to support T-cell differentiation. It is suggested that the cervical thymus branches off from the common thymic anlage before the descent of the thoracic thymus into the chest cavity, and that it matures only after birth [325]. Although much reduced in size, the existence of a second functional thymus in mice may provide an alternative explanation for the ongoing thymopoiesis occasionally observed following traditional thymectomy, although further experiments are necessary to elucidate the true physiological relevance of extrathymic T-cell development from this accessory lymphoid organ [324]. It is worth noting that birds have 14 thymic lobes situated in the neck region [326] – each being functionally equivalent. It therefore seems that the physical location of the thymus does not greatly influence its function, provided there is adequate vascular supply. Indeed full restoration of the immune system can be achieved in athymic animals by transplantation of thymic lobes under the kidney capsule or in subcutaneous fat deposits.

5.2.2 Single Origin Model for Thymus Organogenesis

Although it is well established that the thymic rudiment originates from the third pharyngeal pouch endoderm, the contribution of the third pharyngeal cleft ectoderm had previously

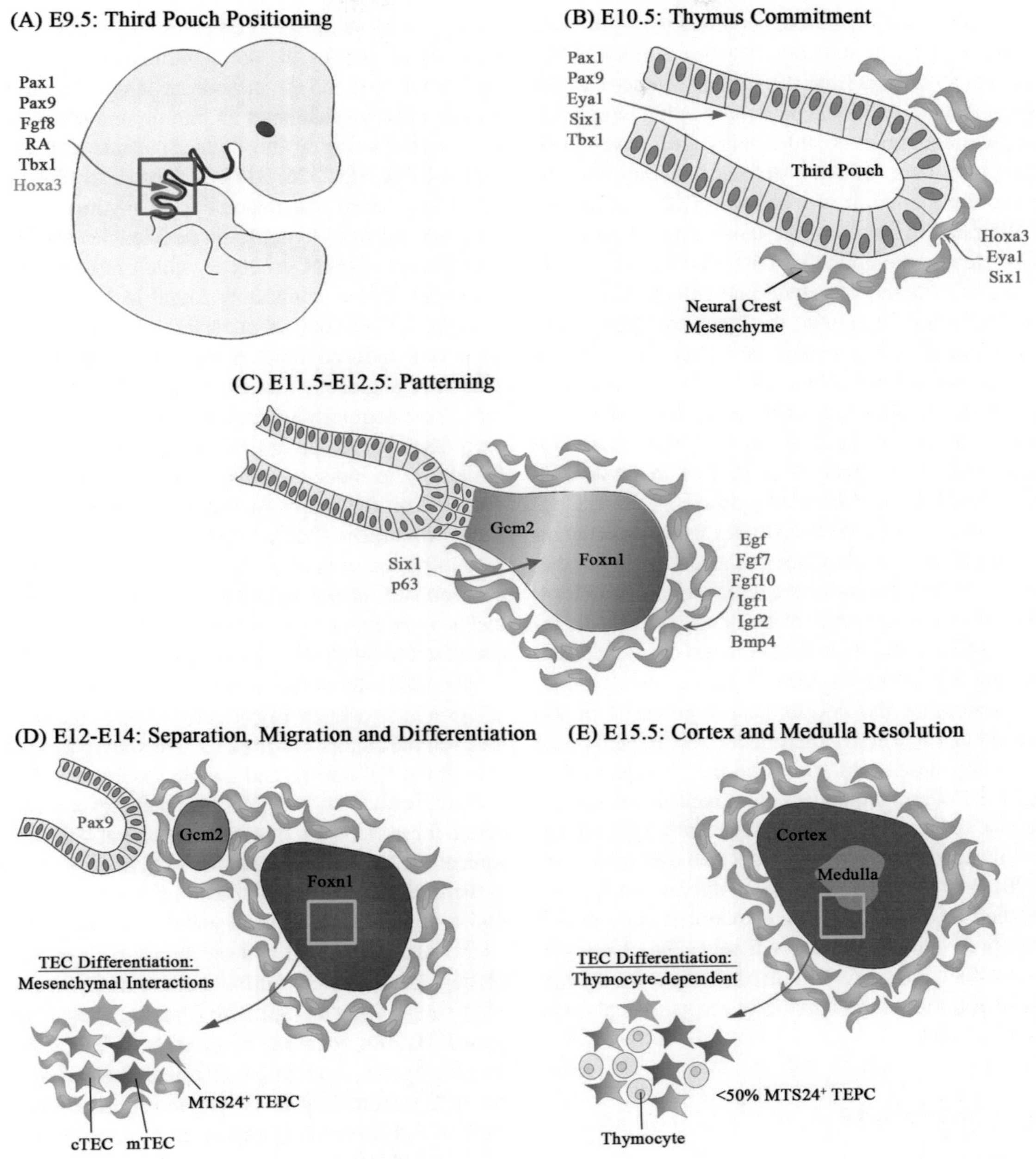

Fig. 3 Molecular control of thymus organogenesis. The thymus and parathyroid glands derive from a common primordium arising from proliferation of third pharyngeal pouch endoderm cells. **(A)** E9.5: Third Pouch Positioning. Initiating signals required for pharyngeal endoderm formation include *Pax1*, *Pax9*, *Fgf8*, *RA*, and *Tbx1*, with *Hoxa3* specifying the third pharyngeal pouch. **(B)** E10.5: Thymus Commitment. Outgrowth of the common thymus-parathyroid rudiment requires expression of *Pax1*, *Pax9*, *Eya1*, *Six1*, and *Tbx1*. A neural crest-derived mesenchymal capsule surrounds and supports the budding rudiment, which is known or postulated to be expressing *Hoxa3*, *Eya1*, and *Six1*. **(C)** E11.5–E12.5: Patterning. The common primordium is patterned into a thymus domain ventrally and a parathyroid domain dorsally, governed by region-specific expression of *Foxn1* and *Gcm2*, respectively. *Six1* may maintain *Foxn1* and *Gcm2* expression to regulate patterning, while *p63* has a proposed role in retaining the proliferative capacity of the thymic epithelium. Communication between the endoderm and mesenchyme continues via soluble factors including Egf, Fgf7, Fgf10, Igf1, Igf2, and Bmp4. **(D)** E12–E14: Separation, Migration, and Differentiation. The common primordium separates from the pharynx, controlled by *Pax9*, resolves into discrete thymus and parathyroid organs and migrates towards the anterior thorax. *Foxn1* is required for ongoing development of thymic epithelial cells (TECs). The early stages of TEC differentiation are influenced by interactions between the epithelium and mesenchyme; cTECs and mTECs differentiate from a bipotent MTS24$^+$ TEPC. **(E)** E15.5: Cortex and Medulla Resolution. Distinct cortical and medullary regions are evident by the later stages of TEC differentiation, which rely on cross-talk between the epithelium and developing thymocytes. Less than 50% of TEC progenitors are MTS24$^+$, reducing the progenitor capacity. cTEC, cortical thymic epithelial cell; mTEC, medullary thymic epithelial cell; TEPC, thymic epithelial progenitor cell (*see* also Color Insert)

been disputed [327, 328]. The more historical view proposed a dual origin model for thymus organogenesis, in which the cortical epithelium derived from the surface ectoderm of the third pharyngeal cleft, while the medullary epithelium originated from the endoderm of the third pharyngeal pouch [329, 330]. Arising from comparative histological studies between wild-type and nude mice, it was believed that the cleft ectoderm proliferated to surround the developing endodermal primordium. Hence, the epithelial defect observed in nude mice was due to a failure of ectodermal cells to contribute normally to the thymic rudiment; the inner medullary compartment subsequently failed to develop owing to the absence of an inductive ectodermal signal [327, 329, 330]. Despite initial support for a physical contribution from the ectoderm, there are now lineage analyses in conjunction with older morphological and functional studies in favor of a single origin paradigm in which all epithelial cells are endodermally derived [331]. Earlier studies using chick-quail chimeras demonstrated that cortical and medullary epithelial cells could be generated from isolated pharyngeal endoderm transplanted into the body cavity of chick embryos [332]. Incredibly, the grafted quail endodermal tissue developed into a functional thymus, which produced T cells of chick origin, thereby demonstrating the developmental potential of the pharyngeal endoderm. These experiments were recapitulated more recently by studies which showed that transplantation of pharyngeal endoderm under the kidney capsule of athymic nude mice was sufficient for development into a functional thymus, with well-organized mature cortical and medullary regions [296]. For confirmation, direct fluorescent lineage tracing revealed no evidence for an ectodermal contribution to the normally developing thymic rudiment [296]. Owing to more stringent data, the single endodermal origin for thymus organogenesis is currently preferentially accepted in place of the dual origin model.

5.2.3 Role of Mesenchyme

Morphogenesis of the thymus is distinguishable by two phases: the early stages of epithelium differentiation are thymocyte-independent and involve interactions with the mesenchyme [53, 297, 333], whereas the later stages depend on a mutual cross-talk between thymocyte-derived signals, together with the mesenchyme and the developing thymic epithelium [298, 299, 334]. Although the initiating signals for thymus organogenesis require further clarification, epithelial-mesenchymal interactions are thought to be essential, with the latter responding to instructive signals from the endoderm to induce the growth and differentiation of the thymic epithelial rudiment [335].

The thymus recruits pharyngeal mesenchymal cells as the organ differentiates, such that at around E10 to E12, the thymus anlage is a proliferating epithelial bud surrounded by mesenchyme [336]. The thymic mesenchyme that encapsulates and invades the epithelial cells originate from both the neural crest and the mesoderm of the pharyngeal arches, and eventually establishes an intrathymic network of fibroblasts, contributing to the thymic capsule and interweaving septae [310, 333, 336, 337]. A functional role for the neural crest–derived cell population in thymus morphogenesis was inferred from hypoplastic deficiencies following physical ablation in chick embryos, which resulted in a range of epithelial defects commonly found in DiGeorge syndrome [333, 338, 339] and an undersized thymus unable to fully support a reduced number of thymocytes [340]. In vitro culture of separately isolated E12 epithelial and mesenchymal tissue demonstrated that the mesenchyme promoted the early stage development of the epithelium as assessed by the ability to induce expression of epithelial surface markers such as the major histocompatibility complex (MHC) I and II antigens [341]. Correspondingly, E12 fetal thymic lobes stripped of their perithymic mesenchymal capsule and subsequently grown in culture exhibited much reduced size and had severely compromised lymphoid development when compared to lobes with intact mesenchyme [333].

The influence of mesenchymal cells on thymic epithelial cell growth is presumably effected through the release of various soluble factors into the ECM, including Egf [342], members of the Fgf family, and insulin-like growth factors-1 and -2 (Igf1, Igf2) [341, 343]. For instance, Egf was found to substitute for mesenchyme in inducing the in vitro lobulation of isolated E12 thymic epithelium [342]. Mice with a deficiency or disruption in signaling of the Fgf receptor, Fgfr2IIIb, possess a thymus with reduced cellularity [344, 345] that still retains mature cortical and medullary cells capable of supporting a qualitatively normal pattern of thymopoiesis [345, 346], suggesting that Fgfr2IIIb signaling may be required for growth but not complete maturation of TECs. In line with this, the ligands Fgf7 and Fgf10 are expressed by mesenchymal cells surrounding the E12 thymus primordium, with their cognate Fgfr2IIIb being expressed by immature TECs [53]. Further evidence was shown by the removal of mesenchyme from E12 thymic lobes which resulted in a loss of TEC proliferation that could be restored by the provision of recombinant Fgf7 and Fgf10 [53]. Hence NC-derived mesenchyme, via its production of Fgf7 and Fgf10, is necessary for the expansion of epithelial cell numbers, and Fgfr2IIIb signaling persists in the postnatal thymus to aid the maintenance of the adult thymic microenvironment [345]. More recently, platelet-derived growth factor receptor-α-positive (Pdgfrα^+) fetal thymic mesenchymal cells were shown to promote the growth of TECs, regulating thymus size and the subsequent development of the intrathymic niche, via the production of Igfs [343]. Similarly, the role of mesenchymal cells in epithelial development was investigated in the thymus anlage of patch (*ph*) mutant mice, which have a primary defect in the mesenchyme caused by the absence of *Pdgfrα* expression

[297]. Addition of intact thymic mesenchymal cells to organ culture induces development of the *ph/ph* thymus anlage, contributing not only to the capsule but also to the fibrous septae that intermingles with the epithelium [297].

Overall these studies signify a compelling role for the mesenchyme in maintaining epithelial function and in fashioning the thymic microenvironment. Nevertheless, the signals provided from cells of mesenchymal origin are necessary but not sufficient for the formation of a regularly structured and appropriately functioning thymus. It is believed that once thymus development has progressed to a stage able to support immature thymocytes, the further differentiation of the epithelial component relies on cross-talk with the developing thymocytes [53]. However, ongoing mesenchymal support in the adult thymus, particularly after damage, is likely [347].

5.2.4 Thymocyte-TEC Cross-talk

Upon entering the thymus, immature lymphoid progenitor cells undergo division and development, requiring tight contact with their associated thymic structural network. Specialized TECs nurture the immature T cells via adhesion contacts and the provision of growth factors and in response the T-lineage committed precursors supply the signals that encourage the maturation of epithelial cells and their differentiation into distinct cortical and medullary compartments. This concept of reciprocal signaling, or "cross-talk," between thymocytes and stromal cells stemmed from observations of disorganized thymic epithelial microenvironments in mutant mouse strains with stage-specific blocks in thymocyte development [300, 318, 348, 349].

For instance, adult CD3εtg26 mice – which show a thymocyte block at the earliest $CD44^{+}CD25^{-}$ double negative (DN1) stage – have abnormal cortical and medullary organization [350]. A similar hypoplastic TEC phenotype was observed in adult recombination-activating gene2/common cytokine receptor γ-chain (*Rag2/γc*)-deficient mice, which are arrested at the $CD44^{+}CD25^{+}$ DN2 stage [298]. These results were supported by studies which indicated that progression from $CD44^{+}CD25^{-}$ towards $CD44^{-}CD25^{+}$ thymocytes induced the three-dimensional organization of epithelial cells in the thymic cortex [300]. In contrast, mice with a later block in T-cell development, resulting in the absence of the T-cell receptor (TCR), displayed a normal cortical microenvironment but defective organization of the medulla, suggesting that medullary TEC development was regulated by the presence of αβ- or γδ-T-lineage committed lymphoid cells [351, 352, 353]. Significantly, the abnormal thymic medullary epithelial organization could be corrected by the introduction of mature receptor-bearing lymphocytes [352, 354]. Lymphostromal interactions are therefore required for maintenance of the thymic epithelial microenvironment, first influencing the arrangement of the cortex then compartmentalizing the medullary architecture.

Insight into whether thymic cross-talk is actually necessary for the differentiation of immature epithelial progenitors into mature stromal cells was gained from analysis of embryonic mouse mutants, which surprisingly differ from adult mice in their dependence on reciprocal interactions.

Mice doubly deficient in the receptor tyrosine kinase c-kit ($ckit^{-/-}$) and the common cytokine receptor γ-chain ($\gamma_c{}^{-/-}$) exhibit complete abrogation of thymocyte development [355], while the aforementioned CD3ε26 transgenic mice have a profound developmental block beyond the earliest stage of T-lineage commitment [350]. Interestingly although the thymic microenvironment lacks organization in these mouse strains, cortical and medullary cells are nonetheless present, disparate to that observed in the adult thymus, suggesting that initial patterning of the epithelial compartment and acquisition of functional competence occurs independently of inductive signals from T-cell precursors in the fetal thymus [298, 300, 334]. Similarly, analysis of embryonic *Rag2/γc*-deficient mice and Ikaros mutant mice (failure of commitment to lymphoid lineage) show retention of proper three-dimensional epithelial organization with mature cortical and medullary cells at E13.5 and E15.5, reiterating that thymocyte-derived signals are not required for the initial stages of TEC development in the embryo [298, 331]. The basis for this difference between the embryonic and adult thymus may be due to the fact that it is only after TECs are appropriately patterned in the embryo that the presence of thymocytes becomes obligatory for the ongoing survival of mature cortical and medullary TECs (cTECs and mTECs) [331]. Although interactions with the surrounding mesenchyme may be the determining factor in the initiation of TEC development, it is believed that thymocyte-derived signals impose on TEC development by E15.5, a timeframe concurrent with the appearance of $CD25^{+}$ thymocytes, and that postnatally thymocyte–TEC interactions are indispensable for maintaining TEC differentiation and organization in the thymus [298].

Various soluble factors, including certain cytokines, and intercellular adhesion molecules are reciprocally expressed by thymic epithelia and thymocytes and are therefore likely to be the effectors of lymphostromal interactions [349, 356]. For instance, signaling through the lymphotoxin-β receptor (LTβR) has been implicated in the control of lymphocyte-dependent development and maintenance of mTECs [357, 358]. The LTβR is expressed on mTECs, while its ligand LTα1β2 is expressed by medullary thymocytes. Impaired thymocyte–mTEC cross-talk in the absence of LTβR results in aberrant differentiation and reduced numbers of medullary epithelial cells [357], highlighting the importance of reciprocal signaling in the organization of proper epithelial architecture. Furthermore, a subset of mTECs expressing high levels of MHCII, express the

Autoimmune Regulator (*Aire*) transcription factor, which directs the intrathymic expression of a set of peripheral antigens, resulting in the purging of T-cell clones reactive to organ-specific, late-onset, and sequestered antigens [331]. Receptor Activator of NF-κB (RANK) signals from lymphoid tissue inducer (LTi) $CD4^{+}CD3^{-}$ cells regulate these *Aire*-expressing mTECs [359] and hence are crucial to intrathymic self-tolerance.

5.2.5 Transcriptional Control and Growth Factors Involved in Thymus Development

Transcription Factors

The transcriptional networks implicated in thymus organogenesis are being unravelled from genetic targeting (selective loss or gain) studies in mice.

The mammalian *Hox* family of 39 known transcription factors are encoded by four gene clusters (A, B, C, and D), organized on four distinct chromosomes in spatial and temporal co-linearity [360–362]. *Hox* genes are identified by their highly conserved 60-amino acid DNA-binding homeodomain and have diverse roles in specifying positional identity [361]. At E9.5, *Hoxa3*, and paralogs *Hoxb3* and *Hoxd3*, are expressed in the NC-derived mesenchyme surrounding the pharyngeal pouches and at E10.5, *Hoxa3*, but not its paralogs, is expressed in the third and fourth pharyngeal endoderm [311]. Homozygous $Hoxa3^{-/-}$ mice displayed athymia and aparathyroidism, while compound heterozygotes $Hoxa3^{+/-}Hoxb3^{-/-}$ $Hoxd3^{-/-}$ displayed ectopic and delayed migration of third pharyngeal rudiments [311, 363], owing to elevated endodermal apoptosis and regression of the postmigratory third NC-mesenchymal arch [362]. *Hoxa3* therefore has intrinsic roles in the interaction of NC-derived mesenchyme and pharyngeal endodermal cells [302]. In addition, combined deletions of *Hoxa1* and *Hoxb1* display disrupted third pharyngeal pouch patterning as a consequential effect of disorganized NC cells [364]. Pbx and Meis/Prep proteins are subfamilies of the TALE-homeodomain superfamily that act as cofactors to several Hox proteins [365]. Phenotypic analysis of $Pbx1^{-/-}$ mice revealed pharyngeal malformations comparable, but varying in severity, to that observed in *Hoxa3* single or compound mutants, suggesting a mediation of *Hox* targets that may either be *Pbx1*-dependent or -independent [366].

The *Pax* family of transcription factors are comprised of nine members, characterized by a conserved paired domain, with some members possessing either 1- or 3-helix homeodomains and/or an octapeptide [367]. *Pax* subgroup I (*Pax1* and *Pax9*) is expressed in foregut endoderm from E8.5 and in the third pharyngeal pouch at around E9.5 (*Pax9*) to E10.5 (*Pax1*) [368, 369]. Undulated $Pax1^{un/un}$ mice display a mild thymic hypoplasia with reduced thymopoiesis [369]. Thymic abnormalities are more drastic in $Pax9^{-/-}$ mice, which have a severely hypoplastic thymus resembling a cystic "polyp-like" structure that fails to detach from the pharynx [370, 371]. *Foxn1* expression and thymocyte infiltration are still evident in $Pax9^{-/-}$ mice, and the decreasing thymic cellularity in $Pax9^{-/-}$ mice correlates with elevated apoptosis [370]. *Pax1* and *Pax9* are placed downstream of *Hoxa3*, as evidenced by the inability of *Hoxa3*-null mice to maintain *Pax* expression [307]. A synergistic and dosage-dependent genetic interaction between *Hoxa3* and *Pax1* was further demonstrated in compound mutants, in which deletions in *Hoxa3* caused a more severe phenotype in *Pax1*-null mice [372, 373]. Differences between *Hoxa3* and *Pax1* compound mutants and *Pax1* single mutants may indicate a redundancy between certain *Pax* genes [303]. *Pax3*, unlike *Pax1/9*, belongs to subgroup III in which members possess an additional homeodomain that confers roles in NC mesenchyme development as evident in Splotch mutant mice [367, 374]. Pharyngeal arch and thymus abnormalities have been identified in these Splotch mutants [375, 376].

The eyes absent (Eya) protein tyrosine phosphatases, historically named after the eye defects in *Drosophila*, have dual roles as a transcription activator and a haloacid dehalogenase phosphatase [377]. At E9.5–E10.5, extensive expression of *Eya1* can be detected in the ectodermal, mesenchymal, and endodermal cells of the pharyngeal clefts, arches, and most notably pouches (second to fourth) [378]. In *Eya1*-null mice, the common thymus and parathyroid primordium fails to form, consistent with the absence of *Foxn1* and *Gcm2* expression [379]. *Eya1* is thought to act independently of *Hoxa3*, *Pax1* and *Pax9* and upstream of *Six1* because $Eya1^{-/-}$ mice express normal levels of *Hox* and *Pax* but are deficient in *Six1* [379]. Eya phosphatases may act to derepress Six proteins, turning them into transcriptional activators [380].

The mammalian *Six* family consists of nine homologs to the *Drosophila* Sineoculis (*So*) genes, characterized by a conserved Six domain and homeodomain [377]. *Six1* and *Six4* are expressed in the third pharyngeal pouch endoderm and surrounding NC cells [381, 382]. Redundancy exists between *Six1* and *Six4* in the initiation of the thymus/parathyroid anlage, while *Six1* exclusively regulates the subsequent patterning of the rudiments by maintaining *Foxn1* and *Gcm2* expression [382]. $Six1^{-/-}Six4^{-/-}$ mice lack the initial thymus/parathyroid primordium (and therefore lack *Foxn1* and *Gcm2* expression), while $Six1^{-/-}$ single mutants exhibit a delayed hypoplastic primordia with reduced *Foxn1* and *Gcm2* expression, and $Six4^{-/-}$ have a normal phenotype [382, 383]. Interestingly, *Pax1* expression is absent in $Six1^{-/-}$ $Six4^{-/-}$ thymus/parathyroid rudiments, suggesting the placement of *Pax1* (but not *Pax9*) downstream of *Six* gene signaling in the pharyngeal endoderm [382].

The T-box transcription factor *TBX1* is associated with the human DiGeorge/velocardiofacial syndrome (DGS/VCFS), also referred to as the chromosome 22q11.2 deletion syndrome [302, 384, 385]. *Tbx1* is extensively expressed in the pharyngeal region from E7.5 to E12.5, with predominant expression in the third and fourth pouches at E10.25, and surrounding non-NC-derived mesenchyme [386–388]. Deletions in murine *Tbx1* impart temporal and gene-dosage effects on pharyngeal development and segmentation, resulting in absent or hypoplastic thymus and parathyroid glands [385]. *Tbx1* may have an additional role in thymus organogenesis, as thymic abnormalities were observed when *Tbx1* was deleted after development of the third pharyngeal pouch at E10.5 [389]. Microarray analysis on E9.5 *Tbx1*-null embryos identified several putative *Tbx1* target genes including *Fgf8* [390], *Fgf10* [391], *Pax9*, *Gcm2*, and *Foxg1* [392].

Foxn1 is a member of the forkhead-box family of transcription factors, defined by the presence of a winged-helix DNA-binding domain [393]. Homozygous deletions in *Foxn1* give rise to the nude phenotype in mice [394, 395], rats [395, 396], and humans [397–399], which is commonly characterized by congenital athymia (resulting in severe immunodeficiency), alopecia, and nail dystrophy. As deduced from the nude phenotype, *Foxn1* is predominantly expressed in the thymus and skin, although there have been a few reports of a widespread epithelial [400] and possibly mesenchymal [401] expression pattern. In the thymus, *Foxn1* can be detected as early as E10.5, although strongest expression is from E11.25 onwards, and is functionally required for the proliferation and differentiation of immature TECs into cortical and medullary TECs as well as subsequent thymocyte infiltration [188, 189]. These roles may also be applicable in the postnatal setting, as Cre-mediated reversion of a nude mouse model has been demonstrated to restore some adult thymic function [402]. However under homeostatic conditions *Foxn1* is not thought to be essential for TEC maintenance, given that *Foxn1*-positive TECs account for only approximately 24% of the total epithelium in two- to four-week-old thymi [403]. The presence of an entopic cluster of thymic stromal progenitor cells in nude mice [188, 189] suggests that *Foxn1* may be more important for initiating terminal differentiation. However, it may still function in lineage specification (of at least a subset of cells) as loss of *Foxn1* results in a respiratory-like phenotype in the thymic anlage [176, 302, 404]. Although the tissue-specificity of *Foxn1* is difficult to discern, owing to the fact that both alternatively spliced forms of the transcript are present in skin, a thymus-specific role is believed to be conferred by the *Foxn1* N-terminal domain, possibly through epithelial-thymocyte cross-talk signals [405]. Molecular and histological comparisons between wild-type and nude thymi have enabled the identification of several putative targets including the Notch Delta-like ligands 1 and 4 (Dll1 and Dll4) [406], PD1 ligand [407], chemokines CCL21 [408] and CCL25 (TECK) [314], and Fgfr2-IIIb [409]. Doubt in the direct regulation of at least two of these targets in the postnatal thymus was raised upon detection of Dll4 and CCL25 proteins in *Foxn1*-negative TECs [403].

The parathyroid-specific equivalent of *Foxn1* is the glial cells missing-2 (*Gcm2*) gene, which demonstrates organ-restricted expression and function [410]. Expression of *Gcm2* begins at E9.5 in the second and third pharyngeal pouches, with expression lost in the second pouch by E10.5 and gradually restricted to the most cranial and dorsal area of the protruding third pharyngeal endoderm, marking the parathyroid domain at E11.5 (Fig. 3C) [309]. *Gcm2*, like *Foxn1*, may also be regulated downstream of the *Hoxa3*, *Pax1/9*, *Eya1,* and *Six1/4* regulatory networks described above [372, 379, 382]. In contradiction to previous observations in *Gcm2*$^{-/-}$ mice [410], a recent publication demonstrated the existence of a prospective *Foxn1*-negative parathyroid region in the common thymus/parathyroid primordium at E11.25, which was lost, possibly due to elevated apoptosis, by E12.5–E13 [411]. This suggests that *Gcm2* may not specify the parathyroid domain as previously thought.

p63, a homolog of the *p53* tumor suppressor gene, has a critical role in the stratification of universal epithelia [412, 413], and is highly expressed in the thymic epithelium from E12 onwards [181]. Until recently, there were contradictions in the specific stages of *p63* control, whether it be in the initiation of differentiation programs or the maintenance of stem cell populations [414, 415]. Comprehensive studies by Senoo et al. [181] postulated a role for *p63* in retaining the proliferative capacity of thymic (and epidermal) epithelial stem/progenitor cells, a functional loss of which results in reduced clonogenicity in vitro and increased apoptotic rates in vivo. Putative downstream targets of *p63* in the thymus include *FgfR2IIIb* and *Jagged-2* [416].

Growth Factors

Regulated Wnt signaling is important for TEC development. Wnt glycoproteins secreted by TECs and thymocytes were found to regulate the expression of *Foxn1* [417]. Kremen1, an inhibitor of the canonical Wnt pathway, is expressed in TECs and certain thymocyte populations and its expression decreases with age [418]. *Kremen1*$^{-/-}$thymi have aberrant architectural defects, increased nonepithelial regions and loss of cortico-medullary delineations despite a relatively normal total thymic size and thymocyte developmental profiles [418]. Similarly, stabilization of β-catenin (by deletions in the adenomatous polyposis coli, *Apc*, gene) in K14-expressing cells (predominantly mTECs) [419] led to an undersized thymus with metaplasia of $K8^{+}K14^{+}$ epidermal-resembling cells and failure of thymopoiesis due to the absence of a supportive epithelial network [420].

Bmp signaling may have roles in epithelial-mesenchymal interactions, inducing Fgf signaling and possibly specifying the thymus primordium. A role for Bmp in thymic development in vivo was evident in *Foxn1*::*Xnoggin* transgenic mice, in which the Bmp antagonist Noggin was expressed in *Foxn1*$^{+}$ TECs, revealing ectopic and severely hypoplastic thymi [421]. The expression of *Bmp4* in the thymic endodermal rudiment and NC-derived mesenchyme infers a prominent role in epithelial-mesenchymal interactions, a deficiency of which may account for the defects observed in the *Foxn1*::*Xnoggin* transgene [421, 422]. In addition, the restricted expression of *Bmp4* in the ventral region and *Noggin* in the dorsal region of the third pharyngeal pouch at E10.5, suggests a role in the respective regulation of *Foxn1* and *Gcm2* expression [422]. However, it remains uncertain whether Bmp is able to mediate direct effects on *Foxn1*, as is evident in the hair follicle setting [253, 423]. In support of a direct interaction, *Foxn1* was upregulated in TEC cultures supplemented with exogenous Bmp4 in vitro [409]. In contrast, the unaltered expression of *Foxn1* in *Foxn1*::*Xnoggin* transgenic mice may suggest an indirect Bmp-Foxn1 interaction [421]. Another interesting outcome of the in vitro studies alluded to the possible induction of *Fgf7* and *Fgf10* by *Bmp4* [409]. The Bmp inhibitor *Chordin* may also have roles in inducing and/or maintaining *Fgf8* and *Tbx1* expression, however this has yet to be shown definitively [302].

Mouse and human TECs secrete Hh proteins (Shh, Ihh and Dhh) and express Hh receptors (Patched1/2, Smoothened) implying a role for Hh signaling in TEC regulation and/or TEC-thymocyte interactions [424, 425]. During embryogenesis, *Shh* was found to upregulate *Gcm2* expression and down-regulate *Bmp4* expression in the third pharyngeal pouch [426]. In *Shh*-deficient mice, the expression of both *Bmp4* and *Foxn1* was found to be up-regulated, occupying the same areas in the thymic and parathyroid primordia (as *Gcm2* expression was absent) [426]. *Shh* may also regulate *Tbx1* expression [427].

Regulated retinoid signaling has been implicated in the proper development of the pharyngeal pouch and its derivatives [428, 429]. Retinoids may elicit its effects through *Hox* genes as retinoic acid (RA) response elements have been identified in the regulatory regions of several *Hox* genes [430]. Inhibition of RA signaling (through antagonism or targeted mutations of RA receptors) resulted in aberrant expression profiles of the *Fgf*, *Hox* and *Pax* (*Pax1* and *Pax9*) genes [429, 431]. Furthermore, perturbed RA signaling may result in phenotypes associated with DiGeorge/velocardiofacial syndrome (DGS/VCFS), such as a deficiency of NC-derived mesenchyme [428, 432, 433]. Notably, mice deficient for retinaldehyde dehydrogenase 2 (*Raldh2*$^{-/-}$), an enzyme required for RA production in the embryo, demonstrate pharyngeal deformities encompassing either athymia or ectopic cervical thymic masses [434].

The mammalian neurotrophin family consists of four members: nerve growth factor (Ngf), brain-derived neurotrophic factor (Bdnf), neutrophin-3 (NT3), and neutrophin-4/5 (NT4/5) that interact with members of the tropomyosine receptor kinase (Trk) family [435]. Rodent and human thymic epithelia express *TrkA*, which interacts with high affinity to Ngf, secreted by lymphocytes [436], and with a lower affinity to NT3 [437–440]. In *TrkA*$^{-/-}$ mice, the postnatal thymus contained multiple cysts lined by endodermal epithelium, and showed a disruption of cortico-medullary borders and reduced thymocyte numbers [441]. *TrkA* is proposed to function in TEC-mesenchyme [442] and TEC-thymocyte interactions [443]. In addition, the differential expression of the three isoforms of *TrkA* (*I/II/III*) in the thymus suggests a predominant role for *TrkAI/II* during embryogenesis and *TrkAIII* in postnatal thymic maintenance [444]. Another hormone associated with the neuroendocrine-immune axis is growth hormone (GH). GH is produced by the pituitary gland but also by TECs and binds to the GH receptor (GHR), with some GH actions being mediated through Igf1 and its receptor. GH production reduces with age and given that GHR is present on many immune cells, including TECs, its declining production has been linked with thymic involution [445].

Cytokines

The expression of transcripts encoding receptors for interleukins (ILs), interferon-γ (IFNγ), Tnf, Tgf, and stem cell factor (SCF) by TECs suggests the involvement of these factors in mediating lymphostromal-associated effects on thymic epithelium [356]. Cytokines in particular may have roles in the proper development and function of TECs. For instance, IL6 serves as an autocrine growth factor for cultured murine TECs [446]. Similarly, supplementation of exogenous IL1 to human TEC cultures induced cells to enter the S phase and exhibit morphological alterations, suggesting a role in TEC proliferation [447]. IL1 may elicit these effects indirectly through the induction of IL6 production by TECs [448]. TEC secretion of IL6 was also induced upon addition of IL4 and IFNγ in human TEC cultures [449]. Collectively IL1, IL4, and IFNγ are some of the cytokines regulating TEC secretory functions and expression profiles [449–451].

6 Stem Cells, Aging, And Regenerative Medicine

Aging is associated with an increased propensity to infections and cancers caused by a diminishing immune system, and a decline in tissue regenerative potential that can lead to the progressive degeneration of tissues and organs. It is not clear whether this arises due to intrinsic aging of the adult stem cells that reside in most mammalian tissues, or whether it is due to an impairment of stem cell function due to an aging tissue environment or niche. These are important considerations in the developing field of regenerative medicine and for the use of stem cells to repair damaged and age-related degenerating or dysfunctional tissue.

6.1 Age-Related Thymic Involution

A paradox of the immune system is that the thymus involutes early in life such that by middle age, the thymus is functioning at only 5% of its maximal capacity [452]. The aged thymus is characterized by a disproportionate loss of TECs, an increase in adipose tissue, enlargement of perivascular spaces, a decline in T-cell development and less distinct cortical and medullary compartments [452]. The mechanisms of aging are not yet clear. Since the thymus requires the continual seeding of bone marrow–derived hematopoietic precursors, alterations can occur at multiple levels. Aging may impact on the structural elements of the thymus – the thymic stromal cell niche, of which TECs are an essential component. Alterations in the thymic niche in turn may impact on the survival, proliferation, differentiation and migration of the developing T-cells. Furthermore, the bone marrow–derived T-cell precursors are ultimately derived from HSCs and these may develop intrinsic defects with age. Alternatively, the bone marrow niche in which the HSCs develop may be altered and subsequently impact on HSC maintenance, proliferation, capacity to differentiate into multiple lineages and mobilization [453].

Thymic atrophy is most profound from puberty onwards in both males and females and has therefore been associated with the increase in production of sex steroids [454]. Consistent with this is the finding that blockade of sex steroids can reverse involution in rodent models within 2–3 weeks [454, 455]. The atrophied aged thymus returns to its young capacity in all respects after sex steroid ablation – thymocyte subsets and stromal cell compartments return in numbers and phenotype, similar to that of a 6-week-old thymus. Repair after damage from cancer treatments, such as after radiation and chemotherapy, is also enhanced with sex steroid ablation [454–457].

This ability of the thymus to dynamically regenerate following sex steroid ablation, and also after GH [458] and Fgf7 [459, 460, 461] treatment, would suggest active participation of epithelial stem or progenitor cells that reside in the thymic tissue throughout adulthood. Why the thymus begins to atrophy so early in life – a unique event among epithelial functional units, remains a topic of physiological and philosophical debate.

6.2 MTS24 as a Marker of Thymic Epithelial Stem or Progenitor Cells

We have developed the monoclonal antibody MTS24 that identifies a putative epithelial progenitor cell in the embryonic mouse thymus. It is expressed in the endodermal layer of the third pharyngeal pouch in the emerging E10.5 thymic anlage. All keratin-positive cells express MTS24 at E13 [183], however, the emergence of MTS24^{-} keratin-positive cells becomes evident from E14. By E15.5, over 50% of TECs have lost expression of MTS24 and associated progenitor capacity. The progenitor capacity of this subset of TECs was based on the finding that as few as 5,000 purified E12 or E15.5 MTS24^{+} cells, but not MTS24^{-} cells, can form a functional thymus that can attract T-cell precursors from the bone marrow and support full T-cell development when reaggregated and grafted under the kidney capsule [184, 462]. However, we subsequently found that if reaggregated in very high numbers (100,000 cells) and in the presence of thymic mesenchyme, E14 and E16 MTS24^{-} cells can also form a thymus [183]. Although putative stem cells may have been included as contaminants, it does indicate additional means of thymic formation and regeneration. The progenitor capacity of the MTS24^{+} cells appears to be lost by late embryogenesis (E18) [183], highlighting changes in the embryonic-adult transitional phase, although they do contribute to the restoration of the epithelium when the thymus regenerates. While an epithelial progenitor cell likely exists in the adult thymus, its phenotype is yet to be precisely defined [463]. More recently, we have identified an adult epithelial progenitor or transit amplifying cell subset in rodents that is activated to replenish epithelial cells after damage by immuno-depleting regimes such as cyclosporine A, cyclophosphamide, and dexamethasone (Fletcher A, *submitted*). However, before this can be clinically relevant, further work is required to identify an equivalent population in humans.

6.3 Clinical Prospects for Thymic Regeneration

In addition to the increased incidence of cancer, autoimmunity and opportunistic infections with the decline in thymic function with age, there are also very stark clinical implications including delayed immune recovery from immunodeficient states induced by cytotoxic Anti-neoplastic treatment or chronic infections such as that caused by the human immunodeficiency virus (HIV). The subsequent increased susceptibility to opportunistic infections can contribute to increased morbidity and mortality. Improving immune-reconstitution by regeneration of thymic tissue rather than expansion of a limited pre-existing T-cell pool, which would not replenish the T-cell receptor repertoire to generate new immune responses, has become important in this new age of regenerative medicine [464]. However, few clinically applicable candidates exist to fulfill these requirements. Building on the concept of sex steroids being at least partly responsible for thymic atrophy, the hormone Luteinizing Hormone Releasing Hormone-Agonist (LHRH-A) is currently being investigated for thymic regeneration clinically. LHRH-A has been used in the clinic for over 25 years to chemically block sex steroid production for the treatment of prostate [465] and breast cancers [466], endometriosis [467], and precocious puberty [468]. Thymic regeneration occurs rapidly in rodents (2–4 weeks), however, patients that have been treated with prostate cancer show a delayed but positive effect of LHRH-A on the increased production of naïve T cells [454]. More recently, a significant enhancement in the regeneration of naïve T cells following both allogeneic and autologous hematopoietic stem cell transplants was observed when LHRH-A was administered prior to transplant [469]. Additional clinical trials are currently ongoing. Other potential growth factors that could impact directly on TEC regeneration include Fgf7 and GH. As previously mentioned, Fgf7 is involved in TEC proliferation and differentiation. Administration of Fgf7 has been shown to enhance thymic recovery from cancer treatments such as cyclophosphamide and radiation and protect TECs from graft-versus-host disease after allogeneic bone marrow transplantation in rodents [459–461]. The clinical impact of Fgf7 treatment is currently under investigation. GH treatment has been used in HIV patients and has shown some success in increasing thymic tissue mass and circulating naïve $CD4^+$ T cells [458], however, toxicity and severe side effects means that the use of GH for immune reconstitution requires further assessment. Whether these growth factors induce the activation of resident epithelial stem cells or merely induce the proliferation of existing mature TECs is not currently clear.

7 Conclusion

Of the many cell types in the body, epithelia are among the best understood. They are characterized by clearly defined structures with relatively easy to identify transitional forms from the stem cells within their stromal niches, through to end-stage functionally mature progeny. It is now clear that epithelial stem cells are able to generate tissues that differ greatly in architecture and function, yet share many growth and differentiation characteristics, including similarity in their stem cell compartment and the array of transcription factors utilized. The generation of diversity is directed by temporal and spatial variations in these generic factors, coupled to as yet poorly defined tissue-specific differentiation signaling. Furthermore, a comparative analysis of the thymus with other epithelial organs and tissues demonstrates the additional importance of intercellular "cross-talk," with the niches providing the scaffold for cellular and soluble growth and differentiation factors operative on stem cells.

Recently, the scientific directions aimed at understanding the developmental sequences defining normal-versus-neoplastic growth have focused on stem cells. The reasoning is that stem cells, because of their inherent ability to self-renew, are likely to accumulate mutations, which collectively and progressively transmit defective growth characteristics onto daughter cells, resulting in tumor formation. Therapeutic strategies are thus being more focused on the conceptual cancer stem cell. Understanding not only whether such cells exist but also how they arise, is in turn dependent on a basic understanding of normal cell growth and differentiation. Our extensive knowledge of epithelial tissues thus represents valuable experimental targets for unraveling the mechanisms underlying normal-versus-pathological, including neoplastic, development.

Arguably, the key question in regenerative medicine is whether residual populations of stem cells exist which can be therapeutic targets. The regular turnover of cells, required for maintaining tissue homeostasis and replenishing cells during injury repair, supports the continued presence of epithelial stem cells in adult tissues. Epithelial stem cells have been in identified in the intestine, corneum, mammary gland, and skin, and are maintained and activated via interactions with resident stromal niche cells, an important component of which is the mesenchyme. Deciphering the molecular footprints for epithelial stem cell maintenance, differentiation, and maturation remains one of the current research challenges. As these factors become elucidated they would be ideally applied ex vivo to a common epithelial stem cell target – if such a cell exists. The recent realization that the amnion is highly enriched in multipotential epithelial stem cells provides a broadly available possibility. While skin burns have been successfully treated with stem cells, restoration of thymus function still requires unequivocal identification of the putative adult stem cell. However, all epithelial tissues may also be restored by self-renewal of pre-existing transit amplifying and mature cell types, not only by the regenerative properties of stem cells. This combination of stem cells and defined molecular signals to chaperone further differentiation along required therapeutic pathways represents the new horizon of cell-based therapy.

Acknowledgments We acknowledge generous research support from the Australian Stem Cell Centre, the National Health and Medical Research Council of Australia and Norwood Immunology.

References

1. Schock F, Perrimon N. Molecular mechanisms of epithelial morphogenesis. Annu Rev Cell Dev Biol. 2002;18:463–93.
2. Slack JM. Stem cells in epithelial tissues. Science. 2000;287 (5457):1431–3.

3. Shook D, Keller R. Mechanisms, mechanics and function of epithelial-mesenchymal transitions in early development. Mech Dev. 2003;120(11):1351–83.
4. Eakin GS, Behringer RR. Gastrulation in other mammals and humans. In: Stern CD, editor. Gastrulation: From cells to embryo. New York: Cold Spring Harbor Laboratory Press; 2004. pp. 275–87.
5. Tam PPL, Gad JM. Gastrulation in the mouse embryo. In: Stern CD, editor. Gastrulation: From cells to embryo. New York: Cold Spring Harbor Laboratory Press; 2004. pp. 233–62.
6. Tam PP, Loebel DA. Gene function in mouse embryogenesis: get set for gastrulation. Nat Rev Genet. 2007;8(5):368–81.
7. Li L, Xie T. Stem cell niche: structure and function. Annu Rev Cell Dev Biol. 2005;21:605–31.
8. Thomson JA, Itskovitz-Eldor J, Shapiro SS, et al. Embryonic stem cell lines derived from human blastocysts. Science. 1998;282(5391):1145–7.
9. Reubinoff BE, Pera MF, Fong CY, Trounson A, Bongso A. Embryonic stem cell lines from human blastocysts: somatic differentiation in vitro. Nat Biotechnol. 2000;18(4):399–404.
10. Spradling A, Drummond-Barbosa D, Kai T. Stem cells find their niche. Nature. 2001;414(6859):98–104.
11. Schofield R. The relationship between the spleen colony-forming cell and the haemopoietic stem cell. Blood Cells. 1978;4(1–2): 7–25.
12. Fuchs E, Tumbar T, Guasch G. Socializing with the neighbors: stem cells and their niche. Cell. 2004;116(6):769–78.
13. Rizvi AZ, Wong MH. Epithelial stem cells and their niche: there's no place like home. Stem Cells. 2005;23(2):150–65.
14. Ohlstein B, Kai T, Decotto E, Spradling A. The stem cell niche: theme and variations. Curr Opin Cell Biol. 2004;16(6):693–9.
15. Mitsiadis TA, Barrandon O, Rochat A, Barrandon Y, De Bari C. Stem cell niches in mammals. Exp Cell Res. 2007;313(16): 3377–85.
16. Moore KA, Lemischka IR. Stem cells and their niches. Science. 2006;311(5769):1880–5.
17. Mackenzie IC. Stem cell properties and epithelial malignancies. Eur J Cancer. 2006;42(9):1204–12.
18. Yamashita YM, Jones DL, Fuller MT. Orientation of asymmetric stem cell division by the APC tumor suppressor and centrosome. Science. 2003;301(5639):1547–50.
19. Perez-Moreno M, Jamora C, Fuchs E. Sticky business: orchestrating cellular signals at adherens junctions. Cell. 2003;112(4): 535–48.
20. Etienne-Manneville S, Hall A. Cell polarity: Par6, aPKC and cytoskeletal crosstalk. Curr Opin Cell Biol. 2003;15(1):67–72.
21. Yamashita YM, Fuller MT, Jones DL. Signaling in stem cell niches: lessons from the Drosophila germline. J Cell Sci. 2005;118(Pt 4):665–72.
22. Betschinger J, Knoblich JA. Dare to be different: asymmetric cell division in Drosophila, C. elegans and vertebrates. Curr Biol. 2004;14(16):R674–85.
23. Marshman E, Booth C, Potten CS. The intestinal epithelial stem cell. Bioessays. 2002;24(1):91–8.
24. Lamprecht J. Symmetric and asymmetric cell division in rat corneal epithelium. Cell Tissue Kinet. 1990;23(3):203–16.
25. Dontu G, Al-Hajj M, Abdallah WM, Clarke MF, Wicha MS. Stem cells in normal breast development and breast cancer. Cell Prolif. 2003;36 Suppl 1:59–72.
26. Claudinot S, Nicolas M, Oshima H, Rochat A, Barrandon Y. Long-term renewal of hair follicles from clonogenic multipotent stem cells. Proc Natl Acad Sci U S A. 2005;102(41):14677–82.
27. Lechler T, Fuchs E. Asymmetric cell divisions promote stratification and differentiation of mammalian skin. Nature. 2005;437(7056):275–80.
28. Watt FM, Hogan BL. Out of Eden: stem cells and their niches. Science. 2000;287(5457):1427–30.
29. Potten CS, Wilson JW, Robert L, Robert L, Joseph V. Adult epithelial tissue stem cells. In: Principles of tissue engineering. 3rd ed. Burlington: Academic Press; 2007. pp. 431–44.
30. Molofsky AV, Pardal R, Morrison SJ. Diverse mechanisms regulate stem cell self-renewal. Curr Opin Cell Biol. 2004;16(6): 700–7.
31. Blanpain C, Horsley V, Fuchs E. Epithelial stem cells: turning over new leaves. Cell. 2007;128(3):445–58.
32. Hanna C, Bicknell DS, O'Brien JE. Cell turnover in the adult human eye. Arch. Ophthalmol. 1961;65:695–8.
33. Potten CS, Saffhill R, Maibach HI. Measurement of the transit time for cells through the epidermis and stratum corneum of the mouse and guinea-pig. Cell Tissue Kinet. 1987;20(5):461–72.
34. Chidgey A, Boyd RL, Hugo P. The thymic niche and thymopoiesis. In: Encyclopedia of life sciences. Chichester: John Wiley & Sons, Ltd; 2007.
35. Gillard GO, Farr AG. Features of medullary thymic epithelium implicate postnatal development in maintaining epithelial heterogeneity and tissue-restricted antigen expression. J Immunol. 2006;176(10):5815–24.
36. Gray DH, Seach N, Ueno T, et al. Developmental kinetics, turnover, and stimulatory capacity of thymic epithelial cells. Blood. 2006;108(12):3777–85.
37. Blanpain C, Lowry WE, Geoghegan A, Polak L, Fuchs E. Self-renewal, multipotency, and the existence of two cell populations within an epithelial stem cell niche. Cell. 2004;118(5):635–48.
38. Hennighausen L, Robinson GW. Information networks in the mammary gland. NatRev. 2005;6(9):715–25.
39. Brawley C, Matunis E. Regeneration of male germline stem cells by spermatogonial dedifferentiation in vivo. Science. 2004;304(5675):1331–4.
40. Kai T, Spradling A. Differentiating germ cells can revert into functional stem cells in Drosophila melanogaster ovaries. Nature. 2004;428(6982):564–9.
41. Potten CS, Grant HK. The relationship between ionizing radiation-induced apoptosis and stem cells in the small and large intestine. Br J Cancer. 1998;78(8):993–1003.
42. Potten CS. Apoptosis in oral mucosa: lessons from the crypt. A commentary. Oral Dis. 2001;7(2):81–5.
43. Barrandon Y, Morgan JR, Mulligan RC, Green H. Restoration of growth potential in paraclones of human keratinocytes by a viral oncogene. Proc Natl Acad Sci U S A. 1989;86(11):4102–6.
44. Ferraris C, Chaloin-Dufau C, Dhouailly D. Transdifferentiation of embryonic and postnatal rabbit corneal epithelial cells. Differentiation. 1994;57(2):89–96.
45. Baba Y, Garrett KP, Kincade PW. Constitutively active beta-catenin confers multilineage differentiation potential on lymphoid and myeloid progenitors. Immunity. 2005;23(6):599–609.
46. Nakagawa T, Nabeshima Y, Yoshida S. Functional identification of the actual and potential stem cell compartments in mouse spermatogenesis. Dev Cell. 2007;12(2):195–206.
47. Park IH, Zhao R, West JA, et al. Reprogramming of human somatic cells to pluripotency with defined factors. Nature. 2008;451(7175):141–6 .
48. Takahashi K, Tanabe K, Ohnuki M, et al. Induction of pluripotent stem cells from adult human fibroblasts by defined factors. Cell. 2007;131(5):861–72.
49. Yu J, Vodyanik MA, Smuga-Otto K, et al. Induced pluripotent stem cell lines derived from human somatic cells. Science. 2007;318(5858):1917–20.
50. Takahashi K, Yamanaka S. Induction of pluripotent stem cells from mouse embryonic and adult fibroblast cultures by defined factors. Cell. 2006;126(4):663–76.

51. Byrne JA, Pedersen DA, Clepper LL, et al. Producing primate embryonic stem cells by somatic cell nuclear transfer. Nature. 2007;450(7169):497–502.
52. Birchmeier C, Birchmeier W. Molecular aspects of mesenchymal-epithelial interactions. Annu Rev Cell Biol. 1993;9:511–40.
53. Jenkinson WE, Jenkinson EJ, Anderson G. Differential requirement for mesenchyme in the proliferation and maturation of thymic epithelial progenitors. J Exp Med. 2003;198(2):325–32.
54. Capdevila J, Izpisua Belmonte JC. Patterning mechanisms controlling vertebrate limb development. Annu Rev Cell Dev Biol. 2001;17:87–132.
55. Rendl M, Lewis L, Fuchs E. Molecular dissection of mesenchymal-epithelial interactions in the hair follicle. PLoS Biol. 2005;3(11):e331.
56. Brittan M, Wright NA. Gastrointestinal stem cells. J Pathol. 2002;197(4):492–509.
57. Marsh MN, Trier JS. Morphology and cell proliferation of subepithelial fibroblasts in adult mouse jejunum. II. Radioautographic studies. Gastroenterology. 1974;67(4):636–45.
58. Cunha GR, Young P, Christov K, et al. Mammary phenotypic expression induced in epidermal cells by embryonic mammary mesenchyme. Acta Anat (Basel). 1995;152(3):195–204.
59. Ferraris C, Chevalier G, Favier B, Jahoda CA, Dhouailly D. Adult corneal epithelium basal cells possess the capacity to activate epidermal, pilosebaceous and sweat gland genetic programs in response to embryonic dermal stimuli. Development. 2000;127(24):5487–95.
60. Taylor RA, Cowin PA, Cunha GR, et al. Formation of human prostate tissue from embryonic stem cells. Nat Methods. 2006;3(3):179–81.
61. Tsai RY. A molecular view of stem cell and cancer cell self-renewal. Int J Biochem Cell Biol. 2004;36(4):684–94.
62. Logan CY, Nusse R. The Wnt signaling pathway in development and disease. Annu Rev Cell Dev Biol. 2004;20:781–810.
63. Chan SD, Karpf DB, Fowlkes ME, et al. Two homologs of the Drosophila polarity gene frizzled (fz) are widely expressed in mammalian tissues. J Biol Chem. 1992;267(35):25202–7.
64. Lowry WE, Richter L. Signaling in adult stem cells. Front Biosci. 2007;12:3911–27.
65. Huber O, Korn R, McLaughlin J, Ohsugi M, Herrmann BG, Kemler R. Nuclear localization of beta-catenin by interaction with transcription factor LEF-1. Mech Dev. 1996;59(1):3–10.
66. Wodarz A, Nusse R. Mechanisms of Wnt signaling in development. Annu Rev Cell Dev Biol. 1998;14:59–88.
67. Wagner TU. Bone morphogenetic protein signaling in stem cells–one signal, many consequences. FEBS J. 2007;274(12):2968–76.
68. Kitisin K, Saha T, Blake T, et al. Tgf-Beta signaling in development. Sci STKE. 2007;2007(399):cm1.
69. Nohe A, Keating E, Knaus P, Petersen NO. Signal transduction of bone morphogenetic protein receptors. Cell Signal. 2004;16(3):291–9.
70. Shi Y, Massague J. Mechanisms of TGF-beta signaling from cell membrane to the nucleus. Cell. 2003;113(6):685–700.
71. Mohr OL. Character changes caused by mutation of an entire region of a chromosome in Drosophila. Genetics. 1919;4(3): 275–82.
72. Carlson ME, Conboy IM. Regulating the Notch pathway in embryonic, adult and old stem cells. Curr Opin Pharmacol. 2007;7(3):303–9.
73. Chiba S. Notch signaling in stem cell systems. Stem Cells. 2006;24(11):2437–47.
74. Fiuza UM, Arias AM. Cell and molecular biology of Notch. J Endocrinol. 2007;194(3):459–74.
75. Iso T, Kedes L, Hamamori Y. HES and HERP families: multiple effectors of the Notch signaling pathway. J Cell Physiol. 2003;194(3):237–55.
76. Gat U, DasGupta R, Degenstein L, Fuchs E. De Novo hair follicle morphogenesis and hair tumors in mice expressing a truncated beta-catenin in skin. Cell. 1998;95(5):605–14.
77. Silva-Vargas V, Lo Celso C, Giangreco A, et al. Beta-catenin and Hedgehog signal strength can specify number and location of hair follicles in adult epidermis without recruitment of bulge stem cells. Dev Cell. 2005;9(1):121–31.
78. Jacob L, Lum L. Hedgehog signaling pathway. Sci STKE. 2007;2007(407):cm6.
79. Riobo NA, Manning DR. Pathways of signal transduction employed by vertebrate Hedgehogs. Biochem J. 2007;403(3): 369–79.
80. Wright NA. Epithelial stem cell repertoire in the gut: clues to the origin of cell lineages, proliferative units and cancer. Int J Exp Pathol. 2000;81(2):117–43.
81. Winton DJ. Stem cells in the epithelium of the small intestine and colon. In: Marshak DR, Gardner RL, Gottlieb D, editors. Stem cell biology. Cold Spring Harbor, NY: Cold Spring Harbor Lab. Press; 2001. pp. 515–36.
82. Potten CS, Loeffler M. Stem cells: attributes, cycles, spirals, pitfalls and uncertainties. Lessons for and from the crypt. Development. 1990;110(4):1001–20.
83. Booth C, Potten CS. Gut instincts: thoughts on intestinal epithelial stem cells. J Clin Invest. 2000;105(11):1493–9.
84. He XC, Zhang J, Tong WG, et al. BMP signaling inhibits intestinal stem cell self-renewal through suppression of Wnt-beta-catenin signaling. Nat Genet. 2004;36(10):1117–21.
85. Potten CS, Owen G, Booth D. Intestinal stem cells protect their genome by selective segregation of template DNA strands. J Cell Sci. 2002;115(Pt 11):2381–8.
86. Bjerknes M, Cheng H. Clonal analysis of mouse intestinal epithelial progenitors. Gastroenterology. 1999;116(1):7–14.
87. Winton DJ, Blount MA, Ponder BA. A clonal marker induced by mutation in mouse intestinal epithelium. Nature. 1988;333(6172):463–6.
88. Sancho E, Batlle E, Clevers H. Live and let die in the intestinal epithelium. Curr Opin Cell Biol. 2003;15(6):763–70.
89. Garabedian EM, Roberts LJ, McNevin MS, Gordon JI. Examining the role of Paneth cells in the small intestine by lineage ablation in transgenic mice. J Biol Chem. 1997;272(38):23729–40.
90. Giannakis M, Stappenbeck TS, Mills JC, et al. Molecular properties of adult mouse gastric and intestinal epithelial progenitors in their niches. J Biol Chem. 2006;281(16):11292–300.
91. Wong MH, Saam JR, Stappenbeck TS, Rexer CH, Gordon JI. Genetic mosaic analysis based on Cre recombinase and navigated laser capture microdissection. Proc Natl Acad Sci U S A. 2000;97(23):12601–6.
92. Potten CS, Owen G, Roberts SA. The temporal and spatial changes in cell proliferation within the irradiated crypts of the murine small intestine. Int J Radiat Biol. 1990;57(1):185–99.
93. Kim KM, Shibata D. Methylation reveals a niche: stem cell succession in human colon crypts. Oncogene. 2002;21(35):5441–9.
94. Paris F, Fuks Z, Kang A, et al. Endothelial apoptosis as the primary lesion initiating intestinal radiation damage in mice. Science. 2001;293(5528):293–7.
95. Karlsson L, Lindahl P, Heath JK, Betsholtz C. Abnormal gastrointestinal development in PDGF-A and PDGFR-(alpha) deficient mice implicates a novel mesenchymal structure with putative instructive properties in villus morphogenesis. Development. 2000;127(16):3457–66.
96. Korinek V, Barker N, Moerer P, et al. Depletion of epithelial stem-cell compartments in the small intestine of mice lacking Tcf-4. Nat Genet. 1998;19(4):379–83.
97. Pinto D, Gregorieff A, Begthel H, Clevers H. Canonical Wnt signals are essential for homeostasis of the intestinal epithelium. Genes Dev. 2003;17(14):1709–13.

98. Kullander K, Klein R. Mechanisms and functions of Eph and ephrin signalling. Nat Rev Mol Cell Biol. 2002;3(7):475–86.
99. Batlle E, Bacani J, Begthel H, et al. EphB receptor activity suppresses colorectal cancer progression. Nature. 2005;435(7045):1126–30.
100. Holmberg J, Genander M, Halford MM, et al. EphB receptors coordinate migration and proliferation in the intestinal stem cell niche. Cell. 2006;125(6):1151–63.
101. Haramis AP, Begthel H, van den Born M, et al. De novo crypt formation and juvenile polyposis on BMP inhibition in mouse intestine. Science. 2004;303(5664):1684–6.
102. Takaku K, Miyoshi H, Matsunaga A, Oshima M, Sasaki N, Taketo MM. Gastric and duodenal polyps in Smad4 (Dpc4) knockout mice. Cancer Res. 1999;59(24):6113–7.
103. Yang Q, Bermingham NA, Finegold MJ, Zoghbi HY. Requirement of Math1 for secretory cell lineage commitment in the mouse intestine. Science. 2001;294(5549):2155–8.
104. Jensen J, Pedersen EE, Galante P, et al. Control of endodermal endocrine development by Hes-1. Nat Genet. 2000;24(1):36–44.
105. Suzuki K, Fukui H, Kayahara T, et al. Hes1-deficient mice show precocious differentiation of Paneth cells in the small intestine. Biochem Biophys Res Commun. 2005;328(1):348–52.
106. Schroder N, Gossler A. Expression of Notch pathway components in fetal and adult mouse small intestine. Gene Expr Patterns. 2002;2(3–4):247–50.
107. Fre S, Huyghe M, Mourikis P, Robine S, Louvard D, Artavanis-Tsakonas S. Notch signals control the fate of immature progenitor cells in the intestine. Nature. 2005;435(7044):964–8.
108. van Es JH, van Gijn ME, Riccio O, et al. Notch/gamma-secretase inhibition turns proliferative cells in intestinal crypts and adenomas into goblet cells. Nature. 2005;435(7044):959–63.
109. Madison BB, Braunstein K, Kuizon E, Portman K, Qiao XT, Gumucio DL. Epithelial hedgehog signals pattern the intestinal crypt-villus axis. Development. 2005;132(2):279–89.
110. Ramalho-Santos M, Melton DA, McMahon AP. Hedgehog signals regulate multiple aspects of gastrointestinal development. Development. 2000;127(12):2763–72.
111. van den Brink GR, Bleuming SA, Hardwick JC, et al. Indian Hedgehog is an antagonist of Wnt signaling in colonic epithelial cell differentiation. Nat Genet. 2004;36(3):277–82.
112. Revoltella RP, Papini S, Rosellini A, Michelini M. Epithelial stem cells of the eye surface. Cell Prolif. 2007;40(4):445–61.
113. Charukamnoetkanok P. Corneal stem cells: bridging the knowledge gap. Semin Ophthalmol. 2006;21(1):1–7.
114. Dua HS, Forrester JV. The corneoscleral limbus in human corneal epithelial wound healing. Am J Ophthalmol. 1990;110(6):646–56.
115. Davanger M, Evensen A. Role of the pericorneal papillary structure in renewal of corneal epithelium. Nature. 1971;229(5286):560–1.
116. Miller SJ, Lavker RM, Sun TT. Keratinocyte stem cells of cornea, skin and hair follicles. In: Potten CS, editor. Stem cells. New York: Academic Press; 1997:331–62.
117. Schlotzer-Schrehardt U, Kruse FE. Identification and characterization of limbal stem cells. Exp Eye Res. 2005;81(3):247–64.
118. Dua HS, Shanmuganathan VA, Powell-Richards AO, Tighe PJ, Joseph A. Limbal epithelial crypts: a novel anatomical structure and a putative limbal stem cell niche. Br JOphthalmol. 2005;89(5):529–32.
119. Dua HS, Joseph A, Shanmuganathan VA, Jones RE. Stem cell differentiation and the effects of deficiency. Eye. 2003;17(8):877–85.
120. Notara M, Daniels JT. Biological principals and clinical potentials of limbal epithelial stem cells. Cell Tissue Res. 2008;331(1):135–43.
121. Li W, Hayashida Y, Chen YT, Tseng SC. Niche regulation of corneal epithelial stem cells at the limbus. Cell Res. 2007;17(1):26–36.
122. Higa K, Shimmura S, Miyashita H, Shimazaki J, Tsubota K. Melanocytes in the corneal limbus interact with K19-positive basal epithelial cells. Exp Eye Res. 2005;81(2):218–23.
123. Narisawa Y, Kohda H, Tanaka T. Three-dimensional demonstration of melanocyte distribution of human hair follicles: special reference to the bulge area. Acta Dermato Venereol 1997;77(2):97–101.
124. Espana EM, Kawakita T, Romano A, et al. Stromal niche controls the plasticity of limbal and corneal epithelial differentiation in a rabbit model of recombined tissue. Invest Ophthalmol Vis Sci. 2003;44(12):5130–5.
125. Kawakita T, Espana EM, He H, Li W, Liu CY, Tseng SC. Intrastromal invasion by limbal epithelial cells is mediated by epithelial-mesenchymal transition activated by air exposure. Am J Pathol. 2005;167(2):381–93.
126. Li DQ, Tseng SC. Three patterns of cytokine expression potentially involved in epithelial-fibroblast interactions of human ocular surface. J Cell Physiol 1995;163(1):61–79.
127. Lovicu FJ, Kao WW, Overbeek PA. Ectopic gland induction by lens-specific expression of keratinocyte growth factor (FGF-7) in transgenic mice. Mech Dev. 1999;88(1):43–53.
128. Touhami A, Grueterich M, Tseng SC. The role of NGF signaling in human limbal epithelium expanded by amniotic membrane culture. Invest Ophthalmol Vis Sci. 2002;43(4):987–94.
129. Qi H, Chuang EY, Yoon KC, et al. Patterned expression of neurotrophic factors and receptors in human limbal and corneal regions. Mol Vis. 2007;13:1934–41.
130. Fokina VM, Frolova EI. Expression patterns of Wnt genes during development of an anterior part of the chicken eye. Dev Dyn. 2006;235(2):496–505.
131. Jin EJ, Burrus LW, Erickson CA. The expression patterns of Wnts and their antagonists during avian eye development. Mech Dev. 2002;116(1–2):173–6.
132. Lyu J, Joo CK. Expression of Wnt and MMP in epithelial cells during corneal wound healing. Cornea. 2006;25(10 Suppl 1):S24–8.
133. Pearton DJ, Yang Y, Dhouailly D. Transdifferentiation of corneal epithelium into epidermis occurs by means of a multistep process triggered by dermal developmental signals. Proc Natl Acad Sci U S A. 2005;102(10):3714–9.
134. Mukhopadhyay M, Gorivodsky M, Shtrom S, et al. Dkk2 plays an essential role in the corneal fate of the ocular surface epithelium. Development. 2006;133(11):2149–54.
135. You L, Kruse FE, Pohl J, Volcker HE. Bone morphogenetic proteins and growth and differentiation factors in the human cornea. Invest Ophthalmol Vis Sci. 1999;40(2):296–311.
136. Chang B, Smith RS, Peters M, et al. Haploinsufficient Bmp4 ocular phenotypes include anterior segment dysgenesis with elevated intraocular pressure. BMC Genet. 2001;2:18.
137. Nicolas M, Wolfer A, Raj K, et al. Notch1 functions as a tumor suppressor in mouse skin. Nat Genet. 2003;33(3):416–21.
138. Umemoto T, Yamato M, Nishida K, et al. Rat limbal epithelial side population cells exhibit a distinct expression of stem cell markers that are lacking in side population cells from the central cornea. FEBS Lett. 2005;579(29):6569–74.
139. Thomas PB, Liu YH, Zhuang FF, et al. Identification of Notch-1 expression in the limbal basal epithelium. Mol Vis. 2007;13:337–44.
140. Ma A, Boulton M, Zhao B, Connon C, Cai J, Albon J. A role for notch signaling in human corneal epithelial cell differentiation and proliferation. Invest Ophthalmol Vis Sci. 2007;48(8):3576–85.

141. Saika S, Muragaki Y, Okada Y, et al. Sonic hedgehog expression and role in healing corneal epithelium. Invest Ophthalmol Vis Sci. 2004;45(8):2577–85.
142. Takabatake T, Ogawa M, Takahashi TC, Mizuno M, Okamoto M, Takeshima K. Hedgehog and patched gene expression in adult ocular tissues. FEBS Lett. 1997;410(2–3):485–9.
143. Richert MM, Schwertfeger KL, Ryder JW, Anderson SM. An atlas of mouse mammary gland development. J Mammary Gland Biol Neoplasia. 2000;5(2):227–41.
144. Kalirai H, Clarke RB. Human breast epithelial stem cells and their regulation. J Pathol. 2006;208(1):7–16.
145. Vaillant F, Asselin-Labat ML, Shackleton M, Lindeman GJ, Visvader JE. The emerging picture of the mouse mammary stem cell. Stem Cell Rev. 2007;3(2):114–23.
146. Williams JM, Daniel CW. Mammary ductal elongation: differentiation of myoepithelium and basal lamina during branching morphogenesis. Dev Biol. 1983;97(2):274–90.
147. Humphreys RC, Krajewska M, Krnacik S, et al. Apoptosis in the terminal endbud of the murine mammary gland: a mechanism of ductal morphogenesis. Development. 1996;122(12):4013–22.
148. Smith GH, Chepko G. Mammary epithelial stem cells. Microsc Res Tech. 2001;52(2):190–203.
149. Chepko G, Smith GH. Three division-competent, structurally-distinct cell populations contribute to murine mammary epithelial renewal. Tissue Cell. 1997;29(2):239–53.
150. Chepko G, Dickson RB. Ultrastructure of the putative stem cell niche in rat mammary epithelium. Tissue Cell. 2003;35(2): 83–93.
151. Smith GH. Experimental mammary epithelial morphogenesis in an in vivo model: evidence for distinct cellular progenitors of the ductal and lobular phenotype. Breast Cancer Res Treat. 1996;39(1):21–31.
152. Alvi AJ, Clayton H, Joshi C, et al. Functional and molecular characterisation of mammary side population cells. Breast Cancer Res. 2003;5(1):R1–8.
153. Welm B, Behbod F, Goodell MA, Rosen JM. Isolation and characterization of functional mammary gland stem cells. Cell Prolif. 2003;36 Suppl 1:17–32.
154. Kordon EC, Smith GH. An entire functional mammary gland may comprise the progeny from a single cell. Development. 1998;125(10):1921–30.
155. Stingl J, Eirew P, Ricketson I, et al. Purification and unique properties of mammary epithelial stem cells. Nature. 2006;439(7079):993–7.
156. Shackleton M, Vaillant F, Simpson KJ, et al. Generation of a functional mammary gland from a single stem cell. Nature. 2006;439(7072):84–8.
157. Dontu G, Abdallah WM, Foley JM, et al. In vitro propagation and transcriptional profiling of human mammary stem/progenitor cells. Genes Dev. 2003;17(10):1253–70.
158. Liao MJ, Zhang CC, Zhou B, et al. Enrichment of a population of mammary gland cells that form mammospheres and have in vivo repopulating activity. Cancer Res. 2007;67(17):8131–8.
159. Parmar H, Cunha GR. Epithelial-stromal interactions in the mouse and human mammary gland in vivo. Endocr-Relat Cancer. 2004;11(3):437–58.
160. Turashvili G, Bouchal J, Burkadze G, Kolar Z. Wnt signaling pathway in mammary gland development and carcinogenesis. Pathobiology. 2006;73(5):213–23.
161. van Genderen C, Okamura RM, Farinas I, et al. Development of several organs that require inductive epithelial-mesenchymal interactions is impaired in LEF-1-deficient mice. Genes Dev. 1994;8(22):2691–703.
162. Andl T, Reddy ST, Gaddapara T, Millar SE. WNT signals are required for the initiation of hair follicle development. Dev Cell. 2002;2(5):643–53.
163. Tepera SB, McCrea PD, Rosen JM. A beta-catenin survival signal is required for normal lobular development in the mammary gland. J Cell Sci. 2003;116(Pt 6):1137–49.
164. Liu BY, McDermott SP, Khwaja SS, Alexander CM. The transforming activity of Wnt effectors correlates with their ability to induce the accumulation of mammary progenitor cells. Proc Natl Acad Sci U S A. 2004;101(12):4158–63.
165. Teuliere J, Faraldo MM, Deugnier MA, et al. Targeted activation of beta-catenin signaling in basal mammary epithelial cells affects mammary development and leads to hyperplasia. Development. 2005;132(2):267–77.
166. Alexander CM, Reichsman F, Hinkes MT, et al. Syndecan-1 is required for Wnt-1-induced mammary tumorigenesis in mice. Nat Genet. 2000;25(3):329–32.
167. Smith GH, Gallahan D, Diella F, Jhappan C, Merlino G, Callahan R. Constitutive expression of a truncated INT3 gene in mouse mammary epithelium impairs differentiation and functional development. Cell Growth Differ. 1995;6(5):563–77.
168. Callahan R, Egan SE. Notch signaling in mammary development and oncogenesis. J Mammary Gland Biol Neoplasia. 2004;9(2):145–63.
169. Gallahan D, Jhappan C, Robinson G, et al. Expression of a truncated Int3 gene in developing secretory mammary epithelium specifically retards lobular differentiation resulting in tumorigenesis. Cancer Res. 1996;56(8):1775–85.
170. Dontu G, Jackson KW, McNicholas E, Kawamura MJ, Abdallah WM, Wicha MS. Role of Notch signaling in cell-fate determination of human mammary stem/progenitor cells. Breast Cancer Res. 2004;6(6):R605–15.
171. Lewis MT. Hedgehog signaling in mouse mammary gland development and neoplasia. J Mammary Gland Biol Neoplasia. 2001;6(1):53–66.
172. Lewis MT, Ross S, Strickland PA, et al. The Gli2 transcription factor is required for normal mouse mammary gland development. Dev Biol. 2001;238(1):133–44.
173. Lewis MT, Ross S, Strickland PA, et al. Defects in mouse mammary gland development caused by conditional haploinsufficiency of Patched-1. Development. 1999;126(22):5181–93.
174. Liu S, Dontu G, Mantle ID, et al. Hedgehog signaling and Bmi-1 regulate self-renewal of normal and malignant human mammary stem cells. Cancer Res. 2006;66(12):6063–71.
175. Kubo M, Nakamura M, Tasaki A, et al. Hedgehog signaling pathway is a new therapeutic target for patients with breast cancer. Cancer Res. 2004;64(17):6071–4.
176. Dooley J, Erickson M, Roelink H, Farr AG. Nude thymic rudiment lacking functional foxn1 resembles respiratory epithelium. Dev Dyn. 2005;233(4):1605–12.
177. Patel DD, Whichard LP, Radcliff G, Denning SM, Haynes BF. Characterization of human thymic epithelial cell surface antigens: phenotypic similarity of thymic epithelial cells to epidermal keratinocytes. J Clin Immunol. 1995;15(2):80–92.
178. Lobach DF, Haynes BF. Ontogeny of the human thymus during fetal development. JClinImmunol. 1987;7(2):81–97.
179. Laster AJ, Itoh T, Palker TJ, Haynes BF. The human thymic microenvironment: thymic epithelium contains specific keratins associated with early and late stages of epidermal keratinocyte maturation. Differentiation. 1986;31(1):67–77.
180. Nishio H, Matsui K, Tsuji H, Tamura A, Suzuki K. Immunolocalization of calcineurin and FKBP12, the FK506-binding protein, in Hassall's corpuscles of human thymus and epidermis. Histochem Cell Biol. 2000;114(1):9–14.
181. Senoo M, Pinto F, Crum CP, McKeon F. p63 Is Essential for the Proliferative Potential of Stem Cells in Stratified Epithelia. Cell. 2007;129(3):523–36.
182. Khlystova ZS, Kalinina, II, Khavinson V. Thymic hormones in human fetal skin epidermis. Bull Exp Biol Med. 2002;133(2):196–8.

183. Rossi SW, Chidgey AP, Parnell SM, et al. Redefining epithelial progenitor potential in the developing thymus. Eur J Immunol. 2007;37(9):2411–8.
184. Gill J, Malin M, Hollander GA, Boyd R. Generation of a complete thymic microenvironment by MTS24(+) thymic epithelial cells. Nat Immunol. 2002;3(7):635–42.
185. Nijhof JG, van Pelt C, Mulder AA, Mitchell DL, Mullenders LH, de Gruijl FR. Epidermal stem and progenitor cells in murine epidermis accumulate UV damage despite NER proficiency. Carcinogenesis. 2007;28(4):792–800.
186. Trempus CS, Morris RJ, Ehinger M, et al. CD34 expression by hair follicle stem cells is required for skin tumor development in mice. Cancer Res. 2007;67(9):4173–81.
187. Nijhof JG, Braun KM, Giangreco A, et al. The cell-surface marker MTS24 identifies a novel population of follicular keratinocytes with characteristics of progenitor cells. Development. 2006;133(15):3027–37.
188. Blackburn CC, Augustine CL, Li R, et al. The nu gene acts cell-autonomously and is required for differentiation of thymic epithelial progenitors. Proc Natl Acad Sci U S A. 1996;93(12):5742–6.
189. Nehls M, Kyewski B, Messerle M, et al. Two genetically separable steps in the differentiation of thymic epithelium. Science. 1996;272(5263):886–9.
190. Mecklenburg L, Tychsen B, Paus R. Learning from nudity: lessons from the nude phenotype. Exp Dermatol. 2005;14(11):797–810.
191. Clark RA, Yamanaka K, Bai M, Dowgiert R, Kupper TS. Human skin cells support thymus-independent T cell development. J Clin Invest. 2005;115(11):3239–49.
192. Braun KM, Prowse DM. Distinct epidermal stem cell compartments are maintained by independent niche microenvironments. Stem Cell Rev. 2006;2(3):221–31.
193. Chuong C-M. Morphogenesis of epithelial appendages: variations on top of a common theme and implications in regeneration. In: Chuong C-M, editor. Molecular basis of epithelial appendage morphogenesis. Austin: R.G. Landes Company; 1998. pp. 3–14.
194. Mikkola ML. Genetic basis of skin appendage development. Semin Cell Dev Biol. 2007;18(2):225–36.
195. Koster MI, Roop DR. Mechanisms regulating epithelial stratification. Annu Rev Cell Dev Biol. 2007;23:93–113.
196. Wilson SI, Rydstrom A, Trimborn T, et al. The status of Wnt signalling regulates neural and epidermal fates in the chick embryo. Nature. 2001;411(6835):325–30.
197. Mack JA, Anand S, Maytin EV. Proliferation and cornification during development of the mammalian epidermis. Birth Defects Res C Embryo Today. 2005;75(4):314–29.
198. Blanpain C, Fuchs E. Epidermal stem cells of the skin. Annu Rev Cell Dev Biol. 2006;22:339–73.
199. Byrne C, Tainsky M, Fuchs E. Programming gene expression in developing epidermis. Development. 1994;120(9):2369–83.
200. Armstrong MT, Turlo K, Elges CJ, Dayton SM, Lee J, Armstrong PB. A novel form of epithelial wound healing of the embryonic epidermis. Exp Cell Res. 2006;312(13):2415–23.
201. M'Boneko V, Merker HJ. Development and morphology of the periderm of mouse embryos (days 9–12 of gestation). Acta Anat. 1988;133(4):325–36.
202. Koster MI, Dai D, Marinari B, et al. p63 induces key target genes required for epidermal morphogenesis. Proc Natl Acad Sci U S A. 2007;104(9):3255–60.
203. Candi E, Schmidt R, Melino G. The cornified envelope: a model of cell death in the skin. Nat Rev. 2005;6(4):328–40.
204. Hardman MJ, Sisi P, Banbury DN, Byrne C. Patterned acquisition of skin barrier function during development. Development. 1998;125(8):1541–52.
205. Segre J. Complex redundancy to build a simple epidermal permeability barrier. Curr Opin Cell Biol. 2003;15(6):776–82.
206. Koster MI, Roop DR. Genetic pathways required for epidermal morphogenesis. Eur J Cell Biol. 2004;83(11–12):625–9.
207. Schmidt-Ullrich R, Paus R. Molecular principles of hair follicle induction and morphogenesis. Bioessays. 2005;27(3):247–61.
208. Waters JM, Richardson GD, Jahoda CA. Hair follicle stem cells. Semin Cell Dev Biol. 2007;18(2):245–54.
209. Fuchs E. Scratching the surface of skin development. Nature. 2007;445(7130):834–42.
210. Millar SE. Molecular mechanisms regulating hair follicle development. J Invest Dermatol. 2002;118(2):216–25.
211. Slominski A, Wortsman J, Plonka PM, Schallreuter KU, Paus R, Tobin DJ. Hair follicle pigmentation. J Invest Dermatol. 2005;124(1):13–21.
212. Barsh G, Cotsarelis G. How hair gets its pigment. Cell. 2007;130(5):779–81.
213. Slominski A, Tobin DJ, Shibahara S, Wortsman J. Melanin pigmentation in mammalian skin and its hormonal regulation. Physiol Rev. 2004;84(4):1155–228.
214. Webb A, Kaur P. Epidermal stem cells. Front Biosci. 2006; 11:1031–41.
215. Potten CS. The epidermal proliferative unit: the possible role of the central basal cell. Cell Tissue Kinet. 1974;7(1):77–88.
216. Potten CS. Cell replacement in epidermis (keratopoiesis) via discrete units of proliferation. Int Rev Cytol. 1981;69:271–318.
217. Braun KM, Niemann C, Jensen UB, Sundberg JP, Silva-Vargas V, Watt FM. Manipulation of stem cell proliferation and lineage commitment: visualisation of label-retaining cells in wholemounts of mouse epidermis. Development. 2003;130(21): 5241–55.
218. Ghazizadeh S, Taichman LB. Organization of stem cells and their progeny in human epidermis. J Invest Dermatol. 2005;124(2):367–72.
219. Lavker RM, Sun TT. Epidermal stem cells: properties, markers, and location. Proc Natl Acad Sci U S A. 2000;97(25):13473–5.
220. Kaur P. Interfollicular epidermal stem cells: identification, challenges, potential. J Invest Dermatol. 2006;126(7):1450–8.
221. Jensen UB, Lowell S, Watt FM. The spatial relationship between stem cells and their progeny in the basal layer of human epidermis: a new view based on whole-mount labelling and lineage analysis. Development. 1999;126(11):2409–18.
222. Jones PH, Harper S, Watt FM. Stem cell patterning and fate in human epidermis. Cell. 1995;80(1):83–93.
223. Lavker RM, Sun TT. Epidermal stem cells. J Invest Dermatol. 1983;81(1 Suppl):121s–7s.
224. Webb A, Li A, Kaur P. Location and phenotype of human adult keratinocyte stem cells of the skin. Differentiation. 2004;72(8):387–95.
225. Wan H, Stone MG, Simpson C, et al. Desmosomal proteins, including desmoglein 3, serve as novel negative markers for epidermal stem cell-containing population of keratinocytes. J Cell Sci. 2003;116(Pt 20):4239–48.
226. Legg J, Jensen UB, Broad S, Leigh I, Watt FM. Role of melanoma chondroitin sulphate proteoglycan in patterning stem cells in human interfollicular epidermis. Development. 2003;130(24): 6049–63.
227. Ito M, Yang Z, Andl T, et al. Wnt-dependent de novo hair follicle regeneration in adult mouse skin after wounding. Nature. 2007;447(7142):316–20.
228. Ghazizadeh S, Taichman LB. Multiple classes of stem cells in cutaneous epithelium: a lineage analysis of adult mouse skin. EMBO J. 2001;20(6):1215–22.
229. Cotsarelis G. Epithelial stem cells: a folliculocentric view. J Invest Dermatol. 2006;126(7):1459–68.
230. Moll I. Differential epithelial outgrowth of plucked and microdissected human hair follicles in explant culture. Arch Dermatol Res. 1996;288(10):604–10.

231. Rochat A, Kobayashi K, Barrandon Y. Location of stem cells of human hair follicles by clonal analysis. Cell. 1994;76(6):1063–73.
232. Cotsarelis G, Sun TT, Lavker RM. Label-retaining cells reside in the bulge area of pilosebaceous unit: implications for follicular stem cells, hair cycle, and skin carcinogenesis. Cell. 1990;61(7):1329–37.
233. Taylor G, Lehrer MS, Jensen PJ, Sun TT, Lavker RM. Involvement of follicular stem cells in forming not only the follicle but also the epidermis. Cell. 2000;102(4):451–61.
234. Oshima H, Rochat A, Kedzia C, Kobayashi K, Barrandon Y. Morphogenesis and renewal of hair follicles from adult multipotent stem cells. Cell. 2001;104(2):233–45.
235. Weinberg WC, Goodman LV, George C, et al. Reconstitution of hair follicle development in vivo: determination of follicle formation, hair growth, and hair quality by dermal cells. J Invest Dermatol. 1993;100(3):229–36.
236. Levy V, Lindon C, Harfe BD, Morgan BA. Distinct stem cell populations regenerate the follicle and interfollicular epidermis. Dev Cell. 2005;9(6):855–61.
237. Ito M, Liu Y, Yang Z, et al. Stem cells in the hair follicle bulge contribute to wound repair but not to homeostasis of the epidermis. Nat Med. 2005;11(12):1351–4.
238. Tumbar T, Guasch G, Greco V, et al. Defining the epithelial stem cell niche in skin. Science. 2004;303(5656):359–63.
239. Li L, Mignone J, Yang M, et al. Nestin expression in hair follicle sheath progenitor cells. Proc Natl Acad Sci U S A. 2003;100(17):9958–61.
240. Amoh Y, Li L, Katsuoka K, Penman S, Hoffman RM. Multipotent nestin-positive, keratin-negative hair-follicle bulge stem cells can form neurons. Proc Natl Acad Sci U S A. 2005;102(15):5530–4.
241. Amoh Y, Li L, Yang M, et al. Nascent blood vessels in the skin arise from nestin-expressing hair-follicle cells. Proc Natl Acad Sci U S A. 2004;101(36):13291–5.
242. Amoh Y, Li L, Campillo R, et al. Implanted hair follicle stem cells form Schwann cells that support repair of severed peripheral nerves. Proc Natl Acad Sci U S A. 2005;102(49):17734–8.
243. Hoffman RM. The pluripotency of hair follicle stem cells. Cell Cycle. 2006;5(3):232–3.
244. Tiede S, Kloepper JE, Bodo E, Tiwari S, Kruse C, Paus R. Hair follicle stem cells: walking the maze. Eur J Cell Biol. 2007;86(7):355–76.
245. Morris RJ, Potten CS. Highly persistent label-retaining cells in the hair follicles of mice and their fate following induction of anagen. J Invest Dermatol. 1999;112(4):470–5.
246. Lyle S, Christofidou-Solomidou M, Liu Y, Elder DE, Albelda S, Cotsarelis G. The C8/144B monoclonal antibody recognizes cytokeratin 15 and defines the location of human hair follicle stem cells. J Cell Sci. 1998;111 (Pt 21):3179–88.
247. Kaur P, Li A, Redvers R, Bertoncello I. Keratinocyte stem cell assays: an evolving science. J Investig Dermatol Symp Proc. 2004;9(3):238–47.
248. Kobayashi K, Rochat A, Barrandon Y. Segregation of keratinocyte colony-forming cells in the bulge of the rat vibrissa. Proc Natl Acad Sci U S A. 1993;90(15):7391–5.
249. Morris RJ, Liu Y, Marles L, et al. Capturing and profiling adult hair follicle stem cells. Nat Biotechnol. 2004;22(4):411–7.
250. Ohyama M, Terunuma A, Tock CL, et al. Characterization and isolation of stem cell-enriched human hair follicle bulge cells. J Clin Invest. 2006;116(1):249–60.
251. Barrandon Y, Green H. Three clonal types of keratinocyte with different capacities for multiplication. Proc Natl Acad Sci U S A. 1987;84(8):2302–6.
252. Finch PW, Rubin JS, Miki T, Ron D, Aaronson SA. Human KGF is FGF-related with properties of a paracrine effector of epithelial cell growth. Science. 1989;245(4919):752–5.
253. Kulessa H, Turk G, Hogan BL. Inhibition of Bmp signaling affects growth and differentiation in the anagen hair follicle. EMBO J. 2000;19(24):6664–74.
254. Botchkarev VA, Botchkareva NV, Roth W, et al. Noggin is a mesenchymally derived stimulator of hair-follicle induction. Nat Cell Biol. 1999;1(3):158–64.
255. Pummila M, Fliniaux I, Jaatinen R, et al. Ectodysplasin has a dual role in ectodermal organogenesis: inhibition of Bmp activity and induction of Shh expression. Development, 2007;134(1):117–25.
256. Atit R, Sgaier SK, Mohamed OA, et al. Beta-catenin activation is necessary and sufficient to specify the dorsal dermal fate in the mouse. Dev Biol. 2006;296(1):164–76.
257. Huelsken J, Vogel R, Erdmann B, Cotsarelis G, Birchmeier W. beta-Catenin controls hair follicle morphogenesis and stem cell differentiation in the skin. Cell. 2001;105(4):533–45.
258. Nguyen H, Rendl M, Fuchs E. Tcf3 governs stem cell features and represses cell fate determination in skin. Cell. 2006;127(1):171–83.
259. Merrill BJ, Gat U, DasGupta R, Fuchs E. Tcf3 and Lef1 regulate lineage differentiation of multipotent stem cells in skin. Genes Dev. 2001;15(13):1688–705.
260. DasGupta R, Fuchs E. Multiple roles for activated LEF/TCF transcription complexes during hair follicle development and differentiation. Development. 1999;126(20):4557–68.
261. Xia J, Urabe K, Moroi Y, et al. beta-Catenin mutation and its nuclear localization are confirmed to be frequent causes of Wnt signaling pathway activation in pilomatricomas. J Dermatol Sci. 2006;41(1):67–75.
262. Lo Celso C, Prowse DM, Watt FM. Transient activation of beta-catenin signalling in adult mouse epidermis is sufficient to induce new hair follicles but continuous activation is required to maintain hair follicle tumours. Development. 2004;131(8):1787–99.
263. Kratochwil K, Galceran J, Tontsch S, Roth W, Grosschedl R. FGF4, a direct target of LEF1 and Wnt signaling, can rescue the arrest of tooth organogenesis in Lef1(-/-) mice. Genes Dev. 2002;16(24):3173–85.
264. Alonso L, Fuchs E. Stem cells in the skin: waste not, Wnt not. Genes Dev. 2003;17(10):1189–200.
265. Botchkarev VA, Sharov AA. BMP signaling in the control of skin development and hair follicle growth. Differentiation. 2004;72(9–10):512–26.
266. Nikaido M, Tada M, Takeda H, Kuroiwa A, Ueno N. In vivo analysis using variants of zebrafish BMPR-IA: range of action and involvement of BMP in ectoderm patterning. Development. 1999;126(1):181–90.
267. Botchkarev VA, Botchkareva NV, Nakamura M, et al. Noggin is required for induction of the hair follicle growth phase in postnatal skin. FASEB J. 2001;15(12):2205–14.
268. Kobielak K, Stokes N, de la Cruz J, Polak L, Fuchs E. Loss of a quiescent niche but not follicle stem cells in the absence of bone morphogenetic protein signaling. Proc Natl Acad Sci U S A. 2007;104(24):10063–8.
269. Andl T, Ahn K, Kairo A, et al. Epithelial Bmpr1a regulates differentiation and proliferation in postnatal hair follicles and is essential for tooth development. Development. 2004;131(10):2257–68.
270. Chiang C, Swan RZ, Grachtchouk M, et al. Essential role for Sonic hedgehog during hair follicle morphogenesis. Dev Biol. 1999;205(1):1–9.
271. Oro AE, Higgins K. Hair cycle regulation of Hedgehog signal reception. Dev Biol. 2003;255(2):238–48.
272. St-Jacques B, Dassule HR, Karavanova I, et al. Sonic hedgehog signaling is essential for hair development. Curr Biol. 1998;8(19):1058–68.
273. Mill P, Mo R, Fu H, et al. Sonic hedgehog-dependent activation of Gli2 is essential for embryonic hair follicle development. Genes Dev. 2003;17(2):282–94.

274. Adolphe C, Narang M, Ellis T, Wicking C, Kaur P, Wainwright B. An in vivo comparative study of sonic, desert and Indian hedgehog reveals that hedgehog pathway activity regulates epidermal stem cell homeostasis. Development. 2004;131(20):5009–19.
275. Niemann C, Unden AB, Lyle S, Zouboulis Ch C, Toftgard R, Watt FM. Indian hedgehog and beta-catenin signaling: role in the sebaceous lineage of normal and neoplastic mammalian epidermis. Proc Natl Acad Sci U S A. 2003;100 Suppl 1:11873–80.
276. Blanpain C, Lowry WE, Pasolli HA, Fuchs E. Canonical notch signaling functions as a commitment switch in the epidermal lineage. Genes Dev. 2006;20(21):3022–35.
277. Rangarajan A, Talora C, Okuyama R, et al. Notch signaling is a direct determinant of keratinocyte growth arrest and entry into differentiation. EMBO J. 2001;20(13):3427–36.
278. Okuyama R, Nguyen BC, Talora C, et al. High commitment of embryonic keratinocytes to terminal differentiation through a Notch1-caspase 3 regulatory mechanism. Dev Cell. 2004;6(4):551–62.
279. Pan Y, Lin MH, Tian X, et al. gamma-secretase functions through Notch signaling to maintain skin appendages but is not required for their patterning or initial morphogenesis. Dev Cell. 2004;7(5):731–43.
280. Vauclair S, Nicolas M, Barrandon Y, Radtke F. Notch1 is essential for postnatal hair follicle development and homeostasis. Dev Biol. 2005;284(1):184–93.
281. Yamamoto N, Tanigaki K, Han H, Hiai H, Honjo T. Notch/RBP-J signaling regulates epidermis/hair fate determination of hair follicular stem cells. Curr Biol. 2003;13(4):333–8.
282. Kopan R, Weintraub H. Mouse notch: expression in hair follicles correlates with cell fate determination. J Cell Biol. 1993;121(3):631–41.
283. Jensen KB, Watt FM. Single-cell expression profiling of human epidermal stem and transit-amplifying cells: Lrig1 is a regulator of stem cell quiescence. Proc Natl Acad Sci U S A. 2006;103(32):11958–63.
284. Rhee H, Polak L, Fuchs E. Lhx2 maintains stem cell character in hair follicles. Science. 2006;312(5782):1946–9.
285. Benitah SA, Frye M, Glogauer M, Watt FM. Stem cell depletion through epidermal deletion of Rac1. Science. 2005;309(5736):933–5.
286. Benitah SA, Watt FM. Epidermal deletion of Rac1 causes stem cell depletion, irrespective of whether deletion occurs during embryogenesis or adulthood. J Invest Dermatol. 2007;127(6): 1555–7.
287. Chrostek A, Wu X, Quondamatteo F, et al. Rac1 is crucial for hair follicle integrity but is not essential for maintenance of the epidermis. Mol Cell Biol. 2006;26(18):6957–70.
288. Castilho RM, Squarize CH, Patel V, et al. Requirement of Rac1 distinguishes follicular from interfollicular epithelial stem cells. Oncogene. 2007;26(35):5078–85.
289. Green H. Terminal differentiation of cultured human epidermal cells. Cell. 1977;11(2):405–16.
290. Rheinwald JG, Green H. Serial cultivation of strains of human epidermal keratinocytes: the formation of keratinizing colonies from single cells. Cell. 1975;6(3):331–43.
291. O'Connor NE, Mulliken JB, Banks-Schlegel S, Kehinde O, Green H. Grafting of burns with cultured epithelium prepared from autologous epidermal cells. Lancet. 1981;1(8211):75–8.
292. Pellegrini G, Ranno R, Stracuzzi G, et al. The control of epidermal stem cells (holoclones) in the treatment of massive full-thickness burns with autologous keratinocytes cultured on fibrin. Transplantation. 1999;68(6):868–79.
293. Ronfard V, Rives JM, Neveux Y, Carsin H, Barrandon Y. Long-term regeneration of human epidermis on third degree burns transplanted with autologous cultured epithelium grown on a fibrin matrix. Transplantation. 2000;70(11):1588–98.
294. Burd A, Ahmed K, Lam S, Ayyappan T, Huang L. Stem cell strategies in burns care. Burns. 2007;33(3):282–91.
295. Boyd RL, Tucek CL, Godfrey DI, et al. The thymic microenvironment. Immunol Today. 1993;14(9):445–59.
296. Gordon J, Wilson VA, Blair NF, et al. Functional evidence for a single endodermal origin for the thymic epithelium. Nat Immunol. 2004;5(5):546–53.
297. Itoi M, Tsukamoto N, Yoshida H, Amagai T. Mesenchymal cells are required for functional development of thymic epithelial cells. Int Immunol. 2007;19(8):953–64.
298. Klug DB, Carter C, Gimenez-Conti IB, Richie ER. Cutting edge: thymocyte-independent and thymocyte-dependent phases of epithelial patterning in the fetal thymus. J Immunol. 2002;169(6):2842–5.
299. Anderson G, Jenkinson EJ. Lymphostromal interactions in thymic development and function. Nat Rev Immunol. 2001;1(1):31–40.
300. van Ewijk W, Hollander G, Terhorst C, Wang B. Stepwise development of thymic microenvironments in vivo is regulated by thymocyte subsets. Development. 2000;127(8):1583–91.
301. Muller SM, Terszowski G, Blum C, et al. Gene targeting of VEGF-A in thymus epithelium disrupts thymus blood vessel architecture. Proc Natl Acad Sci U S A. 2005;102(30):10587–92.
302. Hollander G, Gill J, Zuklys S, Iwanami N, Liu C, Takahama Y. Cellular and molecular events during early thymus development. Immunol Rev. 2006;209:28–46.
303. Nowell CS, Farley AM, Blackburn CC. Thymus organogenesis and development of the thymic stroma. Methods Mol Biol. 2007;380:125–62.
304. Graham A. The development and evolution of the pharyngeal arches. J Anat. 2001;199(Pt 1–2):133–41.
305. Graham A, Smith A. Patterning the pharyngeal arches. Bioessays. 2001;23(1):54–61.
306. Schilling TF. Genetic analysis of craniofacial development in the vertebrate embryo. Bioessays. 1997;19(6):459–68.
307. Manley NR, Capecchi MR. The role of Hoxa-3 in mouse thymus and thyroid development. Development. 1995;121(7):1989–2003.
308. Manley NR. Thymus organogenesis and molecular mechanisms of thymic epithelial cell differentiation. Semin Immunol. 2000;12(5):421–8.
309. Gordon J, Bennett AR, Blackburn CC, Manley NR. Gcm2 and Foxn1 mark early parathyroid- and thymus-specific domains in the developing third pharyngeal pouch. Mech Dev. 2001;103(1–2):141–3.
310. Le Lievre CS, Le Douarin NM. Mesenchymal derivatives of the neural crest: analysis of chimaeric quail and chick embryos. J Embryol Exp Morphol. 1975;34(1):125–54.
311. Manley NR, Capecchi MR. Hox group 3 paralogs regulate the development and migration of the thymus, thyroid, and parathyroid glands. Dev Biol. 1998;195(1):1–15.
312. Blackburn CC, Manley NR. Developing a new paradigm for thymus organogenesis. Nat Rev Immunol. 2004;4(4):278–89.
313. Haynes BF, Heinly CS. Early human T cell development: analysis of the human thymus at the time of initial entry of hematopoietic stem cells into the fetal thymic microenvironment. J Exp Med. 1995;181(4):1445–58.
314. Bleul CC, Boehm T. Chemokines define distinct microenvironments in the developing thymus. Eur J Immunol. 2000;30(12):3371–9.
315. Liu C, Ueno T, Kuse S, et al. The role of CCL21 in recruitment of T-precursor cells to fetal thymi. Blood. 2005;105(1):31–9.
316. Gill J, Malin M, Sutherland J, Gray D, Hollander G, Boyd R. Thymic generation and regeneration. Immunol Rev. 2003;195: 28–50.
317. Amagai T, Itoi M, Kondo Y. Limited development capacity of the earliest embryonic murine thymus. Eur J Immunol. 1995;25(3):757–62.

318. van Ewijk W, Wang B, Hollander G, et al. Thymic microenvironments, 3-D versus 2-D? Semin Immunol. 1999;11(1):57–64.
319. Guidos C. Thymus and T-lymphocyte development: what is new in the 21st century? Immunol Rev. 2006;209:5–9.
320. van Dyke JH. On the origin of accessory thymus tissue, thymus IV: the occurrence in man. Anat Rec. 1941;79:179–209.
321. Tovi F, Mares AJ. The aberrant cervical thymus. Embryology, Pathology, and clinical implications. Am J Surg. 1978;136(5):631–7.
322. Ashour M. Prevalence of ectopic thymic tissue in myasthenia gravis and its clinical significance. J Thorac Cardiovasc Surg. 1995;109(4):632–5.
323. Dooley J, Erickson M, Gillard GO, Farr AG. Cervical thymus in the mouse. J Immunol. 2006;176(11):6484–90.
324. Terszowski G, Muller SM, Bleul CC, et al. Evidence for a functional second thymus in mice. Science. 2006;312(5771):284–7.
325. von Boehmer H. Immunology. Thoracic thymus, exclusive no longer. Science. 2006;312(5771):206–7.
326. Panigraphi D, Waxler GL, Mallman VH. The thymus in the chicken and its anatomical relationship to the thyroid. J Immunol. 1971;107(1):289–92.
327. Manley NR, Blackburn CC. A developmental look at thymus organogenesis: where do the non-hematopoietic cells in the thymus come from? Curr Opin Immunol. 2003;15(2):225–32.
328. Blackburn CC, Manley NR, Palmer DB, Boyd RL, Anderson G, Ritter MA. One for all and all for one: thymic epithelial stem cells and regeneration. Trends Immunol. 2002;23(8):391–5.
329. Cordier AC, Heremans JF. Nude mouse embryo: ectodermal nature of the primordial thymic defect. Scand J Immunol. 1975;4(2):193–6.
330. Cordier AC, Haumont SM. Development of thymus, parathyroids, and ultimo-branchial bodies in NMRI and nude mice. Am J Anat. 1980;157(3):227–63.
331. Anderson G, Jenkinson WE, Jones T, et al. Establishment and functioning of intrathymic microenvironments. Immunol Rev. 2006;209:10–27.
332. Le Douarin NM, Jotereau FV. Tracing of cells of the avian thymus through embryonic life in interspecific chimeras. J Exp Med. 1975;142(1):17–40.
333. Suniara RK, Jenkinson EJ, Owen JJ. An essential role for thymic mesenchyme in early T cell development. J Exp Med. 2000;191(6):1051–6.
334. Jenkinson WE, Rossi SW, Jenkinson EJ, Anderson G. Development of functional thymic epithelial cells occurs independently of lymphostromal interactions. Mech Dev. 2005;122(12):1294–9.
335. Zhang L, Sun L, Zhao Y. Thymic epithelial progenitor cells and thymus regeneration: an update. Cell Res. 2007;17(1):50–5.
336. Jiang X, Rowitch DH, Soriano P, McMahon AP, Sucov HM. Fate of the mammalian cardiac neural crest. Development. 2000;127(8):1607–16.
337. Waldo KL, Kumiski D, Kirby ML. Cardiac neural crest is essential for the persistence rather than the formation of an arch artery. Dev Dyn. 1996;205(3):281–92.
338. Bockman DE, Kirby ML. Dependence of thymus development on derivatives of the neural crest. Science. 1984;223(4635):498–500.
339. Kirby ML, Waldo KL. Role of neural crest in congenital heart disease. Circulation. 1990;82(2):332–40.
340. Kuratani S, Bockman DE. Impaired development of the thymic primordium after neural crest ablation. Anat Rec. 1990;228(2):185–90.
341. Shinohara T, Honjo T. Studies in vitro on the mechanism of the epithelial/mesenchymal interaction in the early fetal thymus. Eur J Immunol. 1997;27(2):522–9.
342. Shinohara T, Honjo T. Epidermal growth factor can replace thymic mesenchyme in induction of embryonic thymus morphogenesis in vitro. Eur J Immunol. 1996;26(4):747–52.
343. Jenkinson WE, Rossi SW, Parnell SM, Jenkinson EJ, Anderson G. PDGFRalpha-expressing mesenchyme regulates thymus growth and the availability of intrathymic niches. Blood. 2007;109(3):954–60.
344. Revest JM, Spencer-Dene B, Kerr K, De Moerlooze L, Rosewell I, Dickson C. Fibroblast growth factor receptor 2-IIIb acts upstream of Shh and Fgf4 and is required for limb bud maintenance but not for the induction of Fgf8, Fgf10, Msx1, or Bmp4. Dev Biol. 2001;231(1):47–62.
345. Dooley J, Erickson M, Larochelle WJ, Gillard GO, Farr AG. FGFR2IIIb signaling regulates thymic epithelial differentiation. Dev Dyn. 2007;236(12):3459–71.
346. Revest JM, Suniara RK, Kerr K, Owen JJ, Dickson C. Development of the thymus requires signaling through the fibroblast growth factor receptor R2-IIIb. J Immunol. 2001;167(4):1954–61.
347. Gray DH, Tull D, Ueno T, et al. A Unique Thymic Fibroblast Population Revealed by the Monoclonal Antibody MTS-15. J Immunol. 2007;178(8):4956–65.
348. van Ewijk W, Shores EW, Singer A. Crosstalk in the mouse thymus. Immunol Today. 1994;15(5):214–7.
349. Ritter MA, Boyd RL. Development in the thymus: it takes two to tango. Immunol Today. 1993;14(9):462–9.
350. Hollander GA, Wang B, Nichogiannopoulou A, et al. Developmental control point in induction of thymic cortex regulated by a subpopulation of prothymocytes. Nature. 1995;373(6512):350–3.
351. Palmer DB, Viney JL, Ritter MA, Hayday AC, Owen MJ. Expression of the alpha beta T-cell receptor is necessary for the generation of the thymic medulla. Dev Immunol. 993;3(3):175–9.
352. Shores EW, Van Ewijk W, Singer A. Disorganization and restoration of thymic medullary epithelial cells in T cell receptor-negative scid mice: evidence that receptor-bearing lymphocytes influence maturation of the thymic microenvironment. Eur J Immunol. 1991;21(7):1657–61.
353. Shores EW, Van Ewijk W, Singer A. Maturation of medullary thymic epithelium requires thymocytes expressing fully assembled CD3-TCR complexes. Int Immunol. 1994;6(9):1393–402.
354. Surh CD, Ernst B, Sprent J. Growth of epithelial cells in the thymic medulla is under the control of mature T cells. J Exp Med. 1992;176(2):611–6.
355. Rodewald HR, Ogawa M, Haller C, Waskow C, DiSanto JP. Prothymocyte expansion by c-kit and the common cytokine receptor gamma chain is essential for repertoire formation. Immunity. 1997;6(3):265–72.
356. Yarilin AA, Belyakov IM. Cytokines in the thymus: production and biological effects. Curr Med Chem. 2004;11(4):447–64.
357. Boehm T, Scheu S, Pfeffer K, Bleul CC. Thymic medullary epithelial cell differentiation, thymocyte emigration, and the control of autoimmunity require lympho-epithelial cross talk via LTbetaR. J Exp Med. 2003;198(5):757–69.
358. Venanzi ES, Gray DH, Benoist C, Mathis D. Lymphotoxin pathway and Aire influences on thymic medullary epithelial cells are unconnected. J Immunol. 2007;179(9):5693–700.
359. Rossi SW, Kim MY, Leibbrandt A, et al. RANK signals from CD4(+)3(-) inducer cells regulate development of Aire-expressing epithelial cells in the thymic medulla. J Exp Med. 2007;204(6):1267–72.
360. Iimura T, Pourquie O. Hox genes in time and space during vertebrate body formation. Dev Growth Differ. 2007;49(4):265–75.
361. Svingen T, Tonissen KF. Hox transcription factors and their elusive mammalian gene targets. Heredity. 2006;97(2):88–96.
362. Chisaka O, Kameda Y. Hoxa3 regulates the proliferation and differentiation of the third pharyngeal arch mesenchyme in mice. Cell Tissue Res. 2005;320(1):77–89.
363. Chisaka O, Capecchi MR. Regionally restricted developmental defects resulting from targeted disruption of the mouse homeobox gene hox-1.5. Nature. 1991;350(6318):473–9.

364. Rossel M, Capecchi MR. Mice mutant for both Hoxa1 and Hoxb1 show extensive remodeling of the hindbrain and defects in craniofacial development. Development. 1999;126(22):5027–40.
365. Mann RS, Affolter M. Hox proteins meet more partners. Curr Opin Genet Dev. 1998;8(4):423–9.
366. Manley NR, Selleri L, Brendolan A, Gordon J, Cleary ML. Abnormalities of caudal pharyngeal pouch development in Pbx1 knockout mice mimic loss of Hox3 paralogs. Dev Biol. 2004;276(2):301–12.
367. Lang D, Powell SK, Plummer RS, Young KP, Ruggeri BA. PAX genes: roles in development, pathophysiology, and cancer. Biochem Pharmacol. 2006;73(1):1–14.
368. Neubuser A, Koseki H, Balling R. Characterization and developmental expression of Pax9, a paired-box-containing gene related to Pax1. Dev Biol. 1995;170(2):701–16.
369. Wallin J, Eibel H, Neubuser A, Wilting J, Koseki H, Balling R. Pax1 is expressed during development of the thymus epithelium and is required for normal T-cell maturation. Development. 1996;122(1):23–30.
370. Hetzer-Egger C, Schorpp M, Haas-Assenbaum A, Balling R, Peters H, Boehm T. Thymopoiesis requires Pax9 function in thymic epithelial cells. Eur J Immunol. 2002;32(4):1175–81.
371. Peters H, Neubuser A, Kratochwil K, Balling R. Pax9-deficient mice lack pharyngeal pouch derivatives and teeth and exhibit craniofacial and limb abnormalities. Genes Dev. 1998;12(17):2735–47.
372. Su D, Ellis S, Napier A, Lee K, Manley NR. Hoxa3 and pax1 regulate epithelial cell death and proliferation during thymus and parathyroid organogenesis. Dev Biol. 2001;236(2):316–29.
373. Su DM, Manley NR. Hoxa3 and pax1 transcription factors regulate the ability of fetal thymic epithelial cells to promote thymocyte development. J Immunol. 2000;164(11):5753–60.
374. Machado AF, Martin LJ, Collins MD. Pax3 and the splotch mutations: structure, function, and relationship to teratogenesis, including gene-chemical interactions. Curr Pharm Des. 2001;7(9):751–85.
375. Conway SJ, Bundy J, Chen J, Dickman E, Rogers R, Will BM. Decreased neural crest stem cell expansion is responsible for the conotruncal heart defects within the splotch (Sp(2H))/Pax3 mouse mutant. Cardiovasc Res. 2000;47(2):314–28.
376. Epstein JA, Li J, Lang D, et al. Migration of cardiac neural crest cells in Splotch embryos. Development. 2000;127(9):1869–78.
377. Jemc J, Rebay I. The eyes absent family of phosphotyrosine phosphatases: properties and roles in developmental regulation of transcription. Annu Rev Biochem. 2007;76:513–38.
378. Kalatzis V, Sahly I, El-Amraoui A, Petit C. Eya1 expression in the developing ear and kidney: towards the understanding of the pathogenesis of Branchio-Oto-Renal (BOR) syndrome. Dev Dyn. 1998;213(4):486–99.
379. Xu PX, Zheng W, Laclef C, et al. Eya1 is required for the morphogenesis of mammalian thymus, parathyroid and thyroid. Development. 2002;129(13):3033–44.
380. Li X, Oghi KA, Zhang J, et al. Eya protein phosphatase activity regulates Six1-Dach-Eya transcriptional effects in mammalian organogenesis. Nature. 2003;426(6964):247–54.
381. Laclef C, Souil E, Demignon J, Maire P. Thymus, kidney and craniofacial abnormalities in Six 1 deficient mice. Mech Dev. 2003;120(6):669–79.
382. Zou D, Silvius D, Davenport J, Grifone R, Maire P, Xu PX. Patterning of the third pharyngeal pouch into thymus/parathyroid by Six and Eya1. Dev Biol. 2006;293(2):499–512.
383. Ozaki H, Watanabe Y, Takahashi K, et al. Six4, a putative myogenin gene regulator, is not essential for mouse embryonal development. Mol Cell Biol. 2001;21(10):3343–50.
384. Lindsay EA. Chromosomal microdeletions: dissecting del22q11 syndrome. Nat Rev Genet. 2001;2(11):858–68.
385. Jerome LA, Papaioannou VE. DiGeorge syndrome phenotype in mice mutant for the T-box gene, Tbx1. Nat Genet. 2001;27(3):286–91.
386. Bachiller D, Klingensmith J, Shneyder N, et al. The role of chordin/Bmp signals in mammalian pharyngeal development and DiGeorge syndrome. Development. 2003;130(15):3567–78.
387. Lindsay EA, Vitelli F, Su H, et al. Tbx1 haploinsufficieny in the DiGeorge syndrome region causes aortic arch defects in mice. Nature. 2001;410(6824):97–101.
388. Chapman DL, Garvey N, Hancock S, et al. Expression of the T-box family genes, Tbx1-Tbx5, during early mouse development. Dev Dyn. 1996;206(4):379–90.
389. Xu H, Cerrato F, Baldini A. Timed mutation and cell-fate mapping reveal reiterated roles of Tbx1 during embryogenesis, and a crucial function during segmentation of the pharyngeal system via regulation of endoderm expansion. Development. 2005;132(19):4387–95.
390. Vitelli F, Taddei I, Morishima M, Meyers EN, Lindsay EA, Baldini A. A genetic link between Tbx1 and fibroblast growth factor signaling. Development. 2002;129(19):4605–11.
391. Xu H, Morishima M, Wylie JN, et al. Tbx1 has a dual role in the morphogenesis of the cardiac outflow tract. Development. 2004;131(13):3217–27.
392. Ivins S, Lammerts van Beuren K, Roberts C, et al. Microarray analysis detects differentially expressed genes in the pharyngeal region of mice lacking Tbx1. Dev Biol. 2005;285(2):554–69.
393. Clark KL, Halay ED, Lai E, Burley SK. Co-crystal structure of the HNF-3/fork head DNA-recognition motif resembles histone H5. Nature. 1993;364(6436):412–20.
394. Cunliffe VT, Furley AJ, Keenan D. Complete rescue of the nude mutant phenotype by a wild-type Foxn1 transgene. Mamm Genome. 2002;13(5):245–52.
395. Nehls M, Pfeifer D, Schorpp M, Hedrich H, Boehm T. New member of the winged-helix protein family disrupted in mouse and rat nude mutations. Nature. 1994;372(6501):103–7.
396. Reth M. Immunodeficiency. Trapping the nude mouse gene. Curr Biol. 1995;5(1):18–20.
397. Pignata C, Gaetaniello L, Masci AM, et al. Human equivalent of the mouse Nude/SCID phenotype: long-term evaluation of immunologic reconstitution after bone marrow transplantation. Blood. 2001;97(4):880–5.
398. Adriani M, Martinez-Mir A, Fusco F, et al. Ancestral founder mutation of the nude (FOXN1) gene in congenital severe combined immunodeficiency associated with alopecia in southern Italy population. Ann Hum Genet. 2004;68(Pt 3):265–8.
399. Auricchio L, Adriani M, Frank J, Busiello R, Christiano A, Pignata C. Nail dystrophy associated with a heterozygous mutation of the nude/SCID human FOXN1 (WHN) gene. Arch Dermatol. 2005;141(5):647–8.
400. Lee D, Prowse DM, Brissette JL. Association between mouse nude gene expression and the initiation of epithelial terminal differentiation. Dev Biol. 1999;208(2):362–74.
401. Ye H, Kelly TF, Samadani U, et al. Hepatocyte nuclear factor 3/fork head homolog 11 is expressed in proliferating epithelial and mesenchymal cells of embryonic and adult tissues. Mol Cell Biol. 1997;17(3):1626–41.
402. Bleul CC, Corbeaux T, Reuter A, Fisch P, Monting JS, Boehm T. Formation of a functional thymus initiated by a postnatal epithelial progenitor cell. Nature. 2006;441(7096):992–6.
403. Itoi M, Tsukamoto N, Amagai T. Expression of Dll4 and CCL25 in Foxn1-negative epithelial cells in the post-natal thymus. Int Immunol. 2007;19(2):127–32.
404. Boehm T, Bleul CC. Thymus-homing precursors and the thymic microenvironment. Trends Immunol. 2006;27(10):477–84.

405. Su DM, Navarre S, Oh WJ, Condie BG, Manley NR. A domain of Foxn1 required for crosstalk-dependent thymic epithelial cell differentiation. Nat Immunol. 2003;4(11):1128–35.
406. Tsukamoto N, Itoi M, Nishikawa M, Amagai T. Lack of Delta like 1 and 4 expressions in nude thymus anlages. Cell Immunol. 2005;234(2):77–80.
407. Bleul CC, Boehm T. Laser capture microdissection-based expression profiling identifies PD1-ligand as a target of the nude locus gene product. Eur J Immunol. 2001;31(8):2497–503.
408. Liu C, Saito F, Liu Z, et al. Coordination between CCR7- and CCR9-mediated chemokine signals in prevascular fetal thymus colonization. Blood. 2006;108(8):2531–9.
409. Tsai PT, Lee RA, Wu H. BMP4 acts upstream of FGF in modulating thymic stroma and regulating thymopoiesis. Blood. 2003;102(12):3947–53.
410. Gunther T, Chen ZF, Kim J, et al. Genetic ablation of parathyroid glands reveals another source of parathyroid hormone. Nature. 2000;406(6792):199–203.
411. Liu Z, Yu S, Manley NR. Gcm2 is required for the differentiation and survival of parathyroid precursor cells in the parathyroid/thymus primordia. Dev Biol. 2007;305(1):333–46.
412. Yang A, Schweitzer R, Sun D, et al. p63 is essential for regenerative proliferation in limb, craniofacial and epithelial development. Nature. 1999;398(6729):714–8.
413. Mills AA, Zheng B, Wang XJ, Vogel H, Roop DR, Bradley A. p63 is a p53 homologue required for limb and epidermal morphogenesis. Nature. 1999;398(6729):708–13.
414. Blanpain C, Fuchs E. p63: revving up epithelial stem-cell potential. Nat Cell Biol. 2007;9(7):731–3.
415. McKeon F. p63 and the epithelial stem cell: more than status quo? Genes Dev. 2004;18(5):465–9.
416. Candi E, Rufini A, Terrinoni A, et al. {Delta}Np63 regulates thymic development through enhanced expression of FgfR2 and Jag2. Proc Natl Acad Sci U S A. 2007;104(29):11999–2004.
417. Balciunaite G, Keller MP, Balciunaite E, et al. Wnt glycoproteins regulate the expression of FoxN1, the gene defective in nude mice. Nat Immunol. 2002;3(11):1102–8.
418. Osada M, Ito E, Fermin HA, et al. The Wnt signaling antagonist Kremen1 is required for development of thymic architecture. Clin Dev Immunol. 2006;13(2–4):299–319.
419. Klug DB, Carter C, Crouch E, Roop D, Conti CJ, Richie ER. Interdependence of cortical thymic epithelial cell differentiation and T-lineage commitment. Proc Natl Acad Sci U S A. 1998;95(20):11822–7.
420. Kuraguchi M, Wang XP, Bronson RT, et al. Adenomatous polyposis coli (APC) is required for normal development of skin and thymus. PLoS Genet. 2006;2(9):e146.
421. Bleul CC, Boehm T. BMP signaling is required for normal thymus development. J Immunol. 2005;175(8):5213–21.
422. Patel SR, Gordon J, Mahbub F, Blackburn CC, Manley NR. Bmp4 and Noggin expression during early thymus and parathyroid organogenesis. Gene Expr Patterns. 2006;6(8):794–9.
423. Ma L, Liu J, Wu T, et al. 'Cyclic alopecia' in Msx2 mutants: defects in hair cycling and hair shaft differentiation. Development. 2003;130(2):379–89.
424. Outram SV, Varas A, Pepicelli CV, Crompton T. Hedgehog signaling regulates differentiation from double-negative to double-positive thymocyte. Immunity. 2000;13(2):187–97.
425. Sacedon R, Varas A, Hernandez-Lopez C, et al. Expression of hedgehog proteins in the human thymus. J Histochem Cytochem. 2003;51(11):1557–66.
426. Moore-Scott BA, Manley NR. Differential expression of Sonic hedgehog along the anterior-posterior axis regulates patterning of pharyngeal pouch endoderm and pharyngeal endoderm-derived organs. Dev Biol. 2005;278(2):323–35.
427. Garg V, Yamagishi C, Hu T, Kathiriya IS, Yamagishi H, Srivastava D. Tbx1, a DiGeorge syndrome candidate gene, is regulated by sonic hedgehog during pharyngeal arch development. Dev Biol. 2001;235(1):62–73.
428. Mulder GB, Manley N, Maggio-Price L. Retinoic acid-induced thymic abnormalities in the mouse are associated with altered pharyngeal morphology, thymocyte maturation defects, and altered expression of Hoxa3 and Pax1. Teratology. 1998;58(6):263–75.
429. Wendling O, Dennefeld C, Chambon P, Mark M. Retinoid signaling is essential for patterning the endoderm of the third and fourth pharyngeal arches. Development. 2000;127(8):1553–62.
430. Daftary GS, Taylor HS. Endocrine regulation of HOX genes. Endocr Rev. 2006;27(4):331–55.
431. Dupe V, Ghyselinck NB, Wendling O, Chambon P, Mark M. Key roles of retinoic acid receptors alpha and beta in the patterning of the caudal hindbrain, pharyngeal arches and otocyst in the mouse. Development. 1999;126(22):5051–9.
432. Lammer EJ, Chen DT, Hoar RM, et al. Retinoic acid embryopathy. N Engl J Med. 1985;313(14):837–41.
433. Mulder GB, Manley N, Grant J, et al. Effects of excess vitamin A on development of cranial neural crest-derived structures: a neonatal and embryologic study. Teratology. 2000;62(4):214–26.
434. Niederreither K, Vermot J, Le Roux I, Schuhbaur B, Chambon P, Dolle P. The regional pattern of retinoic acid synthesis by RALDH2 is essential for the development of posterior pharyngeal arches and the enteric nervous system. Development. 2003;130(11):2525–34.
435. Pezet S, McMahon SB. Neurotrophins: mediators and modulators of pain. Ann Rev Neurosci. 2006;29:507–38.
436. Ehrhard PB, Erb P, Graumann U, Otten U. Expression of nerve growth factor and nerve growth factor receptor tyrosine kinase Trk in activated CD4-positive T-cell clones. Proc Natl Acad Sci U S A. 1993;90(23):10984–8.
437. Hannestad J, Garcia-Suarez O, Huerta JJ, Esteban I, Naves FJ, Vega JA. TrkA neutrophin receptor protein in the rat and human thymus. Anat Rec. 1997;249(3):373–9.
438. Aloe L, Micera A, Bracci-Laudiero L, Vigneti E, Turrini P. Presence of nerve growth factor in the thymus of prenatal, postnatal and pregnant rats. Thymus. 1997;24(4):221–31.
439. Labouyrie E, Parrens M, de Mascarel A, Bloch B, Merlio JP. Distribution of NGF receptors in normal and pathologic human lymphoid tissues. J Neuroimmunol. 1997;77(2):161–73.
440. Parrens M, Labouyrie E, Groppi A, et al. Expression of NGF receptors in normal and pathological human thymus. J Neuroimmunol. 1998;85(1):11–21.
441. Garcia-Suarez O, Germana A, Hannestad J, et al. TrkA is necessary for the normal development of the murine thymus. J Neuroimmunol. 2000;108(1–2):11–21.
442. Vacca A, Di Marcotullio L, Giannini G, et al. Thrombospondin-1 is a mediator of the neurotypic differentiation induced by EGF in thymic epithelial cells. Exp Cell Res. 1999;248(1):79–86.
443. Perez-Pinera P, Garcia-Suarez O, Prieto JG, et al. Thymocyte depletion affects neurotrophin receptor expression in thymic stromal cells. J Anat. 2006;208(2):231–8.
444. Tacconelli A, Farina AR, Cappabianca L, et al. TrkAIII expression in the thymus. J Neuroimmunol. 2007;183(1–2):151–61.
445. Taub DD, Longo DL. Insights into thymic aging and regeneration. Immunol Rev. 2005;205:72–93.
446. Meilin A, Shoham J, Schreiber L, Sharabi Y. The role of thymocytes in regulating thymic epithelial cell growth and function. Scand J Immunol. 1995;42(2):185–90.
447. Galy AH, Hadden EM, Touraine JL, Hadden JW. Effects of cytokines on human thymic epithelial cells in culture: IL1 induces thymic epithelial cell proliferation and change in morphology. Cell Immunol. 1989;124(1):13–27.

448. Galy AH, Dinarello CA, Kupper TS, Kameda A, Hadden JW. Effects of cytokines on human thymic epithelial cells in culture. II. Recombinant IL 1 stimulates thymic epithelial cells to produce IL6 and GM-CSF. Cell Immunol. 1990;129(1):161–75.
449. Galy AH, Spits H. IL-1, IL-4, and IFN-gamma differentially regulate cytokine production and cell surface molecule expression in cultured human thymic epithelial cells. J Immunol. 1991;147(11):3823–30.
450. Shu S, Naylor P, Touraine JL, Hadden JW. IL-1, ICAM-1, LFA-3, and hydrocortisone differentially regulate cytokine secretion by human fetal thymic epithelial cells. Thymus. 1996;24(2):89–99.
451. Lagrota-Candido JM, Villa-Verde DM, Vanderlei FH, Jr., Savino W. Extracellular matrix components of the mouse thymus microenvironment. V. Interferon-gamma modulates thymic epithelial cell/thymocyte interactions via extracellular matrix ligands and receptors. Cell Immunol. 1996;170(2):235–44.
452. Gruver AL, Hudson LL, Sempowski GD. Immunosenescence of ageing. J Pathol. 2007;211(2):144–56.
453. Zediak VP, Bhandoola A. Aging and T cell development: interplay between progenitors and their environment. Semin Immunol. 2005;17(5):337–46.
454. Sutherland JS, Goldberg GL, Hammett MV, et al. Activation of thymic regeneration in mice and humans following androgen blockade. J Immunol. 2005;175(4):2741–53.
455. Heng TS, Goldberg GL, Gray DH, Sutherland JS, Chidgey AP, Boyd RL. Effects of castration on thymocyte development in two different models of thymic involution. J Immunol. 2005;175(5):2982–93.
456. Goldberg GL, Alpdogan O, Muriglan SJ, et al. Enhanced immune reconstitution by sex steroid ablation following allogeneic hemopoietic stem cell transplantation. J Immunol. 2007;178(11):7473–84.
457. Goldberg GL, Sutherland JS, Hammet MV, et al. Sex steroid ablation enhances lymphoid recovery following autologous hematopoietic stem cell transplantation. Transplantation. 2005;80(11):1604–13.
458. Napolitano LA, Lo JC, Gotway MB, et al. Increased thymic mass and circulating naive CD4 T cells in HIV-1-infected adults treated with growth hormone. AIDS. 2002;16(8):1103–11.
459. Alpdogan O, Hubbard VM, Smith OM, et al. Keratinocyte growth factor (KGF) is required for postnatal thymic regeneration. Blood. 2006;107(6):2453–60.
460. Min D, Panoskaltsis-Mortari A, Kuro OM, Hollander GA, Blazar BR, Weinberg KI. Sustained thymopoiesis and improvement in functional immunity induced by exogenous KGF administration in murine models of aging. Blood. 2007;109(6):2529–37.
461. Rossi S, Blazar BR, Farrell CL, et al. Keratinocyte growth factor preserves normal thymopoiesis and thymic microenvironment during experimental graft-versus-host disease. Blood. 2002;100(2):682–91.
462. Bennett AR, Farley A, Blair NF, Gordon J, Sharp L, Blackburn CC. Identification and characterization of thymic epithelial progenitor cells. Immunity. 2002;16(6):803–14.
463. Chidgey AP, Boyd RL. Stemming the tide of thymic aging. Nat Immunol. 2006;7(10):1013–6.
464. Chidgey A, Dudakov J, Seach N, Boyd R. Impact of niche aging on thymic regeneration and immune reconstitution. Semin Immunol. 2007;19(5):331–40.
465. Dondi D, Festuccia C, Piccolella M, Bologna M, Motta M. GnRH agonists and antagonists decrease the metastatic progression of human prostate cancer cell lines by inhibiting the plasminogen activator system. Oncol Rep. 2006;15(2):393–400.
466. Pritchard K. Endocrinology and hormone therapy in breast cancer: endocrine therapy in premenopausal women. Breast Cancer Res. 2005;7(2):70–6.
467. Wang J, Zhou F, Dong M, Wu R, Qian Y. Prolonged gonadotropin-releasing hormone agonist therapy reduced expression of nitric oxide synthase in the endometrium of women with endometriosis and infertility. Fertil Steril. 2006;85(4):1037–44.
468. Vottero A, Pedori S, Verna M, et al. Final height in girls with central idiopathic precocious puberty treated with gonadotropin-releasing hormone analog and oxandrolone. J Clin Endocrinol Metab. 2006;91(4):1284–7.
469. Sutherland JS, Spyroglou LS, Muirhead JL, et al. Enhance immune system regeneration in humans following allogeneic or autologous hemopoietic stem cell transplantation by temporary sex steroid blockade. Clin Cancer Res. 2008;14(4):1138–49.

Hepatic Stem Cells and Liver Development

Nalu Navarro-Alvarez, Alejandro Soto-Gutierrez and Naoya Kobayashi

Abstract The liver consists of many cell types with specialized functions. Hepatocytes are among the main players in the organ, and therefore are the most vulnerable cells to damage. Since they are not everlasting cells, they need to be replenished throughout life. Although the capacity of hepatocytes to contribute to their own maintenance has long been recognized, recent studies have indicated the presence of both intrahepatic and extrahepatic stem/progenitor cell populations that serve to maintain the normal organ and to regenerate damaged parenchyma in response to a variety of insults. The intrahepatic compartment most likely derives primarily from the biliary tree, particularly the most proximal branches, that is, the canals of Hering and smallest ductules. The extrahepatic compartment is at least in part derived from diverse populations of cells from the bone marrow. Embryonic stem cells (ES) are considered as a part of the extrahepatic compartment. Due to their pluripotent capabilities, ES cell–derived cells form a potential future source of hepatocytes, to replace or restore hepatic tissues that have been damaged by disease or injury. Progressing knowledge about stem cells in the liver would allow a better understanding of the mechanisms of hepatic homeostasis and regeneration. Although a human stem cell–derived cell type equivalent to primary hepatocytes does not yet exist, the promising results obtained with extrahepatic stem cells would open the way to cell-based therapy for liver diseases.

Keywords Liver stem cells · Adult stem cells · Embryonic stem cells · Liver development · Intrahepatic stem cells · Extrahepatic stem cells · Liver renewal · Stem cell niche · ES-derived hepatocytes

N. Kobayashi (✉)
Department of Surgery, Okayama University Graduate School of Medicine and Dentistry, 2-5-1 Shikata-cho, Okayama, 700-8558, Japan
e-mail: immortal@md.okayama-u.ac.jp

1 Introduction

For many years, researchers have been trying to understand the body's capacity to renovate and restore the cells and tissues of many organs. After a long journey pursuing the how and why of cell repair mechanisms, scientists have now focused their attention on stem cells. There are no effective treatments for many diseases that shorten lives, but the aim is to find a way to replace what natural processes have taken away. Advances in science have brought us an alternative therapy for different diseases such as diabetes, chronic heart disease, Parkinson's disease, and liver failure, among others, in which organs from one person can be used to replace the diseased organs and tissues of another. This alternative however has been limited by organ donor shortage, which makes unlikely to fully meet the growing demand for organ replacement through organ donation strategies. The use of stem cells to generate replacement tissues for treating such disorders is an option that holds great promise.

Stem cells play a crucial role in the physiological turnover and regenerative responses of many tissues in adult life because it is known that they retain the capacity to generate progeny and renew themselves throughout life. Currently, the evidence suggests that stem cells are present in a variety of organs in order to replenish the multiple mature differentiated cell types and thereby achieve long-term tissue reconstitution. Although, for a long time, it was generally believed that cell loss was reconstituted via stem cells resident within and specific to an organ, this is no longer applicable to all organs and all types of injury. This is mainly due to the fact that recent discoveries have demonstrated that tissue damage may attract migratory stem cells from other compartments to take part into the restoration process. This observation has caused considerable interest in the field of liver diseases by which millions of people of all ages are affected worldwide. While many treatments for liver diseases aim to reduce damage, in some cases it is believed that damage can be reversed by replacing lost cells with new ones derived from cells that can mature into hepatocytes cells,

V.K. Rajasekhar, M.C. Vemuri (eds.), *Regulatory Networks in Stem Cells*, Stem Cell Biology and Regenerative Medicine,
DOI 10.1007/978-1-60327-227-8_34, © Humana Press, a part of Springer Science+Business Media, LLC 2009

derived from any source of stem cells. Therefore, a better understanding of the mechanisms involved in liver regeneration by hepatic stem cells holds great promise for potential development of novel therapeutic approaches to life-threatening liver diseases, although it is certain that there are many research questions to be answered.

New discoveries in stem cell biology and regenerative medicine have expanded our understanding of liver biology and the pathophysiology of various liver diseases such as hepatitis, cirrhosis, and liver cancer and have created hope for the therapeutic potential of such cells in the treatment of hepatic disorders.

At present, liver failure is a catastrophic illness associated with the death of many patients while waiting for transplantation [1]. Considering the vigorous regenerative capacity of the liver [2, 3], some forms of acute liver failure (ALF) can be manageable with several bridging techniques in the waiting time. Nevertheless, the lack of livers to used as a whole for orthotopic liver transplantation (OLT) or to isolate hepatocytes to use as a temporal support while a patient's own liver recovers is one of the major drawbacks [4]. Therefore, there has been a growing interest in liver stem cells (LSC) as an alternative to liver treatments.

The liver has enormous regenerative potential, explained by the mitotic division of hepatocytes and cholangiocytes after injury. For this reason, for decades, the role of stem or progenitor cells in liver regeneration has been controversial. Liver regeneration can be explained as a three-stage cell replacement process. The first stage is characterized by an ability of mature hepatocytes and cholangiocytes achieved after rapid cell division to repopulate the liver in response to certain types of injuries. These cells are the ones that also contribute to normal cell turn over in the liver. The second stage is characterized by the participation of an intra-organ stem cell compartment. This stem cell compartment is believed to be localized in the canals of Hering, which are the smallest, most proximal branches of the biliary tree or in the intralobular bile ducts. The best proof comes from various human analysis and animal models of extensive hepatic damage, where proliferating cells bud from the canals of Hering, and further differentiate towards the biliary and the hepatocytic lineage according to the severity of the disease and the type of mature epithelial cell that is damaged [5–8]. In the third stage, it is believed the participation of a cell source of possible extrahepatic origin, consisting of cells entering from the circulation. The cells are probably of bone marrow origin, although derivation from other sources has not been ruled out. If the cells are from bone marrow origin, it is thought that they enter the circulation through the portal vasculature and establish first next to the ducts in the portal triads when there is marked injury. Thus, the periductular location of these putative liver progenitor cells (LPCs) seems to be of external origin.

A source of controversy surrounds the issue of whether plasticity events are in fact occasions of circulating cells fusing with end-organ cells, such as hepatocytes, leading to the appearance of plasticity where is present. The processes of homing circulating cells to the liver, their engraftment and differentiation into functioning liver parenchymal cells, remain unclear. It is, however, believed that the existence of special factors such as granulocyte colony-stimulating factor (G-CSF) can mobilize stem cells from the bone marrow [9]. The essential role of CXC4 (SDF-1) chemotaxis, as well as the importance of hepatic MMP-9 and HGF expression in recruiting CD34+ stem cells to the liver, has also been confirmed [10].

The theory mentioned above was related to the repair process of the liver itself after injury, however, in terms of sources of hepatocytes used in cell-based therapies, there are additional sources, including stem cells from other adult populations such as bone marrow stromal cells, from fetal liver tissue, or from ex vivo differentiation of embryonic stem cells. The isolation, culture, and expansion ex vivo can generate a large quantity of cells for therapeutic use. Differentiation of mouse embryonic stem (ES) cells into mature hepatocytes has now been readily demonstrated by a number of groups [11–13].

Despite uncertainty surrounding the mechanism underlying the role of stem cells in liver regeneration, there is a great hope with the use of these cells for liver based-therapies. The demonstrated potential of stem cells in other fields [14–16] has increased the enthusiasm in hepatology, because stem cells can be used for the treatment of inherited and acquired end-stage liver diseases. They can also serve as a source of cells for cell transplantation in acquired liver diseases such as acute failure due to toxic or viral injury. Since they can be expanded in vitro to a desired extent, they can be used to populate liver assist devices or artificial livers based on bioengineered matrices. Lastly, they can be used as targets for gene therapies in primary liver diseases or diseases where extra-hepatic manifestations arise from abnormal gene expression or defective protein production by the liver. It is very likely then, that efforts to produce large numbers of transplantable hepatocytes from intra or extra-hepatic LSC will eventually prove to be successful. However, their efficient therapeutic application will demand additional scientific advances, but still has a bright future.

2 Defining a Stem Cell

It is difficult to arrive at a universally applicable definition of a stem cell due to the fact that some of the defined properties for a stem cell can be exhibited by the stem cells in some tissues or organisms but not in others. In spite of that, a generally acceptable consensus defines a *stem cell* as an undifferentiated cell that has the following capacities: self-renewal, production of progeny in at least two lineages, long-term tissue repopulation after transplantation, and serial transplantability. In addition, stem cells exist in a mitotically quiescent form [17] and clonally regenerate all of the different cell types that constitute the tissue in which they exist [18]. They can undergo asymmetric cell division, with production of one differentiated (progenitor) daughter and another daughter that is still a stem cell. The offspring of stem cells are referred to as *progenitor* cells, also called *"transit amplifying cells,"* and therefore cannot be serially transplanted, and are classified as early and late.

The early progenitor or stem/progenitor cells have multilineage potential and similar characteristics to stem cells. The late progenitor cells have differentiated further and produce progeny in only a single lineage. Although they divide rapidly, they are capable of only a short-term tissue reconstitution and they do not self-renew [18]. In order to maintain the pool of stem cells in the adult tissue, some of the cells need to divide without differentiating and others need to undergo asymmetric cell divisions [19]. Tissue stem cells are determined, that is, they lack the biochemical and structural markers of differentiation but are decided to differentiate into a specific cell type. While the size of the stem cell pool remains practically constant in many tissues under steady-state conditions, in some others even under normal circumstances it responds by proliferation and differentiation to replace senescent cells. The skin epithelium is the typical example. The basal cells send daughter cells to replace senescent cells. They can also expand rapidly in response to tissue damage to restore destroyed tissues in pathological conditions [19].

2.1 Stem Cell Niche

It is believed that once postnatal tissues are formed, intraorgan stem cells can only exist in a restricted yet protective microenvironment (stem cell niche), which provides factors that maintain them and excludes factors that induce differentiation [20]. This stem cell niche is an especial compartment of not only stem cells but also diverse gatherings of neighboring differentiated cell types (stem/progenitor cells, stromal cells), which secrete and organize a rich milieu of the extracellular matrix, basement membrane and other factors, whose mysterious interactions modulate the stem/progenitor cell function. Nobody knows yet the precise mechanism occurring in the niche, however studies of the Drosophila ovarian niche and the germ stem cells contained there, have helped us to understand the importance of all the structures contained in the niche. For instance, we know that there are some sort of physical interactions between stem cells and their non-stem cell neighbors contained in the niche that maintain stem cells there and control their relative quiescence or activation.

The evidence suggests that non-stem neighboring cells work as the "molecular glue" that anchors stem cell to their niche mediated by adherence and signaling mechanisms through Notch and WNT pathways. This "molecular glue" has been partially defined in some models, whereas in some others such as the liver still has not been defined yet [21, 22]. Additional factors participating in the retention of stem cells within their niche are integrins, which play an important role mediating cell adhesion to a basal lamina.

In fact, it has been demonstrated that stem cells have high levels of integrins. The niche can retain their stem cells by providing a unique milieu of ECM ligands for the integrin receptors on the surface of them [21]. In the case of HSC, for example, they express $\alpha4\beta1$ and $\alpha5\beta1$ integrins, which bind to fibronectin to promote adhesion to the bone marrow stroma. It has been demonstrated that antibodies against these integrins block hematopoiesis in long-term bone marrow cultures [23]. In the liver, the same could be applied since extensive studies have identified the integrins and basement membrane components predominantly expressed in human biliary epithelium: $\alpha2$, $\alpha3$, $\alpha5$, $\alpha6$ and $\alpha9$, which dimerize with $\beta1$, and laminin and type IV collagen [24]. However, as for the canals of Hering, these studies have not yet been performed. Another important factor involved in the stem cell niche is the innervation. In fact, nerve niche interactions are not yet clear, though some hypothetical mechanism have involved the possibility of direct interaction, via cell-cell junctions with the stem cells and/or stromal cells, or either the secretion of factors into the periniche milieu or autonomic control of vascular supply to these microenvironments. Some of these factors could be stimulatory, but others could be inhibitory of niche activation [25].

3 Liver Development

The definitive endoderm is an epithelial sheet formed at the ventral side of the vertebrate embryo during gastrulation. Invagination of the endoderm at the anterior end of the embryo generates the ventral foregut, which ultimately gives rise to the liver, lung, thyroid and the ventral pancreas [26]. The development of endodermal organs occurs by a coordinated sequence of events. In mice, the liver buds from the foregut endoderm at 7.5–8.5 days post-coitus, and is engrafted by hematopoietic progenitors at 9–10 days. The fetal liver is rich in hematopoietic cells and therefore is the site of proliferating and differentiation of liver-cell precursors. Proliferation of undifferentiated endodermal cells of the ventral foregut is seen around embryonal day (ED)

8.5; these cells then migrate to the septum transversum, and there they come in contact with mesenchymal cells. In the formation of the vertebrate liver is very important the establishment of competence within the foregut endoderm in order to respond to organ-specific signals. This process is followed by liver specification, hepatic bud creation, growth, and differentiation [27].

The specification and development of these areas seems to be controlled by cell-autonomous factors such as transcriptional regulators, as well as by inductive or inhibitory signals from surrounding tissues. Although little is know about the factors that elicit embryonic induction of the liver, extensive research using embryonic tissue explants have provided priceless information on the mechanisms involved in early mouse liver development. Tissue relations and signals from mesoderm are characteristics of endodermal patterning. A combination of positive inductive signals emanating from the cardiac mesoderm, such as fibroblast growth factors (FGFs) FGF1, FGF2, and FGF8, and repressive signals from the trunk mesoderm, specifies a group of primitive pluripotent endodermal stem cells in the ventral foregut to adopt a hepatic fate [28]. It has been hypothesized, however, that the endoderm must first enter a stage in which competence to respond to FGF signaling is established. This finding was based on the observation that portions of dorsal endoderm, which usually does not originate liver, were induced to express albumin (a liver marker) only when those portions were dissected between gestational days 8.5 and 11.5 and cultured with FGF. This competence was lost, however, when the dorsal endoderm was isolated at ED 13.5 or further, indicating that factors required for competence are limited to specific stages of embryonic development [29].

While FGF1 and FGF2 have been shown to induce the expression of hepatic genes at multiple and distinct stages, FGF8 is thought to play a role in liver outgrowth and cell differentiation [28]. Recently, the role of FGF10 has been demonstrated to be a critical factor for liver growth during embryogenesis by enhancing hepatoblast proliferation via β-catenin activation through FGFRs binding [30]. FGFs bind to FGF receptors (FGFRs) in a loose manner, among them the FGFR2-IIIb isoform (FGFR2b), which has been shown to be crucial in the postnatal liver regeneration, based on the findings that adult mice expressing a soluble, dominant-negative form of FGFR2b showed markedly reduced hepatocyte proliferation after partial hepatectomy [31]. The Wnt/β-catenin signaling pathway has also been implicated in the maintenance, survival, proliferation, and cell fate decisions of progenitor cell populations in several organs, including the liver during embryogenesis [32]. It was demonstrated that most of the cells contained in the embryonic liver that displayed the most β-catenin activation pattern were hepatoblasts. It is speculated then that during the hepatogenesis, FGF10 seems to be secreted by the embryonic stellate cells/myofibroblasts, residing around the portal vein. Then, FGF10 is transported to hepatoblasts, which express FGFR2B through the portal vein and the developing sinusoids of the liver. FGF10 then binds to FGFR2B on hepatoblasts and induces the activation of the β-catenin signaling pathway [30]. FGF10 signaling from the adjacent mesenchyme regulates differentiation of the foregut epithelial cell toward hepatic or pancreatic cell lineages in zebrafish, suggesting a significant role for FGF10 in the differentiation of liver precursor cells [33]. FGF10 has been also implicated in the proliferation or differentiation of various stem/progenitor cell populations [34, 35].

Bone morphogenetic protein (BMP) signaling from the septum transversum mesenchyme coordinately works with FGFs to initiate the induction of hepatic gene expression in the endoderm and to exclude a pancreatic fate. BMP seems to affect the levels of the GATA4 transcription factor being also critical for morphogenetic growth of the hepatic endoderm into a liver bud [36]. Furthermore, it has been proposed that the forkhead box A (FoxA) and GATA transcription factors [37, 38] initially facilitate the ventral foregut endodermal cells to go through a stage of competence by opening compacted chromatin structures within liver-specific target genes. Therefore, cells can react to inductive mesodermal signals congregating on a common endodermal domain along the primitive gut tube [37, 39]. By ED 9.0–9.5, after the establishment of the competence process, the cells under the influence of all the above-mentioned factors start to express α-fetoprotein (AFP), and then albumin. The hepatic-specified cells are now considered to be hepatoblast and they have an extraordinarily proliferation capacity. The septum transverse mesenchyme is also invaded by cords of hepatoblast, which will originate stellate cells, and sinusoidal endothelial cells that subsequently will develop into blood vessels [27]. At ED 11 hepatoblasts keep proliferating and they begin to express in addition to AFP and albumin, placental alkaline phosphatase, intermediate filament proteins (CK-14, CK-8 and CK-18), and γ-glutamyl transpeptidase (GGT), later on, they express α-1-anti-trypsin, glutathione-S-transferase (GST)-P, and fetal isoforms of aldolase, lactic dehydrogenase, and pyruvate kinase (M2-PK) [40 , 41]. Immediately before ED 16, hepatoblasts selectively differentiate under the regulation of an array of liver-enriched transcription factors into either hepatocytes or bile duct epithelial cells [42–44].

The Notch signaling is activated in hepatoblast that undergo differentiation along the bile duct epithelial lineage. The expression of the Notch intracellular domain in hepatoblast inhibited hepatic differentiation by reducing the expression of albumin [45]. Notch signaling pathway is antagonized by hepatocyte growth factor, which in combination with oncostatin M promotes hepatocytic differentiation [46].

Recently it was proposed that a gradient of activin/TGFβ signaling is required for the differentiation of hepatoblast towards the cholangyocytes lineage. The inhibition of the activin/TGFβ signaling allows hepatoblast to undergo normal hepatocyte differentiation [47, 48]. In terms of in vitro differentiation towards hepatocyte or bile duct lineage, this point is critical to determine the fate of the desired lineage. This time period is referred to as the differentiation window during the hepatic development [49]. After ED 16, most of the hepatoblast are now committed to either hepatocytic or cholangiocytic lineages, thus they are no longer bipotential, although they continue to proliferate, they become unipotent or late progenitor cells. It is worth mentioning that, in the embryonic liver, some of the LSCs, do not differentiate and progressively go out of the proliferative state and, after ED 16–18, they will just stay in the liver as quiescent potential LSCs [27, 43]. By ED 17, intrahepatic bile ducts are formed surrounding the large portal vein branches. By this time, the essential lobular arrangement of the liver is completed, but the maturation of the hepatic parenchyma will be completely mature several weeks after birth (Fig. 2) [50].

4 Liver Renewal

The liver is both an exocrine and endocrine gland that performs complex functions and has a phenomenal regenerative capacity. This process enables the recovery of lost mass without endangering the viability of the entire organism [2, 3]. Followed by acute injury, stem cells take part of the major role in normal tissue repair and homeostasis in quickly turning over tissues such as the skin or bone marrow [19]. In contrast, liver regeneration after loss of hepatic tissue does not depend on these kinds of cells, but of the proliferation of the existing mature hepatocytes, the parenchymal cells of the organ. In addition, the rest of the hepatic cell types, such as biliary epithelial cells, fenestrated endothelial cells, and Kupffer and Ito cells, proliferate also and contribute to regenerate the lost hepatic tissue [2]. In the case of liver regeneration after toxic damage, it is worth mentioning that another important phenomenon asides from the hepatocyte proliferation: the capacity of the newly formed hepatocytes to adapt themselves to the distinctive three-dimensional architecture of the liver lobules built around portal triads and central veins.

The classical lobule is roughly hexagonal in shape, with groups of hepatocytes arranged in rows that radiate out from the central vein, and defined by loose connective tissue in which the portal canals are found. The portal canals are formed by the portal vein, which is in charge of the blood transport to the liver from the intestine, and hepatic artery with highly oxygenated blood and the bile ducts in charge to carry away the bile. In order to guarantee liver function, the lobule architecture has to assure a generous blood flow from the portal vein through the sinusoids to the central vein within each liver lobule. The rows of hepatocytes in the lobule are one or two cell wide, and are surrounded by sinusoidal capillaries. This arrangement ensures that each hepatocyte is in very close contact with blood flowing through the sinusoids, that is, bathed in blood. Thus, liver regeneration is a complex but precisely defined process [2, 3].

The hepatic acinus has three zones: Zone 1 is high in oxygen and hepatocytes here are the first to receive blood. Zone 2 is lower in oxygen and hepatocytes in this zone are the second cells to receive blood. Zone 3 is the lowest in oxygen and hepatocytes here are the last to receive blood from a branch of the hepatic artery. Thus, the cells with the highest metabolic potential are found in Zone 1 and those with the least are found in Zone 3. Importantly, the cells in Zone 3 are the most susceptible to ischemic conditions due to the already low level of oxygen that reaches them through the blood.

The normal liver has been estimated to be replaced by normal tissue approximately once a year or more [51], since turnover rate of normal liver cells was calculated to be 1 in 20,000–40,000 cells at any given time [52]. Conversely, this slow normal renewal rate differs to the rapid proliferative response to loss of hepatic mass. The liver's self-healing ability was documented since the Greek mythology, and exploited by Zeus to punish Prometheus, a Titan. Zeus, the king of the gods, ordered Prometheus to be chained to a rock in the Caucasian mountains as punishment for stealing the holy fire from Olympia – the home of the gods – and sharing it with mankind. Zeus sent an eagle to the rock to peck away at Prometheus's liver. By night, as the eagle slept, Prometheus's liver grew back so that it was a fresh tasty meal again for the eagle the next morning.

The modern recognition of the liver's self-healing ability is exemplified by experimental partial hepatectomy (PH) [3]. The normal adult liver parenchyma is made up of mitotically quiescent hepatocytes, and cholangiocytes both of them originating from a common endodermal foregut precursor cell. Hepatocytes, as fully differentiated cells, normally turn over very slowly, but have a remarkable ability to re-enter the cell cycle in response to mitotic stimuli. Following two-thirds partial hepatectomy (via removal of the left and median hepatic lobes), nearly all hepatocytes in the adult liver undergo cell division, starting with periportal hepatocytes, as well as those immediately adjacent to the central vein. This is followed by a second round of replication in which about half of the hepatocytes divide again to fully restore the liver of its original mass within a few days with little or almost no evidence of contribution of a liver stem cell [53, 54].

This response to surgical injury is termed "regeneration,"although the term is not precisely accurate since the response involves a compensatory hyperplasia within remaining lobules, not a recreation of the original lobular morphology. Replication of differentiated hepatocytes and biliary epithelial cells accounts for this regeneration. Since hepatocytes are supposed to be terminally differentiated cells, it is believed that they can undergo only one or two rounds of cell proliferation. However, it was demonstrated that this is no longer true, since hepatocytes were able to regenerate the liver after 12 sequential hepatectomies, demonstrating the outstanding proliferative capacity of hepatocytes [55]. Additional experiments performed on a mouse model of tyrosinemia (FAH–/– mice) demonstrated that when hepatocytes from a healthy donor were injected into (FAH–/– mice), these were capable to re-establish the liver mass, rescuing the mice. When hepatocytes isolated from this first generation of rescued mice were serially transplanted, six generations of mice were rescued. Corresponding to approximately a minimal number of 69 cell doublings [56]. This elegant experiment conclusively instituted the ability of mature hepatocytes to repopulate an entire organ and self-renewal. The previous findings imply that hepatocytes could essentially act as their own stem cell, and regenerate the liver. Nonetheless, there is evidence that the replicative activity of hepatocytes decreases in advance cirrhosis in humans and in chronic liver injury in mice, reaching a state of "replicative senescence," probably due to telomere shortening [57]. However, the existence of a common progenitor (hepatoblast) with the ability to give rise to both bile duct epithelial cells and differentiate hepatocytes during embryonic development, and the ability to be responsible of some forms of liver regeneration later in life, suggest the existence of two basic types of liver regeneration: one dependent of mature hepatocytes, and one dependent on the progenitor (stem cells) that may be used when parenchymal hepatocytes are severely damaged and unable to efficiently regenerate the liver [8, 58]. Nevertheless, the proliferation of the different cellular populations in the liver depends mainly on the insult that triggers the process. Mature hepatocytes, for example, respond immediately after chemical injury, such as in the case of carbon tetrachloride. This agent causes hepatocyte necrosis primarily in the periportal areas of the liver lobules. Because hepatocytes in the pericentral areas have much higher expression levels of cytochrome P450 2E1, which is involved in metabolic activation of CCl4, centrolobular necrosis occurs within 36–48 h after administration, and hepatocytes are in charge to restore this damage in a 7-day process [59]. Bile duct structures are, in some cases, stimulated to proliferate, like those seen after bile duct ligation or bile duct necrosis induced by α-naphthyl isothiocyanate (ANIT) or 4,4′-diaminodiphenylmethane (DDPM) [60].

On the other hand, when hepatocyte proliferation is inhibited such as by viral infection or by chemicals, regeneration proceeds from an alternate cells source (oval cells), which respond to the injury by proliferation and differentiation into hepatocytes [8].

Finally, in the liver has been observed the presence of a periductular putative liver progenitor cell (LPC). This is believed to be from a possible extrahepatic origin, the bone marrow [61]. The detail of each of these putative liver stem cells will be discussed in detail in the next section.

5 Liver Stem Cells

5.1 Definition

Fifty years ago, Wilson and co-workers [62] suggested the existence of hepatic stem cells in the adult liver. Nowadays researchers are doing a great effort to characterize, localize, and isolate these cells, although it has been difficult due to the lack of specific cell surface markers. According to the general definition of a stem cell, it is important first to define what liver stem cells (LSCs) are. Hence, a LSC would be one that fulfills the characteristics of being undifferentiated, with a self-renewing ability as well as the ability to produce multilineage (or at least bilineage), and able to repopulate the liver. Those LSCs that do not fulfill all the characteristics are considered as *potential* LSCs. The origin, nomenclature and function of these cells have been a longstanding area of study and debate, since these liver-related stem cell populations will fluctuate accordingly to the liver stage of development and the diverse range of injured circumstances.

5.2 Fetal Liver

The early fetal liver at around ED 12–16 contains two populations of hepatic cells, the fetal hepatic stem cells, and the hepatic progenitor cells (hepatoblasts). Hepatoblasts are bipotent cells derived from endodermal cells; they exhibit many properties expected for hepatic stem cells and they are also known as fetal liver stem/progenitor cells (FLSPC) [58]. During embryonic liver development of rodents, ED 14 in mice and ED 15 in rats, the hepatoblast are located near the vascular spaces, which is going to be the site for portal spaces in later development. These hepatoblast express dual markers of the hepatocyte and biliary lineages, and they are capable of differentiation into either of the two epithelial cell populations of the liver, hepatocytes, and biliary epithelial cells.

The variety of markers expressed by them has been useful for their isolation from the fetal liver. However, during the developmental process, the architecture of the mature liver becomes apparent with the differentiation of the FLSPCs into hepatocytes and sinusoid formation, and subsequently

FLSPCs will express markers only of the committed lineage. In fact, there are several markers used for the identification, proper development and differentiation of FLSPC such as α-fetoprotein (AFP) [37, 63] and albumin, a marker of both hepatoblast and hepatocytes [46]. C/EBPα started to be expressed in endodermal cells on day 9.5 in the liver primordium, and continued to be expressed in FLSPC and hepatocytes throughout development [64]. Cytokeratins (CK)-14 [65], CK-8 [66], and CK-18 [67] are expressed by the embryonic liver diverticulum. Dlk, also known as Pref-1, is highly expressed in the E10.5 liver bud and is a useful marker to enrich highly proliferative FLSPC from fetal liver [68]. E-cadherin was used to isolate fetal liver progenitor cells from ED12.5 mouse livers with high yield and purity [69]. Forkhead Box (Fox) m1b (Foxm1b) is critical for hepatoblast precursor cells to differentiate toward biliary epithelial cell lineage [70]. GCTM-5 is a monoclonal antibody originally derived after immunization of mice with a membrane preparation from a testicular seminoma. Stamp et al. [71] recently discovered this marker to be expressed exclusively in the fetal liver of 7-week human embryos. This marker in the normal and diseased adult and pediatric human liver was expressed in a subpopulation of cells within the biliary epithelium [71]. Gamma-glutamyl transpeptidase (GGT), a major enzyme of glutathione (GSH) homeostasis, is often used as a biliary marker to follow the differentiation of hepatic precursor cells [72]. Hepatocyte nuclear factor-4alpha (HNF4-α) regulates the expression of many genes preferentially in the liver [73] plays a crucial role in early embryonic development [74]. Id3, a negative regulator of helix-loop-helix transcription factors, was demonstrated by Nakayam et al., to be an important regulator of hepatoblast proliferation in the developing chick liver. They demonstrated this marker to be expressed in hepatoblast at early developmental stages, but not in hepatoblasts at later stages [75]. Liv2 a hepatoblast marker [76]. Prox1 is a transcription factor expressed in early embryonic hepatoblasts has been shown to be very important during the liver development. Studies using Prox1 knock-out mice demonstrated that these mice died during early embryogenesis stages, while displaying a very rudimentary liver [77]. Prox1 is still expressed in the adult liver but only by hepatocytes. In addition, it is considered to be one of the earliest markers of liver development together with albumin and AFP [78, 79]. SEK-1 plays a crucial role in hepatoblast proliferation. Studies using mice defective in SEK-1 demonstrated embryonic lethality after ED 12.5 and this was associated with abnormal liver development. The authors also demonstrated in this study that SEK1 is required for phosphorylation and activation of c-jun during the organogenesis of the liver [80]. SMAD, a mediator of BMP signaling, is preferentially expressed in hepatoblast undergoing bile duct morphogenesis in the fetal liver [81]. Sca-1 is a general stem cell marker that is also expressed on murine FLSPC [82]. New cell surface markers were recently demonstrated to be expressed by murine fetal hepatic stem cells, Punc E11 (Nope), and Cd24a [83]. Cd24a was shown to be expressed on the surface of FLSPC but not on mature hepatocytes, whereas (Nope) was only expressed on FLSPC, and was not expressed by hematopoietic stem cells isolated from the adult bone marrow.

Observations of the rat fetal liver by ED 14 showed the presence of three distinct populations: committed immature hepatocytes, expressing AFP and albumin, a second population expressing biliary cell markers such as cytokeratins, and a third population of cell expressing both hepatic and biliary markers. This later small population of bipotent cells, after transplantation, engraft, proliferate, differentiate into hepatocytes and bile duct cells, and stably repopulate normal adult liver [84–86]. After ED 16, the gene expression profile in the liver is remarkably evident, with a more differentiated phenotype and a decrease in the number of bipotent cells [85]. This bipotent population is thought of as the fetal source of hepatic progenitor cells. The bipotential capacity of these cells for liver cell-based therapy have been widely tested, in different liver repopulating models since isolation based in a combination of different markers has been possible despite the controversy regarding this issue. The potential will vary accordingly to the liver stage in which the cells are isolated, since cells isolated in late embryological stages will lose the bipotent capacity, being able to differentiate along only one lineage.

During development, the fetal liver is the main place of hematopoiesis [87], where hematopoietic cells are believed to liberate signals that direct the growth and differentiation of the liver. As the time passes, the hematopoiesis is reallocated in the bone marrow and not any more in the liver. However, during this process we may ask whether any of the transient hematopoietic stem cells stay in the liver to form the hepatic stem cell compartment? This speculation lead investigators to believed that if this was true, then the hepatic progenitor cells could share some cell surface markers associated with hematopoietic stem cells such as CD34 [88] and CD90 Thy-1 [89]. Many surface markers expressed on FLSPC have been reported to be also expressed by progenitor cells in the Adult liver [90]. In the developing fetal liver there is evidence of the existence of a Thy-1$^+$ population [89, 91], and these cells are believed to be located mainly in the portal tracts and express several lineage markers, including co-expression of biliary and hepatocellular proteins [91]. However, not only a Thy-1$^+$ population was found in the fetal liver, but also a Thy-1$^-$ population [92]. A comparison of the properties of these two populations was done in terms of tissue reconstitution, which is one of the characteristic features of stem cells. After isolation and separation of into + and – population of Thy-1 cells from rat ED 14 fetal liver, it was demonstrated that Thy-1$^-$ cells were able to repopulate the normal host liver whereas Thy-1^+ were not. Thy-1^+ cells, however, repopulated the liver of a retrorsine-treated rat. These suggested that Thy-1^+ fetal liver cells at ED 14 represent a population of fetal hepatic progenitor cells that can proliferate and repopulate the liver only after extensive liver injury whereas Thy-1^-fetal liver cells are stem/progenitor that exhibit greater proliferation potential and can repopulate the normal adult liver. New contributions to the field of stem cells are being developed year by year, and in the future will be easier to identify through a combination of several surface markers what best define a liver stem cell.

5.3 Adult Liver

While hepatocytes can be considered conceptually as unipotent stem cells, the presence of true stem or progenitor cells within adult livers has been largely debated.

The restorative reaction of the liver to diverse injuries entails proliferation of cells at different levels in the liver lineage, which consists of stem cells, progenitor cells (transit–amplifying cells), and mature cells. Within the liver, stem cells are thought to reside in a niche composed of cells, extracellular matrix, and soluble factors released by the niche cells that help to maintain the characteristics of the stem cells. Thus, it is believed that the adult liver contains potential LSC that are activated when the regenerative capacity of mature hepatocytes is compromised. The offspring of these potential LSCs are the liver stem progenitor cells (LSPCs) or oval cells. Stem cells in the liver are proposed to be from two origins: endogenous or intrahepatic stem cells and exogenous or extrahepatic stem cells. (Fig. 1). Included in the intrahepatic stem cell compartment are the LSPC, which are greater in number but with a short-term proliferation capacity. LSPCs are thought to be localized within the canals of Hering [93], interlobular bile ducts [94], or in the periductular/intraportal zone of the liver [60]. These cells are called into action when hepatocytes are insufficient or unable to respond [58, 93, 95, 96]. Included in the extrahepatic stem cell compartment are cells derived from bone marrow, and peripheral blood; these cells are usually few, but with long-term proliferation capacity.

5.3.1 Intrahepatic Stem Cells

Oval Cells

During embryonic development, hepatoblast gives rise to the primitive intrahepatic bile ducts, structures that connect parenchymal hepatocytes with the larger segments of the biliary system. These primitive intrahepatic bile ducts correspond to the canals of hering and terminal bile ductules of adult livers, which may constitute the niche for intrahepatic stem cells [50]. Based on this, oval cells are thought to have originated from the cells of the canals of hering in the adult rat liver, thus they may express AFP and some other markers present in both adult and fetal liver cells. However, the extent to which these markers are expressed in a population of proliferating oval cells depends on the agent

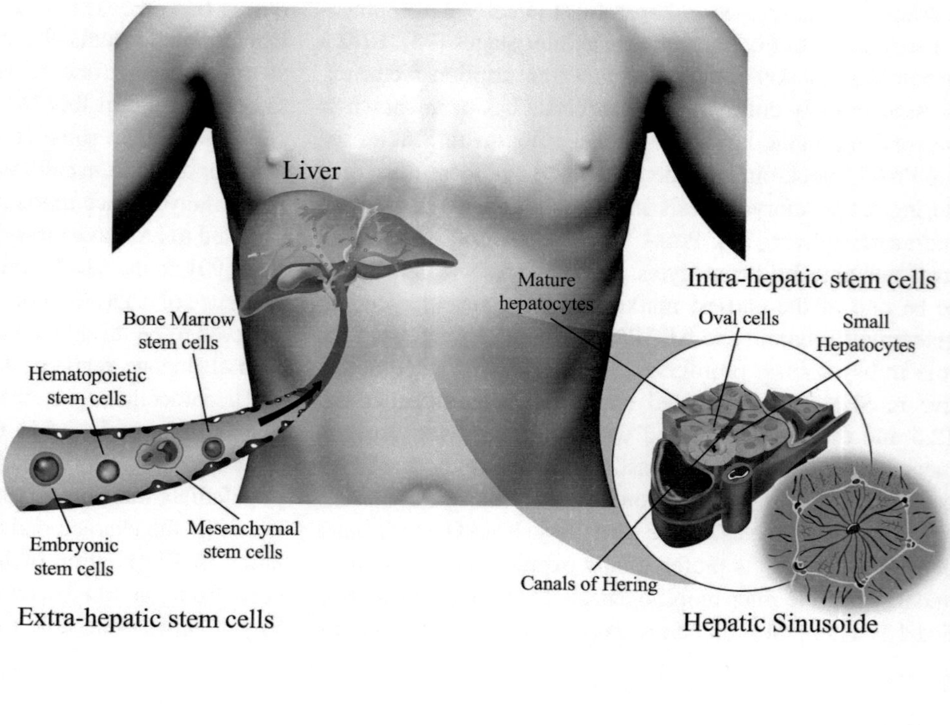

Fig. 1 Possible roles of intra- and extrahepatic stem cells in the repair of hepatic tissue. After tissue injury, hepatocytes act as the first line of defense to replace necrotic cells. When the pool of hepatocytes is exhausted or their capacity is inhibited, stem cells that are intrinsic to the tissue replace necrotic cells. If the pool of endogenous stem cells is exhausted, exogenous circulating stem cells are signaled to replenish the pool and participate in tissue repair. Thus, circulating stem cells may serve as a backup rescue system

that elicited oval cell proliferation. Oval cells are thought to represent a heterogeneous population of transit-amplifying progenitor cells activated to proliferate in several models of hepatic injury, when the regenerative capacity of hepatocytes is inhibited.

The term *oval cell* was first assigned by Farber et al. [8], who observed a population of nonparenchymal cells appearing in the rat liver after treatment with various carcinogenic agents, and described them as small oval cells with scanty lightly basophilic cytoplasm and pale blue-staining nuclei. Oval cells phenotypically resemble fetal hepatoblast since they behave like bipotential progenitors capable of differentiation into mature hepatocytes and biliary epithelial cells [58]. Hepatic oval cells are a heterogeneous population of proliferating cells, with cells having a different capacity and stage of differentiation. Therefore, oval cells express markers associated with immature liver cells, such as a-fetoprotein; mature hepatocytes, such as albumin and γ-glutamyl transferase (GGT); and biliary epithelium, such as cytokeratin 7 and 19, oval cell 6 (OV-6), and OC2 (anti-myeloperoxidase) [97]. In addition, oval cells share some phenotypic characteristics with hematopoietic progenitor cells since they express the hematopoietic stem cell factor and its receptor c-kit tyrosine kinase (c-kit) [98], and the related proteins flt-3 and flt-3 receptor [99]. Oval cells express also CD34 [100], and a marker of early hematopoietic progenitor cells, Thy-1 [101]. In a mouse model of liver injury using 3,5-diethoxycarbonyl-1, 4-dihydro-collidine (DDC), it was shown that proliferating oval cells co-express A6 [102] and the specific marker for hematopoietic stem cells Sca-1, as well as the CD34 and CD45 surface proteins [90]. The leukemia inhibitory factor (LIF) and its receptor are also highly expressed in hepatic oval cells [103].

The expression of these markers suggested their stem cell-like properties. Some of these markers are shared by biliary epithelial cells (GGT, CK19 and CD34). Oval cells have been demonstrated to express also the adenosine triphosphate–binding cassette transporter ABCG2/ BCRP1 (ABCG2), a marker for the bone marrow side population [104]. The localization of these cells still remains controversial, however, they have been identified in the periductular/intraportal zone. This and the fact that oval cells express some hematopoietic stem cell genes, including those found in the bone marrow side population, originated the speculation of the bone marrow origin of oval cells either directly or by transdifferentiation. Regarding this speculation, several researchers have been working continuously in an effort to clarify this possibility. In 1999, Petersen et al. [95] used different approaches to demonstrate that bone marrow cells contribute to liver cells. Bone marrow from a male rat was transplanted into lethally irradiated female animals, followed by treatment with 2-acetylaminofluorene (2-AAF) and CCl4, to simultaneously induce hepatic necrosis and impairment of endogenous hepatocyte proliferation. The appearance of BM-derived oval cells in the liver of these animals was observed. The same group [105] proved the same findings with a different model of liver damage and oval cell activation. Using a model F344 dipeptidyl peptidase IV-deficient (DPPIV(−)) rats plus (2-AAF) and 70% partial hepatectomy (PHx), followed by male F344 DPPIV(+) bone marrow transplantation, the authors concluded that, under certain physiologic conditions, it is possible that a portion of hepatic stem cells arise from the bone marrow and can differentiate into hepatocytes. In addition, X/Y-chromosome analysis revealed that fusion was not contributing to differentiation of donor-derived oval cells [105]. On the other hand, Wang and co-workers [106] demonstrated that mouse liver oval cells are not originated in the bone marrow but in the liver itself by using a fumarylacetoacetate hydrolase (Fah) mice model. Another group reached a similar conclusion by using DPP4$^-$-deficient F344 rats. The authors substituted the BM of lethally irradiated female DPP4$^-$-deficient F344 rats with BM cells from syngeneic normal male F344 rats. Then the recipients were subjected to different models of activation and expansion of oval cells, and they demonstrated that oval cells in the injured liver do not arise through transdifferentiation from BM cells but from the endogenous liver progenitors [107].

Based on the ambiguity of the existing data, the existence of hematopoietic markers in the normal adult liver and after hepatic injury with oval cell proliferation could be interpreted by two possibilities. The first possibility is that a small number of hematopoietic stem cells from the fetal liver remain in the adult liver. If this was the case, then hematopoietic stem cells may be distinct from oval cells, but a component of the oval cell compartment, therefore these cells do not acquire markers of the hepatocyte lineage but they share general stem cell markers with the origin of oval cells. The second possibility is that the hematopoietic cells contained in the adult liver may be pluripotent stem cells, working as the counterpart of embryonic stem cells, able to produce multiple lineages, including the hepatic. If this was the case, hematopoietic stem cells in the liver, thought to be located in the periductular spaces, would by the same stimuli that causes oval cell response differentiate progressively, first into oval cells and finally into hepatocytes [60]. Despite extensive studies, the hematopoietic versus hepatic origin of liver progenitor oval cells remains controversial. It is clear, however, that regardless of the origin, oval cells definitely require certain physiologic conditions and the hepatic niche to proliferate. Oval cell proliferation can be induced in a number of ways, which includes administration of a choline-deficient diet supplemented with ethionine [62, 108]; treatment with other DNA alkylating agents, such as 1,4-bis[N,N'-di(ethylene)-phosphamide] piperazine (Dipin) [109]; feeding 3,5-diethoxycarbonyl-1,4-dihydrocollidine

(DDC) [110]; phenobarbital/cocaine-induced liver injury [111]; and administration of D-galactosamine [112], or treatment with 2-acetylaminofluorene, to block adult hepatocyte proliferation; and then either partial hepatectomy or treatment with carbon tetrachloride, to induce hepatocyte loss and a proliferation signal [95]. In these settings, the failure by adult hepatocytes to respond to growth signals results in activation and rapid proliferation by oval cells [113], which initially appear near bile ductules and later migrate into the hepatic parenchyma.

The use of three different models of oval cell activation in rats, 2-acetylaminofluorene treatment in combination with partial hepatectomy (2-AAF/PH), retrorsine treatment followed by partial hepatectomy (Rs/PH) and D-galactosamine (D-gal) induced liver injury, identified CD133, claudin-7, cadherin 22, mucin-1, ros-1, Gabrp as new surface markers [114]. While in murine adult livers, 3,5-diethoxycarbonyl-1, 4-dihydro-collidine (DDC) identified a population of CD133 expressing oval cells with the gene expression profile and function of primitive, bipotent liver stem cells [115].

These new markers give us a big clue for the isolation of adult progenitor cells, though further research is needed using diverse species in order to confirm the standardization of these markers. Although the characterization and comparison of the oval cell reaction has been tested in several commonly used protocols for stem cell–mediated liver regeneration, it has been demonstrated that the reactions observed among different species vary in several aspects [116]. However, the reasons for these differences are unknown. They could reflect differences in the microenvironment or, alternatively, inherent variations in the endogenous hepatic stem cell compartment.

This suggests that extrapolation of knowledge between mammalian species must be reconsidered, and that further studies are needed for reliable and reproducible experimental models. Nevertheless, the potential of hepatic stem/progenitor cells for use in cell therapy brings great promise for the future treatment of human liver diseases.

Small Hepatocytes

When laboratory animals are exposed to the pyrrolizidine alkaloid retrorsine, which inhibits adult hepatocytes from expanding following a proliferation signal, and undergo partial hepatectomy, liver regeneration cannot proceed normally from adult differentiated hepatocytes, but is initiated by a new type of cell, the small hepatocyte-like progenitor cells (SHPCs) [117–119]. The activation, proliferation, and complete regeneration of normal liver structure from SHPCs have not been recognized in other models of liver injury characterized by impaired hepatocyte replication. Likewise, their precise origin and their defined tissue niche remain controversial. Some investigators have suggested that SH-PCs may represent an intermediate or transitional cell type between oval cells and mature hepatocytes rather than a distinct progenitor cell population [120, 121]. However, the possibility of different cellular origin of SHPCs cannot be excluded. There is evidence suggesting that SHPCs may represent a distinct immature progenitor cell population [117, 118, 122]. Others suggest that SHPCs represent a population of retrorsine-resistant mature hepatocytes [117, 119], while some others have proposed that SHPC simply arise from a subpopulation of hepatocytes that lack the necessary CYP P450 enzymes required to metabolize the vinca alkaloid, retrorsine, and hence are protected from the inhibitory effects of this reagent [118–120]. SHPCs have characteristics of not only mature adult hepatocytes, but also fetal hepatoblasts, and OC. They most closely resemble fully differentiated (but small) hepatocytes morphologically, although they express albumin and transferrin, generate bile canaliculi, and store glycogen. They also express the oval and fetal liver cell markers OC.2, OC.5, and AFP [117], and do not express cytochrome P450 (CYP) genes. During liver repopulation, SHPC tend to form nodules with high proliferative capacity, expressing large amounts of Ki-67 and MCM-2 [119, 120]. These cells also have significant capacity to proliferation in vitro, and following transplantation [123], they can repopulate the liver almost as well as freshly isolated primary hepatocytes (Fig. 1).

5.3.2 Extrahepatic Stem Cells

Bone Marrow and Hematopoietic Stem Cells

The bone marrow (BM) compartment is largely composed of a main stem cell population, hematopoietic stem cells (HSCs), which give rise to all mature blood lineages, and the mesenchymal stem cells (MSCs), which forms stromal tissue.

HSCs in the BM, functionally defined by their ability to reconstitute the BM of a myeloablated host, are contained in a population expressing CD34 (CD34+). In addition, HSCs express CD117 (c-kit), the receptor for stem cell factor (SCF) produced by marrow fibroblast and endothelial cells. Currently, it has been suggested that bone marrow cells possess a broad differentiation potential (plasticity), being able to differentiate into mature cells of various organs [124, 125]. While some groups have attributed this apparent plasticity to transdifferentiation [61, 126], some others however, have suggested that cell fusion could explain these results [127–130]. Suggestive evidence that hematopoietic stem cells may give rise to liver cells has caused considerable interest in the field of liver diseases, where new strategies to restore hepatocyte number, are required. Therefore, many scientists have been trying to define the subpopulations of BM cells capable of generating liver cells, as well as the conversion mechanisms of these cells. So far, two theories have been proposed (*transdifferentiation*: the adoption of a different phenotype by a cell apparently committed to a tissue-specific cell type; and *fusion*), although it is still controversial and not well defined yet (Fig. 1).

Stem Cell Plasticity: Cell Fusion or Transdifferentiation?

The plasticity potential of stem cells has been heavily debated. Yet, experimental research has helped to demonstrate the flexibility of this process. Therefore, stem cells can be instructed by factors of the host's microenvironment to adopt the desired fate. They can transdifferentiate, that is, genetic reprogramming of a differentiated cell under induction of microenvironmental signaling. Alternatively, they can fuse with the recipient's cells, thus leading to cytoplasmic mixing and reprogramming of cell fate. Although it is possible that there might be other pathways to plasticity, these have been so far well documented. The conversion of cells derived from bone marrow into hepatic cells has been suggested in vivo [95, 131–133] and in vitro [134]. Although cell fusion has been proposed to be an alternative mechanism responsible for cell fate changes [127, 128], many other reports have demonstrated conversion without fusion [135, 136].

In the case of humans, it has been possible to suggest the BM origin of hepatocytes by taking advantages of the very useful methods to track hepatocytes that have been developed in the last decades. Such is the case of male recipients of female orthotopic liver transplants, and females who had received bone marrow transplantation (BMT) from male donors [137, 138].On the other hand, by the analysis of biopsy specimens from the liver, gastrointestinal tract, and skin from female patients who had undergone transplantation of HSCs from peripheral blood or BM from male donors, engraftment of donor-derived stem cells was observed, accounting for up until 7% of the cells in histologic sections of the biopsy specimens. With these results, it was demonstrated once again the differentiation potential (plasticity) of circulating stem cells into different types of cells, including their potential to differentiate into hepatocytes [135]. The conversion of cells derived from bone marrow into hepatic cells has been suggested also in animals. The transplantation of bone marrow cells $CD34^{+}$, lin^{-} into myeloablated mice was able to give rise to hepatocytes in the recipient liver up to 2.2% of the total hepatocyte number [132]. In the case of animal models such as fumarylcetoacetate hydrolase (FAH)-deficient mouse, a model of fatal hereditary tyrosinemia type I the demonstration of hepatocyte generation from bone marrow cells, was also possible [131]. FAH–/– mice suffer from severe liver damage as a consequence of accumulation of the hepatotoxic metabolites, fumarylacetoacetate, and its precursor maleylacetoacetate. Due to the deterioration of hepatocytes, FAH-deficient mice cannot survive unless they are treated with the drug 2-(2-nitro-4-trifluoromethylbenzyol)-1,3cyclohexanedione (NTBC), which prevents production of the toxic metabolites. Due to permanent deterioration of hepatocytes, the FAH–/– mice represent an animal model with an extremely high selection pressure for wild-type (i.e., FAH+/– or FAH+/+) hepatocytes. A purified HSCs population (c-kit^{+} $Thy1^{low}$, lin^{-}, and Sca-1^{+}), was transplanted into lethally irradiated FAH–/– mice follow by liver injury by removing the drug NTBC, which is a pharmacological inhibitor of tyrosine catabolism upstream of FAH. As a result, a liver repopulation by HSC was observed, mainly due to the growth advantage that transplanted cells had over endogenous hepatocytes. These data demonstrated the feasibility of correcting a hepatic disease by bone marrow-derived liver-repopulating cells. The mechanism underlying the apparent transdifferentiation of BM-derived cells into liver phenotype in the FAH–/– mouse was years later demonstrated to be caused by cell fusion between the bone marrow-derived transplanted cells and host liver cells. This mechanism appears to be the principal source in liver repopulation models in which there is extensive liver injury and strong selection for survival of transplanted cells [129, 130]. One important remaining question is exactly which cells fuse with the host liver; there is no evidence that it is the stem cells themselves. Instead, it seems more likely that differentiated progeny of the stem cells, such as blood cells known as macrophages, are responsible, because a contribution to the liver is seen only after the donor stem cells have populated the animal's blood system [129, 130]. This issue was clarified further and it was demonstrated that myeloid cells and myeloid progenitor cells were the major source of hepatocyte fusion partners [139, 140]. On the other hand, it was proposed that HSCs could convert into functional hepatocytes without fusion [141, 136]. In spite of continuing efforts, it is still questionable whether bone marrow-derived hepatocytes arise from stem cell "plasticity," "fusion," "transdifferentiation," or another mechanism. Nevertheless, based on the available data, it seems that more than one mechanism is at play, and the liver injury itself is an important ingredient of the response. With all the research performed in the last decades, our understanding of hepatic stem cells has had an outstanding progress. Although we can't deny the fact, that there are still many unanswered questions, nonetheless, it has been clarified that extrahepatic cell reservoirs are capable of contributing to intrahepatic regeneration.

Unfortunately, the essential mechanism to this response (i.e., "plasticity," "transdifferentiation," or "fusion") remains to be fully elucidated. This uncertainty could be elucidated by the great variety of liver injury models used in the different studies. The exploitation possibility of this extrahepatic stem cell reservoir to be translated into cell-based therapies remains to be seen.

6 In Vitro Hepatic Differentiation Potential of Stem Cell

6.1 Adult Stem Cells

Evidence has been accumulated to indicate that certain compartments of bone marrow cells are capable of differentiating into hepatocytes in vitro. All studies performed until now have tried to demonstrate and clarify the confounding issue of cell fusion or transdifferentiation.

6.1.1 Bone Marrow and Hematopoietic Stem Cells

A high level of attention has been paid to the fusion phenomenon in order to explain the plasticity of adult stem cells, however, some studies performed in vitro using bone marrow–derived cells have clearly confirmed that such a phenomenon does not occur. A bone marrow–derived subpopulation enriched for HSCs co-culture with damaged liver tissue was prevented from direct cell-cell contact by the use of transwell plates (which provide the barrier). The minced damaged liver tissue secreted substances into the culture medium that stimulated the hepatocyte differentiation from the marrow-derived cells. A truly direct differentiation potential of bone marrow cells into hepatocytes was demonstrated, ruling out the possibility of cell fusion by genotypic analysis. After just 48 h, albumin and CK18 became detectable in 2–3% of the stem cells. Several liver transcription factors and cytoplasmic proteins expressed during the differentiation of liver (αFP, GATA4, HNF4, HNF3β, HNF1α, and C/EBPα) and in mature hepatocytes (CK18, albumin, fibrinogen, transferrin) were analyzed in the hematopoietic stem cell derived cell population. The expression of all markers increased over time, with the exception of αFP, which initially increased and later decreased, indicating possible maturation. The functionally of these cells was proved by transplantation of the hepatocytes into recipient mice with liver failure induced by CCl4 [141]. Using a similar approach, a special population β2-microglobulinThy1+ cells from bone marrow was co-cultured using a transwell culture system with hepatocytes isolated from cholestatic rat livers (induced by ligation of the common bile duct) in the presence of 5% "cholestatic" serum on matrigel. The β2-microglobulin Thy1+ cells differentiated to a cell type that metabolized ammonia into urea, and expressed albumin, as well as some transcription factors [142]. On the other hand, the adult stem cells' plasticity was also demonstrated when culture of mouse bone marrow cells in the presence and absence of several growth factors showed hepatocyte phenotype. Fibroblast growth factor (bFGF) induced albumin-producing cells as well as the expression of hepatocyte markers and transcription factors such as cytokeratin 18 and albumin HNF1a, HNF3a, HNF3b, HNF4a, GATA4, and GATA6. Although the in vivo function of these cells was not proved, the plasticity process of bone marrow stem cells in this case was demonstrated to be through transdifferentiation [143]. In a similar way, isolated $CD34^+$ bone marrow cells were cultured on collagen-coated plates. After exposure to HGF, EGF, and insulin, these cells showed expression of albumin and CK19 after 28 days, whereas $CD34^-$ cells did not show liver-specific gene expression. The results demonstrated once again the hepatic differentiation potential of adult human bone marrow stem cells [144].

6.1.2 Peripheral Blood

Monocytes from peripheral blood have also been able to show hepatocyte differentiation potential [145, 146]. Zhao and colleagues [145] used a subset of human peripheral blood monocytes that display monocytic and hematopoietic stem cell markers including CD14, CD34, and CD45 to differentiate into liver cells by hepatocyte growth factor. On the other hand, Ruhnke and colleagues [146] treated monocytes with macrophage colony-stimulating factor and interleukin 3 for 6 days, followed by incubation with hepatocyte differentiation media containing FGF4. These programmable cells of monocytic origin were capable of differentiating into neohepatocytes, which closely resemble primary human hepatocytes with respect to morphology, expression of hepatocyte markers, and specific metabolic functions. After transplantation, neohepatocytes were able to integrated well into the liver tissue and showed a morphology and albumin expression similar to that of primary human hepatocytes.

6.1.3 MesenchyMal Cells

Mesenchymal stem cells (MSCs), widely studied over the past decade, are thought to be multipotent cells, present in adult marrow and other tissues that can replicate as undifferentiated cells and that have the potential to differentiate to lineages of mesenchymal tissues, including bone, cartilage, fat, tendon, muscle, and marrow stroma [147, 148]. However, the endodermal differentiation potential of bone marrow or adipose tissue MSCs has just recently been demonstrated. MSCs were isolated from human bone marrow and umbilical cord blood. These cells were serum deprived for 2 days in the presence of EGF and bFGF prior to induction with HGF, bFGF, and nicotinamide for 7 days followed by subsequent exposure to oncostatin M, dexamethasone, and ITS (mixture of insulin, transferrin, and selenium). This procedure resulted in a cell population expressing albumin, α-FP, glucose 6-phosphatase, tyrosine-aminotransferase, CK18, tryptophan 2,3-dioxygenase, and CYP2B6. In addition, cells showed albumin production, urea secretion and uptake of low-density lipoprotein [149]. Lately, using a similar approach, this potential was also confirmed [150]. Isolated MSCs were differentiated in the presence of human hepatocyte growth medium and transplanted in immunodeficient Pfp/Rag2 mice. The resultant cells demonstrated in vitro and in vivo morphological and functional characteristics of hepatocytes. Not only do MSCs from human bone marrow have the potential to differentiate into hepatocytes in vitro and in vivo, but also MSCs derived from adipose tissue [151]. MSCs obtained from of adipose tissue (AT-MSCs) were incubated with several growth factors – hepatocyte growth factor (HGF), FGF1, and FGF4. A special subfraction ($CD105^+$) of these mesenchymal cells exhibited high hepatic differentiation ability in an adherent monoculture condition. The cells revealed several liver-specific markers and functions, such as albumin production, low-density lipoprotein uptake, and ammonia detoxification, and they had the ability to incorporate into the liver parenchyma [151]. The ability to isolate, expand the culture, and direct the differentiation of hMSCs in vitro to particular lineages provides the opportunity to study events associated with hepatocyte commitment and differentiation. It could be used as new therapeutic approaches for the restoration of damaged or diseased liver.

6.1.4 Multipotent Adult Progenitor Cells

Bone marrow–derived stem cells with extensive in vitro expansion ability, termed multipotent adult progenitor cells (MAPCs), have been isolated from mice, rats, and humans [152] and have the capacity to differentiate into cells representing all three germinal layers: ectoderm, mesoderm, and endoderm. By culturing in FGF4 and HGF, MAPCs appear to differentiate into hepatocyte-like cells, which express CK19, AFP, CK18, HepPar-1, and CD26, and produce albumin, urea, and glycogen [134]. Thus, MAPCs, like ES-derived cells, may have potential to develop into a wide spectrum of transplantable cells that could be used to treat a variety of degenerative and inherited diseases. Unlike ES-derived progeny, MAPCs do not develop tumors. Their potential to correct liver disease, however, has not been demonstrated in any animal model.

6.2 Embryonic Stem Cells

Embryonic stem (ES) cells have enormous potential as a source for cell replacement therapies, drug development industry, and as a model for early human development. In general, ES cells have been defined as cell that are self-renewing and pluripotent that can be isolated from the inner cell mass of the blastocyst, proliferate extensively in vitro, differentiate into derivatives of all three germ layers, express a number of characteristic markers like Oct4, SSEA-4, TRA-1-60, and TRA-1-81, and show high levels of telomerase activity [153]. The ability to induce specific differentiation has been demonstrated by the formation of aggregates to form spheroid clumps of cells called embryoid bodies (EBs) leading to a spontaneous differentiation and the production of cells from the three germ layers: ectoderm, mesoderm, and endoderm [154]. In addition, the direct addition of various growth factors to differentiating ES cells, followed by analysis of cell morphology and specific markers expression has been demonstrated. Specific protocols have been developed in order to enrich various cell types during the differentiation of ES cells. The production of neuronal [155], cardiac [156], hematopoietic [157], endothelial [158], pancreatic [159], and hepatic cells [11, 13, 160–170] has been documented.

In general, we can divide the methods of differentiation into spontaneous and directed differentiation. In the protocol of spontaneous differentiation, the cells are grown as EBs for a few days and then usually they are plated on an adherent matrix as a monolayer, either as dissociated cells or as clumps of cells [165, 166, 169, 170]. Direct differentiation has been usually induced by the addition to the culture medium of various kinds of growth factors, cytokines, and extracellular matrices to induce specific gene expression, morphology and most importantly function in the differentiating cells to produce the desired hepatic phenotype [161, 163, 167]. Recently, combinations of both methods, EBs induction in combination with direct application of growth factors and co-cultures have been reported [13, 160, 168].

6.2.1 Endoderm Induction from ES Cells

Since ES cells are derived from and display gene expression and properties characteristic of pluripotent embryonic cells, it is generally believed that directed differentiation of ES cells into specific cell types for therapeutic purposes will necessarily begin by inducing ES cells to form germ layer intermediates. ES cells are in fact a group of undifferentiated cells localized in the epiblast. The epiblast-derived cells give rise to the three principal germ layers and their terminally differentiated tissues through a process called gastrulation [26, 171]. From the three germ cell layers, endoderm is the one that gives rise to hepatic, pancreatic, lung, intestinal, and other therapeutically relevant cell types, yet early endoderm development is not well understood.

The initiation of gastrulation is recognized by the formation of the primitive streak (PS) at the posterior part of the epiblast. Heterotopic transplantation studies have demonstrated that by mid-to-late gastrulation, cells are determined to give rise to endoderm [172]. Several early endodermal transcription factors, including Otx2, Hesx1, Hex, and Cdx2, are regionally expressed prior to the time that organ specific genes are activated [26]. Within the PS, the cells of the mesendoderm regulate the expression of several genes important for the cell-fate differentiation of the definitive endoderm and mesoderm progenitors. Among them, goosecoid (GSC) forkhead box A2, (Foxa 2), chemokine C-X-C motif receptor 4 cxcr4, sex determining region-Y box 17 (Sox17a/b), brachyury, E-cadherin, vascular endothelial growth factor receptor-2, (VEGFR2), VE-cadherin, platelet-derived growth factor receptor-a (PDGFRa), and GATA binding protein 4, (GATA-4) [26, 173]. In addition, extraembryonic endoderm arises at the blastocyst stage and eventually forms two subpopulations: visceral endoderm, the main metabolic component of the visceral yolk sac, and parietal endoderm, which secretes Reichert's membrane and contributes to the transient parietal yolk sac. Extraembryonic endoderm cells share the expression of many genes with definitive endoderm, including the often-analyzed transcription factors Sox17, FoxA1, and FoxA2 [160, 174]. Thus, a better understanding of the genetic pathways that regulate cell fate determination of extraembryonic endoderm, as well as genes that can serve as markers to distinguish definitive and extraembryonic endoderm is needed.

Recent advances have been reported in the gene expression pattern of definitive endoderm and its following enrichment using the markers EpCAM (+), CD38 (–) and Dpp4 (–) [175]. However, a totally pure definitive endoderm population has not been yet purified from ES cells. Several advances have been made in deriving endoderm from ES cells. However, a better understanding of definitive and extraembryonic endoderm is necessary for the field to progress. Recently, several groups have reported differentiation of mouse or human ES cells into definitive endoderm by the combination of Activin A and TGFβ family member, which also binds the same receptors as Nodal (with the exception of the coreceptor Cripto) and a low serum concentration with relative high efficiency. Their analyses clearly indicate the possibility of modulating in vitro a direct differentiation of the ES cells into definitive endoderm derivatives [159, 173, 174, 176]. The common transcriptional machinery in definitive and visceral endoderm implies a similarity in the mechanism of specification of the two tissues. Thus, it is tempting to consider that common signaling events induce Sox17 and the FoxA genes. Therefore, these signaling events confer "endoderm identity." Moreover, selective induction of definitive endoderm from ES cells may require inhibition of visceral endoderm. Thus, factors promoting endoderm formation such as those of the Nodal family [177, 178] should be combined with factors that inhibit induction of the extraembryonic endoderm cassette to specifically induce definitive endoderm. Most of such extraembryonic endoderm-promoting factors are yet unknown, although it has recently suggested an involvement of the FGF signaling pathway [13, 179]. However, a recent work suggested that two conditions are required to induce approximately 70–80% of definitive endoderm from human ES cells: signaling by Activin/Nodal family members and release from inhibitory signals generated by PI3K through insulin/IGF [180].

6.2.2 Hepatic Induction

Growth factor signaling from the cardiac mesoderm and septum transversum mesenchyme specifies the underlying endoderm to adopt a hepatic fate such that by the 6–7 somite stage hepatic gene expression can be detected in the ventral foregut endoderm [28]. Concurrent with these events, the most distal region of the foregut endoderm starts to express pancreatic genes [28, 36]. The growth factors identified were fibroblast growth factors (FGFs) and bone morphogenic proteins (BMPs). Using tissue explants assay, it was demonstrated that FGFs, acid or basic, could substitute for cardiac mesoderm in inducing ventral endoderm to elicit a hepatogenic response [28]. Concomitantly, the same group showed that BMPs secreted from septum transversum mesenchyme are needed in concert with cardiac-derived FGFs to induce the ventral endoderm to adopt a hepatic fate. Usually, the most abundant factors that were used to induce hepatic differentiation are acid FGF, basic FGF, FGF4, BMP4, and BMP2 [12, 13, 161]. Co-cultures of chick cardiac mesoderm were shown recently to induce hepatic differentiation in mouse ES cells, nicely illustrating how embryonic principles help control stem cell differentiation in vitro [168].

Recently, some reports have proven the importance of FGFs and BMPs on mouse ES cells differentiation toward hepatic phenotype. Furthermore, interactions with endothelial cells, a mesodermal derivative in this inductive sequence, are crucial for this early budding phase in hepatic induction [181, 182]. However, the relevant endothelial-signaling molecule is not known. It is not clear how endothelial cells before blood vessel formation appear near the newly specified hepatic endoderm. A recent report of our group used co-culture of ES cells and a combination of liver nonparenchymal cells, including a liver endothelial cell, and it was shown that the interaction of ES cells with the endothelial cell line step up hepatic differentiation [160].

Extracellular matrix plays a key role in the process of differentiation [183]. Generally, collagen or matrigel was chosen as the matrix for growing the cells since the liver bud proliferates and migrates into the septum transversum mesenchyme, which is composed of loose connective tissue containing collagen [160, 161, 163].

Several transcription factors have been identified and proposed as targets of FGFs and BMPs signaling in early hepatic onset. Foxa and Gata genes have been shown by genetic analysis to regulate the competence of foregut endodermal cells to respond to hepatic inductive signals [38, 184]. Gata4 was the first Gata factor to be implicated in development of the ventral foregut. The ability of Gata4, in conjunction with FoxA2, to reposition nucleosomes around this enhancer has lead to the hypothesis that Gata4 potentiates hepatic gene expression [184]. In the endoderm, the onset of Foxa gene expression precedes the induction of the hepatic program by FGF signals. Furthermore, Foxa proteins are able to displace nucleosomes present in the regulatory region of the albumin gene before the gene becomes activated, but other transcription factors that bind to this region are unable to do so [184, 185]. Foxa2 binding can reverse chromatin-mediated repression of alpha-fetoprotein (Afp) gene transcription in vitro [186]. In summary, Foxa1 and Foxa2 are essential for hepatic specification. Foxa proteins function as "pioneer" proteins to open compacted chromatin in regulatory regions of liver-specific genes [37].

HNF4α contributes to regulation of a large fraction of the liver and pancreatic islet transcriptomes by binding directly to almost half of the actively transcribed genes [187]. Moreover, recent observations provided a refined molecular specific hepatic fate characterization, where the transcription factor hepatocyte nuclear factor-6 (HNF6) has a critical role in the proper morphogenesis of both the intra and extrahepatic biliary tree. Furthermore, it would appear that the mechanism by which HNF6 regulates biliary tree development involves the related transcription factor hepatocyte nuclear factor-1 (HNF1) [188].

There is evidence also that there is an uncovered and unexpected relationship between extrahepatic bile duct morphogenesis and pancreas development [189]. Most of the protocols of hepatic differentiation from ES cells used a constitutive expression of a hepatic transcription factor in order to direct and confirm the differentiation towards the endoderm and hepatic lineages.

Some of the nuclear factors that have been used to follow up definitive endoderm and hepatic differentiation are the following. For definitive endoderm development: forkhead box (FOX) transcription factors, FOXA1, FOXA2, goosecoid (GSC), c-Kit and chemokine C-X-C motif receptor 4 (cxcr4), sex determining region-Y box 17 (Sox17a/b), brachyury, E-cadherin, vascular endothelial growth factor receptor-2 (VEGFR2), VE-cadherin, platelet-derived growth factor receptor-a (PDGFRa), and GATA binding protein 4 (GATA-4). For hepatic development: alpha-fetoprotein, albumin (alb), HNF4, HNF6, tryptophan-2,3-dioxygenase (TDO), tyrosine aminotransferase (TAT), the cytochrome P450 enzymes (Cyp7a1, Cyp3a11, Cyp3a4, Cyp3a1), and lucose 6-phosphatase (G6Pase) [12, 13, 174, 176] (Fig. 2).

6.2.3 Hepatic Specification

Hepatocytes and bile duct cells originate from a common precursor, hepatoblasts [190]. Notch signaling promotes hepatoblast differentiation into the biliary epithelial lineage, and HGF does the opposite [45]. Thus, the expression of the Notch intracellular domain in hepatoblasts inhibits their differentiation into hepatocytes. Supporting the idea of HGF as a promoter for the hepatic fate decision, a study found that HGF induces the expression of C/EBPα in albumin-negative fetal liver cells [191]. When C/EBPα activity is blocked through expression of a dominant negative form of C/EBPα, there is no transition of alb– to the alb+ stage. HGF promoted differentiation of alb+ cells from alb– precursors, but inhibited further differentiation of the alb+ cells into biliary cells, suggesting that HGF promotes the establishment of a bipotent state of the hepatoblasts.

Wnt signaling is also involved in regulating biliary epithelial cell fate. The addition of Wnt3A in ex vivo fetal liver culture experiments supports biliary epithelial cell differentiation [192]. Furthermore, the inhibition of β-catenin prevents hepatoblasts from expressing biliary markers [193]. One important factor for activating this differentiation program is HNF6. HNF6 is expressed in hepatoblasts, in the gallbladder primordium, and in biliary epithelial cells of the developing intrahepatic bile ducts [188]. HNF6 knockout mice developed no gallbladder, and the development of the intrahe-patic and extrahepatic bile ducts was abnormal. The intrahe-patic bile ducts had a similar phenotype in conditional HNF1β knockout mice [194]. These results suggest that the effect of HNF1β is downstream of HNF6. Thus, TGFβ, HGF, C/EBPα, and HNF6, in combination with Notch pathway, integrate into a coherent network that controls bipotency and allows further biliary or hepatocytic differentiation (Fig. 2).

6.2.4 Hepatic Maturation

The third important step after the induction of the hepatic fate in endoderm cells and the differentiation into hepatoblasts is the proliferation of these cells. And the mesenchymal

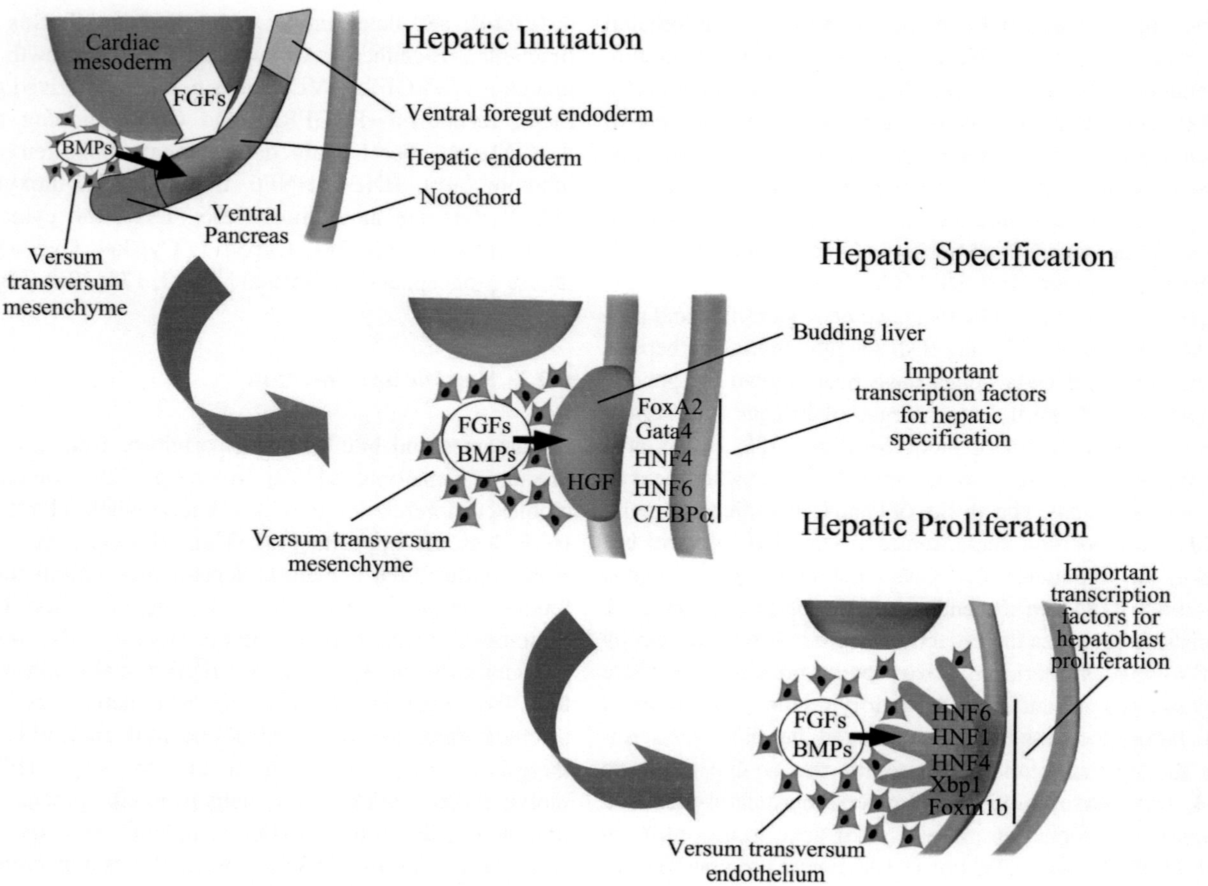

Fig. 2 Signaling that induces hepatic genes in the endoderm. Factors and transcription factors influencing differentiation of the endoderm into liver are shown. Also shown are the location of the cardiac mesoderm and prospective septum transversum mesenchyme cells ("mesenchyme"), both of which signal to the endoderm during this period to promote hepatic induction

component of the liver, derived from the septum transversum mesenchyme, is essential for proliferation of hepatoblasts [195]. Other essential interactions for liver bud growth are the endothelial cells. The requirement for endothelial cells for hepatic endoderm growth could be recapitulated with embryo tissue explants, showing that the effect is independent of oxygen and factors in the bloodstream. These important interactions between endothelial and liver cells appear to persist in the adult liver [196].

HGF is a signaling pathway that controls the proliferation of the fetal liver cells. Genetic studies in mouse embryos showed that the proliferation and outgrowth of the liver bud cells require the interaction of HGF [197, 198] with its receptor, c-met [199]. Either knockout of HGF or c-met showed similar phenotypes and failed to complete the developmental process and die in utero between E 13.5 and 16.5 with multiple abnormalities, including signs of underdeveloped liver. Interestingly, during regeneration of the adult liver, this pathway is important for the proliferation of the hepatocytes, since conditional c-met knockout mice show an inhibition in the proliferation after liver injury where c-met affects primarily hepatocyte survival and tissue remodeling [200]. This is a good example in which pathways for the development of an organism function in a similar way in the adult.

Other transcription factors have been involved in hepatoblast proliferation. The transcription factors Foxm1b and Xbp1 are also required for the liver bud cell proliferation. Foxm1b knockout mice die in utero by E 18.5. The fetal liver shows a 75% reduction in the number of hepatoblasts [70]. Additionally, these animals do not develop intrahepatic bile ducts. Thus, Forkhead Box m1 transcription factor seems to be critical for the differentiation toward the biliary epithelial cell lineage. The Xbp1 knockout mice also show hypoplastic livers and death caused by reduced hematopoiesis, with a reduced growth rate and increased apoptosis of hepatocytes. This is evidence for a link between hematopoiesis and liver development [201].

Researchers have found that Wnt/β-catenin pathway activation plays a role in fetal liver cell proliferation and maturation. Where inhibition of Wnt signaling results in reduced cell proliferation [193].

Hematopoiesis plays an important role in hepatic maturation. After the liver bud emerges from the gut tube, hematopoietic cells migrate there and propagate. The hematopoietic cells secrete oncostatin M (OSM), a growth factor belonging to the interleukin-6 (IL-6) family that includes IL-6, IL-11, leukemia inhibitory factor (LIF), ciliary neurotrophic factor, and cardiotrophin-1. These cytokines often exhibit similar functions since their receptors utilize gp130 as a common signal transducer and the surrounding liver cells express the gp130 receptor subunit OSMR [202]. OSM stimulates the expression of hepatic differentiation markers and induces morphologic changes and multiple liver-specific functions as ammonia clearance, lipid synthesis, glycogen synthesis, detoxification, and cell adhesion. However, OSM also possesses unique functions, for example, growth stimulation of endothelial cells [203] and smooth muscle cells [204]. Oncostatin M not only induces hepatic differentiation but also suppresses fetal liver hematopoiesis. Hepatic cells from E 8.5 support the expansion of hematopoietic stem cells and give rise to myeloid, lymphoid, and erythroid lineages. The addition of OSM and glucocorticoid strongly suppresses this process. In contrast, hepatic cells from E 14.5 no longer support hematopoiesis in co-cultures. However, the hematopoietic cells induce further differentiation of hepatoblasts, and in consequence, the liver stops supporting local hematopoiesis and induces the hematopoietic stem cell switch to the bone marrow [205]. In addition, OSM can induce the down-regulation of cyclins D1 and D2 [206]. This down-regulation is mediated by Stat3, which is activated through OSM and OSM receptor complex interaction. These cyclins are important for the initiation of the cell cycle and therefore for cell proliferation and they are normally down-regulated during liver development.

Glucocorticoids have been also involved in hepatic maturation and found that modulate proliferation and function of adult hepatocytes both in vivo and in vitro. In the fetal liver, physiological concentrations of dexamethasone (Dex), a synthetic glucocorticoid, suppress AFP production and DNA synthesis and up-regulate albumin production [207]. TAT mRNA, which is virtually absent in the early fetal liver, is induced by Dex in primary hepatocytes of late embryonic stage. In contrast, Dex does not regulate TAT levels at earlier stages (mid-gestation: E 12–14), even though these cells are able to express albumin in response to Dex [208].

Recent studies have shown that embryonic stem cells can efficiently be differentiated into hepatocyte-like cells. The transcription factors previously mentioned, their receptors, and the substances related with their stimulation in either way might take part in hepatic maturation and specification from ES cells. A current work reported the co-culture of adult isolated hepatocytes with fetal isolated hepatocytes resulting in the increase of liver-specific gene expression and elevated hepatic functions. Even the effects of the co-culture system were reversible for the fetal hepatocytes; adult hepatocytes provided an appropriate atmosphere for the hepatic maturation of fetal hepatocytes [209]. Inspired by the molecular basis of liver development and regeneration, an elegant work explored the co-culture of liver tissue with hematopoietic stem cells induced hepatic differentiation and maturation [141]. Thus, hepatocytes and liver nonparenchymal cells seem to play a role in liver maturation of stem cells.

Recently, we have combined efforts to generate functional hepatocytes from mouse ES cells. The differentiation protocol is simple, uses defined reagents, and yields to date the most efficiently differentiated hepatocyte-like cells. Starting with a suspension culture system, where early endodermal development is initiated, ES cells are subsequently transferred to plates and cultured in the presence of FGF-2 and Activin A. The predifferentiated cells are then further developed toward hepatocytes in a defined co-culture together with human nonparenchymal liver cells (endothelial cell line, cholangyocyte cell line, and stellate cell line) under the influence of hepatocyte growth factor, dimethyl sulfoxide, and dexamethasone. The improvement of hepatic maturation was observed when co-culture with liver nonparenchymal cell lines was applied. Several cytokines and growth factors were identified in the conditioned medium of the cell lines. Those substances play a key role in liver regeneration [13, 160]. Yet, many questions remain to be further examined before such a protocol can be successfully applied to human ES cells. One of the most particular aspects of ES cell differentiation is whether the cells are homogeneously and specifically differentiated in the desired way.

6.2.5 Current Status of Human ES Cell Differentiation into Hepatocyte-Like Cells

ES cells hold great promise as a source of new hepatocytes, but this potential has proven to be more difficult than expected. Beginning in 2000, papers began to appear [165, 167, 210, 211–213]. Progeny of mouse and human ES cells were reported to express and secrete albumin and even to have CYP enzimes activity. Several approaches have been used to differentiate and to obtain enriched populations. Human hepatic-like cells were isolated and characterized for their phenotype. Through gene manipulation, albumin promoter was used to select the cells, hepatic cells were labeled, and a relative homogenous population of differentiated cell types was demonstrated [165]. However, the cells expressing hepatic phenotype were isolated from EBs, and thus, a considerably low cell number was produced and functionality of the cells was not tested. In one of the few reports on human ES cells, the combination of insulin, DEX, and collagen type I followed by sodium butyrate, led to increased numbers of mature hepatic gene-expressing cells (10–15%). The resultant cells had morphological features similar to that of primary hepatocytes and most of the cells expressed liver-associated proteins [167].

Great care must be taken in defining how closely such cells resemble normal hepatocytes, which are very well characterized in regard to their gene expression, metabolism, growth potential, and secretory functions [214]. We might believe that hepatocyte-like cells derived from stem cells designated for therapeutic replacement must match the extraordinary performance of normal hepatocytes, with their ability to store glycogen, secrete albumin, metabolize drugs, and the other more than 500 different functions performed by the liver [215]. However, it must be recognized that there are developmental pathways found in tissues such fetal pancreas, fetal intestine, and other endodermal derivatives that generate cells that in certain stages of development might express similar hepatic gene pattern and that will never become a true hepatocyte. The lack of success of these early attempts at differentiating human ES cells into functional hepatocytes has focused attention on the fundamentals of normal embryonic development to better understand the early stages of definitive endoderm formation. The difficulty of this approach is the need to produce a directed homogeneous population of definitive endoderm. A recent important contribution is a protocol in which the use of activin A in combination of serum free conditions, resulted in enrichment of definitive endoderm (up to 80%) from human ES cells [176]. Using a modification of this protocol, and a combination of protocols previously reported using mouse ES cells, Cai et al. [161] reported that the addition of FGF, BMP, and HGF can induce the hepatic fate, and the later addition of OSM and Dex to the cell culture induced even more differentiated hepatocyte-like cells in a total time of 18 days. This, in part, is a recapitulation of the events during development. This study also showed that the transplantation of the differentiated cells into mice with drug-induced liver failure incorporated a limited quantity of cells into the liver parenchyma [161]. Transplantation of the differentiated cells is a very important control to confirm that hepatocyte-like cells can function as hepatocytes in vivo. However, adequate animal model of liver failure should be use to evaluate the real functional integration of the resulting hepatocyte-like cells (Fig. 3).

6.2.6 Prospects

The hepatocyte-like cells generated from human ES cells represent an attractive source for the treatment of several hepatic diseases. However, the efficient differentiation of human ES cells into a mature hepatocyte still remains a significant challenge. The differentiation of hepatocyte-like cells from human ES cells has proven to produce a more developmentally heterogeneous population than expected. There is a need to identify reliable markers for hepatocytes and yolk sac tissues. In mice, for example, it was shown that CYP7A1 is expressed in the liver and not expressed in the yolk sac

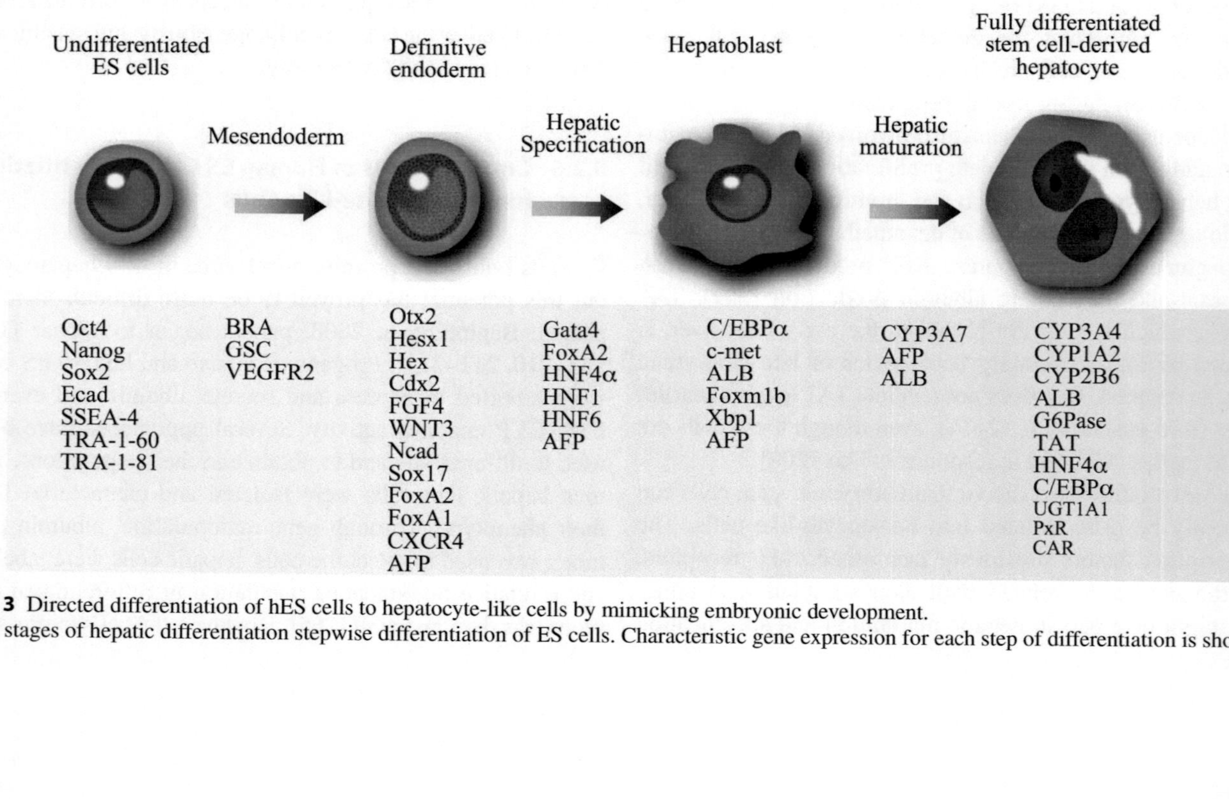

Fig. 3 Directed differentiation of hES cells to hepatocyte-like cells by mimicking embryonic development. Key stages of hepatic differentiation stepwise differentiation of ES cells. Characteristic gene expression for each step of differentiation is shown

tissue, and thus it can be a good marker for hepatocytes [216]. Embryonic, fetal, and adult hepatocytes are different by means of their gene expression and functional activities. Fetal hepatocytes transplanted into the liver cannot replace completely the functional activities of adult hepatocyte since they represent a different developmental stage. Genes such as albumin or AFP are first expressed in early embryos and further on fetal hepatocytes. In the case of AFP, it is expressed very early in embryonic development and later on in the fetus but is turned off. A hepatocyte that had stopped to express AFP can be considered an adult hepatocyte. Thus, their expression cannot tell the state of the differentiated cell unless both markers can be scrutinized. One more hurdle that needs to be overcome is the isolation of pure hepatic cell fractions using systems with clinical significance. To solve that, membrane markers of mature hepatocytes have to be detected. One option is the asialoglycoprotein receptor (ASGPR), which is almost exclusively expressed in hepatocytes [217]. This means cell magnetic sorting would be a consistent purification system. In the future, studies of embryonic stem cells to hepatocyte differentiation can increase our understanding of the molecular basis of liver development. The exact understanding of these developmental processes that lead to a specific cell fate might help us to recapitulate the events in vitro and engineer artificial liver cells and tissue to combat liver diseases. Expectations and hopes are very high, but the difficulty of these approaches remains a challenge. However, with the extraordinary potential of modern science one must remain hopeful that clinical advances will come sooner rather than later.

7 Identifying a Hepatocyte

To date, various ways to cause ES cells to differentiate into hepatic-like cells have been reported. The availability of a homogeneous source of human hepatocytes is considered the most precious tool for toxicity screening. In addition to hepatotoxicity, hES-derived hepatocytes would provide a renewable, cell-based assay to examine other key factors of compound attrition such as the metabolism of xenobiotics by CYP450 enzymes, drug–drug interactions, system for studying hepatic metabolism of xenobiotics, hepatotoxicity, and the activity of drug transporters as well as regenerative medicine. This opens exciting new possibilities for pharmacology and toxicology, as well as for cell therapy. However, the nature of the "hepatocyte-like cells"should be analyzed very carefully under several constrictions and a clear definition of the term hepatocyte has to be implemented.

The expression of hepatocyte markers, such as afp, alb, or CK18, as well as induction of an epithelial phenotype and inducible CYP450 has been reported in several works. Therefore, it is understandable that many scientists gave their stem cell–derived cell types terms such as hepatocyte. Properties such as epithelial morphology and expression of some hepatocyte markers are necessary but not sufficient to consider a cell as a hepatocyte. Albumin expression and CYP450 are an example of this. In fact, hepatocytes are the only cell type that secretes albumin. However, the conclusion that any albumin-secreting or expressing cell necessarily represents a hepatocyte is still premature. For example, it is possible that stem cell–derived cell types express albumin together with a limited number of hepatocyte markers, but this does not mean that they also express the necessary set of hundreds of genes that make up a true hepatocyte. Moreover, CYP enzymes are not exclusively limited to hepatocytes. Indeed, CYP induction was also reported for lung, colon, and small intestine epithelial cells, white adipose tissue, and several other cell types [218–220]. Therefore, it is possible to unequivocally define whether a candidate cell is a hepatocyte or not.

Thus, the definition of hepatocyte should include qualitative studies where the presence/absence of hepatocyte markers are demonstrated together with an enzymatic activity evaluation. We have now moved past the phase of simply detecting the expression of hepatic genes in stem cells and now must find significant measures to judge if stem cell–derived hepatocytes have truly mature hepatic function. It seems reasonable to introduce additional criteria to define whether a cell is a true hepatocyte or only shares several characteristics with a hepatocyte. It is also important to understand what properties a hepatocyte or the substitute hepatocyte-like cell should have. We need to establish a list of mature hepatic functions, which are easily measured. The resulting hepatocyte-like cells should be compared with human fetal and mature human liver and we should define endpoints to measure in stem cell–derived hepatocytes the level of hepatic maturation. Finally, we need fast and easy tests that provide relevant and robust information of the hepatic capacities of the produced stem cell–derived hepatocytes.

From a functional point of view, any candidate hepatocyte-like cell type should represent a minimal set of hepatic functions of a true hepatocyte. Here, we present a battery of relevant studies for the analysis of enzyme activities of stem cell-derived hepatocytes: (a) analysis of expression of genes identified in mature livers; (b) metabolism of xenobiotics and endogenous substances (hormones and ammonia); (c) synthesis and secretion of albumin, clotting factors, complement, transporter proteins, bile, lipids, and lipoproteins; and (d) storage of glucose (glycogen), fat soluble vitamins A, D, E, and K, folate, vitamin B12, copper, and iron.

Finally, a convincing in vivo experiment to prove hepatocellular differentiation is to restore liver function in animal models by means of repopulation assays. However, any repopulation experiment may only evaluate that a certain

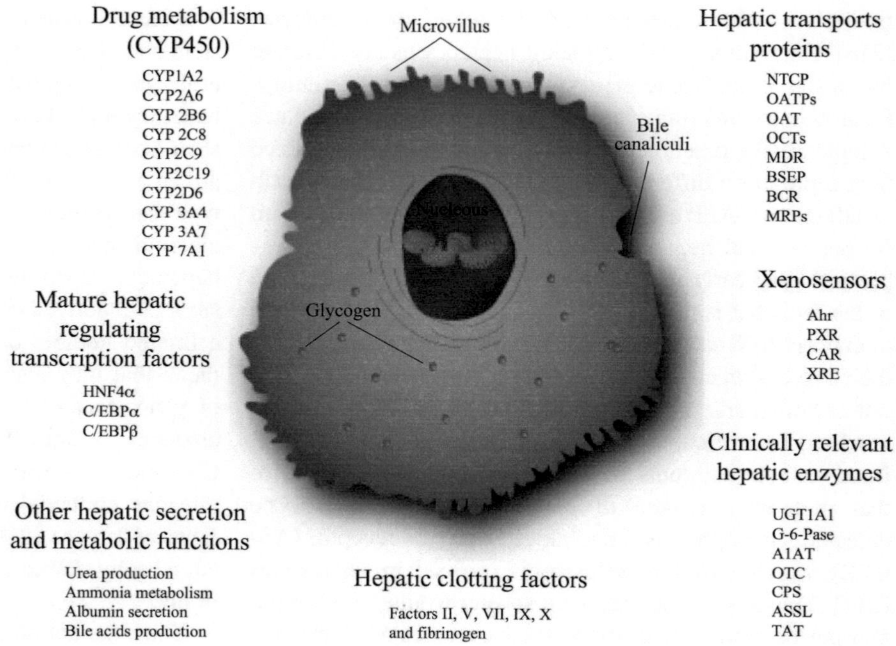

Fig. 4 Distinguishing features of mature hepatocytes Hepatocytes are the chief functional cells of the liver. These cells are involved in protein synthesis, protein storage, transformation of carbohydrates, synthesis of cholesterol, bile salts and phospholipids, and detoxification, modification, and excretion of exogenous and endogenous substances. Roughly 80% of the mass of the liver is contributed by hepatocytes

hepatic cell type has the capacity to generate hepatocytes in vivo. Thus, testing a defined battery of activities and comparing them with primary hepatocytes remains the only feasible option for evaluating the in vitro potential of stem cell–derived hepatocyte cultures as appropriate surrogates for primary human hepatocytes (Fig. 4).

7.1 Hepatocyte Drug Metabolism

The entire hepatic drug-metabolizing enzyme system in an integrated form provides an in vitro model that is a very useful tool for anticipating drug metabolism and drug hepatotoxicity in humans. CYPs are mixed function monooxygenases and the major enzymes in phase I metabolism of xenobiotics. Depending on the nature of the xenobiotic, this oxidative metabolism results in inactivation and facilitated elimination, activation of pro-drugs, or metabolic activation [221]. Evaluation of CYP for specific measurements in stem cell–derived hepatocytes classified as phase 1 metabolism may include CYP1A2, CYP2A6, CYP 2B6, CYP 2C8, CYP2C9, CYP2C19, CYP2D6, CYP 3A4, CYP 3A7, and CYP 7A1. The major site of CYP expression is the liver and CYP3A4 is the most abundant CYP isoenzyme in human adult liver. The enzymes of greatest importance for drug metabolism belong to the families 1–3, responsible for 70–80% of all phase I–dependent metabolism of clinically used drugs [222].

Studies performed in primary human hepatocytes point to CYP 3A4 as an important marker for hepatocytes, as this enzyme is the most abundant CYP enzyme in the human liver. CYP3A4 activity can be measured using 6-beta hydroxytestosterone [223]. It has been reported to be quantitative, sensitive, and specific test for 3A4.

CYP expression and activity present large interindividual variations due to polymorphisms [224]. Moreover, CYPs can be induced several fold or inhibited by specific drugs, resulting in additional, although transient, variability of metabolic activity. Inducibility of CYPS is a mature liver function that must also be observed in useful stem cell–derived hepatocytes. CYPs are inducible by exposure to phenobarbital, rifampicin and to a lesser extent steroid hormones [225].

The CYP2C family also represents a significant proportion of total P450s (2C9, 2C8, 2C19, and 2C18, representing about 20% of the total P450 [226]) and metabolizes many drugs [227], thus making this enzyme subfamily important to monitor. CYP1A2 is a minor enzyme in the liver and only a small number of drugs (4%) are metabolized by this enzyme [227]. However, it is involved in the bioactivation of pro-carcinogens and is therefore considered to be an important enzyme to test [228]. CYP2B6 is emerging as an important enzyme in drug–drug interactions despite a previously reported low abundance in the liver (0.2% of total P450 [226]. Once thought to be of minor importance and uninducible in humans [229], CYP2B6 may actually constitute at least 5% of the total P450, contribute to the metabolism of more than 25% of all pharmaceutical drug metabolism [230] and exhibit high inducibility [231]. CYP2D6 has no known inducer and represents only 2% of the total P450 [226]. However, researchers persist in testing this enzyme because of its major contribution to drug metabolism and its polymorphism [227]. CYP3A7 is mainly expressed in fetal liver even at mid gestation although in rare cases CYP3A7 mRNA has been detected in adults. CYP3A7 activity can be induced by hydroxy progesterone caproate metabolism. The CYP3A forms have demonstrated an equal or reduced metabolic capability for CYP3A5 compared with CYP3A4 and a significantly lower capability for CYP3A7. Thus, active metabolism can be detected for both CYPs 3A7/3A4 [232]. CYP7A1 Cholesterol 7a-hydroxylase is found exclusively in the liver, where it catalyses the first step in the major pathway responsible for the synthesis of bile acids [233]. The expression of this enzyme is subject to feedback regulation by sterols and is thought to be coordinately regulated with enzymes in the cholesterol supply pathways, including the low-density lipoprotein receptor and 3-hydroxy-3-methylglutaryl-coenzyme A reductase and synthase [233].

In summary, we can conclude that gene analysis of the above mentioned CYPs accompanied by induction tests could represent a robust background to follow the maturation of stem cell-derived hepatocytes. In addition, some are expressed at low levels in early development and their increased expression coincides with maturation of hepatic development/function.

7.2 Hepatic Transporters

Hepatic transport proteins and measurement of bile acids can serve as indicators of hepatic maturity as well. However, all hepatic functions do not mature at the same time rate and some hepatic transporters are expressed early in the development and may not be exclusive for liver. There is evidence that they are expressed in the intestine, kidney, brain, and other organs [234]. Some important hepatic transport proteins can be classified as follows: (a) The solute carrier SLC family, comprising among others such as Na+-taurocholate co-transporting polypeptides (NTCP), organic anion-transporting polypeptides (OATPs), organic anion transporters (OATs), organic cation transporters (OCTs). (b) The ATP-binding cassette (ABC) transporter family, including the multidrug resistance proteins (MDR) and bile salt export pump (BSEP, both belonging to the ABCB family), and breast cancer resistance protein (BCRP, belonging to the ABCG or White family). (c) The multidrug resistance associated proteins (MRP), belonging to the ABCC family. MRP1 and MRP2 are involved in biliary excretion of a large variety of structurally unrelated compounds, among others bulky hydrophobic cationic compounds, but also steroid hormones. MRP2 excretes mainly anionic conjugates, among others bilirubin glucuronides, leukotriene C4, and glutathione [225]. The basolateral NTCP transports bile acids from the space of Disse into hepatocytes, and human NTCP accepts most physiological bile acids while, at the canalicular membrane, the efflux of bile acids is by BSEP-mediated concentrative excretion [235]. Sensors like the aryl hydrocarbon receptor (Ahr), pregnane X receptor (PXR), and the constitutive androstane receptor (CAR) are integral to the regulation and induction of the main P450s [229] and their analysis may provide a strong evidence of the maturation state of stem cell-derived hepatocytes due to their up-regulation during liver development. These receptors control the expression of CYP1A (Ahr), CYP2, and CYP3A (PXR and CAR) families. Once activated, the receptors form heterodimers with other factors, such as Arnt (Ahr nuclear translocator) and retinoid X receptor (RXR for both PXR and CAR) and then bind to the target xenobiotic response elements (XRE) located in both the proximal and distal P450 gene promoters, resulting in the transcription of the respective CYP isoform [229].

7.3 Hepatic Transcription Factors, Homeostasis, and Clinically Relevant Hepatic Enzymes

The demonstration of expression of transcription factors regulating hepatic development and maturation is useful (HNF4a, C/EBPα, C/EBPβ). However, they may not be as useful as the CYPs for measuring maturation because they are expressed at near adult liver levels even at mid-gestation. In a very elegant study, Odom et al. have demonstrated the importance of HNF4α for gene regulation in hepatocytes. Microarray data suggest that HNF1α binds to 222 target genes in human hepatocytes, corresponding to 1.6% of the genes assayed. HNF6 bound to 227 (1.7%), and HNF4α bound to 1575 (12%) of the genes, which means that HNF4α bound to nearly half of the active genes in the liver that were tested. In addition, most of the genes bound by HNF1α or HNF6 were also bound by HNF4α, but only a few genes were bound by both HNF1α and HNF6 [187]. The differentiated state of the hepatocytes is regulated by a coordinated interplay of hepatocyte-specific transcriptional factors, including HNF-4 and C/EBPα [236]. HNF-4 is involved in hepatocytespecific expression of serum proteins, such as albumin and transferrin, and of CYP450 proteins. In primary cultures of rat hepatocytes, the expression of C/EBPα is rapidly reduced within a few days of culture, resulting in reduced hepatic functions. Michalopoulos et al. [214] demonstrated that the maintenance of C/EBPα, HNF-4, Nuclear factor-κB (NF-B), and activator protein-1 (AP-1) contributed to the prolonged expression of liver-specific proteins in human hepatocyte cultures. Additionally, analysis of some hepatic clotting factors

(II, V, VII, IX, X, and fibrinogen), albumin production, urea production, or ammonia metabolism and glycogen storage may provide additional robust evidence of an effective hepatic maturation of stem cell-derived hepatocytes.

The presence of hepatic enzymes with clinical implications would be useful in the process of hepatic maturation categorization, for example, UDP-glucuronosyltransferase (UGT1A1), an enzyme of the glucuronidation pathway that transforms small lipophilic molecules, such as steroids, bilirubin, hormones, and drugs, into water-soluble, excretable metabolites [237]. Another important enzyme that is present in mature hepatocytes is glucose-6-phosphatase (G-6-Pase)1, which catalyzes the hydrolysis of G-6-Pase to glucose, and is the terminal step of both hepatic gluconeogenesis and glycogen breakdown [237]. Alpha-1-antitrypsin (A1AT) is another example of a clinically relevant enzyme that is present in a mature stage of hepatocytes. As a member of the serpin superfamily of proteins, A1PI is a potent inhibitor of serine proteases, especially neutrophil elastase, which degrades connective tissue in the lung [238]. The A1AT gene is expressed in cells of several lineages, with expression being highest in hepatocytes [239]. Urea cell cycle related enzymes might be important when hepatic function of stem cell-derived hepatocytes needs to be evaluated. Ornithine transcarbamylase (OTC), the OTC gene is expressed exclusively in liver and intestinal mucosa, is located in the mitochondria, and is part of the urea cycle as well as carbamyl-phosphate synthetase I (CPS) and argininosuccinate synthetase (ASSL) [240] (Fig. 4).

For further progress, it will be important to clearly define activities that closely resemble those of primary hepatocytes and, even more importantly, others that are not hepatocyte-like. For example, basal and rifampicin induced activities of CYP3A4, the most abundant CYP isoform in the human liver. Recapitulating the above-mentioned, gene expression and function of stem cell-derived hepatocytes must be compared to human fetal or adult liver. Mature hepatic characteristics should be demonstrated using drug metabolism detoxification in a gene expression and functional level. Additional characterization can be provided by analyzing hepatic transport proteins, mature hepatic transcription factors, and factors related to homeostasis, albumin secretion, production of bile acids, bilirrubin conjugation assays, ammonia metabolism with the expression of related enzymes, and, finally, the analysis of mature liver gene expression and function in animal models of liver failure after transplantation to the liver or ectopic sites. In addition, a clear definition of nonhepatocyte-like factors is important to identify mechanisms responsible for the lack of activity. Clarification of such mechanisms, for instance, loss of transcription factor expression or modification of signal transducers, is a requirement for further progress. It may be extremely difficult to differentiate stem cells to a cell type that resembles primary hepatocytes in all aspects of drug metabolism. However, promising results have been obtained with extrahepatic stem cells since some previously silent hepatocyte markers become expressed during differentiation and metabolic activities start appearing after new protocols have been reported (Fig. 4).

8 Therapeutic Potential of Stem Cell-Derived Hepatocytes

There are already insufficient donor organs to meet the demand for transplantation. With the worldwide shortage of donor organs likely to increase over the coming decades, research into alternative methods of treatment to whole organ transplantation is essential. Liver cell transplantation and cellular-based therapies are evolving as viable clinical alternatives to whole organ transplantation. Cell therapies provide a better utilization of donor tissue and major surgical procedures can be avoided.

Although liver cell transplants are safe and simple, there are not enough donor organs to spare for a procedure that is still experimental and has not been proven to be effective in the long-term. It would be of great value if an alternative cell source to whole organs could be found for transplantation. Stem cells, whether adult or embryonic derived, offer such a possibility. It is clear that stem cells play a regenerative role in the liver and that different stem cell compartments in the body are activated by different types of physiological or pathological stimuli. Partial hepatectomy leads to regeneration of the mature hepatocyte compartment.

The most elegant demonstration of liver regeneration was shown in a study of serial transplantation of severely immunodeficient, fumarylacetoacetate hydrolase (Fah)-deficient mice. After pretreatment with a urokinase-expressing adenovirus, these animals could be highly engrafted (up to 90%) with human hepatocytes. Furthermore, human cells could be serially transplanted from primary donors and repopulate the liver for at least four sequential rounds, demonstrating the amazing ability of cells within the liver to replenish themselves [241]. Moreover, it is clear that inhibition of the mature hepatocyte compartment through agents such as retrorsine and carbon tetrachloride leads to expansion of the oval cell compartment. Because of this, these cells have been felt to be central to liver repair mechanisms.

Recently, the demonstration in both humans and rodents of the presence of bone marrow-derived cells in diseased livers have suggested that extrahepatic stem cells play at least a minuscule role in liver repair. It is hoped that this mechanism might be employed for therapeutic advantage. There are equally a number of studies that have failed to show disease correction following transplantation of extrahepatic stem cells. However, It is notable that the animal models employed each have different liver injury or pathology. We cannot expect that the same mechanism of stem cell enrollment will be effective for a liver damaged by a metabolic disorder or traumatic injury or chemical toxicity. It is likely that each type of liver pathology will have to be treated as a particular situation that will required an individualized stem cell approach.

Several liver diseases have been identified for the purpose of cell therapeutic options. One of them is fulminant hepatic failure that is characterized by rapid onset of failure of the liver and death of the patient if whole liver replacement does not occur urgently. Cell therapeutic trials for fulminant hepatic failure in the form of liver cell transplants are underway and have shown moderate success [4]. Bioartificial livers that could also theoretically employ stem cells represent an option to treat these kinds of patients as an alternative to cell transplantation, to bridge them to whole organ transplantation or auto-recovery. Chronic liver disease is characterized by simultaneous liver regeneration and development of fibrosis that can finally results in cirrhosis. Patients with this form of liver disease may require treatment of portal hypertension before synthetic failure necessitates whole organ transplantation. Although extensive fibrosis could be an inhibitor of cell engraftment, it is unclear whether liver cells or stem cell–derived cells could provide substantial hepatic support while adequate organ transplantation is performed.

Metabolic liver diseases are characterized by an inherited defect of one hepatic enzyme, including urea cycle defects, bilirubin metabolizing defects, and organic acidaemias. In these disorders, the missing enzyme results in the build-up of toxic metabolites that are harmful to the individual but the rest of the function of the liver is normal. Metabolic liver diseases are ideal targets for development of cell therapeutic programs since only a small number of functional donor cells would effect disease correction through single enzyme replacement. Some metabolic liver diseases such as tyrosineamia type 1 are associated with severe liver injury and therefore engender selective repopulation of the recipient liver with donor cells as observed in animal models. However, most metabolic liver diseases are associated with little or no liver injury, thus cell repopulation remains an issue. However, metabolic liver diseases seem a likely option for donor stem cells to become hepatocyte competent of the native liver parenchyma.

Before the use of stem cells can be effectively translated to clinical practice, both efficacy and safety will need to be demonstrated. At present, transplantation of stem cells from one individual to another would require the recipient to be immunosuppressed particularly for hematopoietic stem cells. However, it is not clear whether this is necessary for all cell types. Lately, several immuno-tolerance techniques have been suggested, for example, donor leucocyte microchimerism has been implicated in liver transplant tolerance [242], and a particular role has been proposed for donor-derived dendritic cell progenitors [242]. This is particularly important since a benefit of transplanting hematopoietic stem cells as an adjuvant to whole organ transplantation to induce tolerance and prevent rejection has been postulated [243]. Not enough is known about the immunogenicity of early lineage stem cells such as those derived from embryos, fetal liver, or liver-derived stem cells at this time, although it is to be hoped that they will be more tolerogenic than adult hepatocytes, more studies need to be done in this area. Regarding the clinical use of differentiated human ES, there are concerns about their potential for tumorigenicity. Embryonic stem cells can provoke the formation of teratomas [244]. Early lineage stem cells from cord blood do not, and have been safely used clinically for many years. The situation with liver progenitor cells is unclear. Since long-term immunosuppression itself is carcinogenic, it remains to be seen whether these risks are clinically relevant and therefore prohibitive.

The development of stem cell therapy is a work in progress. Some of the more speculative and elegant proposals, presently in the research stage, might avoid the problems outlined above for instance in the future possibly specific human ES-equivalent cells could be obtained from patients and thus facilitating immuno-tolerance [245]. It will be important to determine the minimum number of stem cells that can effect disease correction, as has been done for treatment of leukaemias and marrow aplasias. Further work on the homing and engraftment mechanisms of extra-hepatic stem cells in different forms of liver injury will add considerably to optimization of stem cell therapy. Scaled up production of differentiated cells remains a concern as well. The ability of a stem cell population to expand to give clinically relevant numbers of cells might be truncated by terminal differentiation and loss of stem cell function. For instance, primary hepatocytes do not divide well in vitro, and appear to de-differentiate and lose their hepatic potential after prolonged culturing. Growth in culture may also result in loss or change in homing and attractant capacity of stem cells.

Finally, further techniques need to be developed in order to solve the inability to monitor cell function and rejection post-transplant in patients. Current experimental assessment of donor cell engraftment is unreasonable since it is usually obtained postmortem. Since the cell number transplanted is likely to be small, percutaneous liver biopsies may not be representative of graft function within the liver unless selective repopulation of the donor cells occurs. Furthermore, there is no morphological difference between the donor and recipient cells in liver cell transplants and there is unlikely to

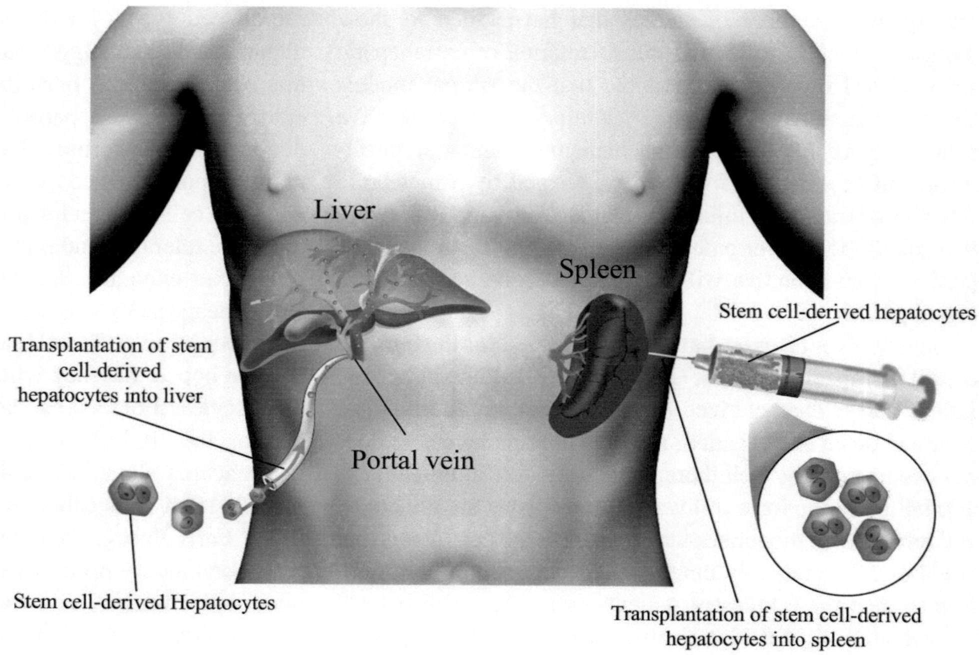

Fig. 5 Therapeutic applications of stem cell–derived hepatocytes. The possible therapeutic applications of stem cell–derived hepatocytes are shown

be a difference in morphology following successfully functioning stem cell transplants. Histology of the recipient liver has been unhelpful in monitoring liver cell graft rejection and this is likely to be the case for stem cell transplants. A recent work reported the pursuit of xenogenic hepatocytes engraftment in the spleen of cynomolgus monkeys by asialoglycoprotein receptor-directed nuclear scanning. This technique may have great impact in future stem cell-derived hepatocytes clinical trials [246] (Fig. 5).

9 Conclusions

Further work is required before we can be confident that stem cells can cure liver disease. We believe that the only conclusive evidence of hepatocyte functionality for stem cells will come from demonstrating disease correction following transplantation. Although there are many promising laboratory studies, there are only a handful of disease models that have been used to test stem cell correction of liver disease and there is an urgent need to develop more clinically relevant models. To date, based on basic studies, we have assembled the concept of a stem cell–derived hepatocyte. However, this conclusion may be premature because the question of whether the produced cells to date are true hepatocytes has not been well addressed. In this case, one should carefully evaluate crucial hepatocyte-defining enzymatic properties. Thus, there is a necessity to establish a standard criterion for defining a true human stem cell–derived hepatocyte. It is essential to understand that the definition of an authentic hepatocyte should not be limited to qualitative assays, but has to include a quantitative analysis of enzymatic activities, which allows direct comparison with primary hepatocytes. Our understanding of the complex nature of liver regeneration and the role of the various stem cell compartments in liver repair has reached new levels. We are only now beginning to understand the biology of hepatic differentiation from stem cells. There are still significant clinical hurdles that will need to be overcome if stem cell therapy is to reach the full potential that basic studies have anticipated. The objective is ambitious and the journey is long, but we have to remain hopeful that stem cell–derived hepatocytes can serve in the near future as a source of cells for transplantation medicine and basic studies related to drug discovery.

References

1. Lee WM. Acute liver failure. N Engl J Med. 1993;329:1862–72.
2. Michalopoulos GK. Liver regeneration. J Cell Physiol. 2007;213(2):286–300.
3. Michalopoulos GK, DeFrances MC. Liver regeneration. Science. 1997;276:60–66.
4. Fisher RA, Strom SC. Human hepatocyte transplantation: worldwide results. Transplantation. 2006;82:441–449.
5. Eleazar JA, Memeo L, Jhang JS, Mansukhani MM, Chin S, Park SM, Lefkowitch JH, et al. Progenitor cell expansion: an important source of hepatocyte regeneration in chronic hepatitis. J Hepatol. 2004;41:983–91.
6. Fotiadu A, Tzioufa V, Vrettou E, Koufogiannis D, Papadimitriou CS, Hytiroglou P. Progenitor cell activation in chronic viralhepatitis. Liver Int. 2004;24:268–74.
7. Paku S, Schnur J, Nagy P, Thorgeirsson SS. Origin and structural evolution of the early proliferating oval cells in rat liver. Am J Pathol. 2001;158:1313–23.

8. Farber E. Similarities in the sequence of early histological changes induced in the liver of the rat by ethionine, 2-acetylamino-fluorene, and 3'-methyl-4-dimethylaminoazobenzene. Cancer Res. 1956;16:142–8.
9. Quintana-Bustamante O, Alvarez-Barrientos A, Kofman AV, Fabregat I, Bueren JA, Theise ND, Segovia JC. Hematopoietic mobilization in mice increases the presence of bone marrow-derived hepatocytes via in vivo cell fusion. Hepatology. 2006;43:108–16.
10. Kollet O, Shivtiel S, Chen YQ, Suriawinata J, Thung SN, Dabeva MD, Kahn J, et al. HGF, SDF-1, and MMP-9 are involved in stress-induced human CD34+ stem cell recruitment to the liver. J Clin Invest. 2003;112:160–9.
11. Soto-Gutierrez A, Navarro-Alvarez N, Rivas-Carrillo JD, Chen Y, Yamatsuji T, Tanaka N, Kobayashi N. Differentiation of human embryonic stem cells to hepatocytes using deleted variant of HGF and poly-amino-urethane-coated nonwoven polytetrafluoroethylene fabric. Cell Transplant. 2006;15:335–41.
12. Gouon-Evans V, Boussemart L, Gadue P, Nierhoff D, Koehler CI, Kubo A, Shafritz DA, et al. BMP-4 is required for hepatic specification of mouse embryonic stem cell-derived definitive endoderm. Nat Biotechnol. 2006;24:1402–11.
13. Soto-Gutierrez A, Kobayashi N, Rivas-Carrillo JD, Navarro-Alvarez N, Zhao D, Okitsu T, Noguchi H, et al. Reversal of mouse hepatic failure using an implanted liver-assist device containing ES cell-derived hepatocytes. Nat Biotechnol. 2006;24:1412–9.
14. Schachinger V, Assmus B, Honold J, Lehmann R, Hofmann WK, Martin H, Dimmeler S, et al. Normalization of coronary blood flow in the infarct-related artery after intracoronary progenitor cell therapy: intracoronary Doppler substudy of the TOPCARE-AMI trial. Clin Res Cardiol. 2006;95:13–22.
15. Assmus B, Schachinger V, Teupe C, Britten M, Lehmann R, Dobert N, Grunwald F, et al. Transplantation of Progenitor Cells and Regeneration Enhancement in Acute Myocardial Infarction (TOPCARE-AMI). Circulation. 2002;106:3009–17.
16. Tse HF, Kwong YL, Chan JK, Lo G, Ho CL, Lau CP. Angiogenesis in ischaemic myocardium by intramyocardial autologous bone marrow mononuclear cell implantation. Lancet. 2003;361:47–9.
17. Morrison SJ, Shah NM, Anderson DJ. Regulatory mechanisms in stem cell biology. Cell. 1997;88:287–98.
18. Potten CS, Loeffler M. Stem cells: attributes, cycles, spirals, pitfalls and uncertainties. Lessons for and from the crypt. Development. 1990;110:1001–20.
19. Lechler T, Fuchs E. Asymmetric cell divisions promote stratification and differentiation of mammalian skin. Nature. 2005;437:275–80.
20. Fuchs E, Tumbar T, Guasch G. Socializing with the neighbors: stem cells and their niche. Cell. 2004;116:769–78.
21. Watt FM, Hogan BL. Out of Eden: stem cells and their niches. Science. 2000;287:1427–30.
22. Wurmser AE, Palmer TD, Gage FH. Neuroscience. Cellular interactions in the stem cell niche. Science. 2004;304:1253–5.
23. Whetton AD, Graham GJ. Homing and mobilization in the stem cell niche. Trends Cell Biol. 1999;9:233–38.
24. Yasoshima M, Tsuneyama K, Harada K, Sasaki M, Gershwin ME, Nakanuma Y. Immunohistochemical analysis of cell-matrix adhesion molecules and their ligands in the portal tracts of primary biliary cirrhosis. J Pathol. 2000;190:93–9.
25. Botchkarev VA, Botchkareva NV, Peters EM, Paus R. Epithelial growth control by neurotrophins: leads and lessons from the hair follicle. Prog Brain Res. 2004;146:493–513.
26. Wells JM, Melton DA. Vertebrate endoderm development. Annu Rev Cell Dev Biol. 1999;15:393–410.
27. Zhao R, Duncan SA. Embryonic development of the liver. Hepatology. 2005;41:956–67.
28. Jung J, Zheng M, Goldfarb M, Zaret KS. Initiation of mammalian liver development from endoderm by fibroblast growth factors. Science. 1999;284:1998–2003.
29. Bossard P, Zaret KS. Repressive and restrictive mesodermal interactions with gut endoderm: possible relation to Meckel's Diverticulum. Development. 2000;127:4915–23.
30. Berg T, Rountree CB, Lee L, Estrada J, Sala FG, Choe A, Veltmaat JM, et al. Fibroblast growth factor 10 is critical for liver growth during embryogenesis and controls hepatoblast survival via beta-catenin activation. Hepatology. 2007;46(4):1187–97.
31. Steiling H, Wustefeld T, Bugnon P, Brauchle M, Fassler R, Teupser D, Thiery J, et al. Fibroblast growth factor receptor signalling is crucial for liver homeostasis and regeneration. Oncogene. 2003;22:4380–8.
32. Micsenyi A, Tan X, Sneddon T, Luo JH, Michalopoulos GK, Monga SP. Beta-catenin is temporally regulated during normal liver development. Gastroenterology. 2004;126:1134–46.
33. Dong PD, Munson CA, Norton W, Crosnier C, Pan X, Gong Z, Neumann CJ, et al. Fgf10 regulates hepatopancreatic ductal system patterning and differentiation. Nat Genet. 2007;39:397–402.
34. Sala FG, Curtis JL, Veltmaat JM, Del Moral PM, Le LT, Fairbanks TJ, Warburton D, et al. Fibroblast growth factor 10 is required for survival and proliferation but not differentiation of intestinal epithelial progenitor cells during murine colon development. Dev Biol. 2006;299:373–85.
35. Nyeng P, Norgaard GA, Kobberup S, Jensen J. FGF10 signaling controls stomach morphogenesis. Dev Biol. 2007;303:295–310.
36. Rossi JD, NR. Hogan, BL. Zaret, KS. Distinct mesodermal signals, including BMPs from the septum transversum mesenquyme, are require in combination for hepatogenesis from the endoderm. Genes Dev. 2001;15:1998–2001.
37. Lee CS, Friedman JR, Fulmer JT, Kaestner KH. The initiation of liver development is dependent on Foxa transcription factors. Nature. 2005;435:944–7.
38. Bossard P, Zaret KS. GATA transcription factors as potentiators of gut endoderm differentiation. Development. 1998;125:4909–17.
39. Holtzinger A, Evans T. Gata4 regulates the formation of multiple organs. Development. 2005;132:4005–14.
40. Brill S, Zvibel I, Reid LM. Maturation-dependent changes in the regulation of liver-specific gene expression in embryonal versus adult primary liver cultures. Differentiation. 1995;59:95–102.
41. Tee LB, Kirilak Y, Huang WH, Smith PG, Morgan RH, Yeoh GC. Dual phenotypic expression of hepatocytes and bile ductular markers in developing and preneoplastic rat liver. Carcinogenesis. 1996;17:251–9.
42. Costa RH, Kalinichenko VV, Holterman AX, Wang X. Transcription factors in liver development, differentiation, and regeneration. Hepatology. 2003;38:1331–47.
43. Zaret KS. Regulatory phases of early liver development: paradigms of organogenesis. Nat Rev Genet. 2002;3:499–512.
44. Lemaigre FP. Development of the biliary tract. Mech Dev. 2003;120:81–7.
45. Kodama Y, Hijikata M, Kageyama R, Shimotohno K, Chiba T. The role of notch signaling in the development of intrahepatic bile ducts. Gastroenterology. 2004;127:1775–86.
46. Tanimizu N, Miyajima A. Notch signaling controls hepatoblast differentiation by altering the expression of liver-enriched transcription factors. J Cell Sci. 2004;117:3165–74.
47. Clotman F, Lemaigre FP. Control of hepatic differentiation by activin/TGFbeta signaling. Cell Cycle. 2006;5:168–71.
48. Clotman F, Jacquemin P, Plumb-Rudewiez N, Pierreux CE, Van der Smissen P, Dietz HC, Courtoy PJ, et al. Control of liver cell fate decision by a gradient of TGF beta signaling modulated by Onecut transcription factors. Genes Dev. 2005;19:1849–54.

49. Blouin MJ, Lamy I, Loranger A, Noel M, Corlu A, Guguen-Guillouzo C, Marceau N. Specialization switch in differentiating embryonic rat liver progenitor cells in response to sodium butyrate. Exp Cell Res. 1995;217:22–30.
50. Van Eyken P, Sciot R, Desmet V. Intrahepatic bile duct development in the rat: a cytokeratin-immunohistochemical study. Lab Invest. 1988;59:52–9.
51. Steiner JW, Perz ZM, Taichman LB. Cell population dynamics in the liver. a review of quantitative morphological techniques applied to the study of physiological and pathological growth. Exp Mol Pathol. 1966;5:146–81.
52. Bucher N. Regeneration of the liver and kidney. Boston: Little, Brown and Co., 1971.
53. Hashimoto M, Sanjo K. Functional capacity of the liver after two-thirds partial hepatectomy in the rat. Surgery. 1997;121:690–7.
54. Grisham J. A morphologic study of deoxyribonucleic acid synthesis and cell proliferation in regenerating rat liver; autoradiography with thymidine-H3. Cancer Res. 1962;22:842–9.
55. Stocker EWH, Brau G. Capacity of regeneration in liver epithelia of juvenile, repeated partially hepatectomized rats. Autoradiographic studies after continous infusion of 3H-thymidine (author's transl). Virchows Archiv. 1973;14:93–103.
56. Overturf K, al-Dhalimy M, Ou CN, Finegold M, Grompe M. Serial transplantation reveals the stem-cell-like regenerative potential of adult mouse hepatocytes. Am J Pathol. 1997;151: 1273–80.
57. Rudolph KL, Chang S, Millard M, Schreiber-Agus N, DePinho RA. Inhibition of experimental liver cirrhosis in mice by telomerase gene delivery. Science. 2000;287:1253–8.
58. Shafritz DA, Oertel M, Menthena A, Nierhoff D, Dabeva MD. Liver stem cells and prospects for liver reconstitution by transplanted cells. Hepatology. 2006;43:S89–98.
59. Hohme S, Hengstler JG, Brulport M, Schafer M, Bauer A, Gebhardt R, Drasdo D. Mathematical modelling of liver regeneration after intoxication with CCl(4). Chem Biol Interact. 2007;168: 74–93.
60. Sell S. Heterogeneity and plasticity of hepatocyte lineage cells. Hepatology. 2001;33:738–50.
61. Krause DS, Theise ND, Collector MI, Henegariu O, Hwang S, Gardner R, Neutzel S, et al. Multi-organ, multi-lineage engraftment by a single bone marrow-derived stem cell. Cell. 2001;105:369–77.
62. Wilson JW, Leduc EH. Role of cholangioles in restoration of the liver of the mouse after dietary injury. J Pathol Bacteriol. 1958;76:441–9.
63. Abelev GI, Eraiser TL. Cellular aspects of alpha-fetoprotein reexpression in tumors. Semin Cancer Biol. 1999;9:95–107.
64. Shiojiri N, Takeshita K, Yamasaki H, Iwata T. Suppression of C/EBP alpha expression in biliary cell differentiation from hepatoblasts during mouse liver development. J Hepatol. 2004;41: 790–8.
65. Rogler LE. Selective bipotential differentiation of mouse embryonic hepatoblasts in vitro. Am J Pathol. 1997;150: 591–602.
66. Suskind DL, Muench MO. Searching for common stem cells of the hepatic and hematopoietic systems in the human fetal liver: CD34+ cytokeratin 7/8+ cells express markers for stellate cells. J Hepatol. 2004;40:261–8.
67. Fiegel HC, Bruns H, Hoper C, Lioznov MV, Kluth D. Cell growth and differentiation of different hepatic cells isolated from fetal rat liver in vitro. Tissue Eng. 2006;12:123–30.
68. Tanimizu N, Nishikawa M, Saito H, Tsujimura T, Miyajima A. Isolation of hepatoblasts based on the expression of Dlk/Pref-1. J Cell Sci. 2003;116:1775–86.
69. Nitou M, Sugiyama Y, Ishikawa K, Shiojiri N. Purification of fetal mouse hepatoblasts by magnetic beads coated with monoclonal anti-e-cadherin antibodies and their in vitro culture. Exp Cell Res. 2002;279:330–43.
70. Krupczak-Hollis K, Wang X, Kalinichenko VV, Gusarova GA, Wang IC, Dennewitz MB, Yoder HM, et al. The mouse Forkhead Box m1 transcription factor is essential for hepatoblast mitosis and development of intrahepatic bile ducts and vessels during liver morphogenesis. Dev Biol. 2004;276:74–88.
71. Stamp L, Crosby HA, Hawes SM, Strain AJ, Pera MF. A novel cell-surface marker found on human embryonic hepatoblasts and a subpopulation of hepatic biliary epithelial cells. Stem Cells. 2005;23:103–12.
72. Holic N, Suzuki T, Corlu A, Couchie D, Chobert MN, Guguen-Guillouzo C, Laperche Y. Differential expression of the rat gamma-glutamyl transpeptidase gene promoters along with differentiation of hepatoblasts into biliary or hepatocytic lineage. Am J Pathol. 2000;157:537–48.
73. Kamiya A, Inoue Y, Gonzalez FJ. Role of the hepatocyte nuclear factor 4alpha in control of the pregnane X receptor during fetal liver development. Hepatology. 2003;37:1375–84.
74. Peiler G, Bockmann B, Nakhei H, Ryffel GU. Inhibitor of the tissue-specific transcription factor HNF4, a potential regulator in early Xenopus development. Mol Cell Biol. 2000;20:8676–83.
75. Nakayama M, Matsumoto K, Tatsumi N, Yanai M, Yokouchi Y. Id3 is important for proliferation and differentiation of the hepatoblasts during the chick liver development. Mech Dev. 2006;123:580–90.
76. Watanabe T, Nakagawa K, Ohata S, Kitagawa D, Nishitai G, Seo J, Tanemura S, et al. SEK1/MKK4-mediated SAPK/JNK signaling participates in embryonic hepatoblast proliferation via a pathway different from NF-kappaB-induced anti-apoptosis. Dev Biol. 2002;250:332–47.
77. Sosa-Pineda B, Wigle JT, Oliver G. Hepatocyte migration during liver development requires Prox1. Nat Genet. 2000;25:254–5.
78. Dudas J, Elmaouhoub A, Mansuroglu T, Batusic D, Tron K, Saile B, Papoutsi M, et al. Prospero-related homeobox 1 (Prox1) is a stable hepatocyte marker during liver development, injury and regeneration, and is absent from "oval cells". Histochem Cell Biol. 2006;126:549–62.
79. Dudas J, Papoutsi M, Hecht M, Elmaouhoub A, Saile B, Christ B, Tomarev SI, et al. The homeobox transcription factor Prox1 is highly conserved in embryonic hepatoblasts and in adult and transformed hepatocytes, but is absent from bile duct epithelium. Anat Embryol (Berl). 2004;208:359–66.
80. Ganiatsas S, Kwee L, Fujiwara Y, Perkins A, Ikeda T, Labow MA, Zon LI. SEK1 deficiency reveals mitogen-activated protein kinase cascade crossregulation and leads to abnormal hepatogenesis. Proc Natl Acad Sci U S A. 1998;95:6881–6.
81. Ader T, Norel R, Levoci L, Rogler LE. Transcriptional profiling implicates TGFbeta/BMP and Notch signaling pathways in ductular differentiation of fetal murine hepatoblasts. Mech Dev. 2006;123:177–94.
82. Nierhoff D, Ogawa A, Oertel M, Chen YQ, Shafritz DA. Purification and characterization of mouse fetal liver epithelial cells with high in vivo repopulation capacity. Hepatology. 2005;42:130–9.
83. Nierhoff D, Levoci L, Schulte S, Goeser T, Rogler LE, Shafritz DA. New cell surface markers for murine fetal hepatic stem cells identified through high density complementary DNA microarrays. Hepatology. 2007;46:535–47.
84. Oertel M, Menthena A, Dabeva MD, Shafritz DA. Cell competition leads to a high level of normal liver reconstitution by transplanted fetal liver stem/progenitor cells. Gastroenterology. 2006;130:507–20; quiz 590.
85. Dabeva MD, Petkov PM, Sandhu J, Oren R, Laconi E, Hurston E, Shafritz DA. Proliferation and differentiation of fetal liver epithelial progenitor cells after transplantation into adult rat liver. Am J Pathol. 2000;156:2017–31.

86. Sandhu JS, Petkov PM, Dabeva MD, Shafritz DA. Stem cell properties and repopulation of the rat liver by fetal liver epithelial progenitor cells. Am J Pathol. 2001;159:1323–34.
87. Zaret KS. Liver specification and early morphogenesis. Mech Dev. 2000;92:83–8.
88. Blakolmer K, Jaskiewicz K, Dunsford HA, Robson SC. Hematopoietic stem cell markers are expressed by ductal plate and bile duct cells in developing human liver. Hepatology. 1995;21:1510–6.
89. Fiegel HC, Park JJ, Lioznov MV, Martin A, Jaeschke-Melli S, Kaufmann PM, Fehse B, et al. Characterization of cell types during rat liver development. Hepatology. 2003;37:148–54.
90. Petersen BE, Grossbard B, Hatch H, Pi L, Deng J, Scott EW. Mouse A6-positive hepatic oval cells also express several hematopoietic stem cell markers. Hepatology. 2003;37:632–40.
91. Masson NM, Currie IS, Terrace JD, Garden OJ, Parks RW, Ross JA. Hepatic progenitor cells in human fetal liver express the oval cell marker Thy-1. Am J Physiol Gastrointest Liver Physiol. 2006;291:G45–54.
92. Oertel M, Menthena A, Chen YQ, Shafritz DA. Comparison of hepatic properties and transplantation of Thy-1(+) and Thy-1(-) cells isolated from embryonic day 14 rat fetal liver. Hepatology. 2007;46(4):1236–45.
93. Theise ND, Saxena R, Portmann BC, Thung SN, Yee H, Chiriboga L, Kumar A, et al. The canals of Hering and hepatic stem cells in humans. Hepatology. 1999;30:1425–33.
94. Baumann U, Crosby HA, Ramani P, Kelly DA, Strain AJ. Expression of the stem cell factor receptor c-kit in normal and diseased pediatric liver: identification of a human hepatic progenitor cell? Hepatology. 1999;30:112–7.
95. Petersen BE, Bowen WC, Patrene KD, Mars WM, Sullivan AK, Murase N, Boggs SS, et al. Bone marrow as a potential source of hepatic oval cells. Science. 1999;284:1168–70.
96. Navarro-Alvarez N. Stem cell-derived hepatocytes. Curr Opin Organ Transplant. 2006;11:659–64.
97. Radaeva S, Steinberg P. Phenotype and differentiation patterns of the oval cell lines OC/CDE 6 and OC/CDE 22 derived from the livers of carcinogen-treated rats. Cancer Res. 1995;55: 1028–1038.
98. Fujio K, Evarts RP, Hu Z, Marsden ER, Thorgeirsson SS. Expression of stem cell factor and its receptor, c-kit, during liver regeneration from putative stem cells in adult rat. Lab Invest. 1994;70:511–6.
99. Omori M, Omori N, Evarts RP, Teramoto T, Thorgeirsson SS. Coexpression of flt-3 ligand/flt-3 and SCF/c-kit signal transduction system in bile-duct-ligated SI and W mice. Am J Pathol. 1997;150:1179–87.
100. Omori N, Omori M, Evarts RP, Teramoto T, Miller MJ, Hoang TN, Thorgeirsson SS. Partial cloning of rat CD34 cDNA and expression during stem cell-dependent liver regeneration in the adult rat. Hepatology. 1997;26:720–7.
101. Petersen BE, Goff JP, Greenberger JS, Michalopoulos GK. Hepatic oval cells express the hematopoietic stem cell marker Thy-1 in the rat. Hepatology. 1998;27:433–45.
102. Engelhardt NV, Factor VM, Medvinsky AL, Baranov VN, Lazareva MN, Poltoranina VS. Common antigen of oval and biliary epithelial cells (A6) is a differentiation marker of epithelial and erythroid cell lineages in early development of the mouse. Differentiation. 1993;55:19–26.
103. Omori N, Evarts RP, Omori M, Hu Z, Marsden ER, Thorgeirsson SS. Expression of leukemia inhibitory factor and its receptor during liver regeneration in the adult rat. Lab Invest. 1996;75:15–24.
104. Shimano K, Satake M, Okaya A, Kitanaka J, Kitanaka N, Takemura M, Sakagami M, et al. Hepatic oval cells have the side population phenotype defined by expression of ATP-binding cassette transporter ABCG2/BCRP1. Am J Pathol. 2003;163:3–9.
105. Oh SH, Witek RP, Bae SH, Zheng D, Jung Y, Piscaglia AC, Petersen BE. Bone marrow-derived hepatic oval cells differentiate into hepatocytes in 2-acetylaminofluorene/partial hepatectomy-induced liver regeneration. Gastroenterology. 2007;132:1077–87.
106. Wang X, Foster M, Al-Dhalimy M, Lagasse E, Finegold M, Grompe M. The origin and liver repopulating capacity of murine oval cells. Proc Natl Acad Sci U S A. 2003;100 Suppl 1: 11881–8.
107. Menthena A, Deb N, Oertel M, Grozdanov PN, Sandhu J, Shah S, Guha C, et al. Bone marrow progenitors are not the source of expanding oval cells in injured liver. Stem Cells. 2004;22: 1049–61.
108. Akhurst B, Croager EJ, Farley-Roche CA, Ong JK, Dumble ML, Knight B, Yeoh GC. A modified choline-deficient, ethionine-supplemented diet protocol effectively induces oval cells in mouse liver. Hepatology. 2001;34:519–22.
109. Factor VM, Radaeva SA, Thorgeirsson SS. Origin and fate of oval cells in dipin-induced hepatocarcinogenesis in the mouse. Am J Pathol. 1994;145:409–22.
110. Preisegger KH, Factor VM, Fuchsbichler A, Stumptner C, Denk H, Thorgeirsson SS. Atypical ductular proliferation and its inhibition by transforming growth factor beta1 in the 3,5-diethoxycarbonyl-1,4-dihydrocollidine mouse model for chronic alcoholic liver disease. Lab Invest. 1999;79:103–9.
111. Rosenberg D, Ilic Z, Yin L, Sell S. Proliferation of hepatic lineage cells of normal C57BL and interleukin-6 knockout mice after cocaine-induced periportal injury. Hepatology. 2000;31:948–55.
112. Lemire JM, Shiojiri N, Fausto N. Oval cell proliferation and the origin of small hepatocytes in liver injury induced by D-galactosamine. Am J Pathol. 1991;139:535–52.
113. Petersen BE, Zajac VF, Michalopoulos GK. Hepatic oval cell activation in response to injury following chemically induced periportal or pericentral damage in rats. Hepatology. 1998;27:1030–8.
114. Yovchev MI, Grozdanov PN, Joseph B, Gupta S, Dabeva MD. Novel hepatic progenitor cell surface markers in the adult rat liver. Hepatology. 2007;45:139–49.
115. Rountree CB, Barsky L, Ge S, Zhu J, Senadheera S, Crooks GM. A CD133-expressing murine liver oval cell population with bilineage potential. Stem Cells. 2007;25(10):2419–29.
116. Jelnes P, Santoni-Rugiu E, Rasmussen M, Friis SL, Nielsen JH, Tygstrup N, Bisgaard HC. Remarkable heterogeneity displayed by oval cells in rat and mouse models of stem cell-mediated liver regeneration. Hepatology. 2007;45:1462–70.
117. Gordon GJ, Coleman WB, Hixson DC, Grisham JW. Liver regeneration in rats with retrorsine-induced hepatocellular injury proceeds through a novel cellular response. Am J Pathol. 2000;156:607–19.
118. Gordon GJ, Coleman WB, Grisham JW. Temporal analysis of hepatocyte differentiation by small hepatocyte-like progenitor cells during liver regeneration in retrorsine-exposed rats. Am J Pathol. 2000;157:771–86.
119. Avril A, Pichard V, Bralet MP, Ferry N. Mature hepatocytes are the source of small hepatocyte-like progenitor cells in the retrorsine model of liver injury. J Hepatol. 2004;41:737–43.
120. Vig P, Russo FP, Edwards RJ, Tadrous PJ, Wright NA, Thomas HC, Alison MR, et al. The sources of parenchymal regeneration after chronic hepatocellular liver injury in mice. Hepatology. 2006;43:316–24.
121. Fausto N. Liver regeneration and repair: hepatocytes, progenitor cells, and stem cells. Hepatology. 2004;39:1477–87.
122. Best DH, Coleman WB. Treatment with 2-AAF blocks the small hepatocyte-like progenitor cell response in retrorsine-exposed rats. J Hepatol. 2007;46:1055–63.
123. Shibata C, Mizuguchi T, Kikkawa Y, Nobuoka T, Oshima H, Kawasaki H, Kawamoto M, et al. Liver repopulation and long-term function of rat small hepatocyte transplantation as

an alternative cell source for hepatocyte transplantation. Liver Transpl. 2006;12:78–87.
124. Morrison SJ. Stem cell potential: can anything make anything? Curr Biol. 2001;11:R7–9.
125. Orkin SH, Zon LI. Hematopoiesis and stem cells: plasticity versus developmental heterogeneity. Nat Immunol. 2002;3:323–8.
126. Orlic D, Kajstura J, Chimenti S, Jakoniuk I, Anderson SM, Li B, Pickel J, et al. Bone marrow cells regenerate infarcted myocardium. Nature. 2001;410:701–5.
127. Terada N, Hamazaki T, Oka M, Hoki M, Mastalerz DM, Nakano Y, Meyer EM, et al. Bone marrow cells adopt the phenotype of other cells by spontaneous cell fusion. Nature. 2002;416:542–5.
128. Ying QL, Nichols J, Evans EP, Smith AG. Changing potency by spontaneous fusion. Nature. 2002;416:545–8.
129. Vassilopoulos G, Wang PR, Russell DW. Transplanted bone marrow regenerates liver by cell fusion. Nature. 2003;422:901–4.
130. Wang X, Willenbring H, Akkari Y, Torimaru Y, Foster M, Al-Dhalimy M, Lagasse E, et al. Cell fusion is the principal source of bone-marrow-derived hepatocytes. Nature. 2003;422: 897–901.
131. Lagasse E, Connors H, Al-Dhalimy M, Reitsma M, Dohse M, Osborne L, Wang X, et al. Purified hematopoietic stem cells can differentiate into hepatocytes in vivo. Nat Med. 2000;6:1229–34.
132. Theise ND, Badve S, Saxena R, Henegariu O, Sell S, Crawford JM, Krause DS. Derivation of hepatocytes from bone marrow cells in mice after radiation-induced myeloablation. Hepatology. 2000;31:235–40.
133. Wang X, Ge S, McNamara G, Hao QL, Crooks GM, Nolta JA. Albumin-expressing hepatocyte-like cells develop in the livers of immune-deficient mice that received transplants of highly purified human hematopoietic stem cells. Blood. 2003;101:4201–8.
134. Schwartz RE, Reyes M, Koodie L, Jiang Y, Blackstad M, Lund T, Lenvik T, et al. Multipotent adult progenitor cells from bone marrow differentiate into functional hepatocyte-like cells. J Clin Invest. 2002;109:1291–302.
135. Korbling M, Katz RL, Khanna A, Ruifrok AC, Rondon G, Albitar M, Champlin RE, et al. Hepatocytes and epithelial cells of donor origin in recipients of peripheral-blood stem cells. N Engl J Med. 2002;346:738–46.
136. Harris RG, Herzog EL, Bruscia EM, Grove JE, Van Arnam JS, Krause DS. Lack of a fusion requirement for development of bone marrow-derived epithelia. Science. 2004;305:90–3.
137. Alison MR, Poulsom R, Jeffery R, Dhillon AP, Quaglia A, Jacob J, Novelli M, et al. Hepatocytes from non-hepatic adult stem cells. Nature. 2000;406:257.
138. Theise ND, Nimmakayalu M, Gardner R, Illei PB, Morgan G, Teperman L, Henegariu O, et al. Liver from bone marrow in humans. Hepatology. 2000;32:11–16.
139. Willenbring H, Bailey AS, Foster M, Akkari Y, Dorrell C, Olson S, Finegold M, et al. Myelomonocytic cells are sufficient for therapeutic cell fusion in liver. Nat Med. 2004;10:744–8.
140. Camargo FD, Finegold M, Goodell MA. Hematopoietic myelomonocytic cells are the major source of hepatocyte fusion partners. J Clin Invest. 2004;113:1266–70.
141. Jang YY, Collector MI, Baylin SB, Diehl AM, Sharkis SJ. Hematopoietic stem cells convert into liver cells within days without fusion. Nat Cell Biol. 2004;6:532–9.
142. Avital I, Inderbitzin D, Aoki T, Tyan DB, Cohen AH, Ferraresso C, Rozga J, et al. Isolation, characterization, and transplantation of bone marrow-derived hepatocyte stem cells. Biochem Biophys Res Commun. 2001;288:156–64.
143. Saji Y, Tamura S, Yoshida Y, Kiso S, Iizuka AS, Matsumoto H, Kawasaki T, et al. Basic fibroblast growth factor promotes the trans-differentiation of mouse bone marrow cells into hepatic lineage cells via multiple liver-enriched transcription factors. J Hepatol. 2004;41:545–50.
144. Fiegel HC, Lioznov MV, Cortes-Dericks L, Lange C, Kluth D, Fehse B, Zander AR. Liver-specific gene expression in cultured human hematopoietic stem cells. Stem Cells. 2003;21: 98–104.
145. Zhao Y, Glesne D, Huberman E. A human peripheral blood monocyte-derived subset acts as pluripotent stem cells. Proc Natl Acad Sci U S A. 2003;100:2426–31.
146. Ruhnke M, Ungefroren H, Nussler A, Martin F, Brulport M, Schormann W, Hengstler JG, et al. Differentiation of in vitro-modified human peripheral blood monocytes into hepatocyte-like and pancreatic islet-like cells. Gastroenterology. 2005;128: 1774–86.
147. Liechty KW, MacKenzie TC, Shaaban AF, Radu A, Moseley AM, Deans R, Marshak DR, et al. Human mesenchymal stem cells engraft and demonstrate site-specific differentiation after in utero transplantation in sheep. Nat Med. 2000;6:1282–86.
148. Pittenger MF, Mackay AM, Beck SC, Jaiswal RK, Douglas R, Mosca JD, Moorman MA, et al. Multilineage potential of adult human mesenchymal stem cells. Science. 1999;284:143–7.
149. Lee KD, Kuo TK, Whang-Peng J, Chung YF, Lin CT, Chou SH, Chen JR, et al. In vitro hepatic differentiation of human mesenchymal stem cells. Hepatology. 2004;40:1275–84.
150. Aurich I, Mueller LP, Aurich H, Luetzkendorf J, Tisljar K, Dollinger MM, Schormann W, et al. Functional integration of hepatocytes derived from human mesenchymal stem cells into mouse livers. Gut. 2007;56:405–15.
151. Banas A, Teratani T, Yamamoto Y, Tokuhara M, Takeshita F, Quinn G, Okochi H, et al. Adipose tissue-derived mesenchymal stem cells as a source of human hepatocytes. Hepatology. 2007;46:219–228.
152. Jiang Y, Jahagirdar BN, Reinhardt RL, Schwartz RE, Keene CD, Ortiz-Gonzalez XR, Reyes M, et al. Pluripotency of mesenchymal stem cells derived from adult marrow. Nature. 2002;418:41–9.
153. Wobus AM, Boheler KR. Embryonic stem cells: prospects for developmental biology and cell therapy. Physiol Rev. 2005;85: 635–78.
154. Itskovitz-Eldor J, Schuldiner M, Karsenti D, Eden A, Yanuka O, Amit M, Soreq H, et al. Differentiation of human embryonic stem cells into embryoid bodies compromising the three embryonic germ layers. Mol Med. 2000;6:88–95.
155. Kim JH, Auerbach JM, Rodriguez-Gomez JA, Velasco I, Gavin D, Lumelsky N, Lee SH, et al. Dopamine neurons derived from embryonic stem cells function in an animal model of Parkinson's disease. Nature. 2002;418:50–6.
156. Kehat I, Khimovich L, Caspi O, Gepstein A, Shofti R, Arbel G, Huber I, et al. Electromechanical integration of cardiomyocytes derived from human embryonic stem cells. Nat Biotechnol. 2004;22:1282–9.
157. Lako M, Lindsay S, Lincoln J, Cairns PM, Armstrong L, Hole N. Characterisation of Wnt gene expression during the differentiation of murine embryonic stem cells in vitro: role of Wnt3 in enhancing haematopoietic differentiation. Mech Dev. 2001;103: 49–59.
158. Suzuki H, Watabe T, Kato M, Miyazawa K, Miyazono K. Roles of vascular endothelial growth factor receptor 3 signaling in differentiation of mouse embryonic stem cell-derived vascular progenitor cells into endothelial cells. Blood 2005;105:2372–9.
159. D'Amour KA, Bang AG, Eliazer S, Kelly OG, Agulnick AD, Smart NG, Moorman MA, et al. Production of pancreatic hormone-expressing endocrine cells from human embryonic stem cells. Nat Biotechnol. 2006;24:1392–401.
160. Soto-Gutierrez A, Navarro-Alvarez N, Zhao D, Rivas-Carrillo JD, Lebkowski J, Tanaka N, Fox IJ, et al. Differentiation of mouse embryonic stem cells to hepatocyte-like cells by co-culture with human liver nonparenchymal cell lines. Nat Protoc. 2007;2: 347–56.

161. Cai J, Zhao Y, Liu Y, Ye F, Song Z, Qin H, Meng S, et al. Directed differentiation of human embryonic stem cells into functional hepatic cells. Hepatology. 2007;45:1229–39.
162. Chen Y, Soto-Gutierrez A, Navarro-Alvarez N, Rivas-Carrillo JD, Yamatsuji T, Shirakawa Y, Tanaka N, et al. Instant hepatic differentiation of human embryonic stem cells using activin A and a deleted variant of HGF. Cell Transplant. 2006;15:865–71.
163. Teratani T, Yamamoto H, Aoyagi K, Sasaki H, Asari A, Quinn G, Sasaki H, et al. Direct hepatic fate specification from mouse embryonic stem cells. Hepatology. 2005;41:836–46.
164. Lavon N, Benvenisty N. Study of hepatocyte differentiation using embryonic stem cells. J Cell Biochem. 2005;96:1193–202.
165. Lavon N, Yanuka O, Benvenisty N. Differentiation and isolation of hepatic-like cells from human embryonic stem cells. Differentiation. 2004;72:230–8.
166. Yamamoto H, Quinn G, Asari A, Yamanokuchi H, Teratani T, Terada M, Ochiya T. Differentiation of embryonic stem cells into hepatocytes: biological functions and therapeutic application. Hepatology. 2003;37:983–93.
167. Rambhatla L, Chiu CP, Kundu P, Peng Y, Carpenter MK. Generation of hepatocyte-like cells from human embryonic stem cells. Cell Transplant. 2003;12:1–11.
168. Fair JH, Cairns BA, Lapaglia M, Wang J, Meyer AA, Kim H, Hatada S, et al. Induction of hepatic differentiation in embryonic stem cells by co-culture with embryonic cardiac mesoderm. Surgery. 2003;134:189–96.
169. Heo J, Factor VM, Uren T, Takahama Y, Lee JS, Major M, Feinstone SM, et al. Hepatic precursors derived from murine embryonic stem cells contribute to regeneration of injured liver. Hepatology. 2006;44:1478–86.
170. Yamada T, Yoshikawa M, Kanda S, Kato Y, Nakajima Y, Ishizaka S, Tsunoda Y. In vitro differentiation of embryonic stem cells into hepatocyte-like cells identified by cellular uptake of indocyanine green. Stem Cells. 2002;20:146–54.
171. Thomson JA, Itskovitz-Eldor J, Shapiro SS, Waknitz MA, Swiergiel JJ, Marshall VS, Jones JM. Embryonic stem cell lines derived from human blastocysts. Science. 1998;282:1145–7.
172. David NB, Rosa FM. Cell autonomous commitment to an endodermal fate and behaviour by activation of Nodal signalling. Development. 2001;128:3937–47.
173. Yasunaga M, Tada S, Torikai-Nishikawa S, Nakano Y, Okada M, Jakt LM, Nishikawa S, et al. Induction and monitoring of definitive and visceral endoderm differentiation of mouse ES cells. Nat Biotechnol. 2005;23:1542–50.
174. Kubo A, Shinozaki K, Shannon JM, Kouskoff V, Kennedy M, Woo S, Fehling HJ, et al. Development of definitive endoderm from embryonic stem cells in culture. Development. 2004;131:1651–62.
175. Sherwood RI, Jitianu C, Cleaver O, Shaywitz DA, Lamenzo JO, Chen AE, Golub TR, et al. Prospective isolation and global gene expression analysis of definitive and visceral endoderm. Dev Biol. 2007;304:541–55.
176. D'Amour KA, Agulnick AD, Eliazer S, Kelly OG, Kroon E, Baetge EE. Efficient differentiation of human embryonic stem cells to definitive endoderm. Nat Biotechnol. 2005;23:1534–41.
177. Takenaga M, Fukumoto M, Hori Y. Regulated Nodal signaling promotes differentiation of the definitive endoderm and mesoderm from ES cells. J Cell Sci. 2007;120:2078–90.
178. Feldman B, Gates MA, Egan ES, Dougan ST, Rennebeck G, Sirotkin HI, Schier AF, et al. Zebrafish organizer development and germ-layer formation require nodal-related signals. Nature. 1998;395:181–5.
179. Chazaud C, Yamanaka Y, Pawson T, Rossant J. Early lineage segregation between epiblast and primitive endoderm in mouse blastocysts through the Grb2-MAPK pathway. Dev Cell. 2006;10:615–24.
180. McLean AB, D'Amour KA, Jones KL, Krishnamoorthy M, Kulik MJ, Reynolds DM, Sheppard AM, et al. Activin a efficiently specifies definitive endoderm from human embryonic stem cells only when phosphatidylinositol 3-kinase signaling is suppressed. Stem Cells. 2007;25:29–38.
181. Cleaver O, Melton DA. Endothelial signaling during development. Nat Med. 2003;9:661–8.
182. Matsumoto K, Yoshitomi H, Rossant J, Zaret KS. Liver organogenesis promoted by endothelial cells prior to vascular function. Science. 2001;294:559–63.
183. Navarro-Alvarez N, Soto-Gutierrez A, Rivas-Carrillo JD, Chen Y, Yamamoto T, Yuasa T, Misawa H, et al. Self-assembling peptide nanofiber as a novel culture system for isolated porcine hepatocytes. Cell Transplant. 2006;15:921–7.
184. Cirillo LA, Lin FR, Cuesta I, Friedman D, Jarnik M, Zaret KS. Opening of compacted chromatin by early developmental transcription factors HNF3 (FoxA) and GATA-4. Mol Cell. 2002;9:279–289.
185. Cirillo LA, McPherson CE, Bossard P, Stevens K, Cherian S, Shim EY, Clark KL, et al. Binding of the winged-helix transcription factor HNF3 to a linker histone site on the nucleosome. EMBO J. 1998;17:244–54.
186. Crowe AJ, Sang L, Li KK, Lee KC, Spear BT, Barton MC. Hepatocyte nuclear factor 3 relieves chromatin-mediated repression of the alpha-fetoprotein gene. J Biol Chem. 1999;274: 25113–20.
187. Odom DT, Zizlsperger N, Gordon DB, Bell GW, Rinaldi NJ, Murray HL, Volkert TL, et al. Control of pancreas and liver gene expression by HNF transcription factors. Science. 2004;303: 1378–81.
188. Clotman F, Lannoy VJ, Reber M, Cereghini S, Cassiman D, Jacquemin P, Roskams T, et al. The onecut transcription factor HNF6 is required for normal development of the biliary tract. Development. 2002;129:1819–28.
189. Offield MF, Jetton TL, Labosky PA, Ray M, Stein RW, Magnuson MA, Hogan BL, et al. PDX-1 is required for pancreatic outgrowth and differentiation of the rostral duodenum. Development. 1996;122:983–95.
190. Germain L, Blouin MJ, Marceau N. Biliary epithelial and hepatocytic cell lineage relationships in embryonic rat liver as determined by the differential expression of cytokeratins, alpha-fetoprotein, albumin, and cell surface-exposed components. Cancer Res. 1988;48:4909–18.
191. Suzuki A, Zheng YW, Kaneko S, Onodera M, Fukao K, Nakauchi H, Taniguchi H. Clonal identification and characterization of self-renewing pluripotent stem cells in the developing liver. J Cell Biol. 2002;156:173–84.
192. Hussain SZ, Sneddon T, Tan X, Micsenyi A, Michalopoulos GK, Monga SP. Wnt impacts growth and differentiation in ex vivo liver development. Exp Cell Res. 2004;292:157–69.
193. Monga SP, Monga HK, Tan X, Mule K, Pediaditakis P, Michalopoulos GK. Beta-catenin antisense studies in embryonic liver cultures: role in proliferation, apoptosis, and lineage specification. Gastroenterology. 2003;124:202–16.
194. Coffinier C, Gresh L, Fiette L, Tronche F, Schutz G, Babinet C, Pontoglio M, et al. Bile system morphogenesis defects and liver dysfunction upon targeted deletion of HNF1beta. Development. 2002;129:1829–38.
195. Gualdi R, Bossard P, Zheng M, Hamada Y, Coleman JR, Zaret KS. Hepatic specification of the gut endoderm in vitro: cell signaling and transcriptional control. Genes Dev. 1996;10: 1670–82.
196. Bhatia SN, Balis UJ, Yarmush ML, Toner M. Effect of cell-cell interactions in preservation of cellular phenotype: cocultivation of hepatocytes and nonparenchymal cells. FASEB J. 1999;13: 1883–900.

197. Schmidt C, Bladt F, Goedecke S, Brinkmann V, Zschiesche W, Sharpe M, Gherardi E, et al. Scatter factor/hepatocyte growth factor is essential for liver development. Nature. 1995;373:699–702.
198. Uehara Y, Minowa O, Mori C, Shiota K, Kuno J, Noda T, Kitamura N. Placental defect and embryonic lethality in mice lacking hepatocyte growth factor/scatter factor. Nature. 1995;373:702–5.
199. Bladt F, Riethmacher D, Isenmann S, Aguzzi A, Birchmeier C. Essential role for the c-met receptor in the migration of myogenic precursor cells into the limb bud. Nature. 1995;376:768–71.
200. Huh CG, Factor VM, Sanchez A, Uchida K, Conner EA, Thorgeirsson SS. Hepatocyte growth factor/c-met signaling pathway is required for efficient liver regeneration and repair. Proc Natl Acad Sci U S A. 2004;101:4477–82.
201. Reimold AM, Etkin A, Clauss I, Perkins A, Friend DS, Zhang J, Horton HF, et al. An essential role in liver development for transcription factor XBP-1. Genes Dev. 2000;14:152–7.
202. Rose TM, Bruce AG. Oncostatin M is a member of a cytokine family that includes leukemia-inhibitory factor, granulocyte colony-stimulating factor, and interleukin 6. Proc Natl Acad Sci U S A. 1991;88:8641–5.
203. Wijelath ES, Carlsen B, Cole T, Chen J, Kothari S, Hammond WP. Oncostatin M induces basic fibroblast growth factor expression in endothelial cells and promotes endothelial cell proliferation, migration and spindle morphology. J Cell Sci. 1997;110 (Pt 7):871–9.
204. Grove RI, Eberhardt C, Abid S, Mazzucco C, Liu J, Kiener P, Todaro G, et al. Oncostatin M is a mitogen for rabbit vascular smooth muscle cells. Proc Natl Acad Sci U S A. 1993;90:823–7.
205. Kinoshita T, Sekiguchi T, Xu MJ, Ito Y, Kamiya A, Tsuji K, Nakahata T, et al. Hepatic differentiation induced by oncostatin M attenuates fetal liver hematopoiesis. Proc Natl Acad Sci U S A. 1999;96:7265–70.
206. Matsui T, Kinoshita T, Hirano T, Yokota T, Miyajima A. STAT3 down-regulates the expression of cyclin D during liver development. J Biol Chem. 2002;277:36167–73.
207. de Juan C, Benito M, Alvarez A, Fabregat I. Differential proliferative response of cultured fetal and regenerating hepatocytes to growth factors and hormones. Exp Cell Res. 1992;202:495–500.
208. Shelly LL, Tynan W, Schmid W, Schutz G, Yeoh GC. Hepatocyte differentiation in vitro: initiation of tyrosine aminotransferase expression in cultured fetal rat hepatocytes. J Cell Biol. 1989;109:3403–10.
209. van Poll D, Sokmensuer C, Ahmad N, Tilles AW, Berthiaume F, Toner M, Yarmush ML. Elevated hepatocyte-specific functions in fetal rat hepatocytes co-cultured with adult rat hepatocytes. Tissue Eng. 2006;12:2965–73.
210. Hamazaki T, Iiboshi Y, Oka M, Papst PJ, Meacham AM, Zon LI, Terada N. Hepatic maturation in differentiating embryonic stem cells in vitro. FEBS Lett. 2001;497:15–9.
211. Abe K, Niwa H, Iwase K, Takiguchi M, Mori M, Abe SI, Abe K, et al. Endoderm-specific gene expression in embryonic stem cells differentiated to embryoid bodies. Exp Cell Res. 1996;229: 27–34.
212. Levinson-Dushnik M, Benvenisty N. Involvement of hepatocyte nuclear factor 3 in endoderm differentiation of embryonic stem cells. Mol Cell Biol. 1997;17:3817–22.
213. Shirahashi H, Wu J, Yamamoto N, Catana A, Wege H, Wager B, Okita K, et al. Differentiation of human and mouse embryonic stem cells along a hepatocyte lineage. Cell Transplant. 2004;13:197–211.
214. Runge D, Runge DM, Jager D, Lubecki KA, Beer Stolz D, Karathanasis S, Kietzmann T, et al. Serum-free, long-term cultures of human hepatocytes: maintenance of cell morphology, transcription factors, and liver-specific functions. Biochem Biophys Res Commun. 2000;269:46–53.
215. Nahmias Y, Berthiaume F, Yarmush ML. Integration of technologies for hepatic tissue engineering. Adv Biochem Eng Biotechnol. 2007;103:309–29.
216. Asahina K, Fujimori H, Shimizu-Saito K, Kumashiro Y, Okamura K, Tanaka Y, Teramoto K, et al. Expression of the liver-specific gene Cyp7a1 reveals hepatic differentiation in embryoid bodies derived from mouse embryonic stem cells. Genes Cells. 2004;9:1297–308.
217. Saunier B, Triyatni M, Ulianich L, Maruvada P, Yen P, Kohn LD. Role of the asialoglycoprotein receptor in binding and entry of hepatitis C virus structural proteins in cultured human hepatocytes. J Virol. 2003;77:546–59.
218. Zhang QY, Dunbar D, Kaminsky LS. Characterization of mouse small intestinal cytochrome P450 expression. Drug Metab Dispos. 2003;31:1346–51.
219. Yoshinari K, Sato T, Okino N, Sugatani J, Miwa M. Expression and induction of cytochromes p450 in rat white adipose tissue. J Pharmacol Exp Ther. 2004;311:147–54.
220. Baijal PK, Fitzpatrick DW, Bird RP. Phenobarbital and 3-methylcholanthrene treatment alters phase I and II enzymes and the sensitivity of the rat colon to the carcinogenic activity of azoxymethane. Food Chem Toxicol. 1997;35:789–98.
221. Bertz RJ, Granneman GR. Use of in vitro and in vivo data to estimate the likelihood of metabolic pharmacokinetic interactions. Clin Pharmacokinet. 1997;32:210–58.
222. Pelkonen O, Maenpaa J, Taavitsainen P, Rautio A, Raunio H. Inhibition and induction of human cytochrome P450 (CYP) enzymes. Xenobiotica. 1998;28:1203–53.
223. Kostrubsky VE, Ramachandran V, Venkataramanan R, Dorko K, Esplen JE, Zhang S, Sinclair JF, et al. The use of human hepatocyte cultures to study the induction of cytochrome P-450. Drug Metab Dispos. 1999;27:887–94.
224. Ingelman-Sundberg M. Genetic polymorphisms of cytochrome P450 2D6 (CYP2D6): clinical consequences, evolutionary aspects and functional diversity. Pharmacogenomics J. 2005;5:6–13.
225. Hewitt NJ, Lechon MJ, Houston JB, Hallifax D, Brown HS, Maurel P, Kenna JG, et al. Primary hepatocytes: current understanding of the regulation of metabolic enzymes and transporter proteins, and pharmaceutical practice for the use of hepatocytes in metabolism, enzyme induction, transporter, clearance, and hepatotoxicity studies. Drug Metab Rev. 2007;39:159–234.
226. Lin JH, Lu AY. Inhibition and induction of cytochrome P450 and the clinical implications. Clin Pharmacokinet. 1998;35:361–90.
227. Zuber R, Anzenbacherova E, Anzenbacher P. Cytochromes P450 and experimental models of drug metabolism. J Cell Mol Med. 2002;6:189–98.
228. Guengerich FP, Shimada T, Iwasaki M, Butler MA, Kadlubar FF. Activation of carcinogens by human liver cytochromes P-450. Basic Life Sci. 1990;53:381–96.
229. Hollenberg PF. Characteristics and common properties of inhibitors, inducers, and activators of CYP enzymes. Drug Metab Rev. 2002;34:17–35.
230. Hesse LM, Venkatakrishnan K, Court MH, von Moltke LL, Duan SX, Shader RI, Greenblatt DJ. CYP2B6 mediates the in vitro hydroxylation of bupropion: potential drug interactions with other antidepressants. Drug Metab Dispos. 2000;28:1176–83.
231. Wang H, Faucette SR, Gilbert D, Jolley SL, Sueyoshi T, Negishi M, LeCluyse EL. Glucocorticoid receptor enhancement of pregnane X receptor-mediated CYP2B6 regulation in primary human hepatocytes. Drug Metab Dispos. 2003;31:620–30.
232. Williams JA, Ring BJ, Cantrell VE, Jones DR, Eckstein J, Ruterbories K, Hamman MA, et al. Comparative metabolic capabilities of CYP3A4, CYP3A5, and CYP3A7. Drug Metab Dispos. 2002;30:883–91.
233. Li YC, Wang DP, Chiang JY. Regulation of cholesterol 7 alpha-hydroxylase in the liver. Cloning, sequencing, and regulation

of cholesterol 7 alpha-hydroxylase mRNA. J Biol Chem. 1990;265:12012–19.
234. van Montfoort JE, Hagenbuch B, Groothuis GM, Koepsell H, Meier PJ, Meijer DK. Drug uptake systems in liver and kidney. Curr Drug Metab. 2003;4:185–211.
235. Mita S, Suzuki H, Akita H, Hayashi H, Onuki R, Hofmann AF, Sugiyama Y. Inhibition of bile acid transport across Na+/taurocholate cotransporting polypeptide (SLC10A1) and bile salt export pump (ABCB 11)-coexpressing LLC-PK1 cells by cholestasis-inducing drugs. Drug Metab Dispos. 2006;34: 1575–81.
236. Duncan SA, Navas MA, Dufort D, Rossant J, Stoffel M. Regulation of a transcription factor network required for differentiation and metabolism. Science. 1998;281:692–5.
237. Rigato I, Cravatari M, Avellini C, Ponte E, Croce SL, Tiribelli C. Drug-induced acute cholestatic liver damage in a patient with mutation of UGT1A1. Nat Clin Pract Gastroenterol Hepatol. 2007;4:403–8.
238. Kolarich D, Turecek PL, Weber A, Mitterer A, Graninger M, Matthiessen P, Nicolaes GA, et al. Biochemical, molecular characterization, and glycoproteomic analyses of alpha(1)-proteinase inhibitor products used for replacement therapy. Transfusion. 2006;46:1959–77.
239. Mulgrew AT, Taggart CC, McElvaney NG. Alpha-1-antitrypsin deficiency: current concepts. Lung. 2007;185:191–201.
240. Saheki T, Kobayashi K, Inoue I. Hereditary disorders of the urea cycle in man: biochemical and molecular approaches. Rev Physiol Biochem Pharmacol. 1987;108:21–68.
241. Azuma H, Paulk N, Ranade A, Dorrell C, Al-Dhalimy M, Ellis E, Strom S, et al. Robust expansion of human hepatocytes in Fah(-/-)/Rag2(-/-)/Il2rg(-/-) mice. Nat Biotechnol. 2007;25: 903–10.
242. Starzl TE, Lakkis FG. The unfinished legacy of liver transplantation: emphasis on immunology. Hepatology. 2006;43: S151–63.
243. Shizuru JA, Negrin RS, Weissman IL. Hematopoietic stem and progenitor cells: clinical and preclinical regeneration of the hematolymphoid system. Annu Rev Med. 2005;56:509–38.
244. Nussbaum J, Minami E, Laflamme MA, Virag JA, Ware CB, Masino A, Muskheli V, et al. Transplantation of undifferentiated murine embryonic stem cells in the heart: teratoma formation and immune response. FASEB J. 2007;21:1345–57.
245. Wernig M, Meissner A, Foreman R, Brambrink T, Ku M, Hochedlinger K, Bernstein BE, et al. In vitro reprogramming of fibroblasts into a pluripotent ES-cell-like state. Nature. 2007;448:318–24.
246. Nagata H, Nishitai R, Shirota C, Zhang JL, Koch CA, Cai J, Awwad M, et al. Prolonged survival of porcine hepatocytes in cynomolgus monkeys. Gastroenterology. 2007;132: 321–9.

Part V
Disease Paradigms and Stem Cell Therapeutics

The Idea and Evidence for the Tumor Stemness Switch

Bikul Das, Rika Tsuchida, Sylvain Baruchel, David Malkin and Herman Yeger

Abstract The maintenance of stemness of normal stem cell is a complex process, where transcription factors like Oct-4, Bmi-1, and signaling pathways such as Wnt/β-catenin play important roles. This molecular set of mechanisms not only expands the population (self-renewal) but also keeps stem cells in a state of "de-differentiation." Thus, stemness and differentiation are mutually exclusive and tightly regulated, where the idea of variation of stemness over time has not been incorporated. However, unlike normal stem cell stemness, tumor stemness may not be tightly regulated, where complexity of tumor microenvironment, especially hypoxic stress may allow for variation in stemness.

In this review, we discuss the stem cell model of tumor growth and the emerging concept of tumor stemness. We also discuss our findings on the expansion of tumor stem cell–like side-population cells following hypoxic and drug-induced stress. We then propose a model of stemness switch, where quiescent TSCs (tumor stem cells) switch to a state of active and self-renewing TSC following stress.

Keywords Tumor stemness · Oxidative stress · Hypoxia · Tumor stem cell · Tumor stem cell niche

1 Stem Cell Model of Tumor Growth

An emerging and provocative concept in tumor biology is that a rare population of "tumor stem cells" (TSC) exists among the heterogeneous population of cells within tumors. The TSC model suggests that proliferative potential and growth patterns of many human tumors may depend upon a small proportion of TSCs that lead to repopulation following cytotoxic therapy [1]. McCulloch and Till proposed that, similar to HSCs (hematopoietic stem cells), a subset of leukemia also has self-renewal property and these leukemic stem cells (LSCs) have unlimited proliferation capacity [2–5]. Subsequently, Bonnet and Dick [6] identified a subset of AML (acute myelogenous leukemia) stem cells (CD 34+ CD38–). Similar to leukemia, which is a "liquid tumor," solid tumor growth may also be maintained by TSCs. In 1977, Hamburger et al. [7] reported that only 1 in 1000 to1 in 5000 lung cancer, ovarian cancer, or neuroblastoma cells formed colonies in soft agar. Therefore, it was hypothesized that, similar to leukemia, solid tumor also contains a small fraction of quiescent stem cells. Similar findings by other investigators [8–18] suggested that, similar to leukemia, solid tumor may also have a stem cell–like hierarchical organization pattern. This stem cell model of solid tumor growth is further supported by the recent isolation of several TSCs from breast, brain, prostate, and lung tumors. These TSCs are characterized by their ability to initiate tumor growth, compared to nontumor stem cells, which are nontumorigenic, even though isolated from the same primary tumors [19–21]. Furthermore, side-population (SP) cells obtained from several solid primary tumors have been found to have TSC-like properties [22, 23]. Normal stem cells have been isolated by their ability to efflux Hoechst 33342 dye and form a distinct population definable by flow cytometry – the SP. SPs have been defined in a number of tissues, and it has been proposed that the SP phenotype is a universal stem cell marker [24–26]. SP cells isolated from bone marrow are about 1000-fold enriched for HSC activity, and most of the HSC-derived SP cells have the Sca-1+/lin neg/low cell surface marker [24]. During 2004–2005, tumor SP cells from primary tumors as well as established cell lines having stem cell–like activities were identified from neuroblastoma, prostate, and other solid tumors.

Established cell lines may not be a good model to study TSCs [20] since questions arise as to whether self-renewal and hierarchical regulatory mechanisms are maintained in the TSCs derived from them. However, substantial evidence suggests that tumor as well as normal stem cell lines retain the hierarchical organization attributed to the normal stem

B. Das (✉)
Stem Cell and Developmental Biology Program, The Hospital for Sick Children, 555 University Avenue, Toronto, Canada. M5G 1X8.
e-mail: bikul.das@utoronto.ca

V.K. Rajasekhar, M.C. Vemuri (eds.), *Regulatory Networks in Stem Cells*, Stem Cell Biology and Regenerative Medicine,
DOI 10.1007/978-1-60327-227-8_35, © Humana Press, a part of Springer Science+Business Media, LLC 2009

cell biology. For example, teratocarcinoma and embryonic stem (ES) cell lines maintain the stem cell self-renewal and hierarchical organization pattern. In the early 1980s, Evans and Martin reported the isolation of teratocarcinoma stem cell lines (EC lines) directly from preimplantation embryos explanted into a culture treated with teratocarcinoma conditioned media [27, 28]. However, when these EC cells were reintroduced into mouse embryos, they contributed widely to tissues of the resulting chimera and subsequently to pass through the germline. These cells were renamed as the ES cells [28, 29]. Thus, the idea and evidence of the existence of ES cells historically came from the research on teratocarcinoma, and knowledge of molecular regulation of ES cells has been historically obtained from studying teratocarcinoma cell lines in their in vitro cultures.

Another example of the maintenance of self-renewal hierarchy in the in vitro culture is the stem cell fraction of normal and malignant epithelial stem cells. When normal keratinocytes are plated at low density, a range of colonies are formed having different morphological attributes. These are classified as holoclones, meroclones, and paraclones, which are derived from, and contain, cells corresponding to stem and early and late transit-amplifying cells, respectively, suggesting that the self-renewal property of stem cells may be intrinsic, and therefore may be maintained in the in vitro cell culture [30]. Lock et al. [31] showed that carcinoma cell lines (CA1 and UK1, cell lines derived from head and neck squamous cell carcinoma) also exhibit holoclone, meroclone, and paraclone patterns of colony formation. Holoclone colonies show higher proliferative and self-renewal capacity than the other types of colonies.

The neuroblastoma cell line (SK-N-BE(2)) comprises neuronal, and Schwannian and intermediate cells. These intermediate cells are considered as the "stem cells" of both the neuronal and Schwannian fractions, which suggests that the cell line maintains the hierarchical organization attributed to stem cells [32]. There is substantial evidence that only a small percentage of cells within tumors can give rise to colonies, suggesting that tumor cell lines may maintain the hierarchical organization, where the stem cell fraction retains unlimited proliferative potential whereas the transit-amplifying cells differentiate and lose their self-renewal and proliferative potential gradually [33].

The most convincing and detailed evidence of the retention of TSCs in the established cell lines have been obtained recently from the work on tumor SP cells. SP cells obtained from established cell lines show TSC-like activities including (i) very high clonogenic and tumorigenic activity (e.g., 100 cells of U373 glioma cell line can form a xenograft within 20 days, whereas non-SP cells of the same cell line do not form a tumor); (ii) high drug resistance; (iii) requirement for growth factor–supplemented serum-free media for maintenance (for example, the C6 glioma cell line requires bFGF and PDGF; without the appropriate growth factors, SP cells rapidly give rise to non-SP cells); and (iv) expression of several embryonal genes including Oct-4 and Rex-1 [22, 23, 34]. We isolated highly tumorigenic SP cells from a diverse group of tumor cells including osteosarcoma (HOS), rhabdomyosarcoma (RH-4), and neuroblastoma (SK-N-BE(2)) that requires appropriate growth factors bFGF, EGF, and PDGF to maintain the SP cells in culture [35]. Only the SP cells of HOS, RH-4, and SK-N-BE(2) showed tumorigenicity whereas non-SP cells did not form tumor even when a large number of non-SP cells were injected subcutaneously to nude mouse, suggesting that SP cells are enriched in the TSC fraction [36–38].

2 Stemness and Tumor Stemness

These previous findings provided substantial experimental evidence to support the concept of a TSC model – a rare population of self-renewing stem cells among the heterogeneous mixture of tumor cells essential for solid tumor growth and progression [19–21]. However, this growing evidence is not by itself convincing without an adequate molecular mechanism explaining how these TSCs maintain their stemness, since there is always a possibility that the so-called solid tumor TSCs may be a highly tumorigenic fraction of the heterogeneous mixture of tumor cells [39]. The lack of knowledge of how TSCs maintain their state of stemness is therefore a major limitation that requires attention in order to improve our understanding on the stem cell model of tumor growth.

Stemness refers to the unique features of resistance to differentiation, self-renewal, and multipotency [40]. One major requirement of maintaining stemness is the inheritance of the epigenetic memory by the daughter stem cells. The daughter stem cells should inherit the same epigenetic status to maintain their characteristics. Thus, it may be predicted that the cellular mechanism that regulates self-renewal may also be involved in maintaining the other two characteristics of stem cells, that is, "resistance to differentiation" and "retaining the multipotent potential." It is therefore not unexpected to find that polyclonal group members such as Bmi-1 are involved in the maintenance of HSC self-renewal and epigenetic memory [41]. The POU home domain transcription factor Oct-4 is an essential mediator of ES cell stemness and implicated in self-renewal as well as lineage-specific differentiation. Oct-4 initiates and maintains the transcriptional activation of genes involved in self-renewal and pluripotency and concomitantly represses genes that facilitate lineage-specific differentiation [42, 43]. Wnt and Notch 1 transcription factors play key roles in HSC self-renewal and lineage differentiation [44]. The maintenance of stemness takes places in the protective

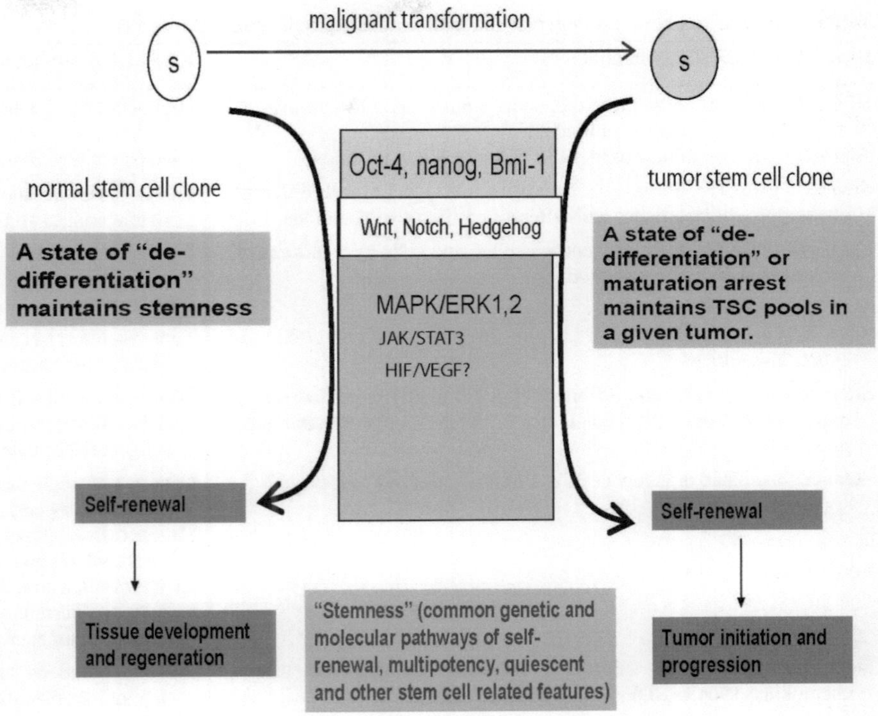

Fig. 1 Normal and tumor stemness. A schematic diagram representing several stemness pathways including Oct-4. Stem cell niche regulates these pathways to keep the stem cells in a state of de-differentiation. Similar pathways may also be involved in maintaining tumor stem cells (TSCs) in their niche in a state of de-differentiation, leading to the persistence of the TSCs in the solid tumors. Hypoxia microenvironmetal signaling such as HIF-1 and VEGF may play an important role in regulating stemness pathways. (Modified from Das Bikul. The role of VEGF autocrine signaling in hypoxia and oxidative stress driven "stemness switch." PhD thesis, University of Toronto, 2007)

microenvironment of the stem cell niche. A complex interaction between the signaling of the stemness process and the niche environment leads to the maintenance of a "de-differentiated state" as well as the expansion of stem cells (Fig. 1). For example, ES cells remain in a relatively hypoxic state where Oct-4 signaling is up-regulated. HIF-2α, a key regulator of hypoxia, also regulates Oct-4 expression in ES cells. In bone marrow, HSCs remain in their hypoxic niche, where microenvironment signaling may help HSCs to maintain their stemness [45].

The first attempt to address TSC stemness was made during 1980s, when Sell et al. [46, 47] proposed the idea of a maturation arrest model to provide a theoretical basis for the persistence of TSCs in teratocarcinoma, leukemia, and epithelial tumors. In this model, the de-differentiation (maturation arrest) state would play the key role in maintaining the hierarchical organization of TSC-mediated growth. The genetic defect would thus occur in a primitive stem cell, but the disease would be manifested in its downstream progenitors, so that the tumor-initiating stem cell would retain its de-differentiated state [46–48].

The major limitation of the maturation arrest model is that it does not explain how the TSCs maintain their stemness state (i.e., preventing "de-differentiation" to maintain the stemness, or the tumorigenic ability). The maturation arrest model can explain the tumor initiation process, however, it does not provide a mechanism for maintaining the stemness of these TSCs once tumor progression from a clone containing a few thousand cells to a rapidly growing tumor occurs. It can be argued, that once a tumor forms, irrespective of cells of origin, clonal evolution would set in, and all the tumor cells would acquire the self-renewal capacity like a TSC, thus losing the hierarchical structure of the tumor cells. However, overwhelming evidence suggests that TSCs are maintained in solid tumors, as discussed earlier in the chapter. Considering the many similarities between normal and tumor stem cells (Table 1), it can be argued that pathways involved in maintaining normal stem cells may also be involved in maintaining TSCs (Fig. 1). Recent evidence suggests that normal stemness signaling pathways such as Oct-4, Wnt/β-catenin, and Bmi-1 may play an essential role in regulating TSC self-renewal and maintenance [41, 49, 50]. Using a mouse/human comparative translational genomics approach, Glinsky et al. identified an 11-gene signature of tumor stemness and showed that developmental signaling pathways such as Bmi-1 may play a key role in enhancing the stemness of the tumor cell population [51]. Recent studies in tumor stemness highlighted the importance of several such signaling pathways in the regulation of tumor stemness.

3 Tumor Stemness–Related Pathways

Studies of human embryonic stem cells (hES) have shown that an intrinsic autocrine loop of Oct-4 and bFGF-2 signaling may maintain hES self-renewal in established hES

Table 1 Comparison between stem cell and tumor initiating cells (tumor stem cells)

Extensive proliferative potential	Extensive proliferative potential
High telomerase activity, therefore, some adult stem cells can live for the entire life span of the individual	High telomerase activity, and therefore immortal
Ability to give rise to new tissue, and take part in regeneration	Give rise to new tissue (tumor), and take part in relapse
Stem cell gives rise to tissue of heterogeneous type i.e. with different phenotypic characteristics and different proliferative potentials	Tumor is composed of heterogeneous cells with different phenotypic characteristics and different proliferative potentials
Stem cell gives rise to heterogeneous tissue and cells by a process of self-renewal and differentiation	Tumorigenic cancer cell may also undergo process of differentiation and self-renewal, even though this process may be aberrant and deregulated
Stem cells are of clonal origin, i.e. one single cell can give rise to an entire tissue/organ	Tumorigenic cells can be viewed as cancer stem cells, where one single transformed cancer stem cell can give rise to the entire tumor
Stem cell self-renewal and differentiation are tightly regulated where transcription factors like Bmi-1, Oct-4, Notch, Wnt/betacatenin take part	Cancer stem cell self-renewal process may be deregulated since Bmi-1, Oct-4, Wnt/beta-catenin and Notch pathways have been found to be deregulated in many cancers
Stem cell mediated organogenesis is a tightly regulated process	Cancer stem cell mediated tumorigenesis may be considered as an aberrant and poorly regulated process of organogenesis Cancer cells display functional and phenotypic attributes of the normal cell from which they are derived, i.e. brain tumor will display the phenotype of brain and not of ovary etc. thus, it can be presumed that brain tumor is an aberrant manifestation of the poorly regulated development of a brain tumor stem cell to an abnormal part of the brain
Stem cells have high migratory and invasive property, where chemokines such as SDF-1/CXCR4 is involved.	Tumorigenic cells are also highly migratory and invasive, where chemokines like SDF-1/CXCR4 is involved in migration and homing
Stem cells are highly drug resistant and expresses drug-pump such as BCRP1, which probably help them to survive hostile environment	Highly tumorigenic side-population cells has been found to express BCRP1
Stem cell has a tendency to grow in hypoxic condition	Tumorigenic cells tend to prefer hypoxic condition

cell lines [52]. A similar mechanism of Oct-4–mediated self-renewal may also operate in teratocarcinoma stem cells [53, 54]. Oct-4 expression is also retained in the neural crest stem cells, even though its down-regulation is essential for the normal development of the endoderm [55]. Recently, Yamaquchi et al. [56] reported that ectopic Oct-4 expression results in dysplastic growth in epithelial tissues. Dysplastic lesions showed stem cell expansion and increased β-catenin transcriptional activity [57]. Neuroblastoma, a pediatric tumor, arises in tissues derived from embryonal neural crest [58] that expresses a high level of Oct-4 [43]. We studied the expression of Oct-4 in MYCN-amplified neuroblastoma cell line SK-N-BE(2) and found that the highly tumorigenic SP fraction expresses Oct-4, nanog, and Rex-1 [36]. Similar findings were observed by Patwarala et al. [22], who showed that high expression of the stemness gene is correlated with high tumorigenicity.

Bmi-1, the polycomb group gene, is usually involved in the self-renewal and proliferation of normal HSCs [50]. Overexpression of Bmi-1 causes neoplastic transformation of lymphocytes and Bmi-1 is widely expressed in all primary myeloid leukemia, human non-small-cell lung cancer, and breast cancer cell lines [50, 51]. Others found that Bmi-1 is highly expressed in tumor stem cell fraction of a diverse group of tumors including breast, prostate, and lung tumors [51, 59–61].

The sonic-hedgehog Sonic hedgehog (Shh) pathway is found to be involved in medulloblastoma initiation and progression. During cerebellar development, the shh/hgg pathway is involved in the maintenance and proliferation of the granular precursor cells (GPCs), which are the neural stem cells that migrate from the dorsal hind brain to the external germinal layer of cerebellum [62]. Medulloblastoma derives from the GPCs and aggressive medulloblastomas show very high activity of the Shh-Ptc1 signaling cascade [62].

The Wnt/β-catenin pathway that regulates the HSCs and neural stem cell self-renewal also plays key role in the EMT (epithelial mesenchymal transition) process, one of the key events in cancer initiation and progression [57]. Wnt/β-catenin signaling plays an important role in the progression of lung, colon, and pancreatic tumors [63, 64].

One of the important questions related to tumor stemness is its relation to tumor progression. Does tumor stemness vary as the tumor progresses from a less aggressive to highly aggressive tumor? Will the TSC acquire new stemness-related property to become more aggressive, drug resistant, and metastatic?

4 Relation Between Tumor Progression and Tumor Stemness: The Hypoxia-Microenvironment Factor

The pathological and clinical criteria of tumor progression are mainly a gradual progression from initial benign growth to the acquisition of malignant characteristics, leading to the development of highly aggressive and metastatic growth [65, 66]. This underlying general behavior of tumors during malignant progression was initially proposed by Leslie Foulds in the early 20th century, who described the tumor as a "dynamic process advancing through stages that are qualitatively different," and progressing from benign stages to the invasive and metastatic spectrum of malignant transformation [65, 66]. Subsequently, Peter Nowell further developed the idea and explained it in the framework of a clonal evolution model, where, as the tumor grows, it acquires and accumulates mutations and genetic instability leading to selection of aggressive tumor cells [15, 67].

Mackillop and Buick [68] first attempted to unite both clonal evolution and TSC model of tumor progression and suggested that deregulation of the genetic pathway involved in normal stem cell self-renewal may result in maturation arrest and the expansion of a particular stem cell clone [68]. The clones with higher self-renewal capacity would have a selective growth advantage and dominate, that is, autonomous growth as the main driving force of clonal selection of self-renewal capacity. This model of the selection of TSCs on the basis of self-renewal capacity suggests the possibility of a varying degree of stemness, because, as the tumor progresses in time, clonal selection may select clones having a higher degree of stemness, whereas the clones having lower stemness may undergo differentiation and apoptosis.

Lessard and Sauvageau [50] showed that leukemia stem cells lacking Bmi-1 lacked self-renewal potential and underwent differentiation and apoptosis. When Bmi-1 was reintroduced to these leukemia stem cells, the self-renewal capacity was rescued [50]. Thus, the higher the expression of Bmi-1 in the leukemia stem cells, the greater their chance of survival and achieving clonal dominance. Considering that tumor cells contain a wide range of genetic and epigenetic alterations, the varying degree of stemness is a possibility that cannot be ignored [39].

In the clonal evolution model of tumor progression, microenvironmental stress such as hypoxia may play a key role in both genetic and epigenetic changes. Graeber et al. [69] showed that solid tumor hypoxia induced apoptosis in p53+/+ cells, whereas p53–/– cells survived. This finding suggested that "hypoxia provides a physiological selective pressure in tumors for the expansion of variants that have lost their apoptotic potential," leading to clonal expansion of a highly resistant and proliferating subclone [69]. Tumor hypoxia activates hypoxia-responsive pathways such as HIF-1α (hypoxia-inducible factor-1α), and HIF-2α, which are transcription factors involved in the regulation of many growth-related pathways of survival and angiogenesis during hypoxia [70]. We reported that tumor hypoxia increases the MAPK/ERK1,2 and VEGF autocrine loop leading to the selection of a highly aggressive, angiogenic, and drug-resistant population [71].

The exposure of TSCs to various stresses such as exposure to drug and also hypoxia/oxidative stress may up-regulate the pathways involved in tumor stemness, leading to increase tumor stemness and, therefore, tumor progression. Studies of bone marrow stem cells suggest that the microenvironmental niche plays a significant role in influencing the status of stemness, that is, the ability to differentiate into different type of progenitors [72]. Bone marrow niche are often hypoxic [45], and hypoxia may also effect stemness related pathways such as Oct-4 [54], Rex-1 [73], Wnt/β-catenin [74, 75], and Notch [76] pathways.

Hypoxia is known to expand normal stem cells in their niche. A key feature of stem cell niche is the regulation of stem cell population. The normal niches provide appropriate signals to the stem cells either to remain quiescent, or to expand in response to injury/stress [77]. In the BM microenvironment, during stress, HSCs migrate from their hypoxic niche near the osteoblasts to the marrow center to expand (223). VEGF signaling contributes to the survival of the HSCs in hypoxia [78]. Brusselmans et al. [79] showed that hES cell growth increases during hypoxia, where VEGF autocrine survival is required to prevent hypoxia-induced apoptosis and growth inhibition.

Hypoxia may influence TSC maintenance in two possible ways: (1) expansion of TSCs, and (2) increasing the "de-differentiation" or the stemness state of TSCs, that is, reverting differentiated TSCs to a "de-differentiated" state. Either way, the net result would be an increase degree of tumor stemness.

5 Expansion of TSCs in Tumor Hypoxic Niche

Tumor hypoxia is in dynamic states of flux with both chronic and intermittent components [80]. The latter state, being an in vivo process of hypoxia-reoxygenation, may mimic "injury/stress" and, therefore, activate/switch quiescent TSCs to self-renewal/expand similar to the expansion of HSCs in their hypoxic niche. Suda et al. [45] showed that HSCs reside in their hypoxic niche, whereas, when required, these HSCs migrate to the angiogenic niche leading to self-renewal and expansion. Whereas during normal physiology such expansion is tightly regulated, we would expect that in tumors an uncontrolled expansion of TSCs would occur during hypoxia-induced stress. We investigated this possibility by using an acute hypoxia-reoxygenation model that we developed recently [71]. In this model, tumor cells are exposed to intense hypoxia (0.1% oxygen) for 24 h and then re-exposed to normoxia for the next 24 h. Using this model, we found that the tumor SP fraction is increased in a diverse group of tumors including the increased fraction of Oct-4-positive SP cells [36, 38, 81]. When this post-hypoxia expanded SP fraction is then cultured under normoxia and subsequently injected into immunocompromised mice, they show a more than 1000-fold increase in tumorigenicity compared to the SP fraction under normoxia [36]. We studied the expression of Oct-4 in MYCN-amplified neuroblastoma cell line SK-N-BE(2) and found that only 10% of SP cells showed expression of Oct-4. Following hypoxia and reoxygenation, the number of SP cells increased significantly, suggesting an expansion of Oct-4 positive TSC fraction. Furthermore, the hypoxic zones of SK-N-BE(2)-derived xenografts showed high expression of Oct-4-positive cells [36]. These results suggest that tumor hypoxia may increase the number of TSCs (Fig. 2a).

6 TSCs May Expand in Response to Injury/Stress

One of the important properties exhibited by stem cells is expansion and then migration to the area of injury for repair and regeneration. Tumor may also be considered an injurious tissue, where hypoxic/necrotic zones may act as the area of injury, and where, in an endless cycle of injury, repair may continue as the tumor progresses. The process of repair and regeneration may play an important role in tumor growth and progression [38]. In such a model of tumor repair and regeneration, TSCs may migrate to the area of injury. Such a property of TSC may be exploited to isolate highly tumorigenic fraction of TSCs. We investigated this possibility by using a Boyden's chamber system, where tumor SP cells were allowed to migrate. Considering that SDF-1α gradient is critical for the migration of normal stem cells to the area of injury zone, we used SDF-1-rich bone marrow stromal cell–conditioned media as a chemokine. We found that migratory SP fraction (SPm) was highly tumorigenic, whereas nonmigratory SP fraction (SPn) did not show tumorigenic activity [36]. Furthermore, hypoxia and reoxygenation increased the

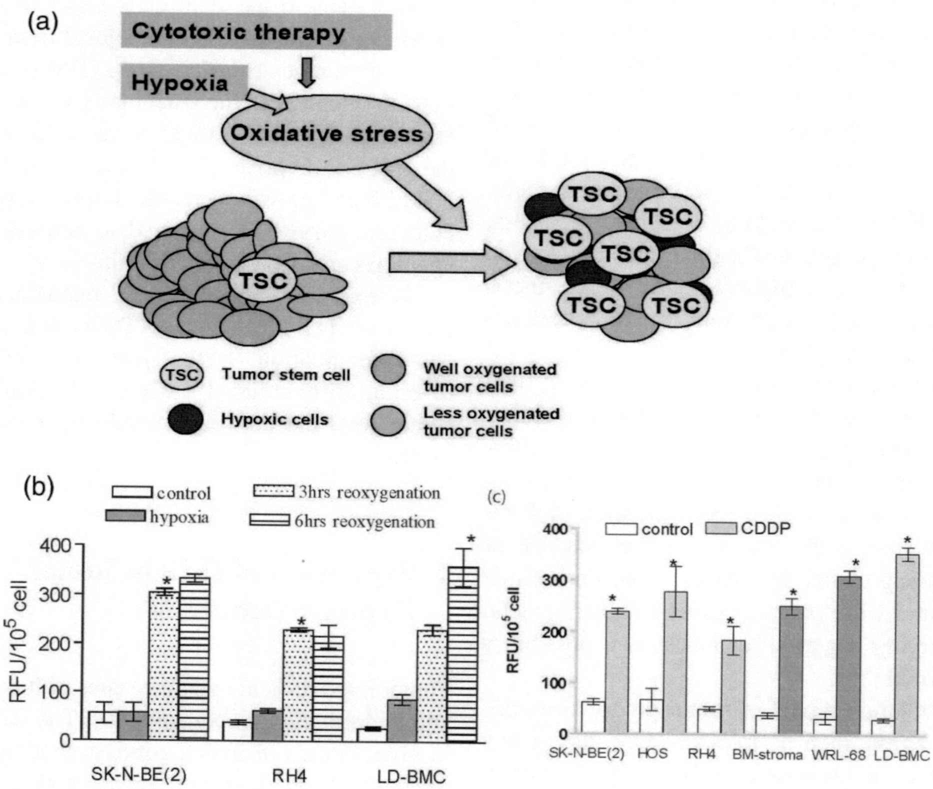

Fig. 2 Hypoxia and drug-induced oxidative stress may expand TSCs **(a)** A schematic diagram to summarize our findings that hypoxia and exposure to cytotoxic therapy such as cisplatin and doxorubicin increases TSC-like SP cells of SK-N-BE(2), RH4, and HOS cells as described in references [81] and [82]. **(b)** Increase production of ROS in tumor cells following hypoxia and reoxygenation. Light-density bone marrow (LD-BMC) was used to represent normal cells. **(c)** Increased production of ROS in tumor cells following exposure to cisplatin (CDDP) 2 μM for 24 h. ROS was measured by DCFH-DA assay as described in [83]. Bone marrow stromal cells (BM stroma), WRL-68, a fetal hepatic cell line, and LD-BMC were used to represent normal cells. Data taken from the PhD thesis of Bikul Das (University of Toronto, 2007) (*see* also Color Insert)

number of SPm fraction by 2–4-fold in neuroblastoma, rhabdomyosarcoma, and small cell carcinoma cell lines [36]. We also found that post-hypoxia SPm fraction (SPm(hox)) cells were highly metastatic and tumorigenic compared to SPm fraction. qPCR studies showed increase expression of Oct-4 and nanog in SPm (hox) versus SPm fraction, suggesting that hypoxia-induced stress may increased the stemness of the SPm fraction [36].

Such expansion and increase of stemness following stress may not be limited to hypoxia-induced stress. We showed that exposure to chemotherapeutic-induced stress may also expand tumorigenic SP fraction with an enhance degree of stemness (the expression of Oct-4, nanog, and Bmi-1 are increases in the TSC fraction following exposure to cisplatin) [35]. In an osteosarcoma cell line HOS, which is nontumorigenic, 3 days' exposure to cisplatin expanded a quiescent but highly tumorigenic SP fraction [35]. Exposure to cisplatin especially increased the SK-N-BE(2) SPm fraction and expanded the highly tumorigenic fraction of HOS and RH-4 (a rhabdomyosarcoma cell line) [35], suggesting that drug-induced stress may expand TSCs (Fig. 2a), similar to the mobilization of bone marrow stem cells following drug-induced stress.

Interestingly, both hypoxia and cisplatin treatment lead to oxidative stress that increases ROS generation (Fig. 2b). Thus, a common underlying mechanism of the expansion and migration of a highly tumorigenic fraction of SP cells may be oxidative stress. The oxidative stress microenvironment plays an important role in the clonal evolution of tumor progression by permitting/potentiating genetic instability, epigenetic modulation of gene expression, and the activation of growth and survival-related signal transduction pathways [82]. Several cellular growth and proliferation-related signal transduction pathways are activated by ROS signaling. Among these are mitogen-activated protein kinase (MAPK) and the redox sensitive kinases [6]. We found that MAPK/ERK1,2 and VEGF/Flt1 autocrine pathways may play significant roles in drug-induced SP cell expansion [35] as well as hypoxia/reoxygenation-induced SPm cell expansion [37, 83]. siRNA inhibition of Flt1 reduced nanog [35] and Oct-4 expression [83], suggesting that stress-induced activation of VEGF/Flt1 and MAPK/ERK1,2 autocrine loop may play an important role in the expansion of TSC fraction.

7 Conversion of Non-TSCs to TSCs: The Process of De-Differentiation

Microenvironmental-generated signaling may play a central role in the conversion of progenitor cell to stem cell. In the normal stem cell niche, Jak/STAT3 signaling has been found to play a key role in the conversion of progenitor cells to germ cells [84]. Recently, mammary fibroblasts were "de-differentiated" to ES cells by transfection with oct-4, nanog, and c-myc [85–88].

Hypoxia has been reported to de-differentiate SK-N-BE(2) cells to a primitive phenotype, resulting in increased tumorigenicity [89]. Hypoxia may also de-differentiate the non-SP to SP phenotype, leading to tumorigenicity. In other words, de-differentiation may play a major role in the tumorigenesis or tumor initiation process by de-differentiating progenitor cells (e.g., non-SP) to TSC like cells (e.g., SP and SPm cells). To investigate this possibility, non-SP cells were exposed to hypoxia with or without treatment with U 0126 and AG 490, inhibitors of MAPK/ERK1,2 and JAK/STAT3 signaling, respectively. We found that a small percentage of cells (2–4%) were converted to SP cells, the process being inhibited by both U 0126 and AG 490 treatment (Fig. 3) [83]. SPm cells isolated from the converted SK-N-BE(2) SP cells formed actively growing tumor in vivo [83], suggesting that even non-SP cells or non-TSC cells can become TSC-like cells following activation of the appropriate signaling pathways under stress.

8 From Dormancy to Tumor Progression: The Tumor Stemness Switch

In the TSC model of tumor growth, dormancy may be explained as a state where TSCs remain in the niche in a quiescent state. Once tumor starts growing, we would expect TSCs number to increase (expand), leading to tumor progression. However, the traditional TSC model suggests that, in a given tumor population (including rapidly growing tumors), only a small fraction will represent true TSCs. This quiescent fraction is often considered to be a small percentage, as low as 1% [39]. During 1980s, when the idea of TSCs began to emerge as an alternative model of tumor growth, a few lines of evidence were suggestive of only a small fraction of tumor cells being tumorigenic (and, therefore, this tumorigenic fraction is TSCs). It was found that a successful radiation treatment could be achieved without killing all the cells [90]. Furthermore, it was found that a relatively large number of tumor cells was required to form an in vivo xenograft [63]. The idea of quiescent TSC achieved greater credibility when the leukemia stem cell was discovered and found to be quiescent, similar to HSCs [91]. However, as the identification of solid tumor TSCs gained momentum, especially during last 5 years, it appears that TSCs may not be as quiescent as first thought. Singh et al. [92] showed that CD 133+ brain TSCs comprised 6–29% of the primary brain tumors, and that a higher number of TSCs correlated with a more aggressive tumor type. Al-Hajj et al. [93] showed that CD 44+, CD 24– breast TSCs comprised

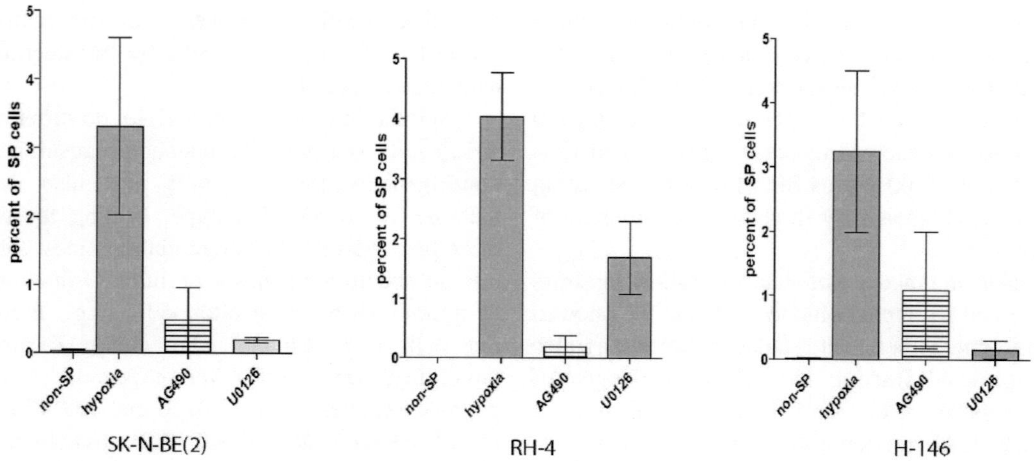

Fig. 3 Conversion of non-SP to SP cells following exposure to hypoxia. Effect of AG490 and U0126 on conversion of non-SP to SP cells following hypoxia in SK-N-BE(2), RH-4, and H-146 cell lines. Freshly sorted non-SP cells were exposed to 24 h hypoxia with and without treatment with 20 μM of AG490 and U0126 and then allowed to expand for next 4 days in normoxia. SP cells were analyzed by FACS. Since U0126 treatment induces cell death of non-SP cells during hypoxia, the surviving cells were mixed with a population of non-SP cells for FACS analysis. Hypoxia treatment showed the reappearance of SP cells (2.4–4.5%). AG940 and U0126 treatment significantly reduced SP cell number. Data taken from the PhD thesis of Bikul Das (University of Toronto, 2007)

11–35% of primary breast cancer. These large variations of number of TSCs suggest that TSCs in a growing tumor are not a limited and quiescent population. This is not surprising and is, in fact, predictable, because, in the tumor microenvironment TSCs would self-renew and expand as a result of microenvironmental stress.

The idea and evidence behind TSCs as a quiescent population were mainly derived from the in vitro culture of clonogenic cells, where it was found that only a few tumor cells harbor tumorigenicity [33, 68, 94]. Usually, to test tumorigenicity, tumor cells are cultured in vitro and then injected in vivo. Therefore, it is obvious that, during the in vitro culture period, in the normoxic environment and also without the appropriate niche environment, TSCs would likely be maintained in a quiescent state while the rest of the population consists mainly of transit-amplifying cells. This notion is supported by evidence that, without appropriate growth factors, TSCs tend to differentiate in vitro [34, 92, 95]. We found that tumor SP cells comprise only a small fraction, 1–2%, in vitro, but when injected in vivo the xenograft contains more than 30% SP cells [35, 36]. Thus, this in vivo expansion of SP cells may be due to the influence of an appropriate microenvironment that switches the quiescent SP to active SP cells, leading to their self-renewal and expansion. This switch from quiescent to active state of stemness of TSC can be referred to as the *stemness switch* (Fig. 4) [36, 83].

The increased number of TSCs in vivo versus in vitro may also be due to the acquisition of TSC-like properties by the resident non-TSC fraction. This phenomenon of de-differentiation, where a more phenotypically differentiated cell acquires stem-like properties and reverts back to a multipotent state, may be a part of the mechanism of the stemness switch.

We found that non-SP derived dormant tumors are very slow growing, lack expression of phospho-MAPK/ERK1,2 (Fig. 5) [83], and contain a very few SP cells (data not shown), suggesting that dormant growth lacked an appropriate microenvironmental stress either for conversion of non-TSC to TSCs or for the expansion of TSCs. Introduction of appropriate microenvironmental stress such as drug-induced oxidative stress may switch the TSCs from dormant to active state leading to tumor growth. We investigated the possibility of a drug-induced switch of dormancy to tumor progression using a nontumorigenic osteosarcoma cell line HOS, which has been shown to became tumorigenic when exposed to platinum agents, but the mechanism was not known [96]. We found that only 3 days of brief exposure of HOS cells to cisplatin leads to increased activation of MAPK/ERK1,2 and VEGF/Flt1 autocrine signaling and subsequent expansion of a SP fraction that expresses Bmi-1 and Nanog. When cisplatin-treated SP cells are injected to mice they form rapidly growing tumors, whereas untreated SP cells formed dormant growth. The number of nanog-positive cells significantly increased in cisplatin-treated, SP-derived xenografts compared to dormant growth. Furthermore, we found that siRNA knockdown of VEGF/Flt1 signaling reduced the tumor growth significantly where nanog positive cells were reduced [35]. We then found that exposure of SP cells to cisplatin increased ERK1,2 phosphorylation, Flt1 expression [35], and nuclear localization of β-catenin in the SP cells but not in the non-SP fraction (Fig. 6) [83], suggesting that microenvironmental stress may specifically target the SP or

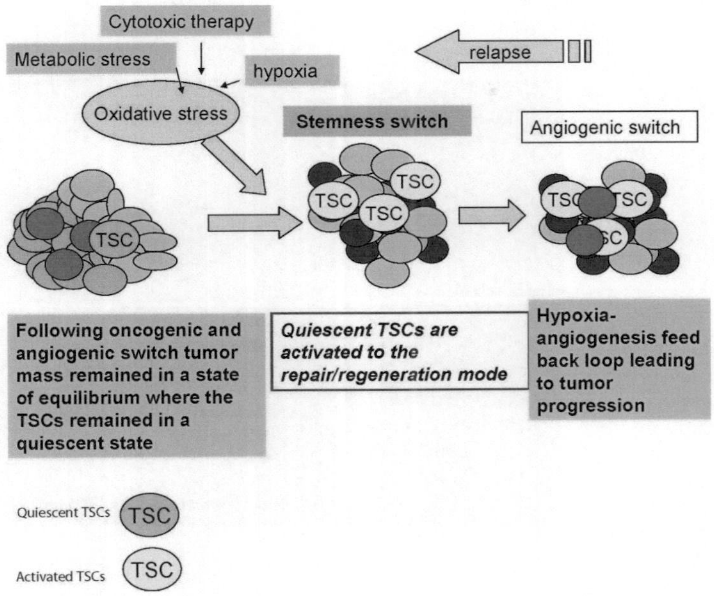

Fig. 4 A stem cell model of tumor repair and regeneration. In this model, a dormant or slowly growing tumor may become aggressive and metastatic when sufficient damage is done to its tissue, leading to the active mobilization of its TSCs. Since the repair regeneration process is de-regulated, zones of hypoxia and angiogenesis are formed in the tumor, leading to the activation of a feedback loop where increasing hypoxia leads to increased repair/regeneration followed by hypoxia, and the process goes on. As a result, TSC fraction is increased significantly and some of the TSCs spread to distant organs for the same purpose of repair/regeneration but then end up forming metastatic clones. Adopted from Das Bikul et al. Stem Cells (May 8, 2008 Online Edition) (*see* also Color Insert)

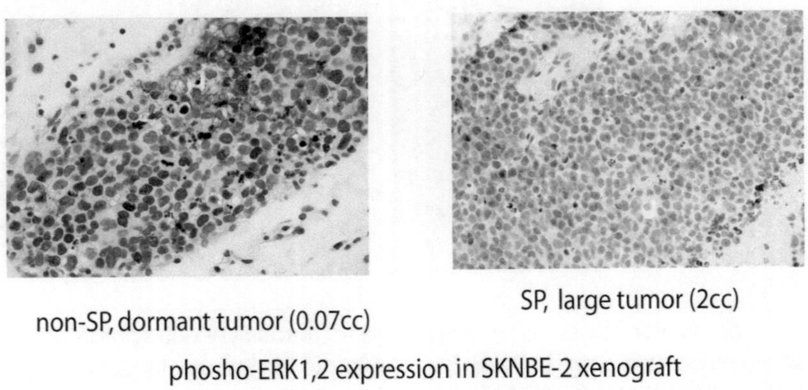

Fig. 5 The expression of phospho ERK 1, 2 in dormant versus rapidly growing tumors. Phospho-ERK 1,2 is highly expressed in SP cell–derived tumor growth, whereas non-SP-derived dormant growth did not show marked expression of phospho ERK1,2. The non-SP growth took 14 weeks to reach the size of 0.07 cc, whereas the SP cell–derived growth took only 7 weeks to reach the size of 2 cc. Data taken from the PhD thesis of Bikul Das (University of Toronto, 2007) (*see* also Color Insert)

TSC fraction by up-regulating stemness signaling pathway such as Wnt/β-catenin, leading to the self-renewal and expansion of highly tumorigenic TSCs. Considering that cisplatin treatment increases ROS (Fig. 2b), it may be presumed that drug-induced oxidative stress may be involved in the up-regulation of stemness signaling pathways. Future research should be directed to find out the effect of oxidative stress on stemness signaling pathways such as Wnt/β-catenin, Oct-4/nanog, Bmi-1, etc.

During the early phase of transition from dormancy to tumor progression, microenvironmental stress such as oxidative stress, may trigger the stemness switch leading to the initial expansion of TSCs. As the tumor grows in size, hypoxic zones would appear, leading to increased population of TSCs, thus initiating a positive feedback loop between the hypoxic zone and the TSC population. During this critical switch from dormancy to rapid tumor growth (equivalent to the period between tumor

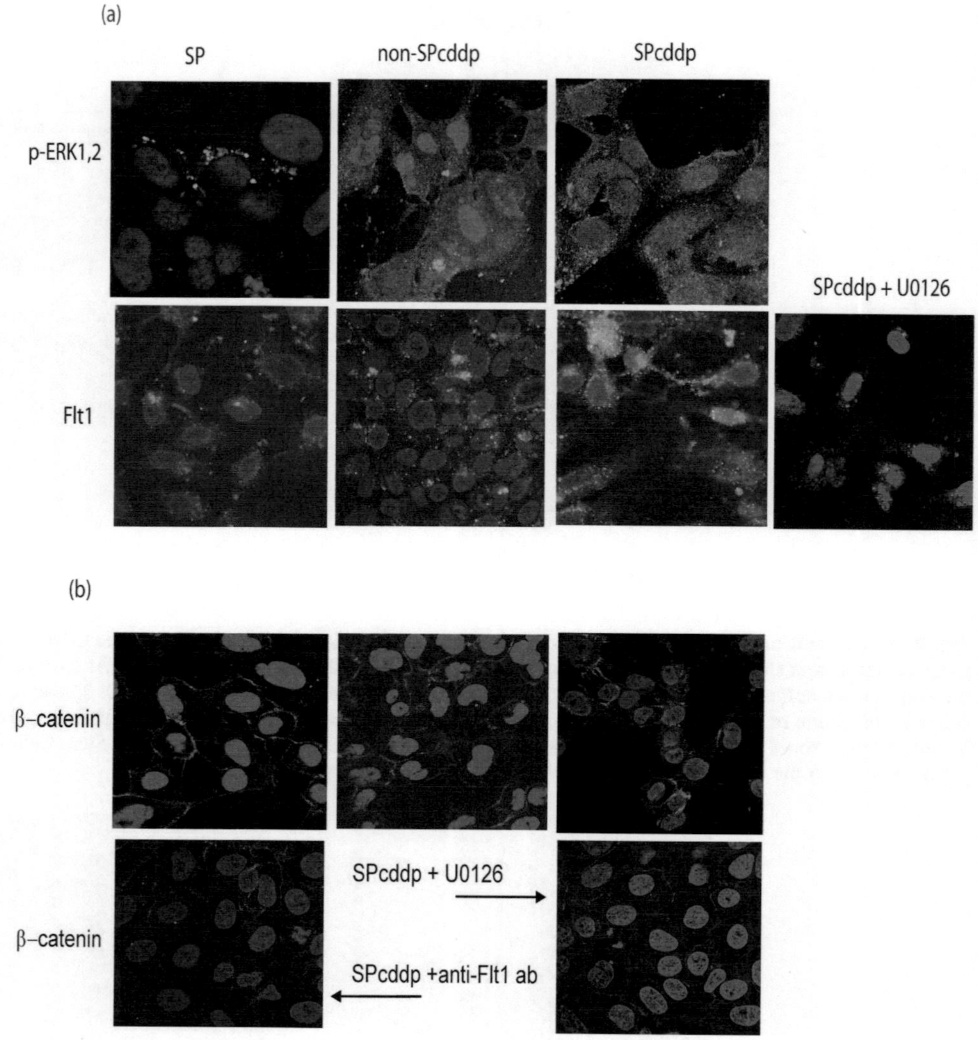

Fig. 6 Cisplatin (CDDP)-mediated activation of MAPK/ERK1,2 and VEGF/Flt1, resulting in the activation of the β-catenin pathway
(a) Immunofluorescence showing marked increase of phosphor-ERK1,2 and Flt1 in SPcddp and non-SPcddp cells. **(b)** Nuclear β-catenin–expressing cells were seen only in the SPcddp group, which was down-regulated following U0126 (a specific inhibitor of MAPK/ERK1,2) and anti-Flt1 treatment. Figure 6(a) is reproduced from Tsuchida et al., Oncogene [35] (*see* also Color Insert)

latency and progression), the tumor mass may enter into a phase of repair/regeneration and may activate key signaling pathways such as MAPK/ERK1,2, JAK/STAT3, Wnt/β-catenin, etc., resulting in an enhanced degree of stemness. Migration and invasion may be an important component of this repair/regeneration process. Interestingly, the EMT (epithelial mesenchymal transition) process, where tumor cells acquire migratory phenotype, plays an important role in tumor progression [57]. In this regard, EMT may be a component of the first phase of the stemness switch, that is, EMT may be considered a transition period between the quiescent (static) and the migratory (repair/regeneration) phenotype switch.

The repeated injury/stress in the tumor microenvironment (mainly due to the presence of the hypoxic zone) may maintain the continuing activation of signaling pathways involved in the mobilization and migration of stem cells during injury. In this respect, stemness switch is a dynamic process, where static (dormant or quiescent) TSC, upon incurring injury, due either to environmental factors (hypoxia/oxidative stress) or drug treatments, would acquire a repair mode ("stemness switch") leading to TSC self-renewal and expansion and the process continued. In this model of tumor stemness switch, the tumor stemness constitutes a larger, dynamic state incorporating TSC, its niche, and its purpose, that is, repair and generation (migration/invasion to injured area, and subsequent angiogenesis/stromal development). The higher the degree of tumor stemness in a tumor, the higher would be the growth, invasion, and metastasis. In this concept, the molecular signaling involved in maintaining the niche would be driven for the sole purpose of preparing the niche for tumor repair and regeneration.

TSCs may inherit the drive for repair/regeneration from the initial trauma of malignant transformation. To explain the malignant transformation of tissue-specific stem cells (TSSCs), Bissel et al. [97] proposed a tissue injury model, where TSSCs, after being damaged either by mutation or another mechanism, modify the microenvironment, which in turn further damages the transformed cells, leading to an escalation in pathological behavior and eventual tumor formation. This pathological behavior is likely to be the "repair/regeneration" response of the damaged TSSC to its surrounding "injured" environment. Oncogene mutation may be linked to the ability of TSCs to maintain and/or increase the degree of tumor repair and regeneration. Recent reports suggest the role of the ras oncogene in the repair process of wound re-epithelization [98] and gastric ulcer [99]. Accordingly, a low-grade tumor would have a lower degree of tumor repair, whereas a high-grade tumor would have a high degree of tumor repair.

It is simplistic to view tumors solely as "autonomous machines of cell proliferation" without any purpose. Instead, the above view of tumor stemness switch proposes a stem cell model of tumor repair and regeneration (SCTR), where the tumor microenvironment is subject to an endless cycle of injury, and repair with further injury leading to a rapid state of tumor progression and metastasis. The SCTR model suggests that the tumors are programmed to create an environment that invites repair/regeneration to "correct" the initial trauma of malignant transformation of the parent stem cell. The SCTR model predicts that metastasis may be linked to establishment of an "injured niche" or "pre-metastatic niche" in the distant organ. Thus, the organ preferences observed during metastasis may be partly explained on the basis of the release of organ-specific toxic factors from a tumor. The SCTR model may also explain why cancer is primarily a disease of aging. As aging progresses, the oxidative stress in the tissues and organs increases, thus increasing the chance of quiescent TSCs to actively switch to the repair mode.

9 Tumor Re-population, Hypoxia, and Stemness Switch

Tumor repopulation addresses the regrowth of a tumor in between the cyclical doses of chemo- and radiotherapy and is a major cause of treatment failure [100]. In 1991, Fowler et al. [101] proposed that well-oxygenated cells proximal to the blood vessels die and are removed following cytotoxic therapy. Consequently, the cells in the hypoxic zones acquire a better blood supply, with the result that the rate of spontaneous cell death decreases and proliferation increases resulting in tumor cell repopulation (Fig.7a, b) [101]. It has been suggested that TSC, being quiescent, would escape from the effects of chemotherapy that target only highly proliferating fraction of tumor cells [100]. However, the opposite may also be true: that TSCs are driven from a quiescent to an active state of self-renewal due to chemotherapy-induced changes to the tumor microenvironment (Fig. 7c) [83]. There is not much known about drug-induced stress on the TSC fraction. Kim and Tannock [100] proposed that, following drug treatment, TSCs may migrate to the angiogenic area leading to accelerated re-population. Interestingly, in our HOS cell model, we found that cisplatin treatment increases VEGF secretion in the SP fraction. We found similar results using rhabdomyosarcoma and neuroblastoma cell lines [35]. Hence, TSCs may not migrate to the angiogenic area. Instead, drugs may activate stress signaling pathways such as MAPK/ERK1,2 and VEGF pathways, leading to the activation and mobilization of TSCs. In this regard, the mobilization or expansion process may be similar to drug-induced mobilization of HSCs [102].

In the stemness switch model of tumor growth, a chemotherapeutic agent will not only select for a TSC but also enhance its stemness, and therefore the repopulating and resistance ability. Thus, tumor drug-resistance may be part of the drug-induced stemness switch. If this is the case, then our current assays and strategies to investigate drug resistance are greatly deficient. Usual drug resistance assays tend to evaluate the number of cells remaining (the greater the LD50 of a drug, the higher is the resistance), which does not accurately reflect the number of TSC increase. In our HOS cell model of drug-induced stemness switch, the parental HOS cells were highly sensitive to cisplatin treatment [35], and a routine drug-resistance assay would not have been able to detect a small fraction of SP cell increase. And yet, that small fraction was the key to the tumorigenic potential of the cell line.

One of the predictions of Nowell's clonal evolution model is that drug resistance is a result of clonal selection, where tumor cells having inherent resistant will escape the drug toxicity and appear as the dominant clone [1–3]. Most of the cytotoxic drugs generate ROS, leading to cellular toxicity, and, therefore, it was found that drug pumps such as Pgp, MRPs, and GSx (glutathione drug pump) would be highly expressed in the drug resistance clone [6]. Laboratory experiments have shown that this is indeed the case. By exposing tumor cell lines to various drugs, clones having high drug pump activity can be selected. In addition, p53 mutations have also been found to constitute a major mechanism of drug-resistance, since mutant p53 protein will reduce drug-induced apoptosis [6]. However, in the clinical setting, so far, inhibitors of the drug pumps have not shown any promising results. In addition, to date, the clinical correlation between high drug pump activity and drug resistance has not been established [36]. The failure of the clinical trials to find direct evidence of the relation between the drug pump expression and resistance may be mostly because of the use of inappropriate assays to measure drug resistance. Hence, in future studies of drug resistance, the degree of stemness should also be measured, which may uncover a strong correlation

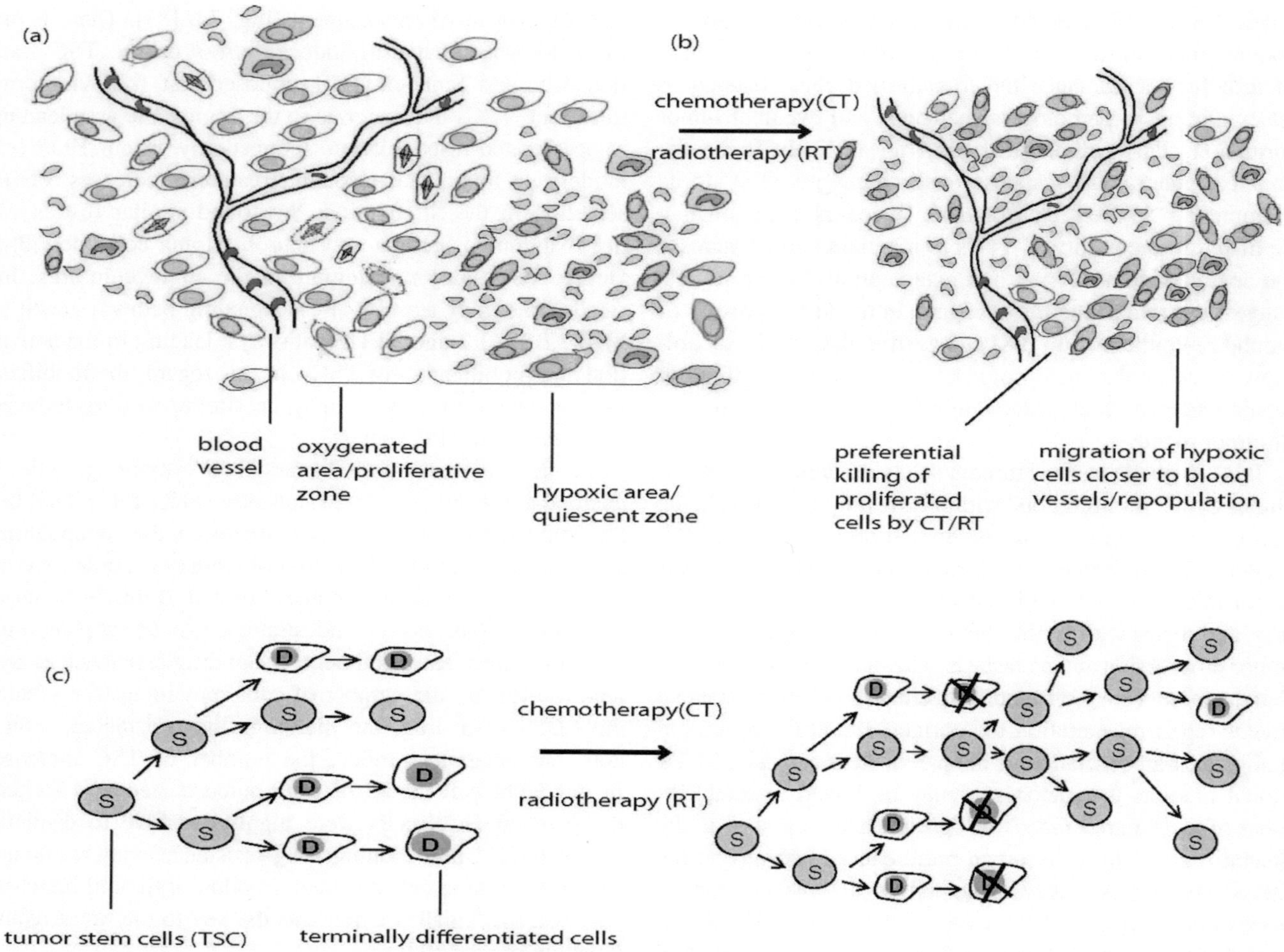

Fig. 7 (a) & (b) Angiogenic model of repopulation. (a) untreated tumor showing angiogenic area (oxygenated zone) and hypoxic area. The tumor cells living in the hypoxic zone tends to die spontaneously. (b) since cytotoxic therapy mainly target the proliferative cells, the angiogenic area experience higher cell death than the hypoxic area. the cells residing in the hypoxic zone migrate closer to blood vessel, and the spontaneous cells death is reduced. Thus, repopulation occurs from the hypoxic cells. Figure (c) stem cell model of repopulation. following cytotoxic therapy, the number of tumor stem cells (TSC) is increased compared with the production of the terminally differentiated cells as a result of stress-induced stemness switch. These figures were modified from Das Bikul: The role of VEGF autocrine signaling in hypoxia and oxidative stress driven "stemness switch". PhD thesis, University of Toronto, 2007.

between expression of the drug pumps and clinical drug resistance.

10 Conclusions

The TSC model has a very strong appeal since it can provide a reasonable explanation of tumor relapse, where TSCs would escape the initial drug toxicity leading to relapse. The idea of TSC mobilization following cytotoxic therapy was initially proposed by Trott and his group in 1991 [56], yet very little has since been done in pursuing this idea. This is mainly because of lack of appropriate assays and models to investigate TSC mobilization/expansion following stress. We found that both hypoxia and drug exposure can switch quiescent SP cells to an active state, leading to their expansion and also an increase in their degree of stemness. In this model of stemness switch, signaling pathways like VEGF and Oct-4 and Wnt/β-catenin may be activated in the TSC fraction and lead to a de-regulated expansion of TSCs. This model of stemness switch also predicts that TSCs are not quiescent and that targeting TSCs may require modulating the microenvironment of TSC niche.

Acknowledgments We thank Dr. John E. Dick, Toronto General Research Institute, University Health Network, University of Toronto, for critical discussion of our work. We also thank Dr. Ernest Cutz, MD, Hospital for Sick Children, Toronto, and Dr. Masabumi Shibuya, Institute of Medical Science, University of Tokyo, for critical review of the manuscript. This work is supported by grants from the National Cancer Institute of Canada. B.D. is supported in part by awards from the Hospital for Sick Children's Research Training Centre, SickKids Foundation.

References

1. Trott KR. Tumour stem cells: the biological concept and its application in cancer treatment. Radiother Oncol. 1994;30(1):1–5.
2. Buick RN, Till JE, McCulloch EA. Colony assay for proliferative blast cells circulating in myeloblastic leukaemia. Lancet. 1977;1(8016):862–3.
3. Iscove NN, Senn JS, Till JE, McCulloch EA. Colony formation by normal and leukemic human marrow cells in culture: effect of conditioned medium from human leukocytes. Blood. 1971;37(1):1–5.
4. McCulloch EA, Buick RN, Till JE. Normal and leukemic hemopoiesis compared. Cancer. 1978;42(2 Suppl):845–53.
5. Till JE, Mak TW, Price GB, Senn JS, McCulloch EA. Cellular subclasses in human leukemic hemopoiesis. Hamatol Bluttransfus. 1976;19:33–45.
6. Bonnet D, Dick JE. Human acute myeloid leukemia is organized as a hierarchy that originates from a primitive hematopoietic cell. Nat Med. 1997;3(7):730–7.
7. Hamburger AW, Salmon SE. Primary bioassay of human tumor stem cells. Science. 1977;197(4302):461–3.
8. Fidler IJ, Hart IR. Biological diversity in metastatic neoplasms: origins and implications. Science. 1982;217(4564):998–1003.
9. Fidler IJ, Kripke ML. Metastasis results from preexisting variant cells within a malignant tumor. Science. 1977;197(4306):893–5.
10. Fogh J, Fogh JM, Orfeo T. One hundred and twenty-seven cultured human tumor cell lines producing tumors in nude mice. J Natl Cancer Inst. 1977;59(1):221–6.
11. Giovanella BC, Stehlin JS, Jr., Williams LJ, Jr., Lee SS, Shepard RC. Heterotransplantation of human cancers into nude mice: a model system for human cancer chemotherapy. Cancer. 1978;42(5):2269–81.
12. Hajdu SI, Lemos LB, Kozakewich H, Helson L, Beattie EJ, Jr. Growth pattern and differentiation of human soft tissue sarcomas in nude mice. Cancer. 1981;47(1):90–8.
13. Hamburger A, Salmon SE. Primary bioassay of human myeloma stem cells. J Clin Invest. 1977;60(4):846–54.
14. Heppner GH. Tumor heterogeneity. Cancer Res. 1984;44(6):2259–65.
15. Nowell PC. Mechanisms of tumor progression. Cancer Res. 1986;46(5):2203–7.
16. Park CH, Bergsagel DE, McCulloch EA. Mouse myeloma tumor stem cells: a primary cell culture assay. J Natl Cancer Inst. 1971;46(2):411–22.
17. Shimosato Y, Kameya T, Nagai K, et al. Transplantation of human tumors in nude mice. J Natl Cancer Inst. 1976;56(6):1251–60.
18. Southam CM, Brunschwig A, Levin AG, Dizon QS. Effect of leukocytes on transplantability of human cancer. Cancer. 1966;19(11):1743–53.
19. Behbod F, Rosen JM. Will cancer stem cells provide new therapeutic targets? Carcinogenesis. 2005;26(4):703–11.
20. Pardal R, Clarke MF, Morrison SJ. Applying the principles of stem-cell biology to cancer. Nat Rev Cancer. 2003;3(12): 895–902.
21. Dick JE. Breast cancer stem cells revealed. Proc Natl Acad Sci U S A. 2003;100(7):3547–9.
22. Patrawala L, Calhoun T, Schneider-Broussard R, Zhou J, Claypool K, Tang DG. Side population is enriched in tumorigenic, stem-like cancer cells, whereas ABCG2+ and ABCG2- cancer cells are similarly tumorigenic. Cancer Res. 2005;65(14): 6207–19.
23. Hirschmann-Jax C, Foster AE, Wulf GG, et al. A distinct "side population" of cells with high drug efflux capacity in human tumor cells. Proc Natl Acad Sci U S A. 2004;101(39): 14228–33.
24. Goodell MA. Multipotential stem cells and 'side population' cells. Cytotherapy. 2002;4(6):507–8.
25. Goodell MA, Brose K, Paradis G, Conner AS, Mulligan RC. Isolation and functional properties of murine hematopoietic stem cells that are replicating in vivo. J Exp Med. 1996;183(4): 1797–806.
26. Smalley MJ, Clarke RB. The mammary gland "side population": a putative stem/progenitor cell marker? J Mammary Gland Biol Neoplasia. 2005;10(1):37–47.
27. Evans MJ, Kaufman MH. Establishment in culture of pluripotential cells from mouse embryos. Nature. 1981;292(5819):154–6.
28. Martin GR. Isolation of a pluripotent cell line from early mouse embryos cultured in medium conditioned by teratocarcinoma stem cells. Proc Natl Acad Sci U S A. 1981;78(12):7634–8.
29. Bradley A, Evans M, Kaufman MH, Robertson E. Formation of germ-line chimaeras from embryo-derived teratocarcinoma cell lines. Nature. 1984;309(5965):255–6.
30. Barrandon Y, Green H. Three clonal types of keratinocyte with different capacities for multiplication. Proc Natl Acad Sci U S A. 1987;84(8):2302–6.
31. Locke M, Heywood M, Fawell S, Mackenzie IC. Retention of intrinsic stem cell hierarchies in carcinoma-derived cell lines. Cancer Res. 2005;65(19):8944–50.
32. Ross RA, Spengler BA, Domenech C, Porubcin M, Rettig WJ, Biedler JL. Human neuroblastoma I-type cells are malignant neural crest stem cells. Cell Growth Differ. 1995;6(4): 449–56.
33. Buick RN, MacKillop WJ. Measurement of self-renewal in culture of clonogenic cells from human ovarian carcinoma. Br J Cancer. 1981;44(3):349–55.
34. Kondo T, Setoguchi T, Taga T. Persistence of a small subpopulation of cancer stem-like cells in the C6 glioma cell line. Proc Natl Acad Sci U S A. 2004;101(3):781–6.
35. Tsuchida R, Das B, Yeger H, et al. Cisplatin treatment increases survival and expansion of a highly tumorigenic side-population fraction by upregulating VEGF/Flt1 autocrine signaling. Oncogene. 2008;27(28):3923–34.
36. Das B, Tsuchida R, Malkin D, Baruchel S, Yeger H. Hypoxia enhances tumor stemness by increasing the invasive and tumorigenic side-population fraction. Stem Cells. 2008;26(7):1818–30.
37. Das B, Tsuchida R, Baruchel S, Malkin D, Yeger H. The role of VEGF/Flt1 signaling in the maintenance of neuroblastoma side-population celll "stemness" during hypoxia. Advances in Neuroblastoma Research meeting, May 2006, Los Angeles.
38. Das B, Tsuchida R, Malkin D, Baruchel S, Yeger H. A stem cell model of tumor repair and regeneration. In: 99th Annual meeting of the American Association of Cancer Research; 2006; Washington DC: American Association of Cancer Research; 2006. p. 1567.
39. Hill RP. Identifying cancer stem cells in solid tumors: case not proven. Cancer Res. 2006;66(4):18911896.
40. Mikkers H, Frisen J. Deconstructing stemness. EMBO J. 2005;24(15):2715–9.
41. Raaphorst FM. Self-renewal of hematopoietic and leukemic stem cells: a central role for the Polycomb-group gene Bmi-1. Trends Immunol. 2003;24(10):522–4.
42. Rao RR, Calhoun JD, Qin X, Rekaya R, Clark JK, Stice SL. Comparative transcriptional profiling of two human embryonic stem cell lines. Biotechnol Bioeng. 2004;88(3):273–86.
43. Lee J, Kim HK, Rho JY, Han YM, Kim J. The human OCT-4 isoforms differ in their ability to confer self-renewal. J Biol Chem. 2006;281(44):33554–65.

44. Jacobsen SE. Defining 'stemness': Notch and Wnt join forces? Nat Immunol. 2005;6(3):234–6.
45. Suda T, Arai F, Shimmura S. Regulation of stem cells in the niche. Cornea. 2005;24(8 Suppl):S12–7.
46. Sell S, Dunsford HA. Evidence for the stem cell origin of hepatocellular carcinoma and cholangiocarcinoma. Am J Pathol. 1989;134(6):1347–63.
47. Sell S, Pierce GB. Maturation arrest of stem cell differentiation is a common pathway for the cellular origin of teratocarcinomas and epithelial cancers. Lab Invest. 1994;70(1):6–22.
48. Sell S. Stem cell origin of cancer and differentiation therapy. Crit Rev Oncol Hematol. 2004;51(1):1–28.
49. Iwama A, Oguro H, Negishi M, Kato Y, Nakauchia H. Epigenetic regulation of hematopoietic stem cell self-renewal by polycomb group genes. Int J Hematol. 2005;81(4):294–300.
50. Lessard J, Sauvageau G. Bmi-1 determines the proliferative capacity of normal and leukaemic stem cells. Nature. 2003;423(6937):255–60.
51. Glinsky GV, Berezovska O, Glinskii AB. Microarray analysis identifies a death-from-cancer signature predicting therapy failure in patients with multiple types of cancer. J Clin Invest. 2005;115(6):1503–21.
52. Armstrong L, Hughes O, Yung S, et al. The role of PI3K/AKT, MAPK/ERK and NFkappabeta signalling in the maintenance of human embryonic stem cell pluripotency and viability highlighted by transcriptional profiling and functional analysis. Hum Mol Genet. 2006;15(11):1894–913.
53. Bhattacharya B, Miura T, Brandenberger R, et al. Gene expression in human embryonic stem cell lines: unique molecular signature. Blood. 2004;103(8):2956–64.
54. Covello KL, Kehler J, Yu H, et al. HIF-2alpha regulates Oct-4: effects of hypoxia on stem cell function, embryonic development, and tumor growth. Genes Dev. 2006;20(5):557–70.
55. Guo Y, Costa R, Ramsey H, et al. The embryonic stem cell transcription factors Oct-4 and FoxD3 interact to regulate endodermal-specific promoter expression. Proc Natl Acad Sci U S A. 2002;99(6):3663–7.
56. Hochedlinger K, Yamada Y, Beard C, Jaenisch R. Ectopic expression of Oct-4 blocks progenitor-cell differentiation and causes dysplasia in epithelial tissues. Cell. 2005;121(3):465–77.
57. Brabletz T, Jung A, Spaderna S, Hlubek F, Kirchner T. Opinion: migrating cancer stem cells - an integrated concept of malignant tumour progression. Nat Rev Cancer. 2005;5(9):744–9.
58. Shimada H, Chatten J, Newton WA, Jr., et al. Histopathologic prognostic factors in neuroblastic tumors: definition of subtypes of ganglioneuroblastoma and an age-linked classification of neuroblastomas. J Natl Cancer Inst. 1984;73(2):405–16.
59. Liu S, Dontu G, Wicha MS. Mammary stem cells, self-renewal pathways, and carcinogenesis. Breast Cancer Res. 2005;7(3):86–95.
60. Hemmati HD, Nakano I, Lazareff JA, et al. Cancerous stem cells can arise from pediatric brain tumors. Proc Natl Acad Sci U S A. 2003;100(25):15178–83.
61. Patrawala L, Calhoun T, Schneider-Broussard R, et al. Highly purified CD44+ prostate cancer cells from xenograft human tumors are enriched in tumorigenic and metastatic progenitor cells. Oncogene. 2006;25(12):1696–708.
62. Wechsler-Reya R, Scott MP. The developmental biology of brain tumors. Annu Rev Neurosci. 2001;24:385–428.
63. Reya T, Morrison SJ, Clarke MF, Weissman IL. Stem cells, cancer, and cancer stem cells. Nature. 2001;414(6859):105–11.
64. Wicha M, Liu S, Dontu G. Cancer stem cells: an old idea-a paradigm shift. Cancer Res. 2006;66:1883–90.
65. Heppner GH, Miller FR. The cellular basis of tumor progression. Int Rev Cytol. 1998;177:1–56.
66. Klein G. Foulds' dangerous idea revisited: the multistep development of tumors 40 years later. Adv Cancer Res. 1998;72:1–23.
67. Nowell PC. The clonal evolution of tumor cell populations. Science. 1976;194(4260):23–8.
68. Mackillop W, Ciampi A, Till J, Buick R. A stem cell model of human tumor growth: implications for tumor cell clonogenic assay. J Natl Cancer Inst. 1983;70(1):9–16.
69. Graeber G, Osmanian C, Jacks T, et al. Hypoxia-mediated selection of cells with diminished apoptotic potential in solid tumors. Nature. 1996;379:88–90.
70. Harris A. Hypoxia- a key regulatory factor in tumor growth. Nat Rev Cancer. 2002;2:38.
71. Das B, Yeger H, Tsuchida R, et al. A hypoxia-driven vascular endothelial growth factor/Flt1 autocrine loop interacts with hypoxia-inducible factor-1alpha through mitogen-activated protein kinase/extracellular signal-regulated kinase 1/2 pathway in neuroblastoma. Cancer Res. 2005;65(16):7267–75.
72. Li L, Xie T. Stem cell niche: structure and function. Annu Rev Cell Dev Biol. 2005;21:605–31.
73. D'Ippolito G, Diabira S, Howard GA, Roos BA, Schiller PC. Low oxygen tension inhibits osteogenic differentiation and enhances stemness of human MIAMI cells. Bone 2006;39(3):513–22.
74. Jiang YG, Luo Y, He DL, et al. Role of Wnt/beta-catenin signaling pathway in epithelial-mesenchymal transition of human prostate cancer induced by hypoxia-inducible factor-1alpha. Int J Urol. 2007;14(11):1034–9.
75. Rich JN. Cancer stem cells in radiation resistance. Cancer Res. 2007;67(19):8980–4.
76. Pear WS, Simon MC. Lasting longer without oxygen: the influence of hypoxia on Notch signaling. Cancer Cell. 2005;8(6):435–7.
77. Fuchs E, Tumbar T, Guasch G. Socializing with the neighbors: stem cells and their niche. Cell. 2004;116(6):769–78.
78. Hattori K, Heissig B, Wu Y, et al. Placental growth factor reconstitutes hematopoiesis by recruiting VEGFR1(+) stem cells from bone-marrow microenvironment. Nat Med. 2002;8(8):841–9.
79. Brusselmans K, Bono F, Collen D, Herbert JM, Carmeliet P, Dewerchin M. A novel role for vascular endothelial growth factor as an autocrine survival factor for embryonic stem cells during hypoxia. J Biol Chem. 2005;280(5):3493–9.
80. Durand R. Intermittent blood flow in solid tumors-an underappreciated source of "drug-resistance". Cancer and Metastasis Rev. 2001;20:57–61.
81. Das B, Tsuchida R, Baruchel S, Yeger H. Idea and evidence of "tumor stemness switch". In: International workshop on cancer stem cell; 2005 November 10–12; Milan, Italy; 2005. p. 34.
82. Cook JA, Gius D, Wink DA, Krishna MC, Russo A, Mitchell JB. Oxidative stress, redox, and the tumor microenvironment. Semin Radiat Oncol. 2004;14(3):259–66.
83. Das B. The role of VEGF autocrine signaling in hypoxia and oxidative stress driven "stemness switch": implications in solid tumor progression and metastasis. PhD thesis: Institute of Medical Sciences, University of Toronto; 2007.
84. Brawley C, Matunis E. Regeneration of male germline stem cells by spermatogonial dedifferentiation in vivo. Science. 2004;304(5675):1331–4.
85. Nakagawa M, Koyanagi M, Tanabe K, et al. Generation of induced pluripotent stem cells without Myc from mouse and human fibroblasts. Nat Biotechnol. 2008;26(1):101–6.
86. Takahashi K, Tanabe K, Ohnuki M, et al. Induction of pluripotent stem cells from adult human fibroblasts by defined factors. Cell. 2007;131(5):861–72.

87. Wernig M, Meissner A, Foreman R, et al. In vitro reprogramming of fibroblasts into a pluripotent ES-cell-like state. Nature. 2007;448(7151):318–24.
88. Takahashi K, Yamanaka S. Induction of pluripotent stem cells from mouse embryonic and adult fibroblast cultures by defined factors. Cell. 2006;126(4):663–76.
89. Jogi A, Ora I, Nilsson H, et al. Hypoxia alters gene expression in human neuroblastoma cells toward an immature and neural crest-like phenotype. Proc Natl Acad Sci U S A. 2002;99(10): 7021–6.
90. Hill RP, Milas L. The proportion of stem cells in murine tumors. Int J Radiat Oncol Biol Phys. 1989;16(2):513–8.
91. Dick JE. Normal and leukemic human stem cells assayed in SCID mice. Semin Immunol. 1996;8(4):197–206.
92. Singh SK, Hawkins C, Clarke ID, et al. Identification of human brain tumour initiating cells. Nature. 2004;432(7015): 396–401.
93. Al-Hajj M, Wicha MS, Benito-Hernandez A, Morrison SJ, Clarke MF. Prospective identification of tumorigenic breast cancer cells. Proc Natl Acad Sci U S A. 2003;100(7): 3983–8.
94. Banerjee TK, Marx JJ, Jr., Spencer SK, Tiffany E. Human tumor stem cell clonogenic assay sensitivity in malignant tumors and clinical correlation. Wis Med J. 1982;81(12):19–22.
95. Kim CF, Jackson EL, Woolfenden AE, et al. Identification of bronchioalveolar stem cells in normal lung and lung cancer. Cell. 2005;121(6):823–35.
96. Miller AC, Mog S, McKinney L, et al. Neoplastic transformation of human osteoblast cells to the tumorigenic phenotype by heavy metal-tungsten alloy particles: induction of genotoxic effects. Carcinogenesis. 2001;22(1):115–25.
97. Bissell MJ, Radisky D. Putting tumours in context. Nat Rev Cancer. 2001;1(1):46–54.
98. Tscharntke M, Pofahl R, Krieg T, Haase I. Ras-induced spreading and wound closure in human epidermal keratinocytes. FASEB J. 2005;19(13):1836–8.
99. Pai R, Jones MK, Tomikawa M, Tarnawski AS. Activation of Raf-1 during experimental gastric ulcer healing is Ras-mediated and protein kinase C-independent. Am J Pathol. 1999;155(5): 1759–66.
100. Kim JJ, Tannock IF. Repopulation of cancer cells during therapy: an important cause of treatment failure. Nat Rev Cancer. 2005;5(7):516–25.
101. Fowler JF. Rapid repopulation in radiotherapy: a debate on mechanism. The phantom of tumor treatment–continually rapid proliferation unmasked. Radiother Oncol. 1991;22(3):156–8.
102. Cottler-Fox MH, Lapidot T, Petit I, et al. Stem cell mobilization. Hematol Am Soc Hematol Educ Prog. 2003:419–37.

The Role of the Tumor Suppressor Fhit in Cancer-Initiating Cells

Hideshi Ishii

Abstract Cancers are genetic and epigenetic disorders with unique tissue compositions. It is proposed that a small proportion of cancer-initiating cells (CICs) give rise to a tumor with extensive heterogeneity and accumulated genomic instability during carcinogenesis. The tumor suppressor Fhit is inactivated in the precancerous or early stages of environmental-carcinogen-induced cancers. Recent studies suggest that Fhit plays a role in modulating the DNA damage checkpoint response, so the dysfunction might lead to the development of CICs with genomic alterations. The possible significance of Fhit in the maintenance of genomic stability is discussed.

Keywords Tumor suppressor · Fhit · Genomic stability · Cancer initiating cells

1 Introduction

Cancer is a genetic and epigenetic disease. Genomic alterations that inactivate tumor suppressors and activate proto-oncogenes accumulate in cancerous cells. Classical tumor-suppressor genes such as retinoblastoma 1 (RB1) and TP53, and oncogenes such as RAS and MYC, have been extensively studied [1, 2] and found to accumulate in a stepwise manner in carcinogenesis, progressing from early stages with a few or several alterations, to advanced stages with numerous alterations leading to aggressive biological behavior. Previous studies have indicated that alterations of common chromosome fragile sites at specific loci, such as active fragile FRA3B/Fhit, can be induced in vitro; such alterations are found frequently in advanced cancer cells but also detectable in precancer or very early stages in vivo [3]. The fragile Fhit gene is one of the first targets for induced DNA damage in precancer as well as in a significant proportion of solid tumors and hematopoietic malignancies. It is suggested that genomic alterations that occur nonrandomly at the early stages might play a role in the progression of carcinogenesis. Moreover, recent studies have shown Fhit to be involved in the surveillance of genome integrity and checkpoint response after exposure to environmental carcinogens or genotoxins. A possible connection between Fhit loss and cancer-initiating cells (CICs) is noted.

H. Ishii (✉)
Center for Molecular Medicine, Jichi Medical University, Yakushiji 3311-1, Shimotsuke, Tochigi, 329-0498, Japan
e-mail: hishii@ms.jichi.ac.jp

2 Tumor Suppressor Fhit from Very Early to Advanced Stages of Cancer

Common chromosome fragile sites are large regions that preferentially exhibit gaps or breaks, visible in metaphase chromosomes, when cells are exposed to replicative stress by carcinogens and DNA synthesis is partially perturbed [3–7]. Fragile site instability in cultured cells is well documented and includes the occurrence of gaps and breaks on metaphase chromosomes, translocation and deletion breakpoints, and sister chromosome exchanges [7]. FRA3B and FRA16D, at chromosome regions 3p14.2 and 16q23.3, respectively, are the two most active or most fragile of the common human fragile sites, and are frequently deleted or altered in cancers associated with environmental carcinogens and also in hematopoietic disorders [3–7]. Genomic sequences at fragile sites [8, 9] and protein signal pathways that monitor their stability (including Atr [10], Brca1 [11], Smc1 [12], Fanconi anemia pathway proteins [13], Chk1 [14], Hus1 [15], and Rad51 [16]) have been characterized [7]. Both homologous or homology-dependent recombination and nonhomologous end-joining repair pathways regulate the stability of the FRA3B and FRA16D fragile sites, indicating that double-strand breaks are formed at common fragile sites as a result of replication perturbation [16]. It has been proposed that the fragile sites are susceptible to delayed DNA replication, which would be monitored by DNA damage checkpoints [6].

V.K. Rajasekhar, M.C. Vemuri (eds.), *Regulatory Networks in Stem Cells*, Stem Cell Biology and Regenerative Medicine, DOI 10.1007/978-1-60327-227-8_36, © Humana Press, a part of Springer Science+Business Media, LLC 2009

The Fhit gene is inactivated or altered in a large proportion of tumors by homo- or hemizygous deletions [3, 17], by translocations [3] and/or by the methylation of 5′ CpG islands [18–22]. These genomic deletions, and possibly the methylation, are involved in Fhit inactivation by environmental carcinogens. Loss of Fhit expression in the early stages of lung cancer is related to smoking [23–28] and a possible correlation between smoking and loss of Fhit expression in epithelial cells has also been shown in early stage esophageal cancer [29]. The urinary bladder is another organ that can be exposed to environmental carcinogens and their metabolites, and Fhit expression has been observed to be altered in 61% [30] and >50% [31] of bladder cancer. Fhit allelic losses have been shown in 22% [32] and >50% [33] of bladder cancer.

3 Fhit Modulates DNA Damage Response

Recent studies have indicated that Fhit has a role in the surveillance of genome integrity [34–37] via the elimination of cells with alterations in checkpoint response. Paradoxically, this suggests that the DNA-damage-susceptible FRA3B/Fhit common chromosome fragile region encodes a protein that is necessary to protect cells from accumulation of this type of damage through its role in the modulation of DNA-damage-response proteins; inactivation of Fhit contributes to the accumulation of abnormal checkpoint phenotypes in cancer development. It was reported that Fhit is involved in the maintenance of genome integrity at the mid-S checkpoint, through the Atr/Chk1 pathway, and that Fhit deficiency altered the response to DNA-damaging genotoxins [34–36]. The introduction of exogenous Fhit into cells in vitro led to modulated expression of the checkpoint proteins Hus1 and Chk1 at the mid-S checkpoint, modulation that led to induction of apoptosis in esophageal cancer cells but not in noncancerous primary cells [37]. Mutation of the conserved Fhit tyrosine 114 within the sequence DSIY114EEL, which can be phosphorylated by Src tyrosine kinase family members [38], resulted in failure of this function.

4 Exposing Genotoxins to Fhit-Deficient Cells in Mice

The significance of Fhit loss in early stages of carcinogenesis has been studied by animal experiments using the carcinogen N-nitrosomethylbenzylamine (NMBA) to induce epithelial tumors and hydroquinone (HQ) relevant to hematopoietic disorders.

4.1 NMBA to Mouse Forestomach

Animal models are useful in the study of heterogeneity of tumor cells. Analogous to the human distal esophagus, the mouse forestomach has a squamous-epithelial lining. The mouse squamo-columnar junction, a transition zone between squamous-epithelial and glandular tissue, corresponds to the human esophago-gastric junction and is commonly studied as a model for the human distal esophagus. The environmental nitrosamine NMBA is the most extensively used carcinogen for the induction of esophageal carcinoma in rodents. On bioactivation, NMBA produces benzaldehyde and an electrophilic agent that methylates DNA resulting in the formation of the promutagenic adduct O^6-methylguanine [39–43]. NMBA was used to investigate the role of the Fhit gene in carcinogen induction of neoplasia [44], indicating that loss of Fhit leads to susceptibility to NMBA-induced epithelial tumor. The introduction of Fhit into epithelial CICs reduced the incidence of tumors, indicating that Fhit-deficient mice are a useful model for carcinogen-induced cancer [44, 45].

4.2 HQ to Mouse Bone Marrow

Exposure to benzene is characterized by multi-site carcinogenicity, that is, tumors develop in various glandular tissues and organs including hematopoietic cells, Zymbal gland, Harderian gland, preputial gland, mammary gland, ovary, and lung [46, 47]. The well-established course of chronic benzene toxicity, resulting from excessive workplace exposure, includes decreases in circulating erythrocytes, leukocytes, and platelets, and, if exposure continues, to pancytopenia, which may accompany bone-marrow aplasia and myelodysplastic syndrome; an alternative outcome is the development of acute myelogenous leukemia [48].

Benzene and its major metabolites are not mutagenic in the Ames Salmonella test [49], but they induce chromosomal aberration in vitro and in vivo [49–51]. The metabolism of benzene to reactive metabolites such as benzene oxide by hepatic enzymes, mainly cytochrome P450-2E1 (CYP2E1), is a prerequisite for cyto- and genotoxicity after benzene exposure [48]. It is likely that HQ and catechol, two major downstream metabolites of benzene, can be metabolically activated, as a result of the peroxidatic action of bone marrow myeloperoxidase or prostaglandin H synthase of bone marrow macrophages, to reactive semiquinone or quinone species that rapidly bind proteins or nucleotides [48]. The metabolic actions are considered to be relevant to the hematological toxicity of benzene, suggesting that benzene metabolites direct benzene-induced disorders.

In vitro colony assay and in vivo transplantation of bone marrow stem and progenitor cells revealed recently that the absence of Fhit expression in hematopoietic cells exposed to

HQ leads to resistance to the induction of cell death. Transplantation of HQ-exposed Fhit-deficient bone-marrow stem and progenitor cells resulted in inhibition of the bone marrow suppression exhibited by wild-type transplanted bone marrow. Comparative immunohistochemical analyses of bone marrow transplanted from Fhit-knockout and wild-type mice showed relative absence of Bax and phospho-p38 expression in Fhit-deficient bone marrow, suggesting resistance to apoptosis and senescence. In vitro assessment of DNA-repair pathways in genotoxin-exposed Fhit-knockout and wild-type cells indicated that error-prone repair was more prevalent in Fhit-deficient cells than wild-type cells. It is proposed that Fhit deficiency would allow unscheduled, long-term survival of genotoxin-exposed Fhit-deficient hematopoietic stem cells carrying deleterious mutations.

5 CIC Hypothesis

Normal stem cells are characterized as having: (1) the capability of self-renewal; (2) the potential to divide and differentiate to generate all functional elements of a particular tissue; and (3) strict control over stem cell numbers [52, 53]. CICs can be defined as cells within the tumor with tumor-initiating potential [53, 54], so they are believed to have no control over cell numbers [53]. CICs are present in very small numbers in whole tumors and are responsible for the growth of the tumor cells [53]. According to a recent model, the involvement of a small fraction of CICs in tumor tissues is an attractive explanation for the observations that it is necessary to inject a high number of cancer cells into an animal model to maximize the probability of injecting cancer stem cells, and that the tumor often shrinks in size in response to treatment only to recur [53]. Treatments targeting whole cancer cells might not be able to target the CICs [53]. Some CICs may play a role in resistance to chemotherapy or radiation therapy and seem be responsible for recurrence after cancer treatment, even when most of the cancer cells appear to have been killed and only a few CICs remain [55]. Heterogeneous tumor tissues as a result of CIC involvement have been observed in leukemia [56–58], head and neck cancer [59], the gastro-intestinal system [60], colon [61, 62], breast [63], and brain [64, 65].

Epithelial CICs and their daughter epithelial tumor cells, including environmental-carcinogen-induced cancers, provide a complex component of epithelial and mesenchymal cells. Epithelial stem cells are specified during development and are controlled by epithelial-mesenchymal interactions [66]. Epithelial tumors contain cellular heterogeneity, which is accounted for by ongoing mutations and accumulating genomic instability. The complex structures can reflect multiple factors, including DNA damage checkpoints, repairs and apoptosis of epithelial cancerous cells or CICs, but also microenvironmental factors such as epithelial-mesenchymal interactions, angiogenesis and infiltrations of lymphocytes, which surround cancer cells. Strategic targeting of the small fraction of CICs would be useful for efficient treatment and for minimizing side effects in therapy.

6 A Role of Fhit in CICs

In vitro studies indicated that loss of Fhit inhibited apoptosis and led to cell survival, whereas the introduction of Fhit stimulated apoptosis in cancer cells during the cell cycle phases S or G2/M [67–70], suggesting a pro-survival effect of Fhit absence. Exposure of Fhit-deficient mice to NMBA resulted in stimulation of proliferation and cell cycling of Fhit-negative epithelium, associated with reduced apoptosis [45]. The study of checkpoint response indicated that introduction of Fhit stimulates the Hus1 pathway and is involved in activating the Atm/Atr pathway, which generally plays a role in DNA damage response after exposure to genotoxins and leads to inhibition or arrest of the cell cycle and the decision of cell fate (repair or apoptosis) [37], suggesting a possible involvement of Fhit in maintaining tumor genomic stability. Furthermore, a study of hematopoietic stem cells indicated that loss of Fhit increased cell survival with reduced apoptosis and p38 senescence marker; hematopoietic stem or progenitor cells were accumulated after exposure to HQ in Fhit-negative cells. Hematopoietic stem cells are slowly cycling or dormant [54]. One might speculate that loss of Fhit contributes to the reduction of apoptosis, leading to the survival and accumulation of stem or progenitor cells carrying deleterious mutations after HQ exposure. Recently, we showed that growth-factor stimulation in a culture of Fhit-deficient cells at dormant phase elicited an acute increase in HQ-induced phosphorylation of both Akt and Chk1, compared with wild-type cells, suggesting that the Fhit-deficient cells might stimulate inappropriate activation of survival and checkpoint kinases. It is noted that Fhit is at least partially involved in regulating a balance between dormant and proliferating cells. Loss of Fhit would disrupt the balance leading to survival and accumulation of damaged stem or progenitor cells. Fhit gene therapy might be a means of correcting the disorders by induction of apoptosis and elimination of surviving damaged cells or CICs.

7 Perspective

Cancer cells accumulate genomic alterations in multi-step carcinogenesis. Loss of the tumor suppressor Fhit in early-stage cancer allows the survival of damaged cells or CICs, which might increase genomic instability in the advanced stages. Therapeutic approaches might allow elimination of CICs carrying deleterious mutations during carcinogenesis.

References

1. Hickman ES, Moroni MC, Helin K. The role of p53 and pRB in apoptosis and cancer. Curr Opin Genet Dev. 2002;12:60–6.
2. Malumbres M, Barbacid M. RAS oncogenes: the first 30 years. Nat Rev Cancer. 2003;3:459–65.
3. Huebner K, Croce CM. FRA3B and other common fragile sites: the weakest link. Nat Rev Cancer. 2001;1:214–21.
4. Glover TW, Berger C, Coyle J, Echo B. DNA polymerase alpha inhibition by aphidicolin induces gaps and breaks at common fragile sites in human chromosomes. Hum Genet. 1984;67:136–42.
5. Glover TW, Stein CK. Chromosome breakage and recombination at fragile sites. Am J Hum Genet. 1988;43:265–73.
6. Zlotorynski E, Rahat A, Skaug J, Ben-Porat N, Ozeri E, Hershberg R, et al. Molecular basis for expression of common and rare fragile sites. Mol Cell Biol. 2003;23:7143–51.
7. Arlt MF, Durkin SG, Ragland RL, Glover TW. Common fragile sites as targets for chromosome rearrangements. DNA Repair (Amst). 2006;5(9–10):1126–35.
8. Inoue H, Ishii H, Alder H, Snyder E, Druck T, Huebner K, et al. Sequence of the FRA3B common fragile region: implications for the mechanism of FHIT deletion. Proc Natl Acad Sci U S A. 1997;94:14584–9.
9. Mimori K, Druck T, Inoue H, Alder H, Berk L, Mori M, et al. Cancer-specific chromosome alterations in the constitutive fragile region FRA3B. Proc Natl Acad Sci U S A. 1999;96:7456–61.
10. Casper AM, Nghiem P, Arlt MF, Glover TW. ATR regulates fragile site stability. Cell. 2002;111:779–89.
11. Arlt MF, Xu B, Durkin SG, Casper AM, Kastan MB, Glover TW. BRCA1 is required for common-fragile-site stability via its G2/M checkpoint function. Mol Cell Biol. 2004;24:6701–09.
12. Musio A, Montagna C, Mariani T, Tilenni M, Focarelli ML, Brait L, et al. SMC1 involvement in fragile site expression. Hum Mol Genet. 2005;14:525–33.
13. Howlett NG, Taniguchi T, Durkin SG, D'Andrea AD, Glover TW. The Fanconi anemia pathway is required for the DNA replication stress response and for the regulation of common fragile site stability. Hum Mol Genet. 2005;14:693–701.
14. Durkin SG, Arlt MF, Howlett NG, Glover TW. Depletion of CHK1, but not CHK2, induces chromosomal instability and breaks at common fragile sites. Oncogene. 2006;25:4381–8.
15. Zhu M, Weiss RS. Increased common fragile site expression, cell proliferation defects, and apoptosis following conditional inactivation of mouse Hus1 in primary cultured cells. Mol Biol Cell. 2007;18:1044–55.
16. Schwartz M, Zlotorynski E, Goldberg M, Ozeri E, Rahat A, le Sage C, et al. Homologous recombination and nonhomologous end-joining repair pathways regulate fragile site stability. Gene Dev. 2005;19:2715–26.
17. Sozzi G, Huebner K, Croce CM. FHIT in human cancer. Adv Cancer Res. 1998;74:141–66.
18. Chang KW, Kao SY, Tzeng RJ, Liu CJ, Cheng AJ, Yang SC, et al. Multiple molecular alterations of FHIT in betel-associated oral carcinoma. J Pathol. 2002;196:300–6.
19. Maruyama R, Toyooka S, Toyooka KO, Virmani AK, Zochbauer-Muller S, Farinas AJ, et al. Aberrant promoter methylation profile of prostate cancers and its relationship to clinicopathological features. Clin Cancer Res. 2002;8:514–9.
20. Maruyama R, Toyooka S, Toyooka KO, Harada K, Virmani AK, Zochbauer-Muller S, et al. Aberrant promoter methylation profile of bladder cancer and its relationship to clinicopathological features. Cancer Res. 2001;61:8659–63.
21. Zochbauer-Muller S, Fong KM, Maitra A, Lam S, Geradts J, Ashfaq R, et al. 5' CpG island methylation of the FHIT gene is correlated with loss of gene expression in lung and breast cancer. Cancer Res. 2001;61:3581–5.
22. Tanaka H, Shimada Y, Harada H, Shinoda M, Hatooka S, Imamura M, et al. Methylation of the 5' CpG island of the FHIT gene is closely associated with transcriptional inactivation in esophageal squamous cell carcinomas. Cancer Res. 1998;58:3429–34.
23. Sozzi G, Pastorino U, Moiraghi L, Tagliabue E, Pezzella F, Ghirelli C, et al. Loss of FHIT function in lung cancer and preinvasive bronchial lesions. Cancer Res. 1998;58:5032–7.
24. Tseng JE, Kemp BL, Khuri FR, Kurie JM, Lee JS, Zhou X, et al. Loss of Fhit is frequent in stage I non-small cell lung cancer and in the lungs of chronic smokers. Cancer Res. 1999;59:4798–803.
25. Tomizawa Y, Nakajima T, Kohno T, Saito R, Yamaguchi N, Yokota J. Clinicopathological significance of Fhit protein expression in stage I non-small cell lung carcinoma. Cancer Res. 1998;58: 5478–83.
26. Sozzi G, Sard L, De Gregorio L, Marchetti A, Musso K, Buttitta F, et al. Association between cigarette smoking and FHIT gene alterations in lung cancer. Cancer Res. 1997;57:2121–3.
27. Nelson HH, Wiencke JK, Gunn L, Wain JC, Christiani DC, Kelsey KT. Chromosome 3p14 alterations in lung cancer: evidence that FHIT exon deletion is a target of tobacco carcinogens and asbestos. Cancer Res. 1998;58:1804–7.
28. Stein CK, Glover TW, Palmer JL, Glisson BS. Direct correlation between FRA3B expression and cigarette smoking. Gene Chromosom Cancer. 2002;34:333–40.
29. Mori M, Mimori K, Shiraishi T, Alder H, Inoue H, Tanaka Y, et al. Altered expression of Fhit in carcinoma and precarcinomatous lesions of the esophagus. Cancer Res. 2000;60:1177–82.
30. Baffa R, Gomella LG, Vecchione A, Bassi P, Mimori K, Sedor J, et al. Loss of FHIT expression in trasitional cell carcinoma of the urinary bladder. Am J Pathol. 2000;156:419–24.
31. Skopelitou AS, Gloustianou G, Bai M, Huebner K. FHIT gene expression in human urinary bladder transitional cell carcinomas. In Vivo. 2001;15:169–73.
32. Wada T, Louhelainen J, Hemminki K, Adolfsson J, Wijkstrom H, Norming U, et al. The prevalence of loss of heterozygosity in chromosome 3, including FHIT, in bladder cancer, using the fluorescent multiplex polymerase chain reaction. BJU Intern. 2001;87:876–81.
33. Louhelainen J, Wijkstrom H, Hemminki K. Multiple regions with allelic loss at chromosome 3 in superficial multifocal bladder tumors. Intern J Oncol. 2001;18:203–10.
34. Ottey M, Han SY, Druck T, Barnoski BL, McCorkell KA, Croce CM, et al. Fhit-deficient normal and cancer cells are mitomycin C and UVC resistant. Br J Cancer. 2004;91:1669–77.
35. Hu B, Han SY, Wang X, Ottey M, Potoczek MB, Dicker A, et al. Involvement of the Fhit gene in the ionizing radiation-activated ATR/CHK1 pathway. J Cell Physiol. 2005;202:518–23.
36. Hu B, Wang H, Wang X, Lu HR, Huang C, Powell SN, et al. Fhit and CHK1 have opposing effects on homologous recombination repair. Cancer Res. 2005;65:8613–6.
37. Ishii H, Mimori K, Inoue H, Inageta T, Ishikawa K, Semba S, et al. Fhit modulates the DNA damage checkpoint response. Cancer Res. 2006;66:11287–92.
38. Pekarsky Y, Garrison PN, Palamarchuk A, Zanesi N, Aqeilan RI, Huebner K, et al. Fhit is a physiological target of the protein kinase Src. Proc Natl Acad Sci U S A. 2004;101:3775–9.
39. Wargovich MJ, Imada O. Esophageal carcinogenesis in the rat: a model for aerodigestive tract cancer. J Cell Biochem Suppl. 1993;17F:91–94.

40. Fong LY, Magee PN. Dietary zinc deficiency enhances esophageal cell proliferation and N-nitrosomethylbenzylamine (NMBA)-induced esophageal tumor incidence in C57BL/6 mouse. Cancer Lett. 1999;143:63–9.
41. Ushida J, Sugie S, Kawabata K, Pham QV, Tanaka T, Fujii K, et al. Chemopreventive effect of curcumin on N-nitrosomethylbenzylamine-induced esophageal carcinogenesis in rats. J J Cancer Res. 2000;91:893–8.
42. Carlton PS, Kresty LA, Siglin JC, Morse MA, Lu J, Morgan C, et al. Inhibition of N-nitrosomethylbenzylamine-induced tumorigenesis in the rat esophagus by dietary freeze-dried strawberries. Carcinogenesis, 2001;22:441–6.
43. Fong LY, Nguyen VT, Farber JL. Esophageal cancer prevention in zinc-deficient rats: rapid induction of apoptosis by replenishing zinc. J Nation Cancer Inst. 2001;93:1525–33.
44. Fong LYY, Fidanza V, Zanesi N, Lock LF, Siracusa LD, Mancini R, et al. Muir-Torre-like syndrome in Fhit-deficient mice. Proc Natl Acad Sci U S A. 2000;97:4742–7.
45. Dumon KR, Ishii H, Fong LYY, Zanesi N, Fidanza V, Mancini R, et al. FHIT gene therapy prevents tumor development in Fhit-deficient mice. Proc Natl Acad Sci U S A. 2001;98:3346–51.
46. Huff JE, Haseman JK, DeMarini DM, Eustis S, Maronpot RR, Peters AC, et al. Multiple-site carcinogenicity of benzene in Fischer 344 rats and B6C3F1 mice. Environ Health Perspect. 1989;82:125–63.
47. Maltoni C, Ciliberti A, Cotti G, Conti B, Belpoggi F. Benzene, an experimental multipotential carcinogen: results of the long-term bioassays performed at the Bologna Institute of Oncology. Environ Health Perspect. 1989;82:109–24.
48. Snyder R. Benzene and leukemia. Crit Rev Toxicol. 2002;32:155–210.
49. Dean BJ. Recent findings on the genetic toxicology of benzene, toluene, xylenes and phenols. Mutat Res. 1985;154:153–81.
50. Yager JW, Eastmond DA, Robertson ML, Paradisin WM, Smith MT. Characterization of micronuclei induced in human lymphocytes by benzene metabolites. Cancer Res. 1990;50:393–9.
51. Wolman SR. Cytologic and cytogenetic effects of benzene. J Toxicol Environ Health Suppl. 1977;2:63–8.
52. Bixby S, Kruger GM, Mosher JT, Joseph NM, Morrison SJ. Cell-intrinsic differences between stem cells from different regions of the peripheral nervous system regulate the generation of neural diversity. Neuron. 2002;35:643–56.
53. Sagar J, Chaib B, Sales K, Winslet M, Seifalian A. Role of stem cells in cancer therapy and cancer stem cells: a review. Cancer Cell Intern. 2007;7:9.
54. Reya T, Morrison SJ, Clarke MF, Weissman IL. Stem cells, cancer, and cancer stem cells. Nature. 2001;414:105–11.
55. Tan BT, Park CY, Ailles LE, Weissman IL. The cancer stem cell hypothesis: a work in progress. Lab Invest. 2006;86:1203–7.
56. Wulf GG, Wang RY, Kuehnle I, Weidner D, Marini F, Brenner MK, et al. A leukemic stem cell with intrinsic drug efflux capacity in acute myeloid leukemia. Blood. 2001;98:1166–73.
57. Lapidot T, Sirard C, Vormoor J, Murdoch B, Hoang T, Caceres-Cortes J, et al. A cell initiating human acute myeloid leukemia after transplantation into SCID mice. Nature. 1994;367:645–8.
58. Bonnet D, Dick JE. Human acute leukemia is organized as a hierarchy that originates from a primitive hematopoietic cell. Nat Med. 1997;3:730–7.
59. Prince ME, Sivanandan R, Kaczorowski A, Wolf GT, Kaplan MJ, Dalerba P, et al. Identification of a subpopulation of cells with cancer stem cell properties in head and neck squamous cell carcinoma. Proc Natl Acad Sci U S A. 2007;104:973–8.
60. Haraguchi N, Utsunomiya T, Inoue H, Tanaka F, Mimori K, Barnard GF, et al. Characterization of a side population of cancer cells from human gastrointestinal system. Stem Cells. 2006;24:506–13.
61. Ricci-Vitiani L, Lombardi DG, Pilozzi E, Biffoni M, Todaro M, Peschle C, De Maria R. Identification and expansion of human colon-cancer-initiating cells. Nature. 2007;445:111–5.
62. O'Brien CA, Pollett A, Gallinger S, Dick JE. A human colon cancer cell capable of initiating tumour growth in immunodeficient mice. Nature. 2007;445:106–10.
63. Al-Hajj M, Wicha MS, Benito-Hernandez A, Morrison SJ, Clarke MF. Prospective identification of tumorigenic breast cancer cells. Proc Natl Acad Sci U S A. 2003;100:3983–8.
64. Piccirillo SG, Reynolds BA, Zanetti N, Lamorte G, Binda E, Broggi G, et al. Bone morphogenetic proteins inhibit the tumorigenic potential of human brain tumour-initiating cells. Nature. 2006;444:761–5.
65. Bao S, Wu Q, McLendon RE, Hao Y, Shi Q, Hjelmeland AB, et al. Glioma stem cells promote radioresistance by preferential activation of the DNA damage response. Nature. 2006;444:756–60.
66. Blanpain C, Horsley V, Fuchs E. Epithelial stem cells: turning over new leaves. Cell. 2007;128:445–58.
67. Siprashvili Z, Sozzi G, Barnes LD, McCue P, Robinson AK, Eryomin V, et al. Replacement of Fhit in cancer cells suppresses tumorigenicity. Proc Natl Acad Sci U S A. 1997;94:13771–6.
68. Ji L, Fang B, Yen N, Fong K, Minna JD, Roth JA. Induction of apoptosis and inhibition of tumorigenicity and tumor growth by adenovirus vector-mediated fragile histidine triad (FHIT) gene overexpression. Cancer Res. 1999;59:3333–9.
69. Dumon KR, Ishii H, Vecchione A, Trapasso F, Baldassarre G, Chakrani F, et al. Fragile histidine triad expression delays tumor development and induces apoptosis in human pancreatic cancer. Cancer Res. 2001;61:4827–36.
70. Ishii H, Dumon KR, Vecchione A, Fong LY, Baffa R, Huebner K, et al. Potential cancer therapy with the fragile histidine triad gene: review of the preclinical studies. JAMA. 2001;286:2441–9.

History of Cancer Stem Cells

Stewart Sell

Abstract It has been hypothesized for over 40 years that cancers contain the same cell populations as normal tissues: stem cells, proliferating transit-amplifying cells, and terminally differentiated (mature cells). The properties of cancer stem cells include the ability to transplant the tumor, the ability to grow in vitro and the ability to resist conventional therapies. The idea that cancer arose from stem cells dates back to the middle of the 1800s as the embryonal rest theory of cancer. Studies on cancers of germinal cells (teratocarcinomas) beginning about 50 years ago established that cancers arise from stem cells and that cancers contain stem cells. For some time teratocarcinomas were believed to be different from other cancers in regard to their origin from stem cells. However, studies on liver cancer, believed to arise from dedifferentiation of mature hepatocytes, revealed that liver cancers also arose from stem cells. More recently cancers of the blood cells (leukemias) were demonstrated to be due to specific blocks to differentiation caused by genetic lesions that allow the leukemic transit-amplifying cells to continue to proliferate and not die (maturation arrest). Conventional chemotherapy, radiotherapy and anti-angiogenic therapies act on the cancer transit-amplifying cells. When these therapies are discontinued the cancer will reform from the therapy resistant cancer stem cells. Re-appreciation of the properties of cancer stem cells allows new directions in differentiation therapy of cancer stem cells by blocking the signals that maintain stemness and allowing the cancer stem cells to differentiate. The goal of this chapter is to introduce the subject of cancer stem cells and present a history of how this concept developed. Two major concepts will be covered: (1) cancers contain stem cells (cancer stem cells); and (2) cancers arise from stem cells. The first concept has significance in designing new stem cell directed therapy for cancer; the second concept is important for directing approaches to prevent cancer.

S. Sell (✉)
Wadsworth Center, Ordway Research Institute and University at Albany, Empire State Plaza, Room C-551; Albany, NY 12201
e-mail: ssell@wadsworth.org

1 Cancer Stem Cells

The cells that take part in normal tissue renewal and in growth of cancer belong to the same populations: stem cells, transit-amplifying cells, and terminally differentiated cells (Fig. 1). The difference between cancer growth and normal tissue renewal is that the number of cells that are produced by cell division in normal tissue essentially equals the number of cells that terminally differentiate in a given time period so that the total number of cells remains the same. In contrast, in cancers, the proliferating transit-amplifying cells do not terminally differentiate. The number of cells in a cancer increases with time because the transit-amplifying cells either continue to proliferate and do not mature, or the mature cells do not die, or both. In both normal tissue renewal and cancer growth there is a small fraction of cells which are not actively proliferating and serve as a reserve cell population. These are the stem cells [1–5]. When a stem cell divides it gives rise to one daughter cell that remains a stem cell and one daughter cell that begins the process of differentiation by becoming a transit amplifying cell. Thus in cancer tissue there is a self-renewing cell that exhibits many of the properties of a normal tissue stem cell.

2 Biological Properties of Cancer Stem Cells

The biological properties of cancer stem cells are the ability to initiate tumor formation when transplanted to a histocompatible or immunodeficient recipient [6], the ability to grow in soft agar [7], and the ability to survive conventional chemo- or radiation therapy [8] In 1952, Green [6] pointed out that embryonic and cancer tissues, but not normal or hyperplastic tissues will grow in immune privileged sites in "alien" animals. Transplantable tumor cells, so-called "tumor initiating cells," were found through dilution studies to comprise from 1 in 30 to 1 in 1000 of the cells in adenocarcinomas and sarcomas [8–10]. For example, to achieve a success rate of 50% required the injection of 300

V.K. Rajasekhar, M.C. Vemuri (eds.), *Regulatory Networks in Stem Cells*, Stem Cell Biology and Regenerative Medicine,
DOI 10.1007/978-1-60327-227-8_37, © Humana Press, a part of Springer Science+Business Media, LLC 2009

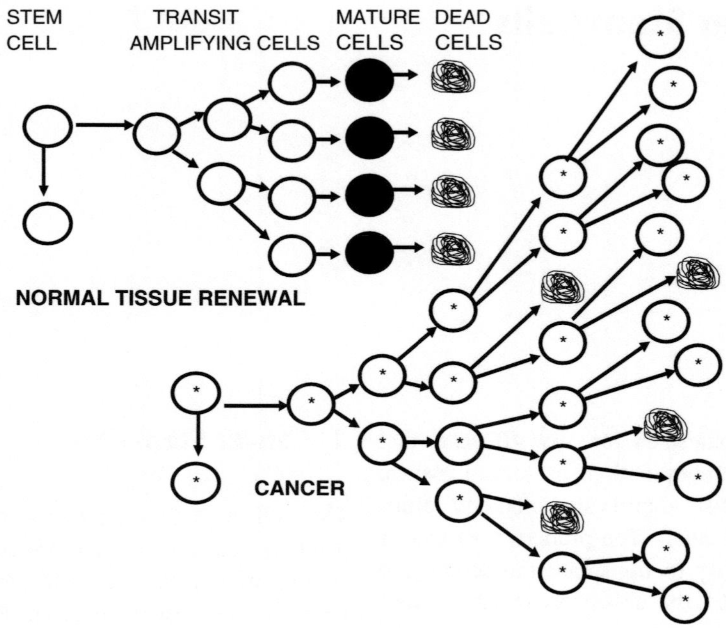

Fig. 1 Scheme of cell populations in normal and cancer tissues. Both normal tissues and cancers are composed of stem cells, transit amplifying cells, terminally differentiated cells and apoptotic cells. Normal tissue renewal and growth of cancer are accomplished by proliferation of the transit amplifying cells. The transit-amplifying cells of normal tissue differentiate to mature cells and then die. The transit-amplifying cells of cancer continue to proliferate; although some do die, there is a continual accumulation of the cancer transit-amplifying cells. The rate of growth of the cancer will depend on the growth fraction of the cancer (i.e., the proportion of cells dividing) and the proportion dying (the death fraction). Both the normal and cancer stem cells divide rarely; and when they do, one daughter cell becomes a transit-amplifying cell and the other remains a stem cell. In this way, the normal and cancer stem cells are self-renewing. The asterisk indicates that the genetic lesion of the cancer is carried in the stem cell

adenocarcinoma cells [11]. The frequency of tumor-initiating cells was later found to be of the same order of magnitude as the frequency of cells that survive after chemo- or radiotherapy and grow in soft agar. Salman [12] found that 1:1000 to 1:100,000 cells from a primary human adenocarcinoma would form colonies in soft agar (called "tumor colony forming units"s). This range of proportions of tumor colony forming units is similar to the proportions recently found for the tumor initiating cells in acute myelogenous leukemia (see below, [13–15]). Finally, as an example of resistance to therapy, highly tumorigenic subpopulations of cancer initiating cells derived from human glioblastomas display resistance to radiation because of increased protection against DNA damage [16], supporting the idea presented above for leukemia that cancer stem cells are resistant to standard therapies [17].

3 Enrichment of Cancer Stem Cells

An approach to partial purification of human breast cancer stem cells using stem cell markers was reported by Al Hajj and co-workers in 2003 [17]. They found that human primary breast cancer cells could be fractionated by cell sorting into two major populations: CD44+CD24–ESA+ and CD44–CD24+ESA–. CD stands for clusters of differentiation, markers for cells at different stages of differentiation; ESA stands for epithelial cell specific antigen. Primary tumors showed a high degree of heterogeneity of expression of these markers by immunohistochemistry, indicating that tumors are composed of at least two types of cells [18]. Using the most highly purified CD44+CD24-ESA+ population, these authors were able to obtain 1/6 transplantation takes with injection of 100 cells. Thus, the frequency of tumor-initiating cells in the purified population was similar to that reported using whole tumor cell populations by Reinhard et al. in 1952 [11], who obtained 6% takes with 18 cells. A higher transplantation frequency could be obtained when CD24–CD44+ESA+^{low} cells were sorted from CD24–CD44+ESA+^{high} cells17. In any case, these results were interpreted by the authors to mean that two types of breast cancer cells may be separated on the basis of these markers. After transplantation of the CD44+CD24–ESA+ breast cancer cells, the growing tumor reconstituted both cell populations, suggesting that the tumor initiating population could produce progeny expressing markers of the nontumor initiating population [17].

Although it is implied that the fractionation of the human breast cancer cells by the CD44+CD24–ESA+ phenotype results in purification of breast cancer stem cells, this may not be the case. So far, the population of the breast cancer cells isolated by these markers consists of a much larger proportion of the cells than those that can transplant the tumor, and much more than is found with other stem cell models. Additional work is required to determine if there really is a stem cell in breast cancer that is different from most of the other cells in the cancer (see below). In addition, the question of the efficiency of transplantability of human cancer cells to immunodeficient mice as compared to that of transplantation of mouse cancer cells to syngeneic mice [19] must be addressed. This may explain why only one of the populations of cells from the human mammary cancers is able to initiate tumors on transplantation to SCID mice. The application of mammary gland stem cell markers, such as endoglin and prion protein, may be able to characterize better the putative breast cancer stem cell [20].

4 Are Cancer Stem Cells a Unique Population?

The concept that cancers contain stem cells has recently attracted much attention [2–5, 21], but this concept has actually been debated for some time. For example, in 1994, contrasting views were published Trott [22] and by Denekamp [23]. Trott [22] reviewed the data that from 0.1% to 100% of all cells from transplantable mouse tumors meet the criteria of a tumor stem cell: that is, "regrowth of the tumour preceded by clonal expansion from a single cell with unlimited proliferative potential." He concluded that tumors contain the same populations of cells as are found in normal tissue, consistent with the proposal of Pierce and Spears for teratocarcinoma (see below [24]). On the other hand, Denekamp [23], considering the same evidence, deduced that the putative cancer stem cells are merely the least differentiated cells in the cancer population, and thus, appear functionally and kinetically different from the mass of tumor cells. She concluded that the cancer stem cell is not as clearly definable as is the normal-tissue stem cell. This debate has now been renewed. For example, Kern and Shibata [25], using mathematical analyses, point out that tumor-initiating capacity might be a varying probabilistic potential for all tumor cells rather than a quantal and deterministic feature of a minority of tumor cells. Identification of tumor initiating cell populations using marker phenotypes may enrich for cells that can transplant tumors, but even with the best purification systems, the so-called nontumorigenic cell population, will contain up to 3% of tumorgeinic cells [26]. In addition, as stated above, the significance of transplantation of human cancer cells into SCID mice as an indicator of a property of cancer stem cells has come into question [19]. In contrast to the finding that only 1 in 250,000 human leukemic cells is transplantable, essentially all of the cells of a mouse B-cell lymphoma will produce tumors when injected into nonirradiated congenic recipients [19], a finding originally reported in 1937 [27] (see below). The possibility is raised that the tissue microenvironment of a SCID mouse may limit the ability of the human leukemic cells to form a tumor and that the low fraction of transplantable cells in human leukemias may be due to an incompatible microenvironment [19]. If this is true, then there may be many more leukemic and breast cancer stem cells than reported.

5 Cancers Arise from Stem Cells

The idea that cancers arise from stem cells has evolved over the last 200 years. In the last 20 years, reappreciation of this idea has resulted in a paradigm change in how we think about cancer [1, 2]. However, the idea that cancer arise from tissue stem cells was first formulated as the embryonal rest theory of cancer in the mid-1800s. The embryonal rest theory of cancer proposes that cancers arise from small collections of embryonic tissue that persisted and did not differentiate in mature tissues with aging [28, 29]. Prior to the 19th century, cancer was believed to arise from and excess of black bile, a view put forth by such luminaries as Hippocrates, Celsus, and Galen; a view consistent with the humoral theory of disease [30]. In the first half of the 19th century, Joseph Claude Anselme Recamier [31] and Robert Remak [32] observed that cancer tissues looked more like embryonic tissues than they resemble adult tissues. This was formalized as the embryonal rest theory of cancer by Franco Durante [28] and Julius Cohnheim [29]. The embryonal theory of cancer served as a forerunner of the stem cell theory of cancer that would be developed in the latter half of the 20th century. However, during the last half of the 19th century, the embryonal rest theory of cancer fell out of favor, to be replaced by the de-differentiation theory of cancer.

6 Dedifferentiation

The dedifferentiation theory of cancer posits that cancers arise from changes in mature differentiated cells so that they now express the properties of cancer cells. A number of observations led to general acceptance of the dedifferentiation theory. As late as 1853, Sir James Paget wrote that cancers came from morbid material in the blood, essentially a variation of the black bile theory [33]. Even the "Father of Pathology," Rudolf Virchow [34], who described embryonic tissues in teratocarcinoma, concluded that cancers arose from connective tissue during chronic inflammation. The earlier observation that chimney sweeps developed cancer of the scrotum [35, 36], was interpreted to mean that cancers were induced in mature epithelial tissues by chemicals. Others considered cancer came from a disequilibrium between connective tissue and epithelium [37], a changed "habit of growth" of normal cells [38, 39], or the loss of restraining influences of the body on displaced tissue cells leading to dedifferentiation of mature cells [40]. Amedee Borrel [41] and a number of other scientists believed infection by parasites caused cancers, and, by 1911, Peyton Rous had identified the Rous sarcoma virus as a cause of cancer in fowl [42]. Finally, Theodore Boveri [43] discovered that sea urchin eggs may have abnormal chromosome content. These observations were each considered support for dedifferentiation as the main mechanism of development of cancer. By 1914, Bainbridge, in his authoritative book *The Cancer Problem* [44], concluded: "The congenital or embryonic theory of the origin of cancer has received no support whatever from the experimental and comparative investigations of recent times." Dedifferentiation remained the dominant theory of the origin of cancer into the 1980s. The role of stem cells in cancer was restored by studies on teratocarcinoma and leukemia, as well as other cancers, that convincingly showed that cancer arose from stem cells and contained cells with stem cell properties.

7 Teratocarcinoma

Teratocarcinoma is a malignant cancer that arises from the germinal cells, usually in the testes or the ovaries of a young adult. This cancer usually grows rapidly, but responds well to therapy, as exemplified by the case of Lance Armstrong. Beginning in the 1950s, Barry Pierce and his colleagues [24, 45, 46], demonstrated that teratocarcinomas contain the same population of cells as is present in normal tissue: stem cells, transit amplifying cells, and differentiated cells In extensive analysis of the cells that make up teratocarcinomas, they demonstrated that well over 99% of the cells of most teratocarcinomas belong to mature differentiated tissues, including derivative cells from all germ layers (skin, glands, connective tissue, vessels, neural tissue, bone marrow, etc.) as well as yolk sac and placental tissue. The only malignant cells of the teratocarcinoma were found in structures in the tumor that resemble early embryos (embryoid bodies). Thus, the embryoid bodies contain the only cells of the teratocarcinoma that can be cultured in vitro or initiate tumors on transplantation [46]. At the same time, when teratocarcinomas were shown to contain a small population of tumor initiating stem cells, it was shown that normal germinal cells or embryonal cells could give rise to teratocarcinomas and that teratocarcinoma stem cells could differentiate into normal tissues.

Leroy Stevens generated teratocarcinomas from normal testicular germ cells [47], tubal mouse eggs [48], or early mouse embryo cells [49], when he transplanted these cells from their natural sites [50] to abnormal tissue sites. Mintz and Illmensee [51, 52] then demonstrated that, when injected into the blastocyst of a normally developing embryo, these transplantable teratocarcinoma stem cells could give rise to normal tissues in mosaic mice. Thus, normal cells could become malignant when removed from their controlling environment, and malignant teratocarcinoma cells could become normal benign cells of various tissues when placed into a controlling and nurturing environment.

From these studies of teratocarcinomas, we learned three critical things about cancer: First, cancers arise from maturation arrest of stem cells, not by dedifferentiation [1]. For many years, the stem cell origin of cancer was considered unique for teratocarcinoma. Dedifferentiation of mature cells was still considered for other cancers, in particular liver cancer. (Because of space limitations, this subject will not be reviewed here; for an extensive recent review see [53].) However, this changed in the 1980s when it was shown that liver cancer also arose from stem cells [54, 55]. Second, we learned that cancers contain the same cell populations as do normal tissues: stem cells, transit-amplifying cells, and terminally differentiated cells [3]. Third, we learned that malignant cells can become benign; later demonstrated directly for squamous cell carcinoma by following the differentiation of pulse-labeled proliferating cancer stem cells [56]. This led to the concept that it may be possible to treat cancer by forcing the transit-amplifying cells to differentiate (differentiation therapy [1–5, 57]). The concepts of maturation arrest of cancer stem cells as a cause of cancer and differentiation therapy were further developed from the study of leukemia [58].

8 Leukemia

Leukemia is a malignant cancer of the cells of the blood and immune system. It is manifested by a tremendous increase in immature white blood cells in the circulation. The presence of so many white blood cells in the blood causes the color of the blood to change from red to creamy or white. The term leukemia means white blood (*leuk* – white; *emia* – blood). Because of the replacement of the normal white blood cells with leukemic cells, patients with leukemia usually die because they are unable to fight off infections. Leukemia is caused by a failure of the transit-amplifying cancer cells in the white blood cell lineage to mature to functional terminally differentiated cells (maturation arrest, Fig. 2).

Cure of leukemia requires elimination of the most primitive leukemic stem cells, which carry the leukemic translocation or mutation, as well as the proliferating leukemic transit-amplifying cells [59]. One of the properties of cancer stem cells is their resistance to chemo- and radiation therapy,

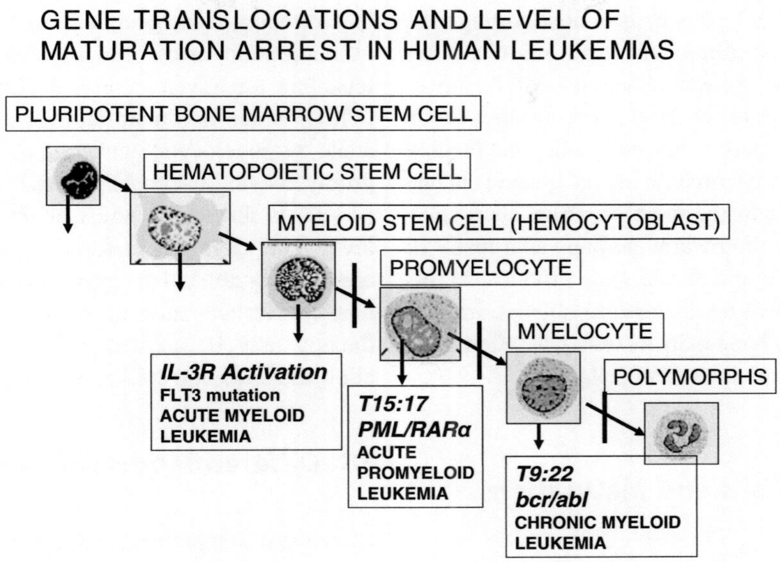

Fig. 2 Stages of maturation arrest in myeloid leukemias are associated with specific fusion gene products of gene translocations. The fusion gene products constitutively activate a signaling pathway that blocks further maturation of the cells. Effective treatment is directed to inhibition of these leukemic signaling pathways. This allows the leukemic cells to mature and die, thus eliminating the leukemic transit amplifying cells (differentiation therapy). However, when the differentiation therapy is discontinued, the leukemia will regrow from the leukemic stem cell. From Reference [58]

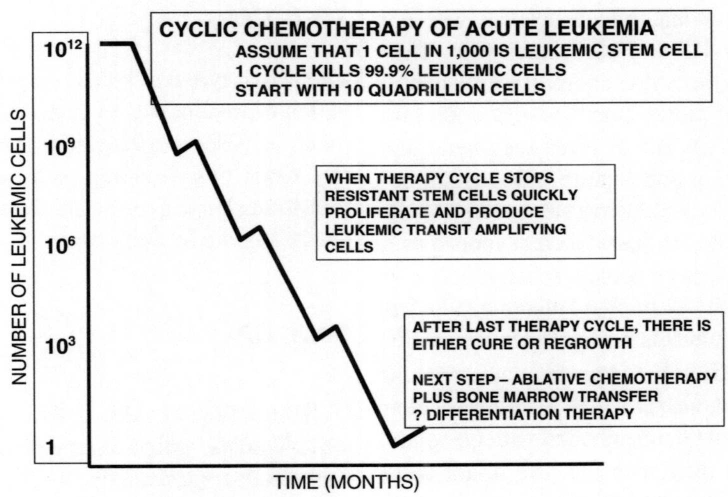

Fig. 3 Kinetics of the decline of the number of leukemic cells in the body during cyclic chemotherapy. Chemotherapy is administered in cycles to allow regeneration of normal hematopoietic transit-amplifying cells. When this is done the leukemia stem cells also proliferate to produce leukemia transit-amplifying cells. Chemotherapy is given again for a total of four cycles. After each cycle the leukemia will begin to regrow. In most cases, it is not possible for chemotherapy alone to eliminate regrowth of the cancer from the cancer stem cells or cancer cells that are in G_0 at the time therapy is administered

as illustrated by the response of AML to cyclic chemotherapy (Fig. 3). When AML is first detected, the tumor load is in the range of 10^{12} cells. In general, chemo- or radiation therapy will be effective in eliminating 99.9% of the AML cells [60]. This kill percentage is consistent either with the assumption that less than 1 in 1000 of the AML cells is a stem cell that is resistant to the therapy, or else the idea that it is ineffective against a leukemia cell that is not dividing when the therapy is administered. Since chemo- and radiation therapies are directed to proliferating cells (the growth fraction) of the AML, the stem or resting (G_0) leukemia cell is not affected by the therapy. When the therapy is discontinued in order to allow the normal hematopoietic cells to recover, the AML stem cells will also regrow and produce more leukemic transit-amplifying cells. Then, a second cycle of therapy is given that will once again eliminate 99.9% of the leukemic cells. Since, the putative AML stem cells or nondividing cells will again be resistant, the tumor will again regrow. After four cycles of therapy (Fig. 3), some leukemia will be cured, suggesting that the most primitive bone marrow stem cell is either not mutated or was caught out of G_0 when the therapy was administered [60]. On the other hand, the genetic change in many AMLs is present in the most primitive stem cells, which are resistant to chemotherapy, perhaps as a result of expression of a multiple drug-resistant gene product. At this point, curative therapy must be changed to ablative irradiation or chemotherapy and bone marrow transplantation in an attempt to eliminate the leukemia stem cell.

9 Leukemia Stem Cells and Maturation Arrest

Leukemia stem cells were first directly identified in 1937 when Furth and Kahn [27] demonstrated that leukemia of mice could be transplanted via a single cell. Later, Makino and Kano concluded that leukemia arises from a progenitor or stem cell [61]. The first definitive demonstration of tissue-specific stem cells for the hematopoietic system was not accomplished until 1961 by Till and McCulloch [62]. They found that 10 to 14 days after intravenous transplantation of small numbers of normal mouse bone marrow cells into heavily irradiated mice, the spleens of these recipient mice contained nodules of maturing and mature blood cells, believed to be a clone of cells derived from a stem cell [63, 64]. More recently, using the nude immunodeficient mouse as a recipient for human cancer cells, an approach first reported in 1969 [65], the transplantability of human leukemic cells has been examined. Of special interest is the finding that only 1 in 250,000 or so leukemic cells is able to transfer acute myeloid leukemia (AML) to a SCID mouse [13–15]. This observation indicates that leukemia cannot be transplanted with the tumor transit-amplifying cells, but only with the true tumor stem cell. Thus, it is concluded that tumor initiation is a property only of the leukemic stem cell. On the other hand, as recently discussed by Kelly et al. [19], the transplantability of human leukemias in SCID mice may be limited by difficulties that human cells have adapting to a foreign (mouse) milieu. In fact, Kelly et al. found, similar to Furth and Kahn [27], that mouse AML could be transplanted to nonirradiated histocompatible recipients by any leukemic cell. As mentioned above, this interpretation has implications for any tumor transplantation studies and may result in an underestimation of the tumor initiating cells in human cancers transplanted to immunodeficient mice. Although there are still inconsistencies in identifying leukemic stem cells, it is clear that leukemias arise from a block in the differentiation of cells in the myeloid pathway, that us, maturation arrest [58].

Examples of the genetic lesions (translocations) resulting in maturation arrest and classification of three types of leukemia are given in Fig. 2. Chronic myeloid leukemia (CML) is due to a maturation arrest at the myelocyte level, acute promyelocytic leukemia is due to an arrest at the promyelocyte level (APL), and acute myeloid leukemia (AML) is due to an arrest at the myeloid progenitor cell level. The stages of maturation arrest for these leukemias are directly related to gene rearrangements that result in constitutive activation of the cells [66, 67]. Differentiation therapy may be applied to reverse maturation arrest and allow the leukemic cells to terminally differentiate.

10 Differentiation Therapy of Leukemia

Effective differentiation therapy of leukemia involves blocking of the constitutive activation signal provided by the fusion products of the specific gene translocations [58]. Such a block reverses the specific maturation arrest at one of the levels shown in Fig. 2.

10.1 CML

The active tyrosine kinase responsible for constitutive activation of myelocytes in CML can be effectively blocked by imatinib (Gleevec, Novartis: formerly called ST1571) [68, 69]. Once this signaling pathway is blocked, the leukemic transit-amplifying cells are free to differentiate, and they eventually die by apoptosis.

10.2 APL

Differentiation of APL cells is induced by treatment with retinoic acids, which react with the fusion gene product and cause its degradation [70, 71]. The fusion gene is the promyelocytic leukemia (PML) protein gene fused with the retinoic acid receptor gene. The fusion product (PML-RARα) inactivates the PML protein, which is required for normal formation of granules, and for maturation of the promyelocyte [70]. Reaction of retinoic acid with the fusion protein results in ubiquitinization and degradation of the protein. The maturation arrest is thereby removed and the leukemic cells are free to differentiate.

10.3 AML

The situation for AML is much more complex than for CML or APL. The complexity arises because there is more than one molecular lesion in AML. The lesions fall into two classes: Class I, proliferative; and Class II, apoptosis inhibitory [72]. Class I or Class II lesions by themselves cause myelodysplasia or chronic lymphoproliferation. Transition to AML occurs when both a Class I lesion and Class II lesion are present, usually because a second mutation occurs in a cell already bearing a lesion. The combination of a proliferative lesion with loss of apoptosis is a double-whammy resulting in AML [72]. Because there are at least two lesions in AML, specific differentiation therapy requires that both lesions be treated. So far, such therapy has met with limited success as agents for only one of the two signals are available in most cases.

From leukemia we have learned that the genetic lesions of cancer occur in the stem cells and the expression of these lesions later in the lineage of the white blood cells determines the stage of maturation arrest of the transit-amplifying cells of the leukemia. Chemo- and radiotherapies act on the leukemic transit amplifying cells; the cancer stem cells are resistant. Newly developed targeted interference with the proliferation signals or apoptosis inhibitors responsible for the maturation arrest allow terminal differentiation of leukemic transit-amplifying cells. However, leukemic stem cells replenish the cancer transit-amplifying cells when chemo-, radio- or differentiation therapy is discontinued and are responsible for regrowth of the leukemia. Clearly what is needed is a way to get at the leukemia stem cell by cancer stem cell directed therapy.

11 Treatment of Therapy-Resistant Cells

Regardless of whether cancers are maintained by stem cells or by a population of cells that are in G_0 at the time of treatment, the properties of these treatment resistant cells can be exploited in efforts to eliminate the therapy-resistant cells. The rationale for this approach is presented in Fig. 4 [2–5, 57, 73]. The nonsurgical therapeutic approaches to cancer include chemotherapy, radiation, anti-angiogenic therapy, and differentiation therapy. The target cells for each of these therapies are the proliferating cancer transit amplifying cells. Each of these therapies may be very effective in reducing the mass of the tumor, and in some cases can actually cure the cancer. However, in most cases therapy must be maintained. If therapy is discontinued, the cancer will grow back from cancer stem cells or cancer cells that are not proliferating at the time of therapy [3, 57]. To prevent regrowth of the cancer, new therapeutic strategies must be directed to the resistant cell population [4, 16, 73, 74]. This may be done by targeting the specific activation signals that maintain stemness [75–77].

12 Conclusions

The stem cell theory of cancer that has evolved in fits and starts over the last 200 years. Two conclusions now accepted

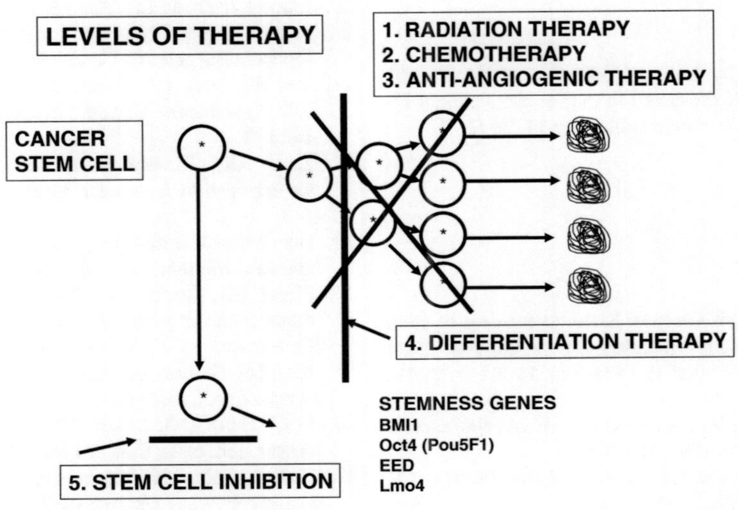

Fig. 4 Differentiation therapy directed to the cancer stem cells. Chemo-, radio-, and anti-angiogenic therapies act on proliferating transit-amplifying cells. Differentiation therapy for leukemia also acts on the transit-amplifying cancer cells. When therapy is discontinued, the cancer re-grows from the therapy-resistant stem cells. If the therapy-resistant cells could be forced to proliferate, then they would become sensitive to conventional therapy. In the future it may be possible to do this by inhibiting the signals that maintain the stemness of the resistant cells

are: (1) cancers arise from maturation arrest of stem cells; and (2) cancers contain the same populations as do normal tissues: stem cells, transit-amplifying cells, and differentiated cells. The biological properties of cancer stem cells are: they can be grown in soft agar; they initiate tumors upon transplantation; and they are resistant to chemotherapy and radiation. Studies done over the last 50 years, and repeated recently, indicate that the number of cancer stem-like cells in tumors varies enormously, from 1 in 1 to 1 in 10^6cells. Some of the cancer transit amplifying cells, such as in CML, can be effectively treated by differentiation therapy, but the cancer will reform from the cancer stem cell when therapy is discontinued.

The concept that cancers contain stem cells, or a subset of cancer cells in G_0, that are resistance to conventional therapy and responsible for long-term survival of the cancer, raises the possibility of inducing the differentiation of the cancer stem cell, or activating the resting G_0 cancer cell, so that it may become susceptible to therapies that are effective on proliferating cancer cells. Given the multiple signaling pathways that have been shown to maintain the stemness of cancer stem cells [75–77], and the identification of many molecules, such as iRNA and small molecular inhibitors, that could be used to effect stem cell differentiation therapy, many new approaches appear possible to "cure" cancer by attacking the cancer stem cell [1–5, 73].

However, it may take a long time to work out how to put the theory into practice. Even if some of these approaches eventually do work out, a word of caution is needed because of the tendency of cancer stem cells to mutate and change characteristics, usually to a more malignant form [78]. This property makes the cancer stem cell a moving target for specific therapy.

Acknowledgments Supported by National Institutes of Health R01 grants ES09494 and CA112481. Presented in part at the National Institutes of Health meeting on cancer stem cells, May 14, 2007.

References

1. Sell S, Pierce GB. Biology of Disease: Maturation arrest of stem cell differentiation is a common pathway for the cellular origin of teratocarcinomas and epithelial cancers. Lab Invest. 1994;70: 6–21.
2. Reya T, Morrison SJ, Clarke MF, Weissman IL. Stem cells, cancer, and cancer stem cells. Nature. 2001;414:105–11.
3. Sell S. Stem cell origin of cancer and differentiation therapy. CR Oncol/Hematol. 2004;51:1–28.
4. Sell S. Potential gene therapy for cancer stem cells. Curr Gene Ther. 2006;6:579–91.
5. Wicha MS, Liu S, Dontu G. Cancer stem cells: an old idea – a paradigm shift. Cancer Res. 2006:66:1883–90.
6. Greene HSN, The significance of the heterologous translatability of human cancer. Cancer. 1952;5:24–44.
7. Puck TT, Marcus PI. A rapid method of viable cell titration and clone production with HeLa cells in tissue culture: use of x-irradiated cells to supply conditioning factors. PNAS. 1955;41:432–7.
8. Buick RN. In vitro clonogenicity of primary human tumor cells: quantitation and relationship to tumor stem cells. In: Salmon S editor. Cloning of humor tumor stem cells. New York: AR Liss; 1980.pp. 15–20.
9. Reinhard MC, Goltz HL, Warner SC. Further studies on the quantitative determination of the growth of a transplantable mouse adenocarcinoma. Cancer Res. 1952;5:102–6.
10. Hewitt HB. Transplantation of mouse sarcomas with small numbers of single cells. Nature. 1952;170:622–3.
11. Reinhard MC, Goltz HL, Warner SG, Mirand EC. Growth rate and percentage takes following inoculation of known numbers of viable mouse tumor cells. Exp Med Surg. 1952;10:254.
12. Salmon S. In vitro effects of drugs on human tumor stem cell assays. In: Salmon S, editors. Cloning of human tumor stem cells. New York: AR Liss; 1952. pp. 197–312.
13. Lapidot T, Sirard C, Bormoor J, Murdoch B, Hoang T, Caceres-Cortes J, Minden M, Paterson B, Caligiuri MA, Dick JE. et al. A cell initiating human acute myeloid leukaemia after transplantation into SCID mice. Nature. 1994;367:645–8.
14. Sutherland HJ, Blair A, Zapf RW. Characterization of a hierarchy in human acute myeloid leukemia progenitor cells. Blood. 1996;87:4754–61.
15. Bonnet D, Dick JE. Human acute myeloid leukemia is organized as a hierarchy that originates from a primitive hematopoietic cell. Nat Med. 1997;3:730–7.
16. Rich JN. Cancer stem cells in radiation resistance. Cancer Res. 2007;67:8980–4.
17. Al-Hajj M, Wicha MS, Benito-Hernandez A, Morrison SJ, Clarke MR. Prospective identification of tumorigenic breast cancer cells. Proc Natl. Acad Sci U S A. 2003;100:3983–8.
18. Shipitsin M, Campbell LL, Argani P, Weremowicz S, Bloushtain-Qimron N, Yao J, et al. Molecular definition of breast tumor heterogeneity. Cancer Cell. 2007;11:259–73.
19. Kelly PN, Dakic A, Adams JM, Nutt SL, Strasser A. Tumor growth need not be driven by rare cancer stem cells. Science. 2007;317:337.
20. Liao M-J, Zhang CC, Zhou B, et al. Enrichment of a population of mammary gland cells that form mammospheres and have in vivo repopulating activity. Cancer Res. 2007;67:8131–8.
21. Tan BT, Park CY, Ailles LE, Weissman IL. The cancer stem cell hypothesis: a work in progress. Lab Investig. 2006;86: 1203–7.
22. Trott KR. Tumour stem cells: the biological concept and its application in cancer treatment. Radiother Oncol. 1994;30: 1–5.
23. Denekamp J. Tumour stem cells: facts, interpretation and consequences. Radiother Oncol. 1994;30:6–10.
24. Pierce GB, Spears WC. Tumors as caricatures of the process of tissue renewal: prospects for therapy by directing differentiation. Cancer Res. 1988;48:1196–204.
25. Kern SE, Shibata D. The fuzzy math of solid tumor stem cells: a perspective. Cancer Res. 2007;67:8985–8.
26. Li C, Heidt DG, Dalerba PK, et al. Identification of pancreatic cancer stem cells. Cancer Res. 2007;67:1030–7.
27. Furth J, Kahn MC. The transmission of leukemia of mice with a single cell. Am J Cancer. 1937;31:276–82.
28. Durante F. Nesso fisio-pathologico tra la struttura dei nei materni e la genesi di alcuni tumori maligni. Arch Memori eed Osservazioni di chirugia practica. 1874;11:217–26.
29. Cohnheim J. Congenitales, quergestreiftes muskelsarkon der nireren. Virchows Arch. 1875;65:64.
30. Shimkin M. Contrary to nature. NIH, USDOH, 1977.

31. Recamier JCA. Recherches sur the traitement du cancer: par la compression methodique simple ou combinee, et sur l'histoire general de la meme maladie.2 vols. Paris: Gabon; 1829.
32. Remak R. Ein beitrag zur entwickelungsgeschichte der krebshaften geschwulste. Deut Klin 1854;6:70–174.
33. Paget J. Lectures on surgical pathology. London; 1853.
34. Virchow R. Dir Krankhoften Geschwulste. Vol II. Onkologie, Pt 1. Berlin; 1863.
35. Pott P. Chirugical observations relative to the cataract, the polypus of the nose, the cancer of the scrotum, the different kinds of ruptures, and the mortification of the toes and feet. London: Hawes, Clarke and Colllins; 1775.
36. Potter M. Percivall Pott's contribution to cancer research. NCI Monograph, 1963;10:1–13.
37. Thiersch, C. Der epithelialkrebs namenlich der Haut. Eine anatomisch kliniche Untersuchung. Mit einen Atlas microscopischer Abbildungen. Lepiz; 1865.
38. Benecke, Zeigler's Beitrage zur Path. Anat., 1892–1893.
39. Adami GA, The principles of pathology. London; 1909.
40. Rippert H. Geschwulstelehre fur Aerzte und Studierende. Bonn; 1904.
41. Borrel A. Le probleme du cancer. Bull Inst Pasteur 1907;5: 497–512, 545, 593–608, 642–662.
42. Rous P. A sarcoma of the fowl transmissible by an agent separable from the tumor cells. J Exp Med. 1911;13:397–411.
43. Boveri R. Zur frage der entstehung maligner tumoren. Jena: Fischer; 1914.
44. Bainbridge WS. The cancer problem. New York: The Macmillan Co;1914.
45. Pierce GB. Dixon FJ. The demonstration of teratogenesis by metamorphosis of multipotential cells. Cancer. 1959;12:573–83.
45a. Pierce GM, Dixon FJ, Verney E. Teratocarcinogenic and tissue forming potentials of the cell types comprising neoplastic embryoid bodies. Lab Invest. 1960;9:583–602.
46. Pierce GB, Shikes R, Fink LM. Cancer: a problem of developmental biology. Englewood Cliffs, NJ; Prentice Hall Inc., 1978.
47. Stevens LC. Origin of testicular teratomas from primordial germ cells in mice. J Natl Cancer Inst. 1967;38:549–52.
48. Stevens LC. The development of teratomas from intratesticular grafts of tubal mouse eggs. J Embryol Exp Morphol. 1968;20: 329–41.
49. Stevens LC. The development of transplantable teratocarcinomas from intratesticular grafts of pre- and post implantation mouse embryos. Dev Biol. 1970;21:364–82.
50. Stevens LC, Experimental production of testicular teratomas in mice. Proc Natl Acad Sci U S A. 1964;52:654–61.
51. Mintz B, Illmensee K. Normal genetically mosaic mice produced from malignant teratocarcinoma cells. Proc Natl Acad Sci U S A. 72;1975:3583–9.
52. Ilmensee K. Reversion of malignancy and normalized differentiation of teratocarcinoma cells in mammals. In: Russel LC, editor. Generic mosaics and chimeras in mammals. New York: Plenum; 1978. pp. 3–25.
53. Sell S. Stem cells in hepatocarcinogenesis. Cell Science Rev.2006;1742–8130.
54. Sell S, Leffert HL. An evaluation of cellular lineages in the pathogenesis of experimental heatocellular carcinoma. Hepatology. 1982;2:77–86.
55. Sell S, Dunsford H. Evidence for the stem cell origin of hepatocellular carcinoma and cholangiocarcinoma. Am J Pathol. 1989;134:1347–63.
56. Pierce GB, Wallace C. Differentiation of malignant to benign cells. Cancer Res. 971;31:127–134.
57. Sell S. Cancer stem cells and differentiation therapy. Tumor Biol. 2006;27:59–70.
58. Sell S. Leukemia: stem cells, maturation arrest and differentiation therapy. Stem Cell Rev. 2005;1:197–205.
59. McCulloch EA. Normal and leukemic hematopoietic stem cells and lineages. In: Sell S, editor. Stem cells handbook. Totowa, NJ: Humana Press. 2004; pp. 119–131.
60. Baird SM. Hepatopoietic stem cells in leukemia and lymphoma. In: Sell S, editor. Stem cells handbook. Totowa, NJ: Humana Press. 2004; pp. 163–175.
61. Makino S, Kano K. Cytologic studies of tumors, XIV. Isolation of single-cell clones from a mixed-cell tumor of the rat. J Natl Cancer Inst. 1955;15:1165–81.
62. Till JE, McCulloch EA. A direct measurement of the radiation sensitivity of normal mouse bone marrow cells. Radiat Res. 1961;14:213–22.
63. Becker AJ, McCulloch E, Till J. Cytological demonstration of the clonal nature of spleen colonies derived from transplanted mouse marrow cells. Nature. 1963;197:452.
64. Till JE, McCulloch EA, Siminovitch L. A stochastic model of stem cell proliferation, based on the growth of spleen colony forming cells. Proc Natl Acad Sci U S A. 1964;51:29–36.
65. Rygaard J, Povlsen CO. Heterotransplantation of a human malignant tumour to "Nude" mice. Acta Pathol Microbiol Scand. 1969;77:758–60.
66. Nowell PC. Diagnostic and prognostic value of chromosome studies in cancer. Ann Clin Lab Sci. 1974;4:234–40.
67. Rowley JD. Nonrandom chromosomal abnormalities in hematologic disorders of man. Proc Natl Acad Sci U S A. 1975;72: 152–6.
68. Drucher BJ, Ralpaz M, Resta DJ, Peng B, Buchdunger E, Ford JM, et al. Efficacy and safety of a specific inhibitor of the BCR-ABL tyrosine kinase in chronic myeloid leukemia. N Engl J Med. 2001;334:1031–7..
69. Drucker BJ, Cuilhot F, O'Brien SG, et al. Five year follow-up of patients receiving Imatinib for chronic myeloid leukemia. N Engl J Med. 2006;355:2408–17.
70. Melnick A, Licht JD. Deconstructing a disease: RARα, its fusion partners, and their roles in the pathogenesis of acute promyelocytic leukemia. Blood. 1999;93:3167–215.
71. Soignet S, Fleischauer A. Pollyak T, Heller G, Warrel Jf RP. All trans retinoic acid significantly increases 5-year survival in patients with acute promyelocytic leukemia: long term follow-up of the New York study. Cancer Chemother Pharmacol. 1997;40: S24–9.
72. Chalandon Y, Schwaller J. Targeting mutated protein tyrosine kinases and their signaling pathways in hematologic malignancies. Haematologica. 2005;90:949–68.
73. Sell S. Cancer and stem cell signaling: a guide to preventive and therapeutic strategies for cancer stem cells. Stem Cell Rev. 2007;3:1–6.
74. Hill RP, Perris R. "Destemming" cancer stem cells. J Natl Cancer Inst. 2007;99:1435–1340.
75. Ivanova N, Dobrin R, Lu R, Kotenko I, Levorse J, DeCoste C, Schafer X, Lun Y, Lemischka. Dissecting self-renewal in stem cells with RNA interference. Nature. 2006;442: 533–8.
76. Sell S. (Ed); Cancer and stem cell signaling. Stem Cell Reviews, vol. 3, Totowa, NJ: Humana Press. 2007; pp. 1–103.
77. Glinsky G. Genomic models of metastatic cancer. Cell Cycle. 2006;5:11–23.
78. Nowell PC. The clonal evolution of tumor cell populations. Science. 1976;194:23–8.

Immune Responses to Stem Cells and Cancer Stem Cells

Xiao-Feng Yang and Hong Wang

Abstract The demonstrated capacity and potential of pluripotent stem cells to repair the damaged tissues holds great promise in development of novel cell replacement therapeutics for treating various chronic and degenerative diseases. However, previous reports show that stem cell therapy, in autologous and allogeneic settings, triggers immune responses to stem cells as shown by lymphocyte infiltration and inflammation. Therefore, an important issue to be addressed is how the host immune system responds to engrafted autologous stem cells or allogenous stem cells. In this chapter, we summarize progress in several related topics in this field, including some of our data, in five sections: (1) mechanisms regarding the immunogenicity of stem cells; (2) methods to inhibit immune rejection to allogeneic stem cells; (3) immune responses to cancer (tumor) stem cells; (4) role of CD4+CD25^{high}Foxp3+ regulatory T cells in regulating anti-stem cell immune reactions; and (5) role of mesenchymal stem cells in regulating anti-stem cell immune responses. Improvement of our understanding on these aspects of immune system-stem cell interaction would definitely facilitate development of stem cell-based therapeutics for regenerative purposes.

Keywords Immunogenicity · Stem cell therapy · Anti-stem cell immunity · Cancer stem cells · Stimulation-responsive splicing model

As aging progresses, the regenerative power of pluripotent stem cells for tissue repair is inadequate to sustain normal tissue function [1]. As a result, the incidence of chronic and degenerative diseases including Parkinson's disease [2], Alzheimer's disease [2], diabetes [3], heart disease [4], leukemia [5], and others [6] has significantly increased in the United States. Over 125 million people suffer from at least one chronic disease and the related medical costs account for 78% of total medical expenses. In the next 20 years, the proportion of populations over 85 years of age in Western countries is anticipated to quadruple to reach 157 millions, placing an unsustainable burden on society [7]. The capacity of pluripotent stem cells to repair the tissues in which stem cells reside has been demonstrated due to various technological advances, which hold great promise as novel cell replacement therapeutics for treating chronic and degenerative diseases [1]. Therefore, the emerging strategies in regenerative medicine will be essential in coping with the challenges outlined above.

Stem cells are defined as clonogenic, self-renewing progenitor cells that can generate one or more specialized cell types. Embryonic stem cell lines (ESC), established in mouse in 1981 [8, 9] and followed by the characterization of human ESC lines in 1998 [10], are derived from the inner cell mass of the blastocyst and are capable of generating all differentiated cell types in the body. To date, there are more than 300 human ES cell lines, but only 22 human ESC lines are commercially available and registered with the NIH (http://stemcells.nih.gov/research/registry/) [6]. Adult (postnatal) stem cells are still pluripotent, but their differentiation ability is restricted to the cell types of a particular tissue, being responsible for organ regeneration. Two general categories of reserve precursor cells exist within the body and are involved in the maintenance and repair of tissue in adults: (a) lineage-committed progenitor cells, and (b) lineage-uncommitted pluripotent stem cells. Young and Black [11] summarized a long list of common characteristics of lineage-committed progenitor stem cells. Progenitor stem cells (progenitor cells) may be committed to one or more specific tissue lineages, which can be further classified into unipotent, bipotent, tripotent, or multipotent, respectively [11]. Each progenitor cell for a particular tissue lineage has a unique profile of cell surface cluster of differentiation (CD) markers [11]. Progenitor cells conform to Hayflick's limit of 50–70 population doublings [12].

Primitive stem cells within the bone marrow niche [13] (hematopoietic stem cell, HSC) possess functional versatility

Xiao-Feng Yang (✉)
Department of Pharmacology, Temple University School of Medicine, 3420 North Broad Street, MRB 300, Philadelphia, PA 19140, USA
e-mail: xfyang@temple.edu

V.K. Rajasekhar, M.C. Vemuri (eds.), *Regulatory Networks in Stem Cells*, Stem Cell Biology and Regenerative Medicine, DOI 10.1007/978-1-60327-227-8_38, © Humana Press, a part of Springer Science+Business Media, LLC 2009

broader than expected, which is termed *trans-differentiation* or *stem cell plasticity*. Stem cell plasticity describes the ability of adult stem cells to cross lineage barriers and to adopt the expression profiles and functional phenotypes of cells unique to other tissues [14]. HSC, expressing markers of the hematopoietic lineage (CD45+) and of hematopoietic stem cells (CD34+, CD133+, and CD117+, Thy-1^{low}, but CD10–, CD14–, CD15–, CD16–, CD19–, and CD20–) [15, 16], are capable of genomic reprogramming upon exposure to a novel environment and give rise to other tissues such as liver, cardiac muscle, or brain [17, 18]. Mouse HSCs' markers are CD117 (c-Kit)+, Sca-1+, and Thy-1^{low} but B220–, CD3–, CD4–, CD8–, Mac-1, Gr-1–, and Ter119– [16]. A defining property of murine hematopoietic stem cells (so-called side population) [19] is low fluorescence after staining with Hoechst 33342 and Rhodamine 123 [20]. The HSCs in mice transplanted at the single cell level gave rise to life-long hematopoiesis, including a steady state of 20 to 1×10^5 HSC and over 1×10^9 blood cells produced daily [16]. In addition to HSC, bone marrow also contains mesenchymal stem cells (MSCs), which have the capacity to proliferative extensively and form colonies of fibroblastic cells (defined as colony-forming units-fibroblastic; CFU-F) [21]. Furthermore, discovery of cancer stem cells in leukemias and solid tumors [22] has added to the complexity of the stem cell field but stimulated great excitement for both stem cell and cancer biologists [23]. Normal ESCs have been used in treating glycemia in a mouse diabetes model [24], generating cardiomyocytes in dystrophic mice [25], improving cardiac function in post-infarcted rats [26], and intervening in the progression of a rat model of Parkinson's disease [27]. Due to the great potential of stem cells in development of novel therapy for chronic and degenerative diseases as well as cancers in autologous and allogeneic settings, several immunological aspects related to stem cell-based cell replacement therapy raise important concerns, as shown in Fig. 1, which are the focus of this brief review.

1 Mechanisms Regarding the Immunogenicity of Stem Cells

One of the important issues regarding stem cell therapy is whether stem cells can be used as immune privileged "BAND-AID" without elicitation of inflammation and immune responses [28]. Transplantation of allogeneic undifferentiated murine ESCs in the heart cause cardiac teratomas, which are immunologically rejected after several weeks in association with increased inflammation and up-regulation of class I and II major histocompatibility complex (MHC) molecules [29]. In addition, in vivo differentiated ESCs transplanted into ischemic myocardium elicit an accelerated immune response as compared with undifferentiated ESCs, suggesting ESC immunogenicity increases upon differentiation [30]. Moreover, immune responses are not limited to transplanted ESCs. Transplantation of neural stem cells also induces immunological responses [31] and lymphocyte infiltration [32]. Furthermore, transplantation of ESCs in the heart elicits infiltration of a few CD3+ T cells even in the syngeneic mouse group, but not in the severe combined immunodeficiency disease (SCID) mouse group [33], the ESCs are not stealthy in the heart. In contrast to these findings, recent studies suggest some immune privilege is associated with human ESC-derived tissues [34, 35, 36]. However, the adaptability of the immune system makes it unlikely that fully differentiated tissues will maintain their immune privilege and permanently evade immune rejection [5]. Generally, an immune-privileged site, such as testis, does not express MHC class I or II molecules [37] and may express FasL to kill attacking lymphocytes [38]. However, a recent report showed that the addition of Fas ligand (FasL) to healthy fetal MSC induces cell death by apoptosis [39], suggesting that stem cell death can be induced by attacking lymphocytes via a Fas/FasL mechanism. Human ES cells express low levels of MHC class I in vitro [40] in their undifferentiated state [41, 42]. The MHC class I expression increases 2–4-fold when the human ES cells are induced to differentiate to embryonic bodies, and an 8–10-fold when induced to differentiate to teratomas [41]. In contrast, other investigators observed MHC class I down-regulation after differentiation induced with retinoic acid on Matrigel or in extended culture [42]. MHC class I expression can be strongly up-regulated after treatment of the ESCs with interferon-γ [43], a potent MHC expression-inducing proinflammatory cytokine known to be released during the course of immune responses [41, 42]. Similarly, Bradley et al. [28] reported that a 4-fold expression of HLA class II molecules in human ESCs upon differentiation in vitro. These results suggest the possibility of upregulation of MHC expression in therapeutic ESCs when their use for treatment of ongoing chronic inflammatory diseases or other pathological interferon-γ and similar cytokine abundant conditions. In addition, our recent report supports the argument that interferon-γ may accelerate processing of T cell-reactive self-tumor antigen epitopes presumably via up-regulation of immunoproteasomes [44], which further emphasizes that stimulations by cytokines, stress, drugs, and other stimuli [45] may up-regulate the immunogenicity of stem cells by promoting self-antigen epitope processing. Dr. Wu's laboratory tested allogeneic undifferentiated mouse ESCs for their ability to trigger allogeneic immune response in a mouse model of myocardial infarction and observed progressive intra-graft infiltration of inflammatory cells mediating both adaptive and innate immune responses [6]. These results suggest that the immunogenicity of mouse

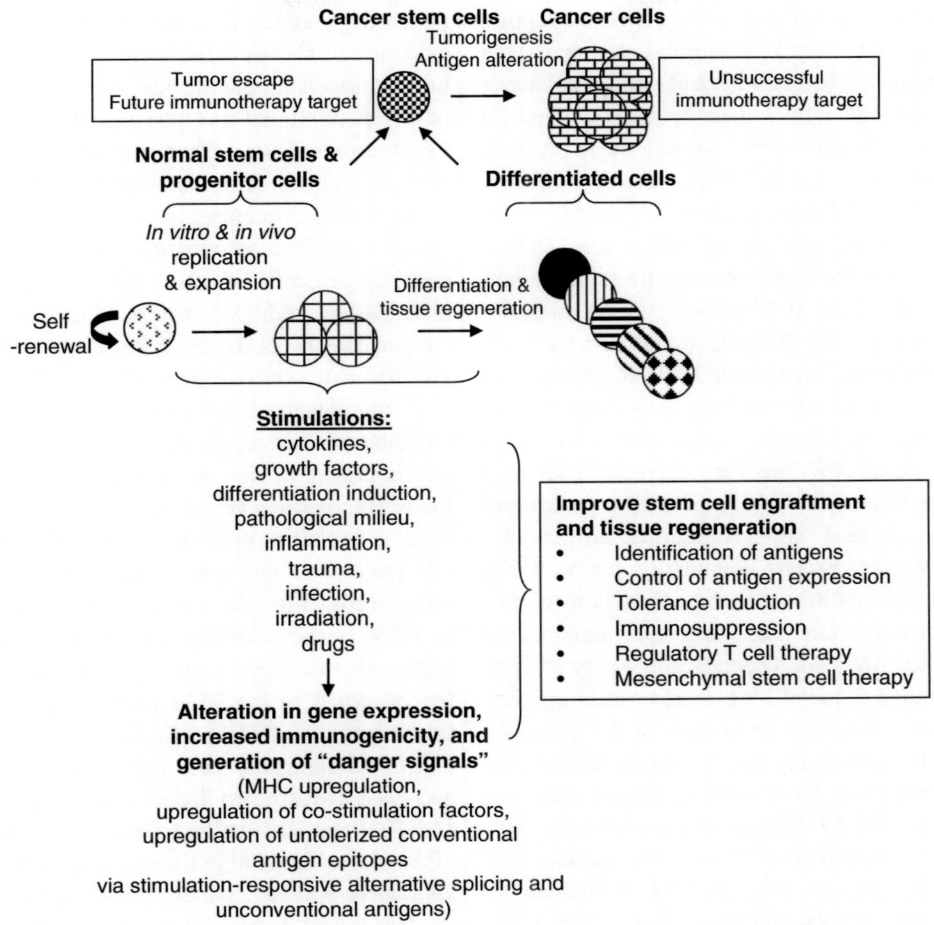

Fig. 1 Immune responses to stem cells and cancer stem cells. To generate enough stem cells for regenerative medicine, stem cells with self-renewal properties need to experience the following two stages: (**a**) in vitro and in vivo replication and expansion; and (**b**) differentiation and tissue regeneration. During these processes, stem cells are stimulated by numerous factors, such as cytokines, growth factors, differentiation induction, pathological milieu, inflammation, trauma, infection, irradiation, and drugs. Under the stimulation, gene expression in stem cells is altered, which leads to increased immunogenicity of stem cells and generation of "danger signals" by up-regulation of MHC molecules, co-stimulation factors, and untolerized conventional antigen epitopes (via the mechanisms of stimulation-responsive alternative splicing and unconventional antigen expression). Therefore, to improve cell engraftment and tissue regeneration, several new strategies have been proposed including identification of stem cell antigens, control of aberrant antigen expression, tolerance induction, immunosuppression, $CD4+CD25^{high}Foxp3+$ regulatory T cell therapy, and mesenchymal stem cell therapy. In addition, normal stem cells and progenitor cells can be developed into cancer (tumor) stem cells due to mutations. Cancer (tumor) stem cells are believed to develop into, via tumorigenesis, cancer (tumor) cells, latter of which have been unsuccessfully targeted by most of current immunotherapies. To enhance the efficacy of future cancer (tumor) immunotherapy, both cancer (tumor) stem cells and cancer cells need to be targeted

ESCs is increased upon their differentiation [6]. Moreover, Mullally and Ritz [46] pointed out that a formerly unappreciated level of "structural variation" within the normal human genome including deletions, duplications, inversions, copy number variants, and single nucleotide polymorphism all increase immunogenicity of transplanted stem cells and affect allogeneic stem cell transplantation. Therefore, allogeneic stem cells may not have reliable immune privilege.

In order to generate sufficient numbers of stem cells for therapeutics, isolated stem cells are often required to expand and induced to differentiate in vitro [47]. For example, to direct autologous adult stem cells into the cardiomyogenic lineage, several strategies have been developed [48] in addition to identification of growth factors and signaling molecules under cell culture conditions [4]. Enucleated cytoplasts generated from human ESC-derived cardiomyocytes could be fused with autologous adult stem cells to generate cytoplasmic hybrids or cybrids. Adult stem cells could also be temporarily permeabilized and exposed to cytoplasmic extracts from these cardiomyocytes. Alternatively, intact cells or enucleated cytoplasts from human ESC-derived cardiomyocytes could be co-cultured with adult stem cells in vitro to provide the cellular contacts and electronic coupling that might enable some degree of trans-differentiation to take place [48]. Long-term in vitro culture and manipulations of ESCs [47] may adversely affect their epigenetic integrity including imprinting. Disruption or inappropriate expression of imprinted genes is associated with several clinically significant syndromes and tumorigenesis in humans. By investigating methylation profiles of CpG sites within the IGF2/H19 IC, Mitalipov [49] demonstrated abnormal hypermethylation within the IGF2/H19 IC in all analyzed ES cell lines consistent with biallelic expression of these genes. Alteration of gene expression leads to changes in antigenic repertoire associated with long-term expansion of autologous stem cells, which may trigger the endogenous "danger signal" sensed by toll-like receptors [50] and activate the host immune system [51]. Since the processing of HLA class I-restricted antigen epitope utilizes the ubiquitination-proteasome protein degradation pathway [52, 53] and non-proteasome pathway [54, 55], then, intracellular antigens cannot escape from presenting their epitopes to the HLA class I pathway [56]. Theoretically, every antigen with epitope structures blanked by proper processing sites [44] encodes T cell antigen epitope despite the potential variations in the HLA presenting alleles and the differences in their immunodominance HLA binding affinity [57] and TAP binding affinity [58, 59] among antigen epitopes [56]. Therefore, cellular overproliferation, tumor formation, and antigenic alteration resulting from in vitro expansion of autologous stem cells are potential problems that must be addressed before clinical trials of ESC-based therapy are initiated.

In addition to above-mentioned immune recognition machinery including the expression of MHC molecules, cytokine function, antigen epitope processing, an important question of how nonmutated self-protein antigens, derived from normal stem cells, other normal cells and tumor cells, gain immunogenicity and trigger immune recognition remain poorly defined [56]. Mutation may be responsible for elevated immunogenicity underlying some tumor-specific antigens generated via mutations (p53 and Ras), chromosome translocations, and abnormalities, such as expression of fusion oncogene Bcr-Abl in chronic myelogenous leukemia (CML) [60–63]. However, the mechanism underlying the immunogenicity of most nonmutated self-tumor antigens is their aberrant overexpression in tumors. Zinkernagel et al. [64] suggested that the overexpression of self-antigens or novel antigenic structure, overcomes the threshold of antigen concentration at which an immune response is initiated [65]. This threshold might be lower for certain untolerized regions of certain antigen epitopes. Overexpressed genes, up to 100-fold, often encode tumor antigens identified by serological identification of self-antigens by screening expression cDNA library with patients' sera (SEREX) [66], which may reflect the inherent methodological bias for the detection of abundant transcript [67]. The overexpression of tumor antigens in tumors can result from transcriptional and post-transcriptional mechanisms. We recently demonstrated that overexpression of tumor antigen CML66L in leukemia cells and tumor cells via alternative splicing is the mechanism for its immunogenicity in patients with tumors [68], which not only illustrated the overexpression of tumor antigen as a principle but also elucidated its molecular mechanism [68]. In addition, expression of one of the major tumor antigen categories, cancer-testis antigens, in tumors has been ascribed to abnormal demethylation [69, 70].

A significant proportion of the SEREX-defined self-tumor antigens are autoantigens [71], for example, CML28 that we identified is also an autoantigen Rrp46p [72]. Beside the overexpression of self-tumor antigens and autoantigens, we also examined the potential mechanisms for nonmutated self-proteins to gain new untolerized structure to trigger immune recognition. We found that alternative splicing occurs in 100% of the autoantigen transcripts. This is significantly higher than the approximately 42% rate of alternative splicing observed in the 9554 randomly selected human gene transcripts ($p < 0.001$). Within the isoform-specific regions of the autoantigens, 92% and 88% encoded MHC class I and class II-restricted T-cell antigen epitopes, respectively, and 70% encoded antibody binding domains. Furthermore, 80% of the autoantigen transcripts undergo noncanonical alternative splicing, which is also significantly higher than the less than 1% rate in randomly selected gene transcripts ($p < 0.001$). These studies suggest that noncanonical alternative splicing may be an important mechanism for the generation of untolerized epitopes that may lead to autoimmunity. Furthermore, the product of a transcript that does not undergo alternative splicing is unlikely to be a target antigen in autoimmunity [73]. To consolidate this finding, we also examined the effect of proinflammatory cytokine tumor necrosis factor-α (TNF-α) on the prototypic alternative splicing factor ASF/SF2 in the splicing machinery. Our results show that TNF-α down-regulates ASF/SF2 expression in cultured muscle cells. This result correlates with our finding of reduced expression of ASF/SF2 in inflamed muscle cells from patients with autoimmune myositis [74]. Based on our and others' data, we recently proposed a new model of stimulation-responsive splicing for the selection of autoantigens and self-tumor antigens [45]. Our new model theorizes that the significantly higher rates of alternative splicing of autoantigen and self-tumor antigen transcripts that occur in response to stimuli could induce extra-thymic expression of untolerized antigen epitopes for elicitation of autoimmune and anti-tumor responses. Of note, our model is not only applied to nonmutated self-tumor antigen-associated tumors and autoantigens associated with various autoimmune diseases, but also applied to composition and expansion of self-antigen repertoire of stem cells. To facilitate the identification of immunogenic isoforms of antigens, we have developed strategies [72, 75–79] using improved SEREX [66] in conjunction with database-mining [73] and immunogenic isoform mapping [68].

However, despite some progress, the detailed definition of antigen repertoire of normal stem cells has not been reported yet. Identification of immunogenic isoforms of autoantigens and self-tumor antigens related to stem cell therapy is very important for the development of novel therapeutics for cell replacement therapy using stem cells [45].

2 Methods to Inhibit Immune Rejection to Allogeneic Stem Cells

Various strategies have been developed to circumvent the immunological barriers and inhibit the rejection of replacement stem cells. Lessons learned from bone marrow transplantation suggest that it is a formidable task to establish a stem cell bank to permit rudimentary matching of tissue [1]. Tissues from MHC–/– mice are rejected by recipients at a rate comparable to their wild-type counterparts [80], suggesting that development of a universal ESC line without expression of its own MHC may not necessarily be beneficial. In addition, transplantation of tissues without expression of MHCs as an immunological surveillance mechanism may create a safe haven for viral infection and malignant transformation of cells [28]. Several approaches have been extensively studied to improve the acceptance of transplanted stem cells. First, use of donated oocytes for somatic nuclear transfer (SNT) to create nuclear transfer ESC lines (ntES cells) [81, 82], which are genetically identical to the recipient in all but their mitochondrial genome remaining the preserve of the oocytes themselves [83]. The question remains whether mitochondrial proteins might act as a source of minor histocompatibility antigens [84, 85] for transplantation rejection antigens in this case although sharing nuclear genes ensure identity of the MHC haplotype. Second, in vitro differentiation of ESCs into desirable cell types for the therapy of diseases followed by purification of cardiomyocytes [25] and neurons [86] has been used to achieve better acceptance of allograft. Third, ESC-derived dendritic cells (esDCs, a professional antigen presenting cell type) are implicated in tolerance induction, which share with therapeutic graft the full repertoire of transplantation antigens, and generation of immature esDCs may polarize responding T cells towards a regulatory phenotype [1]. Harrison's group demonstrated a proof of principle that because T cell tolerance can be induced by presenting antigen on resting antigen-presenting cells (APCs), hematopoietic stem cells engineered to express autoantigen in resting APCs could be used to prevent autoimmune disease [87]. Proinsulin is a major autoantigen associated with pancreatic β cell destruction in humans with type 1 diabetes (T1D) and in autoimmune non-obese-diabetic (NOD) mice. Syngeneic transplantation of hematopoietic stem cells encoding proinsulin transgenically targeted to APCs totally prevents the development of spontaneous autoimmune diabetes in NOD mice. This antigen-specific immunotherapeutic strategy could be applied to prevent T1D and other autoimmune diseases in humans. Fourth, tolerance is induced by establishment of hematopoietic chimerism [6]. Finally, naturally occurring CD4+CD25^{high}Foxp3+ regulatory T cells (Tregs) are differentiated T lymphocytes actively involved in the control and suppression of peripheral immunity [88]. Over the past few years, a number of animal studies have demonstrated the critical role of these regulatory T cells in the outcome of allogeneic hematopoietic stem cell transplantation (HSCT). In these models, Tregs can exert a potent suppressive effect on immune effector cells reactive to host antigens and prevent graft-versus-host disease (GVHD) while preserving the graft-versus-leukemia effect (GVL) [89]. Building on the results from recent studies, a number of therapeutic strategies are being developed to positively modulate Treg pools in vivo and prevent or even correct GVHD. Conversely, clinical interventions can also be envisaged to decrease Treg activity in vivo and enhance the GVL effect [89]. Along this line, our recent data showed that Tregs regulate T cell responses to self-antigen, and that depletion of Tregs via a pro-apoptotic protein Bax-dependent mechanism enhances antigen-specific polyclonal T cell responses. These findings provide support for the idea that stem cell therapy can be improved by therapeutic modulation of survival of Treg cells [90].

Minor histocompatibility antigens (mHA) are allogeneic targets of T cell-mediated graft-versus-tumor (GVT) effects following allogeneic (allo-) stem cell transplantation. Recent research has identified several mHAs as tumor proteins and has also disclosed their unique properties in both the induction and effector phase of GVT reactions. Targeting tumor-specific mHAs by adoptive immunotherapy will prevent tumor tolerance and evoke allo-immune responses, thereby enhancing GVT effects against leukemia and solid tumors [91]. Recently acquired knowledge of the role of donor immunization status, new techniques in the generation of mHA-specific cytotoxic T lymphocytes in vitro, and innovative principles in vaccination will help to design strategies that exploit mHAs in the immunotherapy of cancer. However, the issue of how to control mHA-mediated immune responses and enhance stem cell allograft requires more work.

Of note, pregnant women have been found to tolerate the unborn conceptus expressing a full set of nonmaternal antigens inherited from the father. The exact mechanisms of immune privilege exhibited by fetal tissues remain poorly defined, which may provide useful insights for future tolerance strategies to improve stem cell allograft acceptance [34].

3 Immune Responses to Cancer (Tumor) Stem Cells

The observation of similarities between the self-renewal mechanisms of stem cells and cancer cells has led to the new concept of the cancer stem cell. In 1994, the presence of cancerous stem cells in acute lymphocytic leukemia was documented by cloning such cells and documenting their self-renewing capacity [92]. A self-renewing cancer stem cell population has been identified in solid tumors such as breast [93] and brain [94]. These cancer stem cells represent approximately 1% of the tumor and are the only cells in the tumor generating tumors into nude mice [95]. In cases of multiple myeloma, cells with a high self renewal potential have also been identified [96]. Many researchers now suspect that all cancers are composed of a mixture of stem cells and proliferative cells with a limited lifespan [95]. The implications of this research are far reaching. The relapse of many cancers following therapy could be the result of the survival of the cancer stem cells. Therefore, it is critical to fully characterize the immunological features of these cells and to develop immunotherapeutic approaches to eliminate these cancer stem cells without excessive toxicity to normal stem cells. It has been reported that PTEN dependence distinguishes hematopoietic stem cells from leukemia-initiating cells [97]. In this aspect, molecular characterization of cancer stem cells in comparison to normal stem cells suggests a good start. Since intracellular antigens cannot escape from presenting their epitopes to the HLA class I pathway [56], any differences in proteomic composition between cancer stem cells and normal stem cells can be "translated" into antigenic differences.

Tumor immunosurveillance theory suggests that tumors can be recognized and eliminated as a result of natural anti-tumor immune responses that develop in the host [98–101]. This argument is supported by the discoveries that (i) the immune system can protect the host against the development of spontaneous and chemically induced tumors; (ii) the immunogenicity of a tumor is imprinted on the tumor by the immunological environment; and (iii) individuals with tumor sometimes develop spontaneous reactivity against the antigens of the tumor [98–101]. Many influences either from tumor or environment render a tumor either invisible to the host immune system or resistant to the anti-tumor immune responses. Several situations can lead to this result: (a) the tumor is non-immunogenic, either because it never expressed any tumor antigens or lost them during tumor development, or the tumor acquired defects in the capacity to present tumor antigens to immune cells; (b) the immune system may not be able to recognize or eliminate a tumor because the tumor produces immunosuppressive moieties and induces immunosuppressive responses (also see several excellent reviews in [98–102].

A recent review explores similarities between lymphocytes and cancer cells, and proposes a new model for the genesis of human cancer. This model suggests that the development of cancer requires infection(s) [103] in which determinants from pathogens can mimic self-antigens and co-present to the immune system, leading to breaking T-cell tolerance. However, autoreactive T cells must be eliminated by apoptosis when the immune response is terminated. Some autoreactive T cells suffer genomic damage in this process, but manage to survive. The resulting cancer stem cell still retains some functions of an inflammatory T cell, so it seeks out sites of inflammation inside the body. Due to its defective constitutive production of inflammatory cytokines and other growth factors, a stroma is built at the site of inflammation similar to the temporary stroma built during wound healing. The cancer cells grow inside this stroma, forming a tumor that provides their vascular supply and protects them from cellular immune response. As cancer stem cells have plasticity comparable to normal stem cells, interactions with surrounding normal tissues cause them to give rise to all the various types of cancers, resembling differentiated tissue types. Metastases form at an advanced stage of the disease, with the proliferation of sites of inflammation inside the body following a similar mechanism. Therefore, future development of cancer therapies should provide more support for, rather than antagonizing, the immune system [104].

Substantial antigenic differences have been found between tumors and normal tissues. A milestone in tumor immunology was the cloning of tumor antigen MAGE-1 by Boon's team in 1991 [105–107], and subsequent characterization of the first HLA-restricted T-cell-defined antigenic epitope a year later [108]. In 1995, another breakthrough was reported. Pfreundschuh's team developed a new method of serological cloning approach called SEREX [66, 67, 109, 110]. It allows a systemic and unbiased search for antibody responses against protein antigens expressed by human tumors. More than 2000 tumor antigens have been identified [67] (also see an excellent database http://www.cancerimmunity.org/statics/databases.htm). These advances have led to a renaissance in tumor immunology and studies on anti-tumor immunotherapy [66, 105] (also see our invited reviews [45, 56]).

In addition, studies on identification of HLA-restricted T-cell antigen epitopes of tumor antigens and T-cell-based immunotherapy to tumors have also made significant progress [111]. By 2004, more than 257 HLA class I-, and HLA class II-restricted T cell antigen epitopes have been identified (http://www.istitutotumori.mi.it/INT/AreaProfessionale/Human_Tumor/default.asp?LinkAttivo=17B). Since they are derived from various tumors, these T-cell antigen epitopes are very useful in diagnosis, prognosis, and immunotherapy in treatment of tumors. Furthermore, clinical studies of several formats of active immunization (recombinant viruses, naked DNA, dendritic cells pulsed with peptide, and peptides) in patients with melanoma showed that after two courses of immunization with the gp100, MART-1, or tyrosinase tumor antigens, up to 1–2% of all circulating CD8 T cell had anti-tumor activity, which is several hundred or thousand folds higher than the frequencies of any given antigen-specific T cells in the normal T-cell repertoire.

In identifying tumor antigens associated with cancer stem cells, one needs to bear in mind that there are two major groups of self-tumor antigens. The first group comprises *conventional antigens*, such as proteins encoded by genes with conventional exon-intron organization and translated in the primary open reading frame (ORF) [112]. The conventional tumor-associated antigens include the five groups of tumor antigens above-mentioned [113, 114]. Our reports on tumor antigens associated with chronic myelogenous leukemia (a myeloid stem cell-initiated hematologic malignancy), such as CML66 [68, 76, 90], CML28 [72], PV13 [79], PV65 [79], and others [75] belong to the first group. The second group comprises *unconventional* cryptic peptide antigens, including cryptic antigens encoded in (a) the introns of genes [68] (MPD-associated antigen MPD5) [115]; (b) the exon-intron junctional regions; (c) the alternative reading frames (tumor antigen TRP-1) [116, 117] as opposed to the primary reading frames in mRNAs [118–120]; and (d) the subdominant open reading frames located in the 5′-untranslated region (UTR) or 3′-UTR of the primary open reading frame [112, 115], chromosome rearrangement, and aberrant processing [111]. Recently, we used the SEREX technique to screen a human testis cDNA library with sera from three polycythemia vera (a stem cell-initiated myeloproliferative disease, MPD) patients who responded to interferon-α (IFN-α) and identified a novel unconventional antigen, MPD5. MPD5 belongs to the group of unconventional cryptic antigens without conventional genomic intron/exon structure. MPD5 antigen elicited IgG antibody responses in a subset of polycythemia vera (PV) patients, as well as some patients with chronic myelogenous leukemia or prostate cancer, suggesting that they are broadly immunogenic. Up-regulated expressions of MPD5 in the granulocytes from PV patients after IFN-α [78] or other therapies, might enhance their abilities in elicitation of immune responses in patients. In addition, we recently identified another unconventional antigen MPD6. MPD6 belongs to the group of cryptic Ags without conventional genomic structure and is encoded by a cryptic open reading frame located in the 3′-UTR of myotrophin mRNA. MPD6 elicits IgG antibody responses in a subset of polycythemia vera patients, as well as patients with chronic myelogenous leukemia and prostate cancer, suggesting that it is broadly immunogenic. By using bicistronic reporter constructs, we showed that the translation of MPD6 was mediated by a novel internal ribosome entry site (IRES) upstream of the MPD6 reading frame. Furthermore, the MPD6-IRES-mediated translation, but not myotrophin-MPD6 transcription, was significantly up-regulated in response to IFN-α stimulation [77]. Our findings provide new insights into the mechanism underlying the regulation of the self-antigen repertoire in eliciting anti-tumor immune responses in patients with myeloid stem cell proliferative diseases, and suggest their potential as the targets of novel immunotherapy. What is the significance of identification of unconventional tumor antigens for future immunotherapy? Since these unconventional antigen peptides are not expressed in normal cells and normal stem cells, and are not tolerated by host immune system, they are considered to be tumor-specific or cancer stem cell-specific. These features indicate that these unconventional antigens may be desirable to be targets for future immunotherapy [112].

Despite significant progress in tumor immunology, several important questions remain to be addressed: First, whether there are any differences between the 1% cancer stem cells and the majority of other cancer cells in immunogenicity and antigenic features. Of note, since cancer stem cells represent approximately 1% of the tumor cells [95], tumor antigens highly expressed in cancer stem cells may not be the tumor antigens highly expressed in tumors. The tumor antigens highly expressed in cancer stem cells may have been missed in routine SEREX screening and T cell epitope cloning procedures since immune responses against tumor antigens highly expressed in cancer stem cells are diluted 100 fold during detection. Second, whether any identified tumor antigens are specifically up-regulated in cancer stem cells in comparison to the majority of other cancer cells. Current anti-tumor antigen-specific immune therapies focused on tumor antigens highly expressed in tumor cells are not capable in elicitation of effective anti-cancer stem cell immune responses and inhibiting cancer stem cell growth and cancer relapse after initial treatment. Third, whether there are any ever changing patterns of transient kinetics of tumor antigen expression in cancer stem cells and other cancer cells. Due to this complicated situation, majority cancer antigen-specific immunotherapy may not be able to be effective alone in irradiating tumors, especially in irradiating cancer stem cells [121]. Therefore, future immunotherapy could be in a combinational format, including cancer cell antigen-specific immunotherapy and cancer stem cell antigen-specific immunotherapy as well as anti-tumor immune enhancement therapies including our recently reported promotion of Treg apoptosis [90]. In other words, long-term survival of patients with cancer can only be achieved if effective cytotoxic immune responses against both cancer stem cells and cancer cells are established.

4 Role of CD4+CD25^{high}Foxp3+ Regulatory T Cells in Regulating Anti-stem Cell Immune Reactions

As an active mechanism of immune suppression, distinct subsets of Tregs suppress the activation of autoreactive T cells that have escaped the mechanisms of thymic clonal deletion and T-cell anergy [122]. Several regulatory T-cell subsets [42–48] including CD4+CD25^{high}Foxp3+ T cells, CD4+CD25-PD-1+ T cells [123], CD8+ cells [124], CD8+CD25+ T cells [125, 126], CD8+CD28- T cells(31), TCRγδ+ T cells [124], NKT cells [124], CD3+CD4-CD8-αβTCR+ T cells [127] have been reported, among which CD4+CD25^{high}Foxp3+ Tregs are the best characterized [128–132]. Comprising 5–10% of peripheral CD4+ T cells [56–59], Tregs exhibit potent immunosuppressive functions [133], and play an important role in the regulation of autoimmunity [18, 61], anti-tumor immunity [62–64], pathogenesis of atherosclerosis [65], allergy and asthma [134], transplantation immunity [89, 135], and anti-microbial immunity [69–72]. There are two subsets of Tregs, (a) peripherally adaptive Tregs (aTreg cells, transforming growth factor-β [TGF-β] secreting, and IL-10 secreting) [136, 137], and (b) naturally occurring Tregs (thymic generated, nTregs). After encounter with foreign antigens [138], aTregs can be generated [139] with a standardized protocol [140], which are defined by their cytokine profile, including two subsets: Treg type 1 (T_R 1) cells that secrete high levels of IL-10, and Th3 cells that secrete high levels of TGF-β [141]. nTregs are primarily engaged in the maintenance of self-tolerance and down-regulation of various immune responses via a cell-cell contact manner [78], and aTregs are concerned with ablating an ongoing immune response [142]. nTregs can be defined by cell surface and intracellular marker profile (CD4+, CD25^{high}, intracellular FOXP3+ [143], GITR+ [144], CD62L$^{+/high}$ [82], CD5+, CD27+ [145], CD38+, CD39+ [146], CD45RB^{low} [82], CD45RA^{low}, CD45RO^{high}, CD73+ [146], CD103(integrin αEβ7)+ [147], intracellular CTLA4^{high}, surface CTLA4^{low} [86, 87], HLA-DR^{high}, CD122^{high}, CD127$^{low/-148}$, CD130- [149], CD134(OX40)^{high}, LFA-3^{medium}, CCR4+ [150], CCR7+ [151], CCR8+ [152], TNFR2+ [153]) (also see excellent reviews in [154, 155]). In a recently published review, Roncarolo and Battaglia wrote a very good summary on the difference between mouse and human Tregs [156]. Of note, CD25^{high} has proved to be by far the most useful surface marker for nTregs [97, 98]. Moreover, CD25 is a key component (IL-2 receptor α chain, IL-2Rα) of the high affinity IL-2R so that it is not a mere marker of nTregs but is also functionally essential for nTregs since nTregs are highly dependent on exogenous IL-2 for their survival [157]. nTregs can be activated by self-antigens and non-self-antigens [158]. Once activated, nTregs can suppress T cells in antigen-specific and antigen-nonspecific manners. Interestingly, the suppressive effects of these cells are not restricted to the adaptive immune system including CD4+ T cells, CD8+ T cells [159], NK cells [160], and B cells [161], but can also affect the activation and function of innate immune cells (monocytes, macrophages, dendritic cells) [136] and neutrophils [162]. nTreg suppression in vitro is dependent on cell contact with target cells as shown by experiments utilizing transwells [142]. Blocking antibodies to IL-10 and TGF-β do not affect the nTreg suppression, and nTregs isolated from IL-10- or TGF-β-deficient mice are also functional in vitro [142]. nTregs suppress target cells by four mechanisms: nTregs may kill target cells via a granzyme B-dependent pathway [163], perforin pathway [164], programmed death-1/programmed death-1 ligand pathway [165], or Fas-Fas ligand interaction in an epitope-specific manner [161].

CD4+CD25^{high}Foxp3+ regulatory T cells (Tregs) have potent immunosuppressive function in controlling immune responses against transplanted c-kit+ stem cells [166]. Approaches to suppress transplantation-related immunity have been historically focused on immunosuppressive drugs. However, these drugs have several problems: (a) less effective on activated T cells to abort established immune responses; (b) require continuous use; and (c) suppress immune surveillance for tumor growth [56, 101] and protective anti-microbial immunity [167]. Thus, Treg cells, as a dominant mechanism of immune self-tolerance, can control immune responsiveness to previously untolerized self-antigens, which have opened up exciting opportunities for new therapies in transplantation of autologous and allogeneic stem cells [89]. Several new strategies to effectively inhibit anti-stem cell immune responses have been under intensive study. First, development of new vaccination using antigen-specific Tregs makes the therapies in suppressing anti-stem cell immune responses and other immune responses possible [167]. Second, our lab recently showed that T cell immune responses against CML66, a myeloid progenitor cell antigen [76, 168], are suppressed by Tregs [90], suggesting that Tregs suppress antigen-specific immune responses to stem cells and progenitor cells. Third, we also found that Tregs are highly susceptible to apoptosis, presumably due to higher expression of pro-apoptotic protein Bax in Tregs than in CD4+CD25- T cells. In addition, increased Treg apoptosis via a Bax-dependent pathway enhances anti-self antigen CML66 T cell immune responses [90]. Therefore, in a recent publication, we reviewed the factors affecting apoptosis and homeostasis of Tregs, and proposed that Treg apoptosis pathways are immunotherapeutic targets in modulating immune responses to stem cells and other immune responses [169].

5 Role of Mesenchymal Stem Cells in Regulating Anti-Stem Cell Immune Responses

Mesenchymal stem cells (MSCs) are adherent, fibroblast-like, pluripotent, nonhematopoietic progenitor cells. MSCs are initially isolated from bone marrow, which constitute 0.001–0.01% of the total cell population [170] and have multilineage differentiation potential (i.e., the ability to differentiate into various tissues of mesenchymal and non-mesenchymal origin) [171–173]. MSCs can be easily isolated [174] and found in many different species [171, 172], including humans [175], rodents [176], and primates [177], and in tissues other than bone marrow, including both adult tissues including umbilical cord blood [178], fetal bone marrow, blood, lung, liver and spleen [179], fat [180], hair follicles and scalp subcutaneous tissue [181], and periodontal ligament [182], and prenatal tissues such as placenta [183]. Although there is no agreement on any standardized marker, MSCs are typically defined by a combination of phenotypic and functional characteristics. Using flow cytometry (FACS), human MSCs are negative for hematopoietic markers CD14, CD34, and CD45. Human MSCs are positive in staining for a set of adhesion markers, such as CD44 [184], CD71 [184], CD73, CD90, CD105, and CD166 [21]. Similarly, murine MSCs do not express hematopoietic markers CD45, CD34, and CD11b, while they are positive in surface expression of CD9, Sca-1, and CD44 [184]. The hallmark of MSCs is the trilineage potential in vitro (the ability to differentiate into bone, cartilage and fat upon proper induction) [21]. Human MSCs express HLA class I and can be induced by interferon-γ to express HLA class II. However, in co-culture experiments, human MSCs fail to induce proliferation of allogeneic lymphocytes in vitro, even after provision of a co-stimulatory signal by addition of CD28-stimulating antibodies or transfection of B7-1 or B7-2 co-stimulatory molecules [21]. Several reports showed that MSCs also possess immunoregulatory properties, inasmuch as they can [173]: (1) inhibit the function of mature T cells following their activation by non-specific mitogens [185]; (1) suppress the response of naïve and memory antigen-specific T cells to their cognate peptide in mice [186]; (3) promote the survival of MHC-mismatched skin grafts after infusion in baboons and reduce the incidence of graft-versus-host disease (GVHD) after allogeneic hematopoietic stem cell (HSC) transplantation in humans [187, 188]; (4) cure severe acute GVHD refractory to conventional immunosuppressive therapy [189]; and (5) ameliorate experimental autoimmune encephalomyelitis (EAE) in mice [190]. Therefore, we expect that MSCs may join CD4+CD25^{high}Foxp3+ Tregs in facilitating engraftment of stem cell therapy for regenerative medicine as we proposed [166].

Acknowledgments I am very grateful to Drs. B. Ashby for critical reading, S. Houser, and N. Dun for discussion. This work was partially supported by grants from the National Institutes of Health and the Leukemia & Lymphoma Society and the Myeloproliferative Disorders Foundation.

References

1. Fairchild PJ, Cartland S, Nolan KF, Waldmann H. Embryonic stem cells and the challenge of transplantation tolerance. Trends Immunol. 2004;25(9):465–70.
2. Jordan JD, Ming GL, Song H. Adult neurogenesis as a potential therapy for neurodegenerative diseases. Discov Med. 2006;6(34):144–7.
3. Gangaram-Panday ST, Faas MM, de Vos P. Towards stem-cell therapy in the endocrine pancreas. Trends Mol Med. 2007.
4. Sachinidis A, Fleischmann BK, Kolossov E, Wartenberg M, Sauer H, Hescheler J. Cardiac specific differentiation of mouse embryonic stem cells. Cardiovasc Res. 2003;58(2):278–91.
5. Priddle H, Jones DR, Burridge PW, Patient R. Hematopoiesis from human embryonic stem cells: overcoming the immune barrier in stem cell therapies. Stem Cells. 2006;24(4):815–24.
6. van der Bogt KE, Swijnenburg RJ, Cao F, Wu JC. Molecular imaging of human embryonic stem cells: keeping an eye on differentiation, tumorigenicity and immunogenicity. Cell Cycle. 2006;5(23):2748–52.
7. Faulkner L. Disease management: the new tool for cost containment and quality care: health policy division. NGA Center for Best Practices; 2003.
8. Evans MJ, Kaufman MH. Establishment in culture of pluripotential cells from mouse embryos. Nature. 1981;292(5819):154–6.
9. Martin GR. Isolation of a pluripotent cell line from early mouse embryos cultured in medium conditioned by teratocarcinoma stem cells. Proc Natl Acad Sci U S A. 1981;78(12):7634–8.
10. Thomson JA, Itskovitz-Eldor J, Shapiro SS, et al. Embryonic stem cell lines derived from human blastocysts. Science. 1998;282(5391):1145–7.
11. Young HE, Black AC, Jr. Adult stem cells. Anat Rec A Discov Mol Cell Evol Biol. 2004;276(1):75–102.
12. Hayflick L. The limited in vitro lifetime of human diploid cell strains. Exp Cell Res. 1965;37:614–36.
13. Kiel MJ, Morrison SJ. Maintaining hematopoietic stem cells in the vascular niche. Immunity. 2006;25(6):862–4.
14. Moraleda JM, Blanquer M, Bleda P, et al. Adult stem cell therapy: dream or reality? Transpl Immunol. 2006;17(1):74–7.
15. De Coppi P, Bartsch G, Jr., Siddiqui MM, et al. Isolation of amniotic stem cell lines with potential for therapy. Nat Biotechnol. 2007;25(1):100–6.

16. Shizuru JA, Negrin RS, Weissman IL. Hematopoietic stem and progenitor cells: clinical and preclinical regeneration of the hematolymphoid system. Annu Rev Med. 2005;56:509–38.
17. Cerny J, Quesenberry PJ. Chromatin remodeling and stem cell theory of relativity. J Cell Physiol. 2004;201(1):1–16.
18. Herzog EL, Chai L, Krause DS. Plasticity of marrow-derived stem cells. Blood. 2003;102(10):3483–93.
19. Goodell MA, Rosenzweig M, Kim H, et al. Dye efflux studies suggest that hematopoietic stem cells expressing low or undetectable levels of CD34 antigen exist in multiple species. Nat Med. 1997;3(12):1337–45.
20. Challen GA, Little MH. A side order of stem cells: the SP phenotype. Stem Cells. 2006;24(1):3–12.
21. Le Blanc K, Ringden O. Mesenchymal stem cells: properties and role in clinical bone marrow transplantation. Curr Opin Immunol. 2006;18(5):586–91.
22. Yang ZJ, Wechsler-Reya RJ. Hit em where they live: targeting the cancer stem cell niche. Cancer Cell. 2007;11(1):3–5.
23. Li F, Tiede B, Massague J, Kang Y. Beyond tumorigenesis: cancer stem cells in metastasis. Cell Res. 2007;17(1):3–14.
24. Soria B, Roche E, Berna G, Leon-Quinto T, Reig JA, Martin F. Insulin-secreting cells derived from embryonic stem cells normalize glycemia in streptozotocin-induced diabetic mice. Diabetes. 2000;49(2):157–62.
25. Klug MG, Soonpaa MH, Koh GY, Field LJ. Genetically selected cardiomyocytes from differentiating embryonic stem cells form stable intracardiac grafts. J Clin Invest. 1996;98(1):216–24.
26. Min JY, Yang Y, Converso KL, et al. Transplantation of embryonic stem cells improves cardiac function in postinfarcted rats. J Appl Physiol. 2002;92(1):288–96.
27. Kim JH, Auerbach JM, Rodriguez-Gomez JA, et al. Dopamine neurons derived from embryonic stem cells function in an animal model of Parkinson's disease. Nature. 2002;418(6893):50–6.
28. Bradley JA, Bolton EM, Pedersen RA. Stem cell medicine encounters the immune system. Nat Rev Immunol. 2002;2(11): 859–71.
29. Nussbaum J, Minami E, Laflamme MA, et al. Transplantation of undifferentiated murine embryonic stem cells in the heart: teratoma formation and immune response. FASEB J. 2007;21(7):1345–57.
30. Swijnenburg RJ, Tanaka M, Vogel H, et al. Embryonic stem cell immunogenicity increases upon differentiation after transplantation into ischemic myocardium. Circulation. 2005;112(9 Suppl):I166–72.
31. Modo M, Rezaie P, Heuschling P, Patel S, Male DK, Hodges H. Transplantation of neural stem cells in a rat model of stroke: assessment of short-term graft survival and acute host immunological response. Brain Res. 2002;958(1):70–82.
32. Zheng XS, Yang XF, Liu WG, Pan DS, Hu WW, Li G. Transplantation of neural stem cells into the traumatized brain induces lymphocyte infiltration. Brain Inj. 2007;21(3):275–8.
33. Kofidis T, deBruin JL, Tanaka M, et al. They are not stealthy in the heart: embryonic stem cells trigger cell infiltration, humoral and T-lymphocyte-based host immune response. Eur J Cardiothorac Surg. 2005;28(3):461–6.
34. Fandrich F, Dresske B, Bader M, Schulze M. Embryonic stem cells share immune-privileged features relevant for tolerance induction. J Mol Med. 2002;80(6):343–50.
35. Li L, Baroja ML, Majumdar A, et al. Human embryonic stem cells possess immune-privileged properties. Stem Cells. 2004;22(4):448–56.
36. Burt RK, Verda L, Kim DA, Oyama Y, Luo K, Link C. Embryonic stem cells as an alternate marrow donor source: engraftment without graft-versus-host disease. J Exp Med. 2004;199(7): 895–904.
37. Streilein JW. Unraveling immune privilege. Science. 1995;270(5239):1158–9.
38. O'Connell J, Bennett MW, O'Sullivan GC, Collins JK, Shanahan F. The Fas counterattack: cancer as a site of immune privilege. Immunol Today. 1999;20(1):46–52.
39. Kennea NL, Stratou C, Naparus A, Fisk NM, Mehmet H. Functional intrinsic and extrinsic apoptotic pathways in human fetal mesenchymal stem cells. Cell Death Differ. 2005;12(11): 1439–41.
40. Drukker M, Benvenisty N. The immunogenicity of human embryonic stem-derived cells. Trends Biotechnol. 2004;22(3):136–41.
41. Drukker M, Katz G, Urbach A, et al. Characterization of the expression of MHC proteins in human embryonic stem cells. Proc Natl Acad Sci U S A. 2002;99(15):9864–9.
42. Draper JS, Pigott C, Thomson JA, Andrews PW. Surface antigens of human embryonic stem cells: changes upon differentiation in culture. J Anat. 2002;200(Pt 3):249–58.
43. Magliocca JF, Held IK, Odorico JS. Undifferentiated murine embryonic stem cells cannot induce portal tolerance but may possess immune privilege secondary to reduced major histocompatibility complex antigen expression. Stem Cells Dev. 2006;15(5): 707–17.
44. Yang XF, Mirkovic D, Zhang S, et al. Processing sites are different in the generation of HLA-A2.1-restricted, T cell reactive tumor antigen epitopes and viral epitopes. Int J Immunopathol Pharmacol. 2006;19(4):853–70.
45. Yang F, Chen IH, Xiong Z, Yan Y, Wang H, Yang XF. Model of stimulation-responsive splicing and strategies in identification of immunogenic isoforms of tumor antigens and autoantigens. Clin Immunol. 2006;121(2):121–33.
46. Mullally A, Ritz J. Beyond HLA: the significance of genomic variation for allogeneic hematopoietic stem cell transplantation. Blood. 2007;109(4):1355–62.
47. Oh SK, Kim HS, Park YB, et al. Methods for expansion of human embryonic stem cells. Stem Cells. 2005;23(5):605–9.
48. Heng BC, Haider HK, Sim EK, Cao T, Tong GQ, Ng SC. Comments about possible use of human embryonic stem cell-derived cardiomyocytes to direct autologous adult stem cells into the cardiomyogenic lineage. Acta Cardiol. 2005;60(1):7–12.
49. Mitalipov SM. Genomic imprinting in primate embryos and embryonic stem cells. Reprod Fertil Dev. 2006;18(8):817–21.
50. Miyake K. Innate immune sensing of pathogens and danger signals by cell surface Toll-like receptors. Semin Immunol. 2007;19(1):3–10.
51. Matzinger P. The danger model: a renewed sense of self. Science 2002;296(5566):301–5.
52. Rock KL, York IA, Goldberg AL. Post-proteasomal antigen processing for major histocompatibility complex class I presentation. Nat Immunol. 2004;5(7):670–7.
53. Kloetzel PM. Generation of major histocompatibility complex class I antigens: functional interplay between proteasomes and TPPII. Nat Immunol. 2004;5(7):661–9.
54. Geier E, Pfeifer G, Wilm M, et al. A giant protease with potential to substitute for some functions of the proteasome. Science. 1999;283(5404):978–81.
55. Luckey CJ, King GM, Marto JA, et al. Proteasomes can either generate or destroy MHC class I epitopes: evidence for nonproteasomal epitope generation in the cytosol. J Immunol. 1998;161(1):112–21.
56. Yang F, Yang XF. New concepts in tumor antigens: their significance in future immunotherapies for tumors. Cell Mol Immunol. 2005;2(5):331–41.
57. Altuvia Y, Margalit H. Sequence signals for generation of antigenic peptides by the proteasome: implications for proteasomal cleavage mechanism. J Mol Biol. 2000;295(4):879–90.

58. Uebel S, Kraas W, Kienle S, Wiesmuller KH, Jung G, Tampe R. Recognition principle of the TAP transporter disclosed by combinatorial peptide libraries. Proc Natl Acad Sci U S A. 1997;94(17):8976–81.
59. Daniel S, Brusic V, Caillat-Zucman S, et al. Relationship between peptide selectivities of human transporters associated with antigen processing and HLA class I molecules. J Immunol. 1998;161(2):617–24.
60. Yotnda P, Firat H, Garcia-Pons F, et al. Cytotoxic T cell response against the chimeric p210 BCR-ABL protein in patients with chronic myelogenous leukemia. J Clin Invest. 1998;101(10):2290–6.
61. Pinilla-Ibarz J, Cathcart K, Scheinberg DA. CML vaccines as a paradigm of the specific immunotherapy of cancer. Blood Rev. 2000;14(2):111–20.
62. Zorn E, Orsini E, Wu CJ, Stein B, Chillemi A, Canning C, et al. A CD4+ T cell clone selected from a CML patient after donor lymphocyte infusion recognizes BCR-ABL breakpoint peptides but not tumor cells. Transplantation. 2001;71(8):1131–7.
63. Clark RE, Dodi IA, Hill SC, et al. Direct evidence that leukemic cells present HLA-associated immunogenic peptides derived from the BCR-ABL b3a2 fusion protein. Blood. 2001;98(10):2887–93.
64. Zinkernagel RM, Hengartner H. Regulation of the immune response by antigen. Science. 2001;293(5528):251–3.
65. Shlomchik MJ, Craft JE, Mamula MJ. From T to B and back again: positive feedback in systemic autoimmune disease. Nat Rev Immunol. 2001;1(2):147–53.
66. Sahin U, Tureci O, Schmitt H, et al. Human neoplasms elicit multiple specific immune responses in the autologous host. Proc Natl Acad Sci U S A. 1995;92(25):11810–3.
67. Preuss KD, Zwick C, Bormann C, Neumann F, Pfreundschuh M. Analysis of the B-cell repertoire against antigens expressed by human neoplasms. Immunol Rev. 2002;188:43–50.
68. Yan Y, Phan L, Yang F, et al. A novel mechanism of alternative promoter and splicing regulates the epitope generation of tumor antigen CML66-L. J Immunol. 2004;172(1):651–60.
69. De Smet C, De Backer O, Faraoni I, Lurquin C, Brasseur F, Boon T. The activation of human gene MAGE-1 in tumor cells is correlated with genome-wide demethylation. Proc Natl Acad Sci U S A. 1996;93(14):7149–53.
70. Gure AO, Wei IJ, Old LJ, Chen YT. The SSX gene family: characterization of 9 complete genes. Int J Cancer. 2002;101(5):448–53.
71. Chen Y. SEREX review. Cancer Immunity 2004; http://www.cancerimmunity.org/SEREX/.
72. Yang XF, Wu CJ, Chen L, et al. CML28 is a broadly immunogenic antigen, which is overexpressed in tumor cells. Cancer Res. 2002;62(19):5517–22.
73. Ng B, Yang F, Huston DP, et al. Increased noncanonical splicing of autoantigen transcripts provides the structural basis for expression of untolerized epitopes. J Allergy Clin Immunol. 2004;114(6):1463–70.
74. Xiong Z, Shaibani A, Li YP, et al. Alternative splicing factor ASF/SF2 is down regulated in inflamed muscle. J Clin Pathol. 2006;59(8):855–61.
75. Wu CJ, Yang XF, McLaughlin S, et al. Detection of a potent humoral response associated with immune-induced remission of chronic myelogenous leukemia. J Clin Invest. 2000;106(5):705–14.
76. Yang XF, Wu CJ, McLaughlin S, et al. CML66, a broadly immunogenic tumor antigen, elicits a humoral immune response associated with remission of chronic myelogenous leukemia. Proc Natl Acad Sci U S A. 2001;98(13):7492–7.
77. Xiong Z, Liu E, Yan Y, et al. An unconventional antigen translated by a novel internal ribosome entry site elicits antitumor humoral immune reactions. J Immunol. 2006;177(7):4907–16.
78. Xiong Z, Liu E, Yan Y, Silver RT, Yang F, Chen IH, et al. A novel unconventional antigen, MPD5, elicits anti-tumor humoral immune responses in a subset of patients with polycythemia vera. Int J Immunopathol Pharmacol. 2007;20(2):375–82.
79. Xiong Z, Yan Y, Liu E, et al. Novel tumor antigens elicit anti-tumor humoral immune reactions in a subset of patients with polycythemia vera. Clin Immunol. 2007;122(3):279–87.
80. Grusby MJ, Auchincloss H, Jr., Lee R, et al. Mice lacking major histocompatibility complex class I and class II molecules. Proc Natl Acad Sci U S A. 1993;90(9):3913–7.
81. Munsie MJ, Michalska AE, O'Brien CM, Trounson AO, Pera MF, Mountford PS. Isolation of pluripotent embryonic stem cells from reprogrammed adult mouse somatic cell nuclei. Curr Biol. 2000;10(16):989–92.
82. Hwang WS, Ryu YJ, Park JH, et al. Evidence of a pluripotent human embryonic stem cell line derived from a cloned blastocyst. Science 2004;303(5664):1669–74.
83. Lanza RP, Chung HY, Yoo JJ, et al. Generation of histocompatible tissues using nuclear transplantation. Nat Biotechnol. 2002;20(7):689–96.
84. Morse MC, Bleau G, Dabhi VM, et al. The COI mitochondrial gene encodes a minor histocompatibility antigen presented by H2-M3. J Immunol. 1996;156(9):3301–7.
85. Simpson E, Roopenian D. Minor histocompatibility antigens. Curr Opin Immunol. 1997;9(5):655–61.
86. Li M, Pevny L, Lovell-Badge R, Smith A. Generation of purified neural precursors from embryonic stem cells by lineage selection. Curr Biol. 1998;8(17):971–4.
87. Steptoe RJ, Ritchie JM, Harrison LC. Transfer of hematopoietic stem cells encoding autoantigen prevents autoimmune diabetes. J Clin Invest. 2003;111(9):1357–63.
88. Shevach EM. Certified professionals: CD4(+)CD25(+) suppressor T cells. J Exp Med. 2001;193(11):F41–6.
89. Zorn E. CD4+CD25+ regulatory T cells in human hematopoietic cell transplantation. Semin Cancer Biol. 2006;16(2):150–9.
90. Yan Y, Chen Y, Yang F, et al. HLA-A2.1-restricted T cells react to SEREX-defined tumor antigen CML66L and are suppressed by CD4+CD25+ regulatory T cells. Int J Immunopathol Pharmacol. 2007;20(1):75–89.
91. Hambach L, Goulmy E. Immunotherapy of cancer through targeting of minor histocompatibility antigens. Curr Opin Immunol. 2005;17(2):202–10.
92. Lapidot T, Sirard C, Vormoor J, et al. A cell initiating human acute myeloid leukaemia after transplantation into SCID mice. Nature. 1994;367(6464):645–8.
93. Al-Hajj M, Clarke MF. Self-renewal and solid tumor stem cells. Oncogene. 2004;23(43):7274–82.
94. Hemmati HD, Nakano I, Lazareff JA, et al. Cancerous stem cells can arise from pediatric brain tumors. Proc Natl Acad Sci U S A. 2003;100(25):15178–83.
95. Lou H, Dean M. Targeted therapy for cancer stem cells: the patched pathway and ABC transporters. Oncogene. 2007;26(9):1357–60.
96. Tunici P, Irvin D, Liu G, et al. Brain tumor stem cells: new targets for clinical treatments? Neurosurg Focus. 2006;20(4):E27.
97. Yilmaz OH, Valdez R, Theisen BK, et al. Pten dependence distinguishes haematopoietic stem cells from leukaemia-initiating cells. Nature. 2006;441(7092):475–82.
98. Dunn GP, Old LJ, Schreiber RD. The immunobiology of cancer immunosurveillance and immunoediting. Immunity. 2004;21(2):137–48.
99. Dunn GP, Old LJ, Schreiber RD. The three Es of cancer immunoediting. Annu Rev Immunol. 2004;22:329–60.
100. Dunn GP, Bruce AT, Ikeda H, Old LJ, Schreiber RD. Cancer immunoediting: from immunosurveillance to tumor escape. Nat Immunol. 2002;3(11):991–8.

101. Schreiber RD. Cancer vaccines 2004 opening address: the molecular and cellular basis of cancer immunosurveillance and immunoediting. Cancer Immun. 2005;5 Suppl 1:1.
102. Schreiber H. Tumor immunology. 5th ed. Philadelphia: Lippincott-Raven Publishers; 2003.
103. Ke X, Zhao L, Gao Z, et al. Higher rates of t(11;18) in Chinese patients with transformed type of MALT lymphoma suggest novel pathways for progression of the disease. Leuk Lymphoma. 2007;48(11):2157–66.
104. Grandics P. The cancer stem cell: evidence for its origin as an injured autoreactive T cell. Mol Cancer. 2006;5:6.
105. van der Bruggen P, Traversari C, Chomez P, et al. A gene encoding an antigen recognized by cytolytic T lymphocytes on a human melanoma. Science. 1991;254(5038):1643–7.
106. Boon T, van der Bruggen P. Human tumor antigens recognized by T lymphocytes. J Exp Med. 1996;183(3):725–9.
107. Boon T, Van den Eynde, BJ. Cancer vaccines: Cancer antigens. Shared tumor-specific antigens. In: Rosenberg S, editor. Principles and practice of the biologic therapy of cancer. 3rd ed. Philadelphia,: Lippincott Williams and Wilkins; 2000. pp. 493–504.
108. Traversari C, van der Bruggen P, Luescher IF, et al. A nonapeptide encoded by human gene MAGE-1 is recognized on HLA-A1 by cytolytic T lymphocytes directed against tumor antigen MZ2-E. J Exp Med. 1992;176(5):1453–7.
109. Sahin U, Tureci O, Pfreundschuh M. Serological identification of human tumor antigens. Curr Opin Immunol. 1997;9(5): 709–16.
110. Tureci O, Sahin U, Pfreundschuh M. Serological analysis of human tumor antigens: molecular definition and implications. Mol Med Today. 1997;3(8):342–9.
111. Rosenberg SA. Development of effective immunotherapy for the treatment of patients with cancer. J Am Coll Surg. 2004;198(5):685–96.
112. Shastri N, Schwab S, Serwold T. Producing nature's gene-chips: the generation of peptides for display by MHC class I molecules. Annu Rev Immunol. 2002;20:463–93.
113. Scanlan MJ, Jager D. Challenges to the development of antigen-specific breast cancer vaccines. Breast Cancer Res. 2001;3(2): 95–8.
114. Scanlan MJ, Simpson AJ, Old LJ.. The cancer/testis genes: review, standardization, and commentary. Cancer Immun. 2004;4:1.
115. Xiong Z, Liu E, Yan Y, Silver RT, Zhang S, Yang Y, et al. Novel unconventional and conventional antigens elicit anti-tumor humoral immune reactions in a subset of patients with polycythemia vera. Blood 2005;Submission.
116. Wang RF, Rosenberg SA. Human tumor antigens recognized by T lymphocytes: implications for cancer therapy. J Leukoc Biol. 1996;60(3):296–309.
117. Schirmbeck R, Riedl P, Fissolo N, Lemonnier FA, Bertoletti A, Reimann J. Translation from cryptic reading frames of DNA vaccines generates an extended repertoire of immunogenic, MHC class I-restricted epitopes. J Immunol. 2005;174(8): 4647–56.
118. Wang RF, Johnston SL, Zeng G, Topalian SL, Schwartzentruber DJ, Rosenberg SA. A breast and melanoma-shared tumor antigen: T cell responses to antigenic peptides translated from different open reading frames. J Immunol. 1998;161(7):3598–606.
119. Mandic M, Almunia C, Vicel S, et al. The alternative open reading frame of LAGE-1 gives rise to multiple promiscuous HLA-DR-restricted epitopes recognized by T-helper 1-type tumor-reactive CD4+ T cells. Cancer Res. 2003;63(19):6506–15.
120. Slager EH, Borghi M, van der Minne CE, et al. CD4+ Th2 cell recognition of HLA-DR-restricted epitopes derived from CAMEL: a tumor antigen translated in an alternative open reading frame. J Immunol. 2003;170(3):1490–7.
121. Copland M, Fraser AR, Harrison SJ, Holyoake TL. Targeting the silent minority: emerging immunotherapeutic strategies for eradication of malignant stem cells in chronic myeloid leukaemia. Cancer Immunol Immunother. 2005;54(4):297–306.
122. Miller JF, Basten A. Mechanisms of tolerance to self. Curr Opin Immunol. 1996;8(6):815–21.
123. Totsuka T, Kanai T, Makita S, et al. Regulation of murine chronic colitis by CD4+CD25- programmed death-1+ T cells. Eur J Immunol. 2005;35(6):1773–85.
124. Bach JF. Regulatory T cells under scrutiny. Nat Rev Immunol. 2003;3(3):189–98.
125. Bienvenu B, Martin B, Auffray C, Cordier C, Becourt C, Lucas B. Peripheral CD8+CD25+ T lymphocytes from MHC class II-deficient mice exhibit regulatory activity. J Immunol. 2005;175(1):246–53.
126. Cosmi L, Liotta F, Lazzeri E, et al. Human CD8+CD25+ thymocytes share phenotypic and functional features with CD4+CD25+ regulatory thymocytes. Blood 2003;102(12): 4107–14.
127. Zhang ZX, Young K, Zhang L. CD3+CD4-CD8- alphabeta-TCR+ T cell as immune regulatory cell. J Mol Med. 2001;79(8): 419–27.
128. Schwartz RH. Natural regulatory T cells and self-tolerance. Nat Immunol. 2005;6(4):327–30.
129. Fontenot JD, Rudensky AY. A well adapted regulatory contrivance: regulatory T cell development and the forkhead family transcription factor Foxp3. Nat Immunol. 2005;6(4):331–7.
130. von Boehmer H. Mechanisms of suppression by suppressor T cells. Nat Immunol. 2005;6(4):338–44.
131. Sakaguchi S. Naturally arising Foxp3-expressing CD25+CD4+ regulatory T cells in immunological tolerance to self and non-self. Nat Immunol. 2005;6(4):345–52.
132. Jiang H, Chess L. An integrated view of suppressor T cell subsets in immunoregulation. J Clin Invest. 2004;114(9):1198–208.
133. Horwitz D, Gray JD, Zheng SG. The potential of human regulatory T cells generated ex vivo as a treatment for lupus and other chronic inflammatory diseases. Arthritis Res. 2002;4(4):241–6.
134. Akbari O, Stock P, DeKruyff RH, Umetsu DT. Role of regulatory T cells in allergy and asthma. Curr Opin Immunol. 2003;15(6):627–33.
135. Taylor PA, Lees CJ, Blazar BR. The infusion of ex vivo activated and expanded CD4(+)CD25(+) immune regulatory cells inhibits graft-versus-host disease lethality. Blood. 2002;99(10): 3493–9.
136. Taams LS, Akbar AN. Peripheral generation and function of CD4+CD25+ regulatory T cells. Curr Top Microbiol Immunol. 2005;293:115–31.
137. Wraith DC, Nicolson KS, Whitley NT. Regulatory CD4+ T cells and the control of autoimmune disease. Curr Opin Immunol. 2004;16(6):695–701.
138. Mills KH. Regulatory T cells: friend or foe in immunity to infection? Nat Rev Immunol. 2004;4(11):841–55.
139. Goleva E, Cardona ID, Ou LS, Leung DY. Factors that regulate naturally occurring T regulatory cell-mediated suppression. J Allergy Clin Immunol. 2005;116(5):1094–100.
140. Fantini MC, Dominitzki S, Rizzo A, Neurath MF, Becker C. In vitro generation of CD4+CD25+ regulatory cells from murine naive T cells. Nat Protoc. 2007;2(7):1789–94.
141. Akdis M, Blaser K, Akdis CA. T regulatory cells in allergy: novel concepts in the pathogenesis, prevention, and treatment of allergic diseases. J Allergy Clin Immunol. 2005;116(5):961–8; quiz 9.
142. Wing K, Fehervari Z, Sakaguchi S. Emerging possibilities in the development and function of regulatory T cells. Int Immunol. 2006;18(7):991–1000.
143. Ochs HD, Ziegler SF, Torgerson TR. FOXP3 acts as a rheostat of the immune response. Immunol Rev. 2005;203:156–64.

144. Kanamaru F, Youngnak P, Hashiguchi M, et al. Costimulation via glucocorticoid-induced TNF receptor in both conventional and CD25+ regulatory CD4+ T cells. J Immunol. 2004;172(12):7306–14.
145. Ruprecht CR, Gattorno M, Ferlito F, et al. Coexpression of CD25 and CD27 identifies FoxP3+ regulatory T cells in inflamed synovia. J Exp Med. 2005;201(11):1793–803.
146. Deaglio S, Dwyer KM, Gao W, et al. Adenosine generation catalyzed by CD39 and CD73 expressed on regulatory T cells mediates immune suppression. J Exp Med.;204(6):1257–65.
147. Suffia I, Reckling SK, Salay G, Belkaid Y. A role for CD103 in the retention of CD4+CD25+ Treg and control of Leishmania major infection. J Immunol. 2005;174(9):5444–55.
148. Liu W, Putnam AL, Xu-Yu Z, et al. CD127 expression inversely correlates with FoxP3 and suppressive function of human CD4+ T reg cells. J Exp Med. 2006;203(7):1701–11.
149. Oberg HH, Wesch D, Grussel S, Rose-John S, Kabelitz D. Differential expression of CD126 and CD130 mediates different STAT-3 phosphorylation in CD4+CD25- and CD25high regulatory T cells. Int Immunol. 2006;18(4):555–63.
150. Iellem A, Mariani M, Lang R, et al. Unique chemotactic response profile and specific expression of chemokine receptors CCR4 and CCR8 by CD4(+)CD25(+) regulatory T cells. J Exp Med. 2001;194(6):847–53.
151. Hoffmann P, Eder R, Kunz-Schughart LA, Andreesen R, Edinger M. Large-scale in vitro expansion of polyclonal human CD4(+)CD25high regulatory T cells. Blood. 2004;104(3):895–903.
152. Annunziato F, Cosmi L, Liotta F, et al. Phenotype, localization, and mechanism of suppression of CD4(+)CD25(+) human thymocytes. J Exp Med. 2002;196(3):379–87.
153. Maggi E, Cosmi L, Liotta F, Romagnani P, Romagnani S, Annunziato F. Thymic regulatory T cells. Autoimmun Rev. 2005;4(8):579–86.
154. Holm TL, Nielsen J, Claesson MH. CD4+CD25+ regulatory T cells: I. Phenotype and physiology. Apmis. 2004;112(10):629–41.
155. Yi H, Zhen Y, Jiang L, Zheng J, Zhao Y. The phenotypic characterization of naturally occurring regulatory CD4+CD25+ T cells. Cell Mol Immunol. 2006;3(3):189–95.
156. Roncarolo MG, Battaglia M. Regulatory T-cell immunotherapy for tolerance to self antigens and alloantigens in humans. Nat Rev Immunol. 2007;7(8):585–98.
157. Setoguchi R, Hori S, Takahashi T, Sakaguchi S. Homeostatic maintenance of natural Foxp3(+) CD25(+) CD4(+) regulatory T cells by interleukin (IL)-2 and induction of autoimmune disease by IL-2 neutralization. J Exp Med. 2005;201(5):723–35.
158. Kumar V. Homeostatic control of immunity by TCR peptide-specific Tregs. J Clin Invest. 2004;114(9):1222–6.
159. Camara NO, Sebille F, Lechler RI. Human CD4+CD25+ regulatory cells have marked and sustained effects on CD8+ T cell activation. Eur J Immunol. 2003;33(12):3473–83.
160. Wolf AM, Wolf D, Steurer M, Gastl G, Gunsilius E, Grubeck-Loebenstein B. Increase of regulatory T cells in the peripheral blood of cancer patients. Clin Cancer Res. 2003;9(2):606–12.
161. Janssens W, Carlier V, Wu B, VanderElst L, Jacquemin MG, Saint-Remy JM. CD4+CD25+ T cells lyse antigen-presenting B cells by Fas-Fas ligand interaction in an epitope-specific manner. J Immunol. 2003;171(9):4604–12.
162. Lewkowicz P, Lewkowicz N, Sasiak A, Tchorzewski H. Lipopolysaccharide-activated CD4+CD25+ T regulatory cells inhibit neutrophil function and promote their apoptosis and death. J Immunol. 2006;177(10):7155–63.
163. Gondek DC, Lu LF, Quezada SA, Sakaguchi S, Noelle RJ. Cutting edge: contact-mediated suppression by CD4+CD25+ regulatory cells involves a granzyme B-dependent, perforin-independent mechanism. J Immunol. 2005;174(4):1783–6.
164. Grossman WJ, Verbsky JW, Barchet W, Colonna M, Atkinson JP, Ley TJ. Human T regulatory cells can use the perforin pathway to cause autologous target cell death. Immunity. 2004;21(4):589–601.
165. Kitazawa Y, Fujino M, Wang Q, et al. Involvement of the programmed death-1/programmed death-1 ligand pathway in CD4+CD25+ regulatory T-cell activity to suppress alloimmune responses. Transplantation. 2007;83(6):774–82.
166. Yang XF. Immunology of stem cells and cancer stem cells. Cell Mol Immunol. 2007;4(3):161–71.
167. Bluestone JA, Tang Q. Therapeutic vaccination using CD4+CD25+ antigen-specific regulatory T cells. Proc Natl Acad Sci U S A. 2004;101 Suppl 2:14622–6.
168. Wu CJ, Biernacki M, Kutok JL, et al. Graft-versus-leukemia target antigens in chronic myelogenous leukemia are expressed on myeloid progenitor cells. Clin Cancer Res. 2005;11(12):4504–11.
169. Yang XF. Factors regulating apoptosis and homeostasis of CD4+CD25highFOXP3+ regulatory T cells are new therapeutic targets. Front Biosci. 2008;13:1472–99.
170. Chen X, Armstrong MA, Li G. Mesenchymal stem cells in immunoregulation. Immunol Cell Biol. 2006;84(5):413–21.
171. Pittenger MF, Mackay AM, Beck SC, et al. Multilineage potential of adult human mesenchymal stem cells. Science. 1999;284(5411):143–7.
172. Jiang Y, Jahagirdar BN, Reinhardt RL, et al. Pluripotency of mesenchymal stem cells derived from adult marrow. Nature. 2002;418(6893):41–9.
173. Krampera M, Pasini A, Pizzolo G, Cosmi L, Romagnani S, Annunziato F. Regenerative and immunomodulatory potential of mesenchymal stem cells. Curr Opin Pharmacol. 2006;6(4):435–41.
174. Beyer Nardi N, da Silva Meirelles L. Mesenchymal stem cells: isolation, in vitro expansion and characterization. Handb Exp Pharmacol. 2006(174):249–82.
175. Tremain N, Korkko J, Ibberson D, Kopen GC, DiGirolamo C, Phinney DG. MicroSAGE analysis of 2,353 expressed genes in a single cell-derived colony of undifferentiated human mesenchymal stem cells reveals mRNAs of multiple cell lineages. Stem Cells. 2001;19(5):408–18.
176. Phinney DG, Kopen G, Isaacson RL, Prockop DJ. Plastic adherent stromal cells from the bone marrow of commonly used strains of inbred mice: variations in yield, growth, and differentiation. J Cell Biochem. 1999;72(4):570–85.
177. Devine SM, Bartholomew AM, Mahmud N, et al. Mesenchymal stem cells are capable of homing to the bone marrow of non-human primates following systemic infusion. Exp Hematol. 2001;29(2):244–55.
178. Erices A, Conget P, Minguell JJ. Mesenchymal progenitor cells in human umbilical cord blood. Br J Haematol. 2000;109(1):235–42.
179. in't Anker PS, Noort WA, Scherjon SA, et al. Mesenchymal stem cells in human second-trimester bone marrow, liver, lung, and spleen exhibit a similar immunophenotype but a heterogeneous multilineage differentiation potential. Haematologica. 2003;88(8):845–52.
180. Lee RH, Kim B, Choi I, et al. Characterization and expression analysis of mesenchymal stem cells from human bone marrow and adipose tissue. Cell Physiol Biochem. 2004;14(4–6):311–24.
181. Shih DT, Lee DC, Chen SC, et al. Isolation and characterization of neurogenic mesenchymal stem cells in human scalp tissue. Stem Cells. 2005;23(7):1012–20.
182. Trubiani O, Di Primio R, Traini T, et al. Morphological and cytofluorimetric analysis of adult mesenchymal stem cells expanded ex vivo from periodontal ligament. Int J Immunopathol Pharmacol. 2005;18(2):213–21.

183. In't Anker PS, Scherjon SA, Kleijburg-van der Keur C, et al. Isolation of mesenchymal stem cells of fetal or maternal origin from human placenta. Stem Cells. 2004;22(7):1338–45.
184. Uccelli A, Moretta L, Pistoia V. Immunoregulatory function of mesenchymal stem cells. Eur J Immunol. 2006;36(10):2566–73.
185. Di Nicola M, Carlo-Stella C, Magni M, et al. Human bone marrow stromal cells suppress T-lymphocyte proliferation induced by cellular or nonspecific mitogenic stimuli. Blood. 2002;99(10):3838–43.
186. Krampera M, Glennie S, Dyson J, et al. Bone marrow mesenchymal stem cells inhibit the response of naive and memory antigen-specific T cells to their cognate peptide. Blood. 2003;101(9):3722–9.
187. Bartholomew A, Sturgeon C, Siatskas M, et al. Mesenchymal stem cells suppress lymphocyte proliferation in vitro and prolong skin graft survival in vivo. Exp Hematol. 2002;30(1):42–8.
188. Lazarus HM, Koc ON, Devine SM, et al. Cotransplantation of HLA-identical sibling culture-expanded mesenchymal stem cells and hematopoietic stem cells in hematologic malignancy patients. Biol Blood Marrow Transplant. 2005;11(5):389–98.
189. Le Blanc K, Rasmusson I, Sundberg B, et al. Treatment of severe acute graft-versus-host disease with third party haploidentical mesenchymal stem cells. Lancet. 2004;363(9419):1439–41.
190. Zappia E, Casazza S, Pedemonte E, et al. Mesenchymal stem cells ameliorate experimental autoimmune encephalomyelitis inducing T-cell anergy. Blood. 2005;106(5):1755–61.

Leukemic Stem Cells: New Therapeutic Targets?

Dominique Bonnet

Abstract An emerging concept in cancer biology is that a rare population of cancer stem cells exists among the heterogeneous cell mass that constitutes the tumor. Based on this notion, tumors are thought to be driven by a cellular subpopulation that retains key stem cell features. These include self-renewal, differentiation into the cell types of the original cancer and potent tumor formation. Yet, despite their critical importance, much remains to be learned about the developmental origin of cancer stem cells and the mechanisms responsible for their emergence in the course of the disease. This report will focus more specifically on the blood-related cancer leukemia, which was the first disease where human CSCs, or leukemic stem cells (LSCs), were isolated. We will summarize the recent advance in the study of leukemic stem cells and the potential clinical implications for the development of new therapies.

Keywords Hematopoietic stem cell · Xenotransplantation model · Leukemic stem cell · Cancer stem cell · Self-renewal · Stem cell niche

1 Introduction

The hallmark properties of the stem cells are the ability to balance self-renewal versus differentiation cell fate decisions. The best-studied self-renewing cells are the long-term hematopoietic stem cells, which maintain themselves as a population for the lifetime of an organism. Stem cells divide asymmetrically producing two daughters cells – one is a new stem cell and the other is a progenitor cell that has the ability to differentiate and proliferate, but not to self-renew [1]. In order to assure a persistent pool of regenerating cells without outgrowth of immature cell types, a tight regulation of HSC division is required. Unchecked growth of immature cells is thought to represent a paradigm for malignant outgrowth, at least for acute myeloid leukemia (AML) and chronic myeloid leukemia (CML) [2, 3]. Thus, determining the composition and relationship of the cell types that constitute the human stem cell compartment may help both to identify the cellular and molecular factors that govern normal and leukemic stem cell (LSC) development, and to advance clinical applications of transplantation, gene therapy, stem cell expansion and tumor cell purging. This review will introduce the notion of LSCs, the potential origin of these cells with an emphasis on myeloid leukemia, the molecular mechanisms deregulate in leukemic stem cells, the notion of leukemic stem cell niche and the impacts these discoveries may have clinically and in understanding the organization of cancer of other tissues.

2 Concept of the Leukemia-Initiating Cell

Whereas it was previously believed that most or all cancer cells possess the property to self-renew and replenish new cancer cells, it has recently become clear that cancers, as most normal tissues, are organized in a hierarchical fashion, and that only a small fraction of tumor cells have the ability to reconstitute a new tumor [3]. The existence of such cancer stem cells (CSCs) with self-renewal potential was first documented in leukemias [4, 5], but has now been extended to a wide varieties of solid tumors, such as breast, brain, colon, and prostate cancers raising the possibility that such cells are the apex of all neoplastic systems [6, 7]. As these rare CSCs are both required and sufficient to reconstitute a new tumor, they have immediate and important clinical implications. A better identification and characterization of CSCs should provide a better understanding of tumor developmental biology and the genetic events involved in the transformation process. Recent studies have shown that CSCs, as normal somatic stem cells, express high levels of multidrug resistance (MDR) pumps that efficiently efflux cytotoxic drugs [8], making them particularly difficult to target

D. Bonnet (✉)
Cancer Research UK, London Research Institute, Haematopoietic Stem Cell Laboratory, Cancer Research UK, London Research Institute, 44 Lincoln's Inn Fields, London, WC2A 3PX, UK
e-mail: dominique.bonnet@cancer.org.uk

V.K. Rajasekhar, M.C. Vemuri (eds.), *Regulatory Networks in Stem Cells*, Stem Cell Biology and Regenerative Medicine,
DOI 10.1007/978-1-60327-227-8_39,

with cytotoxic therapies. Similarly, they appear to be less immunogenic than more differentiated cancer cells in the tumor [9], and therefore also more difficult to eradicate with immunotherapy. Thus, it has become evident, that CSCs might prove not only to be the most important but also the most difficult cancer cells to eliminate with conventional therapies, and that a specific monitoring and targeting of these elusive CSCs could become an important tool towards identification and characterization of improved cellular and molecular targets for development of improved cancer therapies. This has opened a new translational research field bridging stem cell and cancer biology [3], which is likely to have considerable impact on the development of clinical oncology.

Based on these recent studies, the paradigm of cancer as a hierarchical disease, sustained by a rare population of CSCs has emerged. Implicit in this model of cancer development is the notion that CSCs are biologically distinct from other cells in the tumor and are able to initiate and sustain tumor growth in vivo whereas the bulk cells are not. Despite of the clear importance of CSCs in the genesis and perpetuation of cancers, little is currently known about the biological and molecular properties that make CSCs distinct from normal stem cells, the developmental/cellular origin of CSCs, the mechanisms responsible for their emergence in the course of the disease, and identification of candidate molecular targets for therapeutic intervention.

3 Leukemic Initiating Cells or Leukemic Stem Cells

The adaptation of the available quantitative assays for normal human stem cells capable of repopulating haematopoiesis in vivo has allowed the identification of leukemic initiating cells. Transplantation of primary AML cells into NOD/SCID mice led to the finding that only rare cells, termed AML-initiating cells (AML-IC), are capable of initiating and sustaining growth of the leukemic clone in vivo, and serial transplantation experiments showed that AML-IC possess high self-renewal capacity, and thus can be considered to be the leukemic stem cells. By LSC, we refer to a cell that has self-renewal and differentiation potential and is able to reinitiate the leukemia when transplanted into NOD/SCID. This definition does not preclude the nature of the cells that is being transformed (i.e., normal HSC, progenitors or mature cells).

Importantly, AML-IC can be prospectively identified and purified as $CD34^{+}/CD38^{-}$cells in AML patient samples, regardless of the phenotype of the bulk blast population, and represented the only AML cells capable of self-renewal [4]. However, subsequently, considerable heterogeneity has been revealed. Using lentiviral gene marking to track the behavior of individual LSCs, following serial transplantation, has revealed heterogeneity in their ability to self-renew, similar to what is seen in the normal HSC compartment [10].

4 Identification/Isolation of the Leukemia-Initiating Cell

The adaptation of xenotransplantation assays to examine the propagation of AML in vivo has allowed the phenotypic identification of the AML-IC. As mentioned above the first demonstration of an in vivo-repopulating AML-IC involved assessment of the phenotype. In this study, it was demonstrated that the only cells capable of engrafting NOD/SCID mice possessed the $CD34^{+}/CD38^{-}$ immunophenotype. This phenotype has been extended via the use of immunodeficient mice to include an absence of CD71, HLA-DR, and CD117, but include expression of CD123 [11–14]. In a recent study, we have extended this phenotype further to include expression of CD33 and CD13 on AML-IC from the vast majority of patients [15]. Hence, the extended immunophenotype of the leukemic stem cell as defined by in vivo propagation is $CD34^{+}/CD38^{-}/CD71^{-}/HLA\text{-}DR^{-}/CD117^{-}/CD33^{+}/CD13^{+}/CD123^{+}$. However, the exclusivity of some markers is debatable. For instance, CD123 is indeed expressed on AML-IC, but is also expressed on the vast majority of AML blasts (unpublished observations) from most patients and hence, could be excluded from the above phenotype of AML-IC.

Apart from immunophenotype, other methods have been utilized to identify the AML-IC. For instance, we have recently published data on the ALDH-1 activity of AML-IC from various AML patients. It seems that in approximately one third of patients, at least a subset of AML-IC possess a high ALDH activity that is detectable using the fluorescent ALDH substrate mentioned previously [16].

5 The Cell of Origin in Leukemic Stem Cells

An important question for a more complete understanding of leukemogenesis is the cellular origin of the leukemic stem cell. Although candidate CSCs have been identified in many hematological as well as solid tumors, their normal cellular origins remains in most cases elusive [17, 18]. The importance of establishing this is not only to better understand how the functional properties (such as self-renewal potential) of the cellular targets of primary and subsequent transforming genetic events (which might occur at the same or distinct developmental and commitment stages) can impact on the biology of the disease, but also on therapeutic strategies. As will be discussed later, there are various pieces of logical and observational evidence to support a stem cell origin for leukemia, but surely the leukemogenic abnormality plays an important role.

The fundamental point is whether the leukemogenic event dictates the cell phenotype or whether the cell type in which this mutation occurs may influence the phenotype of the abnormality. One can envisage different situations where either or both the cell type and mutation may dictate the phenotype of the malignancy. The expression of a particular mutation could be controlled by a cell specific transcription factor. It is also possible that an abnormality only has the opportunity to occur in a certain cell type. Furthermore, it may be that a mutation is molecularly dependent on a cell-specific co-factor. Alternatively, if to cause leukemia, a mutation did not require a cell-specific co-factor or required a co-factor that was expressed in all cell types, then the cell type this abnormality occurred in would be less influential on the cell phenotype of the malignancy. One can imagine how this last "dependency" situation is particularly applicable to mutations that occur in HSC, but are only leukemogenic in certain cell types.

SPI-B is usually expressed in lymphocytes, but can interchange with PU.1 in myeloid development. So the role of PU.1/SPI-B in cells is interchangeable, but specificity of their expression is dependent on cell type [19]. The same may be true for fusion proteins – some may affect the same cell processes in multiple cell types and some may only affect certain cell types due to the lack of expression of co-factors in other cell types.

This question has been addressed in a more direct fashion through the use of retroviral-mediated gene transfer and cell fractionation. Cozzio et al. [20] have tested the leukemogenic potential of the MLL-ENL fusion protein in purified HSC, CMP, GMP and MEP cell populations. Transduction of HSC, CMP and GMP cells by MLL-ENL reportedly produces a similar leukemia as defined by phenotype and leukemogenic latency, whereas transduction of MEP cells did not confer any increase in proliferative or leukemogenic potential. Hence, in the case of MLL-ENL fusion protein, it seems that transformation of stem cells and myeloid progenitors is possible, but transformation of erythroid progenitors is not. Presumably, this is due to the selective expression of a factor that interacts with the MLL-ENL fusion protein in myeloid but not erythroid progenitors. In a similar study by So et al. [21], almost identical results were generated with the MLL-GAS7 fusion protein. HSC, CMP and GMP cell subsets were susceptible to MLL-GAS7–mediated transformation, whereas MEP cell populations could not be transformed. Interestingly, bi-phenotypic leukemia was generated when the MLL-GAS7 fusion protein when HSCs were transformed, whereas transduction of CMP and GMP populations resulted in myeloid leukemia only.

These results confirm that the cell type is important for the leukemogenic effect of certain fusion proteins. However, retroviral-mediated transduction of fusion proteins into various stages of hematopoiesis is a very artificial model, may not represent the leukemogenic event in vivo and hence the leukemia obtained may not behave in a similar fashion. For instance, the phenotype of leukemias generated from HSC, CMP and GMP cell populations is very uniform with no apparent hierarchy [20]. In contrast, there is much greater heterogeneity and hierarchy within human AML samples [4]. Related to this observation, the frequency of AML-IC within the leukemias generated by MLL-ENL fusions is extremely high [20]. Indeed, the transfer of 13 to 46 transduced cells into irradiated mice resulted in high-level engraftment of all five recipients. This suggests a frequency of AML-IC that is completely different to the $1/10^6$ to $1/10^7$ reported for human AML cell populations [22, 23].

A model that is possibly more analogous to the human AML situation is the Cre-loxP mediated translocations described by the group of Terry Rabbitts [24]. In this translocator mouse model, de novo translocations are generated in vivo in adult mice by Cre-mediated excision of LoxP sites. The expression of Cre can be regulated by various lineage-specific promoters to direct the excision of LoxP and hence translocation to a specific cell lineage. These Cre-loxP mediated translocation experiments confirm the ability of the MLL-ENL fusion protein to transform LMO2-expressing HSCs and reveal that the MLL-AF9 fusion protein can also transform HSCs when Cre is under the control of the LMO2 promoter. MLL-ENL translocations may also cause leukemia when the T-cell specific promoter, Lck, controls Cre expression. Interestingly, the MLL-AF9 fusion protein, which usually causes myeloid leukemia, cannot produce any malignancy when Cre is under the control of the Lck promoter [25]. These data seem to be consistent with the known incidence of either lymphoid or myeloid malignancies with each of these fusion proteins: MLL-ENL is associated with both ALL and AML, whereas MLL-AF9 is associated with myeloid leukemias only. These data provide an example that leukemogenesis depends of the cell type and the mutation as some fusion proteins may cause leukemia regardless of cell type (MLL-ENL), whereas some fusion proteins are dependent on cell type (MLL-AF9).

Apart from molecular considerations, there are other pieces of logical evidence that may give a clue as to the cellular origin of AML. As mentioned previously, the concept that AML is organized as a hierarchy, which is maintained by AML-IC, is distinct to the question of the cellular origin of AML. Direct analysis of the leukemogenic potential of certain fusion proteins has determined that it is possible to generate AML from either HSCs or myeloid progenitors [20, 21]. Whether this is possible in vivo and is how AML is generated in humans, still remains to be determined. The immunophenotype and morphology of the bulk of AML cells is very similar to various stages of haematopoietic progenitor development. However, the AML-IC reportedly has a phenotype that is very similar to a subset of non-malignant HSCs. The phenotype that is common between haematopoietic stem cells is: $CD34^{+}CD38^{-}HLA\text{-}DR^{-}CD71^{-}CD123^{+}$ $CD33^{+}$ $CD13^{+}$ [12, 14, 26, 27]. There are some antigens (Thy-1 and c-Kit) that are different between HSCs and AML-IC, but not all HSCs express these antigens [11, 13, 28].

This is indirect evidence that relies in the premise that the phenotype of the AML-IC does not change significantly during leukemogenesis. Some reported observations support this hypothesis. For instance, the immunophenotype of primitive hematopoietic cell does not change significantly when transformed with the AML-ETO fusion protein [29]. However, other studies have described profound alterations in immunophenotype upon MLL-ENL and MLL-GAS7 transformations of HSCs [20, 21]. One consideration however, is that only bulk immunophenotying has been reported, it may be that the AML-IC in these MLL1 generated leukemias has a different immunophenotype to the majority of the leukemia.

A major argument against a progenitor origin for AML is opportunity. The hematopoietic cell turnover is immense; most myeloid cells terminally differentiate and die by apoptosis within a few days. Whereas stem cells are the longest lived cells in the hematopoietic system and, hence, will be exposed to the most damage over time. This is consistent with the multi-hit hypothesis of leukemogenesis. Although certain fusion proteins may directly cause leukemia in mice, it is thought that translocations/inversions and insertions/deletions are the primary events that confers some advantage to mutated cells and provide an opportunity for secondary events and eventual progression to frank malignancy. Indeed, it has been reported that congenital defects are present in neonates and that these do not always progress to frank leukemia, indicating that a pre-leukemic state may exist [30].

Furthermore, there are more similarities between AML-IC and HSC than between AML-IC and progenitors. Hallmark features of both cell types are the ability to self-renew and proliferate extensively. The cellular properties of HSCs are very close to the behaviour of AML-IC and have less changes are required to transform an HSC into an AML-IC.

Another aspect of the multi-hit hypothesis is that the original hit may occur in one cell type, but the final leukemogenic event may occur in a more differentiated progeny of the original cell. In this situation, the first hit could predispose in which cell the second hit may be leukemogenic. For instance, if the first hit was a differentiation block that avoided apoptosis, then the second hit may well occur in the cell type where differentiation was blocked. Alternatively, if the first hit affected self-renewal, then the second hit may occur in HSCs and an HSC-type leukemia could be generated.

6 The Heterogeneity of Leukemia-Initiating Cells

Part of the confusion regarding the origin of the AML-IC may be due to the extreme heterogeneity of AML. Given the various possible routes to AML from a normal hematopoietic cell, it is not surprising that there is great heterogeneity in AML. Indeed, AML may be thought of as a large collection of different diseases that merely share a similar morphology. Indeed, the most effective risk stratification approach so far has been to examine the genetic abnormalities associated with a particular case of AML and compare to previous experience of AML cases with the same abnormality [31, 32]. Although cytogenetic analysis allows the definition of the hierarchical groups with favorable, intermediate, and poor prognosis, the intermediate risk group contains patients with variable outcomes. Assessing the prognosis of this large group of patients is currently difficult.

We have recently reported that the ability of a particular AML to engraft in the NOD/SCID model is related to the prognosis of individual AML cases. Specifically, examination of the follow-up of younger patients with intermediate-risk AML revealed a significant difference in overall survival between NOD/SCID-engrafting and nonengrafting cases. We could not detect a difference between engrafting and nonengrafting cases in various engraftment variables including: homing ability, AML-IC frequency, and immune rejection by the host or alternative tissue sources. Hence, the ability to engraft NOD/SCID recipients seems to be an inherent property of the cells that is directly related to prognosis.

It is extremely interesting to note that although the NOD/SCID model assesses AML independent of the response to chemotherapy, engraftment still correlates with the response to this treatment (prognosis group). A possible explanation is that NOD/SCID engraftment reflects the stem cell nature of each individual AML case. AML cases that engraft in the NOD/SCID assay at six weeks may represent diseases driven by potent leukemia-initiating cells with stem cell-like self-renewal and proliferation abilities whereas non-engrafting AML cases may involve less potent leukemia-initiating cells with more restricted progenitor-type self-renewal and proliferation abilities.

We are not suggesting that there are simply two groups of AML cases that are derived from two different cell types. It is far more likely that there is a great range of cell potential between different AML samples and a large spectrum of cells that each case originated in. The NOD/SCID assay may provide a tool to examine the potential of AML cells by providing a threshold. Below this threshold, AML cases that cannot self-renew or proliferate enough to provide detectable engraftment at six weeks have a favorable prognosis, whereas AML cases that can engraft at six weeks have a worse prognosis.

Alternatively, it may be that certain AML cases require a cytokine or cell–cell interaction that cannot be provided by the NOD/SCID microenvironment. Most cytokines cross-react, but some do not and a lack of a certain cytokine or cellular interaction may be important for the growth of particular AML cases.

7 Signaling Pathways Regulating the Self-Renewal of LSCs

When a hematopoietic stem cell has entered the cell cycle, two fates are possible: self-renewal and differentiation with associated proliferation. The control of self-renewal is under the control of various regulators. Deregulated self-renewal would involve an expansion of primitive cells and is probably very important in leukemogenesis. A detailed description of the molecular pathways responsible for the control of HSC self-renewal is not under the scope of this review. We will only summarize here some of the most important pathways implicated (see reviews in [33–35]).

The hedgehog, Wnt, and Notch pathways have been implicated in promoting normal stem cell self-renewal [36]. The molecular mechanisms by which all these factors interact to regulate stem and progenitor cell self-renewal remain to be elucidated.

The question of whether these signaling molecules play an important role in the regulation of LSCs remains mainly unknown. Nevertheless, Bmi-1 has been reported to play a key role in self-renewal determination in both normal and leukemic murine stem cells [37, 38]. Similarly, the role of Wnt pathway for survival, proliferation, and self-renewal of normal HSCs has raised the hypothesis that aberrant Wnt signaling might also contribute to the pathogenesis of AML/CML. Recently, aberrant activation of Wnt signaling and downstream effectors has been demonstrated in AML as well as in CML [39–42]. Thus, it seems that the pathways that regulate normal commitment/differentiation and self-renewal are not completely abolished in AML-IC.

8 Leukemic Stem Cell Niche

The hematopoietic microenvironment consists of bone cells, stromal cells, macrophages, fat cells, and extracellular matrix. Stem cells and their immediate progeny interact with the hematopoietic microenvironment. Stem/progenitor cells adhere to stromal cells via adhesion molecules. Cytokines and chemokines are also retained in the microenvironment; they bind to extracellular matrix and some are presented on the surface of stromal cells [43]. This co-adhesion of progenitors and cytokines to the microenvironment results in the co-localization of progenitors at a specific stage of differentiation with the appropriate array of cytokines, in niches [43].

Both cellular as well as extracellular matrix components of the stem cell microenvironment or niche are critical in stem cell regulation. By labelling stem cells and re-infusing them, it has been shown that HSCs reside in close proximity to the endosteal bone surface [44, 45]. Using a combinatorial Gata-2 and Sca-1 expression system, Suzuki et al. [46] confirmed these results using in vivo imaging.

The microenvironment has been described to influence survival, proliferation, and differentiation [44, 47]. However, the molecules that regulate stem cell niche interactions and how these may influence the balance between self-renewal and differentiation are just being to be revealed. Indeed, Arai et al. [48] show that the Angiopoietin-1/Tie-2 signaling pathway maintains repopulating quiescent HSC in the niche, presumably via the activation of cell adhesion molecules such as N-cadherin. Two recent studies demonstrate that another osteoblast secreted molecule, osteopontin (Opn), also play a key role in the attraction, retention and regulation of HSC [49, 50].

Although osteoblasts have been shown to have a central role in HSC regulation, other stromal and microenvironmental cell types and extracellular matrix proteins also contribute to this process. Hyaluronic acid as well as the membrane bound form of stem cell factor are also key components of this niche [51, 52]. More recently, it has been shown that calcium sensing receptor (CaR) has a function in retaining HSCs in close physical proximity to the endosteal surface and the regulatory niche components associated with it [53].

Overall, it appears that the regulation of hematopoiesis is the result of multiple processes involving cell–cell and cell–extracellular matrix interactions, the action of specific growth factors and others cytokines as well as intrinsic modulators of haematopoietic development. Currently, the questions whether leukemic stem cells depend on their niche for self-renewal remain unclear. Both normal and LSCs depend on SDF-1/CXCR4 axis for homing [54]. In an AML study, published recently, in vivo administration of anti-CD44 antibody to NOD/SCID mice transplanted with human AML significantly reduced leukemic engraftment. Absence of leukemia in serially transplanted mice demonstrated that AML stem cells have been targeted. The mechanisms underlying this eradication included interference with the homing of these LSC to their niche and alteration of LSC fate. This report thus suggests that LSC require interaction with a niche to maintain their stem cell properties [55]. It has also been well-documented especially for solid tumor that aberrations occur in the functional interactions between cancer cells and cells of the microenvironment, a scenario whereby the normal stem cell niche is replaced by the "tumor microenvironment" [56]. Thus, it is important to understand how cancer stem cells are regulated within the context of their natural tumor microenvironment. In the context of LSC it is unclear to what extend the stem cell niche is or not perturbed by leukemic burden but it is extremely worthwhile to explore this issue in future study.

9 Competition Between Normal and Leukemia Cells

Distinct to the influence of cytokines receptors on the growth of AML cells is the influence of AML cells on normal hematopoiesis. This could have an affect on both the pathogenesis and one of the major symptoms of the disease: bone marrow failure. It is clear that even at presentation of a severe AML, there are normal, nonmalignant HSCs present in the patient's marrow. However, one of the major symptoms of AML is bone marrow failure.

Indeed, the presence of primitive hematopoietic cells has been demonstrated from bone marrow samples of AML patients at presentation [57]. Furthermore, upon chemotherapy, the vast majority of patients recover their non-malignant cell counts to more normal levels. This suggests that HSCs are present in the marrows of AML patients, but these are somehow dormant.

A complex network of cytokines, adhesion molecules, and other cellular interactions controls normal hematopoiesis. The presence of a large "progenitor" population in the form of AML may provide a large negative feedback signal to developing HSCs. This signal could affect developing HSCs and progenitors directly or through changes to the hematopoietic microenvironment. In addition to the presence of a dormant HSC population in AML patients, negative feedback pressure may also explain the absolute lymphocytosis observed in many AML patients. This could be the result of a "lineage switch" to the lymphoid pathway induced by the presence of too many myeloid progenitors (in the form of AML).

If this negative feedback does occur then this may be very important for the pathogenesis of the disease. As mentioned previously, leukemia is thought to be generated by multiple genetic abnormalities in a multi-step process from a slightly abnormal HSC to frank leukemia. It has been reported that although in many people a population of cells with a certain genetic abnormality has been observed, not all go on to develop leukemia [58–60]. This suggests that a pre-leukemic state exists and that in some individuals, this pre-leukemia cell population is prevented from developing to leukemia. The first demonstration of the existence of these pre-leukemic stem cells has been recently reported in the case of TEL-AML1-associated childhood leukemia [61]. The mechanisms that allow the transition from pre-leukemic stage to a full leukemia still remain unclear but should be explored in the future.

10 Concluding Remarks

The insights into the molecular pathogenesis of AML have led recently to the development of more specific targeted agents and have ushered in an exciting new era of anti-leukemia therapy. Such agents include the immunoconjugate gemtuzumab ozogamicin (GO, anti-CD33), multidrug resistance inhibitors, farnesyl transferase inhibitors, histone deacetylase, proteosome inhibitors, Fms-like tyrosine kinase 3 (*FLT3*) inhibitors, and apoptosis inhibitors (for more details, see the review [62]). Nevertheless, none of these treatments have been tested on AML-ICs.

Existing therapies have been developed largely against the bulk blasts population. Thus, the lack of durable response in most cases, suggests that the treatment used can usually ablate the leukemic blasts but may not effective target the AML-IC population. Indeed, the failure of the current therapeutic regimens is likely related to the resistance and persistence of AML-ICs. Attention has turned to the self-renewal potential of AML-ICs as a therapeutic target. Several genes and pathways have been identified that are important for this critical property of self-renewal, including the WNT/β-catenin pathway, Notch, Bmi-1, and Hox family members. Although it is not yet certain how these pathways might be effectively targeted therapeutically, or whether it will be possible to target these pathways without excessive toxicities of the normal hematopoietic system, AML-ICs are ultimately the most important target cells to ablate. Furthermore, combinations of several agents, targeting more than one gene mutation or antigenic determinant, may hold greater promise.

As mentioned above, recent studies in solid tumors indicate that the concept of cancer as a hierarchy that is initiated and maintained by a rare population of stem cells may have broader implications beyond the field of hematopoiesis. The increasing evidence of rare cancer stem cells drive the formation of a number of different tumors types raises the question of whether all cancers originate from and are maintained by cancer stem cells.

Acknowledgments We apologize to authors whose work we may have omitted in an effort to focus on issues that we feel are important. We thank Rachael Swallow for her assistance in the preparation of this manuscript.

References

1. Ho AD. Kinetics and symmetry of divisions of hematopoietic stem cells. Exp Hematol. 2005;33:1–8.
2. Passegue E, Jamieson CH, Ailles LE, Weissman IL. Normal and leukemic hematopoiesis: are leukemias a stem cell disorder or a reacquisition of stem cell characteristics? Proc Natl Acad Sci U S A. 2003;100 Suppl 1:11842–9.
3. Reya T, Morrison SJ, Clarke MF, Weissman IL. Stem cells, cancer, and cancer stem cells. Nature. 2001;414:105–11.
4. Bonnet D, Dick JE. Human acute myeloid leukemia is organized as a hierarchy that originates from a primitive hematopoietic cell. Nat Med. 1997;3:730–7.
5. Lapidot T, Sirard C, Vormoor J, et al. A cell initiating human acute myeloid leukaemia after transplantation into SCID mice. Nature. 1994;367:645–8.
6. Singh SK, Hawkins C, Clarke ID, et al. Identification of human brain tumour initiating cells. Nature. 2004;432:396–401.
7. Al-Hajj M, Wicha MS, Benito-Hernandez A, Morrison SJ, Clarke MF. Prospective identification of tumorigenic breast cancer cells. Proc Natl Acad Sci U S A. 2003;100:3983–8.
8. Zhou S, Schuetz JD, Bunting KD, et al. The ABC transporter Bcrp1/ABCG2 is expressed in a wide variety of stem cells and is a molecular determinant of the side-population phenotype. Nat Med. 2001;7:1028–34.
9. Costello RT, Mallet F, Gaugler B, et al. Human acute myeloid leukemia CD34+/CD38- progenitor cells have decreased sensitivity to chemotherapy and Fas-induced apoptosis, reduced immunogenicity, and impaired dendritic cell transformation capacities. Cancer Res. 2000;60:4403–11.
10. Hope KJ, Jin L, Dick JE. Acute myeloid leukemia originates from a hierarchy of leukemic stem cell classes that differ in self-renewal capacity. Nat Immunol. 2004;5:738–43.
11. Blair A, Hogge DE, Ailles LE, Lansdorp PM, Sutherland HJ. Lack of expression of Thy-1 (CD90) on acute myeloid leukemia cells with long-term proliferative ability in vitro and in vivo. Blood. 1997;89:3104–12.
12. Blair A, Hogge DE, Sutherland HJ. Most acute myeloid leukemia progenitor cells with long-term proliferative ability in vitro and in vivo have the phenotype CD34(+)/CD71(-)/HLA-DR. Blood. 1998;92:4325–35.
13. Blair A, Sutherland HJ. Primitive acute myeloid leukemia cells with long-term proliferative ability in vitro and in vivo lack surface expression of c-kit (CD117). Exp Hematol. 2000;28:660–71.
14. Jordan CT, Upchurch D, Szilvassy SJ, et al. The interleukin-3 receptor alpha chain is a unique marker for human acute myelogenous leukemia stem cells. Leukemia. 2000;14:1777–84.
15. Taussig DC, Pearce DJ, Simpson C, et al. Hematopoietic stem cells express multiple myeloid markers: implications for the origin and targeted therapy of acute myeloid leukemia. Blood. 2005;106:4086–92.
16. Pearce DJ, Taussig D, Simpson C, et al. Characterization of cells with a high aldehyde dehydrogenase activity from cord blood and acute myeloid leukemia samples. Stem Cells. 2005;23:752–60.
17. Fialkow PJ, Denman AM, Jacobson RJ, Lowenthal MN. Chronic myelocytic leukemia. Origin of some lymphocytes from leukemic stem cells. J Clin Invest. 1978;62:815–23.
18. Nilsson L, Astrand-Grundstrom I, Arvidsson I, et al. Isolation and characterization of hematopoietic progenitor/stem cells in 5q-deleted myelodysplastic syndromes: evidence for involvement at the hematopoietic stem cell level. Blood. 2000;96:2012–21.
19. Rosenbauer F, Koschmieder S, Steidl U, Tenen DG. Effect of transcription-factor concentrations on leukemic stem cells. Blood. 2005;106:1519–24.
20. Cozzio A, Passegue E, Ayton PM, Karsunky H, Cleary ML, Weissman IL. Similar MLL-associated leukemias arising from self-renewing stem cells and short-lived myeloid progenitors. Genes Dev. 2003;17:3029–35.
21. So CW, Karsunky H, Passegue E, Cozzio A, Weissman IL, Cleary ML. MLL-GAS7 transforms multipotent hematopoietic progenitors and induces mixed lineage leukemias in mice. Cancer Cell. 2003;3:161–71.
22. Pearce DJ, Taussig D, Zibara K, et al. AML engraftment in the NOD/SCID assay reflects the outcome of AML: implications for our understanding of the heterogeneity of AML. Blood. 2006;107:1166–73.
23. Rombouts WJ, Martens AC, Ploemacher RE. Identification of variables determining the engraftment potential of human acute myeloid leukemia in the immunodeficient NOD/SCID human chimera model. Leukemia. 2000;14:889–97.
24. Rabbitts TH, Appert A, Chung G, et al. Mouse models of human chromosomal translocations and approaches to cancer therapy. Blood Cells Mol Dis. 2001;27:249–59.
25. Drynan LF, Pannell R, Forster A, et al. Mll fusions generated by Cre-loxP-mediated de novo translocations can induce lineage reassignment in tumorigenesis. EMBO J. 2005;24:3136–46.
26. Bhatia M, Wang JC, Kapp U, Bonnet D, Dick JE. Purification of primitive human hematopoietic cells capable of repopulating immune-deficient mice. Proc Natl Acad Sci U S A. 1997;94(10):5320–5.
27. Taussig DC, Pearce DJ, Simpson C, et al. Hematopoietic stem cells express multiple myeloid markers: implications for the origin and targeted therapy of acute myeloid leukemia. Blood. 2005;106:4086–92.
28. Bhatia M, Bonnet D, Murdoch B, Gan OI, Dick JE. A newly discovered class of human hematopoietic cells with SCID-repopulating activity. Nat Med. 1998;4:1038–45.
29. Delaney C, Bernstein ID. Establishment of a pluripotent preleukaemic stem cell line by expression of the AML1-ETO fusion protein in Notch1-immortalized HSCN1cl10 cells. Br J Haematol. 2004;125:353–7.
30. Greaves MF, Maia AT, Wiemels JL, Ford AM. Leukemia in twins: lessons in natural history. Blood. 2003;102:2321–33.
31. Grimwade D, Walker H, Harrison G, et al. The predictive value of hierarchical cytogenetic classification in older adults with acute myeloid leukemia (AML): analysis of 1065 patients entered into the United Kingdom Medical Research Council AML11 trial. Blood. 2001;98:1312–20.
32. Grimwade D, Walker H, Oliver F, et al. The importance of diagnostic cytogenetics on outcome in AML: analysis of 1,612 patients entered into the MRC AML 10 trial. The Medical Research Council Adult and Children's Leukaemia Working Parties. Blood. 1998;92:2322–33.
33. Blank U, Karlsson G, Karlsson S. Signaling pathways governing stem-cell fate. Blood. 2008;111:492–503.
34. Huang X, Cho S, Spangrude GJ. Hematopoietic stem cells: generation and self-renewal. Cell Death Differ. 2007;14:1851–9.
35. Nemeth MJ, Bodine DM. Regulation of hematopoiesis and the hematopoietic stem cell niche by Wnt signaling pathways. Cell Res. 2007;17:746–58.
36. Yilmaz OH, Kiel MJ, Morrison SJ. SLAM family markers are conserved among hematopoietic stem cells from old and reconstituted mice and markedly increase their purity. Blood. 2006;107:924–30.
37. Lessard J, Sauvageau G. Bmi-1 determines the proliferative capacity of normal and leukaemic stem cells. Nature. 2003;423:255–60.
38. Park IK, Qian D, Kiel M, et al. Bmi-1 is required for maintenance of adult self-renewing haematopoietic stem cells. Nature. 2003;423:302–5.

39. Jamieson CH, Weissman IL, Passegue E. Chronic versus acute myelogenous leukemia: a question of self-renewal. Cancer Cell. 2004;6:531–3.
40. Mikesch JH, Steffen B, Berdel WE, Serve H, Muller-Tidow C. The emerging role of Wnt signaling in the pathogenesis of acute myeloid leukemia. Leukemia. 2007;21:1638–47.
41. Muller-Tidow C, Steffen B, Cauvet T, et al. Translocation products in acute myeloid leukemia activate the Wnt signaling pathway in hematopoietic cells. Mol Cell Biol. 2004;24:2890–904.
42. Zhao C, Blum J, Chen A, et al. Loss of beta-catenin impairs the renewal of normal and CML stem cells in vivo. Cancer Cell. 2007;12:528–41.
43. Verfaillie CM. Adhesion receptors as regulators of the hematopoietic process. Blood. 1998;92:2609–12.
44. Haylock DN, Nilsson SK. Stem cell regulation by the hematopoietic stem cell niche. Cell Cycle. 2005;4:1353–5.
45. Potocnik AJ, Brakebusch C, Fassler R. Fetal and adult hematopoietic stem cells require beta1 integrin function for colonizing fetal liver, spleen, and bone marrow. Immunity. 2000;12:653–63.
46. Suzuki N, Ohneda O, Minegishi N, et al. Combinatorial Gata2 and Sca1 expression defines hematopoietic stem cells in the bone marrow niche. Proc Natl Acad Sci U S A. 2006;103:2202–7.
47. Rafii S, Mohle R, Shapiro F, Frey BM, Moore MA. Regulation of hematopoiesis by microvascular endothelium. Leuk Lymphoma. 1997;27:375–86.
48. Arai F, Hirao A, Ohmura M, et al. Tie2/angiopoietin-1 signaling regulates hematopoietic stem cell quiescence in the bone marrow niche. Cell. 2004;118:149–61.
49. Nilsson SK, Johnston HM, Whitty GA, et al. Osteopontin, a key component of the hematopoietic stem cell niche and regulator of primitive hematopoietic progenitor cells. Blood. 2005;106:1232–9.
50. Stier S, Ko Y, Forkert R, et al. Osteopontin is a hematopoietic stem cell niche component that negatively regulates stem cell pool size. J Exp Med. 2005;201:1781–91.
51. Driessen RL, Johnston HM, Nilsson SK. Membrane-bound stem cell factor is a key regulator in the initial lodgment of stem cells within the endosteal marrow region. Exp Hematol. 2003;31:1284–91.
52. Nilsson SK, Haylock DN, Johnston HM, Occhiodoro T, Brown TJ, Simmons PJ. Hyaluronan is synthesized by primitive hemopoietic cells, participates in their lodgment at the endosteum following transplantation, and is involved in the regulation of their proliferation and differentiation in vitro. Blood. 2003;101:856–62.
53. Adams GB, Chabner KT, Alley IR, et al. Stem cell engraftment at the endosteal niche is specified by the calcium-sensing receptor. Nature. 2006;439:599–603.
54. Lapidot T, Kollet O. The essential roles of the chemokine SDF-1 and its receptor CXCR4 in human stem cell homing and repopulation of transplanted immune-deficient NOD/SCID and NOD/SCID/B2m(null) mice. Leukemia. 2002;16:1992–2003.
55. Jin L, Hope KJ, Zhai Q, Smadja-Joffe F, Dick JE. Targeting of CD44 eradicates human acute myeloid leukemic stem cells. Nat Med. 2006;12:1167–74.
56. Clarke MF, Fuller M. Stem cells and cancer: two faces of eve. Cell. 2006;124:1111–5.
57. Guan Y, Ralph S, Hogge DE. Polyclonal normal hematopoietic progenitors in patients with acute myeloid leukemia. Exp Hematol. 2002;30:721–8.
58. Kim-Rouille MH, MacGregor A, Wiedemann LM, Greaves MF, Navarrete C. MLL-AF4 gene fusions in normal newborns. Blood. 1999;93:1107–8.
59. Wiemels JL, Ford AM, Van Wering ER, Postma A, Greaves M. Protracted and variable latency of acute lymphoblastic leukemia after TEL-AML1 gene fusion in utero. Blood. 1999;94:1057–62.
60. Wiemels JL, Xiao Z, Buffler PA, et al. In utero origin of t(8;21) AML1-ETO translocations in childhood acute myeloid leukemia. Blood. 2002;99:3801–5.
61. Hong D, Gupta R, Ancliff P, et al. Initiating and cancer-propagating cells in TEL-AML1-associated childhood leukemia. Science. 2008;319:336–9.
62. Tallman MS, Gilliland DG, Rowe JM. Drug therapy for acute myeloid leukemia. Blood. 2005;106:1154–63.

Solid Tumor Stem Cells – Implications for Cancer Therapy

Tobias Schatton, Natasha Y. Frank and Markus H. Frank

Abstract Cancer stem cells (CSC) that drive tumor initiation and growth through self-renewal and differentiation have been identified in cancers of the hematopoietic lineage and certain solid tumors. Recent findings point to a specific relationship of such tumorigenic minority populations to therapeutic resistance and neoplastic progression. Furthermore, initial proof-of-concept has been established that specific targeting of CSC for cell killing is sufficient to halt experimental tumor growth. These findings indicate that such novel strategies could be suited to improve conventional cytotoxic therapies, which are believed to spare refractory tumor initiators responsible for recurrence and metastasis. Here, we review the currently available scientific evidence in support of the CSC model of tumor development, and discuss the implications of this emerging concept for tumor biology and cancer therapy.

Keywords Cancer stem cells · Tumor-initiating cells · Self-renewal · Differentiation · Tumorigenicity · Cancer progression · Chemoresistance · Targeting · Tumor therapy · ABCB5

1 Introduction

The prevailing stochastic model of cancer growth postulates that all cells within a tumor have an equal capacity for tumor initiation and neoplastic progression. According to this hypothesis, the likelihood of this event at the single cell level is low [1] (Fig. 1A). In contrast, the more recently formulated cancer stem cell (CSC) theory suggests that only a subpopulation of cells within a tumor, the CSC, can proliferate extensively and drive tumorigenic growth [2]. This theory further postulates that CSC give rise to the morphologically diverse malignant cell populations a tumor consists of, including non-tumorigenic tumor bulk components (Fig. 1B).

On the DNA level, tumor development and progression have been associated with cumulative alterations in oncogenes, tumor suppressor genes, and repair/stability genes [3, 4]. On the cellular level, it has been recognized for some time that human cancers consist of phenotypically heterogeneous cell populations [5, 6]. In addition, tumor cells show functional heterogeneity, as evidenced, for example, by differing growth characteristics in proliferation assays in vitro [7] or varying potencies in initiating and sustaining tumorigenic growth in vivo [8]. Ongoing genetic mutations alone can only in part explain this morphological and functional

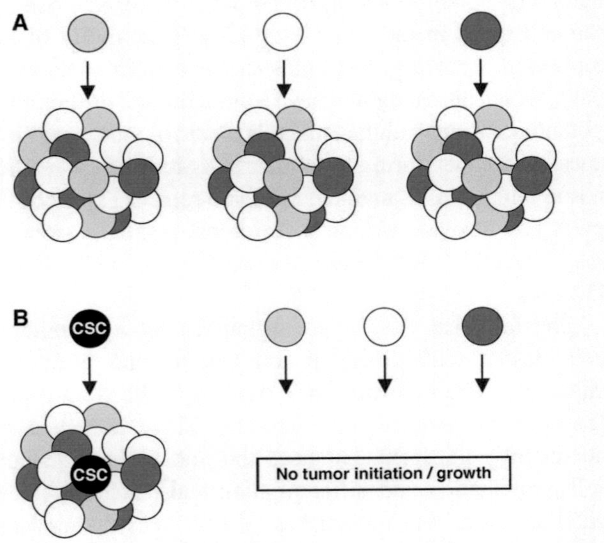

Fig. 1 The stochastic model versus the cancer stem cell (CSC) model of tumor initiation and growth. **(A)** The stochastic model predicts that all cells within a heterogeneous tumor have an equal capacity for tumor formation and neoplastic progression. However, the likelihood of this event at the single cell level is low. **(B)** The CSC model predicts that only a subpopulation within a heterogeneous tumor, the CSC, can proliferate extensively and drive tumor growth. According to this model, the CSC are the cellular source of hierarchically organized and phenotypically diverse tumor cell populations

M.H. Frank (✉)
Transplantation Research Center, Children's Hospital Boston & Brigham and Women's Hospital, Harvard Medical School, 300 Longwood Avenue, Boston, MA 02115, USA.
e-mail: mfrank@rics.bwh.harvard.edu

V.K. Rajasekhar, M.C. Vemuri (eds.), *Regulatory Networks in Stem Cells*, Stem Cell Biology and Regenerative Medicine,
DOI 10.1007/978-1-60327-227-8_40, © Humana Press, a part of Springer Science+Business Media, LLC 2009

heterogeneity, a conclusion that has led to the development of the CSC model of cancer initiation and growth.

In this chapter, we review recent findings that provide increasing evidence in support of the presence and involvement of CSC in the initiation and propagation of many cancers, with a particular emphasis on solid tumors. We analyze these findings in the context of signaling pathways and functional properties ascribed to physiological stem cells. In addition, we will discuss the conceptual implications of the CSC theory for tumor biology and its implications for the development of targeted therapies.

2 Physiological Stem Cells

Physiological stem cells are immature, unspecialized cells that are defined by two distinct properties: firstly, by their ability to create endless copies of themselves through self-renewal, and secondly, by their potential to give rise to more mature, transiently amplifying cell progeny that eventually generates the specialized cells of a particular tissue through a process termed differentiation [2]. In addition, stem cells have the capacity to proliferate extensively [2], for example, in response to injury or during development [9].

There are different kinds of physiological stem cells, including embryonic stem cells (ESC) and adult or somatic stem cells. ESC are derived from the inner cell mass of an early stage embryo and have the potential to give rise to all the cell types in the adult body [10]. This ability of ESC, termed pluripotency, distinguishes them from somatic stem cells, which are thought to have a more limited differentiation potential. Normal adult stem cells reside in most somatic tissues, where they form the cellular basis for lineage-restricted tissue homeostasis, maintenance, and repair [11]. Recent reports now provide evidence that some postnatal stem cells may in addition bear multi-lineage differentiation plasticity [12–14].

Physiological tissues are organized as a hierarchy of cell classes with divergent self-renewal and proliferative capacities. The relatively rare, tissue-specific stem cells are found at the apex of this hierarchy. These immature cells are mainly quiescent, but have the potential for indefinite self-perpetuation and differentiation and hence provide the cellular source that replenishes all cell types that constitute the tissue. Progenitor cells, on the other hand, can proliferate extensively albeit for a limited number of cell divisions. The third and major compartment of physiological tissues is comprised of terminally differentiated cells, which have a more limited lifespan and ultimately undergo apoptosis [9].

Physiological stem cells tend to be more resistant to cytotoxic agents than their differentiated cell progeny, which may relate to their relative quiescence [15], the activation of anti-apoptotic mechanisms [16], or their high expression levels of ABC transporters [17]. The latter class of molecules has in addition been associated with the ability of stem cell populations to efflux the Hoechst dye [17], a feature that allows their isolation using the flow cytometry-based side population (SP) technique [18].

Most importantly, distinct signaling pathways involving BMP (bone morphogenetic proteins), Wnt, Notch, or Hedgehog molecules [19] have been shown to contribute to the self-renewal capacity of physiological stem cells. The interplay between these pathways and the integration of signals from the surrounding microenvironment [20] provide a crucial regulatory balance between self-renewal and differentiation [9]. Tumors demonstrate a loss of such homeostatic mechanisms [21, 22] and, thus, carcinogenesis can be viewed as a dysregulation of self-renewal.

3 Cancer Stem Cells – Definition and Initial Discovery in Tumors of the Hematopoietic Lineage

Striking parallels have been acknowledged for some time between cancer and physiological stem cell development. For instance, many signaling pathways that were first described due to their role in tumor progression were subsequently also found to regulate important functions in stem cell biology, including the regulation of self-renewal [23]. Moreover, both tumorigenic cancer subpopulations and physiological stem cells converge in their capacities to proliferate extensively [5]. In addition, normal tissues and tumors are comprised of morphologically heterogeneous cell populations [6] with varying clonogenic potentials [7], and many of the characteristics of normal stem cells, including stem-cell plasticity, may also be relevant to tumor growth [24].

Taken together, these findings have led to the recent development of the CSC theory, which postulates that only a fraction of the cells in a tumor, the CSC, are essential for its propagation [2]. CSC are defined as a human clinical cancer-derived tumor minority population, prospectively identifiable by a molecular marker or marker combination, which, unlike cancer bulk populations negative for the particular marker or marker combination, can initiate tumor formation and growth in serial xenotransplantation experiments in vivo. Furthermore, CSC possess the exclusive capacity for self-renewal and at the same time give rise to nontumorigenic cancer bulk populations, thereby recapitulating the heterogeneity of the clinical parent tumor [2]. It is important to note that the CSC model does not make any specific assumptions about the relative frequency of tumorigenic CSC populations [2, 25, 26] (Table 1).

Table 1 Defining characteristics of cancer stem cells

1. Increased in vivo tumorigenicity.

Human clinical cancer-derived CSC subsets, prospectively identifiable by a molecular marker of marker combination, unlike tumor bulk components negative for the particular marker or marker combination, can efficiently initiate tumor formation and growth in serial xenotransplantation experiments in vivo.

2. Extensive in vivo self-renewal capacity.

CSC possess the exclusive capacity for extensive self-renewal in serial xenotransplantation experiments in vivo.

3. In vivo differentiation capacity.

CSC can differentiate into phenotypically diverse cancer populations, including nontumorigenic tumor cell subsets, thereby recapitulating the cellular heterogeneity of the original patient tumor in serial xenotransplantation experiments in vivo.

The CSC definition does not make any specific assumptions about the relationship of CSC to physiological stem cells as a potential cell of origin for the cancer or about any observed multipotent transdifferentiation plasticity of such cell subsets. Therefore, investigators in the stem cell field often use the term "tumor-initiating cells" instead. Furthermore, the identification of CSC depends on the prospective purification of tumor-initiating cells, that self-renew and can create a phenocopy of the original patient tumor, irrespective of their frequency.

CSC fulfilling these defining criteria were first identified by the pioneering work of John Dick and others in tumors of the hematopoietic lineage [27, 28]. In 1994, Lapidot et al. [27] isolated $CD34^{+}CD38^{-}$ hematopoietic stem-cell phenotype-expressing tumor cells from human acute myeloid leukemia (AML) patient samples and examined them for their capacity to initiate leukemias in severe combined immunodeficient (SCID) mice. The transplanted cells homed to the recipients' bone marrow where they proliferated extensively, resulting in a repopulation and dissemination pattern similar to that of the parent tumor. $CD34^{+}CD38^{+}$ and $CD34^{-}$ subpopulations failed to initiate leukemias in SCID mice, indicating that the CSC properties were contained exclusively within the $CD34^{+}CD38^{-}$ human AML cell fraction. However, leukemias could not be serially passaged to form secondary neoplasms in this model [27]. Using sublethally irradiated non-obese diabetic mice with severe combined immunodeficiency disease (NOD/SCID), Bonnet et al. [28] confirmed the self-renewal capacity of the $CD34^{+}CD38^{-}$ leukemia initiating cell fraction by isolating these cells from primary and secondary NOD/SCID recipients and injecting them into secondary and tertiary animals, respectively. In line with the earlier finding by Lapidot and colleagues, the leukemia-initiating cell fraction was consistently and exclusively contained within the $CD34^{+}CD38^{-}$subset of cells, which made up 0.02–2% of the total AML population. As few as 5×10^{3} leukemia-initiating cells were found sufficient to initiate and propagate the disease, regenerating the cellular heterogeneity of the original patient tumor in the NOD/SCID model, whereas 5×10^{5} $CD34^{+}CD38^{+}$ AML cells failed to engraft or proliferate. Thus, the study by Bonnet et al. demonstrated that leukemic clones are organized as a hierarchy, whereby leukemia-initiating or stem cells have the exclusive capacity for indefinite self-renewal and differentiation [28].

4 Cancer Stem Cells in Solid Tumors

By applying similar experimental in vivo model systems as the ones established for the identification of leukemic stem cells (LSC), more recent reports have pointed to the existence of CSC in solid tumors, including cancers of the breast [29], medulloblastoma and glioblastoma tumors of the brain [30], colonic cancer [31–33], pancreatic cancer [34, 35], head and neck squamous cell carcinoma (HNSCC) [36], and malignant melanoma [26] (Table 2).

4.1 Cancer Stem Cells in Breast Cancer

A defined population of *solid* tumor stem cells, that could self-renew, differentiate, and had the capacity to proliferative extensively, was first identified by Al-Hajj et al. [29] in cancers of the breast. These tumorigenic human cancer cells, prospectively isolated on the basis of a $CD44^{+}CD24^{-/low}Lineage^{-}$ phenotype, were found to possess the exclusive capacity to form xenograft tumors upon injection into the mammary fat pads of NOD/SCID

Table 2 Cell surface phenotype of cancer stem cells identified in solid tumors

Tumor type	Cell surface marker(s)	Reference
Breast	$CD44^{+}CD24^{-/low}Lineage^{-}ESA^{+}$	[29]
CNS	$CD133^{+}$	[30]
Colon	$CD133^{+}$	[31]
	$CD133^{+}$	[32]
	$ESA^{high}CD44^{+}Lineage^{-}(CD166^{+})$	[33]
HNSCC	$CD44^{+}Lineage^{-}$	[36]
Melanoma	$ABCB5^{+}$	[26]
Pancreas	$CD44^{+}CD24^{+}ESA^{+}$	[35]
	$CD133^{+}$	[34]

recipient mice. Whereas as few as 1×10^2 cells of this phenotype, which were further purified by virtue of their expression of the epithelial specific antigen (ESA) surface marker, were found capable of initiating tumors, as many as 1×10^5 breast cancer cells with alternative phenotypes failed to form tumors in vivo. The breast cancer-initiating subpopulation could be serially passaged into secondary, tertiary, and quaternary mice in these studies, generating new tumors that contained additional $CD44^{+}CD24^{-/low}ESA^{+}Lineage^{-}$tumorigenic cells (2.5–5% of cancer cells), as well as the phenotypically diverse mixed populations of nontumorigenic cells present in the initial tumor. Interestingly, tumorigenic and nontumorigenic breast cancer cells exhibited similar cell cycle distribution patterns, indicating that tumor bulk populations might contribute to tumor growth in the presence of CSC [29].

4.2 Cancer Stem Cells in Tumors of the Brain

In human brain cancer (medulloblastomas and glioblastomas), CSC capable of self-renewal and differentiation were shown to be exclusively contained within tumor cell subsets characterized by expression of the CD133 (Prominin-1/AC133) cell surface antigen [30], which had previously also served as a marker for the isolation of physiological central nervous system (CNS) stem cells [37]. Using the so-called sphere formation approach, which was originally established for physiological stem cells of the brain that selectively grow under these culture conditions whereas differentiated cells rapidly die in response to mitogens present in the medium [38], Singh et al. [39] initially demonstrated that only the CD133-expressing subpopulation of brain tumor cells had the ability to proliferate and form neurospheres in vitro, while their $CD133^{-}$ counterparts failed to proliferate and became adherent. In addition, they demonstrated that enhanced CD133 expression levels correlated with higher tumor grade. Thereafter, Hemmati et al. [40] showed that such cancerous neurospheres could engraft and proliferate in neonatal rat brains. Using an immunodeficient mouse model, Galli and colleagues [41] extended these results by demonstrating that human glioblastoma spheres could be serially passaged in vivo. Subsequently, marker-defined brain tumor-initiating populations were identified on the basis of xenotransplantation experiments with CD133-sorted cancer cell subsets [30]. Intracranial injection of as few as 1×10^2 $CD133^{+}$ human glioblastoma cells from two distinct patients to NOD/SCID mice consistently induced tumor formation, whereas up to 1×10^5 $CD133^{-}$ tumor cells were found to lack tumorigenicity. Tumorigenic $CD133^{+}$ brain tumor cells, which made up 6–29% of the total cancer population, unlike bulk populations, could be serially transplanted into secondary recipients after sorting cells into CSC-enriched populations, and histological analysis of $CD133^{+}$ tumor cell-derived xenografts revealed phenotypic heterogeneity similar to that observed in the respective patient tumors [30]. A subsequent study also found the increased in vivo tumorigenicity of CD133-expressing tumor cell subsets for human ependymomas [42].

4.3 Cancer Stem Cells in Colon Cancer

The CD133 surface marker was also successfully used in two separate studies in order to enrich for highly tumorigenic and self-renewing CSC in human colorectal cancers [31, 32]. Performing limiting dilution xenotransplantation experiments with CD133-sorted cancer populations in immunodeficient mice, both groups independently demonstrated that the capacity to initiate tumor growth was confined to the $CD133^{+}$ colon cancer cell subset. The percent positivity of tumor cells identified by the CD133 marker ranged from 2 to 25% [31] and 1to 6% [32], and as few as 1×10^2 [31] and 1.5×10^3 $CD133^{+}$ human cancer cells [32] were found capable of inducing tumor formation, while 1×10^5 $CD133^{-}$ cancer cells failed to do so, respectively. However, in one study, 1 of 9 mice injected with 2.5×10^5 $CD133^{-}$ tumor cells developed a tumor [31], indicating that cancer bulk components might possess a limited capacity to proliferate. Alternatively, the observed tumorigenic growth could be explained by $CD133^{+}$ contaminants in this cell inoculum. Colon cancer initiating-cells within the CD133-expressing population were able to self-renew as well as differentiate into tumor bulk populations and to initiate tumor growth and re-establish the original patient tumor heterogeneity in serially transplanted secondary, tertiary [31], and quaternary mice [32]. In addition, Ricci-Vitiani et al. [32] showed that $CD133^{+}$ colon cancer cells, unlike $CD133^{-}$ cells, could be maintained in vitro as colon tumor spheres for more than one year, without loosing their in vivo tumorigenic potential.

In a separate study, Dalerba et al. identified a subset of tumorigenic human colon cancer cells based on an epithelial cell adhesion molecule (EpCAM, also known as $ESA)^{high}CD44^{+}Lineage^{-}$ phenotype [33]. The frequency of tumor cells expressing this marker combination ranged from 0.03 to 38% of total live tumor cells. In xenotransplantation experiments to NOD/SCID mice as few as 2×10^2 $ESA^{high}CD44^{+}Lineage^{-}$ cells were capable of forming xenografts that mirrored the morphologic and phenotypic heterogeneity of the respective patient tumors, while 1×10^5 $ESA^{low}CD44^{-}Lineage^{-}$ cells consistently failed to initiate tumor development. However, inoculation of 2×10^5 ESA^{low}CD44$^-$Lineage$^-$ colon cancer cells resulted in tumorigenic growth in one recipient mouse and *serial* xenotransplantation experiments with sorted tumor cell populations for assessment of long-term self-renewal were not performed in this study [33]. In contrast to the findings identifying CD133 as a marker of colon cancer-initiating cells [31, 32], Dalerba et al. described tumors scoring homogeneously negative for CD133 that were capable of forming tumors in NOD/SCID recipients. Some colon cancers, however, displayed high levels of CD133-positivity with only a minority of this cell population coexpressing the CD44 antigen. Interestingly, characterization of the co-surface marker repertoire of ESA^{high}CD44$^-$Lineage$^-$ cells identified the mesenchymal stem cell marker CD166, which was previously also associated with clinical cancer progression [43, 44] and stem-like cancer cells in human malignant melanoma [45], as a candidate marker for the enrichment of colon cancer-initiating cells. Results from initial screening experiments ($n = 2$ mice per experimental group) indicated that tumorigenic cells might be restricted to the ESA^{high}CD44$^+$CD166$^+$Lineage$^-$ colon cancer population [33].

4.4 Cancer Stem Cells in Pancreatic Cancer

In human pancreatic adenocarcinomas, highly tumorigenic cancer cell subsets were identified based on the CD44$^+$CD24$^+$ESA$^+$ marker expression (0.2–0.8% of pancreatic cancer cells) [35], contrary to earlier findings in breast cancer whose CSC subset was described as CD24$^{-/low}$ [29]. Six of 12 NOD/SCID recipients of 1×10^2 CD44$^+$CD24$^+$ESA$^+$ pancreatic cancer cells developed tumor xenografts phenotypically indistinguishable from the original patient's tumor, compared to only 1 of 12 mice that were injected with 1×10^4 tumor cells negative for these markers [35]. No differences in cell cycle distribution could be detected between the respective populations, however, increased expression of the developmental signaling molecule Sonic hedgehog (Shh) was detected by real-time quantitative RT-PCR in CD44$^+$CD24$^+$ESA$^+$ cells compared to tumor bulk components or normal pancreatic epithelium. The latter findings are consistent with the notion that aberrant activation of self-renewal-associated pathways in CSC may at least in part account for their tumorigenic capacity. Results of serial xenotransplantation experiments with sorted pancreatic cell subsets for confirmation of long-term in vivo self-renewal of the tumorigenic subfraction were not described in this study [35].

In a subsequent study, Hermann and colleagues [34] prospectively isolated pancreatic tumor-initiating cells characterized by expression of the CD133 stem cell marker. As already described for tumors of the brain [30] and colon cancer [31, 32], pancreatic cancer cells defined by CD133 expression were exclusively tumorigenic in primary, secondary, and tertiary xenotransplantation experiments, while up to 1×10^6 patient-derived CD133$^-$ pancreatic cancer cells consistently failed to induce tumor formation upon injection into immunocompromised mice [34]. Histologically, CD133$^+$ cell-derived tumor xenografts consistently reproduced the parent tumor heterogeneity. Of note, treatment of pancreatic cell cultures with the chemotherapeutic gemcitabine revealed increased drug-resistance of CD133$^+$ cells compared to their negative counterparts as determined by cell cycle analysis. In addition, increased expression levels of CD133 could be detected in tumor xenografts of gemcitabine-treated compared to vehicle-treated mice. In clinical pancreatic adenocarcinoma specimen, CD133$^+$ cancer cells coexpressing the C-X-C chemokine receptor type 4 (CXCR4) were detected at the invasive front, while CD133$^+$ cells within the tumor bulk were mostly negative for the marker. Most notably, the CXCR4/stromal cell–derived factor 1 (SDF-1) signaling system is known to direct the migration of HSC [46] and the dissemination of metastatic tumor subpopulations [47]. Using a pancreatic tumor cell line, Hermann et al. subsequently compared the ability of CD133$^+$CXCR4$^+$ versus CD133$^+$CXCR4$^-$ cancer cells to metastasize. Following tumor development, circulating CD133$^+$CXCR4$^+$ cancer cells and liver metastases upon splenectomy could only be detected in the CXCR4$^+$ group. Continuous pharmacological inhibition of the CXCR4 receptor significantly reduced the detectable number of migrating cancer cells [34]. These findings point to the possibility that CSC, in addition to their tumor-initiating capacity, may also drive metastatic progression upon acquiring migratory features.

4.5 Cancer Stem Cells in Head and Neck Squamous Cell Carcinoma

In human head and neck squamous cell carcinoma (HNSCC), a subpopulation of cells with CSC properties could be prospectively isolated based on a CD44$^+$Lineage$^-$ phenotype [36]. In this study, 20 of 31 NOD/SCID or Rag2$\gamma^{-/-}$ mice injected with CD44$^+$Lineage$^-$ patient-derived HNSCC cells formed tumors compared to only 1 of 40 mice injected with CD44$^-$ cancer cells. CD44$^+$Lineage$^-$ cells could be serially passaged to form secondary and tertiary tumors, whereas CD44$^-$Lineage$^-$ cells isolated from passaged tumor xenografts didn't initiate tumors. The CD44$^+$ HNSCC cell-derived xenografts showed the original morphologic and phenotypic heterogeneity of the respective patient samples upon histologic examination. Moreover, CD44 expression in HNSCC tumors was co-localized with cytokeratin 5/14 (CK5/14) [36], a marker of normal squamous epithelial stem and progenitor cells [48]. In addition, the stem cell-related gene BMI-1 [49, 50] was expressed at high levels both on the RNA and protein level in the $CD44^+$ subset isolated from three distinct HNSCC patient samples as compared to their $CD44^-$ counterparts or human ESC [36]. Because BMI-1 has also been implicated in LSC proliferation and tumor development [51], the findings by Prince et al. [36] provide initial experimental support for the hypothesis that deregulation of signaling pathways associated with stem cell self-renewal may contribute to the enhanced tumorigenicity of CSC in solid cancers.

4.6 *Cancer Stem Cells in Human Malignant Melanoma*

In human malignant melanomas, CSC capable of self-renewal and differentiation, which are responsible for tumor growth, were prospectively identified recently by our group based on their expression of the chemoresistance mediator [45] ABCB5 [26]. $ABCB5^+$ tumor cells, the frequency of which ranged from 2 to 20% in single cell suspensions prepared from melanoma biopsies, displayed a primitive molecular phenotype and correlated with clinical melanoma progression as determined by immunohistochemical analysis of a progression tissue microarray consisting of 480 distinct patient samples. Subcutaneous injection of $ABCB5^+$ human melanoma cells across a log-fold range to NOD/SCID mice induced tumor xenograft formation in 14 of 23 recipients including all recipients of the highest cell dose, whereas only 1 of 23 mice injected with $ABCB5^-$ tumor cells developed a cancer. Tumorigenic $ABCB5^+$ melanoma cells, unlike bulk populations, could be serially transplanted into secondary recipients after sorting cells into CSC-enriched populations, and histological analysis of $ABCB5^+$ tumor cell-derived xenografts revealed phenotypic heterogeneity similar to that observed in their respective parent tumors [26]. An association of ABCB5 expression to tumorigenesis is supported by findings that revealed the molecule to be closely co-regulated with melanoma tumor antigen p97 (melanotransferrin [MTf]), a known regulator of tumor growth [52].

In an attempt to examine the relative tumor growth contributions of co-injected $ABCB5^+$ and $ABCB5^-$ subpopulations, we tracked the lineage of dividing cancer cells in melanoma xenograft-bearing mice using fluorescent genetic tags. These genetic lineage-tracking experiments confirmed the increased tumorigenic competence of $ABCB5^+$ cancer subsets. In addition, these studies demonstrated the existence of a tumor hierarchy in a solid cancer, with an exclusive capacity of $ABCB5^+$ CSC to self-renew and differentiate, because $ABCB5^+$ cells generated both $ABCB5^+$ and $ABCB5^-$ progeny, while $ABCB5^-$ cells gave rise to $ABCB5^-$ cell progeny only [26]. We furthermore examined whether selective ablation of ABCB5-defined melanoma-initiating cells can inhibit tumor formation and growth, as would be expected in a dynamic tumor environment if the $ABCB5^+$ subset is required for robust tumorigenesis, also of established tumors. Anti-ABCB5 monoclonal antibody, shown to be capable of inducing antibody-dependent cell-mediated cytotoxicity (ADCC), significantly inhibited tumor initiation and halted the growth of established melanoma xenografts in nude mice as compared to controls [26]. These results pointed to a specific relationship between tumorigenic CSC, tumor progression, and chemoresistance in a solid malignancy. Moreover, they provide initial proof-of-principle for the potential therapeutic utility of CSC-targeted strategies.

5 Isolation of Putative CSC Populations

Putative CSC populations were isolated from prostate tumors [53–55], osteosarcomas [56], skin cancers [57, 58], and a series of established tumor cell lines of different etiology [59–64]. While the tumor cell subsets identified in these studies share some of the characteristics of CSC[2], serial in vivo xenotransplantation experiments of surface marker-purified freshly patient-derived cancer cells confirming their self-renewal, differentiation, and tumor-initiating capacity are not yet available [53–59, 61–64].

Collins et al. [53] isolated a putative CSC population from a series of primary and metastatic human patient-derived prostate tumors based on a $CD44^+\alpha_2\beta_1{}^{high}CD133^+$ phenotype. The cellular subset identified by this marker combination, which ranged from 0.1 to 0.3% in the patient samples studied, showed increased in vitro colony-forming and long-term proliferative capacity compared to $CD44^+\alpha_2\beta_1{}^{high}CD133^-$ transiently amplifying populations and tumor bulk components of alternative phenotypes. Moreover, $CD44^+\alpha_2\beta_1{}^{high}CD133^+$ tumor cells were capable of differentiation to a secretory luminal phenotype concomitant with a reduction in CD133 expression in the presence of dihydrotestosterone in vitro [53]. Xenotransplantation experiments to prospectively identify CSC populations according to the aforementioned definition were not performed in this study. Subsequent work by Patrawala and colleagues, however, demonstrated increased tumorigenic capacity of $CD44^+$ compared to $CD44^-$[54] and of $CD44^+\alpha_2\beta_1^{high}$ compared to $CD44^-\alpha_2\beta_1{}^{low}$ cells derived from established prostate cancer cell lines [55].

Gibbs et al. [56] reported the identification of a subpopulation isolated from primary osteosarcoma and chondrosarcoma tumors and an established osteosarcoma cell line, which could be expanded and serially passaged as spherical colonies, termed "sarcospheres," in serum-starved conditions. Sarcospheres showed greater expression levels of the key marker genes of ESC pluripotency Oct3/4 and Nanog [65, 66] as compared to their adherent counterparts that preferentially expressed markers of mesodermal, ectodermal, and endodermal differentiation. Immunohistochemistry of patient biopsies identified bone sarcoma cells expressing the Stro-1, CD105, and CD44 mesenchymal stem cell markers [56]. Purification and xenotransplantation experiments of marker-sorted osteosarcoma cells need to be performed in order to identify a bone tumor stem cell population.

As outlined above, studies by our group led to the identification of CSC in human malignant melanoma, which we prospectively isolated by virtue of their expression of the chemoresistance determinant [45] ABCB5 [26]. Other groups have identified tumorigenic melanoma subpopulations of alternative phenotype [57, 58]. Fang et al. [57] described melanoma cell subsets capable of forming nonadherent spheres when cultured in growth medium for ESC. The sphere-forming subpopulations of human melanoma cells, which differentiated under appropriate conditions into multiple cell lineages and preferentially expressed the CD20 marker, were more tumorigenic when grafted to mice than their adherent counterparts with lesser differentiation plasticity [57]. Sphere-forming melanoma cells persisted after dissociation and replating of in vivo-formed xenografts and in long-term cultures in vitro, indicating their competence to self-renew. Based on these findings, the authors proposed that melanomas contain a subpopulation of stem-like cells that contribute to heterogeneity and tumorigenesis [57]. Although xenografted sphere-derived cells gave rise to significantly larger tumors, the difference in tumor formation capacity compared to adherent populations was modest (3/5 vs. 1/5 tumor xenografts, respectively) [57]. Furthermore, some melanoma spheres lost their ability to differentiate into certain cell lineages over time, indicating potential limitations of the sphere formation assay in efficiently enriching for CSC subsets. It is important to notice, however, that the commonly employed and accepted definition of CSC makes no specific assumptions about the differentiation plasticity of such cell subsets [2]. While the results by Fang and colleagues do not establish the presence of prospectively and molecularly defined melanoma-initiating cells, they provide a rationale to further examine CD20-sorted or other minority populations present in more tumorigenic melanoma spheres in serial xenotransplantation experiments.

As mentioned earlier, tumor cell subsets defined by CD133 enrich for CSC in colon cancers [31, 32], tumors of the brain [30], and pancreatic adenocarcinomas [34]. CD133 expression is likewise detected in a subpopulation of stem-like melanoma cells [45] and Monzani et al. [58] compared the abilities of $CD133^+$ versus $CD133^-$ melanoma cells to initiate tumor formation in vivo. Primary tumor initiating properties were exclusively contained within melanoma cell subsets characterized by expression of CD133, whereas $CD133^-$ melanoma cells were found to lack tumorigenicity [58]. In contrast to the findings by Fang and colleagues [57], more tumorigenic $CD133^+$ melanoma subsets were confined to adherent cell populations, whereas melanoma sphere-associated cells were devoid of CD133 [58]. In tumor xenografts in vivo, and under standard culture conditions in vitro, a melanoma cell line was found to be comprised of a majority of $CD133^+$ cells coexpressing low levels of the side-population determinant ABCG2 [17]. Based on their findings, Monzani et al. proposed that malignant melanoma CSC could be defined by a $CD133^+/ABCG2^+$ phenotype, but serial xenotransplantation of purified populations prospectively identified by these markers was not performed in this study. When others examined ABCG2 as a candidate prospective marker of tumorigenic cancer subpopulations, the molecule was not found to serve such a function [64]. In addition, $CD133^+$ subpopulations may not be consistently present in all primary and metastatic melanomas, according to earlier findings [67]. Hence, more extensive studies, including secondary tumor formation and stringent self-renewal and differentiation experiments are needed to draw more definitive conclusions about CD133 or ABCG2 as candidate CSC markers in melanoma.

Ma and colleagues [63] likewise isolated tumorigenic subpopulations expressing the $CD133^+$ surface phenotype from established hepatocellular carcinoma (HCC) cell lines. In this study, $CD133^+$ cells were found to be more clonogenic in vitro and showed enhanced tumorigenic capacity in immunodeficient mice in vivo when comparing them to their $CD133^-$ counterparts. While 30 of 44 mice injected with $CD133^+$ cells formed primary tumor xenografts, only 5 of 47 transplanted with $CD133^-$ HCC cell line-derived subsets did so. Moreover, $CD133^+$ cells re-isolated from $CD133^+$ cell-derived primary xenografts could be serially passaged to form secondary xenografts and re-established primary tumor heterogeneity with regard to CD133 expression [63].

In addition to immunophenotypic characterization-based methods or using the neurosphere approach, putative CSC populations have also been isolated based on their ability to efflux fluorescent dyes [59–62, 64]. Using fluorescence-activated cell sorting (FACS) approaches, Goodell et al. described the isolation of a subset of cells based on their ability to actively efflux the Hoechst dye, termed side population

(SP). The SP phenotype has been associated with ABC transporter activity [17] and was proposed as a surrogate marker for physiological stem cells [18]. This isolation technique has recently been extended to cancerous tissues to enable the identification of putative CSC. Kondo and colleagues isolated SP cells from the C6 glioma cell line that could be expanded and propagated as floating spheres in vitro. SP cells were in addition found to be more tumorigenic than their non-SP counterparts (tumor growth was detected in 6 of 6 versus 1 of 6 recipient animals, respectively) [62].

A Hoechst SP subset was subsequently also described for freshly patient-derived neuroblastoma samples and established breast cancer, lung cancer, and glioblastoma cell lines [60]. The SP cells displayed a greater capacity to expel the cytotoxic agent mitoxantrone, resulting in better survival as compared to non-SP cells, which may relate to high expression levels of ABC transporters detected in this subset [60]. Similar observations were made for the SP subset of a human malignant melanoma cell line, which was more resistant to Ara-C treatment [68]. Increased chemoresistance to doxorubicin, 5-fluorouracil, and gemcitabine was also detected for SP cells isolated from gastrointestinal cancer cell lines [59]. Accordingly, SP cells from six human lung cancer cell lines, which were enriched in tumor-initiating capability compared with non-SP cells, showed resistance to multiple therapeutic drugs [61]. SP cells could also be detected in clinical cancer samples in this study [61].

Finally, Patrawala et al. [64] systematically analyzed the SP phenotype of multiple established cell lines representing a variety of cancer types and found that only 30% of these cultured human cancer cells possessed a detectable side population. SP cells from a breast cancer and a glioma cell line as well as from a xenograft prostate tumor were more tumorigenic than their non-SP counterparts and could generate non-SP cells in vivo. While ABCG2 expression was found upregulated in the SP fraction, ABCG2$^+$ cells and ABCG2$^-$ cells were similarly tumorigenic [64].

In summary, several recent studies have identified putative CSC populations based on a stem-like phenotype, their ability to grow in floating spheres, the SP isolation technique, or *primary* tumorigenicity experiments. While extensions of these findings might ultimately lead to the molecular identification of CSC in additional cancers, it is important to remember the defining features of CSC so as not to confuse these data with in vivo confirmation of their long-term self-renewal, differentiation, and tumor-initiating capacity.

6 Origins of Cancer Stem Cells

Despite the similarities between cellular hierarchies in physiologic tissues and tumors, it is important to recognize that the term CSC does not imply a specific relationship of CSC to somatic stem cells as a potential cell of origin for the cancer. While definitive experimental evidence for such a putative relationship does not exist in the scientific literature to date, certain characteristics of CSC, however, argue in favor of physiological stem cells as the source of malignant transformation. Perhaps the most obvious argument supporting this hypothesis stems from the fact that restricted progenitor cells and terminally differentiated cells do not possess the capacity for long-term self-renewal. Moreover, the distinct surface markers or marker combinations that identify CSC in AML or brain tumors were found to be identical for multiple subtypes of these cancers [30, 69], indicating that cancerous transformation occurred in undifferentiated tissue stem cells rather than more committed progenitors, which would be expected to molecularly reflect the varying tumor subtypes within the CSC compartment. Hence, it seems plausible that CSC may arise by cumulative genetic alterations from long-lived normal stem cells that are more prone to accumulating mutations and already contain the machinery for indefinite self-renewal [2].

Indirect experimental support of this notion is given by recent studies in model organisms. Using a conditional *K-ras* transgenic mouse model, Kim and colleagues [70] identified epithelial stem cell populations at the bronchioalveolar duct junction termed bronchioalveolar stem cells (BASC). These drug-resistant and otherwise quiescent BASC populations proliferated extensively in response to pharmacologically-induced bronchiolar and alveolar injury to mediate epithelial cell renewal in vivo. BASC could be purified based on a stem cell antigen 1 (Sca-1)$^+$CD45$^-$CD31$^-$CD34$^+$ phenotype and showed increased in vitro clonogenic, multi-lineage differentiation, and self-renewal capacity in serial passages as compared to bulk cultures. Strikingly, induction of the *K-ras* oncogene resulted in a marked and selective expansion of BASC in vitro and in precursors of lung tumors in vivo, with increasing numbers found in more advanced disease [70], raising the possibility that BASC might represent the cellular source of lung adenocarcinomas.

Additional studies focusing on the hematopoietic lineage further suggest that tissue stem cells are indeed the target of malignant transformation. Kato et al. [71] showed that constitutive STAT5 activation in c-kit$^+$Sca-1$^+$CD34$^-$ Lineage$^-$hematopoietic stem cells (HSC) and in CD34$^+$KSL hematopoietic progenitor cells markedly increased the proliferative and self-renewal capacity of both subsets ex vivo. However, only the HSC were capable of inducing myeloproliferative disease (MPD) in an in vivo mouse model, while the progenitor cell population failed to do so [71]. Similarly, Passegue et al. [72] reported that inactivation of the transcriptional regulator JunB in long-term HSC but not at later stages of myelopoiesis could induce MPD in a mouse model of the disease.

However, other studies of the hematopoietic system indicate that CSC might also potentially arise through oncogenic transformation of lineage-restricted progenitor cells. Krivtsov and colleagues [73] showed that LSC could be induced by introduction of a mixed lineage leukemia (MLL)-AF9 fusion protein to committed granulocyte macrophage progenitors (GMP). These generated LSC could be serially passaged to induce leukemia in secondary recipient mice. While the gene expression profile of the MLL-AF9-induced LSC showed a significant overlap as compared to that of normal GMP, a subset of self-renewal-associated genes was specifically activated in the LSC compartment [73]. Cozzio et al. [74] demonstrated that introduction of the leukemia-associated MLL-ENL fusion protein both to HSC and to short-lived myeloid progenitors could induce acute myeloid leukemia when these cells were injected into mice. Similarly, Jaiswal et al. [75] demonstrated that chronic MPD could arise in transgenic mice in which expression of the oncogenic BCR-ABL fusion protein is targeted exclusively to committed progenitors and their myelomonocytic progeny. A subsequent study identified BCR-ABL-expressing GMP from human patients with chronic myelogenous leukemia in blast crisis as candidate leukemic stem cells [76]. Additional results by Huntly et al. [77] showed that transduction of GMP with the MOZ-TIF2 oncogene could confer characteristics of LSC, while BCR-ABL transduction failed to do so, indicating that malignant transformation is dependent on the quality of the oncogenetic stimulus. In an additional study supporting progenitor cells as potential source of CSC, Drynan and colleagues [78] demonstrated that MLL-AF9 translocations caused myeloid neoplasias when targeted to HSC or myeloid progenitors. However, MLL-AF9 fusion protein-expression in the murine T-cell compartment failed to induce leukemia [78], suggesting that carcinogenic transformation is critically dependent on the cell type affected.

Finally, there are alternatives to the hypothesis that physiological stem cells or immature progenitor cells are the main culprits of cancer development. In this regard, several recent studies have demonstrated that terminally differentiated cells, including fibroblasts and epithelial cells, could be transformed by retroviral infection and/or oncogene activation to induce cells with cancerous phenotype and function [79, 80]. In addition, CSC may also evolve through epigenetic transformation of normal stem cells by a dysfunctional environmental niche. In support of this theory, Houghton and colleagues [81] reported that bone marrow-derived cells could give rise to gastric adenocarcinoma when recruited into gastric glands upon chronic injury. Alternatively, it is also conceivable that mutated somatic cells can fuse with normal stem cells, thereby generating malignant CSC [82].

Additional research is required to draw more definitive conclusion about the cellular origin of CSC in human patients. However, many of the features recently described for CSC populations in solid tumors have important implications for tumor resistance, neoplastic progression, and therapy – regardless of the CSC precursor cell.

7 Molecular and Biological Features of Cancer Stem Cells

The prospective identification of CSC in various tumors [26, 29–36] has enabled researchers to begin characterizing their molecular characteristics and biological properties with the ultimate goal to specifically target and eradicate this cell subset for more effective, less toxic therapies for tumor patients (Fig. 2). Some of the conceptual implications for tumor biology emerging from recent findings in CSC populations as well as initial CSC-directed targeting approaches that have been reported in experimental model systems are described below.

7.1 Molecular Signatures and Self-Renewal-Associated Signaling Pathways in CSC – Cues for Differentiation Therapy

The gene expression profile of tumorigenic $CD44^{+}$ $CD24^{-/low}$ breast cancer populations was compared with that of healthy breast epithelium in a recent study by Liu and colleagues [83]. On the basis of this comparison, the authors identified an "invasiveness gene signature," the expression of which correlated with overall survival and metastasis-free survival in patients with breast cancer or other types of

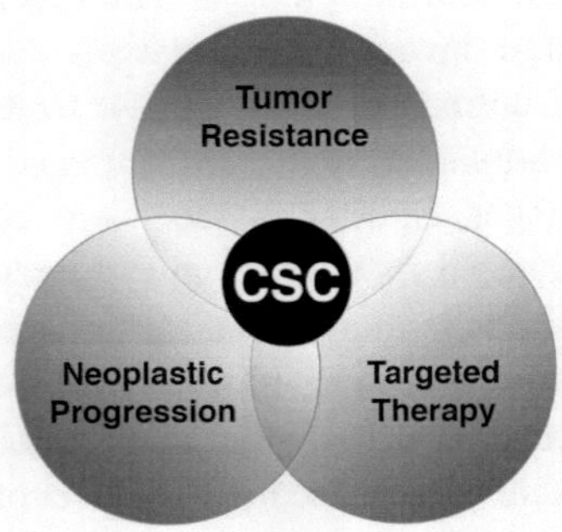

Fig. 2 Implications of the cancer stem cell (CSC) model. Recent studies indicate a specific relationship between tumorigenic CSC, neoplastic progression, and resistance to ionizing radiation and chemotherapeutic agents in solid malignancies. Furthermore, initial proof-of-concept has been established that targeted CSC killing is sufficient to halt experimental tumor growth. These findings indicate that novel CSC-directed strategies might improve conventional cytotoxic therapies, which are believed to spare refractory tumor initiators responsible for recurrence and metastasis

cancer [83]. While this association of a set of genes with disease outcome is important, it remains unclear how these changes in gene expression may relate to nontumorigenic cancer bulk components, which might share several factors of the CSC-related signature [84].

In a separate study, Shipitsin et al. [85] compared the gene expression and genetic profiles of tumorigenic CD44$^+$ versus nontumorigenic CD24$^+$ clinical breast cancer cells. CD44$^+$ cell-specific genes included many known stem-cell determinants and correlated with clinical patient outcome, while CD24$^+$ breast cancer cells were associated with markers of luminal epithelial differentiation. Most notably, distinct signaling pathways were specifically activated in the CD44$^+$ fraction, including the TGF-β pathway. Inhibition of TGF receptors, including TGFBR2, whose expression was found restricted to CD44$^+$ breast tumor cells, led to differentiation of this cell subset in vitro [85], indicating that TGFBR2-directed strategies might prove useful for therapeutic targeting.

In support of this theory, bone morphogenetic proteins (BMP), secreted morphogen members of the TGF-β superfamily, and their receptor (BMPR) transcripts and proteins were found expressed by CD133$^+$ brain tumor-initiating cell-enriched populations [86]. In this study, addition of BMP, in particular BMP4, to human glioblastoma cultures reduced their in vitro proliferation and clonogenicity in response to mitogens concomitant with a significant reduction of the CD133$^+$ pool. Moreover, treated cultures acquired a more differentiated cell morphology as evidenced by increased expression levels of astroglial, neuronal, and oligodendroglial markers. These findings prompted the researchers to investigate the ability of BMP4-treated CD133$^+$ human glioblastoma cells to form tumors in immunodeficient mice. Strikingly, xenotransplantation of glioblastoma-initiating cells transiently exposed to BMP4 resulted in small, delimited lesions with low mitotic indices compared to large, invasive tumor masses with high mitotic activity when untreated tumor cells were injected. In addition, mice implanted with BMP4-treated cells survived significantly longer. Similar results were observed, when BMP4 was delivered in vivo using polyacrylic beads that constantly released the morphogen for one week. Analogous to their in vitro observation, in vivo BMP4-administration resulted in increased differentiation of glioblastoma cells and a concurrent decrease in CD133-expressing tumor cells [86].

BMP4-responsive BMPR1A$^+$ cells, which identify undifferentiated progenitor cells in various physiological stem cell niches [87, 88], also coincide with tumorigenic ABCB5$^+$ populations in human malignant melanoma [26]. Interestingly, Rothhammer et al. [89] revealed a role of BMP4 in melanoma cell-driven vasculogenesis. It has been known for some time that distinct populations of aggressive melanoma cells may contribute to the de novo formation of tumor blood vessels via a process termed 'vasculogenic mimicry' [24, 90]. In the study by Rothhammer and colleagues, BMP4 inhibition resulted in decreased tube formation capacity of melanoma cells in vitro and in reduced tumor growth concomitant with large necrotic areas in an in vivo mouse model [89]. Taken together with the association of the BMP signaling system with tumor-initiating cells [26, 86], these findings suggest that the CSC compartment within a tumor may be responsible for tumor vasculogenesis and that such a potential function of CSC might represent one mechanism by which CSC may drive neoplastic formation and growth [82].

There are several additional signaling pathways that control physiological stem cell function and that have also been implicated in tumorigenesis when deregulated, including Hedgehog [35, 91–95], Wnt [23, 96–98], Notch [99–101], BMI-1 [36, 49–51], and PTEN [102, 103].

In physiological settings, the Hedgehog family, including Shh, regulates stem cell-driven skin regeneration [93]. Notably, tumorigenic pancreatic cancer cells were found to overexpress Shh [35]. In addition, the Hedgehog pathway was specifically linked to CSC maintenance in breast cancer [94], human glioma [92], glioblastoma [91], and multiple myeloma [95].

Another network of signaling molecules that has been implicated both in the regulation of physiological stem cell self-renewal and tumorigenesis is the Wnt family [23]. Overexpression of activated β-catenin, a key effector of the canonical Wnt cascade, impaired differentiation and increased self-perpetuation of HSC [96] and epidermal stem cells [98]. Accordingly, in a mouse model, β-catenin depletion resulted in diminished long-term HSC self-renewal and in a reduced ability of mice to develop chronic myelogenous leukemia (CML), providing a potential link to the regulation of neoplastic stem cell maintenance [97].

A putative association between normal stem cells and CSC is further indicated by common self-renewal mechanisms involving the Notch pathway. In the neural and hematopoietic systems, Notch activation fostered physiological stem cell self-renewal [100, 101]. In medulloblastoma, inhibition of the Notch signaling cascade using γ-secretase inhibitors reduced the CD133-expressing tumor cell fraction in vitro and blocked tumor xenograft formation in mice [99].

Recent studies showed that normal stem cells and CSC might also share the BMI-1 pathway for maintaining themselves. For instance, the polycomb gene BMI-1 is required for the self-perpetuation of neural stem cells [49] and HSC [50], as well as for the proliferation of LSC [51]. In addition, BMI-1 was found up-regulated in the tumorigenic CD44$^+$ cell fraction in HNSCC [36].

Finally, the tumor suppressor protein PTEN was found to drive both the self-maintenance of HSC and leukemogenesis through distinct mechanisms [103]. In a recent study by Yilmaz et al. [103], hematopoietic lineage-restricted loss of PTEN resulted in myeloproliferative disease that subsequently progressed to transplantable leukemias in mice. PTEN deletion also promoted physiological HSC proliferation. However, when transplanted into irradiated mice, PTEN-deficient HSC failed to stably engraft and were incapable of long-term multi-lineage reconstitution. In contrast to leukemia-initiating cells, normal HSC were thus unable to self-renew in the absence of PTEN [103]. Importantly, PTEN can regulate the activity of the mammalian target of rapamycin (mTOR) signaling pathway, an important regulator of cell growth and proliferation [104]. When Yilmaz and colleagues administered the mTOR inhibitor rapamycin to PTEN-deficient mice, they could restore the capacity of HSC to self-renew and to fully reconstitute the hematopoietic system of irradiated recipient mice. In addition, rapamycin treatment could also inhibit leukemogenesis and reduced the frequency of leukemia-initiating cells after the onset of apparent leukemia in this model [103]. These findings demonstrate that it is possible to target signaling pathways that regulate the maintenance and tumor-initiating capacity of CSC without depleting the physiological stem cells pool in an experimental model system.

In summary, studying the signal-transduction machinery that regulates the self-maintenance, differentiation, and proliferation of CSC could provide novel insights into the cellular events responsible for tumorigenic growth and may help identify CSC-specific therapeutic targets.

7.2 The Cancer Stem Cell Niche – A Prominent Regulator of Tumor Growth?

CSC, like normal stem cells, may depend on cellular interactions with differentiated, physiological cell types or on their own nontumorigenic cancer bulk populations to sustain stem cell features [20]. In addition, secreted factors or other niche components might regulate tumor establishment, growth, and progression [20]. Tumorigenicity as determined in xenotransplantation experiments may further vary with the applied experimental conditions, such as the species-specific host milieu, microenvironmental clues at the tissue site of injection, or the activation status of immune effector mechanisms in recipient animals. In support of this hypothesis, Kelly and colleagues [105] showed that a large proportion of mouse tumor cells could sustain myeloid and lymphoid neoplasms in *histocompatible* recipient mice. These findings stand in contrast to those obtained using human to *immunodeficient* mouse xenotransplantation models, which identified a rare subset of human $CD34^+CD38^-$ LSC exclusively capable of tumor development and growth [27, 28].

A pertinent role of the microenvironment was also recognized for the tumorigenic capacity of brain tumor subpopulations [41]. While intracranial injection of these cancer subsets consistently induced tumor formation in mice, only 50% of subcutaneous inocula resulted in neoplastic growth [41]. Likewise, the p53 status of the tumor stroma was found capable of regulating the tumorigenicity of MCF7 breast cancer cells in immunodeficient mice [106]. Recipients of cancer cells reconstituted with fibroblasts from $p53^{-/-}$ mice developed tumors faster than those injected with mixtures of MCF7 cells with wild-type or heterozygous fibroblasts [106]. Calabrese et al. [107] showed that $CD133^+Nestin^+$ brain tumor-initiating cells reside within a perivascular niche, and that endothelial cells within this niche can secrete factors that maintain the CSC subset in a stem-like state. Increasing the number of endothelial cells in brain tumor xenografts expanded the CD133-expressing fraction and promoted cancer development, whereas anti-angiogenic regimen reduced the CSC compartment and arrested tumor growth [107].

On the other hand, CSC might also reciprocally modulate the surrounding niche through the secretion of paracrine factors or direct cell-cell contact. For example, stem-like glioma cells secreted markedly elevated levels of vascular endothelial growth factor (VEGF) in comparison to tumor bulk populations [108]. Accordingly, $CD133^+$ glioma cell-conditioned media had a pro-angiogenic effect on endothelial cells in vitro, which could be abolished by anti-VEGF neutralizing monoclonal antibody treatment. In vivo, VEGF targeting suppressed the growth of stem-like glioma cell-derived xenografts but showed only limited efficacy against matched tumor xenografts derived from $CD133^+$ cell-depleted inocula [108]. Strategies, by which tumorigenic cancer cell subsets might facilitate tumor development, could also include inherent immunomodulatory properties, which have been described both for cancer cells and for physiological stem cells [109–111].

Taken together, these findings highlight that interactions of CSC with their niche environment are critical to tumor initiation and progression. The interpretation of tumorigenicity experiments in xenotransplantation models with sorted, isolated cell subsets may thus prove challenging when pursued outside the context of tumor bulk components and/or requisite human support cells [112]. From a tumor therapy standpoint, a better understanding of the niche factors that regulate CSC self-renewal and differentiation may lead to the identification of novel therapeutic targets within niche-constituent cells that may enable the inhibition of CSC-driven tumor growth [113].

7.3 Cancer Stem Cell-Mediated Resistance to Chemotherapeutic Agents and Radiotherapy

Physiological stem cells and CSC share many functional properties in addition to self-renewal and differentiation capacity. For example, normal stem and progenitor cells are highly resistant to xenobiotics, which may relate to their relative quiescence [15], high expression levels of ATP-binding cassette (ABC) drug efflux pumps [17, 114], or activation of anti-apoptotic mechanisms [16]. In the cancer context, multidrug resistance of tumor cells to therapeutic agents continues to be the leading cause of death [115]. The CSC model may shed new light into the cellular mechanisms of inherent or acquired tumor refractoriness and disease progression, as it predicts that conventional therapies that mainly target cancer bulk components may spare resistant tumor-initiators that subsequently drive neoplastic recurrence [116]. In support of this theory, tumorigenic leukemia subpopulations and LSC were found to overexpress a number of ABC transporters [117] and showed selective resistance to daunorubicin and mitoxantrone treatment [118, 119]. In CML patients, resistance to imatinib (Gleevec^{TM}) was linked to failure of the chemotherapeutic agent to deplete the LSC compartment [120].

Novel insights into CSC-mediated resistance mechanisms in solid tumors are given by recent findings in breast cancers, human malignant melanomas, and brain tumors [26, 121, 122]. In breast cancer patients, tumors from women who had received chemotherapy showed increased levels of $CD44^{+}CD24^{-/low}$ CSC [122], indicating that this subset may be selectively resistant to eradication. In human malignant melanoma, tumor-initiating cells can be prospectively identified based on their expression of the ABCB5 chemoresistance determinant [26], as described above. In a previous study, gene expression levels of ABCB5 across a panel of human cancer cell lines used by the National Cancer Institute for drug screening correlated significantly with chemorefractoriness to 45 out of 119 standard anticancer agents, including doxorubicin, to which clinical melanomas are chemorefractory [45]. Monoclonal antibody-mediated ABCB5 blockade significantly enhanced intracellular drug accumulation and reversed resistance of melanoma cells to doxorubicin [45]. Strikingly, ABCB5 expression, correlates with neoplastic progression in clinical melanomas, thus pointing to a novel, potentially critical link between tumor-initiating cells, cancer progression and chemoresistance in a solid malignancy [26].

In human gliomas, the fraction of $CD133^{+}$ CSC was enriched after ionizing radiation [121]. Mechanistically, this selective survival of CD133-expressing brain cancer cells was linked to preferential activation of the DNA damage checkpoint and increased DNA repair capacity in response to radiation, both in tumor cell cultures in vitro and in the brains of tumor-xenografted immunocompromised mice in vivo. Importantly, the radioresistance of glioma CSC could be reversed through pharmacological inhibition of DNA checkpoint kinases [121]. Similar findings were described for AML stem-like cells, which showed decreased sensitivity to apoptosis induction [118, 123].

Taken together, these findings provide experimental evidence for the hypothesis that conventional therapies directed at the bulk of tumor cells may spare the CSC compartment potentially owing to its inherent resistance mechanisms. Therefore, it seems plausible that CSC might compose the small reservoir of resistant cells that survives chemo- and/or radiation therapy and subsequently drives tumor recurrence and metastatic disease.

8 Cancer Stem Cells – Novel Targets for Tumor Therapy

If tumor-initiating cells are indeed the major culprits of tumor development and progression in human patients, the design of cancer therapies that target markers or signaling pathways specific to the CSC compartment could potentially increase the efficacy of current treatment regimens and might reduce the risk of relapse and metastasis. Two different approaches might prove useful in enhancing responsiveness to systemic therapy in light of the CSC concept: First, quiescent CSC could be induced to differentiate into more mature tumor cell types, either by activation of distinct signaling pathways such as for example through BMP [86], by alteration of gene expression using microRNAs (miRNAs) [122] or by epigenetic differentiation therapy [124]. Second, eradication of the CSC subset by specifically directing anti-tumor agents, including monoclonal antibodies [26] or oncolytic viruses [125, 126], to marker-defined tumor-initiating cells could likewise increase the efficacy of conventional therapies. Recent studies of epigenetic therapy in human cancer patients as well as novel CSC-directed targeting approaches reported in experimental model systems are described below.

8.1 Differentiation Therapy

Epigenetic factors, including the methylation and histone acetylation patterns of specific DNA regions, can maintain the undifferentiated state of physiological stem cells [127]. In addition, regional gains of DNA methylation and histone deacetylation that are associated with inappropriate gene silencing may play fundamental roles in the initiation and progression of human cancer [128]. Thus, the reversal of such epigenetic events by pharmacological inhibition is currently undergoing clinical testing in various types of cancer [129, 130].

In a recent proof-of-principle study of epigenetic therapy for locally advanced breast cancer ($n = 16$ patients), Arce and colleagues [129] added the demethylating agent hydralazine and the histone deacetylase (HDAC) inhibitor magnesium valproate to neoadjuvant doxorubicin and cyclophosphamide to assess their potential biological efficacy. All patients were evaluated for clinical and pathological responses and patient biopsies were subjected to microarray analysis for detection of differences in gene expression before and after epigenetic treatment. Notably, the administration of DNA methylation and HDAC inhibitors tended to increase the efficacy of subsequently administered conventional cytotoxic agents. Genes up-regulated through epigenetic therapy included tumor suppressors and cytochrome P450 family members [129]. Strikingly, the CSC determinant ABCB5 [26] was specifically down-regulated in magnesium valproate plus hydralazine-treated patients [129], indicating that stem-like cancer populations might be driven into differentiation through epigenetic therapy, with improved clinical outcome.

miRNAs represent an additional class of molecules that can influence self-renewal and differentiation through gene regulation [131]. In a recent study, Yu et al. [122] demonstrated a marked reduction in let-7 miRNAs in $CD44^{+}CD24^{-/low}$ breast cancer-initiating cells. Enforced let-7 expression resulted in reduced proliferation and differentiation of this cell subset in vitro, and in inhibition of tumor initiation and metastases in NOD/SCID mice in vivo. These effects were linked to reduced *H-Ras* and HMGA2 (high mobility group AT-hook 2) activities and could be antagonized using antisense oligonucleotides directed at let-7. Of note, silencing *H-Ras* reduced self-renewal, while HMGA2 inhibition caused the cells to differentiate [122].

Development of targeting strategies to specifically deliver epigenetic therapeutics or miRNAs to CSC is needed to potentially translate such novel approaches to the clinic.

8.2 Cancer Stem Cell Ablation Therapy

To date, few studies have addressed the potential efficacy of therapies that target prospective markers of tumor-initiating cells to abrogate CSC-driven tumor growth [26]. Eriksson et al. [125] used oncolytic adenoviruses to kill $CD44^{+}CD24^{-/low}$ breast cancer-initiating cells. Indeed, virus infection could prevent tumor xenograft formation by these cells in immunodeficient mice and exerted tumor-inhibitory effects on established xenografts upon intratumoral injection as compared to control treatment. However, the adenovirus-mediated oncolytic effectiveness was tested on isolated CSC fractions but not on CSC-depleted subsets [125]. It thus remains unclear whether in vivo inhibition of tumor growth resulted from specific CSC depletion. In a separate study, tumor xenografts derived from clinical glioblastoma samples expressing high levels of the CD133 marker were treated with the Delta-24-RGD virus [126]. While viral infection was found to improve the survival of glioblastoma-bearing mice, the oncolytic effect was not specifically directed at CD133-expressing CSC subsets [126].

Most recently, our group investigated whether selective killing of the CSC-enriched melanoma minority population prospectively identified by the ABCB5 marker can inhibit tumor initiation and growth [26]. We administered a monoclonal antibody directed at ABCB5 [45, 114] in a human to nude mouse melanoma xenograft model, because nude as opposed to NOD/SCID mice are capable of tumor cell killing by ADCC [132]. Anti-ABCB5 monoclonal antibody treatment resulted in significantly inhibited tumor formation and could halt the mean tumor growth of established tumors as compared to that determined in control-monoclonal-antibody-treated or untreated mice [26]. Mechanistically, this tumor-inhibitory effect was mediated by murine immune effector populations directed against anti-ABCB5 monoclonal antibody-coated melanoma-initiating cells, as determined both in vitro in an ADCC assay and by histological analyses of anti-ABCB5 or control antibody-treated tumor xenografts [26]. Further studies are required to determine if modulation of ABCB5 function or $ABCB5^{+}$ CSC differentiation might also inhibit tumor growth. Our study provides initial proof-of-principle that targeted ablation of a CSC compartment inhibits tumor growth, thus validating the potential therapeutic utility of the CSC concept.

9 Conclusion and Perspectives

The CSC concept is increasingly being validated in human solid tumors and it has now been established that human solid tumors can be organized as a hierarchy resulting from tumor-initiating CSC. Furthermore, critical links between CSC, tumor progression, and therapy resistance have emerged that underscore the need for novel therapeutic approaches specifically directed at such minority populations. Novel therapeutic strategies might include CSC ablation, sensitization, and differentiation. Studies directed at further understanding the biology of CSC could ultimately help to successfully translate such promising approaches to the clinic in order to improve morbidity and mortality in cancer patients.

References

1. Nowell PC. The clonal evolution of tumor cell populations. Science. 1976;194(4260):23–8.
2. Reya T, Morrison SJ, Clarke MF, Weissman IL. Stem cells, cancer, and cancer stem cells. Nature. 2001;414(6859):105–11.
3. Futreal PA, Coin L, Marshall M, et al. A census of human cancer genes. Nat Rev. 2004;4(3):177–83.
4. Vogelstein B, Kinzler KW. Cancer genes and the pathways they control. Nat Med. 2004;10(8):789–99.
5. Bruce WR, Van Der Gaag H. A quantitative assay for the number of murine lymphoma cells capable of proliferation in vivo. Nature. 1963;199:79–80.
6. Fidler IJ, Hart IR. Biological diversity in metastatic neoplasms: origins and implications. Science. 1982;217(4564):998–1003.
7. Hamburger AW, Salmon SE. Primary bioassay of human tumor stem cells. Science. 1977;197(4302):461–3.
8. Brunschwig A, Southam CM, Levin AG. Host resistance to cancer. Clinical experiments by homotransplants, autotransplants and admixture of autologous leucocytes. Ann Surg. 1965;162(3): 416–25.
9. Morrison SJ, Kimble J. Asymmetric and symmetric stem-cell divisions in development and cancer. Nature. 2006;441(7097):1068–74.
10. Thomson JA, Itskovitz-Eldor J, Shapiro SS, et al. Embryonic stem cell lines derived from human blastocysts. Science. 1998;282(5391):1145–7.
11. Fuchs E. Scratching the surface of skin development. Nature. 2007;445(7130):834–42.
12. Toma JG, Akhavan M, Fernandes KJ, et al. Isolation of multipotent adult stem cells from the dermis of mammalian skin. Nat Cell Biol. 2001;3(9):778–84.
13. Jiang Y, Jahagirdar BN, Reinhardt RL, et al. Pluripotency of mesenchymal stem cells derived from adult marrow. Nature. 2002;418(6893):41–9.
14. Guan K, Nayernia K, Maier LS, et al. Pluripotency of spermatogonial stem cells from adult mouse testis. Nature. 2006;440(7088):1199–203.
15. Harrison DE, Lerner CP. Most primitive hematopoietic stem cells are stimulated to cycle rapidly after treatment with 5-fluorouracil. Blood 1991;78(5):1237–40.
16. Peters R, Leyvraz S, Perey L. Apoptotic regulation in primitive hematopoietic precursors. Blood. 1998;92(6):2041–52.
17. Zhou S, Schuetz JD, Bunting KD, et al. The ABC transporter Bcrp1/ABCG2 is expressed in a wide variety of stem cells and is a molecular determinant of the side-population phenotype. Nature Med. 2001;7(9):1028–34.
18. Goodell MA, Brose K, Paradis G, Conner AS, Mulligan RC. Isolation and functional properties of murine hematopoietic stem cells that are replicating in vivo. J Exp Med. 1996;183(4): 1797–806.
19. Molofsky AV, Pardal R, Morrison SJ. Diverse mechanisms regulate stem cell self-renewal. Curr Opin Cell Biol. 2004;16(6): 700–7.
20. Scadden DT. The stem-cell niche as an entity of action. Nature. 2006;441(7097):1075–9.
21. Leong KG, Karsan A. Recent insights into the role of Notch signaling in tumorigenesis. Blood. 2006;107(6):2223–33.
22. Taipale J, Beachy PA. The Hedgehog and Wnt signalling pathways in cancer. Nature. 2001;411(6835):349–54.
23. Reya T, Clevers H. Wnt signalling in stem cells and cancer. Nature. 2005;434(7035):843–50.
24. Hendrix MJ, Seftor EA, Hess AR, Seftor RE. Molecular plasticity of human melanoma cells. Oncogene. 2003;22(20):3070–5.
25. Kennedy JA, Barabe F, Poeppl AG, Wang JC, Dick JE. Comment on "Tumor growth need not be driven by rare cancer stem cells". Science. 2007;318(5857):1722; author reply
26. Schatton T, Murphy GF, Frank NY, et al. Identification of cells initiating human melanomas. Nature. 2008;451(7176): 345–9.
27. Lapidot T, Sirard C, Vormoor J, et al. A cell initiating human acute myeloid leukaemia after transplantation into SCID mice. Nature. 1994;367(6464):645–8.
28. Bonnet D, Dick JE. Human acute myeloid leukemia is organized as a hierarchy that originates from a primitive hematopoietic cell. Nature. Med. 1997;3(7):730–7.
29. Al-Hajj M, Wicha MS, Benito-Hernandez A, Morrison SJ, Clarke MF. Prospective identification of tumorigenic breast cancer cells. Proc Natl Acad Sci U S A. 2003;100(7):3983–8.
30. Singh SK, Hawkins C, Clarke ID, et al. Identification of human brain tumour initiating cells. Nature. 2004;432(7015): 396–401.
31. O'Brien CA, Pollett A, Gallinger S, Dick JE. A human colon cancer cell capable of initiating tumour growth in immunodeficient mice. Nature. 2007;445(7123):106–10.
32. Ricci-Vitiani L, Lombardi DG, Pilozzi E, et al. Identification and expansion of human colon-cancer-initiating cells. Nature. 2007;445(7123):111–5.
33. Dalerba P, Dylla SJ, Park IK, et al. Phenotypic characterization of human colorectal cancer stem cells. Proc Natl Acad Sci U S A. 2007;104(24):10158–63.
34. Hermann PC, Huber SL, Herrler T, et al. Distinct populations of cancer stem cells determine tumor growth and metastatic activity in human pancreatic cancer. Cell Stem Cell. 2007;1:313–23.
35. Li C, Heidt DG, Dalerba P, et al. Identification of pancreatic cancer stem cells. Cancer Res. 2007;67(3):1030–7.
36. Prince ME, Sivanandan R, Kaczorowski A, et al. Identification of a subpopulation of cells with cancer stem cell properties in head and neck squamous cell carcinoma. Proc Natl Acad Sci U S A. 2007;104(3):973–8.
37. Uchida N, Buck DW, He D, et al. Direct isolation of human central nervous system stem cells. Proc Natl Acad Sci U S A. 2000;97(26):14720–5.
38. Reynolds BA, Weiss S. Generation of neurons and astrocytes from isolated cells of the adult mammalian central nervous system. Science. 1992;255(5052):1707–10.
39. Singh SK, Clarke ID, Terasaki M, et al. Identification of a cancer stem cell in human brain tumors. Cancer Res. 2003;63(18): 5821–8.
40. Hemmati HD, Nakano I, Lazareff JA, et al. Cancerous stem cells can arise from pediatric brain tumors. Proc Natl Acad Sci U S A. 2003;100(25):15178–83.
41. Galli R, Binda E, Orfanelli U, et al. Isolation and characterization of tumorigenic, stem-like neural precursors from human glioblastoma. Cancer Res. 2004;64(19):7011–21.
42. Taylor MD, Poppleton H, Fuller C, et al. Radial glia cells are candidate stem cells of ependymoma. Cancer Cell. 2005;8(4): 323–35.
43. Weichert W, Knosel T, Bellach J, Dietel M, Kristiansen G. ALCAM/CD166 is overexpressed in colorectal carcinoma and correlates with shortened patient survival. J Clin Pathol. 2004;57(11):1160–4.

44. van Kempen LC, van den Oord JJ, van Muijen GN, Weidle UH, Bloemers HP, Swart GW. Activated leukocyte cell adhesion molecule/CD166, a marker of tumor progression in primary malignant melanoma of the skin. Am J Pathol. 2000;156(3):769–74.
45. Frank NY, Margaryan A, Huang Y, et al. ABCB5-mediated doxorubicin transport and chemoresistance in human malignant melanoma. Cancer Res. 2005;65(10):4320–33.
46. Peled A, Petit I, Kollet O, et al. Dependence of human stem cell engraftment and repopulation of NOD/SCID mice on CXCR4. Science. 1999;283(5403):845–8.
47. Muller A, Homey B, Soto H, et al. Involvement of chemokine receptors in breast cancer metastasis. Nature. 2001;410(6824):50–6.
48. Chu PG, Weiss LM. Keratin expression in human tissues and neoplasms. Histopathology. 2002;40(5):403–39.
49. Molofsky AV, Pardal R, Iwashita T, Park IK, Clarke MF, Morrison SJ. Bmi-1 dependence distinguishes neural stem cell self-renewal from progenitor proliferation. Nature. 2003;425(6961):962–7.
50. Park IK, Qian D, Kiel M, et al. Bmi-1 is required for maintenance of adult self-renewing haematopoietic stem cells. Nature. 2003;423(6937):302–5.
51. Lessard J, Sauvageau G. Bmi-1 determines the proliferative capacity of normal and leukaemic stem cells. Nature. 2003;423(6937):255–60.
52. Suryo Rahmanto Y, Dunn LL, Richardson DR. Identification of distinct changes in gene expression after modulation of melanoma tumor antigen p97 (melanotransferrin) in multiple models in vitro and in vivo. Carcinogenesis. 2007;28(10):2172–83.
53. Collins AT, Berry PA, Hyde C, Stower MJ, Maitland NJ. Prospective identification of tumorigenic prostate cancer stem cells. Cancer Res. 2005;65(23):10946–51.
54. Patrawala L, Calhoun T, Schneider-Broussard R, et al. Highly purified CD44+ prostate cancer cells from xenograft human tumors are enriched in tumorigenic and metastatic progenitor cells. Oncogene. 2006;25(12):1696–708.
55. Patrawala L, Calhoun-Davis T, Schneider-Broussard R, Tang DG. Hierarchical organization of prostate cancer cells in xenograft tumors: the CD44+alpha2beta1+ cell population is enriched in tumor-initiating cells. Cancer Res. 2007;67(14):6796–805.
56. Gibbs CP, Kukekov VG, Reith JD, et al. Stem-like cells in bone sarcomas: implications for tumorigenesis. Neoplasia. 2005;7(11):967–76.
57. Fang D, Nguyen TK, Leishear K, et al. A tumorigenic subpopulation with stem cell properties in melanomas. Cancer Res. 2005;65(20):9328–37.
58. Monzani E, Facchetti F, Galmozzi E, et al. Melanoma contains CD133 and ABCG2 positive cells with enhanced tumourigenic potential. Eur J Cancer. 2007;43(5):935–46.
59. Haraguchi N, Utsunomiya T, Inoue H, et al. Characterization of a side population of cancer cells from human gastrointestinal system. Stem Cells. 2006;24(3):506–13.
60. Hirschmann-Jax C, Foster AE, Wulf GG, et al. A distinct "side population" of cells with high drug efflux capacity in human tumor cells. Proc Natl Acad Sci U S A. 2004;101(39):14228–33.
61. Ho MM, Ng AV, Lam S, Hung JY. Side population in human lung cancer cell lines and tumors is enriched with stem-like cancer cells. Cancer Res. 2007;67(10):4827–33.
62. Kondo T, Setoguchi T, Taga T. Persistence of a small subpopulation of cancer stem-like cells in the C6 glioma cell line. Proc Natl Acad Sci U S A. 2004;101(3):781–6.
63. Ma S, Chan KW, Hu L, et al. Identification and characterization of tumorigenic liver cancer stem/progenitor cells. Gastroenterology. 2007;132(7):2542–56.
64. Patrawala L, Calhoun T, Schneider-Broussard R, Zhou J, Claypool K, Tang DG. Side population is enriched in tumorigenic, stem-like cancer cells, whereas ABCG2+ and ABCG2- cancer cells are similarly tumorigenic. Cancer Res. 2005;65(14):6207–19.
65. Chambers I, Colby D, Robertson M, et al. Functional expression cloning of Nanog, a pluripotency sustaining factor in embryonic stem cells. Cell. 2003;113(5):643–55.
66. Mitsui K, Tokuzawa Y, Itoh H, et al. The homeoprotein Nanog is required for maintenance of pluripotency in mouse epiblast and ES cells. Cell. 2003;113(5):631–42.
67. Klein WM, Wu BP, Zhao S, Wu H, Klein-Szanto AJ, Tahan SR. Increased expression of stem cell markers in malignant melanoma. Mod Pathol. 2007;20(1):102–7.
68. Grichnik JM, Burch JA, Schulteis RD, et al. Melanoma, a tumor based on a mutant stem cell? J Invest Dermatol. 2006;126(1):142–53.
69. Hope KJ, Jin L, Dick JE. Acute myeloid leukemia originates from a hierarchy of leukemic stem cell classes that differ in self-renewal capacity. Nature Immunol. 2004;5(7):738–43.
70. Kim CF, Jackson EL, Woolfenden AE, et al. Identification of bronchioalveolar stem cells in normal lung and lung cancer. Cell. 2005;121(6):823–35.
71. Kato Y, Iwama A, Tadokoro Y, et al. Selective activation of STAT5 unveils its role in stem cell self-renewal in normal and leukemic hematopoiesis. JExp Med. 2005;202(1):169–79.
72. Passegue E, Wagner EF, Weissman IL. JunB deficiency leads to a myeloproliferative disorder arising from hematopoietic stem cells. Cell. 2004;119(3):431–43.
73. Krivtsov AV, Twomey D, Feng Z, et al. Transformation from committed progenitor to leukaemia stem cell initiated by MLL-AF9. Nature. 2006;442(7104):818–22.
74. Cozzio A, Passegue E, Ayton PM, Karsunky H, Cleary ML, Weissman IL. Similar MLL-associated leukemias arising from self-renewing stem cells and short-lived myeloid progenitors. Genes Dev. 2003;17(24):3029–35.
75. Jaiswal S, Traver D, Miyamoto T, Akashi K, Lagasse E, Weissman IL. Expression of BCR/ABL and BCL-2 in myeloid progenitors leads to myeloid leukemias. Proc Natl Acad Sci U S A. 2003;100(17):10002–7.
76. Jamieson CH, Ailles LE, Dylla SJ, et al. Granulocyte-macrophage progenitors as candidate leukemic stem cells in blast-crisis CML. N Eng J Med. 2004;351(7):657–67.
77. Huntly BJ, Shigematsu H, Deguchi K, et al. MOZ-TIF2, but not BCR-ABL, confers properties of leukemic stem cells to committed murine hematopoietic progenitors. Cancer Cell. 2004;6(6):587–96.
78. Drynan LF, Pannell R, Forster A, et al. Mll fusions generated by Cre-loxP-mediated de novo translocations can induce lineage reassignment in tumorigenesis. EMBO J. 2005;24(17):3136–46.
79. Fu SL, Huang YJ, Liang FP, et al. Malignant transformation of an epithelial cell by v-Src via tv-a-mediated retroviral infection: a new cell model for studying carcinogenesis. Biochem Biophys Res Commun 2005;338(2):830–8.
80. Sun B, Chen M, Hawks CL, Pereira-Smith OM, Hornsby PJ. The minimal set of genetic alterations required for conversion of primary human fibroblasts to cancer cells in the subrenal capsule assay. Neoplasia. 2005;7(6):585–93.
81. Houghton J, Stoicov C, Nomura S, et al. Gastric cancer originating from bone marrow-derived cells. Science. 2004;306(5701):1568–71.
82. Schatton T, Frank MH. Cancer stem cells and human malignant melanoma. Pigment Cell Melanoma Res. 2008;21(1):39–55.
83. Liu R, Wang X, Chen GY, et al. The prognostic role of a gene signature from tumorigenic breast-cancer cells. N Eng J Med. 2007;356(3):217–26.

84. Massague J. Sorting out breast-cancer gene signatures. N Eng J Med. 2007;356(3):294–7.
85. Shipitsin M, Campbell LL, Argani P, et al. Molecular definition of breast tumor heterogeneity. Cancer Cell. 2007;11(3):259–73.
86. Piccirillo SG, Reynolds BA, Zanetti N, et al. Bone morphogenetic proteins inhibit the tumorigenic potential of human brain tumour-initiating cells. Nature. 2006;444(7120):761–5.
87. Frank NY, Kho AT, Schatton T, et al. Regulation of myogenic progenitor proliferation in human fetal skeletal muscle by BMP4 and its antagonist Gremlin. J Cell Biol. 2006;175(1):99–110.
88. Lim DA, Tramontin AD, Trevejo JM, Herrera DG, Garcia-Verdugo JM, Alvarez-Buylla A. Noggin antagonizes BMP signaling to create a niche for adult neurogenesis. Neuron. 2000;28(3):713–26.
89. Rothhammer T, Bataille F, Spruss T, Eissner G, Bosserhoff AK. Functional implication of BMP4 expression on angiogenesis in malignant melanoma. Oncogene. 2007;26(28): 4158–70.
90. Maniotis AJ, Folberg R, Hess A, et al. Vascular channel formation by human melanoma cells in vivo and in vitro: vasculogenic mimicry. Am J Pathol. 1999;155(3):739–52.
91. Bar EE, Chaudhry A, Lin A, et al. Cyclopamine-mediated hedgehog pathway inhibition depletes stem-like cancer cells in glioblastoma. Stem Cells. 2007;25(10):2524–33.
92. Clement V, Sanchez P, de Tribolet N, Radovanovic I, Ruiz i Altaba A. HEDGEHOG-GLI1 signaling regulates human glioma growth, cancer stem cell self-renewal, and tumorigenicity. Curr Biol. 2007;17(2):165–72.
93. Levy V, Lindon C, Harfe BD, Morgan BA. Distinct stem cell populations regenerate the follicle and interfollicular epidermis. Dev Cell. 2005;9(6):855–61.
94. Liu S, Dontu G, Mantle ID, et al. Hedgehog signaling and Bmi-1 regulate self-renewal of normal and malignant human mammary stem cells. Cancer Res. 2006;66(12):6063–71.
95. Peacock CD, Wang Q, Gesell GS, et al. Hedgehog signaling maintains a tumor stem cell compartment in multiple myeloma. Proc Natl Acad Sci U S A. 2007;104(10):4048–53.
96. Reya T, Duncan AW, Ailles L, et al. A role for Wnt signalling in self-renewal of haematopoietic stem cells. Nature. 2003;423(6938):409–14.
97. Zhao C, Blum J, Chen A, et al. Loss of beta-catenin impairs the renewal of normal and CML stem cells in vivo. Cancer Cell. 2007;12(6):528–41.
98. Zhu AJ, Watt FM. beta-catenin signalling modulates proliferative potential of human epidermal keratinocytes independently of intercellular adhesion. Development. 1999;126(10): 2285–98.
99. Fan X, Matsui W, Khaki L, et al. Notch pathway inhibition depletes stem-like cells and blocks engraftment in embryonal brain tumors. Cancer Res. 2006;66(15):7445–52.
100. Shen Q, Goderie SK, Jin L, et al. Endothelial cells stimulate self-renewal and expand neurogenesis of neural stem cells. Science. 2004;304(5675):1338–40.
101. Varnum-Finney B, Xu L, Brashem-Stein C, et al. Pluripotent, cytokine-dependent, hematopoietic stem cells are immortalized by constitutive Notch1 signaling. Nature Med. 2000;6(11): 1278–81.
102. Wang S, Garcia AJ, Wu M, Lawson DA, Witte ON, Wu H. Pten deletion leads to the expansion of a prostatic stem/progenitor cell subpopulation and tumor initiation. Proc Natl Acad Sci U S A. 2006;103(5):1480–5.
103. Yilmaz OH, Valdez R, Theisen BK, et al. Pten dependence distinguishes haematopoietic stem cells from leukaemia-initiating cells. Nature. 2006;441(7092):475–82.
104. Hay N. The Akt-mTOR tango and its relevance to cancer. Cancer Cell. 2005;8(3):179–83.
105. Kelly JW, Yeatman JM, Regalia C, Mason G, Henham AP. A high incidence of melanoma found in patients with multiple dysplastic naevi by photographic surveillance. Med J Aust. 1997;167(4):191–4.
106. Kiaris H, Chatzistamou I, Trimis G, Frangou-Plemmenou M, Pafiti-Kondi A, Kalofoutis A. Evidence for nonautonomous effect of p53 tumor suppressor in carcinogenesis. Cancer Res. 2005;65(5):1627–30.
107. Calabrese C, Poppleton H, Kocak M, et al. A perivascular niche for brain tumor stem cells. Cancer Cell. 2007;11(1): 69–82.
108. Bao S, Wu Q, Sathornsumetee S, et al. Stem cell-like glioma cells promote tumor angiogenesis through vascular endothelial growth factor. Cancer Res. 2006;66(16):7843–8.
109. Dong H, Strome SE, Salomao DR, et al. Tumor-associated B7-H1 promotes T-cell apoptosis: a potential mechanism of immune evasion. Nature Med. 2002;8(8):793–800.
110. Frank MH, Sayegh MH. Immunomodulatory functions of mesenchymal stem cells. Lancet. 2004;363(9419):1411–2.
111. Le Blanc K, Ringden O. Immunomodulation by mesenchymal stem cells and clinical experience. J intern Med. 2007;262(5):509–25.
112. Kelly PN, Dakic A, Adams JM, Nutt SL, Strasser A. Tumor growth need not be driven by rare cancer stem cells. Science. 2007;317(5836):337.
113. Albini A, Sporn MB. The tumour microenvironment as a target for chemoprevention. Nature Rev. 2007;7(2):139–47.
114. Frank NY, Pendse SS, Lapchak PH, et al. Regulation of progenitor cell fusion by ABCB5 P-glycoprotein, a novel human ATP-binding cassette transporter. J Biol Chem. 2003;278(47): 47156–65.
115. Gottesman MM, Fojo T, Bates SE. Multidrug resistance in cancer: role of ATP-dependent transporters. Nature Rev. 2002;2(1): 48–58.
116. Dean M, Fojo T, Bates S. Tumour stem cells and drug resistance. Nature Rev. 2005;5(4):275–84.
117. de Grouw EP, Raaijmakers MH, Boezeman JB, et al. Preferential expression of a high number of ATP binding cassette transporters in both normal and leukemic CD34+CD38- cells. Leukemia. 2006;20(4):750–4.
118. Costello RT, Mallet F, Gaugler B, et al. Human acute myeloid leukemia CD34+/CD38- progenitor cells have decreased sensitivity to chemotherapy and Fas-induced apoptosis, reduced immunogenicity, and impaired dendritic cell transformation capacities. Cancer Res. 2000;60(16):4403–11.
119. Wulf GG, Wang RY, Kuehnle I, et al. A leukemic stem cell with intrinsic drug efflux capacity in acute myeloid leukemia. Blood. 2001;98(4):1166–73.
120. Michor F, Hughes TP, Iwasa Y, et al. Dynamics of chronic myeloid leukaemia. Nature. 2005;435(7046):1267–70.
121. Bao S, Wu Q, McLendon RE, et al. Glioma stem cells promote radioresistance by preferential activation of the DNA damage response. Nature. 2006;444(7120):756–60.
122. Yu F, Yao H, Zhu P, et al. let-7 regulates self renewal and tumorigenicity of breast cancer cells. Cell. 2007;131(6): 1109–23.
123. Guzman ML, Swiderski CF, Howard DS, et al. Preferential induction of apoptosis for primary human leukemic stem cells. Proc Natl Acad Sci U S A. 2002;99(25):16220–5.
124. Jones PA, Baylin SB. The epigenomics of cancer. Cell. 2007;128(4):683–92.
125. Eriksson M, Guse K, Bauerschmitz G, et al. Oncolytic adenoviruses kill breast cancer initiating CD44(+)CD24(-/Low) Cells. Mol Ther. 2007;15(12):2088–93.
126. Jiang H, Gomez-Manzano C, Aoki H, et al. Examination of the therapeutic potential of Delta-24-RGD in brain tumor

stem cells: role of autophagic cell death. J Natl Cancer Inst. 2007;99(18):1410–4.
127. Boyer LA, Mathur D, Jaenisch R. Molecular control of pluripotency. Curr Opin Genet Dev. 2006;16(5):455–62.
128. Jones PA, Baylin SB. The fundamental role of epigenetic events in cancer. Nat Rev Genet. 2002;3(6):415–28.
129. Arce C, Perez-Plasencia C, Gonzalez-Fierro A, et al. A proof-of-principle study of epigenetic therapy added to neoadjuvant Doxorubicin cyclophosphamide for locally advanced breast cancer. PLoS ONE. 2006;1:e98.
130. Fandy TE, Carraway H, Gore SD. DNA demethylating agents and histone deacetylase inhibitors in hematologic malignancies. Cancer J. 2007;13(1):40–8.
131. Chapman EJ, Carrington JC. Specialization and evolution of endogenous small RNA pathways. Nat Rev Genet. 2007;8(11): 884–96.
132. Kanazawa J, Ohta S, Shitara K, et al. Therapeutic potential of chimeric anti-(ganglioside GD3) antibody KM871: antitumor activity in xenograft model of melanoma and effector function analysis. Cancer Immunol Immunother. 2000;49(4–5): 253–8.

Therapeutic Approaches to Target Cancer Stem Cells

Lisa R. Rogers and Maxs Wicha

Abstract Cancer stem cells (CSCs) have been identified in hematologic malignancies as well as a number of solid tumors. Among solid tumors, the isolation and characterization of the tumorigenicity and signaling pathways of CSCs have been studied most thoroughly in brain and breast cancers. These tumor types share similar normal stem cell and oncogenic regulatory pathways and will be the focus of this review. Efforts to eliminate the CSCs within solid tumors provide a new paradigm for therapy, one that may include standard cytotoxic treatments to target the rapidly proliferating bulk of the tumor and other agents to target the CSCs, which may underlie tumorigenesis and treatment resistance. In addition, because a number of oncogenic pathways may be activated in CSCs, targeting multiple, rather than single, sites of self-renewal and differentiation is likely to be the most efficacious. Proposed treatments designed to target CSCs, singly or in combination, include differentiation therapy, targeting the aberrant signaling pathways in the CSC and the CSC microenvironment ("stem cell niche") and targeting the mechanisms of CSC treatment resistance. Additional research into the biology of CSCs in individual tumors, testing of proposed therapies in animal tumor systems, and carefully designed human clinical trials are needed to demonstrate the efficacy and safety of targeting CSCs.

Keywords CNS Neoplasms · Breast neoplasms · Medulloblastoma · Glioma · Stem cell · Cancer · Neurosphere · Mammosphere

1 Introduction

The stem cell theory of tumorigenesis implies that small subsets of tumor cells are cancer stem cells (CSCs), which have the unique ability to self-renew and to initiate and maintain tumor growth. In this model, disruption of genes involved in the regulation of normal stem cell self-renewal ultimately leads to the generation of CSCs. These CSCs have the capacity to self-renew and thus to expand the CSC pool and also to differentiate into the phenotypically diverse population of nontumorigenic cancer cells that constitute the bulk of the tumor. If tumors are maintained by CSCs, it is clear that therapies such as chemotherapy and radiation, which primarily target actively cycling cells, are unlikely to be effective at eliminating these cells. Conventional treatments may produce transient tumor regression, but the CSCs surviving these treatments will result in tumor recurrence.

Following the identification of CSCs in hematologic malignancies, CSCs have been identified in a number of solid tumors, including breast, prostate, brain, pancreas, head and neck, and colon cancers. In this chapter we present the evidence for a CSC population in two solid tumors, malignant brain tumors (specifically, glioblastoma [GBM] and medulloblastoma [MB]) and breast cancer. We review the techniques used for the isolation and characterization of CSCs and the evidence that CSCs participate in the initiation and maintenance of these tumors. We review the pathways underlying CSC self-renewal and provide the rationale for future therapeutic approaches to target CSCs and to determine treatment efficacy.

2 Evidence of Normal Tissue Stem Cells and Their Relevance to Tumor CSCs in Brain and Breast Cancer

In the CSC model, tumors develop from normal stem cells/progenitors that have disregulated or acquired the ability to self-renew as a result of oncogenic mutations.

L.R. Rogers (✉)
Professor, Department of Neurology, University of Michigan Medical Center, 1500 E. Medical Center Drive, 1920F Taubman Center, Ann Arbor, MI 48109-5316
e-mail: lisaroge@umich.edu

V.K. Rajasekhar, M.C. Vemuri (eds.), *Regulatory Networks in Stem Cells*, Stem Cell Biology and Regenerative Medicine,
DOI 10.1007/978-1-60327-227-8_41, © Humana Press, a part of Springer Science+Business Media, LLC 2009

Several lines of evidence suggest that CSCs in brain and breast cancer arise from normal stem cells/progenitors, including the identification of cell populations within tumors that have cell surface markers identical to normal stem cells. In addition, both normal and malignant stem cells in these tumors have been demonstrated to utilize similar pathways to regulate self-renewal.

2.1 Brain

The presence of neural stem cells (NSCs) in the human brain is well documented. These stem cells exist in restricted regions of the brain, where they are arranged in specific linear hierarchies. The most abundant region of NSCs is the subventricular zone (SVZ), which lines the lateral ventricles.

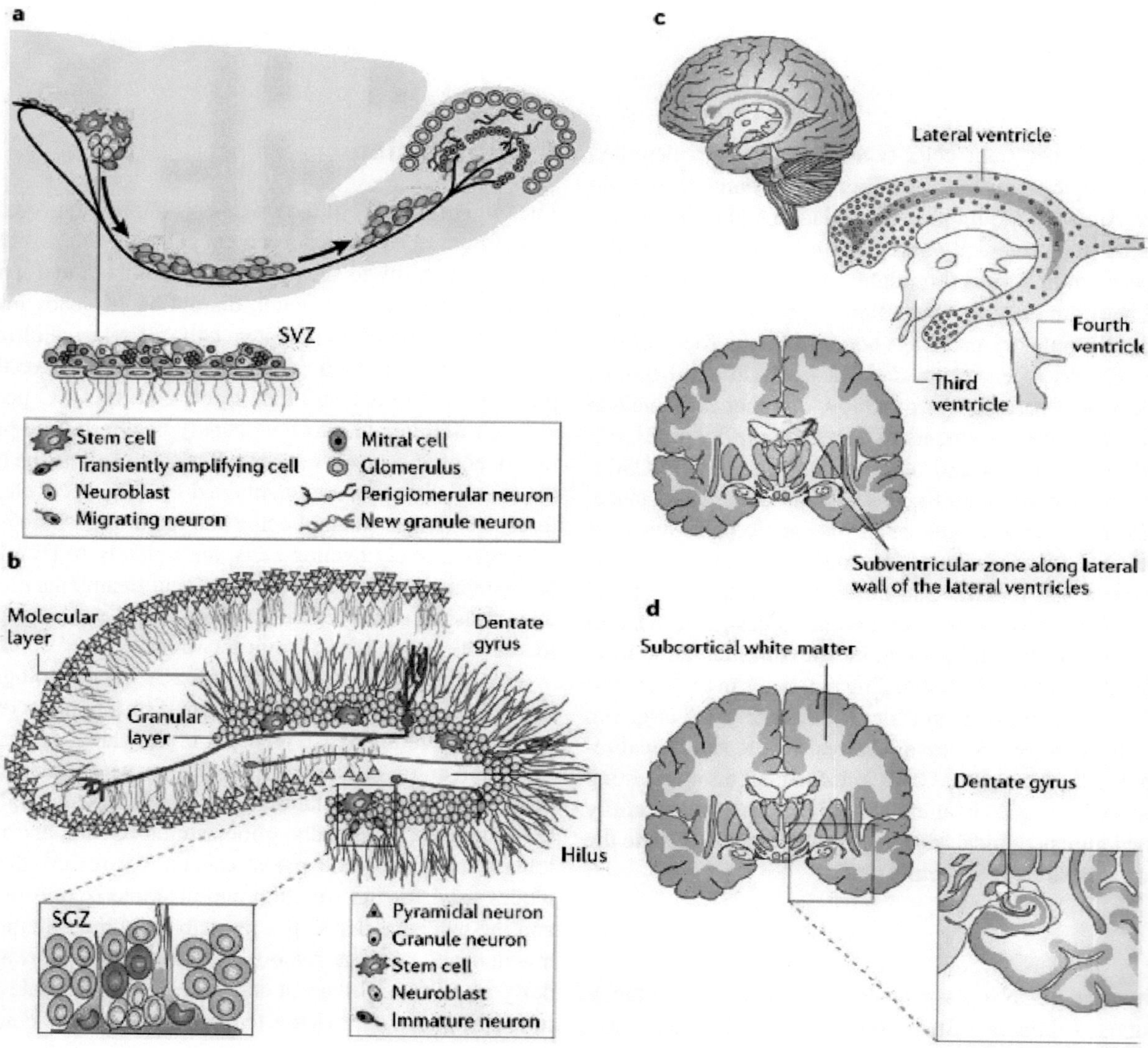

Fig. 1 (**a**) A sagittal section through the lateral ventricle that shows the largest area of adult neurogenesis; that is, the subventricular zone (SVZ). This region lines the lateral ventricles of the forebrain and is comprised of three main cell types. The multipotent, type B astrocytes, that have been identified as the *bona fide* SVZ stem cells, give rise to fast-cycling transiently proliferating precursor cells that are called type C precursors and that, in turn, generate mitotically active type A neuroblasts. The type A cells, while dividing, migrate tangentially towards the olfactory bulbs where they integrate as new interneurons. (**b**) An additional adult neurogenetic region is found in the subgranular zone (SGZ), which is located within the dentate gyrus of the hippocampus. A cellular hierarchy, somewhat similar to that of the SVZ, is seen in the SGZ in which the true stem cell is probably the type B astrocyte, which produces the intermediate type D precursor that eventually gives rise to the type G granule neurons. These neurons integrate functionally into the granule cell layer. (**c**) In the adult human brain, a population of SVZ astrocytes that is organized as a periventricular ribbon has been identified as comprising neural stem cells. In contrast to the rodent SVZ, no signs of tangential neuronal chain migration were detected from the corresponding human area. (**d**) The germinal zone of the adult human hippocampus is located within the dentate gyrus. Neurogenesis in this region has been demonstrated to take place in adult humans. Reproduced with permission from Reference [1]

Astrocytes that line the SVZ are capable of extensive self-renewal, proliferation, and multilineage differentiation. They are capable of generating neurons, astrocytes and oligodendrocytes in vitro. Additional regions that contain NSCs are the subgranular zone of the hippocampus and the subcortical white matter (Fig. 1). It is proposed that most gliomas, including GBM, arise from NSCs.

There is evidence that the precursor cells in MB arise from the granule cell precursors in the external granule layer of the cerebellum [2]. There is also some evidence that cerebellar stem cells may contribute to the development of MB [3].

Nestin and CD133 are expressed in NSCs during development. Stem cells positive for nestin and CD133 have been identified in primary human gliomas, including GBM [4–6]. Medulloblastomas express CD133 and also express CD146, another marker of CSCs [7].

While earlier identification and characterization of NSCs was determined from human fetal and adult brain studies, a more robust method for characterizing NSCs involves cell culture in media which allows for the isolation, expansion, and identification of NSCs. When NSCs are cultured in serum-free media with epidermal growth factor (EGF) and fibroblast growth factor (FGF)2, they proliferate and generate neurospheres, a floating cluster of proliferating cells, of which a small proportion are stem cells. These cells are multipotent, self-renewing, and expandable. This technique was reported in 1992 by Reynolds and Weiss who isolated a small population of NSCs from the adult striatum that could proliferate and generate neurospheres [8].

Singh et al. [9] applied the neurosphere technique to human brain tumor cells and generated tumor spheres (Fig. 2). The majority of the primary tumors studied were MBs. The tumor spheres, like neurospheres, were capable of self-renewal and proliferation. The tumor spheres generated from each of the specimens showed immunoreactivity for nestin and CD133. The sphere-forming population was found to reside in the fraction of primary brain tumor cells that express CD133. The CD133+ cells differentiated into multiple nervous system lineages (neurons, astrocytes, and oligodendrocytes. Other investigators have used the tumor neurosphere assay to further characterize the CSCs in GBM [10–12]. Brain tumor stem cells also express other molecular markers associated with neural stem cells/precursors, including Bmi1, Notch, and Jagged1. In addition, the tumor suppressor gene p53 is expressed in SVZ cells and is frequently deleted or mutated in gliomas [13, 14]. Investigators have subsequently found that the neurosphere assay is a more reliable method of characterizing brain tumors than are primary cell cultures [15, 16].

2.2 Breast

Until recently, the ability to isolate and characterize breast stem cells was limited by lack of suitable markers to identify these cells and systems to maintain them in vitro. The first evidence for stem cells in the breast was generated in rodent models in which the ability to repopulate mouse

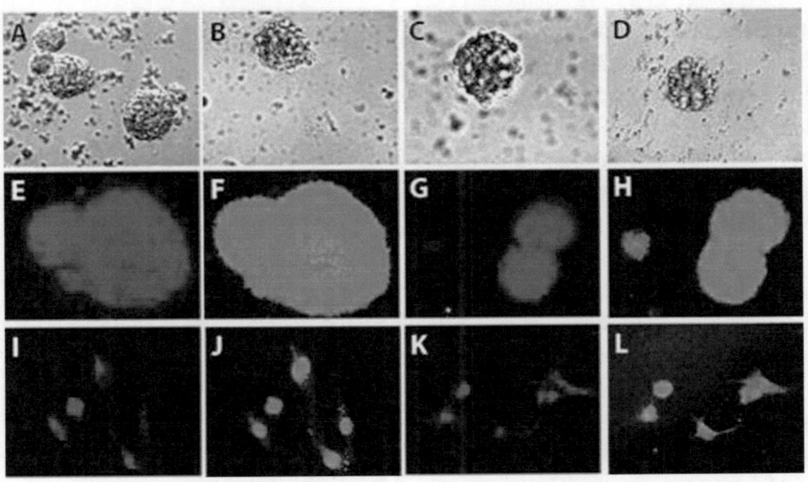

Fig. 2 Primary brain tumors of different phenotypes form neurosphere-like colonies. Photomicrographs of cultured brain tumor cells (magnification ×20) at 24–48 h after plating in TSM, containing EGF and bFGF. Each tumor subtype yielded growth of cells in neurosphere-like clusters, termed tumor spheres. Tumor spheres are shown from a medulloblastoma (*A*), pilocytic astrocytoma (*B*), ependymoma (*C*), and ganglioglioma (*D*). Undifferentiated primary tumor spheres from a medulloblastoma (*E*, *F*, *I*, and *J*) and a pilocytic astrocytoma (*G*, *H*, *K*, and *L*) are immunostained at 4 h for characteristic neural stem cell marker nestin (*E* and *G*) and for CD133 (*F* and *H*). Individual undifferentiated medulloblastoma sphere cells and astrocytoma sphere cells are also shown stained for nestin (*I* and *K*) and CD133 (*J* and *L*). Reproduced with permission from Reference [9]

mammary glands with serial transplantation of retrovirally marked epithelial fragments was demonstrated [17]. More recently, cell surface markers have been identified which can enrich for such cell populations. Cells that express CD29 and/or CD49F (β1 and α6) integrin respectively as well as CD24 display the stem cell properties of self-renewal and multilineage differentiation [18]. A single cell from the $CD29^{high}CD24+$ or $CD49F^{high}CD24+$ population is able to reconstitute a functional mammary gland when this is transplanted into the mouse mammary fat pad. Based on work demonstrating that primitive neural cells could survive in suspension in neurospheres, the development of the mammosphere technique was developed. In this method, primitive mammary cells form floating spherical colonies called mammospheres [19]. These mammosphere-forming cells display stem cell markers including aldehyde dehydrogenase 1 (ALDH1), a marker which is also expressed on NSCs [20].

3 Evidence That CSCs Identified Are Tumorigenic

3.1 Brain

The definitive identification of brain tumor CSCs, as in other tumors, requires the purification of distinct populations and the confirmation of their in vivo tumor-initiating capacity by serial transplantation in animal models. The tumors thus generated should have the same phenotype of the tumor from which it was derived. In 2004, Singh et al. [11] demonstrated that CD133+ cells isolated from human primary GBM could initiate a serially transplantable tumor in the brains of NOD-SCID mice that was a phenocopy of the primary tumor. The CD133- cells engrafted in the brain but did not result in a tumor (Fig. 3). Galli et al. [12] also showed that these CSCs can establish tumors that closely resemble the human disease, even when challenged through serial transplantation in mice.

It is possible that that are other immature cells, apart from CD 133+ cells, that are capable of initiating tumors in the brain. There is an abundant progenitor cell population that can divide and expand in response to CNS injury. These postnatal CNS progenitors are characterized by cell surface neuron-glial (NG) 2 expression and they have been identified in the postnatal SVZ and hippocampus. NG2+ cells are multipotent, differentiating into neurons and astrocytes and, in vivo, differentiating in all CNS lineages and pericytes. NG2+ cells identified in vitro in gliomas are more proliferative but less invasive than NG2− cells. Additional review of the relationship between NSCs and brain tumors is reviewed extensively in a recent publication [21].

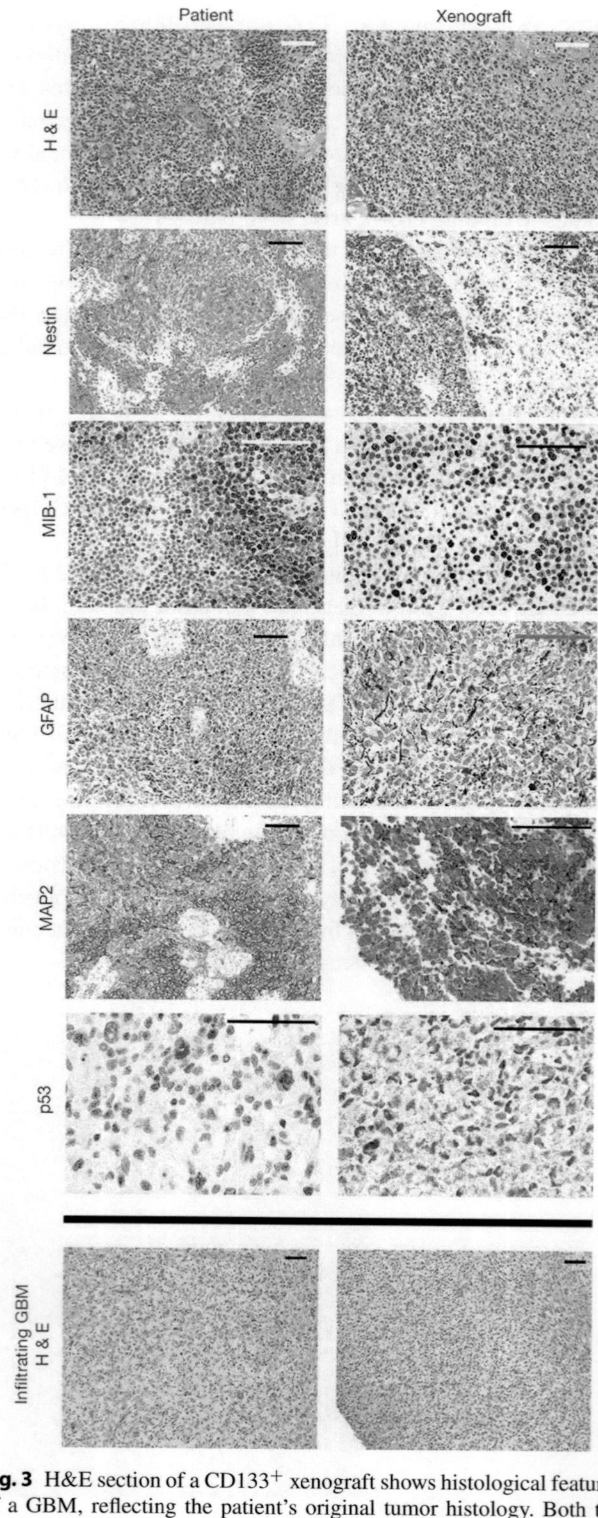

Fig. 3 H&E section of a $CD133^{+}$ xenograft shows histological features of a GBM, reflecting the patient's original tumor histology. Both the $CD133^{+}$ BTIC xenograft and the patient's original tumour show expression of the neural precursor marker nestin, high proliferative indices (MIB-1), p53 expression, and expression of both neuronal and astrocyte differentiated cell markers (MAP2 and GFAP). There are many invading human-specific nestin-positive tumor cells beyond the tumour borders in the xenograft. A diffusely infiltrative GBM is also depicted by H&E (*bottom*). Scale bar, 100 microns. Reproduced with permission from Reference [11]

3.2 Breast

The same markers that have been utilized to isolate and characterize normal rodent and human mammary stem cells also have been useful in identifying and isolating breast cancer stem cells. Cells isolated from human tumor xenografts which display stem cell properties are characterized by the phenotype CD44+CD24^{low}/lin–. As few as 200 of these cells, comprising 1–10% of the total population, were capable of forming tumors when implanted in NOD/SCID mice. In contrast, 20,000 cells that did not express these markers were unable to form tumors. Consistent with a cancer stem cell model, the stem cells were able to generate tumors that recapitulated the phenotypic heterogeneity of the initial tumor [22]. As was the case for normal breast stem cells, breast tumor stem cells also formed mammospheres and also expressed the enzyme ALDH1 [20].

Taken together, these studies demonstrate similar biological characteristics between brain cancer and breast cancer stem cells. Both form spherical colonies in vitro and can be isolated by expression of cell surface markers. Only cells expressing these markers are tumorigenic in NOD/SCID mice and generate tumors upon serial passage which regenerate the phenotypic heterogeneity of the initial tumors. These studies suggest the possibility that similar pathways may be involved in carcinogenesis in both of these tumor systems.

4 Oncogene Expression in Normal and Cancer Stem Cells

Developmental regulatory signal molecules that have been demonstrated to be involved in controlling normal stem cell self-renewal and in regulating lineage fate include Notch, Sonic hedge hog (Shh), Bmi1, and Wingless-type proteins (Wnts).

4.1 Brain

The Notch signaling pathway is required for maintenance of NSCs. Notch receptors and ligands are expressed in primary human GBMs [23]. In addition, Notch receptor and ligand transcripts have been identified in cultures of gliomas [24]. Notch signaling promotes nestin expression and may contribute to GBM CSC potential [25]. Fan et al. [26] have also demonstrated that Notch 1 or 2 transcripts are increased in MBs.

The best known signaling pathway abnormality in human brain tumors is the mutation of multiple sonic hedgehog (Shh) signaling pathways in a subset of MBs. Sonic hedgehog regulates cerebellar development [27]. It is thought that the inability to down-regulate Shh signaling in cerebellar external granule layer cells underlies the development of some MBs. Patched 1, Smo, and SuFu have also been identified. Bmi1 is a target of Shh signaling and is expressed in human MBs [28]. Because Shh plays a critical role in non-neoplastic stem cells, it has been suggested that stem-like neoplastic cells may also be susceptible to pathway blockade. Romer et al. [29] demonstrated the antitumor effect of blocking this pathway in a mouse model of MB by treating with HhAntag, a Smoothened inhibitor. The Wnt-β-catenin pathway has recently been implicated in the regulation of adult neurogenesis and may be involved in the pathogenesis of MBs.

The Shh pathway is also important in growth of the cerebral cortex and NSCs in the adult brain. It plays an important role in neural tube development and is abnormally activated in glioma formation. The transcription factors Gli1, Gli2, and Gli3 are activated by Shh and regulate stem cells by promoting DNA replication and cell-cycle entry. Gli1 is expressed by neuronal progenitors in the SVZ and the Shh-Gli pathways have been shown to maintain the stem cell population and promote the survival and proliferation of stem cell progeny. Ahn and Joyner [30] demonstrated that Shh acts on stem cell niches in the SVZ and SGZ at late embryonic stages. Gli1 is expressed in gliomas of varying grades, including GBM, and the Shh-Gli pathway has been implicated in the initiation and maintenance of gliomas [31]. Bar et al. [32] demonstrated that markers of pathway activity are detected in malignant gliomas and that the pathway appears to regulate the stem cell fraction in GBM cell lines. Sonic hedgehog blockade inhibits the growth of some glioma cell lines.

4.2 Breast

Many of the developmental pathways that have been found to regulate stem cell fate determination of NSCs are also key regulators of breast stem cells. Evidence for the role of Notch signaling in mammary development has been provided by transgenic models. Dontu et al. demonstrated that Notch activation acts as a regulator of asymmetric cell fate decisions in human mammary stem cells by promoting self renewal [33]. In murine models, the targeting of Notch receptors to the mammary gland produces breast cancers [34]. Furthermore, there is evidence for defective Notch signaling in human breast cancers. Up to 40% of human breast cancers have reduced expression of the Notch inhibitor NUMB which results in aberrant Notch signaling [35].

The Shh pathway and the transcription factor Bmi1 also have been found to play an important role in mammary stem cell self-renewal. Transgenic models suggested a role for hedgehog signaling in normal mammary development and carcinogenesis. Furthermore, there is evidence for disregulation of this pathway in a substantial subset of human breast carcinomas. Liu et al. demonstrated that hedgehog signaling regulates self-renewal of both normal and malignant mammary stem cells. This occurs through regulation of the polycomb gene Bmi1 [36]. There is also substantial evidence that Wnt signaling is important in mammary stem cell biology. When Wnt is targeted to the mammary gland, mice develop mammary tumors that display an increased stem cell content [37]. There is also evidence for aberrant Wnt signaling in human mammary cancers [38].

Another important tumor suppressor gene frequently involved in both brain and breast tumors is the PTEN gene. PTEN defects are found in approximately 40% of human breast cancers [39]. PTEN is a lipid phosphotase which regulates PI3 kinase AKT signaling. PTEN has been shown to regulate self-renewal of both hematopoietic and neural stem cells and we have evidence that PTEN has similar effects in normal and malignant breast stem cells (unpublished observation).

5 Growth Factor Signaling Pathways in Normal and Cancer Stem Cells

Genes that encode growth factors and the receptors that control glial cell differentiation are frequently overexpressed in gliomas. Growth factor-signaling pathways play in important role in NSC regulation and gliomagenesis. Epidermal growth factor (EGF)-responsive cells within the SVZ constitute a large population of migratory, rapidly dividing cells. Nearly half of high grade gliomas demonstrate EGFR amplification. Platelet-derived growth factor (PDGF)-responsive cells have also been identified in SVZ stem cells. Excessive PDGF activation induces SVZ NSC proliferation [40]. PDGF overexpression is frequent in malignant gliomas, especially those of oligodendroglial origin. Vascular endothelial growth factor (VEGF) and its receptors are expressed in cultured neurospheres derived from the adult SVZ. Stem cell-like glioma cells secrete markedly elevated levels of VEGF [41]. Vascular endothelial growth factor acts on tumor endothelial cells which express VEGF receptors. Excess of growth factors in tumors leads to autocrine stimulation and increased activity of pathways downstream of these receptors, including those that lead to the Ras and Akt pathways.

6 The Stem Cell Niche (Microenvironment)

The stem cell niche is defined as the local microenvironment of stem cells that functions to maintain stem cell self-renewal and multipotency. It generates extrinsic factors that control stem cell number, proliferation, and fate determination. It is not known if CSCs become independent of the niche. The niche is also a physical anchoring site for stem cells and consists of extracellular matrix (ECM) components and adjacent tissue non-stem cells. Adhesion molecules are involved in the interaction between stem cells and the niche and between stem cells and the ECM.

6.1 Brain

Neural tissue is enriched with several cadherins that facilitate cell to cell adhesion. N-cadherin is expressed by neurons and glial cells [42]. The intracellular regions of cadherins interact with catenins. β-Catenin is involved in the proliferation and differentiation of NSCs [43]. Deregulation of β-catenin signaling is an important event in the formation of a variety of tumors, including MBs, and is a key player in the Wnt signaling pathway. Hypoxia within the niche, similar to hematopoietic stem cells in the bone marrow, is a fundamental condition to promote the survival proliferation of certain populations of NSCs/progenitor cells. Hypoxia may regulate self-renewal versus differentiation via activation of the transcriptional factor hypoxia inducible factor (HIF) 1. In hypoxic conditions, Hif1-a is stabilized and translocates to the nucleus to activate and to regulate downstream targets, including VEGF, VEGF receptor, and matrix metalloproteniases, among others. These factors are strongly involved in the angiogenesis and migration of GBMs [44].

6.2 Breast

There is also evidence that the tumor microenviroment plays a fundamental role in breast carcinogenesis. For example, breast density, which is related to characteristics of breast stromal fibroblasts [45] is an important risk factor for breast cancer development [46]. Growth factors produced by stromal elements may play an important role in regulating mammary stem cell behavior. In addition to mammary fibroblasts, the role of endothelial cells and adiposities in mammary stem cell behavior is currently being investigated.

7 Evidence That CSCs Provide Treatment Resistance

7.1 Brain

In most patients with GBM, the effectiveness of radiation therapy is partial and temporary, which has long suggested that there is inherent radioresistance of a cell population in these tumors. Bao et al. [47] showed that CSCs contribute to glioma radioresistance through preferential activation of the DNA damage checkpoint response and increase in DNA repair capacity. CD133+ glioma cells in cell culture and the brain of immunocompromised mice survived ionizing radiation more than CD133- cells. In addition, the CD133+ cells isolated from human glioma xenografts and primary GBM specimens preferentially activated the DNA damage checkpoint response to radiation and repaired radiation-induced DNA damage more effectively than CD133- tumor cells (Fig. 4).

7.2 Breast

There is also evidence for radiation resistance of breast cancer stem cells compared to more differentiated cells within the tumor. These CD44+CD24– cells showed relative radiation resistance compared to cells without this phenotype [48].

It has been demonstrated that leukemic progenitor cells are less sensitive to chemotherapy than non-stem leukemic cells [49]. It is thought that the resistance of many CSCS may be the high expression of ATP-binding cassette drug transporters, which can protect cells from cytotoxic agents. A wide variety of drug transporters have been demonstrated on normal stem cells [50, 51]. It is possible that such resistance to chemotherapy exists in the CSC of brain and breast cancers.

8 Therapeutic Approaches to Target CSCs

As mentioned previously, according to the CSC theory of cancer, the limitation of standard cytotoxic therapy for solid tumors is that the small subpopulation of CSCs will not be

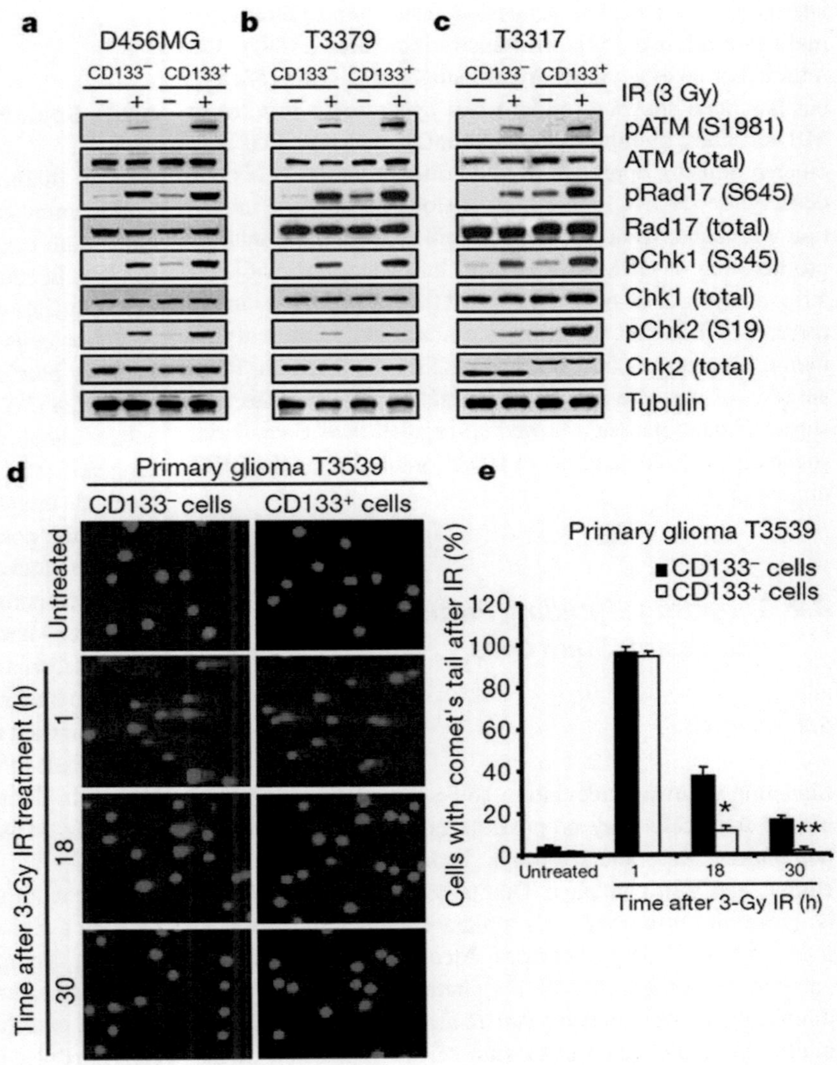

Fig. 4 (**a**–**c**) The activation state of the checkpoint response in matched CD133$^+$ and CD133$^-$ cells from glioma D456MG xenografts (**a**) and glioblastoma specimens (**b**, **c**) was assessed before treatment (–) and 1 h after 3 Gy of IR (+). Whole-cell lysates were immunoblotted for phosphorylated and total amounts of checkpoint proteins (ATM, Rad17, Chk2 and Chk1). (**d**) CD133$^+$ and CD133$^-$ cells derived from primary glioblastoma T3539 were irradiated with 3 Gy of IR. The presence of DNA damage at sequential time points after damage was assessed by single-cell gel electrophoresis assay under alkaline conditions (alkaline comet assay). (**e**) Quantification of the percentages of cells with comet tails at different time points after IR in CD133$^+$ and CD133$^-$ populations. Data are the means ±s.d. ($n = 100$ cells in three trials; $*P < 0.001$; $**P$ 0.002). Reproduced with permission from Reference [47]

targeted and will allow for unchecked expansion of this subpopulation. The most effective method to target these CSCs has not been elucidated, but a number of possibilities exist, including the administration of differentiating agents, targeting the signaling pathways of the CSCs, normal stem cells, or niche, targeting the DNA checkpoint response, and using normal stem cells to home to the region of tumor. It is probable that a combination of one or more of these methods will be required to effectively eradicate the CSC population.

8.1 Differentiating Agents

Because the differentiation of normal stem cells is accompanied by a reduced cell proliferation, it is conceivable that differentiating agents could suppress CSC cell division in a similar way. One such differentiating agent is retinoic acid, which has been studied in malignant gliomas and MBs. Cis-retinoic acid (CRA) shows modest activity when administered after a complete remission from initial therapy for malignant gliomas [52]. All-trans-retinoic acid (ATRA) has minimal or no effect in recurrent malignant gliomas [53, 54], but has been shown to induce cell cycle arrest in vitro in MB cell lines, by inhibition of cyclinD1 or C-myc [55]. Another promising approach to CSC differentiation is histone deacetylase (HDAC) inhibition. Histone deacetylase inhibition with suberoylanilide hydroxamic acid (SAHA) inhibits proliferation of GBM in cells lines and xenografts [56]. Interim analysis of an ongoing Phase 2 trial of SAHA in human recurrent GBM met the primary efficacy endpoint, with a 6 month progression free survival of 23%. In addition, RNA analysis of baseline and post-treatment tumor samples in a subset of these patients showed up-regulation of E-cadherin, suggesting a biologic effect of HDAC inhibition on the GBM tumors [57].

8.2 Targeting Signaling Pathways in Normal and Cancer Stem Cells

8.2.1 Notch

Signaling pathways regulating self-renewal of normal and cancer stem cells may be possible targets against the CSC population. Relevant pathways include the Notch, Shh, Bmi-1, and Wnt pathways. One target for Notch inhibition is presenilin g-secretase, a protease which cleaves and releases Notch from membranes. An elegant model of Notch pathway inhibition of MB by pharmacologic γ-secretase inhibition was reported by Fan et al. In this model, Notch inhibition resulted in suppression of the expression of the pathway target Hes1, led to cell cycle exit, apoptosis, and differentiation of the MB cell lines. In addition, viable populations of better-differentiated cells continued to grow when Notch activation was inhibited but were unable to efficiently form soft-agar colonies or tumor xenografts, suggesting that a cell fraction required for tumor propagation had been depleted [58] (Fig. 5).

As noted previously, almost half of human breast cancers also show defects in the Notch signaling pathway. Furthermore, there is evidence that Notch plays an important role in stem cell self-renewal in these systems [34]. Gamma secretase inhibitors have been shown to be able to effect breast stem cells in vitro and in animal models [59]. Interestingly, gamma secretase inhibitors have been developed for the treatment of Alzheimer's disease. Clinical trials utilizing similar gamma secretase inhibitors in combination with chemotherapy for women with advanced breast cancer are being initiated. Such trials will directly test the hypothesis that targeting breast cancer stem cells improves the therapeutic outcome in these women.

8.2.2 Sonic Hedgehog

Growth inhibition of murine MB by Shh blockade using cyclopamine was shown by Berman et al. [60] in 2002. Cyclopamine can also inhibit the growth of glioma cell lines in vitro. In addition, pretreatment of mice with cyclopamine before injection of human GBM cells intracranially resulted in tumor cells that failed to form tumors. Sonic hedgehog pathway blockade by cyclopamine was also effective in targeting the CSCs in GBM that express Gli1, but not in those cells without evidence of Shh pathway activity, as might be expected (Fig. 6). Bar et al. [32] also identified that Shh blockade targeted some GBM cells not affected by radiation, raising the possibility that combined therapy with radiation and Shh blockade could be effective in GBM. Further work with cyclopamine is needed before it can be tested in humans. It is necessary to determine the maximally tolerated dose, particularly on normal stem cells.

Since only a subset of MBs are driven by the Shh pathway, it is important to seek other targets. Sasai et al. [61] showed that a small-molecule inhibitor of Smoothened inhibited MB growth in Cxcr6 mice, suggesting that this agent may be effective in MBs that lack mutations in known Shh pathway genes.

In light of the evidence sited above for an important role in hedgehog signaling in breast stem cell self-renewal, studies utilizing cyclopamine to inhibit this pathway have been performed in animal models [36]. These studies indicate that, as is the case for brain cancer stem cells, breast cancer stem cells may also be targets for hedgehog inhibitors.

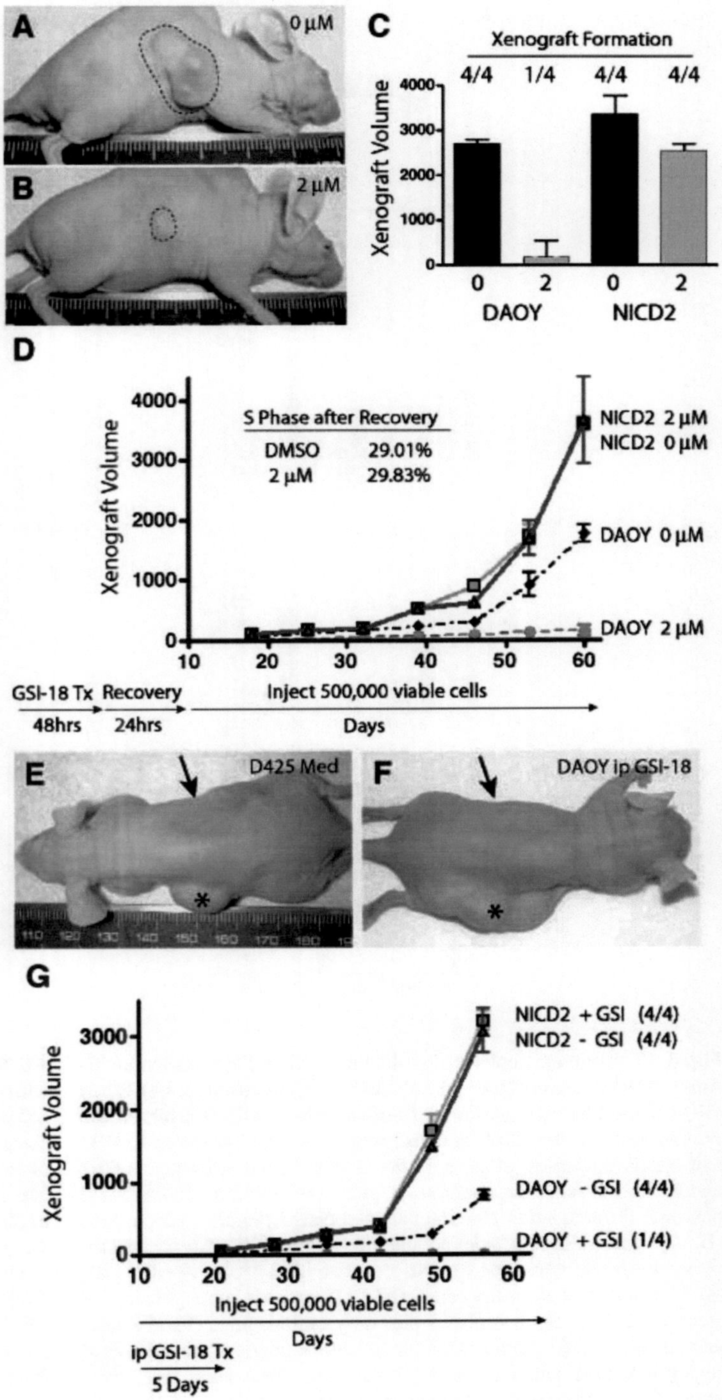

Fig. 5 Notch pathway inhibition blocks xenograft formation. *A* to *C*, bulk xenografts developed at all 12 sites injected with cells in which Notch signaling was active. This included vehicle-treated DAOY ($n = 4$), vehicle-treated DAOY:NICD2 ($n = 4$), and 2 μmol/L GSI-18–treated DAOY:NICD2 ($n = 4$; *A*). Viable cells remaining following Notch blockade with 2 μmol/L GSI-18 formed a tumor mass (*B*) at only one of four injection sites. (*D*) After recovering for 24 h in media containing 10% FBS, GSI-18- and vehicle-treated cultures have identical proliferation rates as determined by flow cytometric analysis of S-phase fraction (*inset*). Nevertheless, cells that had experienced Notch pathway blockade were still unable to efficiently form tumors, with small lesions developing at only two of four injection sites. As before, large xenografts formed at all 12 control sites. (*E*) D425Med cells treated with GSI-18 for 48 h also failed to generate xenografts (*arrow*) whereas vehicle-treated cells always did (*asterisk*). (*F* and *G*) *In vivo* administration of GSI-18 i.p. for 5 days after subcutaneous tumor injection also blocked engraftment of DAOY (*arrow*) but not of DAOY:NICD2 (*asterisk*). Reproduced with permission from Reference [58]

8.3 Targeting Growth Factors

8.3.1 Brain

The autocrine loop of EGFR and its ligands that are present in GBM can be targeted by EGFR tyrosine kinase (TK) inhibitors such as gefitinib and erlotinib. Although EGFR is frequently amplified, overexpressed, or mutated in GBM, only 10–20% of patients show a response to EGFR TK inhibitors. Mellinghoff et al. [62] demonstrated that coexpression of EGFRvIII and PTEN in GBM is associated with responsiveness to gefitinib or erlotinib. Attempts to block PDGF with single agent imatinib in recurrent malignant gliomas demonstrate only minimal activity [63].

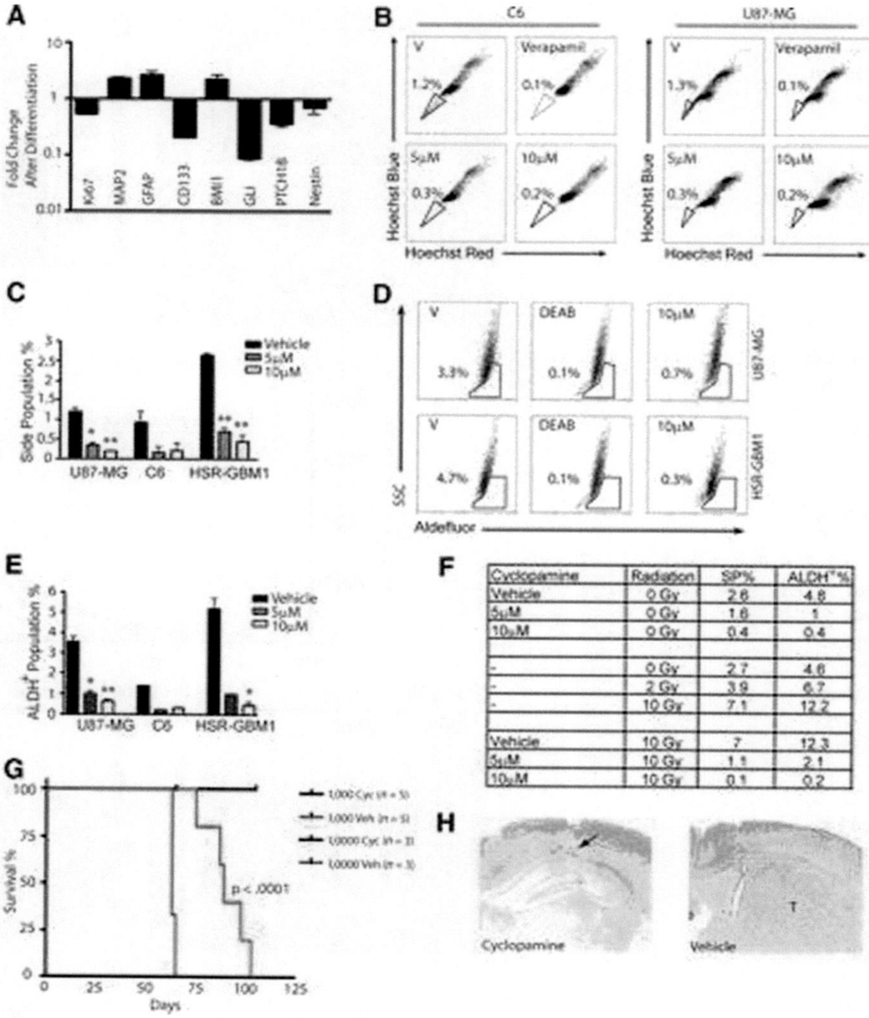

Fig. 6 Cyclopamine depletes stem-like cells from glioblastoma multiforme (GBM) neurospheres and blocks tumor engraftment. (**A**) When GBM neurospheres were forced to differentiate, mRNA levels of the proliferation marker Ki67 and the neural stem-cell markers CD133 and Nestin decreased, along with the Hedgehog pathway targets Gli1 and Ptch1b. In contrast, markers of glial differentiation (GFAP) and neuronal differentiation (microtubule-associated protein 2) increased. (**B**, **C**) The stem-like side population was significantly decreased by cyclopamine (two-sided *t* test; *, $p < .05$; **, $p < .01$). (**D**, **E**) The aldehyde dehydrogenase (ALDH)-positive subpopulation of stem/progenitor cells was also depleted by cyclopamine. (**F**) The side- and ALDH-positive stem/progenitor cell subpopulations were depleted by cyclopamine and increased by radiation when each was given alone. Cotreatment with cyclopamine and radiation depleted side- and ALDH-positive subpopulations. (**G**, **H**) HSR-GBM1 cultures pretreated with vehicle always engrafted when either 1000 or 10,000 viable cells counted at the end of treatment were injected into the athymic mice. In contrast, when the same number of viable cells was injected after 7 days of treatment with 10 μM cyclopamine, no detectable tumors formed (the arrow points to degenerative changes along the injection needle tract). Log-rank analysis of Kaplan-Meier survival curves indicates that the prolongation of survival associated with cyclopamine pretreatment is significant ($p < .0001$). Abbreviations: ALDH, aldehyde dehydrogenase; Cyc, cyclopamine; DEAB, 4-(diethylamino)benzaldehyde; GFAP, glial fibrillary acidic protein; SP, side population; SSC, side scatter; T, *t* test; V, vehicle; Veh, vehicle. Reproduced with permission from Reference [32]

The Wnt/β-catenin pathway is recognized to play a critical role in the malignant transformation of cancer progenitor cells and/or their early progeny during cancer progression through the transcription factors Slug, Snail, and Twist 1.Inhibition of this pathway can be accomplished by a selective Wnt antibody,Wnt protein inhibitors, or repressors disrupting nuclear TLF/β-catenin complexes.

8.3.2 Breast

An important gene that is disregulated in human breast cancers is the HER-2 gene. Approximately 20–25% of human breast cancers display amplification of HER-2 which is a member of the EGF receptor signaling family. Tumors that display increased HER-2 expression are more aggressive and metastatic [39]. Interestingly, these tumors also display increased expression of the stem cell marker ALDH1 [20]. We have recently found that HER-2 overexpression in normal human mammary epithelial cells as well as mammary carcinomas increases the proportion of stem cells as indicated by aldehyde dehydrogenase expression. The monoclonal antibody trastuzumab that blocks HER-2 signaling reduces the stem cell population in trastuzumab sensitive but not resistant breast cancer cell lines [64] . This suggests that HER2 may play a role in mammary carcinogenesis by regulating the stem cell population. The clinical efficacy of inhibitors of this pathway such as trastuzumab may directly relate to the ability of these agents to target breast cancer stem cells.

8.4 Targeting Treatment Resistance

8.4.1 Brain

In a study of radioresistance of CD133+ GBM cells, Bao et al. [47] showed that the radioresistance of the CD133+ cells could be reversed with a specific pharmacologic inhibitor of the Chk1 and Chk2 checkpoint kinases. Therapies to target the checkpoint kinases currently in preclinical and clinical development may provide a method to disrupt the radiation resistance of glioma CSCs and thus improve the therapeutic effectiveness of radiation in treating GBM.

8.5 Other Approaches

Another CSC-targeted approach is to administer normal stem cells to deliver immunotherapy or other treatments and viral therapy. This is a rational approach for brain tumors because NSCs exhibit tropism for tumor cells in the CNS [65]. There is some evidence that immunotherapy can target CSCs in MB [7]. Additionally, flow cytometry analyses of glioma samples reveals that the majority of CD133+ cells do not express detectable MHC I or NK cell activating ligands, which may render them resistant to adaptive and innate immune surveillance. Incubating CD133+ cells in interferon gamma (INF-γ) significantly increased the percentage of CD133+ cells that expressed MHC I and NK cell ligands. Also pretreatment of CD133+ cells with INF-γ rendered them sensitive to NK cell-mediated lysis in vitro. Therefore, CD133+ and CD133– glioma cells may be similarly resistant to immune surveillance, but INF-γ may partially restore their immunogenicity and potentiate their lysis by NK cells [66]. Finally, Jiang et al. [67] recently reported a mouse model of GBM derived from human GBM CSCs is effectively treated that with a virus, Delta-24-RGD.

8.6 Combination Treatments

Clinical experience has demonstrated that the administration of agents targeting only one signaling pathway is not effective in the majority of GBMs (see above). This is because there is likely to be activation of multiple receptor tyrosine kinases (RTKs). Stommel et al. [68] demonstrated that GBM tumor cell lines, xenotransplants, and primary tumors have concomitant activation of multiple RTKs (Fig. 7).

Future therapies will likely include a multitargeted approach, such as combinations of TK inhibitors, differentiating agents, cytotoxic agents or other approaches. Recent examples of effective combination therapy in experimental GBM CSC systems include the combination of RA with okadaic acid (an inhibitor of the N-CoR pathway) [69] and the combination of ATRA and interferon-γ in GBM cell lines [70]. Combination ATRA and paclitaxel-induced differentiation and apoptosis were demonstrated by Karmakar et al. [71] in U87MG xenografts. Loss of the PTEN tumor suppressor gene and amplification of EGFR are common in GBM, resulting in activation of mammalian target of rapamycin (mTOR). Rapamycin, an inhibitor of mTOR, enhances the sensitivity of PTEN-deficient GBM cells to erlotinib [72], suggesting that this may be a rational therapeutic combination to study in GBM patients, especially those with PTEN-deficient tumors. A recent pilot study of combination EGFR (gefitinib or erlotinib) and the mTOR inhibitor sirolimus in recurrent malignant gliomas resulted in a promising 6 month progression free survival [73]. The combination of rapamycin and a nitrosurea appears effective in treating murine xenografts of malignant gliomas [74]. Whereas imatinib alone is not effective in malignant gliomas, the combination of imatinib and hydryoxurea for recurrent malignant gliomas appears promising [75]. Another combination of standard cytotoxic chemotherapy with a TK inhibitor includes a recent dose escalation study of erlotinib and temozolomide in stable or recurrent malignant gliomas [76]. In addition, a trial of radiation therapy, erlotinib, and temozolomide in newly diagnosed GBM is ongoing [77]. Recently, the combination of bevacizumab, an antibody to the VEGF receptor, and irinotecan has shown significant efficacy in recurrent GBM, resulting in a 6 month progression free survival of 46% [78].

9 Timing of Therapy

An important area of investigation is to determine the optimal timing of CSC-targeted therapies with other antineoplastics. Such schedules could include the coadministration of agents for newly diagnosed tumor, after a remission to standard treatment is obtained, or at the time of progression after

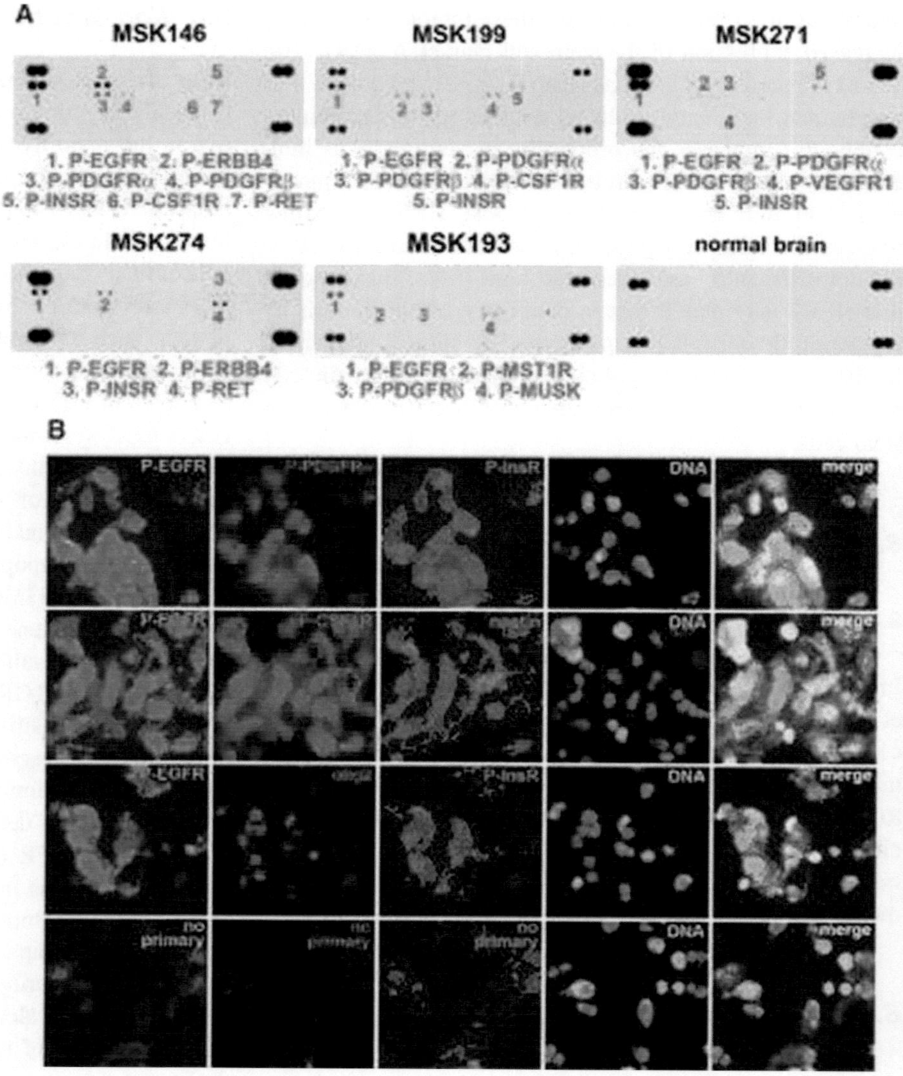

Fig. 7 Multiple RTKs are concomitantly activated in primary GBMs. (**A**) Antibody arrays were performed on protein lysates extracted from snap-frozen primary human gliomas or normal brain autopsy material. (**B**) Coexpression of phospho-RTKs in cells dissociated from primary GBM MSK199. Individual tumor-derived cells were immunofluorescently stained with the phospho-RTK antibodies P-EGFR, P-PDGFR, P-InsR, or P-CSF1R. Each row depicts one field of cells from a slide simultaneously stained with the indicated antibodies. DNA is labeled with Hoechst 33342. Nestin is expressed in neural progenitor cells, tumor endothelial cells, and diffuse gliomas, including astrocytomas and GBMs, and olig2 is expressed in neural progenitors, normal oligodendroglia, and diffuse gliomas. The bottom row depicts cells stained only with secondary antibodies. Reproduced with permission from Reference [68]

standard treatment. In breast cancer, recent studies utilizing a neoadjuvant trial design have permitted the direct testing of the hypothesis that cancer stem cells are relatively resistant to chemotherapy. This trial design permits the assessment of cells with a stem cell phenotype before therapy as well as after therapy. Following adjuvant chemotherapy, there is up to a ten-fold increase in cells which are capable of mammosphere formation as well as with the cell surface phenotype CD44+CD24–, both properties of cancer stem cells [79]. Interestingly, agents that target HER-2, such as lapatinib, were able to decrease this cell population in the neoadjuvant setting [80]. This is consistent with studies suggesting that HER-2 is expressed on breast cancer stem cells. Furthermore, these studies demonstrate that the neoadjuvant trial design is an excellent one to assess the ability of agents to target breast cancer stem cells. Whether these intermediate endpoints correlate with clinically validated outcomes such as patient survival remains to be proven.

The CSC model has an important impact on the identification of future therapeutic targets [81]. By directing expression analyses to enriched populations of CSCs, the identification of novel diagnostic markers and novel therapeutic targets should be more effective than current standard antineoplastic therapies. It remains to be determined whether identification of cell surface markers and profiling of signaling pathways prior to selecting treatment in individual patients is practical. Another challenge in targeting CSC pathways of self renewal is to spare these same pathways in normal stem cells.

10 Methods to Determine the Effectiveness of Therapy Against CSCs

Because CSCs represent only a minority of cells within a tumor and because they are not rapidly dividing, standard imaging response measurements are a poor measure of determining treatment efficacy. In malignant brain tumors, response measures do not have a strong correlation with patient survival and a more meaningful response measure is time to progression, for which reliable endpoints of progression free survival have been established in Phase 2 trials for recurrent GBM [82]. This method of measuring response is applicable to therapy targeting CSCs, as the treatment is designed to prevent repopulation with actively growing cells. There currently exists a window of opportunity also to investigate whether functional imaging techniques, such as diffusion- or perfusion-weighted MRI or thymidine positron emission tomography will be able to identify efficacy early in treatment, as demonstrated thus far in GBM [83]. Manganas et al. [84] recently reported the identification of human hippocampal neural progenitor cells using a specialized signal processing method with proton magnetic resonance spectroscopy. It is not known whether brain tumor CSCs can be imaged similarly. In addition, assessment of the biologic activity of therapy should be sought when targeting identified pathways of self-renewal or differentiation in CSCs [85].

11 Prevention

11.1 Brain

An important area to explore is the targeting of molecular pathways in premalignant or low-grade CNS lesions, in order to prevent transformation into malignant and locally invasive neoplasms. Examples of such lesions include low-grade gliomas, which undergo malignant transformation in up to 50% of cases. Another critical area to explore is whether CSC-targeted therapy will impact the development of malignant tumors in the inherited neurocutaneous disorders with a high propensity to develop CNS tumors, such as neurofibromatosis type I and 2 and von Hippel Lindau disease.

11.2 Breast

The cancer stem cell hypothesis suggests that strategies that target the breast stem cell population may prove effective for cancer prevention and therapy. Agents such as phytoestrogens may have protective effects when administered during adolescent growth periods when stem cell number is determined [86]. Chemopreventive agents such as curcumin from turmeric may function by modulating stem cell self-renewal pathways including Wnt and Notch [87, 88]. In addition, dietary polyphenoids including quercetin and epigallocatechingallate (EGCG), have been shown to modulate Wnt and Notch signaling [89]. Vitamin D3 has also been shown to be involved in stem cell differentiation [90]. The ability of these compounds to affect mammary tumorigenesis by affecting initiated stem cells remains to be proven.

12 Summary

In summary, in two tumor types in which cancer stem cells have been most studied, the brain and the breast, there is increasing evidence to support the cancer stem cell hypothesis. The development of new in vitro and animal systems have been extremely useful to isolate and further characterize these cells. Studies utilizing these systems suggest similar pathways are utilized to regulate stem cell behavior in different organ sites. Pathways which regulate cell behavior may be key targets for the development of innovative prevention and treatment strategies.

References

1. Vescovi AL, Galli R, Reynolds BA. Brain tumour stem cells. Nat Rev Cancer. 2006;6:425–35.
2. Oliver TG, Read TA, Kessler JD, Mehmeti A, Wells JF, Huynh TT, et al. Loss of patched and disruption of granule cell development in a pre-neoplastic stage of medulloblastoma. Development. 2005;132:2425–39.
3. Lee A, Kessler JD, Read TA, Kaiser C, Corbeil D, Huttner WB, Johnson JE, Wechsler-Reya RJ. Isolation of neural stem cells from the postnatal cerebellum. Nat Neurosci. 2005;8:723–9.
4. Dahlstrand J, Collins VP, Lendahl U. Expression of the class VI intermediate filament nestin in human central nervous system tumors. Cancer Res. 1992;52:5334–41.
5. Uchida K, Mukai M, Okano H, Kawase T. Possible oncogenicity of subventricular zone neural stem cells: case report. Neurosurgery. 2004;55:977–8.
6. Strojnik T, Røsland GV, Sakariassen PO, Kavalar R, Lah T. Neural stem cell markers, nestin and musashi proteins, in the progression of human glioma: correlation of nestin with prognosis of patient survival. Surg Neurol. 2007;68:133–44.
7. Castriconi R, Dondero A, Negri F, Bellora F, Nozza P, Carnemolla B, et al. Both CD133^{+} and CD 133^{-} medulloblastoma cell lines express ligands for triggering NK receptors and are susceptible to NK-mediated cytotoxicity. Eur J Immun. 2007;37:3190–6.
8. Reynolds BA, Weiss S. Generation of neurons and astrocytes from isolated cells of the adult mammalian central nervous system. Science. 1992;255:1707–10.
9. Singh SK, Clarke ID, Terasaki M, Bonn VE, Hawkins C, Squire J, et al. Identification of a cancer stem cell in human brain tumors. Cancer Res. 2003;63:5821–8.

10. Hemmati HD, Nakano I, Lazareff JA, Masterman-Smith M, Geschwind DH, Bronner-Fraser M, et al. Cancerous stem cells can arise from pediatric brain tumors. Proc Natl Acad Sci U S A. 2003;100:15178–83.
11. Singh SK, Hawkins C, Clarke ID, Squire JA, Bayani J, Hide T, et al. Identification of human brain tumour initiating cells. Nature. 2004;432:396–401.
12. Galli R, Binda E, Orfanelli U, Cipelletti B, Gritti A, De Vitis S, et al. Isolation and characterization of tumorigenic, stem-like neural precursors from human glioblastoma. Cancer Res.2004;64:7011–21.
13. Barami K. Biology of the subventricular zone in relation to gliomagenesis. J Clin Neurosci. 2007;14:1143–9.
14. Medrano S, Burns-Cusato M, Atienza MB, Rahimi D, Scrable H. Regenerative capacity of neural precursors in the adult mammalian brain is under the control of p53. Neurobiol Aging. 2007 [Epub ahead of print].
15. De Witt Hamer PC, Van Tilborg AAG, Eijk PP, Sminia P, Troost D, Van Noorden CJF, et al. The genomic profile of human malignant glioma is altered early in primary cell culture and preserved in spheroids. Oncogene. 2008;27:2091–6.
16. Lee J, Kotliarova S, Koliarov Y, Li A, Su Q, Donin NM, et al. Tumor stem cells derived from glioblastomas cultured in bFGF and EGF more closely mirror the phenotype and genotype of primary tumors than do serum-cultured cell lines. Cancer Cell. 2006;9:391–403.
17. Kordon EC, Smith GH. An entire functional mammary gland may comprise the progeny from a single cell. Development. 1998;125:1921–30.
18. Stingl J, Eirew P, Ricketson I, et al. Purification and unique properties of mammary epithelial stem cells. Nature. 2006;439:993–7.
19. Dontu G, Abdallah WM, Foley JM, et al. In vitro propagation and transcriptional profiling of human mammary stem/progenitor cells. Genes Dev. 2003;17:1253–70.
20. Ginestier C, Hur M, Charafe-Jauffret E, et al. ALDH1 is a marker of normal and malignant breast stem cells and a predictor of poor clinical outcome. Cell Stem Cell. 2007;1:555–67.
21. Dirks PB. Brain tumour stem cells: the undercurrents of human brain cancer and their relationship to neural stem cells. Phil Trans R Soc B. 2008;363:139–52.
22. Al-Hajj M, Wicha MS, Benito-Hernandez A, Morrison SJ, Clarke MF. Prospective identification of tumorigenic breast cancer cells. Proc Natl Acad Sci U S A. 2003;100:3983–8.
23. Purow BW, Haque RM, Noel MW, Su Q, Burdick MJ, Lee J, et al. Expression of Notch-1 and its ligands, Delta-like-1 and Jagged-1, is critical for glioma cell survival and proliferation. Cancer Res. 2005;65:2353–63.
24. Ignatova TN, Kukekov VG, Laywell ED, Saslov ON, Vrionis FD, Steindler DA. Human cortical glial tumors contain neural stem-like cells expressing astroglial and neuronal markers in vitro. Glia. 2002:39:193–206.
25. Shih AH, Holland EC. Notch signaling enhances nestin expression in gliomas. Neoplasia. 2006;8:1072–82.
26. Fan X, Mikolaenko I, Elhassan I, Ni X, Wang Y, Ball D, Brat DJ, Perry A, Eberhart CG. Notch1 and notch2 have opposite effects on embryonal brain tumor growth. Cancer Res. 2004;64: 7787–93.
27. Dahmane N, Ruiz-iAltaba A. Sonic hedgehog regulates the growth and patterning of the cerebellum. Development 1999;126:3089–3100.
28. Leung C, Lingbeek M, Shakhova O, Liu J, Tanger E, Saremaslani P, et al. Bmi1 is essential for cerebellar development and is overexpressed in human medulloblastomas. Nature. 2004;428: 337–41.
29. Romer JT, Kimura H, Magdaleno S, Sasai K, Fuller C, Baines H, et al. Suppression of the Shh pathway using a small molecule inhibitor eliminates medulloblastoma in Ptc1(+/-)p53(-/-) mice. Cancer Cell. 2004;6:229–40.
30. Ahn S, Joyner AL. *In vivo* analysis of quiescent adult neural stem cells responding to Sonic hedgehog. Nature. 2005;437:894–7.
31. Dahmane N, Sanchez P, Gitton Y, Palma V, Sun T, Beyna M, et al. The Sonic Hedgehog-Gli pathway regulates dorsal brain growth and tumorigenesis. Development. 2001;24:5201–12.
32. Bar EE, Chaudhry A, Lin A, Fan X, Schreck K, Matsui W, et al. Cyclopamine-mediated hedgehog pathway inhibition depletes stem-like cancer cells in glioblastoma. Stem Cells. 2007;25: 2524–33.
33. Dontu G, Jackson KW, McNicholas E, Kawamura MJ, Abdallah WM, Wicha MS. Role of Notch signaling in cell-fate determination of human mammary stem/progenitor cells. Breast Cancer Res. 2004;6:R605–15.
34. Brennan K, Brown AM. Is there a role for Notch signalling in human breast cancer? Breast Cancer Res. 2003;5:69–75.
35. Pece S, Serresi M, Santolini E, et al. Loss of negative regulation by Numb over Notch is relevant to human breast carcinogenesis. J Cell Biol. 2004;167:215–21.
36. Liu S, Dontu G, Mantle ID, et al. Hedgehog signaling and Bmi-1 regulate self-renewal of normal and malignant human mammary stem cells. Cancer Res. 2006;66:6063–71.
37. Li Y, Welm B, Podsypanina K, et al. Evidence that transgenes encoding components of the Wnt signaling pathway preferentially induce mammary cancers from progenitor cells. Proc Natl Acad Sci U S A. 2003;100:15853–8.
38. Wang D, DuBois RN. Cyclooxygenase 2-derived prostaglandin E2 regulates the angiogenic switch. Proc Natl Acad Sci U S A. 2004;101:415–6.
39. Panigrahi AR, Pinder SE, Chan SY, Paish EC, Robertson JF, Ellis IO. The role of PTEN and its signalling pathways, including AKT, in breast cancer; an assessment of relationships with other prognostic factors and with outcome. J Pathol. 2004;204: 93–100.
40. Dai C, Celestino JC, Okada Y, Louis DN, Fuller GN, Holland EC. PDGF autocrine stimulation dedifferentiates cultured astrocytes and induces oligodendrogliomas and oligoastrocytomas from neural progenitors and astrocytes in vivo. Genes Dev. 2001;15: 1913–25.
41. Bao S, Wu Q, Sathornsumetee S, Hao Y, Li Z, Hjelmeland AB, et al. Stem cell-like glioma cells promote tumor angiogenesis through vascular endothelial growth factor. Cancer Res. 2006;66:7843–8.
42. Payne HR, Lemmon V. Glial cells of the O-2A lineage bind preferentially to N-cadherin and develop distinct morphologies. Dev Biol. 1993;159:595–607.
43. Satoh J, Kuroda Y. Beta-catenin expression in human neural cell lines following exposure to cytokines and growth factors. Neuropathology. 2000;20:113–23.
44. Hoelzinger DB, Demuth T, Berens ME. Autocrine factors that sustain glioma invasion and paracrine biology in the brain microenvironment. J Natl Cancer Inst. 2007;99:1583–9.
45. Savarese TM, Low HP, Baik I, Strohsnitter WC, Hsieh CC. Normal breast stem cells, malignant breast stem cells, and the perinatal origin of breast cancer. Stem Cell Rev. 2006;2:103–10.
46. Filip S, Mokry J, English D. Stem cell plasticity and carcinogenesis. Neoplasma/ 2006;53:87–91.
47. Bao S, Wu Q, McLendon RE, Hao Y, Shi Q, Hjelmeland AB, et al. Glioma stem cells promote radioresistance by preferential activation of the DNA damage response. Nature. 2006;444: 756–60.
48. Woodward WA, Chen MS, Behbod F, Alfaro MP, Buchholz TA, Rosen JM. WNT/beta-catenin mediates radiation resistance of mouse mammary progenitor cells. Proc Natl Acad Sci U S A. 2007;104:618–23.

49. Costello RT, Mallet F, Gaugler B, Sainty D, Arnoulet C, Gastaut JA, et al. Human acute myeloid leukemia CD34+/CD38- progenitor cells have decreased sensitivity to chemotherapy and Fas-induced apoptosis, reduced immunogenicity, and impaired dendritic cell transformation capacities. Cancer Res. 2000;60:4403–11.
50. Zhou S, Morris JJ, Barnes Y, Lan L, Schuetz JD, Sorrentino BP. Bcrp1 gene expression is required for normal numbers of side population stem cells in mice, and confers relative protection to mitoxantrone in hematopoietic cells in vivo. Proc Natl Acad Sci U S A. 2002;99:12339–44.
51. Donnenberg VS, Donnenberg AD. Multiple drug resistance in cancer revisited: the cancer stem cell hypothesis. J Clin Pharmacol. 2005;45:872–7.
52. Wismeth C, Hau P, Fabel K, Baumgart U, Hirschmann B, Koch H, et al. Maintenance therapy with 13-*cis* retinoid acid in high-grade glioma at complete response after first-line multimodal therapy – a phase-II study. J Neurooncol. 2004;68:79–86.
53. Phuphanich S, Scott C, Fischbach AJ, Langer C, Yung WKA. All-trans-retinoic acid: a phase II radiation therapy oncology group study (RTOG 91-13) in patients with recurrent malignant astrocytoma. J Neurooncol. 1997;34:193–200.
54. Kaba SE, Kyritsis AP, Conrad C, Gleason MJ, Newman R, Levin VA, et al. The treatment of recurrent cerebral gliomas with all-trans-retinoic acid (tretinoin). N Neurooncol. 1997;34:145–51.
55. Chang Q, Chen Z, You J, McNutt MA, Zhang T, Han Z, et al. All-trans-retinoic acid induces cell growth arrest in a human medulloblastoma cell line. J Neurooncol. 2007;84:263–7.
56. Yin D, Ong JM, Hu J, Desmond JC, Kawamata N, Konda BM, et al. Suberoylanilide hydroxamic acid, a histone deacetylase inhibitor: effects on gene expression and growth of glioma cells in vitro and in vivo. Clin Cancer Res. 2007;13:1045–52.
57. Galanis E, Jaeckle KA, Maurer MJ, Reid JM, Ames MM, Giannini C, et al. N047B: NCCTG phase II trial of vorinostat (suberoylanilide hydroxamic acid) in recurrent glioblastoma multiforme (GBM). J Clin Oncol. 2007;25 suppl:18S.
58. Fan X, Matsui W, Khaki L, Stearns D, Chun J, Li Y-M, et al. Notch pathway inhibition depletes stem-like cells and blocks engraftment in embryonal brain tumors. Cancer Res. 2006;66:7445–52.
59. Nickoloff BJ, Osborne BA, Miele L. Notch signaling as a therapeutic target in cancer: a new approach to the development of cell fate modifying agents. Oncogene. 2003;2:6598–608.
60. Berman DM, Karhadkar SS, Hallahan AR, Pritchard JI, Eberhart CG, Watkins DN, et al. Medulloblastoma growth inhibition by hedgehog pathway blockade. Science. 2002;297:1559–61.
61. Sasai K, Romer JT, Kimura H, Eberhart DE, Rice DS, Curran T. Medulloblastomas derived from *Cxcr6* mutant mice respond to treatment with a smoothened inhibitor. Cancer Res. 2007;67:3871–7.
62. Mellinghoff IK, Wang MY, Vivanco I, Haas-Kogan DA, Zhu S, Dia EQ, et al. Molecular determinants of the response of glioblastomas to EGFR kinase inhibitors. N Engl J Med. 2005;353:2012–24.
63. Wen PY, Yung WKA, Lamborn KR, Dahia PL, Wang Y, Peng B, et al. Phase I/II study of imatinib mesylate for recurrent malignant gliomas: North American Brain Tumor Consortium Study 99-08. Clin Cancer Res. 2006;12:4899–907.
64. Korkaya H, Paulson A, Iovino F, Wicha MS. Her2 regulates the mammary stem/progenitor cell population driving tumorigenesis and invasion. Oncogene 2008 [Epub ahead of print]
65. Shah K, Bureau E, Kim DE, Yang K, Tang Y, Weissleder R, et al. Glioma therapy and real-time imaging of neural precursor cell migration and tumor regression. Ann Neurol. 2005;1:34–41.
66. Wu A, Wiesner S, Xiao J, Ericson K, Chen W, Hall WA, Low WC, Ohlfest JR. Expression of MHC I and NK ligands on human CD133+ glioma cells: possible targets of immunotherapy. J Neurooncol. 2007;83:121–31.
67. Jiang H, Gomez-Manzano C, Aoki H, Alonso MM, Kondo S, McCormick F, et al. Examination of the therapeutic potential of Delta-24-RGD in brain tumor stem cells: role of autophagic cell death. J Natl Cancer Inst. 2007;99:1410–4.
68. Stommel JM, Kimmelman AC, Ying H, Nabioullin R, Ponugoti AH, Wiedemeyer R, et al. Coactivation of receptor tyrosine kinases affects the response of tumor cells to targeted therapies. Science. 2007;318:287–90.
69. Park DM, Li J, Okamoto H, Akeju O, Kim SH, Lubensky I, et al. N-CoR pathway targeting induces glioblastoma derived cancer stem cell differentiation. Cell Cycle. 2007;6:467–70.
70. Haque A, Das A, Hajiaghamohseni LM, Younger A, Banik NL, Ray SK. Induction of apoptosis and immune response by all-*trans* retinoic acid plus interferon-gamma in human malignant glioblastoma T98G and U87MG cells. Cancer Immunol Immunother. 2007;56:615–25.
71. Karmakar S, Banik NL, Ray SK. Combination of all-trans retinoic acid and paclitaxel-induced differentiation and apoptosis in human glioblastoma U87MG xenografts in nude mice. Cancer. 2008;112:596–607.
72. Wang MY, Lu KV, Zhu S, Dia EQ, Vivanco I, Shackleford GM, et al. Mammalian target of rapamycin inhibition promotes response to epidermal growth factor receptor kinase inhibitors in PTEN-deficient and PTEN-intact glioblastoma cells. Cancer Res. 2006;66:7864–9.
73. Doherty L, Gigas DC, Kesari S, Drappatz J, Kim R, Zimmerman J, et al. Pilot study of the combination of EGFR and mTOR inhibitors in recurrent malignant gliomas. Neurology. 2006;67: 156–8.
74. Tanaka K, Sasayama T, Mizukawa K, Kawamura A, Kondoh T, Hosoda K, et al. Specific mTOR inhibitor rapamycin enhances cytotoxicity induced by alkylating agent 1-(4-amino-2-methyl-5-pyrimidinyl)methyl-3-(2-chloroethyl)-3-nitrosourea (ACNU) in human U251 malignant glioma cells. J Neurooncol. 2007;84: 233–44.
75. Desjardins A, Quinn JA, Vredenburgh JJ, Sathornsumetee S, Friedman AH, Herndon JE, et al. Phase II study of imatinib mesylate and hydroxyurea for recurrent grade III malignant gliomas. J Neurooncol. 2007;83:53–60.
76. Prados MD, Lamborn KR, Chang S, Burton E, Butowski N, Malec M, et al. Phase 1 study of erlotinib HCl alone and combined with temozolomide in patients with stable or recurrent malignant glioma. Neuro Oncol. 2006;8:67–78.
77. Krishnan S, Brown PD, Ballman KV, Fiveash JB, Uhm JH, Giannini C, et al. Phase I trial of erlotinib with radiation therapy in patients with glioblastoma multiforme: results of North Central Cancer Treatment Group Protocol N0177. Int J Radiation Oncology Biol Phys. 2006;65:1192–9.
78. Vredenburgh JJ, Desjardins A, Herndon JE 2nd, Marcello J, Reardon DA, Quinn JA, et al. Bevacizumab plus irinotecan in recurrent glioblastoma multiforme. J Clin Oncol. 2007;25: 4722–9.
79. Yu F, Yao H, Zhu P, et al. let-7 Regulates Self Renewal and Tumorigenicity of Breast Cancer Cells. Cell. 2007;131:1109–23.
80. Li X, Creighton C, Wong H, et al. Decrease in tumorigenic breast cancer stem cells in primary breast cancers with neoadjuvant lapatinib. San Antonio Breast Cancer Symposium. San Antonio, TX, 2007.
81. Al-Hajj M, Becker MW, Wicha M, Weissman I, Clarke MF. Therapeutic implications of cancer stem cells. Curr Opin Genet Dev. 2004;14:43–7.
82. Ballman KV, Buckner JC, Brown PD, Giannini C, Flynn PF, LaPlant BR, et al. The relationship between six-month progression-free survival and 12-month overall survival end points for phase II trials in patients with glioblastoma multiforme. Neuro Oncol. 2007;9:29–38.

83. Chen W, Delaloye S, Silverman DH, Geist C, Czernin J, Sayre J, et al. Predicting treatment response of malignant gliomas to bevacizumab and irinotecan by imaging proliferation with [18F] fluorothymidine positron emission tomography: a pilot study. J Clin Oncol. 2007;25:4714–21.
84. Manganas LN, Zhang X, Li Y, Hazel RD, Smith SD, Wagshul ME, et al. Magnetic resonance spectroscopy identifies neural progenitor cells in the live human brain. Science. 2007;318:980–5.
85. Boman BM, Wicha MS, Fields JZ, Runquist OA. Symmetric division of cancer stem cells-a key mechanism in tumor growth that should be targeted in future therapeutic approaches. Clin Pharmacol Ther. 2007;81:893–8.
86. Clarke RB, Anderson E, Howell A, Potten CS. Regulation of human breast epithelial stem cells. Cell Prolif. 2003;36 Suppl 1:45–58.
87. Wang Z, Zhang Y, Banerjee S, Li Y, Sarkar FH. Notch-1 down-regulation by curcumin is associated with the inhibition of cell growth and the induction of apoptosis in pancreatic cancer cells. Cancer. 2006;106:2503–13.
88. Jaiswal AS, Marlow BP, Gupta N, Narayan S. Beta-catenin-mediated transactivation and cell-cell adhesion pathways are important in curcumin (diferuylmethane)-induced growth arrest and apoptosis in colon cancer cells. Oncogene. 2002;21:8414–27.
89. Pahlke G, Ngiewih Y, Kern M, Jakobs S, Marko D, Eisenbrand G. Impact of quercetin and EGCG on key elements of the Wnt pathway in human colon carcinoma cells. J Agric Food Chem. 2006;54:7075–82.
90. Nagler A, Riklis I, Kletter Y, Tatarsky I, Fabian I. Effect of 1,25 dihydroxyvitamin D3 and retinoic acid on normal human pluripotent (CFU-mix), erythroid (BFU-E), and myeloid (CFU-C) progenitor cell growth and differentiation patterns. Exp Hematol. 1986;14:60–5.

Preclinical Evidence for Cellular Therapy as a Treatment for Neurological Disease

Matthew T. Harting, Charles S. Cox and Stephen G. Hall

Abstract There continues to be extraordinary anticipation that stem cells will advance the current therapeutic regimen for both acute and degenerative neurological disease. Given the limited ability of the nervous system for regeneration and repair, combined with the devastating consequences that result from minimal neural tissue damage or cell death, any emerging therapy showing potential for cellular rescue or cell replacement represents a significant opportunity. This chapter will detail the preclinical progress using adult stem cells as a therapy for neurological diseases, both acute and degenerative. The basic adult stem cell types, including bone marrow–derived cells, umbilical cord blood–derived cells, and neural stem cells, will be discussed. The preclinical research using these cells to treat acute neurological diseases, such as traumatic brain injury and cerebrovascular disease, along with degenerative neurological disease, including Parkinson's, Alzheimer's, Huntington's, and multiple sclerosis, will be detailed. Potential mechanisms of action and important limitations of current work will be addressed. Finally, the details of initiating a cellular therapy clinical trial, along with a brief discussion of ongoing clinical trials using cell therapy, with a focus on neurological disease trials, will be provided. Overall, early work using adult stem cell therapy for acute and degenerative neurological disease has provided some promising results, leading the way for the translation to clinical trials. The future of applying stem cell therapy to current treatment regimens of neurological diseases requires continued discovery at the molecular, cellular, tissue, and organism levels, alongside well-designed clinical trials.

Keywords Cellular therapy · Adult stem cells · Neurological disease · Translational research

S.G. Hall (✉)
AlphaGenix, Inc., Department of Regenerative Medicine, 2329 North Career Avenue, Sioux Falls, SD 57107 (605)274-2268
e-mail: hallsg@alphagenix.com

1 Introduction

Cellular therapy is being heavily investigated as a promising therapy in numerous diseases encompassing every organ system. While intense work has focused on cell therapy applications for a wide array of diseases, ranging from cardiac disease to cancer, there continues to be extraordinary anticipation that stem cells will advance the current therapeutic regimen for both acute and degenerative neurologic disease. Much of this excitement stems from the fact that the current therapies for these devastating diseases are limited. Given the limited ability of the nervous system for regeneration and repair, combined with the devastating consequences that result from minimal tissue damage, any emerging therapy showing potential for cellular rescue or cell replacement represents a significant opportunity. Cellular therapy offers the possibility of cellular replacement or microenvironment alteration, preventing cell death, attracting native cells, or minimizing inflammation. This chapter will discuss the basic science and translational preclinical research in the area of adult stem cells as treatment of acute and degenerative brain disease. In the setting of this text, this chapter should bring focus to how the basic stem cell regulatory network discoveries (in vitro) are leading to novel preclinical work (in vitro animal models) and emerging translational/clinical trials.

1.1 Adult Stem Cell Types

There are two main categories of stem cells, embryonic and adult. Embryonic stem cells are derived from the blastocyst of the developing embryo. These cells are considered totipotent, as they are believed to be able to differentiate into any adult cell type. Although very promising preliminary work using embryonic stem cells as therapy for neurologic diseases exists, that work will not be covered in this chapter. Adult stem cells are derived from a niche in a developing or developed organism. These cells are considered multipotent

V.K. Rajasekhar, M.C. Vemuri (eds.), *Regulatory Networks in Stem Cells*, Stem Cell Biology and Regenerative Medicine,
DOI 10.1007/978-1-60327-227-8_42,

or pluripotent because they can differentiate into multiple adult cell types. Some adult stem cells have shown the ability to differentiate into cell types from multiple germ layers. Verfaillie and co-workers first discovered that adult stem cells from the bone marrow had the ability to differentiate down all three germ cell lines (ectoderm, mesoderm, and endoderm) [1]. These cells, termed multipotent adult progenitor cells (MAPC), were differentiated to phenotypic endothelial, hepatic, and neural cells.

Mesenchymal stem cells (MSC) were first discovered in 1970 when Friedenstein and colleagues found that certain cells from the bone marrow adhered to plastic when cultured [2]. More than 20 years later, the ability of the MSC to proliferate extensively and differentiate down multiple lineages was identified [3–5]. The criteria for defining human MSCs include: (1) adherence to plastic in standard culture conditions, (2) expression of CD105, CD73, and CD90 combined with a lack of expression of CD45, CD34, CD14 or CD11b, CD79 or CD19, and HLA-DR surface molecules, and (3) differentiation to osteoblasts, adipocytes, and chondroblasts in vitro [6]. Although controversial, MSCs have also been shown to develop neural phenotypes under specific conditions in vitro [7–9] and in vivo [10, 11].

Neural stem cells (NSC) can be isolated from many parts of the fetal or adult brain, but the most common niche is the subventricular zone of the lateral ventricle [12]. They are multipotent, possessing the ability to differentiate into cells of neural origin including neurons, oligodendrocytes, or astroglia [13, 14]. Selective (rather than specific) immunohistochemical markers of the neural stem/progenitor cell include nestin (an intermediate filament), Musashi1 (an RNA binding protein), Sox1 (a transcription factor), Notch1, $CD133^{+}$, and/or $CD24^{+}$ [15–17]. NSCs grow as neurospheres, spherical collections of a heterogeneous mixture of progenitor and differentiated cells [18]. Although they are more difficult to isolate and culture, these cells hold significant potential as the progenitor cells of the neural lineage.

Umbilical cord blood (UCB) is known to be rich in hematopoietic stem cells and may contain MSCs [19]. The umbilical cord stroma, also known as Wharton's jelly, has been shown to be rich in MSCs [19]. Given the fact that these tissues are a readily available source of cells, they may prove a valuable niche for cell isolation.

Multipotent adult progenitor cells (MAPCs) are actually pluripotent cells, despite their name, derived from the bone marrow and were first fully characterized by Verfaillie and co-workers [1]. MAPCs are able to differentiate, phenotypically, to all three germ cell layers. They have been expanded more than 100 cell doublings and their cultured immunophenotype is $CD34^{-}$, $CD44^{-}$, $CD45^{-}$, c-Kit^{-}, MHC $I\&II^{-}$, $CD13^{+}$, and SSEA-1^{+}. They have previously shown promise in the management of neurological diseases, including stroke and hypoxic-ischemic injury [20, 21].

2 Traumatic Brain Injury

It is estimated that more than 2.5 million Americans live with the devastating consequences of a traumatic brain injury (TBI) [22]. These injuries frequently leave patients with acute and/or chronic deficits in motor, cognitive, behavioral, and/or social function [22]. Acute treatment of TBI is limited to controlling intracranial pressure and optimizing cerebral perfusion pressure to prevent further intracranial damage, while chronic treatment centers on motor, cognitive, and behavioral rehabilitation. No current therapy alters the underlying pathologic processes via cellular salvage, repair, or replacement.

2.1 Bone Marrow–Derived Stem and Progenitor Cells for TBI

A significant amount of research evaluating the therapeutic use of bone marrow derived cells to treat traumatic brain injury has come from the laboratory of Chopp and colleagues. They first reported the use of whole bone marrow transplanted adjacent to the site of TBI in 2001 [10]. They found that some of the cells survived one month, migrated toward the site of injury, developed a phenotypic expression similar to neurons and astrocytes, and even improved motor function. Around that same time they reported intravenous [23, 24] and intraarterial [25] administration of MSCs in a rat model of TBI. Irrespective of route of delivery, they found that the cells survived, migrated to the area of injury, and expressed neural cell markers. Animals treated intravenously were found to have reduced motor and neurological severity score deficits [23]. The administration of bone marrow–derived progenitor cells intra-arterially, as opposed to the intravenous route, has been shown to lead to a significantly greater biodistribution [26]. After the success of the acute therapy, Chopp and co-workers subsequently reported that MSC therapy one week after TBI also led to long-term functional recovery (3 months after treatment) and that the cells remained in the brain at 3 months [27]. They have also shown incremental functional improvement over isolated cell therapy by combining MSCs with atorvastatin [28] or by using a collagen scaffold to aid cell delivery [29].

Human MSCs administered intravenously 24 h after TBI were also shown to migrate to the injured brain, express neural phenotypes, and improve functional outcome [11]. These human MSCs were subsequently shown to remain in the traumatically injured brain for 3 months, affecting long-term functional outcome [30].

2.2 Neural Stem Cells for TBI

More than 10 years ago, McIntosh and co-workers transplanted fetal cortical tissues into the injured cortical areas of adult rats after a fluid percussion injury [31]. Cell transplant survival was seen in 65% of animals and significant improvements in motor and cognitive function were reported. In their study, combining neural transplantation with nerve growth factor infusion also led to functional improvements.

Adult human neurons were shown to survive up to 4 weeks post-transplantation into traumatically injured rat cortex, however no significant improvement in behavioral function was identified [32, 33]. More recently, these postmitotic human neural cells were shown to survive for 12 weeks in immunosuppressed rats after delayed transplantation following TBI, although no improvement in behavioral function was reported [34]. Primed human neural stem cells transplanted into the hippocampus after fluid percussion injury in rats were found to differentiate to neurons, produce neurotropic factors, and mediate cognitive improvement [35].

Human neural stem cells derived from embryos developed astrocytic and neural phenotypes after transplantation into rodent brains after TBI [36–38]. Englund et al. subsequently showed that the neural stem cells established appropriate cortico-thalamic and contralateral hippocampal connections, that they were able to generate action potentials, and that they received functional excitatory and inhibitory synaptic inputs from neighboring cells [39]. This is seminal work, as it is one of the few studies to show functional activity among the transplanted cells, as opposed to the more common (and controversial) phenotypic characteristics evaluated. Additionally, human NSCs have been shown to exert neuroprotective activity by attenuating the number of degenerating neurons in the acute and subacute period after TBI [40]. Human NSCs have been shown to survive in the brain after discontinuation of immunosuppression [41]. This could be particularly important, given the current controversy surrounding the immune response to these cells and whether it may be safe to administer non-autologous cells to patients. One explanation for the possible immuno-privileged nature of human NSCs is the down-regulation of MHC expression [42].

Rodent NSCs have been shown to survive for 14 months, migrate to the area of injury, enhance cognitive recovery, and may offer trophic support [43]. Tate and colleagues reported that NSCs survive in the injured brain and migrate to the fimbria hippocampus within 3 months of implantation, but that in the presence of a fibronectin matrix, increased cell survival and migration was noted [44]. Neural stem cells derived from embryonic or neonatal rodent brains have been conditionally immortalized [45, 46] or altered by retroviral vector-mediated transduction of an oncogene [47] and subsequently shown the ability to develop phenotypic characteristics of native neural cells in vivo. The degree of injury may be important in cell survival, as recent work has revealed that transplanted NSCs may survive better after mild TBI than after severe TBI [48].

2.3 Umbilical Cord Blood

Chopp et al. intravenously infused human umbilical cord blood 24 h after TBI and found that the cells migrated to the parenchyma of the injured brain, expressed neural and astrocytic phenotypes, and significantly reduced motor and neurological deficits [49].

3 Cerebrovascular Disease

Cerebrovascular disease is second only to heart disease as the leading cause of overall death and the fifth leading cause of premature death worldwide [50]. Like traumatic brain injury, the burden of survival is also high, often leading to long-term disability and concomitant emotional and socioeconomic consequences for patients, families, and the health infrastructure overall [51]. In developed countries the vast majority of strokes are the result of an ischemic event (80%), with the rest resulting from a hemorrhagic event (20%) [51, 52]. Acute treatment involves rapid recognition and restoration of circulation via thrombolysis (ischemic stroke) or hemorrhage control and hematoma evacuation (hemorrhagic stroke). If acute treatments are unsuccessful, cell death ensues and loss of function is a common result. In the same fashion as traumatic brain injury, cell therapy may hold the key to cell salvage, repair, or replacement.

3.1 Bone Marrow Derived Stem and Progenitor Cells for Cerebrovascular Disease

In 2001, Chopp and co-workers transplanted whole bone marrow into the ischemic boundary zone of rodents 24 h after MCA occlusion and found that the cells expressed neural and astrocytic phenotypes and that they promoted proliferation of native neural stem cells [53]. The same group found that intrastriatal transplantation of bone marrow derived non-hematopoietic cells 24 h after MCA occlusion also led to neural and astrocytic phenotype expression and functional recovery [54]. Human MSCs intracerebrally transplanted 1 week after MCA occlusion reduced apoptosis in the boundary zone, improved functional outcome, and expressed neural, astrocytic, and glial phenotypes, but the morphologic features were not similar to neurons (few processes), causing the authors to express caution over the ability of these cells to function as neural cells [55]. Chopp and colleagues then administered MSCs intra-arterially (carotid) 24 h after MCA occlusion and again found the development of neural or astrocytic phenotypes [56]. Intra-arterial administration also led to functional improvement. They subsequently found that the intra-arterially infused MSCs facilitate axonal sprouting and remyelination in the cortical ischemic boundary zone and corpus callosum [57].

3.2 Neural Stem Cells for Cerebrovascular Diseease

Human neural stem cells (hNSCs) introduced intravenously or intracerebrally 24 h after MCA occlusion or two-vessel occlusion migrated to the brain, expressed glial or neural phenotypes, and improved motor function recovery [58, 59]. When these cells were intravenously infused in rodents 24 h after a hemorrhagic stroke, they preferentially migrated to the peri-infarct area, expressed the glial or neural phenotypes, and improved motor function recovery [60]. Intracerebral transplantation 4 days after a CCA occlusion showed 8% cell survival, reduced lesion size, neural and glial phenotypes, along with motor and cognitive functional recovery [61]. When hNSCs were intracerebrally transplanted 1 week after MCA occlusion, 30% of the cells had survived at 1 month, glial and neural phenotypes were again noted, but no functional recovery was reported [62].

The use of human neural stem cells in animal models of stroke has been more extensively studied than autologous neural stem cell therapy for animal models of disease. Even when animal neural stem cells are used, they are often from immortalized cell lines which likely have altered growth and therapeutic properties, along with altered risks [63].

In 2001, hippocampus-derived adult NSCs transplanted into the hippocampus of rodents 2 weeks after transient four-vessel occlusion were found to survive (1–3%), differentiate (3–9% of those that survived), and improve cognitive function (in animals where large numbers of cells survived) [64]. Intracisternal transplantation of subventricular zone derived neural stem cells 2 days after MCA occlusion survived, migrated toward the ischemic parenchyma, and improved motor function [65]. Immortalized NSC lines have also shown the ability to survive in vivo, migrate toward the area of cell death, and improve functional recovery [66–68].

3.3 Other Cells That Have Shown Preclinical Promise for Cerebrovascular Disease

Human umbilical cord blood (UCB) is a rich source of hematopoietic stem cells. Human UCB has shown the ability to improve neurologic recovery after intravenous infusion into a rat model of intracerebral hemorrhage [69]. Unique progenitor cells termed human umbilical cord derived neural-like stem cells have been placed intracerebrally 48 h after a thrombotic cortical lesion in a rat model [70]. While these cells migrated to the peri-infarct area and expressed neural and astrocytic phenotypes, few transplanted cells were identified at 1 month, likely due to rejection, highlighting the importance of furthering our understanding of the immunogenicity of these cells. Cells similar to MSCs have also been isolated from the peripheral blood and they showed equal efficacy to bone marrow derived cells, reducing lesion volume, improving regional cerebral blood flow, and improving functional outcome [71]. Peripheral blood isolation of MSCs could prove a valuable technique.

4 Parkinson's Disease

Parkinson's disease (PD) is the second most common neurodegenerative disease and is characterized by tremors, rigidity, and akinesia due to the progressive loss of the dopaminergic neurons and formation of Lewy bodies in the substantia nigra (SN) [72, 73]. The degeneration of dopaminergic neurons in the SN leads to a reduction of striatal dopamine (DA) levels. The rate-limiting step in the biosynthesis of DA is the formation of L-DOPA, which is catalyzed by tyrosine hydroxylase (TH). Cellular transplantation to replace dopaminergic neurons in the SN and dopamine innervation of the striatum with NSCs and MSCs has been attempted in a number of studies. Most of these experiments have involved the use of embryonic stem cells, but a few have used adult NSCs or MSCs. The rat model of PD is the most widely use and is generated by unilateral lesioning of the nigrostriatal dopaminergic pathway with the neurotoxin 6-hydroxydopamine (6-OHDA).

For stem cell therapy to be effective in treating PD, the transplanted stem cells must survive in large numbers and generate enough neurons to sufficiently reinnervate the striatum and release enough dopamine to provide sufficient recovery from PD as well as having multiple implantation sites based on the denervation patterns [74]. Clinical trials involving transplantation of embryonic cells over the last decade have failed to accomplish all of these, but have shown promise [75].

4.1 Bone Marrow Derived Stem and Progenitor Cells for Parkinson's

Ye and colleagues used a medial forebrain bundle lesion rat model of PD to study the effects cells of transplanting MSCs [76]. Two groups were used – one that was transplanted as undifferentiated cells and the second group that was differentiated in vitro and then transplanted. After MSC transplantation, DA content was higher in the lesioned striatum (ST) in the PD model than in the saline control group, suggesting an enhanced dopaminergic neuronal function of the SN in the MSC groups. In addition, asymmetric rotations after apomorphine administration decreased post-implantation, and by 5 months post-implantation were ameliorated in both MSC-treated groups. There were no statistical differences between the non-differentiated MSCs and the pre-differentiated MCSs.

Zhang and colleagues used TH-transfected neuronal stem cells derived from bone marrow stem cells in a rat model of PD [77]. Different transplantation protocols were used, including microinjection into the cerebral ventricles (CV) and the ST. Six weeks after transplantation, rats receiving cell injections in both areas had more movement over a larger area, showed more self-feeding, diminished rotations, and improved locomotor ability compared to the positive control group. Ten weeks after transplantation, the DA levels of the ST and CV groups were restored to 46.6% and 33.0% of control, respectively. Both had significant differences compared to the PD group.

Weiss et al. used umbilical cord matrix stem (UCMS) cells in the 6-OHDA rat model of PD [78]. The rats that received a human UCMS cell transplant showed a significant decrease in the number of rotations compared to sham-transplanted animals. Non-PD model animals with and without an UCMS cell transplant did not rotate following apomorphine treatment. In the PD rats, there was significantly more TH-positive staining in the SN and ventral tegmental area.

4.2 Neural Stem Cells for Parkinson's

Dziewczapolski and colleagues transplanted NSCs into the ST of the 6-OHDA rat model [79]. Their results demonstrated that approximately 60% of the NSCs survive for at least 5 weeks, display a significant radial pattern of migration from the graft site, and integrate into the host brain without deleterious effects. A significant number of the transplanted cells expressed the glial marker, chondroitin sulfate proteoglycan (NG2), and a small proportion expressed the neuronal marker beta-Tubulin III (also known as TuJ1). Retinoic acid (RA) can induce stem cells to differentiate into glial cells (astrocytes and oligodendrocytes) and neurons [80]. In this study, inducing differentiation of the NSCs with RA before transplantation as well as intraperitoneal administration of RA to grafted animals did not promote significant neuronal differentiation in vivo when compared to undifferentiated transplanted NSCs.

5 Alzheimer's

Alzheimer's disease (AD) is the most prevalent cause of dementia and is characterized by the presence of beta-amyloid deposits in the parenchyma (amyloid plaques or senile plaques) and in the cerebral vasculature. While amyloid plaques are associated with cognitive impairment, it is unclear what causes loss of neurons in AD. The amyloidogenic processing of the amyloid precursor protein (APP) takes place through the action of β and γ secretases generating beta-amyloid (Aβ), a protein of 39–43 amino acids in length, which is the likely cause of neuronal damage [81]. The β-secretase cleavage of APP produces a soluble secreted fragment called sAPPβ. The non-amyloidogenic processing of APP by α-secretase cleaves within the Aβ domain preventing Aβ peptide formation thereby releasing soluble sAPPα, which has neurotrophic properties.

While a number of hypotheses have been put forward on the mechanism of damage, none account for all of the deleterious effects Aβ has on neurons. Nonetheless, there is sufficient evidence to support the fact that it does damage to many cells of the CNS [82–85], which makes stem cell treatment for AD a significant challenge. Using stem cell therapy to treat AD is a much greater challenge than PD, because unlike PD, the pathology in AD is diffuse and found throughout the brain. In contrast to the large number of studies investigating the potential use of stem cells for PD, very few studies have been done using them as a cellular therapy for AD.

5.1 Bone Marrow–Derived Stem and Progenitor Cells for Alzheimer's

Chen and colleagues examined the ability of sAPPα to augment the nerve growth factor (NGF)/RA-induced differentiation of MSCs into neural cells with cholinergic neuronal phenotypes [86]. MSCs were isolated from 7-week-old mice and differentiated in vitro followed by transfecting sAPPα to augment the nerve growth factor (NGF)/RA-induced differentiation. In cell culture, there was a significant increase in differentiated MSCs displaying a cholinergic phenotype when compared to the cells differentiated solely with NGF/RA.

The study further examined in vivo effects with transplantation of MSCs into transgenic mice expressing both the chimeric amyloid precursor gene (APP) and human presenilin 1 (PS). Mice carrying both mutant genes (PS/APP) develop AD-like deposits composed of beta-amyloid (Aβ) at an early age [87]. One group was given MSCs that had been pre-differentiated with NGF/RA and another group was given MSCs that been transfected with sAPPα plus differentiated with NGF/RA. Two months after transplantation, immunohistochemistry (IHC) of both groups revealed a small number of differentiated MSCs displaying a cholinergic phenotype and located deep within the brain parenchyma of the PS/APP mice. While sAPPα seems to have the ability to augment the ability of NGF/RS to differentiate MSCs in vitro, there is not any significant data to date that demonstrates its ability to drive MSCs to a cholinergic phenotype in vivo. In addition, the small number of cells displaying the cholinergic phenotype may not be sufficient to ameliorate the effects of PD.

5.2 Neural Stem Cells for Alzheimer's

Using stem cells to treat AD would have to involve replacing cholinergic neurons in the basal forebrain. A very limited number of studies have shown that NSCs implanted into the adult brain can generate cholinergic neurons in situ. One prominent study utilized a rat model that involves transection of the fimbria-fornix (FF) pathway that carries the major cholinergic input to the hippocampus [88], which removes most cholinergic activity in the hippocampus. NSCs obtained from the subventricular zone (SVZ) of postnatal rats were implanted two weeks after lesioning of the FF. After 2 months post transplantation, IHC experiments demonstrated that more NF-200/BrdU, beta-Tubulin-III/BrdU positive neurons were found in the denervated hippocampus, while virtually none were present in the normal side. While this data is encouraging, it presents a challenging problem that is seen in many other studies where there are very few transplanted stem cells that differentiate and subsequently express neuronal markers.

6 Huntington's Disease

Huntington's disease (HD) is a rare autosomal dominant disorder that causes profound neuronal cell death in the striatum and cerebral cortex, which results in chorea, dementia, and death[89]. It is caused by a CAG repeat in the huntingtin gene that results in a polyglutamine expansion in N terminus of the huntingtin protein. HD is a good candidate for stem cell therapy since the inherited disease can be readily diagnosed and during the early stages, the degeneration only occurs in the striatal medium spiny projection neurons. This would allow for homotopic transplantation of stem cells into the striatum, whereas in diseases like PD the distance between the SN and ST precludes the ability to perform a homotopic transplant [90].

6.1 Bone Marrow–Derived Stem and Progenitor Cells for Huntington's

A requirement for the proliferation of stem cells in the SVZ seems to be increased cell death as seen in transgenic mice and lesion rat animal models of HD. In fact, review of many studies shows that both increased progenitor cell proliferation and compensatory striatal neurogenesis is present in many models of neurodegenerative diseases.

The quinolinic acid lesion model of striatal excitotoxicity closely resembles HD. Lescaudron and colleagues demonstrated that cognitive dysfunction in a HD rat quinolinic acid lesion model after striatal lesions can be reduced after intrastriatal transplantation of MSCs [91]. MSCs labeled with PKH26 were observed after 3 weeks to be present in the caudate-putamen, frontoparietal cortex and within the accumbens nucleus/ventral pallidum. However, no MSCs were observed in the hippocampus. At 3 weeks post-transplantation, less than 1% of the transplanted MSCs as determined by staining with beta-tubulin III (Tuj1) and none of the cells were GFAP-positive. The cells displaying a neuronal phenotype were noted to be in small clusters, leaving most of the ST lacking any transplanted neurons. This study showed that the animals treated with MSCs had significantly fewer working memory errors, although there were no significant behavioral differences between the groups (based on visual paw-placement and swim speed tasks). The paper puts forth the plausible hypothesis that the small beneficial effect observed was not due to the transplanted cells differentiating into neurons, but rather due to the transplanted cells providing additional growth factors to the lesioned area. This is explained by the fact that the animals receiving the transplanted cells did not show a reduction in the lesion size as well as by the observation that transplanted cells remained metabolically active for 5 weeks post implantation.

6.2 Neural Stem Cells for Huntington's

Another study involved the use of the HD rat quinolinic acid lesion model where adult rat NSCs were transplanted into the striatum of lesioned animals [92]. Two months after transplantation, approximately 12% of the NSCs had survived, differentiated, and migrated extensively throughout the lesioned striatum. Immunocytochemical analysis of the transplanted BrdU-labeled NSCs revealed that they had differentiated into either astrocytes or mature neurons. A number of the cells also expressed specific markers for striatal medium spiny projection neurons and interneurons. Furthermore, the animals treated with NSCs exhibited significant reductions in asymmetric rotations after apomorphine administration and spontaneous exploratory forelimb use.

Batista and colleagues used the R6/2 transgenic mouse as a model of HD in a set of elegant studies [93]. They used a clonal neurosphere assay in vitro along with IHC and electron microscopy to demonstrate that symptomatic HD R6/2 mice have an increased number of NSCs in the striatal subependyma. The study demonstrates that HD causes the expansion of NSCs and redirected migration of progenitor cells into the ST.

7 Multiple Sclerosis

Multiple Sclerosis (MS) is a T-helper cell (Th1)-mediated autoimmune disease with involvement from B cells as well. Axon damage, including transection of the axon, begins early in MS and correlates with inflammatory activity [94, 95]. The demyelinating diseases such as multiple sclerosis may be particularly amenable to stem cell therapy since they are caused by the sole loss of the oligodendrocyte. Oligodendrocytes are the sole source of myelin in the adult CNS, and their loss or dysfunction is at the heart of a wide variety of diseases of both children and adults. Two primary experimental models are used to study demyelinating diseases-X-irradiated ethidium bromide (X-EB) lesions to the dorsal funiculus [96–98] and experimental allergic encephalomyelitis (EAE), an experimental model of MS [99, 100].

7.1 Bone Marrow–Derived Stem and Progenitor Cells for Multiple Sclerosis

Akiyama and co-workers induced focal demyelinating lesions by X-irradiation and ethidium bromide (X-EB) injection in the spinal cord of adult rats [101]. This experimental lesion model causes nearly complete demyelination and loss of astrocytes and oligodendrocytes in the central region of the dorsal funiculus; however, endogenous remyelination begins after 6–8 weeks post-lesion. In this study MSCs were intravenously injected into the X-EB-lesioned rats, and the spinal cords were studied at 3-week post-lesion induction, which is prior to remyelination. The study found that there was extensive remyelination without obvious ectopic cell differentiation of neurons or other cell types within the lesion. In addition, a subpopulation of remyelinated axons in the MSC-injected rats showed increased conduction velocities. While the endogenous remyelination is likely to play a role in the remyelination and improved conduction velocities, it is almost certain that the MSCs transplanted augment the endogenous repair mechanisms.

Inoue and coworkers also used the X-EB lesion model in the rat and studied the effects of MSC transplantation at 3 weeks post-lesion well in advance of the onset of endogenous repair mechanisms [102]. Injection of as few as 1×10^2 MSCs into the X-EB rats resulted in some remyelination and 1×10^7 MSCs resulted in relatively extensive remyelination.

While no electrophysiological properties were analyzed in this study, the ability of the MSCs to result in remyelination is confirmed in the X-EB lesion model.

While MSCs injected into the X-EB lesion model resulted in remyelination 3 weeks post-lesion, the data may not be relevant in the normal pathology of MS in humans. Metz and co-workers characterized the histopathology of MS lesions in patients who had received autologous stem cell transplantation [103]. Their results demonstrated ongoing acute axonal damage and active demyelination following autologous stem cell transplantation in the absence of substantial lymphocyte infiltration of the lesions. There was also evidence of recent acute axonal damage in areas of macrophage/microglial activation. It is thought that microglia/macrophage-mediated acute axonal damage and active demyelination occurs after autologous stem cell transplantation or the immunosuppression associated with it.

7.2 Neural Stem Cells for Multiple Sclerosis

Pluchino and colleagues used NSCs from adult C57BL/6 mice, that were labeled with a lentiviral vector by which the expression of the *Escherichia coli* β-galactosidase (β-gal) gene (lacZ) to detect nuclear expression [104]. 1×10^6 NSCs were injected into C57BL/6 mice affected by myelin oligodendrocyte glycoprotein (residues 35–55) (MOG(35–55))-induced chronic EAE. Thirty days after injection of the now-differentiated NSCs, many of the donor cells were localized deep within the brain parenchyma with the majority of them located in areas of demyelination and axonal loss. The NSCs containing the β-gal gene showed β-gal expression in proximity with areas of active remyelination.

The study also demonstrated functional recovery in the case of NSC-injected animals by measuring means of motor-evoked potentials. In the case of the NSC-treated mice, the propagation time for the stimulus to travel from the motor cortex to spinal motoneurons was closer to normal than in the sham-treated animals. In addition, the number of oligodendrocyte progenitors derived from NSCs markedly increased thereby actively remyelinating axons. There was a concomitant reduction of reactive astrogliosis and demyelination. In a follow-up paper, the authors have confirmed the persistence of the NSC-derived cells in the inflamed perivascular areas of the CNS for up to 3 months after transplantation in EAE mice [105].

8 Potential Mechanisms of Action

Using cellular therapy as a treatment for neurological disease is an endeavor that remains in its infancy. Even given limited experience, the promise of cellular therapy remains high and the theories surrounding potential mechanisms of action are many. Although recent skepticism has been expressed, differentiation and cell replacement remains a possible mechanism. Neural stem cells, having progressed further down the same lineage as the cells of the brain, may certainly have the ability to replace damaged or non-functioning cells after transplantation, similar to the way they seem to function endogenously when they initiate neurogenesis and migrate toward sites of injury [106–109]. Stem cells may also act as supportive cells, altering the fate of damaged or diseased cells. They may improve cell survival or increase cellular proliferation [110] via direct contact or they may alter the local milieu through growth factor or chemokine secretion. MSCs cultured with TBI brain extracts were noted to produce growth factors [111] and a subsequent in vivo study noted increases in growth factor production with MSC therapy [112]. They may even affect native cells via cell fusion and the transfer of cellular material. Stem cells have been shown to migrate toward sites of inflammation and they may act through the reduction of inflammation, inflammatory mediators, or local edema.

9 Additional Considerations

While the prospect of using stem cells to treat human neurological disorders is good, there are some challenges and concerns that have to be addressed. The ability to culture stem cells for long periods of time is difficult and how this affects their genotype and phenotype is uncertain. Moreover, the ability to reproducibly direct the differentiation of human stem cells into specific neuronal types in sufficient quantities required for effective treatment is not known. To date, attempts to generate clinically significant amounts of dopaminergic, GABAergic nor cholinergic neurons have been unsuccessful. By nature, adult NSCs are more specialized than ESCs and therefore have a more limited differentiation potential. While this may be viewed as a disadvantage, it could be advantageous as the more limited differentiation choices presented to the stem cell may increase the probability of making the right choice based on cues it receives from the target environment.

Little is known about how stem cells will behave following transplantation into humans. There are obvious safety issues related to allogeneic stem cell transplantation even with standard immunosuppression protocols, as the likelihood of introducing genetic mutations or tumorigenic potential is unknown. Moreover, in the case of autologous cell transplantation for genetically based neurodegenerative diseases, using stem cells from the patient may only serve to eventually exacerbate the disease due to genetic predisposition to the disease and/or the age of the individual.

There are some important nuances to critically evaluate when comparing cell-based therapy studies. It is important to consider the cell type used (e.g., MSC, NSC, UCB, etc.), how that cell population is characterized relative to the same cell type in a different study. One must know how many population doublings the cells have gone through and if the genotype and phenotype may have changed. One must determine the number of cells used in the treatment, how they were administered (e.g., IV, IA, IC, etc.) and the time-points that they were given, relative to the injury or disease. The type of cell transplantation (autologous, allogeneic, heterologous, etc.) should be determined.

The type of injury model used is very important (e.g., X-EB vs. EAE for MS; CCI vs. fluid percussion for TBI; MCAo vs. CCAo vs. 4 Vo for Stroke, etc.) and should be suited to approximate the effects of the disease in humans.

The immunological risks involving animals and humans undergoing stem cell transplantation include immune rejection of transplanted stem cells and postoperative infection. The potential risks can be minimized and the results maximized by determining the immuno-competent status of the animal and cells to be used. It is also necessary to determine if immunosuppression was used in the study and to consider the degree the immunosuppression might affect the outcome.

Finally, nearly all of the research discussed here and the vast majority of the published work appears very positive. There are certainly unpublished negative studies. Such publication bias likely skews our perception of the truth.

10 Early Application of These Results to Clinical Practice

10.1 Requirements for Initiating a Clinical Trial

There are a number of requirements for undertaking a cellular based therapeutic intervention clinical trial for a neurologic disease. Prior to embarking on any trial, a clinical infrastructure must exist that allows access to an adequate number of patients of interest. As an example, to conduct a clinical trial evaluating the safety of autologous bone marrow cells in patients with a traumatic brain injury, an American College of Surgeons Level I trauma center is necessary. Further, undertaking an advanced intervention trial also requires that the trauma system be well developed and highly functional prior to initiating a complex trial. In addition, a broad range of medical specialists, experienced and knowledgeable in the management of the specific neurologic disease and in cell-based therapies, is required including (but not necessarily limited to): (1) surgery and trauma surgery, (2) neurosurgery, (3) neurology/neurodevelopmental biology, (4) bone marrow transplantation, (5) critical care, (6) advanced neuroimaging, and (7) neuropsychology. A GMP-certified cell therapeutics laboratory is also required for rapid cell separation, cell expansion, and quality control. A clinical research team that has investigative experience with FDA/IRB regulatory issues is invaluable.

Currently, the CBER section of the FDA requires all cell-based interventions to be conducted under an IND (investigational new drug). While there have been reports of autologous cellular therapeutics administered outside of this regulatory oversight, this is currently unadvisable. The development of the study protocol should focus on cell type, dosing considerations/toxicity monitoring, route of administration, and adverse event monitoring. As with all clinical trials, the trial must be registered with clinicaltrials.gov.

10.2 Current Clinical Trials

Clinical stem cell trials are currently underway in the following areas: heart disease, diabetes, osteogenesis imperfecta, and pediatric traumatic brain injury, to name a few. Cell transplantation has been used for some time in the management of certain cancers and certain autoimmune disorders. The first human neural stem cell trial has been approved and is currently underway for Ceroid Lipofuscinosis [113].

There are approximately 53 clinical trials (as of late 2007) investigating the use of cell therapy for a wide variety of diseases, the vast majority using autologous cells isolated from the bone marrow (Table 1). In addition to the TBI trial discussed below, there are several trials evaluating cell therapy for neurological disease. Two trials are evaluating autologous bone marrow, administered intra-arterially, to treat ischemic stroke patients. Another group is using autologous mesenchymal stem cells, administered intravenously, to treat multiple sclerosis. A third group is implanting human-derived, central nervous system stem cells intracerebrally to treat patients with infantile or late-infantile Neuronal Ceroid Lipofuscinosis (Batten disease), a disease where a reduced amount of, or missing, palmitoyl protein thioesterase 1 (PPT1) enzyme or tripeptidyl peptidase 1 (TPP-I) enzyme leads to excessive accumulation of lipopigments in the body's tissues.

A phase I clinical trial evaluating the use of intravenously infused autologous bone marrow cells (specifically, the mononuclear cell fraction) after isolated traumatic brain injury in children will be completed in early 2009. Pediatric patients (age 5–14) with an isolated traumatic brain injury (Glasgow Coma Score 5–8) are eligible. After obtaining consent, 3 mL/kg bone marrow is harvested, the mononuclear cell fraction isolated, and 6 million cells per kilogram are infused intravenously. The infusion must be complete within 48 h of the time of the injury. Given the complexity of such

Table 1 Approximate number of clinical trials evaluating cell therapy that are enrolling patients or are close to initiating enrollment (www.clinicaltrials.gov, search performed in late 2007; search terms: "stem cell(s)" and/or "nervous system" and/or "brain"). NCL: Neuronal Ceroid Lipofuscinosis (Batten Disease)

Disease	Clinical trials
Myocardial infarction	9
Peripheral vascular disease	9
Coronary artery disease	7
Congestive heart failure	7
Cirrhosis	3
Cardiomyopathy	2
Diabetes	2
Ischemic stroke	2
Tibial fracture	2
Multiple sclerosis	2
Anal fistula	1
Chronic wound	1
Hypercholesterolemia	1
Diabetic neuropathy	1
Myasthenia gravis	1
Osteoarthritis	1
NCL	1
Traumatic brain injury	1

Note – This does not include cancer trials using bone marrow transplantation for recovery after chemotherapy or any trials using 'stem cells' as blood/bone marrow cell replacement. Some of the trials mentioned may be complete at the time of text publication.

a trial and the level of care required, extensive collaboration between multiple specialties, including pediatric surgery, pediatric neurosurgery, hematology, and neuropsychiatry, along with a clinical grade cell therapeutics laboratory and a busy trauma center are all necessary components of this type of clinical trial. The goal of this trial is to critically evaluate the safety of intravenous, autologous mononuclear cell therapy, allowing future studies to more extensively evaluate cell number necessary for possible therapeutic benefit, timing of therapy, route of therapy, and exact cell type.

11 Conclusion

Early work using adult cell therapy for acute and chronic neurologic disease has provided some promising results. The era of translation has now begun, with many clinical trials being initiated. We must continue to corroborate early, positive findings, while advancing our understanding of the basic mechanisms underlying cell therapy. As the fields of cell therapy and stem cell biology grow out of their infancy, patients with devastating neurologic diseases such as traumatic brain injury, stroke, Parkinson's, Alzheimer's, Huntington's, and multiple sclerosis are hopeful that today's promise of cell therapy becomes tomorrow's treatment.

Acknowledgments Supported by National Institutes of Health grant T32 GM008792-06 (MTH).

References

1. Jiang Y, Jahagirdar BN, Reinhardt RL, et al. Pluripotency of mesenchymal stem cells derived from adult marrow. Nature. 2002;418(6893):41–9.
2. Friedenstein AJ, Chailakhjan RK, Lalykina KS. The development of fibroblast colonies in monolayer cultures of guinea-pig bone marrow and spleen cells. Cell Tissue Kinet. 1970;3(4):393–403.
3. Caplan AI. Mesenchymal stem cells. J Orthop Res. 1991;9(5): 641–50.
4. Prockop DJ. Marrow stromal cells as stem cells for non-hematopoietic tissues. Science. 1997;276(5309):71–4.
5. Pittenger MF, Mackay AM, Beck SC, et al. Multilineage potential of adult human mesenchymal stem cells. Science. 1999;284(5411):143–7.
6. Dominici M, Le Blanc K, Mueller I, et al. Minimal criteria for defining multipotent mesenchymal stromal cells. The International Society for Cellular Therapy position statement. Cytotherapy. 2006;8(4):315–7.
7. Tao H, Rao R, Ma DD. Cytokine-induced stable neuronal differentiation of human bone marrow mesenchymal stem cells in a serum/feeder cell-free condition. Dev Growth Differ. 2005;47(6):423–33.
8. Long X, Olszewski M, Huang W, Kletzel M. Neural cell differentiation in vitro from adult human bone marrow mesenchymal stem cells. Stem Cells Devel. 2005;14(1):65–9.
9. Sanchez-Ramos J, Song S, Cardozo-Pelaez F, et al. Adult bone marrow stromal cells differentiate into neural cells in vitro. Exp Neurol. 2000;164(2):247–56.
10. Mahmood A, Lu D, Yi L, Chen JL, Chopp M. Intracranial bone marrow transplantation after traumatic brain injury improving functional outcome in adult rats. J Neurosurg. 2001;94(4): 589–95.
11. Mahmood A, Lu D, Lu M, Chopp M. Treatment of traumatic brain injury in adult rats with intravenous administration of human bone marrow stromal cells. Neurosurgery. 2003;53(3):697–702; discussion-3.
12. Reynolds BA, Weiss S. Generation of neurons and astrocytes from isolated cells of the adult mammalian central nervous system. Science. 1992;255(5052):1707–10.
13. McKay R. Stem cells in the central nervous system. Science. 1997;276(5309):66–71.
14. McKay R, Renfranz P, Cunningham M. Immortalized stem cells from the central nervous system. Comptes rendus de l'Academie des sciences. 1993;316(12):1452–7.
15. Gage FH. Mammalian neural stem cells. Science. 2000; 287(5457):1433–8.
16. Okano H. Stem cell biology of the central nervous system. J Neurosci Res. 2002;69(6):698–707.
17. Murayama A, Matsuzaki Y, Kawaguchi A, Shimazaki T, Okano H. Flow cytometric analysis of neural stem cells in the developing and adult mouse brain. J Neurosci Res. 2002;69(6): 837–47.
18. Xu Y, Kimura K, Matsumoto N, Ide C. Isolation of neural stem cells from the forebrain of deceased early postnatal and adult rats with protracted post-mortem intervals. J Neurosci Res. 2003;74(4):533–40.
19. Secco M, Zucconi E, Vieira NM, et al. Multipotent stem cells from umbilical cord: cord is richer than blood! Stem Cells. 2007.
20. Yasuhara T, Matsukawa N, Yu G, et al. Transplantation of cryopreserved human bone marrow-derived multipotent adult progenitor cells for neonatal hypoxic-ischemic injury: targeting the hippocampus. Rev. Neurosci. 2006;17(1–2):215–25.
21. Tang Y, Yasuhara T, Hara K, et al. Transplantation of bone marrow-derived stem cells: a promising therapy for stroke. Cell Transplant. 2007;16(2):159–69.
22. Consensus conference. Rehabilitation of persons with traumatic brain injury. NIH Consensus Development Panel on Rehabilitation of Persons With Traumatic Brain Injury. JAMA. 1999;282(10):974–83.
23. Mahmood A, Lu D, Wang L, Li Y, Lu M, Chopp M. Treatment of traumatic brain injury in female rats with intravenous administration of bone marrow stromal cells. Neurosurgery. 2001;49(5):1196–203; discussion 203–4.
24. Lu D, Mahmood A, Wang L, Li Y, Lu M, Chopp M. Adult bone marrow stromal cells administered intravenously to rats after traumatic brain injury migrate into brain and improve neurological outcome. Neuroreport. 2001;12(3):559–63.
25. Lu D, Li Y, Wang L, Chen J, Mahmood A, Chopp M. Intraarterial administration of marrow stromal cells in a rat model of traumatic brain injury. J Neurotrauma. 2001;18(8):813–9.
26. Tolar J, O'Shaughnessy MJ, Panoskaltsis-Mortari A, et al. Host factors that impact the biodistribution and persistence of multipotent adult progenitor cells. Blood. 2006;107(10): 4182–8.
27. Mahmood A, Lu D, Qu C, Goussev A, Chopp M. Long-term recovery after bone marrow stromal cell treatment of traumatic brain injury in rats. J Neurosurg. 2006;104(2):272–7.
28. Mahmood A, Lu D, Qu C, Goussev A, Chopp M. Treatment of traumatic brain injury with a combination therapy of marrow stromal cells and atorvastatin in rats. Neurosurgery. 2007;60(3): 546–53; discussion 53–4.

29. Lu D, Mahmood A, Qu C, Hong X, Kaplan D, Chopp M. Collagen scaffolds populated with human marrow stromal cells reduce lesion volume and improve functional outcome after traumatic brain injury. Neurosurgery. 2007;61(3):596–602; discussion-3.
30. Mahmood A, Lu D, Qu C, Goussev A, Chopp M. Human marrow stromal cell treatment provides long-lasting benefit after traumatic brain injury in rats. Neurosurgery. 2005;57(5):1026–31; discussion 1031.
31. Sinson G, Voddi M, McIntosh TK. Combined fetal neural transplantation and nerve growth factor infusion: effects on neurological outcome following fluid-percussion brain injury in the rat. J Neurosurg. 1996;84(4):655–62.
32. Philips MF, Muir JK, Saatman KE, et al. Survival and integration of transplanted postmitotic human neurons following experimental brain injury in immunocompetent rats. J Neurosurg. 1999;90(1):116–24.
33. Muir JK, Raghupathi R, Saatman KE, et al. Terminally differentiated human neurons survive and integrate following transplantation into the traumatically injured rat brain. J Neurotrauma. 1999;16(5):403–14.
34. Zhang C, Saatman KE, Royo NC, et al. Delayed transplantation of human neurons following brain injury in rats: a long-term graft survival and behavior study. J Neurotrauma. 2005;22(12): 1456–74.
35. Gao J, Prough DS, McAdoo DJ, et al. Transplantation of primed human fetal neural stem cells improves cognitive function in rats after traumatic brain injury. Exp Neurol. 2006;201(2):281–92.
36. Englund U, Bjorklund A, Wictorin K. Migration patterns and phenotypic differentiation of long-term expanded human neural progenitor cells after transplantation into the adult rat brain. Brain Res Dev Brain Res. 2002;134(1–2):123–41.
37. Englund U, Fricker-Gates RA, Lundberg C, Bjorklund A, Wictorin K. Transplantation of human neural progenitor cells into the neonatal rat brain: extensive migration and differentiation with long-distance axonal projections. Exp Neurol. 2002;173(1):1–21.
38. Vescovi AL, Gritti A, Galli R, Parati EA. Isolation and intracerebral grafting of nontransformed multipotential embryonic human CNS stem cells. J Neurotrauma. 1999;16(8):689–93.
39. Englund U, Bjorklund A, Wictorin K, Lindvall O, Kokaia M. Grafted neural stem cells develop into functional pyramidal neurons and integrate into host cortical circuitry. Proc Natl Acad Sci U S A. 2002;99(26):17089–94.
40. Hagan M, Wennersten A, Meijer X, Holmin S, Wahlberg L, Mathiesen T. Neuroprotection by human neural progenitor cells after experimental contusion in rats. Neurosci Lett. 2003;351(3): 149–52.
41. Wennersten A, Holmin S, Al Nimer F, Meijer X, Wahlberg LU, Mathiesen T. Sustained survival of xenografted human neural stem/progenitor cells in experimental brain trauma despite discontinuation of immunosuppression. Exp Neurol. 2006;199(2): 339–47.
42. Al Nimer F, Wennersten A, Holmin S, Meijer X, Wahlberg L, Mathiesen T. MHC expression after human neural stem cell transplantation to brain contused rats. Neuroreport. 2004;15(12): 1871–5.
43. Shear DA, Tate MC, Archer DR, et al. Neural progenitor cell transplants promote long-term functional recovery after traumatic brain injury. Brain Res. 2004;1026(1):11–22.
44. Tate MC, Shear DA, Hoffman SW, Stein DG, Archer DR, LaPlaca MC. Fibronectin promotes survival and migration of primary neural stem cells transplanted into the traumatically injured mouse brain. Cell Transplant. 2002;11(3):283–95.
45. Renfranz PJ, Cunningham MG, McKay RD. Region-specific differentiation of the hippocampal stem cell line HiB5 upon implantation into the developing mammalian brain. Cell. 1991;66(4):713–29.
46. Sinden JD, Rashid-Doubell F, Kershaw TR, et al. Recovery of spatial learning by grafts of a conditionally immortalized hippocampal neuroepithelial cell line into the ischaemia-lesioned hippocampus. Neuroscience. 1997;81(3):599–608.
47. Ryder EF, Snyder EY, Cepko CL. Establishment and characterization of multipotent neural cell lines using retrovirus vector-mediated oncogene transfer. J Neurobiol. 1990;21(2):356–75.
48. Shindo T, Matsumoto Y, Wang Q, Kawai N, Tamiya T, Nagao S. Differences in the neuronal stem cells survival, neuronal differentiation and neurological improvement after transplantation of neural stem cells between mild and severe experimental traumatic brain injury. J Med Invest. 2006;53(1–2):42–51.
49. Lu D, Sanberg PR, Mahmood A, et al. Intravenous administration of human umbilical cord blood reduces neurological deficit in the rat after traumatic brain injury. Cell Transplant. 2002;11(3): 275–81.
50. Lopez AD, Mathers CD, Ezzati M, Jamison DT, Murray CJ. Global and regional burden of disease and risk factors, 2001: systematic analysis of population health data. Lancet. 2006;367(9524):1747–57.
51. Feigin VL, Lawes CM, Bennett DA, Anderson CS. Stroke epidemiology: a review of population-based studies of incidence, prevalence, and case-fatality in the late 20th century. Lancet Neurol. 2003;2(1):43–53.
52. van der Worp HB, van Gijn J. Clinical practice. Acute ischemic stroke. New Engl J Med. 2007;357(6):572–9.
53. Li Y, Chen J, Chopp M. Adult bone marrow transplantation after stroke in adult rats. Cell Transplant. 2001;10(1):31–40.
54. Li Y, Chopp M, Chen J, et al. Intrastriatal transplantation of bone marrow nonhematopoietic cells improves functional recovery after stroke in adult mice. J Cereb Blood Flow Metab. 2000;20(9):1311–9.
55. Zhao LR, Duan WM, Reyes M, Keene CD, Verfaillie CM, Low WC. Human bone marrow stem cells exhibit neural phenotypes and ameliorate neurological deficits after grafting into the ischemic brain of rats. Experimental Neurol. 2002;174(1): 11–20.
56. Li Y, Chen J, Wang L, Lu M, Chopp M. Treatment of stroke in rat with intracarotid administration of marrow stromal cells. Neurology. 2001;56(12):1666–72.
57. Shen LH, Li Y, Chen J, et al. Intracarotid transplantation of bone marrow stromal cells increases axon-myelin remodeling after stroke. Neuroscience. 2006;137(2):393–9.
58. Chu K, Kim M, Park KI, et al. Human neural stem cells improve sensorimotor deficits in the adult rat brain with experimental focal ischemia. Brain Res. 2004;1016(2):145–53.
59. Chu K, Kim M, Jeong SW, Kim SU, Yoon BW. Human neural stem cells can migrate, differentiate, and integrate after intravenous transplantation in adult rats with transient forebrain ischemia. Neurosci Lett. 2003;343(2):129–33.
60. Jeong SW, Chu K, Jung KH, Kim SU, Kim M, Roh JK. Human neural stem cell transplantation promotes functional recovery in rats with experimental intracerebral hemorrhage. Stroke. 2003;34(9):2258–63.
61. Ishibashi S, Sakaguchi M, Kuroiwa T, et al. Human neural stem/progenitor cells, expanded in long-term neurosphere culture, promote functional recovery after focal ischemia in Mongolian gerbils. J Neurosci Res. 2004;78(2):215–23.
62. Kelly S, Bliss TM, Shah AK, et al. Transplanted human fetal neural stem cells survive, migrate, and differentiate in ischemic rat cerebral cortex. Proc Natl Acad Sci U S A. 2004;101(32): 11839–44.
63. Mi R, Luo Y, Cai J, Limke TL, Rao MS, Hoke A. Immortalized neural stem cells differ from nonimmortalized cortical neurospheres and cerebellar granule cell progenitors. Exp Neurol. 2005;194(2):301–19.

64. Toda H, Takahashi J, Iwakami N, et al. Grafting neural stem cells improved the impaired spatial recognition in ischemic rats. Neurosci Lett. 2001;316(1):9–12.
65. Zhang ZG, Jiang Q, Zhang R, et al. Magnetic resonance imaging and neurosphere therapy of stroke in rat. Ann Neurol. 2003;53(2):259–63.
66. Wong AM, Hodges H, Horsburgh K. Neural stem cell grafts reduce the extent of neuronal damage in a mouse model of global ischaemia. Brain Res. 2005;1063(2):140–50.
67. Kim DE, Schellingerhout D, Ishii K, Shah K, Weissleder R. Imaging of stem cell recruitment to ischemic infarcts in a murine model. Stroke. 2004;35(4):952–7.
68. Veizovic T, Beech JS, Stroemer RP, Watson WP, Hodges H. Resolution of stroke deficits following contralateral grafts of conditionally immortal neuroepithelial stem cells. Stroke. 2001;32(4):1012–9.
69. Nan Z, Grande A, Sanberg CD, Sanberg PR, Low WC. Infusion of human umbilical cord blood ameliorates neurologic deficits in rats with hemorrhagic brain injury. Ann N Y Acad Sci. 2005;1049: 84–96.
70. Kozlowska H, Jablonka J, Janowski M, Jurga M, Kossut M, Domanska-Janik K. Transplantation of a novel human cord blood-derived neural-like stem cell line in a rat model of cortical infarct. Stem Cells Dev. 2007;16(3):481–8.
71. Ukai R, Honmou O, Harada K, Houkin K, Hamada H, Kocsis JD. Mesenchymal stem cells derived from peripheral blood protects against ischemia. J Neurotrauma. 2007;24(3):508–20.
72. Fahn S. Description of Parkinson's disease as a clinical syndrome. Ann N Y Acad Sci. 2003;991:1–14.
73. Bernheimer H, Birkmayer W, Hornykiewicz O et al. Brain dopamine and the syndromes of Parkinson and Huntington. Clinical, morphological and neurochemical correlations. J Neurol Sci. 1973;20:415–55.
74. Piccini P, Pavese N, Hagell P, Reimer J, Bjorklund A, Oertel WH, Quinn NP, Brooks DJ, Lindvall O.Factors affecting the clinical outcome after neural transplantation in Parkinson's disease. Brain. 2005;128(Pt 12):2977–86.
75. Singh N, Pillay V, Choonara YE. Advances in the treatment of Parkinson's disease. Prog Neurobiol. 2007;81(1):29–44.
76. Ye M, Wang XJ, Zhang YH, Lu GQ, Liang L, Xu JY, Chen SD. Therapeutic effects of differentiated bone marrow stromal cell transplantation on rat models of Parkinson's disease. Parkinsonism Relat Disord. 2007;13(1):44–9.
77. Zhang S, Zou Z, Jiang X, Xu R, Zhang W, Zhou Y, Ke Y. The therapeutic effects of tyrosine hydroxylase gene transfected hematopoetic stem cells in a rat model of Parkinson's disease. Cell Mol Neurobiol. 2008;28(4):529–43.
78. Weiss ML, Medicetty S, Bledsoe AR, Rachakatla RS, Choi M, Merchav S, Luo Y, Rao MS, Velagaleti G, Troyer D. Human umbilical cord matrix stem cells: preliminary characterization and effect of transplantation in a rodent model of Parkinson's disease. Stem Cells. 2006;24(3):781–92.
79. Dziewczapolski G, Lie DC, Ray J, Gage FH, Shults CW. Survival and differentiation of adult rat-derived neural progenitor cells transplanted to the striatum of hemiparkinsonian rats. Exp Neurol. 2003;183(2):653–64.
80. Liu S, Qu Y, Stewart TJ, Howard MJ, Chakrabortty S, Terrence F. Holekamp, TF, McDonald J. W. Embryonic stem cells differentiate into oligodendrocytes and myelinate in culture and after spinal cord transplantation. Proc Natl Acad Sci U S A. 2000;97: 6126–6131.
81. Yankner BA. Mechanisms of neuronal degeneration in Alzheimer's disease. Neuron. 1996;16:921–932.
82. Pike CJ, Walencewicz-Wasserman AJ, Kosmoski J, Cribbs DH, Glabe CG, Cotman CW. Structure-activity analyses of beta-amyloid peptides: contributions of the beta 25–35 region to aggregation and neurotoxicity. J Neurochem. 1995;64: 253–265.
83. Walsh DT, Montero RM, Bresciani LG, Jen AY, Leclercq PD, Saunders D, et al. Amyloid-beta peptide is toxic to neurons in vivo via indirect mechanisms. Neurobiol Dis. 2002;10:20–27.
84. Abramov AY, Canevari L, Duchen MR. Changes in intracellular calcium and glutathione in astrocytes as the primary mechanism of amyloid neurotoxicity. J Neurosci. 2003;23:5088–95.
85. Nagele RG, D'Andrea MR, Lee H, Venkataraman V, Wang HY. Astrocytes accumulate A beta 42 and give rise to astrocytic amyloid plaques in Alzheimer disease brains. Brain Res. 2003;971:197–209.¡img¿¡/img¿
86. Chen CW, Boiteau RM, Lai WF, Barger SW, Cataldo AM. sAPPalpha enhances the transdifferentiation of adult bone marrow progenitor cells to neuronal phenotypes. Curr Alzheimer Res. 2006;3(1):63–70.
87. Holcomb L, Gordon MN, McGowan E, Yu X, Benkovic S, Jantzen P, et al. Accelerated Alzheimer-type phenotype in transgenic mice carrying both mutant amyloid precursor protein and presenilin 1 transgenes. Nat Med. 1998;4(1):97–100.
88. Zhang X, Jin G, Tian M, Qin J, Huang Z. The denervated hippocampus provides proper microenvironment for the survival and differentiation of neural progenitors. Neurosci Lett. 2007;414(2):115–20.
89. Ramaswamy S, Shannon KM, Kordower JH. Huntington's disease: pathological mechanisms and therapeutic strategies. Cell Transplant. 2007;16(3):301–12.
90. Winkler C, Kirik D, Bjorklund A, Dunnett SB. Transplantation in the rat model of Parkinson's disease: ectopic versus homotopic graft placement. Prog Brain Res. 2000;127:233–65.
91. Lescaudron L, Unni D, Dunbar GL. Autologous adult bone marrow stem cell transplantation in an animal model of Huntington's disease: behavioral and morphological outcomes. Int J Neurosci. 2003;113(7):945–56.
92. Vazey EM, Chen K, Hughes SM, Connor B. Transplanted adult neural progenitor cells survive, differentiate and reduce motor function impairment in a rodent model of Huntington's disease. Exp Neurol. 2006;199(2):384–96.
93. Batista CM, Kippin TE, Willaime-Morawek S, Shimabukuro MK, Akamatsu W, van der Kooy D. A progressive and cell non-autonomous increase in striatal neural stem cells in the Huntington's disease R6/2 mouse. J Neurosci. 200611;26(41): 10452–60.
94. Sospedra M, Martin R. Immunology of multiple sclerosis. Annu Rev Immunol. 2005;23:683–747.
95. Hafler DA, Slavik JM, Anderson DE, O'Connor KC, De-Jager P, Baecher-Allan C. Multiple sclerosis. Immunol Rev. 2005;204:208–231.
96. Franklin RJM, Gilson JM, Franceschini IA, Barnett SC. Schwann cell-like myelination following transplantation of an olfactory bulbensheathing cell line into areas of demyelination in the adult CNS. Glia. 1996;17:217–24.
97. Imaizumi T, Lankford KL, Waxman SG, Greer CA, Kocsis JD. Transplanted olfactory ensheathing cells remyelinate and enhance axonal conduction in the demyelinated dorsal columns of the rat spinal cord. J Neurosci. 1998;18:6176–85.
98. Smith PM, Lakatos A, Barnett SC, Jeffery ND, Franklin RJM. Cryopreserved cells isolated from the adult canine olfactory bulb are capable of extensive remyelination following transplantation into the adult rat CNS. Exp Neurol. 2002;176: 402–6.
99. Gold R, Hartung HP, Toyka KV Animal models for autoimmune demyelinating disorders of the nervous system. Mol Med Today. 2000;6:88–91.
100. Steinman L, Zamvil SS. Virtues and pitfalls of EAE for the development of therapies for multiple sclerosis. Trends Immunol. 2005;26:565–71.

101. Akiyama Y, Radtke C, Honmou O, Kocsis JD. Remyelination of the spinal cord following intravenous delivery of bone marrow cells. Glia. 2002;39(3):229–36.
102. Inoue M, Honmou O, Oka S, Houkin K, Hashi K, Kocsis JD. Comparative analysis of remyelinating potential of focal and intravenous administration of autologous bone marrow cells into the rat demyelinated spinal cord. Glia. 2003;44(2):111–8.
103. Metz I, Lucchinetti CF, Openshaw H, Garcia-Merino A, Lassmann H, Freedman MS, et al. Autologous haematopoietic stem cell transplantation fails to stop demyelination and neurodegeneration in multiple sclerosis. Brain. 2007;130(Pt 5):1254–62.
104. Pluchino S, Quattrini A, Brambilla E, Gritti A, Salani G, Dina G, et al. Injection of adult neurospheres induces recovery in a chronic model of multiple sclerosis. Nature. 2003;422(6933):688–94.
105. Pluchino S, Zanotti L, Rossi B, Brambilla E, Ottoboni L, Salani G, et al. Neurosphere-derived multipotent precursors promote neuroprotection by an immunomodulatory mechanism. Nature. 2005;436(7048):266–71.
106. Richardson RM, Sun D, Bullock MR. Neurogenesis after traumatic brain injury. Neurosur Clin N Am. 2007;18(1):169–81, xi.
107. Kernie SG, Erwin TM, Parada LF. Brain remodeling due to neuronal and astrocytic proliferation after controlled cortical injury in mice. J Neurosci Res. 2001;66(3):317–26.
108. Dash PK, Mach SA, Moore AN. Enhanced neurogenesis in the rodent hippocampus following traumatic brain injury. J Neurosci Res. 2001;63(4):313–9.
109. Miles DK, Kernie SG. Activation of neural stem and progenitor cells after brain injury. Prog Brain Res. 2006;157:187–97.
110. Mahmood A, Lu D, Chopp M. Marrow stromal cell transplantation after traumatic brain injury promotes cellular proliferation within the brain. Neurosurgery. 2004;55(5):1185–93.
111. Chen X, Katakowski M, Li Y, et al. Human bone marrow stromal cell cultures conditioned by traumatic brain tissue extracts: growth factor production. J Neurosci Res. 2002;69(5):687–91.
112. Mahmood A, Lu D, Chopp M. Intravenous administration of marrow stromal cells (MSCs) increases the expression of growth factors in rat brain after traumatic brain injury. J Neurotrauma. 2004;21(1):33–9.
113. Bliss T, Guzman R, Daadi M, Steinberg GK. Cell transplantation therapy for stroke. Stroke. 2007;38(2 Suppl):817–26.

Improving Memory with Stem Cell Transplantation

Mathew Blurton-Jones, Tritia R. Yamasaki and Frank M. LaFerla

Abstract Stem cell transplantation is heralded as an exciting and novel approach to treat a wide variety of human brain disorders for which only palliative therapies currently exist. The clinical application of stem cells for the treatment of neurological disorders such as Alzheimer disease, the most common cause of age-related dementia, remains untested. A growing body of work, however, suggests that stem cell transplantation may provide important therapeutic benefits: ameliorating cognition, memory, and locomotor function, albeit via indirect mechanisms. Although neuronal replacement strategies remain a major focus of stem cell research, recent studies indicate that stem cells may provide therapeutic utility not necessarily by replacing dead or dysfunctional cells but by enhancing the survival, health, and activity of existing endogenous neurons. In this chapter, we review the limited regenerative capacity of the brain and the potential role of adult neural stem cells in memory. We also discuss recent advances showing that neural stem cell transplantation provides therapeutic utility for neurological disease in large part via indirect mechanisms. Although neuronal replacement may provide some benefit, indirect mechanisms such as neurotrophic support, immune modulation, and enzyme replacement likely play a far more prominent role in stem cell–mediated functional recovery. Taken together, these studies suggest that at least in the short-term, harnessing or enhancing these alternate mechanisms may provide a more achievable method to treat human disease.

Keywords Neural stem cells · Alzheimer Disease · Neurodegeneration · Neurotrophin · Synaptic Plasticity · Hippocampus · Transgenic · Cognition

F.M. LaFerla (✉)
Department of Neurobiology & Behavior, Institute for Brain Aging and Dementia, 1109 Gillespie Neuroscience Research Facility, University of California, Irvine, Irvine, CA 92697-4545
e-mail: laferla@uci.edu

1 Introduction

Stem cells are undifferentiated cells with the potential to give rise to multiple cell types in the body. By definition, embryonic stem cells are totipotent, capable of giving rise to all embryonic and extraembryonic cell types. In contrast, adult stem cells are multipotent or unipotent, producing a few related cell types or even just a single differentiated cell type. Endogenous adult stem cells contribute to the health and function of most organs, and also influence the regenerative capacity of an organ following an injury or disease process. Under these situations, transplantation of embryonic or adult-derived stem cells may offer a viable approach to improve regeneration and hence treat a variety of human diseases. The potential use of stem cells to repopulate and replace dead or defective cells has focused much of the attention on diseases that afflict organs with very limited regenerative potential, such as the pancreas, heart, and brain. The postmitotic nature of the principle cells of these organs, for example neurons in the brain, renders them particularly vulnerable to injury and disease.

2 The Adult Human Brain Exhibits Limited Capacity for Regeneration

The adult human brain and spinal cord exhibit a very limited capacity for regeneration with little functional recovery observed after serious injury or during neurodegenerative disease. In part, this limited regenerative potential may arise from a lack of sufficient endogenous neural stem cell populations to replenish lost neurons or to disease or age-related processes that impede their induction. Prior to 1965, the dogma concerning adult neurogenesis was summed by the words of Ramon y Cajal: "In the adult centers, the nerve paths are fixed and immutable: everything may die, nothing may be regenerated" [1]. Joseph Altman and Gopal Das [2] directly challenged that concept demonstrating the first evidence of adult neurogenesis in the rat hippocampus. More

V.K. Rajasekhar, M.C. Vemuri (eds.), *Regulatory Networks in Stem Cells*, Stem Cell Biology and Regenerative Medicine,
DOI 10.1007/978-1-60327-227-8_43,

recently, Gage and colleagues found evidence of neurogenesis in the adult human brain [3]. Today, the concept of neurogenesis as a vital and continual post-developmental process, arising from a population of endogenous neural progenitors or neural stem cells, is widely accepted. It is important to recognize, however, that neurogenesis in the adult brain remains extremely limited [4].

There are only two regions of the adult brain where neurogenesis is widely accepted to occur. The first site is the anterior subventricular zone (SVZ), from which new cells move by tangential chain migration along the rostral migratory stream to the olfactory bulb [5]. Once in the olfactory bulb, they differentiate into interneurons, olfactory granule neurons and periglomerular cells [6]. The second commonly accepted site of adult neurogenesis is the subgranular zone (SGZ) of the dentate gyrus in the hippocampus. Studies suggest that up to 10,000 cells/day are generated at this structure in the rat brain [7]. These cells subsequently migrate into the blades of the granule cell layer of the dentate gyrus [8, 9]. Many of these immature cells die before the first month [10], but of those remaining, the majority differentiate into neurons, although a significant fraction ($\sim$20%) become glial cells [8]. These new neurons seem capable of integration into the existing circuitry, sending their dendrites into the molecular layer and axons into the CA3 region around 4 days post-division [11–13]. Once mature, these cells also exhibit electrophysiological properties similar to neighboring granule cells [14, 15]. However, the relative number of new neurons generated in the adult brain remains extremely small. For example, one recent study utilized transgenic mice to inducibly label nestin-expressing stem cells and their progeny with a fluorescent protein and demonstrated that these cells constitute no more than 1% of the total granule cell population within the dentate gyrus of the adult mouse hippocampus [16]. For all other brain regions, except the olfactory bulb, there appears to be no quantifiable contribution of adult neurogenesis.

3 Adult Neurogenesis and Memory

It is interesting to speculate as to why neurons are post-mitotic and why adult neurogenesis remains so restricted. The answer to this question is likely to be complex, but one of the most critical functions of the brain involves the storage and retrieval of experiences that an individual encounters. Differentiated neurons likely remain post-mitotic so as not to interfere with specialized functions like synaptic plasticity and learning and memory. The precise mechanism by which memory is stored in the brain remains unclear. However, evidence suggests that at its most basic level, memories are stored via structural and biochemical changes in the strength and connectivity of neuronal networks [17]. The repeated turnover of neuronal populations through cell death and neurogenesis would likely disrupt the storage and/or retrieval of most aspects of memory.

Despite this, there is some evidence that endogenous hippocampal neural stem cells and their progeny may be involved in at least one specific aspect of memory. The hippocampus is a structure that is strongly implicated in the processing, storage, and retrieval of memories [17]. Although neurogenesis only occurs within a single subfield of the hippocampus, the granule cell layer of the dentate gyrus, these newborn cells may play a specialized role in memory. For example, in one study, administration of a systemic mitotic toxin impaired hippocampal-dependent learning versus saline-treated controls [9]. More evidence for a reciprocal relation between learning and neurogenesis in the hippocampus comes from studies showing that hippocampal-dependent learning actually increases levels of neurogenesis seen at the subgranular zone of the dentate gyrus [18]. Although the precise role of hippocampal neurogenesis in learning and memory remains unclear, Gage and colleagues recently speculated that newborn neurons may aid in the temporal encoding of memories [19]. An interesting property of newborn hippocampal granule neurons is that they exhibit elevated electrophysiological excitability during a brief 1–2-month time window [14]. This increased excitability allows newborn neurons to be preferentially activated in response to environmental stimuli or learning [20–22]. Thus, it has been proposed that these newborn neurons may allow structural and electrophysiological associations to be made between different memories that occur within a similar time window, in essence creating something akin to a "time stamp" on a given set of memories [19]. Despite these intriguing hypotheses, adult neurogenesis remains very restricted.

4 Memory and Neuronal Loss

The major neurologic disorders that afflict humans are typically characterized by a loss of neurons and in many of these diseases memory systems are also affected. For example, Alzheimer's disease (AD) is the most common cause of dementia among the elderly, and afflicts about 5 million individuals in the United States alone. As the disease progresses, it inexorably leads to the destruction of neurons and synapses, and a profound loss of memory, which initially manifests as a loss of episodic memory, followed by impairments in other cognitive domains.

A critical question in the field of neurobiology relates to the efficacy by which learning and memory can be carried out in the brain following the loss of hippocampal neurons. For

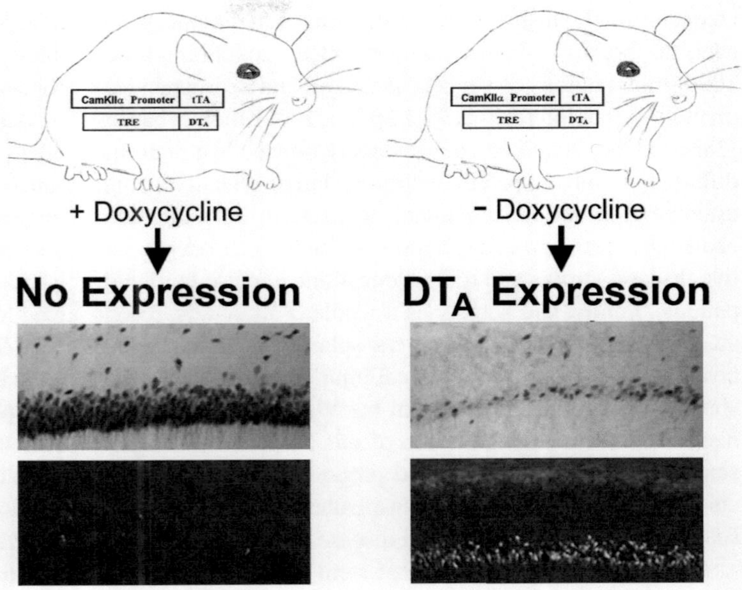

Fig. 1 The inducible *CaM/Tet-DT$_A$* transgenic model utilizes the tetracycline-transactivator (tTA) system and forebrain-restricted CaMKIIα promoter to inducibly express diphtheria toxin (DT$_A$) in the adult mammalian brain. Removal of doxycycline from the diet leads to induction of DT$_A$ expression and ablation of CA1 pyramidal neurons of the hippocampus

example, if a significant number of neurons are lost, as occurs in AD, can memory deficits be rescued? Investigational methods to address this problem have often involved the use of chemical or physical lesion paradigms. Although these experiments have provided important information, interpretation of such studies is limited by a lack of regional and cellular specificity. For example, electrolytic lesions damage not only the target populations, but also surrounding and overlying tissue. Such paradigms also result in a sizable secondary glial response [23].

To better address the precise role of hippocampal CA1 neurons in learning and memory and determine whether memory deficits can be rescued following neuronal loss, we developed an inducible genetic neuronal lesion model [24]. We utilized the tetracycline-regulatable transgenic system in combination with a forebrain-restricted promoter to inducibly express the A chain of diphtheria toxin (DT$_A$) in the adult mammalian brain. As illustrated in Fig. 1, removal of doxycycline (a tetracycline analog) from the diet leads to induction of DT$_A$ expression within a specific set of forebrain neurons. Spatial control is provided by the forebrain-specific CaMKIIα promoter, which expresses most robustly within CA1 pyramidal neurons of the hippocampus. By titrating the duration of doxycycline withdrawal, and thereby transgene expression, we can specifically ablate CA1 neurons alone. In contrast, when doxycycline is removed for longer periods, that is, 30+ days, hippocampal granule neurons and cortical neurons are also ablated.

The establishment and characterization of this novel transgenic model, referred to as *CaM/Tet-DT$_A$* mice, enabled us to examine the impact of CA1 neuronal loss on learning and memory in adult animals. Using a context-dependent novel-object recognition task, which is dependent on the hippocampus, we found that CA1 ablation produces a robust deficit in memory [24]. As expected, however, testing the mice on a cortical-dependent version of object recognition did not reveal any deficits, which attests to the regional specificity of the lesion. Our data clearly demonstrate the utility of this genetic approach for ablating selective neuronal populations in the mammalian brain. Our studies also have implication for age-related neurodegenerative disorders such as AD, in which a profound loss of CA1 neurons occurs. Because this transgenic method of neuronal ablation provides both spatial and temporal control, it can be readily adapted to examine the function of other specific neuronal populations. For example, we are currently utilizing the newborn, neuron-specific promoter doublecortin, to develop a mouse model that allows us to specifically ablate adult born neurons and thereby define the precise role of neurogenesis in memory.

5 Improving Memory Following Transplantation of Stem Cells

Given the importance of hippocampal neurons in learning and memory and the limited neurogenesis that occurs in the adult human brain, cognitive deficits arising from injury or neurodegenerative disease may be extremely difficult to reverse. However, some recent findings from our lab and others suggest that stem cell transplantation may offer a viable approach to prevent or perhaps even reverse memory deficits.

The prospect of using stem cells to replace dead or dysfunctional neurons is extremely appealing, and has important clinical implications as a plethora of brain disorders are all characterized by loss of selective neuronal populations. However, the brain is a highly complex organ, containing up to 100 billion neurons of varying structural and neurochemical phenotype that interconnect via up to a 100 trillion synapses [25]. Coaxing stem cells to effectively replace lost neurons, differentiate into a specific neurotransmitter phenotype, and establish appropriate functional connectivity is a daunting challenge. Yet, congenital brain conditions, neurodegenerative disease, stroke, and trauma constitute a major burden to patients, family, and society as a whole. Fortunately, recent studies suggest that stem cell transplantation may provide benefits for complex neurological injuries and disorders not simply by neuronal replacement but via alternative mechanisms. In the remaining portion of this chapter, we describe some of these key findings and propose four major mechanisms by which stem cell transplantation may aid in the future treatment of neurological disease and injury.

Some of the first evidence that stem cells could improve cognition and memory came from studies in aged rats by Sugaya and colleagues [26]. This group demonstrated that transplantation of human neural stem cells into the lateral ventricle of aged rats led to improved Morris water maze (MWM) performance within one month. A second group found similar results, showing that bone marrow stem cells grafted into the striatum and hippocampus of aged rats improved both motor and cognitive function [27]. It is important to point out, however, that there is minimal neuronal loss that is apparent during normal aging, even in rats that exhibit cognitive deficits [28]. Consequently, it was not possible for these studies to determine whether stem cells might improve function in the context of a more severe dysfunction resulting from neuronal loss. Other groups have begun to examine the impact of stem cell transplantation following more severe paradigms such as traumatic brain injury (TBI), which produces significant neuronal loss. For example, embryonic stem cell transplantation failed to improve cognition despite improved sensorimotor function [29]. In contrast, a second group found that transplantation of partially predifferentiated human fetal neural stem cells improved MWM performance following TBI [30]. Why the discrepancy?

Although TBI leads to neuronal loss, this paradigm produces significant damage to multiple regions and multiple functional domains. Unfortunately, all rodent cognitive testing requires at least some degree of motor activity and performance. Thus, it is difficult to clearly assess whether behavioral improvements following TBI results from cognitive versus simply motor recovery. The inducible *CaM/Tet-DT*$_A$ transgenic model circumvents this limitation, allowing us to examine the impact of neural stem cell transplantation after ablation of CA1 pyramidal neurons. Following transgene induction, these mice exhibit no motor impairments and the focal nature of the lesion allows one to more carefully assess potential cognitive recovery. *CaM/Tet-DT*$_A$ mice were induced for 25 days to elicit a significant ablation of CA1 neurons, which was associated with cognitive dysfunction. Neural stem cells (NSC) derived from GFP transgenic mice were then stereotactically injected into both hippocampi, whereas control mice received vehicle injections. At 1 month and 4.5 months post-transplantation, mice were tested on a hippocampal-dependent memory task. Interestingly, 1 month after transplant, NSC-injected animals exhibited a nonsignificant trend toward improved cognitive function. By 4.5 months, however, NSC-injected animals showed a significant improvement in hippocampal-dependent memory versus vehicle-injected controls. Thus, NSC-injection was able to improve cognitive function despite a massive loss of CA1 hippocampal neurons.

To determine the mechanism by which functional recovery had occurred, we sacrificed the mice and histopathologically examined their brains. As demonstrated in Fig. 2, injected NSCs engrafted well and differentiated into all three lineages: neurons, astrocytes, and oligodendrocytes. Intriguingly, only a small portion (<5%) of these cells differentiated into neurons, suggesting that functional recovery

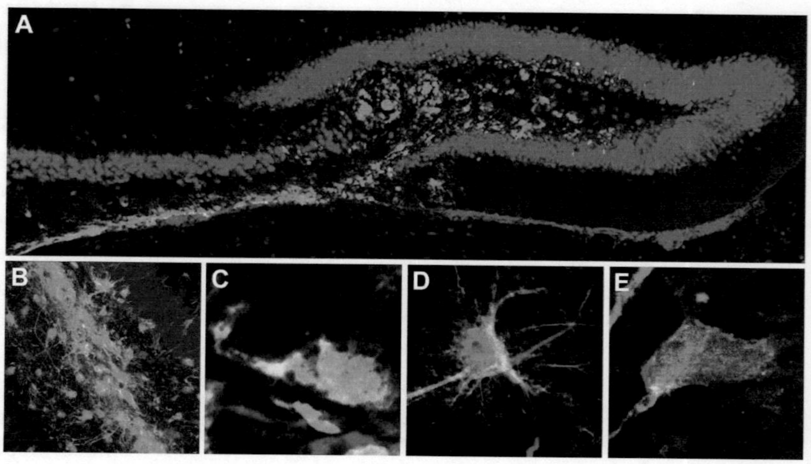

Fig. 2 Neural stem cells derived from GFP transgenic mice (*white*) engraft well into the lesioned adult mouse hippocampus (**A**). Endogenous neurons (NeuN, *gray*) demark the granule cell layer and CA3 of the hippocampus. Engrafted cells differentiate into astrocytes (**B, D**), doublecortin-expressing neurons (**C**), and oligodendrocytes (**E**) (*see* also Color Insert)

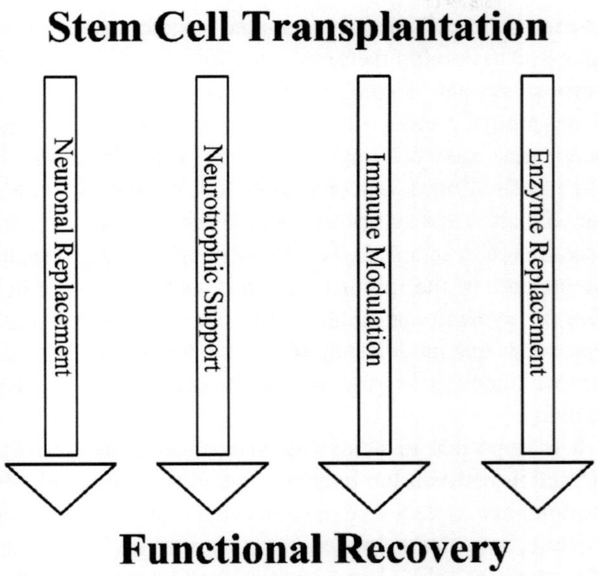

Fig. 3 Stem cell transplantation mediates behavioral recovery via multiple putative mechanisms

was mediated not via neuronal replacement. Hence, we further examined the brains to determine the putative mechanism by which behavioral recovery had occurred. Although the number of functional neurons is clearly important, a second equally vital aspect of brain function is the establishment, maintenance, and modification of appropriate synaptic connectivity, which is referred to as synaptic plasticity. We postulated that NSC transplantation might favorably modulate synaptic plasticity, allowing the brain to compensate for the significant loss of neurons. Using a well-established quantitative analysis, we examined the density of synaptic connections within the region of the hippocampus where CA1 neurons are known to make their connections. As shown in Fig. 3, NSC-injected animals exhibited significantly higher synaptic density versus control-injected animals. These data suggest that the remaining CA1 neurons successfully compensate for the lost cells by increasing their connectivity within the hippocampal circuitry.

The precise mechanism that allows NSCs to promote hippocampal synaptic growth remains unclear. However, preliminary data suggest that NSCs produce large amounts of neurotrophins. Neurotrophins are well-established mediators of synaptic growth and neuronal plasticity. Thus, it is likely that stem cell transplantation improves hippocampal function not by neuronal replacement but via a neurotrophic enhancement of synaptic plasticity and hippocampal connectivity.

The *CaM/Tet-DT$_A$* transgenic model provides a tool by which we can carefully assess the impact of hippocampal neuronal loss and stem cell transplantation on learning and memory. However, in human disease and brain injury, the loss of neurons and occurrence of additional disease-specific pathologies presents a more complex challenge. A key question that remains is whether stem cell transplantation can improve cognition in the context of such aggressive pathologies. Consequently, we have initiated studies to examine the effect of NSC-transplantation in the triple transgenic model of AD (3xTg-AD mice), which are characterized by the age-dependent buildup of both amyloid and neurofibrillary pathology Oddo et al., 2003 [31]. Studies in this and other models of neurodegenerative disease should prove invaluable at further defining the potential of stem cell transplantation to treat human disease.

6 Stem Cells Likely Mediate Improvement via Multiple Mechanisms

Data from the *CaM/Tet-DT$_A$* transgenic mice provide intriguing evidence that stem cell transplantation can ameliorate cognitive dysfunction not simply by neuronal replacement but via alternative mechanisms. Our studies and those of several other groups suggest that at least in the short-term, harnessing or enhancing such alternate mechanisms may provide a more achievable method to treat human disease. Below, we summarize four major mechanisms by which stem cell transplantation may provide functional improvement in the brain (Fig. 3).

6.1 Neuronal Replacement

The original and perhaps ultimate goal of stem cell transplantation in the CNS is to replace dead or dysfunctional neurons. Although the postmitotic nature of neurons is likely critical to normal function, this attribute also makes the brain and spinal cord especially vulnerable to injury and disease. Fortunately, the in vitro differentiation of stem cells into specific subpopulations of neurons continues to advance [32], bringing the goal of neuronal replacement closer to reality. In fact, some studies have demonstrated differentiation and survival of electrophysiologically functional neurons within the brain [32, 33]. However, perhaps the most challenging goal of neuronal replacement is the establishment of appropriate axonal and dendritic growth, synaptic connectivity, and the integration of transplanted cells into a functional neuronal circuit. Regeneration and/or growth of axonal and dendritic processes in the adult CNS are typically very limited (reviewed in [34]). Inhibitory proteins, such as those present within white matter or secreted following an injury, can repulse axonal growth cones, severely limiting regeneration [35, 36]. New strategies are being developed to overcome such inhibition by expression or delivery of growth-promoting factors such as neurotrophins and cyclic-AMP [37]. Likewise establish of neurotrophin gradients of that recapitulate the developmental cues that initially promote synaptic connectivity may hold great promise [38], especially if combined with stem cell transplantation. Neuronal replacement clearly holds great promise, but it is likely that alternative mechanisms may provide more applicable benefits in the near term.

6.2 Neurotrophic Support

One of the most promising mechanism by which stem transplantation may improve cognitive function involves the delivery of growth-promoting neurotrophins. Many neurons remain responsive to neurotrophins for survival, plasticity, and neuronal activity throughout life (reviewed in [39]). Deficits in neurotrophin expression or transport have also been implicated in several neurodegenerative diseases including, AD, Parkinson disease, Huntington disease, and ALS [40–42]. Many stem cell lines have previously been shown to express high levels of various neurotrophins [43–47]. Our own data described above also support the notion that neural stem cells can promote synaptic plasticity via a neurotrophic mechanism. Trophic mechanisms have also implicated in functional recovery in a primate model of Parkinson disease [48] and a mouse model of cerebellar neuronal death [49].

A promising further variation of this approach is to genetically modify stem cells to produce elevated amounts of disease-relevant neurotrophins. For example, work from Clive Svendsen's group has found that overexpression of IGF-1 in neural precursors rescue function in a rat model of Parkinson disease, whereas overexpression of GDNF protects motoneurons from death in a rat model of ALS [50, 51]. The endogenous production of neurotrophins and/or the genetic supplementation of neurotrophin production appear to hold great promise as a therapeutic approach to a number of human diseases. It also is likely that this kind of approach will prove more readily achievable in the near term.

6.3 Immune Modulation

Recent studies have begun to examine the influence of stem cell populations on immune cells and the immune response to brain injury and disease. For example, delivery of neural stem cells in a mouse model of Sandhoff's disease decreases microglial activity, helping to prolong survival [33]. Genetic modification of stem cells to overexpress immune related proteins such as the anti-inflammatory cytokine interleukin-10, also presents an interesting strategy. For example, in one study such an approach effectively decreased both cellular and humoral immune responses in a model of Parkinson's disease [52]. The interactions between stem cells and the immune system are also being harnessed to develop therapies aimed at autoimmune-related nervous system disorders such as multiple sclerosis. In an animal model of this disorder, one recent study found that mesenchymal stem cell transplantation led to a milder disease outcome with fewer relapses, reduced demyelination and axonal loss, and lower T-cell and antibody responses [53]. Interestingly, growing evidence suggests that components of the immune system may also serve critical roles in synaptic plasticity [54]. Thus, stem cell–based approaches that modulate cytokines or chemokines may also improve function in disease that involve altered synaptic plasticity.

It is important to point out the reciprocal nature of the interaction between the immune system and stem cells. For example, just as neural stem cells can modulate the immune response, neuroinflammation can also alter the fate of neural progenitors [55]. Neural stem cells express high levels of receptors to inflammatory chemoattractants such as stromal cell–derived factor 1-alpha, which appears to promote migration of neural stem cells toward areas of brain injury [56]. An interesting extension of this finding is the proposed use of neural stem cells to allow localized delivery of beneficial proteins within the brain [55]. In summary, the reciprocal interactions between stem cells and the immune response may mediate some of the beneficial effects and properties of stem cells that can be therapeutically exploited.

6.4 Enzyme Replacement

Many pediatric neurodegenerative disorders involve deficits in the production or activity of a single specific enzyme. For example, deficits in the enzymes that mediate lysosomal degradation of glycosphingolipids, lead to the development of Sandhoff's disease, Tay-Sachs, or Niemann-Pick [57]. One promising experimental approach to potentially treat these devastating disorders is to decrease levels of the enzymatic substrate that typically accumulates in these disorders. However, a more direct approach, termed enzyme replacement therapy, involves replenishing levels of the missing or defective enzyme [57]. For some disorders, intravenous delivery of enzymes appears to be somewhat effective [58]. However, for diseases that primarily affect the brain, transport of replacement enzymes across the blood brain barrier presents a distinct problem.

More recently, studies have begun to explore the utility of stem cell transplantation as a method to restore brain enzyme activity. In some cases, endogenous production of a missing enzyme by implanted stem cells may be sufficient to at least partially restore enzymatic function and thereby reduce levels of accumulating substrate and disease progression

[33, 49]. For other disorders, genetic modification of stem cells to express and/or secrete elevated levels of the missing enzyme may be required to achieve more significant recovery [59].

In the case of AD, stem cell–mediated enzyme supplementation may also be of value. The disease is thought to be caused by the accumulation of a small toxic peptide termed amyloid-beta (Aβ). Interestingly, a few endogenous enzymes can normally degrade the Aβ peptide, but over the course of many years, the balance between Aβ production and degradation shifts, eventually leading to AD [60]. Recent animal studies suggest that supplementation of these enzymes can slow the progression of AD pathology [61, 62]. It is likely that neural stem cells in particular will be uniquely positioned to provide a novel and more physiologic method to deliver such enzymes.

7 Conclusions and Future Goals

In summary, stem cell transplantation offers an exciting and novel approach to treat a wide variety of human brain disorders for which there are currently only palliative therapies. Although neuronal replacement strategies remains a major focus of stem cell research, our data and that of many other labs suggest that stem cells may provide important therapeutic utility via alternative mechanisms. Nevertheless, a number of further advances must be made before stem cell transplantation can truly offer a viable and widely employed therapy. For example, we propose that a critical goal of stem cell research should be to continue to elucidate and define the precise molecular mechanisms by which stem cell transplantation mediates its beneficial effects within the brain. The four generalized mechanisms described above likely play important roles in the functional recovery afforded by stem cell transplantation for various neurological conditions. However, further study is clearly needed to better define these pathways. For example, if neurotrophic mechanisms are implicated in recovery of cognitive function, identification of the precise neurotrophin(s) involved should be pursued. Such data could in turn lead to the development of improved stem cell–based approaches or even the development of more simple pharmaceutical-based therapies. One of the greatest challenges of this goal is that transplantation of stem cells into the brain likely ameliorates cognitive improvement by multiple mechanisms [33, 48]. Experimental manipulations can be designed to test the importance of a single mechanism without great difficulty. In contrast, when a complex result such as improved cognition involves multiple mechanisms, inhibition of a single pathway may provide limited evidence. Such goals are still worth pursuing, nonetheless.

In conclusion, a growing number of studies support the hope that stem cell transplantation may offer a viable approach to treat a number of devastating neurological disorders and conditions. In the future, these therapies may be mediated via neuronal replacement. However, in the near term researchers should continue to examine additional mechanisms by which stem cell transplantation may provide therapeutic utility far sooner than initially expected.

References

1. Ramon y, Cajal S. Degeneration and regeneration of the nervous system. London: Oxford University Press; 1928.
2. Altman J, Das GD. Autoradiographic and histological evidence of postnatal hippocampal neurogenesis in rats. J Comp Neurol. 1965;124(3):319–35.
3. Eriksson PS, Perfilieva E, Bjork-Eriksson T, et al. Neurogenesis in the adult human hippocampus. Nat Med. 1998;4(11):1313–7.
4. Bhardwaj RD, Curtis MA, Spalding KL, et al. Neocortical neurogenesis in humans is restricted to development. Proc Natl Acad Sci U S A. 2006;103(33):12564–8.
5. Rousselot P, Lois C, Alvarez-Buylla A. Embryonic (PSA) N-CAM reveals chains of migrating neuroblasts between the lateral ventricle and the olfactory bulb of adult mice. J Comp Neurol. 1995;351(1):51–61.
6. Magavi SS, Macklis JD. Manipulation of neural precursors in situ: induction of neurogenesis in the neocortex of adult mice. Neuropsychopharmacology. 2001;25(6):816–35.
7. Cameron HA, McKay RD. Adult neurogenesis produces a large pool of new granule cells in the dentate gyrus. J Comp Neurol. 2001;435(4):406–17.
8. Fuchs E, Gould E. Mini-review: in vivo neurogenesis in the adult brain: regulation and functional implications. Eur J Neurosci. 2000;12(7):2211–4.
9. Shors TJ, Miesegaes G, Beylin A, Zhao M, Rydel T, Gould E. Neurogenesis in the adult is involved in the formation of trace memories. Nature. 2001;410(6826):372–6.
10. Kempermann G, Gast D, Kronenberg G, Yamaguchi M, Gage FH. Early determination and long-term persistence of adult-generated new neurons in the hippocampus of mice. Development. 2003;130(2):391–9.
11. Hastings NB, Gould E. Rapid extension of axons into the CA3 region by adult-generated granule cells. J Comp Neurol. 1999;413(1):146–54.
12. Markakis EA, Gage FH. Adult-generated neurons in the dentate gyrus send axonal projections to field CA3 and are surrounded by synaptic vesicles. J Comp Neurol. 1999;406(4):449–60.
13. Stanfield BB, Trice JE. Evidence that granule cells generated in the dentate gyrus of adult rats extend axonal projections. Exp Brain Res. 1988;72(2):399–406.
14. Ge S, Yang CH, Hsu KS, Ming GL, Song H. A critical period for enhanced synaptic plasticity in newly generated neurons of the adult brain. Neuron. 2007;54(4):559–66.
15. van Praag H, Christie BR, Sejnowski TJ, Gage FH. Running enhances neurogenesis, learning, and long-term potentiation in mice. Proc Natl Acad Sci U S A. 1999;96(23):13427–31.
16. Lagace DC, Whitman MC, Noonan MA, et al. Dynamic contribution of nestin-expressing stem cells to adult neurogenesis. J Neurosci. 2007;27(46):12623–9.
17. Neves G, Cooke SF, Bliss TV. Synaptic plasticity, memory and the hippocampus: a neural network approach to causality. Nat Rev. 2008;9(1):65–75.

18. Gould E, Beylin A, Tanapat P, Reeves A, Shors TJ. Learning enhances adult neurogenesis in the hippocampal formation. Nat Neurosci. 1999;2(3):260–5.
19. Aimone JB, Wiles J, Gage FH. Potential role for adult neurogenesis in the encoding of time in new memories. Nat Neurosci. 2006;9(6):723–7.
20. Kee N, Teixeira CM, Wang AH, Frankland PW. Preferential incorporation of adult-generated granule cells into spatial memory networks in the dentate gyrus. Nat Neurosci. 2007;10(3): 355–62.
21. Ramirez-Amaya V, Marrone DF, Gage FH, Worley PF, Barnes CA. Integration of new neurons into functional neural networks. J Neurosci. 2006;26(47):12237–41.
22. Tashiro A, Makino H, Gage FH. Experience-specific functional modification of the dentate gyrus through adult neurogenesis: a critical period during an immature stage. J Neurosci. 2007;27(12):3252–9.
23. Floyd CL, Lyeth BG. Astroglia: important mediators of traumatic brain injury. Prog Brain Res. 2007;161:61–79.
24. Yamasaki TR, Blurton-Jones M, Morrissette DA, Kitazawa M, Oddo S, LaFerla FM. Neural stem cells improve memory in an inducible mouse model of neuronal loss. J Neurosci. 2007;27(44):11925–33.
25. Kandel ER, Schwartz JH, Jessell TM. Principles of neural science. 4th ed. New York: McGraw-Hill Health Professions Division; 2000.
26. Qu T, Brannen CL, Kim HM, Sugaya K. Human neural stem cells improve cognitive function of aged brain. Neuroreport. 2001;12(6):1127–32.
27. Fernandez CI, Alberti E, Mendoza Y, et al. Motor and cognitive recovery induced by bone marrow stem cells grafted to striatum and hippocampus of impaired aged rats: functional and therapeutic considerations. Ann N Y Acad Sci. 2004;1019:48–52.
28. Rasmussen T, Schliemann T, Sorensen JC, Zimmer J, West MJ. Memory impaired aged rats: no loss of principal hippocampal and subicular neurons. Neurobiol Aging. 1996;17(1):143–7.
29. Hoane MR, Becerra GD, Shank JE, et al. Transplantation of neuronal and glial precursors dramatically improves sensorimotor function but not cognitive function in the traumatically injured brain. J Neurotrauma. 2004;21(2):163–74.
30. Gao J, Prough DS, McAdoo DJ, et al. Transplantation of primed human fetal neural stem cells improves cognitive function in rats after traumatic brain injury. Exp Neurol. 2006;201(2):281–92.
31. Oddo S, Caccamo A, Shepherd JD, Murphy MP, Golde TE, Kayed R, Metherate R. Mattson MP, Akbari Y, and LaFerla FM. Triple-transgenic model of Alzheimer's disease with plaques and tangles: intracellular Abeta and synaptic dysfunction. Neuron 2003;39:409–421.
32. Parish CL, Castelo-Branco G, Rawal N, et al. Wnt5a-treated midbrain neural stem cells improve dopamine cell replacement therapy in parkinsonian mice. J Clin Invest. 2008;118(1):149–60.
33. Lee JP, Jeyakumar M, Gonzalez R, et al. Stem cells act through multiple mechanisms to benefit mice with neurodegenerative metabolic disease. Nat Med. 2007;13(4):439–47.
34. Domeniconi M, Filbin MT. Overcoming inhibitors in myelin to promote axonal regeneration. J Neurol Sci. 2005;233(1–2): 43–7.
35. Buchli AD, Schwab ME. Inhibition of Nogo: a key strategy to increase regeneration, plasticity and functional recovery of the lesioned central nervous system. Ann Med. 2005;37(8): 556–67.
36. Jones LL, Sajed D, Tuszynski MH. Axonal regeneration through regions of chondroitin sulfate proteoglycan deposition after spinal cord injury: a balance of permissiveness and inhibition. J Neurosci. 2003;23(28):9276–88.
37. Lu P, Yang H, Jones LL, Filbin MT, Tuszynski MH. Combinatorial therapy with neurotrophins and cAMP promotes axonal regeneration beyond sites of spinal cord injury. J Neurosci. 2004;24(28):6402–9.
38. Taylor L, Jones L, Tuszynski MH, Blesch A. Neurotrophin-3 gradients established by lentiviral gene delivery promote short-distance axonal bridging beyond cellular grafts in the injured spinal cord. J Neurosci. 2006;26(38):9713–21.
39. Tuszynski MH, Gage FH. Maintaining the neuronal phenotype after injury in the adult CNS. Neurotrophic factors, axonal growth substrates, and gene therapy. Mol Neurobiol. 1995;10(2–3): 151–67.
40. Gauthier LR, Charrin BC, Borrell-Pages M, et al. Huntingtin controls neurotrophic support and survival of neurons by enhancing BDNF vesicular transport along microtubules. Cell. 2004;118(1):127–38.
41. Mufson EJ, Kroin JS, Sendera TJ, Sobreviela T. Distribution and retrograde transport of trophic factors in the central nervous system: functional implications for the treatment of neurodegenerative diseases. Prog Neurobiol. 1999;57(4):451–84.
42. Salehi A, Delcroix JD, Belichenko PV, et al. Increased App expression in a mouse model of Down's syndrome disrupts NGF transport and causes cholinergic neuron degeneration. Neuron. 2006;51(1):29–42.
43. Bentz K, Molcanyi M, Riess P, et al. Embryonic stem cells produce neurotrophins in response to cerebral tissue extract: cell line-dependent differences. J Neurosc Res. 2007;85(5): 1057–64.
44. Kamei N, Tanaka N, Oishi Y, et al. BDNF, NT-3, and NGF released from transplanted neural progenitor cells promote corticospinal axon growth in organotypic cocultures. Spine. 2007;32(12): 1272–8.
45. Pan HC, Cheng FC, Chen CJ, et al. Post-injury regeneration in rat sciatic nerve facilitated by neurotrophic factors secreted by amniotic fluid mesenchymal stem cells. J Clin Neurosci. 2007;14(11):1089–98.
46. Pisati F, Bossolasco P, Meregalli M, et al. Induction of neurotrophin expression via human adult mesenchymal stem cells: implication for cell therapy in neurodegenerative diseases. Cell Transplant. 2007;16(1):41–55.
47. Wei P, Liu J, Zhou HL, et al. Effects of engrafted neural stem cells derived from GFP transgenic mice in Parkinson's diseases rats. Neurosci Lett. 2007;419(1):49–54.
48. Redmond DE, Jr., Bjugstad KB, Teng YD, et al. Behavioral improvement in a primate Parkinson's model is associated with multiple homeostatic effects of human neural stem cells. Proc Natl Acad Sci U S A. 2007;104(29):12175–80.
49. Li J, Imitola J, Snyder EY, Sidman RL. Neural stem cells rescue nervous purkinje neurons by restoring molecular homeostasis of tissue plasminogen activator and downstream targets. J Neurosci. 2006;26(30):7839–48.
50. Ebert AD, Beres AJ, Barber AE, Svendsen CN. Human neural progenitor cells over-expressing IGF-1 protect dopamine neurons and restore function in a rat model of Parkinson's disease. Exp Neurol. 2008;209(1):213–23.
51. Suzuki M, McHugh J, Tork C, et al. GDNF secreting human neural progenitor cells protect dying motor neurons, but not their projection to muscle, in a rat model of familial ALS. PLoS ONE. 2007;2(1):e689.
52. Wang XJ, Liu WG, Zhang YH, Lu GQ, Chen SD. Effect of transplantation of c17.2 cells transfected with interleukin-10 gene on intracerebral immune response in rat model of Parkinson's disease. Neurosci Lett. 2007;423(2):95–9.
53. Gerdoni E, Gallo B, Casazza S, et al. Mesenchymal stem cells effectively modulate pathogenic immune response in experimental

autoimmune encephalomyelitis. Ann Neurol. 2007;61(3): 219–27.
54. Bauer S, Kerr BJ, Patterson PH. The neuropoietic cytokine family in development, plasticity, disease and injury. Nat Rev. 2007;8(3):221–32.
55. Mueller FJ, McKercher SR, Imitola J, et al. At the interface of the immune system and the nervous system: how neuroinflammation modulates the fate of neural progenitors in vivo. Ernst Schering Res Found Workshop. 2005;(53):83–114.
56. Imitola J, Raddassi K, Park KI, et al. Directed migration of neural stem cells to sites of CNS injury by the stromal cell-derived factor 1alpha/CXC chemokine receptor 4 pathway. Proc Natl Acad Sci U S A. 2004;101(52):18117–22.
57. Jeyakumar M, Butters TD, Dwek RA, Platt FM. Glycosphingolipid lysosomal storage diseases: therapy and pathogenesis. Neuropathol Appl Neurobiol. 2002;28(5):343–57.
58. Barton NW, Furbish FS, Murray GJ, Garfield M, Brady RO. Therapeutic response to intravenous infusions of glucocerebrosidase in a patient with Gaucher disease. Proc Natl Acad Sci U S A. 1990;87(5):1913–6.
59. Sidman RL, Li J, Stewart GR, et al. Injection of mouse and human neural stem cells into neonatal Niemann-Pick A model mice. Brain Res. 2007;1140:195–204.
60. Selkoe DJ. Clearing the brain's amyloid cobwebs. Neuron. 2001;32(2):177–80.
61. Hemming ML, Patterson M, Reske-Nielsen C, Lin L, Isacson O, Selkoe DJ. Reducing amyloid plaque burden via ex vivo gene delivery of an Abeta-degrading protease: a novel therapeutic approach to Alzheimer disease. PLoS Med. 2007;4(8):e262.
62. Marr RA, Rockenstein E, Mukherjee A, et al. Neprilysin gene transfer reduces human amyloid pathology in transgenic mice. J Neurosci. 2003;23(6):1992–6.

Index

A

ABCB5 chemoresistance determinant, 538
$ABCB5^+$ tumor cells, 532
Ablation therapy, of CSC, 539
Activin/Nodal/TGFβ signaling pathway, 298–300
Activins, *see* Transforming growth factor (TGF)β family
Acute lymphoid leukemia (ALL), 169
Acute myelogenous leukemia (AML), 473, 490, 496
Acute myeloid leukemia (AML), 169, 172, 500, 519, 529
Acute renal failure (ARF), 379–381, 383
Adeno-associated viruses (AAV), 79
Adenoviruses (Ad), 79–80
Adipose tissue-derived stem cells (ASCs), 257, 262
  DNA methylation in, 262–263
  housekeeping genes, and CpG methylation, 263
Adult stem cells, 285, 561–562
Aequorea, 341
AG 490, 479
Aging
  MTS24, role of, 425
  thymic regeneration, 425–426
  and thymus involution, 425
Ahn, S., 549
Akt–Mer fusion protein, 311
Aldehyde dehydrogenase (ALDH), 350
  expression, 555
Aldehyde dehydrogenase 1 (ALDH1), 548
Al-Hajj, M., 479, 496–497, 529
Allbritton, N.L., 341
Allegrucci, C., 259
Allogenic stem cell transplantations (alloSCTs), 141
Allotransplantation, 384
Alpha-1-antitrypsin (A1AT), 460
ALT-associated PML bodies (APBs), 125
Alternative lengthening of telomeres (ALT), 125
  See also Telomere and telomerase, for regulation of stem cells
Altman, J., 575
Alzheimer's disease (AD), 576
  bone marrow derived cells for treating, 565
  neural stem cells for treating, 566
Ames Salmonella test, 490
AML cells, 500–501
AML-ETO fusion protein, 522
AMLinitiating cells (AML-IC), 520
Amyloid precursor gene (APP), 566
Anagen cycle, in hair follicles, 314
Angiopoietin-1 (Ang-1) signaling, 357
Anti-ABCB5 monoclonal antibody, 532
  treatment, 539
Anti-angiogenic therapy, 501
Antigen-presenting cells (APCs), 509
Anti-inflammatory cytokine interleukin-10, 580
APE/Ref-1 regulator, 70
Aplastic anemia, 123, 128–130
APL cells, 500
Appelhoff, R.J., 219
Ara-C treatment, 534
Arai, F., 166, 523
Arce, C., 539
Ascaris zygote, first divisions
  asymmetry in cell cleavage of, 15–16
ASF/SF2 expression, 508
Asialoglycoprotein receptor-directed nuclear scanning, 462
Asymmetry, in stem cells, 13
  embryonic and germ cells, 14
    first divisions of zygote, 14
    germ cell differentiation, division patterns, 14–17
    neurogenesis, during embryonic development, 16–17
  and immortal strand hypothesis, 20–22
  in postnatal cells, 17–18
    HSCs, 18–19
    skeletal muscle stem cell, 19–20
  role in cell divisions, 13–14
Atm/Atr pathway, 491
ATP-binding cassette (ABC) drug efflux pumps, 538
Avian leukosis virus-A (ALV-A), 78
Axon damage, 567
Azuara, V., 261

B

BAC transgenics, 81–82
Baculoviruses, 80
Bao, S., 551, 555
Bar, E.E., 549, 552
Basic fibroblast growth factor (bFGF), 310, 311
Batista, C.M., 567
Battaglia, M., 512
β-catenin, 303, 312, 314, 364, 409
Bcr-Abl, 508
Beckmann, J., 352
Becskei, A., 89
Benzene, 490–491
Berman, D.M., 552
Bernstein, B.E., 238
Beta-Tubulin-III/BrdU positive neurons, 566
Biglycan, 359
Bioartificial livers, 461
Bissel, M.J., 482
Bisulfite genomic sequencing method, 271

Bivalent chromatin, 236, 238–239, 280
Black, A.C., 505
Black bile theory, 497
Blank, U., 224
Blastocyst, 498
Blue fluorescent protein (BFP), 342
Bmi-1, 475, 476, 481
BMP4-responsive BMPR1A+ cells, 536
Bone marrow (BM), HSCs in, 448
Bone marrow-derived MSCs, 257
Bone marrow transplantation (BMT), 449
Bone morphogenetic protein (BMPs), 15, 299, 321, 330, 416, 442, 452
Bonnet, D., 473, 529
Borrel, A., 498
BOULE, 57–58
Boveri, T., 498
Bradley, J.A., 506
Brain, CSCs in, 546–547
 oncogene expression in, 549
 treatment resistance, 550–551
 tumorigenic, 548
 See also Cancer/Cancer stem cells (CSCs)
Brain tumor stem cells, 547
BrdU-labeled NSCs, 567
Breast, CSCs in, 547–548
 oncogene expression in, 549–550
 treatment resistance, 551
 tumorigenic, 549
 See also Cancer/Cancer stem cells (CSCs)
Brivanlou, A.H., 140
Bromodeoxyuridine (BrdU), 381–382
Bronchioalveolar stem cells (BASCs), 315
Brusselmans, K., 477
Buick, R.A., 477

C
Cadherin-11, 381
Caenorhabditis elegans, asymmetry in development, 14
Cai, J., 456
Cairns, J., 20
Calabrese, C., 537
Calcium sensing receptor (CaR), 523
Calvi, L.M., 357
CaM/Tet-DT$_A$ transgenic model, 577–579
Cancer/Cancer stem cells (CSCs), 140, 315, 401
 ablation therapy of, 539
 anti-mTORC1 therapy in, 314
 biological properties, 495–496
 in brain, 546–547
 in breast, 529–530, 547–548
 cancer stem cell hypothesis, 240
 in colon cancer, 530–531
 concept, 497
 CpG islands, hypermethylation of, 251
 dedifferentiation theory of cancer, 497–498
 definition, 528–529
 and DNA demethylation, 238, 280–281
 enrichment of, 496–497
 growth factor signaling pathways in, 550
 in head and neck squamous cell carcinoma, 531–532
 hematopoietic lineage, 528–529
 in human malignant melanomas, 532
 and idea of cancer development, 497
 model, of tumor initiation and growth, 527
 molecular and biological features of, 535–538
 niche for cancer stem cells, 169
 vs. normal stem cells, 409
 novel targets for tumor therapy, 538–539
 oncogene expression in
  brain, 549
  breast, 549–550
 origins of, 534–535
 in pancreatic cancer, 531
 PcG repressive complex and, 243–244
 PI3K/Akt signaling and, 313–314
 prevention and therapy, 557
 putative populations of, 532–534
 theory, 527
 therapeutic approaches of targeting
  combination therapies, 555
  differentiating agents, 552
  growth factors, 553–555
  signaling pathways, 552
  treatment resistance, 555
 timing of CSC-targeted therapies, 555–556
 treatment efficacy of, 557
 treatment resistance
  brain, 550–551
  breast, 551
 tumorigenic
  brain, 548
  breast, 549
 in tumors of brain, 530
 tumor suppressor gene silencing in, 240–242, 251, 279
Cancer-initiating cells (CICs), 491
 See also Fhit genes
Cancer stem cell (CSC) theory, 527–528, 551
Cancer stem cell hypothesis, 240
Cancer theory, 495, 497
Capecchi, M., 5
Capillary electrophoresis (CE), 339, 341
Cardiogenesis, role of miRNA-1-2 in, 273
Cardiomyocytes and MSCs, 186–187
Carins hypothesis, *see* Immortal strand hypothesis
Carpenter, L., 271
Carrier ChIP (CChIP), 272
β-Catenin, 536
CD133, as cell surface marker, 380–381
CD 44+ brain TSCs, 479
CD 133+ brain TSCs, 479
CD 24–breast TSCs, 479
$CD34^+CD38^+$ AML cells, 529
$CD34^+/CD38^-$cells, 520
$CD44^+CD24^+ESA^+$ pancreatic cancer cells, 531
CD44+CD24-ESA+ population, 496
CD4+CD25^{high}Foxp3+ T cells, 512–513
CD133+ cells, 555
CD133 expression, 533
CD49F^{high}CD24+ population, 548
CD133+ glioma cells, 551
CD29^{high}CD24+, 548
$CD44^+\alpha_2\beta_1$highCD133$^+$ tumor cells, 532
$CD44^+$ HNSCC cell-derived xenografts, 531
CD133 marker, 539
CD8 T cell, 511
Cdx2, 94
Cdx/Hox pathway, 101
Cell cycle control, in pluripotent cells, 31
Cell division and role of asymmetry, 13
 See also Asymmetry, in stem cells
Cell replacement therapy, 3

Cell surface markers, 19
Cellular therapy, for neurological disease, 568
Central polypurine tract (cPPT), 78
Cerberus, 298
Cerebrovascular disease
  bone marrow derived cells for treating, 563–564
Ceroid Lipofuscinosis, 569
C6 glioma cell line, 474, 534
Chadwick, N., 365
Chagraoui, J., 366
Challen, G.A., 381
Chan, T., 384
Chen, Y., 367
ChIP-sequencing technology, 239
Chk1 and Chk2 checkpoint kinases, 555
Cholesterol 7a-hydroxylase (CYP7A1), 459
Chopp, M., 562–564
Chow, D.C., 68
Chow, S., 337
Cho, Y.M., 138
Chromatin
  DNA packaging into, 236
  in ES differentiated cells, 238
  histone modifications and, 236–238, 247
  remodeling, 270
  structure of, 236–237
  subtypes of, 236–237
Chromatin immunoprecipitation (ChIP) assays, 260, 272
Chronic renal failure (CRF), 379–380, 384, 386, 387
Cisplatin, 479, 480
Cited2, 367
CML cells, 500
CML66L, 508
C-myc, 5, 32
Cobblestone area-forming cells (CAFC), 348
Cohnheim, J., 497
Collagen
  type I, 191–194
  type II, 194
Collins, A.T., 532
Conboy, M.J., 22
Corneal epithelium, 411
Cozzio, A., 521, 535
CpG islands, 239, 248, 258, 277
CpG methylation, 258
  in murine ES cells, 259
Cre-loxP mediated translocations, 521
Cripto/Criptic, 298
Curcumin, 557
C-X-C chemokine receptor type 4 (CXCR4), 531
CXCL12-abundant reticular (CAR) cells, 167
CXCR4/stromal cell–derived factor 1 (SDF-1), 531
Cyclin-dependent kinases (CDK), 350
Cyst cell, 15
Cytochrome P450-2E1 (CYP2E1), 490
Cytokeratin 5/14 (CK5/14), 532
Cytotoxic therapy, 473, 478, 483, 484, 551
Czechowicz, A., 353

D

Dao, M.A., 216, 223
Das, B., 481
Das, G., 575
DAZ gene family, members and characteristics, 58
  human *BOULE* gene, 57–58
  phylogenetic tree, 57
DAZL (Deleted in Azoospermia-Like), 58
Dazl protein, 55
DDR receptors, 194
De-differentiation phenomenon, 480, 497–498
Dedifferentiation theory of cancer, 497
Definitive hematopoiesis, 102–103
Dekel, B., 381
Delaney, C., 364
Delta-24-RGD virus, 539
Denekamp, J., 497
Desert Hedgehog, 324
Dexamethasone (Dex), 455
Dialysis, and CRF patients, 379–380
Dicer-1 (Dcr-1), 159, 160
Dick, J., 529
Dick, J.E., 169, 473
Differentiation therapy, 538–539
  of cancer stem cells, 495, 498, 499, 501
  of leukemia, 500–501
DiGeorge syndrome, 420
Ding, 31 [Not found in Reference]
Diseases, from improper imprinting, 271
DNA
  demethylation, 238, 265, 278
  hypermethylation and cancer, 242–244
  methylation, 238, 243, 258, 278–279
    and cancer, 238, 258
    cellular plasticity and, 278
    control of, 277–278
    de novo methylation, 247, 277
    in embryonic cell, 250–252
    maintenance methylation, 247–248
    and pluripotency-associated genes silencing, 239–240
    and replication timing, 264–265
    stem/progenitor cells and, 239, 270–271
  packaging of, 236
DNA methyltransferases (DNMT), 238, 243, 270, 279
  CpG methylation and, 258
  in mammalian cells, 270
DNMT1, 277
  maintenance methylation, 247–248
  UHRF1 and, 248
Dnmt3L, 251
Donehower, L. A., 69
Dontu, G., 549
Doxorubicin, 534
Draper, J.S., 136
Dreesen, O., 140
Dressler, G., 383
Drosophila, 408
  asymmetry in embryonic development of, 14
  ovary, stem cell regulation in
    system for studying stem cell biology, 155–156
    See also Germline stem cells regulation
Drosophila, stem cells and niches, 147
  asymmetric division, translation of extracellular into intracellular polarity, 150
    female GSCs niche, 152
    neuroblasts in, 151
    spindle orientation mechanisms, 151
    tissue aging in fly GSCs, 152
  role of stem cells in tissue homeostasis, 147
    tumorigenesis and tissue aging, 147–148
  signaling in
    blood stem cells, types, 149–150

Drosophila, stem cells and niches (*cont.*)
female germ-line stem cells (GSCs), 148–149
germline stem cells and niche, 149
intestinal and malpighian stem cells, 150
male germline stem cells niche, 149
neuroblasts, 150
Drynan, L.F., 535
Dual-labeling nearest neighbor technique, 271
Dublanche, Y., 89
Duffield, J.S., 380, 382
Durante, F., 497
Dykstra, B., 353
Dyskeratosis Congenita (DKC), 130
Dziewczapolski, G., 565
DZIP (DAZ-Interacting Protein), 58–59

E

Early mesoderm-committed (EM) progenitors
actively dedifferentiated to ES cells by Nanog, 37, 41–42
Early primitive ectoderm-like (EPL) cells, 3, 5
E-cadherin–mediated cell adhesion, 158
ECM molecules, derived fragments
anti-angiogenic fragments
anastellin, 202
Col IV, 203–204
endorepellin, 203
endostatin and restin, 202–203
bioactive fragments and cellular effects, 201
cryptic domains and cellular effects, 200
Fibronectin (FN) role, 200–202
See also Extracellular Matrix (ECM), regulation of MSCs
Ectoderm-like population, pluripotent, 5
Egf signaling pathway, 149
Eiges, R., 76
Electroporation, 76
Embryoid bodies (EBs), 451
Embryonal rest theory, of cancer, 495, 497
Embryonic and germ cells, asymmetry in, 14
first divisions of zygote, 14
germ cell differentiation, pattern for asymmetric divisions, 14–16
neurogenesis, during embryonic development, 16–17
Embryonic development summary, in mouse and human, 28
Embryonic germ cells (EGCs), 56, 310
Embryonic stem cells (ESCs) fate decisions, transcriptional networks regulating, 87–88
genome-wide analyses, 94
ESC maintenance model, 98
microarray expression studies, 94
promoter occupancy studies, 96
shRNA knockdown and overexpression, 95–96
stem cell genes, common, 94–95
systematic integration of expression and promoter occupancy data, 96–98
time courses of differentiation, 95
transcriptional motifs and biological functions
negative and positive autoregulation and feedback, 88–91
transcriptional regulators in, 91
induced pluripotent stem(iPS) cells, 93–94
Nanog, 92
Oct4 and Sox2, 91–92
Sall4, 92–93
transcription factors regulating commitment, 94
Zfp206, 93
Embryonic stem cells (ESCs), in mouse and humans, 55
cell cycle control, in pluripotent cells, 31
derived from inner cell mass (ICM) of blastocysts, 55
phenotypic differences between, 28–29
pluripotency
determinants, genetic and epigenetic, 29–31, 33–34
maintenance, by Nanog, 37–40
-related transcription factors, 31–33
protein expression patterns variations between, 28–29
similarities and differences between and germ cells, 63
Embryonic stem cells (ESCs), self-renewal, 3
hESC self-renewal signaling pathways, 7
induced pluripotent stem (iPS) cells, 8
murine ESC self-renewal signaling pathways, 6–7
GSK3β activity in, 6
LIF role in maintaining mESC, 6
PI3K signaling in, 6
self-renewal by BMP signaling, 6
STAT3-dependent self-renewal in, 6
pluripotent cells from peri-implantation stage, 3–6
roles of LIF, Nanog, and BMP in, 37
STAT3-dependent self-renewal in mESCs, 6
transcriptional networks controling pluripotency, 7–8
Myc transcription factors, 8
Embryonic stem (ES) cells, 238, 248–250, 257, 293, 319, 382, 406, 451, 474, 561
bivalent chromatin domains, 249, 280
DNA methylation pattern in, 250–252, 258–260
demethylation of CpG islands, 251
de novo methylation, 250–251
hypomethylation tolerance, 252
non-CpG methylation, 252
protection of CpGs, from de novo methylation, 251–252
and early development, 293–294
of frog embryo, 295
of human embryo, 296
hyperdynamic chromatin in, 261
insulin-like growth factors and, 303
mapping of histone modifications in, 261
and molecular transcriptional networks, 248–249
PI3K/Akt signaling in, 312–313
pluripotency and chromatin structure, 249–250
post-translational histone modifications, 260
Wnt signaling and, 303
Embryonic theory of origin of cancer, 497–498
Endometrial MSC like cells, use in tissue engineering, 401–402
Endometriosis, and endometrial stem/progenitor cells, 401
Endometrium, 391
bone marrow-derived stem cells and, 399
cancer stem cells and, 401
resident stem/progenitor cells in, 399
stem/ progenitor cells in, 394–395, 398
See also Human endometrium; Mouse endometrium
Endothelial cells (EC), 319
effect on osteogenic differentiation of MSCs, 187–188
endothelial specification and proliferation
BMP signaling, 321
TGFβ signaling, 321–322
TGFβ superfamily and, 320–321
VEGF signaling, 320
Wnt signaling, 322
endothelial stabilization, polarity, and migration, 322–323
ephrins, 323
Hedgehog signaling, 324
Notch signaling, 323
PDGFβ signaling, 323–324
stromal cell–derived factor-1, 324
Endothelial ontogeny, 319
Endothelial progenitor cells (EPC), 321

Endothelium-related cells, AC133 cells, 19
End-replication problem, 123
Englund, U., 563
Environmental signals, in MSC regulation, *see* Mesenchymal progenitor cells, environmental signals regulating
Enzyme replacement, in stem cell therapy, 580–581
Ephrins, 323
Ephrin tyrosine kinase receptors, 410
Epiblast-like stem cells (EpiSCs), 3
Epiblast stem cells (EpiScs), 3, 5, 87
  similarities between hESCs and, 9
Epidermal growth factor receptor (EGFR) signaling, 330
Epidermal proliferative unit, 414
Epigallocatechingallate (EGCG), 557
Epigenetics
  changes during normal development, 237–240
  control, 286
  defined, 235, 243
  modifications, 235–238
  normal development and, 238–240
  regulation, 269–270, 273, 277
Epithelial cells and MSCs, 187
Epithelial CICs, 491
Epithelial mesenchymal transition (EMT) process, 482
Epithelial stem cell niche, 406
  adult niche, 410
    corneal (limbal) niche, 411–412
    intestinal niche, 410–411
    mammary niche, 412–413
  dedifferentiation, 408
  homeostasis, maintenance of, 406–408
  and lineage commitment, 408–410
  mesenchymal cells, role in epithelial cells development, 408–409
  signaling pathways and, 409
    BMP signals, 409
    Hedgehog (Hh) signaling, 410
    Notch signaling, 410
    Wnt signaling pathway, 409
Epithelium, embryonic origins of, 405–406
E-proteins, 366
Eriksson, M., 539
ERK1,2 phosphorylation, 480
Erlotinib, 553, 555
ES cells pluripotency maintenance, by Nanog, *see* Nanog mediated dedifferentiation, of EM progenitors
Escherichia coli β-galactosidase, 567
Escort stem cells (ESCs), 148
ESC pluripotency, transcription factor networks involved, 5
ESC potency, alternate assay for, 5
Estrogen receptor α (ERα), 392, 393
Estrus cycle, murine, 394
Euchromatin, 236, 270
Evans, M., 4
ExGen 500 (Fermentas), 76
Extracellular Matrix (ECM), regulation of MSCs, 191
  collagen types I and II (Col I and Col II), 191
    Col II affect on MSC differentiation, 194
    Col I induce osteogenesis in, 193–194
    ECM domains, effects on cell behavior, 193
    integrin-ECM ligand interactions, 192
    P-15, increase gene expression of BMP-2 and BMP-7, 194
  cryptic domains and bioactive fragments, 200
  regulation of growth factors
    collagen type II role, 198
    effects of binding of growth factors by ECM molecules, 197
    Fibroblast Growth Factor (FGF), 199–200
    Insulin-Like Growth Factor (IGF), 198
    noggin and chordin molecules, 198
    small leucine-rich proteoglycans (SLRPs), 197–198
    transforming growth factor, 196–197
  regulatory mechanisms acting in MSC differentiation, 191
  role in niche and stemness, 194
    EC matrix and matrigel, 194–195
    laminin and laminin-5, 195–196
  See also Mesenchymal stem cells (MSCs)

F

Fabry disease, 386
Facucho-Oliveira, J.M., 137
Fan, 549
Fang, D., 533
Fan, 271 [Not found in Reference]
Fan, X., 551
Farber, E., 447
Feeder cells, 294
Feeder layer culture technique, 27, 29
Feed-forward motifs, *see* Transcriptional networks, regulating ES cells
Feinberg, A.P., 280
Female germ cell development, and asymmetric patterns, 15–16
Fertilized egg, parental genomes and epigenetic asymmetry, 16
Fetal calf serum, 294, 296
Fetal liver stem/progenitor cells (FLSPC), 444
Fgf5, 5, 41, 49, 301
FGF signaling pathway, 301–303, 331–332
Fhit genes, 489
  animal model experiments, 490–491
  apoptosis, 491
  CIC hypothesis and Fhit role in, 491
  DNA damage responses, 490–491
  from very early to advanced stages of cancer, 489–490
Fhit gene therapy, 491
Fibroblast growth factors (FGFs), 301, 442
Fibronectin, 200–202
Flow cytometry, 337–341
Fluorescence-activated cell sorting (FACS), 381, 533
Fluorescence lifetime imaging microscopy (FLIM), 343
Fluorescence resonance energy transfer (FRET), 62, 342–343
5-Fluorouracil, 104, 136, 167, 351, 534
Fms-like tyrosine kinase 3 *(FLT3)* inhibitor, 524
Follicle stem cells (FSCs), 148
Follistatin, 298, 330
Fortunel, N.O., 362
Fowler, J.F., 483
FoxD3, 7
FRA3B/Fhit, 489
Frankel, A.D., 80
Friedenstein, A.J., 176
Frommer, M., 258
FuGENE (Boehringer Mannheim), 76
Fulminant hepatic failure, 461
Furth, J., 500

G

Gadd45a dependent repair pathway, 278
Gage, F.H., 576
G9a, histone methylation by, 253
Galli, R., 530, 548
Gametogenesis and asymmetry in cell division, 17
Gamma-glutamyl transpeptidase (GGT), 445
Gammaitoni, L., 354
Gammaretroviruses, 78
Gap junctions, 158
Gardiner-Garden, M., 258
Gardner, R.L., 14
Gata6, 94
Gefitinib, 553
Geibel, B., 18 [Giebel in Reference]
Gemcitabine, 531, 534
Gene expression, 235, 239
  and DNA methylation, 258
Gene silencing, 235–240, 257, 278–281
  in cancer, 241, 242
  DNA methylation and, 258
  epigenetic modification and, 238
  in ES cells, 33
Genetic and epigenetic reprogramming, 252–254
Genetic manipulation of hESCs, transfection strategies
  by Adeno-associated viruses (AAV), 79
  nonviral transfection of small DNA constructs, 75
Genome Institute of Singapore, 270
Genomic imprinting, 239, 271
  homeostatic regulation of, 279–280
GENSAT library, 81
Germarium, 148, 155–156
Germ cell differentiation, asymmetric divisions in, 14–15
Germ cell markers, 56, 62, 63
Germ cells development and human ESCs, 55–56
  ESC-like properties of, 57
  genes responsible for, DAZ gene family
    BOULE, 57–58
    DAZL (Deleted in Azoospermia-Like), 58
    DZIP gene, 58–59
    NANOG gene, 59
    Nanos gene family, 59
    OCT-4 gene, 59–60
    PUM1 (Pumilio 1) and PUM2 (Pumilio 2), 60
    STELLAR, 60
  telomere and telomerase dynamics in, 126
  in vitro differentiation of hESCS into, 56–57
Germ cells self-renewal, pathway for, 15
Germline stem cells regulation, 155–156
  in *Drosophila* ovary, 155–156, 161
  escort cells and niche for GSC lineage, 159
  factors and pathways for GSC differentiation, 160
  functions of Yb/Piwi-mediated signal, 158
  gap junctions, for maintenance of GSCs, 158
  hedgehog (hh)-lacZ line, 156
  imitation switch (ISWI) and stonewall (Stwl), role in GSC differentiation, 159
  intrinsic factors controlling GSCs self-renewal, 159–160
  molecules essential for GSC renewal and division, 159–160
  niche
    BMP and Piwi-mediated signal from, 157–158
    interactions for stem cell self-renewal, 158
    structural asymmetry, 157
    structure and signaling pathways, 156–157
  signals and nutritional status impact, 158–159
  stem cell aging and, 160
  translational regulators in, 159
Germline stem (GS) cells, 312
Germ plasm, 15
Gerrard, L., 76
Gharwan, H., 78
Gibbs, C.P., 533
Giebel, B., 351
Gilbertson, 169, 172 [Not found in Reference]
Gleevec^{TM}, *see* imatinib
Glial cell line-derived neurotrophic factor (GDNF), 312, 385
Glinsky, G.V., 475
Glioblastoma cells, 172
Glucose-6-phosphatase (G-6-Pase)1, 460
Glycogen synthase kinase 3 (GSK3), 312
Goff, S.P., 79
"Gold standard" assay, 28
Gong, J.K., 356
Gonialblast, 15
Goodell, M.A., 533
Graeber, G., 477
Graft-*versus*-tumor (GVT) effects, 509
Graham, S.M., 172
Granular precursor cells (GPCs), 476
Granulocyte colony-stimulating factor (G-CSF), 440
Green fluorescent protein (GFP), 341
Green, H., 495
Green, M., 80
GSK3β, 6
GSx (glutathione drug pump), 483

H

H4, 260
  acetylation of, 249
Hahn, W.C., 129
Hair follicles (HFs), 408, 414
Hair follicogenesis, embryonic, 414
Hamburger, A.W., 473
Hanna, J., 80
Hayflick limit, 4, 505
Haylock, D.N., 354, 356
HCT-116 colon cancer cells, 243
Hedgehog (Hh), 148
Hedgehog pathway, 324, 410, 523
Helix-loop-helix (HLH), 366
Hematopoiesis, 101–102, 455
  development during early embryonic stage, 213
  stages in vertebrates, 102
Hematopoiesis generated at random (HER), 356
Hematopoietic assays, 103–105
  See also Zebrafish *(Danio rerio),* HSC regulation in
Hematopoietic stem cells (HSCs), 68, 111, 114, 279, 314, 347, 448, 473
  adhesion and extravasation, 355–356
  asymmetric cell division, 352
  asymmetry in HSCs' fate, 18–19
  Bmi-1, role of, 366
  capacity of self renewal, dependence, 68, 69
  cell cycle regulation, 350–353
  Cited2 and CREB-binding protein (CBP), role of, 367
  cytokines, effects of, 354–355
  functional assays for
    in vitro assays, 347–348
    in vivo assays, 348
  Gfi-1 and, 367
  HLF, role of, 368
  homing efficiency of, 349

HOXB4 (*see* HOXB4, role in HSCs regulation)
Hox genes, function of, 368
Cdx4, 370
Cul4A mediated Hox ubiquitination, 369
HoxA9, 368
HoxA10, 368
HoxB4, 368–369
human HSC, 349–350
interaction between HSCs and niche
cancer therapy and, 171–172
ECM contribution to osteoblastic niche, 167
logistic model between osteoblastic and vascular niche, 168
osteoblastic niche, 166–167
vascular niche, 167–168
JAK/STAT pathway and, 365–366
morphogen signaling pathways regulation, 360–361
bone morphogenic protein (BMP), 362–363
Hedgehog (Hh) proteins, 363
Notch receptors and ligands, 364–365
TGF-β, pleiotropic effects of, 360–362
Wnt ligands, 363–364
murine HSC, 349
myeloid and lymphoid lineages, 18
NF-Ya role, 370
niche, regulation, 165–166, 356
bone marrow HSC niches, 357
cancer metastasis mechanism, 171
cancer stem cells evolution, as per microenvironment, 169
connection between CAR cells and, 167
dynamics of two-way interaction, 170
endosteal niche, 356–358
endothelial niche, 358–359
hypoxia and reactive oxygen species (ROS), 359–360
microenvironment disruption and diseases, 168–169
and osteoblastic niche model, 166
reticular cells, role of, 359
role against cancer stem cells, 169–171
signaling pathways between niche and, 170
Nov/CCN3 and, 367
and p21Cip1/Waf1, 350
polycomb group (PcG) proteins and, 366
proliferation and expansion, and cytokine stimulation
c-Kit, 353
Flt3, 354
thrombopoietin (Tpo), 353–354
regulatory pathways for development, 105–107
role of PGE2 in promoting blood cell proliferation, 107
SCL/TAL-1and Id1 proteins, role of, 366–367
and SDF-1/CXCR4 pathway, 355
self-renewal, 351–353
signaling pathways involved, 170–171
targets for gene therapy, 111
thymic engraftment
and enhanced cellularity by HOXB4-GFP in competitive reconstitution assays, 118
in secondary transplants mice, 119–120
zebrafish, gene programs regulating
acute injury recovery assay, schematic of, 104
Cdx/Hox pathway, 105–106
hematopoietic assays, 103–105
Notch pathway, 106–107
primitive and definitive hematopoiesis, 102–103
prostaglandins (PG) signaling, 107
Zfx, role of, 370
See also Hematopoiesis; Stem cells
Hematopoietic stem cell transplantation, allogeneic (HSCT), 509
Hematopoietic Zinc Finger protein (EHZF), 363
Hemmati, H.D., 530
Hemopoietic inductive microenvironments (HIM), 356
Hepatic leukemia factor (HLF), 368
Hepatic transport proteins, 459
Hepatocyte drug metabolism, 458–459
Hepatocyte growth factor (HGF), 451, 453–454, 456
Hepatocyte nuclear factor-4alpha (HNF4-α), 445
Hepatocytes, 443–444, 446, 457–458
drug metabolism, 458–459
hepatic enzymes, 460
hepatic transcription factors, 459–460
hepatic transporters, 459
therapeutic potential of, 460–462
HER-2 gene, 554
Hermann, P.C., 531
Hermitte, F., 216
HESC self-renewal signaling pathways, 7
Heterochromatin, 236–237, 270
Heterochromatin protein 1 (HP1), 253, 260, 261
Heteroplasmy, 136
He, W., 362
Hh signaling pathway, 413
HIF-α
mechanisms regulating, 221
subunit structure-function relationships, 216
tissue and cell distribution of subunits, 216–218
HIF-1α
mechanism of regulation under normoxia and hypoxia, 220
and STAT3 cross-talk in, 224
and targeting of notch-regulated genes, 223
High CpG content promoters (HCPs), 263
High mobility group (HMG) family protein, 248
Hirose, K., 341
Hishikawa, K., 381
Histone code, 237, 270
Histone methyltransferases, euchromatic, 253
Histone modification pulsing model, 261
Histone modifications, 237, 243, 253, 270, 272
HIV-derived lentiviral vectors, 78
H3K27me3, PcG mediated mark, 237, 238
HLA class I-restricted antigen, 508
HMGA2 inhibition, 539
HMG protein Sox2, 5
Hock, H., 367
Hoechst dye, 349, 357, 528, 533
Hoechst 33342 dye, 381, 473
Hoffman, L.M., 272
Holliday, R., 277
Holoclones, 416, 474
Homeobox genes, 105, 370
Homeobox transcription factor (HOXB4), 111
Homeodomain transcription factor, 5
Homologous recombination, in ES cells, 80
Homoplasmy, 136
Houghton, J., 535
HOXB4, role in HSCs regulation
competitive repopulation assay, 115
effects of forced/over expression, 111–112
disturbance to lymphoid-myeloid differentiation in vivo, 114–115
enhances progenitor cell-derived colony formation, 112–113
hematopoietic recovery and thymic engraftment, 116–117
HOXB4-GFP–transduced BMCs, 114

HOXB4, role in HSCs regulation (*cont.*)
immunophenotypic profile of transduced cells, BMCs, 112
increases frequency of Long-Term Culture Initiating Cells (LTC-IC) in vitro, 113–114
leads to long-term hematopoiesis defects, 114
on levels of thymic engraftment and enlargement, 115–116
on progenitor derived CFUs, 113
retrovirus-mediated overexpression of HOXB4, 112
studies of retroviral mediated HOXB4 gene transfer, 111–112
HOXB4 gene in T-cell development, 120
importance in regulating HSCs renewal, 116
role and utility of HOXB4 in expansion, 117
transduces HOXB4 HSCs demonstrating reconstitution and engraftment ability, 120
transduced with GFP-or HOXB4-retroviral vectors, 112
Hox clusters, 105
Hox homeodomain (HD) proteins, 368
H-Ras, 539
HSC/HPC bone marrow niches
oxygen tensions in, 214–215
response to hypoxia in vitro, 215–216
Hsieh, C.L., 270
HTERT, and DNA demethylation, 239
H3, trimethylation of, 249, 260
Hubner, K., 56
Hu-li tai-shao (Hts), 156
Human and mouse ESCs, differences in, 5
Human embryonic stem cells (hESCs), 3, 55, 298, 475
aberrant CpG island methylation, upon hESC differentiation, 260
activin A/Nodal signalling role, 31
and Activin/Nodal/TGFβ signaling, 301
derivation defined by Thomson, 55
DNA methylation in, 259–260
effect of BMPs on differentiation, into germ cells in vitro, 56, 57
factors improving, ES cell survival, 303
FGF/IGF pathway role, 31, 87
FGF signaling pathway in, 302–303
genes playing role in developement of germ cells
DAZ gene family, 57–60
genetic manipulation, transfection strategies
by Adeno-associated viruses (AAV), 79
by Adenoviruses (Ad), 79–80
BAC transgenesis, of hESCs via nucleofection, 81–82
by Baculoviruses, 80
electroporation, 76
episomal maintenance, 82
lipofection, 76
nonviral transfection of small DNA constructs, 75–76
Nucleofection, 76–77
poor clonogenicity, obstacle to genetic manipulation, 75–76
by protein transduction, 80–81
role of environment, 76
targeted transgenesis, by phage phiC31 integrase & ZFNs, 83
transient and stable nucleofection of, 77
by using homologous recombination, 80
and germ cells, 55
conserved protein complex formation in, 60
ESC-like properties of, 57
PUM2 and DAZL, binding specificity of, 61
insulin-like growth factors and, 303
interactions of proteins
BOULE and PUM2, 61–62
DAZ and DZIP, 62
DAZ and PUM2, 60–61
NOS1 and PUM2, 61
PUM2 and DAZL in ESCs, 62
mESCs *vs.*, 298
phenotypic differences between mouse ES cells and, 28–29
self-renewal signaling pathways
and role of Fgf2 and TGFβ family Activin A, 7
and role of PI3K signaling, 7
signaling in, 30–31
signaling pathway, regulating pluripotency, 298–303
similarities between selfrenewal/pluripotency signals in mouse and, 31
telomere and telomerase dynamics in, 126–128
TGFβ family role in pluripotency, 31
transgenesis by
lentiviruses, 78–79
viral transduction, 77–78
up-regulation of ERK signaling and effect on, 31
in vitro differentiation, into germ cells, 56–57
and WNT signaling, 303
See also Embryonic stem cells (ESCs), in mouse and humans
Human endometrium, 400
bone marrow-derived cells and, 399–400
epithelial stem/ progenitor cells in, 395–396
stromal/mesenchymal stem-like cells in, 397–398
structure and function, 392, 393
Human ES cells, derivation of, 27
Human immunodeficiency virus (HIV), 425
Human polycomb repressive complex-1 (PRC1), 286
Human presenilin 1 (PS), 566
Human umbilical vein endothelial cells (HUVEC), 321
Humoral theory of disease, 497
Huntington's disease (HD)
bone marrow derived cells for treating, 566
neural stem cells for treating, 567
Huntly, B.J., 535
Hus1 pathway, 491
Hwang, W.S., 383
Hyaluronan (HA), 355–356
6-Hydroxydopamine (6-OHDA), 564
4-Hydroxynonenal (HNE), 140
Hydryoxurea, 555
Hypoxia, 211–212, 477, 479, 550
cross-talk between HIF-1α, and TGF-β/Smad signaling and, 224
and effects on ESCs in vitro, 212
hematopoietic development in hypoxic conditions, 213
hematopoietic stem/progenitor cells (HSC/HPC) bone marrow niches
oxygen tensions in, 214–215
responses to hypoxia in vitro, 215–216
HIF-α
subunit structure-function relationships, 216
tissue and cell distribution of subunits, 216–218
HIF-1α
mechanism of regulation under normoxia and hypoxia, 220
and STAT3 cross-talk in, 224
and targeting of notch-regulated genes, 223
hypoxia inducible factors (HIFs), and networks regulating response to, 216
cross talks between signalling pathways and, 222
deficiency in vivo and in vitro, 218–219
genes regulating pluripotency, self-renewal or proliferation, 222
regulation mechanisms, 219–221
implantation and placentation during, 212–213
postnatal bone marrow hematopoietic SC niche
hypoxic regions within, 214–215

schematic representation, HIF-1α, HIF-2α and HIF-3α1 subunits., 217
TGF-beta superfamily, 223–224
Wnt and beta catenin, 223
Hypoxia inducible factor (HIF-1), 71, 360, 550
Hypoxia inducible factors (HIFs), 179
and networks regulating response to hypoxia, 216
cross talks between signalling pathways and, 222
deficiency in vivo and in vitro, 218–219
genes regulating pluripotency, self-renewal or proliferation, 222
regulation mechanisms, 219–221
Hypoxia-reoxygenation model, 478

I

ICM-specific markers, Rex1 and Gbx2, 5
Id (Inhibitor of Differentiation) proteins, 297
IGF2/H19 IC, 508
Illmensee, K., 498
Illumina, Inc., 270
Imatinib, 538, 555
Immortal strand hypothesis, 13, 20–22, 67, 352
Immune modulation, in stem cell therapy, 580
Imprinted genes, 271
Imprinting control regions, 272, 280
Induced pluripotent stem (iPS) cells, 3, 8, 93–94, 249, 383, 408
and epigenetic modifications, 34
Ingenuity Pathway Analysis (IPA), 331
Inhibitory proteins, 580
Inner cell mass (ICM), 3, 4, 248
Inside-out signaling, 355
Insulin-like growth factor II (IGF-II), 313
Integrins, 441
Interferon-γ (IFNγ), 424
Interferon-α (IFN-α), 511
Interfollicular epidermis (IFE), 314, 414–417
Interleukins (ILs), 424
Intermediate CpG content promoters (ICPs), 263
Internal ribosome entry site (IRES), 511
International Stem Cell Initiative, 28
Intestinal stem cells (ISCs), 314
Irradiation, 140
Ito, K., 69

J

Jagged1 signaling, 330
Jaiswal, S., 535
Jak/STAT3 signaling, 479
JAK/STAT3 signaling pathways, 482
Jang, Y.Y., 70, 79, 215
Jiang, H., 555
Jordan, 172 [Not found in Reference]
Joyner, A.L., 549
Jumanji family, 34

K

Kahn, M.C., 500
Kale, S.K.A., 380
Karlsson, G., 362
Karmakar, S., 555
Kato, Y., 534
Kelly, J.W., 537
Kelly, P.N., 500
Kern, S.E., 497
Kidney regeneration
from MSCs
nephron construction from hMSCs, by relay culture, 385–386
renal functions, performance of, 386–387
urine production, by modified relay culture system, 386
from stem cells, 384–385
See also Renal stem cells
Kiel, M.J., 352, 358, 359
Kim, C.F., 534
Kim, D., 383
Kim, J.J., 483
Kimura, T., 348
Kinase activity study
single-cell techniques, 337
capillary electrophoresis, 339
flow cytometry, 341
live-cell imaging, 341–343
by traditional biochemical assays, 337, 338
Kirstetter, P., 364
Kitamura, S., 381
Klf4, 32
Kondo, T., 534
Kraynov, V.S., 343
Krivtsov, A.V., 535
Kruppel-like factor 4 (Klf4), 8
Kuang, S., 19
Kunter, U., 380, 383
Kwon, Y.D., 80

L

Label retaining cells (LRC), 396
Lacham-Kaplan, O., 56
Lanza, R.P., 384
Lapidot, T., 167, 529
Larsson, J., 362
Laser-induced fluorescence (LIF) detection, 341
Lck promoter, 521
Leary, A.G., 18
Lefty, 298
Lentiviral vectors, for genetic manipulation, 78
Lentiviruses, 78–79
Lescaudron, L., 566
Lessard, J., 477
Leukemia, 314, 498–500
differentiation therapy of, 500–501
stem cells and maturation arrest, 500
Leukemia inhibitory factor (LIF), 6, 248, 296, 310
Leukemia-initiating cell
cell of origin in, 520–522
concept of, 519–520
heterogeneity of, 522–523
and transplantation of primary AML cells, 520
Leukemic stem cells (LSCs), 473, 520
hematopoietic microenvironment of, 523–524
vs. normal cells, 524
notion of niche in, 523–524
self-renewal and differentiation, 523
Levesque, J.P., 360
Liechty, K.W., 383
Liew, C.G., 76
Li, L., 166, 169
Limbal epithelial crypts, 411
Limbal epithelial stem cells (LESCs), 411
Limbal stroma, 411
LIM domain only 2 (lmo2), 102
Lineage-specific gene expression control, by H3K27 methylation, 261
Lin, F., 380, 382
Lin, H., 165

LIN28, in human ES cells, 249
LipofectAMINE (Life Technologies), 76
Lipofectin (Invitrogen), 76
Liu, B., 354
Liu, R., 535
Liu, S., 550
Live-cell imaging, 339, 341
  FRET, 342–343
  translocation, 341–342
Liver
  development, 441–443
  diseases, 461
  failure, 440
  partial hepatectomy (PH), 443
  regeneration, 440, 443
  renewal, 443–444
  stem cells (*see* Liver stem cells (LSCs))
Liver cell transplantation, 460
Liver progenitor cell (LPC), putative, 444
Liver stem cells (LSCs), 440
  adult liver, 446
    extrahepatic stem cells, 448–450
    intrahepatic stem cells, 446–448
  definition, 444
  fetal liver, 444–446
  in vitro hepatic differentiation potential of, adult stem cell
    bone marrow and hematopoietic stem cell, 450
    mesenchymal stem cells (MSCs), 450–451
    peripheral blood cell, 450
  in vitro hepatic differentiation potential of, ES cell
    endoderm induction, 451–452
    hepatic induction, 452–453
    hepatic maturation, 453–455
    hepatic specification, 453
    into hepatocyte-like cells, 455–457
Liver stem progenitor cells (LSPCs), *see* Oval cells
Lock, 474 [Not found in Reference]
Loewenstein P.M., 80
Lombardo, A., 83
Lonergan, T., 135
Loquacious (Loqs), 159, 160
Lord, B.I., 356
Low CpG promoters (LCPs), 263
Lozenge (Lz), 150
Luteinizing Hormone Releasing Hormone-Agonist (LHRH-A), 426
Lyden, 171 [Not found in Reference]
Lyko, 278 [Not found in Reference]
Lymphotoxin-β receptor (LTβR), 421–422

M

Mackillop, 477 [Not found in Reference]
Mad, 157–159
Madhavan, L., 139
Maeda, Y.T., 90
Maeshima, A., 381
MAGE-1, 510
Magnusson, M., 368
Maloney, M.A., 214
Mammalian target of rapamycin (mTOR) signaling pathway, 537
Mammalian zygote, totipotent, 27
Mammary stem cells (MaSCs), 315
Mancini, S.J., 365
Manganas, L.N., 557
Mangan, S., 91
MAPK/AKT signaling pathways, 299
MAPK/ERK1, 2, 477, 479, 480, 482
Martin, G., 4
Ma, S., 533
Matrosova, V.Y., 167
Matsuzaki, Y., 349
Maturation arrest model, of TSC, 475
Mazurier, F., 348
McCulloch, E., 347, 356
McCulloch, E. A., 473, 500
MCF7 breast cancer cells, 537
McKenzie, J.L., 352
Mechanical stresses, effect MSC differentiation, 180
Medea, 288
Medulloblastoma, 476
Melanocytes, 411
Mellinghoff, I.K, 553
Memory
  adult neurogenesis, 576
  improvement post stem cell transplantation, 577–579
  and neuronal loss, 576–577
Menstrual cycle, 392–393
Meredith, G.D., 341
Mesenchymal progenitor cells, environmental signals regulating
  effects of cytokines/chemokines and growth factors on migration of, 176–177
  functions
    effect of mechanical force on, 180
    oxygen pressure role in regulation of, 179–180
    regulation by Toll-Like Receptor (TLR) ligands, 177
  Hepatocyte growth/scatter factor (HGF/SF) role, 176
  hypoxia as signal, 179
  immune-suppressive activity regulators, 177–178
  inflammatory signals directing MSC to inflicted sites, 176
  interactions with ECM, 178–179
  Reciprocal regulation between MSC and ECM, 178
Mesenchymal stem cells (MSCs), 257, 329, 383, 450–451, 513, 562
  ability for becoming cells of nonmesodermal origin, 185
  autocrine regulation in, 189–190
  cell signaling in, 330
  challenges for, network analysis of, 333
  ECM regulation (*see* Extracellular Matrix (ECM), regulation of MSCs)
  environmental regulation of, 185
    ECM regulation of MSCs, 191–206
    and endochondral ossification, 186
    regulation via cell-cell interactions, 186–188
    regulation via soluble factors, 188–191
  epigenetics of, 262
    histone modifications marking promoters and, 263–264
    promoter DNA methylation, and transcriptional activity, 263
    promoters of lineage-specification genes, and DNA methylation, 262–263
  extracellular signals acting on, 185
  homotypic interactions among
    connexins and gap junctions, 188
    interactions of cadherins, β-catenin, and Wnts, 188
  immunomodulatory activity, 186
  network of all expressed genes of, 331, 332
  properties and nature of environmental influences on, 185
  regulatory networks role
    collagen type X (Col X) role, 205–206
    FN role in, 205
    role of Col II in skeletal development, 205
  signaling networks, compilation of, 330–333
  See also ECM molecules, derived fragments
Meshorer, E., 272
Mesoangioblasts, 19

Metanephros, 384
Methylation events, 33
Methyl-CpG-binding proteins (MBDs), 271
Methyl-DNA immunoprecipitation (MeDIP) assays, 259
Michalopoulos, G.K., 459
Michalska, A.E., 212
Microenvironmental-generated signaling, significance, 479
MicroRNAs (miRNAs), 269, 538
  in transcriptional regulation, 270, 272–273
Migratory SP fraction (SPm), 478
Minor histocompatibility antigens (mHA), 509
Mintz, B., 498
Mitalipov, S.M., 508
Mitochondrial neurogastrointestinal encephalomyopathy (MNGIE), 141
Mitochondria, role in stem cell biology, 135
  aneuploidy and chromosomal alterations in, 136–137
  localization in stem cells, 135–136
  mitochondrial dysfunction and diseases, 140–141
  mtDNA content and affects cells metabolic state, 139
  nutrients conversion to ATP, 135
  and oxygen/oxidative stress role in, 137–139
  role as regulators of apoptosis, 139
  role of ROS in cardiovascular differentiation of stem cells, 139
  roles in processes, 135
  "stemness," 135
Mitogen-activated protein kinase (MAPK), 479
Miyawaki, A., 342
MLL-AF9 fusion protein, 521, 535
(MLL)-eleven-nineteen leukemia (ENL), 169
MLL-ENL fusion protein, 521
MLL-GAS7 fusion protein, 521
MLL-GAS7–mediated transformation, 521
MMPs (matrix metalloproteinases), 171
Mn-superoxide dismutase (MnSOD), 70
Mochizuki, N., 342
Molecular glue, 441
Monzani, E., 533
Morrison, 17, 70 [Not found in Reference]
Mouse embryonic fibroblast cells (MEFs), 55
Mouse embryonic stem cells, pluripotency and signaling pathways
  BMP signaling, 297–298
  derivation of embryonic stem cells, 294–296
  feeder cells activity, effect of LIF on, 296
  FGF signaling, 302
  Id proteins, role of, 298
  LIF function, in early development, 296–297
  STAT3, activation of, 297
Mouse endometrium, 401
  bone marrow-derived cells and, 400
  epithelial stem/ progenitor cells in, 396–397
  stromal/mesenchymal stem-like cells in, 398
  structure and function, 393–394
Mouse epiblast stem cells
  and Activin/Nodal/TGFβ signaling, 300–301
  FGF signaling pathway, function of, 301–302
Mouse pluripotent cells, determinants, 27
MOZ-TIF2 oncogene, 535
MPD5 antigen, 511
MPD6-IRES-mediated translation, 511
MRPs drug pump, 483
Mullally, A., 507
Muller, S., 337
Multidrug resistance associated proteins (MRP), 459
Multiparametric flow cytometry, 333
Multiple Sclerosis (MS)
  bone marrow derived cells for treating, 567
  neural stem cells for treating, 567–568
Multipotent adult progenitor cells (MAPCs), 50, 451, 562
Multipotent stromal cells (MSC), 175
  See also Mesenchymal progenitor cells, environmental signals regulating
Murasawa, S., 139
Murine embryonic stem cells (mESCs), 3
  BMP role in self-renewal, 30
  characterization of steps of differentiation toward mesoderm lineages, 38–40
  cultured on feeders, LIF and serum
    contain committed mesoderm progenitors, 38
  dedifferentiation of mesoderm progenitors, 42
  dependence on LIF, 28–29
  developmental origins, pluripotent cells, 4
  features in culture, 4–5
  feeder layer culture, technique for propagation, 27, 29
  isolations, 4
  and levels of oncogenic factors in, 5
  LIF regulates percentage of *T*-expressing cells in, 39–41
  LIF signaling impact on Oct4 expression in, 9
  LIF/STAT3 signaling for maintenance of, 51
  mechanism for up-regulation, of *Nanog* expression, 42–46
    doxycycline inducible conditional expression system, 42
    role of levels of *T* expression, 40–41
    T controls Nanog-dependent dedifferentiation, 44
  phenotypic differences between human ES cells and, 28–29
  and population of T-expressing cells in, 40
  reversal of mesoderm commitment by LIF, 41
  role of LIF in maintenance of mouse ES cell pluripotency, 49–51
  role of WNT family proteins in, 29
  self-renewal and pluripotency, 37
    depends on expression of POU transcription factor Oct4 and Nanog, 37, 38
    manipulation of, 38
  self-renewal signaling pathways, 6–7
    PI3K/AKT role in, 6
    role in LIF-STAT3, 6–7
  signaling pathways in, 29–30
  similarities between selfrenewal/pluripotency signals in mouse and, 31
  and T-expressing cells in, 40
  in vitro differentiation, into germ cells, 56
  and Wnt-related effects on pluripotency, 30
Murine leukemia virus (MLV), 78
Murine pluripotent cells, developmental origins of, 3–4
Musashi1, 562
Muscle cell population, *see* Skeletal muscle stem cells
MYCN-amplified neuroblastoma, 476
Myegakaryocyte-erythroid progenitor (MEP), 348
Myelodysplastic syndromes (MDS), and telomere length alterations, 127
Myeloproliferative syndromes (MPS), 168

N

Nakamora, T., 16 [Nakamura in Reference]
Nakata, S., 70
Nanog, 5, 7–9, 30–32, 46–51, 239, 248–249, 253, 279, 288, 311–313, 476
Nanog, 38–51, 92
NANOG gene, 59

Nanog mediated dedifferentiation, of EM progenitors
negative feedback mechanism impact, 37, 51
blocks BMP signaling, 48
characterization of steps, 38–40
dedifferentiation of mesoderm progenitors, 42
differentiation-promoting activity of BMPs, 49
EM progenitors, for maintaining mouse ES cells, 50–51
interference with BMP signaling, 47
level of *Nanog* expression in *T-eGFP* ES cells, 41
mesoderm induction blockade by BMPs, 49
mouse ES cells, population of *T*-expressing cells, 40–41
Nanog expression, 38, 44
Nanog regulating dedifferentiation, 42–43
negative feedback and mesoderm differentiation, 49–51
positive regulation of *Nanog e*xpression, 42–46
reversal of mesoderm commitment by LIF, 41
role of LIF in maintenance of mouse ES cell pluripotency, 49–51
self-renewal and pluripotency, 37
significance of T-and STAT3-binding sites, 46
STAT3 and T binding EM enhancer for up-regulation, 45
T controls Nanog-dependent dedifferentiation of EM progenitors, 44
transient EM progenitor population and, 49
transition of EM progenitors to ES cells and, 50–51
up-regulation of *Nanog* expression, 42–46
doxycycline inducible conditional expression system, 42
role of levels of *T* expression, 42–43
T controls Nanog-dependent dedifferentiation, 44
See also Murine embryonic stem cells (mESCs)
NANOS1, 56
Nanos gene family, 59
Nanos protein, 55
Nanos response element (NRE), 61
N-cadherin, 357–358
Negative autoregulatory motif, advantages of, 88
Negative feedback, and negative autoregulation, *see* Transcriptional networks, regulating ES cells
Neo-kidney, from hMSCs, 385–387
Nestin, 381, 562
Neural progenitor cells (NPC), mouse, 238
Neural stem cells (NSC), 546–547, 562
Neuroblastoma, 476
Neuroblastoma cell line (SK-N-BE(2)), 474
Neuroblastoma (SK-N-BE(2)), 474
Neurogenesis, in *Drosophila* model system, 16
asymmetry in neurogenesis of *Drosophila,* 16–18
interactions between Numb and Notch, roles in controlling asymmetry, 17
role of Numb in, 16–17
Neuronal replacement, in stem cell therapy, 579–580
Neuron-glial (NG) 2 expression, 548
Neurotrophic support, in stem cell therapy, 580
Ng, T., 343
Niche, defined, 165, 186
Niche theory, 161
Niehrs, 278 [Not found in Reference]
Niemann-Pick, 580
Nilsson, S.K., 166, 356
Ni, Q., 342
Niwa, H., 34
N-nitrosomethylbenzylamine (NMBA), 490
Nodal, 298
Nodal/Activin signaling, in mouse embryonic stem cells, 300–301
NOD/SCID assay, 522
Nolan, G.P., 339
Nolden, L., 76, 80
Noncoding RNA (ncRNA), 272
See also MicroRNAs (miRNAs)
Non-CpG methylation, in mammal, 252
Nonmigratory SP fraction (SPn), 478
Non-proteasome pathway, 508
Normoxia, 478
defined, 211–212
mechanism of HIF-1α regulation under, 220
Notch intracellular domain (NICD), 410
Notch pathway, 101
Notch signaling pathway, 17, 323, 410–413, 417, 442, 523, 549, 552, 557
Notch 1 transcription factor, 474, 562
Nowell, P., 477
Nuclear transfer ESC lines (ntES cells), 509
Nuclear transplantation technique, 384–385
Nucleofection, 76–77, 82
Nucleosome, 260, 270
Numb proteins, 17, 365
Nystul, T., 148

O

Oct3/4, 7, 8, 32
Oct-4, 56, 60, 218, 221, 222, 302, 398, 474–479, 484
Oct-4, and DNA demethylation, 239
OCT-4 gene, 59–60
Oct-3/4 gene, in ES cells, 248, 252, 311, 313
Oct-4/nanog, 481
Oct4 (Pou5f1), 91–92
Oddo, S., 579
Olguin, H.C., 19
Oliver, J.A., 382
Olwin, B.B., 19
Oncostatin M (OSM), 455
O'Neill, L.P., 272
Open reading frame (ORF), 511
Oral contraceptive pills (OCP), 393
Orimo, A., 324
Ornithine transcarbamylase (OTC), 460
Osafune, K., 385
Osteoblastic niche, 166–167, 214
concept, 357–358
ECM contribution to, 166–167
functional difference between vascular niches and, 168
Osteoblasts, 189
in HSC regulation, 523
and MSCs, 187
Osteopontin, 358
Osteosarcoma (HOS), 474
Oval cells, 446–448
Oxidative stress
and cancer stem cells, 139–140
and neural stem cells, 140
Oxidative stress microenvironment, 479
Oxygen homeostasis, 211
Oxygen pressure, effect on MSC functions, 179–180

P

Pabo, C.O., 80
Palisades of Vogt, 411, 412
Palmitoyl protein thioesterase 1 (PPT1) enzyme, 569

P53 and Akt signaling, 312
Parietal epithelial cells (PEC), 381
Parkinson's disease (PD)
bone marrow derived cells for treating, 565
neural stem cells for treating, 565
Parmar, K., 68, 215, 359
Parthenolide (PTL), 172
Passegue, E., 534
Pathogenassociated molecular patterns (PAMP), 177
Pathological hypoxia, 212
Patrawala, L., 476, 532, 534
Patt, H.M., 214
PcG protein homeostasis, and stem cell identity, 287–288
PDGFβ signaling, 323–324
Pelvic floor prolapse, 402
Percent chimerism, 104
Perez, O.D., 339
Pericytes, 19
Peri-implantation stage mammalian embryos
developmental origins of murine pluripotent cells, 3–4
ectoderm and epiblast stem cells, 5–6
embryonic stem cells from mouse, 4–5
Perry, S.S., 367
Petersen, B.E., 447
PGCs dedifferentiation
and PI3K/Akt signaling
Akt activation, effects of, 311
downstream target of Akt signaling, 312
mitotic activity of PGCs, 311–312
Pten deficiency, effects of, 310–311
into pluripotent stem cells, 310
PGC7/Stella, 16
Pgp drug pump, 483
Phosphatidylinositol 3 kinase (PI3K), 6
Phosphoglycerate kinase *(PGK)* promoter, 78
Phosphoinositide-3 kinase (PI3K), 309
Physiological hypoxia, 212
Physiological stem cells, 528
Piccoli, C., 136
Pierce, B., 498
PI3K/Akt signaling, 309–310
in ES cells
pluripotency, maintenance and regulation, 312–313
Tcl1, and Akt kinase activity, 313
and germ cell dedifferentiation, 310
PGCs dedifferentiation, 310
PI3K/Akt signaling, and PGCs, 310–312
spermatogonial stem cells, 312
and "stemness" of stem cell, 315
in tissue stem cells, 313–315
epithelial stem cells, 314–315
hair follicular and epidermal stem cells, 314
hematopoietic stem cells, 314
tissue stem cells and cancer stem cells, 313–314
Placental growth factor (PlGF), 320
Platelet-derived growth factor (PDGF), 303, 323
Platelet-derived growth factor (PDGF)-responsive cells, 550
Pleotrophin, 303
Plotnikov, E.Y., 141
Pluchino, S., 567
Pluripotency
definition of, 27
and embryonic stem cell, 294
PRCs, involvement of, 261–262
regulators, 87
signaling pathways, regulating mouse ES cells (*see* Mouse embryonic stem cells pluripotency, and signaling pathways)
in somatic cells, 249
transcription factors of, 248
Pluripotency determinants, in mouse and human ES cells
cell cycle control, in pluripotent cells, 31
genetic and epigenetic contributors
epigenetics of PED, 33
role of transcription factors, 33–34
signaling in human ES cells, 30–31, 33–34
signaling in mouse ES cells, 29–30
transcriptional effects of pluripotency-related factors on genes, 33
phenotypic differences between, 28–29
pluripotency-related transcription factors, 31–33
Pluripotent cells
cell cycle control, 31
developmental origins of, 4–5
ectoderm-like population, 5
of embryonic origin, 5
hallmarks of, 3
hESC self-renewal signaling pathways, 7
induced pluripotent stem (iPS) cells, 8
maintenance, 3
murine ESC self-renewal signaling pathways, 6–7
from peri-implantation stage mammalian embryos
developmental origins of murine pluripotent cells, 3–4
ectoderm and epiblast stem cells, 5–6
embryonic stem cells from mouse, 4–5
pluripotency determinants
genetic and epigenetic, 29–31, 33–34
pluripotency-related transcription factors, 31–33
self-renewal pathways for, 6–8
stages of pregastrulation development, 3
states, during early mammalian development, 3
transcriptional networks and factors, 7–8
Pluripotent stem cells, 248
P53 mutations, 483
PNA–FISH method, 126
Polycomb group (PcG) proteins, 238, 243, 261–262, 272, 286, 366
Polycomb repressive complex (PRC), 239, 261–262, 286
Polycomb repressive complex-2 (PRC2), 286–288
Polymerase chain reaction (PCR), 260
Positive feedback, and autoregulation, *see* Transcriptional networks, regulating ES cells
Post-hypoxia SPm fraction (SPm(hox)) cells, 479
Postnatal cells, asymmetry in, 17–18
hematopoietic stem cells (HSCs)
asymmetry in HSCs' fate, 18–19
myeloid and lymphoid lineages, 18
skeletal muscle stem cells
asymmetric division of skeletal muscle satellite cells, 20
and role of asymmetry in, 19–20
Potten, C.S., 22
POU domain transcription factor, Oct3/4, 5
Pou5f1, 32
Preimplantation embryonic development (PED), 27
Pre-implantation (ICM) stage, 5
Primitive ectoderm (PrEct), 4
Oct4+ pluripotent population, 3
Primitive endoderm (PrEn), 4
Primitive hematopoiesis, 102
expression of *gata1* and *pu.1* in ICM cells, 102

Primordial germ cells (PGCs), 14, 55–56, 310
Prince, M.E., 532
Progenitor stem cells, 505
Progesterone receptor A isoform (PR-A), 392, 393
Proinsulin, 509
Proliferating cell nuclear marker (PCNA), 107
Prostatic intraepithelial neoplasia (PIN), 314–315
Protamines, 248
Protein transduction, in hESCs, 80–81
Prox1, in liver development, 445
PSA (prostate-specific antigen), 171
PTEN gene, 550
PTEN protein, 536
Pulse-labeled proliferating cancer stem cells, 498
Pumilio protein, 55
PUM1 (Pumilio 1) and PUM2 (Pumilio 2) gene, 60
PUM2, 56
Putative CSC populations, 532–534

Q
Qi, X., 6
Quercetin, 557

R
Rabbitts, T., 521
Radiation, 501
Ramsahoye, B.H., 271
Rangappa, S., 186
RAS-MAP kinase pathway, 301
Rathjen, J., 49
Reactive oxygen species (ROS), 68
  effects on mitochondrial metabolism, 138
  effects on stem cell types, 138–139
  See also Stem cells
Receptor tyrosine kinases (RTKs), 555
Redox sensitive kinases, 479
Regeneration capacity, of human brain, 575–576
Reinhard, M.C., 496
Remak, R., 497
Renal stem cells
  in adult kidney, 380–381
  in bone marrow, 380
  from ES cells, 382–383
  from iPS cells, 383
  from MSCs, 383–384
  niche, 382
  See also Stem cells
Replication timing, and DNA methylation, 264–265
Retinoblastoma 1 (RB1), 489
Retinoic acid (RA), 565
Retroviral gene therapy trial, 78
Retroviral vectors, 78
Rex-1, 476
Reya, T., 362–364
Reynolds, B.A., 547
Rhabdomyosarcoma (RH-4), 474
Rho-associated kinase (ROCK), 76, 303
Rhyu, M.S., 17
Ricci-Vitiani, L., 530
Riggs, A.D., 277
Ritz, J., 507
RNA-binding protein with multiple splicing (RBPMS), 362
RNA-dependent silencing, in stem cells, 279
Rogers, S., 384, 386
Romer, J.T., 549
Roncarolo, M.G., 512
ROS, and p38/MAPK pathway, 360
Rosenfeld, N., 89
Rothhammer, T., 536
Rrp46p, 508
Ruehr, M.L., 343
Ruhnke, M., 450

S
SAGE (Serial Analysis of Gene Expression), 270
Sagrinati, C., 381
Sall4, 92–93
Salman, S., 496 [Salmon in Reference]
Sandhoff's disease, 580
Santa Maria, L., 189
SAPPβ, 565
Sarcospheres, 533
Sasai, K., 552
Satellite cells, 19, 187
Sato, N., 7
Sauvageau, G., 477
Scadden, D.T., 166, 167
Scaffold attachment region (SAR), 78
Scheller, M., 364
Schlesinger, Y., 243
Schofield, R., 165, 356
Schwarze, S.R., 80
SDF-1/CXCR4 axis, 523
SDF-1/CXCR4 pathway, and HSC, 355
SDF1/CXCR4 signaling axis, 324
Sell, S., 475, 478
Senoo, M., 423
Sensory organ precursor (SOP) cell, 17
SEREX-defined self-tumor antigens, 508, 510
Serpent (Srp), 150
Serrano, L., 89
Sharkis, S.J., 70, 215
Shh pathway, 549, 552
Shibata, D., 497
Shinin, V., 22
Shipitsin, M., 536
Shojaei, F., 365
Side-population (SP) cells, 19, 473
  detection of, 381
  See also Mitochondria, role in stem cell biology
Siemen, H., 76, 77
Signal transducers and activators of transcription 3 (STAT3), 312
Singh Roy, N., 76
Singh, S.K., 479, 530, 547, 548
Single/multi-input motif, *see* Transcriptional networks, regulating ES cells
Sipkins, D.A., 359
Skeletal muscle and MSCs, 187
Skeletal muscle stem cells
  asymmetric division of skeletal muscle satellite cells, 20
  and role of asymmetry in, 19–20
Skin
  epithelia, 413
  organogenesis
    adult stem cells, in skin, 414–416
    epidermis, 414
    hair follicle (HF), 414
    signaling pathways, 416–417
  in regenerative medicine, 417
SK-NBE(2) cells, 479
Smad family, 224
  members in human, 360

Smad pathway, and HSCs, 360–361
Smad2/3 signaling pathway, 300
Small light cells, 412
Smith-Arica, J.R., 80
Smithies, O., 5
Smith, A.G., 297
So, C.W., 521
Soluble factors, regulating MSCs
  autocrine regulation in, 189–190
  endothelial cells, 189
  hypoxia, role in MSC function, 191
  MSC protein expression due to TGF-β stimulation, 190
  osteoblast-secreted factor, role in stemness, 189
  relationship between chondrocytes and, 189
  signaling molecules role
    growth factors, EGF, FGF, PDGF, VEGF, 190–191
    TGF-β superfamily, 190
    Wnt family, 190
  tissue injury and impact on MSC activity, 189
  See also Mesenchymal stem cells (MSCs)
Solvatochromic dyes, 343
Somatic cell nuclear transfer (SCNT), 408
Somatic cyst cells, 159
Somatic nuclear transfer (SNT), 509
Sonic-hedgehog (Shh/hgg) pathway, 476
Sox1, 562
Sox2, 7, 32
Sox2, 91–92
Sox2, as transcription factor, 248–249, 311
Spangrude, G.J., 349
SP-derived xenografts, 480
Spees, J.L., 140
Spemann, H., 14
Sperm entry position (SEP), 14
Sphere-forming melanoma cells, 533
Sphingosine-1-phosphate (S1P), 303
Spindle orientation, 17
  See also Asymmetry, in stem cells
Spradling, A.C., 148, 166
SSEA receptor family, 28
Stamp, L., 445
STELLAR gene, 60
Stem cell antigen 1 (Sca-1)$^{+}$CD45^{-}CD31^{-}CD34^{+} phenotype, 534
Stem cell factor (SCF), 310
Stem cell leukemia (*scl*), 102
Stem cell model, of tumor growth, *see* tumor stem cells (TSC) model
Stem cell model of tumor repair and regeneration (SCTR), 483
Stem cell niche, 550
  concept, 165
  defined as, 147
  signaling in, 148
Stem cells, 3, 69, 439
  asymmetric division and stem cell niche
    translation of extracellular polarity into intracellular, 150–151
  asymmetric *vs.* symmetric cell division, 148
  asymmetry (*see* asymmetry, in stem cells)
  based regenerative therapy, 3
  and chromatin remodeling, 238
  definition of, 440–441
  diseases due to lack of telomere and telomerase activity
    aging and cancer, 129
    Aplastic anemia, 130
    cancer stem cells, 130
    Dyskeratosis Congenita (DKC), 130
  DNA methylation and, 239, 270–271
  downstream targets of p38 MAP kinase in, 69
  effects of telomere length and telomerase activity on, 124–128
  fate decisions, ROS sensors influence on, 69–70
  genomic stability in
    FoxO family, transcription factors, 70
    PI3K/Akt role in, 69
    role of p53, 69
  immune responses
    cancer, 510–512
    role of CD4+CD25^{high}Foxp3+ T cells, 512–513
    role of mesenchymal stem cells, 513
  immunogenicity, 506–509
  interaction between HSC and niche (*see* Hematopoietic stem cells (HSCs))
  mitochondrial diseases and, 140–141
  molecular network regulating, 67–68
  niche, 441
  and niche, 165–166
  niche, asymmetry in, 15, 17, 20, 22
  origin of term, 347
  oxidative stress in
    cancer stem cells, 139–140
    control of self-renewal capacity, 67, 70
    neural stem cells, 140
  oxygen
    critical environmental factor regulating fate of, 211
    and oxidative stress effect on, 137–139
  reactive oxygen species threat to cellular DNA, 68
  rejection of replacement, 509–510
  role in tissue homeostasis, 147
    and tissue aging, 147–148
    and tumorigenesis, 147
  role of mitochondrial DNA mutations and metabolism in biology of, 135
    localization of mitochondria in, 135–136
    MtDNA mutations in, 136–137
  self-renewal, role of telomerase in maintainence of, 123
  signaling in stem cell niche, 148
  source of DNA damage in, 67
  strategies for acceptance of transplanted, 509
  telomere and telomerase role in regulation of, 123–131
  and tissue aging in fly Germ-Line stem cells (GSCs), 152
  tissue-specific stem cells (TSSCs), 482–483
  tolerance induction, 509–510
  types, effects of oxygen and reactive oxygen species (ROS) on, 138
  See also Stem cell therapy; Telomere and telomerase, for regulation of stem cells
Stem cell theory of cancer, 501
Stem cell theory of tumorigenesis, 545
Stem cell therapy
  considerations, 568
  memory improvement
    enzyme replacement, 580–581
    immune modulation, 581
    neuronal replacement, 579–580
    neurotrophic support, 580
Stemness switch, 480, 483
Stet mutations, 159
Stochastic model, of tumor initiation and growth, 527
Stommel, J.M., 555
Stra8, spermatogonial-specific marker, 57
Stroka, D.M., 217
Stromal cell–derived factor-1 (SDF-1), 324
Subventricular zone (SVZ), 546
Suda, T., 18, 477
Sugaya, K., 578

Sugiyama, T., 167, 348, 359
Sulston, J., 14
Superfect (QIAGEN), 76
Surani, M., 33
Suzuki, N., 357, 523
Syngeneic transplantation, of hematopoietic stem cells, 509
SYCP3, 56

T
Takahashi, K., 71, 383
Talks, K.L., 217
Tanimura, A., 342
Tannock, I.F., 483
Targeted transgenesis, 83
Tate, M.C., 563
TAT-HoxB4 fusion protein, 369
Tay-Sachs, 580
T cell leukemia 1 (Tcl1), 313
TEL-AML1-associated childhood leukemia, 524
Telomere and telomerase, for regulation of stem cells, 123–124
  animal models for, 128–129
  diseases due to deficiency in activity, 129
  dynamics in human stem, germ and embryonic cells, 125–126
  length and impact on stem cell behaviors, 126
  maintenance importance, 123
  mechanisms for telomere elongation, 124–126
  role in stem cell self-renewal, 123
  stem cell diseases, 129–130
Telomere and telomerase, for regulation of stem cells (*cont.*)
    aging and cancer, 129–130
    aplastic anemia, 130
    cancer stem cell hypothesis, 130
    Dyskeratosis Congenita (DKC), 130
  structure and elongation mechanism, 124–126
  telomere sister-chromatid exchange (T-SCE), 125
  Tert and Terc core components, 128
Temozolomide, 555
Teratocarcinomas, 294, 474, 498
Terminal filament (TF) cells, 155
Tetracycline-inducible transcriptional activator (rtTA), 42
TGF-beta superfamily, 223–224
TGFβ signaling pathways, 298–300, 320–321, 331
T-helper cell (Th1)-mediated autoimmune disease, 567
Therapeutic intervention clinical trials, 569–570
Therapy-resistant cells, 501
Thiol *N*-acetyll-cysteine (NAC), 69
Thomson, J., 27, 55
Thomson, J.A., 76, 80
Thyagarajan, B., 76, 83
Thymic atrophy, 425
Thymic epithelial cell (TEC), 408, 413, 417–421, 423–424, 426
Thymic epithelial cysts, 413
Thymus, 413
  cytokines, role of, 424
  epithelia, 413
  growth factors, in development, 423–424
  mesenchyme, role of, 420–421
  organogenesis, 417–418
  single origin model, for organogenesis, 418–420
  thymocyte-TEC cross-talk, 421–422
  transcription factors, in development, 422–423
Till, J., 347, 356
Till, J.E., 473
Tissue aging and stem cells, 147–148
Tissue homeostasis, stem cells and SC niches in, 147
Tissue-specific stem cells (TSSCs), 482–483
Tothova, Z., 360
Toyooka, Y., 56
Transcriptional Intermediary Factor1γ (TIF1γ), 362
Transcriptional networks, regulating ES cells
  basic network motifs, 89
  controling pluripotency, 7–8
  core regulators in
    iPS cells, 93–94
    Nanog, 92
    Oct4 and Sox2, 91–92
    Sall4, 92–93
    Zfp206, 93
    Zic3, 93
  ESC-specific transcriptional networks, 92
  factors regulating commitment to lineages, 94
  genome-wide analyses of
    expression analysis, 94
    expression in undifferentiated/differentiated cells, 95
    stem cell genes, 94–95
  integration of data, expression and promoter occupancy, 96–98
  model of ESC maintenance, 97–98
  motifs and biological functions, 87–88
    feed forward, 90
    multi-input motif, 91
    negative and positive autoregulation and feedback, 88–89
    single-input motif, 91
    transcriptional cascade, 91
  promoter occupancy studies, 96
  shRNA knockdown and overexpression, 95–96
  time courses of differentiation, 95
Transcription factors, in mouse/human ES cells
  pluripotency-related, 31–33
    Oct3/4, Sox2, Nanog, Klf4, and c-myc functioning, 31–32
    regulation of Oct3/4 expression, 32
    reprogramming factors, 33
    Zinc Finger Protein 206 (ZFP 206), 32
Transcription, histone modifications, 237
Transforming growth factor (TGF)β family, 298
Transit amplifying (TA) cells, 406–408
Transplantable tumor cells, 495
Transplantation, assay to determine presence of HSCs, 104
Trastuzumab, 555
Traumatic brain injury (TBI)
  bone marrow derived cells for treating, 562–563
  neural stem cells for treating, 563
Treg apoptosis, 513
Trentin, J.J., 356
Tripeptidyl peptidase 1 (TPP-I) enzyme, 569
Trithorax group (Trx) proteins, 262
Trott, K.R., 484, 497
Trumpp, A., 167
Tsai, M.S., 330
Tsuchida, R., 482
TuJ1, 565
Tumor colony forming units, 496
Tumor hypoxia, 477
Tumorigenesis and stem cells, 147
Tumor initiating cells, 495
Tumor necrosis factor (TNF), 351
Tumor stem cells (TSC) model
  abilities, 473
  conversion of non-TSC cells to TSC-like cells, 479
  and de-differentiation phenomenon, 480
  dormancy state, 479–483

expansion in tumor hypoxia, 477–478
hierarchical organization of, 473–474
maintenance of self-renewal hierarchy in *in vitro* culture, 474
maturation arrest model, 475
purpose, 473
repair and regeneration, 478–479
retention rate of, 474
stemness of, 474–475
related pathways, 475–476
relation with tumor progression, 477
switch model, 483–484
tumor repopulation, 483
Tumor suppressor genes, DNA methylation of, 258

U
U 0126, 479
UCMS cell transplant, 565
UDP-glucuronosyltransferase (UGT1A1), 460
U373 glioma cell line, 474
Umbilical cord blood (UCB), 562
Umbilical cord blood transplantation, 107
U87MG xenografts, 555
UPA (urokinase-type plasminogen activator), 171
UTF1, 7

V
Vallier, L., 76
Van Etten, R.A., 169
Van Os, R., 348
Vasa, germline markers, 156
Vascular endothelial growth factor receptor 2 (*vegfr2/flk1*), 102
Vascular endothelial growth factors (VEGFs), 320, 380, 550
Vascular niche, 167–168, 214
Vascular ontogeny, *see* Endothelial cells (EC)
Vasculogenesis, and BMP signaling, 321
VEGF autocrine loop, 477
VEGF/Flt1 autocrine pathway, 479–480
VEGF signaling, 477
Vertebrate hematopoiesis, 101–102
definitive wave, produces long term HSCs, 101–103
primitive and definitive waves, 101–103
See also Hematopoiesis
Vescovi, 172 [Not found in Reference]
Vignearu, C., 383
Villars, F., 187
Viral transduction
by lentiviruses, 78–79
by retroviruses, 77–78
Virchow, R., 497
Vitamin D3, 557

W
Walkley, C.R., 168
Wall, M.E., 91
Walmer, D.K., 394
Wang, X., 447
Wang, Y., 370
Watanabe, K., 75
Weber, M., 263
Weiss, M.L., 547, 565
Weksberg, D.C., 349, 359
Wharton's jelly, 562
Widschwendter, M., 243
Wiesener, M.S., 217
Wilson, J.W., 444
Wnt and beta catenin, 223
Wnt/β-catenin signaling, 312, 416, 442, 454–455
Wnt/β-catenin pathway, 475, 481, 482, 549, 554
Wnt pathway, 523, 550, 557
Wnt transcription factor, 474
Wolf, D., 79
Woodchuck hepatitis virus post-transcriptional regulatory element (WPRE), 78
Woolf, A.S., 384
Wu, M., 352, 364, 365

X
X-chromosome inactivation, 271–272
X-EB lesion model, 567
Xenopus laevis, 321
Xenotransplantation, 384
Xia, X., 79
Xie, T., 166
Xi (inactivated X) chromosome, 239
X-irradiation and ethidium bromide (X-EB) injection, 567
Xu, R.H., 7

Y
Yamanaka, S., 383
Yamaquchi, 476 [Not found in Reference]
Yamashita, Y.M., 15
Ye, M., 565
Yilmaz, O.H., 537
Ying, Q.L., 6
Yoon, J., 187
Young, H.E., 505
Yu, F., 539

Z
Zammit, P.S., 19
Zebrafish *(Danio rerio),* HSC regulation in
caudal genes, 105
Cdx/Hox genes roles, 105
Cdx/Hox pathway, 105–106
for elucidation of genetic pathway, 106–107
hematopoietic assays, 103–105
mindbomb, Delta ligand trafficking, 106
Notch pathway, 106–107
prostaglandins (PG) signaling, 107
role of Notch in definitive HSC induction, 106
stages of hematopoiesis
definitive, 102–103
primitive, 102
primitive myelopoiesis, 102
vertebrate model organism for studying hematopoiesis, 101
See also Hematopoiesis; Hematopoietic stem cells (HSCs)
Zeng, J., 80
Zero population growth (zpg), 158
Zfp206, 93
Zhang, X.B., 369
Zhao, Y., 450
Zic3, 93
Zinc-finger nucleases (ZFNs), 83
Zinkernagel, R.M., 508
Zwaka, T.P., 76, 80
Zygote genome activation (ZGA), 28

MIX
Papier aus verantwortungsvollen Quellen
Paper from responsible sources
FSC® C105338

If you have any concerns about our products,
you can contact us on
ProductSafety@springernature.com

In case Publisher is established outside the EU,
the EU authorized representative is:
Springer Nature Customer Service Center GmbH
Europaplatz 3, 69115 Heidelberg, Germany

Printed by Libri Plureos GmbH
in Hamburg, Germany